STRUCTURAL ENGINEERING IN THE 21ST CENTURY

PROCEEDINGS OF THE 1999 STRUCTURES CONGRESS

April 18–21, 1999
New Orleans, Louisiana

SPONSORED BY
Structural Engineering Institute of ASCE

CO-SPONSORED BY
Structural Engineers Association of Alabama (SEAoAL)
National Council of Structural Engineers Associations (NCSEA)
Florida Structural Engineers Association (FSEA)
Louisiana Section of ASCE
Baton Rouge Branch of ASCE
New Orleans Branch of ASCE

EDITED BY
R. Richard Avent
Mohamed Alawady

Structural Engineering Institute
of the American Society of Civil Engineers

American Society of Civil Engineers
1801 ALEXANDER BELL DRIVE
RESTON, VIRGINIA 20191–4400

Library of Congress Cataloging-in-Publication Data

New Orleans Structures Congress (1999)
   Structural engineering in the 21st century: proceedings of the 1999 New Orleans
Structures Congress / edited by R. Richard Avent, Mohamed Alawady.
      p.   cm.
   Includes bibliographical references and index.
   ISBN 0-7844-0421-6
   1. Structural engineering—Congresses. I. Avent, R. Richard. II. Alawady,
Mohamed. III. Title.
TA630.N497 1999
624'1–dc21   99-18131
                                                               CIP

Any statements expressed in these materials are those of the individual authors and do not necessarily represent the views of ASCE, which takes no responsibility for any statement made herein. No reference made in this publication to any specific method, product, process or service constitutes or implies an endorsement, recommendation, or warranty thereof by ASCE. The materials are for general information only and do not represent a standard of ASCE, nor are they intended as a reference in purchase specifications, contracts, regulations, statutes, or any other legal document.

ASCE makes no representation or warranty of any kind, whether express or implied, concerning the accuracy, completeness, suitability, or utility of any information, apparatus, product, or process discussed in this publication, and assumes no liability therefore. This information should not be used without first securing competent advice with respect to its suitability for any general or specific application. Anyone utilizing this information assumes all liability arising from such use, including but not limited to infringement of any patent or patents.

Photocopies: Authorization to photocopy material for internal or personal use under circumstances not falling within the fair use provisions of the Copyright Act is granted by ASCE to libraries and other users registered with the Copyright Clearance Center (CCC) Transactional Reporting Service, provided that the base fee of $8.00 per article plus $.50 per page is paid directly to CCC, 222 Rosewood Drive, Danvers, MA 01923. The identification for ASCE Books is 0-7844-0421-6/99/ $8.00 + $.50 per page. Requests for special permission or bulk copying should be addressed to Permissions & Copyright Dept., ASCE.

ABSTRACT

This proceedings, Structural Engineering in the 21st Century, contains the papers presented at the 1999 New Orleans Structures Congress held on April 18-21, 1999. Papers are listed by session and sessions are grouped by theme. The themes are: (1) Seismic Effects, (2) Concrete and Masonry Structures, (3) Bridges, (4) Structural Analysis, (5) Structural Design, (6) Wood Engineering, (7) Structural Instrumentation, (8)Steel Structures, (9) Construction, (10) Special Structures, (11) Wind Engineering, and (12)Transmission and Telecommunication Structures. The goal of the congress is to cover the broad range of structural engineering as we move into the 21st century. Topics range from the latest research developments to practical applications of structural engineering principles.

PREFACE

Papers from the 1999 New Orleans Structures Congress are included in this proceedings. There are 69 sessions and three plenary sessions. Sessions address the broad range of structural engineering and range from the latest research developments to practical applications.

We would like to thank the conference organizing committee, sessions organizers, session moderators and authors for their contributions to the conference. The guidance of the Executive committee of the Structural Engineering Institute and the staff at ASCE headquarters is gratefully acknowledged.

Finally, we wish each participant a successful conference experience and hope that you enjoy both the congress and the city of New Orleans.

R. Richard Avent
Mohamed Alawady
Proceedings Editors

Steering Committee

R. Richard Avent
Chairman
Louisiana State University
Baton Rouge, Louisiana

Mohamed Alawady*
Louisiana State University
Baton Rouge, Louisiana

Jasbir S. Arora
The University of Iowa
Iowa City, Iowa

James T. Baylot
U.S. Army Waterways
Experiment Station
Vicksburg, Mississippi

C. Dale Buckner
VMI
Lexington, Virginia

Franklin Y. Cheng
University of Missouri
Rolla, Missouri

James Danner*
Denson Engineers
New Orleans, Louisiana

David Darwin
University of Kansas
Lawrence, Kansas

Om Dixit*
Burk-Kleinpeter, Inc.
New Orleans, Louisiana

Michael H. Dooley*
Sigma Consulting Group
Baton Rouge, Louisiana

W. Sam Easterling
Virginia Tech
Blacksburg, Virginia

Mohamed Elgaaly
Drexel University
Philadelphia, Pennsylvania

Gregory L. Fenves
University of California
Berkley, California

Vijaya K.A. Gopu
Louisiana State University
Baton Rouge, Louisiana

Tim Hogue
Oklahoma State University
Stillwater, Oklahoma

Loren D. Lutes
Texas A & M
College Station, Texas

Norma Jean Mattai*
University of New Orleans
New Orleans, Louisiana

Amir Mirmiran
University of Central
Florida
Orlando, Florida

Gary Parks
Bonneville Power
Administration
Redmond, Oregon

Charles W. Roeder
University of Washington
Seattle, Washington

David V. Rosowsky
Clemson University
Clemson, South Carolina

Richard J. Schmidt
University of Wyoming
Laramie, Wyoming

Art Schultz
University of Minnesota
Minneapolis, Minnesota

George Voyiadjis*
Louisiana State University
Baton Rouge, Louisiana

Patrick Zuraski
U.S. Air Force Academy
Colorado Springs,Colorado

**Local Planning
Committee*

Contents

Wood Engineering

Special Structures

Construction

Indexes

Plenary Sessions

Manuscripts for some presentations were not available at time of publication.

1. *STRUCTURAL ENGINEERING: MANAGING THE RISK*

Lawrence Griffis
Walter P. Moore & Associates, Inc.
Houston, TX

2. *FATIGUE AND FRACTURE BEHAVIOR OF STEEL STRUCTURES:*
THE IMPORTANCE OF DETAILS

Stanley K. Rolfe
University of Kansas
Lawrence, KS

3. *NEW ENGINEERED TIMBER STRUCTURES*

Julius Natterer
Bois Consult Natterer SA
Etoy, Switzerland

Fracture and Fatigue Behavior of Steel Structures - The Importance of Details
Stanley T. Rolfe, Fellow, ASCE[1]

The importance of proper design of structural details to prevent fracture and fatigue problems in steel structures often is not appreciated. Fracture problems in steel structures originally were viewed as a materials problem to be solved by metallurgists. However, this is just not so. Fracture and fatigue problems in steel structures never have been just material problems; they are a result of a combination of factors including loading, design, fabrication, and material properties. Thus, prevention of fracture and fatigue problems depends on the structural engineer using the appropriate design, fabrication, and materials selection for a given loading condition.

In the case of the World War II ship brittle fractures, the initial solution to the fracture problem was to pay greater attention to the structural details and the fabrication. In the 1950's and 60's, as fracture mechanics was being developed, emphasis on material behavior became the primary factor in fracture control. However, several structural failures in the 1960's and early 70's re-emphasized that factors other than the materials were extremely important in assessing, understanding, and preventing brittle fractures in steel structures. In particular, a brittle fracture occurred in the Ingram Barge due to overload at a highly constrained detail even though the steel exhibited a Charpy V-notch impact value of 55 ft-lb at the service temperature.

Analysis of the Point Pleasant and Bryte Bend Bridge fractures as well as others, lead to the development of the *AASHTO Fracture Control Plan* in the mid 70's. In this plan, the steel bridge industry put reasonable controls on all aspects of structural behavior, namely, loading, materials, design, and fabrication. The details were improved and allowable stress ranges established for fatigue loading.

---

[1]Stanley T. Rolfe, A. P. Learned Professor of Civil and Environmental Engineering, The University of Kansas, 2006 Learned Hall, Lawrence, KS 66044-2225

These actions, along with moderate notch toughness requirements, have led to safer structures and the absence of major brittle fracture or fatigue problems in steel bridges. There have been out-of-plane distortion problems from secondary loading in bridges, but these have been more local fatigue problems than critical fracture problems.

In the late 1970's and early 80's, the Kemper Arena and the Hyatt Regency Hotel in Kansas City had major structural failures. In the Kemper Arena, a detail was subjected to fatigue loading which was not anticipated during design. In the Hyatt Regency failure, there were several factors including poor detail design, a mis-use of the laws of statics, and poor communication between the structural engineer and the fabricator. This failure led to the creation of the ASCE document, "Quality in the Constructed Product - A Guide for Owners, Designers and Constructors." In that document the design professional, i.e., the structural engineer, is the one who retains responsibility for the performance of the structure, including the behavior of the connection details. Hence the structural engineer must be more aware of the importance of details in preventing fracture and fatigue failures in structures.

In the 1990's, the TransAlaskan Pipeline Service (TAPS) had cracking problems in oil tankers and the Coast Guard established a special committee to recommend a solution. Once again, it was certain structural details that developed fatigue cracks. A fracture control plan involving all of the contributing factors was developed by the special committee. Considerations of the structural details, fabrication, materials, inspection, and consideration of the loading resulted in an overall solution to this situation. The National Oceanographic Atmospheric Administration (NOAA) wave data were used to predict stress histograms to analyze the fatigue behavior. This fatigue analysis resulted in theoretical stress ranges greater than were measured during actual ship voyages. In this situation, the concept of voyage planning, i.e., simply avoiding the severe storms (although this increased the time for a journey), resulted in a marked reduction in the fatigue loading of the ships.

In 1994, the Northridge earthquake

> "dramatically demonstrated that the pre-qualified, welded beam-to-column moment connections commonly used in the construction of welded steel moment resisting frames (WSMF) in the period 1965-1995, were much more susceptible than previously thought."
> (SAC Report 96-02)

In these connection fractures, there were many factors that are still being studied by the engineering community. However, earthquake loading is one of the most severe loadings to which a building can be subjected. Accordingly, this loading should have resulted in much closer control on design, fabrication, materials, and inspection than was done by that industry. The designs were based on inelastic behavior (whereas most structures are designed for elastic behavior), and resulted in large members, severely constrained details that were difficult to fabricate, no notch toughness specifications, and no upper-bound strength requirements to insure the strong column - weak beam design concept. However, the behavior of the Northbridge moment connections should not have been a surprise to the structural engineering community on the basis of what has been known for years about fracture and fatigue control in other industries.

Constraint can be a very large problem in structural design as has been evidenced in many structural failures. Thus, proper attention to the effect of constraint on the behavior of details should become a significant part of the structural design process. In addition, proper fabrication and inspection must be adhered to, and the existing American Welding Society (AWS) recommended practices should be followed.

Strength, stiffness and ductility generally are the controlling material parameters specified by the structural engineer. However, when the loading is very severe, such as is the case for earthquakes, then the structural engineer must insure that the design details are compatible with the ductility requirements, i.e. minimize the constraint and severe tri-axial states-of-stress. In addition, the structural engineer should consider the need for an additional material property, namely some level of fracture toughness, to help insure satisfactory performance for inelastic behavior, particularly if the possibility of cracking exists.

In summary, fracture and fatigue control must become a part of the structural design process. The structural engineer is the key person who can assess the possible fracture or fatigue behavior of structural details, not the owner, metallurgist, welding engineer, or fabricator. Indeed, it is the structural engineer, as the design professional referred to in the ASCE document, "Quality in the Constructed Project," who must establish the various performance criteria because most factors that affect the performance of structural connections, including the details, are controlled or can be influenced strongly by the structural engineer.

Seismic Effects

Manuscripts for some presentations were not available at time of publication.

1. *RECENT ADVANCES ON THE TORSIONAL EFFECTS OF STRUCTURES*

MODERATOR: Rakesh K. Goel,
California Polytechnic State University
San Luis Obispo, CA

(1) Prediction of Nonlinear Dynamic Response for Asymmetric Building Structures With the Aid of Push-Over Analysis
Vojko Kilar,
University of Ljubljana,
Ljubljana, Slovenia

(2) Inelastic torsional response of simple structures
Jon Sfura, Douglas Foutch
University of Illinois
Champaign, IL

John R. Hayes
U.S. Army Construction Research
Urbana, IL

(3) Torsion in symmetric structures isolated with FPS
J. C. De la Llera and J. L. Almazan,
Pontificia Universidad Catolica da Chile,
Santiago, Chile

(4) Seismic control of Asymmetric Structures
 Rakesh K. Goel,
California Polytechnic State University,
San Luis Obispo, CA

10. *INNOVATIVE SEISMIC RETROFITS OF BUILDINGS*

MODERATOR: John Hayes
US Army Construction Engineering Research Laboratories
Champaign, IL

(1) Innovative Seismic Strengthening - Meeting the Client's Needs
Ted A. Pruess
EQE International
Maryland Heights, MO

(2) Seismic Upgrade of the UCLA Knudsen Hall with Viscous Damping Devices
Farzad Naeim
John A. Martin and Associates, Inc.
Los Angeles, CA

(3) Preliminary Study/Design Concepts for Seismic Retrofit of Stacy Park Water Storage Reservoir
Duane Siegfried and Amir A. Arab
Horner & Shifrin, Inc.
St. Louis, MO

(4) Application of Emerging Seismic Analysis Techniques and Innovative Concepts to Retrofit an Existing High-rise Building
Saiful Islam and Balram Gupta
SAIFUL/BOUQUET, Inc.
Pasadena, CA

(5) National Civil Rights Museum-Exhibit Space Expansion Seismic Retrofit
Richard W. Howe
Stanley D. Lindsey & Associates
Nashville, TN

19. *INNOVATIVE SEISMIC RETROFITS OF BRIDGES IN MID-AMERICA*

MODERATOR: John Hayes
US Army Construction Engineering Laboratories
Champaign, IL

(1) Seismic Retrofit of Elevated Roadways, Part 1
Howard Hill
Wiss, Janney, Elstner Associates, Inc.
Northbrook, IL

(2) Seismic Retrofit of the Poplar Street Interchange in East St. Louis, IL, Part II
Neil Hawkins
University of Illinois at Urbana-Champaign
Urbana, IL

Iraj Kaspar and M. Karshenas
Illinois Department of Transportation
Springfield, IL

(3) Seismic Retrofit of the Poplar Street Bridge
Mark Capron
Sverdrup Civil, Inc.
Maryland Heights, MD

(4) Seismic Retrofit of the I-40 Hernando DeSoto Bridge over the Mississippi River in Memphis, TN
Roy A. Imbsen and Dennis D. Pecchia
Imbsen & Associates, Inc.
Sacramento, CA

**28. *JAPAN-U.S.-CHINA INTERNATIONAL STUDY ON THE
MITIGATION OF EARTHQUAKE HAZARDS***

MODERATOR: Takayuki Shimazu
Hiroshima University
Hiroshima, Japan

Yan Xiao
University of Southern California
Los Angeles, USA

**(1) Overall Program and Its Implementation Among the Three Countries
(USA-CHN-JPN)**
Takayuki Shimazu
Hiroshima University
Hiroshima, Japan

**(2) Current Approaches in Seismic Measures for Building Structures
(US - side features)**
Yan Xiao
University of Southern California
Los Angeles, CA, USA

**(3) Current Approaches in Seismic Measures For Building Structures
(China - side features)**
Yayong Wang
China Academy for Building Research
Beijing, China

**(4) Current Approaches in Seismic Measures for Building Structures
(Japan - side Features)**
Takayuki Shimazu
Hiroshima University
Hiroshima, Japan

**Overall Discussion on Seismic Measures of All the Three Countries
(Free discussion including questions and answers)**
Takayuki Shimazu
Hiroshima University
Hiroshima, Japan

Yan Xiao
University of Southern California
Los Angeles, CA, USA

37. *SEISMIC STRUCTURAL CONTROL WITH SOIL-STRUCTURE INTERACTION*

MODERATOR: Franklin Y. Cheng
University of Missouri-Rolla
Rolla, MO

(1) State-of-the-Art of Control Systems and SSI Effect on Active Control Structures with Embedded Foundation
Franklin Y. Cheng
University of Missouri-Rolla
Rolla, MO

H.P. Jiang
Trus Joist MacMillan
Greenwood Village, CO

Menglin Lou
Tongji University
Shanghai, China

(2) Seismic Performance of Tuned Mass Dampers in reducing the Response of Structures on Soft Soil Medium
Genda Chen and Jingning Wu
University of Missouri-Rolla
Rolla, MO

Menglin Lou
Tongji University
Shanghai, China

(3) Effect of Soil-Structure Interaction on TMD Control for Wind and Seismic Response of Structures
Menglin Lou
Tongji University
Shanghai, China

Jingning Wu and Franklin Y.Cheng
University of Missouri-Rolla
Rolla, MO

(4) Considerations on Soil-Structure Interaction Effects on Vibration Control of Seismically Excited Structures
Raimondo Betti
Columbia University
New York, NY

46. *EFFECTS OF GROUND MOTION CONTENT ON STRUCTURAL RESPONSE*

MODERATOR: Gregory A. MacRae
University of Washington
Seattle, WA

(1) Effects of Near-Fault Ground Motions on the Response of Flexible Structures
Babak Alavi and Helmut Krawinkler
Stanford University
Stanford, CA

**(2) Effects of Vertical Content of Groung Shaking on 2-D and 3-D Inelastically
Responding Steel Frames**
Gregory A. MacRae, Joshua Mattheis
University of Washington
Seattle, WA

David Fields
Skilling Ward Magnusson and Barkshire Inc.
Seattle, WA

(3) Near-Source Effects and the Isolation Provisions of the 1997 UBC
John F. Hall
California Institute of Technology
Pasadena, CA

Keri Ryan
University of California
Berkeley, CA

**(4) Importance of Impulse Ground Motions on Performance-Based Engineering:
Historical and Critical Review of Major Issues and Future Directions**
Vitelmo V. Bertero and Mehrdat Sasani
University of California
Berkeley, CA

James C. Anderson
University of Southern California
Los Angeles, CA

55. RECENT CASE STUDIES IN SEISMIC ISOLATION

MODERATOR: Jagtar S. Khinda
New York City Department of Transportation
New York, NY

(1) Current Progress and Future Challenges in the Application of Seismic Isolation
Ronald L. Mayes, Martin R. Button, and Amarnath Kasalanati
Dynamic Isolation Systems, Inc.
Laffayette, CA

**(2) Seismic Isolation Design of the New International Terminal at San Francisco
International Airport**
Anoop S. Mokha, Peter L. Lee, Xiaofong Wang, and Peter Yu
Skidmore, Owings, & Merrill, LLP
San Francisco, CA

(3) Seismic Isolation of Bridges Using Friction Pendulum Bearings
Victor A. Zayas and Stanley S. Low
Earthquake Protection Systems, Inc.
Richmond, CA

(4) Recent Seismic Evaluations of New York City Bridges
Jagtar S. Khinda, Jay A. Patel, and K. Kishore
New York City Department of Transportation
New York, NY

Feng-Bao Lin
Polytechnic University
Brooklyn, NY

Nonlinear Seismic Response of Asymmetric Systems

Jon Fredric Sfura[1]
Dr. John R. Hayes[2]
Dr. Douglas A. Foutch[1]

Abstract

The purpose of this research project is to investigate the nonlinear, inelastic response of one-story, symmetric- and asymmetric-plan systems to biaxial lateral earthquake ground motions. This combined experimental-analytical investigation seeks to characterize the lateral-torsional response of the system for a range of system parameters.

Background

The complete inelastic response of structures is of great interest because collapse is prevented during earthquakes primarily by the dissipation of energy that occurs after yielding. The current strategy employed in designing structures to withstand large seismic ground motions is one in which the structure is sacrificed to the benefit of its occupants; the preservation of human life is the primary goal. As long as the structure doesn't collapse, it has not failed, and this is achieved by dissipating the energy imparted to the structure by the ground through the inelastic, hysteretic behavior of the structure. In structures with low damping, this hysteretic behavior is the primary mechanism of energy dissipation. Thus, understanding the seismic behavior of structures in the inelastic region is the first step in an effective design.

One mechanism that is of particular interest in seismic behavior of structures is torsion. Torsion is, in essence, the twisting of the structure, and when it is coupled with the lateral response, the forces and displacements of the various resisting elements are different from those experienced when the structure responds only in the planar directions. Torsion is the result when the mass distribution and stiffness distribution of a structure are not

---

[1] Department of Civil & Environmental Engineering, 3129 Newmark Lab, 205 N. Mathews, Urbana, IL
[2] US-Army Construction Engineering Research Laboratory, Champaign, IL

symmetric with respect to both planar axes of motion. Due to architectural constraints, buildings are typically designed with the aforementioned asymmetries. However, even when they are designed symmetrically, in actuality, asymmetries are present as a result of the imprecise nature of construction, the variability of the objects and people occupying the structure, etc. Thus, lateral-torsional coupling, to some degree, is a mechanism always present in the seismic response of a structure.

Therefore, to generate a complete understanding of the response of structures during earthquakes, the inelastic lateral-torsional behavior must be characterized. Initially, the elastic seismic response of structures was studied, and a complete picture of this behavior now exists. However, the elastic response represents an incomplete picture of the overall seismic structural response. As time passed and analysis tools advanced, the focus of study shifted to the inelastic seismic response of structures. A great deal of research has been done in this area, and with good reason; however, until this time, the facilities have not been present in this country to conduct complete three-dimensional earthquake simulations on a structure. Now, this is no longer true; the Triaxial Earthquake and Shock Simulator (TESS) located at USACERL in Champaign, Illinois has full three-dimensional motion capability. An experimental study of this nature is a natural and necessary companion to the analytical work that has been done thus far in the seismic behavior of structures.

Structure

Being that this is the first experimental study of this kind, a simple one-story system is appropriate to lay the groundwork for further experimental studies of more complex structures. The structure used in this study is basically a diaphragm, approximately two and a half meters on a side, supported by four circular steel pipe columns, one and a half meters in length, (Figure 1). The diaphragm is composed of four W12x65 beams on the perimeter, with concrete used to fill the void in the center. Eight octagonal and eight rectangular steel masses are available to be attached to the slab in various configurations in order to provide dead load and mass asymmetry. The pipe columns have base- and top-plates welded to the bottoms and tops, respectively, of the columns to provide attachment points to the shaketable and to the diaphragm. The diaphragm is designed to be used throughout the entire sequence of tests, while the columns, having plastically deformed, will be replaced after each model configuration sequence is complete. Circular pipe columns were chosen because of their complete plan symmetry and their reduced sensitivity to torsional buckling and, thus, greater stability. The focus of this study is the lateral-torsional behavior of the system as a whole, not the local behavior of the columns; non-circular columns would only add unnecessary complexity to the analysis of the system.

Seven different model configurations are studied (Figure 2). The configurations consist of different combinations of varying parameters: one-quarter asymmetric mass, one-half asymmetric mass, symmetric mass, asymmetric and symmetric column strength, and concentric lateral bracing. The mass asymmetries are achieved simply by placing the masses all on one-quarter and one-half of the diaphragm. In both the asymmetrical and symmetrical cases, half of the masses are attached above and half below the diaphragm to keep the vertical mass center reasonably near the center of the diaphragm. Three different types of pipe columns are used: four-inch extra strong, five-inch standard, and four-inch double extra strong. The strength asymmetry in the structure is achieved by pairing two sets of columns which have similar stiffness, but different yield strengths, which in this case are the five-inch standard and four-inch double extra strong columns. The stiffness asymmetry is achieved through the use of asymmetrical lateral bracing. Brackets were designed to allow the attachment of lateral bracing to the structure. The inclusion of each of these different types of asymmetry will allow the scope of the results to be more general in nature.

Experimental Work

Following construction, the model was placed on the shaking table for the simulations. Approximately 72 channels of data were recorded in each test. Response quantities measured included accelerations and displacements of the table and slab in both planar directions, and strains in the columns.

The test sequences for each model configuration were nearly identical. Preliminary tests were performed before any earthquake simulations in order to determine the natural frequencies and damping characteristics of the structure. First, the structure was subjected to white noise base motions, in order to determine the uncoupled torsional and lateral frequencies. Second, the structure was excited sinusoidally at its measured natural frequency and allowed to decay to rest, in order to determine the equivalent viscous-damping ratio. Third, the structure was subjected to sinusoidal motions ranging in frequency from 1-20 Hertz. The response of the structure to the sine sweep tests was used to confirm the natural frequencies of the structure. Each of the aforementioned tests were performed once in each planar axis, and once in the yaw-axis, for a total of nine preliminary tests for each model configuration.

Following the preliminary tests, the structure was subjected to earthquake simulations. The 230 [X] and 140 [Y] degree acceleration components from the 1979 Imperial Valley earthquake recorded at Bonds Corner were chosen as the base earthquake motions based on the large response spectrum magnitudes near the natural frequency of the structure. The goal in choosing the motions was to achieve a ductility in the response of the structure in the neighborhood of four to five.

The structure was first subjected to low-level earthquake tests, typically at 10-25 percent of the reference. The low-level tests were performed using first the x-axis input motion, then the y-axis input motion, followed by both axes simultaneously. Next, the structure was subjected to the full-scale reference accelerograms, followed by white noise tests to analyze any changes in the natural frequencies of the structure. This general sequence of preliminary tests and earthquake simulations was followed for each of the seven different model configurations.

Acknowledgments

This study is supported by a grant from the US-Army Construction Engineering Research Laboratory, Champaign, IL. All findings, opinions and conclusions expressed in this paper are those of the authors and do not necessarily represent those of the sponsors.

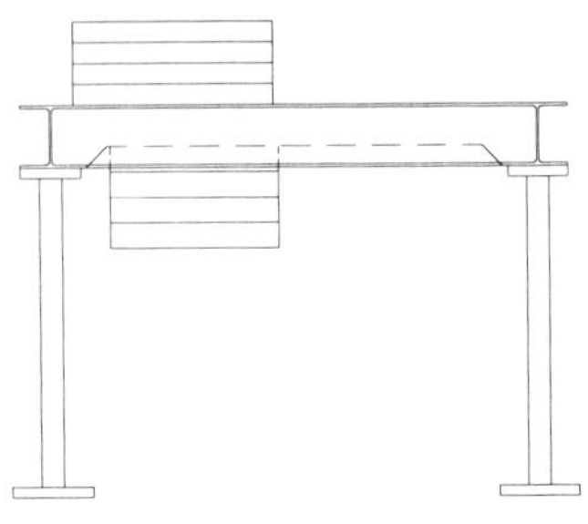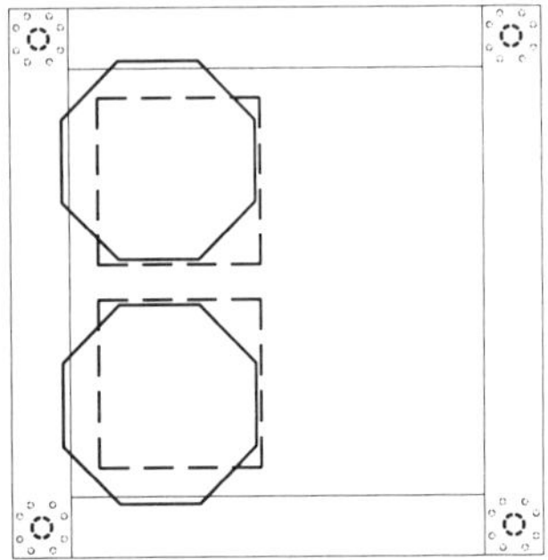

Figure 1. Plan View and Elevation View of Test Structure - Configuration 4 (Typical)

Test Configuration	Input Motion	Mass Distribution	Stiffness Distribution	Strength Distribution	Notes
1	Uniaxial	Symmetric	Symmetric	Symmetric	4 - 4" Extra Strong Columns
2	Biaxial	Symmetric	Symmetric	Symmetric	4 - 4" Extra Strong Columns
3	Biaxial	1/4 Asymmetric	Symmetric	Symmetric	4 - 4" Extra Strong Columns
4	Biaxial	1/2 Asymmetric	Symmetric	Symmetric	4 - 4" Extra Strong Columns
5	Biaxial	1/2 Asymmetric	Symmetric	Symmetric [X] Asymmetric [Y]	2 - 5" Standard Columns 2 - 4" Double Extra Strong Columns
6	Biaxial	1/2 Asymmetric	Symmetric [X] Asymmetric [Y]	Symmetric	4 - 4" Extra Strong Columns Asymmetric Lateral Bracing [Y]
7	Biaxial	1/2 Asymmetric	Symmetric [X] Asymmetric [Y]	Symmetric [X] Asymmetric [Y]	2 - 5" Standard Columns 2 - 4" Double Extra Strong Columns Asymmetric Lateral Bracing [Y]

Figure 2. Test Configuration Sequence

Seismic Control of Asymmetric Structures

Rakesh K. Goel[1], M.ASCE

Abstract

This paper presents results of an investigation on how supplemental viscous damping can be used to control excessive deformations in asymmetric-plan buildings. It is shown that symmetric distribution of supplemental damping devices in the building plan is not necessarily the best way to control excessive deformations in an asymmetric-plan building; a value of the damping eccentricity equal to the structural eccentricity in magnitude but opposite in algebraic sign leads to higher reduction.

Introduction

The performance of structures during past earthquakes has shown that asymmetric-plan buildings are especially vulnerable to earthquake damage. Therefore, numerous investigations in the past have investigated the earthquake behavior of asymmetric-plan buildings. As a result, procedures to account for undesirable effects of plan asymmetry, such as increased force and ductility demands on lateral load-resisting elements, have been developed and incorporated into seismic codes of many countries. However, there remains a need for additional research to develop techniques that will control excessive earthquake-induced deformations in asymmetric-plan buildings.

Although, the control of earthquake-induced vibrations in symmetric-plan buildings through the use of supplemental damping has been a subject of numerous recent studies (e.g., Aiken and Kelly, 1990; Hanson, 1993; Reinhorn et. al., 1995), there has been a lack of efforts toward developing a fundamental understanding of how these devices and their plan-wise distribution influence the lateral-torsional coupling in asymmetric-plan systems. Therefore, the objectives of this investigation were to (1) to identify the system parameters that control the seismic response of

---

[1] Assistant Professor, Department of Civil and Environmental Engineering, California Polytechnic State University, San Luis Obispo, CA 93047.

asymmetric-plan buildings with fluid viscous dampers; and (2) to investigate the effects of the controlling parameters on edge deformations in asymmetric-plan buildings. This paper focuses on the effects of one of the most important parameters related to the supplemental damping, namely the damping eccentricity.

System and Ground Motion

The system considered was the idealized one-story building of Figure 1 consisting of a rigid deck supported by structural elements (wall, columns, moment-frames, braced-frames, etc.) in each of the two orthogonal directions. Supplemental damping is provided by fluid viscous dampers in the building's bracing system. The mass is assumed to be distributed symmetrically in the system plan. Therefore, center of mass (CM) is located at the geometric centroid of the plan area. The stiffness eccentricity is denoted by the distance, e, between the CM and the center of rigidity (CR), and the damping eccentricity is denoted by the distance, e_{sd}, between the CM and the center of supplemental damping (CSD).

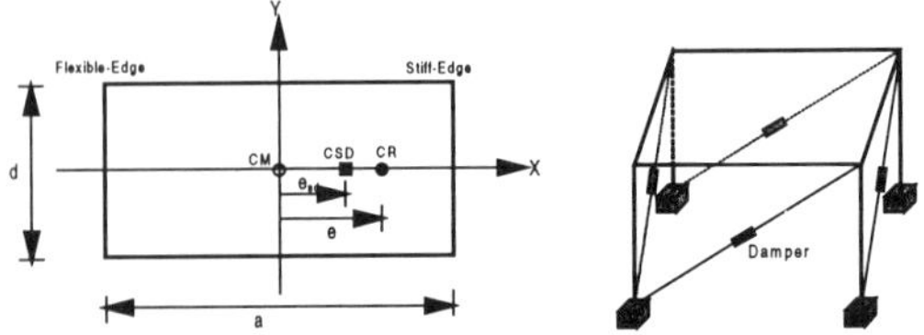

Figure 1. System considered.

The ground motion considered is the North-South (360°) component recorded at the Sylmar County Hospital parking lot during the 1994 Northridge earthquake. The peak values of the ground acceleration, velocity, and displacement recorded at this site were $826.8\,\mathrm{cm/s^2}$, $128.9\,\mathrm{cm/s}$, and $32.55\,\mathrm{cm}$, respectively.

System Parameters and Response Quantities

The parameters that control the system response are (Goel, 1998): (1) transverse vibration period, $T_y = 2\pi/\omega_y$ (ω_y = transverse vibration frequency) of the reference symmetric building; (2) normalized stiffness eccentricity, $\bar{e}$; (3) ratio of the torsional and transverse frequencies, Ω_θ; (4) aspect ratio, α; (5) mass and stiffness proportional constants, a_0 and a_1, which in turn depend on the natural damping ratio in the two vibration modes of the system; (6) supplemental damping ratio, ζ_{sd}; (7) normalized supplemental damping eccentricity, $\bar{e}_{sd}$; and (8) normalized supplemental damping radius of gyration, $\bar{\rho}_{sd}$. In this study, the following system parameters were fixed: $\Omega_\theta = 1$, $\bar{e} = 0.2$, $\alpha = 2$, and constants a_0 and a_1 were selected such that $\zeta_1 = \zeta_2 = \zeta = 5\%$.

The response quantities selected in this investigation were the deformations of the flexible and stiff edges in asymmetric-plan building normalized by the deformation of the reference symmetric building, $\bar{u}_f = u_f \div u_o$ and $\bar{u}_s = u_s \div u_o$. A value of the normalized edge deformation by more than one indicates a larger edge deformation in the asymmetric-plan building as compared to the reference symmetric building; conversely, a value of normalized edge deformation smaller than one implies a smaller edge deformation in the asymmetric-plan building.

Effects of Supplemental Damping

This paper summarizes the effects of one of the three important system parameters related to the supplemental damping, namely the damping eccentricity, $\bar{e}_{sd}$. Presented in Figure 2 are the normalized edge deformations computed for a range of $\bar{e}_{sd}$ values between −0.5 and 0.5 for buildings with $T_y = 1$ s; the extreme values of $\bar{e}_{sd} = -0.5$ and 0.5 correspond to all dampers located either at the flexible or at the stiff edge, respectively.

The presented results (Figure 2) show that the supplemental damping eccentricity significantly affects the edge deformations. In particular, deformation of the flexible edge decreases and that of the stiff edge increases as $\bar{e}_{sd}$ varies from 0.5 to −0.5, i.e., the CSD moves from the right to the left of the building plan (Figure 1). These results also show that $\bar{u}_f$ is the smallest for $\bar{e}_{sd} = -0.5$ indicating that the largest reduction in deformation of the flexible edge would be obtained by concentrating all dampers at the flexible edge. The stiff edge deformation, on the other hand, is the smallest for $\bar{e}_{sd} = 0.5$, implying that the largest reduction would be obtained by locating all dampers at the stiff edge. Furthermore, asymmetric distribution of supplemental damping results in much higher reduction in the edge deformation compared to the symmetric distribution as apparent from deformation of the flexible edge being much smaller for $\bar{e}_{sd} = -0.5$ compared to $\bar{e}_{sd} = 0$.

It is apparent from the presented results that the same distribution of dampers does not lead to the most reduction in deformations of both edges: the distribution that results in the largest reduction in the flexible edge deformation leads to the smallest reduction in the stiff edge deformation and vice versa. For asymmetric-plan buildings, the flexible edge is generally the most critical edge because of higher earthquake-induced deformations. Therefore, dampers should be distributed such that the CSD is as far away from the CM, on the side opposite to the CR, as physically possible -- a distribution that leads to the largest reduction in deformation of the flexible edge. Although this distribution does not lead to the largest possible reduction in deformation of the stiff edge, it none the less reduces deformations as compared to deformation of the same edge in buildings without dampers.

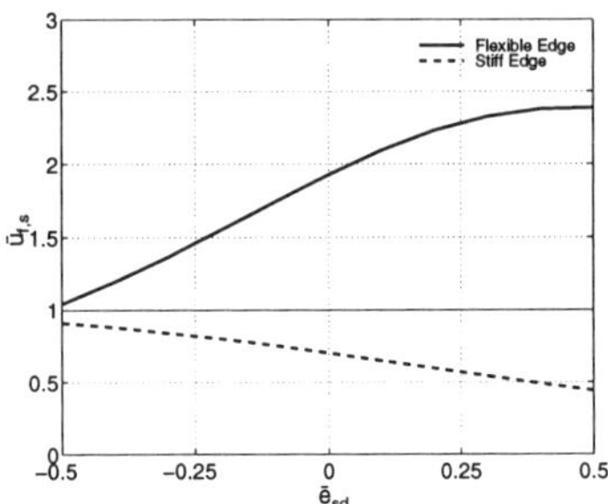

Figure 2. Response of asymmetric-plan buildings with supplemental damping: $\bar{e} = 0.2; \Omega_\theta = 1; \alpha = 2; \zeta = 5\%; T_y = 1\,\text{s}$; and $\zeta_{sd} = 10\%$.

Conclusions

It is shown that the response reduction in asymmetric buildings strongly depends on the plan-wise distribution of the supplemental damping. In particular, it was found that (1) asymmetric distribution of the supplemental damping led to a higher reduction in edge deformations as compared to symmetric distribution; and (2) the largest reduction in the critical edge, i.e., flexible edge, deformation occurred when the CSD was as far away as physically possible from the CM and on the side opposite to CR.

References

Aiken, I. D., and J. M. Kelly. (1990). Earthquake Simulator Testing and Analytical Studies of Two Energy-Absorbing Systems for Multistory Structures, *Report UCB/EERC-90/03*, Earthquake Engineering Research Center, University of California, Berkeley, CA.

Goel, R. K. (1998). Effects of Supplemental Viscous Damping on Seismic Response of Asymmetric Systems, *Earthquake Engineering and Structural Dynamics*, 27, 125-141.

Hanson, R. D. (1993). Supplemental Damping for Improved Seismic Performance, *Earthquake Spectra*, 9(3), 319-334.

Reinhorn, A. M., C. Li, and M. C. Constantinou. (1995). Experimental and Analytical Investigation of Seismic Retrofit of Structures with Supplemental Damping: Part 1 - Fluid Viscous Damping Devices, *Report No. NCEER-95-0001*, National Center for Earthquake Engineering Research, Buffalo, NY.

Innovative Seismic Strengthening – Meeting the Client's Needs

By Theodore A. Pruess[1], P.E., Member ASCE
& Elizabeth M. Dausman[2], Member ASCE

Abstract: Innovative seismic strengthening can be defined in several ways, depending on your perspective. As consulting engineers, we will examine the concept of innovation from the viewpoint of the client. The client's goals are achieving the expected performance with respect to his constraints. Case studies will be presented to illustrate this concept of innovative seismic strengthening.

Introduction:

Innovative seismic strengthening design can be defined in several ways, depending on your perspective. The researcher often equates innovation with state-of-the-art high-tech concepts. The structural engineer often thinks of innovation as providing the client with the best design for the available professional fee. The client in a seismic strengthening project usually doesn't think in terms of innovation. His primary goal is to pay for a project which will achieve his performance expectations within his constraints.

We will examine innovative design with the client's perspective in mind. The client in a seismic strengthening project usually has a number of constraints which are imposed on the design team. The client's constraints include:

* Budget – There is only a finite amount of money to achieve the goal
* Operations – Often the facility remains in operation during the construction and locations for construction are limited
* Architectural – The strengthening solution must accommodate the architectural layout

---

[1] Project Manager, EQE-Theiss, St. Louis, MO
[2] Lead Engineer, EQE-Theiss, St.Louis, MO

❖ Construction Methods – Limitations on noise, vibration, dust, water, and construction locations

These constraints drive the engineering solution. The definition of innovative seismic strengthening which we propose is as follows: The structural engineer uses his experience, judgement, and creativity to provide the client with the best design which meets both the client's performance expectations and constraints. This definition will usually preclude the use of the state-of-the-art solutions because of the budget constraint.

Several case studies are presented to examine how this definition of innovative seismic strengthening applies to real projects.

Bevo Building

The Bevo building was constructed in 1917. The structure is on the National Register of Historic Places, and is a landmark building. The structure is an eight story reinforced concrete frame building with a brick facade. Expansion joints separate the building into three structures. The resulting composite building, which is approximately 602 feet long at its west side and 542 feet long at its east side, has a skewed south wall and is approximately 252 feet wide. There is a high bay area in the upper levels of the building. The building encloses approximately 1.1 million square feet. The building houses bottling and packaging operations for the brewery. It is the largest multi-story bottling plant in the world.

The client's performance objective was to minimize disruption to the facility's operations after the design level seismic event. The motive was to minimize business disruptions.

The vulnerability analysis revealed several significant deficiencies in the building's lateral force-resisting system. A large number of columns and beam-column joints at the 6th and 7th floor mezzanines, and at the roof were determined to require strengthening.

The client's constraints included:
❖ The facility operates 24 hours a day, and must remain in operation during construction of the strengthening elements
❖ No debris, including water and dust, could fall on the bottling and packaging processes
Engineering challenges included attaching to the old concrete columns with spiral reinforcing and weak concrete.

The following strengthening solutions were considered:
- ❖ Base Isolation
  - ➢ Advantages:
    - ▪ Minimize strengthening on the upper levels of the building used for the bottling and packaging processes
  - ➢ Disadvantages:
    - ▪ Cost – approximately double the conventional strengthening
    - ▪ Life-cycle considerations of isolators
- ❖ Concrete Shear Walls
  - ➢ Advantages:
    - ▪ Compatible with existing construction
  - ➢ Disadvantages:
    - ▪ Disruption of operations during construction and after completion
- ❖ Steel Braced Frames
  - ➢ Advantages:
    - ▪ Least disruptive during and after construction
  - ➢ Disadvantages
    - ▪ Connection to existing concrete construction

The final design was driven by the cost and operations constraints. This design used steel bracing which was installed while the bottling and packaging processes remained in operation. The connection between the steel bracing and concrete columns was accomplished using steel collars fabricated to surround the columns. This solution minimized the interference between new anchors and the spiral reinforcing and avoided disturbing the weak concrete in some areas of the existing columns.

Southeast Missouri Central Dial Offices

The local telephone utility had a network of small central dial offices throughout southeast Missouri. Many of these buildings were within the boundaries of the New Madrid Seismic Zone. These buildings were typically one story unreinforced masonry bearing wall buildings with steel joist roofs. The typical footprint was 2000 to 5000 square feet.

The owner's performance expectation was continued operations immediately after the design level seismic event. The motive was that these facilities would be required for the post-earthquake recovery effort.

The owner's constraints included:
- ❖ The facility operates 24 hours a day, and must remain in operation during construction of the strengthening elements
- ❖ The telephone equipment was sensitive to vibration and dust

- ❖ The existing roofing could not be removed to strengthen the diaphragm
- ❖ There could be no penetrations through the roof or walls
- ❖ Work inside the building was prohibited
- ❖ Foundation work outside the building was limited by the telephone grounding grid

The following strengthening solutions were considered:
- ❖ Base Isolation
  - ➢ Advantages:
    - Reduce loads on existing URM walls and steel deck diaphragm
  - ➢ Disadvantages:
    - Cost
    - Lead time to obtain isolators
    - Life-cycle considerations of isolators
    - Limitation on work inside building
- ❖ Exterior Concrete Buttresses
  - ➢ Advantages:
    - No work inside building
    - Aesthetics
  - ➢ Disadvantages:
    - Foundation work could disrupt the grounding grid
    - Appearance of the building changed
- ❖ Exterior Steel Braced Frames
  - ➢ Advantages:
    - No work inside building
    - Foundations can avoid grid
  - ➢ Disadvantages
    - Aesthetics / change appearance of building

The final design was driven by the cost and construction limitation constraints. This design used the exterior steel braced frames and was chosen with the approval of the owner's architects.

Application of Emerging Seismic Analysis Techniques and
Innovative Concepts to Retrofit an Existing High-rise Building

Saiful Islam, Ph.D., S.E.[1] and Balram Gupta, Ph.D.[2]

Abstract

This paper discusses the application of emerging seismic analysis techniques to analyze a non-ductile concrete high-rise building in an active seismic region and the use of innovative concepts to strengthen the building. The subject building is one of the three high-rise concrete buildings of a residential/commercial development constructed circa early 1960's. The present study emphasizes the importance of considering significant strength and stiffness degradation exhibited by non-ductile elements under reversed cyclic loading. It is shown that some of the acceptance criterion for individual elements recommended in FEMA-273 (1997) are too stringent and warrant a second look.

Introduction

The traditional approach to seismic evaluation and retrofit is founded upon new building design where limited damage control and collapse prevention objectives are presumably assured through satisfaction of prescriptive provisions of the codes. These provisions include providing the minimum code strength and pre-yield stiffness, and enhancing component and system ductility. It provides little or no insight into the post-yield behavior of the building that actually dictates its survivability in case of a major earthquake. The rational approach to seismic evaluation and rehabilitation of existing buildings requires that the design criteria be performance-based with emphasis on story drifts.

General Building Information

The building considered in this study is one of the three high-rise reinforced concrete buildings of a residential/commercial development constructed circa early 1960's. It has a rectangular plan (66'x274') with 22 stories above grade and one basement. Total building height above the ground level is 218'. The vertical load resisting system consists of pan-joist system supported on RC walls. The specified

---

[1] President, SAIFUL/BOUQUET Inc., Ph. (626) 304 2616, Fax (626) 304 2676, email: saifbouq@sbise.com
[2] Senior Analyst, SAIFUL/BOUQUET Inc., Ph. (626) 304 2616, Fax (626) 304 2676, email: BGupta@sbise.com

concrete strength for all elements except columns is 3,000 psi. For columns, it varies from 4,500 psi at lower levels to 3,000 psi at higher levels. The lateral load resistance is provided exclusively by a series of concrete shear walls. In the transverse direction (EW), the building is braced by ten 6" thick interior walls and a series of 12" thick walls and columns interconnected by 12"x 40"x36" long coupling beams along the north and south faces. Below 3rd floor level, number of shear walls either terminate or offset. These include five walls in western half of the tower. Only the transverse building direction is discussed in this paper.

The subject building possesses certain features that could be critical from seismic standpoint. These include – (a) possible excessive axial force demands on columns supporting transverse walls below 3rd level, (b) possible shear yielding of walls, and (c) short span coupling beams.

Selected Performance Objectives & Element Limit States

The primary performance objective selected for the building was "life-safety" for a seismic hazard having a 10% probability of being exceeded in 50 years. However, damage control was also, initially, a secondary consideration. The performance limit states were selected as – (a) maximum concrete compressive strain limited to 0.003, (b) drift angle for coupling beams less than 1%, (c) maximum inter-story drift less than 1.25% with weak stories avoided, and (d) cracks in beams and shear walls should be repairable using epoxy (secondary consideration).

Analyses Performed

 Linear as well as nonlinear analyses were performed to understand the building behavior. Elastic analyses included 3D response-spectrum analysis and 3D time-history analysis. Nonlinear analyses included 2D pushover analysis as well as 2D time-history analyses (for 3 ground motions – Bonds Corner, Taft, and Tabas). Results of only the nonlinear time-history analyses are presented in this paper. Expected material strengths were used to compute the stiffness and strength properties of various elements. The initial stiffness properties were computed corresponding to the secant stiffness at first significant yield.

Primary objective of performing elastic analyses was to obtain preliminary insight into the expected force and deformation demands on various members. Nonlinear analyses were performed to (a) study the effects of material non-linearity and (b) study the effect of hysteretic degradation on local and global response.

Nonlinear Modeling and Analysis Assumptions

Two levels of material non-linearity (moderate and severe) were considered for time-history analyses. Building's mathematical model included five interior frames and two end frames. P-Δ effects were also included.

Non-linear Time History Analysis Results

The fundamental period of vibration was computed as 1.93 second. A summary of displacement and drift response is presented in Table 1. Figures 1 and 2 show story displacement and inter-story drift envelopes for the existing (unstrengthened) building configuration. The inadequacy of elastic methods to predict inter-story drifts and their location can be readily seen from Figure 2. The excessive drifts at the lower level predicted by inelastic analyses are result of shear yielding of walls at that level.

Table 1. Summary of Displacement & Drift Response

Analysis Type	Ground Motion	Overall Degradation	δ_{roof} (inch)	Max. Drift (%)	
				Value	Bet. Levels
Elastic	Resp. Spectrum	-	17.62	0.87	17 & 16
	Time-History	-	17.25	0.98	18 & 17
Inelastic (Time-History)	Bonds Corner	Moderate	15.58	1.27	4 & 3
		Severe	15.64	1.59	4 & 3
	Taft	Moderate	15.25	1.24	4 & 3
		Severe	15.26	2.70	4 & 3
	Tabas	Moderate	15.53	3.46	4 & 3
		Severe	13.68	5.20	4 & 3

Effects of Degrading Hysteresis

The effect of severe degradation was negligible on the global response (roof displacement) while the same had substantial effect on inter-story drift at lower level (local response). These drifts are much higher than permissible drift of 0.6% per FEMA-273 (1997) and certainly pose a "life-safety" concern. This indicates that, as a minimum, some retrofit is required between 3rd and 4th levels to control inter-story drifts. It was however observed that shear strengthening of only one level was not enough to control the excessive inter-story drifts. Iterative studies had to be carried out to optimize additional shear capacity required, number of levels to be retrofitted, and number of such walls. Shear capacity of existing walls was enhanced by using fiber-wrap. Wall retrofit of total 10 levels was proposed. Drift profile of retrofitted structure is also shown in Figure 2.

Time-history analyses indicated that most of the coupling beams would experience shear yielding. The computed chord rotation demands were found much higher than those permissible per FEMA-273 (1997) indicating life-safety concerns. However, these concerns are not realistic due to two reasons. First, the experimental study of a typical coupling beam suggested satisfactory behavior up to drifts substantially higher than those permitted by FEMA-273 (1997). Second, severe shear cracking of many coupling beams would certainly be a damageability issue but not a life-safety concern. Based on these findings, the initially selected performance levels for coupling beams were reconsidered. Since the number of coupling beams needing repair would be excessive and many would not be repairable using epoxy, it was decided to ignore the contribution of these beams by assuming hinges at their ends. This reduced the overall structure stiffness considerably (fundamental period changed from 1.93 seconds to 2.55 seconds). Retrofit measures were then designed for the new flexible structure.

Most of the columns supporting discontinuous walls were found elastic considering all gravity and seismic demands. Two corner columns on end frames, though, had excessive axial force as well as moment demands and were retrofitted by concrete jacketing.

Conclusions

This study has shown that conventional force-based methods are not adequate for seismic evaluation of buildings. Degrading hysteretic characteristics must be accounted for in the nonlinear analyses. The global effect of severe degradation was found minimal but its local effect was substantial and must not be ignored. Some acceptability criteria for "life-safety" of shear-critical coupling beams, as recommended by FEMA-273 (1997), were found to be too stringent and warrant a second look.

References

FEMA-273, (1997). *NEHRP Guidelines for Seismic Rehabilitation of Buildings.* Building Seismic Safety Council, Washington, D.C.

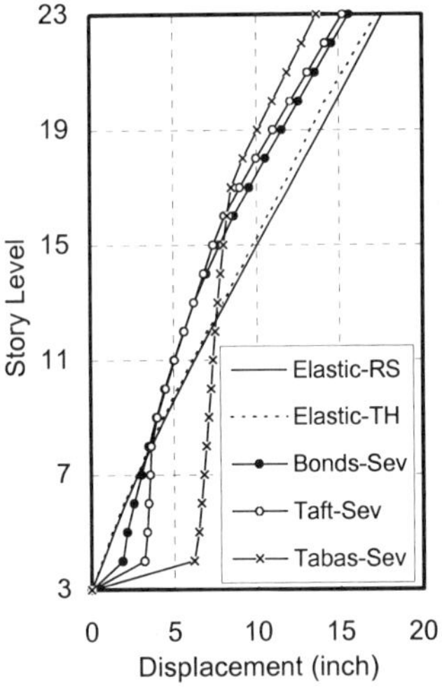

Figure 1. Displacement Profiles

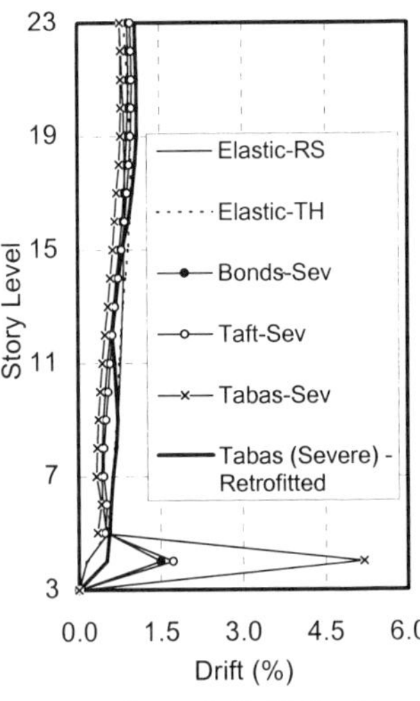

Figure 2. Drift Profiles

National Civil Rights Museum – Exhibit Space Expansion Seismic Retrofit

Richard W. Howe, Stanley D. Lindsey & Associates, Nashville, TN

Background

The Exhibit Space expansion of the National Civil Rights Museum is being designed by Looney Ricks Kiss, Architects, Inc. and associate Self-Tucker, Architects, of Memphis, TN. Exhibit space consultation is being provided by Ralph Applebaum Associates, New York City, NY (designers of The Holocaust Museum in Washington, DC).

The exhibit space expansion will be located in existing historic buildings located in city block adjacent to the existing National Civil Rights Museum in downtown Memphis. The existing and proposed structures feature two stories with a full basement will be connected to the museum by an underground tunnel and adjacent plaza.

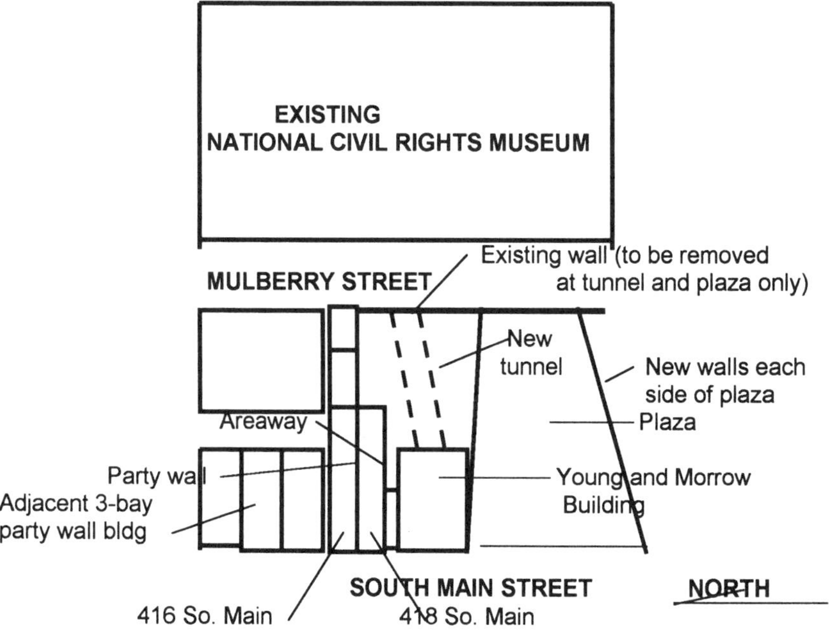

Seismic considerations and challenges

Addressing seismic provisions for the Exhibit Space expansion of the National Civil Rights Museum presents as fascinating challenge to the design team for it requires consideration of a broad spectrum of issues.

- Historic preservation issues from two perspectives. The structure(s) under consideration are in the National Register's South Main Historic District. Internally, certain areas of the structure(s) are of interest from the perspective of the National Civil Rights Museum (i.e., the crime scene comprised of the area in which James Earl Ray slept and from which he shot Martin Luther King). Exterior (historic) walls and certain walls in the crime scene are required to be preserved. Further, general finishes and partition walls in the crime scene area must be preserved in place while establishing new seismic-resistant construction in that area.
- Combination of new and preserved historic construction. The general interior structure of the existing buildings – deteriorated wood frame construction inadequate for public assembly (museum) occupancy - is to be demolished, while retaining the existing historic exterior walls and existing construction in the crime scene area.
- Involvement of multiple structures in the analysis. An existing structure (the Young & Morrow Building) is to be joined to the existing 416/418 South Main Building.
- Impact of multiple ownership. 416/418 South Main is a party wall structure, only half of which is owned by the NCRM and subject to modification and new construction. The remaining half is privately owned and occupied and, while not subject to modifications or new construction, must be preserved and not subject to any potentially adverse impact of the adjacent NCRM project.
- Impact of and on adjacent buildings. 416/418 South Main and the Young & Morrow Building are separated by a ¼-inch separation joint from an adjacent similar URM, 3-bay party wall building.
- Interpretation of applicable building code requirements (i.e., City of Memphis and Shelby County Joint Building Code – SBC 1994). It should be noted that the Code provides discretion to the Building Official regarding the degree of conformance of historic buildings to the Code provisions for new construction.
- Determination of appropriate design standards meeting Code requirements and NCRM objectives. (SBC 1994 and referenced standards – e.g., 1991 NEHRP *Provisions*; current SBC 1997 and referenced standards – again, 1991 NEHRP *Provisions*; or "state-of-the art" standards, such as 1997 NEHRP *Provisions* and/or FEMA 273 *Seismic Rehabilitation of Existing Buildings*)

- Do provisions for "new buildings" or "existing buildings" apply or do both and to what degree? Or is this a "new addition" to an "existing building"? Or should this be considered an "existing building" in spite of comprehensive replacement of primary structure and the building be analyzed as "existing" under the provisions of FEMA 273 and its two-level design provisions.
- Selection of appropriate seismic objectives and levels of ground motion. What is the hazard/risk level for this Central US site and how should it be presented to the NCRM Board for decision making purposes?
- Evaluation of existing wood frame and URM load-bearing construction.
- Unreinforced masonry (URM) wall construction considerations apply. Can existing multiple-wythe URM (brick) walls serve as primary structure or are they to be considered strictly as a URM veneer?
- Involvement of multiple seismic lateral force-resisting schemes/systems (reinforced concrete - shotcrete - shear walls in one primary direction; welded steel moment-resisting frames on story-deep cast-in-place concrete foundation girder walls in the other).
- Application of multiple technologies for strengthening and anchoring URM walls (shotcrete; center coring; polymer/epoxy resin bonding; and/or more conventional mechanical anchoring techniques).
- Relative timing or sequence of property acquisition and development of and commitment to project concept relative to selection and engagement of consultants (specifically, structural).
- And as always, budget constraints. Pursuit of supplemental funding through programs such as FEMA's *Hazard Mitigation Grant Program (HMGP)*.

Structural framing options

A variety of potential structural options were considered ranging from minimum to maximum scope of (structural) work. Preliminary cost estimates were developed for the purpose of comparing options. The potential seismic performance of the existing buildings was determined to be very poor when exposed to ground motions associated with a City of Memphis Code design basis earthquake (Av=0.20 – 10%/50 yrs). Options 1 through 4 provided little or no enhancement in that regard. Estimated cost of viable options (with no improvement in seismic performance) ranged from $150-250,000.

Options 5, 6, and 7 provide progressively improved seismic performance and are presented below. ***Option 7 has been selected in order to meet NRCM seismic objectives.***

<u>Option 5</u> - **Comprehensive replacement of floor and roof systems utilizing existing load-bearing masonry walls as vertical (and lateral) load supporting system. (I.e., save exterior brick walls; rebuild all else.) Results in two separate structures as currently exist with connectors. See following sketch.**

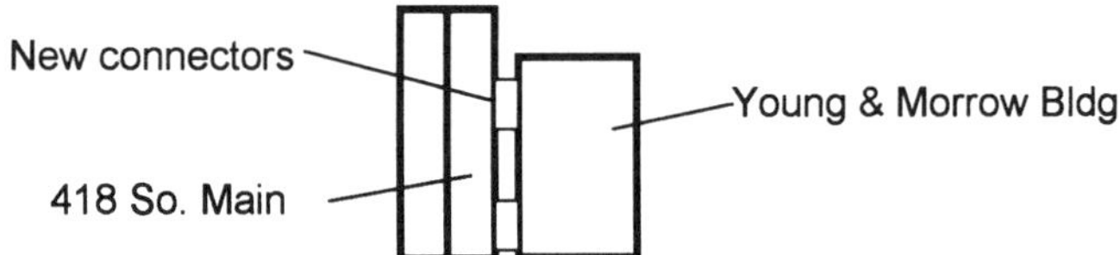

Primary floor and roof replacement framing options include light gage steel systems (C-joists) and open web steel joists (or perhaps steel beams) with concrete on steel deck floor systems. Provisions would be made for effective floor diaphragm load transfer thus introducing some improvement in seismic performance.

Estimated cost: $225-300,000 including temporary extrerior wall stablization.

<u>Option 6</u> - **Comprehensive replacement of floor and roof systems utilizing *project perimeter* existing load-bearing masonry walls as vertical (and lateral) load supporting system. (I.e., save *project perimeter/exterior* brick walls; rebuild all else.) Results in one structure of approximately 7500 sf (footprint) – including 416 South Main.**

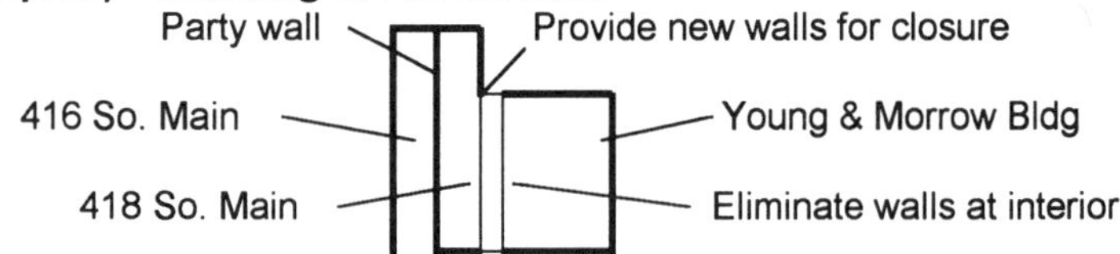

Primary floor and roof replacement framing options as above.

This option will provide somewhat improved seismic performance due to improved floor and roof diaphragms and connections to perimeter shear walls (existing masonry walls). It *will not*, however, meet the intent of Code provisions for new construction due primarily to unreinforced nature of existing masonry walls. Further, existing N-S masonry walls certainly inadequate to accommodate forces from adjacent building (416 South Main) due to party wall (and impact of adjacent 3-bay party wall through grossly inadequate separation joint).

Estimated cost: $225-250,000 including temporary exterior wall stabilization.

<u>Option 7</u> - Comprehensive replacement of floor and roof systems with new structure, including perimeter CMU load-bearing/shear walls. Maintain existing *project perimeter* masonry walls as (historic) exterior brick wall system (regard as veneer only). (I.e., save *project perimeter/exterior* brick walls; rebuild all else.) Results in one structure of approximately 7500 sf (footprint) – including 416 South Main.

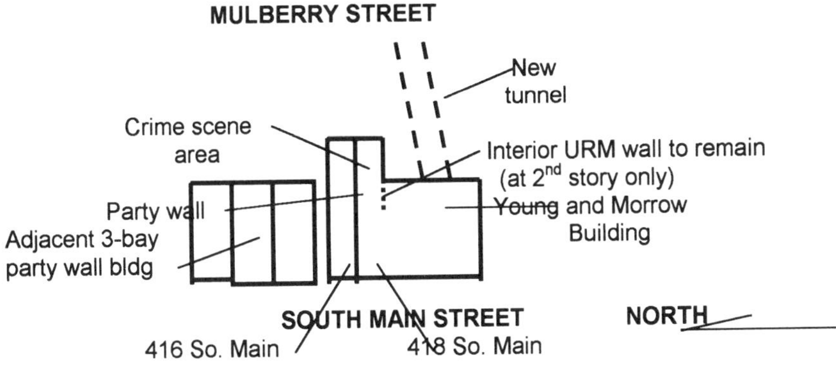

At the north and south walls, applied reinforced concrete walls (cast-in-place or more likely shotcrete) are recommended as load carrying and shear walls as well as providing out-of-plane lateral support for existing URM (brick) walls. Drilled-in masonry veneer retrofit anchors would be used to tie veneer to new structural wall designed to provide lateral support for veneer as well as serving as primary shear walls.

Welded steel moment-resisting frames are recommended at the east and west walls due to extensive openings that must be honored from a historical perspective *as well as the requirement that modifications to 418/Young & Morrow not adversely impact 416 South Main.* The flexibility of the steel MR-frames must emulate the relatively flexible behavior of the existing buildings. All buildings on this city block *possess little stiffness in the N-S direction.* Steel MR-frames will be designed to provide little stiffness up to the point of elastic failure but then prevent collapse of the NCRM expansion component of the city block. Provisions to bond or anchor potentially failed fascia/veneer will be made, potentially through a relatively flexible polymer resin interface.

Primary floor and roof replacement framing options are similar to Option 5.

It is anticipated that this option can be designed to meet City of Memphis Code seismic provisions for DBE ground motion of 0.20 Av (10%/50 criteria).

Estimated cost: $350-450,000

SEISMIC RETROFIT OF ELEVATED ROADWAYS, PART 1

Howard J. Hill[1]

Abstract

In East St. Louis, Illinois, a system of elevated roadways provides access to the Poplar Street Bridge that crosses the Mississippi River. The elevated system was designed in the 1960's with no significant seismic performance requirements. Recent seismic evaluations of the structures indicated that current AASHTO seismic demands would cause extensive damage and long-term loss of use.

After replacing the system's two main roadways, the Illinois Department of Transportation (IDOT) began a project to improve the seismic resistance of remaining sections. Wiss, Janney, Elstner Associates, Inc. (WJE) developed seismic retrofit methods that could both provide performance that exceeds current AASHTO seismic design requirements, and be implemented for a small fraction of the cost of replacement. The WJE design included cutting of main column base dowel bars to force inelastic deformations to occur in easily accessible and therefore, easily modified elements of each pier. In these areas, reliable inelastic deformation capabilities were established by providing external confinement for anticipated hinge locations. Other modifications included bumpers, ties and bracing designed to engage, and fully develop, the capacities of currently inactive or underutilized piers.

Introduction

In East St. Louis, Illinois, a system of elevated roadways serves the eastern end of the Poplar Street Bridge over the Mississippi River. The roadways were designed and built in the 1960's. A typical roadway comprises several sections of one-to-four span plate girder superstructures, supported on reinforced concrete piers with pile foundations. Figure 1 shows typical roadway structural elements.

---

[1] Unit Manager, Wiss, Janney, Elstner Assoc., Inc. 330 Pfingsten Road, Northbrook, IL 60062

"

Figure 1. Typical Structural Elements

In the past few years IDOT has commissioned seismic evaluations of the 1960's vintage structures. These efforts showed that, under current AASHTO seismic demands, the original elevated roadway system would sustain extensive damage, and would be out of service for many months. Based upon these findings and other considerations, IDOT decided to replace the system's two main roadways, and to investigate methods for reducing the seismic vulnerability of the remaining structures.

In 1994, WJE was retained to perform detailed evaluations of the primary original roadways that were still in service. Following the evaluation, WJE was asked to design modifications that would enable the subject roadways to sustain current AASHTO level seismic demands with minimal impact on serviceability.

Characteristic Weaknesses

In the existing structures, inertial forces originating in the concrete deck and steel girder superstructures must follow load paths that include: various bearing assemblies, pier columns, foundation walls, pile caps and piles. These paths are identical to the load paths that are mobilized to sustain wind loads and other design forces that are applied to the superstructure. Unlike these other design forces however, the specified seismic demands exceed the ultimate strengths of the existing load paths. While this situation is not uncommon (seismic resistant structures built today typically have an ultimate strength that is between one sixth and one third of the corresponding peak seismic demand), it creates a situation where survival depends heavily on the ability to sustain significant, repeated, inelastic deformations. In other words, if a structure is unable to sustain specified seismic demands elastically (or nearly elastically), ductility is essential. Providing a threshold level of elastic strength, in combination with a minimum level of ductility, is the most common formula used in seismic resistant design today.

The evaluation showed that, lacking the ultimate strength necessary to remain essentially elastic given the specified seismic demands, the subject structures would experience significant ductility demands. Therefore, in each load path, the ultimate strength of the weakest link would be exceeded, and subject to large inelastic deformations. Unfortunately, the load paths did not include any significantly ductile links, weak or otherwise.

In this system a typical load path consists of: deck, deck/pier connection (i.e., bearings), pier columns, foundation wall, pile cap, and piles. Examples of all but the pile cap and piles can be seen in Fig. 1. A given load path has only one weak link (e.g., flexure at the bases of the columns or bearing strength) that limits its capacity and sustains inelastic deformations. Examples of common weak links included: bearing capacity, column base flexure, foundation wall flexure, pile cap flexure, and pile capacity.

Retrofit Approach

Retrofit options included strengthening the load paths to eliminate ductility demands. While effective, such an approach would require accessing, and greatly strengthening, almost every component of more than half the piers on the project (most expansion joint piers have elastomeric bearings that effectively isolate them from significant seismic demands). The associated costs would be prohibitive.

Another retrofit approach included modifications to the load paths to ensure adequate ductility in the weakest links. For example, if the pile cap represents the weakest link, it could be modified to be able to deform in a ductile fashion. A variation of this approach would be to strengthen the pile cap (forcing the weak link elsewhere), and modify the new weak link to achieve the desired ductility. In either case, excavation and extensive below-grade work would be required.

The WJE design team realized that below-grade modifications would be costly. Therefore, it was decided to eliminate weak links in the substructure by weakening above ground elements of affected piers. In other words, the need for below-grade work could be eliminated by manufacturing a weak link above grade. The following approach was implemented:

- Eliminate below-grade weak links by manufacturing flexural weak links at bases of pier columns. This would be done by cutting foundation wall dowel bars that develop primary flexural reinforcing at the base of each column.

- Strengthen deck bearings as required to develop ultimate strengths of modified columns.

- Confine bases of modified columns to provide adequate ductility in the weak link.

The resulting structural systems will have the two features necessary to ensure ductile response to large seismic demands: 1) a ductile element in each load path, and 2) load paths proportioned so that the ductile elements are also the weakest links in their respective load paths (Hill).

The design of bar cutting patterns and bearing strengthening measures involves straightforward, conventional structural engineering principles. In contrast, the design of retrofit confinement for column hinges that include lap splices was based on relatively new structural engineering concepts. To date, the 1995 Federal Highway Administration (FHWA) publication, "Seismic Retrofitting Manual for Highway Bridges," (FHWA-RD-94-052) comprises one of the few published documents to include an applicable formal design procedure. Much of the research work that lead to the development of the FHWA-RD-94-052 retrofit confinement design procedures was conducted at the University of California, San Diego (UCSD), under the supervision of M.J.N. Priestly.

In 1994, the University of Illinois (U of I) began a series of research efforts related to the external wrapping of column hinges that included lap splices. These efforts included cyclic load testing of both original and modified columns that were part of the East St. Louis, Poplar Street elevated roadway system. The U of I tests demonstrated: 1) the original columns had very little ductility, 2) loss of lap splice integrity in the hinge region was a common and brittle mode of failure, and 3) external wrapping could enable columns to maintain full flexural strengths, through several cycles of displacement, at large ductility levels (Shkurti, Lin, Gamble, Hawkins). The U of I tests gave all of the involved parties a great deal of confidence that column hinge wrapping could be used effectively on this project. The retrofit design documents allow a variety of external confining systems, including: prestressing strand hoops, glass fiber reinforced resin jackets, and carbon fiber reinforced resin jackets.

The modified structures will have the ability to deform in a ductile fashion via formation of plastic hinges at the bases of the pier columns. Given the specified level of seismic demand, maximum deck displacements are expected to be kept within serviceable limits.

References

1)   Hill, H.J., *Proportioning: Missing Step in Model Building Code Seismic Design,* ASCE Practice Periodical on Structural Design and Construction, Vol. 3, No. 1, February, 1998

2)   Shkurti, F.P.; Lin, Y.; Gamble, W.L.; Hawkins, N.M., *Testing of Bridge Piers: Poplar Street Bridge Approach, East St. Louis, Illinois,* a Report to Illinois Department of Transportation on Project IHR-330, University of Illinois at Urbana-Champaign, Department of Civil Engineering, June 1995

Seismic Retrofit of the Poplar Street Bridge

Mark R. Capron, P.E., Member ASCE[1]

Abstract

In work for the Missouri Department of Transportation (MoDOT) Sverdrup Civil, Inc. performed a seismic evaluation and developed a seismic retrofit strategy for the Poplar Street Bridge over the Mississippi River at St. Louis. The 660 meter (2,165 foot) structure consists of two parallel five span continuous roadways with orthotropic steel plate deck and variable depth steel box girders. The seismic evaluation considered three levels of design earthquakes and identified deficiencies in the bearings, reinforcement splices in the columns and piers, and one foundation. The retrofit strategy includes adding force transmitters or dampers to the existing expansion bearing piers, adding transverse shear blocks to the beam seats, jacketing the column splices, adding rock anchors to one foundation, and adding supplemental reinforcement to address lap splice deficiencies at the base of the piers.

Introduction

Sverdrup Civil, Inc. has supported MoDOT in evaluating and retrofitting several major bridges and bridge complexes throughout the state. Evaluation of these structures using current guidelines (FHWA 1983, 1995) has identified significant seismic deficiencies in critical bearing, column, and foundation elements, in spite of design acceleration coefficients as low as 0.12g. This paper presents the application of the seismic retrofitting guidelines to the Poplar Street Bridge over the Mississippi River at St. Louis. Design of seismic retrofits for this structure is being developed in two phases. Phase 1, consisting of seismic evaluation and development of retrofit strategy has been completed. Phase 2, consisting of additional studies and final design of selected alternatives is scheduled for completion in 1999. The following paragraphs present a description of the bridge, the approaches and results of the seismic evaluation, the retrofit strategy, and conclusions from this project.

---

[1] Project Manager, Sverdrup Civil, Inc., 13723 Riverport Drive, Maryland Heights, Missouri 63043, Phone (314)-770-4520

Description of the Bridge

The Poplar Street Bridge carries more than 130,000 vehicles per day across the Mississippi River at St. Louis via U.S. 40/I-64 and I-55/70. The 660 meter (2,165 foot) structure consists of two parallel five span continuous roadways, each providing four lanes of traffic. Span lengths from west to east are 91, 152, 183, 152, and 81 meters (300, 500, 600, 500, and 265 feet). Each roadway is supported by two variable depth steel box girders. The deck is of orthotropic steel plate construction consisting of a deck plate, trapezoidal longitudinal ribs and transverse floor beams. The deck plate acts as the top flange of the transverse floor beams, the trapezoidal ribs, and the main longitudinal box girders. The reinforced concrete substructure from west to east consists of a hollow shear wall type structure at Pier 1, solid shafts with rectangular columns and continuous cap beams at Piers 2, 3, 4, and 5, and a hollow shear wall structure at Pier 6. The substructure is founded on six foot diameter caissons to rock at Piers 1, 3, 4, 5, and 6, and a spread footing on rock at Pier 2. The superstructure is supported on the substructure by expansion bearings that allow rotation using spherical bearings and longitudinal translation through guided rollers at Piers 1, 2, 4, 5, and 6, and fixed bearings that allow only rotation with spherical bearings at Pier 3. The bridge was originally designed by Sverdrup and Parcel and Associates in 1963. The pier reinforcing steel is spliced near the base of the columns with lap splices of 24 bar diameters in length and confined with #5 ties spaced at 30 cm (12 inches), as typical of bridges designed in this timeframe. There are also similar splices in the reinforcing steel where the base of the pier shafts connect to the footings.

Seismic Evaluation

The analytical model used in the seismic evaluation consisted of over 15,000 degrees of freedom and included representation of the deck, box girders, and Piers 1 and 6 with shell elements, while the bearings, cross frames, and Piers 2, 3, 4, and 5 were modeled with beam elements. Linear spring elements were included in the model to represent the stiffness of the soil/foundation system, and mass was applied to account for unmodeled components such as parapets, major sign structures, and one lane of traffic live load per roadway.

Dynamic analysis was performed using the linear elastic response spectrum method and spectra based on the parameters and return periods shown in Table 1. In addition to the three design spectra, the analysis considered cases with completely rigid soil springs, with linear soil springs, with existing bearings, and with expansion bearings supplemented by "rigid" restrainers.

The results of the dynamic analyses were evaluated using the Capacity /Demand (C/D) ratio method presented in the guidelines (FHWA 1983, 1995). In the context of the guidelines, C/D ratios approximate the percentage of the design earthquake at which a particular component can be expected to fail; therefore, ratios less than 1.0 indicate insufficient capacity. The significance of individual C/D ratios

Table 1. Design Response Spectra Parameters

Acceleration Coef. (A)	Soil Type	Return Period (yr.)	Reference
0.12	2	475	(AASHTO 1996)
0.23	S_2	2,500	(FEMA 1991)
0.50	2	---	Maximum Credible Earthquake (MCE) - estimated from unpublished literature review

depends on the consequences of failure of each particular component, or groups of components. Table 2 summarizes the C/D ratios computed for the existing structure at the various design levels.

Table 2. C/D Ratios for Existing Structure

Component	C/D Ratios			Remarks
	Design Level			
	AASHTO 475 yr.	FEMA 2500 yr.	MCE	
Bearing Displacement	8.9	5.2	1.7	Unrestrained bearings at Piers 1 and 6, Method 2, Method 1 0.59 all levels
Bearing Force (longitudinal)	0.3	0.3	0.1	Pier 3 only point of longitudinal seismic resistance
Bearing Force (transverse)	0.8	0.5	0.2	All Piers similar
Cap Beam Yielding	0.5 - 0.7	0.4 – 1.2	*	No collapse mechanisms at AASHTO Pier 3 near collapse mechanism at FEMA
Splice Details at base of Columns	0.2 – 0.6	0.1 – 0.4	*	* Collapse mechanisms Piers 3, 4, and 5 at MCE
Splice Details at base of Piers	0.2 – 0.9	---	---	Longitudinal collapse mechanism at AASHTO, ratios not computed for FEMA 2500 or MCE, results < AASHTO
Foundation Rotation	0.9 - 10.8	0.7 – 8.8	---	Ratios not computed for MCE
Cross Frames	1.2 – 3.7	0.6 – 2.0	0.3 – 0.9	

The C/D ratios for the existing structure identified significant deficiencies in the bearings, the splices at the base of the columns, the splices at the base of the Piers at the AASHTO design level. Evaluation of the structure with engaged restrainers at expansion Piers 1, 2, 4, and 6 showed notable improvement in the critical C/D ratios of Pier 3 at the AASHTO design level, but indicated the need to increase the overturning capacity of Pier 1. Comparison of the AASHTO 475 year and FEMA 2500 year design levels showed similar types of retrofits would be required for both levels, while the MCE design level would require major changes in the structural response, which could conceptually be provided by isolation bearings, and/or major increases in capacity.

Seismic Retrofit Strategy

Based on the findings of the evaluation a retrofit strategy was adopted that involved adding force transmitter type longitudinal restrainers to expansion Piers 1, 2, 4, 5, and 6; increased transverse force capacity at Piers 1, 4, 5, and 6 using reinforced concrete shear blocks; strengthening the longitudinal and transverse capacity of the Pier 3 bearings; adding rock anchors to the Pier 1 foundations; confinement of the splices in the reinforcement at the base of the columns; and reinforcement of the splices at the base of Piers 2, 3, 4, and 5. The construction cost of these retrofits was estimated to be on the order of $7.2 million dollars. In addition, MoDOT selected the AASHTO 475 year spectrum as the design level for the seismic retrofit.

During Phase 2, the retrofit strategy will be further refined by evaluating damper versus force transmitter restrainers at all piers, and at Piers 2, 4, and 5 with bearing seat extensions at Piers 1 and 6; development of criteria for the maximum capacity to be developed by restrainers for earthquakes greater than design level; investigation of composite wrapping versus steel jacket confinement of the splices in the column reinforcement; and investigation of alternatives to reinforcing the splices at the base of the piers. Dynamic analysis of the structure with dampers will be based on the time history method using AASHTO spectra compatible synthetic ground motions and damper characteristics developed from recommendations from damper manufacturers.

Conclusion

Seismic evaluation of this structure using current guidelines identified deficiencies in the bearings, piers, and foundations, in spite of a design acceleration of only 0.12g. A retrofit strategy was developed for the specified design level that involved strengthening critical bearing and substructure components.

Acknowledgment

The Author wishes to express appreciation to the Missouri Department of Transportation for their support in the preparation of this paper.

Appendix. References

American Association of State Highway Transportation Officials (AASHTO). 1996. "Standard Specifications for Highway Bridges." Sixteenth Edition. Washington, D.C.
Federal Emergency Management Agency (FEMA). (1991). "NEHRP Recommended Provisions for Seismic Regulations for New Buildings." No. 222. Washington, D.C.
Federal Highway Administration (FHWA). (1983). "Seismic Retrofitting Guidelines for Highway Bridges." Report No. FHWA/RD-83/007, McLean, VA.
FHWA. (1995). "Seismic Retrofitting Manual for Highway Bridges." Publication No. FHWA-RD-052, McLean, VA.

The I-40 Mississippi River Bridge Seismic Retrofit

Roy A. Imbsen[1], Dennis D. Pecchia[2], Gerald V. Davis[3], and Ging Song Chang[3]

Abstract and Introduction

The I-40 Mississippi River Bridge (Figure 1) is a vital transportation, commerce, and defense link, being one of only two crossings of the Mississippi River in the Memphis, Tennessee area. The I-40 bridge is situated at the southeastern edge of the New Madrid Seismic Zone, where three of the largest earthquakes in the Central United States occurred in the early 1800's. Considering the potential for another major earthquake and closure of the two bridge crossings due to earthquake damage, significant impacts to national/regional transportation and economics would result. Recognizing this consequence, the Arkansas and Tennessee Departments of Transportation have given priority status to the seismic evaluation and retrofit of the I-40 bridge. The Departments of Transportation have contracted Imbsen & Associates, Inc. (IAI) to conduct a seismic evaluation and prepare retrofit design for the I-40 bridge. Retrofitting the I-40 bridge, significantly enhances public safety and reduces the potential for transportation and economic loss.

Figure 1. Steel Box Girder Spans and Steel Tied Arch Spans

[1] President, Imbsen & Associates, Inc., 9912 Business Park Drive, Suite 130, Sacramento, CA 95827
[2] Project Engineer, Imbsen & Associates, Inc. 9912 Business Park Drive, Suite 130, Sacramento, CA 95827
[3] Senior Engineer, Imbsen & Associates, Inc. 9912 Business Park Drive, Suite 130, Sacramento, CA 95827

The I-40 bridge, built in the late 1960s, has a total length of 5.3 km (3.3 miles), including main channel spans, approaches, and ramps. The bridge is comprised of 164 spans, 160 piers, and 10 abutments. The main channel spans consist of five steel box girder spans (2 @ 100.6 m/3 @ 121.9 m (2 @ 330'/3 @ 400')) and two steel through tied arch truss spans (2 @ 274.3 m (2 @ 900')). The west approach to the channel spans is primarily precast prestressed concrete I girder spans, except for twelve welded steel plate girder spans; while the east approach and connecting ramps are entirely welded steel plate girder spans.

IAI is currently preparing final design, plans, specifications, and estimates (PS&E) for the main channel spans. PS&E for these spans will be finished and construction is expected to start in the Spring of 1999. Features of the main channel spans retrofit include:
- Serviceable performance following the maximum probable "Contingency Level Earthquake" (2500 year return period).
- Three dimensional, nonlinear, time history analysis.
- Friction pendulum isolation bearings.
- Additional piles and strengthening of footings, columns, and webwalls.
- Replacement and strengthening of selected superstructure members.

Construction of the remaining I-40 approaches and ramps will commence once construction of the main channel spans has been finished.

Seismic Hazard and Site Specific Ground Motion

The I-40 bridge is situated at the southeastern edge of the New Madrid Seismic Zone, one of the most active zones in Central United States. In the winter of 1811-1812, three of the largest earthquakes (all three had magnitudes greater than M_w=8.0) in the Central United States occurred within this seismic zone. The site condition is characterized by deep soil deposits, up to 762 m (2500') deep, that result in dominant long period motion in the ground input motion. In addition, the potential for liquefaction is currently being evaluated and may exist at certain locations along the bridge. Spatially varying ground motion time histories were developed at the multiple supports. The maximum probable "Contingency Level Earthquake" (2% chance to occur in 50 years, 2500 year return period) of magnitude m_b 7 from the New Madrid Seismic Zone was the primary seismic event investigated for the project. The earthquake is assumed to occur 64.3 km (40 miles) from the bridge site at a depth of 19.3 km (12 miles).

Computer Model and Analysis

A three dimensional analytical model of the box girder portion and the tied arch portion of the bridge was developed using the IAI-NEABS (IAI-Nonlinear Earthquake Analysis of Bridges Systems, 1991) program. The model also included soil-structure interaction, foundation radiation damping, and foundation rocking. See Figure 2 for the cross section of the tied arch center pier and analytical idealization for caisson-soil interaction.

Seismic Performance Criteria

Seismic performance goals were established, in conjunction with the Departments of Transportation, that the bridge must remain "operational/serviceable" following the maximum probable "Contingency Level Earthquake" (2500 year return period). The bridge is a vital defense and commerce transportation link. Bridge closures are expected to last no longer than several days. Any damage is expected to be relatively minor and repairable under traffic.

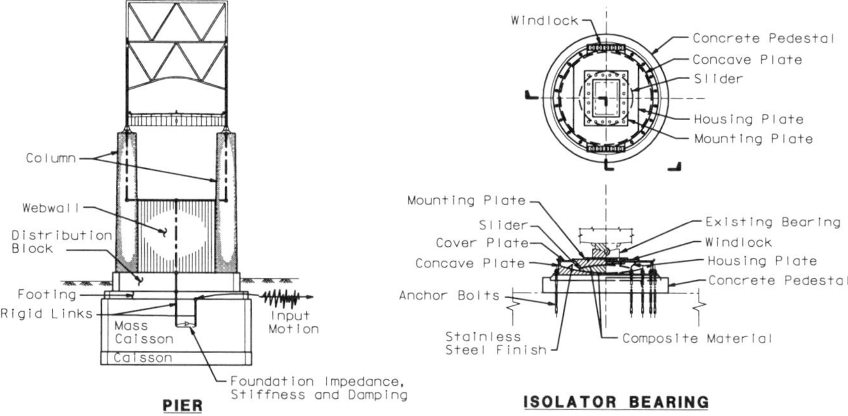

Figure 2. Analytical Pier Model   Figure 3. Friction Pendulum Bearing

Existing Structure Vulnerabilities

Originally, the bridge was designed primarily for transverse wind forces, not earthquakes. The original design only considered transverse seismic forces of between 2% to 2.5% of gravity. Consequently, there are many seismic deficiencies with the existing bridge. The primary vulnerabilities of the existing box girder spans and tied arch spans include:

- Factors of safety against overturning in the longitudinal direction are marginal at piers.
- Inadequate footing reinforcement.
- Piers would experience excessive damage at the base of the columns and webwalls.
- The existing tall steel bearings would probably topple or displace laterally, allowing the bridge spans to drop.
- Cross frames and lateral bracing systems are deficient.
- Inadequate shear transfer system between the concrete deck, stringers, and floorbeams.
- Finger type expansion joints are not adequate for expected seismic displacements.
- Each portal frame of the truss, cause is extremely overstressed.

Initial Retrofit Strategy—Maintain Existing Bearing Strategy

During the early development of seismic retrofit philosophies, a combination of "strength and ductility" was a common approach to provide maximum seismic resistance. For the I-40 bridge, this strength approach (utilizing existing bearings) would have required retrofitting nearly all superstructure and substructure members and connections. The estimated cost of this initial retrofit strategy escalated quickly, totaling more than $43,200,000 for the box girder and tied arch spans. At this point, a review of the retrofit strategy seemed appropriate based upon the extent of retrofit work that would be needed for the "Existing Bearing Strategy".

Final Retrofit Strategy and Design—Isolation Bearing Strategy

With recent advances in isolation bearing technology and understanding, isolation has become a viable retrofit strategy for many bridges. Isolation provides a means to limit structure forces; however larger displacements often occur. A retrofit strategy incorporating isolation bearings was investigated for the I-40 bridge. Figure 3 shows typical details of a friction pendulum bearing. Comparing the "Isolation Bearing Strategy" to the "Existing Bearing Strategy", significant reductions of force due to isolation result in major retrofit cost savings. The following retrofit is currently being prepared for the box girder spans and tied arch spans:

- Cofferdams required at all piers except one. Each cofferdam could cost up to $1 million dollars.
- Footing strengthenings and top reinforcement mats. Tie down anchors.
- Minimal webwall modifications to tie the tops of the webwalls to the columns.
- Strengthening of the columns.
- Column cap enlargement to accommodate isolation bearings.
- Jacks and temporary supports with 2,340 tonnes (5,150 kips) total lifting capacity for raising the bridge to replace bearings typically required at piers. 5,440 tonnes (12,000 kips) total lifting capacity required at the tied arch center pier.
- Replacement of existing bearings with friction pendulum isolator bearings.
- Replacement of some bottom lateral braces and strengthening of others.
- Strengthening of steel cross frames at all piers.
- Deck of the box girder spans is to considered as a horizontal diaphragm spanning between piers.
- Replacement of the existing expansion finger joints with modular swivel expansion joints having ±510 mm (20") of longitudinal movement and ±610 mm (24") of transverse movement.
- Truss retrofit, add truss members that laterally brace the portal frame posts.

Overall construction costs and the extent of retrofit work were significantly reduced by switching from the "Existing Bearing Strategy" to the "Isolation Bearing Strategy". This cost savings was accomplished while maintaining a high level of structural safety and serviceable performance criteria.

Conclusions

The estimated cost of the "Existing Bearing Strategy" escalated quickly, totaling more than $43,200,000. The "Isolation Bearing Strategy" reduced estimated construction cost to $26,700,000 – a 40% reduction and $16,500,000 savings. Recognize that these costs are for the main channel spans only, consisting of the box girder and tied arch spans. The entire I-40 retrofit project, including these channel spans and the remaining west approach, east approach, and ramp spans is currently projected to cost more than $105,000,000. The selected isolation strategies offer structurally safe and economic solutions. Solutions that are, in the case of the I-40 bridge, 40% less costly.

The interest and guidance provided by Mr. Edward Wasserman, the Tennessee Department of Transportation Civil Engineering Director, Structures, is highly appreciated.

Japan-U.S.-China International Joint Study on the Mitigation of Earthquake Hazards (part 1):

Overall Program and Its Implementation Among the Three Countries

Takayuki Shimazu[1], Yayong Wang[2] and M. J. Nigel Priestley[3]

Abstract

Joint research team outline, main item of the program and its realization as well as motivations and objectives of the joint study are presented herein. Finally the brief comparison on historical background of code modification and major earthquake are explained among the three countries as the first step for parts two to four of above title..

Introduction

After the occurrence of the Kobe Earthquake (January 17, 1995) and the damage it had made to structures, the importance of reforms and fundamental revision of the current code became clear. Besides, both seismiç safety evaluation and retrofitting of existing buildings appeared to be very essential. On the other hand, the Tangshan earthquake that happened in China (July 28, 1976) had claimed an unprecedented structural damage and human life losses, while in the US, the Northridge earthquake (January 17, 1994) caused severe damage to residential houses.

In each of these three countries, different approaches were taken as measures against earthquake damage and so were the reforms that were forced into the design   codes. According to the unique experience of each of these three countries, some convergence and divergence is observed in their approach.

An international committee was initiated in 1995 by Prof. Takayuki Shimazu of Hiroshima University with the collaboration of Prof. M. J. N. Priestley, the University of California at San Diego, USA, and Prof. Yayong Wang, China Academy for Building Research, China. The project is mainly funded by the Ministry of Science, Culture, and Education of Japan (Grant no. 08044152).

---

[1] Professor, Dept. of Arch./Structural Eng., Hiroshima Univ., Higashi-Hiroshima 739-8527, Japan

[2] Professor, IEE, China Academy for Building Research, Beijing 100013, P. R. of China

[3] Professor, Dept. of Civil Eng., Univ. of California, San Diego, California 92093-0085, USA

The objective of this joint research is to exchange the experiences of each country in this field through seminars and field studies. The research comprises the study of performance of building structures along with the changes of codes and their updates, aspects of the seismic safety evaluation of structures, retrofitting problems of existing structures and the evaluation of current isolation system. The duration is three years during which about three rounds have been held in each year.

In this paper the overall results of the last two and half years are presented mainly focusing on the historical background of design code modification and the state of retrofitting of existing buildings in the three countries, although several number of items have been also investigated

Joint research-team outline

Table 1 shows a list by the members participating in the study. The committee consists initially of 14 members, and the coordinators for each country's team are indicated in the table, and namely are Prof. Priestley, Prof. Wang and Prof. Shimazu. The latter is the overall chairman of the committee. However, according to necessity, more participants or presenters have been assembled.

Table 1   List by the participants of each country's side

USA	China	Japan
Prof. M.J.Nigel Priestley Univ. of California, San Diego	**Prof. Yayong Wang** Director of I.E.E. China Building Research Academy (CABR)	**Prof. Takayuki Shimazu** Hiroshima University Faculty of Engineering
Prof. Stephen A. Mahin Univ. of California, Berkeley	Prof. Wei Chengji Vice Director of I.E.E. (CABR)	Prof. Koji Tominaga Hiroshima Univ. (International Development & Cooperation IDEC)
Prof. Yan Xiao University of Southern California	Miss Liu Wen Engineer at I.E.E. (CABR)	Assoc. Prof. Hideo Araki Hiroshima Univ.. (Faculty of Eng.)
Prof. Gary C. Hart University of California, Los Angeles	Assoc. Prof. Ye Liaoyuan Yunnan Polytechnic Uni.	Assoc. Prof. Haruyuki Yamamoto, Hiroshima Univ. (IDEC)
	Prof. Gong Sili (CABR)	Res. Assoc. Aiman Alawa Hiroshima Univ. (Faculty of Eng.)
	Prof. Zhou Xiyuan (CABR)	Res. Assoc. Kenji Kabayama* Hiroshima Univ. (Faculty of Eng.)
	Mr. Jiao Jun* Senior Engineer at The Earthquake Resistance Office of Yunnan Province	Res. Assoc. Ali Bakhshi* Hiroshima Univ. (Faculty of Eng.)
	Mr. Sha An* Research Engineer at I.E.E. (CABR)	

*) Cooperative members joined the project at midway

Main items of the program and its realization

The main items of the program are shown in Table 2. In 1996, the first round were held in Japan, Hiroshima and Kobe cities, the second round in China, Beijing Tangshan,

Kumming, Lijiang and Shanghai cities and the third in the U.S.A., San Diego, Los Angeles and San Francisco cities, in which earthquake history, earthquake effects on human society, mitigation, effects of code revisions and technological developments as well as possibility of codes harmonization were discussed. In 1997 three rounds, focusing on the seismic safety evaluation and retrofitting of buildings have been held in each of the three countries. On the other hand, the round of 1998 will conclude the meetings of recent two years studies

Table 2   Main items of the program

1996	1997	1998
1. Earthquake history with geological features	1. Strong motion accelerogragh network	1. Possibility of Codes harmonization
2. Code revision history with history of damage of buildings	2. Seismic safety evaluation and retrofitting of buildings	2. Acceleration method of implementation of retrofitting of buildings
3. Technical development for mitigation of earthquakes hazard	3. Implementation of retrofitting of buildings	3. Overall conclusions of the three years (seeds and needs for the future)

Historical background of code modifications

Table 3 shows a brief history of code modifications along with the occurrence of major earthquakes in the three countries [1]. It can be found from the Table that US and Japan have about one century long history of measures against earthquake while China has less than half a century. This might be attributed partially to the fact that both countries, the US and Japan, lie directly on the seismic zone of the Pacific belt, and also may be attributed to the difference in the frequency of earthquake occurrence at the same location.

As each country has its own geological features, earthquake history and design philosophy, there seems to be conspicuous differences among the three codes. However, there are also similarities on the items treated although the design steps to be taken and the values treated are different. Both China and Japan codes stipulate the use of dual steps (the first level design and the second level design, etc.), while the US codes uses a single step using ductility factor (R_W), directly into the second level. To compare the size of seismic design level among three countries it should be noted that load combinations should also be considered

Conclusions and Recommendations

Overall trends for code developments was explained. Each side features of the three countries will be presented in the next parts of 2, 3, and 4.

Acknowledgements

A series of these four papers is mainly based on a research project sponsored by the Ministry of Science, Culture, and Education of Japan, which needs to be acknowledged. The cooperation provided by all the persons related with this project is highly appreciated.

Table 3 Historical background of code modifications as well as major earthquakes in the three countries

Year	U.S.A. major earthquakes and seismic regulation establishment	Year	China severe earthquakes and seismic regulation establishment	Year	Japan great earthquakes and seismic regulation establishment
1906	San Francisco Earthquake S. F. rebuilt to 1.16 kg/m^2 lateral load	before-1964	Only required by the local authorities	1891 1919 1923	Nobi Earthquake (M=8.4) Structural design regulations Kanto catastrophic Earthquake (M 7.9)
1927	UBC (Uniform Building Code): F=CW C=0.075 for firm soil C=0.10 for soft soil ; W=DL+0.5LL	1964	Draft for Seismic Resistant Design Code	1924	seismic design regulation: F=kW (k=0.1)
1933	Long Beach Earthquake (M 6.3)	1966	Xingtai Earthquakes (M 6.5 and M 7.2)	1933	Sanriku Earthquake (M 8.3)
1940	El Centro Earthquake (M 6.7)	1974	Seismic Design Code for Industrial and Civil Buildings (TJ 11-74)	1947	Japan architectural Standard: k=0.20 for firm soil site k=0.30 for wooden building on soft soil
1952	Taft Earthquake (M 7.7)	1975	Haicheng Earthquake (M 7.3)	1950	Building Standard Law founded
1966	SEAOC $$C = \frac{0.05}{\sqrt[3]{T}} \quad \&$$ $$T = \frac{0.03\,h_n}{\sqrt{D}}$$ (T= period, h_n= height, and D=width of the building)	1976	Great Tangshan Earthquake (M 7.8) (242,829 people died and 4,332,700 rooms collapsed)	1968	Tokachi (Hokkaido) Earthquake (M 7.9)
1971	San Fernando Earthquake (M 6.6)	1979	Seismic Design Code (TJ 11-78)	1970	height limitation nullified
1976	UBC: V=ZIKCSW: $$C = \frac{1}{15\sqrt{T}}$$	1993	Seismic Design Code for Building and Structures (GBJ 11-89) $$F_{EK} = \alpha_1 G_{eq}$$ $$\alpha_1 = \frac{1}{R} k\beta$$ $$\& \beta = (\frac{T_g}{T})^{0.9} k\beta_{max}$$ k: Zones factor R: Ductility factor equal to 2.8 β: Design elastic response spectra T_g: Factor dependent on Site-specific coefficient	1978 1981	Miyagi offshore Earthquake (M 7.4) current resistant-design Code: V=ZR$_t$A$_i$C$_0$W C$_0$=0.2 (moderate) to 1.0 (severe earthquake) Z: Zone factor R$_t$= Factor based on dynamic response and site soil type A$_i$: Shear coefficient distribution factor along height of structure
1991	UBC: $$V = \frac{ZIC}{R_w}W \quad ;$$ $$C = \frac{1.25S}{T^{2/3}}$$ Z: Zone factor I: Occupancy importance factor S: Site specific coefficient				
1989	Loma Prieta Earthquake (M 6.9)				
1994	Northridge Earthquake (M 6.7)	1996	Lijiang Earthquake (M 7.0)	1994	Hokkaido (s-w area) Earthquake (M 7.8)
1997	Current UBC (Zone 4)			1995	Hyogoken-Nanbu Earthquake (M 7.2)

Japan-U.S.-China International Joint Study on the Mitigation
of Earthquake Hazards (part 2):

Current Approaches in Seismic Measures for Building Structures (US-side features)

Yan Xiao[1], M. J. Nigel Priestley[2], Stephen A. Mahin[3] and Gary C. Hart[4]

Abstract

This paper presents the design philosophy of retrofitting of buildings and its implementation in U.S. as well as the code development process to a next code. At first, the basic design philosophy on displacement-based seismic assessment of existing reinforced concrete structures are introduced and actual state of implementation of retrofitting of buildings briefly explained. Then, code development process, moving towards a unified national code, that is , international building code, as well as new trend on the code and etc. are presented.

Introduction

The deficiency in expected seismic performance of existing unreinforced masonry buildings has long been recognized, with cities such as Los Angeles and Wellington requiring owners to strengthen or demolish deficient structures within a given time frame. Typically these buildings were designed before the advent of the first seismic design codes in the early 1930's. More recently, a second class of earthquake risk buildings (ERBs) has been identified – reinforced concrete buildings designed between 1930, and about 1975 when design codes were implemented containing seismic provisions more or less equivalent to those currently in practice.

In this paper the design philosophy to be established for retrofitting of existing buildings from now on are first explained together with the actual state of the implementation of retrofitting in U.S. Second, code development process towards a unified international building code in U.S. has been pointed out.

---

[1] Assist. Prof., Dept. of Civil Eng., Univ. of Southern California, Los Angeles, CA 90089-2531, USA

[2] Professor, Dept. of Civil Eng., Univ. of California, San Diego, California 92093-0085, USA

[3] Professor, Dept. of Civil Eng., Univ. of California, Berkeley, California 94720, USA

[4] Professor, Dept. of Civil & Envir. Eng., Univ. of California in L. A. , Los Angeles 90095-1593, USA

On the retrofitting of existing buildings in U.S.A. (Priestley, 1996a and Xiao, 1997)

Deficiency of seismic performance is generally a consequence of lack of ductility rather than inadequate lateral strength. Seismic design coefficients in current codes generally imply dependable inelastic cyclic response to significant levels of ductility. In older buildings, the ductility deficiency is a consequence of two major fallings in the original design process – poor detailing of reinforcement, and the lack of a capacity design philosophy. In this paper, a more meaningful system approach to the assessment of existing frame buildings, that is, displacement-based seismic assessment than traditional force-based seismic assessment is introduced.

A two-level seismic assessment procedure has been outlined intended to determine the risk, in terms of annual probability of exceedance, associated with both serviceability and ultimate limit states. Determination of the ultimate limit state involves an attempts to identify the most critical collapse mechanism, and calculation of its associated strength and ductility in system response terms. Strength and structural ductility were combined to provide an equivalent elastic response force level, which, by comparison with the design elastic response spectrum could be used to determine annual probability of exceedance corresponding to development of structural capacity. The basis for identifying the critical collapse mechanism is a modified form of capacity design, which permitted local element failure provided overall structural integrity has not been jeopardized.

In the U.S.A,. the west coast, especially California, is leading the research and implementation for retrofit policy and practice. The retrofit practice is different for different construction materials. For example, there is a mandate for retrofit of all unreinforced masonry buildings (UMB's) in the state California, while no such policy exist for wood or concrete framed buildings. Policy on steel moment frames is still up in the air, pending the final research results from the collaboration of SEAOC (Structural Engineers Association of California), CUREe (California Universities for Research in Earthquake Engineering) and ATC (Applied Technology Council).

Recently, ATC published guidelines for the retrofit of nonstructural elements. FEMA (Federal Emergency Management Agency) published "NEHRP Guidelines for the Seismic Rehabilitation of Buildings". In the meantime, most retrofits are governed by local modifications to the UBC (Uniform Building Code). Most of the older public school buildings have been retrofitted in the 1970's, and just about all of the older school buildings have been retrofitted. Most of the State owned buildings will be reviewed. Private owners typically do not voluntarily spend money to retrofit their buildings, except high-tech firms. High profile high-tech firms (with more to lose during a seismic event) are interested in retrofit using base isolation, viscous dampers, etc. Very recently, a special report has been also published by EERI (Earthquake Engineering Research Institute), which mainly deals with owners' concern.

Code development in U.S.A. (Priestley, 1996b)

a) Model Code Agencies develop codes to be adopted by local cities and countries. The followings are typical model codes,
    i.   Uniform Building Code (western U.S.)
    ii. National Building Code (northeastern U.S.)
    iii. Standard Building Code (southeastern U.S.)
    iv. Building Official and Code Administrator Code (mid-western U.S.)

Codes indicate that they are essentially to provide minimum standard to safeguard life or limb, health, property and public welfare by regulating and controlling the design, construction, quality of materials, use and occupancy, location and maintenance of buildings.

Model code agencies are made up of building officials who are often not engineers, which means that anyone can submit suggestions for code changes. It should be noted that however, SEAOC has historically developed recommendations for inclusion in the Uniform Building Code. Furthermore, adoption of provisions by consensus balloting of member building officials is usually underway.

b) Towards a Unified National Code

The National Earthquake Hazard Reduction Program (NEHRP) has resulted in tentative provisions for seismic design (development funded by FEMA). The main features are as follows,

    i.   Developed by committees of engineers, manufacturers, and building officials.

    ii.   Developed on the basis of life safety

    iii. More rational mapping of seismic hazards

However, since the provisions have no legal authority, they are not enforced in practice.

Model code agencies in U.S. are merging into one agency and will produce one model building code in the U.S. for the year 2000. Code will be called the "International Building Code, i.e. IBC". In this new code, seismic provisions will be based largely on and be consistent with format of the NEHRP provisions.

c) Near-term Trend

    (1) Uniform Building Code:

      i.   For the year 2000, the major model building codes will merge into a single national code called: **International Building Code**

      ii.   Uniform Building Code will no longer exist

      iii. Seismic provisions will be based on current activities within the NEHRP program

    (2) NEHRP Provisions:

      i.   Continuing to be updated

      ii.   Major effort to improve seismic provisions

         a.   To include several new materials approaches (e.g. composite construction, isolation, etc.)

         b.   FEMA has several parallel investigation to support development (steel structures, rehabilitation, force reduction factors, mapping, etc)

d) Other Trends for U.S. codes

    1. Steel Buildings

      i.   Special code is being developed for steel buildings including new construction, repair of earthquake damage, evaluation of existing buildings

    2. Existing buildings

      i   FEMA 272 and 273 cover the analysis and evaluation of existing buildings.

      ii.   Relies heavily on nonlinear analysis (static pushover analysis)

      iii. Based on reduced reliability that required for new buildings.

3. Performance Based Design

Many efforts are now underway to develop performance-based criteria and procedures for seismic design, these include,

- i.   Vision 2000 – SEAOC
- ii. Action Plan for Performance based Seismic Design – FEMA
- iii. Efforts are focused mainly on criteria for performance and on acceptance criteria
- iv. Having an important impact on thinking and design of individual projects
- v.   Research/ knowledge base

e) Recent activities to better characterize seismic risk in the U.S. and related design criteria are based on 1997 NEHRP Provisions with the following purposes:

- i.   Utilize latest techniques for seismic hazard mapping, including
    - a.   New fault data regarding activity rate
    - b.   New attenuation relations
- ii. Provide design spectra representative of seismic environment around nation
- iii. Recognize the large ground motions may occur in the near source region
- iv. Provide uniform risk of failure for building structures

f) Uniform Risk of Failure

- i.   Current NEHRP provisions proportion all structures based on the smaller of :
    - a.   475 year return period (10 % exceedance over 50 years) motion
    - b.   Zone 4 design ground motion
- ii. There are two problems, however,
    - a.   In zones of high risk, this motion represents a much more frequent event than 475 years
    - b.   In zones of lower risk, the 475 year event is not a good estimate of motion from very rare, but large events
- iii. Solution approach taken by NEHRP : Proportion all structures for 1/1.5 times the ground motion resulting from a Maximum Considered Earthquake (MCE)
    - a.   1/1.5 expresses the presumed inherent margin in current structural design

in which MCE defined as: Probabilistic 2500 year (2 % / 50 years) motion, but not larger than the larger of :   1. Current zone 4 design basis

2. 1.5 times the median motion resulting from a characteristic event on any known fault

Conclusions and Recommendations

Measures for the acceleration of implementation of retrofitting of existing buildings in U.S.A. is quite urgent. The next code is expected to be certainly suitable not only for inside the U.S. but also for "International" applications.

References

Xiao, Y. and Chang, H., Seismic evaluation and retrofit practice in the U.S., presented in the second round meeting, Los Angeles., 1997.
Priestley, M.J.N., Displacement-based seismic assessment of Reinforced Concrete buildings, presented in the first round meeting, Hiroshima, 1996a.
Priestley, M.J.N., Moving towards International building code, presented in the first round meeting, Hiroshima, 1996b.

Japan-U.S.-China International Joint Study on the Mitigation of Earthquake Hazards (part 3):

Current Approaches in Seismic Measures for Building Structures
(China-side features)

Yayong Wang[1], W. Chengii[2] L. Win[2], Y. Liaoyuan[3], G. Silii[2], Zh. Xiyuan[2], J. Jun[4], and, Sh. An[2]

Abstract

This paper presents the overall trend of building strengthening in China, as well as the various aspects to be considered for a new round of revision to the next code. In the former part, seismic zone, achievement and target of strengthening, standard and regulation for seismic strengthening and retrofitting, three progressive stages of strengthening and retrofitting in China, and efficiency analysis and investment optimization are presented, while in the latter part, the opinion for a new code replacing the current code for aseismic design of buildings are introduced.

Introduction

Since the 1976 great Tangshan earthquake, which brought tremendous damage to building structures in China, the code TJ11-74, the first national standard for aseismic design code for industrial structures and civil buildings in China had been revised to code TJ11-78 and then the current code GBJ11-89.

However, it has been widely recognized in China that the lessons learnt from the 1994 Northridge and the 1995 Kobe earthquakes should be introduced into the revisions on GBJ11-89 code, due to the lack of knowledge and experience of earthquake damages to modern building and civil engineering facilities in China. In this paper, the actual state of the building strengthening in China as well as the key

---

[1] Professor and Director, IEE, China Academy for Building Research, Beijing 100013, P.R. China.

[2] Senior Engineer or Researcher, IEE, China Academy for Building Research, Beijing 100013, China.

[3] Professor & Director. Yunnan Institute of Earthquake Eng., Kyunming, Yunnan 650051, P.R. China.

[4] Vice-Director, Aseismic Design Office of Construction Dept. of Yunnan Province, Kyunming, Yunnan 650032, P.R. China.

points on the revision to the next code from the current stage of practice are presented.

Overall trend of building strengthening in China (Wang, 1997 and Jun, 1997)

1) Classification of Seismic Zones
60 % of the continental area are seismic zones

Table 1. Seismic Intensity Distribution

Intensity	Area (10^4 km^2)	Percentage
5 and below	384.5	40.10
6	263.5	27.40
7	206.4	21.55
8	71.3	7.37
9	72.6	2.46
10 and above	10.7	1.12

2) Achievement and Target of Strengthening
a. 300 million square meters of buildings have been strengthened since the 1976 Tangshan Earthquake that cost about 5 billion RMB
b. Category of strengthening targets
    i.   Public buildings such as school, hospital, office, theater, library, etc.
    ii. Residence : masonry apartment, wood-mud houses, old existing houses, etc.
    iii. Life line facilities : railway station, control of center communication system, transform
    iv. Station, water supply, gas and heating system, highway and bridge engineering, etc.

3) Standard and Regulation for Seismic Strengthening and Retrofitting
a. Seismic Evaluation Standard for Existing Buildings GB50023-95, state level
b. Technical Code for Seismic Strengthening of Buildings, department level
c. Classification Standard for Earthquake Damaged Buildings, department level
d. Standard for Classification of Seismic Protection of Buildings GB 50223-95, state level
e. Aseismic Technical Specification for Multistory Masonry Buildings with Reinforced Concrete Tie Column JGJ/T 13-94, department level
f. Seismic Evaluation Standard for Industrial Facilities, state level
g. Seismic Evaluation Standard for Industry Special Structures, state level
h. Seismic Evaluation Standard for Water Supply, Water Drain Away Facilities Outdoor, state level
i. Seismic Evaluation Standard for Water transport and Hydraulic Engineering, department level
j. Regulation of Seismic Strengthening and Acceptance for Communication facilities, department level

4) Three Progressive Stages of Strengthening and Retrofitting in China
a. Primary Stage : Emergent-Temporary Measures and Simple Strengthening Technology
    i.   Emergent repair and temporary support after earthquake
    ii. "Sandwich" concrete

      iii. Added tie-column
      iv. Added tie beam and steel tensile bar
b.   Advance Stage : Pre-event Strengthening and Post-event Retrofitting
      i.   High pressure grouting technology
      ii. R.C. enveloping and steel binding technology for R.C. frame structure
      iii. Added wind wall of shear wall to R.C. frame structure
      iv. Added steel brace to spacious R.C. workshop
c.   Comprehensive Stage : in addition to seismic strengthening and retrofitting
      i.   Extension of existing building
      ii. Decoration
      iii. Performance improvement
      iv. High-tech utilization (base isolator, energy dissipation facility, TMD, etc.)

5) Efficiency Analysis and Investment Optimization

a.   Direct loss and Induced loss
b.   Cost for and Benefit from Strengthening
c.   Efficiency Ratio of J:

$$J = 2.Z' / Z$$

Where Z' is the benefit (reduction of loss), Z is the cost for the pre-event strengthening, it is assumed that the cost of reconstruction equal to the earthquake loss. It is obvious that the efficiency ratio J depends on the regional seismic risk (zone intensity), the original value of the building, acceptable damage or failure to the object building and the cost of reconstruction. The ratio J is generally ranged in 20 – 100.

Important aspects to be considered for the next code in China (Wang, 1996)

Since the lack of knowledge and experience of earthquake damages to modern building structures and engineering facilities in China, lessons learnt from the 1994 Northridge and Kobe earthquakes should be introduced into the provisions on GBJI1-89 code.

It has been more than twenty years since the 1976 great Tangshan earthquake. There have not occurred any severe earthquake events in the metropolitan cities of China where lots of modern buildings and infrastructures have been constructed since then. Most of earthquake damages and failures were created and found in traditional buildings such as unreinforced multi-story masonry structures, RC frame structures, single-story factory buildings, single-story spacious buildings, adobe, wood, and stone structures from recent earthquake disasters.

We have little, or poorly understood the behavior and response of the modern buildings and civil engineering facilities, such as RC high-rise buildings, composite structures (RC + SRC), bound masonry structures, steel structures, large-span structures (bridge, huge gymnasium and etc.) which are designed and constructed in the ways more or less different from those adopted in USA or Japan. On the other hand, the size, height, configuration, structural layout of buildings are often beyond the scopes and limitation specified by the current code, i.e. GBJ11-89. The questions and worries about the safety of these buildings for earthquakes are raised recently.

The Chinese expert professionals are considering to revise the GBJ11-89 code on the basis of the following aspects:

1. The design parameters which describe long-period contents of the strong ground motions such as the PGV and PGD or long-period response spectrum and the effects of the near-source.
2. The applicable and practical methodologies for the elastic and inelastic time-history analysis of structures, including the principle of input accelerograms, the hysteric model of the relationship between forces and displacement, the yielding model of moment and axial force of elements of beam and column, and etc.
3. The criterion of failure and collapse of structures in terms of inter-story drift index.
4. The ductile details of sections of element and connections.
5. The base-isolator and energy dissipater devices for bridge multi-story building structures.
6. Seismic design for new type of buildings and structures such as steel-frame structures, composite structures (RC+SRC & RC+S), reinforced masonry structures.
7. The seismic design for non-structural elements and equipment on floors of multi-story buildings.
8. The evaluation for the seismic safety of the construction site.

Conclusions and Recommendations

Urgent measures for acceleration of retrofitting of existing buildings have been strongly recognized in China. The model code of year 2000 of seismic design for buildings will perform as a guideline of seismic provision for infrastructures and engineering facilities. The tasks for short-term action as well as long-term action should be figured out. A concrete cooperation among all aspects of earthquake engineering societies in home and abroad is also required.

References

Jun, J., Introduction of the Technology and some relevant problems of building's aseismic fortification in Yunnan province, presented in the second round meeting, 1997.

Wang, Y., History of revision and questionnaires on seismic design in China, presented in "International Joint Study on the Mitigation of Earthquake Hazards" the third round meeting, 1996

Wang, Y., Overall trend of building strengthening in China, presented in "International Joint Study on the Mitigation of Earthquake Hazards" the second round meeting, 1997.

Japan-U.S.-China International Joint Study on the Mitigation
of Earthquake Hazards (part 4):
Current Approaches in Seismic Measures for Building Structures
(Japan-side features)

Takayuki Shimazu [1], Koji Tominaga[2] Hideo Araki[3], Haruyuki Yamamoto[4],
Kenji Kabayama[5], Ali Bakhshi[5],

Abstract

The method of determining the priority of retrofitting of existing buildings in Japan, the actual state of the implementation of such retrofitting process as well as the method of prediction of earthquake damages in each prefecture in Japan are presented The recommendation of utilization of base isolation systems is also introduced together with the basic concept as fundamental approach towards the next code.

Introduction

The 1995 Kobe (Hyogoken Nanbu) earthquake brought catastrophic disaster in Kobe and the adjacent area. Most of prefectures in Japan have started to conduct "prediction of earthquake just after the earthquake to estimate human, building and economical losses caused by probable major or extremely strong earthquake to be anticipated to their regions in the near future. Those predictions revealed that taking appropriate seismic countermeasures was an urgent matter of consideration, particularly on the improvement of seismic capacity of the existing buildings as well as on the introduction of site specific characteristics into the revision of a new provision for the next code. Based on those predictions, the utilization of high technological devices such as base-isolation systems has also been recommended.

---

[1] Professor, Dept. of Arch./Structural Eng., Hiroshima Univ., Higashi-Hiroshima, Japan.

[2] Professor, International Development and Cooperation (IDEC), Hiroshima Univ., Japan.

[3] Associate Prof., Dept. of Arch./Structural Eng., Hiroshima Univ., Higashi-Hiroshima, Japan

[4] Associate Prof., International Development and Cooperation (IDEC), Hiroshima Univ., Japan.

[5] Research Associate, Dept. of Arch./Structural Eng., Hiroshima Univ., Higashi-Hiroshima, Japan

Countermeasures to earthquake damages in Japan (Shimazu, 1997a)

There are four steps in the prediction of earthquake damages. In step 1, 'Investigation of natural condition', probable earthquake were determined based on the locations of dangerous faults and historical earthquakes, while ground condition map of all over the prefecture was produced based on several thousands of existing boring data. In the step 2, 'Prediction of Natural Phenomena', various seismic factors such as acceleration on the bed rock surface, acceleration of the ground, risk of liquefaction, landslide and tsunami were estimated on each 500 m mesh in regard to each probable earthquake.

In the step 3, 'Prediction of Damages', damage of buildings, transportation, lifeline, human and so on were estimated based on the results of the step 2. Building damages were estimated on each structural type according to the average deteriorating level of buildings and response spectra on each 500 m mesh including the influence of liquefaction. Transportation was including roads, harbors, railways and airports. Damages of lifelines such as water supply, gas supply, electric power supply, sewerage and telephone networks were estimated based on intensity of ground motion, liquefaction and spreading of fire. In the human damages, number of death, injured and refugee were estimated applying proposed equations based on damages during past major earthquakes.

In Step 4, 'General Estimation', the estimation was produced including attentive points which should be solved or improved immediately.

These predictions revealed that improvement of seismic capacities of existing buildings was an urgent task in the whole country. 'The Network Committee on Seismic Safety Evaluation and Retrofitting of Existing Buildings' was organized according to the law established of late 1995, and branches of the committee have been opened in most of prefectures. Enterprises of the committee also include training specialist, serving manuals, enlightening owners of buildings and so on. The committee reported that hundred thousands buildings have been seismic-safety-evaluated and more than 10% of them have been retrofitted in Japan after Hyogoken-Nanbu Earthquake.

In the implementation of retrofitting of existing buildings, the ranking on priority of retrofitting of buildings has been made. This ranking is made based on both the social and seismic characteristics of the buildings. Social ranking depends on the degree of building function to be required during and after earthquakes, while seismic ranking is determined based on ground acceleration level, building strength and risk of liquefaction. In the case of Hiroshima city with dense population in Hiroshima prefecture, one of the ten ordinance designated cities in Japan, ranking has been made on more than two thousands public buildings by combining the above two elements.

Evaluation of actual base isolated buildings under recent earthquakes (Shimazu, 1997a)

During 1995 Hyogoken-Nanbu earthquake some unexpected cases had been revealed on so called earthquake resistant buildings. One is the unexpected first-story collapse of the multi-story buildings having slits set between every outer columns and low walls with the expectation of overall collapse mechanism, due to ineffectiveness of the slits, in spite of the fact that it has slits set, that is the formation of weak beam-strong column frame collapse type under earthquake loading. Second is also an unexpected collapse, different from the proposed overall collapse mechanism in design, due to the exceeding of beam strength to that of adjacent columns which was caused by unexpected high axial force occurring in beams.

Third is again an unexpected excessive wide cracks which occurred not only over each story beams but also over beam-column joints and each floor slabs even though weak beam-strong column collapse mechanism was fully realized. This building was finally determined to be demolished due to high cost of repair.

The above three phenomena explain that so-called earthquake resistant designed building structures have high possibility to behave in completely different way from a introductory designed way when subjected to strong ground motions with falling into vibration resonance.

It can be concluded from this fact that utilization of base isolation systems would be recommended to building structures constructed on the site locations having high possibility of high intensity ground motions, because of significant reduction in vibration resonant tendency due to complete shift of natural period of buildings from predominant period of site-soil conditions with high energy dissipation due to added dampers in addition to reliably simple behavior of isolated structures.

The utilization of current base isolation systems to buildings in Japan has been initiated in 1985. The number of the base-isolated buildings approved on their construction by the steering committee, Building Center of Japan under the Ministry of Construction has increased considerably since the 1995 Hyogoken-Nanbu earthquake, so that it is exceeding six-hundred as of late 1998.

According to the results of a research project granted from the Building Center of Japan, investigating actual behavior of the base-isolated buildings when subjected to earthquake-induced ground motions, the amplification factors for base isolated buildings reaches three times larger than input peak acceleration for some cases up to 10 gal of ground motion acceleration level in both longitudinal and transverse directions, while these values decrease considerably to be less than 50% for 100 gal or larger of peak ground acceleration.

It was also found out from the recorded values at both top and bottom levels of building structures during earthquakes that base-isolated buildings move like a rigid body during earthquakes, featuring the third merit of seismic isolation in addition to both desirable characteristics of non-resonance and high damping effects against strong earthquakes. In contrast to base-isolated buildings, earthquake-resistant-designed buildings usually behave very complicatedly during earthquakes. Finally, it should be added that the recorded response factors for base-isolated buildings against vertical ground motions range from 0.8 to 2.0 regardless of level of input peak acceleration, which is consistent with those of non-isolated buildings.

During recent earthquakes such as 1994 Northridge and 1995 Hyogoken-Nanbu events, considerably strong ground motions with maximum velocity of more than 100 cm/sec have been recorded. Non-linear dynamic analyses have been made using a single-degree-of-freedom system representing the base isolated building subjected to these recorded strong motions. The natural periods were set from 1 sec to 10 sec. The restoring force characteristic of the isolator was assumed to be linear while dampers to be bilinear. The additional stiffness and yield strength of dampers were selected as the optimum based on minimizing lateral shear force as well as bearing displacement during calculation.

It was found out that more than 5 sec is required for fundamental period of base-isolated buildings in order to control peak acceleration responses within 200 gal, with maximum bearing displacement responses of 60 cm.

Towards a new code

Based on the recognition of necessity of introduction of site-specific characteristic as well as diversity of building product levels into seismic buildings design levels, performance-based design building code has been planned to be established in 2000 in Japan.

Recently buildings are becoming more intelligent, that is, having sensitive non-structural elements or variety of controlling instruments mounted on floors. During and after the Kobe earthquake, the panic caused by the stopping operation of non-structural systems amongst the citizens was extremely high. That inconvenience had been continued for a long time since them. The utilization of new high technological devices such as base isolation systems has been recommended to solve these recent facts regarding earthquake damage.

In the 2000 code in Japan, the combination among the seismic objects (Frame, member, equipment, utensil and soil) and the limitation level (serviceability, reparability and safe) will be established.

Conclusions and Recommendations

Urgent measures for the acceleration of implementation of retrofitting of existing buildings have been strongly recognized. Design procedure of a new code will be based on that of current approach of base-isolation systems, in addition to the combination matrix among seismic objects and limitation levels.

References

Shimazu, T., Prediction of earthquake damages in Japan, presented in the second round meeting, Beijing, 1997a.

Shimazu, T., Evaluation of responses of actual base-isolated buildings under recent earthquakes, presented in the first round meeting, San Diego, 1997b..

STATE-OF-THE-ART OF CONTROL SYSTEMS AND SSI EFFECT ON ACTIVE CONTROL STRUCTURES WITH EMBEDDED FOUNDATION

Franklin Y. Cheng[1], Fellow, ASCE
H.P. Jiang[2], Member, ASCE, Menglin Lou[3]

Abstract

This paper reviews recent developments of structural control systems with focus on hybrid control currently in vogue. The presentation also includes the effect of soil-structure interaction (SSI) on active controlled seismic-resistant structures with embedded foundation. Numerical results reveal that SSI requires much larger control force than that for the same controlled structure without SSI.

Review of Structural Control Systems

Structural control implies that performance and serviceability of a structure are controlled to maintain prescribed limits during the application of environmental loads. Structural control is achieved in several ways: with passive or active control devices or with semi-active or hybrid systems. Passive devices utilize the fact that energy-dissipating mechanisms can be activated by the motion of the structure itself. Examples are 1) friction devices, 2) hysteretic devices, 3) viscoelastic devices, and 4) base-isolation.[8,9,15] Semi-active devices are based on a passive device and improved by installing certain performance-adjusting functions such as semi-active stiffness, and semi-active vibration absorbers.[11,14] Active control devices require external energy for their operation. Examples are 1) active mass damper, 2) active tendons or bracing, and 3) appendages.[1,2,4,11]

---

[1]Curators' Professor of Civil Engineering and Senior Investigator with Intelligent Systems Center, University of Missouri-Rolla, Rolla, MO 65409-0030; [2]Structural Engineer, TrusJoist MacMillan, Greenwood Village, CO 80111 (former Ph.D. student at UMR); [3]Professor and Director, Institute of Structural Theory, Tongji University, China, Visiting Scholar at UMR

In order to gain the advantages of both passive and active control devices, hybrid systems have recently been developed. Types of hybrid may be categorized as active mass damper, active base isolation, and hybrid damper-actuator bracing system.[7,10,11] The former two systems are limited in their application. For large structures and earthquakes of high intensity, an active mass damper needs an extensive amount of energy to achieve control. For tall buildings, active base isolation tends to lose effectiveness. The larger overturning moments of tall buildings can cause the uplift bearing to stop functioning. Furthermore, the control friction force on a sliding surface may fail to be modeled on actual three-dimensional ground motion since earthquake characteristics differ in each direction. In view of the advantages and disadvantages of these two systems, hybrid systems are developed to overcome the weaknesses just discussed and offer strengths such as capability of controlling a wide range of earthquake intensity and multicomponent seismic response. UMR's work has been focused on hybrid damper-actuator bracing system.[6,7]

The hybrid damper-actuator bracing control system is based on a tube fluid damper (passive device) with an active controlled piston mounted on a structural brace as shown in Fig. 1. Passive, active, and hybrid devices can be separately installed at different floors in order to provide both economical and effective control for a structure. The generalized optimal active control (GOAC) algorithm was developed and implemented for structural response simulation with both analytical and shaking table tests.[3,5] This algorithm overcomes the sensitivity problem of time-increment with the optimum instantaneous control algorithm previously developed. Studies showed that for controlling the same amount of allowable displacement of a structure, the active control force for a hybrid control system can be reduced significantly compared to that of an active control system. The hybrid system can be further developed as an intelligent hybrid control system in which active control begins to operate

Fig. 1 Hybrid Control System on Shaking Table

once maximum structural response exceeds the threshold value.[5] Thus structural response due to mild or moderate earthquakes can be mitigated by the passive control component while response due to major earthquakes can be controlled by the hybrid system through an intelligent control algorithm.

Effect of Soil-Structure Interaction (SSI) on Structural Control

Most studies of control of seismic-resistant structures were based on the assumption that structures are fixed-supported. This assumption is valid when structures are built on rock or are not subjected to ground motion such as wind forces or mechanical vibration. It is well-known that ground motion induces SSI which can significantly influence response of a superstructure built on soft soil. Thus to consider SSI modeling of a controlled structure requires more d.o.f. than a fixed-base structure. These additional d.o.f. will relate to the state-space vector and consequently affect control action and even stability. Inclusion of SSI is particularly important in seismic zones such as near-field where the dynamic response of soils can drastically alter structural response, as pointed out in the current Uniform Building Code.[12]

Researchers have studied the effect of SSI on active-controlled structures considering frequency domain with shallow foundation and passive-controlled structure with grouping pile foundation.[13,16] This paper presents a case study of SSI on active-controlled tall buildings with embedded foundation by using boundary element and frequency domain in half-space formulation which represents a practical construction in an urban area with a basement having several floors below ground level.

Sample Response Behavior and Concluding Remarks

A dynamic system of a 10-story one-bay shear structure is equipped with an active tendon system attached to its foundation and second floor. The structure sits on a 6.0 × 6.0 m rigid square foundation having its 3.0-meter depth embedded into a half-plane. Structural properties are: each floor mass (m_i) = 10 ton; each floor mass moment of inertia = 20.83 ton-sq.m; foundation mass = 28.75 ton; foundation mass moment of inertia = 172.5 ton-sq.m; floor translational stiffness = 1244 MN/m; proportional damping ratios = 0.02 for the first mode and 0.10 for the second mode; and floor height = 3.00 m for every floor. Half-plane properties are: shear modulus of elasticity = 7.2 MN/sq.m; soil density = 2.0 gram/cu.cm; Poisson's ratio = 0.33; hysteretic damping ratio = 0.04; and shear wave velocity is 60 m/sec. The first 20-second 1940 El Centro north-south component is applied at the ground system's rigid interface.

Influences of the control parameter s1/r on structural response with consideration of fixed-support and half-space are shown in Figs. 2 and 3. For the FIX-model with control (see Fig. 2), an increase of s1/r ratio results in an increase of maximum control force and then a decrease of maximum top floor displacement relative to footing. At the optimal point, maximum displacement is reduced from 0.053 m (without control) to 0.016 m with maximum control force of 8.68 MN. For the SSI-model with control (see Fig. 3), an increase of s1/r ratio also results in the same pattern as the case of the FIX-model. But the change in maximum top floor displacement relative to footing excluding rotational effect (DISP W/O ROT) is small. This also indicates that a decrease of displacement relative to footing is mainly the result of a decrease of footing rotation. At

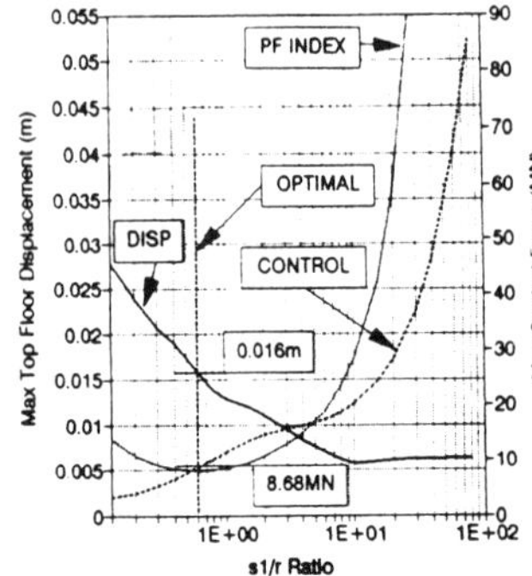

Fig. 2 Top Floor Displacement,
          FIX-Model

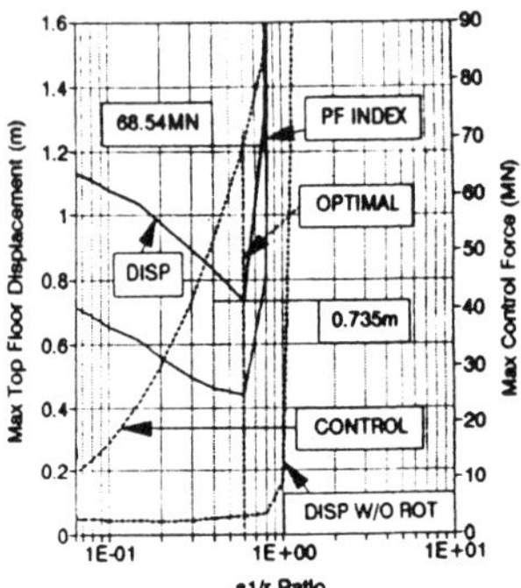

Fig. 3 Top Floor Displacement,
          SSI-Model

the optimal point, maximum displacement is reduced from 1.219 m (without control) to 0.735 m with maximum control force of 68.54 MN. Footing rotation is reduced from 0.035 radian (without control) to 0.022 radian. Responses of these two models are different and require different control forces. The control force required in the SSI-model is greater than that in the FIX-model. An additional amount of control force is needed to limit the movement of foundation. The control force set by the FIX-model's gain matrix is dictated only by floor translations. By using the SSI-model's gain matrix, a control force is determined as the function of both floor and foundation responses.

Acknowledgments

This paper is resulting from the partial support from the National Science Foundation, National Center for Earthquake Engineering Research Center at Buffalo, MRTC and ISC at UMR. The support is gratefully acknowledged.

References

1.   Abdel-Rohman, M. (1984), "Optimal Control of Tall Buildings by Appendages," *ASCE J. Structural Eng.*, Vol. 110, pp. 937-946.

2.   Cheng, F.Y. and Pantelides, C.P. (1986), "Optimum Seismic Structural Design with Tendon and Mass Damper Controls and Random Process," *Proc. ASCE Structures Congress, Recent Developments in Structural Optimization*, ed. F.Y. Cheng, pp. 40-53.

3.   Cheng, F.Y., Tian, P., and Suthiwong, S. (1992) "Generalized Optimal Active Control Algorithm of Seismic Structures and Related Soil-Structure Formulation," *Computational Mechanics in Structural Engineering: Recent Developments and Future Trends*, eds. F.Y. Cheng and Fu Zizhi, Elsevier Science Ltd., pp. 49-62.

4. Cheng, F.Y. and Tian, P. (1994), "Generalized Optimal Active and Hybrid Control for Seismic Structures," *Proc. 1st World Conf. Structural Control*, Vol. 2, TP1, pp. 21-30.

5. Cheng, F.Y. and Tian, P. (1995), "Assessment of Algorithms, Material Nonlinearity, and Foundation Effects on Structural Control of Seismic Structures," *Proc. Intl. Symp. Public Infrastrucure System Research*, pp. 211-226.

6. Cheng, F.Y., Tian, P., Rao, V., Martin, K., Liou, F. and Yeh, J.H. (1996), "Theoretical and Experimental Studies on Hybrid Control of Seismic Structures," *Proc. 12th ASCE Conf. Analysis and Computation*, pp. 322-328.

7. Cheng, F.Y. and Jiang, H. (1998), "Optimum Control of a Hybrid System for Seismic Excitations with State Observer Technique," *J. Smart Materials and Structures*, Issue 5, Vol. 7, pp. 654.663.

8. Earthquake Eng. Research Inst. (EERI) (1993), Theme Issue: Passive Energy Dissipation, *Earthquake Spectra*, Vol. 9, No. 3.

9. EERI (1990), Theme Issue: Seismic Isolation, *Earthquake Spectra*, Vol. 6. No. 2.

10. Feng, Q.M. (1993), "Application of Hybrid Isolation System to Buildings," *ASCE J. Eng. Mechanics*, Vol. 119, No. 10, pp. 2090-2108.

11. Housner, G.W. and Masri, S.F., eds. (1994), *Proc. 1st World Conf. Structural Control*, Vols. 1, 2, and 3.

12. ICBO (1997), *Uniform Building Code*, Vol. 2, Whittier, Calif.

13. Lou, M.L., Wu, J. and Cheng, F.Y. (1999), "The Effect of Soil-Structure Interaction on TMD Control for Wind and Seismic Response of Structure," *Proc. ASCE Structures Congress XVI*.

14. Patten, W.N., Kuo, C.C., He, Q., Liu, L. and Sack, R.L. (1994), "Seismic Structural Control via Hydraulic Semi-active Vibration Dampers (SAVD), *Proc. 1st World Conf. Structural Control*, Vol. 3, FA1, FA2, Los Angeles, pp. 30-38, 83-89.

15. Soong, T.T., and Dargush, G.F. (1997), *Passive Energy Dissipation System in Structural Engineering*, John Wiley & Sons.

16. Wu, W.H. and Smith A.H. (1994), *Optimal Structural Control Considering Soil-Structure Interaction Effects*, Report No. 112, C.E. Dept., Stanford University.

SEISMIC PERFORMANCE OF TUNED MASS DAMPERS
IN REDUCING THE RESPONSE OF STRUCTURES ON SOFT SOIL MEDIUM

Genda Chen[1], Jingning Wu [2], and Menglin Lou[3]

Abstract

Parametric studies in this paper have shown that soil-structure interaction significantly affects the tuned-mass-damper's performance. As the soil medium is softening, the damper becomes less effective in reducing the maximum responses of the structure due to high damping characteristics of the soil-structure system.

Introduction

Seismic effectiveness of a tuned mass damper (TMD) in reducing the maximum responses of a structure depends on good understanding of the structural dynamics. Since soil-structure interaction may significantly modify the dynamic characteristics of a structure and adopt the seismic input to the structure, it will generally affect the seismic performance of a damper device. Previous studies by Samali *et al* (1992) and Gao *et al* (1996) lead to inconsistent conclusions. The objective of the present study is to better understand the effect of soil-structure interaction on TMDs' performance by conducting extensive parametric studies on a building structure.

Modeling of Soil-Structure-TMD Systems

The soil-structure-TMD system under consideration consists of a shear-type of building structure, a tuned mass damper, a rigid mat footing and a half-space soil mass. Soil-structure interaction of the system can be characterized by impedance functions of the rigid foundation. These functions can be expressed into $K_F + i\omega C_F$, in which K_F and C_F are the effective stiffness and damping matrices of the foundation.

---

[1]Assistant Professor and   [2]Ph.D. Candidate, Department of Civil Engineering, University of Missouri-Rolla, 1870 Miner Circle, Rolla, MO 65409-0030.
[3]Professor, Institute of Structural Theory, Tongji University, Shanghai 200092, China.

The motion equation of the soil-structure-TMD system can be written as

$$\begin{bmatrix} M & 0 \\ 0 & M_F \end{bmatrix}\begin{Bmatrix} \ddot{X} \\ \ddot{X}_F \end{Bmatrix} + \begin{bmatrix} C & -C\Gamma \\ -\Gamma^T C & \Gamma^T C\Gamma + C_F \end{bmatrix}\begin{Bmatrix} \dot{X} \\ \dot{X}_F \end{Bmatrix} + \begin{bmatrix} K & -K\Gamma \\ -\Gamma^T K & \Gamma^T K\Gamma + K_F \end{bmatrix}\begin{Bmatrix} X \\ X_F \end{Bmatrix} = -\begin{bmatrix} M & 0 \\ 0 & M_F \end{bmatrix}\begin{bmatrix} \Gamma \\ I \end{bmatrix}\begin{Bmatrix} \ddot{x}_g \\ \theta_g \end{Bmatrix} \quad (1)$$

in which M, C and K are the mass, damping, and stiffness matrices of the structure-TMD system; X and X_F respectively denote the displacement vectors of the structure-TMD system and the rigid footing relative to the free field ground motion; M_F is the mass matrix of the footing; and Γ defines the transformation from the free field motion to that of the superstructure. The quantities $\ddot{x}_g$ and θ_g are the translational and rocking components of the free field acceleration, and θ_g is neglected in this study.

Random Response Analysis

The ground acceleration $\ddot{x}_g$ in Eq.(1) is considered as a stationary stochastic process here. Its power spectral density function can be described by the modified Kanai-Tajimi spectrum. The normalized root-mean-square (RMS) response R_r can then be determined by

$$R_r = \sqrt{E[r^2]/E[r_0^2]} \quad (2)$$

in which $E[r^2]$ and $E[r_0^2]$ represent the mean-square responses of any quantity r of engineering interest of a controlled and uncontrolled structure, respectively. When $R_r < 1.0$, the damper can mitigate the seismic response r. The smaller the value R_r, the more effectively the damper performs. In what follows, the relative displacement x_1 and the absolute acceleration $\ddot{y}_1$ at the top floor of the structure and the base shear force Q_b are discussed in details.

A 12-story shear building resting on viscoelastic soil is considered here. The 16m×16m mat footing of 9.8×10^5 kg is assumed rigid. Five shear wave velocities ranging from 100m/s to infinity are used to simulate various soil conditions. The 45m tall superstructure has a floor mass of 5.3×10^5 kg and an internal stiffness of 8.4×10^5 kN/m with a damping ratio of 1% or 5% for all vibration modes. The tuned mass damper atop the structure is of 8% damping ratio and has a mass equal to 1% of the total mass of the superstructure.

The fundamental frequency and damping ratio of the example structure without dampers are listed in Table 1 for five soil conditions. As one can see, the frequency is reduced by 45% when the shear wave velocity V_s varies from infinity to 100m/s while the damping ratio increases significantly. Due to the material hysteresis and radiation in soil medium, damping of the soil-structure system is insensitive to structural damping when V_s=100m/s.

The normalized RMS responses are calculated for various soil shear wave velocities and plotted in Figs. 1 and 2 for two structural dampings. It can be seen that a damper performs the best when tuned into the fundamental frequency of the soil-structure system. The damper reaches its ultimate performance for a lightly-damped structure resting on stiff soil. The RMS displacement and base shear in this case are reduced by 30%~50%. The damper's performance degrades as the soil medium is softening. Comparisons between Figs.1 and 2 indicate that the damper's performance is more sensitive to structural damping when a structure is supported on stiff soil. When the soil medium becomes very soft, little change in performance is observed regardless of structural damping.

An interaction factor α_r is defined as the response ratio of a structure with and without soil-structure interaction effect. Figure 3 presents the interaction factors corresponding to 1% structural damping ratio. The solid and dotted lines respectively show the interaction factors for an optimally-controlled and uncontrolled structure while the dash line for a structure controlled with a damper tuned into the fundamental frequency of the structure alone (mistuned). It can be observed that the interaction factors, $\alpha_{x_1}, \alpha_{y_1}$ and α_{Q_b}, of a controlled structure are always smaller than those of an uncontrolled structure and generally closer to unity for various shear wave velocities. This indicates the weaker soil-structure interaction due to the presence of dampers. It is also observed that all interaction factors are not particularly sensitive to mistuning of a damper for various shear wave velocities.

Concluding Remarks

Based on this study, the damper's effectiveness rapidly decreases as the soil medium gets softer due to significant contribution to the damping of soil-structure system from soil material hysteresis and radiation effect. For structures resting on very soft soil, soil-structure interaction can make a damper on the structure totally ineffective.

References

Gao, H., Samali, B. and Kwok, K.C.S. 1996. "Structural vibration control by passive dampers considering soil-structure interaction." *Proc. 2nd Int. Workshop on Structural Control*, HKUST, Hong Kong, 174-185.
Samali, B., Kwok, K.C.S. and Geoghegan, P.K. 1992. "Soil-structure-damper interaction under earthquake loading." *Proc. 1st Int. Conf. MOVIC*, Japan, 7-11.

Table 1 Fundamental frequency and damping ratio

	Shear Wave Velocity V_s (m/s)	100	150	200	350	∞
	ω_i (rad/s)	2.7	3.5	4.1	4.6	5.0
ξ_i (%)	1% Struct. damping	7.9	5.7	4.1	2.2	1.0
	5% Struct. damping	8.5	7.1	6.3	5.4	5.0

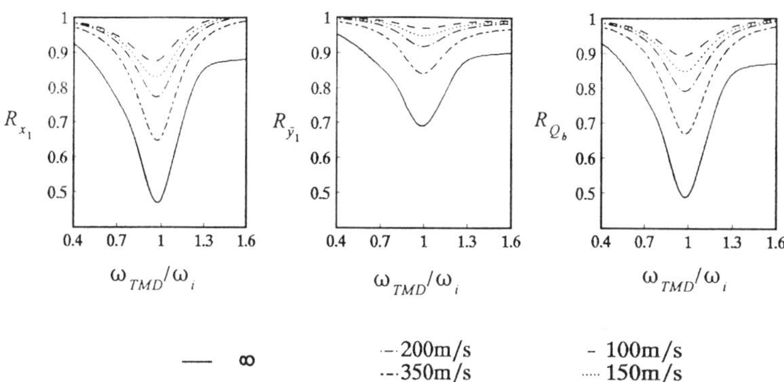

Figure 1 Responses of Flexible-Base Structures: ξ=1%

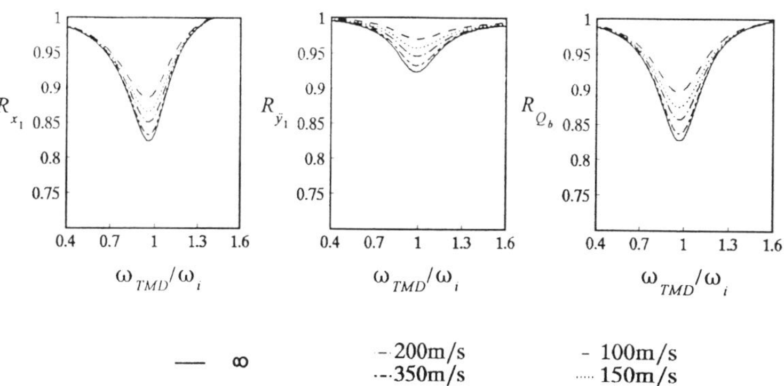

Figure 2 Responses of Flexible-Base Structures: ξ=5%

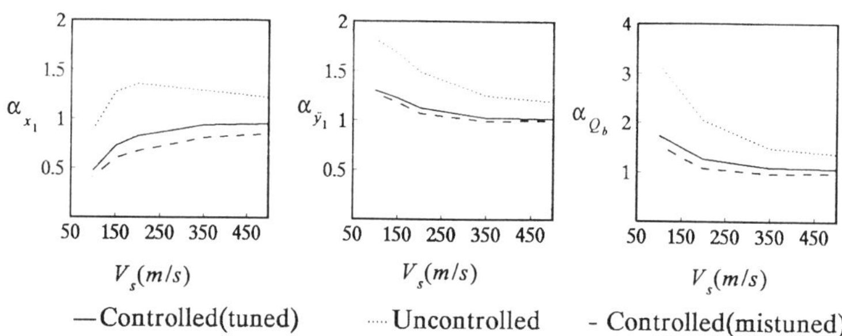

Figure 3 Interaction Factors

EFFECT OF SOIL-STRUCTURE INTERACTION ON TMD CONTROL
FOR WIND AND SEISMIC RESPONSE OF STRUCTURES

Menglin Lou[1], Jingning Wu[2] and
Franklin Y. Cheng[3], Fellow, ASCE

Abstract

Soil-structure interaction (SSI) effect on TMD performance in suppressing wind-induced responses of structures is studied by using random vibration method. Differences in how SSI affects TMD control for mitigation of wind and seismic response of structures are discussed.

Introduction

It is known that the efficiency of structural control is dependent significantly such characteristics as fundamental frequency and damping of a controlled structure. It is also known that natural frequencies and vibration damping of a structural system are modified due to soil-structure interaction (SSI). These facts become particularly important for a controlled structure built on soft soil. The study of the effect of SSI on structural control has arisen in recent years. Several papers have been published on the effect of SSI on active and passive seismic response control of various structures (Wong and Luco, 1991, Kawano et al., 1992, Samali et al., 1992, Sato and Toki, 1992, Alam and Baba, 1993, Cheng et al., 1994, Smith et al., 1994, Lou and Wu, 1997). These

---

[1]Professor and Director, Institute of Structural Theory, Tongji University, Shanghai 200092, China; Visiting Scholar, Dept. of Civil Engineering, University of Missouri -Rolla, Rolla, MO 65409

[2]Ph.D. Student, Dept. of Civil Engineering, University of Missouri-Rolla, MO 65409

[3]Curators' Professor of Civil Engineering and Senior Investigator with Intelligent System Center, University of Missouri-Rolla, Rolla, MO 65409

publications show that SSI affects control process, algorithm and performance efficiency. Some work has also been done on the effect of SSI on TMD wind response control of structures (Xu and Kwok, 1992, Gao et al., 1996). In the research just mentioned, only shallow or embedded foundations are considered and the soil is usually assumed as to be an elastic medium.

In this paper, the effect of SSI on TMD performance in suppressing wind-induced structural responses is studied under more complex conditions: those of group-pile foundations and viscoelastic soil medium. Moreover, differences in the effects of SSI on TMD control for wind-induced vs. earthquake-induced responses is discussed.

Equations of motion

1. Structure-TMD system

When the soil is assumed to be rigid, the equations of motion for a structure-TMD system under wind or seismic excitation can be written in time domain as

$$M\ddot{x}(t) \ + \ C\dot{x}(t) \ + \ Kx(t) \ = \ p(t) \tag{1}$$

where M, C and K are mass, damping and stiffness matrices for structure-TMD system; x(t) is displacement vector with respect to the structure's base and a dot represents the derivative with respect to time; p(t) is dynamic load vector induced by wind or earthquake.

Applying Fourier transformation, Eq.(1) can be written in frequency domain as

$$(K \ - \ \omega^2 M \ + \ i\omega C)X(\omega) \ = \ P(\omega) \tag{2}$$

Using random vibration analysis, the power spectrum of the structural response can be expressed by

$$S_{xx}(\omega) \ = \ H^{*}(\omega)S_{pp}(\omega)H^{T}(\omega) \tag{3}$$

where $H(\omega)$ is transfer function; $S_{pp}(\omega)$ is power spectrum of the random dynamic load which is assumed as a stationary stochastic process. The root-mean-square (RMS) responses of the structure can be obtained from $S_{xx}(\omega)$.

2. Soil-structure-TMD system

Utilizing the substructure method, equations of motion for a soil-structure-TMD system can be expressed in time domain as

$$\begin{bmatrix} M & 0 \\ 0 & M_f \end{bmatrix} \begin{Bmatrix} \ddot{x}(t) \\ \ddot{x}_f(t) \end{Bmatrix} + \begin{bmatrix} C & -CE \\ -E^{T}C & E^{T}CE \end{bmatrix} \begin{Bmatrix} \dot{x}(t) \\ \dot{x}_f(t) \end{Bmatrix} + \begin{bmatrix} K & -KE \\ -E^{T}K & E^{T}KE \end{bmatrix} \begin{Bmatrix} x(t) \\ x_f(t) \end{Bmatrix} = \begin{Bmatrix} p(t) \\ p_f(t) \end{Bmatrix} \tag{4}$$

where $x_f(t)$ is structure's base displacement vector; $p_f(t)$ is interaction force vector between structure and soil. Usually, the interaction force is expressed in frequency domain as

$$-P_f(\omega) = S_f(\omega)X_f(\omega) = K_f(\omega)X_f(\omega) + i\omega C_f(\omega)X_f(\omega) \tag{5}$$

Applying Fourier transformation and substituting Eq.(5), Eq.(4) can be written as

$$\left(\begin{bmatrix} K & -KE \\ -E^TK & E^TKE+K_f(\omega) \end{bmatrix} - \omega^2 \begin{bmatrix} M & 0 \\ 0 & M_f \end{bmatrix} + i\omega \begin{bmatrix} C & -CE \\ -E^TC & E^TCE+C_f(\omega) \end{bmatrix} \right) \begin{Bmatrix} X(\omega) \\ X_f(\omega) \end{Bmatrix} = \begin{Bmatrix} P(\omega) \\ 0 \end{Bmatrix} \tag{6}$$

Formulations similar to those of Eq.(2) and Eq.(3) can be obtained to evaluate the wind or seismic RMS response of the structure. In this paper, the spectrum of longitudinal turbulence, including the space correlation, is used as the along-wind load spectrum (V_{20} = 20m/sec). The modified Kanai spectrum serves as seismic power spectrum. Maximum acceleration of ground motion is equal to 0.2g.

Efficiency of TMD performance

When wind or seismic RMS responses $\bar{s}_x$ and s_x are computed for controlled and uncontrolled structures, the efficiency ratio R_x is defined for TMD performance as: $R_x = \bar{s}_x/s_x$. If $R_x = 1$, it indicates that TMD does not perform its function. If $R_x < 1$, it indicates that TMD mitigates structural response. The smaller the ratio, the more efficient the TMD performance. If $R_x > 1$, it indicates that TMD behaves contrarily and dynamic responses of controlled structures increase compared to those of uncontrolled structures.

A numerical example of a building with group-pile foundation and different soil properties is presented for illustration purpose. A 24-story building with 16m square section is 96m tall. The group-pile foundation consists of 64 concrete piles and a 16m square mat. Each pile has a circular cross-section of 0.4m diameter. Five shear wave velocities of the soil are considered: V_s = 100, 150, 200, 350 and ∞ m/sec. $V_s = \infty$ m/sec represents rigid soil. Hysterestic damping ratio of the soil medium is chosen as 0.05 for wind excitation and 0.125 for seismic excitation. In consequence, fundamental frequency of the structure is equal to 0.28, 0.31, 0.33, 0.36 and 0.41Hz, respectively. Figs. 1 and 2 show the numerical results of the ratios R_a and R_Q for wind-induced horizontal accerelation $a(t)$ at the top of the building and total shear force $Q(t)$ acting on the building base. Correspondingly, Fig. 3 depicts for seismic-induced response. In these figures, ω_0 is the fundamental frequency of the building considering SSI effect.

Conclusions

1. Numerical results show that soil-structure interaction has some influence on the

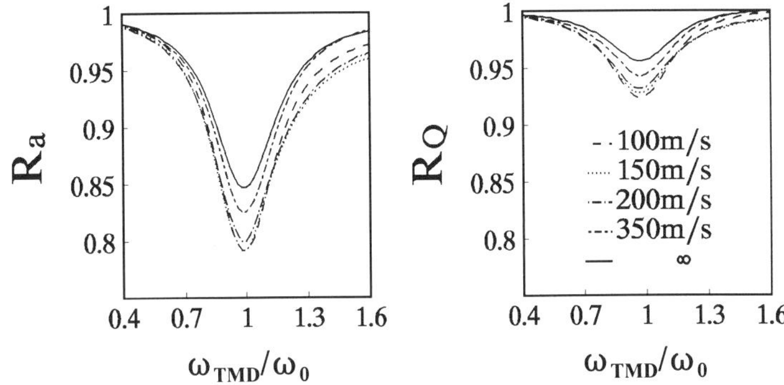

Figure 1 TMD Performance Efficiency (wind excitation, structural damping 5%)

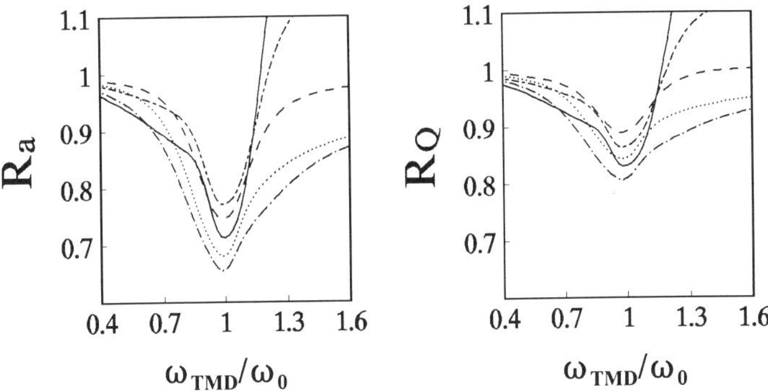

Figure 2 TMD Performance Efficiency (wind excitation, structural damping 1%)

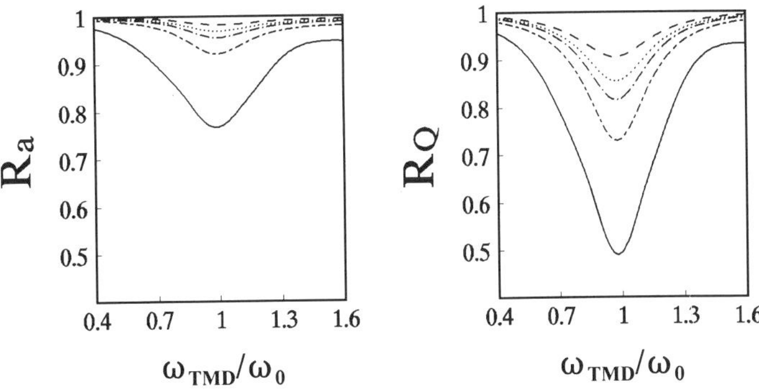

Figure 3 TMD Performance Efficiency (seismic excitation, structural damping 1%)

efficiency of TMD performance to mitigate the wind response of structures. But the effect is not as strong as on TMD performance to mitigate the seismic response of controlled structures.

2. Structural damping and shear wave velocity in a soil medium are the most influential factors to possibly defeat the TMD performance in wind and seismic applications.

3. To minimize wind and seismic response of a structure, TMD must be tuned to the fundamental frequency of the soil-structure system instead of the structure alone. This is especially important for TMD application to mitigate wind-induced response of structures with low structural damping.

4. The effect of SSI on wind- and seismic-induced responses of controlled structures should be studied before designing TMD.

Acknowledgments

This research is sponsored by National Natural Science Foundation of China through Grant No. 59778027 and State Education Committee of China through Grant No. 9524713. This support is gratefully acknowledged.

References

Alam, S. M. and Baba, S. (1993), "Robust Active Optimal Control Scheme Including Soil-Structure Interaction." *J. Struct. Engrg.*, ASCE, 119(9), 2533-2551

Cheng, F. Y., Suthiwong, S. and Tian, P. (1994), "Generalized Optimal Active Control with Embedded and Half-Space Soil-Structure Interaction." *Proc. 11th Conf. on Analysis and Computation* (ed., F. Y. Cheng), 337-346

Gao, H., Samali, B. and Kwok, K. C. S. (1996), "Structural Vibration Control by Passive Dampers Considering Soil-Structure Interaction." *Proc. 2nd International Workshop on Structural Control,* Hong Kong, 174-185

Kawano, K., Venkataramana, K. and Fnrukawa, K. (1992), "Seismic Response of Offshore Platform with TMD." *Proc. 10th WCEE*, 2240-2246

Lou, M. and Wu, J. (1997), "Effects of Soil-Structure Interaction on Structural Vibration Control." *Dynamic Soil-Structure Interaction* (eds., C. Zhang and J. P. Wolf), 188-201

Samali, B., Kwok, K. C. S. and Geoghegan, P. K. (1992), "Soil-Structure-Damper Interaction under Earthquake Loading." *Proc. 1st Int. Conf. on Motion and Vibration Control*, Japan, 164-169

Sato, T. and Toki, K. (1992), "Predictive Control of Seismic Response Structure Taking into Account the Soil-Structure Interaction." *Proc. 1st Eur. Conf. on Smart Structures and Materials*, Scotland, 245-250

Smith, H. A., Wu, W. and Borja, R. (1994), "Structural Control Considering Soil-Structure Interaction Effects." *Earthq. Engrg. and Struct. Dyn.*, 23(6), 609-626

Wong, J. P. and Luco, J. E. (1991), "Structural Control Including Soil-Structure Interaction Effects." *J. Engrg. Mech.*, ASCE, 117(10), 2237-2250

Xu, Y. L. and Kwok, K. C. S. (1992), "Wind-Induced Response of Soil-Structure-Damper System." *J. Wind Engrg. and Industrial Aerodyn.*, 41, 2047-2068

Considerations on Soil-Structure Interaction Effects
on Vibration Control of Seismically Excited Structures

Raimondo Betti[1], Member

Abstract

In this paper, some considerations on the effects of the complex phenomenon
of soil-structure interaction on the design and analysis of vibration control systems
for structures subjected to earthquake excitation are presented.

Introduction

In recent years, there has been increasing interest in studying the effects of
dynamic soil-structure interaction (SSI) on structural systems that are equipped with
vibration control systems and that are subjected to earthquake excitation. Such
interest has sprung from the recognition that soil-structure interaction plays a
fundamental role in the seismic response of large structures, structures that are now
becoming equipped with vibration control devices, either active, passive or both.

These research initiatives have mainly focused on active control devices and
have followed two main approaches (Luco (1998)). In the first approach,
investigations are based on the analysis of soil-structure interaction effects on the
seismic response of structures with control systems that have been designed without
accounting for the SSI effects. Such control devices are designed by using different
control methodologies (mainly optimal control theory) on fixed-base structural
models and then their efficiency is tested on similar structural models that include
SSI. This two-stage process could lead to erroneous results since, in this case, it is
quite cumbersome to guarantee the structural stability. The second approach directly
considers the inclusion of the soil-structure interaction effects into the design of the
control systems. In this case, stable control algorithms that include this interaction
from the beginning have to be determined.

---

[1] Associate Professor, Department of Civil Engineering and Engineering Mechanics,
Columbia University, New York, NY 10027.

Nonlinear vs. Linear Analyses

The effectiveness of vibration control systems in reducing the structural response is mainly measured on their performance during strong ground shaking. In this type of event, the behavior of the soil-foundation-superstructure system is strongly nonlinear. The soil has mechanical properties that, in the case of large deformations, cannot be considered linear. Soil liquefaction could be present and play an important role in the soil response. Similarly, rapid variation of the soil morphology and topography could lead to a complex pattern of the spatial distribution of the ground motion. Nonlinearities in the form of lost contact and/or uplifting could occur at the interface between the surrounding soil and the foundation. In addition, for the case of strong ground motion, it is unrealistic to expect the control system to be able to keep the structure within the linear range, unless a large control force is provided. In the case that the structure presents a nonlinear behavior, control gains, obtained by using classical control theory on an equivalent linearized structural model, can bring the structure to final collapse because of instability problems (Huang (1998)). Hence, because of all these sources of nonlinearities, we can conclude that a realistic design of vibration control systems for structures with included SSI effects should be placed within a nonlinear framework.

To capture the essence of the soil-structure interaction on the response of actively controlled structures, linear soil-foundation-superstructure models have been used (i.e. Wong and Luco (1990), Smith and Wu (1997) and Luco (1998)). In these studies, the soil has been modeled as an infinite, isotropic, homogeneous half-space while the structural models are two-dimensional single or multiple degree-of-freedom systems and shear beams. Although different approaches are used, all the studies converge on the determination of an "equivalent" or "modified" structural system that includes both the effects of the soil-structure interaction and those of the control action. The results show that a control algorithm that considers SSI effects is more effective in reducing the structural response but requires control forces larger than those required for the corresponding case of a fixed-based model.

Frequency Domain Approach vs. Time Domain Approach

One of the main difficulties in the analysis of SSI effects on the response of actively controlled structures is that, traditionally, the SSI problem is analyzed in the frequency domain while the classical control analysis is performed in the time domain. As an example of such a duality, it is noteworthy that the foundation impedances are frequency-dependent functions and that, in optimal control theory, the cost functional contains an integral over the duration of the event.

To circumvent such differences, one approach would be to obtain the impedance functions in the time domain from the corresponding counterparts in the frequency domain. A formulation of such an approach has been presented by Yoshida (1996), using the Hilbert transform and the causal inverse fast Fourier transform to derive the impulse function of the dynamic stiffness of the soil.

An alternative approach, used in the previously mentioned studies, consists in obtaining an " equivalent" or "replacement" model supported on a rigid soil and characterized by a modified natural frequency and damping ratio. These values of the frequencies and damping ratios are adjusted to account for the frequency dependent nature of the impedance functions and for the associated radiation damping. Once such a "new" structural system has been defined, then a "normal" control methodology is followed for determining the control gains.

A third approach is to assume approximate frequency-independent foundation impedances and formulate the whole SSI/control problem directly in a time domain framework. Although it is always possible to introduce an observer to compensate for the inaccurate representation of the soil-structure interaction effects, possible observation spillover associated with the uncontrolled modes is a cause of major concern.

Three Dimensional Analysis – Torsional Effects

In performing two-dimensional analyses in SSI problems, emphasis is placed on translational and rocking components of the Foundation Input Motion. However, previous studies done by the author on the three-dimensional, dynamic soil-structure interaction of cable-supported bridges (Betti *et al.*(1993)) have shown that there is a strong coupling between the in-plane and the out-of-plane vibrational modes. In fact, because of the foundation presence, even the simplest representation of the earthquake action results in a complex three-dimensional response of the foundation and, consequently, of the superstructure. The three-dimensionality of such a response is also emphasized by asymmetry of the foundation geometry and of the structural mass distributions. The importance of the lateral and torsional structural responses was also emphasized in the work by Hahn and Liu (1996) who investigated the structural responses to vertically propagating, incoherent seismic motion. To visualize the three dimensional character of the structural response and its effects on the vibration control systems, let us consider the model of a SDOF system with active tendon subjected to arbitrarily inclined SH-waves, as shown in Fig. 1.

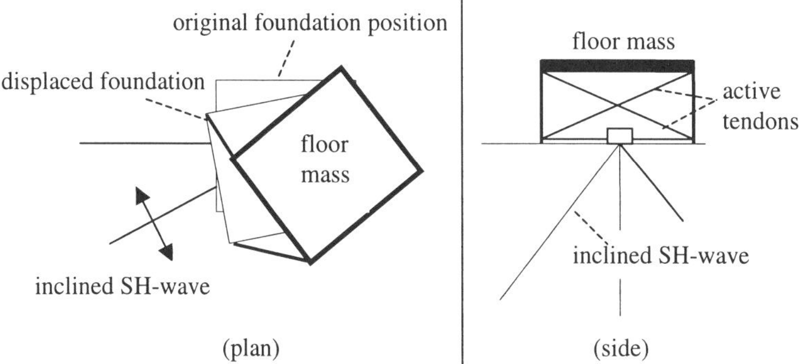

Figure 1: Effects of Torsional Motion on Vibration Control Systems

It is immediately evident how the torsional component of the structural response imposes additional requirements to the control device, requirements that are completely neglected in a current control analysis. This will lead to an underestimation of the required control action and to a reduction of the effectiveness of the vibration control system.

Conclusions

From this brief review of some of the most important effects of the dynamic soil-structure interaction on the efficiency of vibration control systems for structural applications in seismic zones, it appears that a thorough analysis of such effects must be three-dimensional and must include some elements of structural and soil nonlinearities. Although the present studies on such a topic represent an excellent starting point, more research is needed in the area of vibration control of nonlinear, hysteretic structures with soil-structure interaction.

References

- Betti, R., Abdel-Ghaffar, A.M. and Niazy, A.S.: "Kinematic Soil-Structure Interaction for Long-Span Cable Supported Bridges", *Earthquake eng. struct. dyn.*, Vol. 22, No. 5, 415-430 (1993).
- Hahn, G.D., and Liu,X.:"Assessment of Lateral and Torsional Structural Responses Induced by Incoherent Ground Motions", *Proceedings of the ASCE Engineering Mechanics Specialty Conference*, Fort Lauderdale, FL., 188-191, (1996).
- Huang, K, and Betti, R.:"Instantaneous Optimal Control of Nonlinear Hysteretic Structures Subjected to Earthquake Excitation", submitted to *Earthquake eng. struct. dyn.* (1998)
- Luco, J.E.:"A Simple Model for Structural Control Including Soil-Structure Interaction Effects", *Earthquake eng. struct. dyn.*, Vol. 27, 225-242 (1998).
- Smith, H.A., and Wu, W.-H.:"Effective Optimal Structural Control of Soil-Structure Interaction Systems", *Earthquake eng. struct. dyn.*, Vol. 26, 549-570, (1997).
- Wong, H.L., and Luco, J.E.:"Active Control of the Seismic Response of Structures in the Presence of Soil-Structure Interaction Effects", *Proceedings of the U.S. National Workshop on Structural Control Research*, Los Angeles, CA, 231-235, (1990).
- Yoshida, N.:"Dynamic Sub-Structure Method in Time Domain Using Analytical Representation of Dynamic Stiffness of Soil", *Proceedings of the ASCE Engineering Mechanics Specialty Conference*, Fort Lauderdale, FL., 184-187, (1996).

EFFECT OF NEAR-FAULT GROUND MOTIONS ON THE RESPONSE OF FRAME STRUCTURES

Babak Alavi[1] and Helmut Krawinkler[2]

Abstract

The results of the study summarized here are intended to shed light on some of the important issues that affect the response of frame structures to near-fault ground motions. Examples are presented that indicate that such motions may cause very high story ductility demands in structures designed to present code requirements.

Introduction

Recordings from recent earthquakes have provided much evidence that ground shaking near a fault rupture is characterized by a pulse with very high energy input. This holds true particularly in the "forward" direction, where the propagation of the fault rupture towards a site causes most of the seismic energy from the rupture to arrive in a single large pulse of motion that occurs at the beginning of the record (Somerville, 1998).

Velocity spectra of the fault-normal and fault-parallel components of a typical near-fault record are shown in Fig. 1. The figure shows a large difference between the two components, and indicates in the velocity spectrum

Table 1. Near-Fault Ground Motions whose Characteristics are Illustrated in Figs. 2 and 3

Designation	Earthquake	Station	Magnitude	Distance
LP89lgpc	Loma Prieta, 1989	Los Gatos	7.0	3.5
LP89lex	Loma Prieta, 1989	Lexington	7.0	6.3
EZ92erzi	Erzincan, 1992	Erzincan	6.7	2.0
LN92lucr	Landers, 1992	Lucerne	7.3	1.1
NR94rrs	Nothridge, 1994	Rinaldi	6.7	7.5
NR94sylm	Nothridge, 1994	Olive View	6.7	6.4
KB95kobj	Kobe, 1995	JMA	6.9	3.4
KB95tato	Kobe, 1995	Takatori	6.9	4.3

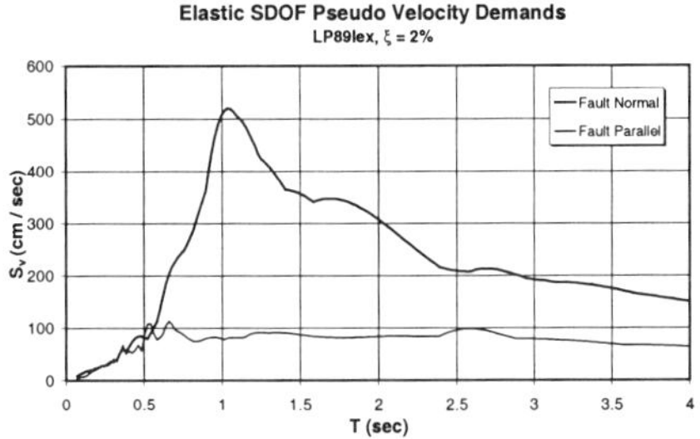

Figure 1. Velocity Spectra of a Near-Fault Record, Fault-Normal and Fault-Parallel Comp.

[1] Research Assistant, Dept. of Civil & Env. Engineering, Stanford U., Stanford, CA 94305-4020
[2] Professor, Dept. of Civil & Env. Engineering, Stanford U., Stanford, CA 94305-4020

of the fault-normal component a dominant peak at a well defined period, which is the period of the pulse contained in the record. The large intensity and particular nature of the fault-normal component deserves great attention in the design of structures located near seismic sources. This paper summarizes a few of the findings of two related studies on the effects of near-fault ground motions on the story ductility demands of frame structures. The findings are based on analytical studies of single-bay frames, which are subjected to a set of fault-normal components of near-fault records and to simple pulses that represent the effects of these record components.

Ductility Demands for Recorded Near-Fault Ground Motions

The velocity spectra of several records (see Table 1) are shown in Fig. 2. Superimposed is the mean spectrum from a set of reference ground motions that are scaled to match the 1997 UBC soil profile S_D ground motion spectrum disregarding near-fault effects.

The ductility demands imposed by these ground motions on frames with a fundamental period of 2.0 sec. and of two different strengths are shown in Fig. 3. The base shear strength of the frame structures is defined by the seismic coefficient $\gamma = V_y/W$. For the strong structure ($\gamma = 0.4$, Fig. 3a), the ductility demands are relatively high in the upper stories, but for the weak structure ($\gamma = 0.15$, Fig. 3b), the ductility demand

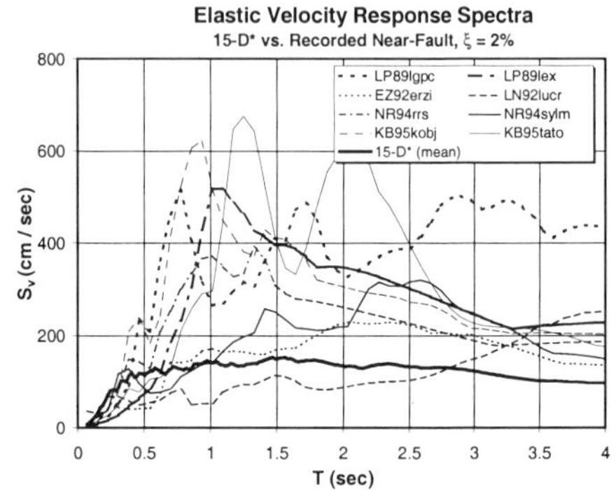

Figure 2.   Velocity Spectra of Near-Fault Records and Reference Ground Motions

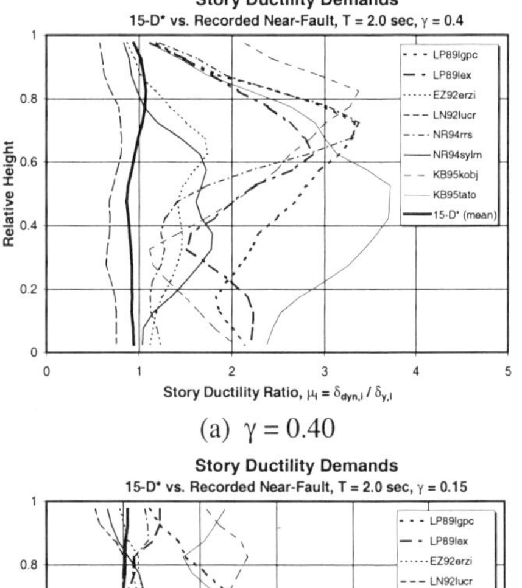

(a) $\gamma = 0.40$

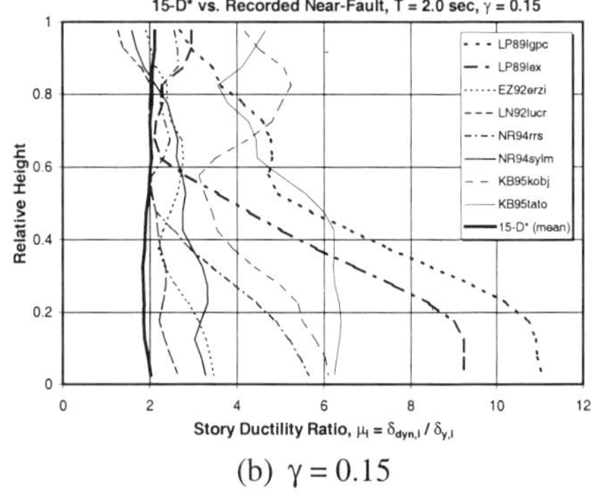

(b) $\gamma = 0.15$

Figure 3. Story Ductility Demands for Several Near-Fault Records, T = 2.0 sec.

is high in the lower stories. This migration of maximum story ductility demands from the upper to the lower portion of structures, with a decrease in strength, is a fundamental characteristic of the response of frames to near-fault records. So are the very large demands imposed by these records compared to the demands imposed by "standard" design ground motions (see curves for 15-D*).

Representation of Near-Fault Records by Equivalent Pulses

Studies by the authors (Alavi & Krawinkler, 1997, Krawinkler & Alavi, 1998) have shown that the effects of the fault-normal component of near-fault records can be represented by the effects of equivalent pulses. Fig. 4 shows one of the three pulses investigated. The pulse is defined by a pulse period T_p and a maximum ground acceleration $a_{g,max}$. The response of frame structures to this pulse shows the same characteristics as the response to near-fault records (compare Figs. 3 and 5).

Elastic and inelastic SDOF strength demand spectra for this pulse are shown in Fig. 6. If MDOF effects are explicitly accounted for, the relationships between story ductility demand and base shear strength shown in Fig. 7 are obtained. If the pulse period T_p and effective ground acceleration $a_{g,max}$ of the pulse that represents the near-fault ground motion are known, this figure can be used directly for design.

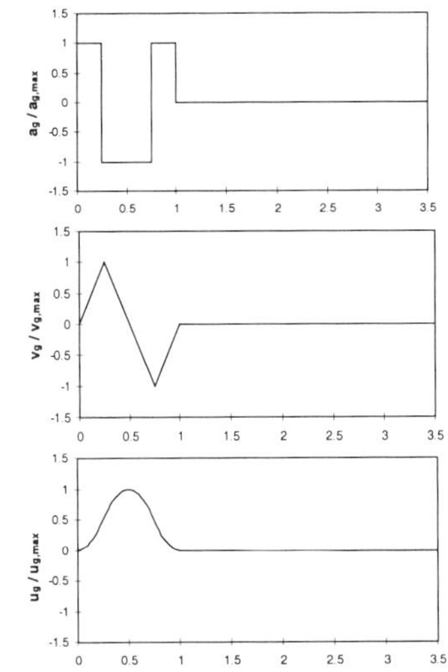

Figure 4. Acceleration, Velocity, and Displacement Time Histories of Pulse P2

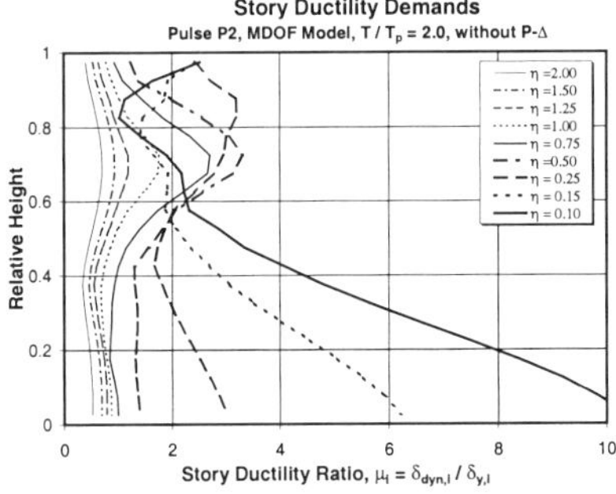

Figure 5. Story Ductility Demands for Pulse P2, Various Values of η, $T/T_p = 2.0$

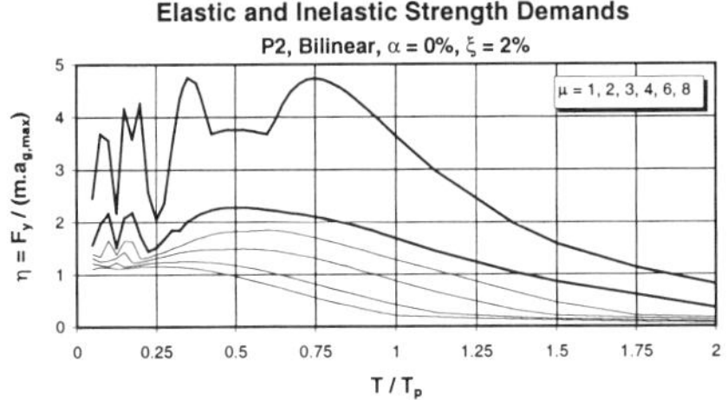

Figure 6. Elastic and Inelastic Strength Demand Spectra for Pulse P2

Design Implications

A preliminary model that relates period and PGV of equivalent pulses to earthquake magnitude and distance is proposed in Somerville, 1998. Design implications can be investigated by using this model and the story ductility demand spectra in Fig. 7.

For instance, for frame structures located 3 km from a seismic source, the base shear strength demands shown in Fig. 8 are obtained for a magnitude 7.0 event. Also shown are estimates of the strength of structures designed according to the 1997 UBC, without and with the appropriate near-fault factor. An R factor of 8 and an overstrength of 2 are assumed, thus, the UBC spectral values are divided by 4. As can be seen, in certain period ranges the ductility demands for code designed structures are very large. But it must be considered that these demands are obtained for the fault-normal component with forward directivity.

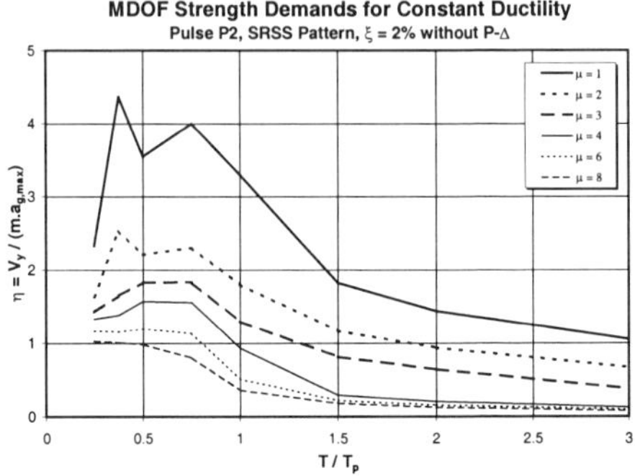

Figure 7. MDOF Base Shear Strength Demands for Target Maximum Story Ductilities

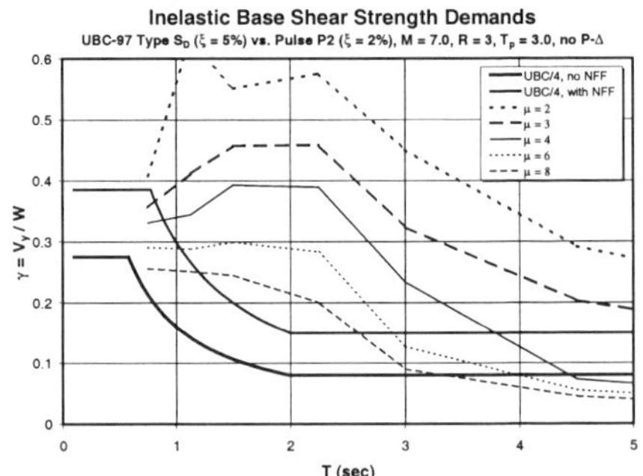

Figure 8. Base Shear Strength Demands for Equivalent Pulse P2 at M = 7.0 and R = 3 km

Acknowledgments

This research is supported by grants from the California Strong Motion Instrumentation Program and from the CUREe/Kajima research program.

References

Alavi, B. and Krawinkler, H., 1997. "Structural Design Implications of Near-Field Ground Motion," Kajima-CUREe Research Report 1997.9.

Krawinkler, H., and Alavi, B., "Development of Improved Design Procedures for Near-Fault Ground Motions," *SMIP98 Seminar on Utilization of Strong-Motion Data*, Oakland, CA.

Somerville, P.G., 1998. "Development of an Improved Ground Motion Representation for Near Fault Ground Motions," *SMIP98 Seminar on Utilization of Strong-Motion Data*, Oakland, CA.

Effects of Vertical Content of Ground Shaking
on 2-D and 3-D Inelastically Responding Steel Frames

by Gregory A. MacRae[1], David Fields[2] and Josh Mattheis[3]

Introduction

Structures in seismic regions are subject to vertical as well as horizontal components of ground shaking. In the past, vertical ground motions have generally been ignored in design. However, recent earthquakes have shown that large vertical accelerations may occur at the same time as large horizontal accelerations. While a significant number of studies have been conducted to understand the effect of vertical shaking on elastically responding structures, as summarized by MacRae (1998), very little work has been carried out to evaluate vertical shaking effects on the response of inelastically responding 2-D or 3-D structures. This paper examines the behavior of steel moment resisting frames to vertical and horizontal shaking from near-fault ground motions.

Structures Analyzed and Records Used

Two-dimensional frames of 3, 9 and 20 story moment resisting structures, designed for Los Angeles by SAC (MacRae, 1998), referred to as LA3, LA9 and LA20 respectively, as well as a 3 story 3-D building frame similar to the LA3 structure, were analyzed with SAC near-fault ground motion records using the programs DRAIN-2DX and DRAIN-3DX. Centerline models of the frame were used and members were provided with 3% strain hardening. Rayleigh damping of 2% was used for the first mode and at a period of 0.2s. *P*-delta effects were considered. Horizontal and vertical masses were lumped at joints. Horizontal fundamental periods for the structure were 1.03s, 2.34s and 3.98s for the LA3, LA9 and LA20 frames respectively. Vertical periods, found by setting the horizontal mass to zero, were 0.058s, 0.097s and 0.173s respectively. An analysis time step of 0.001s was used. Smaller time steps showed identical behavior. However, the acceleration record time steps of 0.01s or 0.02s may be too large to adequately represent the record characteristics for some of these short period structures.

---

[1] Asst. Prof., Dept. of Civil & Env. Eng., Univ. of WA, Seattle, WA98195-2700
[2] Skilling Ward Magnusson Barkshire, 1301 5th Ave, Suite 3200, Seattle, WA98101
[3] Former Grad. Student., Civil & Env. Eng., Univ. of WA, Seattle, WA98195-2700

Frame Response

Vertical ground accelerations primarily affected column axial loads. For external seismic columns, "median" axial compression forces were increased by less than 5% due to vertical shaking. LA9 frame external column median peak compressive axial forces for the 20 strike-normal records (*NF-SN*) are compared with the strike-normal values and vertical records together (*NF-SN-V Actual*) in Figure 1. Also shown is the *gravity* load without horizontal shaking, the axial load assuming a *mechanism* in which all beams yield at their ends simultaneously without strain-hardening due to lateral and gravity loading, and predictions of the likely response using the square-root-of-the-sum-of-the-squares (SRSS) and the sum-of-absolute-values (SAV) methods. Predictions were based on analyses with *gravity and horizontal shaking* together, and *vertical shaking* alone, rather than on independent analyses with *gravity load* alone, *horizontal shaking* alone, and *vertical shaking* alone, to show vertical shaking effects on the response. The value of 15% of the squash load, $0.15f'_yA$, is also shown.

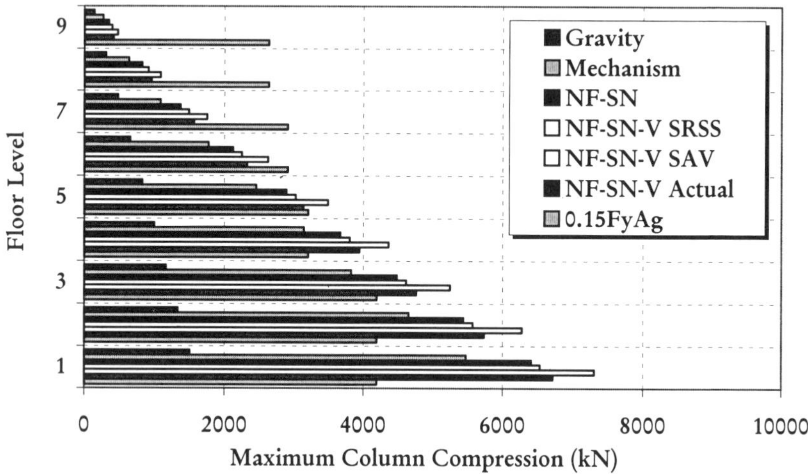

Figure 1. Median peak external column compression from near-fault strike normal horizontal and vertical accelerations for LA9

It may be seen that the *NF-SN* force is significantly larger than the gravity axial force. It is also larger than the *mechanism* response as a result of strain hardening in the beams. The SAV estimate is always greater than the *NF-SN-V Actual* axial force. The SRSS method generally underestimates the actual response by about 5% because SRSS is based on the assumption of elastic behavior. If the response is inelastic due to beam yielding then the axial force due to horizontal shaking is limited as shown in Figure 2 and the possibility of axial loads from horizontal and vertical shaking being near their peaks simultaneously is increased.

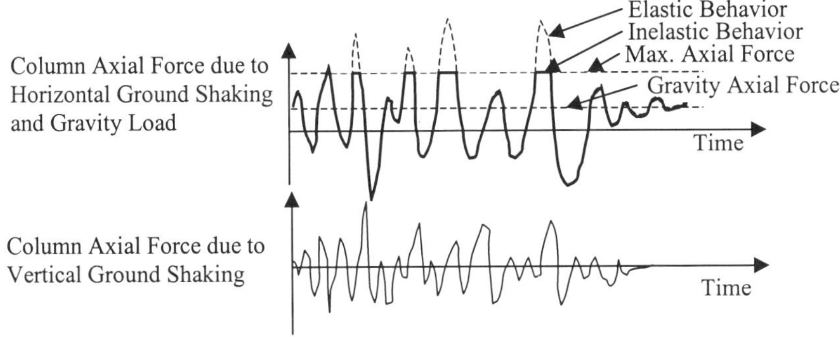

Figure 2. Effect of Horizontal and Vertical Shaking on Column Response

Internal seismic column axial forces did not change with lateral shaking since the seismic shears in the beams either side of these columns cancelled out. However, they did change by 63% in the first story and by 96% in the top story of the LA9 frame due to vertical shaking. Each story had greater percentage increases than the story below it. SDOF oscillators with periods equal to the fundamental period of the internal column or of the frame had median spectral accelerations greater than 1.0g implying an increase in axial force of more than 100% for an elastic SDOF oscillator. The column increases in axial force were less than 100% since the mass was distributed over the frame height. The internal columns achieved tension in few analyses, one out of twenty possible instances for LA3, 3/20 for LA9, and 2/16 for LA20.

Table 1 shows that modeling a frame with the nodal masses distributed at quarter points along the beam length (*member mass*), tends to estimate similar or slightly smaller axial forces than if rather than mass concentrated at the joints, (*joint mass*).

Table 1. Joint and Member Vertical Mass Modeling Effect
on Level 1 Column Compressive Force in LA3, Record NF1_2V

Column	Gravity Axial Force (kN)	Column Axial Force (kN)	
		Joint Mass	Member Mass
1	485	1878	1882
2	832	1749	1749
3	837	1678	1589
4	819	2559	2394
5	490	948	846

In the 3-D frame, gravity columns were pinned at the base. Column axial forces were not affected by lateral shaking but vertical shaking increased their axial forces from $0.30f_yA$ to $0.495f_yA$. This also increased the likelihood of column yielding as shown in the strong-axis moment versus axial load (M-P) interaction plots at the top of the first story column in Figure 3. No tension occurred in any columns of the 3 story 3-D frame due to vertical shaking.

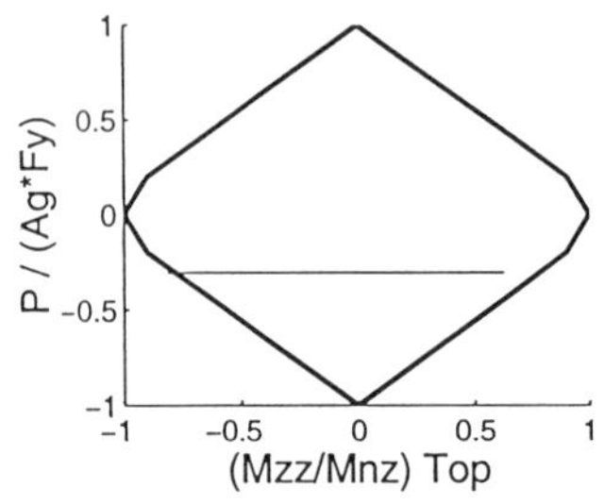

(a) Horizontal Shaking Alone

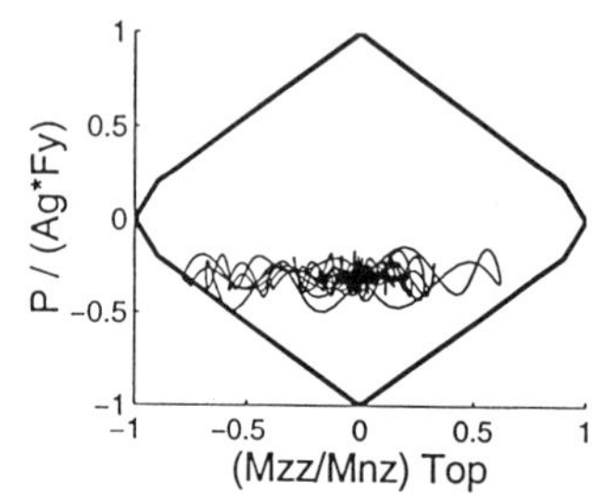

(b) Horizontal and Vertical Shaking

Figure 3. Gravity Column at Top of Story 1 *M-P* Interaction Plot, Record NF1718v90

Conclusions

Analyses were carried out of 2-D and 3-D steel frames to evaluate the effects of vertical shaking. It was shown that:

1. Since fundamental vertical response periods of frames may be less than 0.06s, ground motion records with normal time-steps may not contain sufficient information to allow an accurate understanding of structural response to be obtained.

2. Vertical ground accelerations primarily affected column axial loads. Internal seismic columns were more affected than external seismic columns.

3. The SRSS prediction of median peak external seismic column compression due to vertical and horizontal shaking was about 5% non-conservative due to frame plasticity. The SAV method was always conservative.

4. Distributing the vertical masses along quarter points of the beam caused similar or slightly smaller column axial loads compared to lumped joint mass loading.

5. Vertical shaking in the gravity columns of the 3 story 3-D frame increased the axial forces from $0.30f_yA$ to $0.495f_yA$. Flexural yield was reached in these columns.

Acknowledgements

The authors gratefully acknowledge Federal Emergency Management Agency (FEMA) who partially funded this work through the SAC Steel Project. The opinions expressed in this paper are those of the authors and are not necessarily those of the sponsor.

References

MacRae G. A., "Parametric Study of the Effect of Seismic Demands of Ground Motion Intensity and Dynamic Characteristics", Draft Report, SAC Steel Program, June 1998.

NEAR-SOURCE EFFECTS AND THE ISOLATION PROVISIONS OF THE 1997 UBC

John F. Hall[1] and Keri L. Ryan[2]

Abstract

Demands placed on base-isolated buildings by near-source ground motions are evaluated through time-history analyses of 6-story buildings designed according to the 1997 Uniform Building Code. Results are presented for actual ground motion recordings containing near-source effects as well as for a suite of simple pulse motions.

Introduction

Near-source factors for base-isolated buildings increased significantly in the 1997 Uniform Building Code compared to the 1994 edition. This reflects greater understanding of the potential for large, rapid ground displacements in the near-fault region to place severe demands on long-period structures. Such demands can be assessed through computer simulation as presented here. The isolated building studied is six stories in height and meets the near-source requirements of the 1997 UBC.

Building Design

Dimensions of the structure are shown in Figure 1. Dead loads of 3.83 *kPa* roof, 4.55 *kPa* floors 2 to 6, 6.22 *kPa* first floor, and 1.68 *kPa* cladding give a total dead weight W = 31.14 *MN*. Asuming Zone 4 and soil S_D, source and fault distance are chosen to give the highest near-source factor N_v of 2 (Case N2) or an intermediate value of 1.6 (Case N1.6).

Design of the isolation system opts for squat rubber bearings without supplemental damping or restraint devices, as advocated by Kelly (1997). Short bearings and a large design displacement mean high design shear strains. Beyond about 200%, stiffening in the rubber provides natural restraint which is viewed as an advantage. Selection of the code damping factor β_D as 10% and the nominal design period T_D as 2.6 seconds produces a bearing design displacement D_D of 69 *cm* (Case N2) or 55 *cm* (Case N1.6, a reduction in D_D of 20%). Accounting for torsion and maximum earthquake increases this displacement (D_{TM} in the code) by about 50%, at which a stability check is required. With these considerations, the bearing diameter is taken as 114 *cm* (Case N2) or 91 *cm* (Case N1.6).

The 20 columns are supported by identical bearings. To achieve T_D = 2.6 seconds, the secant shear modulus of the rubber at displacement D_D is taken as G_{SEC} = 344 *kPa*, which is about as low as is currently practical, and this gives a rubber height of 38 *cm* (Case N2) or 30 *cm* (Case N1.6). Thus, the design displacement D_D corresponds to a shear strain of 181%, and D_{TM} to 271%. The horizontal secant stiffness of a bearing is K_{SEC} = 927 *kN/m*. Each bearing is fully bonded to the structure and foundation (not a dowelled connection).

The base shear V_S for superstructure design is the force carried by all bearings at displacement D_D, divided by R_I = 2. Thus, V_S/W = 0.205 for Case N2 or 0.164 for Case N1.6.

[1]Professor of Civil Engineering, Caltech, Mail Code 104-44, Pasadena, CA 91125
[2]Graduate Student in Structural Engineering, University of California, Berkeley

Superstructure design is considered for gravity loads and for lateral seismic loads in the short direction (N-S) of the building. Torsion is included by an increase in V_S of 6%. The seismic design is actually controlled by the story drift limit of 0.5%, which means that extra strength above the code base shear value will be present. A major consideration, evident under dynamic analysis, is handling uplift (Ryan and Hall, 1998). In this regard, in order to distribute more of the overturning to interior columns, all five N-S frames are designed as moment frames; exterior frames on the east and west faces are made less stiff than interior ones; girders in north and south bays are made less stiff than center-bay girders; and simple connections are used at the northern and sourthern-most columns from floor 2 to the roof. The completed design for Case N2 (Fig. 1) is carried out using A36 steel and static planar analysis of the N-S frames with rigid diaphragm coupling. Design for Case N1.6 is accomplished by reducing all beam and column plate dimensions in the E-W direction by 20%.

The superstructures exceed the fixed-base requirements in the UBC for similar design parameters, owing to different design objectives . A fixed-base design is expected to yield considerably under the design earthquake, while an isolated design is intended to remain nearly elastic, as indicated by the R_I of 2 which is essentially an over-strength factor.

Computer Model for Dynamic Analysis

The five N-S frames are modelled as planar beam-column frames with rigid diaphragm coupling; P-Δ effects are included; and yielding is represented by plastic hinging (yield strength of steel at 290 *MPa*). Foundation is assumed rigid. Live load is present as gravity load (0.72 kPa) and as building mass (0.48 *kPa*). Viscous damping is provided by dashpots that resist horizontal interstory velocities. The force carried by all the dashpots in a story is capped at a level equal to the story shear force produced by lateral design forces corresponding to $V_S/W = 0.02$. This cap is reached at an interstory velocity of 25 *cm/sec*.

Each bearing follows a nonlinear horizontal force-deflection relation (Fig. 2) which is bilinear up to a shear strain of 200% and stiffens beyond. A ratio of secondary to primary stiffness of 0.3 is used for the bilinear part, and this gives damping $\beta_D = 10\%$ for a cycle at D_D. A vertical bearing stiffness of $K_V = 1250$ *MN/m* is used for both compression and tension and is not a function of the lateral displacement. The moment which develops in the bearing as it shears is included in the analysis as a moment load applied to the frame.

The model for Case N2 produces a fundamental period of the superstructure of 0.98 seconds, computed assuming infinitely stiff bearings in horizontal and vertical directions. Lateral push-over strength of this superstructure in terms of the dead weight is 0.41W (with weight of first floor still included in W). The computed period of the Case N2 isolated building, using K_{SEC} and K_V of the bearings, is 2.84 seconds. Corresponding values for Case N1.6 are 1.10 seconds, 0.33W and 2.88 seconds, respectively.

Results

Earthquake responses are computed for a set of recorded accelerograms containing strong near-source effects as well as for a suite of simple displacement pulses. The recorded motions are 1978 Tabas (TAB), 1989 Loma Prieta at Los Gatos Presentation Center, 1992 Landers at Lucerne Valley (LUC), 1994 Northridge at Sylmar County Hospital free-field (SYL) and at Rinaldi Receiving Station (RRS), and 1995 Kobe at Takatori (TAK). All of these are horizontal components in the direction of maximum peak-to-peak ground velocity (Hall, 1997). The simple motion (Fig. 3) consists of a forward and back displacement pulse which is characteristic of the fault-normal component of near-source ground motion. Duration T_P of the pulse is varied from 0.2 seconds to 5.0 seconds with the peak velocity held constant at 128.6 *cm/sec*, except below $T_P = 1.05$ seconds where the peak acceleration is held constant at 0.80g. This produces a peak displacement increasing from 1.1 *cm* at $T_P = 0.2$ seconds to 188 *cm* at $T_P = 5.0$ seconds. Magnitudes of the earthquakes which could create such motions would be greater for the larger displacements.

Results for peak values of story drift, horizontal bearing displacement, and vertical tensile stress in a bearing are shown below (format: Case N2/Case N1.6) for the recorded motions. As expected, in terms of story drift, Case N2 provides a higher degree of isolation

	TAB	LGP	LUC	SYL	RRS	TAK
Story drift (%)	0.77/0.83	0.69/1.38	0.49/0.64	0.69/0.88	0.54/0.68	0.85/1.22
Bearing disp (*cm*)	61/67	55/71	39/43	59/63	43/44	76/69
Bearing tension (*MPa*)	0.70/0.53	0.36/1.45	----/----	0.37/0.80	----/----	1.02/0.93

than does Case N1.6. In fact, none of the Case N2 story drifts exceed 1%, indicating minimal yielding in the frame. For Case N1.6, several of the actual earthquake records push the rubber bearings into their stiffening range (horizontal bearing displacement of 61 *cm* for Case N1.6), and when this happens, the higher force transferred up into the building is effective in increasing the drifts: over 1% for LGP and TAK. This behavior is especially evident in the results for the suite of simple displacement pulses (Figure 4) where large bearing displacements and story drifts occur over a considerable range of T_P above 2 seconds. These drifts, reaching 3% for Case N2 and exceeding 5% for Case N1.6, demonstrate some vulnerability of base-isolated buildings to ground motions containing large, rapid displacements. The possibility of actual ground motions containing such pulses, or even several cycles of them, is an important research topic. Regarding the issue of uplift, the peak bearing tensile stresses are 1.45 *MPa* for LGP and 1.92 *MPa* from Figure 4, values which do not seem excessive. Thus, the design measures taken to reduce uplift have met with some success, although vertical ground motion should also be assessed.

Final comments

Because the UBC requires a dynamic design procedure using site-specific ground motion spectra when considering near-source effects, the results shown here are not intended as an assessment of the code provisions, although they are relevant. At the least, they indicate the extent of measures needed to deal with strong near-source effects and the possible consequences when the ground displacements are large and rapid. An important design consideration is control of uplift, which is especially true for taller buildings.

Use of the large shear strain capability of rubber bearings, as done here and as advocated by Kelly (1997), is a departure from past practice and should be subjected to additional laboratory testing. Such tests should be done at full scale and include application of vertical tensile and compressive load simultaneous with maximum shear. The benefit of rubber stiffening at large strains as a natural restraining device may be an improvement over sudden impact with a surrounding moat wall as considered by Hall et.al. (1995), but it still significantly increases the force transferred up into the superstructure and so can cause large yielding excursions. Possible advantages of friction isolation devices in accommodating large displacements at smaller shear force levels need to be assessed.

References

Hall, J. F., T. H. Heaton, M. W. Halling and D. J. Wald, "Near-Source Ground Motion and Its Effects on Flexible Buildings," Earthquake Spectra, EERI, Vol. 11, No. 4, Nov. 1995.
Hall, J. F., "Seismic Response of Steel Frame Buildings to Near-Source Ground Motions," Report No. EERL 97-05, Caltech, Pasadena, California, 1997.
Kelly, J. M., *Earthquake-Resistant Design with Rubber* , Springer-Verlag, 2nd Ed., 1997.
Ryan, K. and J. F. Hall, "Aspects of Building Response to Near-Source Ground Motions," *Proc. of Struc. Engr. World Cngrs.*, San Fran., 1998.

	Exterior Frame		Interior Frame	
	Column	Girder	Column	Girder
1	W14X193	W30X148	*W30X211*	W30X261
2	W14X145	W24X104	*W30X173*	W24X192
3	W14X109	W24X76	*W30X116*	W24X131
4	W30X211	W24X57	B24/1.50	W21X101
5	W30X173	W30X191	B24/1.12	W30X391
6	W30X116	W24X146	B24/0.75	W24X279
7		W24X104		W24X192
8		W21X083		W21X147

Figure 1. Data and sketch of Case N2 building. B = box with dimensions in inches. Weak axis in italics.

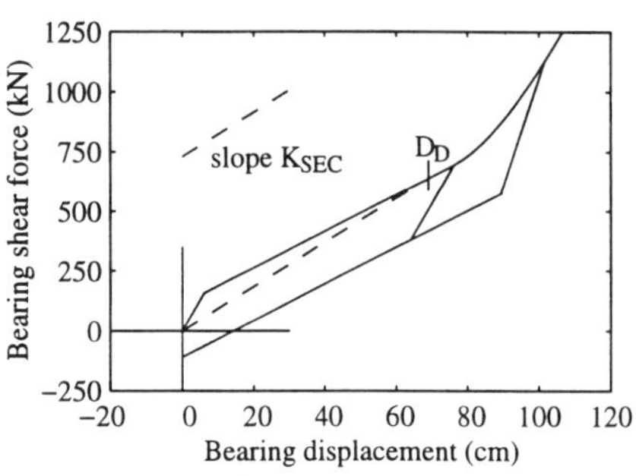

Figure 2. Horzontal force-deflection relation for a Case N2 bearing. Left half (not shown) is similar.

Fig. 3. Displacement-pulse ground motion.

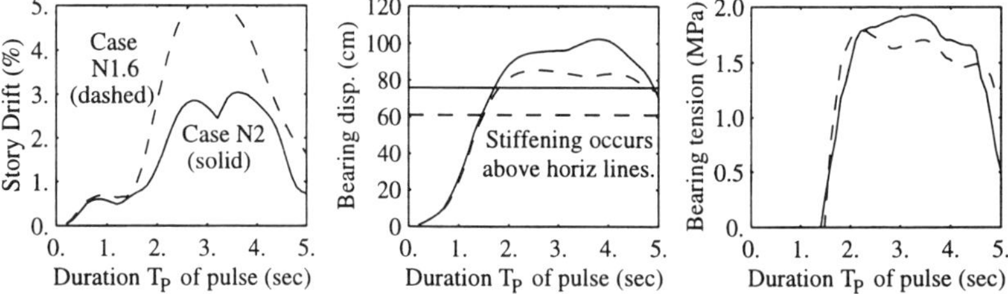

Figure 4. Peak building responses to suite of displacement-pulse ground motions.

IMPULSE EQGMs: A HISTORICAL AND CRITICAL REVIEW

Vitelmo V. Bertero (FM)[1], James C. Anderson (M)[2], and Mehrdad Sasani[3]

Abstract

This paper presents a historical and critical review of recorded EQ ground motions (EQGMs) that contain severe acceleration pulses and the effects of these motions on the performance of buildings in order to identify the major issues and directions for their solution. Results of some previous studies are summarized.

Introduction

EQ ground motions generally consist of random vibrations, however, their possible effect on structures has been visualized and analyzed by assuming that they are periodic or harmonic motions. In spite of the fact that analyses of accelerograms recorded during EQs which have occurred since the 1950s show that EQGMs can also be of impulse type, only recently has the possible occurrence of EQGMs with one or more severe acceleration pulses and their importance for the EQ-resistant design of civil engineering structures been recognized and introduced into the seismic provisions of building codes through the introduction of the near-source factor (UBC 1991, 1977). The main objective of this paper is to present a historical and critical review of recorded EQGMs that contain severe acceleration pulses and of the studies regarding the effects of these pulses on the seismic performance of buildings. The critical review focuses on identifying the following: the main results obtained from the studies conducted; the reasons for the delay in recognizing the importance of these results and incorporating them into the building code seismic provisions; the reasons for limiting the possible occurrence of EQGMs having severe acceleration pulses to building sites located less than 15 km from the seismic source when recorded data has indicated that severe acceleration pulses can occur at sites located at epicentral distances larger than 150 km; the reliability of applying the recommended near source factor to only the required strength (i.e. the seismic base shear).

Historical and Critical Review

The record obtained during the Port Hueneme EQ of March 18, 1957 was shown to be essentially a single pulse of energy (Housner and Hudson, 1958). In spite of its small

---

[1] Nishkian Emeritus Professor and Research Engineer, Earthquake Engineering Research Center, University of California, Berkeley
[2] Professor, Department of Civil Engineering, University of Southern California, Los Angeles
[3] Graduate Student, University of California, Berkeley

Richter Magnitude (M) of 4.7 and low peak ground acceleration of 0.08g, it caused exceptional damage. Destructive single-shock EQGMs associated with moderate M (5.4 to 6.2) and shallow foci (<30 km) occurred in Agadir in 1960 (Despeyroux, 1960), Libya in 1963 (Minami, 1965), Skopje in 1963 (Ambraseys, 1964), and San Salvador in 1965 (Rosenblueth and Prince, 1965).

A relatively small M=5.6 shallow EQ occurred on the San Andreas fault at Parkfield on June 27, 1966. An EQGM record obtained at a distance of two hundred feet from the fault shows that the strong phase of shaking was very short, about 1.5 seconds, and consisted of four successive acceleration pulses, one of triangular shape which had a very high PGA of 0.5g but a short duration of only 0.3 seconds. The four acceleration pulses resulted in one displacement pulse of about 23 cm and a duration of 1.3 seconds. This EQGM did relatively little damage because of the very short duration of the pulses and of the total phase of the strong ground motions (Cloud, 1967). For this reason it did not receive much attention in the engineering community.

Perhaps the first destructive EQ in the United States with recorded accelerograms that contained acceleration pulses with very severe damage potential was the February 9, 1971 San Fernando EQ with M=6.2. The derived records from the recorded accelerations at the top of the Pacoima Dam and at the Van Norman Dam have acceleration pulses that resulted in very large incremental velocities of 158 cm/sec at Pacoima and 172 cm/sec for Van Norman. The senior author and his researchers carried out several studies on the effects of these recorded and derived EQGMs on buildings and compared the effects with those that can be expected from the design EQs specified in the seismic codes. In one paper (Bertero, et al., 1976), they offered a series of conclusions and recommendations among which the following are considered relevant for this review: (1) When safety rather than serviceability controls the design, large but controllable inelastic deformations can be tolerated; (2) The use of inelastic design response spectra (IDRS) derived directly from recommended linear elastic design response spectra (LEDRS) through displacement ductility factors as suggested by present methods, does not appear to be conservative for buildings located in the immediate area of causative faults; (3) The main drawback of present methods is that direct derivation of IDRS from LEDRS has been shown to be basically untenable due to the fact that the dynamic characteristics of the critical ground excitations for elastic and inelastic responses are completely different. While periodic pulses having frequencies equal to those of the predominant modes of vibration of the building constitute the critical ground motions for linear-elastic systems, a critical inelastic response can be induced by accelerations of equal to or greater value than the effective seismic resistance coefficient of the structure; (4) Near fault records of the San Fernando EQ, such as those for Pacoima and Van Norman Dams contain severe long acceleration pulses which resulted in large velocity increments. These have been found to be characteristic of near fault motions. Results of an analytical study of a building near the fault zone using the Pacoima record correlated well with the observed damage. This damage appears to have been the result of only a few large displacement excursions rather than of numerous oscillations; (5) Unusually large ground velocities may be developed at near fault sites. Methods for constructing elastic and inelastic design response spectra should reflect the larger values recorded at such sites. Additional research is needed to establish bounds on the different parameters that define the characteristics of severe long pulses, i.e. the largest incremental velocity and associated effective acceleration that can be developed according to the mechanical (dynamic) characteristics of the soil present at a site. These values will enable the design engineer to determine an upper bound on the energy that can be transmitted to the foundation of the structure so that the structure can be designed accordingly. It is also necessary to know the number of long severe pulses that can occur at the site since repeated pulses can lead to an incremental (crawling) type of collapse. Furthermore, the above authors describe a procedure that could be implemented for establishing design EQs for near fault sites. In 1975 in discussing the magnitudes, aftershocks and fault dynamics of the 1971 San Fernando EQ, Bolt gave a detailed explanation for the resultant large horizontal displacement pulse obtained from the recorded EQGMs on the abutment of Pacoima Dam. He called the ground velocity and displacement pulses the source "fling." Bolt also emphasized that this pulse-like ground displacement seemed likely to be the normal condition because it had a theoretical justification in the elastic rebound mechanics of a rupturing fault.

In 1979 the Imperial Valley EQ produced several records of EQGMs which contained severe acceleration pulses resulting in severe velocity and displacement pulses. The record obtained at the James Road station (approximately 2.5 miles from the fault trace) had a long duration acceleration pulse having a PGA of 0.36g which resulted in an incremental velocity of 160 cm/sec. As reported by Singh in 1981 and in more detail in 1985, the presence of these severe pulses had a theoretical justification which was that given by Bolt in 1975. Anderson and Bertero investigated the effects of several of these recorded EQGMs on flexible steel frame buildings. The results of this study are summarized in an ASCE paper (Anderson and Bertero, 1987) which contained the following observations which are relevant to this critical review: (1) In certain geological regions, such as the near-fault region, structures may be subjected to impulse-type EQGMs that can be particularly damaging to the lower floors of buildings. This part of a structure is more critical than the upper floors due to the high axial loads carried by the columns. Under an impulse-type load, the deformation tends to be concentrated in the lower floors, thereby increasing the interaction effects between lateral displacement and axial load (P-delta); (2) The peak ground acceleration is not a very accurate means of classifying the severity of strong ground motion with regard to structural damage potential. In order to obtain a more meaningful classification, the use of other parameters such as incremental velocity and peak ground displacement may be required; (3) Current design response spectra do not recognize two important factors that are the result of impulse-type ground motions: (a) the increase in spectral response in the long-period region (1-4 seconds) that results in significantly higher accelerations, velocities and displacements that should be included in spectra used to design structures in close proximity to active faults, and (b) the increase in the ductility requirement for more rigid structures which is due to the increase in the ratio of the pulse duration to the period of the structure; (4) This study has shown that EQGMs resulting from the same EQ may have significant variations in dynamic characteristics due to such factors as local geological conditions, directionality of fault propagation and distance from the fault plane. In addition, the nonlinear structural response is sensitive to the duration of the acceleration pulse relative to the fundamental period of the structure and to the yield resistance seismic coefficient of the designed structure. These factors combine to create a large uncertainty in the establishment of a design EQGM which must be taken into account in the design process; (5) These studies have shown that a structure in the near-fault region may experience a dynamic response that is twice that of a similar structure located at some distance from a fault even though the peak ground accelerations are about the same; (6) The El Centro record used for evaluating the response of structures has very little damage potential when compared with the more recent Imperial Valley (1979) records. This illustrates the importance of directionality effects associated with the direction of fault rupture propagation.

Although the presence of severe pulses creating impulsive types of EQGMs might be mainly a characteristic of near-source records, they have also been recorded at sites far from the EQ source such as the Bucharest accelerogram which was recorded at a distance of 150 km from the epicenter of the 1977 Vrancea (Romania) EQ, resulting in severe damage to many modern buildings in Bucharest. Furthermore, stations located on the lake bed zone of Mexico City during the 1985 Michoacan EQs, with an epicentral distance of about 350 km, caused severe damage to low-rise modern buildings. The senior author made these observations: Although the recorded EQGMs were periodical (practically harmonic) with periods higher than 1.5 seconds and with PGAs up to 0.23g, from the point of view of their effect to buildings with relatively low yield strength capacities, these EQGMs were of the impulsive type, i.e. a series of reversal acceleration pulses.

In view of the results obtained in the above studies and their publication and oral presentation to the EQ community, it has been very surprising that until 1991, the effects of impulsive types of EQGMs have not been incorporated in seismic code provisions for the practical analysis and design of civil engineering structures, even for those located in the near-field of the source (exceptions have been for the design of special facilities such as dams (Bolt, 1997). The 1991 UBC EQ requirements for seismic isolated structures introduced near source factors related to the proximity of the building or structure to known faults with established magnitudes and slip rates in determining the seismic coefficient.

For the response analysis as well as the design of conventional building structural systems it took additional dramatic records with severe pulses that were obtained in the Northridge EQ (1994) and the Kobe EQ (1995) to convince the experts (researchers and practitioners) that in the very near field, strong shaking and pulses were more likely to be the rule rather than the exception (Jennings, 1997; Bolt, 1997). Thus the 1997 UBC introduced in its EQ design regulations for buildings using conventional structural systems, near-source factors. Although the introduction of these factors in the code seismic regulation has been a step forward toward the recognition of the importance of the effects of impulsive type of EQGMs, there are some concerns about the reliability of the required procedures, particularly in the following two areas: (1) That the occurrence of these pulses has been limited to sites located within 15 Km of the known seismic source and (2) that it is doubtful that the effects of these pulses on the building response can be obtained by simply increasing the required base shear which is determined using as a base a LEDRS.

Anoop S. Mokha[1], Ph.D., S.E.; Peter L. Lee[1], S.E.; Xiaofeng Wang[2], D.Sc., P.E.; Peter Yu[1], P.E.

Abstract

This paper presents the seismic isolation design of the new International Terminal at San Francisco International Airport, which is the centerpiece of the Airport's $2.4 billion expansion and modernization program. In response to the critical nature of the functions of a large strategically located airport with both international and domestic traffic, the San Francisco International Airport addressed seismic performance goals with the stated desire to further limit potential damage and disruption to the new facilities following a major earthquake. Linear dynamic time history analyses were performed to evaluate the performance of various structural systems including seismic isolation system. Furthermore, the design parameters for the isolation system design and the superstructure response to different levels of earthquake is also presented. The New International Terminal Building will be the largest base isolated building in the world with over 1.2 million square feet of floor area and over 22 million cubic feet of interior volume.

Introduction

The San Francisco International Airport (SFIA) expansion program consists of three major structures: the central new International Terminal, the shoulder building with boarding gates and the access roadway bridge structure. The geological conditions at the expansion site are not ideal from an earthquake point of view. The site predominantly consists of bay mud under-lain by sand, silt, clay and soft rock with depth of bedrock varying from 90 feet to 150 feet below the ground level. The original site-specific spectra developed in 1993 for the project were subsequently upgraded following the 1994 Northridge earthquake. The Airport Commission established a goal to ensure minimum disruption of the Airport's infrastructure and a desire for continued operation. Since the Main International Terminal spans over the existing approach roadways to the domestic terminal, it was required that the New Terminal structural system be designed to the highest standards that would ensure no collapse and continued operation with minimal damage.

New International Terminal

The configuration of the new terminal is dominated by the glass enclosed "great hall" – which is 215 m (705 ft) long, 64 m (210 ft) wide and about 25 m (83 ft) high. The roof

---

[1] Associate, Skidmore, Owings & Merrill LLP (SOM)
[2] Skidmore, Owings & Merrill LLP (SOM)

structure of the great hall consists of five lines of tubular trusses which are 8.8 m (29 ft) deep. Below the great hall floor are two levels one each for arrival and baggage handling. Rising above the great hall floor is an office tower consisting of 2 levels and a roof (see Fig. 1). Four different structural systems were evaluated for their elastic behavior (see Table 1).

Site-Specific Ground Motions

40 to 50 feet of Soft Bay Mud underlie the International Terminal site. The project geotechnical-engineering consultants developed site-specific seismic hazard evaluations for use in the analysis and design of the new International Terminal. Three earthquake levels, A, B & D, with peak ground accelerations of 0.34g, 0.6g, and 0.62g, respectively, were considered. The probabilities of exceedance of earthquake levels A, B and D were 50%, 10% & 10% in 50, 50 & 100 years, respectively. Fig. 2 shows the respective site-specific spectra.

Seismic Performance Objectives

The New International Terminal structure, responding to the architectural desire for a "great hall" enclosure with a tall glass window wall, the highest performance levels. For Earthquake D (a probability exceedance of 10% in 100 years), no structural damage to the main roof and window wall were expected. Minor architectural damage was acceptable with no breakage of glass in the window wall system. The Architects established a limit of 11 inches for the maximum relative displacement between the top and bottom of the window wall to ensure no breakage of glass. This limit is consistent with acceptable performance drift criteria of 1% commonly used to define the operational limit.

Table 1: Performance of Structural Schemes under Elastic Conditions

Structural Schemes	Braced Frame and Moment Frame			Moment Frame			Moment Frame and Fluid Dampers			Moment Frame and Braced Frame Plus Seismic Isolation		
Earthquake	EQ-A	EQ-B	EQ-D	EQ-A	EQ-B	EQ-D	EQ-A	EQ-B	EQ-D	EQ-A	EQ-B	EQ-D
Base Shear to Total Wt. Ratio	0.49	0.95	1.28	0.22	0.72	1.00	0.12	0.46	0.53	0.11	0.21	0.30
Expected Structural Damage	None to min.	Mod. repair-able	Severe no collapse	None	Mod. Repair-able	Severe, no collapse	None	None to min.	Mod repair	None	None	None to min.
Expected Non Struct. Damage	Min.	Severe	Severe	Mod.	Severe	Severe	None	None to min.	Mod.	None	None	None to min.
Roof Shear to Roof Wt. Ratio	0.48	1.34	2.06	0.50	2.16	2.36	0.32	1.29	1.50	0.33	0.46	0.78
Roof Displ. Rel. to 3rd Level, (in.)	5	21	23	8	32	43	8	23	28	4	5	10

Performance Evaluation of the New International Terminal

Linear dynamic time history analyses were performed to assess the performance of different structural schemes. The linear analysis was performed on a three-dimensional model (see Fig. 3). The program ETABS v6.10 was used to perform time history analysis for the four structural schemes. The structural elements were modeled as elastic elements except that fluid dampers and base isolators were explicitly modeled for their nonlinearity. The results of the time history analysis for the three levels of earthquakes are summarized both qualitatively and quantitatively in Table 1. From Table 1 it can be concluded that from a performance point of view in terms of roof displacement and expected damage, the moment frame and braced frame combination is preferred rather than just the moment frame. Also the reliability of a pure moment frame structural scheme is questionable from the point of view of ductility. The addition of dampers reduces the base shear and expected damage but not sufficient enough to reduce the roof displacement to less than 11 inches. The scheme with Seismic isolation yields the best overall results in terms of reduced base shear, ductility demand, drift, and expected damage and roof displacement, thus ensuring a continued safe operation of the new Terminal structure after a major earthquake.

Evaluation of Different Isolations Systems

Three isolation systems: Friction Pendulum Bearings, Lead-Rubber bearings and High damping rubber bearings were identified as practical and designed for the same level of structural shear. For the elastomeric bearings, the steel frame member sizes were increased to resist the additional eccentric gravity load moments (i.e. P-Δ moments) that occur during seismic movements. Lower bound and upper bound limits on parameters affecting the design of these isolation systems were established to account for aging effects. Friction Pendulum bearing isolation system was selected on a competitive bid of the super structure steel and the isolation systems. The selection of Friction Pendulum bearings resulted in savings of about 680 tons of structural steel as compared to that for the steel frame required for the elastomeric bearings.

Implementation of Friction Pendulum Bearings

Figure 4 shows the construction details of the Friction Pendulum bearings under the columns. A total of 267 such bearings support over 1.2 million square feet making it the largest base isolated building in the world. Table 2 slows the design parameters and the resulting global response of the superstructure for earthquake D.

Table 2: Isolated Superstructure's Response for Earthquake D (PGA= 0.62g)

Max. Coeff. of Friction (%)	Isolation Period (Sec.)	Max Structural Shear		Max. Base Shear		Max Resultant Bearing Displ. (inches)	Max Floor Acc. (g)	
		X	Y	X	Y		X	Y
5.0[1]	3.0	0.21w	0.14w	0.28w	0.18w	19.23[5]	0.43	0.45
5.8[2]	3.0	0.20w	0.14w	0.27w	0.17w	17.17[5]	0.45	0.53
6.0[3]	3.0	0.20w	0.15w	0.27w	0.17w	16.92[5]	0.45	0.55
7.0[4]	3.0	0.20w	0.15w	0.25w	0.18w	15.74[5]	0.50	0.65

[1] Lower bound limit
[2] Average of Production Bearing Test
[3] Average of lower and upper bound
[4] Upper bound limit
[5] Bearing displacement capacity 20 in.
w = seismic weight of structure

Conclusion

The linear evaluation of different structural schemes for the expected performance of the new International Terminal dictated a structural scheme that utilizes seismic isolation. The seismic design of the architecturally complex New International Terminal with long spans and tall curtain walls was accomplished with the use of seismic isolation.

Figure 1: Transverse section through New Terminal "great hall" and access roadways

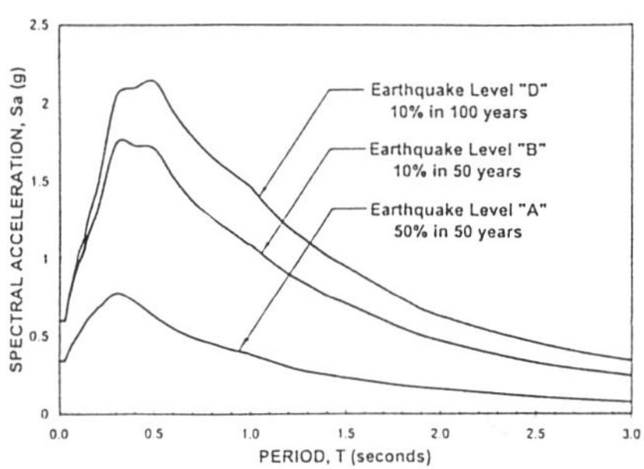

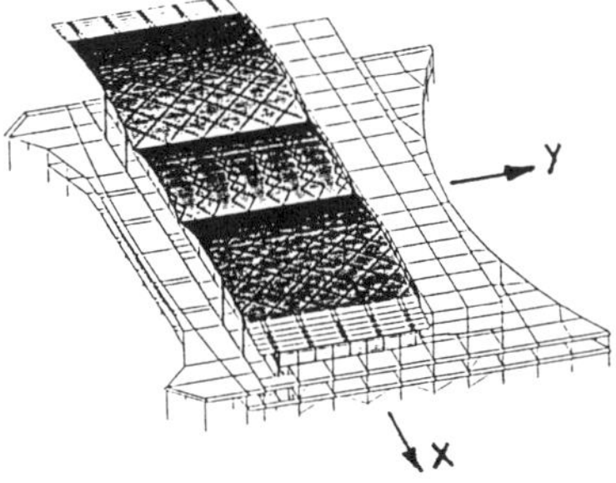

Figure 2: SFIA Site Specific Spectra   **Figure 3**: 3-D Structure Model

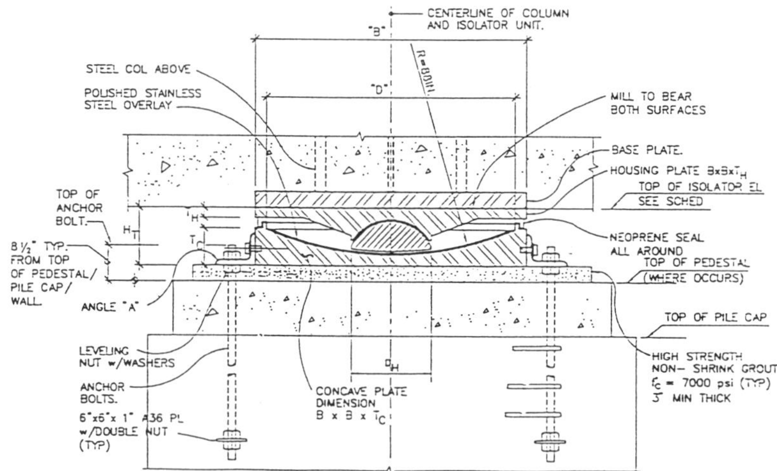

Figure 4: Friction Pendulum Bearing Under Columns

Seismic Isolation of Bridges using
Friction Pendulum Bearings

Victor A. Zayas[1], M. ASCE and Stanley S. Low[2]

Abstract

This paper presents recent examples of bridges seismically isolated using Friction Pendulum seismic isolation bearings. The Friction Pendulum bearings were used to achieve an elastic, no-damage seismic design, and provided the lowest total construction cost for this performance level.

Benicia-Martinez Bridge

The Benicia-Martinez bridge is a 1,894 meter long bridge; has 10 steel truss spans supported by concrete piers; carries 6 lanes of traffic and averages 100,000 vehicles a day (Fig. 1). It is one of the largest bridges to date to undertake a seismic isolation retrofit. The design earthquake included very strong near-fault ground motions, with near-fault fling and deep soil site effects resulting in design spectra accelerations of up to 8g (Fig. 2). The Friction Pendulum bearings for this project have a period of 5 seconds; a lateral displacement capacity of 1.3 meters (53 inches); a design vertical load of 22 million N (5 million lb.); measure 3.9 meters (13 feet) in diameter, and weigh 178,000 N (40,000 lbs.) (Fig. 3). They are the largest seismic isolation bearings ever manufactured. The 22 such bearings are installed at the tops of the concrete piers, directly under the roadway trusses, 2 bearings per pier. The seismic retrofit will be completed while maintaining traffic flow across this important San Francisco Bay transportation link.

----

[1] President & [2] Structural Engineer, Earthquake Protection Systems, Inc., 2801 Giant Hwy., Bldg. A, Richmond, CA 94806 ; Tel: (510) 232-5993; Fax (510) 232-6577

Figure 1: Benicia-Martinez Bridge

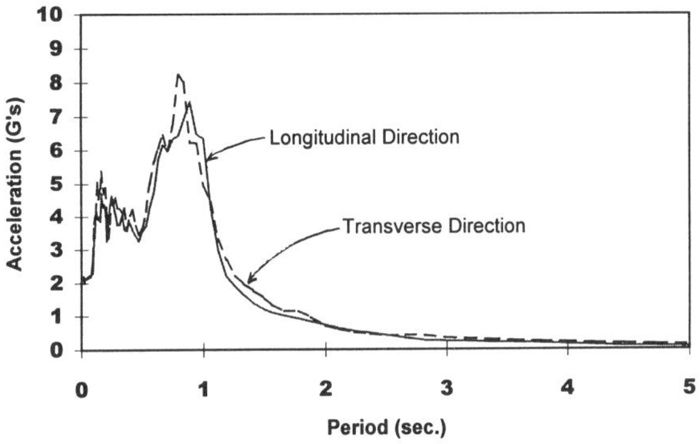

Figure 2: Ground Motion Spectra at Pier 6 Footing

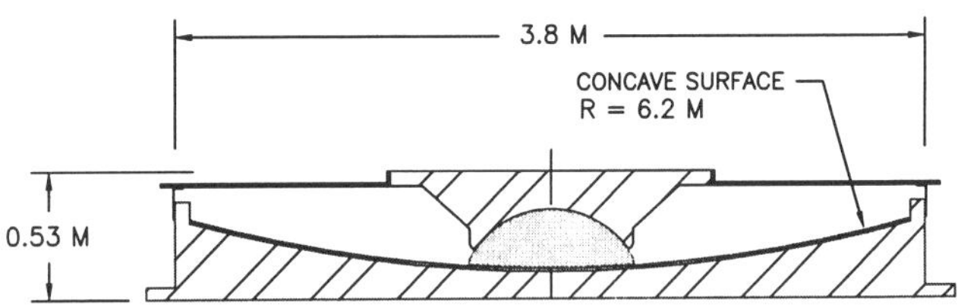

Figure 3: Section of Benicia-Martinez Bridge Bearing

American River Bridge

The 690 meter long bridge across the American River at Lake Natoma, Folsom, California (Fig. 4), uses Friction Pendulum seismic isolation bearings to protect it from damaging earthquake motions (Fig. 5). It is one of the largest new bridges in

Figure 4: American River Bridge, Folsom, California

Figure 5: Installed Bearing, American River Bridge

the U.S. to use seismic isolation. Use of seismic isolation reduces the earthquake lateral forces on the bridge foundation and piers, and this saved $1 million in construction costs, as compared to the non-isolated bridge design. Seismic isolation also allowed the bridge to elastically resist the design earthquake with no damage anticipated, whereas damage would have been expected for the non-isolated design.

The bridge structure consists of two post-tensioned concrete box frames supported by 48 Friction Pendulum bearings. The bearings were installed pre-displaced to accommodate post-tension and shrinkage movements.

White River Bridge

Friction Pendulum bearings were used in the construction of the new White River Bridge located in the Yukon, Canada (Fig. 6). The 180 meter long steel girder bridge is supported by 9 Friction Pendulum bearings, and consists of 2 spans and carries 2 lanes of traffic. The Friction Pendulum bearings were tested at temperatures as cold as -110°C, and demonstrated desirable and stable performance without incurring bearing damage. Moreover, the bearings maintained the project specified design stiffness and damping properties over a temperature range of -70°C to + 60°C. The isolation bearings achieve an elastic structure response for the design level earthquake (0.2g peak ground acceleration), at a substantially lower cost than would have been possible without isolation.

Figure 6: White River Bridge, Yukon, Canada

Recent Seismic Evaluations of New York City Bridges

Jagtar S. Khinda[1], P.E., Fellow ASCE; Jay A. Patel[2], P.E., Member ASCE;
Kamal Kishore[2], P.E., Member ASCE; and Feng-Bao Lin[3], P.E., Member ASCE

Abstract
New York City (NYC) has a total of 2,102 bridges. New York City Department of Transportation
(NYCDOT) owns 765 of them. Of these, about 20 fall in the Bridge Importance Category of "Critical",
245 "Essential", and 500 "Other", from the seismic point of view. The paper discusses recent seismic
evaluations of such bridges under the new seismic hard rock motions recommended by a consensus of a panel
of national experts specializing in NYC region's geology and seismicity.

Introduction
Soon after the Loma Prieta earthquake of 10/17/89, New York State like many states in the central and
eastern United States adopted provisions for the seismic design of highway bridges (Khinda 1993, 1996). The
Northridge Earthquake of 1/17/94 and the Kobe Earthquake of 1/17/95 further highlighted the importance
of considering local site effects in the seismic design of bridges.

To this end, three local seismic evaluation projects were started: Queensboro Bridge (NYCDOT); Bronx-
Whitestone Bridge (MTA Bridges and Tunnels); and JFK International Airport (Port Authority of NY & NJ).
Their consultants recommended rock motion required response spectra (RRS) that differed from each other
and from the (AASHTO 1996) RRS and the new 1996 USGS seismic hazard maps significantly (see Fig. 1
& 2). Since the RRS chosen has a big impact on the retrofit cost of a bridge, NYCDOT, with help from New
York State Department of Transportation (NYSDOT) and Federal Highway Administration (FHWA),
selected a panel of national experts familiar with the NYC local geology and seismicity, to review above
project seismic reports, resolve the differences and recommend hard rock motion RRS which all projects in
the entire NYC region could use. This study has just been finalized (Weidlinger 1998) and is based on new
soil classes (A, B, C, D, E, and F) adopted in (NEHRP 1997). The new seismic guidelines take into account
the east-west United States differences as well as local NYC site effects. Revisions to NYSDOT and
AASHTO Seismic Specifications have also been recommended.

The study serves as a good example for other big cities in the eastern and central US to follow and
FHWA is actively encouraging the distribution of this report to other states.

NYCDOT New Seismic Design Criteria Guidelines
Two-level (Lower and Upper) Approach is used for "critical" bridges:
- *Lower level-functional evaluation*: (500-year return period or 10% probability of exceedance in 50 years),
  the bridge shall provide *immediate* access to normal traffic, with *minimal* damage and no collapse.
- *Upper level-safety evaluation:* (2500-year return period or 2% probability of exceedance in 50 years),
  the bridge shall provide *limited* access (reduced lanes, light emergency traffic) within 48 hours and full
  access within months, with *repairable* damage and no collapse.

Site-specific ground motions at the bridge are computed from the panel supplied hard rock motion time
histories by proper wave propagation and soil-structure analyses. Multi-mode linear/nonlinear dynamic

[1]Seismic Projects Engineer, NYCDOT Div. of Bridges, 2 Rector Street, New York, NY 10006
[2]Deputy Chief Engineer, NYCDOT Div. of Bridges, 2 Rector Street, New York, NY 10006
[3]Professor of Civil Engineering, Polytechnic Univ., 6 Metrotech Center, Brooklyn, NY11201

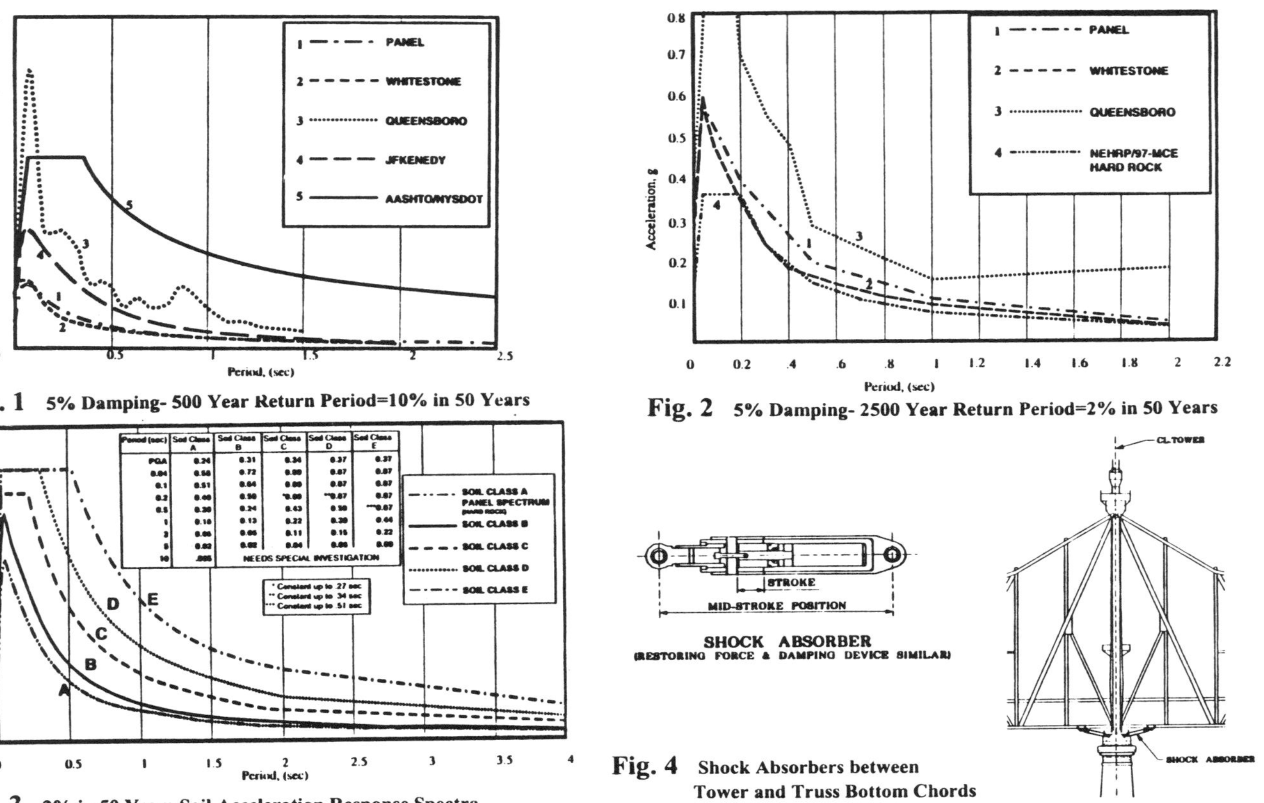

Fig. 1   5% Damping- 500 Year Return Period=10% in 50 Years

Fig. 2   5% Damping- 2500 Year Return Period=2% in 50 Years

Fig. 3   2% in 50 Years Soil Acceleration Response Spectra

Fig. 4   Shock Absorbers between Tower and Truss Bottom Chords

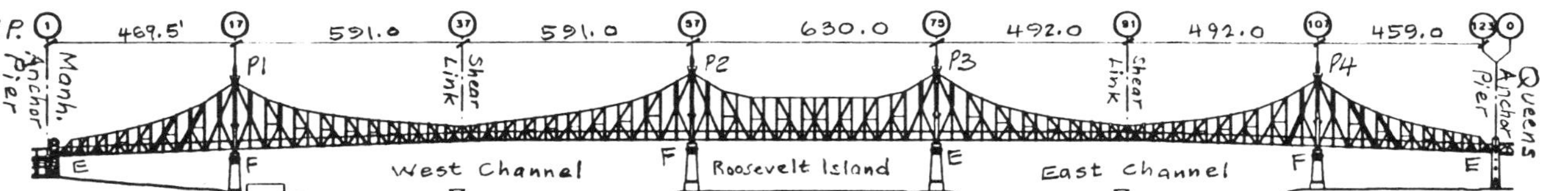

Queensboro Bridge is a two-level, double cantilever, truss structure with extensive approaches and approach ramps. Steinman prepared seismic evaluation study and retrofit concepts. NCEER, Lamont Doherty Earth Observatory and other consultants performed geotechnical studies and developed the site-specific ground motions. For the Steinman study, two levels of site-specific earthquakes (the SSE and OBE) and AASHTO type earthquake (see Fig. 1&2) were used. Computer modeling and analysis used SAP90 software package. The Main Bridge, a long-span heavy truss structure, is supported on six unreinfoced masonry piers, which are weak in resisting longitudinal forces. Towers near their base and fixed bearings are subjected to excessive stresses. The Queens Approach is located over layers of soft soil, including a layer of very soft organic soil (peat) which is vulnerable to liquefaction.

Summary of significant study results is given below:

Bridge Portion	Vulnerable Components under the SSE event	Recommended Retrofit Measures	Cost Estimate
Main Bridge	a. Most piers and towers. b. Most truss bearings. c. Most upper lateral system members.	• Strengthen piers and/or towers. • Retrofit fixed bearings; Install seismic isolation devices between superstructure and piers and/or distribute seismic loads to 6 rather than the current 3 piers resisting the longitudinal forces; Install shock absorbers at Pier 3 (see Fig. 4) and /or install restoring force and damping devices. • Strengthen top lateral system.	$ 30 Million
Manhattan And Queens Approaches & Ramps	d. Many foundations, abutments and retaining walls. e. Most bearings; most superstructure members; most members of towers and bents; Manhattan Approach - many masonry elements; Queens Off Ramp and Approach Ramps A, B, C. f. Most column base anchorages. g. Several column to floorbeam connections and other superstructure items. h. General liquefaction.	• Provide vertical reinforcement in the footings. • Provide seismic isolation between truss structure at the bents and towers or strengthen members of the framing system connecting columns of bents and towers; provide dampers between various portions of the truss. • Strengthen column to footing connections. • Strengthen columns, floorbeam connections and bracing members of Approach Ramps. • Provide new drilled pile supports for the footings in the liquefaction prone soil areas; install stone columns/drains or use some other ground improvement scheme to prevent liquefaction.	$ 25 Million

Fig. 5 Queensboro Bridge and its Seismic Evaluation

analysis of the bridge subjected to site-specific uniform and/or spatially varying ground motions is performed to determine the Capacity/Demand (C/D) ratios for all members for all modes and retrofit is recommended if the minimum C/D ratios are less than 1 (FHWA 1995).

One-level Approach is used for "Essential" and "Other" bridges:

- *One level-safety evaluation:* [2/3(2% probability of exceedance in 50 years) = 2/3 MCE]--
  The "Essential" bridge shall provide *limited* access (reduced lanes, light emergency traffic) within 3 days and full access within months, with *repairable* damage and no collapse. The "Other" bridge shall have no collapse. *Significant* damage is allowed -- damage should occur in visible and controlled locations only. It may require closure to repair. Soil RRS curves (Fig. 3) x2/3 are used for these bridges.

Recent Seismic Evaluations of "Critical" Bridges— a few case studies:

East River Bridges:

Name of Bridge	Bridge Type	Status of Rehabilitation/Seismic Evaluation
Queensboro Bridge	Truss	(Steinman 1996) reports study results for the OBE, SSE and AASHTO type earthquake (see Fig. 1,2,4 & 5). This will be re-evaluated in view of the above new seismic criteria guidelines.
Williamsburg Bridge	Suspension	Seismic studies are in progress for Side and Main Spans. This is presently under major reconstruction and rehabilitation.
Manhattan Bridge	Suspension	Rehabilitation is ongoing. For Suspended Spans, seismic and modeling studies are in progress.
Brooklyn Bridge	Suspension	Proposal for seismic evaluation study is being prepared.

Movable Bridges over Harlem River:

Madison Avenue	Swing	Under construction.
Third Avenue	Swing	Seismic and modeling studies are in progress.
145th Street	Swing	Seismic and modeling studies are in progress.
Willis Avenue	Swing	Seismic and modeling studies are in progress.
Macombs Dam	Swing	Seismic and modeling studies are in progress.

Note: Modifications to Performance Criteria for Movable Bridges to that enumerated above, in general, may be needed to solve mechanical machinery operability problems under earthquake forces as follows:

- To avoid damage to mechanical moving machinery, bridge supporting structure should be isolated from the bridge moving machinery by making use of seismic isolation devices in all new designs.

Conclusions

New seismic design criteria guidelines and some recent seismic evaluations have been presented.

Acknowledgments

The authors would like to thank Satinder P.S. Puri, Chair, ASCE Subcommittee on Seismic Performance of Bridges, for reviewing the paper and making helpful comments. Especial thanks are due to Ms. Connie Crawford, Past Chief Engineer, Division of Bridges, for her encouragement and advice.

References

AASHTO (1996). "Standard Specifications for Highway Bridges." 16th Edition, Washington, D.C.

FHWA (1995). "Seismic Retrofitting Manual for Highway Bridges." Report No. FHWA-RD-94-052.

Khinda, J. S., Avison, R.L.and Deitch, I. (1993). " Rehabilitation and Seismic Retrofit of Bridges in New York City." 10th Annual International Bridge Conference, Pittsburgh, PA., June 14-16.

Khinda, J. S., and Lin, F. B (1996). "Seismic Isolation of Bridges in New York City." Proceedings of ASCE Structures Congress, Chicago, Illinois, April 15-18.

NEHRP (1997). "Recommended Provisions for Seismic Regulations for New Buildings and Other Structures." Vol.1 (Provisions), Vol. 2 (Commentary).

Steinman (1996). " Queensboro Bridge: Seismic Condition Assessment and Retrofit Recommendations." Report to NYCDOT by Steinman Engineers, New York, NY.

Weidlinger (1998). "New York City Seismic Hazard Study and its Engineering Application." Report to NYCDOT by Weidlinger Associates, New York, NY.

Concrete and Masonry Structures

Manuscripts for some presentations were not available at time of publication.

2. *ADVANCES IN ENGINEERED MASONRY STRUCTURES*

MODERATOR: Arturo E. Schultz,
University of Minnesota,
Minneapolis, MN

(1) Update of Masonry Building codes and Specification, 1995 to 1998 Editions
Max L. Porter
Iowa State University
Ames, IA

J. Gregg Borchelt
Brick Industry Association
Reston, VA

(2) Advanced Composites in Construction and Retrofit of Masonry Structures
Ahmad Hamid
Drexel University
Philadelphia, PA

(3) Out-of-Plane Seismic Behavior of Brick Masonry Walls of Turn-of-the-Century North American Residential Buildings
Andre Filiatrault
Ecole Polytechnique
Montreal, Canada

Michel Bruneau and Jocelyn Paquette
University of Ottawa
Ottawa, Canada

(4) Stability of URM Compression Members under Out-of -Plane Lateral Loads
Arturo Schultz and James Mueffelman
University of Minnesota
Minneapolis, MN

11. *ADVANCES IN PRESTRESSED MASONRY*

MODERATOR: Trey Hamilton
University of Wyoming
Laramie, WY

(1) An Introduction to the New Prestressed Masonry Code Provisions
Arturo Schultz
University of Minnesota
Minneapolis, MN

2) Prestressed Masonry in the U.S.
David Biggs
Ryan-Biggs Associates, P.C.
Troy, NY

(3) European Perspective of Prestressed Masonry
Malcolm Phipps
University of Manchester Institute of Science and Technology
Manchester, UK

Hans R. Ganz
VSL International
Paris, France

(4) The Moduloc Post-Tension Approach for Strengthening Existing Masonry Buildings
Michael Schuller
Atkinson-Noland & Associates, Inc.
Boulder, CO

(5) Unbonded vs. Bonded Prestressing: Which is better for Masonry?
Trey Hamilton
University of Wyoming
Laramie, WY

20. DESIGN OF COMPOSITE JOISTS AND JOIST GIRDER SYSTEMS

MODERATOR: Roberto T. Leon
Georgia Institute of Technology
Atlanta, GA

(1) Composite Joist Behavior and Design Requirements
W. Samuel Easterling
Virginia Tech
Blacksburg, VA

(2) Floor Vibrations: Ultra-Long Span Joist Floors
Barry S. Band Jr.
Stanley D. Lindsey and Assoc.
Nashville, TN

Thomas M. Murray
Virginia Tech
Blacksburg, VA

(3) Application of the Composite Joist Floor System in Office Building Construction
Kurt D. Swensson
Stanley D. Lindsey and Associates
Atlanta, GA

(4) Composite Joist Case Histories
David L. Samuelson
Nucor Research and Development
Norfolk, NE

29. *CONCRETE: FROM MATERIAL MODELING TO STRUCTURAL DESIGN*
I. *Structural Modeling*

MODERATOR: E. Spacone and P. B. Shing
University of Colorado
Boulder, CO

(1) Reliability Assessment of Reinforced Concrete Structures with Respect to Material Degradation
W. B. Kraetzig, Y. S. Petryna, M. Roik, and F. Stangenberg
Ruhr-University Bochum
Bochum, Germany

(2) Snap-Through Behavior and Stability of Composite Concrete/Masonry Structures
J. ``, C.M. Frissen, P.H. Feenstra
TNO Building and Construction Research
Delft, The Netherlands

(3) Assessment of Ultimate Load Analyses of Reinforced Concrete Plates and Shells by Means of Adaptive Methods
R. Lackner and H. Mang
TU-Vienna, Karlsplatz
Wien, Austria

(4) Modeling of Shear Behavior of RC Columns
B. Rose, Shing, and E. Spacone
University of Colorado
Boulder, CO

38. *CONCRETE: FROM MATERIAL MODELING TO STRUCTURAL DESIGN*
II. *Structural Retrofits*

MODERATORS: G. Meschke
Ruhr-University Bochum
Bochum, Germany

F. Ulm
Massachusetts Institute of Technology
Cambridge, MA

1) Retrofitting of a RC Cooling Tower: From Concrete Modeling to Structural Design
G. Meschke
Ruhr-University Bochum
Bochum Germany

H.A. Mang
TU-Vienna
Wien, Austria

M.N. Fardis and S.N. Bousias
University of Patras
Patras, Greece

(2) Finite Element Investigation of Concrete Slabs Post Strengthened by Fiber Reinforced Polymers
M. Hormann, H. Menrath, E. Ramm
University of Stuttgart
Stuttgart, Germany

F. Seible
University of California, San Diego
La Jolla, CA

(3) Finite Element Modeling of Reinforced Concrete in Bridge Seismic Retrofit Design Practice
R.A. Dameron and Y.R. Rashid
ANATECH Corp.
San Diego, CA

(4) FRC Containers Design to Ensure Impact Performance
F. Toutlemonde
Laboratoire Central des Ponts et Chaussees (LCPC)
Paris, France

J. Sercombe
TU-Vienna
Wien, Austria

47. CONCRETE: FROM MATERIAL MODELING TO STRUCTURAL DESIGN
III: Seismic Performance

MODERATOR: F. Ulm
LCPC
Paris, France

G. Meschke
Ruhr-University
Bochum, Germany

(1) Simplified Modeling Strategies for RC Structures under Seismic Loading using Damage Concepts
J. Mazars, F. Raguenau, and Shahrokh Ghavamian
Laboratoire de Mechanique et Technologie
Cachan Cedex, France

(2) Reinforced Concrete Building Prediction Analyses
R. K. Dowell and L. Zhang
ANATECH Corp.
San Diego, CA

(3) Collapse Analysis of RC Structures including Shear
A. S. Elnashai, A. M. Mwafy, and D. Lee
Imperial College
London, UK

(4) Simulation of Hysteretic Behavior of Reinforced Concrete Members
J. Lee, F. C. Filippou, and G. L. Fenves
University of California, Berkeley
San Francisco, CA

56. *CONCRETE: FROM MATERIAL MODELING TO STRUCTURAL DESIGN*
IV: Constitutive Modeling

MODERATOR: K. Willam
University of Colorado
Boulder, CO

(1) Durability Scaling of HPC-Structures Using the Probability Crack Approach
F. Ulm, P. Rossi
Laboratoire Central des Ponts et Chaussees
Paris, France

I. Schaller
Service d'Etude Techniques des Routes et Autoroutes
Bagneux, France

D. Chauvel
Electricite de France
Villeurbanne, France

(2) Modeling confined concrete under monotonic and cyclic loading
M. Kwon, E. Spacone, and T. A. Balan
University of Colorado
Boulder, CO

(3) Constitutive modeling of concrete in numerical simulation of anchoring technology
J. Nienstedt, R. Mattner, and J. Wiesbaum
Hilti AG.
Schaan, Principality of Lichtenstein

(4) Evaluation of a Rate -Sensitive Material Model for Concrete

H. D. Kang, K. J. William, and Y. Xi
University of Colorado
Boulder, CO

(5) Nonlinear Analysis of Composite Steel-Concrete Structures

A. Haufe, H. Menrath, and E. Ramm
University of Stuttgart
Germany

**UPDATE OF MASONRY BUIILDING CODE AND SPECIFICATION,
1995 TO 1998 EDITIONS**

Max L. Porter, Ph.D., P.E., FASCE[1] and J. Gregg Borchelt, P.E., MASCE[2]

Introduction

The Masonry Standards Joint Committee (MSJC), sponsored by American Concrete
Institute (ACI), the American Society of Civil Engineers (ASCE), and The Masonry
Society (TMS), has finished its work on the 1998 editions of the *Building Code
Requirements for Masonry Structures (ACI 530-98/ASCE 5-98/TMS 402-98)* and the
Specification for Masonry Structures (ACI 530.1-98/ASCE 6-98/TMS 602-98). The
changes made to the 1995 editions of the Code and Specification are expected to
increase their usefulness to the design and construction community.

The purpose of the Code continues to be minimum requirements for the design and
construction of masonry. It is to be used as part of legally adopted building codes.
Therefore, it contains requirements with respect to the preparation of contract
documents, but the Code is not intended to become part of those documents. This
distinction is clarified by the addition of the following new statement in the Code
Commentary: "This Code is not intended to be made a part of the contract
documents. The contractor should not be asked through contract documents to
assume responsibility regarding design (Code) requirements, unless the construction
entity is acting in a design-build capacity..."

The Specification continues to set minimum requirements for constructing masonry
designed in accordance with the Code. These include requirements for materials and
construction procedures. The Specification is primarily directed to the contractor,
although the design professional must review it and modify project specifications as
appropriate. Specification checklists provide guidance to the specifier in this regard.

Background to the content of the Code and Specification is found in the
Commentaries for these standards.

---

[1]Professor of Civil Engineering, Department of Civil and Construction Engineering,
Iowa State University, Ames, IA and Chairman, Masonry Standards Joint Committee

[2]Vice President, Engineering and Research, Brick Industry Association, Reston, VA
and Secretary, Masonry Standards Joint Committee

New Format

Throughout both standards and their related commentaries many changes have been made to increase readability and to clarify the intent of the provisions, but perhaps the most significant editorial change is that the Code and Code Commentary have been completely reorganized. Since 1988 when the Code was first published, new chapters and sections have been added as appropriate. New chapters were located sequentially in the Code as they were developed. With the addition of a new chapter on design of prestressed masonry and the impending addition of a new design method using limit states design, the MSJC felt that it was time to reformat the entire document to follow a more logical progression in the order of Code provisions.

This change does not entail any substantive changes to the Code but does result in many changes to adjust the cross references and to reorganize sections to match the new organization. All general Code requirements have been moved into Chapter 1. This includes all of the old Chapters 1, 2, 3, 4, and 10, and most of Chapters 5 and 8. The remainder of the Code is now divided up by design method as follows:

- Chapter 2 covers Allowable Stress Design (ASD). Section 2.1 is general ASD requirements (formerly Chapter 5), 2.2 is unreinforced ASD requirements (formerly Chapter 6), and Section 2.3 is reinforced ASD requirements (formerly Chapter 7).
- Chapter 3 is reserved for limit states design, which the Committee expects to finalize for the next edition of the Code.
- Chapter 4, a new portion of the Code, covers design of prestressed masonry.
- Chapter 5 addresses empirical design of masonry (formerly Chapter 9).
- Chapter 6 on design of veneers (formerly Chapter 12) incorporates a new section on design of adhered veneer.
- Chapter 7 is design of glass unit masonry (formerly Chapter 11).

The Committee believes that these changes will result in a document that is easier to use.

TECHNICAL CHANGES

Several substantive changes have been made to the Code, and Code Commentary. These include establishment of minimum criteria for quality assurance, addition of design provisions for prestressed masonry and for adhered veneer. The Committee also made changes to the Specification and Specification Commentary as needed to be consistent with these Code revisions. Finally, the hot weather construction provisions of the Specification and Specification Commentary were substantially modified. These subjects are discussed in more detail in this article.

Additional technical changes include:

- A new method of determining the modulus of elasticity of masonry. The use of tables based on unit strength and mortar type has been replaced with equations that use a multiplier of the specified compressive strength of masonry.
- Permitting the use of mortar cement (ASTM C 1329). This product has bond strength requirements established to provide comparable flexural tensile bond strength to that achieved using portland cement-lime mortar. Masonry built with mortar cement mortars has the same allowable flexural tensile stress values as unreinforced masonry built with portland cement-line mortars.
- Empirical design requirements to have net uplift forces resisted by anchors and the introduction of foundation piers.
- Verification of the compressive strength of masonry under ASTM C 1314. This standard describes prism dimensions, construction, handling, and curing. Clay and concrete masonry now use the same prism slenderness reduction factors on which to base compressive strength.

Quality Assurance

The Masonry Standards Joint Committee has reaffirmed several times that the Code is based on inspected masonry work. Similar to requirements in previous editions of the Code, it now states, "A quality assurance program shall be used to ensure that the constructed masonry is in compliance with the contract documents." However, the Code was not clear about the content of the quality assurance program or how it was established. The 1998 Code establishes minimum criteria for three levels of quality assurance, based on whether the facility is an essential or non-essential facility (as defined in the general building code or ASCE 7) and the design procedure used. For each level of quality assurance, minimum tests and submittals are identified and minimum inspection before and during masonry construction is set. In general, more testing and inspection are required for essential facilities and engineered design than for non-essential facilities designed using empirical or prescriptive design procedures.

The quality assurance program must also set forth:

- Procedures for reporting and review.
- Procedures for resolution of noncompliance's.
- Qualifications for the testing laboratories and inspection agencies.

The Specification repeats the tables that establish these minimum requirements. It lists services and duties of the testing agency, inspection agency, and contractor. The submittals portion of the Specification was revised to coordinate with the quality assurance requirements and industry practice.

Prestressed Masonry
Code provisions for the design of prestressed masonry focus on post-tensioned
masonry, since that is the most common application. The prestressing tendons may
be bonded or unbonded, and, if unbonded, lateral restrained or unrestrained. Design
requirements are based on allowable stress design procedures except that the design
of prestressed masonry with laterally-restrained prestressing tendons is designed
using a combination of allowable stress design and strength design. For simplicity,
the design methods for unreinforced masonry have been applied to prestressed
masonry wherever possible. The Specification contains requirements for
prestressing materials, tendon installation, and stressing procedures.

Adhered Veneer
The chapter on veneer was reorganized with the addition of adhered veneer. General
requirements are given first. Anchored veneer and adhered veneer requirements
follow and are separated. The Code retains the traditional definition of masonry
veneer as an element that is not relied upon to add to the load-resisting capacity of a
wall system. The adhered veneer design provisions are prescriptive, and are similar
to those currently included in the Uniform Building Code. Alternative design
procedures using the principles of mechanics are permitted, with conditions outlined.
This is the same process used for anchored veneer.

Hot Weather Construction
Hot weather construction requirements of the Specification have been revised to
conform to current recommended practices. The new provisions include preparation
requirements to be accomplished prior to conducting masonry work, construction
requirements to be followed while work is in progress, and protection requirements
for the newly constructed masonry. The preparation and construction requirements
are indexed to ambient air temperatures and wind velocities, while the protection
requirements are indexed to mean daily temperatures and wind velocities.

Closing
The proposed documents have completed public review. It is anticipated that the
documents will be published during the spring of 1999. The 1998 edition of the
Code has been submitted to the International Code Council for consideration in the
International Building Code, expected in the year 2000.

This brief overview is by no means a complete listing of the changes to the MSJC
Code, Code Commentary, Specification, and Specification Commentary. The
Committee has worked long and hard to make the documents more user-friendly and
update them to incorporate current masonry design, materials, and. The Committee
continues to examine the documents, looking for ways to improve their provisions
for the design and construction of masonry.

**Out-of-Plane Seismic Behavior of Brick Masonry walls of
Typical Turn-of-the-Century North American Residential Buildings**

Michel Bruneau[1], M. ASCE, André Filiatrault[2], M. ASCE, and Jocelyn Paquette[3]

Abstract

Sample brick masonry walls only tied with nails to their timber backing were extracted
from an existing building and tested on a shake table. Out-of-plane resistance of this
type of construction widely used at the turn-of-the-century was measured, and
compared with performance when retrofitted using anchors or fiberglass strips.

Introduction

Three masonry walls with their wood backing were extracted from an existing three-
story residential building in Montréal, Canada, scheduled for demolition. These
specimens are representative of a type of construction widely used in North America at
the beginning of the century (and before), in which a single wythe exterior masonry wall
was added to wood buildings as a fire-propagation prevention measure in dense urban
areas. The masonry walls typically resisted their own weight and were tied only with
nails to the timber structure, and an irregular gap was present between the masonry and
timber walls. The timber structure consisted of stacked 3"x10" rough-cut timber
laterally supported by vertical post at 12 feet spacing.

To seismically retrofit such buildings, special seismic-resistant anchorage of the walls
would be required at every floors at a minimum. Questions remained however as to the
out-of-plane resistance of the remaining walls spanning between floors, recognizing the

[1] Professor and Deputy Director of Multidisciplinary Center for Earthquake Engineering
Research, Department of Civil, Structural and Environmental Engineering, 130 Ketter Hall,
State University of New York at Buffalo, Buffalo, NY 14262.
[2] Professor, Department of Applied Mechanics and Engineering Science, University of
California, San Diego, 9500 Gilman Drive, Mail Code 0085, La Jolla, CA 92093.
[3] Graduate Research Assistant, Ottawa Carleton Earthquake Engineering Research Centre,
Department of Civil Engineering, University of Ottawa, 161 Louis Pasteur, Ottawa,
Ontario, Canada, K1N 6N5.

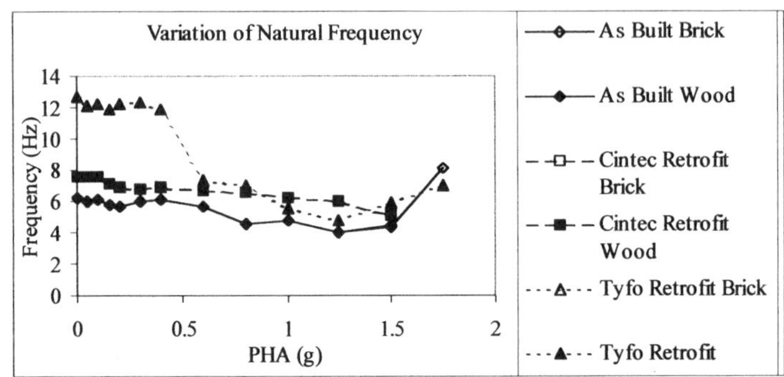

Figure 1 Variation of wall vibration frequency as a function of peak
horizontal acceleration reached in prior test.

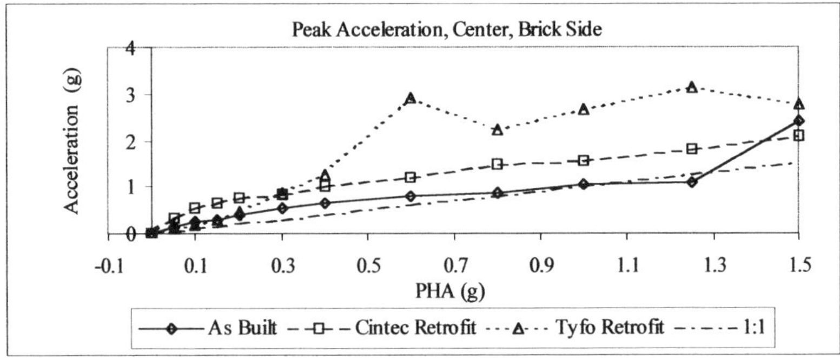

Figure 2 Peak acceleration measured at middle of wall (brick side) as a function of
peak horizontal acceleration in respective tests.

possible interaction with the timber backing, and as to the best way to strengthen these
walls having access to the masonry surface from only one side.

To partly answer these questions, the three specimens extracted from the existing
building were tested on the Earthquake Simulator Facility at École Polytechnique in
Montréal, Canada. The wall specimens were subjected to earthquakes of progressively
larger intensity until structural failure. All specimens were 4'x6' panels tested in their
out-of-plane direction. A rigid steel A-frame and a concrete base were used to tie the
top and bottom of the wall rigidly to the shake table; this was to ensure that both the top
and bottom of the walls would receive the same seismic excitation, as predicted for
unreinforced masonry buildings having wood floor diaphragms per the commonly
accepted practice for the seismic evaluation and retrofit of such these buildings (see the
Uniform Code for Building Conservation, for example).

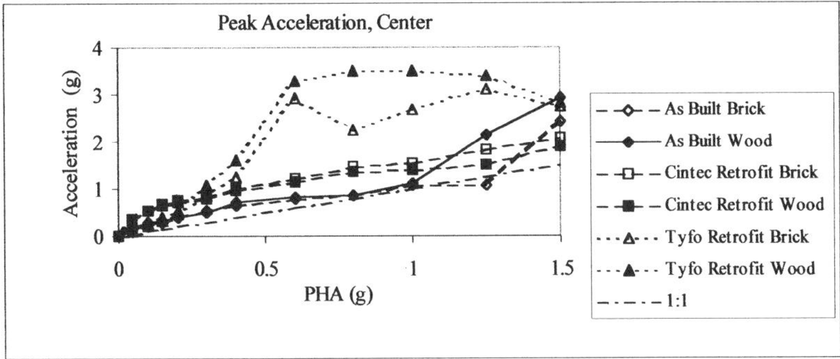

Figure 3 Comparison of peak acceleration measured at middle of brick and wood sides of wall as a function of peak horizontal acceleration in respective tests.

Two of the three walls were retrofitted prior to testing. First, the h/t ratio of the unsupported wall was reduced by bolting the masonry wall to the timber backing using through thickness bolts. Cintec bolts with sleeves filled with grout were used to ensure a bearing surface between the two walls over the expanded sleeve surface in the gap between the two walls. Second, fiberglass strips were added to one side of the wall. Although double sided strips are recommended based on prior testing, access to both sides of the walls is not economically possible for these type of structures. The retrofit design expectation, however, was that the fiberglass strip would serve to prevent dynamic out-of-plane instability in the outward direction, and that the wood backing would preclude the same in the inward direction, in spite of the small gap present between the two walls. Such a design assumption can only be verified experimentally using shake-table testing.

Figs. 1 to 4 illustrate the behavior observed during these tests. As shown in Fig. 1, the existing unreinforced masonry wall was able to resist significantly large peak ground accelerations (PGA) without substantial degradation in natural frequency. Large PGA at mid-wall were also necessary to make the

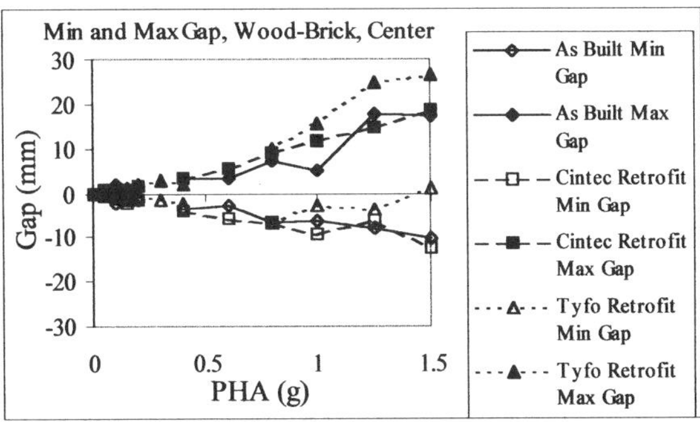

Figure 4 Size of gap between brick and wood at wall center

wall collapse out-of-plane (Figs. 2 and 3). Although bricks at the edge of the wall started to fall as PGA approached 1.0g, the existing walls certainly behaved beyond expectations. Although maximum gaps observed between the wood and masonry

Figure 5: Left to right, top row: As built wall before test and after collapse, and wall retrofitted with Cintec anchors at mid-height; Bottom row: Failure mode of wall retrofitted with Cintec anchors, wall retrofitted with Tyfo fiberglass stips, and failure mode of wall retrofitted with fiberglass strips.

walls in this assembly increased significantly as PGA increased during these tests, time histories are being studied to determine the extent to which the loosely connected wood and masonry walls behaved as a unit.

Conclusions

Data collected during this test series is being analyzed to explain the observed behaviors, assess effectiveness of the proposed retrofits, and compare results with predictions from existing seismic evaluation methodologies.   In particular, normalization of the findings will be required to extrapolate these results on 6' tall panels to walls of full story heights.

Stability of URM Compression Members under Out-of-Plane Lateral Loads

A. E. Schultz[1], A. M. and J. Mueffelman[2]

Abstract

Numerical solutions for the buckling capacity of URM compression members subjected to out-of-plane lateral loads are presented. The impact of bending on buckling capacity is illustrated, and the interaction between critical load and out-of-plane bending is discussed.

Introduction

Bending arising from out-of-plane lateral loads has a dramatic impact on the stability of unreinforced masonry (URM) compression members (Fig. 1). Due to the low tensile strength of masonry, flexural tension translates into cracking, and it reduces the effective portion of the cross-section. This renders the member more flexible, augments lateral deflection due to out-of-plane bending, and gives rise to second-order (P-Δ)

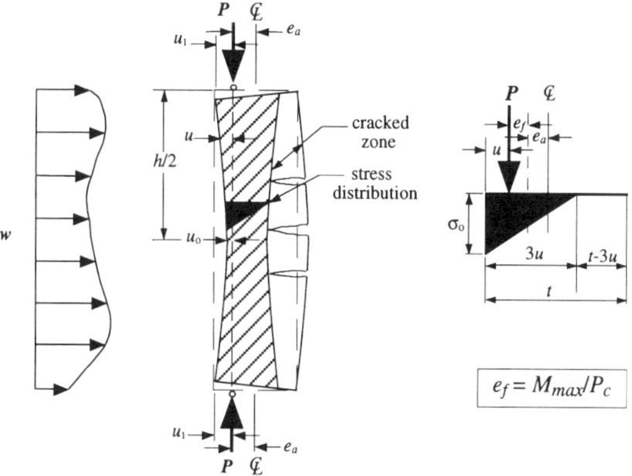

Figure 1. Influence of Lateral Load on Buckling of URM Members

[1]Assoc. Prof., Dept. of Civil Engrg., Univ. Of Minnesota, Minneapolis, MN, 55455.
[2]Struct. Engineer, Bonestroo, Rosene, Anderlik & Assoc., Roseville, MN, 55113.

moments. These moments generate more tension which further reduces the cross-section, and this sequence of events is repeated until equilibrium of the deflected member is established, or until it fails due to instability (i.e. buckling). Prestress forces in masonry walls that are post-tensioned with unbonded and unrestrained tendons can be viewed in a similar manner to external vertical loading in URM compression members.

The first source of out-of-plane bending that was recognized to affect the stability of URM compression members is eccentricity of axial load. The effect of axial load eccentricity was analyzed extensively from the 1950's to the 1970's (Angervo, 1954; Sahlin, 1961; Yokel, 1971; Colville, 1979). This body of research led to the development of design code provisions that include eccentricity as a factor in the buckling capacity of URM walls. In the U.S., a check on the buckling strength of URM walls including the effects of axial load eccentricity is required (MSJC,1995).

Surprisingly, the existing design approach for URM compression members in the U.S. has not led to problems, even though it does not address out-of-plane bending. Most of the existing URM building stock in the U.S. was designed according to older codes that are overconservative in terms of allowable compression stresses and wall slenderness. Yet, new masonry construction, even if it is unreinforced, often features members that are more slender and more highly stressed than those in older buildings. This trend comes at a time when design standards (*Minimum*, 1995) and model codes (NEHRP, 1998) have recently adopted significant increases in wind and seismic loads.

Influence of Axial Load Eccentricity

Linear analysis techniques have been used to solve the governing differential equation for the lateral deflection of URM walls under eccentric axial loading (Angervo, 1954; Sahlin, 1961). Solutions for deflection, as a function of a axial load, were used to define critical loads. Two U.S. researchers, working independently, developed equivalent solutions for the eccentric buckling strength of URM compression members (Yokel, 1971; Colville, 1979). Yokel (1971) illustrated the accuracy of this approach for predicting the buckling strength of walls tested over a wide range of variables.

The impact of nonlinearity in the masonry stress-strain relation on the eccentric buckling capacity of URM walls has generated interest in the past decade (La Mendola et al., 1995; Ganduscio and Romano, 1997). A variety of solutions have been obtained, but these have not found their way to design practice because they typically involve coupled systems of implicit nonlinear equations that must be solved iteratively. Furthermore, the importance of the nonlinearity of masonry in compression has not been assessed. In U.S. practice, some of this nonlinearity is accounted for by using a secant modulus of elasticity.

Bending from Out-of-Plane Lateral Loads

La Mendola et al. (1995) and Ganduscio and Romano (1997) extended the nonlinear analysis of eccentric buckling in URM compression members to include lateral loading. They obtained analytical solutions for cantilever walls with uniform loading, but these solutions are still cumbersome. Sahlin (1961) had earlier considered the influence of out-of-plane bending on the buckling capacity of a solid wall made using a linear, elastic material with no tensile strength but, this approach did not yield a practical solution.

Yokel's second-order formulation for the deflection of a linear, elastic URM wall, with solid cross-section and no tensile strength can be extended to include the effect of out-of-plane bending (Schultz and Mueffelman, 1998). An additional variable eccentricity e_f which is defined at a section as the ratio of moment M to axial load P (Fig. 1). Differential equations were defined for moment distributions arising from four different combinations of support conditions and lateral loading. These include 1) equal end-moments at top and bottom of a simply-supported wall, 2) uniformly distributed lateral load on a simply-supported wall, 3) concentrated lateral load on top of a cantilever wall, and 4) uniformly-distributed lateral load on a cantilever wall.

Impact of Bending on Buckling Capacity

Schultz and Mueffelman (1998) obtained numerical solutions for the lateral displacement of URM compression members under a given axial load. These solutions were found to describe axial load-displacement relationships like that shown in Fig. 2 for a simply-supported wall with a uniform load at ultimate w_u and which produces maximum first-order moment M_{max} at mid-height. From these solutions, it is evident that the axial force-lateral displacement response of an URM compression member exhibits a strong dependance on bending magnitude. The relative maxima for these curves represent points of impending instability, and the corresponding axial load is the buckling strength P_c.

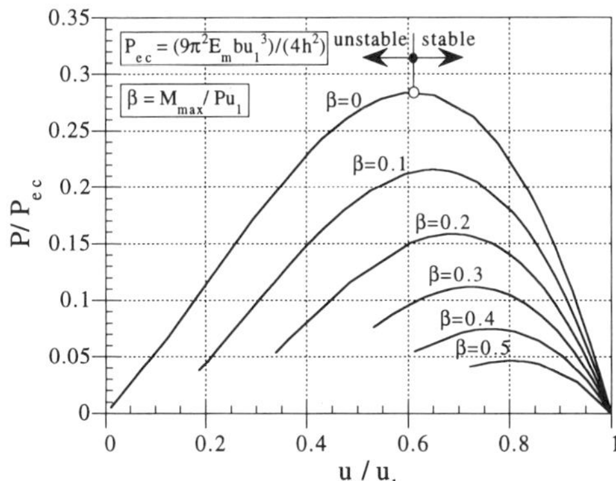

Figure 2. Stability Curves for Simply-Supported Uniformly-Loaded URM Member

The buckling load solutions obtained by Schultz and Mueffelman (1998) also describe a highly nonlinear interaction between buckling load P_c and bending moment M_{max}. In Fig. 3, normalized buckling load is plotted against normalized moment for all four load cases considered. As noted earlier, the magnitude of bending has a dramatic impact on critical axial load. Moreover, for a given value of moment, there are two values of axial load that produce instability for a given compression member. One of these axial loads ($Ph^2/\pi^2 EI < 0.4$) represents a regime for which increases in axial load are beneficial since they reduce flexural tension, but, for the other axial load ($Ph^2/\pi^2 EI > 0.4$) the reduction in flexural tension does not offset the increase in instability. It is also noted, as a result of the zero tensile strength assumption, that bending capacity vanishes when no axial load is present.

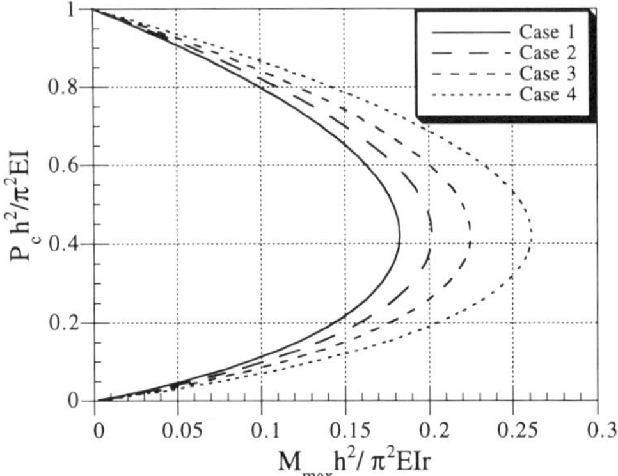

Figure 3. Moment-Buckling Load Interaction Diagrams for URM Members

Conclusions

Numerical solutions to the differential equations for flexure of URM members demonstrate the deleterious effect of out-of-plane bending on buckling strength, and the nonlinear interaction between moment and axial load. This phenomenon appears to affect post-tensioned members with unrestrained tendons as well. The implications for URM design are troubling because bending effects are currently not addressed by U.S. building code provisions for buckling strength. It is concluded that the influence of bending on the buckling instability of URM members poses a grave and far-reaching problem in view of the exposure of masonry structures to lateral loads from wind and earthquakes in the U.S.

References

Angervo, K. (1954). *Uber die knickung und tragfahigkeit eines excentrisch gedrueckten pfeilers*, Staaliche Tecnische Forschungsansalt, Helsinki.

Colville, J. (1979). "Stress reduction factors for masonry walls." *Proc.*, ASCE, 105(ST10), 2035-2051.

Ganduscio, S., and Romano, F. (1997). "FEM and analytical solutions for buckling of nonlinear masonry members." *J. Struct. Engrg.* ASCE, 123(1), 104-111.

Masonry Standards Joint Committee (MSJC). (1995). *Building Code Requirements for Masonry Structures.* ACI 530-95, ASCE 5-95, TMS 402-95.

La Mendola, L., Papia, M., and Zingone, G. (1995). "Stability of masonry walls subjected to transverse forces." *J. Struct. Engrg.* ASCE 121(11), 1581-1587.

Minimum Design Loads for Buildings and Other Structures. (1995). ASCE 7-95, Reston, VA, 232 pp.

NEHRP Recommended Provisions for Seismic Regulations for New Buildings. (1994). FEMA 222A. Federal Emergency Management Agency. Washington, DC.

Sahlin, S. (1961). "Transversely loaded compression members made of materials having no tensile strength." *Proc.*, Int. Assoc. for Bridge and Struct. Engrg., 21, 243-253.

Schultz, A. E. and Mueffelman, J. (1998). "Interaction of bending and compression in masonry members." Dept. of Civil Engrg., Univ. of Minnesota, Minneapolis.

Yokel, F. Y. (1971). "Stability and capacity of members with no tensile strength." *J. Struct. Div.*, ASCE, 97(7), 1913-1926.

An Introduction to the New Prestressed Masonry Code Provisions

Arturo E. Schultz[1], A.M.

Abstract

A brief history of prestressed masonry construction is presented, as well as a summary of ongoing efforts to develop design provisions for prestressed masonry in the US. A brief overview of the proposed Masonry Standards Joint Committee (MSJC) design provisions for prestressed masonry is also given.

A Short History of Prestressed Masonry

Prestressed masonry has emerged as a bona fide loadbearing system only in the last two decades of the 20th century. Yet, for more than one century, engineers and builders experimented with the use of prestressed metal tendons to erect masonry structures (Schultz and Scolforo, 1991). The earliest documented use of prestressed masonry occurred in the United Kingdom (UK) during the Victorian era. Precompression stresses of unknown magnitude were imparted to brick masonry in tunnel construction by placing and restraining heated wrought-iron tie rods in the masonry. In the United States (US), many attempts were made between the 1880's and 1930's to develop masonry roof and floor systems that relied on metal rods and nuts to hold masonry units in place. It was not until 1935, however, that Anderegg and Dalzell made an explicit and rational attempt to use predefined precompression stresses to offset flexural tension stress in masonry floor beams. It took another two decades before Haller, from Switzerland, implemented prestressed masonry floor systems in construction practice.

Development of prestressed masonry walls did not begin until the 1960's, with pioneering work in the UK, Australia and New Zealand (Schultz and Scolforo, 1991). However, it was the UK which experienced expanded use and sustained

---

[1]Assoc. Professor, Dept. of Civil Engrg., Univ. of Minnesota, 500 Pillsbury Dr. SE, Minneapolis, MN 55455

development of prestressed masonry construction (Curtin et al., 1988). Structural design code provisions were drafted initially in 1978 (Haseltine) and were approved in 1985 by the British Standards Institute (*Code*). Prestressing has been used in the UK to increase the resistance of masonry walls subjected to lateral loads from high winds, earth pressure, differential settlement, blast, and accidental vehicular impact. Most applications have utilized vertical post-tensioning in the walls of low-rise structures which often feature open floor plans. The walls in these applications usually have hollow, non-rectangular built-up brickwork sections with either internal diaphragms or external fins (deep pilasters).

During the past decade, a significant effort has been expended in Switzerland to develop single-wythe, prestressed masonry walls (Ganz, 1990). Hollow unit walls are post-tensioned with monostrand tendons that are placed in protective ducts, and the cavities containing the ducts are grouted. Additional development of prestressed masonry during the past decade has been reported in Canada, Australia and New Zealand. A more complete literature review on prestressed masonry can be found elsewhere (Schultz and Scolforo, 1991).

Development of Prestressed Masonry in the US

In the US, the development of prestressed masonry has been slower, and more convoluted (Schultz and Scolforo, 1991). In the early 1980's, Ng and Cerny reported on the construction and testing of prototype, post-tensioned, prefabricated concrete masonry beams. However, the earliest modern use of prestressed masonry in the US has been for the construction of residential basements, and light commercial construction, in southwestern states. More recently, the VSL Corporation designed and implemented retrofit systems incorporating post-tensioned tendons for two historic unreinforced masonry buildings damaged during the 1989 Loma Prieta earthquake in California.

During the past decade, a persistent effort has been made to develop design provisions for prestressed masonry structures in the US. This effort began shortly after the release of the 1988 version of the Masonry Standards Joint Committee (MSJC) document (ACI 530, 1988) as a comparison of US design practice for reinforced and unreinforced masonry with British design practice for prestressed masonry (Schultz and Scolforo, 1992). Shortly after release of the 1992 edition of the MSJC code document, a subcommittee of the MSJC was created to draft design provisions for prestressed masonry.

During the past six years, the Prestressed Subcommittee of the MSJC has worked arduously on a set of code provisions which can be used to design cost-effective prestressed masonry structures which perform well under expected loads and actions. As of the writing of this paper, the proposed prestressed masonry design provisions for the MSJC code document have been reviewed and approved by The Masonry Society (TMS) and the American Society of Civil Engineers (ASCE). Final approval from the American Concrete Institute (ACI) is pending.

With the expected publication of the prestressed masonry design provisions in the MSJC code document, engineers and builders will be in a better position to obtain approval for the design and construction of masonry buildings that incorporate prestressed tendons. Many applications have been envisioned in the US including prefabricated lintels and veneer panels. But, the most common application is likely to be the tall single-wythe hollow unit wall for low-rise commercial and industrial buildings. In such systems, low vertical stresses and large slenderness ratios conspire to aggravate flexural tension stress demand. The use of strategically placed tendons can offset these effects while retaining a significant degree of the economy of unreinforced masonry.

Proposed MSJC Code Provisions for Prestressed Masonry

The greatest benefit of prestressing is that tensile stresses in masonry can be reduced or eliminated allowing for longer beams and more slender walls and columns than in unreinforced masonry. Since the prestressing tendons can be spaced farther apart than reinforcing bars in reinforced masonry, prestressed masonry requires less grouting. Also, member sections remain uncracked, so prestressed members are stiffer than comparable reinforced members. However, these features come at the expense of the prestressing hardware and tendon placement and alignment. Furthermore, the effective prestress that can be relied upon during design is that magnitude of stress that is likely to be present after all losses. The MSJC code document recognizes the same mechanisms for loss of prestress that affect prestressed concrete (ACI 318, 1995).

The proposed MSJC code provisions for prestressed masonry address structures that are prestressed with either bonded or unbonded tendons, and unbonded tendons can be either unrestrained or laterally-restrained. A restrained tendon is not free to move laterally with respect to the adjacent masonry, whereas an unrestrained tendon can do so. Prestressing tendons can comprise steel strand, steel bar or steel wire. As with the bulk of the MSJC document, design of prestressed masonry follows the working stress design procedure, and elastic analysis is utilized for the determination of stresses.

Masonry can be either pre-tensioned, in which case it is erected around an externally stressed tendon which is released after the masonry has achieved the minimum required strength (i.e. transfer). Or, it can be post-tensioned, in which case the tendon is stressed against the masonry once the latter has achieved the required minimum strength. The proposed MSJC provisions for prestressed masonry control the amount of stress in the tendons at jacking and after transfer for pre-tensioned masonry, and at the time of prestress application for post-tensioned masonry. Stresses in the masonry of prestressed members have to satisfy the provisions for allowable stresses applicable to unreinforced masonry. And, the combined effects of prestressing force and dead load cannot generate flexural tension stress in the masonry. Furthermore, a moment strength check is required for sections with laterally-restrained tendons.

The moment strength check utilizes a rectangular stress block similar to that used for reinforced concrete in the ACI 318-95 code document, and a formula is provided for estimating the increase in tendon stress at ultimate moment conditions for members with laterally-restrained but unbonded tendons. For bonded tendons, the stress at ultimate must be obtained from strain compatibility considerations. Shear stress in masonry members is limited to the stresses allowed in an unreinforced masonry section according to the Coulomb-type friction expression. In addition, two other alternate allowable shear stress expressions are included to limit the maximum principal tensile and compressive stresses in masonry members that resist large shear and axial forces. Net axial tension in a masonry section must be resisted by either reinforcing bars, prestressing tendons or both.

Closing Statement

The MSJC proposed provisions, should they be included in the 1998 MSJC document are but a temporary respite in a long saga. These provisions will continue to evolve as the design and construction sectors of the masonry building industry gain experience with its use. Changes can be expected regarding several topics, including a moment strength check for sections with unbonded tendons, a stability check for members with unbonded and unrestrained tendons and out-of-plane moment, and more accurate parameters for seismic design. In addition, as the entire MSJC code document will adopt a Limit States Design format in the near future, so the prestressed masonry provisions will inevitably follow suit. All of these changes will require additional research and continuous evaluation of the research results.

References

Building Code Requirements for Structural Concrete. (1998). ACI 318-95. American Concrete Institute, Farmington Hills, MI.

Building Code Requirements for Masonry Strucrtures. (1988). ACI 530-88/ASCE 5-88. American Concrete Institute, Detroit MI, American Society of Civil Engineers, New York, NY.

Code of Practice for Use of Masonry. (1985). BS 5628. British Standards Institution. London.

Curtin, W. G., Shaw, G., and Beck, J. K. (1988). *Design of Reinforced and Prestressed Masonry Structures*. Thomas Telford, Ltd. London, 1988.

Ganz, H. R. (1990). "Post-Tensioned Masonry Structures." *VSL Report Series*, VSL International, Ltd., Berne, Switzerland, 35 pp.

Haseltine, B. A. (1982). "Codification of Reinforced and Prestressed Masonry." *Reinforced and Prestressed Masonry*. Thomas Telford, Ltd., London.

Schultz, A. E. and Scolforo, M. J. (1991). "An Overview of Prestressed Masonry." *TMS Journal*, 10(10, 6-21.

Schultz, A. E. and Scolforo, M. J. (1992). "Engineering Design Provisions for Prestressed Masonry." *TMS Journal*, 10(2), 29-64.

Prestressed Masonry in the United States

David T. Biggs, FASCE[1]

Abstract

Prestressed masonry has been used in the United States for many years. Most applications have utilized proprietary systems. This paper describes the types of systems and the uses for prestressed masonry.

Introduction

Masonry design in the United State has generally fallen into one of two categories, Empirical and Rational methods. The Rational methods are further divided into categories for Unreinforced and Reinforced Masonry.

The Empirical techniques are developed around ratios, such as the height to thickness of a wall. These ratios are based upon historical data for completed masonry structures.

The Rational methods in the United States are based upon analytical techniques using either Working Stress or Strength Design. All building codes have procedures for designing with Working Stress, whereas only the Uniform Building Code (UBC) has a Strength Design option.

----

[1]Principal, Ryan-Biggs Associates PC, 291 River Street, Troy, NY 12180-3278

The Unreinforced Masonry procedure is based upon evaluating both strength and stability. Strength is determined by checking allowable stresses in the masonry for both compression and tension under service loads. Stability is determined by checking wall and column elements for buckling.

The Reinforced Masonry procedure has similar strength and stability checks, using allowable stresses and service loads. However, in this method, reinforcing is introduced to take all tensile stresses.

The Strength Design method is based upon ultimate strength techniques which were supported by field test data. This method is currently part of the UBC and is useful for tall-wall type buildings such as warehouses, gymnasiums, commercial businesses, and schools.

Prestressed masonry offers another alternative to masonry designers. This proposed 1998 Masonry Standards Joint Committee (MSJC) code is generally based upon the Working Stress procedure, with a supplemental Strength Design check when laterally restrained tendons are utilized. Proprietary prestressing systems are based upon either Working Stress or Limit States procedures.

Although prestressed masonry has not been codified in the United States, there have been numerous projects constructed which utilize tendons which are prestressed or post-tensioned. These have included both new construction and renovation/repair projects.

Prestressed Masonry Systems

Prior to 1998, the use of prestressed masonry was achieved by either proprietary systems or single-user systems. The proprietary systems developed and marketed by large companies are intended to reinforce the masonry to resist the loads on it. The single-user systems are developed by individual masonry suppliers or contractors for use on a more regional basis. Thus, it is more difficult to identify these users. One known user prestresses the masonry for secondary effects such as crack control.

There are two primary types of proprietary systems; one uses rods for tendons and the other uses strands. The rods vary from 414,000 TO 690,000 kPa (60 to 100 ksi) and utilize threaded couplers. Rods can be installed in short lengths or full length. Tension is often measured using a torque wrench. One rod system has been approved by IBCO and uses patented concrete masonry units.

The strands are high-strength. The technology is taken from prestressed concrete and modified for use with masonry. Patented fittings are used with full-length strands. Tension is applied and measured using hydraulic jacks.

Each proprietary system has unique anchorage systems. Generally, prestressing tendon loads are transferred to the masonry by bearing plates on the masonry.

Of the known single-user systems, it appears the prestressing has been useful for prefabricated masonry panels. Reinforced brick is prefabricated with prestressing to minimize cracking during handling. The mild reinforcing is intended to carry the actual loads.

Uses

Prestressing is intended to compensate for tensile stresses in the masonry. There are numerous applications which have been constructed. These include:
   a. Residential foundation walls: the prestressing counteracts earth pressures (Figure 1).

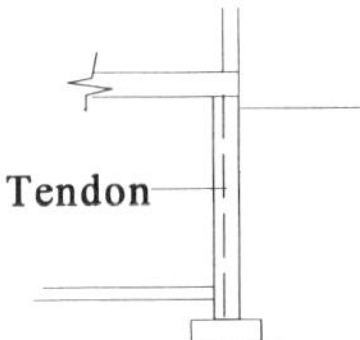

Figure 1 - Basement Wall

   b. Screen walls: prestressing allows these walls to function as cantilevers to resist wind and seismic loads (Figure 2).

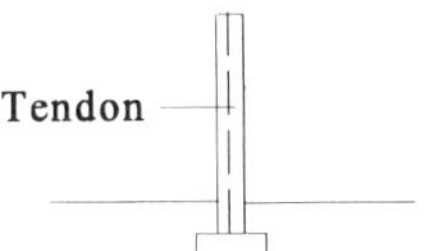

Figure 2 - Screen Wall

c. Building walls: prestressing allows taller, thinner walls to be constructed (Figure 3).

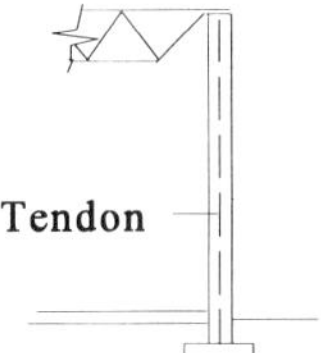

Figure 3 - Tall Wall

d. Seismic upgrading: prestressing provides stabilizing forces which improve the seismic resistance of walls for both out-of-plane flexure and in-plane shear.

e. Crack control: prestressing places the masonry of prefabricated wall panels and beams into compression and thereby minimizes cracking due to handling.

f. Interior partition walls which may be cantilevered off the floors or braced at the underside of the structure above.

Summary

Prestressed masonry offers new opportunities for designers. Performance and economics of the systems will dictate how often they are used. Based upon the benefits of the systems, applications for prestressed masonry will tend toward use on one-or two-story buildings, tall single-story walls, basement foundation walls, partition walls, retaining walls, bridge abutments, and prefabricated wall panels.

European Perspective of Prestressed Masonry

M E Phipps[1] and H R Ganz[2]

Abstract

The paper describes the design concepts and codes for prestressed masonry used in Europe. Five projects built with prestressed masonry in Britain, Switzerland and Germany are then described to show the range of applications to date.

Introduction

The structural performance of masonry can be enhanced by the application of compressive prestress. This is because although masonry has a high compressive strength its tensile strength is low. If sufficient prestress is applied the tensile stresses in the masonry can be eliminated entirely under working conditions so that the compressive strength of the whole masonry cross-section can be utilised. In prestressed members, unlike reinforced members, cracking need not take place under working loads.

Development

In the nineteenth century masonry was sometimes prestressed with tightened iron or steel rods. The first recorded application of prestressed masonry using modern high tensile steel prestressing tendons was in the 1950's and since then there has been a steady increase in its usage. The most widespread use of prestressed masonry in Europe today is for the post-tensioning of vertically spanning walls. In Britain high tensile steel bar tendons are most commonly used and the structural performance of the masonry is often enhanced by constructing the masonry member with a geometric cross-section, i.e., a cross-section with the structural units arranged to give greater structural efficiency than that obtained from the equivalent section in a solid rectangular wall or column or a cavity wall. In Switzerland post-tensioning tendons are placed inside large cores of bricks and blocks, typically in the centre of a wall. The tendons are made of individually greased and sheathed 7-wire prestressing strand, monostrand, placed inside a steel or plastic duct, grouted into the cores of the brick and block. [VSL International Ltd., 1990]. The main application of prestressed masonry in Switzerland is for relatively slender walls exposed to wind with low axial load.

---

[1]UMIST, PO Box 88, Manchester M60 1QD, United Kingdom.
[2]. VSL Management, 41 Avenue du Centre, 78067 St Quentin-Yvelines Cedex, France.

Research

There is a considerable body of test evidence on the performance of prestressed masonry to support the increasing use in practice [Phipps, 1999]. Significant work has been carried out at the University of Manchester Institute of Science and Technology [Phipps, Montague 1987, Roumani, Phipps 1988],and at the Institute of Structural Engineering, ETH Zurich [Ganz, Thürlimann 1984, Ganz, Thürlimann 1985].

Codes

The first code which covered prestressed masonry was published in 1985 as British Standard Code of Practice BS 5628: Part 2: 1985. This was revised in 1995 and a new edition is expected to be published in 1999. In Switzerland prestressed masonry was first addressed in the 1995 revision of the Swiss Masonry Code SIA V177. The developing European code for masonry, DD ENV 1996 Eurocode 6, incorporates a section on prestressed masonry. DD ENV 1996 is a limit state code in SI units. Unlike a working stress code, such as the MSJC code ACI 530 / ASCE 5 / TMS 402, a limit state code considers loads which will cause collapse (ultimate limit state) as well as service loads (serviceability limit state).

Prestressed masonry members under the serviceability limit state.

The safety factor for load, γ_f, is taken as 1.0 at transfer of prestress and under the design loads. In BS 5628 the material safety factors, γ_m, are replaced by compression and tension stress limitations. For example, stresses in the masonry at transfer must be $<0.4f_{ki}$ and >0 and under design loads $<0.33f_k$ and >0 where f_{ki} is the characteristic strength of the masonry at transfer and f_k is the specified characteristic strength.

The analysis of a section at the serviceability limit state assumes that plane sections in the masonry remain plane after bending, stress is proportional to strain and the effective prestressing force does not change after all prestress losses have taken place. DD ENV 1996 suggests that although flexural cracking and crushing will be satisfied by the above, a deflection check may need to be carried out. SIA V177 does not require a verification of the maximum compression stresses in service conditions and instead of tensile stress, the nominal crack width is limited on the tension face to 0.20mm and 0.05mm for normal and superior performance requirements, respectively.

Prestressed members under the ultimate limit state

At ultimate DD ENV 1996 requires γ_f for a permanent action such as earth to be taken as 1.35 if the effect is unfavourable and 1.0 if it is favourable. The safety factors for the prestressing force, γ_p, are 0.9 if the effect is favourable and 1.2 if the effect is unfavourable. The safety factor for masonry, γ_m, is 1.7 and for steel, γ_s , is 1.15. The tendon stresses are limited to 70% of their characteristic strength, f_{kp}. SIA V177 considers prestress as resistance and therefore, does not apply a factor to the prestressing force. However, $\gamma_s = 1.2$ is applied, as for reinforcing steel.

The flexural behaviour of unbonded post-tensioned masonry at the ultimate limit state is similar to reinforced masonry. However, the lack of bond causes differential longitudinal movement between the tendons and the adjacent masonry and, when the voids in the masonry are large, there can be differential lateral movement also. The analytical treatment of the general case, i.e., the case for any shape of masonry cross-section, involves an interactive method which can easily be carried out by computer. Many members, however, particularly walls, have rectangular compression zones and unbonded tendons and for these cases an empirical method is given in BS 5628. SIA V177 limits the depth of the compression zone, x, to one quarter of the wall thickness to assure sufficient ductility.

The Eurocode requires axially loaded prestressed members to be designed as unreinforced members and highlights the possible need to limit the prestress force to that which the member can carry without buckling due to slenderness

In prestressed masonry shear failure is due to the principal tensile stresses reaching the diagonal tensile strength of the masonry so forming a diagonal crack. BS 5628 sets out a design method in terms of applied shear stress and shear strength which caters for this and also satisfies DD ENV 1996. The Swiss Code requires the shear strength of walls to be verified with strut-and-tie models neglecting any tensile strength of masonry.

Loss of prestress with time
The sources of prestress loss are, relaxation of tendons, elastic deformation of the masonry, moisture movement of masonry, creep of masonry, anchorage seating, friction effects and thermal effects. As far as post-tensioned masonry is concerned prestress loss will normally stem only from tendon relaxation, masonry moisture content changes and creep in the masonry.

Road bridge abutments, England
5.8m high post-tensioned brickwork road bridge abutments were built on the Glinton bypass highway in Cambridgeshire. The walls were hollow box sections 1.57m wide and were post-tensioned with 25mm and 50mm diameter high tensile steel bars at 390mm centres.

New support walls for a 14th century roof, England
Post-tensioned brickwork walls were used to carry the large outward roof thrusts from a repositioned 14th century roof in the New Guesten Hall at Worcester Cathedral. The walls incorporate hollow triangular shaped sections of brickwork 0.89mm deep.

Fire Station, England
The two longitudinal supporting walls of a fire station were constructed of post-tensioned brickwork. One wall was 9.0m high and one was 5.0m high. Both walls had hollow box, or diaphragm, sections with overall depths of 0.44m. The post-tensioning gives the walls sufficient stiffness to provide lateral stability to the building without the use of additional cross-walls.

Industrial Centre, Altendorf, Switzerland
A new warehouse was added adjacent to an existing building. Two 13.8m tall walls facing the existing building and exposed to full wind load were designed in 8.2m post-tensioned masonry, sitting on a 5.6m reinforced concrete wall. Post-

tensioning tendons consisted of 15mm monostrands spaced at 0.57m to 0.95m depending on actual exposure.

German Technical Museum, Berlin, Germany

The Swiss post-tensioned masonry technique was recently introduced in Germany. The four storey addition to the museum presently under construction in Berlin consists of a structural steel frame, concrete floors, and double wythe masonry curtain walls. The exterior wall is exposed to full wind load, has axial load due to self weight only, and span up to 8.8m. They were therefore, axially prestressed at the centre of the inner wythe of the wall by 34 15mm monostrand tendons spaced at 1.75m. The tendons were each 26.5m long and anchored at the bottom in a reinforced concrete basement wall, see Figure 1.

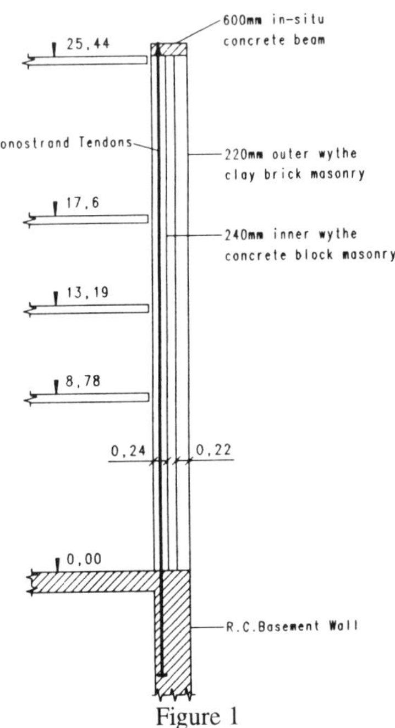

Figure 1

References

Ganz H R, Thürlimann B 1984 Versuche an Mauerwerksscheiben unter Normalkraft und Querkraft (Tests on Masonry Walls Loaded by Normal Force and Shear), *Test Report No. 7502-4*, Institute for Structural Engineering, ETH Zurich, Birkhäuser Publisher, Basel.

Ganz H R, Thürlimann B 1985 Shear Design of Masonry Walls. *Proceedings of the ASCE Structures Congress* 1985 "New Analysis Techniques for Structural Masonry".ASCE/Chicago, IL.

Phipps M E, Montague T I 1987 The testing of plain and prestressed concrete blockwork beams and walls of geometric cross-section. *Masonry International* Vol 1 No 3 1987: 96-99

Phipps M E Prestressed Masonry *Progress in Structural Engineering and Materials* Vol2, Issue 3, 1999

Roumani N A, Phipps M E 1988 The ultimate shear strength of unbonded prestressed brickwork I and T section members. *Proceedings of the British Masonry Society* Masonry 2: 82-84

VSL International Ltd. 1990 Post-tensioned Masonry Structures, *Technical Report*, Berne, 35pp.

Unbonded vs. Bonded Tendons: Which is Better for Masonry?

H. R. Hamilton III[1]

1. INTRODUCTION

The design of prestressed masonry has recently been introduced into the model building code by the Masonry Standards Joint Committee. It is anticipated that post-tensioning hollow unit walls that carry small gravity loads will be the most economical and popular method of application. Presumably a bonded system will give a higher strength and certainly a more ductile system than that of an unbonded, laterally restrained system. In addition, it would be expected that the increased level of effort required to do a moment-curvature analysis of the cross section would provide some benefit in the way of larger capacity than the assumption of yield stress in the capacity calculation. A practical problem using concrete masonry flexural design is examined to determine the benefit gained by using bonded vs. unbonded systems. In addition, the advantage of determining the flexural capacity using strain compatibility is examined.

2. TENDON RESTRAINT

There are two terms that describe how stress and force are transferred between the tendons and the masonry. A *bonded* tendon is a tendon that is encapsulated by prestressing grout in a corrugated duct. The duct is then bonded to the surrounding masonry through grouting (see Figure 1). This is analogous to post-tensioned concrete systems in which the duct is cast in the concrete and the tendon is subsequently placed in the duct and grouted. This grout provides a continuous bond between the masonry and tendon. A *laterally restrained* tendon is defined as a tendon that is not free to move laterally within the cross section of the member (see Figure 2). This normally

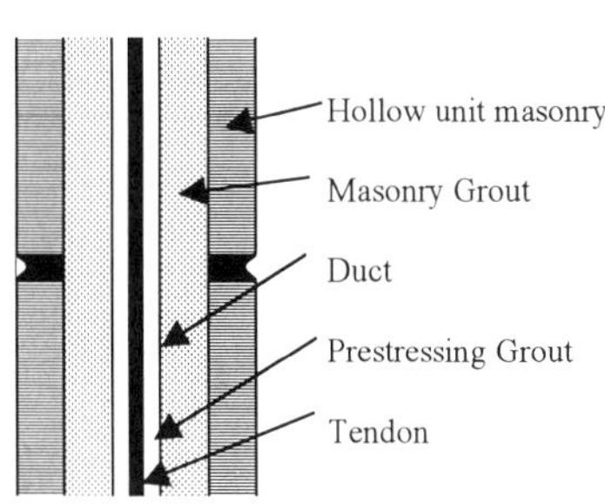

Figure 1. Bonded Tendon

applies to unbonded tendons. Note that a bonded tendon is considered a restrained tendon but a restrained tendon is not necessarily a bonded tendon.

3. FLEXURAL CAPACITY CHECK

The proposed provisions of the MSJC utilize a working stress philosophy. In all cases, the designer is required to check the flexural and axial stresses under service loads against the allowable compressive and tensile stresses given for unreinforced masonry. If the tendon is laterally unrestrained then the designer must check to ensure that the buckling capacity is sufficient under service axial loads *and* loads imposed by the prestressing force. When the tendon is laterally restrained then the designer must decide whether the system is to be bonded or unbonded. If the tendon is unbonded then the designer must check the flexural capacity using the equations given in the code. If the tendon is bonded, then the designer may calculate the moment capacity assuming the prestressing steel has yielded (using the yield strength for the calculation) or they may use strain compatibility and the basic constitutive relationships for the masonry and prestressing steel.

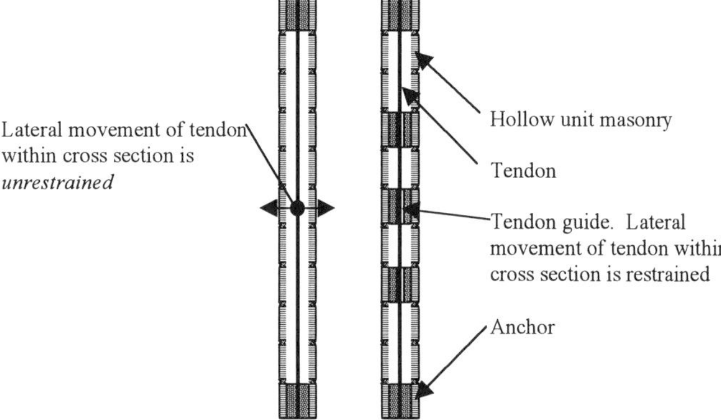

Figure 2. Restrained and Unrestrained Unbonded Tendons.

4. EXAMPLE

The example chosen for this parametric study is that of a 20-ft. tall masonry wall with no axial load. Prestressing losses were assumed to be zero for comparison purposes. The assumption is that the wall will be constructed from 8"cmu and that face shell bedding will be used. The objective of the parametric study is to study the advantages and disadvantages of bonded vs. unbonded tendons. The flexural capacity of the bonded system was calculated in the two ways allowed in the proposed provisions. The first method was to assume that the prestressing steel had yielded and use the "yield" strength of the steel. The second method was using strain compatibility and the constitutive relationship for the prestressing steel. Note that the yield point in high strength steels is not as well-defined as that of conventional reinforcing. To complete the parametric study, the constitutive relationships for three common prestressing steels were simplified to one of a bilinear elastic-plastic relationship that models the actual behavior of the prestressing steel (see Figure 3). The three steels used in this analysis were 270 ksi strand (243/270), 167 ksi bar (150/167), and 95 ksi bar (75/95).

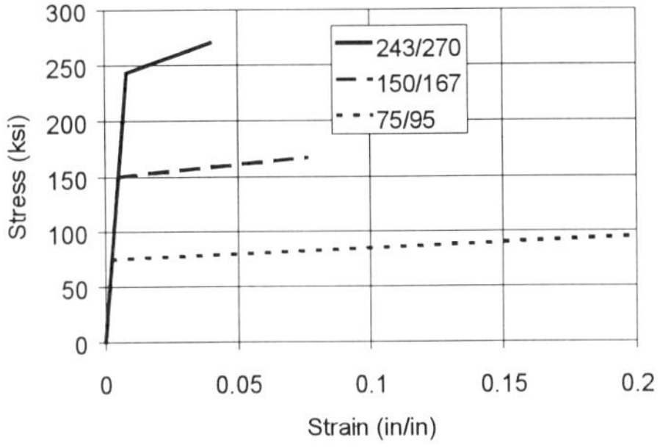

Figure 3. Constitutive Relationship for Prestressing Steel.

Three flexural capacities were calculated for each configuration of prestressing quantity and strength. The unbonded flexural strength was calculated using the equations given in proposed provisions. The bonded flexural capacity was calculated in two ways. The first was to assume an elastic perfectly plastic constitutive relationship for the prestressing steel and the whitney stress block with a limiting masonry strain capacity of 0.003. The second method made use of the bilinear relationships shown in Figure 3. This method required iterative calculations in order to balance the internal forces on the cross-section.

5. COMPARISON OF METHODS

MSJC places a limit on the size of the equivalent stress block (a/d<0.425) to ensure a ductile failure mode for the bonded systems and to keep the amount of prestressing to a reasonable level in unbonded conditions. The reinforcing limits (for this example) associated with the a/d limits are shown in Table 1.

Table 1 – Limiting reinforcing ratios for a/d<0.425

steel	bonded	unbonded
243/270	0.00297	0.00381
150/167	0.00482	0.00616
75/95	0.00963	0.0124

Figure 4 shows a sample graph of the moment capacity for bonded and unbonded systems using the 75/95 prestressing steel. There is very little difference between the moments calculated using the two methods for bonded systems. This is true for the full range of reinforcing ratios for each of the three prestressing steels (The plots for other prestressing steels were not included due to limited space). For this particular example, the designer does not gain much capacity with the investment of additional design time necessary for strain compatibility analysis. The calculated strength of the bonded system is approximately 25% higher than that of the unbonded system.

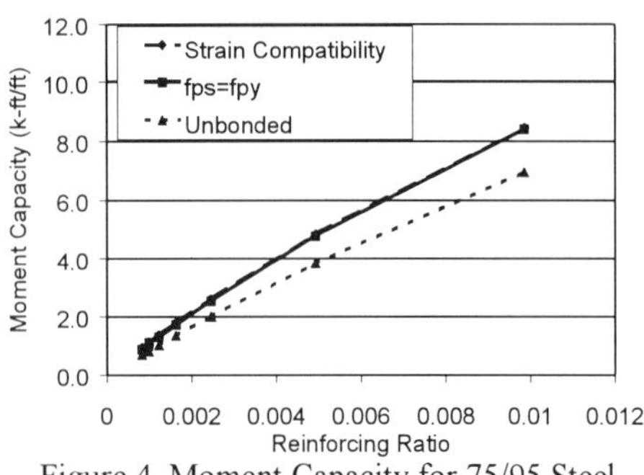

Figure 4. Moment Capacity for 75/95 Steel

Figure 5 shows the moment capacity of the sample wall using each of the three prestressing steels. Note that the plot also includes the effective prestress level as a function of the moment strength. Using the limiting reinforcing ratios from Table 1 the maximum prestress level for each of the prestressing steels is approximately 800 psi. This value is much larger than would be allowed under allowable stress design for concrete masonry.

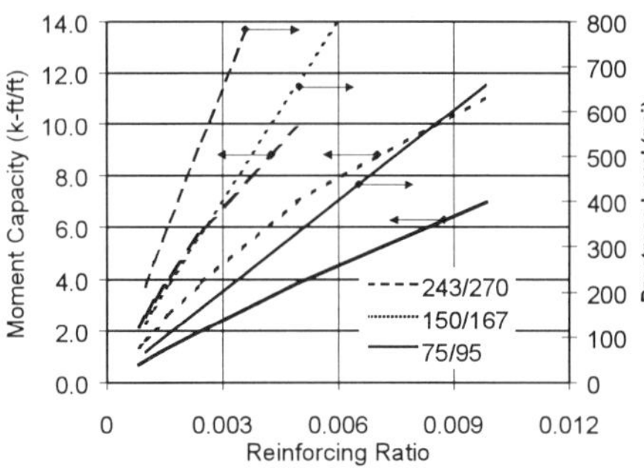

Figure 5. Unbonded System

6. CONCLUSIONS

Based on the sample prestressed masonry design problems presented, the following conclusions can be drawn:

1. There is no significant benefit in using strain compatibility to calculate the moment capacity.
2. The moment capacity of the unbonded sample wall was calculated at approximately 25% less than that of the bonded sample wall.

Composite Joist Behavior and Design Requirements

W. Samuel Easterling,[1] P.E.

Introduction

Composite joists or trusses are structural members that have become increasingly popular in the last several years due to economical and functional considerations. These members consist of a steel joist and composite slab that are connected by welded headed shear studs. The joists are specifically designed for the composite application and are not selected from standard Steel Joist Institute (SJI) load tables. Composite slabs are constructed using steel deck, topped with normal or light weight concrete. No distinction is made in this paper between "joists" and "trusses," given that joist is an accepted industry description of a truss.

The benefits of open, essentially column free floor spaces in buildings are well recognized. Such a configuration provides maximum flexibility in leasable space arrangements and thus gives an owner the ability to easily accommodate the requests of new tenants. Composite joists are one structural system that provides these large open areas. In addition to the benefits of column free space, the open web configuration of the joists permits easy access for mechanical and service systems, without necessarily increasing the floor-to-floor heights in the building. Both of these characteristics are significant benefits and make composite joists an economical structural system.

This paper is part of a four part session focusing on the design of composite joist and joist girder systems. Two of the papers deal with behavioral and design issues (this paper and one by Band and Murray), while the other two papers (Swensson and Samuelson) focus on various case studies that used composite joist floor systems. The general behavior of composite joists will be reviewed in this paper, followed by a brief introduction to a proposed design specification developed by an ASCE task committee (ASCE 1996.)

---

[1] Assoc. Prof. and Asst. Dept. Head, The Charles E. Via, Jr. Dept. of Civ. and Env. Engr., Virginia Polytechnic Institute and State University, Blacksburg, VA, 24061

Composite Joist Behavior

The overall behavior of a composite joist, subjected to gravity load, is a reflection of the behavior of individual components. Specifically, the behavior of the bottom chord, top chord, web members, shear connectors and composite slab dictate the overall composite joist behavior. Although not a requirement in current steel specifications (*Load and* 1993, *Standard Specifications* 1996), it is generally desirable to have a vertical load carrying truss respond to load in a ductile manner. In the case of composite joists this means that the bottom chord should experience general yielding of the cross section while the top chord, web members, shear connectors and composite slab continue to carry load. If none of the components of the composite joist reach their maximum capacity prior to the occurrence of bottom chord yielding, then the composite joist will attain a well-defined yield plateau, in terms of a load vs. deflection response.

The flexural strength of the composite joist is obtained from a couple that consists of the slab being in compression and the bottom chord being in tension. Shears are resisted by the web members. This "bending" analogy is frequently made when considering trusses using hand computations.

If any of the elements in the composite joist, other than the bottom chord, reach their maximum capacity prior to the bottom chord reaching a general cross sectional yielding, then a brittle failure will likely occur. A "brittle failure" is characterized by the attainment of a peak load, followed by a rapid decrease in the load carrying capacity, with no evidence of a yield plateau. The designer must be aware of the behavioral characteristics of the components and of the typical values of the actual yield stress of the bottom chord material when proportioning composite joist components.

The fundamental idea in proportioning the components of the composite joist is that no element should reach it's maximum capacity prior to the bottom chord reaching it's yield load which is given by:

$$P_y = A_s F_y \qquad\qquad (1)$$

where P_y = yield load, A_s = gross area of bottom chord, F_y = yield stress in bottom chord. The value of the yield stress that is used for determining the design strength is the nominal yield stress. The strength of the bottom chord is obviously related linearly to the yield stress. Other elements can then be sized so that the load on the joist that causes P_y to be reached in the bottom chord is smaller than the load that will cause these elements to reach their maximum capacity. However, when the compression webs are designed based on this approach, the designer must account for the actual yield stress of the bottom chord.

Composite joists are commonly fabricated using steel shapes that have an actual yield stress approximately 10% greater than the nominal value. Thus the load that will be required to achieve bottom chord yielding will be approximately 10%

higher. The strength of a compression web is not linearly related to the yield stress. The closer the member slenderness approaches the transition between inelastic and elastic buckling, the less dependent the strength is on the yield stress. Thus, the compression webs will not likely be 10% stronger as a result of the actual yield stress being 10% greater than the nominal value. Design of the tension webs is not affected by the difference between nominal and actual yield stress values, because increased strength in those members will parallel the increase in the bottom chord.

The top chord, although a compression member under construction loads, need not receive the same attention in the design process as the compression webs. The top chord is generally ignored in the composite section calculations due to the proximity of the top chord neutral axis to the plastic neutral axis of the cross section. The only requirement of the top chord during the composite stage is that the base metal thickness satisfy the minimum requirements for stud welding.

Shear studs in composite joist applications behave as they do when used for composite hot-rolled beam applications. The proportioning requirements given in the proposed specification (ASCE 1996) consider a strength reduction factor for the shear studs and recognize the typical 10% increase in actual vs. nominal yield stress of the bottom chord.

Proposed Design Specification for Composite Joists

The *Proposed Specification and Commentary for Composite Joists and Composite Trusses* (ASCE 1996) was written by and ASCE task committee and addresses the design of simply supported composite joists used in one way floor systems. The specification relies heavily on the American Institute of Steel Construction (AISC) *Load and Resistance Factor Design Specification for Structural Steel Buildings* (1993) and the SJI *Standard Specifications, Load tables and Weight Tables for Steel Joists and Joist Girders (1996)*. The fundamental difference between the way one would design a "truss" using either one of these documents and the proposed composite joist specifications is the requirement of ductility.

The required ductility, for which the compression web member and shear stud behaviors are described above, results in slightly different design requirements (ASCE 1996). Compression webs are designed using a strength reduction factor, ϕ, of 0.75 instead of the usual 0.85. The shear studs are designed with a requirement of

$$NQ_n = 1.3\ A_s\ F_y \qquad\qquad (2)$$

where N = number of shear studs between point of maximum moment and support and Q_n = shear capacity of a single shear connector. The coefficient of 1.3 reflects the use of a strength reduction factor of 0.85 for the studs and the 10% difference in the nominal vs. actual yield stress in the bottom chord.

More complete design provisions, including a section on serviceability concerns, and a detailed example problem are given in the referenced paper (ASCE

1996.) Note that the paper by Band and Murray (1999), which is part of the same session as this paper, address in detail the floor vibration checks required for composite joists. The reader is also referred to the AISC Design Guide No. 11 (Murray, et al, 1997).

Summary

Behavioral characteristics of composite joists have been presented in this paper, with attention being given to those characteristics that differ from typical truss design. In particular the influence of the difference between nominal and actual yield stress in the bottom chord has been highlighted. The applications for composite joists will continue to grow, as engineers become more familiar with their benefits both functionally and economically.

References

ASCE Task Committee on Design Criteria for Composite Structures in Steel and Concrete (Darwin, D., Donahey, R. C., Clawson, W. C., Deierlein, G. G., Easterling, W. S. and Leon, R. T.) (1996). "Proposed Specification and Commentary for Composite Joists and Composite Trusses." *Journal of Structural Engineering*, ASCE, 122(4), 350-358.

Band, B. S., Jr. and Murray, T. M. (1999). "Floor Vibrations: Ultra-Long Span Joist Floors." *Proceedings of the ASCE Structures Congress*, New Orleans, *to appear*.

Load and Resistance Factor Design Specification for Structural Steel Buildings. American Institute of Steel Construction. Chicago, Illinois. September, 1993.

Murray, T. M., Allen, D. E. and Unger, E. E. (1997). *Floor Vibrations Due to Human Activity*, AISC/CISC Steel Design Guide Series – No. 11, American Institute of Steel Construction, Chicago.

Standard Specifications, Load tables and Weight Tables for Steel Joists and Joist Girders, Steel Joist Institute, Myrtle Beach, SC, 1996.

FLOOR VIBRATIONS: ULTRA-LONG SPAN JOIST FLOORS

Barry S. Band, Jr.[1] and Thomas M. Murray[2], P.E., Ph.D.

Abstract

Tests and analyses for annoying floor vibrations of three ultra-long span floor systems are presented.

Introduction

With the development of composite joist design, ultra-long span floor systems can be economically designed and built. "Ultra-long span" refers to floor systems having a span greater than 12.2 m (40 ft). A significant design concern for such floors is occupant induced floor vibrations. The new AISC/CISI Design Guide 11 *Floor Vibrations due to Human Activity* (Murray, Allen and Ungar 1997) has criteria to evaluate floors subject to walking and rhythmic excitations and for floors supporting sensitive equipment. Also included are design methods to account for shear deformation and the effect of joist seats when computing the fundamental frequency of the floor. The purpose of this paper is to show that the criterion for walking excitations and the joist frequency calculation methods are applicable to ultra-long span floor systems.

The walking excitation criterion in the Design Guide states that the floor system is satisfactory if the peak acceleration, a_p, due to walking excitation as a fraction of the acceleration of gravity, g, determined from

$$\frac{a_p}{g} = \frac{P_o \exp\left(-0.35 f_n\right)}{\beta W} \tag{1}$$

where P_o = a constant force representing the excitation, f_n = fundamental natural frequency, β = modal damping ratio, and W = effective weight supported does not exceed the acceleration limit, a_o/g for the occupancy. Recommended values of P_o, β, and a_o/g limits for several occupancies are given in the Design Guide.

The most important parameter for the above criteria is the first natural frequency of the floor system, f_n. The first natural frequency of the floor system is estimated from

$$f_n = 0.18 \sqrt{\frac{g}{\left(\Delta_j + \Delta_g\right)}} \tag{2}$$

where g is the acceleration of gravity, and Δ_j and Δ_g are the beam or joist and girder deflections due to the weight supported, respectively. The supported weight, w, used in calculating Δ_j and Δ_g must be estimated carefully. The actual dead and live loads, not the design dead and live loads, should be used in the calculations. The Design Guide suggests the live load be taken as 0.5 kN/m^2 (11 psf) for office floors.

---

[1] Structural Engineer, Stanley D. Lindsey and Associates, Nashville, TN.
[2] Montague-Betts Professor of Structural Steel Design, Virginia Tech, Blacksburg, VA.

In calculating the fundamental natural frequency, the transformed moment of inertia is to be used if the slab (or deck) is attached to the supporting member. This assumption is to be applied even if structural shear connectors are not used, because the required minimal shear forces at the slab/member interface are resisted by deck-to-member spot welds or by friction between the concrete and metal surfaces. If the supporting member is separated from the slab, overhanging beams or joist/girders supporting joists with seats, full composite behavior should not be assumed.

The effects of web shear deformations and eccentricity at web member joints and joist seats must considered in calculating the effective moment of inertia of joists and joist girders. Traditionally, the effective moment of inertia of parallel chord trusses has been taken as 0.85 times the moment of inertia of the chords alone. Physical tests have shown that the 0.85 rule is a good predictor of actual joist behavior if the span-to-depth ratio, L/D, is greater than about 18. For ratios less than 18, the rule becomes unconservative, and very unconservative, below about 13.   To account for shear deformation and joint eccentricity, the Design Guide based on studies by Kitterman and Murray (1994) and by Band and Murray (1996) recommends that the effective moment of inertia of joist or joist girders be estimated using

$$I_{mod} = C_r I_{chords} \tag{3}$$

where, for joists or joist girders with single or double angle web members,

$$C_r = 0.90(1 - e^{-0.28(L/D)})^{2.8} \tag{4}$$

with $6 \leq L/D \leq 24$, L = span length, D = nominal depth of the joist, and I_{chords} = moment of inertia of the chords. Equation (3) predicts the effective moment of inertia of the non-composite member. The effective moment of inertia of a composite joist or joist girder can be calculated from

$$I_{eff} = \cfrac{1}{\cfrac{\gamma}{I_{chords}} + \cfrac{1}{I_{comp}}} \tag{5}$$

where $\gamma = 1/C_r - 1$ and I_{comp} = full composite moment of inertia of the joist chords and concrete slab.

In a joist floor system, the girder and slab are coupled by joist seats and the degree of composite action between the girder and slab depends on the capacity of the joist seats to transfer the forces from the top chord of the joist girder to the overlying slab. Allen and Murray (1993) proposed that the girder or joist girder may develop only partial composite action and recommended using a moment of inertia less than the full composite moment of inertia. Their recommendation was based on engineering judgement without experimental verification and is

$$I_g = I_{nc} + (I_c - I_{nc})/4 \tag{6}$$

where I_{nc} and I_c are the non-composite and fully composite moments of inertia, respectively. For joist girders, I_{nc} is the modified moment of inertia calculated using Equation (3) and I_c is the effective composite moment of inertia calculated using Equation (5).

Results of Tests and Analyses

Since the tolerance criterion in the Design Guide, as well as older criteria, were calibrated using floors with spans less than 12.2 m (40 ft), four long span floors were tested and analyzed by Band and Murray (1996). The joist spans in these buildings ranged from

13.7 m (45 ft) to 35.9 m (117 ft 9 in.). For each building, field measurements were made and subjective evaluation obtained. Acceleration ratio predictions using Equation (1) were calculated and compared to the tolerance acceleration for the occupancy, 0.5%g, and the subjective evaluations.

Building 1. Building 1 is a multistory office building constructed during 1995-96 in the Southeast. The structural system consists of 36LH series composite joists at 4.3 m (14 ft). on center which frame directly into the supporting joist girders or concrete columns. The composite joist girders are 30G-2N and frame into concrete columns. The girder spans are 8.5 m (28 ft) and the joist spans are 13.9 m (45 ft 8 in.). The slab is 165 mm (6 ½ in.) deep of 20.7 MPa (3000-psi) lightweight concrete supported by a 76mm (3 in.) deep metal deck. At the time the measurements were taken on the third floor, a few full depth partitions had been constructed between the second and third floors.

The floor was analyzed assuming 0.15 kN/m^2 (3 psf) of additional dead load due to various construction materials on the floor and no live load. The assumed modal damping was 0.03 to model that expected in the furnished building. The measured and predicted frequencies, the predicted peak acceleration, and the criterion evaluation for four center-of-bay locations are shown in Table 1. At the first three locations, the measured frequencies were greater than both the predicted joist and system frequencies. At the fourth location the measured frequency was less than the predicted joist frequency but greater than the predicted system frequency. The Design Guide criterion for walking excitation, 0.5%g, was satisfied at all locations. The subjective response of the four-member measurement team was that the floor did not produce discernable vibrations. Furthermore, no complaints have been received from the occupants of the finished building.

Building 2. The second floor of this mid-west building was built over an existing one-story building. Composite joists, 32LH series, were used to span between new columns erected outside of the existing structure. The joists were spaced at 1.75 m (5 ft 9 in.) and supported by W21x50 beams.   The joist spans are 16.6 m (54 ft 5 in.) and the girder spans are 7 m (23 ft). The floor deck is 102 mm (4 in.), 27.6 MPa (4000-psi) concrete on 38mm (1 ½ in.) deep metal deck.

The floor was analyzed assuming 0.2 kN/m^2 (4 psf) of additional dead load and 0.5 kN/m^2 (11 psf) of live load. The assumed modal damping was 0.03 since the building was occupied and full height partitions were not present. The measured and predicted frequencies, the predicted peak acceleration, and the criterion evaluation for two locations are shown in Table 1. At both locations, the measured frequencies were greater than both the predicted joist and system frequencies. The Design Guide criterion for walking excitation, 0.5%g, was satisfied at both locations. The subjective response of the four-member measurement team and the occupants was that were no noticeable vibrations.

Building 3. The second and third floors of Building 2 were constructed over an existing single story building with no interior columns. Long span composite joists were designed to span 35.9 m (117 ft 9 in.) on each of the two floors. The second floor is supported by 2.0 m (80 in.) deep joists and the third floor by 1.5 m (60 in.) deep joists. The joist spacing varies between 2.1 m (7 ft 0 in.) and 3.0 m (9 ft 9 in.). The composite joists are supported by 72.5G joist girders on the second floor and 52.5G joist girders on the third floor, both are designed to be fully composite. The joists support a 127 mm (5-in.), 27.6 MPa (4000-psi), normal weight concrete slab on a 50 mm (2-in.) deep metal deck.

The floors were analyzed assuming 0.2 kN/m^2 (4 psf) of additional dead load and 0.5 kN/m^2 (11 psf) of live load. The assumed modal damping was 0.03 since the building was occupied and full height partitions were not present. The measured and predicted frequencies, the predicted peak acceleration, and the criterion evaluation for five center of

Table 1- Summary of Test and Analysis Results

Location	Measured Frequency (Hz)	Predicted Frequency Of Joist (Hz)	Predicted Frequency Of System (Hz)	Predicted Peak Acceleration %g	Criterion Evaluation
Building 1					
2	6.25	5.7	4.9	0.44	Satisfied
5	6.5	5.8	4.9	0.44	Satisfied
7	7.0	5.8	4.9	0.44	Satisfied
9	9.5	10.5	6.5	0.26	Satisfied
Building 2					
1	5.75	4.98	4.54	0.47	Satisfied
2	5.25	4.98	4.54	0.47	Satisfied
Building 3					
1	3.25	2.38	2.32	0.23	Satisfied
2	4.00	2.38	2.32	0.23	Satisfied
3	4.25	2.77	2.71	0.24	Satisfied
4	3.5	2.65	2.33	0.25	Satisfied
5	2.50	2.65	2.33	0.25	Satisfied

bay locations are shown in Table 1. At the first four locations, the measured frequencies were greater than both the predicted joist and system frequencies. At the fifth location the measured frequency was less than the predicted joist frequency but greater than the predicted system frequency. The Design Guide criterion for walking excitation, 0.5%g, was satisfied at all locations. The subjective response of the four-member measurement team and the occupants was that the floor was generally "satisfactory" but that large impacts were felt. The criterion predicted that the floor would be satisfactory by a large margin.

Conclusions and Acknowledgements

From the above data, it can be concluded: (1) that the procedures recommended in the Design Guide under estimate the frequency of long span floors; (2) that for the aspect ratios of the tested floors, the floors tend to vibrate at the natural frequency of the joists rather than that of the system; and (3) that the tolerance criteria given in the Design Guide is applicable to long span joist supported floor systems.

The recommendations and results presented in this paper were developed in research projects at Virginia Tech sponsored by Nucor Research and Development.

References

Allen, D. E. and Murray, T. M. (1993) "Design Criterion for Vibrations due to Walking," Engineering Journal, American Institute of Steel Construction, 4th Qtr, pp. 117-129.

Band, B. S. and Murray, T. M. (1996) "Vibration Characteristics of Joist and Joist-Girder Members," *Research Report CE/VPI-ST 96/07*, Department of Civil Engineering, Virginia Polytechnic Institute and State University, Blacksburg, VA.

Kitterman, S. and Murray, T. M. (1994) "Investigation of Several Aspects of the Vibration Characteristics of Steel Member Supported Floors," Research Report CE/VPI/ST-94/11, Virginia Polytechnic Institute and State University, Blacksburg, Virginia.

Murray, T. M., Allen, D. E. and Ungar, E. E. (1997) *Floor Vibrations Due to Human Activity*, AISC/CISC Steel Design Guide Series 11, American Institute of Steel Construction, Chicago, IL.

Composite Joist Case Histories

David Samuelson,[1] P.E.

Abstract: Composite joist supported floor systems are growing in popularity within the United States. Four projects containing Vulcraft open web steel composite joists are discussed. Physical dimensions of these floors are described along with advantages offered by such a floor system.

Reed Arena, Texas A & M University, College Station, TX

The $35 million Reed Arena completed in May 1998 provides a new arena and special events center on the College Station campus of Texas A & M University. This facility provides 12,500 seats for student convocations, NCAA basketball, other sports, concerts, circuses, rodeos, conferences, and special events.

Lockwood, Andrews, and Newman from Houston, Texas were project architects. Walter P. Moore and Associates, Inc. and D. Y. Davis Associates, both from Houston, Texas, provided structural engineering services. The 1,115 m^2 (12,000 ft^2) four floor level conference and meeting room area on the end of the ovular shaped arena were designed to be supported by composite joists. Spanning 25.3 m (83 ft) and spaced on 2.44 m (8 ft) centers, the shallow 838 mm (33 in) deep composite joists with a span / depth ratio of 30.2 minimized the floor to floor heights. The concrete slab having an overall depth of 159 mm (6.25 in) is supported by 51 mm (2 in) high composite steel floor deck with a 1.06 mm (19 Ga.) design thickness. Reinforcing steel, consisting of one layer of 9.5 mm (0.375 in) diameter bars at 305 mm (12 in) centers each way, was placed in the concrete slab. The floor was designed to support a dead load of 3.26 KPa (68 psf) plus a live load of 4.79 KPa (100 psf).   Lightweight concrete was specified with $f_c' = 27.6$ MPa (4 ksi). Top and bottom chords for the joists typically were 2L –102 mm x 102 mm x 11.13 mm (2L- 4 in x 4 in x .438 in) and 2L – 152 mm x 152 mm x 19 mm (2L- 6 in x 6 in x .750 in), respectively.

---

[1] Structural Research Engineer, Nucor Research and Development, 1601 W. Omaha Ave., Norfolk, NE 68702

Three rows of bolted diagonal erection bridging were installed to stabilize the joists. Hoisting cables were not released until one end of each joist was welded to its support and all erection bridging completely installed. Shear transfer between the concrete slab and underlying composite joists was accomplished by welding 19 mm (0.75 in) diameter x 127 mm (5 in) long shear studs through the metal deck.

Primary advantages for using the composite joists included a cost savings of $50,000 compared to alternate floor systems, ability to span the 25.3 m (83 ft) with a shallower depth, and ease of erection.

General contractor for this project was Huber, Hunt, and Nichols, Indianapolis, IN; steel fabrication was done by Hirshfeld Steel, San Angelo, Texas; and steel erection performed by Irwin Steel, Trophy Club, Texas.

Advanced Communication and Information Technology Center, Blacksburg, VA

Construction began in the fall of 1998 to build a new $21,315,000 Information Technology Center for Virginia Polytechnic Institute and State University. This center will provide additional classrooms, offices, and lecture halls.

Sherertz Franklin Crawford Shaffner, Inc., Project Architect of Record and Structural Engineers for this project suggested that composite joists be utilized for this 13,955 m^2 (150,209 ft ^{2}) project. Esocoff & Associates, Washington DC were design architects. The Technology center has a length of 96.6 m (317 ft) and width of 33.2 m (109 ft) plus an offset floor area measuring 11.3 m (37 ft) wide x 18.3 m (60 ft). The second and third level floors supported by composite joists will provide additional classrooms, offices, and lecture halls. Stated advantages for using composite joists on this project include 1) Cost savings 2) Greater flexibility to accommodate future changes in floor usage 3) Large column free bay sizes, and 4) more room for heating, ventilating, and air-conditioning equipment.

The 660 mm (26 in) deep composite joists spanned 12.06 m (39 ft 7 in) with a 2.13 m (7 ft) cantilever. Joists were spaced at 3.05 m (10 ft) centers and supporting 76 mm (3 in) high galvanized metal deck having a thickness of 1.20 mm (18 Ga.). The total slab depth was 152 mm (6 in) of normal weight concrete with $f_c' = 27.6$ MPa (4 ksi). The slab was reinforced with one layer of 152 x 152 –MW9 x MW9 (6 x 6 –W1.4 x W1.4) welded wire fabric. Additional reinforcement was placed in the slab over the bottom chord bearing cantilever to resist the applied negative moment. Rows of diagonal bridging were installed at the location of the cantilever support and at the end of the cantilever plus one row of diagonal bridging at the joist midspan. Design loads on the floor included 3.59 KPa (75 psf) dead load and 4.79 KPa (100 psf) live load.

Steel shear studs having a diameter of 19 mm (0.75 in) and length of 127 mm (5 in) were welded through the metal deck to the underlying steel joist. Quantity of studs varied from 18 – 58 studs / joist depending on a given joists floor loading and span.

Sherertz Franklin Crawford Shaffner, Inc. estimated that the composite joists provided steel savings of 24.4 kg/m^2 (5 psf) compared to noncomposite joists and 17 kg/m^2 (3.5 psf) steel savings compared to composite beam construction.

Sherertz Franklin Crawford Shaffner indicated that "with the ability to space the joists at 3.05 m (10 ft) centers and span 12.2 m (40 ft), it really opened up the plan and provided lots of space for mechanical, electrical, and plumbing."

General contractor for this project was Branch & Associates, Roanoke, VA with steel fabrication done by Banker Steel, Lynchburg, VA.

Associated Wholesale Grocers, Kansas City, KS

When Associated Wholesale Grocers rapidly expanding business required extra office space they looked at the possibility of adding a second level to their existing office in Kansas City, Kansas. Working closely with George Butler and Associates, Project Architects, Kansas City, KS and A. T. Renczarski and Co., Inc, Structural Engineers, Kansas City, MO several solutions were considered. One of the critical design requirements was that construction activities could not in any way interfere with the computer center located on the ground floor , which was used on a constant basis for processing customer orders. Heating and ventilating units located on the existing roof could not be shut down during the construction process.

The new second and third floor office addition had a footprint of 35.9 m (117 ft 9 in) x 56.4 m (185 ft) providing an additional 4,048 m^2 (43,570 ft^2) of office space. Composite joists were selected to simply span 35.9 m (117 ft 9 in) over the entire first floor level thereby not disturbing any office activities. Steel columns supported by augercast pilings were erected immediately adjacent to the perimeter of the existing building. The composite joists and noncomposite joist girders provided a moment resisting frame for the building.

A joist depth of 2032 mm (80 in) was selected for the second story floor with these deeper joists straddling the roof top HAV equipment. Shallower 305 mm (12 in) deep noncomposite joists, supported by transverse headers, spanned over the existing HAV equipment. Design loads for the floor consisted of a 2.92 KPa (61 psf) noncomposite dead load, 2.39 KPa (50 psf) composite live load, and 1.15 KPa (24 psf) composite dead load [0.19 KPa + 0.96 KPa (4 psf mechanical + 20 psf partitions)]. Large file rooms were designed to support a load of 5.99 KPa (125 psf). To minimize any problems with differential live loading deflections, the maximum differential deflection between adjacent composite joists was specified not to exceed 12.7 mm (0.5 in). Composite joists were cambered for 100% of the noncomposite dead load deflection plus 25% of the composite live load deflection.

Top and bottom chords for the second story composite joists typically were 2L – 102 mm x 102 mm x 12.7 mm (2L – 4.0 x 4.0 x 0.500) and 2L – 152 mm x 152 mm x 15.9 mm (2L – 6.0 x 6.0 x 0.625) angles, respectively with a minimum yield of 345 MPa (50 ksi). The angle webs of the joists were arranged in a modified warren configuration with a 1.52 m (5 ft) horizontal half panel length. Shear transfer between the overlying concrete slab and steel joists was provided by welding 46 to 88 19mm (0.75") diameter x 102 mm (4 in) long studs through the metal deck depending on exact load combinations required at each composite joist.

The joists were spaced on 2.64 m (8 ft 8 in) centers and supported 51 mm (2 in) high composite steel deck with a thickness of 0.909 mm (20 Ga.). A total concrete slab depth of 127 mm (5 in) was supported by the metal deck. Design f_c' for the normal weight concrete was specified to be 27.6MPa (4 ksi).

When designing such long floor and roof joists special consideration needs to be given to the following items:

- Transverse differential deflections between adjacent joists and / or points, which do not move vertically, need to be carefully examined.

- When placing the concrete for such long span floors, the concrete should be placed at a constant thickness versus trying to establish a constant surface elevation for the slab. Ideally the concrete placement starts near the center of the span and progresses toward either end of the joists. If one attempts to place the concrete at a preset elevation, initial concrete thicknesses will be less than the specified minimum slab thickness until the majority of the joist camber has been removed under full noncomposite load of the wet concrete. Likewise slab surface elevations will continue to move downward until the full span has had all the concrete placed on it.

- Partitions, piping, risers, electrical conduits, and HAV ducts between floors ideally should be framed to allow differential vertical movement in the likely case that floors or the roof do not have equal live load applied to them.

- When bridging such long span joists, it is suggested that diagonal bridging not be utilized in the last exterior joist space. When the more heavily loaded interior span is loaded the diagonal bridging will also pull down the spandrel joist, which may have more restrictive deflection limitations due to window and cladding serviceability requirements. For the last joist space it is suggested that horizontal bridging be utilized or a vertical slip connection be provided between the bridging and the spandrel joists.

General contractor for this project was Bohnert Construction Co., Inc., Kansas City, Kansas; steel fabrication was done by Sunbelt Steel, Tulsa, OK, and steel erection was performed by Lawson Erectors, Belton, MO.

Sound Track, Thornton, CO

Ultimate Electronics in Wheatridge, CO constructed a new $14,000,000 Soundtrack retail showroom, office, and warehouse in Thornton, CO utilizing composite joists. Intergroup, Inc., Denver, CO was the project architect with Martin/Martin Consulting Engineers, Wheatridge, CO providing the structural design of this project.

The second story of this building measuring 25.6 m (84 ft) x 76.2 m (250 ft) was supported using composite joists spanning 12.8 m (42 ft). Joists had a depth of 559 mm (22 in) and were spaced 1.829 m (6 ft 0 in) on center. Steel composite deck having a height of 51 mm (2 in) supported the 127 mm (5 in) deep concrete slab. Normal weight concrete was utilized with $f_c' = 24.1$ MPa (3.5 ksi).

The joists in the office area were designed to support a dead load of 3.59 KPa (75 psf) plus a live load of 3.35 KPa (70 psf). Certain areas of the office floor were designed to carry a high-density storage loading of 9.58 KPa (200 psf). With the open web composite joists, it was possible to run the electrical and mechanical lines through the webs thereby reducing the need for beam penetrations and reducing the required overhead ceiling space.

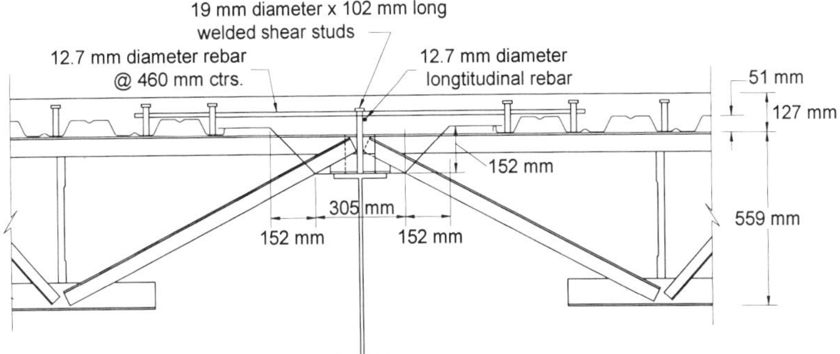

FIG. 1. Haunch Detail

One unique feature for this floor was that a trapezoidal shaped concrete haunch as shown in Fig. 1 was formed over each girder making it possible to develop composite action in the wide flange girders. The composite joists with standard 127 mm (5 in) joist seats rested directly on top of the wide flange girders which allowed very fast joist erection. Adjacent joists were staggered 267 mm (10.5 in) from one another so that the concrete totally surrounded each joist bearing. Sealing of the concrete around the joist bearing seats, top chord, and end web required the installation of special formed sheet metal closures which added extra time to the deck laying.

Shear studs 19 mm (0.75 in) diameter x 102 mm (4 in) long were welded through the metal deck to the underlying steel joists. Where only 1 stud was required per deck rib, the studs were staggered from one top chord angle to the other to assure a uniform transfer of shear load into the joist top chord. Shear studs were located on the side of the deck stiffening rib closest to the ends of the joists to develop the maximum shear capacity from the studs.

Welded shear studs 19 mm (0.75 in) diameter x 203 mm (8 in) long were welded to the top flange of the girder transferring shear forces from the girder to the concrete haunch. Transverse and longitudinal 12.7 mm (0.5 in) diameter reinforcing bars were located in the haunch to transfer compressive forces into the concrete adjacent to the haunch.

General contractor for this project was Saunders Construction, Parker, CO with the steel fabrication performed by Ripsam Mfg. and Fabricating, Loveland, CO.

Conclusions

Vulcraft composite joists described in the above projects provided an efficient and economical longspan floor system. Floor to floor height can often be reduced when the mechanical systems can be routed through the joist open webs. Large column free areas provide increased flexibility for laying out floor plans. Reductions in building weight as the result of the efficient composite design resulted in direct savings to the building owner.

Reliability assessment of reinforced concrete structures with respect to material degradation

F. Stangenberg[1], W. B. Krätzig[2], Y. S. Petryna[3] & M. Roik[4]

ABSTRACT: A special research center („SFB") consisting of some 15 projects sponsored by the German Science Foundation (DFG) is established at Bochum university. The main object is to develop a basis for long-term prognoses of the structural behavior with respect to changes of the material properties and of their scattering. E.g., mechanical damage accumulation models and further deteriorative prediction models are generated. The time histories of these influences are of special importance, as different categories of design life times are considered.- In this contribution, special reference is given to the structural design for different working lives, with respect to probabilistics.

A reliability assessment concept based on nonlinear material models is developed for reinforced concrete structures. Thus, a basis is obtained for calibrating partial safety factors, as they are preferably used in engineering practice.

Principal elements of the nonlinear deterministic and probabilistic assessment are demonstrated. Actual problems and necessary future developments are also referred to.

1 INTRODUCTION

This contribution refers to influences of long-term changes of material properties on the structural behavior of reinforced concrete structures. Not only these changing properties, but also their scattering including the time-dependent distributional changes are significant. In order to obtain reliability-based results for the expected service life, it is necessary to predict these changes. The users of buildings have different life-time expectations according to the specific serving purposes of the building.

The building codes, however, do not reflect life-time classifications, in general. Current research, e.g. the German special research program (SFB) since 1996, refers to this deficiency. The SFB-objectives are: life-time related design of concrete, steel structures and their foundations, with emphasis on the structural components, as far as they are significant for the resistance against actions and for the corresponding serviceability. Solutions are being achieved for:

A: Models for quantifying the influences relevant for life time,

B: Methods for evaluating life-time relevant influences in structures,

C: Future, life-time oriented design strategies.

2 GENERAL ASSESSMENT CONCEPT

The reliability (Fig. 1) can be calculated by the difference of action and resistance. This reliability and its time dependence must be a design criterion, which anticipates the influences of all phases of the erection and service time of the building.

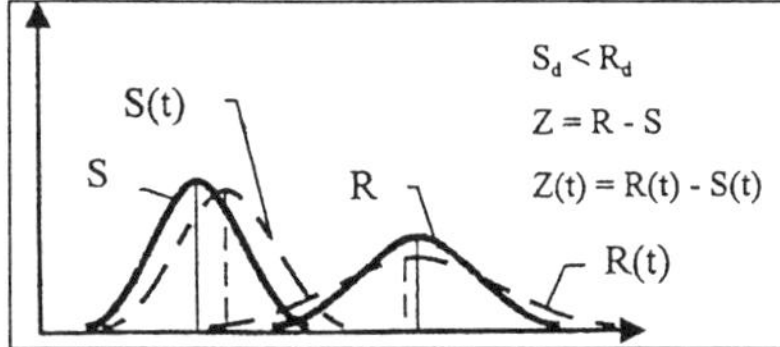

Figure 1: R/S dependence of reliability

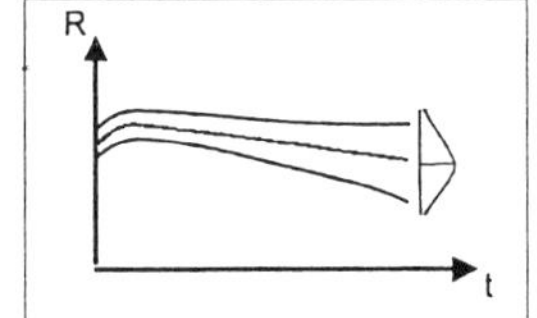

Figure 2: Time-dependent resistance of a RC structure

---

[1] Univ.-Prof., Inst. for Reinforced and Prestressed Concrete,

[2] em. Prof., Inst. for Statics and Dynamics,

[3] Dr.-Ing., Inst. for Statics and Dynamics,

[4] Dipl.-Ing., Inst. for Reinforced and Prestressed Concrete,

Civil Engineering Department, Ruhr-University Bochum, Germany

Both, the action S and the resistance R, are time dependent, and with them, the reliability Z.

Corresponding time-dependent scattering distributions result from the damage evolution, as exemplarily shown in Fig. 2. After an initial rise of the resistance - due to the continuous hardening process of the concrete -, a permanent decrease of the resistance, with a corresponding change (probably: increase) of the scattering is expected.

The required reliability [Z(t) = R(t) - S(t)] is to be fixed in the conceptual phase. It must be sufficient, in order to obtain an acceptable safety at the end of the design service life.

Simplified approaches are necessary for actions and resistances, with account for their probabilistic behavior and time-dependent changes including their scatterings.

3 MATERIAL DEGRADATION MODELS

Load-independent deteriorations (in case: due to shrinkage; chemical aging alterations and loss of safety against corrosion etc.) are not treated, so far.

Here, emphasis is laid on degradations due to fatigue, particularly high-cycle fatigue, as e.g.: decrease of strength, micro-cracks development, and macro-cracking. The material formulations for these damage evolution effects are converted into macro-level models appropriate for finite-element implementation. This can be achieved by the principle of the so-called „elastic" degradation where an integral substitute elasticity modulus changing with the strain history expresses the time-dependent resulting degradation - or damage evolution, respectively - of a structural element (Fig. 3). Introducing the upper limits (suffix: max) and lower limits (suffix: min) of stresses and strains as well as introducing the standard ultimate strength and standard ultimate strain (suffix: u), the Fig. 4 is obtained.

On this basis, macro-level formulations for the degradation behavior of concrete are generated.

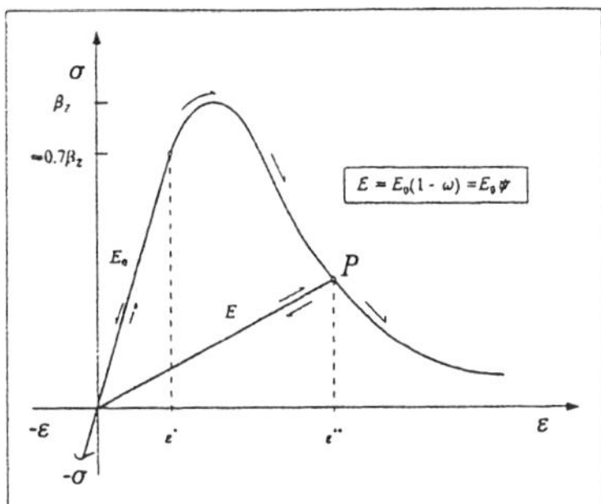

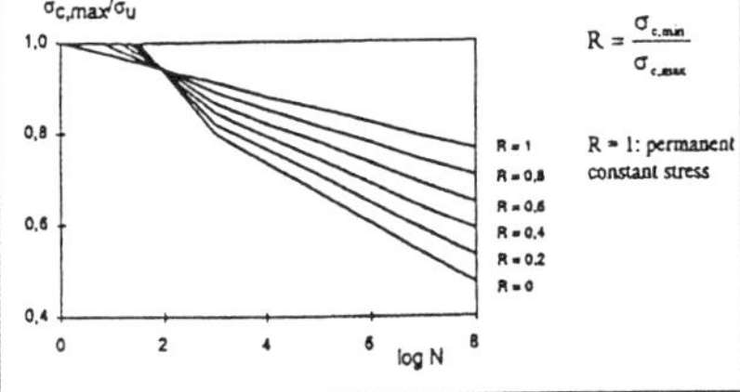

Figure 3: Principle of „elastic" degradation

Figure 4: Fatigue of concrete under compression according to [5], on the basis of [3]

4 NONLINEAR STRUCTURAL ANALYSIS

4.1. Material model of reinforced concrete

Reinforced concrete has an extremely nonlinear material behavior. For its simulation, the plastic-fracturing theory by Bazant/Kim is used, which, originally, has been developed for stress-strain states in the compression-compression domain. It was extended into the compression-tension and tension-tension domains, and proved by numerical experiments to be satisfactory. To determine the initiation of tensile cracking, a principal stress criterion is used. Structural crack formation is not described by discrete discontinuities, but by the concept of „smeared cracks". The crack growth over the thickness is considered just by means of the

layered discretization, while cracking conditions are checked separately in each layer of a multi-layered shell continuum (Fig. 5).
For the reinforcing steel, an elasto-plastic constitutive law has been applied including the Bauschinger effect. Reinforcing bars are modelled as smeared steel layers with only uniaxial stiffness and strength properties (Fig. 6).

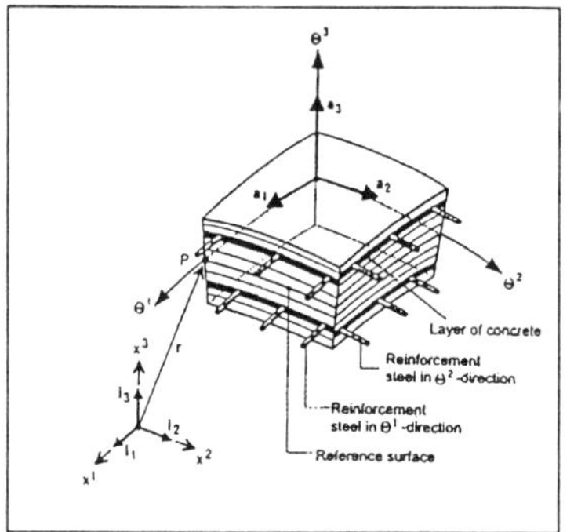

Figure 5: Multi-layered shell continuum.

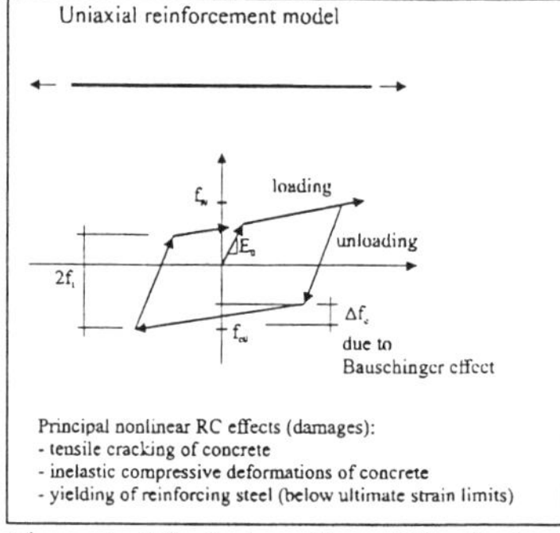

Figure 6: Principal nonlinear RC effects

The bond between steel bars and surrounding concrete is considered as stiff, assuming strains in steel bars and adjacent concrete as equal. Tension stiffening is attributed to the steel stiffness. For this path-dependent concrete model, the incremental constitutive equations relating differential stress $\Delta\tau$ to strain $\Delta\gamma$ increments are defined. The material tensor C^{ijkm} includes the elastic, plastic and micro-fracturing material constituents.

4.2. Finite element formulation

Using the above mentioned incremental material model and applying the displacement-related finite element formulation, the so-called tangential governing equations are obtained. The static version is:

$$K_T(V)\Delta V = \Delta\lambda P, \tag{1}$$

the dynamic version is:

$$M\Delta\ddot{V} + C\Delta\dot{V} + K_T(V)\Delta V = \lambda P(t + \Delta t), \tag{2}$$

where ΔV is the unknown vector of nodal displacement increments; K_T is the tangential stiffness matrix, depending on the current stress-strain state V; M,C are mass and damping matrices correspondingly; P is the vector of external loads and λ is a load scaling parameter. The FEMAS-software [1] allows to calculate nonlinearly and step-by-step the evolution of arbitrary structural responses under varying loads.

5 RELIABILITY ASSESSMENT

Starting from a deterministic approach, time-invariant probabilistic analyses were performed using Monte-Carlo simulations. Then time-dependent effects on the design parameters distinguishing between load-dependent and load-independent effects were studied. The data obtained are used for a time-variant reliability assessment.

6 BASIC VARIABLES

Investigations concerning significant input data were performed. On the one hand, this refers to actions (permanent, non-permanent and exceptional loads for ordinary and industrial buildings) and their simplified modeling, e.g. by substitute combined scenarios. On the other hand, characteristic material parameters as well as their time variance are studied and also brought in a simplified form, Fig. 2.
Additionally, sensitivity studies will help to find out what degree of significance special influences have.

7 SAMPLE APPLICATIONS

Numerical studies were performed for some practical applications, for instance for a slab bridge near Wuppertal, Germany. The author refers to the publication [4] where a detailed presentation is given.

The selected random variables used for a typical probabilistic assessment are given in Fig. 7 and Table 1.

One cross-section of the obtained response surface [2] is given in Fig. 8. The convergence of the calculated failure probability is shown in Fig. 9.

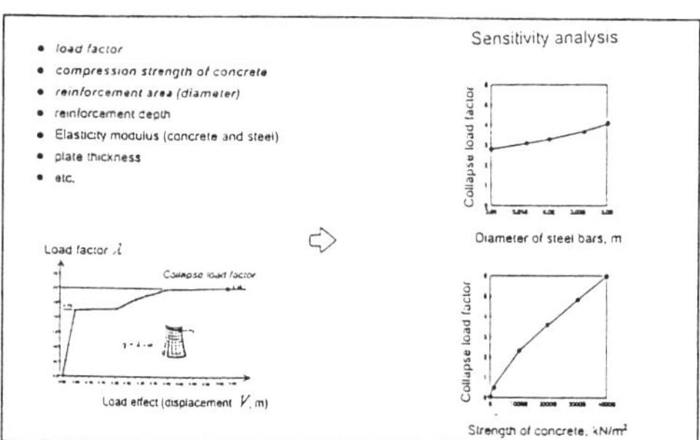

Figure 7: Selected random variables

	λ [ ]	f_c [kN/m²]	d_s [mm]
mean value	1.0	27000	26
standard deviation	0.3	4200	4

Table 1: Parameters of normal distributions (load factor λ, compressive strength f_c, steel diameter d_s)

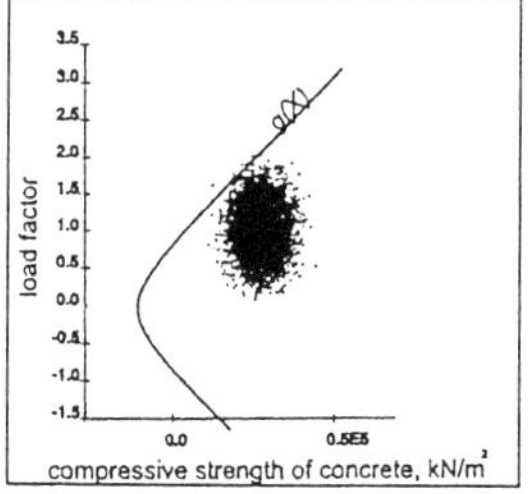

Figure 8: Response surface

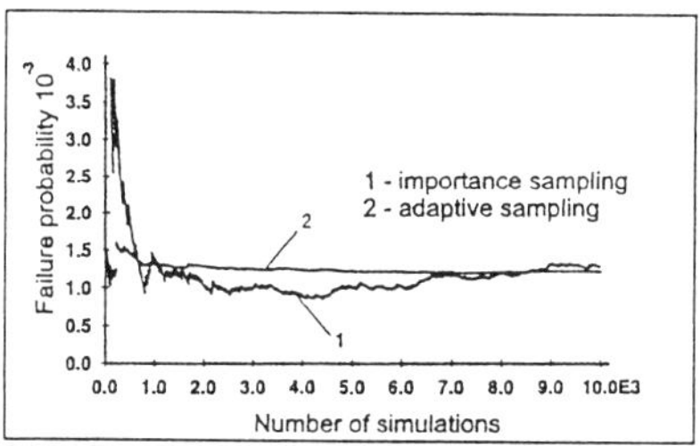

Figure 9: Convergence of results

[1] **Beem, H.; Könke, C.; Montag, U.; Zahlten, W.:** FEMAS-2000 Users Manual. Release 3.0. Bochum, Germany: Institute for Statics and Dynamics, Ruhr-University Bochum, 1996.

[2] **COSSAN :** COSSAN User's Manual. Stand-Alone Toolbox. Innsbruck, Austria: Institute of Engineering Mechanics, 1996.

[3] **Hsu, T. T. C.:** Fatigue of Plain Concrete, ACI-Journal 1981, Vol. 78, page 292 - 305

[4] **Krätzig, W. B.; Petryna, Y. S.:** Probabilistic reliability assessment of concrete structures, ESREL '98 European Safety and Reliability Conference, 16 - 19 June 1998, Trondheim, Norway. Balkema, Rotterdam 1998, page 1007 - 1012

[5] **Sippel, T. M.; Eligehausen, R.:** Trag- und Verformungsverhalten von Stahlbetontragwerken unter Betriebsbedingungen *(Bearing and Deformation Behavior of Reinforced Concrete Structures Due to Service Conditions)* , Berichte des Abschlußkolloquiums vom 16./17.06.1997 zum DFG-Schwerpunktprogramm „Bewehrte Betonbauteile unter Betriebsbedingungen", Stuttgart 1997

Towards design rules based on snap-through analysis
of composite concrete/masonry structures

Jan G. Rots[1]
Chantal M. Frissen[2]
Peter H. Feenstra[2]

Abstract

The relevance of computational fracture mechanics rationalizing design rules for stacked concrete/masonry structures is demonstrated. Interface softening models for tension and shear are used to simulate the initiation, propagation and snap-through of cracks and slip planes in these unreinforced structures. Two examples are included, for the serviceability limit state (cracking under restrained shrinkage) and the ultimate limit state (stability of the buildings under wind load).

Two design problems

Buildings and houses often consist of masonry walls, masonry piers and concrete slabs. The masonry walls and piers can be either load-bearing, in the form of calcium-silicate or concrete blockwork, or non load-bearing, referring to e.g. clay brickwork as "cladding". For the design two aspects should be checked:
- The cracking of the masonry, in the Serviceability Limit State (SLS).
  Practical significance: the design of movement-joint schemes for crack prevention and crack control.
- The stability of the total system, in the Ultimate Limit State (ULS).
  Practical significance: can the structure withstand wind load?

Approach with softening lines and softening hinges

As masonry structures in Europe are often unreinforced, cracking and slip will localize in a few lines. Furthermore, the connections between the prefab concrete parts and masonry parts are weak, with little cohesion or even dry friction. It is generally not difficult for an engineer to judge on the potential lines (or planes) of

---

[1] Senior Research Fellow, TNO Building and Construction Research / Delft University of Technology, P.O. Box 49, 2600 AA Delft, The Netherlands.
[2] Research Engineer, TNO Building and Construction Research.

failure in these structures. As a consequence, the model does not have to be predictive with regard to the *location* of failure. Instead, potential softening lines or hinges can be pre-assumed in the mesh. All nonlinearity is lumped into interface elements, so that possible convergence problems with continuum softening solutions is circumvented. The constitutive model covers three modes: mode-I discrete cracking with softening on the tensile strength, non-associated mode-II Coulomb friction with softening on the cohesion, and a compression cap or compression traction-displacement relation with softening to simulate crushing of the continuum material.

SLS: Cracking under restrained shrinkage

Figure 1 shows a masonry wall subjected to (hygral or thermal) shrinkage, which is restrained by a concrete foundation. The restraint generates tensile stresses which may lead to cracking. The potential primary vertical crack in the middle of the wall is modelled with interface elements and tension softening. Upon increasing shrinkage, the crack initiates, propagates and eventually explosively snaps through to the top surface. The snap-through complies with practical experience of a dynamic propagation upon such cracking. By varying the length of the wall and collecting the critical shrinkage values at which snap-through occurs, we arrive at design rules for movement joint spacing. For a given design value of shrinkage, an engineer can find the required movement joint spacing. If he would not install the movement joint, the wall would crack. An example of such fracture mechanics based rule is included in Figure 1 and compared with three traditional, strength-based rules that show mutual scatter. This research is relevant as crack formation in walls due to restrained shrinkage is number one in the top-10 of damage events in the Netherlands. Details of the analyses are given by Rots (1997).

ULS: Stability of buildings

The frame of masonry walls and concrete floors is often stabilized via masonry piers. The resistance against horizontal wind load depends on the strength of the pier and on the connection between pier and wall. Figure 2 shows three possible mechanisms: (a) tilt (turn-over), (b) compression failure of the pier, or (c) sliding failure of the vertical pier/wall joint, which is often a glued joint between calcium-silicate units. First, 2D analyses on a single pier-wall connection were carried out (Rots 1997). The Coulomb friction model with cohesion softening appeared to correctly reproduce experiments on the third mechanism, whereby a sharp snap-back behaviour was found indicating brittle response of this relatively large unreinforced structure. Subsequently, the step towards complete 3D analyses of buildings has been made. Figure 2 shows examples of deformed meshes, whereby gapping in horizontal joints and sliding in the vertical joint can be recognized. Crushing near the toe of the pier has been successfully modelled by lumping compression-softening in the interface underneath the pier. The analyses include geometrically nonlinear behaviour. The maximum loads predicted are used to rationalize quasi-nonlinear design approaches for stability. Results prove that there is significant reserve in these structures and that current Dutch design approaches are conservative (Frissen and Wijte 1999). The research helps the brick/block industry in getting a larger market share in (high-rise) structural masonry.

Acknowledgement

The research is supported by STW (Netherland's Technology Foundation), brick/block industry and CUR.

References

Rots J.G. (Ed.), Structural masonry - An experimental/numerical basis for practical design rules. A.A. Balkema, Rotterdam, The Netherlands, 152 pp. (1997)

Frissen C.M., Wijte S.N.M., Stability of masonry structures. TNO report (in preparation, in Dutch), TNO, Rijswijk, The Netherlands (1999)

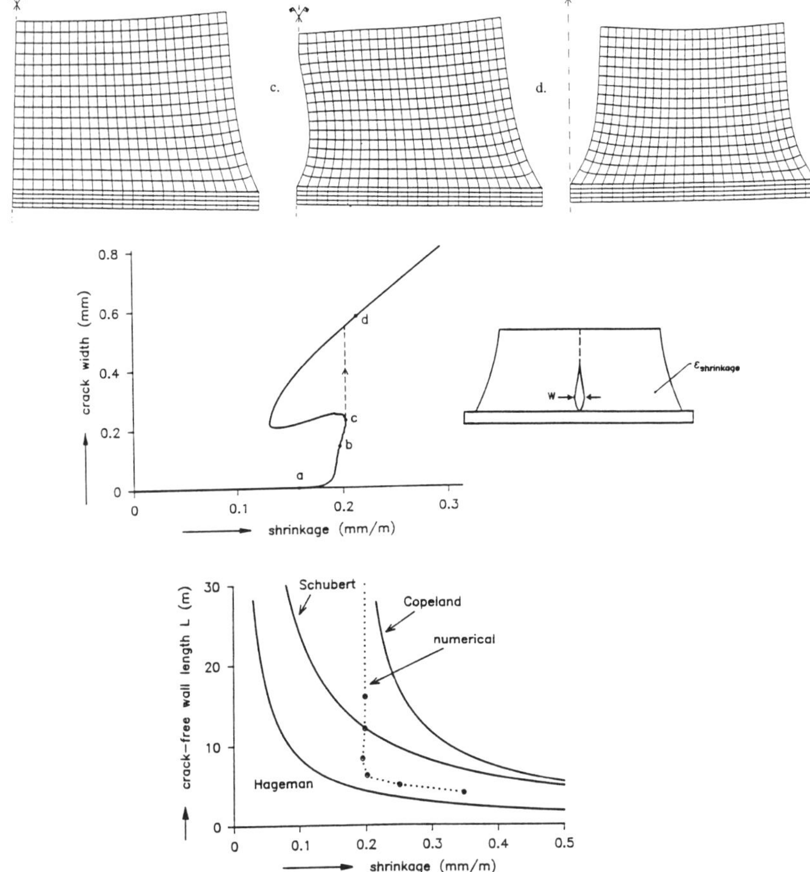

Figure 1. Cracking in walls due to restrained shrinkage.
Deformed meshes and snap-through shrinkage-crack width behaviour (top).
Design rules for crack-free wall length i.e. movement joint spacing (bottom).

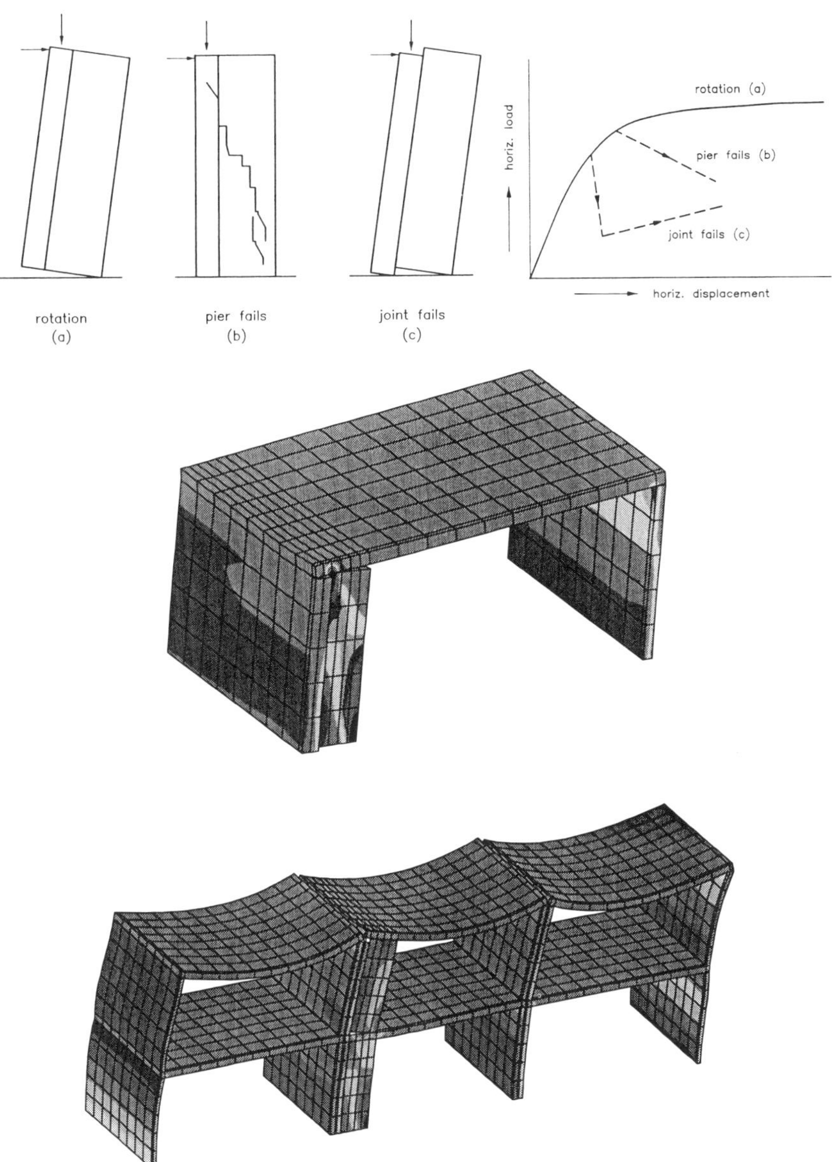

Figure 2. Stability of masonry structures.
Three possible failure mechanisms of pier-wall connection (top).
Examples of 3D simulations of buildings (bottom).

Assessment of Ultimate Load Analyses
of Reinforced Concrete Plates and Shells
by Means of Adaptive Methods

Roman Lackner[1] and Herbert A. Mang[2]

Abstract

This paper reports on adaptive methods for ultimate load analyses
of reinforced concrete (RC) surface structures (plates and shells). The finite
element (FE) mesh is adapted according to the distribution of an "estimated"
error of the FE results obtained from incremental-iterative analysis. For this
purpose, the recovery technique of Zienkiewicz and Zhu (1992) is extended
to elastoplastic material behavior considering hardening as well as softening.
The developed adaptive scheme is applied to a reinforced shear wall panel,
making use of the layer concept.

Introduction

Increasing experience in nonlinear analyses of RC structures together
with improvements of the respective computational algorithms has paved the
way for the application of realistic material models. In general, these models
are evaluated on the level of integration points in the form of recalculations
of experimental data. Applications in the form of simulations of the mechanical behavior of real-life structures including cracking of concrete and yielding
of the reinforcement are notable exceptions. Such simulations are performed
within the framework of the finite element method (FEM) appearing to be
the best tool for this kind of nonlinear analyses. Obviously, this numerical

[1]University Assistant, Institute for Strength of Materials, Vienna University of Technology, Karlsplatz 13/202, A-1040 Vienna, Austria

[2]Professor, Institute for Strength of Materials, Vienna University of Technology,
Karlsplatz 13/202, A-1040 Vienna, Austria

approach leads to a discretization error. As a consequence of this error, the structural stiffness is often overestimated. If the discretization error is controlled at all, this is generally done by means of uniform mesh refinement. The required series of calculations is expensive as regards time and computer resources. Moreover, this approach results in a nonuniform distribution of the discretization error. This situation has been the motivation for the development of adaptive methods. Instead of refining the mesh uniformly, it is refined according to the distribution of an "estimated" error. Originally, the adaptations of the discretization were restricted to problems of the linear theory of elasticity. Later on, adaptivity was extended to ideal plasticity and strain-hardening plasticity. Extension of adaptivity to ultimate load analyses of RC surface structures is part of ongoing research at the Vienna University of Technology.

In this paper, uniform as well as adaptive mesh refinement is applied to an RC panel. In the following, the considered material models and the definition of the error estimator will be described briefly.

Plasticity Model for Reinforced Concrete

The different behavior of concrete in tension and compression is taken into account by the multi-surface plasticity concept. The Drucker-Prager yield criterion is used for the simulation of the ductile response in compression. The Rankine criterion serves for the description of cracking.

For the description of the behavior of steel, a 1D model is used. It refers to the strain component in the direction of the steel bar.

In reinforced concrete, usually several cracks develop until a stabilized crack pattern is formed. The distribution of the cracks depends on the geometric properties (bar diameter, concrete cover, etc.). Herein, the average crack spacing is computed from the distribution of bond slip along the reinforcement bar. The increase in the released strain energy because of bond slip and the development of secondary cracks is considered within the fracture energy concept (for details see (Lackner 1999)).

Error Estimator for Elastoplasticity

In case of localization, the analytical solution for the stress distribution contains discontinuities. Accordingly, for the estimation of the error the recovery technique by Zienkiewicz and Zhu (1992) was adapted to allow for stress jumps parallel to the element edges. Using the recovered total and incremental stresses, σ^* and $\Delta\sigma^*$, the following absolute error measure is defined:

$$\Delta(e^2) = \int_V \sum_{i,j=1}^{3} |(\sigma_{ij}^* - \sigma_{ij}^h) \sum_{k,l=1}^{3} [D_{ijkl}(\Delta\sigma_{kl}^* - \Delta\sigma_{kl}^h)]| dV$$

$$+ \int_V \sum_{i,j=1}^{3} |(\sigma_{ij}^* - \sigma_{ij}^h)(\Delta\varepsilon_{ij}^{p,*} - \Delta\varepsilon_{ij}^{p,h})]| dV \; , \tag{1}$$

where the error in the incremental plastic strains, $\Delta\varepsilon^{p,*} - \Delta\varepsilon^{p,h}$, is computed from the recovered stress increments, $\Delta\sigma^*$ (for details see (Lackner 1999)). In Equation (1), the superscript "h" refers to the FEM solution; D_{ijkl} is the compliance tensor. A relative error measure η is obtained by relating $\Delta(e^2)$ to the reference quantity $\Delta(u^{h,2})$: $\eta^2 = \Delta(e^2)/\Delta(u^{h,2})$.

As regards the extension of the recovery scheme to the layer concept, stress resultants such as the axial force $\mathbf{n}^h$ and the bending moment $\mathbf{m}^h$ are smoothed, yielding $\mathbf{n}^*$ and $\mathbf{m}^*$. The stresses required for the evaluation of the error are computed from $\mathbf{n}^*$ and $\mathbf{m}^*$ by means of an extreme value problem with a constraint condition. This problem is formulated for each stress component at each integration point, using the Lagrangian multipliers λ_n and λ_m:

$$\mathcal{F}(\hat{\sigma}_{ij}^{1,*}, ... \hat{\sigma}_{ij}^{n,*}; \lambda_n, \lambda_m) = \sum_{\ell=1}^{n} \left(\hat{\sigma}_{ij}^{\ell,*} - \hat{\sigma}_{ij}^{\ell,h}\right)^2 + \lambda_n \left(\hat{n}_{ij}^* - \frac{h}{2}\sum_{\ell=1}^{n} \hat{\sigma}_{ij}^{\ell,*}\Delta\xi^\ell\right)$$

$$+ \lambda_m \left(\hat{m}_{ij}^* + \frac{h^2}{4}\sum_{\ell=1}^{n} \hat{\sigma}_{ij}^{\ell,*}\xi^\ell\Delta\xi^\ell\right) \quad \rightarrow \quad \text{stationary} \; . \tag{2}$$

Values at the considered integration point are indicated by the symbol "$\hat{\ }$". Analogous to the situation for one layer, the difference between the recovered stresses and the stresses obtained from the FE analysis is minimized. In Equation (2), n denotes the number of layers, h is the thickness and ξ is a normalized coordinate orthogonal to the middle surface of the shell.

Numerical Example. Shear Wall Panel

Experimental data of a shear wall panel, subjected first to the vertical load p_v and then to the horizontal displacement $\bar{u}$ (see Figure 2(left)), are taken from (Maier and Thürlimann 1985). The support and loading beam of the experimental setup are not considered for the error estimation. Figure 1 contains load-displacement curves obtained from calculations based on uniform meshes as well as from adaptive analysis. Convergence of the ultimate load for both refinement strategies is observed. Nevertheless, the experimentally observed ductile deformation capacity and, consequently, the failure load (P_u=392 kN) are underestimated. This is mainly caused by the assumption of plane stress in the concrete layer and the two steel layers of the element. Hence, the confinement of concrete in consequence of the reinforcement is not taken into account.

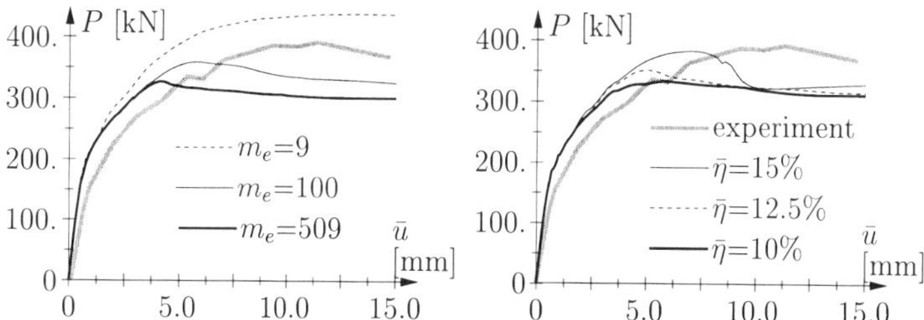

Figure 1: Shear wall panel V4: load-displacement curves obtained from (left) uniformly refined meshes (m_e: number of elements) and (right) adaptive calculations

The final meshes generated during the course of the adaptive calculations are illustrated in Figure 2. High mesh density is observed in the left lower corner where the concrete fails in compression. This is also reflected by the plotted vertical stress component in the concrete layer. The location of the maximum compressive stress has moved inside the panel. The reason for this is strain softening of concrete under compression. Tensile failure of concrete is initiated at the right lower corner. The reinforcement favors the distribution of the crack. Hence, cracks have developed in the entire right side of the panel.

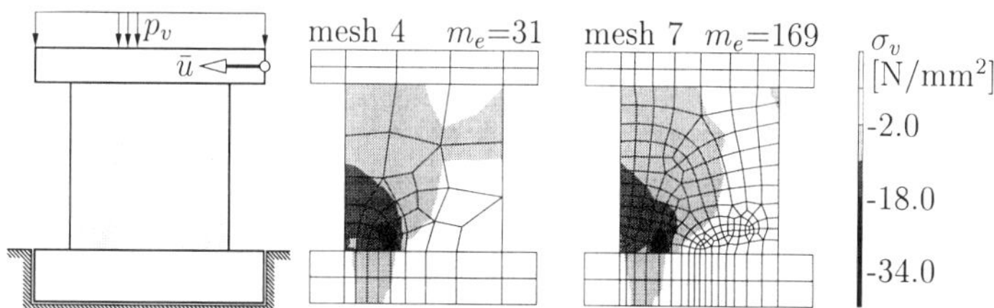

Figure 2: Shear wall panel V4: final meshes showing the vertical stress in the concrete layer for the adaptive calculation for $\bar{\eta}=15\%$ (center) and $\bar{\eta}=10\%$ (right) at $\bar{u}=15$ mm

References

Lackner, R. (1999). *Adaptive finite element analysis of reinforced concrete plates and shells*. Dissertation, Vienna University of Technology. In preparation.

Maier, J. and Thürlimann, B. (1985). Bruchversuche an Stahlbetonscheiben. Technical Report 8003-1, Institut für Baustatik und Konstruktion, Eidgenössische Technische Hochschule Zürich. In German.

Zienkiewicz, O. and Zhu, J. (1992). The superconvergent patch recovery and *a posteriori* error estimates. Part 1: The recovery technique. *International Journal for Numerical Methods in Engineering*, 33:1331–1364.

Modeling of Shear Behavior of RC Columns

Brian Rose[1], P. Benson Shing[2], Member ASCE, and
Enrico Spacone[3], Associate Member ASCE

Abstract

A constitutive model for reinforced concrete based on the Modified Compression Field Theory (MCFT) is presented. The model differs from the MCFT in that cracks do not rotate continuously and the kinematics incorporate crack interface sliding. The model and its validation are summarized here.

Introduction

A newly developed constitutive model, capable of simulating the shear behavior of orthogonally reinforced concrete, is presented in this paper. This research is part of a project to simulate and study the load-deflection behavior of reinforced concrete bridge columns subjected to seismic loads. The model is based on the Modified Compression Field Theory (MCFT) proposed by Vecchio and Collins (1986) but includes several major modifications as outlined in this paper.

Kinematics

Experimental tests of orthogonally reinforced panels subject to pure shear show that cracks are parallel, and crack spacing is more or less uniform. Accordingly, the idealized panel model adopted here consists of an assembly of equal width struts inclined at an angle α (Figure 1). Hence, two regions of interest arise: the strut and the crack interface. The proposed model assumes that the struts separate and slide past each other at the crack interface. Additionally, the struts can elongate, contract and shear. The crack opening and sliding are assumed to be the same from crack to crack. Similarly, the deformations are assumed the same from one strut to another.

---

[1] Res. Asst, Dept. of Civ. Engrg., Univ. of Colorado, Boulder, CO 80309
[2] Prof., Dept. of Civ. Engrg., Univ. of Colorado, Boulder, CO 80309
[3] Asst. Prof., Dept. of Civ. Engrg., Univ. of Colorado, Boulder, CO 80309

Consider the 1-2 coordinate system shown in Figure 1. Equivalent overall strains in the 1-2 coordinate system are defined as, $\varepsilon_1 = \varepsilon_1^c + w/s$, $\varepsilon_2 = \varepsilon_2^c$ and $\gamma_{12} = \gamma_{12}^c + v/s$, where , ε_1^c, ε_2^c, and γ_{12}^c are concrete strains, w v, and s are crack opening, sliding, and strut width. This model differs from the MCFT in three aspects: *i*) there exist shear stresses and strains in the strut in the 1-2 coordinate system *ii*) the struts are allowed to slide at crack interfaces, and *iii*) the crack angle does not always rotate with the principal stress or strain direction. The crack angle rotates to coincide with the principal stress direction only if the first principal stress is greater than the cracking strength of concrete. The crack angle remains constant otherwise.

Equilibrium

The panel is subjected to uniform normal and shear stresses at its boundaries (Figure 1). As in the MCFT, the equilibrium of the panel can be best described by free body diagrams. A free-body diagram results from "slicing" a panel horizontally (or vertically) as shown in Figure 2. The slice exposes concrete stresses and stresses in rebars located within the strut. Another free-body diagram (Figure 3) results from "slicing" along a crack interface. This slice exposes the normal and shear stresses on the crack interface and localized stresses of the rebars at the crack interface.

Material Laws

Constitutive laws tie the kinematics to equilibrium. The required material models are the concrete and reinforcing steel stress-strain laws and crack interface stress-displacement models. All three models are cyclic.

A simple orthotropic law is used to model the concrete in the 1-2 directions. A formula proposed by Saenz (1964) defines the compression behavior. The stress-strain behavior in the 1 and 2 directions are coupled; tensile strain in the 1 direction reduces the compressive strength in the 2 direction (Vecchio and Collins 1986). The shear stress-strain relation in the 1-2 directions is linearly elastic.

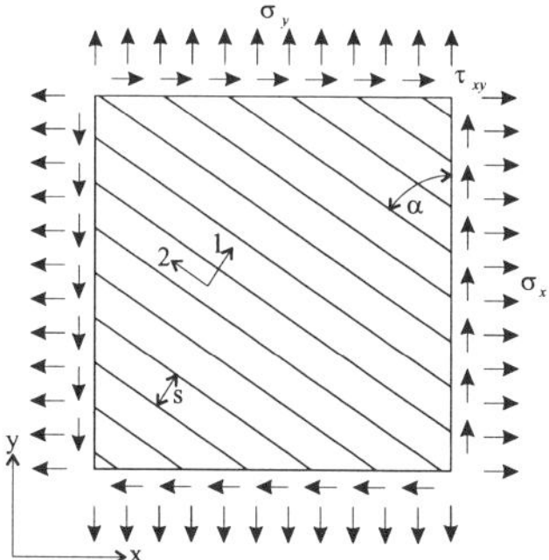

Figure 1: Reinforced concrete panel

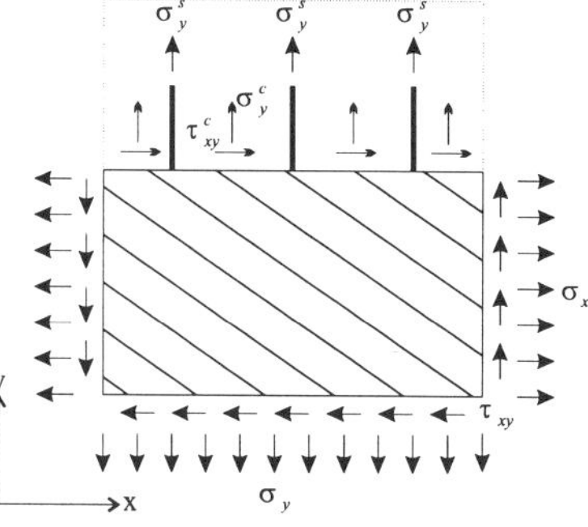

Figure 2: Equilibrium of strut

The Menegotto-Pinto law, modified by Filippou (1983), models the stress-strain behavior of the reinforcing bars within the strut and at the crack interface. This model incorporates strain hardening and the Bauschinger effect.

An aggregate interlock law proposed by Walraven (1981) is used. It describes the relationship between the crack opening and sliding to the normal and shear stresses on the crack face (Figure 4). A crack interface is modeled as an assembly of rigid-perfectly plastic cement matrix and rigid spherical aggregate particles. The model incorporates both variable aggregate size and aggregate embedment depths with respect to the crack face.

Comparison of Experimental and Simulated Results

Simulation and experimental results, for three panels loaded in pure shear from the University of Toronto, are presented here (Stevens, *et al*, 1987 and Vecchio & Collins, 1982). Figure 5 shows the shear behavior of two panels subjected to monotonically increasing loads. Panel PV 18 and PV 22 have reinforcing ratios of 1.78/1.52% and 1.78/0.32%, respectively. Panel PV 18 shows low shear strength; the simulation reveals that rebars in the weaker reinforcement direction yield, allowing significant crack openings. This limits the interface shear stress capacity, thereby never allowing the concrete struts to develop the full compressive strength. Panel PV 22 shows higher shear strength; the large amount of rebars limits the crack opening, thereby allowing higher interface shear strength. In turn, the struts develop the full compressive strength.

Figure 6 shows results for a panel (SE 8) loaded cyclically in pure shear. The reinforcing ratios are 3.0/1.0%. Like panel PV 18, the weak direction rebar yields, limiting interface shear stress, and never allowing the struts to develop full compressive strength. The interface sustains a major portion of the total shear deformation. All three panels show good agreement between simulation and experimental results.

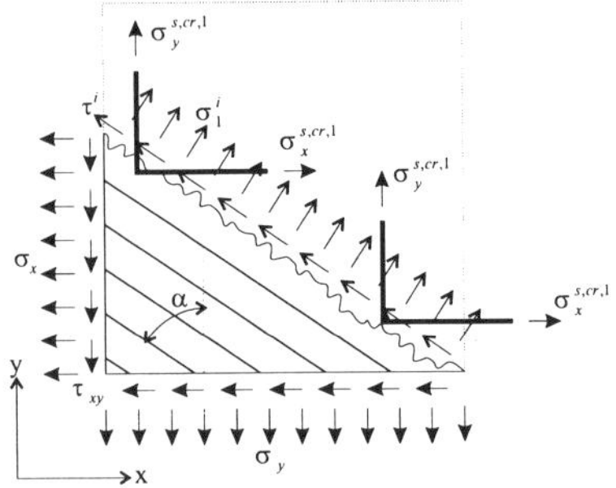

Figure 3: Equilibrium of crack interface

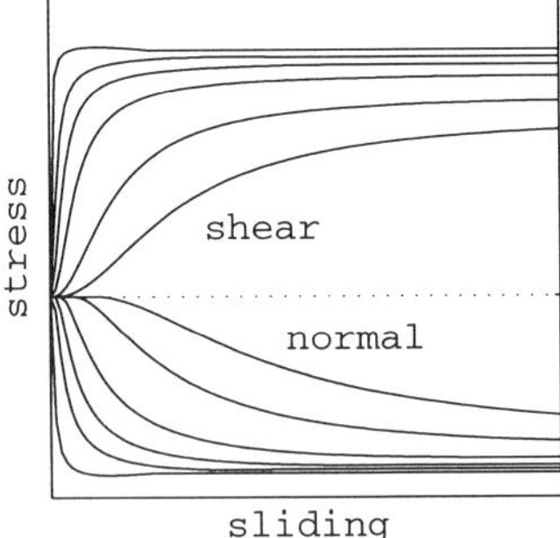

Figure 4: Aggregate interlock law

Conclusions

A new constitutive model for orthogonally reinforced concrete is presented. The model is based on the Modified Compression Field Theory. However, the crack does not rotate continuously and the crack interface is permitted to slide. Simulation and experimental results, for panels loaded both monotonically and cyclically in pure shear, show good agreement. This model will be implemented in a beam-column element to analyze reinforced concrete columns.

Acknowledgments

This study is sponsored by the NSF under Grant No. CMS-9622940. However, opinions expressed in this paper are those of the writers and do not necessarily reflect those of the sponsor.

References

Filippou, F. C., Popov, E. P., and Bertero, V. V. (1983). "Effects of bond deterioration on hysteretic behavior of reinforced concrete joints." EERC Report 83/19, Earthquake Engineering Research Center, University of California, Berkeley.

Saenz, L. P. (1964). "Discussion of equation for the stress-strain curve of concrete by Desayi and Krishman." J. Am. Concr. Inst., 61 (9), 1229 - 1235.

Stevens, N. J., Uzumeri, S. M., and Collins, M. P. (1987). "Analytical modeling of reinforced concrete subjected to monotonic and reversed loadings." Report No. 87-1, Univ. of Toronto, Toronto, Canada.

Vecchio, F. J., and Collins, M. P. (1982)."The response of reinforced concrete to in-plane shear and normal stresses." Report No. 82-03, Univ. of Toronto, Canada.

Walraven, J. C. (1981). "Fundamental analysis of aggregate interlock." J. Struct. Div. ASCE, 107 (ST11), 2245 -2270.

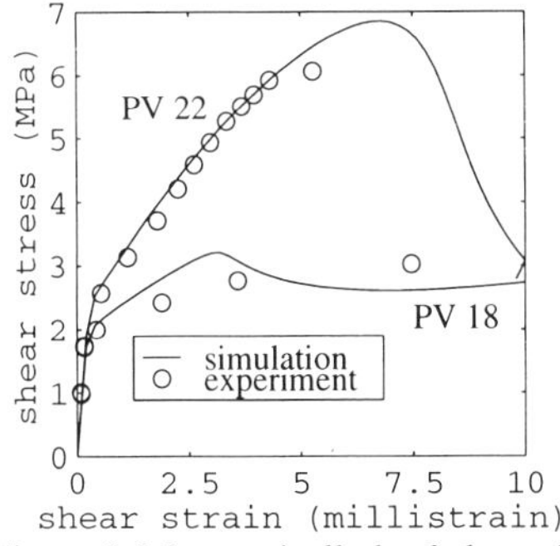

Figure 5: Monotonically loaded panels

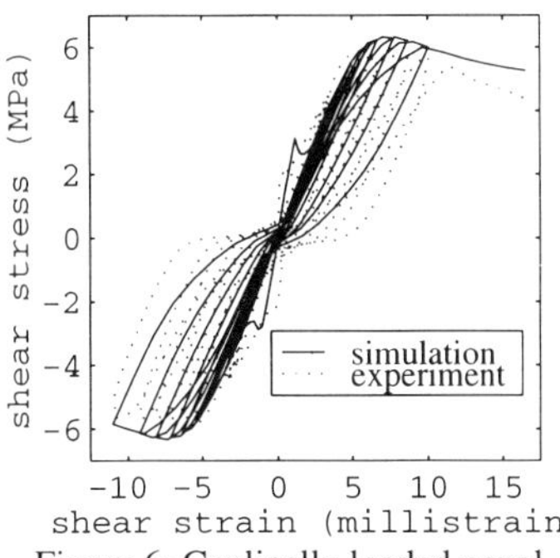

Figure 6: Cyclically loaded panel

Retrofitting of a RC cooling tower:
from concrete modelling to structural design

G. Meschke[1], H.A. Mang[2], M.N. Fardis[3] and S.N. Bousias[3]

Abstract

The paper addresses the computer-aided retrofitting of a damaged cooling tower shell. Structural details concerning the construction of the stiffening rings and consequences drawn from the numerical results for decisions on the mode of construction of the stiffening rings are described.

Introduction

In the 50's and 60's, a large number of RC structures were erected. Many of them are now showing signs of damage. Hence, retrofitting of partially damaged engineering structures is attracting increasing attention [7, 1]. The present condition of a hyperbolic RC shell (height 82 m, shell thickness 10 cm), built in 1964, is characterized by a nonsymmetric distribution of thermally

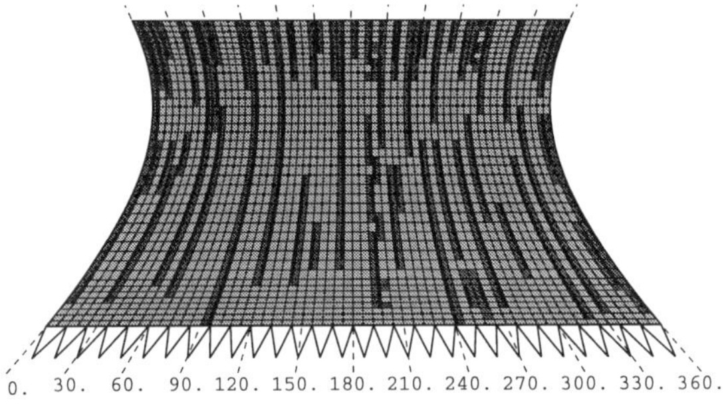

Figure 1: View of the Cracked Shell and Representation of Initial Cracks on the Outside Shell Surface (Dark (Light) Grey: Cracked (Intact) Elements)

[1]Institute for Structural Mechanics, Ruhr-University Bochum, D-44780, Germany
[2]Institute for Strength of Materials, Vienna University of Technology, A-1040, Austria
[3]Department of Cvil Engineering, University of Patras, GR-26500, Greece

induced meridional cracks (Fig.1). Except for the lower part of the shell, the structure is reinforced by only one layer each of meridional and circumferential reinforcement located in the middle of the shell. Corrosion of the reinforcement in the vicinity of the cracks results in a reduction of the diameter ranging from 11 % to 30 %. The repair includes two *in-situ* concrete stiffening rings and the strengthening of the cornice.

Computational Model of the Damaged Cooling Tower Shell

Cracks are modelled by means of the fixed-orthogonal-crack model within the framework of the smeared-crack concept [3], which is considered a suitable approach to replicate the existing crack pattern on both faces of the shell in the analyses [4]. The descending part of the load displacement-curve of the localized crack-zones, averaged over the respective part of the isoparametric thick-shell finite element is considered by a fracture-energy based formulation, which depends on the size of the finite element and on the crack direction [6]. The residual interface shear transfer across cracks is considered. Tension stiffening is taken into account by increasing the fracture energy G_f for a single crack by a crack space-dependent factor which depends on the diameter of the reinforcement bars, the concrete cover and the orientation of the crack with respect to the reinforcement [2]. The opening of the existing meridional cracks is triggered by the application of a thermal load history prior to applying the wind load in the most critical direction [4]. Corrosion is taken into account with respect to the end of the residual lifetime in the year 2018.

Numerical Assessment of a Repair of the Cracked Shell

The coefficient of safety against structural collapse of the undamaged shell according to the design is $\lambda_u = 1.35$. For the unstiffened cooling tower shell this coefficient is obtained for the year 2018 as $\lambda_u = 0.9325$. Out of 5 alternatives, the optimal location of the two stiffening rings was found as $z = 43.0$ m and $z = 73.0$ m, resulting in $\lambda_u = 1.61$ as the value for the safety coefficient.

For levels of the upper stiffening ring not less than 73.0 m, large stresses in the meridional reinforcement located at the windward meridian between the lower and the upper stiffening ring lead to plasticizing of the reinforcement steel and eventually to the collapse of the shell. For lower levels of the upper stiffening ring, however, structural collapse is triggered by the plasticizing of the circumferential reinforcement located between the upper stiffening ring and the cornice. The chosen level of 73.0 m provides an optimal "balance" between the two failure mechanisms. For load levels $\lambda > 1.20$, the structural behavior of the stiffened RC shell is characterized by two localized failure zones located between the upper stiffening ring and the cornice of the shell (Fig. 2b). The location of these zones coincides with the location of the maximum external suction caused by the wind load. In these zones the shell is severely weakened by the existing cracks (Fig.1b).

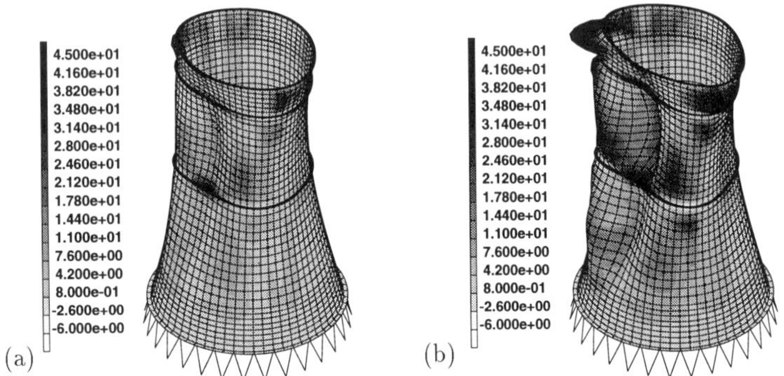

Figure 2: Stress Distribution in the Circumferential Reinforcement of the Stiffened Shell: (a) $\lambda= 1.30$; (b) $\lambda= \lambda_u$ (20-fold Magnification of Displacements)

Stresses Induced by the Stiffening Rings During Construction

A 2D nonlinear FE analysis of the ring-shell connection, neglecting the effect of a closed ring, was performed to compute the stresses induced during construction. A fracture-energy-based fixed-crack approach in conjunction

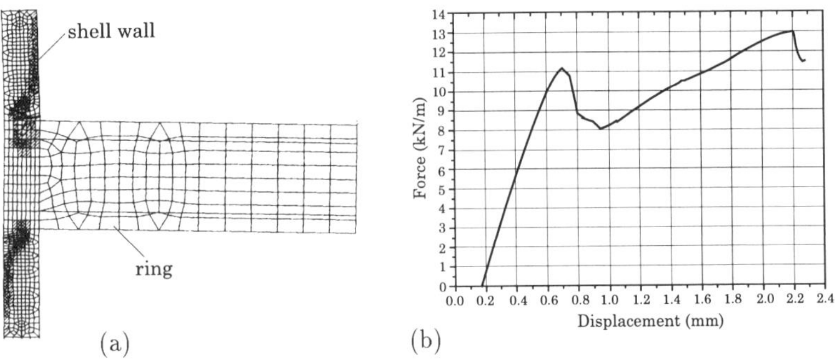

Figure 3: (a) Crack Distribution at the Ultimate State of the Stiffening Ring Subjected to Life Load during Erection and (b) Load-Displacement Relation

with an orthotropic $\sigma - \epsilon$ law was adopted for the cracked concrete. A peak value of $P= 11$ kN/m after the opening of horizontal macrocracks at both sides of the shell in the vicinity of the ring (Fig. 3) is observed prior to the yielding of the reinforcing steel at $P_{ult}= 13$ kN/m. Each ring is cast in a single day all around the perimenter. This mode of construction prevents large vertical tensile stresses in the shell concrete that would be superimposed to those resulting from subsequent wind loading. It avoids lapping of circumgerential bars at the same location and cold joints in the ring concrete. Both stiffening rings (Fig. 4) are attached to the shell by means of L-shaped dowels drilled

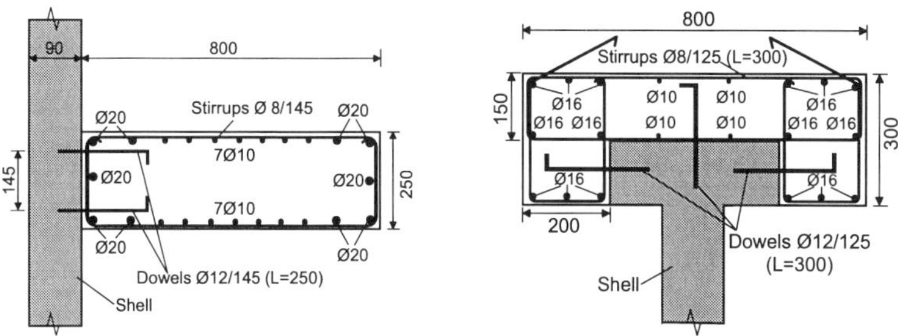

Figure 4: Cross-Section of Stiffening Rings and the Strengthened Top Cornice

into the shell and injected by a two-component hybrid adhesive (Hilti HY-150 system).

Summary and Concluding Remarks

The retrofitting of a cracked RC cooling tower shell, considering the state of damage at the end of the anticipated lifetime, was described. The repair was based on two cast in-situ stiffening rings and on the strengthening of the cornice. The coefficient of structural safety of the restored RC shell was increased from $\lambda = 0.93$ to $\lambda = 1.61$. The shell-ring-connection was investigated to determine construction-induced stresses. It was shown that the structural behavior of the stiffened cooling tower is dominated by two localized failure zones. A fine discretization was necessary to detect this mode of failure.

References

[1] P.B. Bosman, I.G. Strickland, and R.P. Prukl. Strengthening of natural draught cooling tower shells with stiffening rings. In *Natural Draught Cooling Towers*, 293–301. Balkema, 1996.

[2] P. Feenstra and R. De Borst. Constitutive model for reinforced concrete. *J. Eng. Mech. (ASCE)*, 121(5):587–595, 1995.

[3] G. Hofstetter and H.A. Mang. *Computational Mechanics of Reinforced and Prestressed Concrete Structures*. Vieweg, Braunschweig, 1995.

[4] G. Meschke, T. Huemer, and H.A. Mang. Computer-aided retrofitting of a damaged cooling tower. *J. Struct. Eng. (ASCE)*, 1999. in press.

[5] G. Meschke, J. Macht, and R. Lackner. A damage-plasticity model for concrete accounting for fracture-induced anisotropy. In R. de Borst, N. Bicanic, H. Mang, G. Meschke, editors, *Proc. Int. Conf. Comput. Modelling of Concrete Structures, Badgastein, Austria*, 199–208, 1998, Balkema.

[6] J. Oliver. A consistent characteristic length for smeared cracking models. *Int. J. Num. Meth. Eng.*, 28:461–474, 1989.

[7] Symposium San Francisco 1995. *Extending the Lifespan of Structures*, ETH Hönggerberg, Zürich, Switzerland, 1995. IABSE-AIPC-IVBH.

Finite Element Investigation of Concrete Slabs Post Strengthened
by Fiber Reinforced Polymers

M. Hörmann[1], H. Menrath[1], E. Ramm[1], F. Seible[2]

Abstract

The present work deals with the post strengthening of concrete slabs and the numerical calculation of their load carrying behavior. The used strengthening materials are Fiber Reinforced Polymers (FRP) which are of increasing interest in the civil engineering applications, i.e. textile reinforced concrete tubes, cables of cable–stayed bridges or even entire bridges. In this work, the used FRPs consist of carbon fibers in a epoxy–matrix or glass fibers in a vinylester–matrix. The reinforced polymers were applied in different ways to concrete slabs; the strengthened structures have been tested to failure. Afterwards, nonlinear finite element analyses are performed gaining more informations about modelling and prediction of the structural response of strengthenend constructions.

1. Introduction

It is widely accepted that concrete structures can be strengthened with steel plates externally bonded to the tension side. Great efforts have been undertaken to exchange the steel plates by Fiber Reinforced Polymers, a light weigth and corrosion resistant material. Adding FRP to a steel reinforced concrete structure leads to a completely changed structural response and failure mode. The structure fails in a brittle manner due to the elastic behavior of FRP up to failure and the following sudden peeling away of the strips or sudden rupture of the fibers. In order to study this phenomenon, a test series of 13 scale model and one full scale slab was undertaken at the University of California, San Diego .The slabs were strengthened using carbon FRP (CFRP) strips, CFRP fabrics and sprayed glass fiber reinforced composite in different designs of the composite materials.

[1] Institute of Structural Mechanics, University of Stuttgart, Germany
[2] Division of Structural Engineering, University of California, San Diego, USA

The investigated designs and strengthening systems indicaed not only different failure loads but also different failure modes, which are described in detail in Hörmann 1997. Fig. 1 shows therefore only the geometry and the test setup of the strengthened slabs.

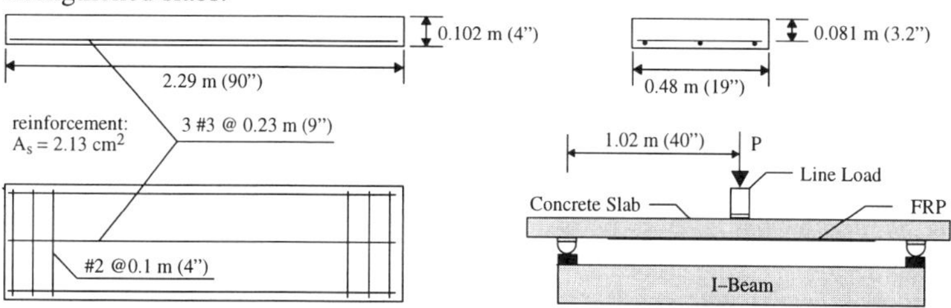

Figure 1. Geometry and Test Setup of Strengthened Slabs

In addition, a finite element investigation is performed at the University of Stuttgart. For this, the beam like slabs are idealized by a two–dimensional (2D) model for the concrete and the FRP assuming a plane stress condition. A softening plasticity law for concrete with fracture energies G_f and G_c as controlling parameters is used. For the reinforcement perfect bond is assumed and its 1D constitutive law is based on an elastoplastic model with isotropic hardening.

Compared to the test results the finite element analyses show stiffer load displacement responses due to the simplified 2D model of the slabs. The calculation for the CFRP fabric strengthened slab indicates the best agreement to the test results due to the fact, that the fabric covers the width of the slab completely. On the other hand, the CFRP strips are not uniformly distributed across the width of the slab. Consequently the calculated load displacement response based on the simplified 2D model differs from the test results.

2.   Finite Element Analysis

In the finite element analysis the slabs are modelled in a 2D design space, i.e the slabs are idealized as beam like structures. Linear kinematics are assumed. The used material model for reinforced concrete was developed by Menrath *et al.* 1998.

The structure is discretized by fully integrated 9–noded plane stress elements for the concrete and the FRP. Three different element meshes, namely 10x20 (coarse mesh–1), 10x40 (mesh–2) and 10x80 (fine mesh–3) are used, each with ten elements across the height. The inner steel reinforcement is distributed over three elements in height in order to avoid big stiffness changes and to provide a gradual structure behavior. For each strengthening material perfect bond to the concrete is assumed, i.e. no adhesive layer is introduced.

The material parameters such as tensile strength, Young's modulus, fracture energies of concrete and Young's modulus and hardening of reinforcement are chosen according to Eurocode 2 1992 and CEB–FIP Model code 1990. The tensile strength of the concrete f_t is set to 2.0 MN/m^2 (29 psi) and the concrete compression fracture energy G_c to 500 kN/m (2.86 kip/in). Young's modulus of concrete is chosen as 30000 MN/m^2 (4350 ksi) and the compression strength is 33.16 MN/m^2 (4.8 ksi). Young's modulus of steel is taken as 200000 MN/m^2 (29000 ksi), the hardening modulus as 10000 MN/m^2 (1450 ksi) and the tensile stress as 478 MN/m^2 (69 ksi). f_t and G_f are varied from 0.0 to 3.3 MN/m^2 (0.0 to 47.9 psi) and 0.02 to 0.25 kN/m (0.114 to 1.428 lbf/in.) in a preliminary investigation in order to calibrate these non predefined material parameters.

The calculation of the *CFRP strip strengthened slabs* (Fig. 2) shows a stiffer load displacement response than the test result due to following reasons. Firstly, the strips are modelled with perfect bond to the concrete surface which does not reflect the real situation, since the tests showed debonding in the center. Therefore, the stiffness of the slab is in reality lower resulting in a less stiff load displacement response in part II and III since debonding is correlated to crack openings. Secondly, the cross section of the CFRP strip strengthened slabs is not uniformly distributed across the width since the strips covered only 20 % of the slab width. Consequently, the strips have to bridge (crackspanning) significantly bigger cracks compared to strengthening systems with fully covered slab width. Moreover, the forces on the tension side are concentrated nearby the strips due to the high stiffness of the strips. Consequently, a three dimensional stress state appears in the surrounding area of the strips. Due to these reasons it is apparent that the calculated load displacement response based on the simplified 2D model differs from the test results. In order to avoid the above mentioned drawbacks, a 3D model of the slabs is necessary.

The peeling away of the strips is modelled by a negative hardening modulus of the strips. For the onset of the peeling away, a substitute tensile strength of the strip is introduced, which is assumed to be 0.65 % times Young's modulus of the strips. The value of 0.65 % is approximately the ultimate tensile strain of the adhesive and was observed in the tests as a failure criterion.

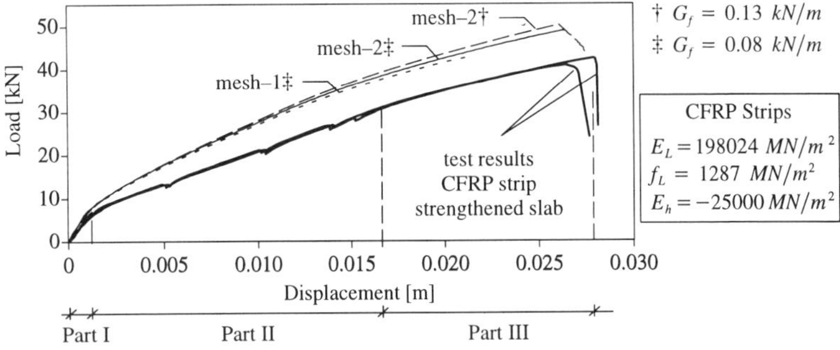

Figure 2. Load Displacement Responses of CFRP Strip Strengthened Slabs

The ultimate displacement (calculated with mesh–2) is predicted very well while the ultimate load is exceeded by about 18 %. The influence of the tensile fracture energy G_f on load displacement response and on ultimate load is small. The difference in ultimate loads is about 2.1 %.

The calculation of the *CFRP fabric strengthened slab* (Fig. 3) shows the best agreement to the test result due to the fact, that the fabric covers the entire slab width.

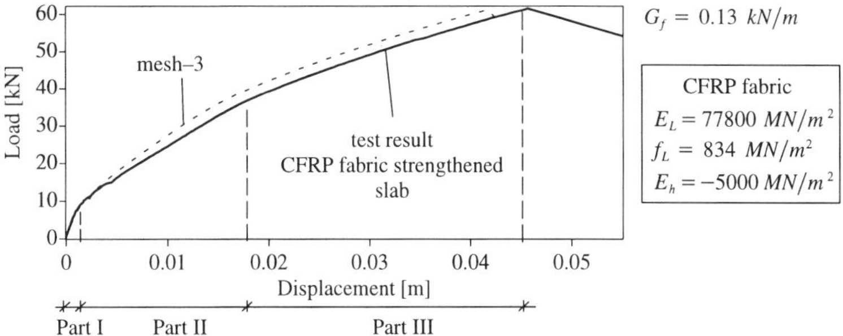

Figure 3. Load Displacement Responses of CFRP Fabric Strengthened Slab

The cracks in the concrete along the cross section are completely bridged by the fabric. Therefore the stresses in the cross section are uniformly distributed and a 2D model for the finite element calculation is justifiable. Again, the peeling away is modelled by a negative hardening modulus of the fabric. The ultimate tensile strength of the fabric identified in a tension test is used for the onset of the peeling away.

The calculation of the *Glass Fiber strengthened slab* (Fig. 4) leads to a slightly stiffer load displacement response compared to the test result.

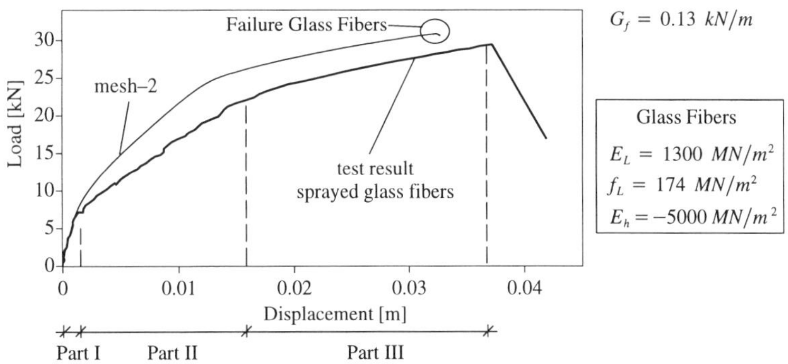

Figure 4. Load Displacement Responses of Glass Fiber Strengthened Slab

The failure load differs by 4.8 % while the displacement at ultimate load differs by 12.9 %. Moreover, the load at onset of steel yielding is well predicted. The ultimate tensile stress of the glass fibers is used to model the fracture of the glass fibers.

3.   Conclusions

The finite element analyses of the *CFRP fabric strengthened slabs* lead to a remarkable agreement with the results of the tests (Fig. 3). Not only the first crack load is well predicted but also the tensile failure of the carbon fabric. Due to the fact that the fabric covers the complete width and that it provides an almost uniformly distributed stress state along the width, a 2D plane stress finite element model is justified. The difference of the load displacement response in part II and III can be explained by the fact that perfect bond between fabric and concrete is assumed.

The results of the analyses for the *CFRP strip strengthened slabs* are not in such an ideal agreement with those observed in the tests (Fig. 2). The first crack is well predicted while the difference of the load displacement response in part II and III can be explained by the onset of debonding of the strips. Since the failure is governed by debonding, the adhesive layer should be considered in the finite element model. Moreover, an interface element, similar to what is known from the numerical simulation of nonlinear behavior of composite steel–concrete structures should be used to simulate debonding of the strips, see Menrath *et al.* 1998. Also, a 3D finite element model should be used allowing to model the varying response in the width direction, the stress concentrations at the cracks and therefore a more realistic simulation of the failure mode.

References

1. CEB–FIP – Model code 1990. Bulletin d'information CEB.

2. Eurocode 2 (1992) – *Planung von Stahlbeton– und Spannbetontragwerken* – DIN V 18932 (10.91); DIN ENV 1992–1–1 (06.92)

3. Hörmann, M. (1997) *Post Strengthening of Concrete Structures by Externally Bonded Fiber Reinforced Polymers (engineering diploma thesis)*, Institute of Structural Mechanics, University of Stuttgart, Stuttgart (in cooperation with the University of California at San Diego, UCSD).

4. Menrath, H., Haufe, A., Ramm, E. (1998) A Model for Composite Steel–Concrete Structures, in R. de Borst, N. Bicanic, H. Mang, G. Meschke (eds.), *Proceedings of the EURO–C 1998 Conference on Computational Modelling of Concrete Structures (Badgastein/Austria)*, A. A. Balkema Publishers, Rotterdam, Volume 1, pp. 33–42.

Finite Element Modeling of Reinforced Concrete
in Bridge Seismic Retrofit Design Practice

Robert A. Dameron[1], Yusef R. Rashid[1]

Abstract

This paper summarizes the current methods for concrete bridge seismic analysis used in California and the advances made in recent years in those methods. The paper describes an analysis approach in which a three-tiered analytical hierarchy is established for modeling complex phenomena and the corresponding analytical tools needed for each analysis level. In this hierarchy, the highest level of detail, continuum cracking analysis with discrete representation of reinforcement, is reserved for structure component evaluation, such as the prediction of deep-beam or column shear failures, rebar bond-slip, and torsion effects. Stick models are more appropriately utilized in the analysis of global behavior of the structure. While this hierarchical procedure has now been applied on many of the major bridges in California, space only allows illustration of the methods for one example, the San Diego-Coronado Bay Bridge Seismic Retrofit project, in which all three levels of concrete modeling detail were used.

Introduction

Since the Loma Prieta Earthquake of 1989, the authors have worked closely with the California Dept. of Transportation (Caltrans) and many of its design consultants to improve the analysis methods for concrete bridges. Through this experience, a hierarchy of methods involving various levels of sophistication and complexity were developed and applied to the study of the behavioral issues involved. The analysis techniques developed have been demonstrated for cyclically loaded concrete bridge components and have been validated by prediction analyses and comparisons to Caltrans-sponsored laboratory tests.

After the Loma Prieta Earthquake, a California Governor's Board (Housner, 1990) called for the performance of "comprehensive earthquake vulnerability analyses and evaluation of important transportation structures....using state-or-the-art methods in

---

[1] ANATECH Corp., 5435 Oberlin Drive, San Diego, CA 92121

earthquake engineering." Caltrans has now conducted such seismic vulnerability assessments of all the State's toll bridges, and retrofit designs have been completed. One such toll bridge is the San Diego-Coronado Bay Bridge, which extends 1.6 miles across San Diego Bay. The bridge consists of three long channel spans of orthotropic box girders and twenty-six shorter spans of girder and composite lightweight concrete deck construction, and a 90-degree curve on the western spans. One of the bridge's most distinguishing features are the reinforced concrete pier supports, ranging from 20 to nearly 200 feet in height, and supported by tall, 54"-diameter prestressed concrete piles. Seismic evaluation and retrofit of this support system provides a comprehensive example of the range of levels of concrete analysis detail.

Concrete Modeling Hierarchy

Seismic analysis procedures for bridges have evolved over the last decade to three types with increasing levels of complexity and sophistication as follows:

1. Linear elastic "stick" models analyzed statically for vertical loads, and with linear response spectrum analysis for seismic loads.
2. Nonlinear "stick" and shell model analysis for vertical loads, and nonlinear quasi-static "pushover" analysis for seismic loads.
3. Two-dimensional and three-dimensional nonlinear continuum analysis.

Prior to 1990, the only procedure in widespread use in California bridge analysis was Level-1. The Loma Prieta Earthquake changed this for California. Use of linear analysis in seismic design relies on the concept of ductility demand because elastic analysis for strong ground motion spectra typically produce member forces greatly in excess of strength capacities. If components can be shown to be ductile then allowable force or moment overload factors are assigned. This method is clearly limited in evaluating modes of damage (which usually depend on sequences of nonlinear events in the structure's response), and the accuracy of structure displacement prediction is poor. Nevertheless, elastic, Level 1 analysis is an important first step in the initial assessment of the seismic vulnerabilities of a structure. Such analysis can provide initial indication of "hot spots" to plan a retrofit strategy, categorization of members by D/C ratio, and an enveloping of peak response quantities to a design spectrum. Level 2 analysis may utilize the same model as a Level 1 analysis, but it invokes multi-support time history input (usually by direct time integration) and introduces nonlinearities, preferably those nonlinearities that are essential to accurate response prediction. Such nonlinear behaviors as expansion hinge contact, material yielding, and P-Delta effects can be characterized. The drawbacks of Level 2 over Level 1 analysis are the difficulty in selecting which nonlinearities are important and how to characterize them, and the limitations caused by the use of a discrete time history to design for a range of possible seismic events. For these reasons, the authors still recommend the use of both Level 1 and Level 2 analyses for thorough seismic evaluations.

Level 3 analysis is applied to the capacity side (both force and deformation-ductility) of the demand vs. capacity design equation. The distinguishing feature of Level 3 is the choice and discretization of the elements. Stick (or beam element) models provide poor representation of shear and torsion, especially in combination with large axial loads and biaxial bending. The most detailed methodology now used extensively for California bridges evolved through the authors' research in the behavior of nuclear safety structures (Rashid, 1968). Using detailed constitutive subroutines for the concrete and steel which are called by general purpose finite element programs (Rashid, et al, 1996) the concrete stress-strain behavior is handled separately from (but in close coupling with) rebar and prestressing subelements. The concrete model handles 3-dimensional cracking, crushing, and post-cracking shear retention, and the steel element models handle plasticity, strain hardening, and bond-slip behavior.

Application to the San Diego-Coronado Bay Bridge

Analyses performed for the seismic vulnerability and retrofit studies included:
1. Three-dimensional linear (Level 1) and nonlinear (Level 2) modeling of the entire bridge, including accounting for soil-structure interaction. These models used a linear response spectrum and nonlinear time history inputs.
2. Component models (Level 3) for many of the most critical components, including typical piers (short, medium, and tall piers), and prestressed concrete piles.

Balancing the level of detail in analysis involved making tradeoffs between the additional level of insight gained from increasingly complex models and the reality of schedule constraints. The Level 1 and Level 2 models utilized are shown in Figure 1, and an example of a Level 3 model is shown in Figure 2. The selection of nonlinearities used in the Level 2 global demand evaluation are based on experience and, in large part, on the retrofit strategy and damage criteria that is adopted for the project. For Coronado, the strategy was to replace many of the bearings with high damping rubber isolation bearings, so these were important nonlinear elements to include. The damage acceptance criteria allowed considerable yielding in the tall concrete pier legs, so material nonlinearity and P-delta effects were important for the pier modeling. Since primarily elastic behavior was observed in the superstructure (on the basis of Level 1 analysis), this was represented with elastic beam elements. Together, the use of all modeling levels provided comprehensive assessment of force and deformation demands and capacities and allowed the use of a deformation-based retrofit design approach.

References

Housner, G. W., et al., "Competing Against Time," Governor's Board of Inquiry Report on the 1989 Loma Prieta Earthquake, State of California, May 31, 1990.
Rashid, Y. R, "Ultimate Strength Analysis of Prestressed Concrete Pressure Vessels," Nuclear Engineering and Design, July, 1968, pp. 334-344.
Rashid, Y. R., et al., "Constitutive Modeling of Reinforced Concrete and Steel," ANATECH Corp., San Diego, CA, September 1996.

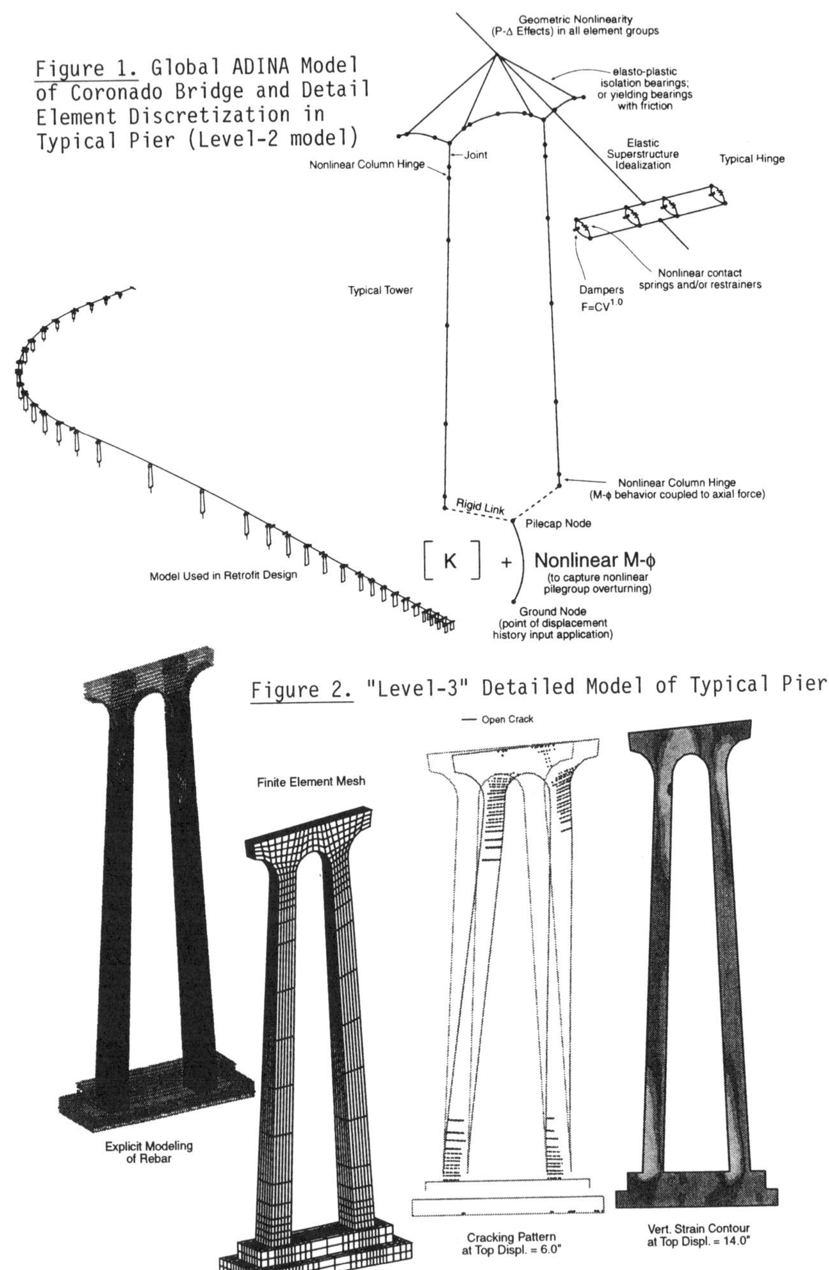

Figure 1. Global ADINA Model of Coronado Bridge and Detail Element Discretization in Typical Pier (Level-2 model)

Figure 2. "Level-3" Detailed Model of Typical Pier

FRC Containers Design to ensure Impact Performance

François Toutlemonde[1]
Jérôme Sercombe[2]

Abstract

The 3D computational modeling of 5-m drop tests of 'high integrity' fiber-reinforced concrete (FRC) containers used for the storage of radioactive waste is presented. Attention is focused on the preliminary design stage of Reactive Powder Concrete (RPC) containers, using an adapted material law for concrete in high rate dynamics.

Industrial problem and requirements

'High integrity' interim storage containers are required for certain types of radioactive waste. They shall resist, among other physico-chemical requirements, a 5-m high drop test without loosing integrity, i.e. waste widespreading. Presently available fiber reinforced concrete (FRC) containers do not satisfy this requirement without a second (expensive) external envelope. High integrity containers made of Reactive Powder Concrete (RPC) are thus under study (Torrenti, 1996). Appropriate numerical computations shall help the preliminary design stage of such containers to ensure impact performance. Computations must discriminate the materials considered (correct accounting for concrete 'rate effects' is thus necessary), predict the failure mode of the containers (local crushing, cracking, etc.) and the critical fall height at which general failure occurs. They should give correct estimations of the impact parameters (impact force and duration) and of the structural damage : amount of scabbing material, crack opening and length. So most of the prototyping shall be done numerically, reducing the costs of experimental development.

[1] Struct. Eng. Dep., LCPC, 58 bd Lefebvre 75732 Paris Cedex 15 France
[2] Concrete Lab., CEA, Saclay, Bât. 158, BP2, 91191 Gif s/Yvette France

Material characterization and constitutive modeling

The model is based on the experimental and theoretical knowledge concerning concrete behaviour at high strain rates developed at LCPC since 1990 (Toutlemonde, 1995). It is chosen in the frame of continuum mechanics to favour 3D computation and global analysis for the designer. Irreversible damage and cracking of the material is represented using plasticity, which is admissible for a material which exhibits progressive crack opening, such as FRC. Locally (around the impact location), concrete is reduced to powder. It implies an important loss of stiffness which must be accounted for using a damage variable. Accounting for rate effects (Young's modulus and concrete strength increase, especially in tension, related to loading rate) has been achieved by introducing a non-plastic hardening variable, which has the dimension and physical significance of the deformation of the viscous fluid (water) occupying concrete porous space (Sercombe, 1998[b]). Solving the mechanical problem (coupled material non-linearity) in dynamics with a precise control of iterations concerning plastic and non-plastic variables has implied the development of specific algorithms within the general F.E. code CESAR-LCPC used for the problem (Sercombe, 1997). Validating the material constitutive modelling and data identification has been carried out using experimental results at different rates : direct tension and compression on cylinders, bending of plain concrete prisms, flexural tests of RC slabs using a shock tube (Toutlemonde, 1995).

Parameters required to describe the RPC behaviour according to this model were identified following the guidelines of the recommendations for design of FRC structures (Casanova, 1997). Parameters characterizing RPC compressive and multiaxial behaviour in compression were derived from existing results. However, since the tensile behaviour of FRC greatly depends on casting conditions and fiber orientation, and since rate effects on this ultra-high performance material had not been previously characterized, a specific important program of direct tensile tests at loading rates ranging from 0.05 to 50000 MPa/s was performed, using the techniques of modified Split Hopkinson Bars and a consistent quasi-static direct tensile testing process. Cylindrical specimens were drilled from representative L-shaped elements cast under realistic conditions relatively to future containers. Detailed results of this characterization can be found in (Toutlemonde, 1998). Since as a first approach, anisotropy was not accounted for in the computations, only the average value of tensile strength has been considered. Due to RPC hardening behaviour in tension, up to a stress peak corresponding to strain localization, and due to the smooth softening slope in the post-cracking regime, two sets of data have been defined. The first one corresponds to a 'macroscopically sound' (uncracked) state, with crack openings lower than 0.3 mm. The yield limit and its evolution with strain rate are derived from the maximum stress obtained experimentally on unnotched specimens. The other one is related to a 'cracking admissible state', conventionnally defined for crack openings lower than 1 mm (which is considered as a limit for the integrity of the containers). The yield limit is then computed as the equivalent stress leading to the same consumed energy as experimentally obtained, up to this limit of 1 mm. The average experimental values have then been reduced to include a safety margin, leading to following specific input data for RPC modelling under high strain rates (Tab. 1, after Sercombe 1998[a]).

Table 1 : RPC characteristics for computations using viscous hardening plasticity

	Static value (MPa)	Rate effect (abs. trend MPa/log unit)
Tensile Strength	8.	0.8
Equiv. Plast. Stress	7.	0.5
Young's modulus	52,000.	500.

Drop test computational modeling, results and validation

The container represented as a cylinder is dropped on the lid in an inclined position, such that the center of gravity is aligned with the impact point. 3D modelling is required, but because of the symmetry only half of the container may be discretized. A 1:3 scale RPC prototype container is modelled, for drop tests results on such prototypes are available. The container is filled with fine sand, assumed as elastic and perfectly bonded, representing the waste material. Only one node at the point of impact is restrained initially, an initial speed depending on the height of fall is prescribed to all other nodes. Requirements concerning the containers consider the hypothetical case of its drop on an infinitely rigid surface. The contact modelling may then only consist in a geometrical verification (position of the impactor's nodes relatively to the plane limiting the impacted body). In a first approach, contact is assumed without friction and non-interpenetration is ensured by a classical penalty method. No specific implicit algorithm was developed to check simultaneously plasticity and contact conditions, but it was verified by comparison with reference solutions that a satisfactory solution was obtained, provided a refined mesh of the contact zone is used as well as a compatible time discretization. The computation is continued until contact is lost (after about 3 ms). Computed plastic strains are translated in terms of crack openings, by muliplying them by twice the length of the fibers, which corresponds to the possible extension of the 'process zone' where deformations are concentrated due to fiber anchoring on both sides of a crack (Sercombe, 1997).

Simulations have been carried out for falls from drop heights of 5 m (required performance) and 12 m (critical height where crack openings of about 1 mm have been observed on a 1:3 scale container, without loss of integrity). Macroscopic damage is represented by the extension of zones with a high principal plastic strain DP1 (maximal extension). Detailed representation of the results in terms of plastic strain isovalues and possible crack analysis can be found in Sercombe, 1998[a]. Clearly for the 5-m high fall, the results show that damage remains concentrated around the impact point, which is consistent with the (limited in space) spalled and crushed material observed experimentally. Locally the material is submitted to very high triaxial compressive stresses, therefore it seems important to use the modified 4-parameters Willam-Warnke criterion (Sercombe, 1997). On the opposite for the 12-m high fall important irreversible strains extend to the whole structure. They may be considered as representing a crack in the container shell, extending up to half of the container height, and making an angle of about 40° with the lid. The crack opening may reach 1 mm near the lid, which is close to experimental observation. Using data from instrumented drop tests, it could be verified that the mean deceleration level during the shock, and the total contact time were correctly estimated. A parametric study was carried out to confirm (for a container at scale 1:1) the critical height corresponding to the development of cracking overall in the container beyond the impact zone. The effects of the material characteristic strength, of concrete rate sensitivity, and the necessity of accounting for damage under high triaxial compressive stresses have been emphasized.

Conclusion

The methodology presented here, consisting in an appropriate constitutive modelling based on viscous hardening plasticity and relying on an adapted FRC characterization, has proven to be reliable to describe the major relevant macroscopic phenomena for an effective preliminary design of high integrity containers to ensure their impact performance. More detailed prototyping can still be continued using this validated tool.

References

Casanova P., Rossi P. (1997) "Analysis and design of steel fiber-reinforced concrete beams", *ACI Struct. Journal*, 94(5), 595-602.
Sercombe J. (1997) "Modelling concrete behaviour under high strain rates. Application to the design of high integrity containers", Ph. D. Thesis, 288 p. (in French, English abstract), ENPC Paris (F).
Sercombe J., Toutlemonde F., Ulm F.-J. (1998[a]) "From material behaviour to structural design for concrete in high rate dynamics", in *Computational modelling of concrete structures*, proc. EURO-C '98 conf., Balkema, 633-642.
Sercombe J., Toutlemonde F., Ulm F.-J. (1998[b]) "Viscous hardening plasticity for concrete in high-rate dynamics", *J. Engrg. Mech.*, ASCE, 124(9), 1050-1057.
Torrenti J.-M. *et al.* (1996) "High integrity containers for interim storage of nuclear wastes using reactive powder concrete", in *High strength/High performance concrete*, proc. BHP'96 Symp., Presses ENPC, Paris (F), 1407-1413.
Toutlemonde F. (1995) "Shock strength of concrete structures. From material behaviour to structural design", (348 p., in French, English abstract), *Res. Report hors coll.*, LCPC, Paris (F).
Toutlemonde F. *et al.* (1998) "Characterization of reactive powder concrete (RPC) in direct tension at medium to high loading rates", in *Concrete under severe Conditions 2*, proc. CONSEC'98 conf., E&FN SPON, 887-896.

Simplified Modelling Strategies for RC Structures under Seismic Loading using
Damage Concepts

Jacky Mazars[1], Frédéric Ragueneau[1] and Shahrokh Ghavamian[1]

Abstract

Continuum damage mechanics allows a realistic description of concrete like materials degradation. Used in combination with a simplified finite element code based on multilayered beam kinematics assumptions, unilateral local damage models become an efficient tool to describe the response of RC structures under seismic loading.

Introduction

The response of a structure submitted to severe loadings, depends on strong interaction between "material" (local non-linearities), "structural" (geometry, mass distribution, joints) and "environmental" (interaction of the structure with its support) effects (Pauley and Priestley 1992). For concrete structures, the local non-linearities are mainly the consequences of, partly the opening and reclosure of cracks, and for the rest, bond slip and materials reinforcements. A good description of those phenomena is required for describing changes in stiffness of the structure and for understanding the mechanism up to failure. Discretisation process of a structure is of high importance, since a compromise should be obtained between the need of a fine modelling in order to allow a good positioning of masses, and giving access to localised damage, and a sufficiently coarse mesh so that the calculation can feasibly be performed. The additional flexibilities and non-linearities are also necessary (eg. soil-structure interaction) since their effect, specially on eigenfrequencies of the structure are very strong.

This presentation shows how are solved these problems through "simplified finite element methods" and the efficiency of this strategy through different national and international experimental seismic programs (Mazars et al. 1998, Ghavamian and Mazars 1996, Ragueneau and Mazars 1998).

---

[1] Laboratoire de Mécanique et Technologie, Ens de cachan
61, avenue du Président Wilson, 94235 Cachan Cedex, France

"Simplified" approach strategies: code EFiCoS

Two levels have to be considered: the f.e. discretisation of the structure and the mechanical behaviour of the materials. Assuming simplifications for the structural kinematics allow to consider refine constitutive relationships at the local level whithout jeopardising the computation efficiency. Concerning the materials constitutive laws, the use of damage mechanics introduces physical features like local stiffness degradation or unilateral behaviour of a crack. Based on irreversible thermodynamics processes, a Gibbs free energy can be defined and the resulting constitutive equation is obtained by derivation (see figure 1 for the response of the model):

$$\varepsilon = \frac{\partial \chi}{\partial \sigma} = \frac{\sigma_+}{E(1-D_1)} + \frac{\sigma_-}{E(1-D_2)} + \frac{\nu}{E}\left(\sigma - \mathrm{Tr}(\sigma)\mathbb{1}\right) + \frac{\beta_1 D_1}{E(1-D_1)}\frac{\partial f(\sigma)}{\partial \sigma} + \frac{\beta_2 D_2}{E(1-D_2)}\mathbb{1}$$

E is the initial Young's modulus and ν the Poisson ratio, D_1 and D_2 are respectively the damage variables for traction and compression, β_1 and β_2 are material constants governing the inelastic strains evolution, $f(\sigma)$ and σ_f are the crack closure function and the crack closure stress respectively.

The constitutive equation for steel here concerned is a standard plasticity model with a nonlinear kinematic hardening (see figure 1).

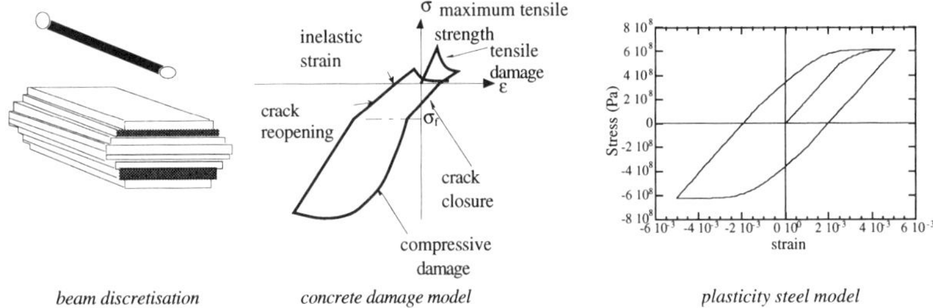

Figure 1. Beam discretisation and local behaviour for concrete and steel

The choice of using a multilayered f.e. configuration combines the advantage of using beam type finite elements with the simplicity of uniaxial behaviour (or uniaxial behaviour enhanced to include shear as one can see below). Each finite element is a beam which is discretized into several layers (see figure 1). Equivalent formulation can also be found in (Spacone et al. 1996). The basic assumption is that plane sections remain plane (Bernouilli's kinematic) allowing to consider a uniaxial behaviour of each layer. This is no longer satisfactory when shear strains take a major role (Mazars and Ghavamian 1996). In that particular case the shear strains have to be introduced in the model and the layer behaves now bi-axially. Cross sectional distortion is introduced through a Timoshenko's kinematic assuming a parabolic distribution of the shear strains over a rectangular cross section (see figure 2 for an example). The stress/strain relation considered is :

$$\begin{pmatrix} \varepsilon_{11}^{e} \\ \varepsilon_{12}^{e} \end{pmatrix} = \begin{bmatrix} \dfrac{1}{E_D} & 0 \\ 0 & \dfrac{1}{E_D} + \dfrac{\upsilon}{E_0} \end{bmatrix} \begin{pmatrix} \sigma_{11} \\ \sigma_{12} \end{pmatrix} \text{ With } E_D = E_0\left(1 - D_I\right) \text{ when } \sigma = \sigma^{+}$$

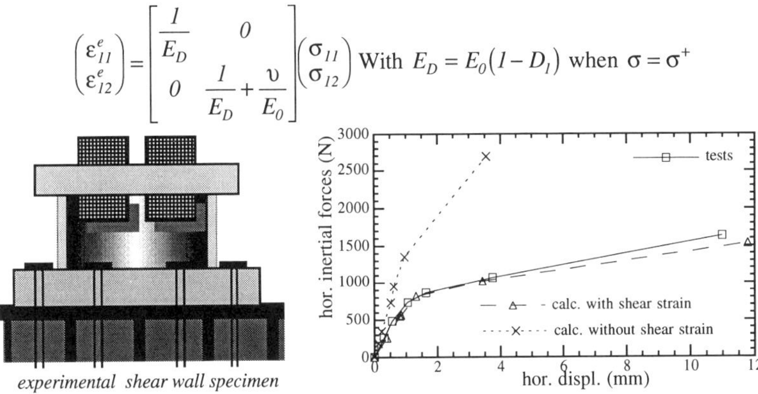

Figure 2. Shear strain effects on a shear wall submitted to a shaking table test (maximum values for 5 different earthquake levels are plotted. Computations are performed with the code EFiCoS)

Application

The application presented concerns a structure subject to earthquake at high levels. The main purpose of this experimental program (CAMUS) is to demonstrate the ability of reinforced concrete bearing walls to resist during seismic loading. To reach this goal, a 1/3rd scaled model has been tested on the shaking table of C.E.A (France). This mock-up is composed of two parallel walls (0.06mx5mx1.6m) linked by 6 square slabs (1.6m²). A highly reinforced footing allows the anchorage to the shaking table. The equations of motion are solved thanks to the implicit Newmark algorithm. The Rayleigh viscous damping matrix allows to introduce in the computations the damping forces. Due to a particular behaviour of the contact and to the flexibility of the table, special boundary conditions have been introduced (Mazars et al. 1998). The accelerograms are imposed at the basis of the structure with increasing levels of maximum accelerations. The figures below present comparisons between "EFiCoS" modelling and experimental results in terms of horizontal top displacement and frequency analysis of the responses.

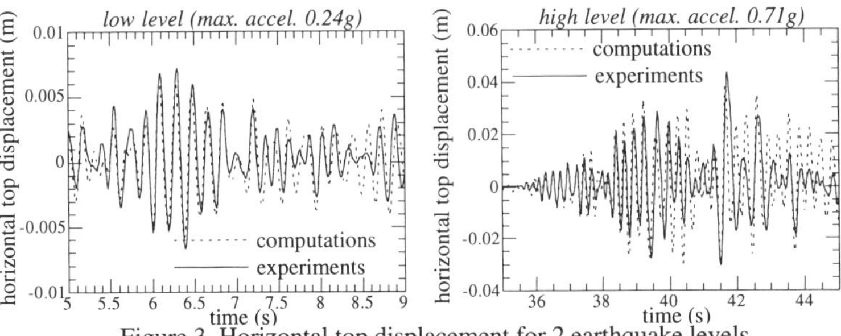

Figure 3. Horizontal top displacement for 2 earthquake levels

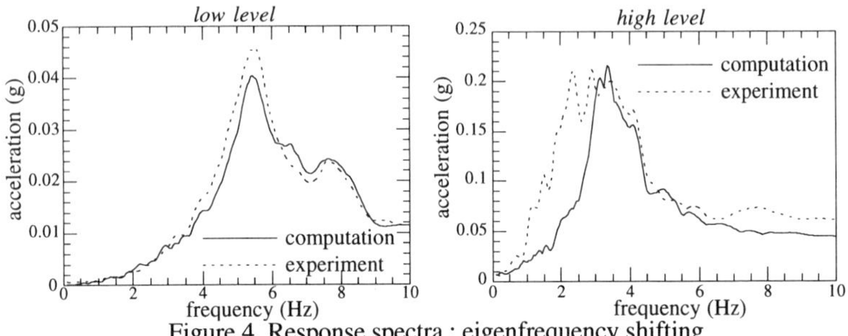

Figure 4. Response spectra : eigenfrequency shifting

For the highest level, the critical cross sections are severely loaded : concrete is cracked and the steel reinforcement bars are buckled or have failed under tension. We can observe that the model is able to simulate the displacement time history in a good agreement with the experiment. The decrease of eigenfrequencies due to cracks opening is obtained thanks to the local damage model used for concrete whose main implication remains in the degradation of the Young's modulus.

Conclusions

Comparisons with experiments of different kind of structures enlighten the relevance of this approach of simplified structural analysis combined with a refine description of the local behaviours. An international benchmarck (AFPS 1998) pointed out the benefits of using such a formulation with regards to the results obtained for 2D or 3D models. Future developments concern the way to take into account in a physical manner the dissipative phenomena (Ragueneau et al. 1998). The aim is to decrease the importance of the global Rayleigh matrix by shifting a part of it at a local level accounting for crack lips frictional sliding. Moreover, enriched local behaviour should help us to improve the determination of an ultimate failure criteria which is strongly influenced by the failure mechanism at a micro-level (crack opening, bond slip, ...)

References

Ghavamian, Sh. and Mazars, J. (1996). "The ISP Shear Wall - Analysis of the non-linear behaviour using multi-layered shear beam elements", *OCED-NUPEC comparisons report*, Yokohama Japon.

Mazars, J., Ghavamian, Sh., and Dubé, J.F.(1998). "Tools to Analyse the Scaling effects on Concrete Mock-up under Dynamic Loading", *Proc. EURO-C 1998, Ed. Balkema*, Vol. 2, pp. 703-711, Badgastein, Austria.

Pauley, T. and Priestley, M.J.N. (1992). *Seismic design of reinforced concrete and masonry buildings*. John Wiley & Sons, inc. USA.

Ragueneau, F., La Borderie, Ch. and Mazars, J. (1998) "Damage Model for Concrete Including Residual Hysteretic Loops: Application to Seismic and Dynamic Loading", *Proc. FRAMCOS-3*, Fracture Mechanics on Concrete Structures., pp. 685-696, Gifu, Japon, 12-16 Octobre 1998.

Spacone, E., Filippou, F.C. and Taucer, F.F. (1996). "Fibre Beam-column Model for Non-linear Analysis of R/C Frames: Part I. formulation", *Earth. Engng. & Struct. Dyn., Vol. 25* , No. 7.

AFPS (1998). *CAMUS International Benchmark-Synthesis Report*, Workshop CAMUS, 11st ECEE, Paris, 6-11 Septembre 1998.

Reinforced Concrete Building Prediction Analyses

Robert K. Dowell[1], Liping Zhang[1]

Abstract

As part of the CAMUS International Benchmark in France, three-dimensional nonlinear time-history prediction analyses have been conducted to assess the response of a 1/3-scale, six-floor reinforced concrete building tested on a shaking table. The paper presents predicted time-history responses and compares these results to the reported measured results from the benchmark organizers. A pushover analysis also is discussed.

Introduction

Nonlinear dynamic time-history analysis of reinforced concrete structures has been the subject of many investigations over the past three decades. Important nonlinear effects to be considered include cracking of concrete in tension and strain-softening in compression, as well as a reduction in shear stresses at wide open flexural cracks. Because such analyses can be extremely complicated, few investigations have shown good agreement between prediction and test results, especially at large deformations. As part of an experimental and analytical effort, organizers of the CAMUS International Benchmark (Mazars, 1997) invited participants to perform blind prediction analyses of a 1/3-scale, six-floor reinforced concrete building which was tested with a series of three increasingly large earthquake records. Benchmark organizers requested that participants also perform a pushover analysis and assessment of the final failure mode of the structure.

This paper presents prediction analyses of the 1/3-scale building tested on the shaking table using nonlinear concrete and rebar modeling techniques. Results are compared with measured test data, which became available after the prediction analyses were completed and submitted to the benchmark organizers.

---

[1] ANATECH Corp., 5435 Oberlin Drive, San Diego, CA 92121

Description of Test Specimen

The 1/3-scale test specimen consists of two reinforced concrete cantilever shear walls, linked by six square floors, supported on a heavily reinforced concrete footing and tied down to a shaking table, as indicated in Figure 1 (one-half of the structure shown -- symmetry model). The walls are 5.1 m high, 1.7 m wide and 6 cm thick, reinforced by three groups of vertical bars. Vertical reinforcement is terminated in stages, 10 cm below each floor, with the center group of bars completely eliminated by the top floor. Construction joints are created just above the middle of each floor by casting the walls in separate pours. The structure is loaded in the plane of the walls, with transverse bracing provided to increase the stiffness in the out-of-plane direction. Concrete and steel mass blocks have been added to the floors to correct for scale effects. A detailed description of the test specimen is given elsewhere (Mazars, 1997).

Finite Element Model, Material Properties and Analysis Procedures

A detailed three-dimensional finite element model was developed for one-half of the building, due to symmetry of the structure and loading. The concrete is modeled by 8-node continuum elements and the transverse steel bracing is modeled by beam elements. All reinforcing bars are explicitly modeled using truss type sub-elements embedded in the concrete continuum elements. Rocking of the shaking table was included by modeling it explicitly, including its mass and tie-downs.

The concrete material model is the ANATECH smeared cracking model that allows the concrete to crack in tension and strain-soften in compression (Rashid, 1996), while the steel reinforcement material model is based on the classical plasticity approach. Included in the concrete model is shear shedding, which allows the shear stresses to be reduced at an open crack based on the magnitude of the normal tensile strain. Directed cracks have also been provided at construction joints, reducing the vertical tensile capacity at these locations to zero.

Analyses were conducted with the ANATECH material models, in conjunction with the general purpose nonlinear finite element program ABAQUS (Hibbitt, 1996). Prior to conducting time-history analyses, modal and pushover analyses were performed to determine the fundamental frequency and mode shape of the structure as well as its static capacity and final failure mode. In the dynamic analyses, the structure was subjected to three earthquake motions of increasing intensity with peak accelerations of 0.24g, 0.40g and 0.71g, respectively. Implicit time integration is conducted with numerical damping to remove any high frequency noise.

Preliminary Analysis

The modal analysis shows that the structure has a fundamental mode of 8.28 Hz, which is in reasonable agreement with the measured first mode frequency of

7.24 Hz. The pushover analysis indicates that the final failure mode is a combination of flexural and shear behavior of the walls. As Figure 2 shows, the region of damage is not at the base of the building, as might be expected, but just below the third floor. This is a consequence of reinforcement being terminated at each floor level (flexural demand versus capacity is not the greatest at the base of the wall) and a large tension shift effect. From the pushover analysis, it was predicted that the primary longitudinal reinforcement will fail just below floor three when the maximum top displacement approaches about 5 cm. The predicted failure mode is in good agreement with test results (Figure 2).

Dynamic Analysis

As described previously, the test specimen was subjected to three seismic motions of increasing magnitude. Time-history responses of the relative displacement at the top floor, base shear and base moment are shown in Figure 3 for the largest earthquake record applied. Predictions for all three quantities are in good agreement with measured results. Computed rebar strains just below floor three exceed 0.055 (reported strain capacity of the large diameter longitudinal steel) at approximately 11 seconds, indicating rebar rupture. This prediction was confirmed following the test, which showed that the structure had failed at about 11 seconds due to rebar rupture just below the third floor, resulting in a large diagonal crack (Figure 2).

Conclusions

Good agreement between the blind prediction analyses and measured test results for the six-floor building demonstrates that it is possible to provide an accurate assessment of a complicated reinforced concrete structure, subjected to large ground motions, using three-dimensional nonlinear time-history analysis. The pushover analysis proved useful in providing locations of high strain, or "hot-spots", so that these local regions could be monitored throughout the dynamic analyses, allowing the correct prediction of failure.

References

Hibbitt, D., et al., "ABAQUS/Standard User's and Theory Manuals," Ver. 5.6, Hibbitt, Karlsson & Sorensen, Pawtucket, RI, 1996.

Mazars, J., et al., "Mock-up and Loading Characteristics Specifications for the Participants Report," Report 1, CAMUS International Benchmark, France, 1997.

Rashid, Y. R., et al., "Constitutive Modeling of Reinforced Concrete and Steel," ANATECH Corp., San Diego, CA, September 1996.

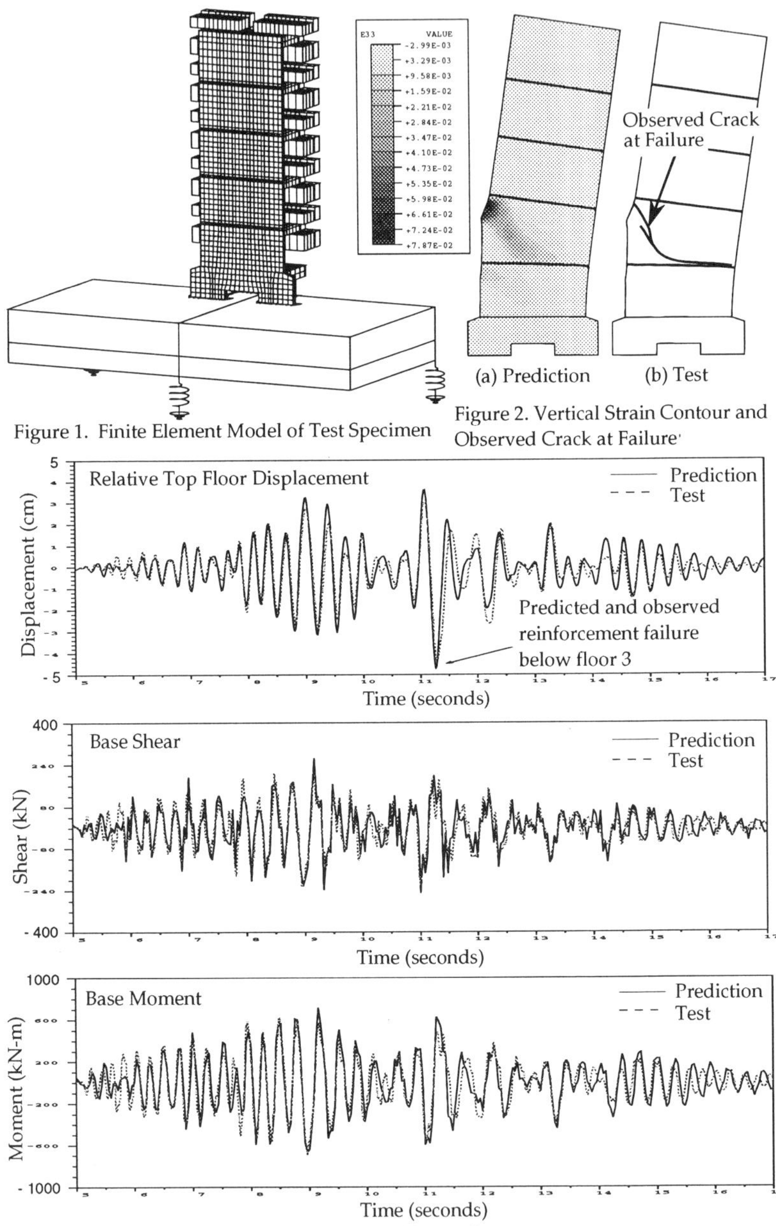

Figure 1. Finite Element Model of Test Specimen

Figure 2. Vertical Strain Contour and Observed Crack at Failure

Figure 3. Time History Responses

Collapse Analysis of RC Structures including Shear

A.S.Elnashai[1], FASCE A.M.Mwafy[2] and D.Lee[2]

Abstract

New trends in performance based seismic design, alongside requirements for capacity design of members and connections, dictate stringent requirements on analytical tools for analysis of RC structures. Shear stiffness, strength and post-peak response remain a major challenge especially under variable amplitude reversed actions. This paper describes briefly two developments, on the member and panel levels, to allow for assessment of structural collapse by shear failure, implemented in the nonlinear dynamic analysis program ADAPTIC, developed at Imperial College.

Introduction

The concept of capacity design, as applied to members, requires that the flexural strength is lower than the shear strength, in order to avoid shear failure. Whereas the flexural strength of members is relatively straightforward to assess, even by section analysis and hand calculations, the shear component of deformation and strength is rather more taxing. Moreover, since actions on a structure are distributed according to the distribution of stiffness (be it elastic or inelastic), it follows that ignoring shear stiffness, yielding and capacity would potentially lead to erroneous distribution of actions.

ADAPTIC, a program developed at Imperial College over the past twelve years (Izzuddin and Elnashai, 1989), has recently been modified in two ways to allow shear representations. The first, member level, approach utilises an empirical model for shear capacity to signal failure; this model is implemented in a step-wise fashion to allow for shear-axial interaction. The second more refined approach, possibly used in a second stage analysis to ascertain and sharpen results from the first model, is a novel implementation of the modified compression field theory. The developments, and a glimpse of the applications are outlined below.

---

[1]   Professor of Earthquake Engineering, Civil Engineering Dept., Imperial College, London, UK.
[2]   PhD candidate, Civil Engineering Dept., Imperial College, London, UK.

Member Level Representation of Shear Failure

To evaluate the inelastic seismic performance of RC buildings designed with different ductility classes and to obtain accurate analytical predictions of the response modification factors, a number of response criteria are needed. A shear-based criterion has been utilized amongst other limit state criteria to member failure. A shear model capable of providing an experimentally verifiable estimate of shear supply in RC members was proposed by Priestley et al. (1994) and has been adopted in the current study. In this model, shear strength is considered to be the sum of the following terms: (i) the inherent concrete shear resisting mechanism, V_c, the magnitude of which accounts for the influence of flexural ductility, (ii) the truss mechanism, V_s, and (iii) strength resulting from the inclined compression strut, V_p. The model is implemented as a post-processsor attached to ADAPTIC. Extensive analysis of RC buildings to collapse was undertaken, as below.

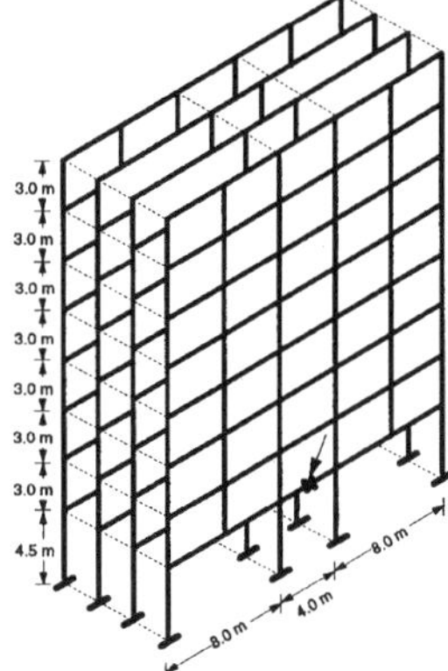

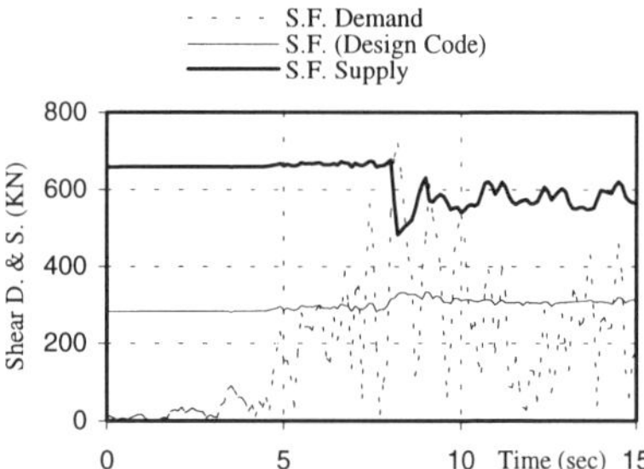

Figure 1: ADAPTIC Model of Irregular Frame

Figure 2: Shear Supply and Demand-Time

Twelve RC buildings designed and detailed in accordance with Eurocodes 2 and 8 are investigated. The buildings are considered to represent different types of structural systems, structural ductility class, design ground accelerations and height. Figure 1 depicts the ADAPTIC model for one of the buildings. Results of the application of the shear failure criterion on a low ductile building under Loma Prieta 1989 record at Aloha Ave. station, scaled up to collapse, are shown in Figure 2. Moreover, results of collapse analyses for four irregular buildings (IF) with three different ductility classes (High, Medium and Low) and designed for two design acceleration levels (0.15g and 0.30 g) were obtained. These indicate that values of collapse load and deformations are sensitive to the inclusion of shear failure, which may precede other limit states. This is particularly true if near-field effects are pronounced, particularly vertical earthquake motion (Elnashai and Papazoglou, 1996; Papazoglou and Elnashai, 1997). It is noted that shear deformation is not included in the analysis above, an effect included in the analysis described below.

Detailed Modelling of Shear Stiffness and Strength

An axial-shear interaction model was proposed and implemented in the fibre element of ADAPTIC (Lee, 1999). The basic concept of the model is that the stiffness in the current time step is calculated by introducing smooth transitions corresponding to the current level of axial force between series of envelop curves derived for various axial force values. To describe the inelastic shear response of the model, a refined version of the hysteretic rules proposed by Ozcebe and Saatcioglu(1989) is adopted for investigating loading and unloading (total and partial). The shear envelop curve is based on the modified compression field approach (Vecchio and Collins, 1986). In order to verify the implementation of the proposed model, the analytical results obtained with ADAPTIC incorporating the new model is compared with experimental results described by Maruyama, Ramirez and Jirsa (1984). The test specimen is a short square column the response of which is dominated by shear. As a representative case, analytical prediction is given for specimen '120C-U' which is subjected to constant compressive axial load. Table 1 gives the peak strength values at each ductility level obtained from the experiment and ADAPTIC analysis with shear. Figures 3 and 4 show a comparison of hysteresis loops (first cycles only) with and without shear, respectively.

Peak strength(kN)		
$\mu=1$	Exp	185
	An	207
$\mu=2$	Exp	269
	An	266
$\mu=3$	Exp	258
	An	275

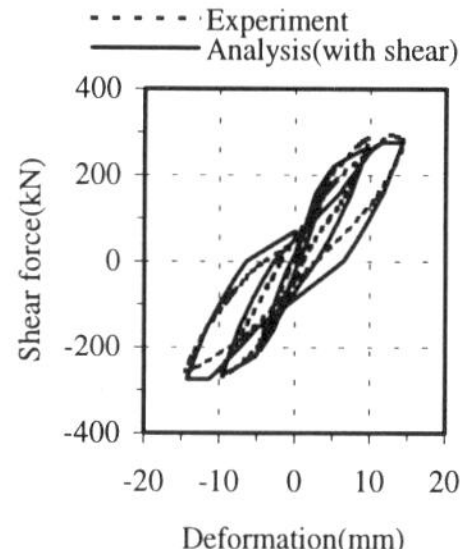

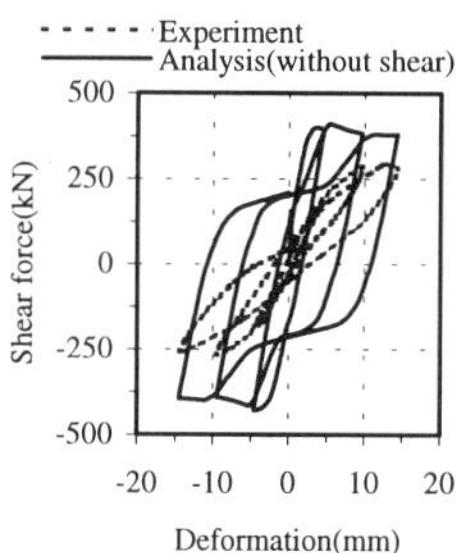

Table 1: Comparison of Force Resistance for Different Ductility

Figure 3: Comparison of the specimen 120C-U with shear

Figure 4: Comparison of the specimen 120C-U without shear

As illustrated in Figures 3 and 4, whereas analytical results for flexure only bare little resemblance to the experimental measurements, the ADAPTIC analysis with shear is adequately accurate. This emphasises that use of FE models based on the flexure-axial interaction is inadequate to predict the response of reinforced concrete members of low shear ratios. Further investigations were conducted of a real reinforced concrete bridge structure, which is Collector-Distributor 36 of I-10 Santa Monica; La Cienega-Venice Boulevard undercrossing. This structure was extensively damaged during the Northridge earthquake of January 1994 (Broderick et al, 1994), especially pier 8, for which results are shown. Figure 5 displays the analytical results of displacement time-histories with and without shear for the pier 8 under three components of the earthquake input. As illustrated in Figure 5, the analytical prediction with shear is significantly greater than that without shear, which correlates well with observed. Moreover, observations of the analytical response of the three piers supporting the Collector-Distributor structure show that shear failure would have occurred much more extensively, and only when all components of ground motion were included. Depicted in Figure 6 is the hysteretic component of shear response in which the effect of fluctuation in stiffness due to axial-shear interaction is clearly demonstrated.

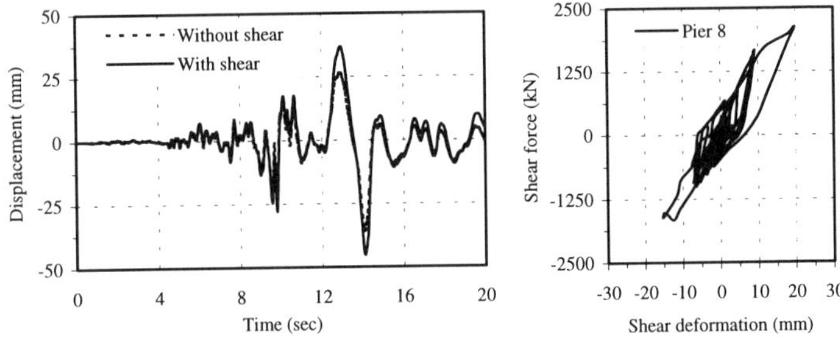

Figure 5: Displacement time histories of pier 8   Figure 6: Shear hysteresis loops of pier 8

It is clear from the above discussion, and brief presentation of results, that modelling of shear in seismic analysis is essential. It affects significantly the conclusions drawn regarding failure mode control and the success or otherwise of capacity design recommendations. In some applications, representation of shear failure only may be adequate. However, representing shear deformation and varying stiffness is necessary where shear-dominated response is anticipated.

References

Broderick et al, 1994, "The Northridge earthquake of 17 January 1994: Observation, strong-motion and correlative analyses", ESEE Report 94/4, Imperial College.

Elnashai A.S. and Papazoglou, A.J., 1997, "Procedures and Spectra for the Evaluation of Vertical Earthquake Forces", Journal of Earthquake Engineering, Vol. 1, No.1, pp.121-156.

Izzuddin, B.A. and Elnashai, A.S., 1989, "ADAPTIC A program for static and dynamic analysis of structures by adaptive mesh refinement, user manual", ESEE Report 89/7, Imperial College.

Lee, D., 1999, "Inelastic seismic analysis and behaviour of RC bridges", PhD thesis, University of London, to be submitted.

Maruyama, K., Ramirez, H. and Jirsa, J.O., 1984, "Short RC columns under bilateral load histories", ASCE, Vol. 110, No.1, pp. 120-137.

Ozcebe, G. and Saatcioglou, M., 1989, "Hysteretic shear model for reinforced concrete members", ASCE, Vol. 115, No.1, pp. 132-148

Papazoglou, A.J. and Elnashai, A.S., 1996, "Analytical and field evidence of the damaging effect of vertical earthquake ground motion", Earthquake Engineering and Structural Dynamics, Vol. 25, pp.1109-1137.

Priestley, M.J.N., Verma, R. and Xiao, Y., 1994, "Seismic shear strength of reinforced concrete columns", ASCE, 120(ST8), pp. 2310-2329.

Vecchio, F.J. and Collins, M.P., 1986, "The modified compression field theory for reinforced concrete elements subjected to shear", ACI Journal, 83(2), 219-231.

Simulation of Hysteretic Behavior
of Reinforced Concrete Members

Jeeho Lee, Filip C. Filippou and Gregory L. Fenves[1]

Abstract

A plastic-damage concrete constitutive model is used with reinforcing steel and bond-slip models for simulating the hysteretic behavior of reinforced concrete members. The numerical simulation using the present reinforced concrete model agrees well with the data from cyclic testing of a reinforced concrete column.

Introduction

The numerical simulation of damage in reinforced concrete elements is of paramount importance in the evaluation of existing structures and in the design of economic measures of strengthening of seismically deficient structures. In reinforced concrete structures the effect of shear and bond plays an important role in the damage evolution of the member and its eventual failure mode. This paper presents an analytical model of the damage process in reinforced concrete members under the substantial amount of inelastic cyclic displacements. The model consists of plastic-damage concrete elements, truss elements for reinforcing steel and bond link elements simulating the bond-slip of reinforcement. The proposed model is validated by comparing the numerical simulation of a RC column under cyclic loads with the experimental result.

Plastic-Damage Model for Concrete

In this study, cyclic concrete behavior is represented by the plastic-damage model developed by Lee and Fenves (1998a). The plastic-damage model is derived from the Barcelona model (Lubliner *et al.* 1989) by introducing two damage variables, one for tensile damage, κ_t, and the other for compressive damage, κ_c. The evolution equation for the damage variable vector, $\boldsymbol{\kappa} = [\kappa_t \ \kappa_c]^{\mathrm{T}}$, is obtained by factoring strength functions into two parts to represent the effective stress $\bar{\sigma}$ and degradation of elastic stiffness (degradation damage).

The damage index is defined based on the ratio of dissipated plastic energy to the energy capacity per unit volume of materials. The internal energy capacity is reduced as the material resists tensile cracking and compressive crushing. To maintain objective results at the structural level, the characteristic length, which is the crack bandwidth along which the energy is dissipated, is specified as a material property.

For modeling the cyclic behavior of concrete, which has very different tensile and compressive yield strengths, it is necessary to use two cohesion variables in the yield function: a tensile cohesion variable c_t, and a compressive cohesion variable c_c.

---

[1]Department of Civil and Environmental Engineering, University of California, Berkeley, CA 94720

The yield function in Lubliner *et al.* (1989), only models isotropic hardening behavior in the classical plasticity sense, is modified to include the two cohesion variables as follows:

$$F(\bar{\sigma}, \kappa) = \frac{1}{1 - \alpha}[\alpha I_1 + \sqrt{3J_2} + \beta(\kappa)\langle\hat{\sigma}_{max}\rangle] - c(\kappa) \tag{1}$$

where I_1 is the first stress invariant, J_2 is the second deviatoric stress invariant, $\hat{\sigma}_{max}$ denotes the algebraically maximum principal stress, and α is a parameter which is evaluated by the initial ratio of uniaxial compressive strength to biaxial compressive strength. The evolution of the yield function is determined by defining $\beta = (1 - \alpha)(c_c(\kappa))/(c_t(\kappa)) - (1 + \alpha)$ and the cohesion parameter $c = c_c(\kappa)$. The plastic strain ε^p is evaluated by the non-associative flow rule using a Drucker-Prager type function as the plastic potential function.

The mechanism of stiffness degradation under cyclic loading is complicated because of the opening and closing of microcracks. The crack opening/closing behavior can be modeled as elastic stiffness recovery during elastic unloading from a tensile state to a compressive state. Using a multiplicative parameter, $0 \leq s \leq 1$, on the tensile degradation variable D_t, the degradation damage variable is defined as $D = 1 - (1 - D_c(\kappa))(1 - sD_t(\kappa))$, where D_c is the compressive degradation variable. Accordingly, the total stress σ is written as:

$$\sigma = (1 - D_c(\kappa))(1 - sD_t(\kappa))\mathbf{E}_0:(\varepsilon - \varepsilon^p) \tag{2}$$

where $\mathbf{E}_0$ is the initial elastic stiffness tensor. A modified evolution relation (Lee and Fenves 1998b) is used to simulate large crack opening/closing process in the continuum context. It is assumed that the microcracks are joined to construct a discrete crack if $\kappa_t \geq \kappa_{cr}$, where κ_{cr} is an empirical value near unity. At that tensile damage level, the evolution of the plastic strain caused by the tensile damage is stopped.

Representing the softening behavior with a model based on rate-independent plasticity makes the governing initial-boundary value problem become ill-posed such that a unique solution may not exist (Needleman 1988). In the present model, a rate-independent version of the plastic-damage model is regularized by introducing a viscoplastic concept into the inelastic strain and degradation variable to ensure a well-posed structural system.

Reinforced Concrete Members

In finite element analysis a reinforced concrete member can be modeled as a composite of three components: concrete, reinforcing steel, and bond-slip link. The concrete model plays a crucial role because it must represent initiation and localization of cracking and crushing. The plastic-damage model is used to represent nonlinear concrete behavior including material hardening and softening, stiffness degradation, and stiffness recovery on crack closing. The constitutive relation is embedded in a four-node quadrilateral isoparametric element with 2-by-2 Gauss integration.

The reinforcing steel bar is modeled by truss elements. To represent steel behavior under cyclic loading, a steel model should include the constitutive relations of the isotropic and kinematic hardening, and the Baushinger effect. In this study, the stress-strain relation proposed by Filippou and *et al.* (1983), which is a modified version of the Monegotto and Pinto model (1973) to incorporate the isotropic hardening effect, is used for longitudinal and transverse steel bars.

The reinforcing steel elements are connected to the surrounding concrete elements through zero-length bond-slip link elements. Slip is assumed to take place only along the longitudinal direction of a reinforcing bar, and transverse directional slip is constrained by imposing large stiffness for that direction. A combination of multi-linear and exponential functions is used for the smooth representation of bond stress loading curves.The large amount of slip during unloading is included in the model.

Numerical Simulation of RC Column

A cyclically loaded reinforcing concrete column is tested to demonstrate performance of the present plastic-damage concrete model with the other RC members. The present RC members are implemented in the finite element program FEAP. The geometry and finite element mesh of the column are illustrated in Figures 1(a) and (b). Cyclic horizontal displacement and constant vertical force loads are applied.

Figure 2 shows the numerical simulation of cyclic load-displacement curve at the top of the column compared with the experimental response (Bousias *et al.* 1995). Four observation times are marked as A, B, C and D in Figure 2. The numerical result agrees well with the experimental data. It shows outstanding performance of the present model in reproducing the pinching effect during unloading and reloading following the substantial displacement loading path. The numerical model realistically simulates the cyclic response of the column with large inelastic displacements.

Contour plots for the tensile damage evolution at the four observation times are shown in Figure 3. In Figure 4, the simulation of crack opening and closing is demonstrated as evolution contours of the degradation variable. At time C and D the stiffness at the cracked bottom corners is partially recovered as the cracks close.

Conclusions

The proposed concrete constitutive model can realistically simulate the nonlinear load-displacement behavior of RC structures under cyclic loading. The crack opening and closing is represented through the stiffness degradation and recovery relation. The results also show the ability of the model to simulate the reduced energy dissipation capacity of reinforced concrete elements under cyclic bond deterioration.

References

Bousias, S. N., Verzeletti, G., Fardis, M. N., and Guitierrez, E. (1995). "Load-Path Effects in Column Biaxial Bending and Axial Force." *Journal of Engineering Mechanics*, ASCE, Vol. 125, No. 5, pp. 596-605.

Filippou, F. C., Popov, E. P., and Bertero, V. V. (1983). "Effects of Bond Deterioration on Hysteretic Behavior of Reinforced Concrete Joints."*EERC Report 83-19*, Earthquake Engineering Research Center, Berkeley, California.

Lee, J., and Fenves, G. L. (1998a). "Plastic-Damage Model for Cyclic Loading of Concrete Structures." *Journal of Engineering Mechanics,* ASCE, Vol. 124, No. 8, pp. 892-900.

Lee, J., and Fenves, G. L. (1998b). "A Plastic-Damage Concrete Model for Earthquake Analysis of Dams." *Earthquake Engineering and Structural Dynamics*, Vol. 27, pp. 937-956.

Lubliner, J., Oliver, J., Oller, S., and Onate, E. (1989). "A Plastic-damage Model for Concrete." *International Journal of Solids and Structures*, Vol. 25, No. 3, pp. 299-326.

Monegotto, M., and Pinto, P. E. (1973). "Method of Analysis for Cyclically Loaded Reinforced Concrete Plane Frames Including Changes in Geometry and Non-Elastic Behavior of Elements under Combined Normal Force and Bending." *Proceedings, IABSE Symposium on Resistance and Ultimate Deformability of Structures Acted on by Well Defined Repeated Loads*, Lisbon, pp. 15-22.

Needleman, A. (1988). "Material Rate Dependence and Mesh Sensitivity in Localization Problems." *Computer Methods in Applied Mechanics and Engineering*, Vol. 67, pp. 69-85.

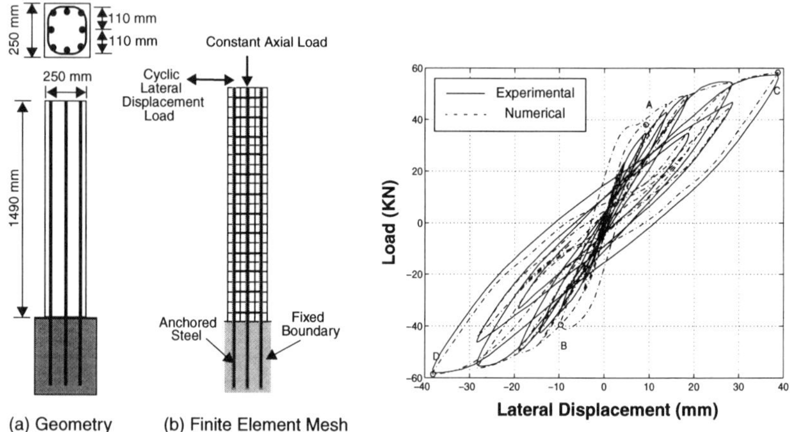

FIG. 1. Geometry and Finite Element Mesh

FIG. 2. Numerical Simulation of Experimental Load-Displacement Relationship

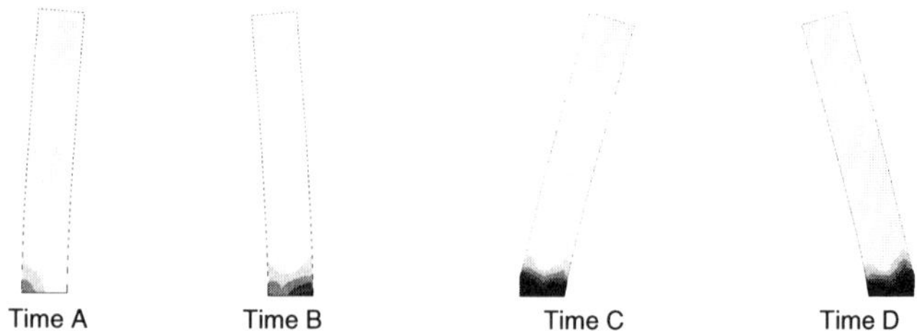

FIG. 3. Evolution of Tensile Damage Variable (κ_t) for Concrete

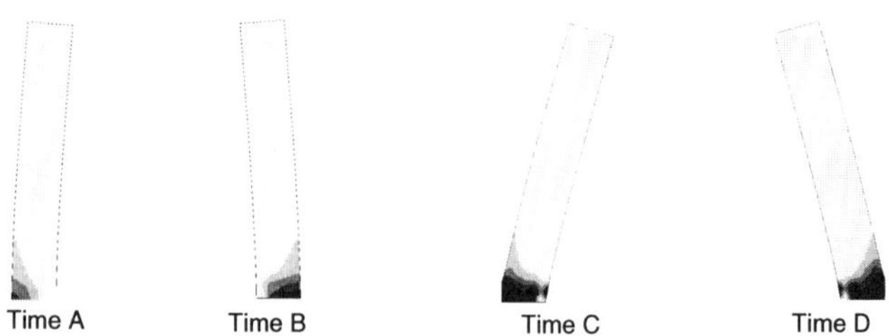

FIG. 4. Evolution of Tensile Damage Variable (D) for Concrete

Durability Scaling of HPC-structures using the
Probabilistic Crack Approach

F.-J. Ulm[1], P. Rossi[2], I. Schaller[3], D. Chauvel[4]

Abstract

The paper summarizes some results of a combined experimental-theoretical
research program on the durability of cooling towers with minimum reinforcement
made of normal and high strength concrete, when subject to hygral and thermal
gradients (Ulm et al., 1998). The study confirmed that the structural durability
performance is governed by the evaporable water content of the concrete: it scales
both the magnitude and the time scale of drying, and thus the crack opening and
its long term propagation.

Introduction: Probabilistic Crack Approach with Diffusion Couplings

Concrete cracking is a major cause of life span limitation of concrete structures. Aside from external actions, this cracking can be induced by phenomena
involving hygro-mechanical couplings. Such phenomena involve scale effects related to local or global gradients of the involved physical quantities, as well as
size effects on cracking in concrete structures. These size and scale effects are
quite important when aiming at predicting the structural durability performance
of concrete structures. The scale effects related to drying are suitably taken into
account by the nonlinear diffusion theory of evaporable water content C:

$$\frac{\partial C}{\partial t} = \mathrm{div}(D(C)\mathrm{grad}C) \tag{1}$$

with $D(C) =$ the nonlinear apparent diffusivity of concrete.

As for size effects for concrete cracking we use the probabilistic crack approach
(Rossi et al., 1996): the random distribution of constituents and initial defects is
taken into account in finite element analysis, by assigning randomly distributed

---

[1]Res. Eng., Laboratoire Central des Ponts et Chaussées, 75732 Paris Cedex 15, France;
after Jan. 1, 1999, Assoc. Prof., Massachusetts Institute of Technology, Dpt. of Civil and
Environmental Engineering, Cambridge, MA 02139; ulm@mit.edu.

[2]Res. Dir., Laboratoire Central des Ponts et Chaussées, 75732 Paris Cedex 15, France.

[3]Sen. Eng. (former LCPC), SETRA, 46, av. Aristide Briand, BP 100, 92223 Bagneux.

[4]Sen. Eng., EDF- SEPTEN, 12–14, Av. Dutriévoz, 69628 Villeurbanne Cedex, France.

203

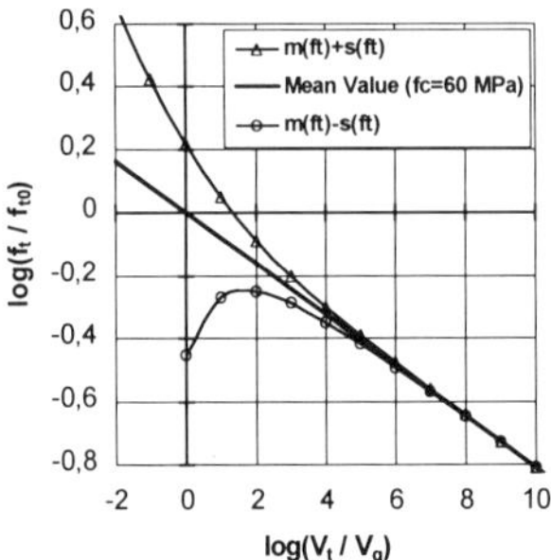

Figure 1: Probabilistic tensile strength domain.

material properties to both the solid elements and the contact elements interfacing the former (discrete crack approach). The crack-opening $[[u]]$ is governed by the strength criteria:

$$f = \max(f_\sigma, f_\tau); \quad f_\sigma = \sigma_n - f_t \leq 0; \quad f_\tau = |\tau|/k - f_t \leq 0 \tag{2}$$

where $\sigma_n = \boldsymbol{n} \cdot \boldsymbol{T}$ and $\tau = \boldsymbol{t} \cdot \boldsymbol{T}$ = the normal and the tangential components of the stress-vector $\boldsymbol{T} = \boldsymbol{\sigma} \cdot \boldsymbol{n}$ in the local base $(\boldsymbol{n}, \boldsymbol{t})$ of the contact element; $k \approx 5 =$ a constant which relates the cohesion $c = k f_t$ to the local –randomly distributed– tensile strength f_t. $\boldsymbol{\sigma}$ denotes the stress tensor which is updated according to:

$$d\boldsymbol{\sigma} = \frac{E}{1+\nu} d\boldsymbol{e} + \frac{E}{3(1-2\nu)}(d\epsilon - 3\kappa_s dC)\boldsymbol{1} \tag{3}$$

with E = randomly distributed Young's modulus, and ν = Poisson ratio assumed constant; $\boldsymbol{e} = \boldsymbol{\varepsilon} - \frac{1}{3}\epsilon\boldsymbol{1}$ = deviator part of strain tensor $\boldsymbol{\varepsilon}$, and $\epsilon = \mathrm{tr}\boldsymbol{\varepsilon}$ = average volume variation; $\kappa_s > 0$ = the uniaxial empirical coefficient of hygral dilatation.

The statistical distribution functions (mean values $m(x)$ and standard deviation $s(x)$) of tensile strength f_t and Young's modulus E were obtained experimentally (Rossi et al., 1994), and depend only on the compressive strength f_c, representing the quality of the concrete matrix, and on the volume ratio V_t/V_g of the sample V_t to the coarsest aggregate V_g, refering to the *elementary material heterogeneity*. Figure 1 shows, in a double logarithmic space, the tensile strength domain as a function of the volume ratio V_t/V_g for a concrete of compressive strength $f_c = 60$ MPa, used in the probabilistic crack approach.

The case of Cooling Towers

Cooling towers are thin reinforced concrete shells which are subject to severe loading of physical-chemical origin, in particular due to the great difference in

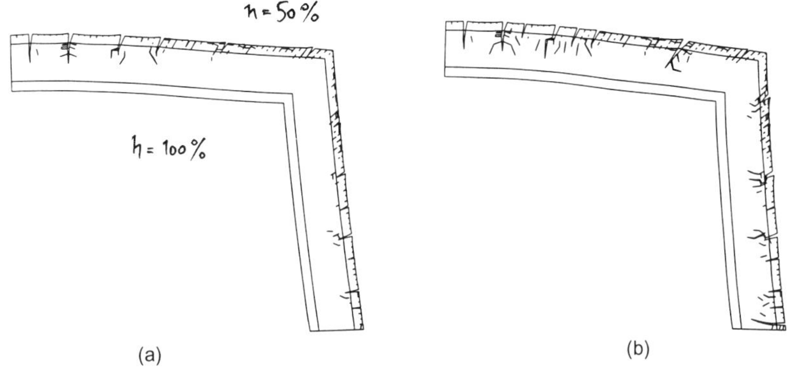

Figure 2: Probabilistic crack patterns: (a) after 1 year (b) after 12 years of drying

temperature and relative humidity between the inside and the outside. The first generation of nuclear cooling towers was designed in normal strength concrete; it is foreseen that the next one will be built in high performance concrete, provided an increase in structural durability performance. This is the aim of the durability analysis (Ulm et al., 1998). The task involved comparing the durability performance of nuclear cooling towers carried out in normal strength concrete (characteristic compressive strength $f_c = 30$ MPa) and in high strength concrete ($f_c = 60$ MPa), when subject to hygral and thermal gradients. The durability is characterized in terms of crack opening (width and depth) and crack spacing. To ensure the validity of the chosen approach, and thus the relevance of the results, a reduced 1:2-scale experiment was carried out at LCPC, simulating the hygral and thermal loading of nuclear cooling towers. The experimental data was compared with the numerical results obtained from finite element analysis, which allowed for the validation of the model. Finally, the numerical analysis was applied to the scale-1 structure. The figures 2 show two probabilistic crack patterns after 1 year and 12 years of drying.

Conclusion

The main results of this study may be summarized as follows:

1. Scale effects in hygro-mechanical problems result from the diffusion problem and from concrete cracking. Both should be taken into account in durability analysis of concrete cracking due to drying, the first scales the time scale of the crack propagation, the second the crack width, and thus the structural durability performance.

2. High strength concrete (HSC) is not equal to high performance concrete (HPC) as far as the structural durability performance with respect to drying is concerned. The governing parameter for the long-term performance of concrete

structures subject to drying is the evaporable water content. It scales the drying magnitude (and thus the drying shrinkage) and the time scale of drying governing the crack propagation. The higher the evaporable water content, the faster the crack propagation, and thus the lower the structural durability performance.

3. A higher structural durability performance in terms of maximum crack opening can be obtained by reducing the evaporable water content in the mix composition, which allows the shrinkage magnitude and the drying time to be scaled, and hence the crack propagation. In other words, "High Performance" means here a concrete with a lower w/c-ratio than the one considered in this study. In turn, we should note that the crack opening is not the only parameter governing the steel corrosion. At equal crack openings, reinforced HSC is probably more durable, as the corrosion process involves the exchange of ions through the concrete porosity (and not just through the cracks alone); the porosity decreases as the concrete strength increases.

4. More reinforcement does not necessarily imply a durable protection against cracking induced by restrained shrinkage. It seems that an optimal minimum reinforcement ratio exists that provides an optimal functioning of the steel-concrete composite during cracking due to drying. For standard design situations this minimum reinforcement should be calculated on account of the matrix quality (tensile strength including size effects) *and* governing parameters of drying relative to the mix composition of the concrete.

5. For prototype studies or for RC-structures with particular hygral and thermal loading and required durability performance, advanced numerical tools should be used, which provide the relevant information of cracking at the time and space scale of the structure (crack opening, crack spacing and crack depth), with a minimum of material parameters of clear physical significance and accessible by standard material tests. An example of such a FE-durability analysis using the probabilistic crack approach, applied to nuclear cooling towers, is shown in the paper (Ulm et al., 1998).

Acknowledgement: This work was undertaken at the Laboratoire Central des Ponts et Chaussées on behalf of Electricité de France under research contract EDF/SEPTEN ND 1494/I–II. Their financial support is gratefully acknowledged. The numerical analyses were achieved using the finite element program CESAR-LCPC.

References

Rossi P., Wu, X., and Le Maou, F. (1994) "Scale effects on concrete in tension". *Mater. & Struct.*, RILEM, 27, 437–444.

Rossi P., Ulm, F.-J., and Hachi, F. (1996) "Compressive behavior of concrete: physical mechanisms and modeling". *J. Engn. Mech.*, ASCE, 122(11), 1038–1043.

Ulm, F.-J., Rossi, P., Schaller, I., Chauvel, D. (1998) "Durability scaling of cracking in HPC-structures subject to hygral gradients". submitted to *J. Struct. Engng.*, ASCE, July 98.

A Hypoelastic Model for Cyclic Response of Concrete Structures

M. Kwon[1], E. Spacone[1] Associate Member, ASCE,
and T. Balan[2], Member, ASCE

Abstract

This paper discusses enhancements to an existing concrete constitutive model for the 3-D finite element analysis of reinforced concrete structures. The model is based on the concept of equivalent uniaxial strain that allows the use of a uniaxial constitutive law for concrete. The main features of the constitutive law are presented. The focus is on the cyclic uniaxial constitutive law and on the post-peak behavior necessary to describe the response of concrete under triaxial states of stress. Unconfined and confined compression tests on concrete specimens are used to validate the material law under different levels of confinement.

Introduction

Concrete is a complex material whose response depends heavily on the applied stress and strain histories. In particular, concrete is characterized by a brittle behavior whose post-peak response depends on the confinement level. For low confinement stresses, specimens loaded in compression show vertical cracks that eventually lead to the specimen failure. If the lateral confinement is increased, failure occurs along a shear band. Eventually, for very high confinement stresses, concrete exhibits a ductile response. Still much remains to be learned on the behavior of concrete, mainly because experimental tests on brittle materials are difficult to conduct since the post-peak softening behavior is difficult to control.

From a modeling standpoint, a number of constitutive laws have been developed in recent years based on nonlinear elastic theories or various plasticity approaches, some including damage and fracture mechanics concepts. Plasticity-based laws are definitely elegant, but they are also more complex from a theoretical standpoint and are computationally demanding due to iterations needed to find converged solutions within the constitutive driver.

---

[1] Dept. of Civil Environmental and Architectural Engineering, University Colorado, Boulder, CO 80309-0428, Tel. (303) 492-7607, Fax (303) 492-7317.
[2] Weidlinger Associates, Applied Science Division, 4410 El Camino Real #110, Los Altos, CA 94022

The model discussed herein is based on the equivalent uniaxial strain concept (Darwin and Pechnold 1977, Elwi and Murray 1979). The model builds on the 3-D law described in Balan et al. (1997) and focuses on the nonlinear cyclic uniaxial constitutive law and on the post-peak response under different levels of confinement.

Model Description

The incremental stress-strain relation of concrete with respect to the orthotropic axes is written

$$
\begin{Bmatrix} d\sigma_1 \\ d\sigma_2 \\ d\sigma_3 \\ d\tau_{12} \\ d\tau_{23} \\ d\tau_{31} \end{Bmatrix} = \frac{1}{\phi} \begin{bmatrix} E_1(1-v_{23}v_{32}) & E_1(v_{21}+v_{23}v_{31}) & E_1(v_{31}+v_{21}v_{32}) & 0 & 0 & 0 \\ E_2(v_{12}+v_{13}v_{32}) & E_2(1-v_{13}v_{31}) & E_2(v_{32}+v_{12}v_{31}) & 0 & 0 & 0 \\ E_3(v_{13}+v_{12}v_{23}) & E_1(v_{23}+v_{13}v_{21}) & E_3(1-v_{12}v_{21}) & 0 & 0 & 0 \\ 0 & 0 & 0 & G_{12}\phi & 0 & 0 \\ 0 & 0 & 0 & 0 & G_{23}\phi & 0 \\ 0 & 0 & 0 & 0 & 0 & G_{31}\phi \end{bmatrix} \begin{Bmatrix} d\varepsilon_1 \\ d\varepsilon_2 \\ d\varepsilon_3 \\ d\gamma_{12} \\ d\gamma_{23} \\ d\gamma_{31} \end{Bmatrix} \tag{1}
$$

where v_{ij} are the mixed Poisson's ratios, E_i is initial modulus of elasticity along the orthotropic axis i, G_{ij} is the shear modulus in the ij plane (Balan et al. 1997), and $\phi = 1 - v_{21}v_{12} - v_{31}v_{13} - v_{32}v_{23} - v_{12}v_{23}v_{31} - v_{21}v_{32}v_{13}$. The orthotropic axes are parallel to the current principal stress axes. Principal stress and principal strain axes are in general not coaxial. Eqn (1) can be modified in the following manner

$$
\begin{Bmatrix} d\sigma_1 \\ d\sigma_2 \\ d\sigma_3 \\ d\tau_{12} \\ d\tau_{23} \\ d\tau_{31} \end{Bmatrix} = \begin{bmatrix} E_1 & 0 & 0 & 0 & 0 & 0 \\ 0 & E_2 & 0 & 0 & 0 & 0 \\ 0 & 0 & E_3 & 0 & 0 & 0 \\ 0 & 0 & 0 & G_{12} & 0 & 0 \\ 0 & 0 & 0 & 0 & G_{23} & 0 \\ 0 & 0 & 0 & 0 & 0 & G_{31} \end{bmatrix} \begin{bmatrix} B_{11} & B_{21} & B_{31} & 0 & 0 & 0 \\ B_{12} & B_{22} & B_{32} & 0 & 0 & 0 \\ B_{13} & B_{23} & B_{33} & 0 & 0 & 0 \\ 0 & 0 & 0 & 1 & 0 & 0 \\ 0 & 0 & 0 & 0 & 1 & 0 \\ 0 & 0 & 0 & 0 & 0 & 1 \end{bmatrix} \begin{Bmatrix} d\varepsilon_1 \\ d\varepsilon_2 \\ d\varepsilon_3 \\ d\gamma_{12} \\ d\gamma_{23} \\ d\gamma_{31} \end{Bmatrix} \tag{2}
$$

where the B_{ij} terms can be obtained from Eqn. (1). The above equation can be rewritten in the form:

$$
d\sigma_i = E_i d\varepsilon_{ui} \qquad (i = 1,2,3) \tag{3}
$$

with $d\varepsilon_{ui} = B_{ij}d\varepsilon_j$. The incremental strain $d\varepsilon_{ui}$ defines the increment of the equivalent uniaxial strain ε_{ui} (Darwin and Pecknold, 1977). Eqn. (3) requires a uniaxial constitutive law. In this work, the compression envelope proposed by Saenz (1964) and modified by Bashur and Darwin (1978) is used:

$$
\sigma_i = \frac{E_o \varepsilon_{ui}}{1 + A_i\left(\dfrac{\varepsilon_{ui}}{\varepsilon_{ci}}\right) + B_i\left(\dfrac{\varepsilon_{ui}}{\varepsilon_{ci}}\right)^2 + C_i\left(\dfrac{\varepsilon_{ui}}{\varepsilon_{ci}}\right)^3} \qquad (i = 1,2,3) \tag{4}
$$

where E_0 is the initial modulus of elasticity, A, B and C are parameters that depend on the concrete strength f_{ci}', the corresponding strain ε_{ci}, and on a point on the post-peak descending branch f_{fi}, ε_{fi} (Figure 1). Three different uniaxial curves $\sigma_i - \varepsilon_{ui}$ are defined. The concrete strength in each direction is obtained by projecting a straight line from the origin to the ultimate surface through the current stress point $(\sigma_1, \sigma_2, \sigma_3)$ in the principal stress space. The ultimate curve is defined by the five-parameter surface proposed by Willam and Warnke (1975) with the modifications proposed by Balan et al. (1997). A cap model is added to avoid unrealistically high strengths for loading along or near the hydrostatic axis.

The point f_{fi}, ε_{fi} is important because it determines the slope of the softening branch. In order to account for the effect of the lateral confinement on the post-peak behavior, this point is found using:

$$\varepsilon_{fi} = 4\varepsilon_{ci} \qquad f_{fi} = f_{ci}\left(5 - \frac{f_{ci}'}{f_c'}\right), \qquad \frac{f_{ci}'}{f_c'} \le 4.3 \tag{5}$$

where f_c' is the unconfined concrete compression strength. Since f_{ci}' depends on the confinement level as determined by the ultimate curve, the descending point indirectly depends on the level of lateral confinement. The above curve was calibrated using available experimental data on concrete specimens loaded with different lateral confinement stresses (Smith et. al, 1989).

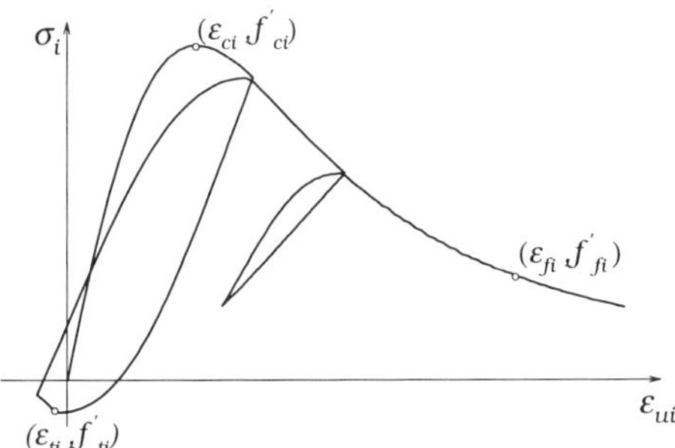

Figure 1. Cyclic uniaxial constitutive model

The original curve by Saenz (1964) was extended to describe loading in both compression and tension and unloading/reloading through simple shifting of the curve each time a load reversal occurs. This way, a unique curve described by Eqn. (4) represents the entire constitutive law.

Examples

Two confined monotonic and cyclic tests conducted by Smith et. al(1989) and Hurlbut (1985) are used to validate the modifications to the concrete model. The results in Figure 2 and Figure 3 show close agreement between experimental and analytical data. The effects of confinement on the specimen strength and post-peak behavior are well traced by the constitutive model. As the confinement stresses increase, the concrete migrates from a brittle to a ductile response (Figure 2). In Figure 3 the unloading slope of the analytical curve is smaller than in the experimental results. The experimental tests unload elastically, while a degrading stiffness was selected in the uniaxial constitutive law.

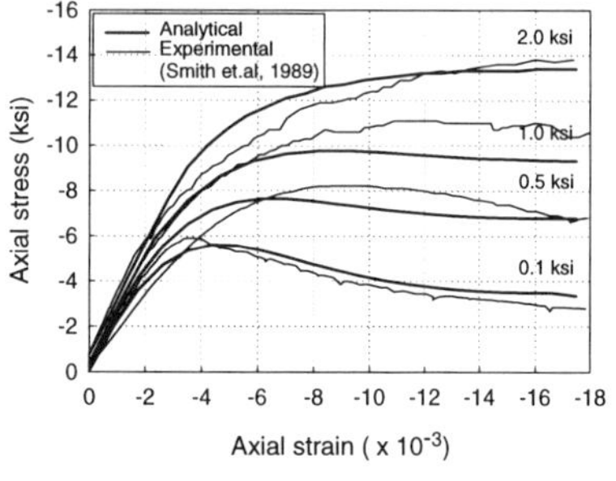

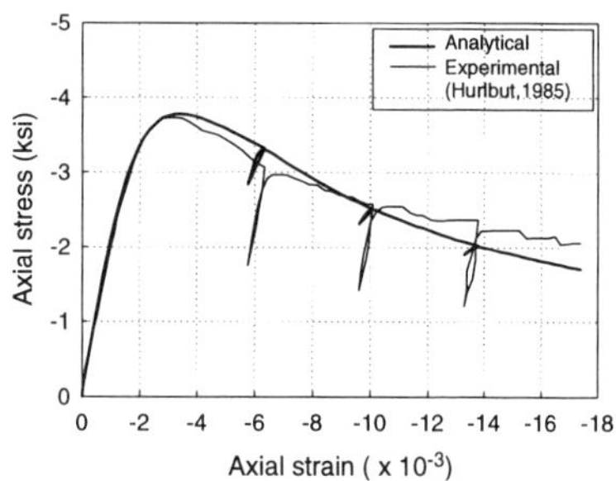

Figure 2. Confined compression tests

Figure 3. Cyclic test (σ_{conf} =0.1 ksi)

Conclusions

Recent enhancements to an existing 3D hypoelastic constitutive law for the analysis of three-dimensional concrete structures are discussed in the paper. The constitutive law is generalized to the full cyclic nonlinear range. A discussion is presented on how to account for the transition from brittle to ductile response in the post-peak branch as the lateral confinement increases. Ongoing work is addressing other issues such as Poisson ratios and dilatancy in view of the model's application to the finite element analysis of reinforced concrete structures

References

Balan, T.A., Filippou, F.C., Popov, E.P. (1997). "Constitutive model for 3D cyclic analysis of concrete structures." *Journal of Eng. Mech., ASCE*, 123(2), 143-153.

Bashur, F.K and Darwin, D.(1978). "Nonlinear biaxial law for concrete." *Journal of Struct. Eng., ASCE*, 104(1), 157-170.

Darwin, D., and Pecknold, D.A. (1977). "Nonlinear biaxial stress-strain law for concrete." *Journal of Eng. Mech., ASCE*, 103(2), 229-241.

Elwi, A.A., and Murray, D.W. (1979). "A 3D hypoelastic concrete constitutive relationship." *Journal of Eng. Mech., ASCE*, 105(4), 623-641.

Hurlbut, B.J. (1985). "Experimental and computational investigation of strain-softening in concrete." *M.S Thesis*, CEAE Dept., Univ. of Colorado, Boulder.

Saenz, L.P. (1964). " Discussion of 'Equation for the stress-strain curve of concrete' by P. Desayi and S. Krishan." *ACI Journal*, 61(2), 195-211.

Smith, S.S., Willam, K.J., Gestle, K.H., and Sture, S. (1989). "Concrete over the top, or: is there life after peak?", *ACI Journal*, 86(5), 491-497.

Willam, K.J., and Warnke, E. P. (1975). "Constitutive model for the triaxial behavior of concrete." *IABSE Proceedings*, 19, 1-3.

Constitutive Modelling of Concrete
in Numerical Simulation of Anchoring Technology

Jürgen Nienstedt[1], Richard Mattner[1], Johannes Wiesbaum[1]

Abstract

Developing new and innovative products require an in-depth understanding of the physical processes forming the basis of the anchoring mechanisms. One essential ingredient to simulate fastening systems utilizing anchors is a sophisticated material model of the base material, i.e. the concrete. The practicability of the utilized model based on the smeared crack concept is shown for a HKD flush anchor.

Introduction

The development process for new and innovative products in fastening technology has at least two essential requirements. The first one is a sophisticated material model. A realistic description of the physical processes is the basis to get an insight into the structural behavior. The second one is a numerically stable model which is easy to handle for the development engineer and applicable to all anchoring systems.

The material model implemented in an implicit finite element code, developed at Hilti, is an uniaxial model based on the smeared crack approach. The softening behavior in the tension regime is driven by the fracture energy. The compressive domain is described by a nonlinear hardening behavior.

The applicability of the model described is shown by the simulation of the setting procedure and the pull-out of a HKD flush anchor. This anchor is modelled axisymmetrically and as a three dimensional structure.

[1] Hilti AG, Corporate Research, FL-9494 Schaan, Principality of Liechtenstein

Constitutive modelling

The failure of an anchoring system is often determined by the tension domain, i.e. cracking of the concrete material. Hence the constitutive modelling described by a stress/strain dependence is of main interest when simulating the structural behavior.

The constitutive modelling of concrete material is done with a uniaxial model. In the tension domain this model is based on the well known smeared crack approach. The softening behaviour of the rotating smeared crack model is defined in terms of fracture energy. The qualitative shape of this softening branch is a bilinear curve.

The stress/strain relation is completed in compression domain with a nonlinear hardening function up to the compressive strength of the concrete material. The postpeak behaviour, in general the softening behaviour, is a function of the actual stress state, here described by the first stress invariant.

Numerical simulation

The numerical example presented in this chapter is a Hilti HKD flush anchor. This anchor is modelled both axisymmetrically, as well as a three dimensional structure. Both relevant loading cases, the setting of the anchor and the pull-out have been considered. Further applications for different anchors (axisymmetrical modelling) can be seen in Jussel et al. (1994) and Nienstedt et al. (1995).

Axisymmetrical modelling

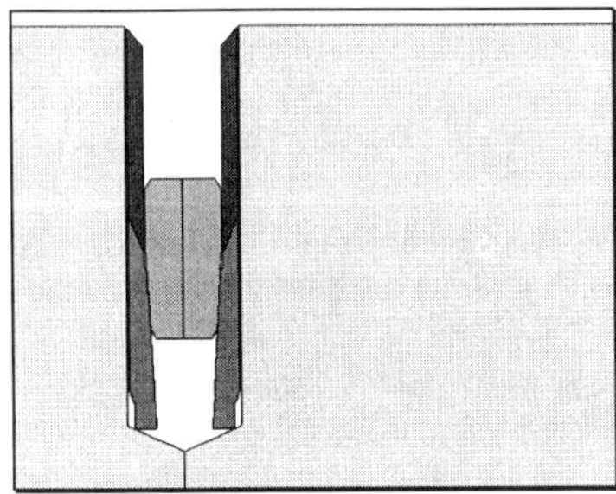 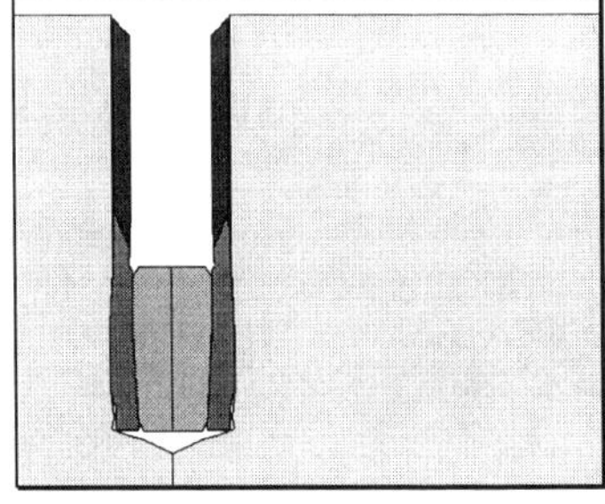

Fig. 1 Axisymmetrical modelling   Fig. 2 Displacement at the end
         of the HKD flush anchor   of the setting procedure

The geometry of the anchor and a part of the base material is shown in fig.1 for the axisymmetric model. The anchor consists of a cone and an expansion sleeve. The setting procedure of the anchor is performed by a hammering of the cone until the end of the expansion sleeve reaches the bottom of the borehole (s. fig. 2). This is done utilizing a special setting tool.

The stress state occuring in the base material at the end of the setting procedure is shown in fig. 3. Here the minimum principal stresses can be seen using a logarithmic scale. The main loading effects are caused by compression and a circumferential tension at the bottom of the borehole

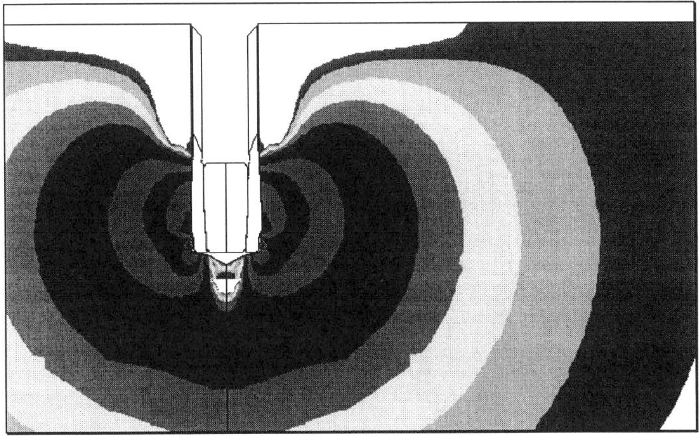

Fig. 3 Minimum principal stresses in the base material

Three dimensional modelling

Considering the existing symmetry of the flush anchor set in the middle of a concrete plate it is sufficient to model one quarter of the complete structure. Fig. 4 shows the anchor and a part of the base material at the initial state.

The three dimensional modelling allows the calculation of effects like e.g. influence of edge distances, distances between axes of multiple anchor fastenings. Fig. 5 shows the beginning of radial cracking during the pull-out simulation of the anchor. In fig.5 the base material is reduced by damaged or partly damaged elements.

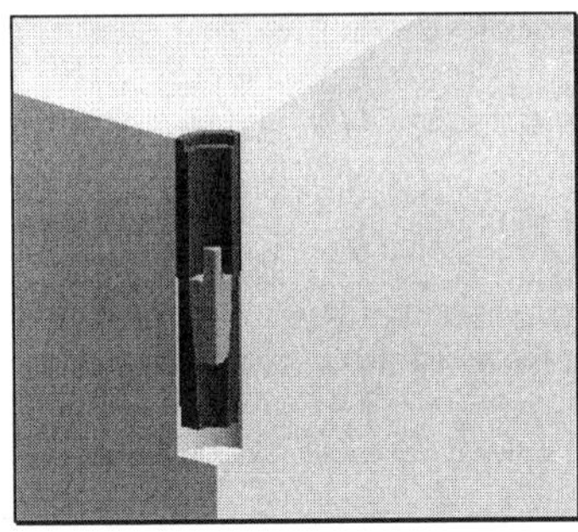 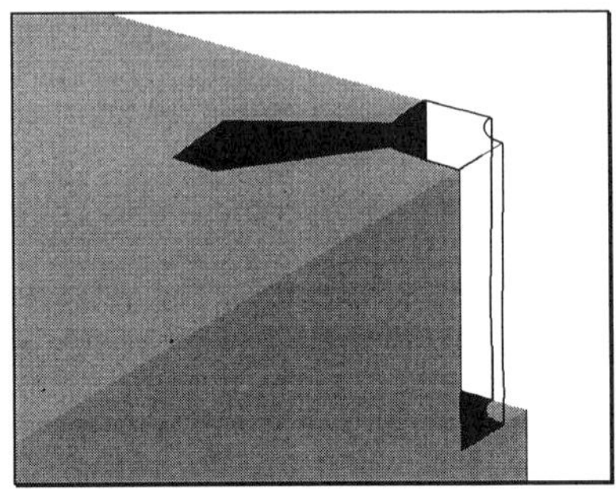

Fig. 4 Three dimensional modelling   Fig. 5 Damaged elements during
        of the HKD flush anchor   pull-out loading

Summary

The described uniaxial material model for concrete based on the smeared crack approach proved to be a suitable tool when analysing anchoring systems. The implementation of this model in a finite element program provides the basis for the understanding of the physical processes guaranteeing the performance of the fastening.

Acknowledgement

The authors would like to thank Prof. Dr. W.J. Huppmann, head of Corporate Research, Hilti AG, for his support and continuous interest in this research project.

References

Jussel, P., Wall, F.J. and Bourgund, U. (1994): „Application of axisymmetric finite element analysis for anchors in concrete", in Computational modelling of structures (eds H. Mang, N. Bicanic and R. de Borst), Pineridge Press, Swansea, U.K., 1047-1056.

Nienstedt, J. and Dietrich, C. (1995): „Application of the finite element method to anchoring technology in concrete", in Fracture mechanics of concrete structures (ed F.H. Wittmann), Aedificatio Publishers, Freiburg/Breisgau, Germany, 1909-1914.

Evaluation of a Rate-Sensitive Material Model for Concrete

Hong D. Kang[1], Kaspar J. Willam[2] and Yunping Xi[3]

Abstract

Unlike the performance under static (infinitely slow) loading, increasing loading rates normally enhance the strength and reduce the ductility of engineering materials. In concrete the failure properties, i.e. the triaxial strength and the failure mode, depend not only on the load path, but also on the loading rate especially at high speed impact. In this paper an effort is undertaken to evaluate the performance of a comprehensive triaxial concrete model and its extension from rate-independent elastoplasticity to rate-dependent viscoplasticity. To this end the Duvaut-Lions overstress formulation is adopted to explore the dynamic strength enhancement in tension, shear and compression in terms of the underlying viscosity or the equivalent relaxation time.

Introduction

The loading rate mobilizes the micromechanical features of concrete materials in different ways because of the fundamental difference of the tangential and normal interface bond among aggregate particles and the hardened cement paste. Thereby, the heterogeneity of the different mass densities makes it difficult to separate inertia from material rate effects at very high loading rates. In spite of these complications, the different failure mechanisms of particle interaction and the interface layers determine the surface roughness which is a measure of fracture energy release during failure.

The large number of material models for rate effects may be classified into viscoelastic without, and viscoplastic with an explicit loading condition. Among the latter the Perzyna-model (1966) is the most prominent extension of classical Bingham viscoplasticity, while the formulation by Duvaut-Lions (1972) and the 'fully consistent' viscoplastic overstress formulation by Wang et al. (1997) and

---

[1] Postdoctoral Fellow of CEAE Dept. in University of Colorado, Boulder CO 80309.
[2] Professor of CEAE Dept. in University of Colorado, Boulder CO 80309.
[3] Assistant Professor of CEAE Dept. in University of Colorado, Boulder CO 80309.

Etse et al. (1997) have been recently proposed to improve the asymptotic behavior of viscoplastic formulations.

To compare the predictions of rate independent with rate dependent concrete behavior under different loading rates the triaxial elastoplastic concrete formulation by Kang (1997) is adopted which has been implemented in the 3-D FE program FEAP. Thereby, a single finite element is used to study the difference of viscous effects in uniaxial tension, shear, and uniaxial compression. For shear loading, the equibiaxial tension-compression test (T-C) is used to evaluate the dynamic strength enhancement of concrete subject to in-plane shearing under plane stress. A literature survey shows that the increase of strength of concrete is not very significant for loading rates up to $\dot{\epsilon} = 1 \times 10^{-2}\,[mm/mm/sec]$. However, for loading rates higher than $\dot{\epsilon} = 1 \times 10^{-1}\,[mm/mm/sec]$ the dynamic strength enhancement may be very significant.

Viscoplastic Formulation of Concrete Model

The concrete failure criterion by Kang (1997) delimits the triaxial strength in stress space and describes the loading surface for isotropic hardening/softening in a smooth fashion. The curvilinear loading surface $F(\boldsymbol{\sigma}, \boldsymbol{q}) = 0$ which is C^1-continuous except at the apex in equi-triaxial tension, was calibrated with the aid of conventional triaxial compression tests, Willam et al. (1986).

Analogous to infinitesimal elastoplasticity, the total strain rate is decomposed into an elastic and into a viscoplastic part, $\dot{\boldsymbol{\epsilon}} = \dot{\boldsymbol{\epsilon}}_e + \dot{\boldsymbol{\epsilon}}_{vp}$. Considering linear elasticity it follows that

$$\dot{\boldsymbol{\sigma}} = \boldsymbol{\mathcal{E}} : [\dot{\boldsymbol{\epsilon}} - \dot{\boldsymbol{\epsilon}}_{vp}] \tag{1}$$

In the Duvaut-Lions formulation the viscoplastic strain rate and the rate of state variables are defined in the form of the linear overstress model

$$\dot{\boldsymbol{\epsilon}}_{vp} = \frac{1}{\tau}\boldsymbol{\mathcal{E}}^{-1} : [\boldsymbol{\sigma} - \bar{\boldsymbol{\sigma}}]; \qquad \dot{\boldsymbol{q}} = \frac{1}{\tau}[\boldsymbol{q} - \bar{\boldsymbol{q}}] \tag{2}$$

whereby τ designates the relaxation time, and $(\bar{\boldsymbol{\sigma}}, \bar{\boldsymbol{q}})$ stand for the 'backbone' stress and the set of internal state variables associated with the elastoplastic problem. Thus, the differential stress-strain relationship (1) expands into the linear overstress format

$$\dot{\boldsymbol{\sigma}} = \boldsymbol{\mathcal{E}} : \dot{\boldsymbol{\epsilon}} - \frac{1}{\tau}[\boldsymbol{\sigma} - \bar{\boldsymbol{\sigma}}] \tag{3}$$

The backward Euler strategy of time integration transforms the differential equation into algebraic form and advances the solution in the time step $\Delta t = t_{n+1} - t_n$:

$$\Delta\boldsymbol{\sigma}_{n+1} = \boldsymbol{\mathcal{E}} : \Delta\boldsymbol{\epsilon}_{n+1} - \frac{\Delta t}{\tau}[\boldsymbol{\sigma}_{n+1} - \bar{\boldsymbol{\sigma}}_{n+1}] \tag{4}$$

Tangential linearization for the iterative Newton-Raphson solution of the nonlinear Backward Euler equations leads to the algorithmic elastic-viscoplastic tangent operator of Duvaut-Lions,

$$\left[\boldsymbol{\mathcal{E}}_{vp}^{alg}\right]^{DL} = \frac{d\boldsymbol{\sigma}}{d\boldsymbol{\epsilon}} = \left[\frac{\tau}{\tau + \Delta t}\boldsymbol{\mathcal{E}} + \frac{\Delta t}{\tau + \Delta t}\boldsymbol{\mathcal{E}}_{ep}^{alg}\right] \tag{5}$$

whereby the subscript $t = t_{n+1}$ is omitted for the sake of clarity. Note, this expression involves the algorithmic elastic-plastic tangent operator $\mathcal{E}_{ep}^{alg}$ which determines the backbone stress $\bar{\sigma}_{n+1}$. Moreover, the limiting condition $\frac{\Delta t}{\tau} \to 0$ results in instantaneous elasticity $d\sigma = \mathcal{E} : d\epsilon$, while $\frac{\Delta t}{\tau} \to \infty$ results in instantaneous elastoplasticity $d\sigma = \mathcal{E}_{ep}^{alg} : d\epsilon$.

Performance of Viscoplastic Concrete Model

The rate-dependence of the viscoplastic concrete formulation exhibits localization properties which are shown in Figures 1, 3 and 5. The three diagrams illustrate the variation of the normalized algorithmic localization indicator $\det(Q_{vp})/\det(Q_e)$ of the viscoplastic Duvaut-Lions model as a function of the angle of inclination θ between the normal to the discontinuity surface with the axis of minor stress. In the case of inviscid elastoplasticity, the underlying tangent stiffness exhibits localization in uniaxial tension and shear (T-C), but not in uniaxial compression. In contrast, the rate-dependent viscoplastic results in Figures 1, 3 and 5 exhibit regular behavior $(\det(Q_{vp}) > 0)$ suppressing localization even in tension and shear (T-C). The figures demonstrate that the parameters τ and Δt, which control the amount of inelastic degradation, result in large differences along the three load paths. Note, localization analyses in this study were performed at the elastoplastic localization point.

The results in Figures 2, 4, and 6 illustrate the dynamic strength enhancement when the Duvaut-Lions extension of the triaxial concrete model is subject to uniaxial tension, shear (T-C) and uniaxial compression at different loading rates. The tensile response in Figure 2 illustrates a modest strength enhancement with increasing strain rate $\dot{\epsilon} = 10^{-3}\,to\,10^{-1}$. However at $\dot{\epsilon} = 10^0$ to 10^1 for $\tau = 10^{-1}\,to\,10^{-4}\,sec$, it shows a drastic increase of 3-4 times the static tensile strength. The shear (T-C) results in Figure 4 illustrate a similar increase of shear strength however at a lower rate of loading, $\dot{\epsilon} = 10^{-3}$ to 10^{-2} for $\tau = 10^0 sec$. Finally the results of uniaxial compression in Figure 6 exhibit only a very modest increase of 8 % in compressive strength at $\dot{\epsilon} = 10^{-1}$ to 10^0 for $\tau = 10^{-1} sec$. Unlike tension and shear(T-C), very fast loading in uniaxial compression causes numerical convergence problems of the Duvaut-Lions model due to the high degree of nonlinearity enforcing lateral dilatation.

Concluding Remarks

The viscoplastic extension regularizes the localization properties of the non-associated strain softening concrete model and retrofits the loss of ellipticity of rate-independent inviscid material descriptions. At the same time the rate-sensitivity of the simple Duvaut-Lions overstress model enhances the static strength properties much more in tension and shear than in compression.

Acknowledgments

The authors wish to thank Maj. Mike Chipley and AFOSR, for the partial support of this effort under grant F 49620-98-1-0159 to the University of Colorado, Boulder.

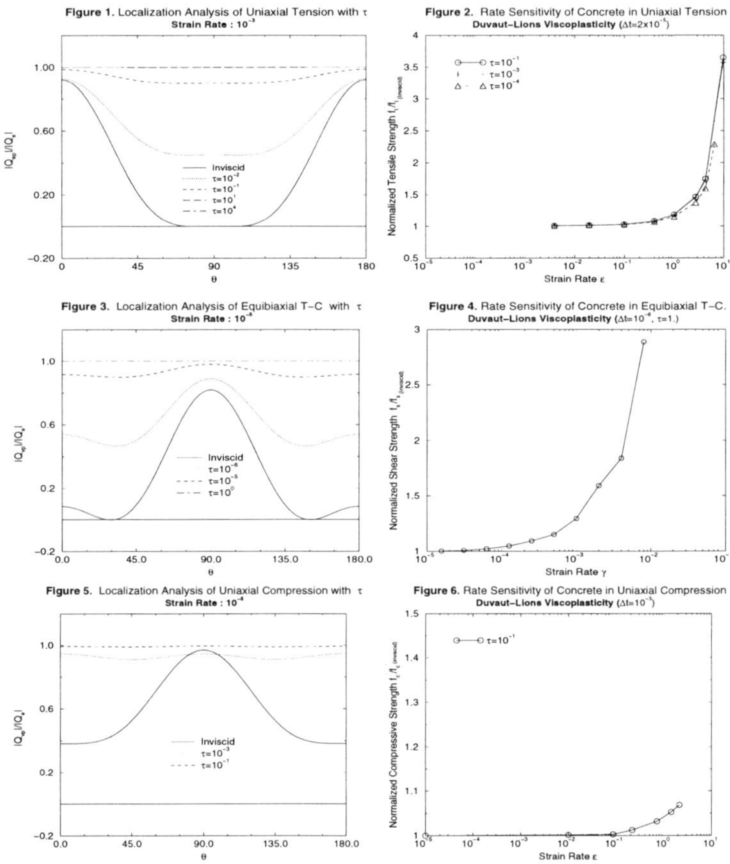

Figure 1. Localization Analysis of Uniaxial Tension with τ Strain Rate : 10^{-3}

Figure 2. Rate Sensitivity of Concrete in Uniaxial Tension Duvaut–Lions Viscoplasticity ($\Delta t=2\times10^{-5}$)

Figure 3. Localization Analysis of Equibiaxial T–C with τ Strain Rate : 10^{-5}

Figure 4. Rate Sensitivity of Concrete in Equibiaxial T–C. Duvaut–Lions Viscoplasticity ($\Delta t=10^{-4}$, $\tau=1$.)

Figure 5. Localization Analysis of Uniaxial Compression with τ Strain Rate : 10^{-4}

Figure 6. Rate Sensitivity of Concrete in Uniaxial Compression Duvaut–Lions Viscoplasticity ($\Delta t=10^{-3}$)

References

Duvaut, G. and Lions, J.L. (1972) *Les Inéquations en Méchanique et en Physique*, Dunod, Paris, France.

Etse, G., Carosio, A., and Willam, K. (1997) "Limit Point and Localization Analysis of Elastoviscoplastic Materials Models", *Int. J. Mech. Coh.-Frict. Matls.*, submitted for publication.

Kang, H.D., (1997) *Triaxial Constitutive Model for Plain and Reinforced Concrete Behavior*, Ph.D. Dissertation, University of Colorado Boulder.

Perzyna, P. (1966) "Fundamental Problems in Viscoplasticity." *Advances in Applied Mechanics*, Academic Press, New York, N.Y., 9, 244-368.

Wang, M.W., Sluys, L.J., and de Borst, R. (1997) "Viscoplasticity for Instabilities due to Strain Softening and Strain-Rate Softening." *I. J. Num. Meth. Eng.*, 40, 3839-3864.

Willam, K., Hurlbut, B., and Sture, S. (1986) "Experimental and Constitutive Aspects of Concrete Failure" *Proc. Japan-US Symp. on Finite Element Analysis of Reinforced Concrete Structures*, C. Meyer and H. Okamura (eds.), ASCE, New York, 226-254.

Nonlinear Analysis of Composite Steel–Concrete Structures

A. Haufe[1], H. Menrath[1], E. Ramm[1]

Abstract

An efficient model for the material nonlinear behaviour of composite steel–concrete structures is presented. In a first approach a two–dimensional discretization is used for the implementation of the 3D constitutive description of each material or their interaction. Within the framework of linear kinematics the special properties of concrete are presented in this paper. As a low number of parameters was a strong design criterion for the approach presented all of them can be identified using the EC2– and EC4–standards or the CEB–FIP model code [3]. The numerical simulation of 'classical' steel–concrete beams, which are composed of an upper concrete slab and a lower steel beam, connected at the interface by studs transmitting shear and normal forces, is mainly adressed.

1   General approach

Rate–independent elastoplasticity for small strains is applied individually to each of the three material components (steel, concrete and interface layer). The numerical analysis based on a *Return–Backward–Euler* algorithm for multisurface yield functions is enhanced by consistent linearization of the stress–strain relation within the framework of a Newton–Raphson scheme on the structural level [8]. For the *steel beams* classical von Mises–elastoplasticity with linear kinematic and isotropic work–hardening is applied (see e. g. [12]) while the algorithmic constitutive law for the *reinforced–concrete* combines several features commonly accepted for modelling plain concrete, reinforcement and tension–stiffening as so–called interaction stress contribution.

The concrete is modelled by softening plasticity with the fracture–energy G_f as controlling parameter. Feenstra's 2D–Model ([5]–[7]) is expanded to a 3D–multisurface yield criterion wherein four parameters describe the combined yield function with two Drucker–Prager regions and a spherical cap. Due to the underlying approach mesh–independent results can be obtained with respect to the postcritical regime.

---

[1]*Institute of Structural Mechanics, University of Stuttgart, Germany*

For the reinforcement perfect bond is assumed; tension–stiffening is considered as an additional stress in rebar direction. The constitutive law for reinforcement is based on an elastoplastic model with hardening which is described in [5]. An interface layer is introduced to model the bond behaviour and the geometrical discontinuities in a smeared approach using a Coulomb friction law [11]. A more detailed discussion of the latter subjects can also be found in [10].

2   Constitutive model for concrete

The first Drucker–Prager yield function Φ_1 controls the tensile region and the mixed parts in the principal stress space, whereas the second one Φ_2 and the spherical cap Φ_4 control the compressive region (Fig. 1a). Stress–strain relationships are approximated by an exponential function for the tensile region that takes the rebar diameter into account and a compression–softening model defined by a biparabolic function for the equivalent values of stress and strain, respectively (Fig. 1b).

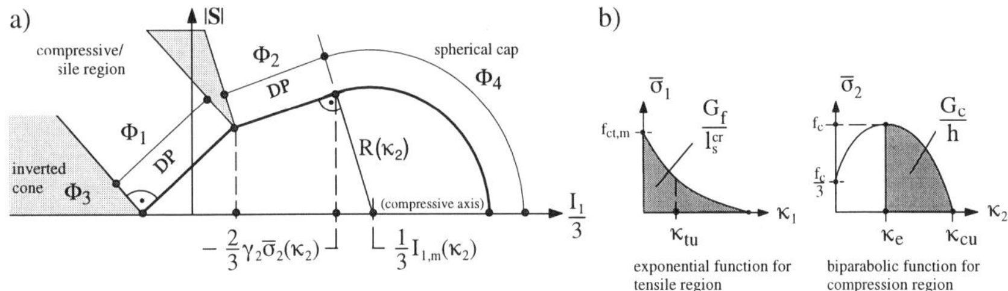

Fig. 1: a) Concrete yield surface in invariant stress space; b) Equivalent stress–equivalent strain diagrams

The total stress $\boldsymbol{\sigma}$ in a cracked state can be defined as sum of the stress contributions of plain concrete $\boldsymbol{\sigma}_c$, reinforcement $\boldsymbol{\sigma}_s$ and additional stress due to tension stiffening $\boldsymbol{\sigma}_{ia}$ (1). Assuming small strains the strain rate can be divided into an elastic, reversible part $\dot{\boldsymbol{\epsilon}}^e$ and an inelastic part $\dot{\boldsymbol{\epsilon}}^{pl}$ (1).

$$\boldsymbol{\sigma} = \boldsymbol{\sigma}_c + \boldsymbol{\sigma}_s + \boldsymbol{\sigma}_{ia} \qquad \dot{\boldsymbol{\epsilon}} = \dot{\boldsymbol{\epsilon}}^e + \dot{\boldsymbol{\epsilon}}^{pl} \qquad \dot{\boldsymbol{\epsilon}}^{pl} = \sum_{i=1}^{n} \dot{\lambda}_i \, \partial_\sigma g_i \qquad \dot{\lambda}_i \geq 0 \qquad (1)$$

Associated plasticity is assumed, i. e. the plastic potential functions g_i are identical to the yield functions Φ_i. The evolution of the inelastic part is evaluated by Koiter's rule [9], i. e. according to (1) each active yield function Φ_i contributes to the plastic strain rate $\dot{\boldsymbol{\epsilon}}^{pl}$. The loading/unloading condition is given by the standard Kuhn–Tucker conditions.

The rate independent yield function depends on the first invariant I_1 of the stress tensor $\boldsymbol{\sigma}$, the norm of the deviatoric stress tensor $\mathbf{S}$ and internal parameters κ_i describing an isotropic hardening/softening behaviour. In order to reduce the number of discontinuities the tensile stress region and the mixed parts in the principal stress space are controlled by only one Drucker–Prager yield function Φ_1.

$$\phi_i(\mathbf{S}, I_1, \kappa_i) = |\mathbf{S}| + \alpha_i I_1 - \sqrt{\frac{2}{3}}\, \bar{f}_i(\kappa_i); \quad \bar{f}_i(\kappa_i) = \beta_i \bar{\sigma}_i(\kappa_i); \qquad\qquad i = 1,2 \tag{2}$$

$$\text{with parameters for } \phi_1: \qquad \alpha_1 = \sqrt{\frac{2}{3}}\,\frac{\gamma_1 f_c - f_{ct,m}}{\gamma_1 f_c + f_{ct,m}}; \quad \beta_1 = \frac{2\,\gamma_1\,f_c}{\gamma_1 f_c + f_{ct,m}} \tag{3}$$

$$\text{and for } \phi_2: \qquad \alpha_2 = \sqrt{\frac{2}{3}}\,\frac{\gamma_2 - 1}{2\gamma_2 - 1}; \quad \beta_2 = \frac{\gamma_2}{2\gamma_2 - 1} \tag{4}$$

The factors α_i and β_i in (2) depend on four parameters, namely the uniaxial tensile stress $f_{ct,m}$, the uniaxial compressive stress f_c and two fitting parameters γ_1 and γ_2 defining the contour of the failure surface. They are calibrated by experimental data of Kupfer and Gerstle [10]. The obvious problem of the singularity at the apex of Φ_1 is overcome by defining a yield function Φ_3 as an inverted cone and assuming of the plastic potential $g_3 = \Phi_1$. The yield function for the spherical cap Φ_4 (5), defined by the midpoint $I_{1,m}$ and the radius R, is chosen to provide a C_1–continuous surface with Φ_2.

$$\phi_4(\mathbf{S}, I_1, \kappa_2) = \sqrt{|\mathbf{S}|^2 + \frac{1}{9}(I_1 - I_{1,m}(\kappa_2))^2} - R(\kappa_2) \tag{5}$$

$$\text{with:} \qquad I_{1,m}(\kappa_2) = -(\sqrt{54}\,\alpha_2 + 2)\,\gamma_2\,\bar{\sigma}_2(\kappa_2) \qquad R(\kappa_2) = \sqrt{\frac{2}{3} + 6\alpha_2^2}\,\gamma_2\,\bar{\sigma}_2(\kappa_2)$$

The nonlinear behaviour of concrete due to cracking and crushing can be represented by the internal parameters κ_1 in tension and κ_2 in compression. Assuming that the equivalent stresses are related to energy dissipation via a work–hardening hypothesis, the evolution of the internal variables can be determined based on an uncoupled damage behaviour

$$\dot{W}_i^{pl} = \boldsymbol{\sigma}^T \lambda_i\, \partial_\sigma g_i \equiv \bar{\sigma}(\kappa_i)\,\dot{\kappa}_i \qquad \text{where} \qquad \dot{\lambda}_i = \dot{\kappa}_i\ . \tag{6},(7)$$

The maximum dissipated energy in tension and compression, respectively, is given by

$$G_f = l_s^{cr} \int_{t=0}^{t=\infty} \dot{W}_1^{pl} d\tau \quad \text{and} \quad G_c = h \int_{t=0}^{t=\infty} \dot{W}_2^{pl} d\tau, \quad \text{where} \quad l_s^{cr} = \min\{l_s, h\} \tag{8},(9),(10)$$

is the effective average crack spacing.

The equivalent stress–strain relations are defined as follows (Fig. 1b; see also [7]).

$$\bar{\sigma}_1(\kappa_1) = f_{ct,m}\exp(-\frac{\kappa_1}{\kappa_{tu}}) \qquad\qquad \text{where:} \quad \kappa_{tu} = \frac{G_f}{l_s^{cr} f_{ct,m}} \tag{11}$$

$$\bar{\sigma}_2(\kappa_2) = \begin{cases} \dfrac{f_c}{3}\,(1 + 4\dfrac{\kappa_2}{\kappa_e} - 2(\dfrac{\kappa_2}{\kappa_e})^2) & \text{if } \kappa_2 < \kappa_e \\[2mm] f_c\,(1 - (\dfrac{\kappa_2 - \kappa_e}{\kappa_{cu} - \kappa_e})^2) & \text{if } \kappa_e \le \kappa_2 < \kappa_{cu} \end{cases} \quad \begin{array}{l} \text{where:} \quad \kappa_e = \dfrac{4\,f_c}{3\,E_c} \\[3mm] \kappa_{cu} = \dfrac{3\,G_c}{2\,h\,f_c} + \kappa_e \end{array} \tag{12}$$

3   Example

Composite steel–concrete beams with an opening in the web region have been tested at the University of Kaiserslautern [1]. Beam *A2* of the test series (Fig. 2) has been analyzed using the proposed material model.

Since it was already shown in [2],[10] that perfect bond between steel beam and concrete slab leads to an unacceptable stiff result, the used stud connectors had to be discretized by the aforementioned interface layer. A parameter study was performed to calibrate the interface parameters according to a given load–slip curve. The results are given in Fig. 2. Especially the pre–peak performance and the maximum load show good agreement with the experiment.

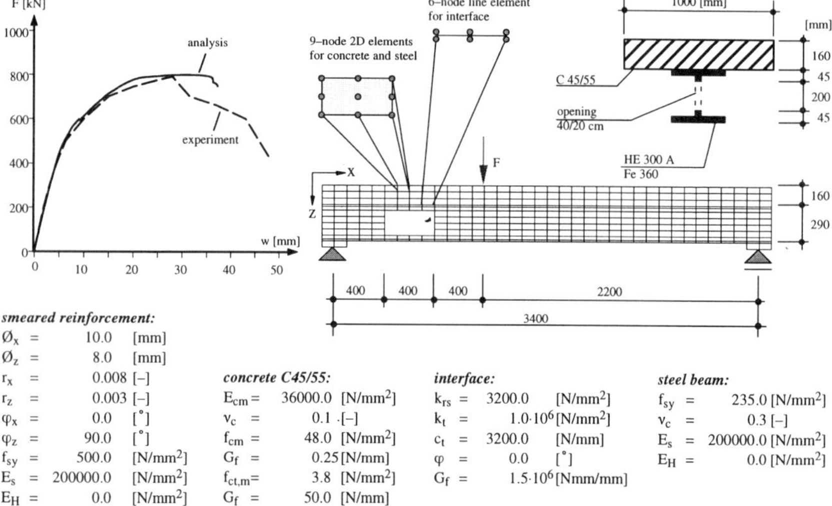

smeared reinforcement:

$\varnothing_x$	=	10.0	[mm]
$\varnothing_z$	=	8.0	[mm]
r_x	=	0.008	[–]
r_z	=	0.003	[–]
φ_x	=	0.0	[°]
φ_z	=	90.0	[°]
f_{sy}	=	500.0	[N/mm²]
E_s	=	200000.0	[N/mm²]
E_H	=	0.0	[N/mm²]

concrete C45/55:

E_{cm}	=	36000.0	[N/mm²]
ν_c	=	0.1 ·[–]	
f_{cm}	=	48.0	[N/mm²]
G_f	=	0.25	[N/mm]
$f_{ct,m}$	=	3.8	[N/mm²]
G_f	=	50.0	[N/mm]

interface:

k_{rs}	=	3200.0	[N/mm²]
k_t	=	$1.0 \cdot 10^6$	[N/mm²]
c_t	=	3200.0	[N/mm]
φ	=	0.0	[°]
G_f	=	$1.5 \cdot 10^6$	[Nmm/mm]

steel beam:

f_{sy}	=	235.0	[N/mm²]
ν_c	=	0.3	[–]
E_s	=	200000.0	[N/mm²]
E_H	=	0.0	[N/mm²]

Fig. 2: Load–displacement curve, discretization, loading and material parameters of beam *A2*

4   Conclusions

The present approach represents a simple but effective and robust method to model composite steel–concrete structures. All involved constitutive laws are based on classical plasticity using fracture energy to define softening/hardening characteristics. The number of parameters has been kept as low as possible. The physical parameters for concrete (f_c, $f_{ct,m}$, G_f, G_c) can be derived directly from EC2–standard or CEB–FIP rules.

The Coulomb friction law embedded for the interface elements uses three parameters, the friction angle φ, the cohesion c_t and the fracture energy G_f, which again must be verified by experimental results.

Advantageous to the used material models is the fact that they can be extended in a straight forward manner to describe a more sophisticated material behaviour, e. g. triaxial compression, simple shear and modification of the yield function in the π–plane. The present research concentrates on extending the formulation to plate and shell–type problems.

References

[1] Bode, H., Stengel, J., Künzel, R.: *'Stahlverbundträger mit großen Stegausschnitten'*, Stahlbau, Vol. 63/1 and 2, Ernst & Sohn (1994).

[2] Bode, H., Ramm, W., Zhou, D.: *'Physikalisch nichtlineare Berechnungen von Verbundträgern mit Stegöffnung mit dem Programmpaket ANSYS'*, Fachtagung Verbundkonstruktionen, Juli 1997, DFG–Forschergruppe Verbundbau, Universität Kaiserslautern (1997).

[3] CEB–FIP – Model code 1990, Bulletin d'information CEB.

[4] Eurocode 2 (1992): *'Planung von Stahlbeton– und Spannbetontragwerken'*, DIN V 18932 (10.91), DIN ENV 1992–1–1 (06.92).

[5] Feenstra, P. H.: *'Computational Aspects of Biaxial Stress in Plain and Reinforced Concrete'*, Dissertation, Delft University of Technology Press (1993).

[6] Feenstra, P. H., De Borst, R.: *'A Plasticity Model and Algorithm for Mode–I Cracking in Concrete'*, International Journal for Numerical Methods in Engineering, Vol. 38 (1995), pp. 2509–2529.

[7] Feenstra, P. H., De Borst, R.: *'A Composite Plasticity Model for Concrete'*, International Journal for Solids & Structures, Vol. 33 (1996), pp. 707–730.

[8] Hofstetter, G., Mang, H. A.: *'Computational Mechanics of Reinforced Concrete Structures'*, Vieweg & Sohn, Braunschweig/Wiesbaden (1995).

[9] Koiter, W. T.: *'Stress–strain relations, uniqueness and variational theorems for elastic–plastic materials with a singular yield surface'*, Quart. of Appl. Mech., Vol. 11 (1953), pp. 350–354.

[10] Menrath, H., Haufe, A., Ramm, E.: *'A Model for Composite Steel–Concrete Structures'*, Proceedings of the EURO–C 1998 Conference on Computational Modelling of Concrete Structures, Eds.: De Borst, R., Bicanic, N., Mang, H. A., Meschke, G.; Badgastein/Austria, A. A. Balkema Publishers, Rotterdam, Volume 1, (1998) pp. 33–42.

[11] Schellekens, J. C. J.: *'Interface Elements in Finite Element Analysis'*, TU–Delft, Report Nr. 25-2-90-5-17 (1990).

[12] Simo, J. C., Hughes, T. J. R.: *'Computational Inelasticity'*, Springer–Verlag New York, Inc. (1998).

Bridges

Manuscripts for some presentations were not available at time of publication.

3. *HORIZONTALLY CURVED BRIDGES*

MODERATOR: A. Zureick,
Georgia Institute of Technology,
Atlanta, GA

(1) Measured vs. Computed Stresses in a Curved Steel Bridge
Jerome F. Hajjar and Theodore V. Galambos
University of Minnesota
Minneapolis, MN

Wen-Hsen Huang
Virginia Dept. of Transportation
Richmond, VA

Brian E. Pulver
Elstner Associates
Northbrook, IL

Roberto T. Leon
Georgia Institute of Technology
Atlanta, GA

Brian J. Rudie
Minnesota Dept. of Transportation
Roseville, MN

(2) Top Flange Bracing Systems for Box Girders
Zhanfei Fan and Todd A. Helwig
University of Houston
Houston, TX

(3) Verification Modeling for Stability Analysis of Curved Girder Bridges
Michael D. Simpson and Peter C. Birkemoe
University of Toronto
Toronto, Canada

(4) FHWA Experimental Studies of Curved Steel Bridge Behavior During Construction
D. Linzell, A. Zurcick and R.T. Leon, M. Grubb
Georgia Institute of Technology
Atlanta, GA

12. *ADVANCES IN BRIDGE SAFETY AND RELIABILITY*

MODERATORS: Dan Frangopol
University of Colorado
Boulder, CO

Gongkang Fu
Wayne State University
Detroit, MI

1)Life-Cycle Cost Design of Deteriorating Bridges Using Genetic Algorithm
Hitoshi Furuta and Makiko Saito
Kansai University
Takatsuki, Osaka, Japan

D.M. Frangopol
University of Colorado
Boulder, CO

(2)Performance Based Reliability Assessment and Calibration for Seismic Highway Design
Gongkang Fu and A.G. Moosa
Wayne State University
Detroit, MI

(3) Bridge Safety Management Based on Reliability and Cost
Dan M. Frangopol, Michael P. Enright, and Emhaidy Gharaibeh
University of Colorado
Boulder, CO

(4)Fatigue Reliability of Steel Girder Bridges
Andrzej S. Nowak and Maria M. Szerszen
University of Michigan
Detroit, MI

21. *EVALUATION AND REHABILITATION OF HISTORIC BRIDGES*

MODERATOR: J. Michael Stallings
Gottlieb Associate Professor
Auburn University
Auburn, AL

(1) Rehabilitation of a Nineteenth Century Cast and Wrought Iron Bridge
Perry S. Green
University of Florida
Gainesville, FL

Robert J. Connor
Lehigh University
Bethlehem, PA

Christopher C. Higgins
Clarkson University
Potsdam, NY

(2) Restoring Historic Bridges Using Modern Methods
Joseph J. Pullaro
A.G. Lichtenstein & Associates
Paramus, NJ

(3) Seismic Retrofit of Historical Kolekole Bridge
Harold Hamada
University of Hawaii
Honolulu, HI

David Fujiwara and Chad Nakamoto
KSF, Inc.
Honolulu, HI

(4) Diagnostic Testing of a Unique Historic Bridge
Leon Lung-Yang Lai
Specialty Engineering, Inc.
Morrisville, PA

30. *RECENT CABLE-SUPPORTED BRIDGES AND DESIGN OF CABLES*

MODERATOR: Gerard F. Fox
HNTB Corp. (Ret.)
Gardon City, NY

(1) The Sidney Lanier Cable-Stayed Bridge
Man-Chung Tang
T. Y. Lin International
San Francisco, CA

(2) The Charles River Cable-Stayed Bridge
Raymond J. McCabe
HNTB Corp.
Fairfield, NJ

(3) Behavior and Design of Bridge Stay Cables
Khaled Shawwaf
DSI-USA Inc.
Boilingbrook, IL

(4) The New San Francisco Oakland Bay Suspension Bridge
Brian H. Maroney
Caltrans Engineering Service Center
Sacramento, CA

39. *NEW TECHNOLOGIES FOR LOCAL EVALUATION OF BRIDGE COMPONENTS*

MODERATOR: Steven B. Chase
Federal Highway Administration
Mclean, VA

(1) Coating Tolerant Thermography Inspection System
Jon Lesniak and Daniel J. Bazile
Stress Photonics, Inc.
Madison, WI

(2) Health Monitoring of Suspension Cables
Jack Elliot
Pure Technologies
Calgary, Alberta, Canada

(3) Tomographic Imaging of Bridge Decks Using Radar
Steven B. Chase
Federal Highway Administration
Mclean, VA

(4) Evaluation of Prestressed Concrete Girders Using Magnetic Flux Leakage
Al Ghorbanpoor
University of Wisconsin,
Madison, WI

(5) Damage Assessment of Steel and Concrete Bridge Components Using Acoustic Emission
Gregory Muravin
Margon Physical Diagnostics, Ltd.
Netanyo, Israel

48. *BRIDGE EVALUATION USING LOAD AND RESISTANCE FACTOR PHILOSOPHY*

MODERATOR: Charles M. Minervino
Lichtenstein Engineering Associates
New York, NY

(1) Load Rating and Permit Review Using Load and Resistance Factor Philosophy
Bala Sivakumar
Lichtenstein & Associates, Inc.
Paramus, NJ

(2) Calibration of Load Factors for Load and Resistance Factor Evaluation
Fred Moses
University of Pittsburgh
Pittsburg, PA

(3) Fatigue Evaluation Procedures in the Proposed Manual for Condition Evaluation and Load and Resistance Factor Rating of Highway Bridges
Dennis Mertz
University of Delaware
Newark, DE

(4) The Latest Developments in the AASHTO-LRFD Bridge Design Specifications
Wagdy Wassef and John M. Kulicki
Modjeski and Masters, Inc.
Harrisburg, PA

(5) Implementing the LRF Bridge Evaluation Manual
Charles Minervino
A. G. Lichenstein & Associates, Inc.
Paramus, NJ

57. BRIDGE FOUNDATION IDENTIFICATION AND EVALUATION

MODERATORS: Masoud Sanayei,
Tufts University
Medford, MA

Sreenivas Alampalli
New York State Department of Transportation
Albany, NY

(1)Nondestructive Determination of Unknown Bridge Foundation
Larry Olson, Ming Liu, and Marwan F. Aouad
Olson Engineering
Wheat Ridge, CO

(2) Nondestructive Testing for Quality Control and Length Determination of Deep Foundations for Bridges
Stephen Borg,
New York State Department of Transportation,
Albany, NY

(3) Bridge Foundation Stiffness Identification Using Static Field Measurements
Masoud Sanayei
Tufts University
Medford, MA

Kenneth R. Maser
Infrasense
Arlington, MA

(4) Bridge foundation stiffness identification using dynamic field measurements
Erin M. Santini and Masoud Sanayei,
Tufts University,
Medford, MA

Ming Liu and Larry Olson
Olson Engineering
Wheat Ridge, CO

(5) Foundation stiffness design charts for seismic analyses of bridges
Zia Zafir and Mansour Tabatabaie,
Kleinfelder, Inc.,
Pleasanton, CA

64. *FIELD TESTING OF STRUCTURES*

MODERATOR: Andrzej S. Nowak,
University of Michigan
Ann Arbor, MI

(1) Dynamic Testing of Cable-Stayed Bridges
Juan R. Casas,
University of Catalunya
Barcelona, Spain

(2) Forced and Ambient Vibration Tests of Hakucho Suspension Bridge
Yozo Fujino, Hajime Shibuya, Shunichi Nakamura, Masashi Sato,
Masato Yanagihara, and Yoshifumi Sakamoto
Tokyo, Japan

(3) Laser Based Testing and Monitoring of Large Bridges
Jan Bien and Pawel Rawa,
Wroclaw University of Technology
Wroclaw, Poland

(4)Load Distribution of Damaged Prestressed Bridges
Francesco M. Russo
Iowa DOT
Ames, IA

F.W. Klaiber and T.J. Wipf,
Iowa State University
Ames, IA

(5) Noncontact Nondestructive Testing for Structure Health Monitoring
Gongkang Fu and Adil G. Moosa
Wayne State University
Detroit, MI

Measured vs. Computed Stresses in a Curved Steel Bridge

Jerome F. Hajjar[1], Member, ASCE, Theodore V. Galambos[2], Honorary Member, ASCE, Wen-Hsen Huang[3], Associate Member, ASCE, Brian E. Pulver[4], Roberto T. Leon[5], Member, ASCE, and Brian J. Rudie[6], Associate Member, ASCE

Abstract

This research investigates the behavior of the steel superstructure of a curved steel I-girder bridge system during all phases of construction, and ascertains whether the actual stresses in the bridge are represented well by linear elastic analysis software developed for this project and typical of that used for design. Sixty vibrating wire strain gages were applied to a two-span, four-girder bridge, and elevation measurements were taken by a surveyor's level. The resulting stresses and deflections were compared to computed results for the full construction sequence of the bridge, as well as for live loading from up to nine 50 kip trucks. The analyses correlated well with the field measurements, especially for the primary flexural stresses. Stresses due to lateral bending and restraint of warping induced in the girders, and the stresses in the crossframes, were more erratic, but generally showed reasonable correlation. The results also show that analyses in which composite behavior is assumed in the negative moment region (where no shear connectors were used) yield better correlation than analyses in which just the bare steel girders are used.

Introduction

Composite, I-shaped, steel curved girder bridges may be quite flexible and potentially susceptible to stability problems during construction, prior to their stabilization after installation of all diaphragms and hardening of the concrete deck. In order to insure safe design, it is vital that the stresses and deflections resulting from analyses used

---

[1] Assoc. Prof., Dept. of Civil Engrg., Univ. of Minnesota, Minneapolis, MN 55455
[2] Prof. Emeritus, Dept. of Civil Engrg., Univ. of Minnesota, Minneapolis, MN 55455
[3] Struc. Eng., Div. of Struc. and Bridges, VA Dept. of Transp., Richmond, VA 23219
[4] Struc. Eng., Wiss, Janney, Elstner Associates, Northbrook, Illinois 60062
[5] Prof., Sch. of Civil and Envir. Engrg., Georgia Inst. of Tech., Atlanta, GA 30332
[6] Struc. Eng., Off. of Bridges and Struc., MN Dept. of Transp., Roseville, MN 55113

during design be representative of the service-level stress state in the actual bridge structure. The primary objective of this research was to monitor the strains in the steel superstructure of a two-span curved I-girder bridge during its entire construction process, and to compare these field measurements with results obtained from the University of Minnesota Steel Curved Girder Bridge System Analysis Program (a.k.a. "UM program").   While the UM program is similar to linear elastic software used commonly for design of curved girder bridges, it permits detailed specification of loading and assessment of stress states (Huang 1996). Details of this research are presented in (Galambos et al. 1996; Huang 1996; Galambos et al. 1999). To date there have been few measurements of actual stresses recorded during construction in curved steel I-girder bridges (Zureick et al. 1993).

Minnesota Department of Transportation (Mn/DOT) Bridge No. 27998, selected for this project, includes four concentric I-girders, each of differing depth ranging from 50 to 70 inches. The bridge girders are divided into three segments over two spans, with one central support. The spans range from 139 to 155 feet each, and the girders are continuous over the center support. The in-plane radius of curvature of the bridge varies from approximately 270 feet to 300 feet. Two of the three supports have substantial skews. Crossframes consisting of a bottom chord (a WT section), a double angle top chord, and double angle X-brace diagonals, are welded to gusset plates that are in turn bolted to the I-girder transverse stiffeners. Stiff I-shaped end diaphragms are used at the two end abutments of the bridge in lieu of crossframes.

Finite Element Model

The Grillage Method, a stiffness-based finite element formulation that uses a two-dimensional planar grid model to simulate the three-dimensional effects of bridge superstructures, is used by the UM program. The curved I-girders were represented by a three-dimensional, two-node curved beam element having four degrees-of-freedom at each node (Galambos et al. 1996; Huang 1996). Shear connectors were not supplied in the negative moment region of all four girders (i.e., in the region of the center pier). Once the concrete deck hardened, composite action of the concrete deck with the four girders was considered in the positive moment region (i.e., in regions with shear studs). For the negative moment region, there was some speculation that friction and adhesion may induce some level of composite action between the concrete deck and the four curved steel girders. Thus, analyses were conducted both with and without composite action in the negative moment region of the bridge (i.e., including the added stiffness due to the longitudinal reinforcing bars in the deck). For all composite action, the effective slab width was taken as twelve times the thickness of the structural slab. The ratio of the steel to concrete modulus was taken as either $N = 6$ or $N = 8$.

Gage Placement

In order to monitor the strains and stresses induced during the successive construction phases, and to assess the ability of the UM program to model actual behavior, strains were measured during all phases of erection up to the completion of construction. In addition, strains were also measured during two field tests using up to nine trucks with known weight and axle configuration. Sixty gages were attached to the steel superstructure of the bridge. Twenty-four gages, six per girder, were placed along a section near the midspan of one span (span 1) in order to determine the member stresses occurring in the positive moment region of that span. For each girder, four gages, oriented along the longitudinal axis of the girder, were placed close to the flange tips, with two gages attached to the bottom surface of the top flange and two gages attached to the top surface of the bottom flange. The last two gages were affixed to the girder web, also oriented along the longitudinal axis of the girder approximately 1.5 inches away from each flange. These web gages are most appropriate for tracking the predominant strong axis flexural strains in the girders, while the gages at the flange tips also track straining due to lateral bending and resistance to warping. Twenty-four gages were also placed along a section parallel to the skewed middle pier to determine the member stresses occurring in the negative moment region of the girders. The final twelve gages were placed on three crossframes spanning the width of the bridge approximately at the midspan of span 1. A gage was attached to diagonal and chord member of the crossframe, oriented longitudinally along each member, to monitor the axial strain present in each of the members. *Geokon VK-4100* vibrating wire strain gages were selected for use.

Comparison of Measurements and Finite Element Results

Field measurements were taken at all key stages of construction, including eight key dead load stages (e.g., after erection of each span, after placement of deck reinforcing bar, continuous readings during pouring of the concrete deck, after placement of the parapet and overlay, etc.) [see (Galambos et al. 1996; Huang 1996) for details]. In addition, within two years after the bridge was opened for traffic, the bridge was twice closed for a night and loaded with trucks weighing approximately 50 kips each. A total of eleven static loading cases of significance were measured, including nine trucks (arranged in three back-to-back groups of three side-by-side trucks) at six locations along the bridge, three side-by-side trucks at four locations along the bridge, and four trucks placed simultaneously at midspan of each of the two spans. Stresses were obtained from the strains assuming linear elastic behavior.

Percent errors between computed and measured stresses for each gage were calculated for all key stages of loading, as were errors between computed and measured deflections. Each gage was assigned an overall rating indicating its correlation between measured and computed stresses.

Conclusions

A summary of the conclusions from the results from this research follows:

1. This bridge was shored in the early stages of construction of the steel superstructure, and the bridge design was controlled by stiffness, not strength. The stresses were well below the yield stress throughout construction.
2. Computational results generally matched qualitatively and often quantitatively with measured results, particularly in the girders' positive moment region. The bridge behavior was generally predictable at all stages, particularly in the girders.
3. The primary difference between measured and computed results was due mainly to the erratic effects of warping restraint and minor axis bending on the measured results, and to the less predictable behavior seen in the measured results of the crossframes. In addition, the correlation between measured and computed results in the negative moment region due to live truck loading improved substantially if composite action was assumed over the center pier. As there were no shear connectors in the negative moment region of the bridge, composite action in this region was due to friction and adhesion. The unreliable nature of these mechanisms of force transfer justify the traditional practice of neglecting composite action in the negative moment region of these types of bridges. Nevertheless, the magnitude of stresses predicted in the analyses impacts design calculations, particularly for fatigue design. Fit-up stresses were seen in the measured results, especially in the crossframes, but they dissipated as the construction progressed, and they remained below 6 ksi.
4. Further research is required to determine how best to model the bridge over interior piers, where the effects of composite action are more ambiguous, and also in the crossframes, which show less predictable behavior at all stages of dead and live loading, and often produce unconservative results in the analyses.

Acknowledgments

Funding for this project was provided by the Minnesota Department of Transportation and the Center for Transportation Studies at the University of Minnesota.

References

Galambos, T. V., Hajjar, J. F., Leon, R. T., Huang, W.-H., Pulver, B. E., and Rudie, B. J. (1996). "Stresses in Steel Curved Girder Bridges." Rep. No. MN/RC-96/28, Minn. Dept. of Transportation, St. Paul, MN.

Galambos, T. V., Hajjar, J. F., Huang, W.-H., Pulver, B. E., Leon, R. T., and Rudie, B. J. (1999). "Comparison of Measured and Computed Stresses in a Steel Curved Girder Bridge," *Journal of Bridge Engineering*, ASCE, submitted for publication.

Huang, W.-H. (1996). "Curved I-Girder Systems." Ph.D. diss., Dept. of Civil Engrg., Univ. of Minn., Minneapolis, MN.

Zureick, A., Naqib, R., and Yadlosky, J. M. (1993). "Curved Steel Bridge Research Project. Interim Report I: Synthesis." FHWA-RD-93-129, FHWA, McLean, VA.

Top Flange Bracing Systems for Box Girders
By Zhanfei Fan[1] and Todd A. Helwig[2]

ABSTRACT: A critical design stage for steel girders occurs during casting of the concrete bridge deck, when the non-composite steel section must support the wet concrete and the entire construction load. Although a composite box girder has a high torsional stiffness in the completed bridge, during construction the open section is relatively flexible in torsion. A horizontal truss system is usually installed at the top flange level to increase the torsional stiffness. Finite element results presented in this paper show that large forces can develop in the horizontal truss system due to vertical bending of the box girder. Many current design methods and computer programs do not consider the truss forces induced from bending. Expressions to estimate the forces in the lateral truss system due to bending are presented.

INTRODUCTION

Composite box girder bridges usually consist of steel girders of trapezoidal cross-section with two top flanges and a concrete slab. The closed cross-section of the box in the completed bridge has a torsional stiffness that may be one hundred to more than one thousand times the stiffness of a comparable I-girder section. The large torsional stiffness makes box girders attractive for application in horizontally curved bridges in which the bridge geometry may result in large torques on the girders. There also are several other structural, maintenance and aesthetic advantages that make box girders attractive for use in both curved and straight bridges.

Although the torsional stiffness of a composite box girder is large in the completed bridge, during transport, erection and construction the girder consists of an open section with relatively low torsional stiffness. This poses a major problem during the early stages of bridge construction when the steel section may be subjected to large torques. Bracing is typically provided to control the torsional behavior of the steel section. Internal cross-frames and diaphragms are provided to prevent distortion of the cross-section. Although external diaphragms between adjacent girders can be used to increase the torsional stiffness of the bridge, they are mainly used only at supports due to aesthetic and fatigue concerns. A horizontal truss fastened to the box near the top flanges is commonly used to increase the

---

[1] Graduate Research Assistant, University of Houston, 4800 Calhoun, Houston, TX. 77204-4791.
[2] Assistant Professor, University of Houston, 4800 Calhoun, Houston, TX. 77204-4791.

torsional stiffness of the steel section, thereby forming a quasi-closed box girder. The top flange truss may have a single diagonal (SD-type) or have two diagonals, therefore forming an X (X-type).

This paper focuses on the design and behavior of the top flange horizontal truss that is used to increase the torsional stiffness of the steel section during the girder erection and bridge construction. In particular, the effects of vertical bending of the box girder on the truss forces will be addressed. Most current design methods neglect the effects of girder bending stresses on the horizontal truss behavior, however in some cases truss forces in the maximum bending region may actually exceed the maximum member forces caused by the torsional moments. In addition, depending on the type of truss system that is used, large lateral bending stresses may result in the top flange. These large brace forces and girder stresses are often not predicted by computer programs for box girder analysis. An analytical method to evaluate truss forces and girder stresses is therefore warranted.

DESIGN and BEHAVIOR OF TOP FLANGE TRUSS

The torsional analysis of a quasi-closed box girder is usually performed using the Equivalent Plate Method (EPM) developed by Kollbrunner and Basler (1969). Two types of truss systems usually considered consist of a single diagonal (SD-type) system and an X-system (X-type). While the EPM does an excellent job of predicting the truss forces due to torsion, the method usually significantly underestimates truss forces for girder subjected to combined torsion and bending. The lateral struts of the truss are usually only designed for the horizontal component of the top flange loading that results from the shear change in the sloping webs of a trapezoidal section.

The box girder shown in Figure 1 was analyzed with a three-dimensional finite element model using ANSYS (1994). The radius of curvature of the girder is 291 m. (955 ft.), which results in a 19.7° subtended angle of curvature for every 100-m. (328 ft.) girder length. The girder is non-prismatic, with two cross-sections, Section P and N, used in the respective positive and negative moment regions as shown in Fig. 1. The horizontal top truss was an X-type system with 64

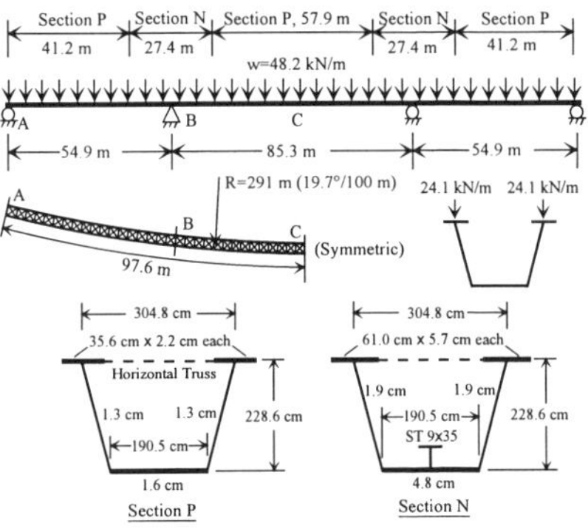

Figure 1: Curved Box Girder Considered

panels along the length of the bridge and a panel size of approximately 3 m. (10 ft.) The distributed load of 24.1 kN/m (1.65 kips/ft) on each of the top flanges, simulates the gravity load from wet concrete as well as other construction loads. The torsional moments were caused by the horizontal curvature of the girder.

The resulting moment and torsion diagrams are shown in Fig. 2. Since the girder is symmetrical about point C, the diagrams for only half the girder length is shown. The bending moment diagram was determined by neglecting the curvature and analyzing a straight bridge, followed by applying the M/R method (Tung and Fountain 1970) to determine the torsional moments. The top flange truss consisted of WT6x13 sections for the diagonals and L3x3x1/4 members for the struts. These members were selected using the EPM based on the maximum torque of 836 kN-m that occurs to the left of the support at point B.

Figure 3 shows a graph of the member forces along the girder length predicted by the EPM and the results from the finite element analysis (FEA). X1 and X2 indicate the two different diagonals in the X-type truss. Negative values indicate compression. The graphs show that the results predicted using the EPM have poor agreement with the FEA results. The EPM significantly underestimates the member forces at several locations along the length. There are a number of locations in which the EPM predicts tension in members that are actually in compression from the FEA results which therefore may lead to problems with buckling of the brace.

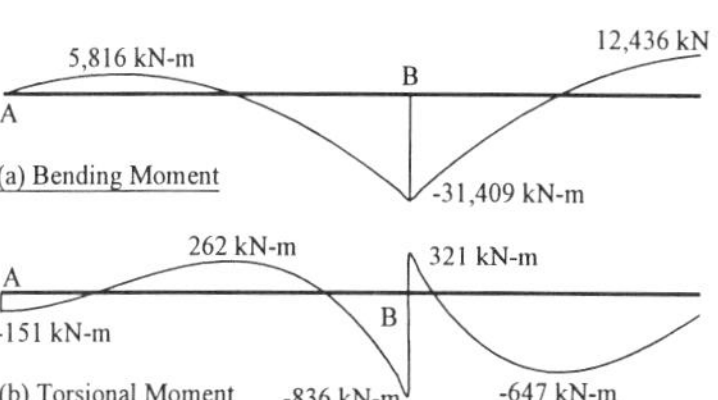

Figure 2: Bending and Torsional Moments

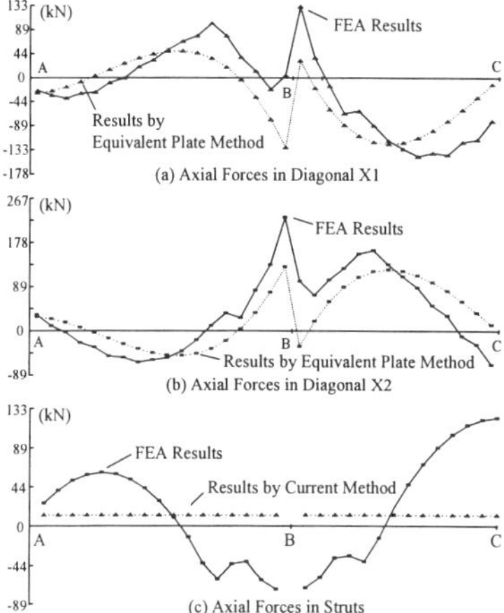

Figure 3: FEA and EPM results

The reason that the FEA results and the EPM have such poor agreement is due to vertical bending of the box girder. The truss is connected to the girders near the top flange, which is a region of large bending stress. Due to strain compatibility the truss must experience the same strain as the girder at the point of connection. This therefore leads to compression in the truss in a positive moment region, and tension in the negative moment region. The following equations were derived to account for the effects of box girder bending on the force in the diagonal, D, and the strut force, S:

SD-Type Truss

$$D = \frac{f\, s \cos\alpha}{K_1} \qquad (1a)$$

$$S = -D \sin\alpha \qquad (1b)$$

X-Type Truss:

$$D = \frac{f\,s\cos\alpha}{K_2} \qquad (2a)$$

$$S = -2D\sin\alpha \qquad (2b)$$

in which K_1 and K_2 are parameters defined by

$$K_1 = \frac{d}{A_d} + \frac{b}{A_s}\sin^2\alpha + \frac{s^3}{2b_f^{\,3}t_f}\sin^2\alpha \qquad (3)$$

$$K_2 = \frac{d}{A_d} + \frac{2b\sin^2\alpha}{A_s} \qquad (4)$$

where, f is the girder bending stress at the depth of the truss connection, s is the panel length, α is the angle between the top flange longitudinal axis and the truss diagonal, d is the diagonal length, b is the spacing between the top flanges, A_d and A_s are the respective areas of the diagonals and the struts, and b_f and t_f are the respective width and thickness of a top flange. Unlike the axial forces due to torsion, for the X-type truss, the two diagonal forces have the same magnitude and sign. Figure 4 shows a comparison between the above equations and the FEA results for the girder previously shown in Fig. 1. The forces due to bending were computed with the above equations and added to the forces from torsion which were computed with the EPM. The proposed equations provide excellent agreement with the FEA results.

SUMMARY

Vertical bending of box girders can result in significant forces in top flange lateral truss systems which are usually only designed for torsional loading. Equations have be presented to compute the component due to vertical bending, which can be directly added to the torsional component computed using the EPM. The forces in the truss also lead to significant lateral bending stresses in the top flange, particularly for SD-type truss systems.

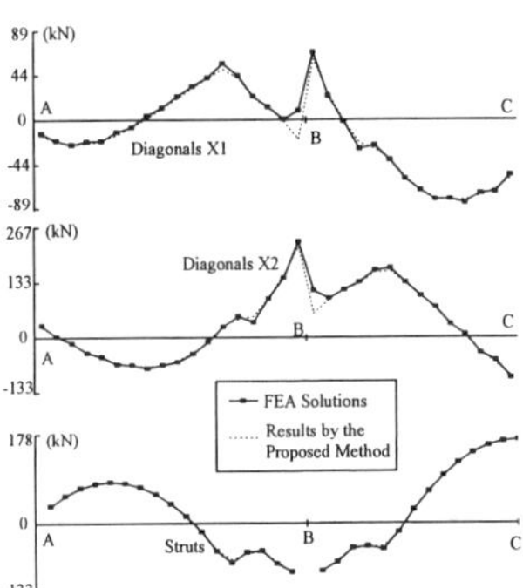

Figure 4: FEA Results and Proposed Equations

Equations are presented in a different paper for the top flange lateral bending stresses induced by forces in the truss.

REFERENCES

ANSYS, Finite element program users manual, Version 5.3. (1996), ANSYS, Inc.

Kollbrunner, C. F., and Basler, K., (1969), "Torsion in Structures --- An Engineering Approach", Springer-Verlag, New York.

Tung, D. H. H., and Fountain, R. S., (1970), "Approximate Torsional Analysis of Curved Box Girders by the M/R Method", *AISC Eng. Jour.*, V. 7, No. 3, pp. 65 - 74.

FHWA EXPERIMENTAL STUDIES OF CURVED STEEL BRIDGE BEHAVIOR DURING CONSTRUCTION

D. Linzell[1] S. M. ASCE, A. Zureick[2] M. ASCE, R. T. Leon[2] M. ASCE

Abstract

An overview of a series of full-scale tests of a curved steel bridge structure during construction is presented. The tests were completed as a part of the Federal Highway Administration's (FHWA) Curved Steel Bridge Research Project (CSBRP). The behavior of single girders and twin and three girder systems were examined within six different framing plans.

Manuscript was withdrawn immediately prior to publication.

---

1. Graduate Research Assistant, Georgia Institute of Technology, School of Civil Engineering, Atlanta, GA 30332
2. Professor, Georgia Institute of Technology, School of Civil Engineering

Life-Cycle Cost Design of Deteriorating Bridges
Using Genetic Algorithm

Hitoshi Furuta[1], Dan M. Frangopol[2], and Makiko Saito[3]

Abstract

The purpose of life-cycle cost bridge optimization is to implement an optimal inspection / repair strategy for the minimum expected total life-cycle cost that includes initial, preventive maintenance, inspection, repair and failure cost while the bridge maintains the target reliability. This paper presents an optimization methodology for the life-cycle cost design of reinforced concrete T-girders that deteriorate over time. An attempt is made here to extend and improve the previous work by Frangopol et al. (1997) so as to determine the time, quality and number of inspections and repairs by applying Genetic Algorithm. A numerical example is presented to demonstrate the applicability of the proposed method.

Introduction

Highway bridges follow the life cycle that consists of design, construction, inspection, repair and failure. At present, it is desirable to develop an optimal strategy for the bridge management through the lifetime in order to reduce the overall cost.

An efficient method that provides adequate inspection / repair strategies was proposed by Frangopol et al (1997). This method can determine how many inspections are appropriate for lifetime, and at what time inspections and repairs should

---

[1] Professor, Department of Informatics, Kansai University, 2-1-1 Ryozenji-cho, Takatsuki, Osaka 569-1095, Japan

[2] Professor, Department of Civil, Architectural, and Environmental Engineering, University of Colorado, Boulder, CO 80309-0428, USA

[3] Graduate Student, Department of Informatics, Kansai University, 2-1-1 Ryozenji-cho, Takatsuki, Osaka 569-1095, Japan

be done, while taking into account all bridge repair possibilities based on an event tree. However, if the number of design variables increases, it is difficult to solve the problem.

Therefore, an attempt is made in this paper to extend and improve the previous work by Frangopol et al. (1997) using Genetic Algorithm.   A numerical example is presented to demonstrate the applicability of the proposed method.

Application of Genetic Algorithm To Life-Cycle Cost Design

Life-cycle cost optimization is a nonlinear problem that includes integer and discrete variables.   Therefore, it is necessary to apply a combinatorial optimization method to solve it.   The purpose of this study is to find the most economical plan for inspection / repair.   It is evident that a nouniform interval of inspection / repair strategy is more economical than a uniform one (Frangopol et al. 1997).   It is easily understood that the combination of inspection techniques with different detection capabilities in a strategy is more economic.   Discrete variables are useful in determining when and how inspections and repairs have to be performed and what methods have to be used. Genetic Algorithm (GA) is a representative algorithm of combinatorial optimization methods.   Using GA, it is possible to decide the number of lifetime inspections, the time of each inspection, and which inspection has to be used.   Then, the time of repair is decided based on an event tree analysis.

Numerical Example

The life-cycle cost optimization is reduced to the following mathematical programming problem.

$$\text{Minimum} \quad C_{ET} \quad \text{subject to} \quad P_{f.life} \leqq P_{f.life}^{*} \tag{1}$$

where C_{ET} is the expected total cost, $P_{f.life}$ is the lifetime probability of failure, and $P_{f.life}^{*}$ is the maximum acceptable lifetime probability of failure.   As an example, a prefabricated reinforced concrete T-girder bridge shown in Fig.1 is considered for the inspection / repair maintenance analysis.   Two lanes of HS-20 trucks provide the loading.   Girder spacing S is 2.44m, total width of the bridge is 7.32m, and the span L is 18.30m.

The interior girder in Fig.1 was designed in Lin and Frangopol (1996).   The design shown in Fig.2 was based on the reliability and optimization according to the constraints specified by the American Association of State Highway and Transportation Officials (AASHTO) (Standard 1992).

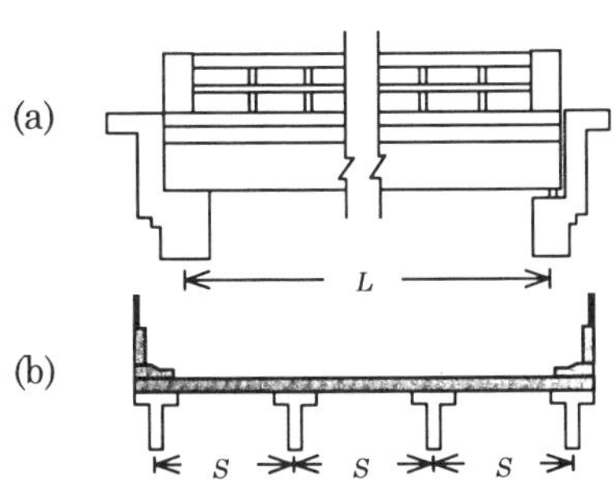

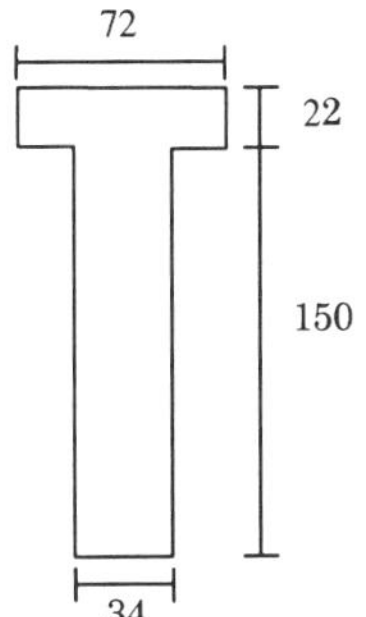

Fig. 1 Reinforced Concrete T-Girder Bridge: **Fig. 2** Optimum Design
(a) Elevation, (b) Cross Section

Assuming a lifetime of 75 years, the maximum number of inspections is considered to be 10.   The initial reliability index is β=3.76 (probability of failure : P_f = 0.000085), the target reliability level is β^*=2.0 (probability of failure : P_f^*=0.02275), the uniform corrosion rate of bending reinforcement is assumed to be 0.0089 cm/year, and three types of inspection methods are considered as follows: high quality (i.e. $\eta_{0.5}$ =0.05), medium quality (i.e. $\eta_{0.5}$=0.1), low quality (i.e. $\eta_{0.5}$=0.15) (see Frangopol et al. 1997).

The numerical computation was implemented by changing GA parameters several times. The best solution is presented in Table1.   The optimal strategy has 5 lifetime inspections at 30, 43, 51, 64 and 65 years, repairs are performed after every inspection, and the calculated expected total cost is 1,427.1.   The medium and low quality inspections are used at 30 and 43, 51, 64, 65 years, respectively.   The GA used 50 individuals, 5000 generations, a rate of crossover of 0.6, and a rate of mutation of 0.1.

Table 1 Optimal Solution

Inspection Time (year)	30	43	51	64	65
Inspection Quality ($\eta_{0.5}$)	0.1	0.15	0.15	0.15	0.15

Fig. 3 shows the change of reliability index β when using the optimal strategy in Table1, and Fig. 4 shows the change of the expected total cost (fitness value).

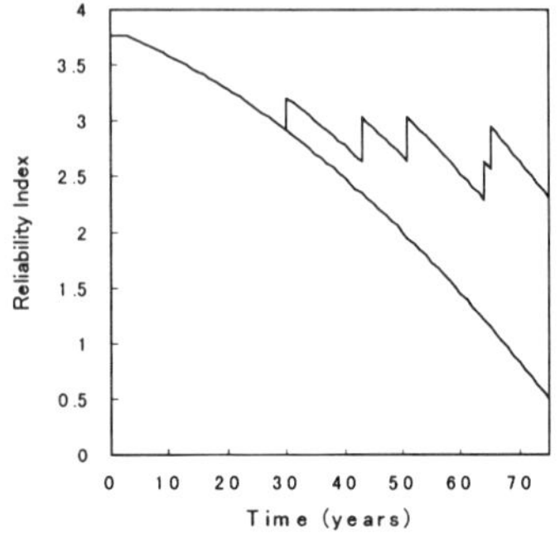
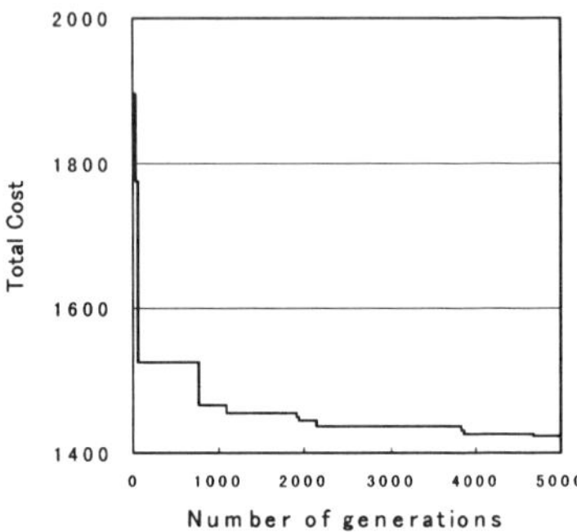

Fig. 3 The Change of Reliability Index

Fig. 4 The Change of Fitness

Conclusions

Several conclusions can be drawn from this study. First, it is more economical to combine inspection methods with different capabilities in achieving an optimal lifetime inspection / repair strategy.   Second, Genetic Algorithm is effectively applicable to the life-cycle cost problem of deteriorating bridges.   Third, it is desirable to consider the combination of inspection and repair methods with different detection capabilities and strength improvement, in order to obtain a more economical and practical lifetime strategy.   Finally, GA is a promising technique to solve practical life-cycle optimization problems for civil infrastructure systems including highway bridges, because it can deal with discrete and continuous variables very effectively (Goldberg 1989).

Acknowledgement

The financial support of the Ministry of Education, Japan through Grant 10044185 is acknowledged.

References

Frangopol, D. M, Lin, K-Y, and Estes, A. C. (1997), "Life-Cycle Cost Design of Deteriorating Structures", *Journal of Structural Engineering, ASCE*, 123(10) 1390-1401.

Goldberg, D. E. (1989), "Genetic Algorithms in Search, Optimization and Machine Learning", *Addison-Wesley Publishing Company, Inc.*

Lin, K.-Y., and Frangopol, D. M. (1996), "Reliability-Based Optimum Design of Reinforced Concrete Girders.", *Struct. Safety*, 18(2/3), 239-258.

Standard Specifications for Highway Bridges. (1992), 15th Ed., American Association of State Highway and Transportation Officials, Washington, D.C.

Performance Based Reliability Assessment and Calibration
for Seismic Highway Bridge Design

Gongkang Fu[1] and Adil G. Moosa[2]

Abstract

This paper discusses on performance based highway bridge seismic design and related reliability assessment and calibration. Span falling-off and column failure modes are addressed. Performance requirements are considered for different intensities of seismic event. System reliability at these excitation levels is assessed for a typical highway bridge designed according to current AASHTO seismic design code.

Introduction

Recent seismic damage to critical bridges has shown costly lessons. It impaired post-earthquake emergency response, causing significant additional loss. This also alludes to inadequacy of today's US seismic bridge design code. It includes: (1) the designer is not required to directly address bridge's seismic performance, and (2) associated uncertainty is not explicitly covered. On the other hand, performance based design has been suggested to the profession. FEMA guidelines [1997] pioneer the concept for retrofitting buildings. Several design codes have explicitly covered uncertainties by reliability calibration. In this respect, US seismic design for bridges has not reached a compatible stage. This paper addresses this issue, and presents a system reliability model for highway bridges under various performance requirements.

Performance Based Seismic Design

Performance-based design should be addressed as an integral and critical part of performance based engineering, whose goal is to have the final product of engineering (including design and construction) meet pre-determined objectives. FEMA 273 [1997] represents a significant step in this direction. In order for the concept to be completely implemented, the requirements need to be included in design codes, mandating the

[1]Associate Professor [2]Graduate Research Student, Department of Civil and Environmental Engineering, Wayne State University, Detroit, MI48202

designer to ensure the structure's performance. Of course, these requirements have to be consistent with state of the knowledge, in terms of both theory and practicing tools.

System Reliability of Highway Bridges under Seismic Load

Structural reliability can be one type of performance based requirements for seismic design of highway bridges. This is rationalized by the following facts. A) The theory of structural reliability has been well established. B) Several design codes have adopted the concept. C) Current state of knowledge about strong earthquakes is probability based. To that end, a bridge system reliability model is presented below.

For highway bridges, span falling and column failure have been commonly observed critical seismic failure modes. System reliability is assessed here by considering them in series. Let respective safety margins g_d and g_f be:

$$g_d = d_C - d_D \tag{1}$$
$$g_f = F_C - F_D \tag{2}$$

Displacement capacity d_C is given by the design code. d_D is defined here by the seismic displacement spectrum as $AC_e/(2\pi/T)^2$. A is constant for the ground acceleration levels, whose uncertainty is covered in C_e, the elastic acceleration response spectrum. T is the period of the bridge system. F_C and F_D in Eq.2 are respectively the force capacity and demand for the column. Depending on the performance requirement and seismic intensity, this force failure mode may take the form of yielding onset (for response within the elastic range) for relatively frequent seismic events, or ductility failure (for response in the non-elastic range) for rare and occasional seismic events. The safety indices for these cases, β_f and β_d indicates the probability of failure given that an earthquake occurs at intensity levels covered.

For elastic response, F_C is provided according to current design code. For non-elastic response under stronger ground motions, a ductility failure model is appropriate according to [Ghosn & Chen 1991] [Priestley & Park 1987]:

$$F_C = [\ 1 + 3\ (1 + 5.4\ \alpha)\ (L_p/L)\ (2 - L_p/L)\ ]\ \lambda \tag{3}$$

F_C is now the ductility capacity. α is the ratio of provided- to required-reinforcement volumetric ratios. L_p/L is the ratio of plastic hinge length to the distance between the contraflexure and the maximum moment section. λ is a modeling factor, to cover the bias and variation of the estimated value from reality. The ductility demand μ_D is modeled according to [Ghosn & Chen 1991; Priestley & Park 1987]. β_f and β_d for respective failure modes can be obtained using first order reliability method. The system reliability assessment here is performed by bounding, which is conditioned on occurrence of an earthquake at the level covered.

Example, Discussion, and Conclusions

Consider a typical US highway bridge, given as the design example in the AASHTO specifications [1991]. For design, this bridge was subject to displacement check at the abutment supports, and force check (including ductility check) at the column supports. Three levels of seismic events are considered here: 1. frequent events with a return period of 50 years; 2. occasional events with a return period of 500 years; and 3. rare events with a return period of 2500 years. For a bridge life of 75 years, corresponding performance requirements are defined as: 1. responses must be kept within elastic range, for being operational when frequent events occur; 2. span collapse must be prevented, for being operational without immediate need for repair when an occasional earthquake occurs; and 3. span collapse must be prevented, for being operational after quick repair upon occurrence of a rare event. Three levels of ground acceleration are accordingly assumed: A= 0.1g, 0.4g, and 0.6g, respectively. Assume that the bridge was designed for a ground acceleration level of A_{design} = 0.4g, according to current design code. The code requires only one level of excitation be designed for, and does not explicitly mandate assurance of performance, although performance requirements were included as development principles for the provisions.

In Eq.1, d_C = 12 + 0.03L_t + 0.12 H = 26 (in.), where L_t= 376 ft and H = 25 ft. d_C is assumed to be deterministic for its low uncertainty, compared with other variables involved. C_e in the response spectra is assumed to be a lognormal random variable. Its mean and standard variation are given as follows [Ghosn & Chen 1991]:

$$\mu_{Ce} = 6.54 - 18.77\,T_n + 27.48\,T_n^2 - 17.87\,T_n^3 + 4.19\,T_n^4 \quad (T_n >0.3) \qquad (4)$$
$$\sigma_{Ce} = 3.33 - 13.49\,T_n + 23.92\,T_n^2 - 17.77\,T_n^3 + 4.70\,T_n^4 \quad (T_n >0.3) \qquad (5)$$

where T_n is the nominal period of the bridge, taken to be 0.6 sec for this example. T is the period of the bridge, assumed to be a lognormal random variable with mean equal to 0.648 sec (=1.08 T_n) and coefficient of variation (COV) equal to 20%. As discussed earlier, three As are used for respective levels of earthquake intensity.

For elastic response in Eq.2, F_C = 1.13 $\eta\gamma$ (1.2) A_{design} S / (ϕ $RT_n^{2/3}$) + (1.13 $\eta\gamma$ / ϕ −1) (B L_b^2)/ (H L_t), where γ and ϕ are load and resistance factors for design, taken to be 1.0 and 0.9 respectively. A_{design} is the ground acceleration coefficient used to design the bridge, which is 0.4g. S is the factor to cover effects of the soil, taken to be 1.2 for Type II soil profile. R is the response modification factor for design, permitting non-elastic response (for the design earthquake). It is set to be 5 here for multiple columns. T_n is equal to 0.6 sec, being the nominal period of the bridge. All of the above factors are specified in the design code. η is a modeling factor covering random variation in the capacity, assumed to be a lognormal random variable, with mean equal to 1.0 and COV equal to 13%. B is a deterministic coefficient for dead load moment in the column, and found to be 0.0327 for this example. L_b is equal to 127 ft, being the span length influencing the dead load moment. H and L_t are the height of the columns and the total length of the bridge, respectively, as given above. For ductility consideration under higher seismic loads (occasional and rare events), F_C in Eq.3 is found, for this example, to have α = 1.25. L_p and λ are assumed to be lognormally distributed independent random variables. L_p/L has mean of 0.17 and COV of 21%. The mean and COV of λ are 1.54 and 32% [Ghosn & Chen 1991].

Considering three levels of excitation intensity, Table 1 shows the resulting safety indices for the displacement and force failure modes, and the system. Note that β_f for the case of frequent events is calculated for yielding onset. For the other two cases it is assessed considering ductility. The safety indices for the displacement failure mode are generally higher than those for the force failure mode. For code calibration, recorded performance data need to be collected to serve as a benchmark compared with these safety indices. The force failure safety index appears to be low, raising a question about the level of current code's assurance for preventing (possibly small) damage under relatively low earthquakes. This is a first step towards a fully calibrated performance based design code. The numerical example shows that the span falling mode has much higher reliability than the column failure mode. It is desirable if the column failure does not lead to structural collapse, because the displacement failure mode (span falling) is brittle and more likely to cause additional life loss.

Acknowledgments

This work was funded by CULMA of Wayne State University and FHWA Project DTFH61-97-P00549, which is gratefully appreciated. However, the authors' views presented are not necessarily those of the sponsors.

References

[1] AASHTO Standard Specifications for Seismic Design of Highway Bridges, 1991
[2] Buckle,I.G. and Friedland,I.M. "Seismic Retrofitting Manual for Highway Bridges", FHWA-RD-94-052, May 1995
[3] FEMA 273 "NEHRP Guidelines for Seismic Rehabilitation of Buildings", 1997
[4] Fu,G. (1998) "Elastically Supported Cantilever Beam Subjected to Nonstationary and Colored Seismic Excitation" Earthq.Engg & Struc.Dyn., **27**, p.977
[5] Fu,G. (1995) "Response Statistics of SDOF System to Exponentially Decaying Seismic Input: An Explicit Solution", Earthq.Engg. & Struct.Dyn.,**24**, p.1355
[6] Ghosn,M. and Chen,G. (1991) "Reliability of Highway Bridge Columns under Earthquake Loading", Proc. 3rd US Conf. on Lifeline Earthq.Engg, LA, CA, p.692
[7] Priestley,M.J.N. and Park,R. (1987) "Strength and Ductility of Concrete Bridge Columns under Seismic Loading", ACI Structural Engg Journal, Jan.-Feb. 1987

Table 1   Numerical Results of the Example

EQ Frequency (Max Ground Acc. A)	Rare (0.6g)	Occasional (0.4g)	Frequent (0.1g)
β_d	3.73	4.40	6.73
β_f	1.97	2.65	1.25
β_{sys}	1.96	2.65	1.25

LIFETIME RELIABILITY OF CONCRETE BRIDGES

Michael P. Enright, [1] Associate Member, ASCE,
and Dan M. Frangopol, [2] Fellow, ASCE

Abstract

The need for using time-variant reliability methods for bridge service-life prediction is becoming increasingly recognized in the U.S. This study presents recent progress in the analysis of deteriorating concrete bridges using a time-variant system reliability approach. The results can be used for lifetime bridge system reliability prediction and for the development of whole life reliability-based maintenance planning strategies for bridges.

Introduction

Although the U.S. bridge design code (*LRFD* 1994) is based on reliability concepts, it is largely limited to component-level safety requirements and does not generally consider the behavior of bridges modeled as structural systems. For nonredundant bridges, the component-level approach is underconservative. For redundant bridges, this approach can be very conservative, particularly for structures exhibiting significant ductility and/or post-failure strength. When the redundancy of the bridge needs to be considered in the overall safety evaluation, fail-safe system reliability methods have to be used.

This study illustrates recent progress in the lifetime reliability assessment of concrete bridges. It is based on several recent studies at the University of Colorado (Enright 1998; Enright and Frangopol 1998a,b,c). Time-variant system reliability estimates are shown for an existing bridge in Colorado. The influence of post-failure material behavior on bridge system failure probability is included. A comparison of failure probabilities for several system failure models is also presented.

Reliability of Bridges Modeled as First-Failure Systems

For bridges modeled as first-failure (i.e., weakest-link) systems consisting of

---

[1] Graduate Research Assistant, Department of Civil, Environmental, and Architectural Engineering, University of Colorado, Campus Box 428, Boulder, CO 80309-0428.

[2] Professor, Department of Civil, Environmental, and Architectural Engineering, University of Colorado, Campus Box 428, Boulder, CO 80309-0428.

elements with uncorrelated strengths, the cumulative-time system failure probability increases as the number of elements in the system increases. To illustrate this effect, the cumulative-time failure probability of a deteriorating reinforced concrete T-beam highway bridge is evaluated using the RELTSYS computer program (Enright 1998). Weakest-link systems consisting of one, three, and five girders are considered. The girders are subjected to linear strength degradation (e.g., $g(t) = 1 - 0.005t$, where $g(t) = R(t)/R_o$ = resistance degradation function, R_o = initial resistance, $R(t)$ = resistance at time t, and t = elapsed time since damage initiation) and the initial girder strengths are uncorrelated ($\rho_R = 0.0$). A complete description of the bridge, including main descriptors of initial resistance R_o, of corrosion initiation time T_I, and of dead (DL) and live (LL) load effects, is given in Enright and Frangopol (1998a).

Cumulative-time failure probabilities associated with three first-failure systems are shown in Fig. 1. The effects of the coefficient of variation of live load, $V(LL)$, and of the number of elements on the cumulative-time failure probability are shown in Fig. 1. A detailed investigation of these and other effects is presented in Enright (1998) and Enright and Frangopol (1998a).

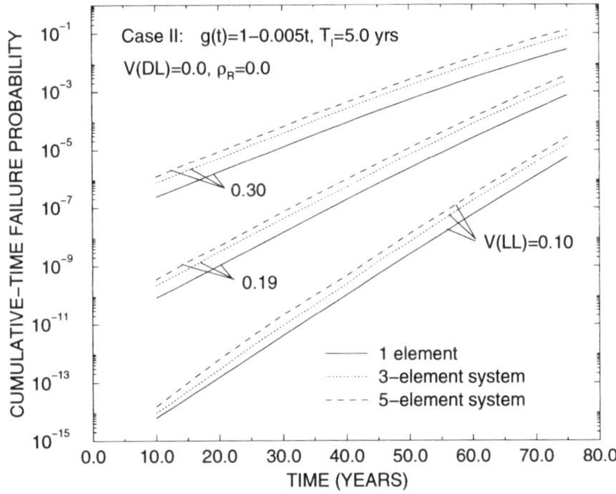

Figure 1: Effect of Number of Components and Live Load Coefficient of Variation, V(LL), on Failure Probability of First-Failure Systems

Reliability of Fail-Safe Systems

Many bridges do not collapse when a single member (e.g., girder, diaphragm) fails. Failed members may also continue to carry a portion of the load, depending on the post-failure behavior (represented by the post-failure behavior coefficient of the member η). For example, if the post-failure behavior is perfectly brittle

$(\eta = 0)$, the failed member carries no load, and the load originally carried by the failed member is completely redistributed to the other members in the system. For perfectly ductile behavior $(\eta = 1)$, the failed member continues to support a load equal to its resistance.

The cumulative-time failure probability of a deteriorating five-member fail-safe system is shown in Fig. 2 (see Enright and Frangopol 1998b for details). The effect of the post-failure behavior coefficient η is illustrated. As η increases from 0 to 1, the cumulative-time failure probability of the system decreases. This is due to the increasing ability of the deteriorating structure to redistribute load.

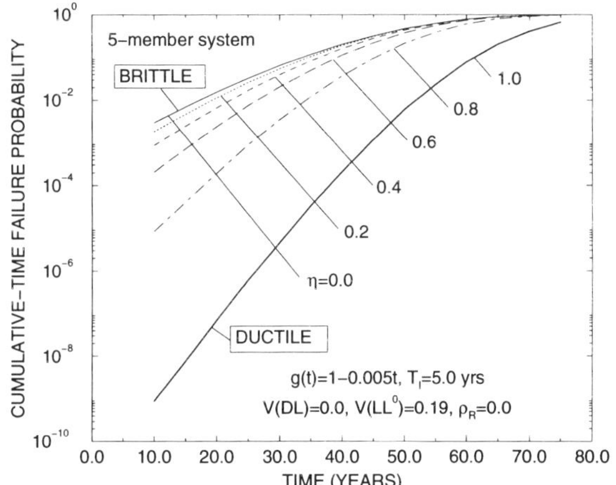

Figure 2: Effect of Post-Failure Behavior Coefficient η on Failure Probability of a Deteriorating Five-Member Fail-Safe System

A variety of system models have been proposed for the reliability analysis of girder bridges, in which failure of a specified number of girders is represented as a combination of series and parallel systems. Four system failure criteria (i.e., system models) are considered for an existing five-girder bridge, Colorado L-18-BG. Complete details regarding the bridge geometry, strength degradation, loads, post-failure material behavior, and system failure modes are given in Enright and Frangopol (1998c).

A comparison of the shear reliabilities associated with four different system failure criteria is shown in Fig. 3. Differences in the failure probabilities associated with the four system failure criteria could be substantial. Failure probabilities associated with system I (i.e., shear failure of any single girder) and system IV (i.e., shear failure of any four adjacent girders) are the highest and lowest, respectively. As shown, the influence of post-failure material behavior (i.e., amount of load redistributed after failure) may also be significant.

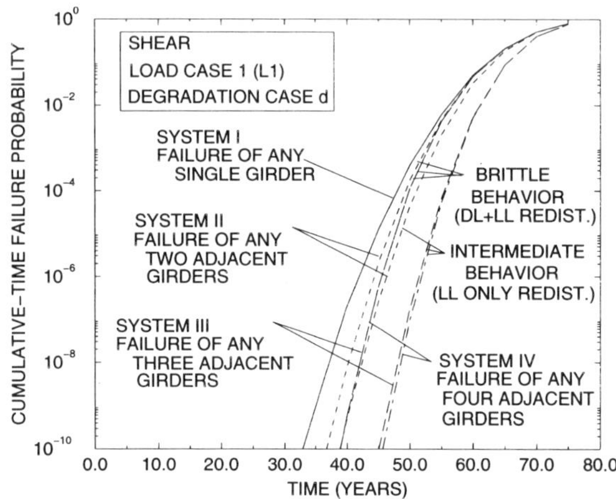

Figure 3: Comparison of Time-Variant Failure Probability Predictions For System Failure Criteria Models I, II, III, and IV

Concluding Remarks

As illustrated in this study, the time-variant failure probability of deteriorating bridges is sensitive to many factors such as load variabilities, load case, degradation rate, number of members in the system, post-failure member behavior, and system failure criterion, to name a few. The work presented provides an improved understanding of the effects of these factors on the cumulative-time failure probability of deteriorating concrete bridges. The results can be used for lifetime bridge system reliability predictions and for the development of whole life reliability-based maintenance planning strategies for bridges.

References

Enright, M.P. (1998). "Time-variant reliability of reinforced concrete bridges under environmental attack." PhD Thesis, Dept. of Civ., Envir., and Arch. Engr., University of Colorado, Boulder, Colo.

Enright, M.P., and Frangopol, D.M. (1998a). "Service-life prediction of deteriorating concrete bridges." *J. Struct. Engrg.*, ASCE, 124 (3), 309-317.

Enright, M.P., and Frangopol, D.M. (1998b). "Failure time prediction of deteriorating fail-safe structures." *J. Struct. Engrg.*, ASCE, 124 (12), in press.

Enright, M.P., and Frangopol, D.M. (1998c). "Reliability-based condition assessment of deteriorating concrete bridges: A case study." submitted for publication.

LRFD bridge design specifications (1994). 1st Ed., American Association of State Highway and Transportation Officials, Washington, D.C.

Fatigue Reliability of Steel Girder Bridges

Andrzej S. Nowak[1] and Maria M. Szerszen[2]

Abstract

Reliability analysis is performed for fatigue limit state in steel bridges. The load and resistance parameters are treated as random values. The fatigue load model includes the magnitude and frequency of occurrance. Fatigue resistance is based on the available test data, in particular S-N curves. Reliability indices are calculated as a function of time.

Introduction

Repeated application of live load may lead to failure of material even when the load level is lower than for the ultimate limit states. In their service life, bridges are exposed to traffic loads, sometimes very heavy, specially on the high volume roads. Multiple application of dynamic load may lead to fatigue-specific changes in the structural materials. The number of trucks on the slow lane of highways can be very high, in some cases 5,000 per day was observed in Michigan. Usually, a real number of load cycles during the service life of the bridge is greater than it is assumed in design codes.

Load and resistance parameters are random variables. Therefore, structural performance can be measured in terms of reliability. Traditional analytical models have been developed for the ultimate limit states, such as bending capacity (moment) and shear. However, fatigue loads require a special approach, because they are determined not only by magnitude but also by frequency of occurrence. Fatigue resistance must be considered in relation to load (magnitude and frequency).

Analysis of fatigue performance involves the determination of loads and material strength. Material response has been studied by many researchers. For example Fisher (1974) developed S-N curves for various categories of details in steel structures. The distribution of the number of cycles to failure can be approximated as normal, with the coefficient of variation decreasing for decreasing stress levels.

----

[1]Professor of Civil Eng., University of Michigan, Ann Arbor, MI 48109-2125
[2]Associate Research Scientist, University of Michigan, Ann Arbor, MI 48109-2125

Fisher's work demonstrated the importance of load level, particularly magnitude and frequency of occurrence.

Fatigue Load Model

The most important fatigue parameters are amplitude and frequency of loading. To investigate fatigue of bridges loaded with heavy trucks, it is convenient to use the load model based on weigh-in-motion (WIM) measurements (as for example Laman and Nowak 1996). The field data can be collected and recorded including stress histograms for the girders and other components. The rainflow method counts the number, n, of cycles in each predetermined stress range, σ_i, for a given stress history (as for example shown by Nowak et al 1992; Frank 1992). The results indicate that magnitude and frequency of truck loading are strongly site-specific and stress/strain values are strongly component-specific.

The equivalent stress, σ_{eq}, needed to find the number of cycles to failure from the S-N relationship can be calculated for each girder using the root mean cube (RMC) formula (Moses et al. 1987):

$$\sigma_{eq} = \sqrt[3]{\sum \left(\sigma_i^3 \times p_i \right)} \tag{1}$$

where σ_i = midpoint of the stress interval i and p_i = the relative frequency of cycle counts for interval i. The stress, σ_i, is calculated as a product of the strain and modulus of elasticity of steel. This formula is originally based on Miner's rule (Miner 1945), and his assumption that the damage in material (steel) is accumulating in a linear way according to the applied number of load cycles. The sequence of applied cycles is not included in the Miner's rule. Cube root in Eq. 1 gives enough accuracy in estimating the equivalent stress.

Resistance Model

In the fatigue analysis, resistance is the ability of the structure to resist cyclic loads. For each girder, the load amplitude varies. Most of the available material tests were performed for a constant load amplitude. For example, Fisher et al. (1974) presented results of tests to determine the number of load cycles to failure for steel beams. The example of resulting cumulative distribution functions (CDF) for one of tested cross-sections is shown in Fig. 1 on the normal probability paper. The horizontal axis represents the number of cycles to failure and different curves correspond to different stress levels.

The distribution of the number of cycles to failure can be approximated as lognormal. However, the number of tested specimens was not very large and therefore, there is a need for further verification of the results.

Reliability Analysis

In recent studies on the ultimate limit states, the structural performance was measured in terms of the reliability index (Nowak 1995). There are several procedures available as described for example by Thoft-Christensen and Baker (1982) or Melchers (1987).

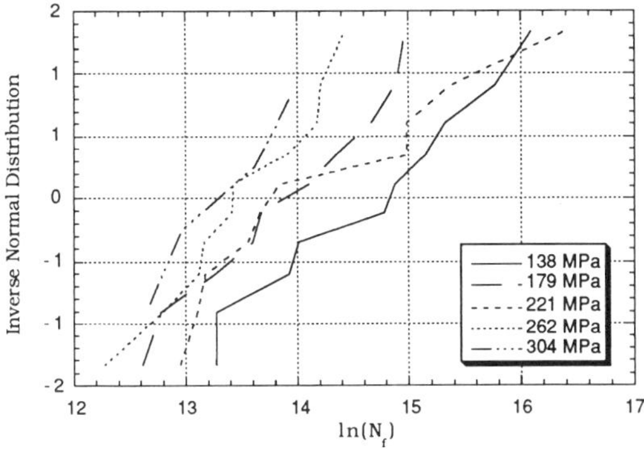

Fig. 1. CDFs of Number of Cycles to Failure for Tested Steel Beams, Depth 330 mm.

It is further assumed that the resistance and load parameters (N_F and N_n) are lognormal random variables. Therefore, the reliability index, β, is

$$\beta = \frac{\ln\left(m_{NF}/m_{Nn}\right)}{\sqrt{V_{NF}^2 + V_{Nn}^2}} \tag{2}$$

where m_{NF} = mean number of cycles to failure, V_{NF} = coefficient of variation of the number of cycles to failure, m_{Nn} = mean number of cycles applied, V_{Nn} = coefficient of variation of the number of cycles applied. The example of reliability index analysis for an existing bridge is shown in Fig. 2.

Conclusion

Reliability analysis is performed for fatigue limit state in steel bridges. The parameters of load (number of cycles applied) and resistance (number of cycles to failure) are derived from the available test data, e.g. load spectra and S-N relationship. Component stress spectra can be presented in a form of CDFs, and basing on these data the equivalent stresses can be calculated. Live load stress spectra are strongly component-specific. The number of cycles to failure (resistance) for a given load stress level is a subject of a considerable variation. The reliability index is calculated for an example bridge, which is carrying a motorway. The live load used in this analysis is the standard fatigue vehicle according to BS5400.

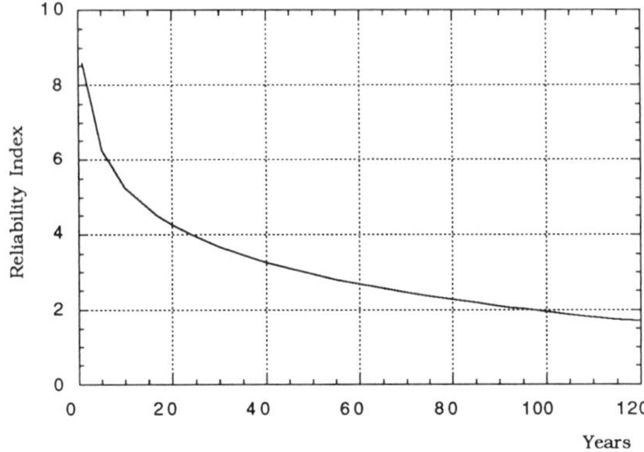

Fig. 2. Reliability Analysis for a Two Lane Steel Bridge, 28 Years Old
with the Span Length of 40 m.

References

1. Frank, K.H., (1992), Using Measured Stress Histories to Evaluate the Remaining Fatigue Life of Bridges, 3rd International Workshop on Bridge Rehabilitation, Darmstadt, Germany, June 1992, pp. 529.
2. Fisher, J.W., et al. (1974), Fatigue Strength of Steel Beams with Welded Steffeners and Attachments, Report 147, Transportation Research Board, National Council, Washington, DC.
3. Laman, J.A., and Nowak, A.S., (1996), "Fatigue Load Models for Girder Bridges", ASCE Journal of Structural Engineering, Vol. 122, No. 7, pp. 726-733.
4. Melchers, R.E., (1987), "Structural Reliability Analysis and Prediction", Ellis Horwood Limited, Chichester, England.
5. Miner M.A., (1945), "Cumulative Damage in Fatigue", Transactions of the American Society of Mechanical Engineers, Vol. 67.
6. Moses, F., Schilling, C., and Raju, S., (1987), "Fatigue Evaluation Procedures for Steel Bridges," NCHRP Report 299, Transportation Research Board, Washington, D.C.
7. Nowak, A.S., (1995), "Calibration of LRFD Bridge Code", ASCE Journal of Structural Engineering, Vol. 121, No. 8, pp. 1245-1251.
8. Thoft-Christensen, P. and Baker, M.J., (1982), "Structural Reliability Theory and Its Applications", Springer-Verlag.

Rehabilitation of a Nineteenth Century Cast and Wrought Iron Bridge

Perry S. Green[1], Robert J. Connor[2], and Christopher Higgins[3]

Abstract

A cast and wrought iron bridge, constructed circa 1860, has undergone a complete rehabilitation and has been placed back into service as a pedestrian bridge at a new location adjacent to its previous site on Walnut Street in Hellertown, PA. As part of this rehabilitation, the material properties of the antiquated members were established. Both the cast and wrought iron demonstrated strength and ductility consistent with the period with the wrought iron exhibiting significant tensile strength and ductility as well as good impact toughness. The material test results verified that member strengths were adequate, indicated that connections between castings required reinforcement to ensure ductile behavior, and provided reference data for other historic bridge restoration projects.

Introduction and Background

The Walnut Street Bridge is a 16.8 m long single-span, cast and wrought iron, Pratt through-truss bridge. The structure has been designated PA-206 as part of the Historic American Engineering Record (HAER) collection and is featured in <u>Landmark American Bridges</u> (DeLony 1992).

The bridge was constructed circa 1860 by Charles N. Beckel, whose iron foundry and machine shop, located on Sand Island in Bethlehem, PA, cast many of the bridge members. The bridge employs several patented structural details of well-known engineer Francis C. Lowthorp of Trenton, NJ. These details permitted the attachment of tension diagonals, bracing, and bottom chord members to the panel point connections without the need for expensive forging or machining which was typical of the period. The bridge makes efficient use of both wrought iron and cast iron materials. Round wrought iron bars were used for the tension diagonals, bottom chord, and sway bracing. Cast iron was used for the compression and bending members including the top chord, verticals, endposts, lateral struts, and floorbeams. All the castings are remarkable in craftsmanship and structural detail. The bridge makes use of several unusual design features including modular connections, prefabricated construction practices, and optimally detailed cross-sections which include variable beam depth, variable flange widths, and integrally cast vertical and diagonal stiffeners.

In 1970, the bridge was removed from vehicular service due to structural and functional deficiencies. During construction of the new bridge, the old iron span was lifted from its bearings and placed alongside the creek where it had spanned. Spared from demolition by a group of concerned citizens, the bridge remained abandoned.

Condition Prior to Restoration

During the spring of 1994, a cursory visual inspection of the bridge was performed by a group of Lehigh University graduate students and the interesting structural and historical details were "rediscovered". Subsequently, a detailed visual inspection was conducted during that summer to assess the existing structural condition of the bridge, document overall structure geometry and member dimensions, and

---

[1]University of Florida, Gainesville, FL; [2]Lehigh University, Bethlehem, PA; [3]Clarkson University, Potsdam, NY

determine the potential for restoration. The majority of the cast iron members exhibited only minor corrosion damage and appeared structurally sound. The historically significant floorbeams were undamaged. Three of the eight cast iron verticals were damaged having large cracks or pieces missing near the bottom chord connection. One of these verticals was sufficiently damaged to warrant replacement. All of the wrought iron members exhibited significant section loss due to corrosion at connection locations and required replacement. After presenting several restoration alternatives to the bridge owners, a decision was made to erect the truss as a pedestrian bridge in a local historical park and a restoration effort was initiated.

Material Properties of Antiquated Engineering Materials

Prior to developing plans for the rehabilitation, it was deemed necessary to evaluate the engineering properties of the cast and wrought iron materials. To characterize the mechanical behavior of the cast iron, material tests in compression, flexure, tension, and Charpy impact were conducted. A sample of extraneous cast iron was removed from one of the end posts to permit fabrication of the required test specimens. Although the wrought iron bars were to be replaced as part of the bridge rehabilitation, these materials were also tested to determine their tensile behavior and toughness. Wrought iron was taken from the 32 mm diameter bottom chord members.

Tensile specimens were fabricated and tested according to ASTM E8. Average strain was measured over a 51 mm gage length with an extensometer and stress was calculated from the measured initial cross-sectional properties and applied load. The stress-strain behavior for the wrought iron and cast iron specimens are shown in Figs. 1 and 2, respectively. Figure 1 shows that the wrought iron exhibited significant tensile strength and ductility, with a distinct yield plateau before strain hardening, necking, and eventual fracture. Tensile properties of the wrought iron including yield stress, ultimate stress, and maximum strain are given in Fig. 1. In comparison, the stress-strain behavior of the cast iron (Fig. 2) exhibited minimal tensile strength and ductility, with no indication of imminent failure prior to fracture. Tensile properties of the cast iron including ultimate stress and maximum strain are shown in Fig. 2.

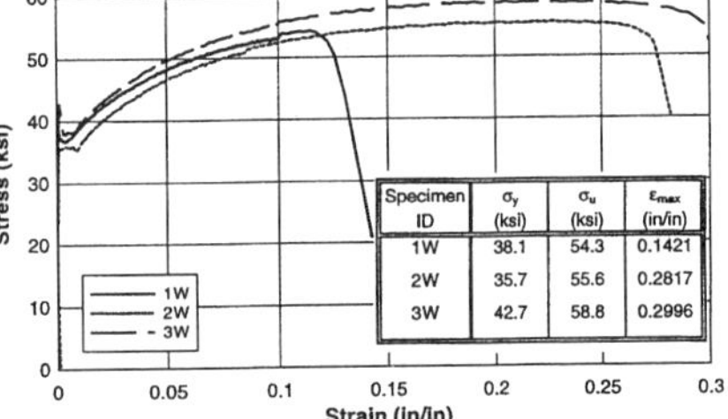

Specimen ID	σ_y (ksi)	σ_u (ksi)	ε_{max} (in/in)
1W	38.1	54.3	0.1421
2W	35.7	55.6	0.2817
3W	42.7	58.8	0.2996

Figure 1. Wrought Iron Tensile Behavior

Two specimens were fabricated from the cast iron and subjected to three-point bending according to ASTM E290. The nominally 13 mm x 25 mm rectangular samples were placed in a loading fixture with a 102 mm span and tested about the strong axis. Flexural strains were measured by a strain gage located on the bottom surface of the specimens directly under the load point. Flexural stresses were computed based on measured cross-sectional properties and the magnitude of applied load. The stress-strain response of the three-point bend specimens is shown in Fig. 3. The material exhibited a nonlinear behavior from the onset of loading and continued to soften with increasing load. There was no unloading branch as the material fractured without warning. The peak flexural tension strain for

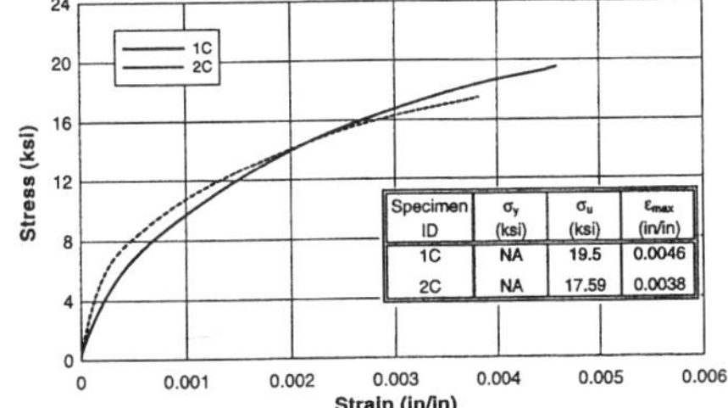

Specimen ID	σ_y (ksi)	σ_u (ksi)	ε_{max} (in/in)
1C	NA	19.5	0.0046
2C	NA	17.59	0.0038

Figure 2. Cast Iron Tensile Behavior

both specimens was approximately 0.006 mm/mm and the peak flexural stresses were 269 kN and 275 kN for specimens 1C and 2C, respectively. Figure 3 shows that specimen 1C was unloaded at a strain of approximately 0.0045 mm/mm and subsequently reloaded. The reloading branch was approximately linear and exhibited higher stiffness than the secant stiffness of the original loading curve at the point of unloading. This observation could potentially be used to identify the past load history of a cast iron member. It is important to note that the material from which these specimens were fabricated was extraneous to the casting and could not have been subjected to any significant loading history.

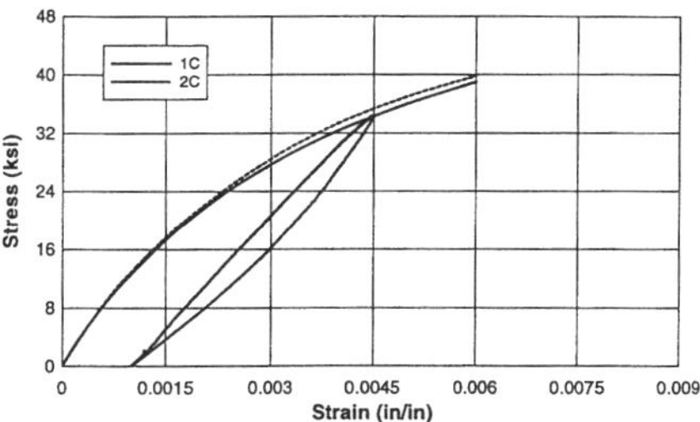

Figure 3.   Cast Iron Flexural Behavior

To characterize the compression behavior of the cast iron, three specimens were fabricated 13 mm in diameter and 38 mm long and tested according to ASTM E9. Strains were measured by both strain gages and displacement transducers attached to the specimens and stress was computed from the measured initial cross-sectional properties and applied load. The compressive stress-strain behavior of the cast iron specimens is shown in Fig. 4. The material exhibited an initial linear stiffness up to approximately 241 kN before becoming nonlinear. The cast iron specimens demonstrated compressive strengths greater than 552 kN and had a ductile behavior with a distinct unloading branch. Specimen 2C was unloaded at a strain of approximately 0.008 mm/mm and subsequently reloaded. The reloading branch again exhibited a higher stiffness than the secant stiffness of the original loading curve and was approximately linear.

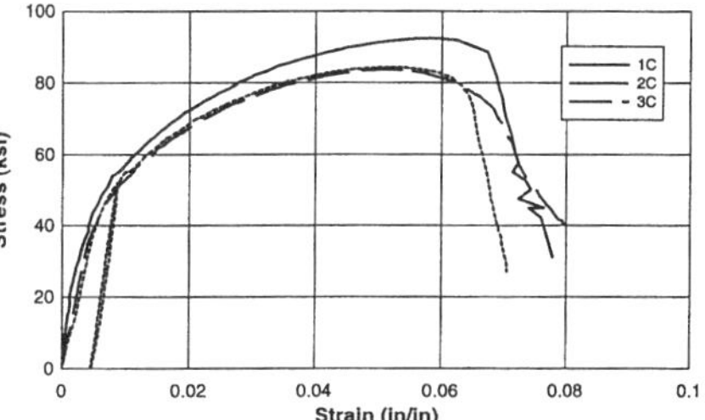

Figure 4. Cast Iron Compressive Behavior

To characterize the notch toughness of the materials, Charpy impact specimens were fabricated and tested for both the cast and wrought iron materials according to ASTM E23. Tests were performed over a temperature range of 20 to 98 °C. Measured absorbed energy and percent of shear fracture over the failure surface for the wrought iron are shown in Table 1. The wrought iron exhibited similar toughness to typical modern hot-rolled mild steel while the cast iron exhibited no measurable toughness. This finding was not unexpected given the documented inadequacies of cast iron in the literature.

Results of material tests indicated that the cast iron members were sufficient for use in a pedestrian bridge application. The notch toughness of the cast iron was minimal and thus it was important to ensure no significant stress concentrations were introduced during rehabilitation and any connections that might be subjected to tensile stress concentrations or flexure were reinforced with modern ductile materials.

Rehabilitation

A new location for the bridge was found in a local historical park near the original site on Walnut Street where it once stood. The restored bridge would span the millrace coming from a recently restored grist mill located nearby. Since the flow of water under the bridge would be controlled at the mill, this would alleviate any future periodic flooding or scour and also reduced the permitting requirements for the bridge.

A rehabilitation program was established which preserved the historically significant cast iron members and replaced the corrosion damaged wrought iron elements. Prior to disassembly, the individual members were tagged to ensure proper placement during reconstruction of the bridge. The bridge was taken down during the period September 23 - 25, 1994. Following disassembly, the individual castings were inspected for damage, sandblasted, primed, painted, and stored until reconstruction of the bridge at the new site. A foundry was located which produced three replicas of the original vertical cast iron members. Wrought iron members were replaced with modern A36 round steel bars of corresponding diameters. Connections between castings were reinforced with steel pipe or reinforcement plates to ensure a ductile connection between the members. Other components were fabricated from new materials as required.

The bridge was reconstructed on temporary steel falsework girders with an initial camber set equal to the dead load deflection of the structure and built in at the floorbeam locations. After all members were erected, the diagonals were tensioned and the falsework was lowered. The bridge is being redecked with solid sawn white oak timbers and rubble masonry will be used to cover the concrete abutments and wing walls.

Table 1. Charpy impact test results

Wrought Iron

Specimen Number	Test Temp. (°F)	Energy Absorbed (ft-lbs)	Percent Shear
A1	68	7	10
B1	68	15	20
C1	68	7	5
A5	104.7	36	80
B5	102.8	27	65
C5	102.3	7	10
A6	136.3	44	85
B6	135.3	22	40
C6	132.5	28	80
A3	163.5	47	100
B3	158	34	80
C3	155.5	18	35
A4	176.8	45	90
B4	177.4	59	90
C4	175.3	57	100
A2	208.4	57	100
B2	208.4	49	95
C2	208.1	59	100

Cast Iron

Specimen Number	Test Temp. (°F)	Energy Absorbed (ft-lbs)	Percent Shear
2	68.2	<1	N/A
6	68.9	<1	N/A
13	68.1	<1	N/A
4	208.6	<1	N/A
9	208.5	<1	N/A
8	208.5	<1	N/A

1 joule = 1.3558 ft-lbs

1 °C = (1 °F - 32) / 1.8

Summary and Conclusions

A historic cast and wrought iron bridge has recently undergone restoration. The mechanical properties of the antiquated materials were determined as part of the restoration process. The tests indicated the cast iron possesses significant compressive strength and ductility, limited flexural strength, and negligible tensile strength and notch toughness. The wrought iron exhibited significant tensile strength and ductility and notch toughness. Evaluation of the test results indicated that the connections between cast iron members required reinforcement to ensure ductile behavior.

References

DeLony, E. (1992). Landmark American Bridges, ASCE, New York, NY.

ASTM (1994), Metals---Mechanical Testing; Elevated and Low-Temperature Tests; Metallography, Annual Book of ASTM Standards, Vol. 03.01, Section 3; E8-94a, pp 60 - 80; E9-89a, pp 101 - 108; E23-94a, pp 140 - 160; E290-92, pp 335 - 338.

Restoring Historic Bridges Using Modern Methods

Joseph J. Pullaro[1]

Abstract

During the past twenty (20) years there has become an awareness of our history as seen through bridges with added emphasis being placed on ways to rehabilitate them for increased loads. Modern methods including finite element analyses, strain gauging and material testing has allowed us to better understand these structures and with innovative strengthening many can be kept in productive use. The following two (2) projects describe some of these methods.

Shelby Street Bridge - Nashville, Tennessee

The Shelby Street Bridge spans more than 900m across the Cumberland River in downtown Nashville. The bridge is composed of 48 spans, which are mostly concrete T beams and metal trusses, and two very unique reinforced concrete truss spans with a span of 30m and a rise of 4.5m. (See Figure 1) The bridge was designed by Howard M. Jones, a noted engineer, and erected in 1907. It is currently owned by the Metropolitan-Davidson County government. The bridge provides a 12.2m wide roadway with 2.7m sidewalks on each side.

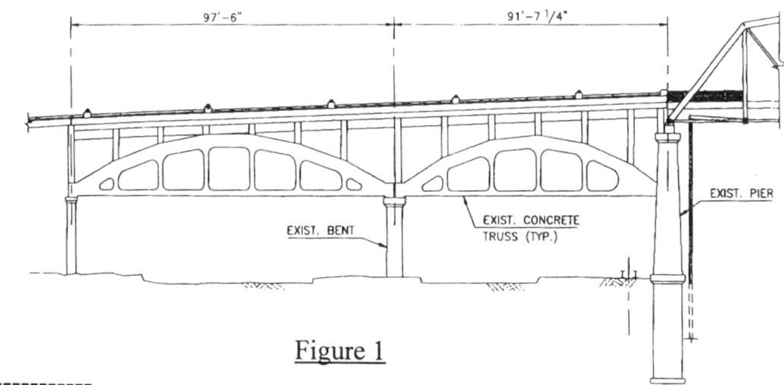

Figure 1

[1]Principal, A.G. Lichtenstein & Associates, Inc, Paramus, NJ

A new football stadium for the Tennessee Oilers is being constructed on the east side of the river with completion scheduled for the beginning of the 1999 season. The bridge which today carries vehicular traffic will be rehabilitated and its use changed to that of pedestrians. Architectural details such as lookouts at the river, lighting, decorative concrete and a replication of the original railing are part of the construction to make this crossing an enhancement to the stadium and the city. The bridge is on the National Register of Historic Places with the most interesting feature being the concrete trusses which are at the west approach spanning over an abandoned railroad.

Mr. Jones considered many options for the span over the railroad including a steel truss and a concrete arch but they were eliminated because of head room, dead load or effect of corrosive sulfurous smoke from the exhaust on exposed steel. He settled on the concrete trusses as the structure of choice even though there was no known American precedent at the time. He chose a bow string type truss but without diagonal members. From a structural standpoint the bottom chords act as ties to take the horizontal component of the end thrusts of the arched top chords. Thus, these two spans are technically trusses and partially function as trusses, but they also distribute the forces within the spans as an arch does. The chords are rectangular and include up to 18-54mm steel rods as reinforcement.

Mr. Jones apparently was aware of the limitations inherent in using concrete for truss construction, especially its low tensile strength. He sought to compensate for this weakness by trying to induce pre-compression into the bottom chords by torquing the nuts at the end of the rebars after the concrete set. This may well have been the first application of prestressing in bridge construction in this country. It was however not entirely successful as evidenced by the extensive deterioration of the bottom chord concrete.

The condition of the trusses was poor for many years as evidenced by spalls and areas of delamination. The bottom chords were found to be the most deteriorated probably due to the tension forces in the chord. Concrete coring indicated strength of 24.8MPa.

In order to be consistent with the design assumptions and intended behavior of the concrete truss, it was decided to perform a 3-dimensional finite element elastic analysis using uncracked section properties for the concrete truss members. However, it was realized that true truss type behavior cannot be achieved due to the rigid connections used. Hence, it was decided to use space frame members in the modeling of concrete truss members; space frame members are capable of carrying both axial force and moments.

The entire cross sections of all bottom chord members are in tension, under the uncracked section assumption. The maximum tensile stress is greater than 4.8MPa well above the modulus of rupture. Concrete in the bottom chord members would have cracked at these stress levels, in the absence of prestress forces. The excessive deterioration observed in the bottom chord leads one to believe that the original prestress was not effective in inducing sufficient compression in the bottom chords or that much of the original prestress has been dissipated by losses. In any case, the bottom chord effectively acts as a cracked concrete member, which means that the entire tensile forces are carried mostly by the embedded steel rods. All top chord members are in compression. The maximum compressive stress of 7.6 Mpa is within the allowable compressive stress for concrete as measured by the cores.

The truss members that exhibit the most concrete deterioration are the bottom chord and vertical members. These members also happen to be subjected to high levels of tensile stresses, direct tensile loads in the bottom chord, and bending stresses in the verticals. It can be surmised that the stress levels in these members have as much to do with the observed concrete deterioration as with the quality of concrete. Even good quality/strong concrete will experience premature cracking and deterioration at these high stress levels. Conventional removal of deteriorated gunnite and original concrete and patching will not provide a long term solution as the existing tensile stresses in the concrete must be substantially reduced. Hence, a long term solution to the truss concrete deterioration problem should consider ways of providing sufficient prestress to the concrete so that the tensile stresses never exceed the modulus of rupture of the concrete.

In order to restore these spans to full use, removal of deteriorated concrete and installation of new prestressing and recasting of concrete was proposed. 3-15mm pretensioning strands were installed at either side of the bottom chord. 100mm concrete encasement was installed and the strands jacked after the concrete attained a strength of 34.5 MPa. All this work was accomplished after the floorbeams and deck were removed and the bridge span placed on new bearings. By this method prestressing was introduced into the new and existing concrete.

Mr. Jones had a vision to construct something unique and long lasting. His attempt to post tension a bridge would not be practical until the 1950's but nevertheless he tried to do it with the means available to him. By adding post tensioning his bridge is now complete and should last many years.

<u>John Mack Bridge - Wichita, Kansas</u>

The John Mack Bridge constructed in 1929 spans the Big Arkansas River and was designed by James Marsh who received a patent for this design in 1912. There are a few of this type remaining in the U.S. and it is eligible for listing on the National Register of Historic Places.

The eight span 243.8m long structure is composed of composite steel and concrete tied arches with each 30.5m. long arch rising above the deck at the curb line. The spans are constructed on a 30° skew with reinforced concrete floorbeams and sidewalk brackets perpendicular to the arches at each panel point. The road width is 9.1m with two (2) 1.7m sidewalks. The slab is 230mm thick. (See Figure 2)

Figure 2

The arch is actually composed of steel angles and plates with concrete being placed around the arch after full dead load of the superstructure was applied. The ribs consist of four steel angles embedded in the 760mm x 1000mm deep concrete. The hangers transfer the loads from the deck to the arches in tension and are composed of four steel angles. The bottom tie beam is composed of two steel channels.

The bridge is generally in good condition except for the arch hangers which exhibit extensive cracking of the concrete with cracks up to 3mm wide. It is the condition of the hangers which caused concern for the structural capacity of the bridge.

The structure is very stiff and complicated due to the steel lattice and skew. It was decided that a three-dimensional analysis would be appropriate to model the stresses and find the cause of the cracks. In the three-dimensional analysis the influence lines for the live loads exhibited significant longitudinal distribution of the unit loads on a floorbeam, thereby reducing the influence coefficients for axial loads and moments at the hangers and floorbeams. Additionally, the stiffness of the deck and tiebeams also contributed to the reduced influence coefficients for moments in the arch ribs. The axial loads in the arches were not noticeably affected by the deck stiffness. The three-dimensional analysis also revealed the presence of transverse bending moments transverse to the arches in the arch hangers, most likely due to the skewed geometry of the bridge. The bending moments when combined with the tensile loads significantly increased the stresses in the hanger steel. These stresses are considered very conservative as the concrete will have a stiffening effect on the hangers under bending moments. The transverse bending moments in the hangers may also have played a role in initiating the cracking of concrete encasements.

The maximum stresses in the arch rib based on an 325 KN truck were 7.9 Mpa for the concrete and 41.8 Mpa for the steel. The stress in the steel of the arch tie is

100.8 Mpa, while the arch hanger stress was measured at 156.4 Mpa.

Concrete cores were obtained and tested at 26.9 Mpa minimum compressive strength resulting in an allowable stress of 8.3 to 11.0 Mpa, which is greater than the 7.9 Mpa stress in the rib. Steel samples were not taken, but structural steel of this age has an allowable stress of from 110 Mpa to 124 Mpa.

In order to verify the analytical results a diagnostic load test was performed utilizing a 23 metric crane. Strain gauges were attached to the slab, floorbeams, exposed hanger angles and concrete encasement of the arch. The strains in the upper and lower rib were substantially less than what the model predicted. It is felt this was due to the almost total participation of the skewed deck slab in carrying the load. Actual strains in the hanger steel were at least 50% lower than the model indicating that the concrete is participating in carrying the load.

In summary it is generally expected that load tests of this nature, especially on concrete bridges, produce strain readings lower than computed values, showing that many structures have greater load carrying capacity than that predicted by calculations. The thick skewed deck appears to substantially participate in the load transfer and the hanger concrete is working compositely with the steel. Having established that the bridge had a structural capacity far in excess of its requirements, the rehabilitation was only concerned with repairs to the hangers.

The method of repair chosen was to attach full-height high strength rods 35mm in diameter to all hangers. These rods reduce the stress on the existing hangers to within allowable levels, based on analytical results. The installation of the rods required the removal of all concrete encasements on the east and west faces of the hangers for the full length. The newly installed rods and the exposed angles were then encased in a new concrete cover constructed to match the original architectural treatment. The new cover is 50 mm thicker than the original cover but blends in with the hangers so it cannot be noticed.

It is evident that a structure of this magnitude requires a 3-D analysis to understand how it is behaving and that a load test is a good way to understand the shared behavior of stresses between the steel lattice and concrete encasement.

In this case, the concrete is participating in carrying the tension loads of the hanger, a conclusion which could not be supported without this type of test. Also, it was established that the very thick skewed deck participates in carrying the loads to a much greater extent than first believed. With the true behavior of the bridge understood, a simple solution for retrofit could be designed.

Seismic Retrofit of Historical Kolekole Bridge

*Harold Hamada[1], David Fujiwara[2] and Chad Nakamoto[3]

ABSTRACT

The Kolekole Bridge is a multi-span steel girder, steel truss bridge supported on concrete piers. The bridge embodies four plate girder spans and two steel truss spans with its total length in excess of 500 feet. It is one of many bridges selected for the Hawaii State Department of Transportation Seismic Retrofit Program. Located in Hilo, Hawaii, Kolekole Bridge provides the only major access road between Hilo and neighboring Hamakua communities and destruction of the roadway would sever transportation of goods and emergency services to large portions of the community. The retrofit scope required the bridge to remain serviceable after the seismic event, however, design strategies and alternatives were severely inhibited by environmental and historical concerns.

INTRODUCTION

The Kolekole Bridge was initially used for transport of sugar cane trains and later widened to support vehicular traffic. The original steel trestles were destroyed following a tsunami on April 1, 1946 and were rebuilt using major portions of the salvaged plate girders and steel truss members supporting a 7-1/2 inch thick concrete pavement. The bridge is supported by five concrete piers generally built in the shape of the letter "H" varying in height according to the land topography. Column cross-sections are square (5x5 feet) with square connecting beams (4x4 feet). The foundation footing is buried deep to prevent scouring and concrete pedestals extend between the pier columns and footings.

Spans 2, 5 and 6 comprise of 78-inch deep plate girders measuring 72, 72 and 73 feet, respectively. Steel trusses extend 132 feet over bays 3 and 4. The legs of the tallest piers, 2, 3 and 4 are inclined with a 2 to 12 batter. The rocker bearing

1) Professor, Civil Engineering, University of Hawaii, Honolulu, HI, 2) Principal, and 3) Engineer, KSF, Inc., 615 Piikoi Street, Suite 300, Honolulu, HI 96814.

"

Connection between the truss and pier permits translation and rotation and is considered to be the most seismically vulnerable due to its large vertical dimension.

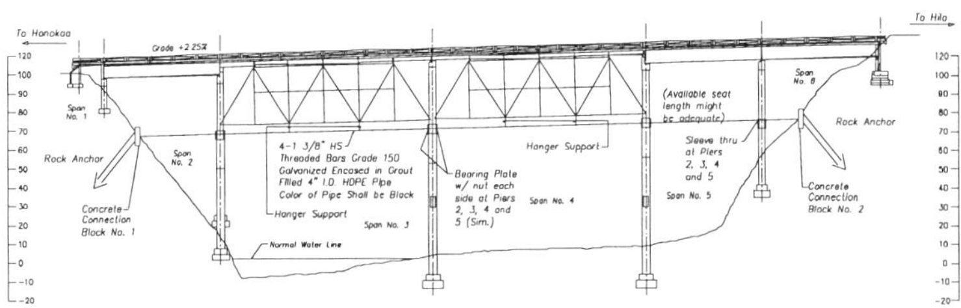

Fig. 1 Elevation Kolekole Bridge

SEISMIC ENVIRONMENT

The Hawaiian Islands have a recorded history (since 1834) of moderate seismic activity with numerous shocks causing damage in the VII-VIII Modified Mercalli range (Algermissen 1983). The seismic activity centers on the island of Hawaii and much of this activity is associated with volcanic processes, however, the stronger shocks are generally of tectonic origin. The greatest known earthquake occurred in April 1868 measuring at Richter 7.2. Other earthquakes causing significant damage occurred in 1938, 1951 and 1975. AASHTO specifications for the Kolekole bridge is category D, assigning it as a region with maximum seismic ground acceleration. The seismic retrofit was performed for a maximum ground acceleration of 0.42 g's, consistent with the seismic zoning also reported in the I.C.B.O. Uniform Building Code (UBC 1994). The shape of the spectrum associated with ground motion utilized in our retrofit scheme is one recommended by the American Association of State Highway and Transportation Officials (AASHTO 1996).

ENVIRONMENTAL AND HISTORIC CONCERNS

Kolekole Bridge is located within a few hundred yards from the coast and is directly exposed to the persistent northeasterly trade winds and salt spray. In addition to the effects of the marine environment, the prevailing high humidity and warm temperatures contribute to the corrosive environment. The bridge is also sanctioned under the State of Hawaii historic register, requiring that the visual impact of the seismic retrofit be minimized to preserve the bridge's original appearance. A decision was made to avoid working or excavating in the stream near the base of the footing to elude obtaining permits for environmental concerns. Design discretion was further retarded by existing air and water pollution regulations pertaining to lead levels which dissuaded the removal of red lead paint primer from the plate girders and steel truss members at the superstructure level. Corrosion susceptibility and the aforementioned constraints provided additional adversity and formidable impediments to the design and retrofit of the bridge.

STRUCTURAL SYSTEM

Inspection of the bridge structural system revealed the required strengthening for lateral seismic forces in the longitudinal direction. The structural system selected, within the confines of environmental and historical boundaries, was a cable system (high strength threaded bars) extending longitudinally beneath the bottom chord of the truss and plate girders connecting each pier and fastened with rock anchors at the bridge extremities. The design appropriates the existing structural system to carry the dead loads and the daily vehicular traffic and the cable to resist the lateral seismic forces in the longitudinal direction. The high strength bars are tensioned to minimize the dead weight catenary deflection and stress when seismic loads are applied to the bridge. In the transverse direction, the bridge incorporates wide concrete piers denoting structural integrity. A "lollipop" model was used, with each pier carrying its share of the transverse seismic load.

LATERAL FORCE RESISTING SYSTEM

Four 1-3/8-inch diameter high strength threaded bars encased in a 4-inch diameter high-density polyethylene pipe were selected as the lateral force resisting system (See Fig. 1). Each high strength threaded bar weighs 5.56-lb/ft, its ultimate stress is 150-ksi, and is capable of resisting an axial force of 189.6-kips at yield and 237-kips at ultimate. The encasement system is grout filled weighing 17.6-lb/ft and designed to retard corrosion of the threaded bar. The high strength rod will be stressed to 45.6-kip initially, 20 percent of its ultimate axial strength, producing a catenary deflection of 3-inch per 132-ft span between piers 2 and 3.

NONLINEAR ANALYSIS

In the transverse direction, a nonlinear static (pushover) analysis (Priestley et al. 1996) was performed using SC-Push3D. Developed by SC Solutions in Santa Clara, California, the computer program is capable of applying horizontal forces to frames in an incremental manner and predicts collapse by instability or by large deflections defined by the user. The beams and columns were modeled in the customary way in the elastic range and by using classical yield surfaces in the inelastic region. For concrete columns an interaction diagram was used to describe the concrete column behavior.

The interaction diagram was determined by a computer program developed by the State of California Transportation Department(CALTRANS). The stress-strain curves for the concrete and reinforcing steel were input variables. The steel-strain curve simulates strain hardening. Mander's equation is used to simulate the concrete stress-strain curve, where it is possible to simulate the lateral ties in the column. With the cross section and reinforcing scheme inputted, the program gives a moment-curvature diagram for a given axial column load. By varying the axial load, the moment- curvature diagram is constructed.

RETROFIT MEASURES FOR BEARINGS, SEATS & EXPANSION JOINTS

At the abutments a bearing seat extender was designed to accommodate the anticipated large longitudinal displacement and to prevent steel girders from falling from its supports. A concrete platform was designed to be supported by the abutment footing and fastened to the existing bearing support by drilling and epoxy mild steel reinforcement to insure shear transfer across the plane of the old and new concrete. The procedure suggested by the Federal Highway Administration Manual (Buckle and Friedland, 1995) was followed.

At the pier caps, similar bearing seat extenders and concrete fascia were designed for longitudinal and transverse displacement, respectively. In addition, in the areas where plate girders sat on the pier caps, cable restrainers were design to further restrict the plate girders from falling off their supports. In the transverse direction the lateral bracing of the plate girders required retrofit to accommodate the large forces anticipated in the girders bumping into the concrete fascia.

CONCLUSION

No remedial measures were made to the beam and column members constituting a relatively simple and economical retrofit design. The existing foundation was determined adequate for the day to day operations of carrying the dead loads and the traffic loads. The separation of the two loads, vertical and horizontal, enabled the cable solution to fully comply with the environmental and historical restrictions placed on the retrofit design within the budget of the project.

REFERENCES

(1) Algermissen, S.T. (1983), *An Introduction to the Seismicity of the United States*, Earthquake Engineering Research Institute, El Cerrito, California, 95-98.
(2) International Conference of Building Officials, *Uniform Building Code*, ICBO, Whittier, California, 1994.
(3) American Association of State Highway and Transportation Officials, *Standard Specifications For Highway Bridges, Sixteenth Edition*, Washington, D.C., 1996.
4) Priestley, M.J.N., Kowalsky, M.J., Ranzo, G. and Benzoni, G., "Preliminary Development of Direct Displacement-Based Design for Multi-Degree of Freedom System," *Proceedings 1996 Convention Structural Engineers Association of California*, Maui, Hawaii.
(5) *SC-Push3D, 3D nonlinear Pushover Analysis Program, User's Guide*, SC Solutions, Santa Clara, California, 1996.
(6) Buckle, I.G. and Friedland, I.M. (Editors), *Seismic Retrofitting Manual for Highway Bridges*, Federal Highway Administration, 6300 Georgetown Pike, McLean, VA, FHWA-RD-94-052, May, 1995.

Diagnostic Testing of A Unique Historic Bridge

Leon Lung-Yang Lai[1], Ph.D., P.E., S.E.

Introduction

A dual Whipple+Baltimore truss bridge was closed to traffic due to severe section losses to many truss members. The bridge was originally a two-span, wrought iron double intersection Whipple truss structure constructed in 1871. In 1937 Baltimore trusses were added to the outside of the original structure. Two sidewalk timber stringers sat on the floorbeams and were close to the bottom chords of the Baltimore trusses. Water contained in the timber and trapped in the gap between the two stringers provided a high moisture environment. Consequently, the horizontal legs of the Baltimore truss bottom chord angles were completely holed through at a number of panel points. Other structural defects were also identified through an in-depth inspection program. Major defects included: loose Whipple truss diagonal members, severe section losses to the hanger bars and other truss members.

In order to develop a rehabilitation program so that the bridge could be reopened to car traffic and the owner specified emergency vehicles, a diagnostic load testing was conducted. Surprisingly, although the Baltimore trusses had severe section losses to the bottom chords, they carried more than 48% of the test truck load. Three-dimensional finite element models were also developed to diagnosis the testing results and were further used to develop rehabilitation procedures. This paper reports results of the testing program and diagnostic analyses.

Bridge Description

The bridge is a two-span, through truss structure which consists of a 1937 Baltimore truss constructed outside the original 1871 wrought iron double intersection Whipple truss (Figure 1). Both trusses support floorbeams through hangers. Single U-bolts are used on the Whipple trusses and straight partially threaded bars (4 at each panel

---

[1] President, Specialty Engineering, Inc., Morrisville, PA

point) are used at the Baltimore truss panel points. Each span is 39,724 mm for an overall length of 80,467 mm.

The bridge carries a 5,791 mm roadway and is on a 90-degree skew angle. The deck is comprised of a 127 mm deep open steel grid deck that is welded directly to the stringer top flanges.

All Baltimore truss members are comprised of two angle sections. The top chords of Whipple trusses

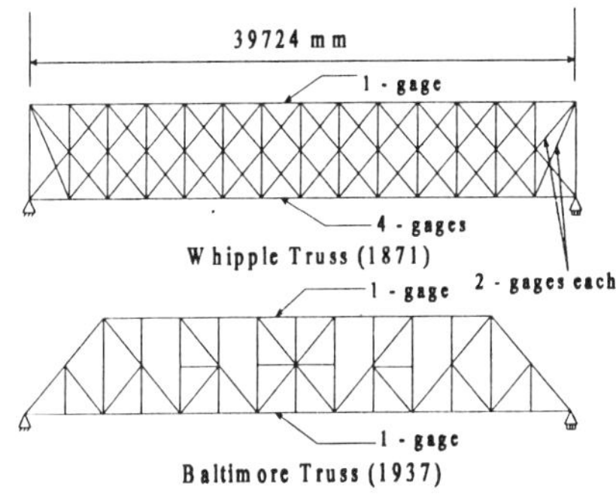

Figure 1 Bridge Elevation and Gage Locations

are comprised of two channels topped with one or two cover plates. All Whipple truss diagonal members consist of one or two rectangular bars and all bottom chords consist of two or four eye bars.

In addition to section losses to many bottom chord members, many vertical and diagonal members in the Whipple trusses also exhibit collision damage. Many diagonals of the Whipple trusses are loose under dead load. The floorbeam hanger rods of the Baltimore trusses typically exhibit severe section losses. At some locations, the hanger assemblies have broken away completely.

Load Testing and Analysis

Load-deformation monitoring was accomplished using fifty (50) electrical resistance type strain gages installed on selected Baltimore and Whipple truss members. The gage locations are shown in Figure 1.

The test vehicle was a dump truck. The gross vehicle weight of the truck was measured to be 111.6 kN with the front axle weighed 48.0 kN and rear axle weighed 63.6 kN. The spacing between both axles is 4,140 mm and the gage distance is 1,498 mm. The truck was placed in a number of positions, with the rear (heavier) axle at a panel point and in various transverse positions. For the east span, a total of 19 load cases were investigated. Nine locations were in both the south lane and north lane. One location was at the center of the roadway. Fifteen load locations were investigated for the west span. Seven of them were in the south lane and seven in the north lane. One was at the center of the roadway. The truck faced west on all 34 locations and moved from the east end toward the west end in both spans. The centerline of the outer rear wheel was approximately 762 mm from the inside face of the curbs when the test truck was in either the south lane or the north lane.

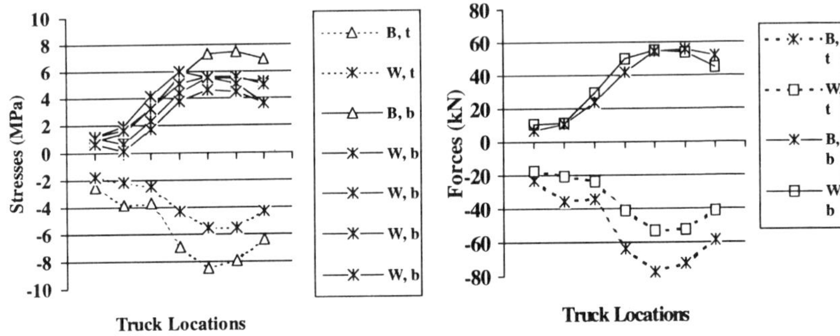

Figure 2 Measured Stresses   **Figure 3 Measured Member Forces**

Figure 2 shows the stresses measured on the south side trusses in the east span under the test truck loading assuming the modulus of elasticity to be 207×10^3 MPa. Sign convention is positive (+) for tension and negative (-) for compression. The symbols 'B' and 'W' in the legend box represent Baltimore and Whipple trusses, respectively; and 't' and 'b' represent top and bottom chords, respectively.

The test shows that for members containing more than one element (e.g. two eye-bars) the forces were distributed unevenly among the components. In some cases, the force in one element was more than twice of that in another component in the same member. Therefore, forces instead of stresses were used to determine the load sharing between the Whipple and Baltimore trusses. By combining the member stresses in Figure 2, Figure 3 shows the measured forces in the top and bottom chords of the south trusses in the east span. Based upon the forces measured in the top and bottom chords of the Whipple and Baltimore trusses, the load sharing between the two trusses was computed. Table 1 shows percentage of the test truck load carried by the Baltimore trusses for various truck locations. Even though the Baltimore trusses have severe section losses at a number of panel points, the Baltimore trusses carries more than 48% of the live load.

Table 1 Percetage of Load Carried by the Baltimore Trusses

	Test Truck in		
	North Lane	Center	South Lane
N.E. Truss	0.57	0.56	0.51
S. E. Truss	0.54	0.48	0.52

Figure 4 shows the forces in the end panel diagonal members of the Whipple trusses in the east span. The symbols 'S' and 'N' represent the south and north Whipple trusses, respectively. Analytically, each of the diagonal members in the first panel (U0L1 and U0L2) of a double intersection Whipple truss alternately carries forces when the load is at different panel points. This behavior can be seen when the rear axle of the truck is located at panel point 1. However, it becomes less obvious as the truck is away from panel point 1.

The STAAD-III computer program and three-dimensional finite element models were used to analyze the trusses so that the analytical results can be compared with the load testing results. The models include both Whipple and Baltimore trusses, floorbeams, and lateral bracings. Major defects, such as loose members in the Whipple trusses and broken floorbeam hangers in the Baltimore trusses, were also included in the models to simulate the existing condition of the bridges. The results

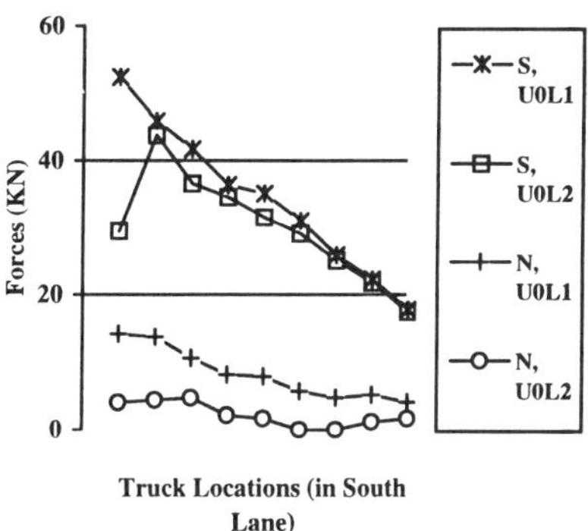

Figure 4 Measured Diagonal Member Forces

were compared with the measured forces in the gaged members as shown in Figure 5 for the south trusses of the east span. The symbols 'a' and 't' represent analytical and test results, respectively. The analytical and measured forces match very well at most truck locations.

Conclusions

Several findings were observed from this diagnostic testing:

1. By careful modelling and taking into account the major defects, the analytical results compare reasonably well with the test results. Therefore, the finite element models can be used for other loading to determine force or stress distribution in the trusses.
2. Although the Baltimore trusses have severe section losses to the bottom chords, the Baltimore trusses shared more than 48% of the load under the test truck.
3. Truss members containing more than one element have uneven force distribution among the elements. In some cases, the difference is more than 100%.

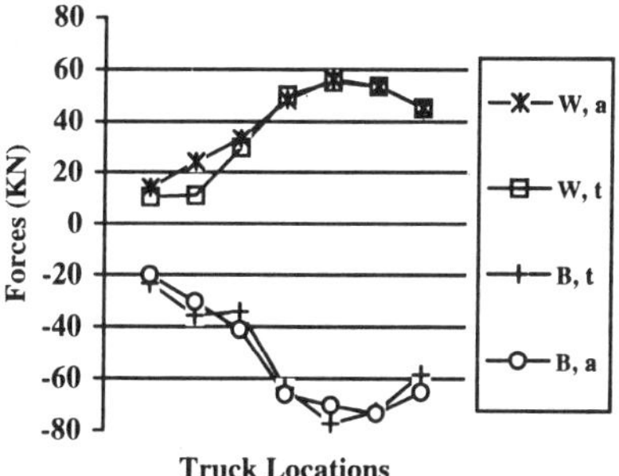

Figure 5 Test and Analysis Comparision

Coating tolerant thermography inspection system

Jon R. Lesniak[a], Daniel J. Bazile[a]

Introduction

Thermal methods correlate structural integrity with thermal diffusivity. If the molecular structure is altered, impairing transfer of forces, then the conduction of heat energy is also impeded (Cramer,92)(Osiander,96). Coating Tolerant Thermography projects a pattern of slowly translating thermal stripes to force conduction across cracks thereby, optimizing the measurable thermal gradient (Lesniak,95,96,97,97). Heat travels from "hot" stripe to "cool" stripe as the stripes slowly comb the structure for cracks. The in-plane heat flow is impeded by a structural flaw, such as a crack, creating a gradient in the thermal image, which clearly defines the crack. The fundamental principle behind Coating Tolerant Thermography is that only the thermal spatial derivative of a true structural anomaly will change sign with opposing heat flow. When the heat is flowing from the left, the gradient is positive (as defined) because the heat builds up behind the crack on the left. When heat is flowing from the right, the gradient changes sign becoming negative because now heat builds up behind the crack on the right side. Only a true structural flaw has this characteristic.

Projector development

The projector utilizes readily available, inexpensive and safe incandescent light sources. The source consumes about 1000W at 120 volts and emits most of its energy in the near infrared below 3μm. The light energy does not significantly conflict with the sensitivity range of typical infrared cameras (3-5μm and 8-12μm). The projector parses the radiation from the line source into 3 light paths via a gold plated elliptical light separator (Lesniak,98) and channels (Fig. 3).

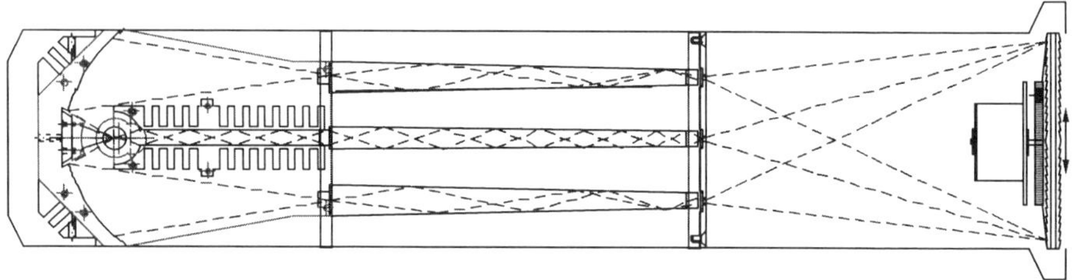

Fig. 3 Top view of projector layout (horizontal light management)[a]

[a] Stress Photonics, Inc. 3002 Progress Rd. Madison WI 53716

The lightweight Fresnel projector lens is translated in order to shift the position of the stripes. The dimensions of the projection cell are 380mm x 75mm x 150mm. The final system will incorporate several of the line sources so that higher line counts can be achieved. The final configuration will be determined through lab and field testing of actual bridge components.

<u>Infrared cameras</u>

In order to control cost of the final system, an uncooled bolometer has been chosen as the sensor engine. Table 1 compares the attributes of the bolometer to the Sterling cooled InSb.

	ICC Bolometer	**SBFP camera**
Thermal resolution (1s ave)	50mK	10mK
Spatial Resolution	320 x 240	128 x 128
Wavelength sensitivity	8-12µm	3-5µm
Cost	$15k	$50k

Table 1 Comparison of bolometer camera vs. SBFP

Figure 4 compares the ability of a bolometer and a Sterling cooled InSb camera operated at similar sampling times to resolve low magnitude thermal stripes.   As can be seen the InSb array camera has 3-4 times the thermal resolution. If a better signal to noise ratio is required, more energy can be projected, which is cheaper than a more expensive camera.

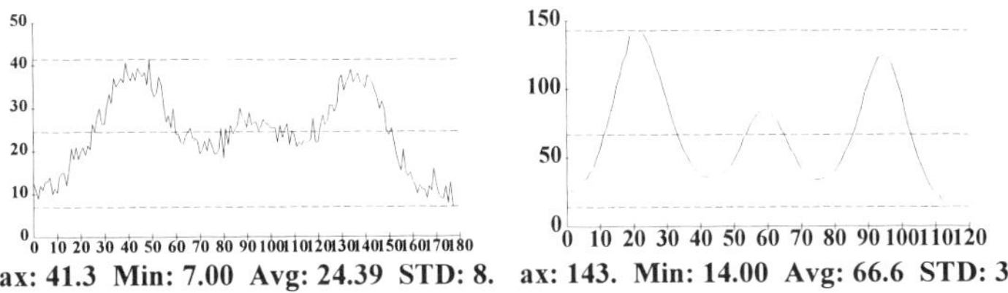

ax: 41.3 Min: 7.00 Avg: 24.39 STD: 8.   ax: 143. Min: 14.00 Avg: 66.6 STD: 3

a) Bolometer image of hot stripes   b) InSb camera image of hot stripes

Fig. 4 Thermal resolution of bolometer vs. InSb

Processing Algorithm

Each projected thermal stripe yields a positive and negative gradient zone as heat flows away from the stripe. The stripe is moved through four or more steps to accomplish complete coverage. The final challenge is applying the CTT algorithm to the plethora of thermal images with their respective gradient zones . Because the system is intended to be used on simple but varying geometries the location of the stripes in the image area is not certain. An automatic identification algorithm has been developed (Fig. 7). To accomplish total coverage several positions will be sampled.

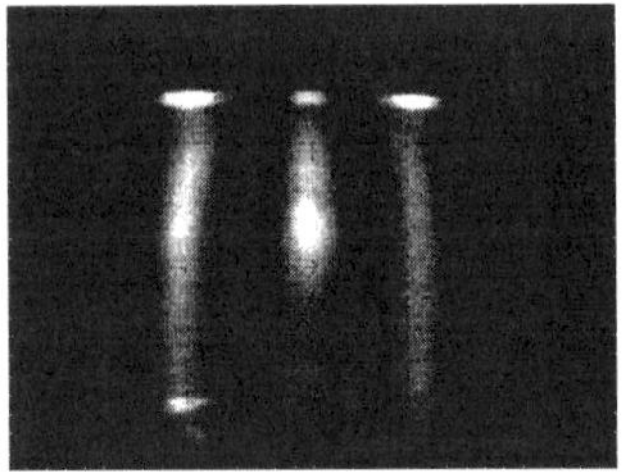
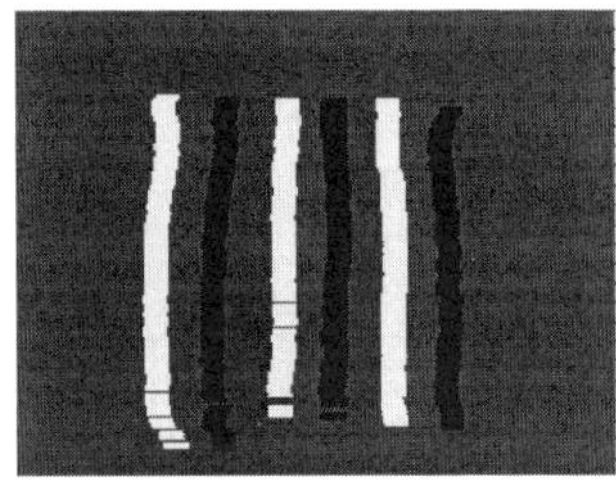

a) Thermal image   b) controlled width stripe

Fig. 7 Gradient mapping algorithm

Conclusion

Coating Tolerant Thermography is specifically designed for inspection of large, steel structures that are heavily and nonuniformly coated with paints and perhaps debris. Emissivity variances caused by such nonuniformities are completely eliminated in the final result. There is a clear path to an inexpensive hand held system that will be portable, robust, rapid and widely reproducible. Several issues remain to be resolved including final compilation algorithms.

Acknowledgments

Stress Photonics would like to acknowledge the supporters of this research, which included the Federal Highway Administration and the National Aeronautics and Space Administration. There are several individuals without whom this project would not have been possible: Steve Chase of the FHWA, Elliott Cramer of NASA LaRC and Phil Fish of the Wisconsin DOT.

REFERENCES

1. Cramer, K. E. and Winfree, W. P., "Thermographic imaging of cracks in thin metal sheets," Thermosense XIV, Jan., K. Eklund, Editor, Proc. SPIE Vol. 1682, pp. 162-170,1992.

2. Osiander, R., Spicer, J. W. M. and Murphy, J. C., "Analysis methods for full-field time-resolved infrared radiometry," Thermosense XVII, April, D. D. Burleigh and J. W. M. Spicer, Editors, Proc. SPIE Vol. 2766, pp. 218-227, 1996.

3. Lesniak, J. R. and Boyce, B. R., "Forced-Diffusion Thermography," Thermosense XVII, Sharon A. Semanovich, Editor, Proc. SPIE Vol. 2473, pp. 179-189, Orlando April 1995.

4. Lesniak, J. R. and Bazile, D. J., "Forced-Diffusion Thermography technique and projector design," Thermosense XVIII, D. Burleigh and J. Spicer, Editors, Proc. SPIE Vol. 2766, pp. 210-217, Orlando, April 1996.

5. Lesniak, J. R., Bazile, D. J. and Zickel, M. J., "Structural Integrity Assessment via Coating Tolerant Forced Diffusion Thermography," ASCE Structures Congress XV, Leon Kempner , Jr. and Colin A Brown, Editors, pp. 924-928, April 13-16, 1997, Portland, OR.

6. Lesniak, J. R., Bazile, D. J. and Zickel, M. J., "Coating tolerant thermography for the detection of cracks in structures," Themosense XIX, R. N. Wurzbach and D. Burleigh, Editors, Proc. SPIE Vol. 3056 pp.235-241 Orlando, FL, April 1997.

7. Lesniak, J. R., Bazile, D. J. and Zickel, M. J., "Theory and application of coating tolerant thermography," Themosense XX, John R. Snell,Jr. , R. N. Wurzbach, Editors, Proc. SPIE Vol. 3361 pp.325-330, Orlando, FL, April 1998.

Tomographic Imaging of Bridge Decks Using Radar

Steven B. Chase[1]

Abstract

Rapid, accurate, nondestructive and non-disruptive inspection of reinforced concrete bridge decks is needed to effectively manage the Nation's network of highway bridges. A new approach to the application of ground penetrating radar for bridge deck inspection has been developed to meet this need. This new approach utilizes an array of very low power impulse radar to create a synthetic aperture which interrogates a 2 meter width of bridge deck, up to 30 cm deep. Inspection speeds of up to 90 kilometer per hour have been obtained. The data from the array, along with spatial registration data is digitized, stored and post processed to produce the tomograms. A time-focused ,multi-frequency diffraction tomography imaging technique is used on the coherent backward propagation of the received reflected wavefield to form a spatial image of the scattering interfaces within the bridge deck. Two radar tomography imaging systems have been developed. The first is a slow speed, very high resolution, single element system and the second is a high speed, medium resolution array based system.

Introduction

Rapid, accurate, quantitative, nondestructive and non-disruptive inspection of reinforced concrete bridge decks is essential to effectively manage the Nation's network of highway bridges. The importance of bridge deck condition assessment is emphasized by the information presented in Table 1, which summarizes national bridge deficiencies by contributing factors. Highway bridges can be classified as structurally deficient for one or more of the five reasons listed under Table 1. Of the approximately 98,000 structurally deficient bridges in the National Bridge Inventory (NBI), 22 percent have poor or worse deck condition ratings. If relative proportions are based upon cumulative deck area or cumulative average daily traffic, the percentages for bad decks are 28 and 37 percent respectively. Highway bridges can

----

[1] Program Manager for Nondestructive Evaluation Research and Development, Federal Highway Administration, 6300 Georgetown Pike, McLean, Virginia 22101

also be classified as functionally obsolete for five reasons as shown in Table 1. Of the approximately 81,000 functionally obsolete bridges in the NBI, 60 percent of them have substandard deck widths. The relative percentages are also large when considering cumulative deck

Table 1. Why Bridges are Classified as Deficient (relative percentage)[*]

	Structurally Deficient Bridges					Functionally Obsolete Bridges				
Reason	DK	SU	SB	SA	WW	DW	CL	AL	SA	WW
By Number	22	21	25	28	4	60	16	9	11	3
By Area	28	24	26	14	8	45	42	4	8	1
By ADT	37	22	25	6	9	43	50	2	5	<1

[*]based upon an analysis of the 1996 National Bridge Inventory Database

Reasons for Structural Deficiency
DK : Deck condition rating
SU : Superstructure condition rating
SB : Substructure condition rating
SA : Structural appraisal rating
WW : Waterway appraisal rating

Reasons for Functional Obsolescence
DW : Substandard deck width
CL : Vertical or horizontal clearance
AL : Substandard approach alignment
SA : Structural appraisal rating
WW : Substandard waterway appraisal

area and cumulative ADT. This is significant from a bridge deck condition assessment perspective because very often, in order to determine if it is more prudent to widen a narrow bridge or replace the deck, the condition of the existing deck must be accurately determined.   The need for better bridge deck condition assessment methods is also supported by frequent cost over-runs and delays on bridge deck rehabilitation projects.

There has been much previous research and development and many nondestructive methods to inspect bridge decks, especially asphalt covered bridge decks, have been developed and used. These methods range from point measurements using impact echo and pulse echo stress wave methods, to full field infrared thermography and continuous strip ground penetrating radar. However, all of these previous methods have significant shortcomings. The acoustic or stress wave based methods such as impact echo, chain drag and pulse echo are essentially point measurements which require physical contact with the deck. These methods are inherently slow and are less reliable when an asphalt concrete layer is present. Infrared thermography full field and more rapid, with maximum inspection speeds of approximately 40 kilometers per hour (kph), but is often confounded by surface emissivity variations, provides no depth information for thermal anomalies and is much less sensitive when an asphalt concrete layer is present. The current ground penetrating radar systems provide high speed inspections, with inspection speeds of 90 kph. With up to four antennas, these systems cover multiple strips of bridge deck at a time. However, the spatial resolution is rather course with resolution on the order of 30 cm and data interpretation is based upon pattern recognition of amplitudes and phases of the radar echoes.   The synthetic aperture, planar array based radar tomography system overcomes most of these limitations. The high speed system uses an array of 64 bi-static, ultra-wide band, very low power radar to provide 3 cm cross-range resolution for a 2 meter wide section of bridge deck with a maximum inspection speed of 90 kph. The most unique feature of this system, however, is the application of time-domain focusing and multi-frequency diffraction tomography digital signal processing to produce tomographic images of the interior of reinforced concrete bridge decks. The major advantages to utilizing radar tomography are increased inspection speed, reduced cost, better spatial resolution and imaging capabilities.

Radar Tomography

The basic active element of the radar tomography system is a low power impulse radar. The radar is driven by a very short voltage impulse with an effective frequency bandwidth of .5 to 4 GHz . The pulse is converted to an electromagnetic wave and launched into the air as a roughly planar wavefront via a specially designed wide-beam, horn antenna. The wave passes through the air and enters the bridge deck. Scattering reflections occur at all dielectric interfaces. Typical interfaces include the air and the bridge deck, the asphalt and Portland cement concrete interface, reinforcing bars and concrete, delaminations, and the bottom interface between the Portland cement concrete and the air. The   interaction between the planar electromagnetic wave and the inhomogeneous, anisotropic, dispersive, and conductive three-dimensional solid reinforced concrete bridge deck is complex. The resulting reflections are received by a separate high gain antenna which converts the reflected electromagnetic waves to voltages. These voltages are amplified, digitized and stored. A synthetic aperture is formed by moving the antenna to a new position and repeating the process. A large, 2–3 meter wide, synthetic aperture is formed by positioning the antenna at many locations. The stored data is processed using a hybrid time-domain focusing and frequency-domain diffraction tomography methods to produce a three dimensional image of the scattering interfaces. The process is slow when a single antenna element is used. To attain practical inspection speeds a physical array of radar elements are moved over the bridge deck.

Radar tomography systems and results.

The status of the radar tomography systems are summarized in Figures 1 through 4. Figure 1 shows the low speed, high resolution radar tomography system. It consists of a single radar element on a self propelled scanning cart. The antenna sweeps back and forth as the cart slowly moves forward. Figure 2 shows a planar reconstruction of the interior of a test slab with an embedded Styrofoam target at 5 cm below the surface. The #5 reinforcement and the foam insert are clearly depicted. Figure 3 shows the high speed, array based, radar tomography system (HERMES) during a recent field test. Figure 4 shows a 2 meter by 20 meter planar tomographic image of the top layer of reinforcement using data acquired by the array based system.

Figure 1. High resolution, low speed system

Figure 2. High resolution image

Figure 3. High speed, array based radar system

Figure 4. Planar radar tomograph of top layer of reinforcement form bridge deck.

Evaluation of Prestressed Concrete Girders Using Magnetic Flux Leakage

Al Ghorbanpoor[1], FASCE

Abstract

A non-destructive evaluation system is developed based on the magnetic flux leakage concept to assess the condition of steel within prestressed concrete girders. Corrosion and/or fracture of prestressing steel could compromise the safe load carrying capability of prestressed concrete girders and could now be detected using the new system. The system can scan the length of a girder and is controlled through wireless communication that is integrated with a remote notebook computer. The system is capable of detecting approximately 5 percent loss of cross section in the prestressing steel and is easy to install on the girder within a few minutes.

Introduction

Prestressed concrete girders have been used in bridges and other structures since the 1950's. These members have generally performed well. However, with aging of these members, evidence of corrosion and deterioration of prestressing steel has been observed in some cases throughout the world. Since the prestressing steel is the primary load-carrying component of prestressed concrete girders, it must remain free of corrosion or other deterioration. Highway bridges, including those built with prestressed concrete girders are inspected routinely in the U.S. on a scheduled basis to determine their condition. The evaluation procedure has been based on visual inspection in most cases. Corrosion of prestressing steel in both the prestressed and post-tensioned concrete members may not be seen through corrosion stain and concrete cracking during the early stages. Normally, when visual signs of corrosion of prestressing steel are present in these members, it can be concluded that the corrosion activities have been in progress for some time and extensive damage could have resulted. In order to better and more reliably assess the condition of prestressed concrete members, it is desirable to develop a non-destructive evaluation (NDE) capability that is field worthy and effective.

---

[1] Professor, Department of Civil Engineering and Mechanics, University of Wisconsin-Milwaukee, P.O. Box 784, Milwaukee, WI 53201

The concept of magnetic flux leakage (MFL) has shown promise for assessing the condition of steel in prestressed concrete members. A study, supported by the U.S. Department of Transportation, the Federal Highway Administration (FHWA), was undertaken that has lead to the development of a NDE system based on the MFL concept. The system is capable of detecting approximately 5 percent loss of cross section in steel, within concrete, due to corrosion.

Principle of Magnetic Flux Leakage

The magnetic based NDE concept for assessing the condition of steel in concrete structures utilizes the ferromagnetic property of the steel to detect perturbance of an externally applied magnetic field due to the presence of flaws in the steel. A more comprehensive discussion of the concept and its application to bridge structures can be found in the published literature by the author and others[1-4].

If the direction of the applied magnetic field is set to be collinear with the longitudinal prestressing steel, the flux lines will also be collinear with the steel elements within concrete. Any change in the cross sectional area of the steel at any point will cause a flux leakage or fringing, where the continuous flux lines within the steel will be forced into the medium surrounding the flaw. Since the effect of concrete on the magnetic field is negligible, the field leakage may be measured by sensors in the air near the surface of the concrete. If the magnetic source and sensors are moved along the length of a concrete member, the changes in the field, due to the presence of flaws in steel, can be recorded as a continuous signature in terms of time and the field amplitude. The signature is then analyzed to obtain information relevant to the location and extent of the flaw in the steel (Fig. 1). Signal features such as the amplitude (A) and peak-to-peak separation (p), as shown in the figure, are used to characterize the flaw. The flux leakage usually results in a three dimensional perturbance of the magnetic field in the vicinity of the flaw. The perturbance of the field may be measured by using an array of sensors and is shown in Fig. 2. The amplitude of the perturbance is normally associated with the flaw size. Other factors also have significant influence on the field perturbance and should be taken into consideration. These include the strength of the magnetic source, adjacent steels including stirrups, the distance between the magnet and the steel, and the distance between detecting sensors and the steel.

A two- or three-dimensional analysis method may be used to evaluate the recorded MFL signals. A two-dimensional method includes a simple subtraction technique where point-by-point subtraction is performed on two signals recorded by two different sensors. The effects from most ferromagnetic artifacts will be minimized in the resulting signal after the subtraction. Another two-dimensional method of signal analysis is a technique based on the correlation concept. Here, a recorded signal is compared, point-by-point, with a mathematically defined ideal flaw signal. The ideal signal is defined based on the data obtained from laboratory

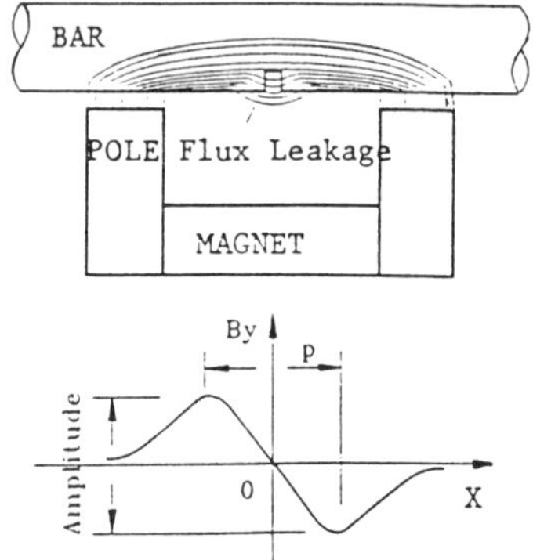

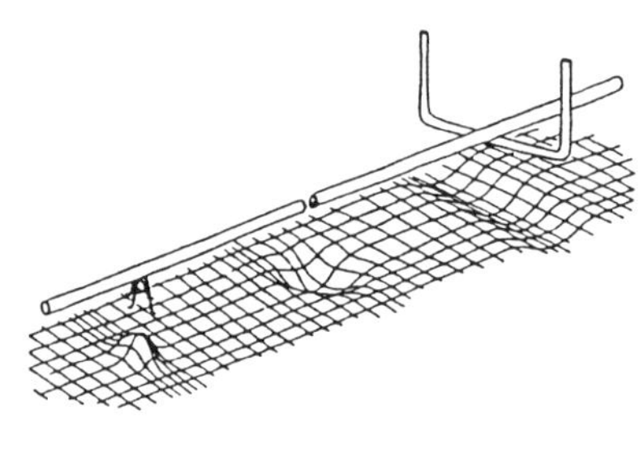

Figure 1. Magnetic flux leakage and signal[3]

Figure 2. Three-dimensional variation of magnetic field[3]

and field tests of components with various flaws. A perfect match is represented by a correlation factor of +1.00 and a total mismatch is indicated by a value of 0.00. The subtraction and correlation techniques may be combined to enhance the MFL data interpretation capability, Fig. 3. In the figure, the correlation values corresponding to different percentages of loss of section are also shown.

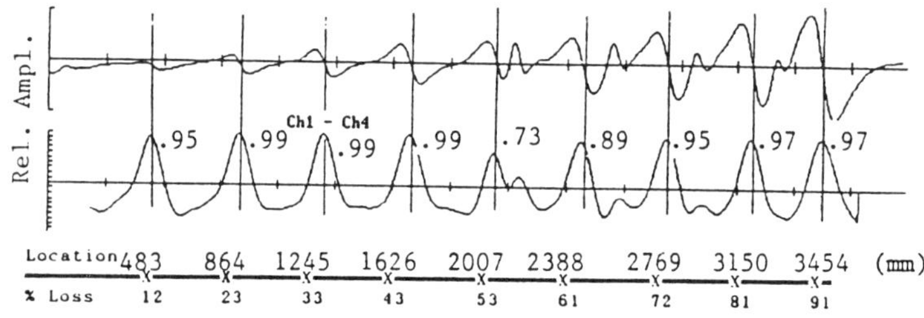

Fig. 3. Magnetic field signal analysis (data subtraction & correlation techniques)[3]

In the three-dimensional analysis method, multiple records of signals with varying amplitudes and shapes are obtained from a series of sensors located near the concrete member's surface. A graphical view of the data, shown in Fig. 2, can be used to detect the presence of discontinuities such as flaws in the steel within concrete. The method offers superior analysis capability where flaws in a steel bar or cable can be easily differentiated from the effects of ferromagnetic artifacts in concrete.

The MFL System

The MFL system developed in this study is comprised of two major components, a scanning device that is installed on a prestressed girder, and a remote computer that controls the scanning device and records and analyzes the MFL data. The device is easy to install and can scan the entire length of the girder under its own power. It consists of a magnet, a series of Hall effect sensors, relevant electronic circuits, a wireless data transmitter, and a series of mechanical and electrical devices for its operation, Fig. 4. It has an effective scan width of 200 mm (8 in.) in each direction and the data from different scans can be combined to get an overall MFL record for the entire width of the girder under test. The remote computer is a notebook computer that consists of a wireless data receiver, and software for the control of the scanning device and for the acquisition and analysis of the MFL data. The use of the system has been demonstrated in both the laboratory and field.

Figure 4. The MFL Scanning Device Installed on a Prestressed Girder

Acknowledgments

Support for the conduct of this study was provided by the U.S. Department of Transportation, the Federal Highway Administration.

References

1. A. Ghorbanpoor, G.R. Steber, and T.E. Shew, "Evaluation of steel in concrete bridges: *The Magnetic Field Disturbance (MFD) System*," Final Report No. FHWA-SA-91-026, Federal Highway Administration, Washington, D.C., 1991.

2. F.N. Kusenberger and R.S. Birkelbach, "Detection of flaws in reinforcing steel in concrete bridge members," Final Report No. FHWA/RD/-83/081, Federal Highway Administration, Washington, D.C., 1983.

3. A. Ghorbanpoor and S. Shi, "Assessment of corrosion of steel in concrete structures by magnetic based NDE techniques," <u>Techniques to Assess the Corrosion Activity of Steel Reinforced Concrete Structures, ASTM STP 1276</u>, Neal S. Berke, Edward Escalante, Charles K. Nmai, and David Whiting, Eds., American Society for Testing and Materials, Philadelphia, 1995.

4. D.E. Bray and R.K. Stanley, *Nondestructive Evaluation*, pp 173-364, McGraw-Hill, New York, 1989.

LOAD RATING AND PERMIT REVIEW USING LOAD AND RESISTANCE FACTOR PHILOSOPHY

Bala Sivakumar[1], P.E.

ABSTRACT

The objective of NCHRP Project 12-46 is to develop a manual with supporting commentary and illustrative examples for the evaluation of highway bridges that will be consistent with the AASHTO LRFD Specifications in using a reliability based limit states philosophy. The primary goal of this project is to produce a practical evaluation manual that is easy to understand and use, which will receive broad support for adoption and distribution by the AASHTO Subcommittee on Bridges and Structures. The reliability aspects will remain invisible to the evaluation engineer through the use of calibrated load and resistance factors, as was done in the LRFD Specifications.

INTRODUCTION

The 1994 LRFD Specifications for highway bridge design introduce a limit states design philosophy, based on structural reliability methods, to achieve a more uniform level of safety (reliability) throughout the system. The current AASHTO Manual for Condition Evaluation of Bridges (MCE) contains provisions for only the conventional (deterministic) Allowable Stress and Load Factor methods of load rating existing bridges. Hence, there is an acute need for a comprehensive new evaluation manual that will not only be consistent in philosophy and approach with the LRFD Specifications but will also be technologically current. NCHRP Project 12-46 was initiated in March 1997 to develop a new AASHTO Load and Resistance Factor Evaluation Manual for highway bridges that will introduce reliability based evaluation methods. A Lichtenstein led Research Team was awarded the contract for the project. Mr. Charles Minervino is the Principal Investigator and Mr. Bala Sivakumar is the Co-Principal Investigator. Dr. Fred Moses and Dr. Dennis Mertz

[1] Associate, A.G. Lichtenstein & Associates, Inc., 45 Eisenhower Drive, Paramus, New Jersey 07652

serve as consultants. A Draft Manual was completed and submitted for review in August 1998. Once adopted the new Manual will replace the existing MCE.

Major revisions to the existing MCE will include:
- A new section on load rating bridges using the Load and Resistance Factor philosophy.
- Load rating procedures that are contingent upon the live load model and the intended use of evaluation results.
- Procedures to determine site-specific load factors for load rating.
- Customized load factors by permit type for overload permit review.
- A new section on fatigue evaluation of steel bridges.
- A new section on non-destructive load testing of bridges.
- Parallel commentary with several illustrative load rating examples.

RELIABILITY BASED ASSESSMENT

Upgrading an in-service bridge is far more costly than incorporating extra capacity at the design stage. Therefore, a more refined approach to evaluating the load capacity of existing bridge can be economically justified. Structural reliability methods contain the necessary ingredients to provide a more rational, a more flexible, and a more powerful evaluation strategy for existing bridges. In a reliability based approach the evaluation could provide a uniform safety level, without resorting to traffic restrictions or strengthening, by reducing uncertainties. This can be done by obtaining improved resistance data, site-specific traffic data, and improved load distribution analysis. In evaluation, uncertainties can be reduced based upon site-specific considerations that were unavailable to the designer.

DESIGN VS. EVALUATION:

Bridge design and evaluation, though similar in overall approach, differ in important aspects. In the design stage there is greater uncertainty in loading over the life of the bridge whereas in evaluation there is greater uncertainty on the resistance side, especially in the case of degraded bridges. In the LRFD Specifications much of the emphasis and calibration of design factors was based on multi-girder steel and concrete bridges that were considered representative of current and future trends in bridge design. In developing a compatible evaluation manual it is important to recognize that the existing inventory of bridges is comprised of a wide array of bridge types, structural systems, material types, and physical conditions. Additionally, the evaluation criteria pertaining to reliability indices, limit states, and load and resistance models could be different from that used for new bridge design in the LRFD Specifications.

KEY TECHNICAL ISSUES PERTINENT TO EVALUATION

Reliability Indices

In LRFD code calibrations, the target reliability index (beta) of 3.5 for the strength limit state was selected as it was considered indicative (average) of the reliability indices inherent in girder bridges designed by the AASHTO Standard Specifications. The level of reliability in existing multi-girder bridges designed to the AASHTO Standard Specifications was found to vary from a low of 2.0 to a beta as high as 4.5.

For design, a relatively high target reliability index was chosen as the cost of compliance is only marginal. For evaluation, a lower bound on acceptable reliability is more appropriate as the cost impact due to bridge strengthening or traffic restrictions could be quite significant. The reliability index for evaluation was selected based on calibrating with existing practice, which generally allows up to operating stress levels to be used for redundant bridges. The operating level (beta = 2.5) has served as an acceptable safety level for posting and permit decisions for redundant systems in the past and was selected as an acceptable basis for the new Manual calibrations.

Limit States

The new LRFD Specifications introduce the philosophy of limit states design, which in essence is a systematic way of considering all applicable design criteria by grouping them into limit states. This philosophy has now been extended to the evaluation of existing bridges. Differences exist between the application of limit states to design vs. evaluation. As in the case of reliability index, there may be a high cost penalty for imposing certain non-strength related limit states in evaluation compared to design, where the cost impact may be negligible.

The very nature of evaluation calls for different limit states to be applied. If the evaluator is concerned about accumulating excessive fatigue damage during normal usage, the fatigue-and-fracture limit state should be considered for evaluation. On the other hand, if the evaluator is only interested in precluding damage from more limited, perhaps single, passages of permit vehicles, the service and strength limit states would only be considered, and not the fatigue-and-fracture limit state.

Live Load Models

As a result of the exclusions permitted by the grandfathered rights the HS loading does not bear a uniform relationship (for varying span lengths) to many of the vehicles allowed on the roads. In developing the LRFD design specifications it was determined that if the objective of developing a new specification was a more uniform and consistent safety of bridges, a new live load model (designated as HL93 loading) would be necessary. While this notional load provides a convenient and uniform basis for design, it bears no resemblance or correlation to any vehicle type on the roads. Practical difficulties are therefore bound to arise in using HL93 rating results for load posting. Additionally, the local live load environment is often less severe than the prescribed design loading.

The AASHTO legal loads (TYPE 3, TYPE 3S2, TYPE 3-3) model three portions of the Federal Bridge Formula which control short, medium, and long spans. Therefore, the combined use of these three AASHTO legal loads results in uniform reliability over all span lengths, as was achieved with the HL93 notional load model. These vehicles are presently widely used for load rating and load posting purposes. They are based on present legal load limits. These AASHTO vehicles model much of the configurations of present truck traffic. They are appropriate for use as rating vehicles as they satisfy the major aim of providing uniform reliability over all span lengths.

PERMIT REVIEW

Bridge owners usually have established procedures which allow the passage of vehicles above the legally established weight limitations on the highway system. These procedures involve the issuance of a permit which describes the features of the vehicle, its load, and route of travel. Permits are granted for a single trip or for multiple trips valid up to one year or more. Single trip permits for excessively heavy loads may require the use of escorts and the restriction of all other traffic on the bridge being crossed.

The new Manual provides procedures for checking overweight trucks that are analogous to load rating for legal loads except that load factors are selected based upon the permit type. The actual permit vehicle will be the live load used in the evaluation. The reliability level for permit crossings is established as the same level as for legal loads. Load superposition is modeled by the live load factor which is sensitive to the permit type and the number of crossings. For routine permits with unlimited crossings the critical load case will be a situation with a permit in one lane and normal truck traffic in the other lane. For heavy special permits (>125 K) with limited number of crossings the likelihood of another heavy vehicle alongside the permit truck during a bridge crossing is small. The probability of this occurring increases with ADTT and the number of permit crossings. To maintain the target bridge reliability higher live load factors are specified for higher ADTT and increasing number of total crossings. A Table of permit load factors suitable for most commonly encountered permit situations will be provided for easy application

Acknowledgment

This project is funded by the Transportation Research Board with David Beal as Program Manager. The research work leading to this paper was performed jointly by the Research Team.

Calibration of Load Factors for Load and Resistance Factor Evaluation

Fred Moses[1] , M. ASCE

Abstract

This paper outlines the derivations of the live load factors in the proposed AASHTO Condition Evaluation Manual (MCE). Load factors are derived for bridge rating under legal loadings, routine and special permits, incorporation of site specific traffic data and weight posting curves.

Introduction

There has been considerable research and data gathering in connection with recent LRFD - based bridge codes. These codes contain load and resistance factors which provide consistent target reliability levels for design of components. The steps in a structural design code calibration include the following activities:

- Assemble a data base on the various load and resistance variables.

- Define a calculation procedure for the safety index, β.

- Select a target safety index.

- Formulate a design model. For example, in LRFD practice, a component strength is typically checked by an equation of the form:

$$\phi\, R_n = \gamma_d\, D + \gamma_L\, L_n \qquad (1)$$

where ϕ is the resistance factor, R_n is the nominal component resistance computed by a prescribed formula, γ_d is the dead load factor, D is the nominal dead load effect, γ_L is the live load factor and L_n is the nominal live loading effect prescribed by a loading model such as HS20, HL 93 or some other vehicle and/or uniform loading model.

- Select partial load and resistance factors ϕ, γ_d and γ_L to meet the target betas.

A major concern for the proposed CE Manual is the selection of the load and resistance factors for a wide range of applications, that is, different traffic and live loading environments as well as possible cases of deteriorated spans. A significant random variable is the projected maximum truck traffic live loading, discussed herein.

---

[1]   Professor , Dept. of Civil and Environmental Engineering, University of Pittsburgh, Pittsburgh, PA 15261

Derivation of Live Load Factors

The derivation of live load factors included the following steps(Moses, 1998). First, a description of a truck weight spectra is given to make the evaluation methodology in the AASHTO CE Manual consistent with calibration of the AASHTO LRFD Design Specification (Nowak, 1993). An evaluation should utilize a site specific truck population and it is necessary to have a basis for comparison of the site traffic with a reference traffic for calibrating the respective rating live load factors. The Nowak truck weight data was determined equivalent to weight parameters for the AASHTO legal vehicles. After trial and error, these truck data(top 1/5th of weight spectrum) were best matched by a 3S2 population with a normal distribution and a mean of 303kN(68 kips) and a standard deviation of 80kN(18 kips).

The next step was to review traffic models for assessments of the side by side occurrences of heavy trucks. These multiple presence events usually control the maximum live loading on a span. Site truck volumes based on recorded ADTT values should be used to select a more optimum load factor for rating and evaluation.

The basis for calibration of live load factors in the MCE Manual is consistent with the safety indices generated for the AASHTO design specifications. The Ontario data was used by Nowak to calibrate the AASHTO LRFD Specification to a safety index of 3.5. At least for the time the data was taken, the 3.5 target beta is higher at most Interstate sites. The calibrated loading model is used as a reference in the MCE Manual for rating, postings and permit checking.

Evaluation Live load model

The proposed MCE adopts the AASHTO legal vehicles as the basis of the calculation of ratings. These vehicles are familiar to rating agencies and used to check posting requirements and to determine if a bridge is satisfactory for legal loads. It was also shown in the AASHTO Guide Specification for evaluating steel and concrete bridges that uniform reliability could be achieved by using the legal vehicles as the nominal load effect calculation model. The AASHTO legal vehicles have the added benefit that ratings can be converted to tons and reported in a familiar format.

Although the AASHTO LRFD Specification adopted the HL93 load model in place of the HS20 load model, it was noted that the AASHTO legal vehicles have a consistent bias to the HL93. The nominal HL93 model closely averages 1.77 times the effect of a 3S2 vehicle with weight equal to 320kN(72 kips) over a range of simple spans. The factored HL93 design load effect is then 1.77 times the load factored (1.75) 3S2, which is equivalent to 992kN(223 kips!) This margin creates considerable flexibility in the rating process to achieve safety for typical site traffic situations.

Calibration

The recommended approach for calibration is suitable for an evaluation code which requires greater flexibility in specifying live load factors for different site traffic and permits and amount of site traffic data retrieved. The issues of marginal costs and allowable target risks are considered but not in an explicit format. Since a wide variety of engineers in different rating organizations must use the same criteria, the evaluation format must be clear, unambiguous and lead to similar results by different investigators. Furthermore, the AASHTO evaluation criteria must relate back to the

AASHTO design specification and provide a clear relationship of the reliability and design margins in design with those in evaluation.

To simplify, the calibration was based on similar data used for the AASHTO-LRFD design rules, namely the dead and live load factors and resistance factors (Nowak, 1993). The design load factor of 1.75 and target index of 3.5 applies to a very severe traffic loading case. Adjustments were made to consistently reflect both site realities and economic considerations of evaluation criteria vs. design criteria.

The ADTT value at a site which is often known was used to project the expected maximum load effect. The required live load factor depends directly on this expected load effect as follows:

$$\text{live load factor} = \frac{\text{expected mean load effect}}{\text{Most severe load effect}(=240 \text{ kips})} \times \text{most severe live load factor} \quad (2)$$

For the most severe traffic case, the equivalent expected mean load effect was projected as 1068kN(240 kips) or equal to two 120 kip vehicles moving side by side. The corresponding evaluation load factors for different traffic cases were derived consistent with the statistical live load data used in calibrating the AASHTO-LRFD Specification (Nowak,1993). The AASHTO design live load factor of 1.75 is adjusted in an evaluation to account for the following considerations:

-reduce the 1.75 to 1.6 since the latter is acceptable though the factor of 1.75 was adopted to be "conservative". Using 1.6 reduces the live load factor by a ratio of 0.91 (i.e., 1.6/1.75). Note, the target safety index is an average and not a bound.

- reduce the reliability level from the design level of 3.5 to the operating level of 2.5, corresponding to the working stress factor operating methods. Several parametric analyses indicated this reduction in beta corresponds to a reduced load factor ratio of about 0.76.

-reduce the live load factor to account for a five year instead of a 75 year exposure. This leads to a reduction in load factors by about a ratio of 0.94.

-Comparing the HL 93 bending effects with the three AASHTO legal vehicles gives a quite uniform ratio of about 1.73.

To compare the live load factors with the extreme live load factor in the AASHTO Guide Specification which is 1.8, it must be considered that the Guide used a target for beta corresponding to load factor rating (LRF) of 2.3. This further reduces the live load factors by a ratio of 1.80 / 1.95 = 0.92.(See NCHRP 301 Report and AASHTO Guide Specification for comparisons).

Balancing these different considerations means the design level load transforms in the rating of bridges for legal loadings to:

1.75 HL 93 =1.75 x 0.91x 0.76 x 0.94 x 1.73 x 0.92 x 3S2 = 1.81 x 3S2   (3)

Thus, the value recommended for the CE Manual for the most severe traffic case is 1.80 . For an ADTT of 1000, it was shown that the maximum expected load effect acting in both lanes projected from the Ontario data is only 966kN(217 kips). The load factor for this traffic case case is from eqn. 2:

$$\gamma \,(ADTT= 1000) = 1.8 \times 217/240 = 1.62, \text{ use } 1.60 \qquad (4)$$

Similarly, for ADTT= 100, the expected load effect is 774kN(174 kips) or 86.5 kips of load per lane. The ratio of 174/240 leads to a load factor of 1.30 for a two lane distribution. However, in this case a one lane load case may govern. By raising the live load factor to 1.4 instead of 1.3 helps to compensate and provide adequate safety.

POSTING: To reach a lower acceptable bound on the rating factor, assume that a bridge is closed if the posted load falls to 3 tons as determined by the computed R.F. value using the methods in the CE Manual. For a 3 Ton Type 3 vehicle (legal weight 25 tons), modifying the rating to provide a design level of reliability (a ratio of 1.3), increased dynamic allowance (a ratio of 1.1), a load shifting allowance(a ratio of 1.1), and an additional 44.5kN(10000 lb) arbitrary load overweight allowance (an additional live load ratio of 1.78) gives the following expression:

$$R.F. = [3 \text{ Ton}/25 \text{ Ton }] \times 1.78 \times 1.1 \times 1.1 \times 1.3 = 0.34, \qquad (5)$$

The suggested lower limit for the rating factor before a bridge closing should be considered is given in the proposed manual as 0.3.

PERMITS: The analysis of routine permits was done with three categories of site traffic, namely ADTT of 5000, 1000 and 100. Site truck weight statistics were taken as above with the same multiple presence modeling used for the three ADTT categories used for routine legal traffic rating, namely a probability of a truck alongside the permit was 1/15, 0.01 and 0.005 respectively. The number of permits was taken as 10 and 100 permits per day to cover the likely range. A range of permit weights of 356kN(80 kips), 556kN(125 kips), 668kN(150 kips) and 890kN (200 kips) were examined. For each case, compute the expected number of multiple presences, the corresponding weight fractile, and the expected maximum weight of the alongside vehicle. This load effect was added to that of the permit weight. Using eqn. 2 with the same assumptions given for calculating the live load factor for rating legal vehicles, the live load factor for routine permits was calculated to attain the target reliability level(operating level). For the special permit case, use a one lane distribution analysis and a given number of permit crossings. This analysis of permit factors assumes the target reliability is satisfied by adjusting the ratio of the expected maximum load effect to that of the most severe traffic load case as shown in Eqn.2 and that the uncertainties for single and two lane loadings are the same. These live load factors are summarized in the proposed MCE Manual.

Acknowledgements

The work described herein was performed as part of NCHRP Project 12-46, being directed by the firm of Lichtenstein and Associates including Charles Minervino and Bala Sivakumar. The opinions given, however, are those of the author

References
Moses, F. (In preparation)."Load Factor Derivations for Evaluation", Final Report, NCHRP 12- 46.
Nowak, A. (1993), "Calibration of LRFD Bridge Design Code", Final Report, NCHRP 12-33.

Fatigue Evaluation Procedures in the Proposed *Manual for Condition Evaluation and Load and Resistance Factor Rating of Highway Bridges*

Dennis R. Mertz,[1] Member, ASCE

Abstract

A new *Manual for Condition Evaluation and Load and Resistance Factor Rating of Highway Bridges* is being developed. New fatigue-life evaluation procedures for existing steel bridges combine aspects of the *Guide Specifications for Fatigue Evaluation of Existing Steel Bridges* and the fatigue design procedures of the *LRFD Bridge Design Specifications*. Highlights are described herein.

Introduction

A new *Manual for Condition Evaluation and Load and Resistance Factor Rating of Highway Bridges* is being developed as the National Cooperative Hgihway Research Program (NCHRP) Project 12-46. Upon completion, the Manual may be considered by the American Association of State Highway and Transportation Officials (AASHTO) Highway Subcommittee on Bridges and Structures for adoption as a replacement document for the *Manual for Condition Evaluation of Bridges* (AASHTO 1994).

New fatigue-life evaluation procedures for existing steel bridges are specified in Section 7 of the Manual which combine aspects of the *Guide Specifications for Fatigue Evaluation of Existing Steel Bridges* (AASHTO 1990) and the fatigue design procedures of the *LRFD Bridge Design Specifications* (AASHTO 1998). Highlights of these fatigue-life evaluation procedures are described herein.

The load-induced fatigue-life evaluation procedures are divided into two levels: the infinite-life check of Article 7.2.4 of the Manual and the finite fatigue life

---

[1]Associate Professor of Civil Engineering, University of Delaware, Newark, DE 19716

estimate of Article 7.2.5 of the Manual. Both levels use estimated stress ranges specified in Article 7.2.2 of the Manual.

Table 1 - Partial Load Factors; R_{sa}, R_{st} and R_s

column 1	column 2	column 3	column 4
Fatigue-life Evaluation Methods	Analysis Partial Load Factor, R_{sa}	Truck-weight Partial Load Factor, R_{st}	Stress-range Estimate Partial Load Factor, R_s (In general, R_s equals R_{sa} times R_{st})
For Evaluation or Minimum Fatigue Life			
stress range by simplified analysis, and truck weight per Article 3.6.1.4 of the *LRFD Specifications*	1.0	1.0	1.0
stress range by simplified analysis, and truck weight estimated through weight-in-motion study	1.0	0.95	0.95
stress range by refined analysis, and truck weight per Article 3.6.1.4 of the *LRFD Specifications*	0.95	1.0	0.95
stress range by refined analysis, and truck weight by weight-in-motion study	0.95	0.95	0.90
stress range by field-measured strains	na	na	0.85
For Mean Fatigue Life			
all methods	na	na	1.00

<u>Estimating Stress Ranges</u>

The estimate of stress range for use in the fatigue-life evaluation procedures utilize partial load factors which are a function of the uncertainty of the analysis procedure and the estimate of the effective fatigue truck weight. The effective stress range is estimated by:

$$(\Delta f)_{eff} = R_s(\Delta f)$$

where R_s is given in column 4 of Table 1 and, in general represents the product of columns 2 and 3, and Δf is the calculated or measured stress range.

Infinite-Life Check

A detail is assumed to have infinite fatigue life if the maximum stress range, defined as twice the effective stress range, is less than the constant-amplitude fatigue threshold of the *LRFD Specifications*.

Estimating Finite Fatigue Life

The finite fatigue life of details on existing steel bridges which fail the infinite-life check is calculated using variables defined in the *LRFD Specifications* and in Table 7.2.5.1-1 of the Manual, given as Table 2, herein. A new evaluation fatigue life has been defined, and the fatigue life is calculated as:

$$Y = R_R A/[365n(ADTT)_{SL}((\Delta f)_{eff})^3]$$

where R_R is given in Table 2, with the remainder of the variables are given in the fatigue provisions of the *LRFD Specifications*.

Table 2 - Resistance Factor for Evaluation, Minimum or Mean Fatigue Life, R_R

Detail Category (from Table 6.6.1.2.3-1 and Figure 6.6.1.2.3-1 of the *LRFD Specifications*)	R_R		
	Evaluation Life	Minimum Life	Mean Life
A	1.7	1.0	2.8
B	1.4	1.0	2.0
B'	1.5	1.0	2.4
C	1.2	1.0	1.3
C'	1.2	1.0	1.3
D	1.3	1.0	1.6
E	1.3	1.0	1.6
E'	1.6	1.0	2.5

<u>Summary</u>

The principles of reliability and uncertainty employed in the *LRFD Specifications* for design have been extended to fatigue-life evaluation in the new Manual and presented briefly herein.

<u>References</u>

American Association of State and Highway Officials (1998). *LRFD Bridge Design Specifications*, 2nd Edition. AASHTO, Washington, D.C.

American Association of State and Highway Officials (1994). *Manual for Condition Evaluation of Bridges*. AASHTO, Washington, D.C.

American Association of State and Highway Officials (1990). *Guide Specifications for Fatigue Evaluation of Existing Steel Bridges*. AASHTO, Washington, D.C.

<u>Acknowledgment</u>

The development of the proposed Manual, and the fatigue evaluation procedures therein, is sponsored by the AASHTO, in cooperation with the Federal Highway Administration, and is conducted in the NCHRP which is administered by the Transportation Research Board of the National Research Council.

IMPLEMENTING THE LRF BRIDGE EVALUATION MANUAL

Charles M. Minervino[1]

The 1994 AASHTO *Load and Resistance Factor Design (LRFD) Specifications for Highway Bridge Design* introduced a limit states design philosophy, based on structural reliability methods, to achieve a more uniform level of safety (reliability) throughout the system. The current AASHTO *Manual for Condition Evaluation of Bridges* (MCE) contains provisions for only the conventional (deterministic) Allowable Stress and Load Factor methods of load rating existing bridges. Hence there is an acute need for a comprehensive new evaluation manual that will not only be consistent in philosophy and approach with the LRFD Specifications but will also be technologically current.

NCHRP Project 12-46 was initiated in March 1997 to develop the new AASHTO *Load and Resistance Factor Evaluation (LRFE) Manual for Highway Bridges* that will introduce reliability based evaluation methods. Once adopted by AASHTO, the new Manual will replace the existing *Manual for Condition Evaluation of Bridges.* It is critical that the new LRFE procedures be integrated into practice upon adoption so that consistency with the Design Code LRFE procedures can be initiated. The proposed plan for implementing the new LRFE Code is the subject of this paper.

The key to implementation of the new Manual is acceptance by the bridge engineering community of the LRFE concept and its benefits. The general LRF philosophy is now more widely understood as use of the LRFD design specification is becoming more widespread. It is critical that the importance of extending these principles to evaluation be conveyed to bridge engineers. The preliminary Implementation Plan involves a five part approach to introducing the LRFE procedures.

----

[1]Principal, A.G. Lichtenstein & Associates, Inc, Paramus, NJ

Provide Sufficient Explanatory Technical Material in the New Manual to Enable Users to Better Understand the Principles and Procedures.

The proposed method of load and resistance factor evaluation of bridges is highly consistent with the LRFD Bridge Design Specifications. On the other hand, the procedures are considerably different than the current bridge rating methods contained in the MCE. However, the new LRFE approach is more an extension of the Load Factor method of rating than a radical new procedure.

The new LRFE Manual will contain considerable commentary that expands on the fundamental concepts behind LRFE and clarifies the procedures described in the primary text. In addition, the Final Report for NCHRP 12-46 will contain a summary document which will identify the differences from current practice, explain the intent of each, and provide rational discussion of the development of the new procedures. Much of the discussion material will be background technical information that describes the limitations of existing rating procedures and the advantages and flexibility of the new methods.

The new Manual will contain 8 to 10 fully worked example ratings of typical bridges prepared by the research team in developing the Manual. These examples will be selected to demonstrate the basic procedures of LRFE as well as the flexibility contained in the procedures. To expand on the limited number of examples, the project panel has authorized a testing phase in order to get a more widespread evaluation of the procedures over a broader range of bridges. The AASHTO Subcommittee on Bridges and Structures through Technical Committee T-18 will encourage volunteer state DOT's to participate within the Research Team by rating a series of bridges completely, using the new Manual under the technical guidance of the Research Team. The result of this testing will be a compendium of completed example bridge ratings that will be a part of the project Final Report, and may be made available as a workbook or in CD format.

The practical testing will allow the raters and research team to identify areas of the Manual which are missing, vague, or inappropriate. The Final Draft of the Manual will be reversed accordingly.

Gain Confidence of State Bridge Engineers in the Procedures by Actively Involving this End User Group in the Research Phase.

The Research Team has maintained coordination with the AASHTO State Bridge Engineers via a presentation of the Research Team's LRFE approach to AASHTO Technical Committee T-18 at the 1997 AASHTO Annual Bridge Committee meeting and a progress report at the 1998 meeting. An additional presentation is scheduled for the 1999 meeting. In the interim, the T-18 Committee has met twice with the research team to discuss and review the proposed LRFE procedures in detail. The

T-18 Committee has endorsed the testing phase of the project, and a number of states will participate. We believe that this ongoing discourse with the T-18 Committee will be the foundation for its endorsement of the LRFE procedures leading to approval by the full AASHTO Bridge Committee.

Educate the Bridge Engineering Community in Advance of Publication of the Manual.

A valuable lesson learned by observing the successful transition to the LRFD bridge design code is the importance of educating the general bridge engineering community in advance . The research team members are attending various bridge conferences over the course of the project in order to present papers that describe the LRFE concepts and promote the benefits of LRFE. In this way bridge engineers will be introduced to the upcoming Manual and its procedures, and will recognize LRFE as a logical extension of current bridge rating procedures. The bridge engineering community will follow the lead of the bridge owners, but awareness and education will further aide the implementation process.

Coordinate with Simultaneous Development of AASHTO's LRFE Bridge Evaluation Software

One major impediment to full and immediate implementation of the LRFD design code has been the lack of available design software. LRFD bridge design software, entitled OPIS, is being developed by AASHTO while some state DOT's and consultants have independent design programs in various stages of completion. Bridge rating programs have been in common use for a number of years. AASHTO is developing load factor based bridge rating software, entitled VERTIS, for release in 1999.

The research team is coordinating with the VERTIS contractor to ensure that the VERTIS software is prepared in a format that allows updating to accommodate the final version of the LRFE procedures when approved by AASHTO. This continuing effort should result in an LRFE version of VERTIS bridge rating software being available at about the same time that the new Manual is printed and released.

The research team believes that the availability of VERTIS LRFE software will be a major step toward timely implementation of the new Manual. Without available software, bridge engineers would continue to use the load factor version of VERTIS and other software until such time as LRFE software became available.

Train Bridge Rating Engineers to Use the Manual and Appreciate its Flexibility.

The final step of implementation proposes training bridge engineers to use the new Manual. The research team believes that a two or three day seminar would be sufficient to train experienced bridge rating engineers to apply the LRFE principles contained in the Manual. The content and duration of these seminars would be better defined after conducting the testing described above. Training bridge engineers in the use of the LRFE version of VERTIS software should be considered as part of the training seminar. We believe that State DOT sponsored training sessions to which public and private sector bridge rating engineers are invited could be held around the country and would be sufficient to train a significant part of the bridge engineering community.

SUMMARY

The research team's proposed implementation plan for LRFE includes providing extensive technical support in the Manual and project documents, taking active steps toward gaining the confidence and acceptance of bridge owners, coordinating with LRFE software development, educating bridge engineers in the concepts and benefits of LRFE, and training bridge rating engineers to use the Manual,. The research team will continue to refine and update the Implementation Plan with the project panel and Technical Committee T-18 as the project progresses.

The Latest Developments in the
AASHTO-LRFD Bridge Design Specifications

Wagdy G. Wassef[1]
John M. Kulicki[2]

Abstract

Since it was published in 1994, the AASHTO-LRFD Bridge Design Specifications has been used in parallel to the AASHTO Standard Specifications for Highway Bridges. Both specifications are updated annually and these updates are published in an annual interim specification issued by AASHTO. This paper presents an overview of the changes to the LRFD specification as a result of the annual interims since its introduction.

Introduction

The AASHTO-LRFD Bridge design Specification was adopted by AASHTO in 1993 and was published in 1994 as a design specification parallel to the AASHTO Standard Specifications. In 1997, the AASHTO Highway Subcommittee on Bridges and Structures voted to archive the Standard Specifications after implementing the ballot items that will be presented in the 1999 AASHTO meeting. After that, the Standard Specifications will be available for use and AASHTO will continue to correct any errors in the Standard Specifications. However, no enhancements involving new research, new materials or technologies will be introduced. This will leave the LRFD specification as the only bridge design specification actively maintained by AASHTO.

After publishing updating information in the 1996 and 1997 interims to the LRFD specifications, AASHTO published the second edition of the LRFD Specification in 1998. The first interim to the second edition is scheduled for release in early summer of 1999. These interims are necessary to implement new information, to clarify ambiguities and to correct any typing errors discovered in the specification or in the original research on which the specification was based. Following is a brief description of the more significant changes to the specification since it was first published in 1994. The changes are listed by the specification section. A missing section indicates minor or no changes to this section to-date.

---

[1] Associate, Modjeski and Masters Inc., Harrisburg, PA 17055

[2] President and Chief Engineer, Modjeski and Masters Inc., Harrisburg, PA 17055

Changes to Section 3, Loads and Load Factors

- Changes to the application of wind loads and some wind load factors were introduced to produce forces comparable to the forces used successfully in the past.
- Modifications to the method used to determine the tire contact area were implemented to include new research results and to simplify the calculations for the design truck and design tandem tires.
- Modifications to the application of temperature changes and maps for the seasonal temperature variations were introduced. In addition, research results consistently indicated that the negative temperature gradient in the specification was overly conservative. Therefore, a reduced value for the negative temperature gradient was introduced.
- Additional information pertaining to the calculations of earth pressure was added
- Changes to the calculations of ice loads were implemented

Changes to Section 4, Structural Analysis and Evaluation

- Refinements to some of the live load distribution factors were introduced
- New provisions for the design of the lateral components (diaphragms and cross-frames) in seismic areas were implemented
- The uniform load method of calculating seismic forces was added
- A reinforcement table for typical concrete deck slabs was introduced as an Appendix

Changes to Section 5, Concrete Structures

- The second edition of the AASHTO Guide Specification for Segmental Concrete Bridges was integrated into the LRFD Specifications
- New provisions were introduced for welded and mechanical splices of reinforcement and for the investigation of hollow rectangular concrete sections
- The provisions pertaining to the control of cracking of concrete structures were modified to ensure consistent application of these provisions
- Modified stress limits for prestressing strand and prestressed concrete were introduced along with modifications to the calculations of the elastic shortening losses of prestress. The elastic shortening determination and verification are now the responsibility of the contractor performing the prestressing operation.
- Many additions and modifications to the sections pertaining to the design of the anchorage zones were implemented
- Some of the reinforcement detailing requirements were enhanced
- The interaction equation for members under combined axial load and moment was updated

Changes to Section 6, Steel Structures

- The articles on flexural member design have undergone a major editorial reorganization. Many article numbers were changed and additional commentary was added to facilitate the use of the specification and simplify cross-referencing. However, the content of the articles was essentially unchanged by the reorganization.
- The provisions for bolted splice design were updated. The new provisions provide some relief in the design of web splices for members with shear capacity significantly larger than the applied shear. New provisions for calculating the net area of members and cross-section components in tensions were introduced.
- Resistance factors for ASTM A307 bolts in shear and tension were added
- The fatigue category for eye bars was defined
- Based on new information, the provisions pertaining to the design of longitudinally stiffened webs were updated
- Updated provisions for determining the moment capacity, the ductility requirements and slenderness interaction of compact sections were implemented
- The bolt shear capacity equations were updated to implement new research results and new provisions for the design of ASTM A307 bolts were added

Changes to Section 7, Aluminum Structures

- The material properties for aluminum alloys were updated for both the base metal and near welds. New alloys were also added.
- New provisions for the buckling stress for different components of the cross section were implemented

Changes to Section 9, Decks and Deck Systems

- Additional requirements for cross-frames and diaphragms were added to the conditions to be satisfied for the application of the concrete deck empirical design procedures. These provisions are for both I-girder and box girder bridges.
- Limits on the elastic deflection of stay-in-place formwork were introduced

Changes to Section 10, Foundations

The modifications to the articles on foundation design are more extensive than for other sections of the specification. However, many of the modifications were the result of adding information that is readily available in the literature. This information was added upon the request of many bridge engineers. Following is a brief description of the changes in the foundation design articles.

- The provisions for calculating the settlement of footings were expanded
- Resistance factors for piles were modified and additional factors, dependent on the sophistication of the method of analysis, were added.
- Additional information on the design of footings on rock was implemented. These additional provisions include rock properties, settlement analysis and compressive strength.
- Additional details were added to the procedures to estimate the load capacity of footings on cohesive and cohesionless soils
- More details were added to the procedures to estimate the required pile length and pile group lateral load capacity were introduced

Changes to Section 12, Buried Structures and Tunnel Liners

- Extensive design provisions were added to the long-span structural plate structures. This reflects the growing popularity of these systems.
- Equations to determine the moments in structural plate box structures were added
- Modified and additional provisions for the design of concrete pipe structures were implemented
- New provisions dealing with the design of precast reinforced concrete three-sided structures were introduced

Changes to Section 13, Railings

The National Cooperative Highway Research Program (NCHRP) Report 350 "Recommended Procedures for the Safety Performance Evaluation of Highway Features" was implemented. This resulted in seven Test Levels replacing the Three Performance Levels originally in the LRFD Specification.

Expected future changes

Two NCHRP projects related to the LRFD specification are being developed. The first will produce provisions for the design of curved bridges. The second will attempt to develop simplified provisions for shear design of concrete structures. The retaining wall section is being reviewed by an independent group following a major rewrite of the wall provisions in the Standard Specification.

References

American Association of State Highway and Transportation Officials, "AASHTO LRFD Bridge Design Specifications," First Edition, 1994, Second Edition, 1998, 1996 and 1997 Interim Specifications, and 1998 Ballot items.

Nondestructive Determination of Unknown Bridge Foundation

Ming Liu, Marwan F. Aouad and Larry D. Olson[1]

Abstract

This paper presents a brief description of three nondestructive testing (NDT) methods used by the authors to investigate the problem of determining unknown depths of bridge foundations. This was part of a research project sponsored by the National Cooperative Highway Research Program (NCHRP) project 21-5 entitled "Determination of Unknown Subsurface Bridge Foundations". The three methods discussed in this paper are the surface method of Ultraseismic testing and the borehole methods of Parallel Seismic and Induction Field testing. A case study is presented herein to illustrate the use of the Borehole Induction Field method.

Introduction

The unknown depths of bridge foundations pose significant problems to State DOTs when evaluating the susceptibility of bridges to scour and earthquake events (Elias, 1992). The National Cooperative Highway Research Program (NCHRP) 21-5 research project "Determination of Unknown Subsurface Bridge Foundations" (Olson et al, 1995) was conceived to address these urgent concerns and to find accurate, cost-effective nondestructive testing (NDT) methods to determine unknown foundation conditions. Nine NDT technologies were evaluated in this research work and were divided into two categories: 1) Surface methods and 2) Borehole methods. The surface techniques require access to the exposed parts of the bridge substructure elements while the borehole methods require access through a nearby boring.

Based on the results of the NCHRP research project 21-5, the Parallel Seismic test was found to have the broadest applications for determining the bottom depths of unknown bridge foundations. The Ultraseismic surface method has applications for determining depths of certain types of unknown bridge foundations such as wall piers,

Project Engineer, Project Manager and President, respectively
Olson Engineering, Inc., 5191 Ward Rd, Wheat Ridge, CO 80033

but provides no information on piles below massive substructures. The Induction Field method is the electromagnetic analog to the Parallel Seismic method. However, its application is limited to steel piles and reinforced concrete piles with exposed steel.

Parallel Seismic Method

The Parallel Seismic (PS) method was researched and developed specifically to determine the depths of unknown foundations by the CEBTP research organization headquartered in Paris, France. The PS test method is based on the principle that an impact to the exposed structure generates wave energy that travels down the foundation. The response of the foundation is monitored by receivers in a nearby parallel boring to determine when the signal weakens and slows down. This indicates the receiver has gone beyond the bottom of the foundation, and the depth is therefore determined. Figure 1 shows a schematic of the PS method.

The PS test involves impacting the side or top of exposed bridge substructures with a 1.4 kg (3 lb) or 5.6 kg (12 lb) hammer to generate the wave energy that travels down the foundation and the surrounding soil. The wave arrival is tracked by a hydrophone receiver suspended in a water-filled cased borehole or by a clamped, 3-component geophone receiver in a grouted and cased borehole (or in an open non-caving borehole). A hydrophone receiver is sensitive to pressure changes in the water-filled tube when the wave arrives, but is subject to tube wave energy contamination. A clamped 3-component geophone receiver in cement-bentonite, bentonite, or sand-backfield, 10 cm (4 in.) ID, PVC cased borings was also used to better examine the wave propagation behavior with reduced tube wave energy noise. The boring is drilled typically within 1 to 1.5 m (3 to 5 ft) of the foundation edge and should extend at least 3 m (10 ft) deeper than the anticipated and/or minimum required foundation depth. By using a 3-component geophone receiver in cased borings with good contact between the casing and soils, improved quality of PS results can be obtained at field sites with variable soil velocity conditions.

Ultraseismic Method

The Ultraseismic (US) method was researched and developed during the NCHRP research project 21-5 for determination of the depth of unknown bridge foundations. The US method relies on geophysical digital data processing techniques to analyze the propagation of induced compression and flexural waves as they reflect from foundation substructure boundaries. The US method uses multi-channel, 3-component (vertical and two perpendicular horizontal accelerometer receivers, i.e. tri-axial receiver) recording of reflected seismic waves up to 4-5 kHz. The seismic processing can greatly enhance data quality by identifying and clarifying reflection events that are from the foundation bottom and minimizing the effects of undesired wave reflections from the foundation top and attached superstructure.

The US test involves striking the side or top of exposed bridge foundation with a 0.09 kg (0.2 lb) to 5.6 kg (12 lb) hammer to generate both compression and flexural wave energy which travels down the foundation and reflect back from the bottom of the foundation or any existing boundary. The tri-axial receivers are mounted on the surface or side of the accessible bridge foundation at intervals of 1 ft or less. The reflected wave arrival is tracked by the tri-axial receivers. The recorded receiver outputs from many receiver locations are stacked together in order to identify the reflected wave from the bottom of the foundation. Figure 2 shows a schematic of the US method.

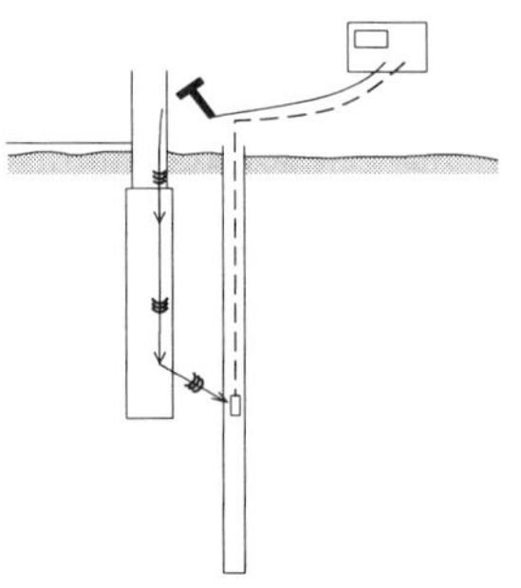

Figure 1 Schematic of the PS method

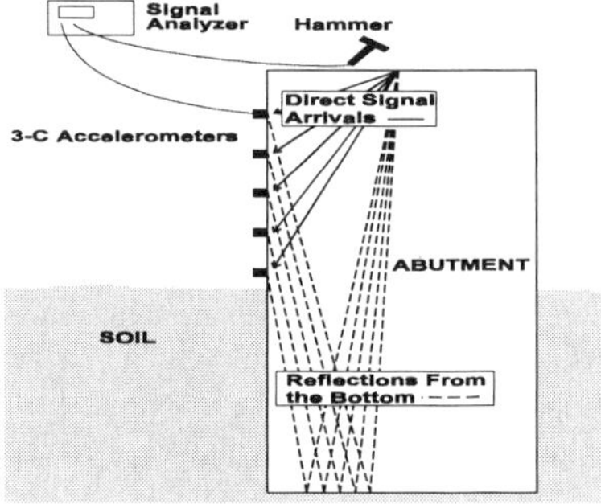

Figure 2 Schematic of the US method

Induction Field Method

The Induction Field (IF) method was researched and developed in New Zealand for determining the depths of reinforced concrete piles and steel piles. In the IF method, an AC current flow is impressed into a steel pile or steel rebars in a reinforced concrete pile from which the current couples into the subsurface and finally to a return electrode. A receiver coil which is suspended in a nearby boring is then used as a sensor of the magnetic field induced by the alternating current flow in the pile. By plotting the magnitude of the induced voltage versus the depth of the receiver coil, an indication of the bottom of the pile foundation is provided. Once the receiver coil is below the bottom of the foundation, the measured induced voltage tends to stabilize at a low value because of the residual conductivity of the soil or bedrock.

Case Study

A case study is presented in this paper to illustrate the use of the borehole Induction Field (IF) method. Figure 3 shows a schematic of the IF method.

An Induction Field data set collected on a bridge located on US 287 near Longmont, Colorado is presented below. The piers of this bridge consist of concrete beams supported by steel H-piles. The IF tests were conducted at intervals of 0.3 m (1 ft) using a 3-axial magnetometer to measure the strength of the induced magnetic field by an AC current flow through the piles. The AC current was generated by a function generator operating at a frequency of 5 kHz, and then amplified by an audio power amplifier. Figure 4 shows the voltage versus depth of the receiver coil. The predicted depth was 8.23 m (27 ft), where a significant drop in voltage of the receiver coil occurred. The actual length of the tested pile was 9.07 m (29.8 ft), a good agreement within a 10% accuracy.

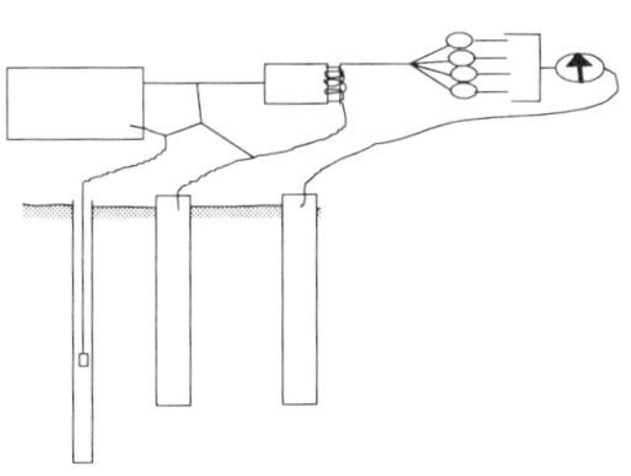

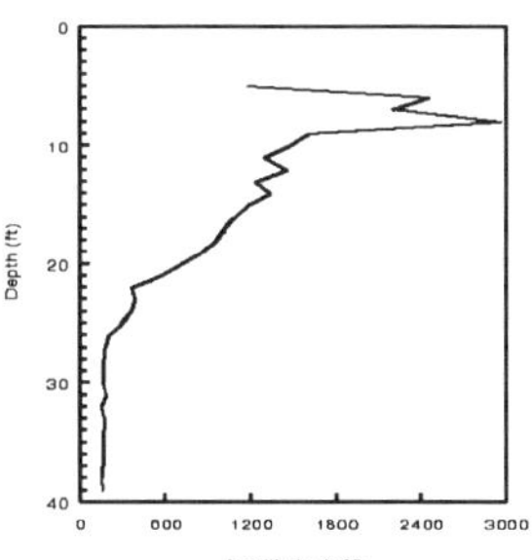

Figure 3 Schematic of the IF method

Figure 4 Results of the IF tests

Conclusions

The PS and US methods have the broadest applications for determining depths of unknown foundations. Limited research with the IF method showed its promise.

Acknowledgments

This research study was sponsored by the National Cooperative Highway Research Program (NCHRP). Olson Engineering would like to thank NCHRP and the research panel members for their support.

References

Elias, V.A., Strategies for Managing Unknown Bridge Foundations. *Report FHWA-RD092-030*, Washington, D.C., January, 1992.

Olson, L.D., Jalinoos F. and Aouad, M.F., Determination of Unknown Subsurface Bridge Foundations. *NCHRP Project 21-5, Final Report*, Transportation Research Board, National Research Council, Washington, D.C., 1995.

Nondestructive Testing for Quality Control and
Length Determination of Deep Foundations for Bridges

Stephen L. Borg[1]

Abstract

This subject of this paper is the practice of New York State Department of Transportation (NYSDOT) using nondestructive testing (NDT) methods for quality control of deep foundations. Two NDT methods that are used by NYSDOT on new foundations are discussed. Determining the pile length of existing structures is a challenge when detailed construction records are not available. One NDT method used by NYSDOT to determine pile lengths for existing structures is presented.

Introduction

The standard pile test method used by New York State Department of Transportation (NYSDOT) is dynamic pile testing (DPT) also referred to as the Case Method. NYSDOT uses DPT on 20 percent of newly constructed bridges that are supported by driven piles.

For large diameter deep foundations, such as drilled shafts, the NYSDOT policy is to test every shaft on all projects using an ultra sonic pulse NDT method referred to as crosshole sonic logging (CSL). Data from a CSL test aids in evaluating the quality of the shaft concrete. When CSL is coupled with concise inspection records the quality control of the shaft improves, and a better judgement can be made on the applicability of the shaft construction method for a given site.

----

[1]Civil Engineer II, New York State Department of Transportation, Bldg. 7, 1220 Washington Avenue, Albany, NY 12232

During the design phase of a structure replacement or rehabilitation, the need to determine the length of existing deep foundations occasionally arises. In these cases the pile is usually not physically accessible, nor is it generally practical to employ a high strain NDT method such as DPT. In this case, NYSDOT uses a low strain NDT method referred to as parallel seismic (PS).

Dynamic Pile Testing (DPT)

DPT is a high strain method for evaluating a driven foundation element, most commonly steel pipe piles, H-piles, prestressed concrete and wood piles. The dynamic response to the impact of the pile driving hammer and resistance to soil is measured at the pile top. Strain and acceleration measurements are taken during the initial installation of the pile, and to account for soil strength gain they are taken some time after the pile is driven. DPT is used to:

- Calculate pile capacity
- Monitor hammer performance
- Measure pile stresses
- Detect pile damage

The DPT program is based in the Main Office in Albany and all tests are run by NYSDOT personnel. Two pile driving analyzers (PDAs) are used to cover work for the entire state. NYSDOT has used this technology since 1975 and has owned and operated PDAs since 1978. Testing is exclusively conducted by properly trained engineers under the guidance of a staff who have maintained a continuity of PDA experience for over 20 years.

NYSDOT has a pay item for DPT in its Standard Specifications. The contractor's cost to NYSDOT essentially reflect's down time from pile driving production work, which is generally estimated to be one-hour. On an as needed basis, the test requirement is written into the Foundation Design Report for the project. A note is placed in the Contract Documents stating the frequency and circumstances for each test, always with the stipulation that additional tests may be ordered by the project engineer.

Some common soil conditions and design circumstances that necessitate DPT fall into the following categories:

- Insufficient subsurface information
- Questionable density of sand and gravel
- Cohesive and Silty Soil (loss of soil strength during installation)
- Limited foundation redundancy
- Pile type subject to damage (prestressed concrete - tensile failure)

Over the last three years, the average contract bid price for each DPT has been 2,000 dollars. Cost data shows that there is no relationship to the item cost per test and number of tests on the contract. If NYSDOT did not run its own testing program, the contractor would be required to hire a firm, and the cost of the contract item would increase significantly.

Crosshole Sonic Logging (CSL)

NYSDOT has a contract item for CSL testing of drilled shafts, which is covered under a special specification. The contractor hires a testing firm, and submits the credentials of the prospective firm for approval by NYSDOT.

To provide access for transmitting and receiving hydrophones (transducers sensitive to change in water pressure), the contractor casts plastic or steel access pipes into the shaft concrete, then fills them with water. Once the concrete cures for the stipulated period, the testing can be conducted. Hydrophones are lowered to the bottom of the shaft and a low end ultra-sonic pulse is transmitted between pairs of pipes. The larger the diameter of the shaft the more access pipes are used.

A profile of the pulse arrival time is logged for the full length of the shaft. The shorter the pulse travel time is between the hydrophones, the stronger the concrete. Should the travel time be excessively long or significant anomalies (degradation or loss of signal) appear in the data, a defect in the concrete could be present.

If a defect is suspected, NYSDOT requires coring to determine strength and quality of the concrete core samples and to detect possible voids or soil inclusions. NYSDOT has used CSL as a contract requirement for the past three years on 151 shafts for 7 projects. To date CSL test data has lead to coring three shafts, one of which was reconstructed.

A typical NYSDOT drilled shaft design utilizes permanent casing installed to bedrock, and a rock socket is drilled below the casing. This type of installation is generally less susceptible to concrete defects than shafts installed with temporary casing and/or drilling mud. CSL has replaced the more costly and time consuming practice of coring to verify the quality of concrete placed under water.

The average contract bid price to test each shaft is 1,300 dollars. The cost to the contractor to provide CSL should decrease as the number of shafts tested for each mobilization by the testing firm increases.

Parallel Seismic (PS)

PS testing requires that casing be installed adjacent to the existing pile to some depth below the anticipated pile toe. At the time of testing the casing is water filled and free from debris. A hydrophone is lowered to the bottom of the casing and raised incrementally while the pile cap or preferably the pile is struck with a hand held hammer. The hammer impact induces a low strain pulse to travel down the pile and into the surrounding soil, thus triggering the hydrophone.

The pulse travel time between each impact and triggering of the hydrophone is plotted against the coinciding depth of the hydrophone in the casing. A change in the slope of this curve represents the location of the pile toe. This determination is possible when the input wave is sufficiently strong to trigger the hydrophone and the speed of sound in the pile material and soil medium is sufficiently different.

NYSDOT has employed the services of a testing firm to determine the length of unknown deep foundations for ten bridges. PS is used on existing bridges to evaluate:

- Susceptibility to stream bed and tidal scour
- Susceptibility to seismic soil liquefaction
- Effect of increased loads on foundations
- Feasibility of incorporating existing piles into new foundations

PS data presents itself to direct interpretation and is generally considered to be the most accurate low strain NDT method for pile length determination. It can be most cost effective when it is considered during the subsurface exploration phase of a project. Drill holes can be located close to the substructure in question, allowing for casing to be installed for later testing.

In NYSDOT contracts where drill holes were progressed for the sole purpose of installing casing for PS testing, the cost of the PS testing was 15 percent of the drilling costs.

Conclusions

Dynamic Pile Testing (DPT) is a cost effective method of ensuring the quality of impact driven deep foundations.

Parallel Seismic Testing (PS) is a definitive method in determining pile lengths, and can be highly cost effective when incorporated into the subsurface exploration program.

Crosshole Sonic Logging (CSL) can be a cost effective method of increasing the quality control of drilled shafts.

Bridge Foundation Stiffness Identification Using Static Field Measurements

Masoud Sanayei[1] Kenneth R. Maser[2]

Abstract

A simple low-cost nondestructive testing (NDT) method is presented for identifying unknown bridge foundations for vulnerability assessment to scour, seismic, and other sources of foundation degradation. The proposed new method is based on direct foundation stiffness estimation using bridge field measurements subjected to live loads. Foundation stiffness coefficients of a bridge were calculated from field measurements on the exposed portions of the structure. Measured patterns in foundation stiffness coefficients could provide a signature for identifying different foundation types.

Introduction

This paper quantifies a bridge pier foundation stiffness using measurements of strain, rotation, and displacements for direct foundation stiffness estimation. Static measurements are made in response to a 20-ton load truck rolling across the bridge at a crawl speed. Each foundation is characterized by a soil-substructure superelement (SSS) called (K_{sss}) matrix (1) (Sanayei et. al. 1998). The symmetric K_{sss} accounts for off-diagonal terms, such as $K_{H\theta}$ denoting the interaction of the horizontal and rotational forces (Poulos and Davis 1980). K_{HV} and $K_{V\theta}$, which are zero for symmetric foundations, become nonzero for unsymmetric foundations. Maser et al. (1998) showed a significant difference in the foundation stiffness ratios of deep and shallow foundations.

$$[K_{sss}] = \begin{bmatrix} K_{HH} & K_{HV} & K_{H\theta} \\ K_{VH} & K_{VV} & K_{V\theta} \\ K_{\theta H} & K_{\theta V} & K_{\theta\theta} \end{bmatrix} \tag{1}$$

---

[1] Associate Professor, Dept of Civil & Env. Engineering, Tufts University, Medford, MA 02155
[2] President, INFRASENSE, Inc., 14 Kensington Road, Arlington, MA 02476

The proposed NDT method involves collection of strain, rotation, and displacement data on the pier and the adjacent superstructure. Bending moment, horizontal and vertical forces in the pier near the foundation level are calculated from strain measurements. As the load truck moves across the deck, a series of forces, moments, displacements, and rotations are recorded at predetermined positions on the deck. Assuming that data is recorded at N locations, then:

$$[F] = [K_{sss}] [U] \tag{2}$$

where $[F]$ and $[U]$ are 3 x N matrices in which each column of $[F]$ has two forces and one moment, and each column of $[U]$ represents the corresponding displacements and rotation at a given time. Using the least-square fit solution $[K_{sss}]$ is computed as:

$$[K_{sss}] = [F][U]^{T} ([U][U]^{T})^{-1} \tag{3}$$

Eq. (3) represents a direct calculation of the foundation stiffness elements from the field data. However, there may be numerical problems in implementing this method depending on the conditioning of the matrix $([U][U]^{T})^{-1}$. One simplification is to uncouple the vertical stiffness from the remaining stiffnesses (i.e., assume K_{HV} and $K_{V\theta}$ are zero and directly calculate K_{VV}). This results in the inversion of a 2x2 matrix for the remaining three unknown parameters. Another level of simplification is to assume that all off diagonal terms are zero and directly divide the measured internal forces by the measured displacements to evaluate only the diagonal stiffness terms of $[K_{sss}]$. This procedure assumes no coupling between the stiffnesses.

Nondestructive Field Tests

The above methodology was implemented and field-tested in April and May 1997. Two bridges in Eastern Massachusetts were tested: one with a spread footing, carrying Fruit Street over I-495, and the second with a pile foundation carrying Winter Street over I-90. Details of these bridges are presented by Maser et al. (1998).

(a) Center Span

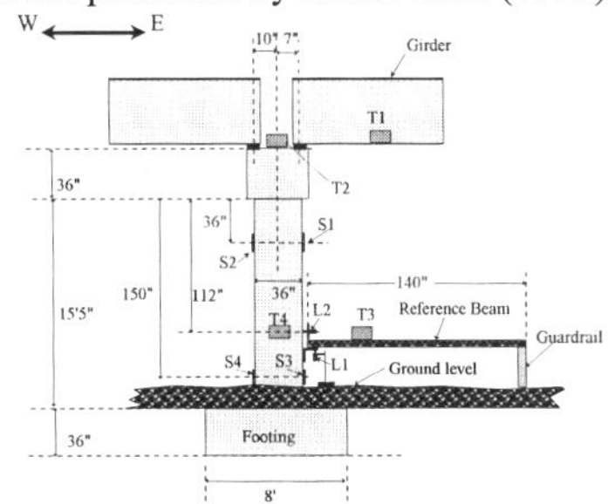

(b) Instrumentation of East Pier

Figure 1 - Fruit Street Bridge Test

Figure 1 is the center span of the Fruit Street bridge with the instrumentation used for the east pier NDT. Similar setup was used for the Winter Street Bridge. The instrumentation shown are tiltmeters (T1-T4), strain gages (S1-S4), and LVDT's (L1 and L2). Measurements were made as a 20-ton truck traveled across the deck. Typical data shown in Figure 2 are (a) rotation vs. time recorded directly with the tiltmeters, and (b) axial force vs. time computed from four strain gages on the column. The marks on the bottom are placed at reference truck locations on the deck.

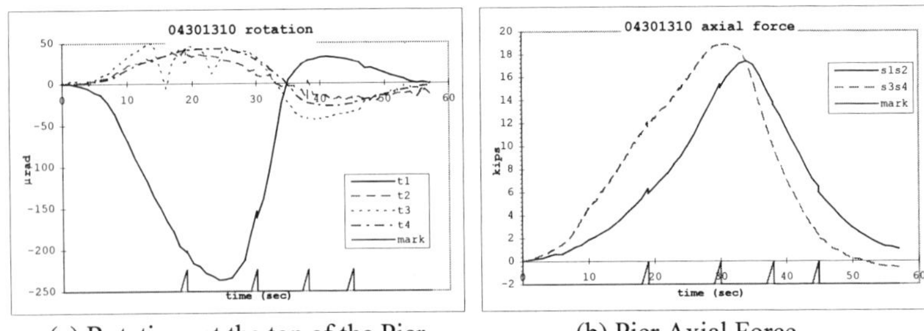

(a) Rotations at the top of the Pier   (b) Pier Axial Force

Figure 2 - Typical East Pier Measurements from Fruit Street Test

Bridge Modeling and Data Analysis

A 2-D FEM of the Fruit Street Bridge was loaded with a two-axle truck similar to the test truck. It was made to replicate the observed field behavior and explain some of the less familiar phenomenon. For example, it was observed in the field that the top of the pier always rotates away from the loaded span. This is explained by the fact that the neutral axis of the deck/girder is almost 5 feet (1.52 m) above the bearing, thus resulting in a horizontal pier motion and end rotation away from the truck. The FEM incorporated this neutral axis offset and the field behavior was verified. The field data for both bridges was analyzed using the direct foundation stiffness estimation. In order to deal with computational issues, it was assumed that the vertical stiffness was uncoupled from the other stiffnesses. Table 1 shows the diagonal stiffness coefficients calculated from the NDT field data.

Table 1 - Summary of Averaged Foundation Stiffnesses Data

Bridge Test	$K_{\theta\theta}$		K_{VV}		K_{HH}	
	(ft.kips/rad)	(kN.m/rad)	(kips/ft)	(kN/m)	(kips/ft)	(KN/m)
Fruit St.	891,666	1,208,877	1,656	24,166	41,640	607,660
Winter St.	386,407	523,871	24,600	358,992	8,131	118,657

In order to evaluate these results, study of foundation stiffness coefficients was carried out using a 2-D finite element foundation model. Models were formulated for both spread footing and pile foundations, and several different soil types were considered (Maser et al. 1998). The ratios of the diagonal stiffness elements of the foundation model were compared with those calculated from field data. Table 2 shows that the pattern of differences between the Fruit Street and Winter Street foundation stiffnesses determined from field data is similar to the FEM prediction. For example, the ratio K_{VV}/K_{HH} that was shown in theory to be much larger for the pile foundation than for the spread footing actually turned out that way when computed from the field data. A similar pattern was observed for the ratio $K_{\theta\theta}/K_{HH}$ as predicted by theory.

Table 2 - Diagonal Stiffness Ratios - Theory versus Field Measurement

Bridge	K_{VV}/K_{HH}		$K_{\theta\theta}/K_{HH}$	
Test	Theory	Measured	Theory	Measured
Fruit Street	1.2	.04	19.9	21.4
Winter Street	13.8	3.02	136.7	47.52

Conclusions

A new method has been proposed for direct estimation of unknown bridge foundations through calculation of foundation stiffness coefficients from static testing. Foundation stiffness coefficients have been determined from field test data, and the differences in the ratios of the diagonal stiffness coefficients between spread footing and pile foundations are shown to have similar patterns as demonstrated in theory.

Acknowledgements

The authors would like to acknowledge the Federal Highway Administration and Dr. Steven B. Chase for sponsoring this research under SBIR grant number DTRS57-96-C-00087. The following contributors are also acknowledged: Abba Lichtenstein, Bridge Consultant; Laura McGrath, Data Analyst; Bruce Ambuter, Data Acquisition Consultant; Gerard Grippo, Bridge Engineer from A.G. Lichtenstein and Associates; Amy Stern and Chitra Javdekar, Graduate Students at Tufts University.

References

Maser, K.R, Sanayei, M., Lichtenstein, A., and Chase, S.B. (1998). "Determination of Bridge Foundation Type from Structural Response Measurements." *Nondestructive Evaluation Techniques for Aging Infrastructure & Manufacturing*, SPIE, 3400, 55-67.

Poulos, H.G., and Davis E.H., (1980). *Pile Foundation Analysis and Design*, John Wiley & Sons, New York.

Sanayei, M., McClain, J. A. S., Wadia-Fascetti, S., Gornshteyn, I., and Santini, E.M. (1998). "Structural Parameter Estimation Using Modal Responses and Incorporating Boundary Conditions." *Proc. Structural Engineering World Congress*, ASCE, T118-2.

Bridge Foundation Stiffness Identification Using Dynamic Field Measurements

Erin M. Santini[1] Masoud Sanayei[2], Ming Liu[3], and Larry D. Olson[4]

Abstract

A dynamic field test was conducted on Bent 12 of Trinity Bridge located west of Liberty County, Texas. Foundation stiffness estimation was studied using 2D FEM computer simulations. The most feasible foundation stiffness parameters were selected for estimation using dynamic field data. Olson Engineering, Inc. performed the dynamic field tests while Tufts University performed the foundation stiffness estimation using a computer program, PARameter Identification System (PARIS).

Introduction

The focus of this paper is to quantify the foundation stiffness for Trinity Bridge using dynamic field data and parameter estimation. Bent 12 is a concrete frame with four piers/piles. The single pile under each pier is depicted by a soil-substructure superelement (SSS) (Sanayei et. al. 1998). The symmetric stiffness (K_{sss}) and mass (M_{sss}) matrices are presented in equation (1), where K_{sss} accounts for off-diagonal terms, such as $K_{H\theta}$ denoting the interaction of the horizontal and rotational forces (Poulos and Davis 1980). K_{HV} and $K_{V\theta}$, which are zero for symmetric foundations, become nonzero for unsymmetric foundations. Maser et al. (1998) shows a significant difference in the stiffness ratios of deep and shallow foundations.

$$K_{sss} = \begin{bmatrix} K_{HH} & K_{HV} & K_{H\theta} \\ K_{VH} & K_{VV} & K_{V\theta} \\ K_{\theta H} & K_{\theta V} & K_{\theta\theta} \end{bmatrix} \qquad M_{sss} = \begin{bmatrix} M_{HH} & 0 & 0 \\ 0 & M_{VV} & 0 \\ 0 & 0 & M_{\theta\theta} \end{bmatrix} \tag{1}$$

---

[1] Doctoral Student, Dept of Civil & Env Engineering, Tufts University, Medford, MA 02155
[2] Associate Professor, Dept of Civil & Env Engineering, Tufts University, Medford, MA 02155
[3] Project Engineer, Olson Engineering, Inc., Wheat Ridge, CO 80033-1905
[4] President and Principal Engineer, Olson Engineering, Inc., Wheat Ridge, CO 80033-1905

Trinity Bridge- Nondestructive Test and Foundation Stiffness Estimation

Figure 1 shows the 2D FEM for Bent 12 of Trinity Bridge. K_s denotes the horizontal linear springs for the lateral stiffness of the slab. The 2D FEM was adjusted to mimic the 3D FEM response by adjusting the tributary slab mass and K_s (Santini, 1998). A specially designed truck, Vibrosies (Bay et al. 1995) was used to apply a dynamic force at Node 5, while the 11 nodes shown as "●" in figure 1 were measured in both the vertical and horizontal direction, i.e. a total of 22 measurements were recorded for Bent 12. Figure 2 shows the mode shapes at 11.657 Hz and 14.143 Hz from the 2D FEM based on field geometric measurements, blueprints and analytical foundation stiffness values (Poulus and Davis, 1980).

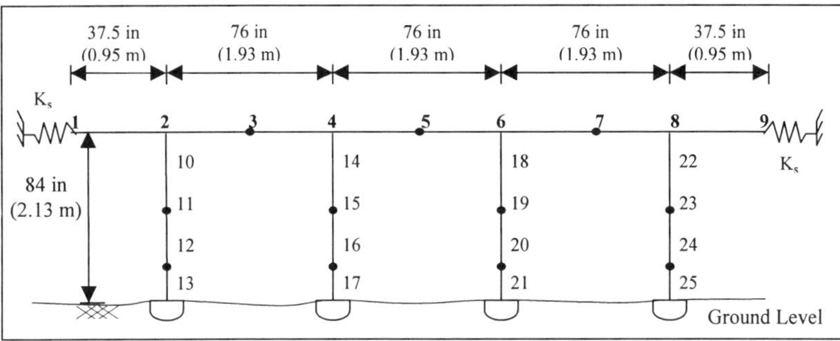

Figure 1. Finite Element Model for Trinity Bridge in Liberty County, Texas.

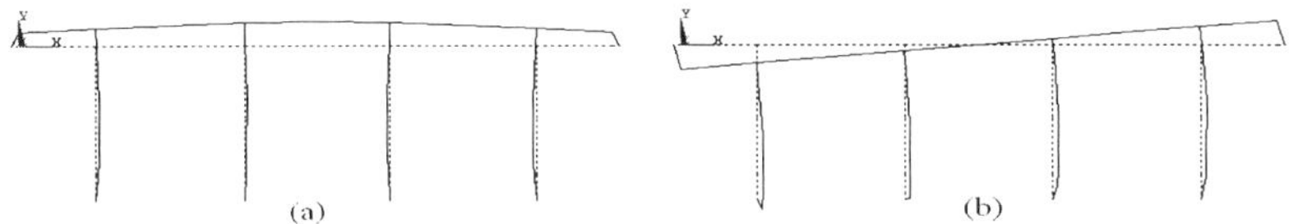

Figure 2. Simulated Mode Shapes-(a) Mode 1 (11.657 Hz), (b) Mode 2 (14.143 Hz).

Bent 12 was tested under (a) undamaged, (b) scour, (c) damaged foundation conditions for the far right pile. Figure 3 compares the vertical frequency response functions (FRF) at nodes 3 and 23. The experimental mode shapes for each test were extracted from the FRFs at each node (Aouad et al. 1998). Only 11 vertical measurment were used to extract the mode shape similar to figure 2(a) while both vertical and horizontal measurements were used to extract the mode shape simliar to figure 2(b). The FRF for node 23 (Figure 3b) which is closer to the damaged pile shows a bigger difference in magnitude compared to the FRF for node 3 (Figure 3a).

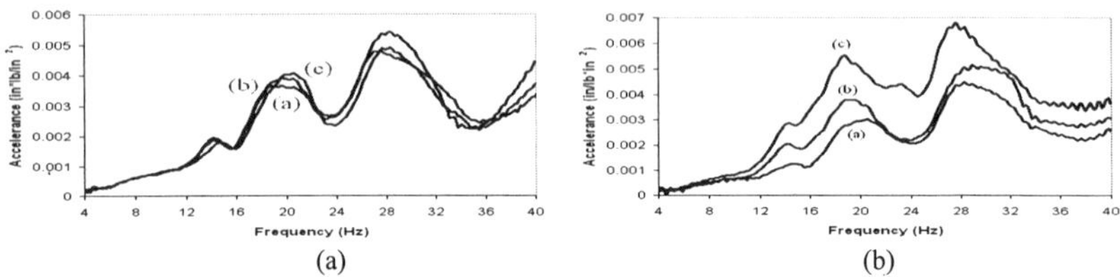

Figure 3. The FRFs for (a) Node 3 and (b) Node 23 for Trinity Bridge.

PARIS was developed to quantify foundation stiffness using field data. The characteristic equation for dynamic loading can be partitioned in terms of mode shape values at measured and unmeasured DOF for each selected natural frequency (Sanayei et al. 1998). By condensing out the unmeasured mode shapes, the "modal error function" is minimized to estimate the stiffness parameters. This equation uses mode shapes only at measured DOF. The error magnification in the foundation stiffness estimates depends on both test data error and the error tolerance of PARIS.

Using modal shape data from the 2D FEM simulations, several PARIS runs were conducted using the first two modes of vibration to determine the most noise-tolerant cases, in terms of number and location of unknown parameters. For the cases in Table 1, in addition to the four foundation stiffnesses the moments of inertia for the beam and piers were considered unknown. Using simulated data, all converged simulations found the exact stiffness values. The most noise-tolerant foundation stiffness parameter was K_{VV}. PARIS runs used the dynamic field data from three foundation conditions. Table 2 shows the K_{VV} values from PARIS estimates while Table 3 compares the frequencies from the initial and adjusted 2D FEM to the field data for mode 2. The mode shapes corresponding to Table 3 are shown in Figure 4.

Table 1. Simulated Cases for Parameter Estimation of Trinity Bridge

Measured Modes	No. of Meas.	Unknown Foundation Elements	Iterations	Comments
1	11 (V)	$4(K_{VV})$	4	Converged
2	11(H+V)	$4(K_{SSS})$	NA	Rank Deficiency
2	11(H+V)	$4(K_{HH}, K_{VV}, K_{\theta\theta})$	4	Converged

Table 2. K_{VV} Estimates Using Dynamic Data from Mode 2 of the Undamaged Case

Analytical K_{VV} N/m (lb/in)	K_{VV}^1 N/m (lb/in)	K_{VV}^2 N/m (lb/in)	K_{VV}^3 N/m (lb/in)	K_{VV}^4 N/m (lb/in)
11e7 (6e5)	65e7 (37e5)	81e7 (46e5)	21e7 (12e5)	8.4e7 (4.8e5)

Table 3. Comparison of Frequencies for Undamaged Foundation Conditions

Initial FEM	Adjusted FEM by PARIS	Dynamic Field Data
14.143	21.817	21.882

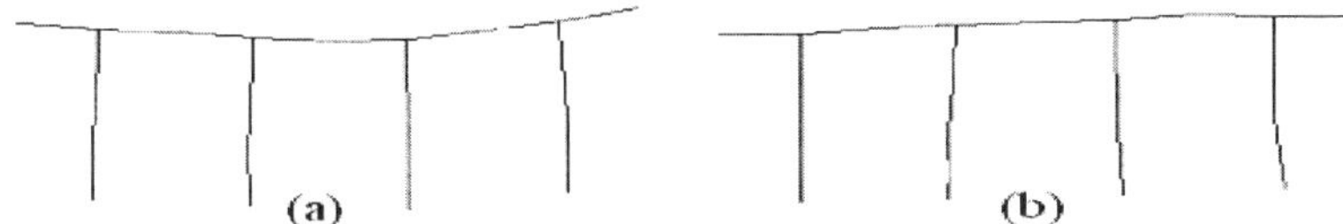

Figure 4. Mode 2 (a) using parameter estimates, (b) using dynamic field data

Although the frequencies of the adjusted FEM and field data match closely, the mode shape (figure 4a) does not fully reflect the shape from field data (figure 4b). This could be due to the fact that only the contributing mass and lateral stiffness of the slab are represented in the 2D FEM but not the bending stiffness of the slab.

Conclusions

Due to the complex nature of the inverse solution of the eigenvalue problem, PARIS requires relatively clean data. The level of uncertainty in the data was higher than what PARIS can tolerate. As a result, only K_{vv}'s were identifiable for Bent 12.

Acknowledgements

This research is sponsored by FHWA grant number DTFH61-96-C-00030 with Mr. Michael Adams as the project technical coordinator.

References

Aouad, M. F., Olson, L. D., and Liu, M. (1998). "Dynamic Bridge Substructure Evaluation and Monitoring System." *Proc., Nondestructive Evaluation Techniques for Aging Infrastructure & Manufacturing,* SPIE 3400, 44-54.

Bay, J. A., Stokoe II, K. H., and Jackson, J. D. (1995). "Development and Preliminary Investigation of Rolling Dynamic Delfectometer." *Transportation Research Record,* 1473, 43-54.

Maser, K.R, Sanayei, M., Lichtenstein, A., and Chase, S.B. (1998). "Determination of Bridge Foundation Type from Structural Response Measurements." *Nondestructive Evaluation Techniques for Aging Infrastructure & Manufacturing,* SPIE, 3400, 55-67.

Poulos, H.G., and Davis E.H., (1980). *Pile Foundation Analysis and Design.* John Wiley & Sons, New York.

Sanayei, M., McClain, J. A. S., Wadia-Fascetti, S., Gornshetyn, I., and Santini, E. M. (1998). "Structural Parameter Estimation Using Modal Responses and Incorporating Boundary Conditions." *Proc., Structural Engineering World Congress,* ASCE, T118-2.

Santini, E.M. (1998). "Bridge Foundation Stiffness Estimation Using Dynamic Nondestructive Test Data." MS Thesis 5322, Tufts University, Medford, MA.

Dynamic Testing of Cable-Stayed Bridges

Joan R. Casas[1]

Abstract

The paper shows the different dynamic tests performed in two-cable-stayed bridges: the Alamillo in Spain and the General Belgrano (Chaco-Corrientes) in Argentina. The first is a newly constructed bridge and the dynamic tests performed just after completion of the construction were justified to update the mathematical and scaled models used in the wind tunnel tests. The second corresponds to an existing cable-stayed bridge where a repair and strengthening work had to be undertaken. In this case, the model updating and correlation was mandatory to know some of the mst important parameters involved in the design of the most effective repair (deck properties, cable forces,...). It is shown how with a minimum instrumentation and recording set-up, and with very simple excitation techniques, it is possible to perform reliable and helpful dynamic field tests.

Description of the bridges and dynamic tests

The Alamillo bridge is a cable-stayed bridge located in Sevilla (Spain). It is used as an access to the Island of La Cartuja, seat of the Expo '92 Universal Exhibition (Fig. 1).The deck of the bridge, with a 200 m span, is supported every 12 m by a pair of stays. Further information is available in Aparicio and Casas (1997). The objectives of the dynamic test were: 1) to check the agreement between frequencies of the real bridge and mathematical and wind-tunnel models, 2) to check if the damping in the real bridge is greater than or equal to the damping in the wind-tunnel model, 3) to determine the final cable forces after completion of the bridge and 4) to evaluate the dynamic increment (impact factor) due to traffic.

The equipment consisted of 4 accelerometers and 1 displacement transducer (Fig. 2). The dynamic excitation was provided by two 2-axle trucks of 196 kN crossing the bridge at different speeds from 5 to 15 m/s. To excite mainly bending or torsional modes in the deck and transverse and longitudinal modes in the tower, different crossing arrangements were chosen (symmetric and eccentric load). Several passages were performed over the undisturbed pavement and other, with the trucks passing over a plank placed at midspan. A complete description of the tests can be found in Casas 1995. A dynamic test in the cables was also used to measure the final forces in the cables after construction according to the vibrating chord theory. An accelerometer was attached to the lower end of the cable-sheating pipe. The excitation was achieved by releasing a hanging weight in a rhythm similar to the natural frequency of the cable. The test results and process of derivation of the cable forces and the problems encountered are fully described in Casas 1994.

----

[1]Professor, Tech.Univ. of Catalunya, Civil Eng. Dept.Gran Capitan s/n. Modulo C1. 08034 Barcelona (Spain)

Figure 1. View of the Alamillo cable-stayed bridge (Sevilla, Spain)

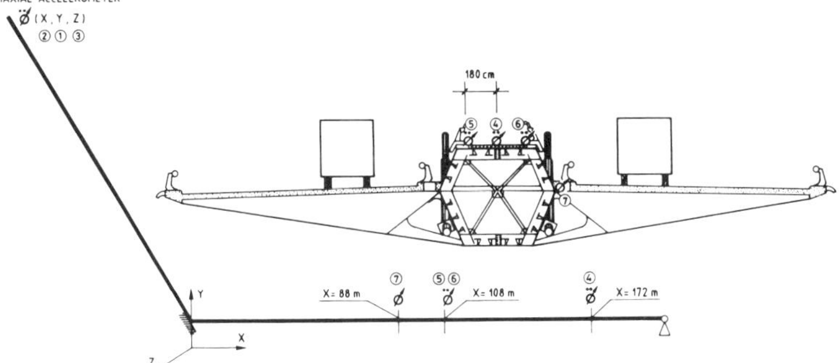

Figure 2. Cross-section, instrumentation used and location in the Alamillo bridge

The Chaco-Corrientes cable-stayed bridge is the central link of a major roadway connection across the Paraná River in Argentina. Inaugurated in 1973, its structure consists of two main longitudinal box girders of segmental precast prestressed concrete. Pylons are cast-in-place concrete space frame rigidly connected to the box girders at deck level. A general view is given in Figure 3. The central span measured at centerlines of pylons is 220 m. The original cables were of locked-coil type with external layers of galvanized wires. In this case, the objective of the dynamic test was the reconstruction of the current state of stress in the different elements of the bridge. That was necessary to decide and design the repair and strengthening works to be carried out. The works consist basically in the replacement of all cables due to the damage present as well as in the restitution of the correct profile of the bridge deck that showed a large and unpleasant deflection after some years of operation due to concrete creep and cables relaxation. The updating was possible by direct measurement of cable forces and vertical deck accelerations recorded over time. Determination of the current stiffness of the concrete deck was accomplished through interpretation of ambient vibration records of the bridge deck under traffic and wind forces. Measured natural frequencies are used to calibrate the effective stiffness of a numerical model, which in turn is used to determine the additional forces in cables and girders associated with restitution of the longitudinal profile. The ambient vibration records consist of a series of vertical acceleration time histories of three points.   Sensor 1: Cross section passing through the intersection of the longitudinal axis of the girder and the resultant force of the groups of 6 stays. Sensor 2: Cross section passing through the intersection of the auxiliary stays (not yet installed when the records were made) and the longitudinal axis of the girder. Sensor 3: Cross section passing through the intersection of the longitudinal axis of the girder and the resultant force of the groups of 4 stays. These sets of 3 simultaneous records, made along both main girders in the two half bridges, were repeated several times at each location to provide a variety of forcing functions during intervals with large oscillations induced by traffic and wind. Vibrations of the cables were also induced in this case to derive the actual forces in the cables to be replaced. Because the technology of stays and anchorages in this bridge was much older and simpler than in the case of the Alamillo bridge, the actual forces were easily derived from the records without requiring additional information and correction procedures. More information can be found in Prato et al. 1997.

Fig 3. View of the Chaco-Corrientes (General Belgrano) cable-stayed bridge

Results

The FFT technique was used to identify frequencies and mode shapes. In the Alamillo, the frequency resolution achieved with the part of the total recorded signals suitable for post-processing was 0.025 Hz. The main results concerning the dynamic parameters of the bridge and their comparison with theoretical ones are summarized in table 1. The range in the values of percentage of critical damping (ζ) indicates the maximum and minimum values obtained from records with different maximum vibration amplitudes depending on the test. Only the transverse vibrations of the pylon were measured. As deduced from the table, the agreement between dynamic parameters of the real bridge and theoretical and scaled models (tested in wind-tunnel) was completely satisfactory. Also in the table is shown how, in spite of the scatter, the actual measured damping ratio is always greater than damping in the aerolastic model. Therefore the conclusion was that the vibration level because of vortex shedding will not derive in unpleasant or dangerous (from the fatigue point of view) vibrations. In the General Belgrano bridge, to get longer records in time and to achieve a better resolution in frequency, the individual records acquied in the same sensor location in different times due to wind and traffic, were added. Results are shown in table 2.

Conclusions

The results have confirmed the possibility of performing dynamic tests in long-span bridges even with relatively low excitation means consisting of two trucks if the measurement set-up is accurate enough. Therefore, this excitation technique (passage of trucks over obstacle) becomes a useful alternative to the ambient vibration test (by wind or traffic). The correct dynamic behaviour of the bridge in response to traffic and wind (vortex shedding, flutter, etc.) can be

Vibration mode	Theoretical model	Aerolastic model			Actual bridge (acceleration)		Actual bridge (displacement)	
	f (Hz)	f (Hz)	ζ (%)		f (Hz)	ζ (%)*	f (Hz)	ζ (%)
Transverse pylon 1	0.292	0.30	0.41		0.30	--	--	--
Longitudinal (pylon + deck) 1	0.373	0.39	0.21		0.40	1.9-4	0.40	1.9
Longitudinal (pylon + deck) 2	0.610	0.65	0.37		0.66	1.1-4	0.65	1.6
Transverse deck 1	1.088	1.20	0.72		--	--	--	--
Longitudinal (pylon + deck) 3	1.191	1.19	--		1.205	0.6-2.6	1.20	--
Torsion deck 1	1.235	1.11	0.25		1.155	0.5-1.5	1.16	--
Transverse pylon 2	1.583	1.67	--		1.537	--	--	--
Longitudinal (pylon + deck) 4	2.196	1.97	--		2.155	0.6-3.7	2.06	--
Torsion deck 2	2.298	2.19	--		2.295	0.5-0.8	--	--
Longitudinal (pylon + deck) 5	2.312	--	--		2.78	--	--	--
Transverse deck 2	3.244	3.4	--		--	--	--	--

Table 1. Summary of results in the dynamic test of the Alamillo bridge

Record	Mode				
	4 (Bending)	9 (Bending)	12 (Torsion)	14 (Bending)	28 (Bending)
CN	0.64	1.64	-	4.09	13.15
CN'	0.51	1.59	-	3.99	13.20
CS	0.62	-	-	4.07	13.20
CS'	0.50	1.58	-	3.88	13.03
HN	0.65	1.62	-	4.04	13.18
HN'	0.65	1.62	-	-	12.99
HS	0.62	1.69	-	4.00	13.10
HS'	0.47	1.57	-	3.87	13.11
TOR1	0.58	1.67	3.72	4.03	13.09
Mean	0.58	1.62	3.72	4.00	13.12
Standard dev.	0.067	0.040	-	0.076	0.069
Numerical model	0.57	1.616	3.80	4.071	13.048

Table 2. General Belgrano bridge. Theoretical and experimental frequencies derived by FFT.

deduced. The measurements in the cables permits the updating of their dynamic model, being the basis of future inspections of the cable structural performance based on the vibrating method. In the case of the Chaco-Corrientes bridge, the ambient vibration technique has shown as very useful in its application to cable-stayed bridges, where wind actions are of importance. Actual forces in the cables to be replaced were also found in the dynamic tests. Only in this way the correct force to be introduced in the new cables can be obtained.

References

Aparicio, A.C. and Casas, J.R. (1997). "The Alamillo cable-stayed bridge: special issues faced in the analysis and construction". *Proc. of the Institution of Civil Engineers, Structures and Buildings,* 122, 432-450.

Casas, J.R. (1995). "Full-scale dynamic testing of the Alamillo cable-stayed bridge in Sevilla (Spain)". *Earthquake Engineeering and Structural Dynamics*, 24, 35-51.

Casas, J.R. (1994). "A combined method for measuring cable forces: the cable-stayed Alamillo bridge, Spain". *Structural Engineering International*, Vol. 4, N. 4, 235-240.

Prato, C.A.; Ceballos, M.A.; Casas, J.R. and Aparicio, A.C. (1997). "Interpretation of ambient vibration records for restitution of deck profile of the Chaco-Corrientes cable-stayed bridge". Proceedings of Structural Faults and Repair-97, 387-394, Edinburgh.

Forced and Ambient Vibration Tests of Hakucho Suspension Bridge

Yozo Fujino[1], Shun-Ichi Nakamura[2], Hajime Shibuya[3],
Masashi Sato[3],Masato Yanagihara[3] and Yoshifumi Sakamoto[4]

Abstract

Full-scale forced and ambient dynamic tests were carried out on the Hakucho Suspension Bridge to study its dynamic behavior. The emphasis is placed on measurement of natural frequencies as well as vibration shapes of high modes of the bridge using dense instrumental array. The natural frequencies and mode shapes obtained from the forced and ambient vibration tests agreed well with those calculated from three dimensional finite element model..

Introduction

Dynamic testing of full-scale structures is the only reliable method to verify an analytical dynamic model and it can determine dynamic parameters such as natural frequencies, mode shapes and damping[1,2]. This can give useful information to structural monitoring in order to detect structural damage. Full-scale tests of long-span bridges have been conducted on the Honshu-Shikoku project in Japan, but these tests mainly aimed to examine the wind design criteria and hence only low frequency vibration of stiffening girders were studied[3]. This paper describes two types of full-scale dynamic tests, namely forced and ambient vibration tests, conducted on Hakucho Suspension Bridge. Taking responses and possible damage caused by winds, earthquakes and traffic into consideration, high frequency modes of girders as well as towers were studied using dense array measurement system.

Description of Hakucho Bridge and dynamic tests

[1]Professor of Civil Eng., University of Tokyo, Tokyo, Japan,   [2]Professor of Civil Eng., Tokai University,   [3]Hokkaido Development Bureau,   [4]Nippon Steel Corporation

Hakucho Bridge is a suspension road bridge with main-span of 720m and two side-spans of 330m each. The girder is a streamlined steel box one with width of 23.0m and maximum web height of 2.5m. The tower s are made of steel with full welding, which makes the damping very low; indeed the logarithmic decrement of the tower during free standing erection stage was found to be as low as 0.005. Active vibration control system was successfully applied to this tower to suppress the wind-induced vibration[4]. The bridge crosses the Muroran Bay and receives frequent attack of strong winds in winter and typhoon seasons, and is located in a seismically active zone. The monitoring system consisting of servo-type accelerometers and ultrasonic anemometers has been installed to observe dynamic behavior of the bridge to assure the safety of the structure and the vehicle passage.

Dynamic tests were carried out from March to May in 1998 just before the bridge opening in June. A number of accelerometers were installed for the tests in addition to the monitoring system.

Figure 1.   Hakucho Bridge

Forced vibration tests

The bridge was excited by two vibrating machines with capacity of 0.6-23.5kN each and the frequency range of 0.1-2.0Hz. The machines were set at span-center or quarter position depending on the modes to be excited, and the two machines were operated in-phase for vertical and lateral modes while in 180° anti-phase for torsional modes. 16 accelerometers were attached to the girder, 12 on the tower and 8 on the cables to collect the data. All the data were stored in 3 digital recorders. The girder was excited with a 0.001Hz interval to find peak frequency. Damping was measured from the free vibration which was generated by abruptly stopping the vibrating machines.

Ambient vibration tests

Ambient vibration tests

In the ambient tests 40 accelerometers were placed in the half main-span girder (30m interval) and in the side-span(55m interval), and 29 were in the towers. The data was recorded for at least 30 minutes in one measurement. Totally the measurement was made over 35 hours under various levels of the wind speed up to15m/s. Time interval for recording was 0.01sec and the total length of each measurement was 409.6sec, providing frequency resolution of 0.00244Hz..

Results obtained by both vibration tests

Table 1 shows natural frequencies obtained from the forced and ambient tests and those calculated by three dimensional finite element model. Those from the two tests have good agreement, and also agreed well with the calculated values. Tower modes were obtained by exciting the girder with the natural frequencies of the tower. Logarithmic decrement obtained from the free vibration is also shown in the table. Most of the modes have logarithmic decrement ranging from 0.02 to 0.05, but the 1st lateral mode has a far large value while the tower 1st in-plane mode has a very small value.

Table 1. Dynamic Tests Results

Vibration Mode			Natural Frequency (Hz)			Logarithmic Decrement
			Forced	Ambient	Analysis	
Girder Vertical Mode	Symmetric	1	0.129	0.131	0.123	0.040
		2	0.218	0.223	0.217	0.027
		3	0.435	0.440	0.436	0.025
		4	0.719	0.724	0.720	0.035
	Anti-Symmetric	1	0.149	0.152	0.149	0.015
		2	0.317	0.318	0.320	0.020
		3	0.568	0.571	0.567	0.036
		4	0.906	0.902	0.893	0.050
Girder Torsional Mode	Symmetric	1	0.487	0.494	0.472	0.037
		2	1.164	1.160	1.128	0.054
		3	1.866	1.860	1.852	0.032
	Anti-Symmetric	1	0.772	0.778	0.761	0.032
		2	1.502	1.530	1.490	0.022
Girder Lateral Mode	Symmetric	1	0.099	0.093	0.101	0.270
		2	0.560	0.566	0.541	0.039
Tower In-Plane Mode		1	0.599	0.598	0.624	0.010

Figure 2 compares two examples of the mode shapes calculated by FEM analysis and the mode shapes obtained from the forced and ambient tests. It is clearly shown that the both test results are very close to the calculated mode shapes.

Wind speed was measured by ultrasonic anemometers at mid-span and tower top. Ambient vibrations were measured for wind speed up to 15m/s, and the collected data were analyzed by Fourier transform. It is found that, as the wind speed increased, acceleration responses of girder and tower increased, and that Fourier spectrum shape became sharp and dominant frequency shifted to a lower value. GPS system was also applied to directly measure quasi-static displacements of girder; this was found to be successful in obtaining reasonably accurate static lateral displacements due to strong winds[5].

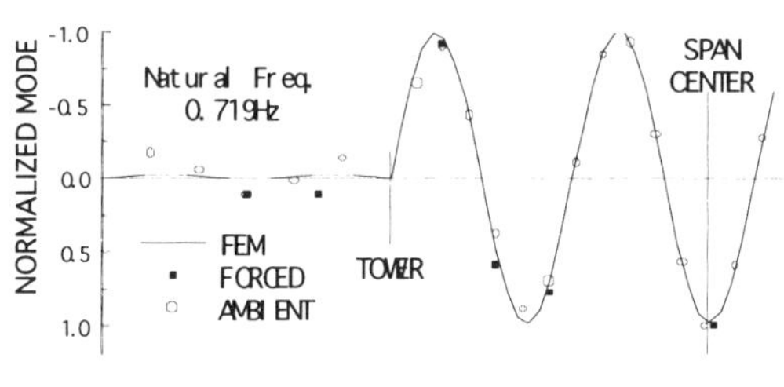

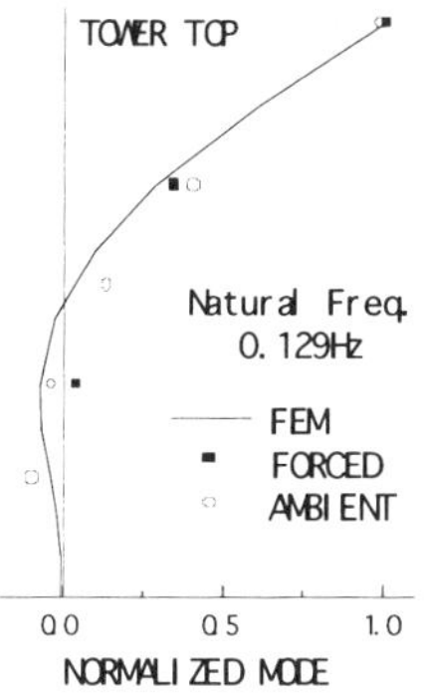

Figure 2. Comparison of Mode Shapes between Analysis and Measured Data for Girder Vertical Symmetric 4th and Tower 1st Out-of-plane Vibration Modes

Conclusions

Full-scale forced and ambient dynamic tests were conducted on the Hakucho Suspension Bridge to study the dynamic behavior. In the test, not only low modes but also high modes of girders and towers were carefully studied using dense array measurement system. The natural frequencies and mode shapes obtained from the ambient and forced vibration tests agreed well with those calculated by three dimensional FEM. Damping in most of the modes has logarithmic decrement ranging 0.02-0.05, but the 1st lateral mode has far large value while the tower 1st in-plane mode has very small value.

References

1. Ahmed M. Abdel-Ghaffar and George W. Housner: Ambient Vibration tests of suspension bridge, *Journal of Engineering Mechanics*, ASCE, Vol.104, EM5, 1978, pp.983-999.

2. Ahmed M. Abdel-Ghaffar and Robert H. Scanlan: Ambient Vibration Stidies of Golden Gate Bridge I. Suspended Structure, II. Pier-Tower Structure *Journal of Engineering Mechanics*, ASCE, Vol.111, EM4, 1985, pp.463-499,

3. Y. Fujino, T.T. Soong, B.F. Spencer Jr.: Structural Control, Basic Concepts and Applications, Proc. of the XIV ASCE Structural Congress, Chicago, 1996, pp.1277-1287.

4. Shun-Ichi Nakamura ital: Monitoring of Wind-Induced Displacements of Suspension Bridges using GPS, to be presented, *IABSE Symposium in Rio*, 1999.

5. Isao Okauchi ital: Vibration Test of Ohnaruto Bridge to Confirm Wind-Proofness, Proc. *IABSE Symposium in Tokyo*, 1986.

Laser-Based Testing and Monitoring of Large Bridges

Jan Bień[1]
Paweł Rawa[1]

Abstract

In the report there were presented experiments with laser-based system appliances for measuring the displacement of large bridge structures during the proof loads as well as for the structure displacement monitoring during operating. The presented measurement results carried from the distance of up to 150 m by means of laser-based equipment were compared with the results achieved by other measuring methods and the values calculated theoretically.

Measurement system

To the tests there was applied the measurement system PSM200 produced by the NOPTEL OY company in Finland. The system consists of a laser transmitter and the position-sensitive receiver connected to a laptop PC by means of a serial port (Fig. 1). The laser beam forms a reference line from the transmitter to the position- -sensitive detector. When the receiver placed on the structure moves, the position of the laser beam on the sensitive screen is continuously recorded (x and y coordinates) and stored in the PC. The laser beam acts as a reference line for the measurement, which requires a steady solid base for the transmitter. The transmitter as well as the receiver are separate units functioning on a 12 volt DC supply. The resolution of the receiver is 0.1 mm at a sampling rate of 500 Hz for the standard receiver diameter of 200 mm.

Proof loads

According to the regulations in Poland, all road bridges of the span over 20 m have to be submitted to the proof load tests before being accepted for use. In the case

----

[1] Wrocław University of Technology (WUT), Institute of Civil Engineering (ICE),
   Bridge Group, Wyb. Wyspiańskiego 27, 50-370 Wrocław, Poland

of the structures of big span length (over 100 m) situated over the water or deep valley, there occur some problems with thorough measurement of displacements by using traditional measurement methods. The example of one such structure is the

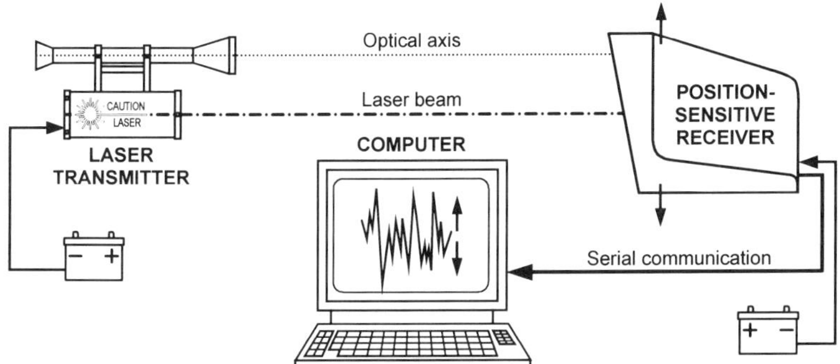

Figure 1. Configuration of the laser-based system PSM200

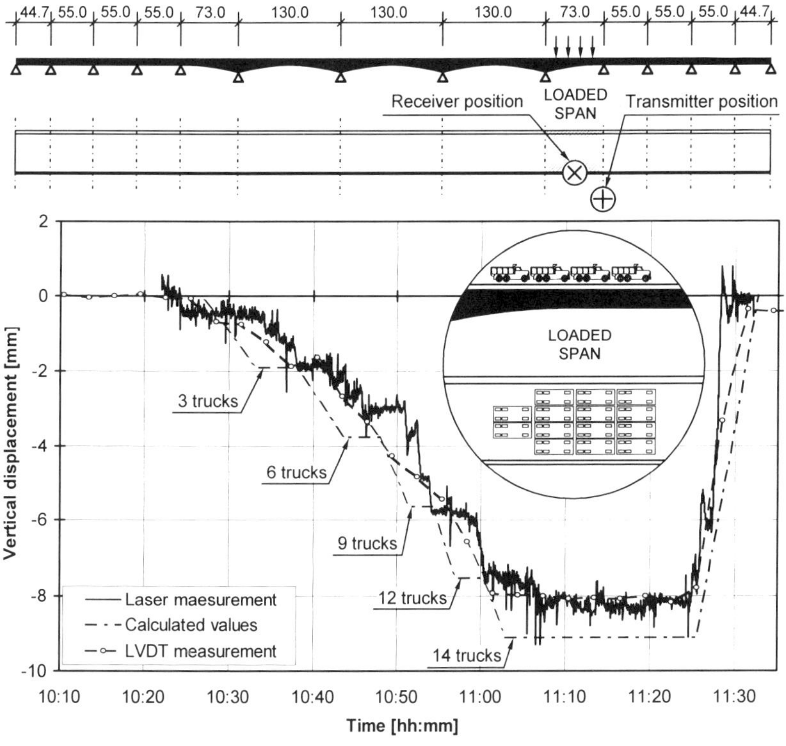

Figure 2. Measured and calculated vertical displacement of the highway bridge

bridge over Vistula River near Toruń belonging to highway A1. The construction of the bridge was accomplished in June 1998. The object is a 13-span box structure of prestressed concrete of the total length 955.40 m (Fig. 2, top part). The center spans the length of 130 m are situated directly over the Vistula current. Because of the innovative solutions applied during the bridge construction (the first bridge in Poland constructed by the cantilever method) and because of the longest prestressed concrete span in Poland, Wrocław University of Technology (WUT) has made comprehensive proof loads of the structure. During the measurement lasting over a week the strains, angles of rotation at selected sections inside the structure as well as displacements of spans and supports were measured. Because of the land configuration, only the span displacements over the flood lands were measured by classical methods, i.e. by using inductive displacement detectors. The over-water spans - which could not be reached from the ground level - were measured by the laser system. To check the correctness of the laser equipment a preliminary measurements have been compared with the other measurement methods results and calculated values. Fig. 2 presents the test measurement results of vertical displacement in the middle of the span of 73 m long loaded with 14 cars (14 x 300 kN = 4200 kN). Displacement changes for each group of loading cars were measured by the laser system and LVDT measurement system. After the conformability of the results achieved by various methods had been stated, the loads of the over-water spans were conducted. The example results are shown in Fig. 3 where the measured displacements in the middle of the span were compared with those calculated theoretically. Detail results of proof loads and computer simulation of the structure are presented in reports (Bień et al., 1997, Bień et al., 1998). The achieved results confirmed the correctness of the design and the construction as well as the usefulness of the applied equipment for large bridge testing.

Monitoring

Similar problems concerning large bridge displacement measurements occur not only during proof loads but during monitoring of such structures while being used as well. The laser equipment was also used for testing two bridges over the Odra River in Wrocław: a one-span hanging steel structure of the length of 100 m and a three-span box structure of prestressed concrete of span lengths: 33.1 m + 52.5 m + 33.1 m. Both structures are situated over the current of the river. The measurements were made from various posts and the distance between the transmitter and the receiver was 70 m to 100 m The applied method confirmed its usefulness and enabled to estimate condition of the bridge structures which were always difficult to be monitored by classical measuring methods.

Conclusions

The laser-based system displacement measurement technology is relatively new (Tervaskanto, 1998), but confirms its usefulness fully on the field measurements

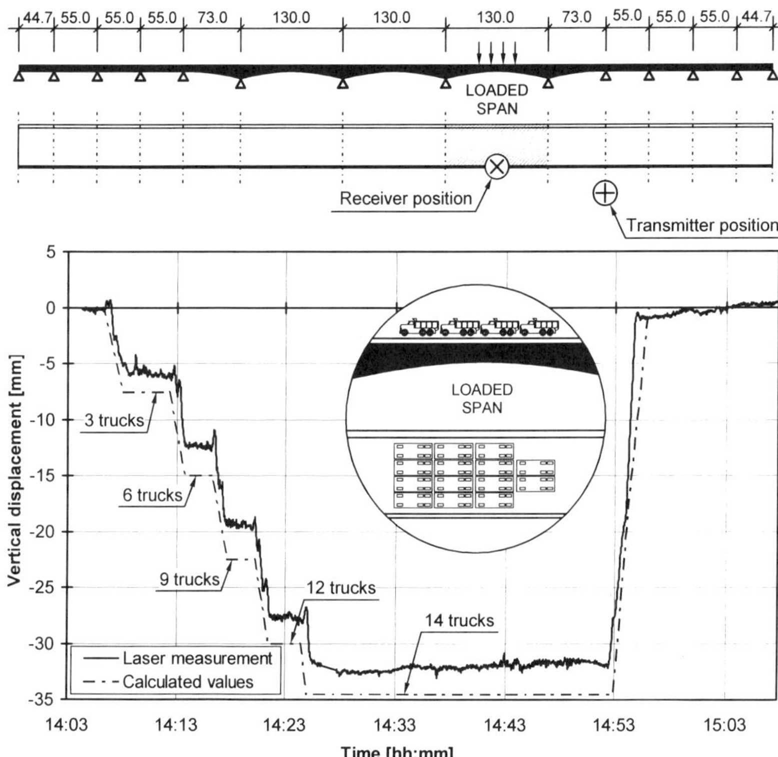

Figure 3. Measured and calculated vertical displacement of the highway bridge

of large structures. At measurements from the distance up to 150 m the method ensures the precision efficient for technical purposes in both static and dynamic tests. The set of measuring equipment is relatively easy in fitting and operation as well as resistant to environmental influences (sun, temperature changes, precipitation, etc.).

References

Bień J., Kmita J., Rawa P., Zwolski J.: Proof loads project of the southern lane of the road bridge across the Vistula River near Toruń, WUT, Report ICE, No. SPR-77/97, 1997.

Bień J., Kmita J., Maliszkiewicz P., Rawa P., Zwolski J., Szymkowski J.: Results of the proof loads of the southern structure of the road bridge across the Vistula River near Toruń, WUT, Report ICE, No. SPR-53/98, 1998.

Tervaskanto M.: A laser based displacement measurement technology for monitoring and testing the dynamic behaviour of large structures, Symposium of Geodesy for Geotechnical and Structural Engineering, Eisenstadt, Austria, 1998.

Testing of a Damaged Prestressed Concrete Bridge

Francesco M. Russo[1], F. Wayne Klaiber[2], Terry J. Wipf [2]

Abstract

The use of diagnostic load testing to determine live load distribution in an undamaged, damaged, and repaired prestressed concrete I-beam bridge is described herein. Brief results from a research project are presented. The effects of lateral impact damage from overheight vehicles is quantified in terms of its effect on the live load distribution in the damaged structure. The replacement of the damaged beams appears to have restored the original live load distribution pattern.

Introduction

The objective of this paper is to briefly describe the diagnostic load testing of prestressed concrete I-beam bridges to assess the load distribution characteristics. The results of three field tests are described in which load testing of a damaged bridge, later repaired by beam replacement, were compared to a complementary undamaged parallel structure. The objective of the testing was to determine if accidental damage due to overheight vehicles has a measurable effect on the load distribution in prestressed concrete bridges. The load tests are part of a larger research project that is assessing load distribution in damaged bridges, remaining strength of damaged beams removed from service, and the use of CFRP for strength and stiffness restoration.

The two bridges tested in this project carry I-680 eastbound and westbound over county road L34 in Pottawattamie County, Iowa. The bridges are three-span asymmetric bridges with 11 beam lines. The spans are 13 145mm, 17 145mm, and 14 415mm from east to west. Beams 11W through 5W are straight with a beam spacing of 1525mm. Beams 4W through 1W are flared with the beam spacing varying from 1065mm to 1525mm. The bridge was designed with a 150mm nominal slab thickness and has since been overlaid to a total thickness in excess of 230mm.

[1] Iowa Department of Transportation, 800 Lincoln Way, Ames, IA 50010
[2] Iowa State University, Department of Civil Engineering, Ames, IA 50011

Figure 1 Collision Damage to Beam 1W

The impetus for this research project was the collision of an unknown vehicle with the north three beams, beams 1W, 2W and 3W, of the westbound structure in July 1996. Damage was centered ± 1.5m west of the mid-span diaphragm of the center span. Approximately 1.8m of the bottom flange spalled from the north fascia stringer, beam 1W, exposing the bottom two layers of 12.7mm diameter strands (see Figure 1). Several of the strands on beam 1W seem to be lax but no strands were severed during the 1996 collision. There was a pre-existing severed strand from a 1993 collision. Cracking of the bottom flange and web as well as fracturing of the core concrete was present in the first two beams, but to a lesser extent on beam 2W. Cracking appears to have been arrested by the presence of the cast-in-place concrete diaphragm. The second interior beam, beam 3W, was also damaged, but not as severely, with the damage consisting of the spalling of a patch installed following prior collisions with the bridge.

Prior to the initiation of this research project, the Iowa DOT decided that beams 1W and 2W would be replaced due to uncertainties concerning their remaining strength, effect of damage on load distribution, and long-term serviceability of the damaged beams. By supporting this research, the Iowa DOT wanted to develop a more refined criteria through which the effects of damage on the behavior and strength of damaged prestressed concrete bridges can be more accurately assessed. One of the objectives of the project is to present guidelines for the simulation of damage in an analytical model for predicting live load distribution in damaged prestressed concrete bridges.

Figure 2 Test Trucks, Multiple Trucks in Same Lane Shown

Field Testing of the I-680 Beebeetown Bridges

The load tests used two Iowa DOT maintenance vehicles of known axle loads and axle spacing. In the 1997 baseline tests, trucks weighing an average of 265kN were used. Approximately 25% of this load is distributed to the front axle with the remainder carried by the rear tandem. In the 1998 tests on the repaired westbound bridge, slightly lighter trucks with an average weight of 232kN and similar axle load distribution were used. To date, no attempt has been made to scale the 1998 test results so as to simulate the somewhat heavier trucks (approximately 14%) used in the 1997 tests. A majority of the load tests were conducted with a single truck at various longitudinal and transverse positions. Several tests were conducted with trucks side-by-side in adjacent lanes and with multiple trucks in the same lane, the latter of which is shown in Figure 2.

Figure 3 is a plot of several transverse deflected shapes taken at midspan of the damaged center span of the westbound bridge and the complementary location in the eastbound bridge. L1W is a test of the damaged westbound bridge, L1W(R) is a test of the repaired westbound bridge, and L1E is a test of the undamaged eastbound bridge. For all tests, the rear tandem of a single test truck is centered at midspan with the wheels located as close as practicable to the gutter line over the damaged beam lines. This location has been defined as Lane 1, i.e. L1, in the test description.

Figure 3 shows comparable deflections and therefore load distribution in the undamaged eastbound and repaired westbound bridges. Both bridges are stiffer (deflect less) than the damaged westbound structure. The data tends to support a

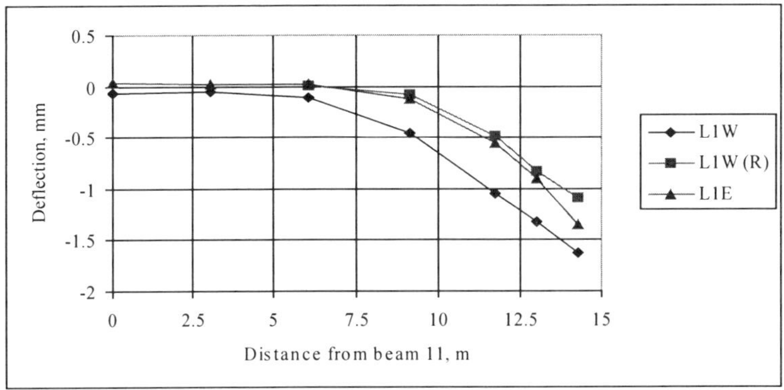

Figure 3 Transverse Deflected Shapes, Lane 1 Loaded

conclusion that the westbound bridge, in lieu of damage, should have performed roughly the same as the undamaged eastbound bridge. The difference in measured deflection in the westbound damaged and repaired tests can be attributed to the replacement of damaged beams 1W and 2W.

Other transverse deflected shapes, not shown here due to space limitations, tend to concur with those results presented in Figure 3. The repaired bridge behaves more like the undamaged eastbound bridge. However, because of the inherent longitudinal and transverse stiffness of the two test bridges, the actual difference in deflections from undamaged, damaged, and repaired tests is very small.

Analytical Investigations

At this time, a series of analytical models have been developed using the grillage analogy. Additional models are being developed using the ANSYS finite element program to help understand the experimental data. The intent of the analytical investigation is to develop some guidelines that describe how damage to a longitudinal beam in a prestressed concrete bridge affects live load distribution in the bridge.

Summary

This paper has briefly described an ongoing investigation into the effect of lateral impact damage on prestressed concrete I-beam bridges. At this time, dual bridges have been field-tested and a preliminary evaluation of that data completed. Tentative conclusions indicate that the eastbound and westbound bridges are behaving similarly now that the damaged beams have been replaced. The measured deflections under known loads are small, the maximum being less than 2.5mm under a total load in excess of 530kN.

Noncontact Nondestructive Testing for Structure Health Monitoring

Gongkang Fu[1], M.ASCE, Adil G.Moosa[2], Juan Peng[2]

Abstract

This paper presents a study on nondestructive testing for bridge inspection using advanced data acquisition capabilities. They are able to quickly obtain a large amount of data for low cost. A probability based approach is proposed to process the measurement data for diagnosing an "after" state with reference to a "before" state. For illustration, a lab test was conducted using a laser device for deformation measurement, showing that the concept is promising and further field experiment is warranted.

Introduction

Current bridge inspection largely relies on visual observation by trained personnel. It is labor intensive and subjective. Improvement is desired for higher cost-effectiveness. Research on nondestructive testing has been one of the efforts in this direction. A number of advanced data acquisition concepts and devices have been made available for this purpose. Compared with traditional approaches, they can obtain much large amount of data for less time and lower cost. For data obtained by these devices, this paper proposes a new diagnosis algorithm, named probabilistic advancing cross-diagnosis (PAC) method.

An example of such devices is the recently developed coherent laser radar system (CLRS) [CRC 1996]. It is a portable, light-weight, non-contact measurement system. It can be easily setup by two people in minutes to measure bridge deflection at multiple points from 2 to 30 m away. It is applicable to surface of any color or texture, under any lighting condition. Its laser is certified as unconditionally eye safe. For field application, the system can be transported to the site by a small van. The scanner head has a horizontal (Az) range of $\pm 200^O$ and a vertical (El) $\pm 60^O$ [CRC 1996]. In this study, this system was used in the structures lab of Wayne State University to obtain deformation data, for which the proposed algorithm was applied for diagnosis.

---

[1]Associate Professor [2] Graduate Research Assistant, Department of Civil and Environmental Engineering, Wayne State University, Detroit, MI48202

PAC Method

Let H+1 sets of deformation data be available, for $L=D$ as follows:

$$[B]^{1L} = (B^{1L}_1, B^{1L}_2, B^{1L}_3, B^{1L}_4, ..., B^{1L}_M)$$

$$..........................$$

$$[B]^{HL} = (B^{HL}_1, B^{HL}_2, B^{HL}_3, B^{HL}_4, ..., B^{HL}_M)$$

$$[A]^L = (A^L_1, A^L_2, A^L_3, A^L_4, ..., A^L_M) \qquad (L=D,S,C,C_2) \qquad (1)$$

where $[B]^{1D}$ to $[B]^{HD}$ are H replicate matrices of deformation for the "before" state of the structure. $[A]^D$ is deformation for the "after" state of the structure to be diagnosed. Each data vector in Eq.1 includes J replicates. Each replicate covers interested areas of the structure by a grid of M data points. This grid is pre-selected according to the need of bridge inspection. Further, the deformation data can be used to numerically obtain other relevant physical quantities or features, indicated by L: $D=$ deformation, $S =$ slope, $C=$ curvature, and $C_2 =$ curvature squared.

For two data vectors s and t, a likelihood factor is defined as $LF_{s,t} = LF_{t,s}$, $=Ln(P_{s,t}*P_{t,s})$, where

$$P_{s,t} = \int_{min(s)}^{max(s)} f_t(t)\, dt \qquad\qquad P_{t,s} = \int_{min(t)}^{max(t)} f_s(s)\, ds \qquad (2)$$

where $max(\ )$ and $min(\ )$ are the maximum and minimum of the data set. $P_{s,t}$ is the probability that data elements in s belong to the distribution of t, $f_t(t)$. Normally distributed $f_t(t)$ and $f_s(s)$ are used here. Their mean and standard deviation are estimated using s and t respectively. A lower LF_{st} means that more likely s and t do not belong to each other's population, or there is damage from state s to state t or vice versa.

Using Eq.2, a single index of comprehensive likelihood factor (*CLF*) covering all features for cross-diagnosis is defined as follows:

$$CLF = \Sigma_{L=D,S,C,C_2}\, w_L\, [LF_{A^L, B^{1L}\&...\&B^{HL}} - (LF_{B^{1L}, B^{1L}\&...\&B^{HL}} + ... + LF_{B^{HL}, B^{1L}\&...\&B^{HL}})/H] \qquad (3)$$

where all *LF*s refer to a data point. Subscript $B^{1L}_{\&...\&} B^{2L}$ is combination of B^{1L} to B^{HL} having HJ replicates. *CLF* is a weighed sum of signals for each and every feature. Each weight w_L depends on the feature's sensitivity to damage or deterioration, possibly affected by measurement noise. CLF close to zero indicates low likelihood of deterioration, and a lower *CLF* signals higher likelihood of deterioration. CLF is based on comparison of data for each point. Its definition can be extended for neighborhoods, each including several points. The more points are included in a neighborhood, the less vulnerable to noise CLF becomes, and the lower resolution CLF will have for diagnosis. It need to be noted that the dot product of two data sets is proposed to be used to quantify their correlation. It is then converted to a normal variable through the Fisher Transfer, to be used in Eq.2 for computing CLF [Fu et al 1998].

It is seen that the PAC method has several merits. 1) It needs no assumption of whether and where there is deterioration. 2) It uses a probability based approach to explicitly dealing with possible noise. 3) It uses multiple features for reliable cross-diagnosis. 4) It is flexible to particular application. 5) The diagnosis process is mechanical, thus it can be implemented in computer software.

Laboratory Experiment

The CLRS was used in this study to measure deformation data for a simply supported steel beam modeling bridges. The laser scanner was set approximately 2.5 m away from the beam, which is apparently not optimal distance for CLRS. The data grid was selected to consist of two lines on the bottom flange, one being at the center and the other close to the front edge. Each line included 70 equally spaced data points (M=2*70=140). A concentrated load was applied at the center of the beam. For the "before" state, deformation was read by the CLRS for 2*15 replicates (H=2,J=15). For each replicate, it scanned the entire grid of 140 points for 47 sec. in a rate of 3 points per sec. This was a computer controlled automatic process. Then two cuts were introduced to the beam by a grinder with a 3.2 mm (0.125 in.) thick blade, as shown in Fig.1. One was between Points 42 and 43 and 25.4 mm (1 in.) long, the other betwwen Points 51 and 52 and 12.7 mm (0.5 in.) long. They respectively reduced stiffness by 7.8 and 3.8%. Then 15 replicates of deformation data were obtained by the CLRS for this "after" state.

The data were first treated for noise reduction, mainly for the derivatives (S,C,C_2). Fig.2 shows resulting CLF, which is viewed as a condition map for the beam for diagnosing the "after" state. $w_D = 0.1$; $w_S = 0.2$; $w_C = 0.3$; $w_{C_2} = 0.4$ were used in Eq.3. Lower CLF values clearly show the two areas of possible deterioration: one around Points 45 to 46, and the other around 49 to 52. It was also found that the beam was not symmetric longitudinally and transversely, as shown in Fig.2. More simulated cases were experienced in this study, and consistent effectiveness was seen for the proposed diagnosis algorithm [Fu et al 1998].

Conclusions

Advanced noncontact data acquisition is now a reality for structural testing. Using a non-optimal application condition for CLRS, the proposed method is able to identify multiple small damages, as well as respective locations. The diagnosis resolution depends on data grid's resolution, and data quality. Noise treatment for measured data by CLRS is necessary, to have more reliable numerical derivatives of deformation. Field experiment is warranted based on these lab results, where bridge self weight may be used as the load.

Acknowledgments

Funding for this work, by FHWA Research DTFH61- 97P00549 and CULMA of Wayne State University, is gratefully acknowledged. Special thanks are due to Dr. Steven Chase, the program director, for his inspiration and encouragement. Dr. Paul Fuchs with FHWA assisted in the lab testing. Many fellow faculty and students at Wayne State University assisted in testing. Their efforts are appreciated.

References

[1] CRC "Coherent Laser Bridge Measurement", Final Report to FHWA, Feb. 1996
[2] Fu,G., Peng,J., and Moosa,A.G. "Probabilistic Pattern Recognition Using Coherent Laser Radar System for Bridge Inspection", Final Report to FHWA, Department of Civil & Environmental Engineering, Wayne State University, Detroit, MI, July 1998
[3] Fu,G. "Bridge Inspection: Modal Testing for Global Diagnosis", (in press) 7th Int'l Conference on Structural Safety and Reliability, Kyoto, Japan, Nov. 24-28, 1997

Fig.1   Grinder Cuts to the Model Structure

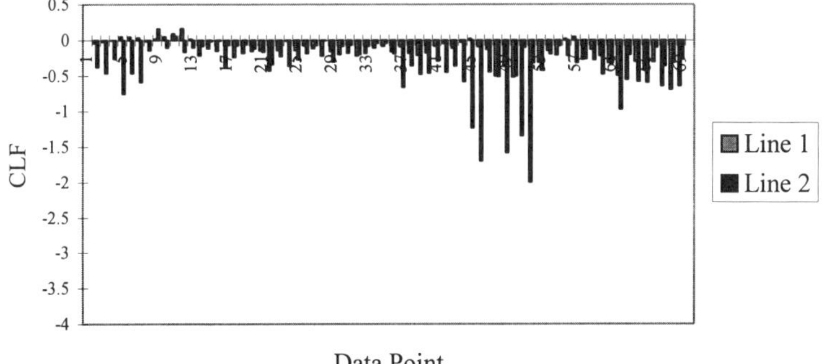

Fig.2   Condition Map using Comprehensive Likelihood Factor (CLF)

Structural Analysis

Manuscripts for some presentations were not available at time of publication.

4. *STRUCTURAL ASSESSMENT REPAIR, RETROFIT, AND CONTROL*

MODERATOR: Ronald Sack
National Science Foundation
Arlington, VA

(1) Monitoring the Health of Civil Infrastructure
Darryll J. Pines
University of Maryland
College Park, MD

(2) Damage Assessment of Jacketed Columns
Maria Q. Feng and Eun Young Bahng
University of California
Irvine, CA

(3) Seismic Hazard Mitigation Using Multiple Magnetorheological Devices
S. J. Dyke, F. Yi
Washington University
St. Louis, MO

J.D. Carlson
Lord Corporation
Cary, NC

(4) Dynamic Process Simulation Model of Cast-in-Place Concrete Process
E. Sarah Slaughter
Massachusetts Institute of Technology
Cambridge, MA

Michael N. Carr
Sellen Construction Company
Seattle, WA

(5) RC Columns with Lap Splices Subjected to Earthquakes
Olga Reyes and José A. Pincheira
University of Wisconsin
Madison, WI

13. *PERFORMANCE-BASED ENGINEERING OF STRUCTURAL FIRE PROTECTION*

MODERATOR: James A. Milke
University of Maryland
College Park, MD

(1) Estimating Fire Exposures for the Purpose of Evaluating the Expected Performance of Structural Elements
Morgan J. Hurley
Society of Fire Protection Engineers
Bethesda, MD

(2) Estimating the Fire Peformance of Steel Structural Members
James A. Milke
University of Maryland
College Park, MD

(3) Estimating the Fire Performance of Concrete and Masonry Structural Members
James A. Milke
University of Maryland
College Park, MD

22. *STRUCTURAL DAMAGE RESULTING FROM SHIP IMPACTS*

MODERATOR: Timothy J. Dickson
Stuart K. Jacobson & Associates, Ltd.
Northbrook, IL

(1) The Consequences of Vessel Impacts on the Mississippi River Bridges in New Orleans
Zolan Prucz and William B. Conway
Modjeski and Masters
New Orleans, LA

(2) Lessons Learned from Historical Ship/Barge Accidents with Bridges
Michael A. Knott
Moffatt & Nichol Engineers

(3) Vessel/Bridge Collisions Investigated By the National Transportation Safety Board
Ron Weber
National Transportation Safety Board

31. *RECENT ADVANCES IN NONLINEAR ANALYSIS OF BUILDING STRUCTURAL SYSTEMS (I)*

MODERATORS: Sherif El-Tawil
University of Central Florida

Donald White
Georgia Institute of Technology
Atlanta, GA

(1) Nonlinear Analysis of Buildings with PRC Connections
Bulent N. Alemdar, J. Taylor, D. W. White, and R. T. Leon
Georgia Institute of Technology
Atlanta, GA

(2) Earthquake Response Analysis of Steel Building Frames Considering Brittle Fractures at Member-ends
Koji Uetani and Hiroshi Tagawa
Kyoto University
Kyoto, Japan

(3) Pushover Analysis of Isolated Flexural Walls
B. Gupta and S. K. Kunnath
University of Central Florida
Orlando, FL

(4) Modeling the Seismic Behavior of Planer Moment Frames using Fishbone Models
K. Inoue
Kyoto University
Kyoto, Japan

K. Ogawa
Kumamoto University
Japan

H. Kamura
NKK Corporation
Japan

(5) Localization Issues in Nonlinear Frame Analysis
Enrico Spacone and J. Coleman
University of Colorado
Boulder, CO

40. *RECENT ADVANCES IN NONLINEAR ANALYSIS OF BUILDING STRUCTURAL SYSTEMS (II)*

MODERATORS: E.M. Lui
Syracuse University
Syracuse, NY

W.F. Chen
Purdue University
West Lafayette, IN

(1)Practical Design Method for Flexibly Jointed Frames with LRFD
N. Kishi, M. Komuro, K.G. Matsuoka
Muroran Institute of Technology
Muroran, Japan

W.F. Chen
Purdue University
West Lafayette, IN

(2) Local Buckling Analysis of Steel Bridge Piers under Cyclic Loading
Eiki Yamaguchi, Masataka Hayashi, Yoshinobu Kubo
Kyushu Institute of Technology
Japan

Yoshiaki Goto
Nagoya Institute of Technology
Japan

(3) Inelastic Critical Loads by Eigenvalue Analysis
Ronald D. Ziemian
Bucknell University
Lewisburg, Pennsylvania

(4) Behavior and Design of Laterally Braced Inelastic Columns
E.M. Lui and J. Lee
Syracuse University
Syracuse, NY

49. *ANALYTICAL AND NUMERICAL TECHNIQUES FOR PRACTICING STRUCTURAL ENGINEERS*

MODERATOR: Sreenivas Alampalli and Osman Hag-Elsafi
NYSDOT
Albany, NY

(1) Analysis of Bridges: What Do The Numbers Really Mean?
Dennis Mertz
University of Deleware
Newark, DE

(2) Estimating Suspension Cable Strength
Robert J. Perry
New York State Department of Transportation
Albany, NY

(3) Strength Evaluation of A Complex Bridge
William X. Zhang
Vanasse Hangen Brustlin, Inc.
Watertown, MA

(4) Comparisons of Structural Linear and Nonlinear Analysis of Steel Moment Frames with Supplemental Damping
Anna Yu
Structural Design Engineers
San Francisco, CA

Wenshen Pong
San Francisco State University
San Francisco, CA

(5) Easy, Concise Calculations for Eccentrically Loaded Welded Connections
Thomas W. Hartmann
TwHartmann, Inc.
Fort Collins, CO

(6) Simplified Procedure for the Analysis of End/Base Plates of Traffic Support Structures
Osman Hag-Elsafi, Frank Owens, and Sreenivas Alampalli
New York State Department of Transportation
Albany, NY

58. NEW TRENDS IN STRUCTURAL RELIABILITY ASSESSMENT

MODERATORS: Pavel Marek,
ITAM Academy of Sciences of the Czech Republic,
Prague, Czech Republic·

Andrzej S. Nowak
University of Michigan,
Ann Arbor, MI

(1) Issues Related to Achieving Qualitative Improvements in Reliability Assessment in Structural Design
Pavel Marek
ITAM Academy of Sciences of the Czech Republic
Prague, Czech Republic

Le-Wu Lu
Lehigh University
Bethlehem, PA

(2) The JCSS Probabilistic Code for New and Existing Structures
Dimitris Diamantidis
Fachhochschule Regensburg
Regensburg, Germany

(3) Probabilistic Assessment of the Safety of HSC Columns
Sofia M. C. Diniz
Departamento de Engenharia de Estruturas, EE UFMG
Belo Horizonte, Brazil

Dan M. Frangopol
University of Colorado
Bolder, CO

(4) Deterioration Models for Corrosion in Concrete Using Monte Carlo Simulation
J. Edward Gannon
Sear and Brown Engineers
State College, PA

Paul J. Tikalsky
The Pennsylvania State University
University Park, PA

(5) Critical Review of Fully Probabilistic Design for Seismic Loadings
J. Huh, A. Mehrabian, and A. Haldar
University of Arizona
Tucson, AZ

A.R. Salazar
University of Sinaloa
Mexico

65. *USE OF GENETIC ALGORITHMS IN STRUCTURAL OPTIMIZATION*

MODERATOR: Shahram Pezeshk
The University of Memphis
Memphis, TN

(1) Genetic Algorithm for Multiobjective Optimization and Life-Cycle Cost
Franklin Y. Cheng
University of Missouri-Rolla
Rolla, MO

Dan Li
Ludwing Buildings, Inc.
Harahan, LA

A.H-S. Ang
University of California
Irvine, CA

J. H. Lee
University of California
Irvine, CA

(2) Optimal Design of Transmission Line Structures for Earthquake Loads Using a Genetic Algorithm
Fatma Y. Kocer and Jasbir S. Arora
University of Iowa

(3) Composite Frame Design Using a Genetic Algorithms
Charles V. Camp, Jifei Li, and Shahram Pezeshk
University of Mempis
Memphis, TN

(4)Genetic Algorithim for Design of Nonlinear Framed Structures
Shahram Pezeshk, Charles Camp
University of Memphis
Memphis, Tennessee

Debin Chen
Green Mountain Geophysics
Boulder, CO

Monitoring the Health of Civil Infrastructure

Darryll J. Pines[1]
Department of Aerospace Engineering
University of Maryland, College Park, MD 20742-3015
Email: djpterp@eng.umd.edu

Abstract

Recent advances in smart materials and structures embedded sensor technology offer many unique opportunities to assess the structural integrity of civil structures. However, the remote operational environment of large civil structures such as highways, buildings, and bridges, makes condition-based health monitoring for damage assessment difficult in the event of a natural disaster. During such disasters, electrical power is lost and phone lines are under heavy usage. This limits the retrieval of very important sensor data. However, recent rulings by the Federal Communication Commission coupled with advances in wireless communication products has now made it possible to circumvent existing wired and cellular infrastructure to retrieve data from smart sensors remotely and more economically. This paper discusses an ongoing effort to develop an inexpensive wireless monitoring system for civil infrastructure.

1.0 Introduction
1.1 The Need for Monitoring the Health of Large Civil Structures

The motivation for investigating the application of smart materials and structures technology to large civil structures arises from the desire to mitigate potential hazards to the general public. These potential hazards include natural disasters such as hurricanes, typhoons, tornadoes, and earthquakes. Unfortunately, current inspection approaches are either passive or occur after the natural disaster has caused a catastrophic structural failure. Embedded smart structures technology (actuators and sensors) offers the unique ability to assess damage on demand to deliver the current condition of the structure prior, during and after a natural disaster has occurred. Advances in wireless communication technology permits this structural health monitoring to be performed remotely.

1.2 State of Art in Condition-Based Health Monitoring Instrumentation

There are essentially four main elements of condition-based health monitoring systems of large civil structures. They include *sensors and actuators, data acquisition, data retrieval*, and *data processing* for condition monitoring and damage assessment. *Sensors* used for condition-based health monitoring typically involve measurements of strain, acceleration, displacement, temperature, wind speed and in some instances angular rotation. More recently, fiber optic sensors (Kodindouma and Idriss 96) that are adaptive and self-calibrating have found use as strain sensors for monitoring the deformation of concrete poured highway structures. On the other

---

[1] Assistant Professor, Senior Member of AIAA, Member of AHS

hand, *actuators* have not found widespread use because of the large forces required to excite large civil structures. Most sensed data is obtained from ambient excitations from ground motion or moving vehicles. Nevertheless, there have been a few examples of actuators implemented on bridges to obtain modal response data for dynamic damage detection algorithms. *Acquisition* of embedded or externally mounted sensor data has received probably the most attention over the past five years. *Data acquisition* system design has been driven by number of sensors and dynamic range as opposed to sampling frequency. Because of the desire to be both sensitive to in-service and adverse loading conditions, it is not uncommon to find data acquisition systems for health monitoring with 16 bit or higher A/D converter resolution. *Data retrieval* from sensors on large civil structures have primarily relied on wired connections to local portable data acquisition systems. However, there has been increasing interest on monitoring large civil structures remotely, so that data can be retrieved and analyzed at some central or mobile processing center. Some examples of remote health monitoring systems have been implemented via wired connections on various in-service bridge structures. More recently, remote monitoring of large civil structures has been performed via a wired connection across the internet (Fuhr et. al. 96, Ballard and Chen 96). Wireless data retrieval and health monitoring has been implemented using the existing cellular telephony infrastructure to transfer data via modems. Finally, once the raw data has been retrieved either locally or remotely, *data processing* is performed to assess the condition of the structure. This processing can take the form of threshold level indication, to damage localization and identification (Doebling et al. 96). Most current in-service systems are based on threshold level monitoring. If an adverse loading condition arises, a warning system is triggered to alert operators that the structure should be taken out-of-service for possible visual inspection. The future of condition-based health monitoring may involve the modular monitoring strategy under development by Straser and Kiremidjian (1996, 1998). This scheme proposes the use of local microprocessors coupled with local area networks to transmitt data back to a central processing center for damage assessment.

To date there have been many examples of condition-based health monitoring systems (Abdel-Ghaffar et al. 95, Chang and Kim 96, Robison 96, Shahawy and Arocklasamy 96) which have been installed on in-service structures involving to some degree all of the four main elements mentioned above. Table 1 gives an abbreviated list of health monitoring systems which have been installed on in-service structures. It is evident from this list that most monitoring systems have been installed on bridge structures. These structures are of interest because they affect the economic viability of a region, since bridges are used to transport people, goods and services.

1.3 Remote Monitoring of Civil Infrastructure

This paper describes the components of a system for remote wireless condition-based health monitoring (Pines 98) of static and vibratory loads on large civil structures using wireless communication technology, data communication software, smart sensors and custom damage detection software. The uniqueness about this system is that it avoids the cellular infrastructure which can be costly and at times unreliable during real-life natural disasters. The system is mobile, self-contained, inexpensive and can be programmed to actively excite the structure to retrieve sensor data, or to passively retrieve sensor data generated from ambient excitations.

2.0 Local Structural Health Monitoring System

A more economical approach to wireless data transfer for infrastructure health monitoring applications has evolved from a recent FCC ruling to allocate a significant portion of the 900 MHz frequency band license-free to digital data transfer over short distances. This ruling has resulted in the development of a number of new products including digital home telephones, security systems and wireless modems. The digital home telephones and security systems typically operate at 900 MHz frequencies with ranges up to 500 ft. Similarly, the wireless modems operate in the 902 to 928 MHz frequency range using spread spectrum technology and are made by several manufacturers with data throughputs greater than 28.8 Kbps. The advantages of spread spectrum technology used in wireless modems include system flexibility, interference immunity, error-free communication and real-time data throughput. In addition, these wireless modems can be configured to have two-way capability along with the ability to transfer voice data. However, the most unique feature about the wireless modems is that the FCC has permitted them to be operated over greater distances (approximately 3 to 5 miles) with minor modifications in the transmitting antenna. Even greater distances are possible by using them as transceivers involving *multiple wireless hops* before arriving back at a some central or mobile processing center. Interference from other similar wireless modems can be minimized by customizing the communication software of the modems which handles error detection, packett sequencing and flow control. Therefore, the existing communications infrastructure for cellular telephony can be avoided altogether by effectively using spread spectrum wireless data transfer to individual nodes.

3.0 Components of Remote Wireless Health Monitoring System

Figure 1 displays a conceptual schematic of a simple and inexpensive remote health monitoring system for retrieving data from large civil structures before, during and after an adverse loading conditions. The system consist of two wireless modems, a notebook computer, a data acquisition system and data communication software for uplinking commands and downloading sensor data. It is assumed that sensors, mounted either internally or externally to the structure, are monitored by a ruggedized data acquisition system in close proximity to the civil structure of interest. The system can be powered by battery in case of loss of electrical power. Commands from a notebook computer are uplinked via the wireless modems to instruct the data acquisition system to record sensor measurements. Upon retrieving this data, additional commands are issued to download diagnostic information about the presence of a potential fault. Low data rates are required since damage assessment can be either performed at the local node. The use of the notebook computer permits the operator to be mobile or stationary.

4.0 Summary and Conclusions

The future of structural health monitoring of civil infrastructure will involve multiple sensors with potentially a telepresence for inferring diagnostics about critical structural elements. This local information about the structure's health will help to infer damage and schedule proper maintenance. This paper has presented the stages of a very simple approach involving wireless nodes. Future systems will have to rely on embedded sensor systems for detecting incipient damage. More impotantly these systems must be cheap since many 100's of sensors may be used along with several independent nodes.

5.0 Acknowledgments

This work was supported by the National Science Foundation, contract no. CMS9625004, with Dr. William Anderson serving as contract monitor.

6.0 References

Abdel-Ghaffar A M, Masari S F and Nigbor R L, 1995, "Preliminary Report on the Vincent Thomas Bridge Monitoring Test", Report No. M9510, Civil Engineering Dept. University of Southern California.

Alampalli S and Fu G, 1994, "remote Bridge Monitoring System for Bridge Condition", Report 70, Engineering Research and Development Bureau, New York State Department of Transportation, Albany, New York.

Ballard C M, and Chen S S, 1996, "Automated Remote Monitoring of Structural Behavior via the internet", *SPIE 3rd Annual Smart Structures and Materials   Conference* (San Diego, CA, February 26-29)

Chang S-P and Kim S, 1996, "Online structural monitoring of a cable-stayed bridge", *SPIE 3rd Annual Smart Structures and Materials. Conference* (San Diego, CA,   February 26-29)

Doebling S W, Farrar C R, Prome M B and Shevitz D W, 1996, "Damage Identification   and Health Monitoring of Structural and Mechanical Systems from Changes in Their Vibration Characteristics: A Literature Review", Report No. LA-13070-MS, Los Alamos National Laboratory (Los Alamos, NM, May)

Fuhr P L, Huston D R, Ambrose T P, and Mowat E F, 1995., :An Internet Observatory: Remote Monitoring Instrumented Civil Structures Using the Information Superhighway", *Smart Materials and Structures* 4, 14-19.

Kodindouma M B and Idriss R L, 1996, "An integrated sensing system for highway bridge monitoring", *SPIE 3rd Annual Smart Structures and Materials Conference* (San Diego, CA, February 26-29)

Pines D J and Lovell, P., 1998, "Conceptual Framework of a Remote Monitoring System", *Smart Materials and Structures*, Vol. 7, No. 5, 627-636.

Robison R, 1996, "Saving Scotland's Busiest Bridge", Civil Engineering, (January)

Shahawy M A and Arocklasay M, 1996, "Field Instrumentation to Study the Time-Dependent Behavior in Sunshine Skyway Bridge", ASCE J. Bridge Eng., Vol. 1, No. 2.

Straser E g, and Kiremidjian A S, 1996., "A Modular, Visual Approach to Damage Monitoring for Civil Structures", *2nd International Workshop on Structural Control, The Hong University of Science and Technology* (Hong Kong, December 18-21)

Table 1: Examples of Condition-Based Health Monitoring Systems

Structure	Researchers	Year	Location	Sensors and Actuators	Data Acquisition	Data Retrieval	Processing
Many highway bridges	Kodindouma and Idriss	8/96	*	fiber optic sensors	FLS3000 (Electrophontonic Corp.)	Locally	Strain Based
Vincent Thomas Bridge	Abel-Ghaffar et al.	12/95	Los Angeles	*	19 Bit A/D	Cellular	FEM based

HaengJu Bridge	Chang and Kim	8/96	Seoul, Korea	strain gauges, displacment, thermometers, accelerometers, etc. (65 total)	PC-Based Custom made	Local and Remote Wireless Telemetry, Wired Transmission, Wired Modems	Sensor Level Monitoring
Sunshine Bridge	Shahawy and Arocklasamy	5/96	Florida	strain, temperature, dispacement	*	Locally and remotely by wired modem	Sensor Level Monitoring
Kingston Bridge	Robison	1/1996	Scotland	strain, temperature, displacment, wind speed	*	Remotely via wire	Sensor Level Monitoring (Alarms)
I490/Conrail Bridge	Alampalli et al.	8/94	New York State	*	*	Remotely via wire	Modal parameter analysis
Winooski One Dam	Fuhr, Dryve et al.	8/96	Vermont	fiber optic sensors	Microcomputer (74 sensors)	Remote via wire (Internet)	Static Strains
Light Pole	Ballard and Chen	8/96	Orchid Park, NY	strain, accelerometers, temperature, wind speed	PC-Based	Remotely via wire (Internet)	Sensor Level Monitoring

*-not known

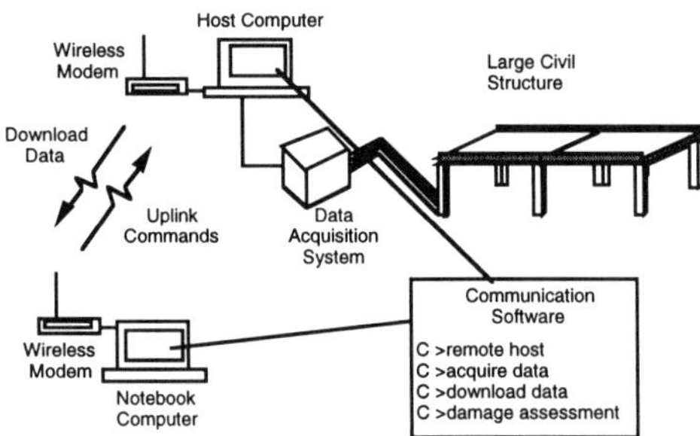

Figure 1: Schematic of Wireless Structural Health Monitoring System

Damage Assessment of Jacketed Columns

Maria Q. Feng [1] and Eun Young Bahng [2]

Abstract

This paper presents a vibration testing method using neural network-based system identification techniques for assessing the extent and location of seismic damage on jacketed bridge columns. The degradation of both flexural and shear stiffnesses of the damaged columns were correlated to their vibration characteristics through experimental and analytical studies. The applicability of this method to an entire bridge including its superstructure is also studied.

Introduction

There is an urgent need to develop post-earthquake damage detection technologies for highway bridges retrofitted with steel and composite jackets. The authors of this paper are currently developing a method assessing damage in jacketed columns by taking advantage of change in the bridge vibration characteristics including the natural frequencies and mode shapes before and after damage. To this end, this study experimentally and analytically develops a fundamental knowledge base to correlate vibrational characteristics and damage described by stiffness degradation of jacketed column. A back-propagation neural network technique was employed for identifying the extent and location of damage, without expensive and cumbersome search processes as required in a conventional system identification technique (Szewczyk and Hajela 1994). Finally, the damage assessment method was applied to an entire bridge involving a superstructure by a simulation study.

----

[1] Associate Professor, Dept. of Civil and Environmental Engineering, University of California, Irvine, E4130 Engineering Gateway, Irvine, CA 92697-2175
[2] Visiting Researcher, Dept. of Civil and Environmental Engineering, University of California, Irvine, E4130 Engineering Gateway, Irvine, CA 92697-2175

Experiments

A half-scale bridge column was built and tested in the Structures Laboratory of University of California, Irvine. It represents the existing California bridge columns designed using older (Pre 1971) specifications. In order to compensate for the insufficient lap splice length and reinforcement confinement, the column was retrofitted with carbon fiber composite jackets. Cyclic horizontal loads were applied on top of the column by an actuator fixed on the strong wall to introduce damage on the column. The details of cyclic loading tests are given in the reference (Haroun and Feng, 1997). Up to ductility factor of three, no apparent damage was observed on the jackets or the column by visual inspection. At ductility factor of seven, severe failure occurred at the lap splice region. Crushed concrete was observed in the 1-inch gap area (between the jackets and the footing) of the column, while no cracks were visible on the jackets and other parts of the column. Vibration tests using a shaker installed on top of the column (see Fig.1) were also performed for the jacketed/undamaged, moderately damaged (ductility two), and severely damaged (ductility seven) cases. The natural frequencies and mode shapes in these cases were obtained. Measurable shifts in the frequencies resulting from damage were clearly demonstrated. Especially at ductility two where no damage was observed by visual inspection, the frequencies of the column did change clearly.

Figure 1. Vibration Test

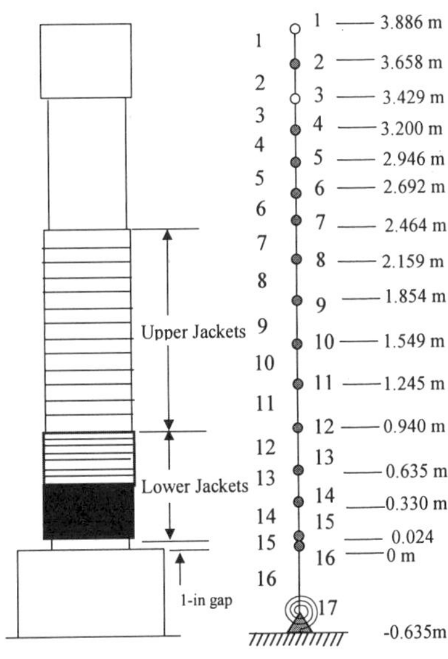

Figure 2. Finite Element Model

Damage Assessment

The test column was modeled by a finite element model as shown in Fig 2. The footing of the column was tensioned on the strong floor with six high strength rods during the cyclic loading and vibration tests. This support was modeled by a zero length torsional spring with a specified stiffness. Using the neural network approach based on the measured vibration data of the jacketed/undamaged case, the moment of inertia of the column and the torsional stiffness of the support spring was estimated. Using the estimated stiffness parameters, the lower two frequencies were calculated to be 13.97 Hz and 75.17 Hz, very close to the measured 13.84 Hz and 75.26 Hz.

In order to investigate the effectiveness of the vibration testing method in detecting invisible damage of a jacketed column, vibration test was performed on the carbon fiber-jacketed column at ductility two induced by the cyclic loading test. The stiffness degradation (representative of damage) was identified based on the measured vibration data. It was found that the moment of inertia and the effective shear area of the one-inch unwrapped gap portion were decreased by 98.4% and 99.7% due to damage, compared with the stiffness of the jacketed/undamaged case. Using the estimated damage, the lower two frequencies were calculated to be 8.68 Hz and 56.74 Hz, very close to the measured 8.66 Hz and 56.02 Hz. Figure 3 shows that the calculated (estimated) and measured mode shapes also achieved excellent agreement. The shear deformation of the gap portion (Element 15) due to the damage made an important contribution to the second mode shape. The damage at ductility factor of seven was also assessed. The damage severity in the jacketed column parts was slightly increased, compared with the moderately damaged case. The effective shear area of Element 15 was reduced by only a small value from the moderately damaged case (Feng and Bahng 1999).

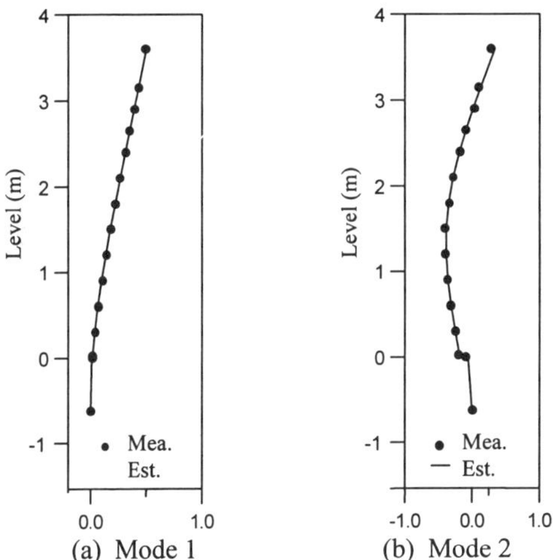

Figure 3.   Estimated and measured mode shapes

Application to an entire bridge

A three-span planar bridge structure was selected for example analyses to investigate the applicability of the proposed method to an entire bridge system including a superstructure and jacketed columns, as shown in Fig. 4. Based on the results of the column test, the unwrapped gap parts of the column are most vulnerable to the external loading. Therefore, the moments of inertia and shear areas of those parts of the columns were assumed to be unknown and changed due to damage. After a neural network was trained using 1000 training patterns, the average estimation errors for 100 testing patterns were obtained. The estimated results are judged to be very reasonable even in the input case with only first mode information. It means that as long as the first mode information can be obtained from the vibration test, it is sufficient for assessing the damage in the gap portion of the jacketed columns.

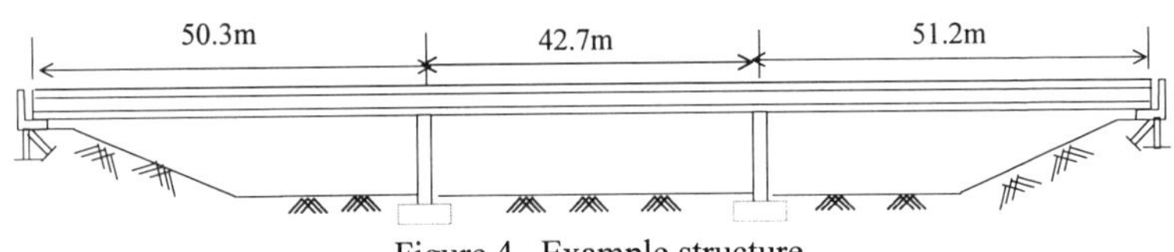

Figure 4. Example structure

Conclusions

The neural network technique was effective in correlating the stiffness (representative of damage) with the measured vibration characteristics. Degradation of the shear stiffness as well as the flexural stiffness in the column, especially in the unwrapped gap portion, was a key parameter to represent the damage. Finally, it is demonstrated that the proposed vibration testing method can be effectively applied to an entire bridge for assessing the damage on its jacketed columns.

Acknowledgment

This work was supported by the National Science Foundation under Grant CMS-9501796 and CMS-9812585.

References

Haroun, M. A. and Feng, M.Q. (1997). "Lap splice and shear enhancements in composite-jacketed bridge columns." *Proc. 3rd US-Japan Bridge Workshop.* Tsububa: Japan.

Szewczyk, Z. and Hajela, P. (1994.) "Neural network based damage detection in structures." *J. Comput. & Civil Engrg.* ASCE. 8(2): 163-178.

Feng, M. Q. and Bahng E. Y. (1999) "Damage assessment of jacketed RC columns using vibration tests." *J. Struct. Engrg.* ASCE. Accepted for publication.

SEISMIC HAZARD MITIGATION
USING MULTIPLE MAGNETORHEOLOGICAL DEVICES

S.J. Dyke[1], F. Yi[2], and J.D. Carlson[3]

Abstract

This paper discusses the results of a series of experiments conducted to evaluate the performance of magnetorheological devices for earthquake hazard mitigation. Shear-mode MR dampers are used to control the responses of a six story building subjected to ground motion. A multiple-input configuration is used in which two dampers are rigidly attached between the base and first floor, and between the first and second floors of the test structure. A clipped optimal control algorithm that is based on acceleration feedback is used to determine the control action.

Introduction

The application of modern control techniques to mitigate the effects of seismic loads on civil engineering structures offers a promising alternative to traditional earthquake resistant design approaches. In particular, semi-active control devices appear to present the best opportunity for near-term acceptance by the civil engineering community. Devices in this class offer the ability to dynamically vary their properties, indicating that they will be effective for a variety of loading conditions. They typically have low power requirements, eliminating the need for a large external power source. Furthermore, semi-active devices are considered to be stable because they do not have the ability to input energy into the structural system (including the control device and the structure).

A variety of semi-active devices have been proposed, including variable orifice dampers, variable friction devices, adjustable tuned liquid dampers, variable stiffness dampers, and controllable fluid dampers. Magnetorheological (MR) dampers are classified as controllable fluid devices. Prior studies using MR dampers have indicated that these devices have a great deal of promise for civil engineering applications (Dyke et al 1996, 1998; Yi et al, 1998). Experimental and analytical studies have demonstrated that the performance of this semi-active system is superior to that of comparable passive systems. This paper summarizes the results of a recent experiment conducted in the Washington University Structural Control and Earthquake Engineering Laboratory (*www.seas.wustl.edu/research/quake/*) to demonstrate the efficacy of multiple MR dampers for seismic response reduction.

---

1. Assistant Prof., Dept. of Civil Engrg., Washington University, St. Louis, Missouri 63130.
2. Doctoral Cand., Dept. of Civil Engrg., Washington University, St. Louis. Missouri 63130.
3. Mechanical Products Division, Lord Corporation, Cary, North Carolina 27511.

Magnetorheological Dampers

Magnetorheological (MR) fluids have the ability to change from a liquid to a semi-solid upon application of a magnetic field. Because they have this capability, MR fluids are an attractive choice for the development of controllable fluid dampers. MR dampers also have many features that indicate that they are an appropriate choice for civil engineering applications. They require minimal power, are able to respond quickly to changes in the con-

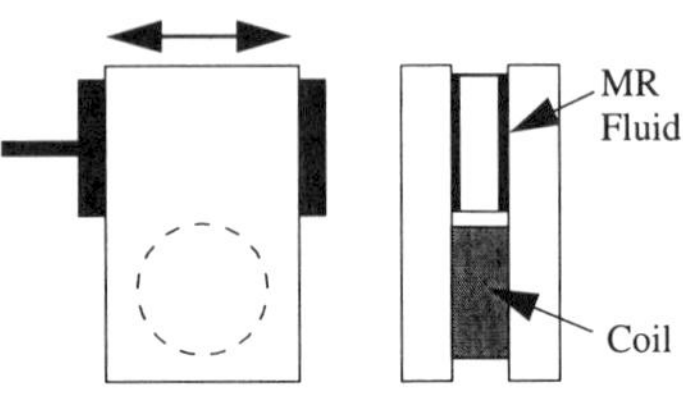

Figure 1: Schematic Diagram of a Shear Mode MR Damper.

trol input, are expected to be relatively inexpensive, and are considered to be quite reliable because they have few moving parts. Moreover, recent tests on a 20-ton MR damper at the University of Notre Dame have demonstrated that these devices can provide forces of the magnitude required for full-scale structural control applications (Carlson and Spencer, 1996; Spencer *et al.*, 1997).

The MR dampers employed in this experiment are prototype devices, shown schematically in Fig. 1, obtained from the Lord Corporation for testing and evaluation (see: *www.mrfluid.com*). The device consists of two steel plates placed parallel to each other. The dimensions of the damper are 4.45×1.9 ×2.5 cm (1.75×0.75×1.0 in). The magnetic field produced in the device is generated by an electromagnet consisting of a coil located at one end of the device. Forces are generated when the moving plate, coated with an MR fluid saturated foam, slides between the two parallel plates.

Experimental Studies

To experimentally demonstrate the efficacy of the MR damper, the device was applied to control a six story test structure. A clipped optimal control algorithm based on acceleration feedback (Dyke and Spencer, 1996; Dyke *et al.*, 1996ab, 1998) was used to determine the control action to apply to the MR damper. Prior analytical and experimental studies have demonstrated that the MR damper, used in conjunction with this clipped optimal control algorithm, was effective for controlling a multi-story structure with a single MR damper (Dyke *et al.*, 1996b, 1998). Dyke and Spencer (1996) then extended the control algorithm to control multiple MR devices, and performed a numerical study demonstrating the efficacy of the control algorithm.

To evaluate the efficacy of the controller harmonic tests near the first two modes of the structure were conducted to study the performance of the control algorithm in a worst case scenario. In the first set of tests the excitation frequency was chosen to be 1.38 Hz, which is approximately the first mode of the structure. To avoid damaging the structure, an uncontrolled test was not conducted with this excitation. However, the performance of the semi-actively controlled structure was compared to that of the passive-on configuration, in which both dampers are supplied with a constant 5V command input. Here, the semi-actively control system achieved an additional 22% reduction in sixth floor absolute acceleration over that of the passive-on system. Fig. 5 shows the resulting absolute acceleration responses of the sixth floor of the structure.

In the second set of tests the excitation frequency was chosen to be 4.2 Hz, which is the second natural frequency of the structure. In these tests the controlled system achieved a 62% reduction in the sixth floor absolute acceleration over the uncontrolled

Figure 3: Photo of
Experimental Structure.

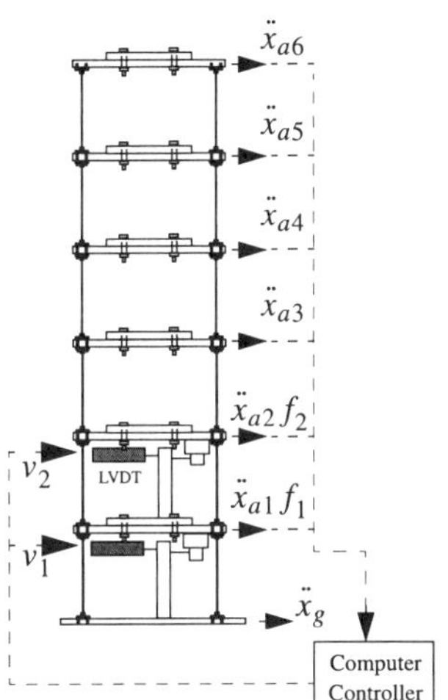

Figure 4: Diagram of MR
Controller Implementation.

system (no MR devices present). This represents an improvement of 30% over the passive-on response at this frequency. Furthermore, these results demonstrate that the semi-actively controlled structure has the ability to significantly reduce the structural responses in more than one mode of vibration. Fig. 6 shows the experimentally measured absolute acceleration responses of the sixth floor of the structure.

Summary

The efficacy of magnetorheological dampers for the control of civil engineering structures has been demonstrated through a series of experiments conducted in the Washington University Structural Control and Earthquake Engineering Laboratory. Each of the control devices is capable of supplying a maximum force of 20 kN (approximately 1.36% the weight of the structure). The semi-active controller achieved significantly better results than that of the passive-on control system at the first two modes of vibration. The results demonstrate that multiple MR dampers can effectively be used to control a six story test structure using a clipped optimal control algorithm based on acceleration feedback.

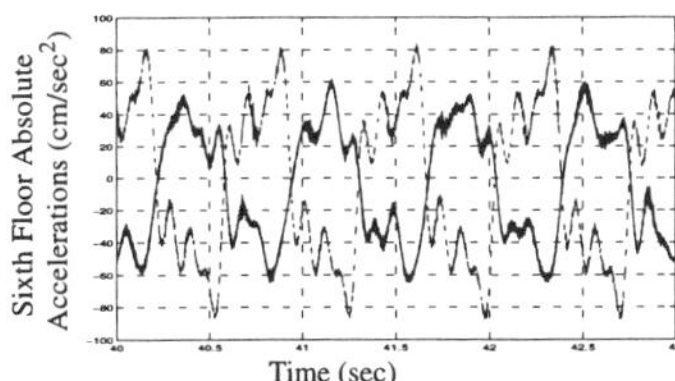

Figure 5: Passive-On and Semi-Active
System Responses to 1.38 Hz
Sine Wave (First Mode).

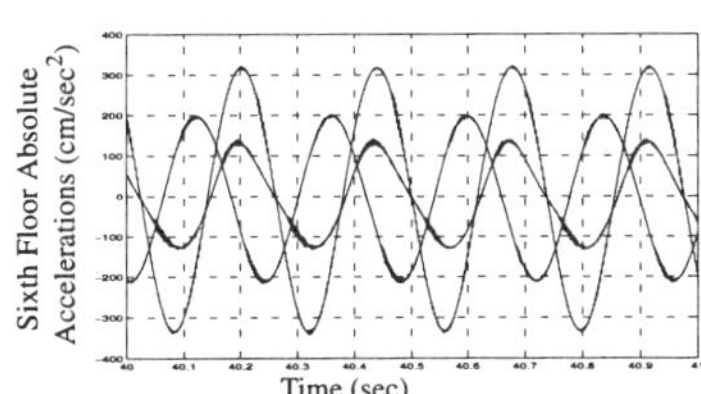

Figure 6: Uncontrolled, Passive-On, and
Semi-Active System Responses to 4.2 Hz
Sine Wave (Second Mode).

Future studies will consider the performance of this control system for more general loading conditions including random and historical earthquakes. Video clips that visually document the results of these experiments will soon be available at *www.seas.wustl.edu/research/quake/*. Further information can be obtained by contacting Dr. S. J. Dyke at *sdyke@cive.wustl.edu*.

Acknowledgment

This research is partially supported by National Science Foundation Grant No. CMS 97-33272. The authors from Washington University would also like to thank Lord Corporation for providing the prototype MR dampers used for this study.

References

Carlson, J.D. and Spencer Jr., B.F. (1996). "Magneto-Rheological Fluid Dampers for Semi-Active Seismic Control," *Proc. 3rd Int. Conf. on Motion and Vibr. Control*, Chiba, Japan, Vol. III, pp. 35–40.

Carlson, J.D., Catanzarite, D.M., and St. Clair, K.A. (1996). "Commercial Magneto-Rheological Fluid Devices," *International Journal of Modern Physics*, Vol. 10, Nos. 23 & 24, pp. 2857–2865.

Dyke, S.J., and Spencer Jr., B.F. (1996). "Seismic Response Control Using Multiple MR Dampers." *Proc. of the 2nd International Workshop on Struc. Control*, Hong Kong, December.

Dyke, S.J., Spencer Jr., B.F., Quast, P. and Sain, M.K. (1995). "Role of Control-Structure Interaction in Protective System Design," *J. of Engrg. Mech, ASCE*, Vol. 121, No. 2, pp. 322–38.

Dyke, S.J., B.F. Spencer, Jr., M.K. Sain and J.D. Carlson (1996a). "Modeling and Control of Magnetorheological Dampers for Seismic Response Reduction," *Smart Materials and Structures*, Vol. 5, pp. 565–575.

Dyke, S.J., B.F. Spencer, Jr., M.K. Sain and J.D. Carlson (1996b). "Experimental Verification of Semi-Active Structural Control Strategies Using Acceleration Feedback," *Proc., 3rd Int. Conf. on Motion and Vibr. Control*, Chiba, Japan, Vol. III, pp. 291–296.

Dyke, S.J., B.F. Spencer, Jr., M.K. Sain and J.D. Carlson (1998). "An Experimental Study of MR Dampers for Seismic Protection," *Smart Materials and Structures: Special Issue on Large Civil Structures*.

Dyke, S.J. (1999). "Performance and Stability of Controlled Civil Engineering Structures Using MR Dampers," *Proc. of the 1999 Struc. Cong.*, New Orleans, LA.

Dyke, S.J., Spencer Jr., B.F., Sain, M.K. and Carlson, J.D. (1998). "An Experimental Study of MR Dampers for Seismic Protection," *Smart Materials and Structures: Special Issue on Large Civil Structures*, Vol. 7, pp. 69.3–703, 1998.

Spencer Jr., B.F (1996). "Recent Trends in Vibration Control in the U.S.A.," *Proc. of the 3rd Intl. Conf. on Motion and Vibr. Control*, Chiba, Japan, Vol. II, pp. K1–K6.

Spencer Jr., B.F., Dyke, S.J., Sain, M.K. and Carlson, J.D. (1996). "Phenomenological Model of a Magnetorheological Damper." *J. Engrg. Mech., ASCE*, Vol. 123, No. 3, pp. 230–238.

Spencer Jr., B.F. and Sain, M.K. (1997) "Controlling Buildings: A New Frontier in Feedback," *IEEE Control Systems Magazine: Special Issue on Emerging Technologies* (Tariq Samad Guest Ed.), Vol. 17, No. 6, pp. 19–35.

Yi, F., Dyke, S.J., Frech, S., and Carlson, JD. (1998). "Investigation of Magnetorheological Dampers for Earthquake Hazard Mitigation" *Proc. of the 2nd Intl. Conf. on Structural Control*, Kyoto, JAPAN, June 30–July 2.

Dynamic Process Simulation Model of Cast-in-Place Concrete Process

E. Sarah Slaughter[1] , Associate Member, and Michael N. Carr[2]

Abstract

Introduction of innovations into constructed facilities can create many primary, secondary, and tertiary impacts throughout the specific system being changed and among the multiple systems constituting a facility. New research has developed dynamic process simulation models as a means through which to directly assess the time, cost, and safety impacts of particular innovations for specific projects.

Introduction

Constructed facilities are complex, multi-system products that require dynamic processes to assemble and transform the materials to create a completed and functioning unit. Introduction of new designs and technologies in constructed facilities creates impacts which can spread from a single point throughout a system, and into other systems and processes as well. Current techniques to evaluate the impacts of design and technology alternatives cannot fully assess these impacts because they often concentrate on the innovation itself and its immediate region of impact.

Research at MIT is creating a set of computer-based dynamic simulation models of construction activities to accurately represent the processes to analyze the system and inter-system impacts of innovations (Slaughter 1999). The models incorporate the process-specific experiential data associated with construction processes into the simulation models, which are tailored to respond to the characteristics of projects which change (i.e., the design specifics, the resources and their related production rates, and the process management strategies). The models represent the performance of work on-site at the level at which it is actually performed-- unit by unit. For instance, for the skilled trades involved in erecting a cast-in-place concrete structure, the work consists of creating the reinforcing steel and formwork elements, and then creating new units (i.e., columns, walls, slabs) through the placing of the concrete around those elements. These activities are performed for each structural unit of the frame, and the process can be modified to incorporate the particular attributes of each unit in a specific spatial location over time. Many different activities can be performed simultaneously throughout the site, and the resources can be immediately reassigned to different tasks as they are needed.

This new approach using process simulation is a dramatically different way to assess the project cost, duration, and worker safety for a specific project. Current

---

[1] Assistant Professor, Department of Civil and Environmental Engineering, Massachusetts Institute of Technology, 77 Massachusetts Ave., Cambridge, MA 02139
[2] Field Engineer, Sellen Construction Company, 228 9th Ave. North, Seattle, WA 98109

scheduling and project performance approaches (e.g., critical path method scheduling, queuing-based simulation models) are generally at too high a level of aggregation of the physical systems to clearly represent the simultaneity of task performance, and most of these approaches assume that resources are assigned for the duration of an activity, and often are specialized to a particular cycle of tasks.

In contrast, the dynamic process simulation uses as data inputs the number and type of each unit (e.g., columns, windows, or pipe runs), and can be easily modified in the number of resources and their production rates, and the site specifics which influence the organization and sequence of activities. In many ways, this process simulation approach is more similar to a quantity takeoff cost estimating program than to current scheduling techniques, since it is completely responsive to the number of units and resources for each project. However, unlike cost estimation programs, it has imbedded process information, such as the logical progression of tasks and standard resources with associated production rates, that allow the designer and planner to clearly assess different scenarios to organize and effectively realize a complex construction process.

Ten system and material-specific models are completed or in process (structural steel erection, cast-in-place concrete, glass/metal curtainwall erection, precast concrete exterior panel erection, plumbing installation, fire protection installation, HVAC installation, electrical wiring installation, interior wall partition installation, and suspended ceiling installation). In addition, a meta-model of the whole building being developed will provide a means through which designers and construction planners can assess the impacts of design and technology innovations for a whole project.

This paper will describe a specific model, cast-in-place concrete, to demonstrate the structure of the model and its application to a particular system and material-specific construction process (Carr 1998). It will also discuss the results of the analyses of several innovations.

Cast-in-Place Concrete Model Structure

Three key sets of activities are performed within the process of cast-in-place reinforced concrete construction, which are associated with the formwork, the reinforcing steel bars, and the placement, curing, and finishing of the concrete (Figure 1).

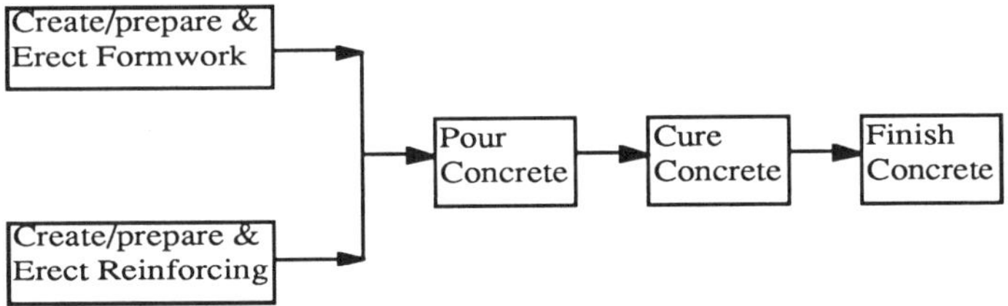

Figure 1: Major Subprocesses in Cast-in-Place Concrete Construction

Each of these subprocesses can be further broken down into their own subprocesses and specific tasks. The order in which the specific tasks are performed follows a specific logical progression, and the tasks can be repeated for the number of units being processed, with specific resources being used for a task, but shared among different subprocesses and the process overall. The relationship among the subprocesses, such as formwork placement and reinforcing steel installation, depends upon the nature of the element being created; for instance, a floor slab always has the floor formwork placed first, followed by the reinforcing steel, while a column has the reinforcing steel set first, followed by the formwork.

Therefore, the simulation model distinguishes among vertical and horizontal cast-in-place concrete members. Because of the logical progression for concrete superstructure, the vertical elements precede the horizontal elements (Figure 2). (This precedence relationship, however, can be easily modified to apply the process model to include substructure elements.) Within the stages of the construction of the vertical and horizontal element are the specific subprocesses of making/preparing and erecting the formwork and reinforcing steel, and the placement of the concrete and its curing and finishing. The latter stages of the processes by element can overlap with the early stages of the subsequent element processes; that is, the column reinforcing and formwork can be set while the slabs are still curing to their final strength and being finished.

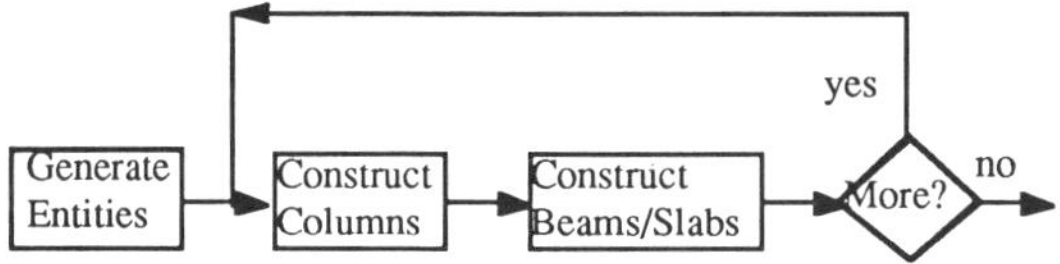

Figure 2: Simulation Model of Cast-in-Place Concrete

The entities which pass through the model are the specific physical components which are being worked upon. For the creation of the formwork and reinforcing steel, the raw materials are often brought onto site and cut, bent, and otherwise transformed into new elements which are then further aggregated to create a section (e.g., part of a formwork group for a specific structural element such as a column). The tasks in the simulation model are at this fine level of detail, and can therefore capture the specific resources and production rates necessary to perform this task, and their relationship to other tasks and subprocesses. On most sites, these subprocesses can be performed for multiple elements at different locations on the site; for instance, the column reinforcing steel may be prepared for a column in one place, while elsewhere a slab is poured or a beam is stripped.

Innovations in cast-in-place concrete are currently being developed at a high rate, due in large part to government research programs, but these innovations are not necessarily being accepted into the industry practice because of the risks they pose. The perceived risks can include technical uncertainty in being able to achieve required performance levels, and also uncertainty about how the innovation will affect other construction activities. The dynamic process simulation models provide an environment in which several scenarios can be examined, and the impacts of the innovation on time, cost, and worker safety for a specific project can be directly assessed.

Cast-in-Place Concrete Innovations

An innovation can be defined as a significant improvement in a component, system or process which is actually used and is novel to company using it (Slaughter 1998). Three specific innovations in cast-in-place concrete construction were directly assessed to demonstrate the functions of the simulation model (Carr 1998). The innovations are the Talon2 360 Rebar Crosstie System (a machine which automatically wires together reinforcing steel bars), Self-Compacting High Performance Concrete (a novel mixture of standard concrete materials which eliminates the vibration activities of the placement subprocess), and precast concrete stay-in-place forms (which eliminates completely the subprocesses associated with creating and later stripping the forms).

To assess the impacts of the innovations in terms of time, cost, and worker safety, a specific prototype building was designed (Carr 1998). The building is 5 stories, with a floor to floor height of 3 m (10 ft), and consisting of 20 bays per floor, with each bay spanning 7.6 m (25 ft). The structure is cast-in-place reinforced concrete columns with 2-way slabs that are 200 mm (8 in.) thick.

Using empirical data (collected through numerous site visits and interviews with specialty contractors), the common standard procedures are characterized and incorporated into the model, such as the specific resources needed for each task and the related production rates. (These rates and the resources themselves can be modified to reflect the actual availability of resources and their productive capacity.) The results of the model were evaluated by industry professionals, and found to accurately represent the duration, cost and worker safety for this size and type of project using standard processes and resources. The duration of the project, including completion of curing, was 55 working days (of 8 hours per day), and the total cost was $217,491. The workers' exposure to dangerous conditions was an index of 947 (Slaughter and Eraso 1997).

The innovations each influenced a different facet of the cast-in-place concrete process. The Talon Rebar System decreased the time needed to assemble the reinforcing steel (although the inclusion of only 5 pieces of equipment reduced its impact on overall CIP concrete duration to a negligible amount), and reduced the costs by 6% and reduced worker exposure to dangerous conditions by 7% (Table 1). The self-compacting concrete eliminated a set of activities during concrete placement, and reduced the process duration by 7%, the construction costs by 3%, and the worker exposure by 7%. The precast concrete forms also eliminated several stages, and thereby reduced duration by 2%, reduced cost by 39%, and reduced worker exposure by 42%. To explore the complementary nature of innovations that often occurs in construction processes, due to their system and inter-system level impacts, the combination of the Talon Rebar System and the Stay-in-Place was analyzed, and indeed the combination provided significant benefits, reducing duration by 9%, costs by 44%, and improving worker safety by 49%.

Innovation	Cost (% Reduction)	Duration (% Reduction)	Worker Danger (% Reduction)
Standard Method	$38,416	55 days	948
Talon Rebar System	(6%)	(0%)	(7%)
Self-Compacting Concrete	(3%)	(7%)	(7%)
Stay-in-Place Forms	(39%)	(2%)	(42%)
Talon/Stay-in-Place Forms Combination	(44%)	(9%)	(49%)

Acknowledgements: This research is sponsored by a grant from the NSF, CMS-9596227, under the CAREER program, and has been aided significantly through the contributions of the construction industry and many specialty companies.

References:
Carr, M. N. (1998). "Simulation to Assess Cast-in-Place Concrete Construction Innovation," Master of Science, Massachusetts Institute of Technology, Cambridge, MA.
Slaughter, E. S. (1998). "Models of Construction Innovation." *Journal of Construction Engineering and Management*, 124(2).
Slaughter, E. S. (1999). "Assessment of Construction Processes and Innovations Through Simulation." *Construction Management and Economics*, Forthcoming.
Slaughter, E. S., and Eraso, M. (1997). "Simulation of Structural Steel Erection to Assess Innovations." *IEEE Transactions on Engineering Management*, 44(2), 196-207.

RC Columns with Lap Splices Subjected to Earthquakes

Olga Reyes and José A. Pincheira[1]

Abstract

A nonlinear model and analysis procedure for reinforced concrete columns with short lap splices is presented. The model is based on local bond stress and slip relations to estimate the strength and deformation capacity of the spliced bars. A comparison of the calculated and measured response showed good agreement in terms of the lateral strength, deformation, and failure mode.

Introduction

Reinforced concrete columns with short, lightly confined splices are often found at critical sections in older frame construction. These splices are usually located immediately above the floor level, a region where large inelastic deformations may occur during a severe earthquake. The behavior of these splices is nonductile and is characterized by strength and stiffness decay with loading cycles and increasing deformation amplitude.

In past studies (Pincheira and Jirsa, 1992), various efforts have been made to develop realistic models that represent the nonductile response of splices for the evaluation of older buildings. More recently, new experimental data have become available (Aboutaha, 1994; Lynn, et al., 1996) that permit verification and correlation of the proposed models. In this paper, a nonlinear analysis procedure for estimating the seismic response of reinforced concrete columns with nonconforming lap splices is presented. The success and accuracy of the procedure are evaluated by comparing the calculated response with test data of isolated building columns.

Modeling Approach

Past studies have yielded valuable information on the failure mechanisms of lap splices, but have focused primarily on strength estimates. Local force and slip relations of lap spliced bars have not been established. In lap splices, the bars interact with each

---

[1] Graduate Research Assistant and Assistant Professor, The University of Wisconsin, Madison, WI 53706

other to form a complex force transfer mechanism. The cracking and splitting behavior of splices, however, have been found to be similar to those of a single anchored bar. For simplicity and in the absence of appropriate relations for lap spliced bars, it is postulated that the local bond stress and slip relations obtained for a single bar may be used to simulate the behavior mechanism of lap spliced bars. Since the splices found in older construction typically have modest amounts of transverse reinforcement, the effect of this reinforcement on the behavior of the splice may be ignored. Based on these assumptions, the local bond stress and slip relation for a single bar in unconfined concrete proposed by Eligehausen et al. (1983) was used in this study.

Three resistance mechanisms are considered to describe the response of a reinforced concrete column: flexure, shear and anchorage slip of the lap spliced bars. A single prismatic element is used to model the hysteretic response in flexure and shear. The flexural characteristics are obtained from moment-curvature relations, while the shear response is estimated from the Modified Compression Field Theory. To simulate the rigid-body rotations due to bond slip of the spliced bars, a nonlinear rotational spring is used at the element end where the lap splices are located. The moment-rotation characteristics of this spring are calculated as follows. A one-dimensional model with discrete uniaxial springs, whose properties are based on the local bond stress and slip relation mentioned above, is used to obtain the pullout force and slip relation for each spliced bar under consideration. The latter relation is then used to compute the moment for a given bar force and the corresponding rigid-body rotation using the calculated slip.

Comparison of Analytical Model and Experimental Observations

Details of the column chosen for this presentation are shown in Figure 1. This column was designed according to past standards and had lap splices of 24 bar diameters at the base. The column was subjected to unidirectional reversed cyclic loading, using prescribed displacement increments until failure.

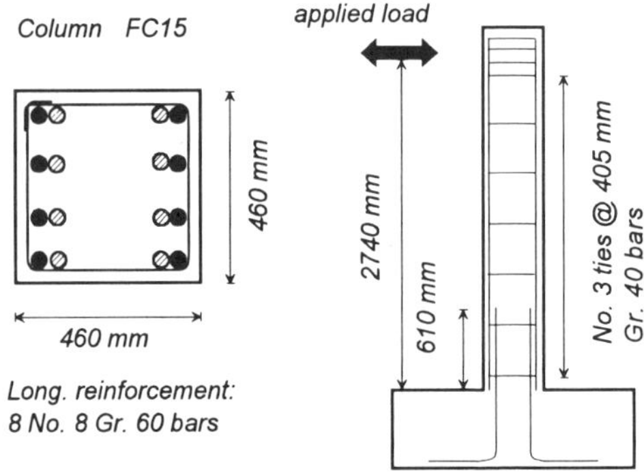

Figure 1 Column dimensions and reinforcing details.

The response of the column was calculated using a modified version of the program DRAIN-2D (Pincheira et al., 1998). In this version of the program, nonlinear behavior in flexure and shear can be considered, including strength and stiffness decay with increasing deformation amplitude. Modeling parameters were based on the measured material properties and recognized theories of reinforced concrete.

The calculated pullout force and slip relation for the spliced bars is presented in Figure 2. As shown, the calculated failure mode is due to bar pullout before the development of its yield strength. This result is consistent with the experimental observations which showed that none of the bars yielded (Aboutaha, 1994).

The measured and calculated load and deformation responses of the column are compared in Figure 3. In this figure, the calculated response was obtained by applying the same displacement increments used in the test. It may be noted that the calculated load and lateral displacement corresponding to splice failure are underestimated by about 8% and 23%, respectively. They provide, however, a reasonable and conservative estimate of the measured values. Post peak behavior is not represented as well by the model. The calculated response shows an abrupt and total loss of strength upon reaching splice failure. This result is consistent with the bond stress-slip model and the imposed displacement history used in the calculations. The measured response also shows a sudden loss of strength immediately after reaching the peak load-carrying capacity. During the test, however, the column was unloaded before allowing a complete loss of strength and reloaded to the same deformation amplitude. As a result, a more gradual rate of strength decay appears to have occurred following splice failure. If the calculated post-peak response is compared with the envelope curve defined by the points

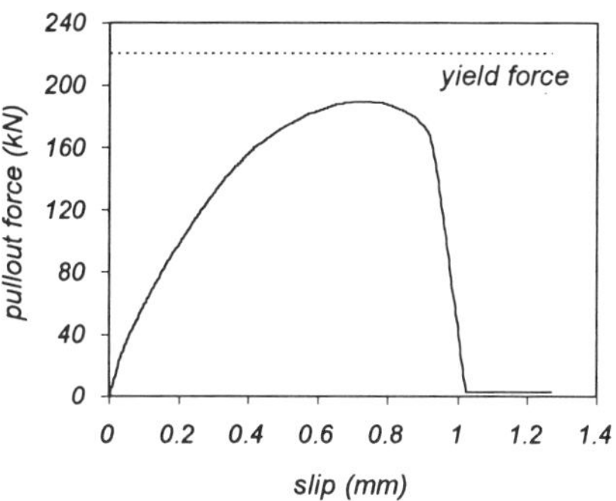

Figure 2 Pullout force and slip relation of spliced bars.

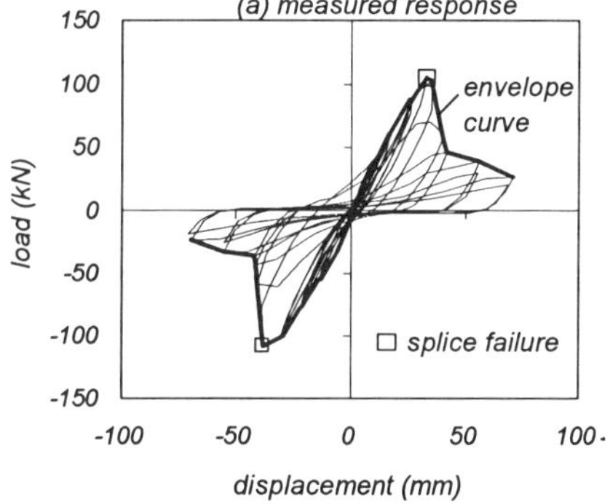

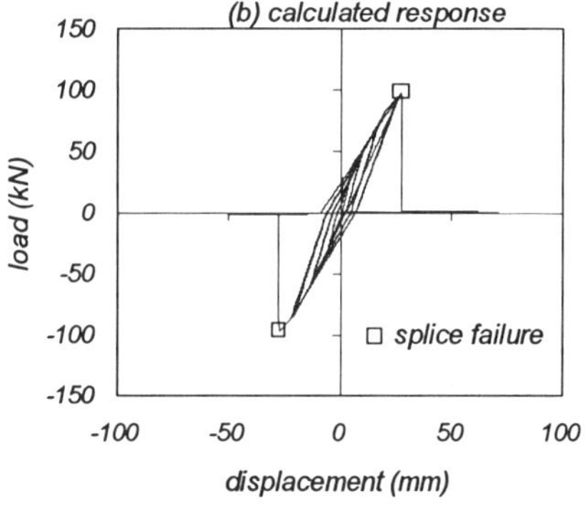

Figure 3 Calculated and measured response.

of peak response in each loading cycle, a better correlation is obtained. The test column shows, however, a residual strength at large displacement amplitudes of about one third of the maximum measured strength, a result that is not captured by the model.

Conclusions

A nonlinear modeling approach for the analysis of reinforced concrete columns with short lap splices subjected to earthquake loading is presented. A comparison of the calculated and measured responses of a column with nonconforming lap splices shows that the proposed procedure provides a reasonable, yet a conservative estimate of the measured lateral strength and deformation of the column. The calculated failure mode was also in good agreement with that observed in the test. Post-peak behavior is not represented as well, but it provides a conservative estimate of the expected response of the column.

Acknowledgements

The work described in this paper was sponsored by the National Science Foundation under Grant No. CMS-9624801. The authors thank Drs. Riyad Aboutaha and Michael D. Engelhardt for providing the test data. The opinions, findings, and conclusions expressed in the paper are solely those of the authors and do not necessarily represent the views of the sponsors or those of the individuals mentioned here.

References

Aboutaha, R.S. (1994) "Seismic Retrofit of Non-Ductile Reinforced Concrete Columns using Rectangular Steel Jackets," Ph.D. Dissertation, Department of Civil Engineering, The University of Texas at Austin, Austin, Texas, 373 pp.

Eligehausen, R., Popov, E.P., and Bertero, V.V. (1983) "Local Bond Stress-Slip Relationships of Deformed Bars under Generalized Excitations," Report UCB/EERC-83/23, University of California, Berkeley, California, 162 pp.

Lynn, A.C., Moehle, J.P., Mahin, S.A., and Holmes, W.T. (1996) "Seismic Evaluation of Existing Concrete Building Columns," *Earthquake Spectra*, Vol.12, No. 4, pp. 715-739.

Pincheira, J.A. and Jirsa, J.O. (1992) "Seismic Strengthening of Reinforced Concrete Frames using Post-Tensioned Bracing Systems," PMFSEL Report No. 92-3, Department of Civil Engineering, The University of Texas at Austin, Texas, 250 pp.

Pincheira, J.A., Dotiwala, F.S., D'Souza, J.T. (1998) "Seismic Analysis of Older Reinforced Concrete Columns," accepted for publication in *Earthquake Spectra*.

Estimating Fire Exposures for the Purpose of Evaluating the Expected
Performance of Structural Elements

Morgan J. Hurley, P.E.[1]

Abstract

The Society of Fire Protection Engineers is developing a "pre-standard" on fire
exposures that will be integrated into an ASCE standard on fire resistance of
building assemblies. This is part of a collaborative effort between ASCE, SFPE
and representatives of the masonry, timber, steel and concrete industries.

This paper describes the plan for creation of a performance-based standard on fire
exposures and summarizes the literature that will be used to develop the standard.

Introduction

Historically, fire resistance of structural elements has been determined by
subjecting individual components, including any protection provided, to a furnace
test (ASTM, 1995). Approximately 10 years ago, ASCE and SFPE formed a
partnership to develop a series of standards on fire resistance of structural
elements. The first effort of this partnership, a standardized set of calculation
methods based on standardized time-temperature curves, is almost complete.

The next standard in this series will be "performance oriented," and will give
guidance on estimating temperature and heat flux boundary conditions of
anticipated fire exposures. Correlations for the thermal response and reactions of
structural elements given these boundary conditions will also be included in the
standard. This effort follows a "white paper" outlining the development of a
performance-based standard on fire resistance of building assemblies (Milke &
Hill, 1997).

---

[1] Technical Director, Society of Fire Protection Engineers, 7315 Wisconsin Ave.,
Suite 1225W, Bethesda, MD 20814-3202

As of the writing of this paper, the SFPE working group formed to complete this project has agreed on a format for the portion of the pre-standard on fire exposures and surveyed applicable literature.

The portion of the pre-standard on fire exposures will be divided into four sections: fully developed enclosure fires, fire plumes and window flames. The fourth section will provide guidance on determining the appropriate inputs to the calculation methods.

Fully Developed Enclosure Fires

Several methods have been identified by the group during their literature search in the area of fully developed enclosure fires. These include methods by Babrauskas (Walton & Thomas, 1995), Law (Walton & Thomas, 1995), Magnusson and Thelanderson (Walton & Thomas, 1995), (Magnusson &Thelanderson, 1970), Lie (Lie, 1995), small-scale investigations by Thomas et al. (Thomas, 1970), (Thomas & Heseldsen, 1972), (Thomas, P. H. & Nilsson, 1973) and a computer model by Babrauskas (Babrauskas, 1975).

These correlations require input factors such as room dimensions, dimensions of ventilation openings, fire heat release rates, enclosure thermal properties such as specific heat, conductivity and density, and fire load. The output from these correlations is typically temperature rise above ambient, while assumptions may be made for heat flux parameters to structural elements.

Fire Plumes

Beyler (Beyler, 1986) summarizes correlations for fire plumes, ceiling jets, flames and ceiling heat transfer. The paper compares related correlations and provides examples that demonstrate solving practical fire protection problems.

The correlations require as input burning rate, height above the fuel, and fuel bed dimensions. The correlations yield temperature rise above ambient as a function of height and radial distance from the plume center line. Heat flux parameters may be determined from knowledge of the fuel and the structural element.

Window Flames

Early attempts to assess external fire exposures involved experiments utilizing ASTM E119 furnace tests with a window opening in a furnace wall (Prior, 1965). However, these tests were determined to be less severe than actual building fire exposures (Underwriter Laboratories, 1969). Experimental fires were later analyzed to obtain code approvals for specific buildings (Ove Arup, 1977), which later led to a design approach that was ultimately accepted by the model building code organizations (AISI, 1979).

These methods require the dimensions of the room or enclosure, dimensions of opening(s), and the fire load in the enclosure as input. The fire load is expressed as "wood equivalent" weight, where the weight of materials other than wood are adjusted based on the ratio of their calorific value to the calorific value of "wood."

These correlations yield the burning rate, the fire temperature, the dimensions of the window flame, and the temperature of the window flame at the point of impingement of any structural members. Given the fire effects, calculation methods are given to determine the heat transfer from the compartment to structural elements, the heat transfer from the window flames to the structural elements, and the heat losses from the structural elements to the surrounding air and other elements. The heat transfer correlations given are limited to steel elements. Also, correlations are given to determine maximum steel temperatures in the structural elements.

Future

For each calculation method, guidance will be developed that will include a summary of the calculation method, data requirements and sources, assumptions, limitations and comparison with published data. The final product will be balloted through the ASCE process and published as an ASCE/SFPE standard.

References

American Iron and Steel Institute, *Design Guide for Structural Steel*, Washington, November 1979.

ASTM, 1995, "Standard Test Methods for Fire Tests of Building Construction Materials," ASTM E119, Philadelphia, American Society of Testing and Materials.

Babrauskas, V., COMPF: A Program for Calculating Post-flashover Fire Temperatures (UCB FRG 75-2). Fire Research Group, University of California, Berkeley (1975).

Beyler, C., L., "Fire Plumes and Ceiling Jets," *Fire Safety Journal*, Vol 11 Nos. 1 & 2, Elsevier, The Netherlands, July/September 1986.

Lie, T., T., "Fire Time Temperature Relations," *SFPE Handbook of Fire Protection Engineering*, 2nd Ed., P. J. Dinenno (ed), Quincy, NFPA, 1995.

Magnusson and Thelanderson, "Temperature-Time Curves of Complete Process of Fire Development. Theoretical Study of Wood Fuel Fires in Enclosed Spaces," Acta Polytechnica Scandinavia, *Civil Engineering and Building Construction Series No. 65*, Stockholm, 1970

Milke, J & Hill, S. "Development of a Performance-Based Fire Protection Standard on Construction, September, 1997 (unpublished).

Pryor, A. J., Fire Exposure of Exterior Structural Members, Southwest Research Institute, San Antonio, TX, July 1965.

Ove Arup and Partners, Design Guide for Fire Safety of Bare Exterior Structural Steel – Technical Reports and Designers Manual. Report for American Iron and Steel Institute, Ove Arup and Partners, London, January 1977.

Thomas, P. H., "The Fire Resistance Required to Survive a Burn Out," *Fire Research Note No. 901*, Fire Research Station, Borehamwood, UK, November 1970.

Thomas, P. H. & Heseldsen, A. J. M., "Fully Developed Fires in Single Compartments, A Cooperative Research Program of the Conseil International du Batiment (CIB Report No. 20), *Fire Research Note No. 923*, Fire Research Station, Borehamwood, UK, August 1972.

Thomas, P. H. & Nilsson, L., "Fully Developed Compartment Fires: New Correlations of Burning Rates, *Fire Research Note No. 979*, Fire Research Station, Borehamwood, UK, August 1973.

Underwriters Laboratories, Inc., "Fire Severity at the Exterior of a Burning Building and its Effect on Exposed Structural Members," January 9, 1969.

Walton, W. D. & Thomas, P.H., "Estimating Temperatures in Compartment Fires," *SFPE Handbook of Fire Protection Engineering*, 2nd Ed., P. J. Dinenno (ed.), Quincy, NFPA, 1995.

Estimating the Fire Performance of Steel Structural Members

James A. Milke, Ph.D., P.E.[1]

Abstract

A fire resistance analysis of steel members for performance-based design involves a determination of the structural integrity of the heated steel member. As a prerequisite to conducting the structural evaluation, the temperature rise of the fire-exposed steel member needs to be assessed to evaluate the material properties and thermal strains. In some cases, the structural analysis can be conducted using the same elementary equations from mechanics that are used in ambient temperature design.

Introduction

With the development of performance-based building codes, there will be a need to develop engineering standards that document the supporting analytical methods. Formulation of an engineering standard to assess the fire resistance of steel assemblies is underway through the joint efforts of ASCE and SFPE, with the support of industry groups, including the AISI and AISC. Analysis of the response of fire-exposed steel members requires consideration of fire exposure, material properties at elevated temperatures, thermal response of the structure and structural response of the heated assembly [Milke and Hill, 1997], as reflected in Figure 1. Hurley addresses the means of selecting and describing the fire conditions [1999].

Material Properties at Elevated Temperature

Thermophysical properties include the conductivity, k, specific heat, c, and density, ρ, while the mechanical properties are the strengths, σ, modulus of elasticity, E, and coefficient of thermal expansion, α. Thermophysical and mechanical properties of structural steel have been reported at elevated temperatures, such as presented in Figure 2. [ECCS, 1985] [Lie, 1992] [Milke, 1995] [Eurocode 3, 1995].

---

[1] Associate Professor, Department of Fire Protection Engineering, University of Maryland, College Park, MD 20742

Figure 1. Process of Performance-Based Analysis of Fire Resistance

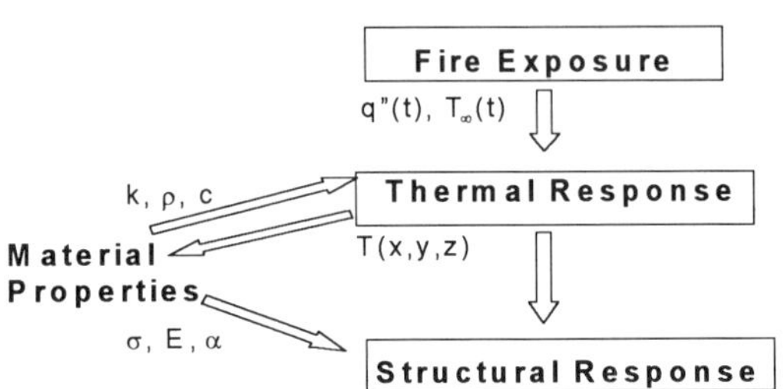

Thermal Response

A thermal response analysis involves analyzing the conduction heat transfer for the fire-exposed assembly to determine the temperature rise within the assembly. The conditions considered for the thermal response analysis depends on the type of assembly being examined. The fire exposure is assumed to only be on one side for walls and floor/ceiling assemblies, while interior columns are assumed to be exposed around their entire perimeter. In most thermal response analysis, any protection system is assumed to remain in place. Calculation methods range from applying simple algebraic equations to multi-dimensional, time-dependent computer models.

One algebraic-equation based method is the lumped heat capacity approach (LHCA) which assumes that any incident heat from the fire causes a uniform temperature increase throughout the exposed steel member [Malhotra, 1982]. The LHCA can be applied to steel members that range from being completely unprotected to completely protected. For the case of time-varying exposure conditions, as typically experienced in fires, the algebraic equation needs to be applied repeatedly, dividing the fire exposure into several small time steps.

Predictions from the LHCA versus measurements of the temperature rise of a steel column are presented in Figure 3 [Berger, 1989]. Predictions of temperature rise in steel beams by the LHCA are prone to be inherently less accurate than those for steel columns. A better estimate can be achieved with a relatively simple model [Milke, 1997] or a finite element model [Jeanes, 1984].

Figure 2. Yield Stress and Modulus of Elasticity of Structural Steel at Elevated Temperature [Lie, 1992]

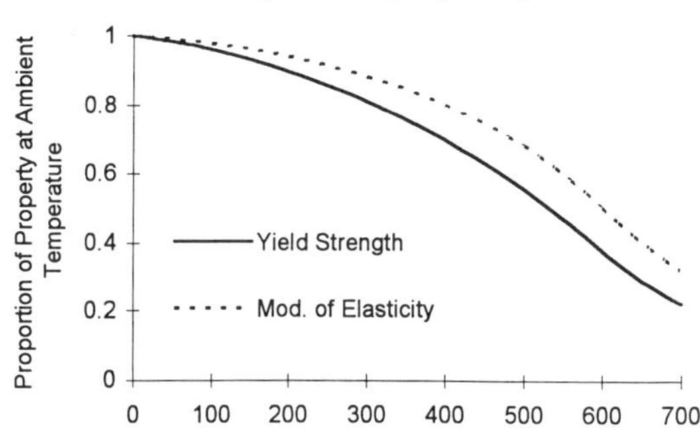

Figure 3. LHCA Prediction for W10X49 Protected Steel Column

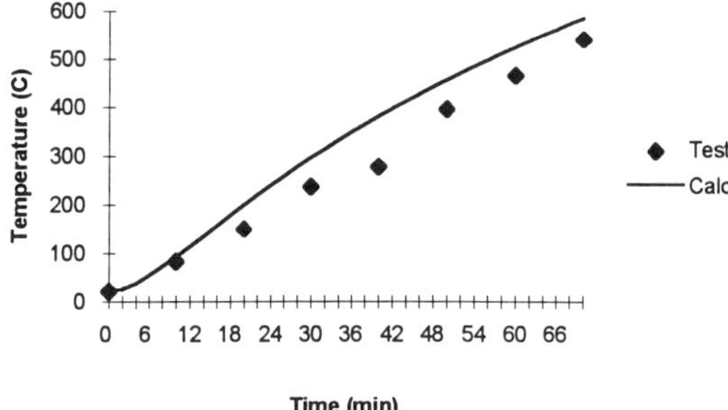

Structural analysis

The structural response analyses, such as column buckling and beam moment capacity analyses, assesses the impact of reductions in material properties and thermally-induced stresses. As an example, an elastic buckling analysis can be conducted by applying Euler's buckling equation for columns with a uniform, elevated temperature (such as estimated by LHCA), The column buckling analysis reduces the modulus of elasticity to account for the elevated temperature. The influence of temperature on the critical stress for elastic buckling is indicated in Figure 4. The structural response analyses may involve elementary calculations conducted with hand calculators or spreadsheets. Finite difference and finite element models can be applied to address situations involving non-uniform temperatures, arbitrary load conditions or adjoining structural members [Milke and Hill, 1997].

Figure 4. Elastic Buckling Stress of Heated Columns

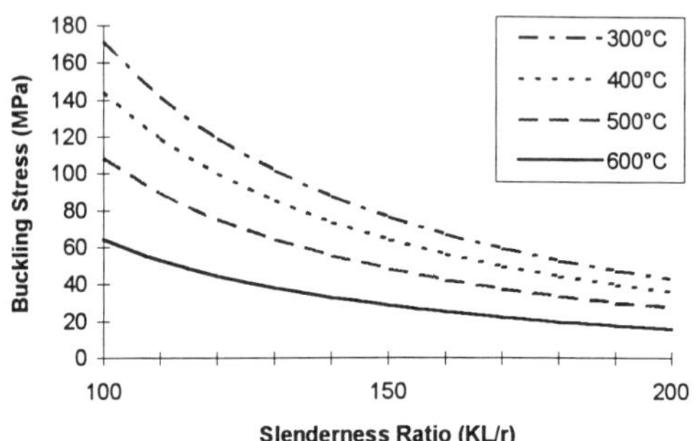

Summary

A performance-based fire resistance analysis, including thermal response and structural response analyses, can be conducted using a range of methods. A standard is being developed to document the general methodology and provide a detailed description of the simplified methods. One of the principal constraints is the availability of material property data at elevated materials, especially for protection materials.

References

Berger, G., "Estimating the Temperature Response of Wide Flange Steel Columns in the ASTM E119 Test," College Park: University of Maryland, 1989.

European Convention for Constructional Steelwork, 1985, *Design Manual on the European Recommendations for the Fire Safety of Steel Structures*, TC 3.

Eurocode 3, 1995, "Design of Steel Structures-Part 1-2, General Rules – Structural Fire Design," CEN.

Hurley, 1999, "Estimating Fire Exposures for the Purpose of Evaluating the Expected Performance of Structural Elements" ASCE Structure Congress.

Jeanes, D.C., 1984, "Predicting Temperature Rise in Fire Protected Structural Steel Beams," Technical Report 84-1, SFPE.

Lie, T.T., ed., 1992, *Structural Fire Protection*, ASCE.

Malhotra, H.L., *Design of Fire-Resisting Structures*, N.Y.: Chapman and Hall, 1982.

Milke, J.A., 1995, "Analytical Methods for Determining Fire Resistance of Steel Members," *SFPE Handbook of Fire Protection Engineering*, 2nd edition, NFPA.

Milke, J.A., 1997, "A Simplified Model for Estimating the Thermal Response of Steel Beam/Concrete Slab Ceiling Assemblies," ICFRE2.

Milke, J.A., and Hill, S.M., 1997, "Development of a Performance-Based Fire Protection Standard on Construction," ASCE.

Estimating the Fire Performance of Concrete and Masonry Structural Members

James A. Milke, Ph.D., P.E.[1]

Abstract

A fire resistance analysis of concrete members for performance-based design involves a determination of the structural integrity of the heated concrete member. As a prerequisite to conducting the structural evaluation, the temperature rise within the fire-exposed concrete member needs to be assessed to evaluate the material properties and thermal strains. In some cases, the structural analysis can be conducted using the same equations that are used for ambient temperature design.

Introduction

With the development of performance-based building codes, there will be a need to develop engineering standards, which document the supporting analytical methods. Formulation of an engineering standard to assess the fire resistance of concrete assemblies is underway through the joint efforts of ASCE and SFPE, with the support of the concrete industry. Analysis of the response of fire-exposed concrete members considers the fire exposure, elevated temperature material properties, and thermal response and structural response of the heated assembly [Milke and Hill, 1997]. The relationship of these four issues in a performance-based fire resistance analysis is reflected in Figure 1. Hurley addresses the means of selecting and describing the fire exposure conditions [1999].

Material Properties at Elevated Temperature

Temperature-dependent thermophysical (thermal conductivity, k, density ρ, and specific heat, c) and mechanical (strengths, σ, modulus of elasticity, E, and coefficient of thermal expansion, α) material properties influence the thermal and structural response analyses. Elevated temperature properties for various concretes are available [Schneider, 1988].

---

[1] Associate Professor, Department of Fire Protection Engineering, University of Maryland, College Park, MD 20742

Figure 1. Process of Performance-Based Analysis of Fire Resistance

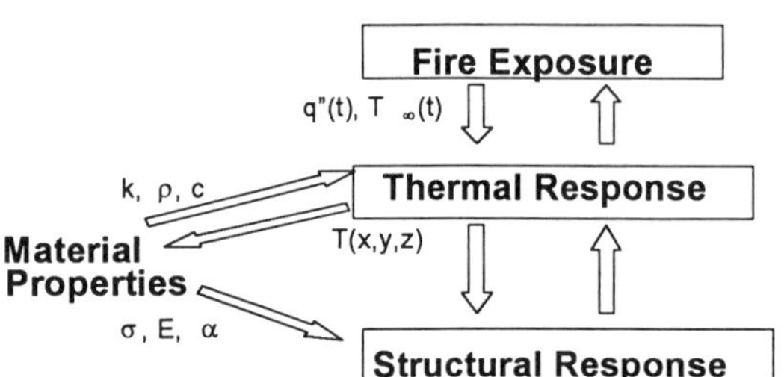

Because of the strong dependence of concrete properties on aggregate, each type of concrete has its own unique properties [Harmathy, 1993]. Concrete undergoes changes in composition upon heating and may spall [Harmathy, 1993]. These material related effects are generally included implicitly in the reported material properties. Alternatively, Ahmed and Hurst model pore pressure build-up to provide an improved understanding of the phenomenon of spalling [1995]. Some high performance concretes are available with high strength or high temperature capabilities, though the properties for these concretes are not as well known as traditionally used types of concrete.

Elevated temperature material properties for reinforcing and pre-stressing steel are also widely available. However, little information is available on the elevated temperature characteristics of other reinforcement materials such as glass fibers.

Thermal Response

A conduction heat transfer analysis for the assembly is applied to determine the temperature rise within the assembly and assess thermal strains and material properties. The exposure conditions considered depend on the type of assembly, with the fire exposure assumed to only be on one side for walls and floor/ceiling assemblies and interior columns are exposed around their entire perimeter.

For engineering purposes, the thermal response analysis of masonry or concrete assemblies assumes that the assembly is comprised only of concrete. This assumption is based on the negligible impact of reinforcing steel on heat transfer within a concrete assembly, given the ratios of concrete to steel volume and steel to concrete thermal conductivity. Graphical solutions of the temperature rise within concrete beams and slabs exposed to the standard time-temperature exposure are available [Gustaferro and Martin, 1989] [Fleichmann, 1995].

Alternatively, for a unique time-temperature exposure, the temperature rise within a concrete element is determined by a finite difference or finite element heat transfer analysis. An example is the one-dimensional, finite difference model of the coupled heat and mass transfer developed by Ahmed and Hurst for carbonate and siliceous aggregate concrete slabs [1995]. Sullivan, *et al.* [1994] provide a review of five finite element models to evaluate the thermal response from fire.

Structural analysis

The structural response analysis addresses the impact of reductions in material property values and thermally-induced stresses on the integrity of load-carrying members. The structural response analyses include column buckling and beam moment capacity analyses.

Moment capacity analyses formulated for fire-exposed reinforced and prestressed concrete flexural members are adapted from analysis methods used in room temperature concrete design adapted for fire resistance analyses supported by large-scale test data from the Portland Cement Association [Fleischmann, 1995]. The moment capacity of a concrete beam or slab is given as:

$$M_{cap} = \sigma \, A_s (d - a/2)$$

where M_{cap} is the moment capacity (N-mm), σ is the critical stress in steel (MPa), A_s is the cross-sectional area of the steel (mm^2), d is the from extreme compression fiber to centroid of steel (mm), and a is the depth of the equivalent rectangular stress block (mm), given by:

$$a = \frac{A_s \sigma}{0.85 \sigma_c b}$$

σ_c is the strength of concrete (MPa) and b is the breadth of the slab (mm).

When considering the positive moment capacity, the strength of the concrete in a slab is assumed to remain unchanged from that at ambient temperature. However, the strength of the steel is reduced commensurate with the temperature at the position of the reinforcement. Results of the analysis for one set of concrete slabs are illustrated in Figure 2.

Negative moment analyses are appreciably more complex. If the temperature is in excess of the threshold value, that portion of concrete on the lower edge of the beam or slab is neglected to reduce the negative moment arm [Gustaferro and Martin, 1989] [CRSI, 1980]. The strength of the concrete and reinforcing steel is reduced given the temperatures at the centroid of the equivalent stress block and reinforcement location. In some analysis methods, any portion of concrete which attains a threshold temperature of at least 650°C for siliceous aggregates or 760°C for carbonate aggregates is neglected [Gustaferro and Martin, 1977] [CRSI, 1980]. Development lengths of steel reinforcement need to be evaluated, accounting for the redistribution of moments. The thrust due to the expansion is evaluated to determine its effect on adjoining structural members and continuity of the assembly.

The structural response analyses described above involve relative elementary calculations. Computer-based finite difference and finite element models are also available [Sullivan, *etal.*, 1994][Milke and Hill, 1997]

Summary

A performance-based fire resistance analysis, including thermal response and structural response analyses, can be conducted using a range of methods. A standard is being developed to document the general methodology and provide a detailed description of the simplified methods. One of the principal constraints is the precision of thrust analysis for concrete beams and slabs exposed to fire exposures other than the standard time-temperature exposure.

Selected References

Ahmed, G.N. and Hurst, J.P., 1995, "Modeling the Thermal Behavior of Concrete Slabs Subjected to the ASTM E119 Standard Fire Condition," *J. of Fire Protection Engineering*, 7, 4, 125-132.

CRSI, 1980, *Reinforced Concrete Fire Resistance*, Chicago, Concrete Reinforcing Steel Institute.

Fleischmann, C., 1995, "Analytical Methods for Determining Fire Resistance of Concrete Members," *SFPE Handbook of Fire Protection Engineering*, 2nd edition, NFPA.

Gustaferro, A.H., and Martin, L.D., 1989, *Design for Fire Resistance of Precast Prestressed Concrete*, Chicago, Prestressed Concrete Institute.

Harmathy, T.Z., 1993, *Fire Safety Design & Concrete*, N.Y., John Wiley.

Hurley, 1999, "Estimating Fire Exposures for the Purpose of Evaluating the Expected Performance of Structural Elements" ASCE Structure Congress.

Lie, T.T., ed., 1992, *Structural Fire Protection*, NY, ASCE.

Milke, J.A., and Hill, S.M., 1997, *Development of a Performance-Based Fire Protection Standard on Construction*, ASCE.

Schneider, U., 1988, "Concrete at Elevated Temperature - A Critical Review," *Fire Safety J.*, 13, 55-68.

Sullivan, P.J.E., Terro, M.J., and Morris, W.A., 1994, "Critical Review of Fire Dedicated Thermal and Structural Computer Programs," *J of Applied Fire Science*, 3, 2, 113-135.

The Consequences of Vessel Impacts on the Mississippi River Bridges in New Orleans

Zolan Prucz[1] and William B. Conway[2]

Introduction

The Mississippi River crossings in the New Orleans area include the Crescent City Connection Bridges No. 1 and 2, the Huey P. Long Bridge and the Hale Boggs Bridge. These are high level bridge crossings of significant economical and societal importance.

The relatively large number of recent events of ship power and steering losses, groundings and collisions, which culminated with the December 14, 1996 collision of the Bright Field with the Riverwalk have drawn public attention and concern over the possibility of vessel collision with the bridge structures in New Orleans. The stretch of the Mississippi River in the New Orleans area is at particular risk for vessel collision accidents due to the combination of high vessel traffic density, high river currents and sharp bends in the river.

In 1996 the State of Louisiana, Department of Transportation and Development commissioned an extensive statewide vessel collision vulnerability study, which also includes all the Mississippi River bridges in New Orleans. This paper is based on information obtained during the LaDOTD study and investigations of past collision accidents. It reviews vessel collision risk factors, historical vessel collision data, bridge capacity levels in relation to calculated collision loads and discusses possible consequences of vessel impacts.

Bridge Evaluation for Vessel Collision

The evaluation of a bridge for vessel collision needs to address both the probability of occurrence of a collision and its consequences. The probability of occurrence of a collision is determined by factors related to the bridge characteristics, waterway characteristics and vessel traffic characteristics. The consequences of a vessel collision depend on the vessel type, size speed, loading condition and type of cargo carried, the bridge geometry, strength, stiffness, ductility and redundancy characteristics, and the effectiveness of the existing protection system. The location, direction and magnitude of the impact loads and their consequences are also affected by the river stage and current conditions.

The commonly accepted bridge performance criteria with respect to vessel collision is to prevent bridge collapse. Damage of structural members after a major vessel collision is usually accepted provided it does not result in bridge collapse or long term bridge service interruptions. Due to the prohibitive costs of designing a bridge for the worst case scenario involving the largest projected vessel hitting a bridge element head-on and at high speed, a certain amount of risk is considered acceptable.

Collision Risk Models

Collision risk models are used to evaluate bridges in relation to risk acceptance criteria, which is usually specified as a function of the importance of the bridge. In general, the occurrence of a collision is separated into three events; (a) a vessel approaching the bridge becomes aberrant, (b) the aberrant vessel hits a bridge element, and (c) the bridge element that is hit fails. Collision risk models consider the effects of the vessel traffic, the navigation conditions, the bridge geometry with respect to the waterway, and the bridge element strength with respect to the impact loads, and are commonly expressed in the following form (AASHTO, 1994):

$$AF = (N)(PA)(PG)(PC) \tag{1}$$

where, AF = annual frequency of collapse of a bridge element due to a given N, N = annual number of vessels classified by type, size, and loading condition which can strike a bridge element, PA = probability of vessel aberrancy, which defines the likelihood of a vessel losing control in the vicinity of a bridge, PG = geometric

---

[1]Senior Associate, Modjeski and Masters, Inc., New Orleans, Louisiana
[2]Chairman and CEO, Modjeski and Masters, Inc., New Orleans, Louisiana

probability of a collision between an aberrant vessel and a bridge pier or span, PC = probability of bridge collapse due to a collision with an aberrant vessel. The annual frequency of collapse for the entire bridge is taken as the summation of AF over all bridge elements exposed to collision and all the vessel groups considered.

Bridge, Waterway and Vessel Characteristics

Crescent City Connection Bridges No. 1 and 2

The Crescent City Connection Bridges No. 1 and 2 are twin high level steel cantilever through truss structures (see Figure 1). They cross the Mississippi River at River Mile 95.7 above Head of Passes. The Crescent City Connection Bridge No. 1 was opened to traffic in 1958, while the construction of the main span of the Crescent City Connection Bridge No. 2, which is located 400 feet downstream of the Crescent City Connection No. 1, was completed in 1985. The center span is 1,575 feet long and the vertical clearance is 150 feet from high water level, which corresponds to river stage +20.0. Pier II of both bridges are the only piers exposed to deep draft vessel impact. Each pier has two solid concrete shafts supported on caissons founded at elevation -180.0 (see Figure 2). The subshaft extends to elevation is at +30.0. A fender system that consists of horizontal timbers and timber posts is connected to the pier.

The navigation channel is 750 wide and the design draft is 45 feet. Deepening of the channel to 55 feet has been authorized, but it is uncertain whether it will be constructed. The mean water elevation and the mean high water elevation are +6.0 feet and +14.5 feet, respectively. The mean current velocity at river stage +14.5 is about 7 feet per second. A sharp bend of approximately 100 degrees in the river is located about 6,000 feet downstream from the Crescent City Connection Bridge No. 2 at Algiers Point (see Figure 3). The Gouldsboro Bend, which is a fairly gradual 70 degree bend, is located only 600 feet upstream from the Crescent City Connection Bridge No. 1.

The sharp turn and strong currents at Algiers Point and the turn at the Gouldsboro Bend make vessel navigation near the Crescent City Connection Bridges difficult. In order to position themselves for the turn at Algiers Point, ships coming downstream have to swing towards the East Bank as they approach the Crescent City Connection Bridges. The river traffic in the New Orleans area is monitored and guided by the Gretna and the Governor Nicholls stations operated by the U.S. Coast Guard.

The vessel movements in the New Orleans stretch of the Mississippi River include the main river traffic, which is composed of deep draft ships and barge tows, and local traffic composed of ferries, river boats and floating equipment. Both the river and the local traffic densities are very high. Also, there is significant activity at the wharves and the fleet mooring facilities near all bridges. The annual number of trips of vessels exceeding 150,000 DWT is over 50, and that figure is expected to increase significantly in the future, especially if the channel is deepened to 55 feet. The high river traffic density and the significant local vessel activity in the area add to the difficulty of navigation.

Huey P. Long Bridge

The Huey P. Long Bridge crosses the Mississippi River at River Mile 106.1. The main bridge is a steel cantilever through truss structure (see Figure 1). The center span is 790 feet long and its vertical clearance is 135 feet at river stage +18.0. The piers exposed to deep draft ship impact include Piers I through IV. These piers have two solid concrete shafts supported on caissons founded at elevations of about -170.0 (see Figure 2). Fender systems consisting of timber and rubber units are attached to Piers I and II.

The navigation channel is 500 feet wide. The mean and the mean high water elevations are 6 feet and 15.1 feet, respectively. Carrollton Bend, which is a sharp bend in the river of approximately 120 degrees is located 6,000 feet downstream from the bridge (see Figure 3). Another river bend is located about 3 miles upstream from the bridge at Twelvemile Point.

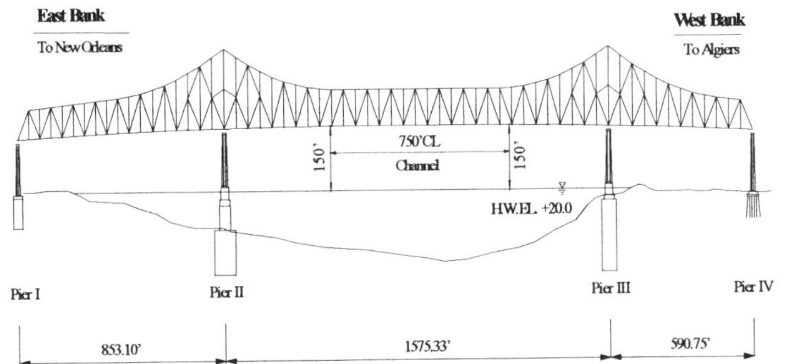

Crescent City Connection Bridges: Upstream Elevation of the Main Bridge (CCC1 shown)

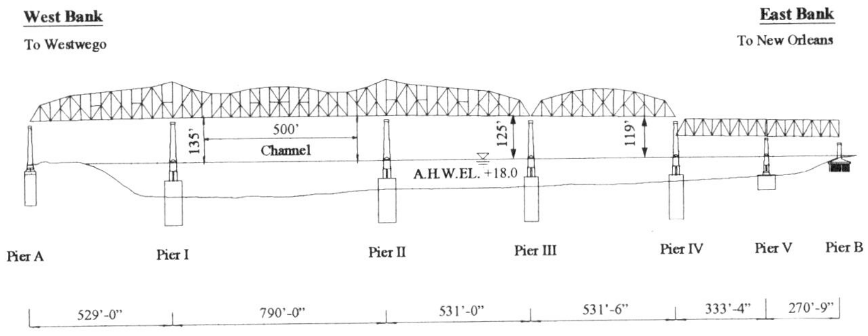

Huey P. Long Bridge: Downstream Elevation of the Main Bridge

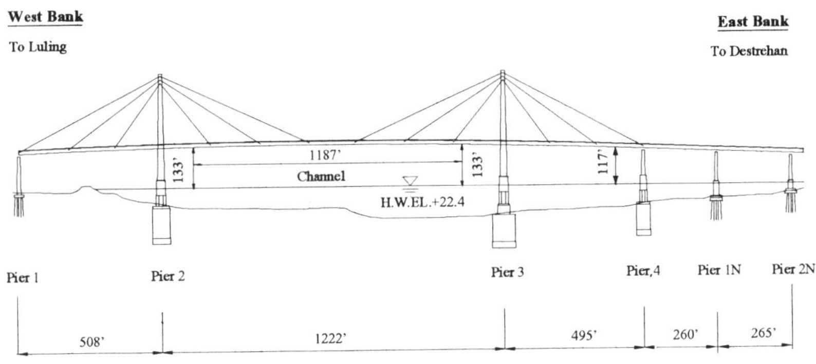

Hale Boggs Bridge: Downstream Elevation of the Main Bridge

Figure 1 - Elevation Views of the Mississippi river Bridges in New Orleans

Crescent City Connection Bridge No. 1

Huey P. Long Bridge

Hale Boggs Bridge

Figure 2   -   View of Channel Piers

The navigation conditions at the Huey P. Long Bridge are relatively difficult mainly during medium to high river stages. Ship traffic usually maintains to the centerline of the navigation channel while barge tows also use the adjacent spans, especially when passing each other under the bridge.

Hale Boggs Bridge

The Hale Boggs Bridge crosses the Mississippi River at River Mile 121.7. The main bridge has a cable-stayed steel box girder superstructure with an orthotropic steel deck supported by steel towers (see Figrure 1). The channel span is 1,222 feet long and its vertical clearance is 133 feet from high water level, which corresponds to river stage +22.4. The span to the east of the main span is also navigable. It has an horizontal clearance of 460 feet and a vertical clearance of 117 feet from high water level. The piers exposed to deep draft ship impact include Piers 2, 3, 4, 1N and 2N. Piers 2, 3 and 4 are supported by caissons founded at elevations ranging from -150.0 to -185.0. The top of the pier subshaft elevation is at +50.0 (see Figure 2). A timber fender system is connected to Piers 2, 3 and 4.

The mean and the mean high water elevations are 8 feet and 18.6 feet, respectively. The mean current velocity at river stage +18.6 is about 7 feet per second. A bend of approximately 55 degrees in the river is located about 6,000 feet upstream from the bridge at 26 Mile Point (see Figure 3).

The navigation conditions at the Hale Boggs Bridge are not very difficult. Ship traffic maintains to the centerline of the navigation channel, while barge tows also use the adjacent east span. Ships occasionally meet under the bridge and barge tows routinely pass each other mostly under different spans.

Table 1 includes a summary of some of the relevant bridge data described above.

History of Vessel Collisions

Crescent City Connection Bridge No. 1

There have been a total of seven recorded vessel collisions with the Crescent City Connection Bridge No. 1. The reported causes of the accidents were ship steering failures, vessel operator error, loss of control of barge tows, and strong winds and currents.

Five out of the seven collisions involved barge tows. In most cases the barge tow was headed downstream and the river stage was high. The consequences of the barge collisions included damage to the Pier II fender system and damage to the barge tow. The damage to the barges ranged from minor indentations to crushing. In one case, two barges of a twenty five coal barge tow sank in front of the pier. The barge tow broke apart upon impact with the pier.

Two of the collisions with the bridge involved deep draft ships. In 1961, a Japanese freighter, the Kyuo Maru, hit Pier II while trying to turn downstream after leaving a nearby wharf. The bow of the freighter broke open as a result of the impact, but the pier itself was not damaged, except for its fender system. Pier II was struck again in 1980 by a Filipino freighter, the Antipolo-I, which was headed upstream. The ship bow hit one of the pier shafts about 10-15 feet above the fender, leaving two scrape marks (see Figure 4). The fender system had only minor damage, but the damage to the ship was more serious. Witnesses of the accident noted a strong rolling of the ship as a result of the impact.

Crescent City Connection Bridge No. 2

There has been only one reported vessel collision with the Crescent City Connection Bridge No. 2 since its construction, but evidence of another collision was discovered during a 1991 bridge inspection. The reported accident involved a tow of six empty barges headed upstream. The accident occurred during strong winds and high river stage. It was attributed to vessel operator error. There was no apparent damage to the bridge, except for scraping of the timbers of the Pier II fender system.

Huey P. Long Bridge

There have been 42 vessel collisions with the Huey P. Long Bridge reported since the bridge was

Crescent City Connection Bridges

Huey P. Long Bridge

Hale Boggs Bridge

Figure 3 - Channel Layout

View of the Damaged Antipolo-I

View of the Damage to the Pier II Shaft and Fender System Caused by Antipolo-I

Figure 4   -   August 30, 1980 Collision of the Antipolo-I with the Crescent City Connection Bridge No. 1

opened to traffic in 1936, which is high relative to the other Mississippi River bridges. The reported causes of accidents include vessel operator's failure to ensure adequate vertical clearance, barge breakaways from upstream facilities during high river stages, loss of control of barge tows and mechanical failures.

The substructure was hit in 25 of the reported accidents. The majority of these accidents involved barge tows headed downstream, however, in some cases the bridge piers were hit by deep draft ships. The 1988 M/V Turpial accident is quite typical. The 580-foot Venezuelan tanker Turpial that was headed downstream hit the Pier II shaft and fender system after losing steerage. The pier shaft had only surface scrapes while the ship suffered an indentation in its port bow (see Figure 5). Witnesses said that the ship rolled to port during the impact. The Huey P. Long Bridge has four piers in the river and a relatively short main navigation span of 790 feet, which is probably the reason for the large number of substructure hits.

The superstructure was hit in 17 of the reported accidents. Most of these accidents were caused by vessels headed upstream. The vessels involved included ships with high cargo booms, masts or antennas and large crane barges. In one case, the cargo boom of a ship that was preparing to dock upstream of the bridge was lifted too soon. In most cases, however, the vessel operators misjudged the vessel height or vertical clearance of the bridge, which is 17 feet lower than the vertical clearance of the Crescent City Connection Bridges 1 and 2, located 10.4 miles downstream. Typical consequences of such accidents are shown in Figure 6, which includes photographs of a damaged cantilevered roadway section after the 1982 M/V Nova Gorica collision.

Hale Boggs Bridge

There have been only two reported vessel collisions with the Hale Boggs Bridge in the 15 years since its construction. One accident involved a barge tow that hit Pier 4 and damaged its fender, and the other accident was caused by several breakaway barges during an unusually high river stage. In both cases there was no damage to the bridge piers.

Ship Collision Loads vs. Bridge Capacities

The estimation of the load on a bridge pier during a ship collision is a very complex and uncertain problem. The actual force is time dependent, and, among other factors, it depends on the size and construction of the vessel, its velocity, loading condition and degree of water ballast, the location and direction of impact, as well as the geometry, strength and stiffness characteristics of the pier. There is very large scatter among the collision force values recommended in various vessel collision guidelines or used in various bridge projects. The ship collision loads used for the design of new bridges in the United States are defined in AASHTO, 1994 as a function of the vessel size and speed. These loads assume a head on collision scenario with a perfectly rigid body, where all the energy is absorbed by plastic deformation of the ship structure. The loads associated with a more yielding structure or a vessel impact at an angle could be significantly lower.

The capacity of a pier to withstand ship collision loads is determined by its overall strength and stability and by the local strength of the bridge elements exposed to direct contact with any portion of the ship's hull or bow. A concentrated collision load applied at mean high water level is usually used to calculate the overall pier capacity. For localized impact evaluation, collision loads are assumed uniform along the depth of the ship's bow. Possible failure mechanisms that need to be considered in a caisson supported pier include foundation sliding and overturning and structural failure at critical sections of the subshaft below vessel reach, such as construction joints and transition sections of different geometry. In addition to the pier characteristics, the governing failure mode for a pier is also affected by the ship size and its loading condition, and by the river stage. Collision Load/ Pier Capacity ratios calculated for a given ship size and river stage may be used to identify governing failure modes and to facilitate comparison between different piers or bridges. Table 2 illustrates channel pier Load/Capacity ratios corresponding to a 100,000 DWT ship travelling 13 mph and at mean water level. Load/Capacity ratios of over 1.0 indicate that the design collision load is larger than the pier capacity. Note that of the four bridges presented, only the Huey P. Long Bridge has a global Load/Capacity ratio greater than 1.0, and only the twin CCC Bridges have a local Load/Capacity ratio greater than 1.0. The significance of these ratios needs to be assessed in relation to the

View of the Damaged Starboard of the Turpial

Area of Contact with the Turpial and the Resulting Damage to the Pier II Shaft and Fender

Figure 5   -   January 10, 1988 Collision of the Turpial with the Huey P. Long Bridge

View of Eastbound Roadway

Note damage is mainly confined to one panel.

General View of Damage

Note bent roadway stringer and floorbeam bracket.

Figure 6 - June 8, 1982 Collision of the Nova Gorica with the Huey P. Long Bridge

risk of occurrence of a collision. Values of the probability of vessel aberrancy, PA, and the annual frequency of bridge collapse, AF_B, which are defined in Equation 1, are also included in Table 2, for reference. PA reflects the risk of a ship losing control in the vicinity of a bridge, while AF_B accounts for both the risk of collision with a bridge element and the capacity of the bridge element to resist that collision. The river stage can have a significant effect on AF_B, especially when the local pier capacity governs. The AF_B values shown in Table 2 correspond to mean water level which is considered to be closer representation of the actual conditions . The mean high water levels are usually used as reference in design. Acceptable AF_B values for design are 10^{-3} for "regular" bridges and 10^{-4} for "critical" bridges.

Table 1   –   Summary of Bridge and Vessel Collision Data

BRIDGE	LOCATION	LENGTH OF CHANNEL SPAN (FEET)	VERTICAL CLEARANCE (FEET)	NO. OF PIERS EXPOSED	AGE	SUBSTRUCTURE HITS	SUPERSTRUCTURE HITS
CCC No. 2	Mile 95.7	1,575	150	1	13	1	0
CCC No. 1	Mile 95.7	1,575	150	1	40	7	0
Huey P. Long	Mile 106.1	790	135	5	62	25	17
Hale Boggs	Mile 121.7	1,222	133	5	15	2	0

Table 2   –   Load/Capacity Ratios for a Channel Pier, Probability of Vessel Aberrancy and Annual Frequency of Bridge Collapse Data

BRIDGE	LOAD/CAPACITY GLOBAL		LOAD/CAPACITY LOCAL		PA		AF
	LONGITUDINAL DIRECTION (*)	TRANSVERSE DIRECTION	LONGITUDINAL DIRECTION (*)	TRANSVERSE DIRECTION	UPSTREAM	DOWNSTREAM	
CCC No. 2	0.5	0.4	1.4	0.7	1.0×10^{-4}	2.2×10^{-4}	0.4×10^{-3}
CCC No. 1	0.9	0.9	1.6	0.8	1.0×10^{-4}	2.2×10^{-4}	2.0×10^{-3}
Huey P. Long	1.3	1.5	0.3	0.2	0.8×10^{-4}	1.0×10^{-4}	2.0×10^{-3}
Hale Boggs	0.6	1.0	0.1	0.1	0.8×10^{-4}	1.0×10^{-4}	1.0×10^{-3}

(*) Longitudinal Direction is Along the Channel

Summary and Conclusions

The Mississippi River bridges in the New Orleans area were designed prior to the development of the AASHTO vessel collision criteria. Nevertheless, vessel collision loads were considered in their design based on the state of the art knowledge and understanding of their impact. The steady increase in vessel sizes and traffic density has put the older bridge structures at relatively greater risk for vessel collision damage.

The type of fender systems attached to the piers of the New Orleans bridges are mainly intended for minor vessel collisions. They prevent surface damage to the piers and reduce the formation of sparks during a collision. The fenders do not have the strength and energy absorption capacity to affect the consequences of major vessel collisions. These collisions must be resisted by the piers, while energy is absorbed through plastic deformations of the ship structure.

Review of past vessel collisions has shown that the bridge piers were able to withstand both barge and ship collisions with only minor surface damage. The fenders were damaged, but they were relatively easy to repair or replace. The collisions with the superstructure of the Huey P. Long Bridge, although quite frequent, have only caused localized damage in the cantilevered roadways, and the truss bottom chords and laterals. However, in some cases, the deformations of the roadway stringers, floorbeam brackets and deck, and the damage to the truss bottom chord were significant, resulting in traffic interruptions during repairs. Also, the roadway damages that occurred during these collisions created serious traffic hazards.

The risk of barge tow collisions is quite high, however, barge tows were found not to pose a threat to the integrity of the bridge crossings in New Orleans. Nevertheless, the consequences of barge collisions involving hazardous spills could be very severe in the densely populated area of New Orleans. The probability of a ship causing bridge collapse is remote. However, there is no guarantee that the New Orleans bridges could survive a head-on collision under worst case conditions, such as those involving a large vessel travelling at high speed and during high river stages.

References

AASHTO. 1994. *LRFD Bridge Design Specifications and Commentary*. American Association of State Highway and Transportation Officials, Washington D.C.

VESSEL/BRIDGE COLLISIONS INVESTIGATED BY THE NATIONAL TRANSPORTATION SAFETY BOARD

Ronald A. Weber, P.E.
National Transportation Safety Board

Forward

The National Transportation Safety Board is an independent Federal agency that determines the "probable cause" of transportation accidents through the investigative process and promotes transportation safety by making recommendations based on its investigations. The Safety Board has no regulatory authority. The Safety Board also conducts safety studies, and evaluates the effectiveness of other government agencies' transportation safety programs. The Safety Board conducts accident investigations and safety studies in the aviation, railroad, pipeline, marine and highway modes.

The Safety Board is completely independent of the United States Department of Transportation because Congress determined that "no Federal agency can properly perform such functions unless it is totally separate and independent from any other...agency of the United States."

Under its accident selection process, the Board's investigative response will depend on:

- the need for independent investigative oversight to insure public confidence in the transportation system
- the need to concentrate on the most significant and life-threatening safety issues

Safety Board investigations include the participation of modal agencies and other parties who can provide technical expertise not readily available at the Board. For example, in bridge collapses the owner of a collapsed highway bridge will normally be a party as well as others with a special interest in the accident. In the collapse of two spans of the Judge William Seeber bridge in New Orleans from the impact of a tow (barge and towboat), the Louisiana Department of Transportation and Development was a party. Other parties participating in the investigation were the Federal Highway Administration, the United States Coast Guard, the American Association of State Highway and Transportation Officials, the United States Army Corps of Engineers, and the owner of the towboat.

Since 1972 the Safety Board has investigated 16 vessel/bridge accidents, including the Sunshine Skyway Bridge. This paper will primarily discuss four of the vessel/bridge accidents that we have investigated since 1988. Two of the accidents involved a highway bridge, and two accidents involved a railroad bridge.

The Collisions

<u>Ramming of the CSXT Railroad Bridge by the Cyprian Bulk Carrier M/V
PONTOKRATIS, Calumet River, Chicago, Illinois, May 6, 1988</u>

On May 6, 1988, the 590-foot-long Cyprian Bulk Carrier PONTAKRATIS was
proceeding outbound in the Calumet River under the control of a Canadian pilot with the
assistance of two harbor tugs. While transiting the CSXT bridgedraw, the navigation
bridge of the PONTOKRATIS struck the leaf of the raised bascule bridge, collapsing the
leaf atop the vessel's wheelhouse. (The channel angle is 50 degrees to the bridge
centerline). The pilot, the master, and crewmembers exited the wheelhouse and ran onto
the stern of the vessel. No one was injured but the bridge was a total loss, estimated
between $10 and $12 million. The ship's navigation bridge was crushed, and damage to
the vessel was estimated to be about $2.5 million.

The safety issues discussed in the report include: the adequacy of the bridgeleaf chords'
vertical channel clearances for the safe transit of vessels having high superstructure; the
performances of the master and the pilot, and the tug operators; the performance of the
bridgetender; the accuracy of the bridge controls; the adequacy of federally published
information concerning vertical channel clearance for open bascule bridges; the adequacy
of U.S. Coast Guard administration of bridge construction and the issuing of bridge
permits as they affect safe vessel navigation; the failure of the Coast Guard to monitor full
implementation of prior Safety Board safety recommendations relative to bridge controls,
clearance surveys, and protective bridge fenders.

The Safety Board determined that the probable cause of this accident was the lack of
sufficient data on navigation charts and publications to permit mariners to determine if a
vessel can safely transit the bridgedraw.

<u>U.S. Towboat Chris Collision with the Judge William Seeber Bridge, New Orleans,
Louisiana, May 28, 1993</u>

About 3:30 p.m. on May 28, 1993, the towboat CHRIS, pushing an empty 202 long tons,
35-foot by 195-foot barge, with 18 inches of draft, drifted into a 14WF136 steel column
of the eastern approach span of the Judge Seeber Bridge, known locally as the Clairborne
Avenue bridge. This bridge carries Highway Route 39 over the New Orleans Inner Harbor
Navigation Canal. The severing of the column collapsed the two-column bent and two
simply supported spans (about 145 feet of bridge deck). The superstructure fell onto the
barge and into the shallow waters of the canal.

Two automobiles carrying three people fell with the four-lane bridge deck, resulting in one
death and serious injuries to the other two people. The bridge was closed to vehicle traffic
for 2 months and the canal was closed to navigation traffic for 2 days. The collision

resulted in approximately $2 million in damage to the bridge and $7 thousand in damage to the barge.

The Safety Board determined the probable causes of the collapse to be (1) the towboat operator's poor judgment in leaving the wheelhouse of the unsecured towboat unattended and (2) the failure of various Federal, State, and local government agencies to institute an effective program for managing risks to this bridge.

Some of the safety issues identified were: adequacy of operator performance, vulnerability of the Claiborne Avenue bridge to vessel collision and collapse, and vulnerability of existing bridges nationwide to vessel collision and collapse.

The Safety Board made recommendations on these issues to the Federal Highway Administration (FHWA), the U.S. Coast Guard, the U.S. Army Corps of Engineers, the Louisiana Department of Transportation and Development, the American Association of State Highway and Transportation Officials (AASHTO), and the Board of Commissioners of the Port of New Orleans. The recommendation to the FHWA and the AASHTO asked them to "...broaden the application of risk-assessment and management programs to existing highways. Such programs should include, among other things, a formal assessment of the vulnerability of bridges to vessel collision and collapse."

<u>Derailment of Amtrak Train No. 2 on the CSXT Big Bayou Canot Bridge near Mobile, Alabama, September 22, 1993</u>

On a foggy September 22, 1993, at 2:45 a.m., barges that were being pushed by the towboat MAUVILLA struck a pier and struck and displaced the Big Bayou Canot through-plate girder span of a CSXT railroad bridge near Mobile, Alabama. At 2:53 a.m., National Railroad Passenger Corporation (Amtrak) train 2, the Sunset Limited, en route from Los Angeles, California, to Miami, Florida, with 220 persons on board, struck a displaced girder and derailed. The three locomotives, the baggage and dormitory cars, and two of the six passenger cars fell into the water. The fuel tanks on the locomotive units ruptured, and the locomotives and the baggage and dormitory cars caught fire. Forty-two passengers and 5 crewmembers were killed; 103 passengers were injured. The towboat's four crewmembers were not injured.

The Safety Board determined that the probable causes of the derailment were the displacement of the bridge when it was struck by the MAUVILLA and tow as a result of the pilot becoming lost and disoriented in the dense fog because of (1) the pilot's lack of radar navigation competency; (2) the pilot's employer failure to ensure that the pilot was competent to use radar to navigate his tow during periods of reduced visibility; and (3) the U.S. Coast Guard's failure to establish higher standards for inland towing vessel operator licensing. Contributing to the accident was the lack of a national risk assessment program to determine bridge vulnerability to marine vessel collision.

Safety issues discussed in the accident report include towboat operator training and evaluation, bridge risk assessment, bridge identification, emergency response and evacuation procedures, and event recorder crashworthiness.

The Safety Board made recommendations addressing these issues to the U.S. Department of Transportation; the U.S. Army Corps of Engineers; the U.S. Coast Guard; Amtrak; the Federal Management Agency; The American Waterways Operators, Inc.; the Warrior & Gulf Navigation Company; the Association of American Railroads; and the American Short Line Railroad Association. The recommendation to the U.S. Department of Transportation stated:

> Convene an intermodal task force that includes the Coast Guard, the Federal Railroad Administration, the Federal Highway Administration, and the U.S. Army Corps of Engineers to develop a standard methodology for determining the vulnerability of the Nation's highway and railroad bridges to collisions from marine vessels, to formulate a ranking system for identifying bridges at greatest risk, and to provide guidance on the effectiveness and appropriateness of protective measures.

<u>The Ramming of the Portland-South Portland (Million Dollar) Bridge at Portland, Maine, by the Liberian Tankship *JULIE N* on September 27, 1996</u>

The 560-foot-long Liberian tankship *JULIE N*, carrying a cargo of heating oil, collided with the south bascule fender, pier, and live-load beam of the Million Dollar Bridge in Portland, Maine at 11:05 a.m. on September 27, 1996. The vessel had just passed between the piers of the new Portland-South Portland bridge (Casco Bay Bridge) and was en route to a terminal about 1.2 miles upstream. The vessel was under the pilotage of a State-licensed docking master. On the approach to the bridge, the pilot had issued three orders for port rudder to swing the bow to the left; he then attended to order the rudder to hard starboard and to increase the engine speed from slow to half ahead to stop the ship's swing and to align the vessel for passage through the drawspan. However, the pilot inadvertently ordered the rudder to hard port instead of hard starboard. He recognized his error within seconds and ordered the rudder to hard starboard. The shifting of the rudder occurred too late to avoid the collision. (The horizontal clearance between fenders was 98.4 feet; the width of the *JULIE N* was 86.3 feet).

The collision penetrated the tankship's single hull. A 33 foot longitudinal tear, 13 feet wide, allowed 2,000 barrels of fuel oil and slightly more tan 2,000 barrels of heating oil into the Fore River. The response to the spill was immediate and lasted until November 14, 1997; it cost approximately $43 million. The cleanup operation resulted in the recovery of 78 percent of the spilled oil. There were no injuries.

The cost of repairs to restore the bridge to normal operation was approximately $32,000. MDOT also replaced the damaged approach fender system with a steel structure that provided extra shielding for the pier. The cost of the new fender system was about $200,000.

This bridge had been hit many times before. A 1986 MDOT report to the Coast Guard listed 46 vessel collisions between January 1976 and May 1986 which resulted in bridge damage. Two cases were recorded in 1987 and one in 1988. From 1989 through 1996, 22 collisions with the bridge or fender system were recorded. The pilot of the *JULIE N* had been involved in two previous collisions with the bridge while navigating tugs and barges through it.

The Safety Board determined that the probable cause of the collision was the pilot's inadvertent order to port rudder instead of starboard rudder. Contributing to the accident was the narrow horizontal clearance of the bridge drawspan, which afforded little leeway for human error. Contributing to the severity of the damage to the vessel and to the amount of oil spilled was a corner of the bridge pier that was not adequately shielded by the timber fender system.

The new Casco Bay Bridge, which replaced the Million Dollar Bridge, has a horizontal clearance of 196.85 feet and unlimited vertical clearance; it opened to highway traffic on September 1, 1997.

Nonlinear Analysis of Buildings with PRC Connections

Bulent N. Alemdar[1], Joshua Taylor[1], Donald W. White[1] and Roberto T. Leon[1]

Abstract

This paper summarizes one approach for inelastic time-history analysis of steel building structures with partially-restrained composite (PRC) connections. It outlines several new elements capable of directly and efficiently modeling composite actions under cyclic load reversals. These analysis models are targeted at inelastic time-history analysis of three-dimensional building structural systems with PRC connections. The targeted computing platforms are mid- to high-end desktop computers.

Introduction

Although steel moment frames with composite floors are a common type of structural system used in low and mid-rise construction today, very little is known about the ultimate strength, ductility and toughness of these systems. However, the recent Northridge earthquake clearly demonstrated that "simple" connections can provide a significant amount of stiffness and strength to a frame if they act together with the floor system. In fact, many structures which suffered severe damage to welded moment connections survived the earthquake satisfactorily due to the resistance provided by flexible and weak connections. Prior experimental and analytical work has shown that "simple" connections may be coupled with composite slabs, leading to a structural system commonly referred to as partially-restrained composite (PRC) construction. When carefully detailed, this type of structure can provide good ductility and energy dissipation capacity. Moreover, if a large number of PRC connections are present, such as is the case in many modern designs where the ratio of the number of gravity columns to those resisting lateral loads is six or greater, the aggregate effect of these weak and flexible connections can be the same order of magnitude as that of the conventional moment-resisting ones.

This paper summarizes one approach to modeling of the complex behavior of PRC building structures.

---

[1] School of Civil and Environmental Engineering, Georgia Institute of Technology, Atlanta, GA 30332-0355

Beam Modeling

The floor girders are modeled using beam-type kinematics. Both fully-composite and partially-composite interaction are addressed. In the case of partially-composite interaction, the interface slip between the steel girder and the composite decking is modeled explicitly within the beam element formulation. These beam elements are force based. The partially-composite beam element is based largely on Salari et al. (1998). These elements offer significant advantages in coarse-element accuracy over conventional displacement-based finite element models, due to the fact that their force fields satisfy the governing differential equations of equilibrium at the section level along the length of the member. A continuous load-slip relationship is used to model the partial shear interaction in the latter of the two elements. The fully-composite beam element utilizes a stress-resultant moment-curvature constitutive model, whereas the partially-composite element is currently based on a fiber idealization of the slab and beam cross-sections. The modeling of the slab in these elements is based on an effective width idealization. Variation of the effective width along the member length is accounted for in the beam element formulations.

Connection Modeling

The behavior in the vicinity of the beam-to-column connections is modeled using a component-based approach. That is, the force-deformation behavior of each of the structural components that provide the force transfer between the floor system and the steel column is modeled directly, and the different force-deformation models are combined to represent the complete behavior associated with this force transfer. This model may be used to obtain a basic connection moment-rotation (M-θ) curve for analysis with the full-composite beam element, or alternately, the M-θ curve may be specified directly for use with this element. However, for the partially-composite beam element, the connection component model and the beam and beam-column models are combined as illustrated in Fig. 1. This figure represents the idealization of a PRC joint composed of a structural T or seat angle attached to the beam bottom flanges, double-sided web angles, and a composite concrete slab. The model of the slab within the connection region is composed of two fundamental types of spring elements: load introduction and load redirection springs. These springs are targeted at representation of the force transfer mechanism illustrated in Fig. 2 for a PRC connection under negative or sidesway moments. The load introduction spring models the slab force transfer in tension to the opposite side of the column, whereas the load redirection spring models the transfer of the slab force to the column through bearing. The basic layout of the model shown in Fig. 1 is the similar to that developed by Tschemmernegg et al. (1994), except that the slip resistance between the steel beam and the slab is modeled directly within the beam element in this work, as opposed to the use of a discrete slip spring component in Tschemmernegg's research. Also, the strut and tie force transfer mechanism illustrated in Fig. 2 is handled within the beam element formulation.

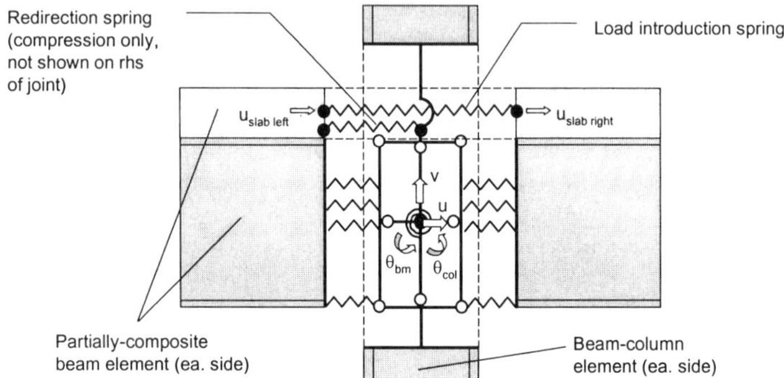

Figure 1. Model of an interior PRC joint in the strong direction.

The component representation of a T-stub connection to the beam bottom flange is outlined in Fig. 3. In Fig. 1, this connection is represented by the single spring between the column face and the bottom flange of the beam. In general, this spring can be broken down into a sub-group of spring components, connected in series, which account for the following deformations: (1) column web transverse deformation due to the load introduction from the T, (2) bolt elongation and shortening plus column flange bending, (3) bolt elongation and shortening plus deformation of the T flange, (4) axial straining of the T-stem, (5) shear displacement within the connection to the beam flange, including slip and bearing deformations, and (6) localized (non-planar) deformations in the beam flange and web due to the transfer of force from the T stem. A similar idealization is used for the web angles, which are illustrated by the springs between the beam webs and the column flanges in Fig. 1.

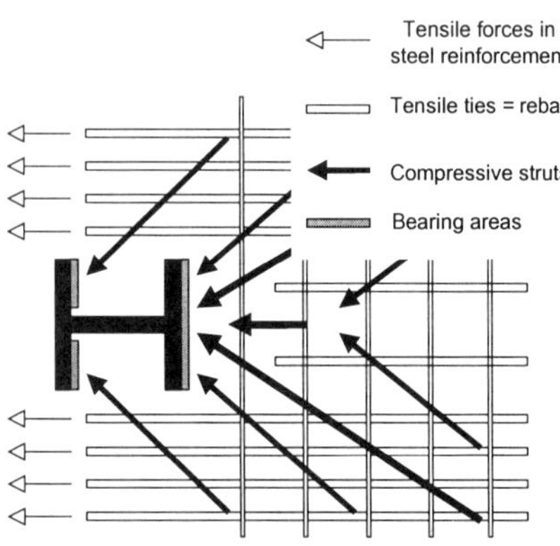

Figure 2. Slab force transfer mechanism for PRC connections under negative or sidesway moments.

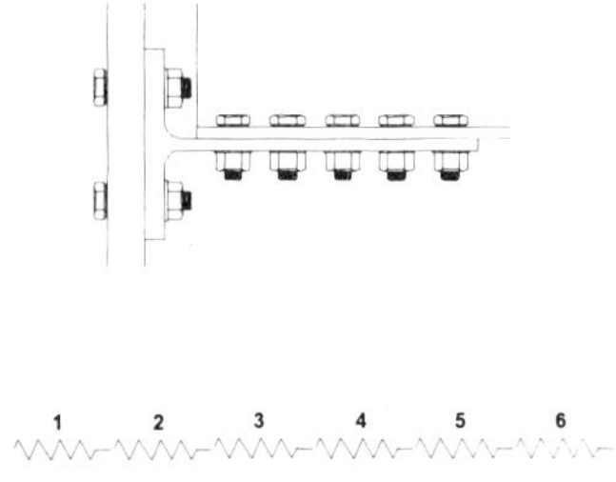

Figure 3. Component representation of a T-stub connection.

Panel Zone Modeling

The finite size and deformability of the column panel zone for strong-axis connections is modeled using a shear-panel constraint (see Fig. 1). This constraint idealizes the deformation of the panel zone as a single uniform shear-racking distortion. The rotations within the plane of strong-axis bending at the column faces on each side of the panel are constrained to be equal, and the in-plane rotations at the top and bottom of the panel, at the connection to the beam-column elements are equal, but different than the column face rotations.

Hysteretic Models

The hysteretic behavior of the various components is represented by a suite of models, which include the following attributes: (1) unsymmetric behavior in tension & compression of some steel components (due to stability effects), (2) stable symmetric hysteretic behavior of certain steel components, (3) abrupt stiffening of the force-deformation curve due to contact between components, e.g., contact in compression between the T-stub flange and the column flange in modeling the effects of bending in these components, (4) slip within shear connections, (5) stiffness and strength degradation and pinching behavior in the slab concrete response.

Beam-Column Model

The steel columns are modeled by a second-order three-dimensional beam-column finite element that is capable of tracking inelastic torsional-flexural actions, including cross-section warping (Nukala and White, 1998). It has 14 global degrees of freedom, three translation and three rotation dofs at each end, and two warping dofs at each end. This element has a mixed formulation, which provides improvements in accuracy for general inelastic analysis using one or only a few elements per member.

Acknowledgements

Funding for this research is provided by the National Science Foundation under the US-Japan Cooperative Earthquake Research Program, Composite and Hybrid Structures. The program director is Dr. Shih-Chi Liu.

References

Nukala, P.K.V.V. and White, D.W. (1998). "A Mixed Finite Element Formulation for Three-Dimensional Nonlinear Analysis of Frames," *Computational Methods in Applied Mechanics and Engineering*, (under review).

Salari, M.R., Spacone, E., Shing, P.B., and Frangopol, D.M. (1998). "Nonlinear Analysis of Composite Beams with Deformable Shear Connectors," *Journal of Structural Engineering*, ASCE, 124(10), 1148-1158.

Tschemmernegg, F., Brugger, R. Hittenberter, R, Wiesholzer, J., Huter, M, Schaur, B.C., and Badran, M.Z. (1994). "Zur Nachgiebigkeit von Verbundknoten," *Stahlbau*, 63, 3-19.

Earthquake Response Analysis of Steel Building Frames
Considering Brittle Fractures at Member-ends

Koji Uetani[1] and Hiroshi Tagawa[2]

Abstract

Two methods, simple one and accurate one, are presented for numerical response analysis of steel frames including brittle fractures at member-ends. The characteristics of these methods are shown through the comparison of the results of the same analysis problem. The earthquake response analyses of the multistory frames are performed in order to investigate the effects of brittle fractures at beam-ends.

Introduction

After the 1994 Northridge Earthquake and the 1995 Hyogoken Nambu Earth-quake, brittle fractures at member-ends near welded connections were observed in many steel building frames. Such kind of damages had been completely outside the designer's expectation and is by no means allowed to occur in the earthquake-resistant design, because the ductile plastic deformation of steel structures is believed to guar-antee sufficient large capacity for absorbing input energy. It is necessary to investigate the seismic response of steel buildings including brittle fractures for enhancing the confidence in the earthquake-resistant design. In the present paper, we propose two response analysis methods; one is based on the lumped-mass assumption but has higher computational efficiency, and the other takes into accounts the distributed-mass effect. By using the proposed method, the earthquake response analyses of 9-story frames with different fracture models are performed, and the analysis results are discussed. In this study, it is assumed that fracture occurs at a member-end when bending moment, rotation angle or cumulative rotation angle of a plastic hinge at the member-end reaches a critical value, and then a mechanical hinge forms at that end (see Fig.1). Two analy-sis methods presented here are different in the way of avoiding difficulties due to the disappearance of bending moment at a cross-section where fracture occurs.

[1]Professor, Dept. of Architecture, Kyoto University, Kyoto 606, Japan
[2]Research Associate, Dept. of Architecture, Kyoto University, Kyoto 606, Japan

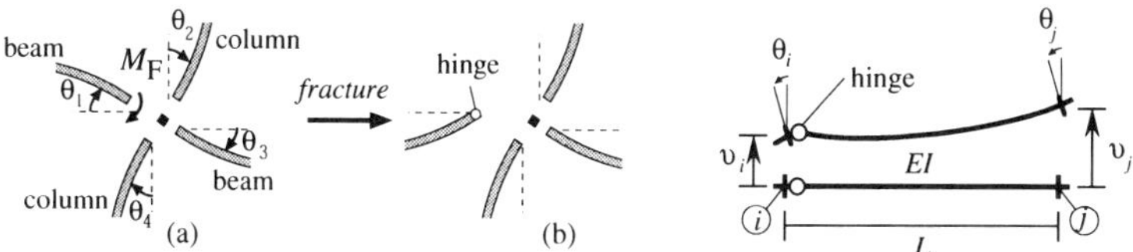

<table>
<tr><td>Figure 1 Fracture Model</td><td>Figure 2 Beam Geometry with a Hinge</td></tr>
</table>

Method-A for simplified analysis

Concentrated plasticity theories are applied. In this section, to make an explanation of the method simpler, consider a case in which fracture occurs at a beam-end reaching the elastic limit. The stiffness equations of the beam segment without hinges and with a hinge at i-th side (see Fig.2) are given by

$$\begin{Bmatrix} V_i \\ M_i \\ V_j \\ M_j \end{Bmatrix} = \frac{2EI}{L^3} \begin{vmatrix} 6 & 3L & -6 & 3L \\ 3L & 2L^2 & -3L & L^2 \\ -6 & -3L & 6 & -3L \\ 3L & L^2 & -3L & 2L^2 \end{vmatrix} \begin{Bmatrix} \upsilon_i \\ \theta_i \\ \upsilon_j \\ \theta_j \end{Bmatrix}, \quad \begin{Bmatrix} V_i \\ M_i \\ V_j \\ M_j \end{Bmatrix} = \frac{EI}{L^3} \begin{vmatrix} 3 & 0 & -3 & 3L \\ 0 & 0 & 0 & 0 \\ -3 & 0 & 3 & -3L \\ 3L & 0 & -3L & 3L^2 \end{vmatrix} \begin{Bmatrix} \upsilon_i \\ \theta_i \\ \upsilon_j \\ \theta_j \end{Bmatrix} \quad (1a,b)$$

respectively, where EI and L are the bending stiffness and the member length, respectively; V_i, M_i, υ_i and θ_i are the shear force, bending moment, vertical displacement and rotation angle at the i-th nodal coordinate of the beam segment, respectively. Condensing vertical and rotational degrees of freedom by the static condensation method and introducing an assumption that the horizontal displacements of all nodes on each floor are equal, the equations of motion for the frame in free vibration can be written as

$$[\mathbf{M}]\{\ddot{u}\} + [\mathbf{C}]\{\dot{u}\} + [\mathbf{K}]\{u\} = \{0\} \tag{2}$$

where $\{\mathbf{u}\}$ is a vector of story horizontal displacements; $[\mathbf{M}]$, $[\mathbf{C}]$ and $[\mathbf{K}]$ denote the mass, damping and reduced stiffness matrices, respectively. When fracture occurs at the beam-end of i-th side, eq.(1a) is replaced by eq.(1b) and the equations of motion, eq.(2), are to be updated. And then, the vectors of nodal displacement and velocity in vertical and rotational directions and the vector of story horizontal acceleration are discontinuous. No difficulty arises in the analysis by *Method A,* because the Nigam-Jennings method (Nigam *et al.* 1964) is used for numerical integration of equations of motion and the response in each time interval can be obtained with vectors of initial horizontal displacements and velocities of each time step. Geometrical nonlinearity is considered of conventional first-order approximation.

Method B for accurate analysis

Finite element methods are applied. The compatibility conditions of the member-end rotation (see Fig.1(a)) are written as $\theta_1=\theta_2=\theta_3=\theta_4 (=\theta_n)$, where θ_n denotes the nodal rotation. After fracture occurs (see Fig.1(b)), these conditions are changed to $\theta_2=\theta_3=\theta_4 (=\theta_n)$. The rotation angle, θ_1, is considered as an independent degree of freedom. $[\mathbf{M}]$, $[\mathbf{C}]$ and $[\mathbf{K}]$ need to be generated for corresponding to new degrees of freedom. Since inertial forces of all degrees of freedom are considered, the rotational acceleration is discontinuous at the instant of fracture. The Newmark-beta average acceleration method, which uses the initial acceleration vector of each time step, is applied for the numerical integration of equations of motion. In order to get the initial acceleration vector just after the fracture, the discontinuous value $\{\Delta\ddot{\mathbf{d}}\}$ of accelerative vector are calculated by eq.(3a,b) and added to the accelerations evaluated at the end of the time step just before fracture.

$$\{\Delta\ddot{\mathbf{d}}\} = -[\mathbf{M}]^{-1}\{\mathbf{R}\} \ , \quad \{\mathbf{R}\} = \{ \ 0 \cdots 0 \ R_F \ 0 \cdots 0 \ -R_F\}^T \qquad (3a,b)$$

R_F denotes the released bending moment acting on fracture section just before fracture. In the vector $\{\mathbf{R}\}$, R_F corresponds to the nodal rotation angle θ_n, and $-R_F$, to the beam-end rotation angle θ_1 which is the newly generated degree of freedom. Geometrical nonlinearity is considered with the local coordinate system rotating with the element.

Comparison of Analysis Methods

The two analysis methods are compared through the analyses for the same problem. The 2-story 2-span frame as shown in Fig.3 is considered. The fundamental natural period is 0.692 sec. The number of degrees of freedom corresponding to the motion for *Method A* and *Method B* are 2 and 138, respectively. The damping ratio of the lowest-mode vibration is chosen as 0.02 and the damping matrix is assumed to be proportional to the initial stiffness matrix. The time-step interval for numerical integration is set to 0.001 seconds. Fracture occurs when the bending moment at the left-side-end of 'Beam A' reaches a specified value. The frame is subjected to the initial velocity distributed in the fundamental eigenmode and scaled so that the horizontal velocity of node 5 is equal to 59.54 cm/s. The free vibration during the first 0.7 seconds is traced. The CPU time spent by *Method A* and *Method B* are 0.52 sec and 86.04 sec, respectively. The computational time with *Method A* is much shorter. Fig.4 shows the definitions of the two rotation angles at node 1, and their time histories are shown in Fig.5. Large disagreement are observed just after fracture points. In the case of *Method B*, a high frequency vibration appears but converges to the curve obtained by *Method A* rapidly due to the effects of viscous dumping.

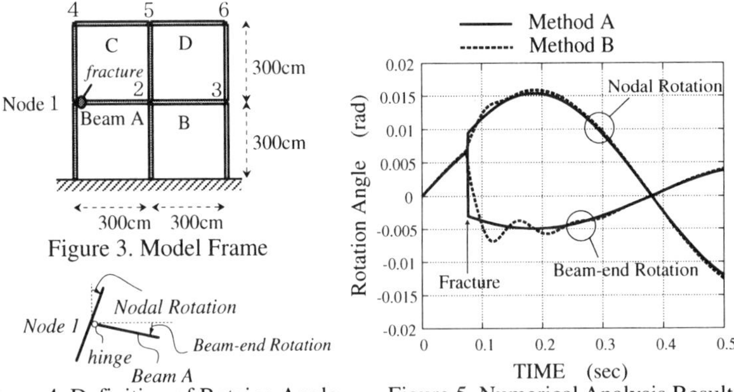

Figure 3. Model Frame

Figure 4. Definition of Rotaion Angle

Figure 5. Numerical Analysis Results

Table 1 List of Fracture Models

Model	Fracture Condtion; Rotation Angle of Plastic Hinges	Closing Cracks
A	Maximum value = 0.01rad	considered
B	Cumulative value =0.02rad	considered
C	Cumulative value =0.02rad	none

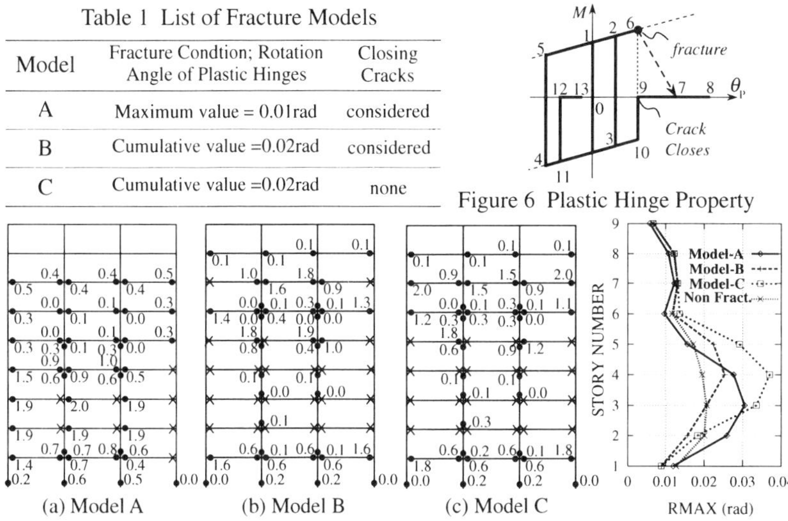

Figure 6 Plastic Hinge Property

(a) Model A   (b) Model B   (c) Model C

Figure 7 Locations of Plastic Hinges and Fractures

Figure 8 Story Drift

Seismic Response Analysis of 9-story Frames with Analysis Method A

Consider three fracture models listed in Table 1. Fracture condition is given so that the maximum or cumulative rotation angle of plastic hinges at the beam-ends reaches a specified value. Fig.6 shows the plastic hinge property of *Model A*. In *Model A* and *B*, fracture occurs only at bottom flanges and the bending moment is applied while a crack closes. Figs.7 show the locations of plastic hinges, ●, and fractures, ×, of the frames subjected to the JMA-Kobe 1995 NS scaled so that the maximum velocity is equal to 80 cm/sec. The maximum rotation angle is indicated in unit 10^{-2} rad at each plastic hinge in Fig.7(a), and the cumulative one, in Figs.7(b,c). Fig.8 shows the maximum story rotation angles. Although many fractures occur in the case of the *Model B*, the story drifts are not so large in comparison with the non-fracture case. Strong effect of fracture model properties on the seismic response of frames is observed.

Concluding remarks

Two numerical analysis methods have been presented. One is simple and effective for analyzing large scale frames. The other is more accurate and can trace the high frequency vibration caused by fractures. The effect of fracture properties on the frame response has been revealed through seismic response analyses of the 9-story frames.

References

Nigam, N.C. and Jennings, P.C. (1964). "Calculation of response spectra from strong motion earthquake records", *BSSA*, 59(2), 909-922.

Pushover Analysis of Isolated Flexural Reinforced Concrete Walls

Balram Gupta[1] and Sashi Kunnath[2], Associate Members ASCE

Abstract

The application of static pushover analyses for seismic evaluation of isolated flexural walls is investigated. The objective is to verify the adequacy of the uniform and modal lateral load patterns in predicting the inelastic behavior of flexural walls under seismic loads and propose a new nonlinear static procedure which overcomes the limitation of existing pushover procedures. Predictions based on different pushover procedures are compared to results obtained from detailed nonlinear time history analyses. Ground motions corresponding to 2% probability of being exceeded in 50 years for Los Angeles area are considered in the evaluation. It is shown that traditional pushover analyses, such as those recommended in FEMA-273 (1997), are incapable of reproducing computed behavior using nonlinear time-history evaluations and that the new pushover procedure represents a significant improvement in current analytical capabilities of nonlinear static methods to predict nonlinear behavior of structural walls under seismic loads.

Introduction

There are two critical issues in the inelastic behavior of walls under seismic loads: amplification of base shear demands and progressive yielding. It has been shown that the inelastic base shear demands on isolated walls can be much higher than the base shear strength for which the walls have been designed, and yielding in the wall can progress to higher floors even after a plastic hinge has formed at the base (Seneviratna, 1995). The implication of base shear amplification is that a shear failure is possible due to the effect of higher modes. A nonlinear time-history analysis would provide information about this phenomenon. The second issue of progressive yielding is important from the viewpoint of simplified analysis. Since the isolated walls are vertical cantilevers, theoretically speaking, the walls should rotate as a rigid body once a plastic hinge forms at the base if the post-yield stiffness is

---

[1] Senior Analyst, Saiful/Bouquet Consulting Structural Engineers Inc., Pasadena, CA 91105
[2] Assoc. Prof., Dept. of Civil Engrg., University of Central Florida, Orlando, FL 32816-2450

negligible. Nonlinear time-history analyses, however, suggest that yielding above the base is possible even after plastic hinging at the base. A simplified static analysis using uniform or modal load patterns would never indicate yielding at other levels once a plastic hinge has formed at the base. It is shown that a new adaptive modal pushover procedure is able to reasonably capture these effects.

Proposed Pushover Procedure

The main difference between the traditional pushover methods and the proposed method is that the applied load pattern in the proposed method continually changes depending on the instantaneous dynamic properties of the system. The load pattern can consider as many modes as deemed important during the course of the analysis. Modal base shears are calculated using an elastic response spectrum of the site-specific ground motion. These shears are combined using SRSS to compute the building base shear. Then a static analysis of the structure is carried out for story forces corresponding to each mode independently. This means that for modes other than the fundamental mode, the structure will be pushed and pulled simultaneously. Whenever some element(s) yields, a new structure is created by changing the stiffness of the yielded element(s) and the response spectral analysis is repeated. Details of the procedure can be found in Gupta (1998).

Description of the Wall Structures Used in the Study

Four isolated walls with 8, 12, 16, and 20-stories are considered. These walls are modeled as vertical cantilevers. The yield moment at the base corresponds to the overturning moment for a base shear of 15% of the seismic weight applied per UBC-1988 load pattern. A constant flexural strength is assumed along the full height of the walls. Table 1 gives the dynamic properties of the wall structures.

Table 1. Dynamic Properties of Isolated Wall Structures

Mode	Parameter	8 Story	12 Story	16 Story	20 Story
1	T_1	0.80	1.10	1.40	1.70
	$\%M^1$	65.31	63.94	63.27	62.87
2	T_1/T_2	6.32	6.29	6.28	6.28
	$\%M$	19.98	19.61	19.41	19.30
3	T_1/T_3	17.82	17.68	17.62	17.60
	$\%M$	6.86	6.74	6.67	6.63

[1] Participating Mass (%)

Ground Motions Used in the Study

The nonlinear response of structures is strongly influenced by ground motion characteristics. As part of the SAC Phase-2 steel project, response spectra and time-histories have been developed for use in various structural investigations (SAC Draft Report, 1997). Fifteen time-histories corresponding to 2% probability of exceedance in 50 years (designated as 2%/50 hereafter) from SAC records are chosen for the present study. The basic SAC records were scaled to the median (geometric mean) spectral acceleration at the fundamental frequency of the respective structure. This scaling was done to reduce the dispersion of response parameters (Shome et al., 1997) and to study the effect of higher modes. Figure 1 shows the scaled mean 5% damped elastic response spectra for 2%/50 records. The response spectrum corresponding to Los Angeles area and stiff soil conditions suggested by FEMA-273 is also shown for comparison. In the period range of interest, the building-specific (2%/50) spectra are extremely severe as compared to the FEMA spectrum.

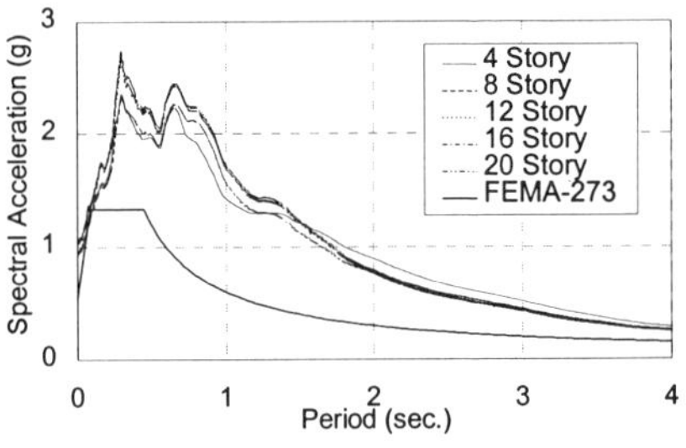

Figure 1. Scaled Spectra for 2%/50 Records

Base Shear Amplification

Figure 2 shows the variation of the ratio of mean base shear obtained from pushover analyses to that obtained from nonlinear time-history analyses for all three pushover procedures at a roof displacement *equal* to that predicted by nonlinear time-history analyses. The adaptive modal pushover procedure predicts the base shear to within 75% of the dynamic base shear while the traditional procedures predict shears much less than the dynamic base shear. The reason for this behavior is as follows. For low post-yield stiffness, as soon as a plastic hinge is formed at the base, its rotation increases rapidly, not allowing any significant increase in the base shear for traditional lateral load patterns. The inclusion of higher modes in the adaptive pushover procedure is able to capture this behavior more accurately.

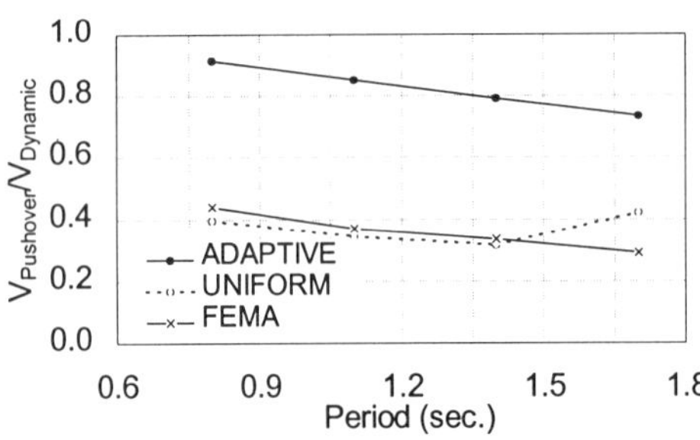

Figure 2. Ratio of Pushover to Dynamic Base Shear

Additionally, the contribution of higher modes to base shear amplification can be better understood by considering the global force-deformation behavior as shown in Figure 3. The displacement increment for a given base shear increment is much higher for the fundamental mode than if the first two modes are considered for the same increment in base shear.   It should also be noted that the rate of change of fundamental period is generally much higher than that for periods in the higher modes. This means that as the structure yields, the story forces, which are directly proportional to spectral acceleration and modal participation factor, are relatively higher for the second mode as compared to those for the first mode.

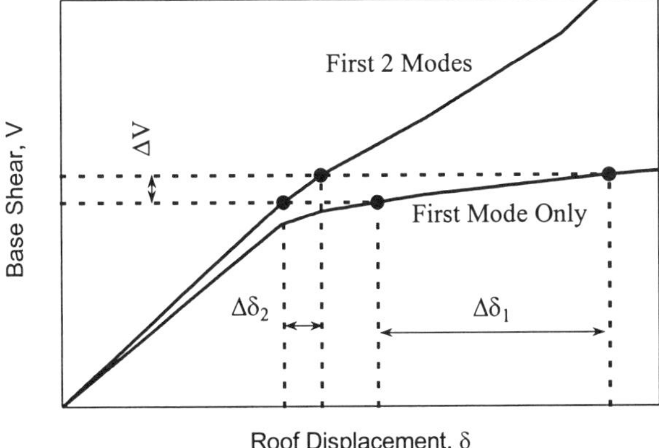

Figure 3. Variation of Roof Displacement for Various Modes

Progressive Yielding

Progressive yielding refers to the spread of plastic hinging to higher levels after the base has yielded. Due to reasons discussed in the previous section, the uniform and modal load patterns can not predict this phenomenon. The adaptive modal procedure, however, is able to capture this since the force and displacement increments are computed by a SRSS combination of the corresponding modal values. Thus, although, at some levels, the story forces corresponding to higher modes act in the opposite direction to those for the fundamental mode, the net effect is a monotonic increase in the forces because of a SRSS combination. Thus, the moments at some higher levels can also reach their yield levels before the roof displacement attains a value predicted by a nonlinear time-history analysis.

References

FEMA-273, (1997). *NEHRP Guidelines for the Seismic Rehabilitation of Buildings.* Building Seismic Safety Council, Washington, D.C.

Gupta, B. (1998). "Enhanced Pushover Procedure and Inelastic Demand Estimation for Performance-Based Seismic Evaluation of Buildings".*Ph.D. Dissertation,* Dept. of Civil and Env. Engrg., Univ. of Central Florida, Orlando, FL.

Seneviratna, G.D.P.K. 1995. "Evaluation of Inelastic MDOF Effects for Seismic Design". *Ph.D. Dissertation,* Dept. of Civil Engineering, Stanford University.

Shome, N., Cornell, C.A., Bazzurro, P., and Carballo, J.E. (1997). "Earthquakes, records, and nonlinear MDOF responses". *Report No. RMS-29,* Department of Civil Engineering, Stanford University, June 1997.

Modeling the Seismic Behavior of Planar Moment Frames Using Fishbone Models

Kazuo Inoue[1], Koji Ogawa[2] and Hisaya Kamura[3]

Abstract

A method for simulating the seismic behavior of planar moment frames using the fishbone model is presented. Good correlation is achieved between the seismic response calculated from the simplified fishbone model and results from more detailed analyses.

Introduction

Rotation demand of members of moment frames subjected to severe earthquake depends on various structural factors such as stiffness, strength, collapse mechanism, etc. This suggests that the rotation demand at the beam ends is affected by the beam-to-column strength and stiffness ratio. To obtain the rotation demand of members of steel moment frames, it is necessary to calculate the earthquake response considering a wide range of structural factors and many input ground motions. For this purpose, simple and reasonable modeling of steel moment frames is required. A simulation method based on the fishbone model is presented here.

Fishbone-shaped Frame and Hysteretic Characteristics of Fishbone-members-

Fig. 2 (b) shows a fishbone-shaped model representing a planar moment frame. To distinguish the beams and columns of the original moment frame shown in Fig. 2 (a), the beam and the column of the fishbone-shaped frame are called a fishbone-beam and a fishbone-column , respectively. The mass of each floor is concentrated at the intersection of the fishbone-beam and the fishbone-column. In the calculations, the fishbone-beam is treated as the rotational spring as shown in Fig. 2 (c). The main assumptions of modeling the fishbone-shaped frame are as follows :

[1]Professor, Dr. Eng., Dept. of Architecture and Environmental Design, Kyoto University.
[2]Professor, Dr. Eng., Dept of Architecture and Civil Engineering, Kumamoto University.
[3]Senior Research Engineer, M. Eng. Applied Technology Research Center, NKK Corp.

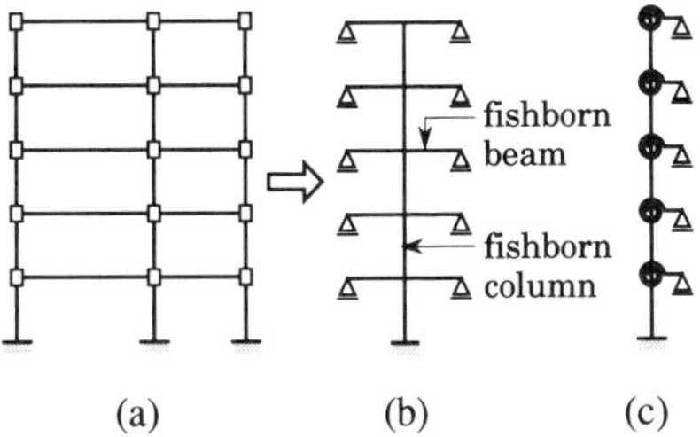

Fig. 1 Modeling to fishbone-shaped frame;
(a) Original Frame, (b) Fishbone-shaped
frame, (c) Numerical model.

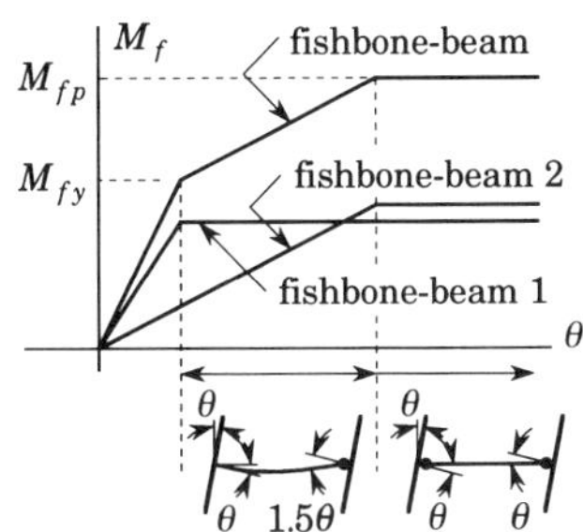

Fig.2 Tri-linear model of fishbone-
beam and decomposition into two
perfectly elastic-plastic relations.

[1] Horizontal joint displacement and rotation on the same floor is assumed to be the
     same for all joints.
[2] Hysteretic characteristic of the fishbone-column is a perfectly elastic-plastic.
[3] Hysteretic characteristic of the fishbone-beam is a tri-linear type (see Fig. 2).

The hysteretic behavior of fishbone members is determined from 1) elastic
stiffness, 2) plastic moment. In addition, for the fishbone beams, 3) the initial yield
moment and 4) the second slope are required. From assumption [1], elastic rotational
stiffness of the fishbone-beam and the fishbone-column are obtained from the summation
of the elastic stiffnesses of the beams and columns of the original frame. Reduction of
story stiffness due to the axial deformation of columns is reflected in the stiffness of
fishbone-beams.

The plastic moment of the fishbone-column at the ith floor can simply be obtained
by summation of the plastic moments of columns at the same floor. To make the collapse
load and collapse mechanism of the fishbone-shaped frame coincide with those of the
original frame at any static horizontal loads, we introduce the concept of "floor plastic
moment". The floor moment M_f is defined as the sum of the moments at the beam ends
on the same floor. The plastic moment of the fishbone-beam of ith floor is equated to
the floor plastic moment which is defined as the smaller of the sum of the plastic moments
of the beam ends on the same floor and the sum of the plastic moments of the columns
connected that floor.

The initial yield moment of the fishbone-beam is the floor moment at which the
first plastic moment is reached at a beam end. This is obtained from the results of
elastic analysis of the original frame.

The second slope, after the initial yield moment of the tri-linear model, is constant
whereas the slope decreases gradually in the original frame. It is appropriate to regard
that the slope gives the average value when plastic hinges occurred at one end of all the
beams. In this case, the second slope becomes a quarter of the initial slope.

Thus, the hysteretic characteristics of the fishbone-column and the fishbone-
beam can be determined by the results of elastic analysis and the plastic moments of
individual members. The shear deformation of members, stiffness and strength of beam-
to-column joint panels and PΔ-effect can be taken into account in the fishbone-model.

These factors are included in the numerical analysis described below.

Cumulative Plastic Rotation of Fishbone-beam

It is expected that a pre-formed plastic hinge is subjected to the maximum plastic rotation. The plastic rotation of pre-formed hinges does not correspond to the plastic component of tri-linear relation. Then, the tri-linear model of the fishbone-beam is decomposed into two perfectly elastic-plastic relations as shown in Fig. 2. Fishbone-beam 1 indicates the moment-rotation relation of pre-yielding beams, and fishbone-beam 2 indicates that of post-yielding beams.

When the plastic hinge occurs at one side of the beam end as shown in Fig. 2, the increment of plastic hinge rotation is 1.5 times that of joint rotation. Then the cumulative plastic rotation of pre-formed hinges is obtained by the following equation:

$$\Sigma \theta_p = 1.5 \left(\Sigma \theta_{p1} - \Sigma \theta_{p2} \right) + \Sigma \theta_{p2} \tag{1}$$

in which, $\Sigma\theta_{p1}$ and $\Sigma\theta_{p2}$ are the cumulative plastic rotation of the fishbone-beam 1 and 2, respectively.

Verification by Earthquake Response Analysis

The 8-story moment frames shown in Fig. 3 are used in the earthquake response analysis. The frames were designed according to the Japanese seismic-resistant code, and are of the weak beam type. The results obtained by using the fishbone-model (FM) are compared with those of the detailed model (DM). In the DM, both material and geometrical nonlinearities are considered (Ogawa and Tada). The shear force-shear deformation relation of the joint panel is assumed to be bi-linear. The first and second natural periods of FM and DM are listed in Table 1.

Two ground acceleration records shown in Table 2 were used, namely the 1940 El Centro earthquake record (N-S component), and 1995 Kobe Meteorological Agency (JMA) record. The intensity of the input ground motions was selected as shown in Table 2. The viscous damping ratio of the system was assumed to be 0.02.

Fig. 4 shows the maximum story drift angle R_{max} for both sides, and Fig. 5 shows the cumulative plastic rotation $\Sigma\theta_p$ of the fishbone-beam and the beams of the

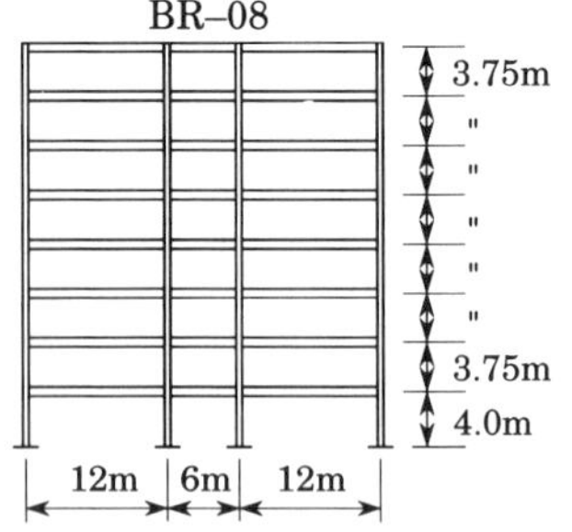

Fig. 3 Dimensions of 8-story frames for analysis

Table 1 Natural period (s)

	DM	FM
T1st	1.16	1.15
T2nd	0.41	0.41

Table 2 Input earthquake motions

Input motion	Max. V.	Max. Acc.	Duration
El Centro NS	1.0m/sec	10.22m/sec²	20sec
JMA Kobe NS	1.0m/sec	9.40m/sec²	30sec

original frames. In Fig. 4, a dashed line and a solid line show the results of FM and DM, respectively. The mark $\Diamond$ in Fig. 5 indicates the $\Sigma\theta_p$ of each beam end obtained from DM. The $R_{\max}$ distribution by FM agrees well with that of DM. This implies that the good estimation of the maximum amplitudes of plastic hinge rotation of beams can be obtained by FM. In regard to $\Sigma\theta_p$, the results of FM are close to the maximum value of the results of DM as expected.

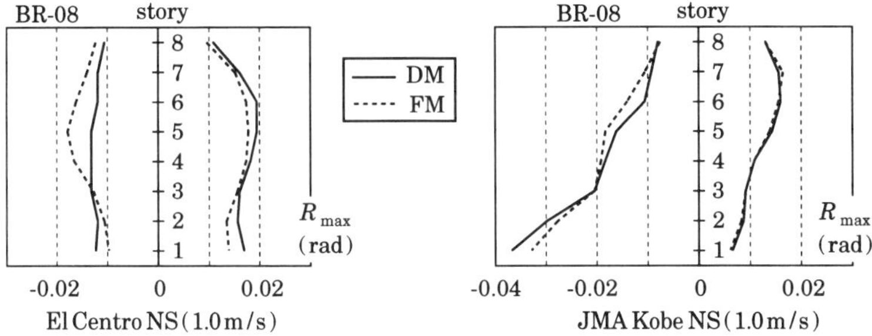

Fig. 11 Maximum story drift angle $R_{\max}$

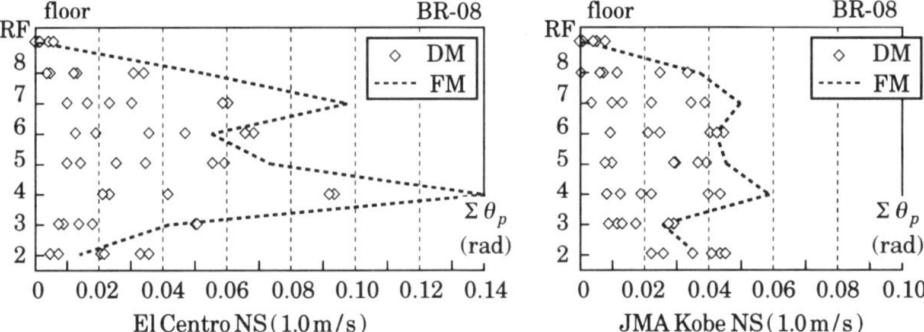

Fig. 12 Cumulative plastic rotation $\Sigma\theta_p$ of fishbone-beam and beams of original frames

Conclusion

A method for simulating the seismic behavior of planar steel moment frames using a fishbone-shaped frame model has been described. Good accuracy was achieved in calculating the maximum story angle and the plastic deformation of weak members. One special benefit of using the fishbone-shaped frame is that it is easy to use since only the results of elastic analysis and the plastic moments of the individual members of the original frame are required.

References

Ogawa, K and Tada, M (1994) : "Computer Program for Static and Dynamic Analysis of Steel Frames Considering the Deformation of Joint Panel," *Proc. of the 17th Symposium on Computer Technology of Information, Systems and Applications,* 79-84.

Localization Issues in Nonlinear Frame Analysis

E. Spacone[1] Associate Member, ASCE, and J Coleman

Abstract

Nonlinear finite element analyses of brittle structures show mesh dependent results due to strain concentrations in localized regions or bands. The problem has been studied mainly for solid finite elements and appropriate remedies have been suggested to eliminate or alleviate the problem, especially for concrete in tension regions. This paper reports on an ongoing study on strain localization issues in frame analysis. The topic is of primary importance for the use of pushover and nonlinear dynamic analyses in performance-based engineering of new and existing structures. The discussion focuses on the bending response of concrete and steel specimens followed by simple tension and compression tests of concrete specimens. The study relies on the use of a force-based beam element that yields stable results even under element softening.

Introduction

Strain localization problems have been extensively studied in the last few years, especially in tension regions of concrete structures. Localization implies that the post-peak behavior of the stress-strain response of concrete depends on the specimen size (in experimental tests) or on the mesh discretization (in finite element analyses). In particular, in finite element analyses of softening structures such as concrete specimens, strains localize in bands whose size depends on the size of the finite elements used in the region where strain-softening occurs. The smaller the finite element size, the larger the strains needed to obtain the same nodal displacements. In order to alleviate the problem, solutions have been proposed that rely on a stress-displacement approach rather than on a stress-strain measure. A solution widely accepted in the finite element community utilizes the fracture energy concept (Bazant and Planas, 1997). In this approach the post-peak behavior of the stress-strain curve is rescaled depending on the element size in order to retain the specimen's known fracture energy. A different solution, more popular in the reinforced concrete design community, relies on the concept of plastic hinge length. The finite elements should be scaled in such a way that plasticity concentrates in a

---

[1] Dept. of CEAE, University Colorado, Boulder, CO 80309-0428.

region whose length should match the experimentally measured plastic hinge dimensions.

This paper reports on the early stages of an ongoing study on the effects of deformation localization in frame analysis. The problem is of primary importance and should be addressed in view of the application of nonlinear frame analysis to performance-based design engineering of buildings and bridges (FEMA,1997). This study illustrates how frame elements suffer from localization problems similar to those in studies involving solid finite elements. The study relies on the robustness of a force-based beam element (Spacone et al. 1996) that provides a stable response even when the material yields or softens after reaching the peak stress (as in the case of concrete). The element uses a Gauss-Lobatto integration scheme. Though less precise than Gauss integration, the above scheme has the first and last extreme integration points at the end sections of the element and allows for a more accurate prediction of the element ultimate load. A fiber model was used for the section response. All numerical tests are conducted under applied displacements.

Global vs Local Response in Frame Members

The first case is that of a cantilever beam made of elastic-perfectly plastic steel (L = 2000 mm, A = 93000 mm^2 ,f$_y$ = 414 MPa). Figure 1 shows the response of the cantilever under an applied tip displacement. One beam element with different number of integration points (NIP) is used. Figure 1a shows the element response, which is basically unaffected by the number of integration points. Figure 1b, on the other hand, shows how the response of the first integration point (at the fixed end) varies as NIP is increased. Unloading at the end of the analysis was prescribed to distinguish the response for different values of NIP. The reason for this different response is due to the curvature localization at the first integration point. This is the only integration point that yields. As NIP increases, the length of the first integration point (the "plastic hinge" length) decreases. For an equal applied tip displacement, the curvature of the first integration point must therefore increase.

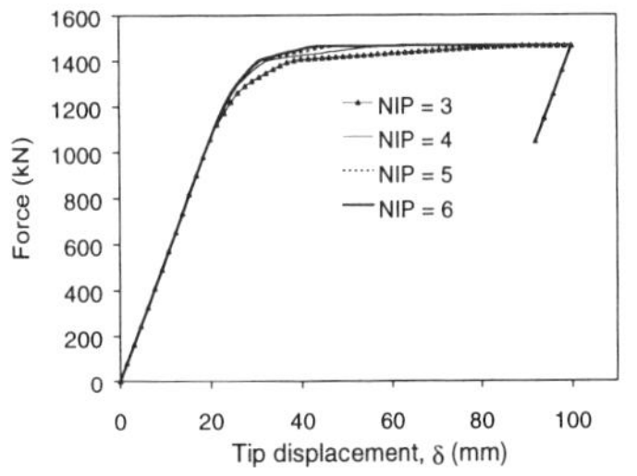

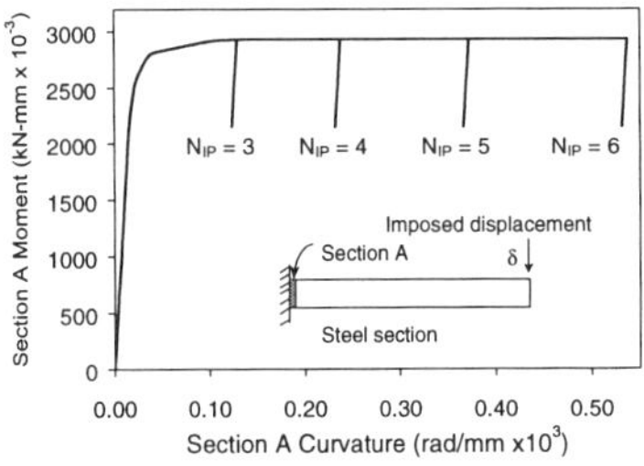

a) lateral load-displacement response   b) first integration point response

Figure 1 Response of cantilever beam made of elastic-perfectly plastic steel

The second example extends the study to a softening element. In this case, a reinforced concrete column under a large axial load was selected. The response is shown in Figure 2. The effect of localization is shown at both the section and the element level. As NIP increases, the length of the first integration decreases and the post-peak softening behavior becomes steeper. The response will eventually snap-back if NIP is further increased. The response of the first section (Figure 2b) does not change in shape, only the magnitude of the curvature increases.

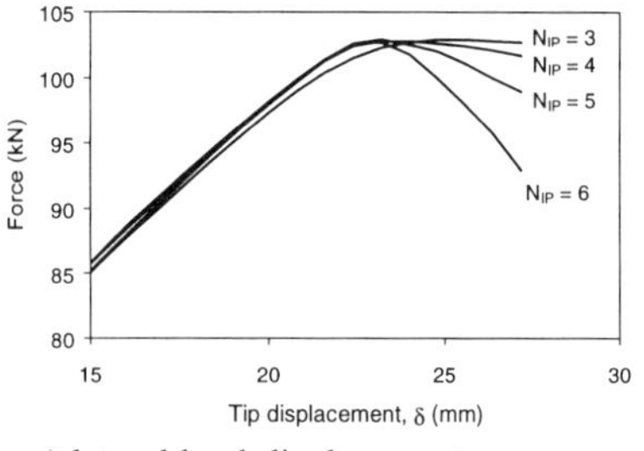

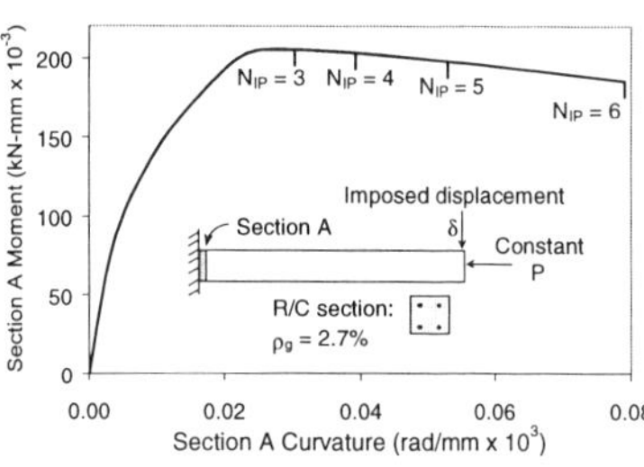

a) lateral load-displacement response   b) first integration point response

Figure 2 Response of a RC column under constant compressive axial force

Localization in Tension and Compression Tests

Most of the research in concrete localization problems has focused on tension regions. More recently, attention has shifted to localization in compression (Ulfkjaer 1998) to investigate if similar problems exist in this case. Two extreme examples have been selected to illustrate the problem in a concrete element: one is a specimen under pure compression, the other is the same specimen under pure tension. The concrete law in compression follows Kent and Park (1971) law. This law has a linear post-peak softening curve until a stress of 20% f_c'. From this point on, the stress remains constant. In tension, the concrete law is linear up to f_t', then decreases linearly until zero stress is reached.

The response of the specimen in pure tension is shown in Figure 3. A weak link has been introduced (by lowering f_t' at one of the integration points) to induce localization at the corresponding integration point. The overall response reflects the localization phenomenon. The force-based beam element used for the study is a series system in which the axial load is constant throughout the element. This implies that if one section has a lower f_t' (due for example to a numerical round-off), this section is the only one to reach f_t' and then lose strength, while all other sections unload elastically. As NIP is increased, the response loses objectivity after the peak as the strain localizes in the weak integration point.

The same behavior is observed in a pure compression test. In this case, f_c was lowered at one of the integration points. Figure 4 shows that while the pre-peak behavior is basically unaffected by NIP, the post-peak behavior depends on the number of integration points. As NIP increases, the post peak slope increases. If the number of integration points were further increased, the response would eventually snap back in both pure tension and pure compression tests. In the pure compression test, only the weak section reaches f_c' and then softens. All other sections unload elastically. The numerical error needed in f_c' (or in f_t') to cause localization is a function of the step size used to drive the tip displacement. The smaller the step increment, the smaller the numerical error that may cause localization.

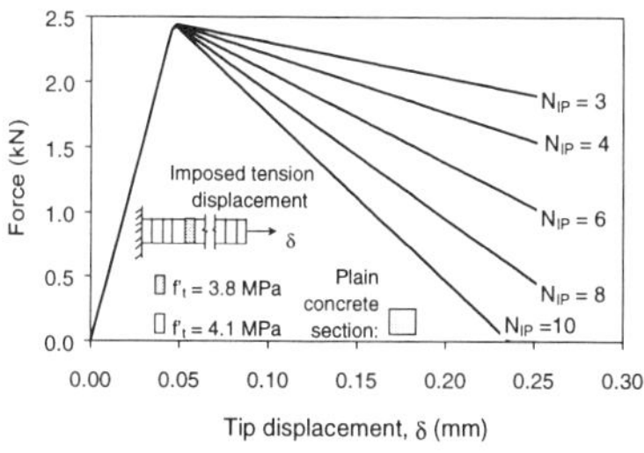

Figure 3 Pure tension test

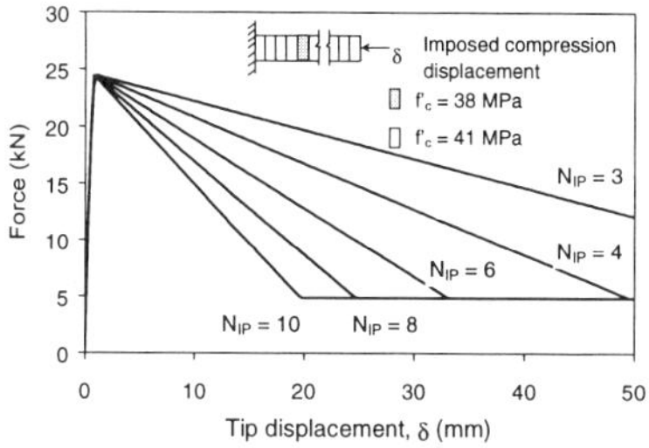

Figure 4 Pure compression test

Summary and Future Work

This paper reports on an ongoing study on deformation localization issues in frame elements. The problem is central to the application of nonlinear frame analysis to performance-based engineering. As long as strains localize in mesh-dependent bands, the local and global responses of the structure are not unique and strain and curvature demands cannot be predicted. A few simple examples were presented to illustrate the nature of deformation localization in frame elements. Future work will focus on searching for problem solutions based on the experience from strain localization in tension in solid finite elements.

References

FEMA. (1997). *NEHRP guidelines for the seismic rehabilitation of buildings.* Report No. FEMA-273. Washington, D.C.

Bazant, Z.P. and Planas, J. (1997). *Fracture and size effect in concrete and other quasibrittle materials.* CRC Press, Boca Raton, Florida.

Kent, D.C., and Park, R. (1971). "Flexural members with confined concrete." *Journal of structural division, ASCE,* 97(7), 1964-1990.

Spacone, E., Filippou, F.C., and Taucer, F.F. (1996). "Fiber beam-column model for nonlinear analysis of R/C frames. I: Formulation. II: Applications" *Earthquake engineering and structural dynamics,* 25(7), 711-742.

Ulfkjaer, J.P. (1998). "Experimental investigation of over-reinforced concrete beams of three different types of concrete and at two different size scales." *FRAMCOS-3,* Gifu, Japan, October, 1253-1260.

PRACTICAL DESIGN METHOD FOR
FLEXIBLY JOINTED FRAMES WITH LRFD

Norimitsu Kishi [1], *M. ASCE, Wai-Fah Chen* [2], *M. ASCE*
Masato Komuro [3], *M. JSCE and Ken-ichi G. Matsuoka* [4], *M. JSCE*

ABSTRACT

To establish a practical design method for flexibly jointed frames as well as fully rigid frames based on the AISC-LRFD specification, a simplified method for the estimation of column design moments is proposed and its applicability is numerically discussed. In this paper, in order to convert a nonlinear connection stiffness into a linear one, the beam-line method is applied. Two connection stiffnesses are taken, one of which is a secant connection stiffness for first-order elastic analysis of frames, and the other is a tangent one for estimation of the column K-factor. The column end moments are estimated by using only B_2 factor (P-Δ effect). Comparing the results with those obtained from the second-order elastic analysis for a flexibly jointed frame example, it can be shown that proposed simplified method gives reasonable values and will be useful for a practical frame design.

1. INTRODUCTION

In AISC-LRFD specification (1994), two basic types of construction and associated design assumptions are permissible which are Type FR (fully restrained) and Type PR (partially restrained). The former type commonly designates as rigid frames. The latter type refers to frame with beam to column connections those have insufficient rigidity to maintain the angle between intersecting members. Type FR construction can be designed by using LRFD Specification. However, Type PR construction has been limited to simple framing design because no design procedure considering connection flexibility based on LRFD is provided in the specification.

In order to establish a design method of the PR construction (flexibly jointed frames) based on the LRFD specification, many researchers have been studied so far. The authors also have been working on the issue. To this effect, they developed a data-base of beam-to-column connections and investigated the applicability of three-parameter power model to evaluate the nonlinear stiffness of semi-rigid connections. They also provided a method to find column effective length factor in flexibly jointed frames by means of an alignment chart.

In this paper, in order to evaluate column design moment M_u for flexibly jointed frames considering the nonlinear connection stiffness and second-order

---

[1] Prof., Civ. Engrg., Muroran Inst. of Tech., Muroran, 050-8585 Japan.
[2] Prof. and Head of Struct., School of Civ. Engrg., Purdue Univ., West Lafayette, IN 47907.
[3] Res. Asso., Civ. Engrg., Muroran Inst. of Tech., Muroran, 050-8585 Japan.
[4] prof., Civ. Engrg., Muroran Inst. of Tech., Muroran, 050-8585 Japan.

effects, a simplified B_1/B_2 method is applied and its applicability is numerically discussed. Here, beam-line method is used to accommodate nonlinear connection stiffness. An applicability of the proposed method is investigated comparing the results with those obtained from a second-order elastic analysis directly considering nonlinear connection stiffness (Goto and Chen, 1987).

2. B_1/B_2 METHOD

According to AISC-LRFD specification, column design end moment M_u for rigid frames is obtained by using first-order column end moments as follows:

$$M_u = B_1 \, M_{nt} + B_2 \, M_{lt} \tag{1}$$

where,

M_{nt}, M_{lt} : first-order column end moment due to no lateral and lateral translation loads, respectively;

B_1 : $P-\delta$ moment amplification factor;

$$B_1 = \frac{C_m}{1 - P_u/P_{ek}} \geq 1.0 \tag{2}$$

B_2 : $P-\Delta$ moment amplification factor;

$$B_2 = \frac{1}{1 - \sum P_u/\sum P_{ek}} \quad or \quad B_2 = \frac{1}{1 - \sum P_u \Delta_{oh}/\sum HL} \tag{3), (4}$$

C_m : $0.6 - 0.4(M_1/M_2)$, where M_1 and M_2 are smaller and larger end moments in column, respectively;

P_{ek} : Euler buckling load;

$\sum P_u$: total required axial forces surcharged to all columns in the story;

Δ_{oh} : translation of the story under consideration;

$\sum H$: sum of all story horizontal forces producing Δ_{oh}.

It is seen that Eq. (1) is formulated by using a superposition principle for two divided loading conditions (no lateral translation and lateral translation loads). However, this principle does not logically fit with the first-order elastic analysis considering nonlinear connection stiffness. Here, to easily estimate column design moments in flexibly jointed frames, it is proposed that only B_2 factor is used for the frames such as:

$$M_u = B_2 \, M_{fu} \tag{5}$$

where a suffix "f" in right-hand side means first-order elastic analysis, and M_u is column bending moment of unconstrained flexibly jointed frames under given factored loads. This method is formulated based on the concepts as mentioned below:

1) In elastic analysis of frames under no lateral translation loads, B_1 factor will be unity (1.0) because the columns usually bent in double curvature.

2) Then, the proposed B_2 method will provide a conservative design comparing with the conventional B_1/B_2 method following Eq. (1).

3) The design procedure using this proposed method is simpler than that using the conventional method. Because only one time of first-order elastic analysis for estimation of the column bending moments is demanded without dividing two loading systems in case of rigid frames.

In this paper, Eq. (3) is used to estimate B_2 factor, in which Euler buckling load P_{ek} is obtained by using an alignment chart with modified relative stiffness factors to accommodate nonlinear connection stiffness.

3. OUTLINE OF THE PROPOSED ANALYSIS PROCEDURE

Having developed the appropriate estimation method on column design moments in flexibly jointed frames, the analysis procedure can now be outlined in the following steps.

1. Determine moment-rotations curve for semi-rigid connections used in frames.

2. Determine the linearized connection stiffnesses R_{ks} and R_{kt} by using beam-line method under factored loads.

3. Determine the column moments M_{fu} of flexibly jointed frames under factored loads applying first-order elastic analysis method incorporated with the linearized connection stiffness R_{ks}.

4. Determine the value of Euler buckling load P_{ek} for each column by using alignment chart, in which the modified relative stiffness factors G' considering the linearized connection stiffness R_{kt} is used in place of the relative stiffness factors G for rigid frame cases.

5. Evaluate the amplification factor B_2 according to Eq. (3).

6. The column design moments in flexibly jointed frames considering the nonlinear connection stiffness and second-order effects are determined by using Eq. (5).

Determining column design moment M_u and Euler buckling load for each column, the column stability in flexibly jointed frames by using an interaction equation H1-1 in AISC-LRFD specification is checked.

4. NUMERICAL STUDY

4.1. A semi-rigid frame example

In this paper, a 1-bay 3-story frame example is used for numerical investigation as shown in Fig. 1, in which member sections, story heights and width are shown in this figure. Element and node numbers are shown in boxes and circles, respectively. The intensities of applied load are listed in Table 1. The load combination for factored load is chosen for 1.2D + 0.5L + 1.3W as per AISC-LRFD specification. The frame spacing is taken as 7.62 m (300 in). Then, the distributed

Table 1 List of load intensities (kPa).

	dead load (D)	live load (L)	wind load (W)
roof (R)	0.958	0.958	0.958
floor (F)	3.257	1.916	

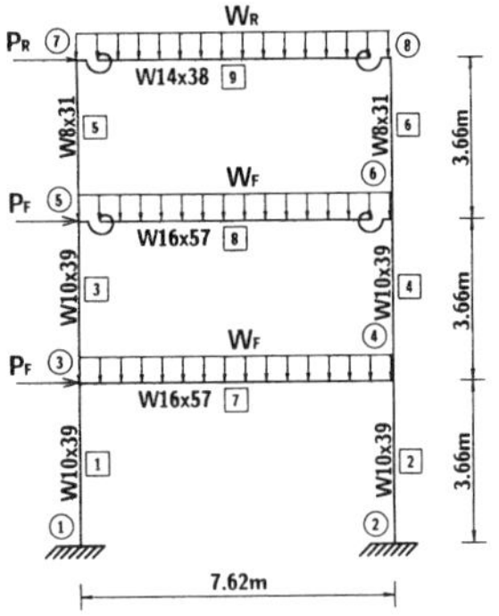

Figure 1. 1-bay 3-story frame

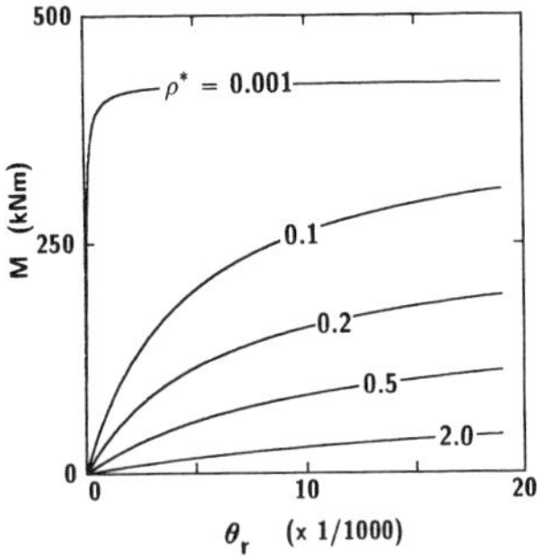

Figure 2. $M - \theta_r$ curves for numerical analysis (W16×57)

load (W_R, W_F) and the concentrated loads (P_R, P_F) become as follows:
W_R = 12.41 kN/m, W_F = 37.08 kN/m, P_R = 17.37 kN and P_F = 34.73 kN

4.2. Applied moment-rotation curves of connections

Nonlinear moment-rotation curves of semi-rigid connections are evaluated by using three-parameter power model (Kishi and Chen, 1990). The model is composed of initial connection stiffness R_{ki}, ultimate moment capacity of connection M_c and shape parameter n. In this study, the ultimate moment capacity of connection M_c is varied from $0.2M_p$ to $1.0M_p$ according to the level of R_{ki}, in which M_p is the beam plastic moment (M_p = 250.2 kNm for W14×38 and 427.2 kNm for W16×57). The shape parameter n is kept constant $n = 1$. The initial connection stiffness R_{ki} is nondimensionalized with reference to the beam bending stiffness (EI_b/L_b) as follows:

$$\rho^* = \frac{EI_b}{L_b R_{ki}} \tag{7}$$

Figure 2 shows five moment-rotation curves considered here in case connected to a W16×57 beam, in which ρ^* is varied ranging from nearly rigid connection (ρ^* = 0.001) to nearly pinned connection (ρ^* = 2.0).

4.3. Numerical results

Figure 3 shows the maximum end moment for each column obtained by the proposed method and the distribution of its nondimensionalized value M^* with reference to the value obtained from second-order elastic analysis for each flexibly jointed frame. Since, generally, windward and leeward columns will be designed using the same sectional forces, an applicability of proposed method is discussed using the absolute maximum column end moment for each story. In this frame, the end moments for leeward columns (column number: 2, 4 and 6) are considered.

From this figure, it is seen that all the leeward column end moments m^* are almost equal to or greater than unity for all semi-rigid connection considered here. This result implies that the proposed method gives conservative column end moments for flexibly jointed frames with arbitrary connection stiffness.

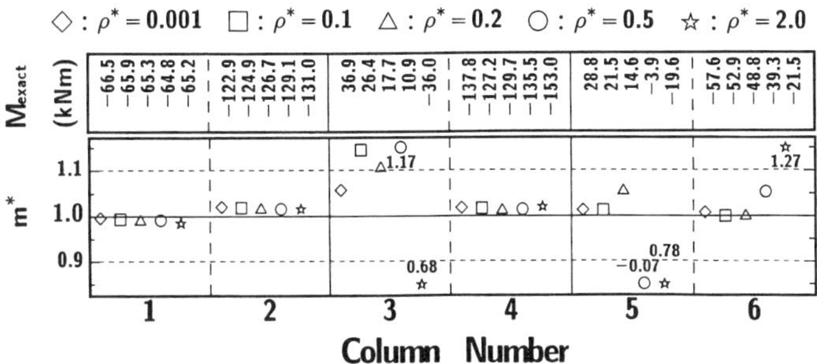

Figure 3. Distribution of column end moment m^*

5. CONCLUSION

In order to easily evaluate column design moments for flexibly jointed frames, the simplified method using only B_2 factor in B_1/B_2 method is numerically discussed. From this study, it is seen that the proposed method can be used for practical frame design.

Local Buckling Analysis of Steel Bridge Pier under Cyclic Loading

E. Yamaguchi,[1] Y. Goto,[2] M. ASCE, M. Hayashi,[3] and Y. Kubo[4]

Abstract

After the 1995 Hyogo-ken Nanbu Earthquake, much research is conducted in Japan so as to improve the seismic performance of steel bridge piers. In the present study, we deal with a steel bridge box pier with round corners, which is often employed in an urban area from an aesthetic viewpoint. In particular, we conduct the large deformation analysis of this bridge pier by the finite element method. For constitutive relationships, the three-surface model as well as the conventional J_2-plasticity model is utilized. The analyses reproduce the local buckling near the lower end of the pier. The comparison of the numerical results with experimental data confirms the validity of the three-surface model for the analysis of a steel bridge box pier with round corners.

Introduction

Quite a few steel bridge piers were damaged during the 1995 Hyogo-ken Nanbu Earthquake, Japan. Since then, much research effort has been made to improve behaviors of steel bridge piers during a severe earthquake. Since the damage modes observed in the aftermath of the earthquake are reproducible under the loading condition of constant vertical load and cyclic horizontal load, many experiments of steel bridge piers have been carried out under such loading condition.

Although experimental data is valuable, experiments are so expensive that the number of tests has to be limited. Because of this, the finite element analysis of steel bridge piers has been also conducted, and good agreement between numerical results and experimental data are reported for bridge piers with a box section and a circular

---

[1] Assco. Prof., Dept. of Civ. Engrg., Kyushu Inst. of Tech., Kitakyushu 804-8550, Japan

[2] Prof. Dept. of Civil Engrg., Nagoya Inst. of Tech., Nagoya 466-8555, Japan

[3] Grad. Student, Dept. of Civ. Engrg., Kyushu Inst. of Tech., Japan

[4] Prof., Dept. of Civ. Engrg., Kyushu Inst. of Tech., Japan

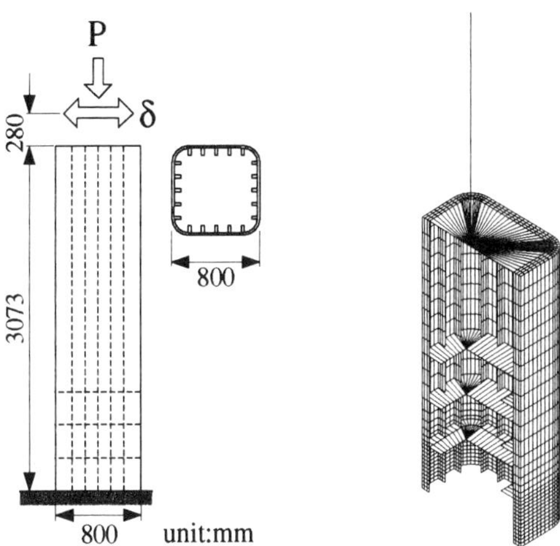

Figure 1. Steel Bridge Pier   Figure 2. Finite Element Mesh

section (e.g. Goto et al. 1998).

The corners of the box section of a bridge pier are often rounded in an urban area in Japan from an aesthetic viewpoint. The structural characteristics of a box beam-column with round corners have been investigated to some extent in the past (Watanabe et al. 1992). However, the performance of this type of bridge pier under cyclic loading does not seem to be known adequately, so that the prediction of its performance during a severe earthquake is still a difficult task.

In the present study a bridge box pier with round corners subjected to constant vertical load and cyclic horizontal load is analyzed. The finite element method is employed and the bridge pier is modeled by shell and beam elements. The effects of large deformation and material nonlinearity are included, and two constitutive models are used. The numerical results thus obtained are compared with experimental data (Yoshizaki et al. 1997).

<u>Bridge Pier and Loading Condition</u>

The steel bridge pier (Fig. 1) employed in the experiment (Yoshizaki et al. 1997) is considered herein. The corners of the cross section are rounded with a radius of 142 mm. The plate thickneeses of a flange and a web are both 7 mm, and the thickness of each stiffener is 8 mm. The axial compressive force P is 15% of the nominal squash load of the cross section, and the horizontal load H is applied to trace the predetermined path of the horizontal displacement δ at the loading point: $0 \rightarrow +\delta_y \rightarrow -\delta_y \rightarrow +2\delta_y \rightarrow -2\delta_y \rightarrow \cdots$, where δ_y is the displacement at the initial yielding due to the horizontal force.

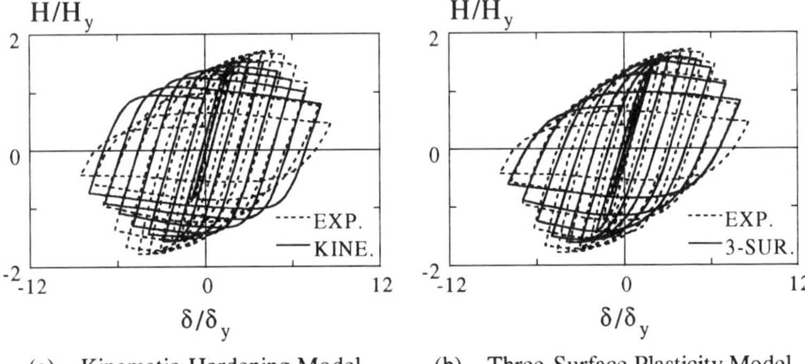

(a)   Kinematic-Hardening Model   (b)   Three-Surface Plasticity Model

Figure 3. Horizontal Load-Displacement Curves

<u>Analysis Model</u>

Due to symmetry, only a half of the bridge pier needs be considered. Using 2966 shell elements and 33 beam elements, the bridge pier is modeled as shown in Fig. 2. Finer mesh is applied to the lower portion of the bridge pier.

The material behavior is assumed to be described by the plasticity theory of von Mises type. We consider two plasticity models in this study: the conventional kinematic-hardening model and the three-surface model.

The three-surface model has been proposed by Goto et al. (1998). The model is an extension of the two-surface model (Dafalias and Popov 1976) and can express the material features of structural steel such as the yield plateau and the reduction of elastic region. The bounding surface, the yield surface and the discontinuous surface are defined to this end. The yield surface changes its size as it moves within the bounding surface. The discontinuous surface lies between the bounding and yield surfaces, and it controls the hardening modulus. Through various numerical studies, the values of some material parameters have been identified, so that in practice all the material parameters can be determined only by the result of a uniaxial material test.

<u>Numerical Results</u>

The numerical result of the horizontal load-displacement relationship at the loading point is presented together with the experimental data (Yoshizaki et al. 1997) in Fig. 3, where H_y is the horizontal force corresponding to δ_y. In the case of the kinematic-hardening model, the effect of the constant elastic region is clearly observed: the structural stiffness in the reversed loading paths is overestimated. By the three-surface model, much better hysteresis loops are obtained: the shape of each loop is quite similar to that of the experiment and the discrepancy of the ultimate strength is only about 7% whereas the kinematic-hardening model underestimates the strength by 11%.

The numerical result due to the three-surface model indicates that local buckling

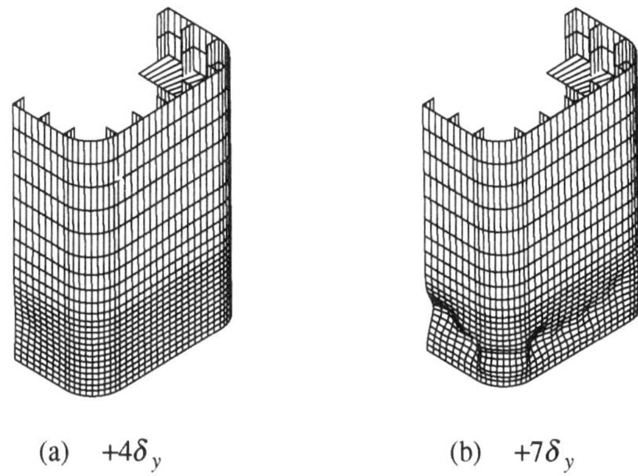

(a) $+4\delta_y$ (b) $+7\delta_y$

Figure 4. Deformed Configuration

appears at the loading-point displacement of around $+4\delta_y$. The out-of-plane displacement across the stiffened plates occurs at around $+7\delta_y$. Figure 4 shows the deformed configuration of the specimen at these stages. These deformation states compare very well with the experimental observations.

Concluding Remarks

We have conducted the finite element analysis of a bridge box pier with round corners under cyclic horizontal load. Different structural responses are obtained for the different plasticity models, and the superiority of the three-surface model over the other plasticity model is confirmed. It may be stated then that the finite element analysis with the three-surface model is a reliable means to evaluate the structural behavior of a bridge box pier with round corners under seismic loading.

References

Dafalias, Y.E. and Popov, E.P. (1976). "Plastic internal variables formalism of cyclic plasticity." *J. Applied Mech.*, 43, 645-651.

Goto, Y, Wang, Q., Takahashi, N. and Obata, M. (1998). "Three surface cyclic plasticity model for FEM analysis of steel bridge piers subjected to seismic loading." *J. Struct. Mech. Earthquake Eng.*, JSCE, 591/I-43, 189-206.

Watanabe, E., Sugiura, K., Mori, T. and Suzuki, I. (1992). "Strength and ductility of short stiffened box beam-column with round corners." *J. Struct. Eng.*, JSCE, 38A,143-154.

Yoshizaki, N., Murayama, T., Yasunami, H., Natori, T. and Tuji, H. (1997). "An experimental study on the cyclic elasto-plastic behavior of steel box column with round corners." *Proc. of Nonlinear Numerical Analysis and Seismic Design of Steel Bridge Piers*, JSCE, 339-346.

INELASTIC CRITICAL LOADS BY EIGENVALUE ANALYSIS

Ronald D. Ziemian,[1] Associate Member, ASCE

Abstract

This paper presents the details of an iterative procedure for calculating inelastic critical loads by performing a series of eigenvalue analyses. An example is also provided that compares results obtained from this approach with those of an elastic critical load analysis.

Introduction

As an alternative to performing a complete nonlinear analysis, a great deal of insight into the limit state behavior of many frame and truss systems can be obtained from a critical load analysis (McGuire, et al. 1999). By employing a finite element approach, the elastic critical load, often referred to as the buckling or bifurcation load, is computed by solving the eigenvalue problem that corresponds to the equations of equilibrium at the critical state

$$[[K_{E,ff}] + \lambda[K_{G,ff}]]\{\Delta_f\} = \{0\} \tag{1}$$

where $[K_{E,ff}]$ is the elastic stiffness matrix for the structure's unsupported degrees of freedom; and $[K_{G,ff}]$ is the corresponding geometric stiffness matrix and is based on the distribution of element forces computed for a given applied load $\{P\}$. After solving (1), the minimum positive eigenvalue λ_{min} is the elastic critical load ratio ($\beta_E = \lambda_{min}$ with an elastic buckling load of $\beta_E\{P\}$) and its associated eigenvector $\{\Delta_f\}$ is the buckled shape.

This approach neglects material nonlinear behavior and assumes the relative distribution of element forces is the same at all ratios of applied load. Hence, (1) will calculate the critical load of an Euler pinned end column as $P_{cr} = \pi^2 EI/L^2$ regardless of whether the column is long and slender or short and stocky. It is generally accepted that there is a difference in the stability of these two types of columns and it can be explained through the use of tangent modulus theory. Since it is known that a short, stocky column will partially yield due to the combination of

---

[1] Associate Professor, Department of Civil Engineering, Bucknell University, Lewisburg, PA 17837

430

resisting higher compressive loads and the presence of residual stresses from the manufacturing process, the theory postulates that bifurcation will occur at a load of $P_{cri} = \pi^2 E_t I/L^2$. In this case, the flexural stiffness of the column is a function of the axial force it resists. The stiffness is elastic for an axial force below some proportional limit with $E_t = E$. Above this point the column is partially yielded or inelastic with a gradually decreasing stiffness represented by the tangent modulus with $E_t < E$. An empirical equation often used to represent the performance of steel columns in the inelastic range, and given in the *Guide to Stability Design Criteria for Metal Structures* (1998), is

$$E_t = E \qquad\qquad \text{for } P/P_y \leq 0.5$$

$$E_t = 4E\left(1 - \frac{P}{P_y}\right)\frac{P}{P_y} \qquad \text{for } 0.5 < P/P_y \leq 1.0 \tag{2}$$

where P/Py is the ratio of the axial force to the squash load ($\sigma_y A$).

The proceeding eigenvalue approach for determining elastic critical loads can be modified to include the material nonlinear behavior defined in (2). To account for the reduction in material stiffness at the critical load, $[K_{E,ff}]$ is replaced by the inelastic stiffness of the structure $[K_{I,ff}]$. Since the distribution of element forces, and hence $[K_{I,ff}]$ and $[K_{G,ff}]$, are not linear functions of the given applied load $\{P\}$, the objective of an inelastic critical load analysis is to determine the minimum load ratio β that will result in the following equation having an eigenvalue of $\lambda = 1$

$$\left[\left[K_{I,ff}(\beta P)\right] + \lambda\left[K_{G,ff}(\beta P)\right]\right]\left\{\Delta_f\right\} = \{0\} \tag{3}$$

The product of the applied load $\{P\}$ and the minimum value for β satisfying the requirements of (3) is the inelastic critical load, $\beta_I\{P\}$. The eigenvector corresponding to $\lambda = 1$ represents the buckled configuration. The following section contains a detailed procedure for systematically prescribing different values of β and performing the associated eigenvalue analyses until an inelastic critical load ratio β_I is calculated to within a user-defined tolerance. Two essential parts of this algorithm are the use of an iterative nonlinear analysis to determine the force distribution for calculating $[K_{I,ff}]$ and $[K_{G,ff}]$, and an interpolation scheme for predicting the inelastic critical load ratio.

Procedure

a. For the given applied load $\{P\}$, calculate element end forces by performing a first-order elastic analysis. Retain the first-order elastic global stiffness matrix for the unsupported degrees of freedom $[K_{E,ff}]$.

b. For all elements, determine the magnitude of the largest ratio of compressive axial force to squash load: $\qquad \alpha_{max} = \left| \max_{1 \leq ele \leq n}(P_{ele}/P_{y,ele}) \right|$.

Note that if no elements are in compression, the procedure should stop and a message should be returned indicating that no critical load will be calculated.

c. Using the element forces calculated in step *a*, assemble the global geometric stiffness matrix for the unsupported degrees of freedom, $[K_{G,ff}]$.

d. Determine the elastic critical load ratio β_E, which is the minimum positive eigenvalue λ_{min} that satisfies equation (1).

   If $\beta_E \leq 0.5/\alpha_{max}$, elastic critical buckling controls. In this case, the procedure should stop and an appropriate message reported along with the elastic critical load ratio β_E and the buckled shape, which is the eigenvector $\{\Delta_f\}$ corresponding to λ_{min}.

e. Estimate the lower bound on the inelastic critical load ratio as $\beta_I^{low} = 0.5/\alpha_{max}$. Note that for a factored applied load of $\beta_I^{low}\{P_f\}$, the corresponding minimum eigenvalue λ_{min}^{low} will equal β_E/β_I^{low}, a value that exceeds unity. Estimate the upper bound on the inelastic critical load ratio as $\beta_I^{up} = \min(\beta_E, 1.0/\alpha_{max})$.

f. Set the applied load ratio β_I equal to β_I^{up}.

g. For an applied load of $\beta_I\{P_f\}$, calculate element end forces by performing an iterative first-order inelastic analysis. Note that this analysis does not include the use of geometric stiffness matrices and hence, only includes material stiffness reduction according to the tangent modulus of each element, see equation (2). From the last iteration, retain the global tangent stiffness matrix for the unsupported degrees of freedom $[K_{I,ff}]$.

h. Using the element forces calculated in step g, assemble the global geometric stiffness matrix, $[K_{G,ff}]$. Note that this may be different from a previously calculated $[K_{G,ff}]$ because the internal force distribution may have changed.

i. Determine the minimum positive eigenvalue λ_{min} that satisfies equation (3).

   If λ_{min} equals unity or is within the acceptable tolerance, the inelastic critical load ratio has been determined. In this case, the procedure should stop and an appropriate message returned along with the inelastic critical load ratio β_I and the buckled shape, which will be the eigenvector $\{\Delta_f\}$ corresponding to λ_{min}.

j. If λ_{min} calculated in step i is less than unity by more than the acceptable tolerance, set λ_{min}^{up} equal to λ_{min}. In this case the inelastic critical load ratio is now known to be bounded by β_I^{low} and β_I^{up}.

   If λ_{min} exceeds unity by more than the acceptable tolerance, then β_I, which is equivalent to β_I^{up}, did not provide a true upper bound on the inelastic critical load ratio. In this case, steps e through i should be repeated with β_I set equal to the elastic critical load ratio β_E. Since this will provide a λ_{min} that must be less than unity, set λ_{min}^{up} equal to this λ_{min}. The inelastic critical load ratio is now bounded by β_I^{low} and $\beta_I^{up} = \beta_E$.

k. Using the data pairs $(\beta_I^{low}, \lambda_{min}^{low})$ and $(\beta_I^{up}, \lambda_{min}^{up})$, use linear interpolation to determine a load ratio β_I that should provide a λ_{min} that equals unity.

l. Repeat steps g through i.

m. If λ_{min} is not within the acceptable tolerance of unity, update the bounds on the inelastic critical buckling load ratio accordingly. That is, if λ_{min} is less than unity, refine the upper bound pair to $\beta_I^{up} = \beta_I$ and $\lambda_{min}^{up} = \lambda_{min}$, otherwise refine the low bound pair to $\beta_I^{low} = \beta_I$ and $\lambda_{min}^{low} = \lambda_{min}$.

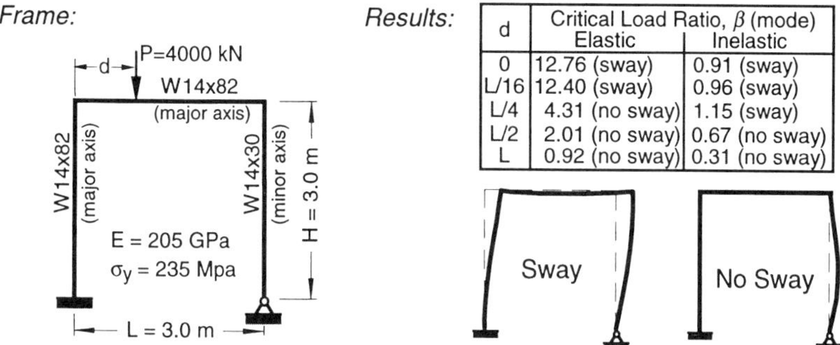

d	Critical Load Ratio, β (mode)	
	Elastic	Inelastic
0	12.76 (sway)	0.91 (sway)
L/16	12.40 (sway)	0.96 (sway)
L/4	4.31 (no sway)	1.15 (sway)
L/2	2.01 (no sway)	0.67 (no sway)
L	0.92 (no sway)	0.31 (no sway)

Fig. 1. Example of critical load analyses

n. Using data pairs (β_I, λ_{min}) from the current and previous iterations, use the secant method of linear interpolation/extrapolation (Chapra and Canale, 1998) to determine a load ratio β_I that should provide a λ_{min} that equals unity. In no case should the newly calculated β_I be permitted to exceed the bounds of β_I^{low} or β_I^{up}. If the predicted β_I attempts to exceed a bound, this indicates that the relationship between β and λ_{min} is highly nonlinear and it is suggested that the secant method be re-performed with the data pairs mapped into $\log_{10}$ space.

o. Repeat steps *l* through *n*, which includes steps *g* through *i*, until either the inelastic critical buckling load ratio is computed and reported in step *i* or a maximum number of iterations has been performed. If the latter occurs, the user should be appropriately notified and no ratio reported.

Example and Summary

The above procedure is implemented within the structural analysis computer program MASTRAN2. This software is included with the manuscript by McGuire et al. (1999). Figure 1 provides an example of the results that may be obtained. All members are assumed fully restrained against out-of-plane behavior. The beam and left column are oriented with their webs in the plane of the frame and the right column with its web perpendicular to the plane. All connections are rigid and members are discretized into four elements. It is interesting to note that the buckled configuration depends not only on the location of the load but also on the type of analysis employed. In all cases, the inelastic critical load was determined in five or less iterations with an analysis time of approximately thirty seconds.

References

McGuire, W., Gallagher, R.H., and Ziemian, R.D. (1999). *Matrix Structural Analysis*, Second Edition, John Wiley and Sons, Inc., New York, N.Y.

Guide to Stability Design Criteria for Metal Structures (1998). Fifth Edition, T.V. Galambos, John Wiley and Sons, Inc., New York, N.Y.

Chapra, S.C., and Canale, R.P. (1998). *Numerical Methods for Engineers*, Third Edition, McGraw-Hill, New York, N.Y.

Behavior and Design of Laterally Braced Inelastic Columns

E.M. Lui[1] and J. Lee[2]

Abstract

This paper presents a numerical study of the behavior and load-carrying capacity of laterally braced geometrically imperfect inelastic columns under uniform compression. Although laterally braced compression members are commonly used in civil engineering structures, the study of the behavior of these members is quite limited in that most of the reported studies are based on elastic behavior (Winter, 1958; Timoshenko and Gere, 1961; Plaut, 1993, Plaut and Yang, 1993, Zhang, et al, 1993) In the event that inelasticity is considered, the investigation is limited to cases with equally spaced braced points (Ales and Yura, 1993; Clarke and Bridge, 1993).

The study reported in this paper employs the pseudo load method of inelastic analysis. The compression member to be analyzed is pinned at both ends and is supported at some intermediate point by a spring brace. The location of the spring brace can be anywhere within the length of the member and is not necessarily at midspan. Although the spring brace is assumed to be elastic, the column can experience inelasticity. The inelastic behavior of this braced compression member as well as the lateral bracing requirements and the effect of brace location on the ultimate strength of the column are investigated.

The load-carrying capacity of a compression member is very much affected by its slenderness ratio. To investigate the slenderness effect on compression strength, different values of slenderness parameter $\lambda = (KL/r\pi) \sqrt{(F_y/E)}$ (where K is the effective length factor, taken as unity in this study, L is the member length, r is the radius of gyration, F_y is the yield stress and E is the modulus of elasticity) are used in this study. The load-carrying capacity of each member is obtained as the peak point of the nonlinear load-deflection curve generated using the pseudo load method of inelastic member analysis. Based on the results of this study, observations in regard to the inelastic response of braced compression members are made.

---

[1] Associate Professor, Department of Civil & Environmental Engineering, Syracuse University, Syracuse, NY 13244-1190
[2] Graduate Assistant, Department of Civil & Environmental Engineering, Syracuse University, Syracuse, NY 13244-1190

Analysis Model

The model used for the numerical analysis is shown in Figure 1. The compression member, pinned at both ends, is braced at an intermediate point by a spring brace. The stiffness of the brace is denoted as k_s. To simulate geometrical imperfections that are unquestionably present in a real member, an initial crookedness in the form of a half sine curve is introduced to the model. The maximum out-of-straightness of the member is taken as $0.001L$ at midspan. The initial deflection from the member chord due to member crookedness is denoted as d_o, and any additional deflection due to member buckling is denoted as d.

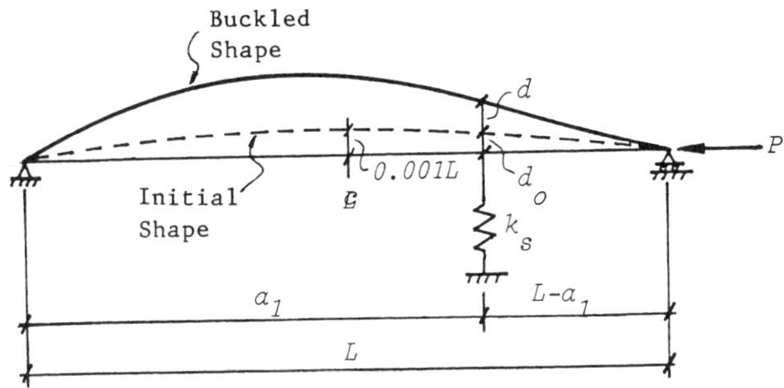

Figure 1. Analysis Model

Analysis Method

A load-deflection analysis using the pseudo load method of inelastic analysis (Lui and Zhang, 1990) is employed to obtain ultimate compressive strengths P_u of the member shown in Figure 1 for different values of brace stiffness k_s, slenderness parameter λ, and brace location a_1/L. An incremental load control method is used to trace the nonlinear load-deflection curve and P_u is obtained as the peak point of this curve. The nonlinear behavior of the system is attributed to both geometrical and material effects. In the pseudo load method of inelastic analysis, geometrical nonlinearity is accounted for by the application of pseudo in-span and member end loads to the structure and material nonlinearity is accounted for by the use of a predetermined cross-section moment-curvature-thrust relationship in forming the member stiffness matrix and in calculating the pseudo loads.

Numerical Studies

All analyses reported herein were carried out using the following material properties: E=29,000ksi (200×10^3MPa) , and F_y=36ksi (248MPa). Figures 2a and 2b show the variation of P_u/P_e (where $P_e=\pi^2EI/L^2$) with brace location a_1/L and brace stiffness k_sL/P_e for two different values λ. λ=1 represents the case when the column experiences more inelasticity than that for λ=2 when P_u is reached. The following observations can be made: (1) The ultimate strength of a braced column increases as the brace is moved towards midspan indicating that the brace is more effective when placed at the column midspan. However, this increase in effectiveness is less pronounced when the column becomes more inelastic. (2) Regardless of whether $\lambda=1$ or 2, a stationary value for P_u exists for a centrally braced column once the brace stiffness exceeds some values. The brace stiffness at which no increase in P_u will occur is referred to as the ideal brace stiffness k_{id}. Note that k_{id} only exists when the brace is placed at locations corresponding to the natural stationary points of the braced member's higher instability modes. The value of the ideal brace stiffness decreases when the column becomes more inelastic. The effect of the brace having an ideal stiffness is identical to that of a rigid support as far as instability of the braced member is concerned.

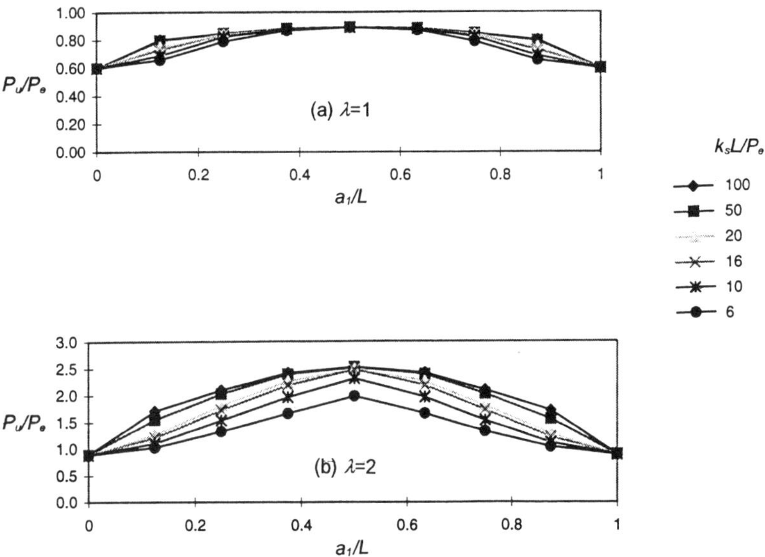

Figure 2. Effects of Bracing Location and Bracing Stiffness on Member Strength

Figures 3a to 3d show the variation of the deflection ratio d_o/d as a function of the k_sL/P_e for various values of P_u/P_{rigid} (where P_{rigid} is the ultimate strength of the compression member when the brace stiffness is infinite) for four cases ($\lambda=1$, $a_1/L=0.25$ and 0.50, and $\lambda=2$, $a_1/L=0.25$ and 0.50). As can be seen from these graphs, d_o/d varies as a linear function of k_sL/P_e for a given value of P_u/P_{rigid}. These graphs can be used to determine the brace force ($F_s=k_sd$) for given values of λ, k_s, and a desired value of P_u/P_{rigid}.

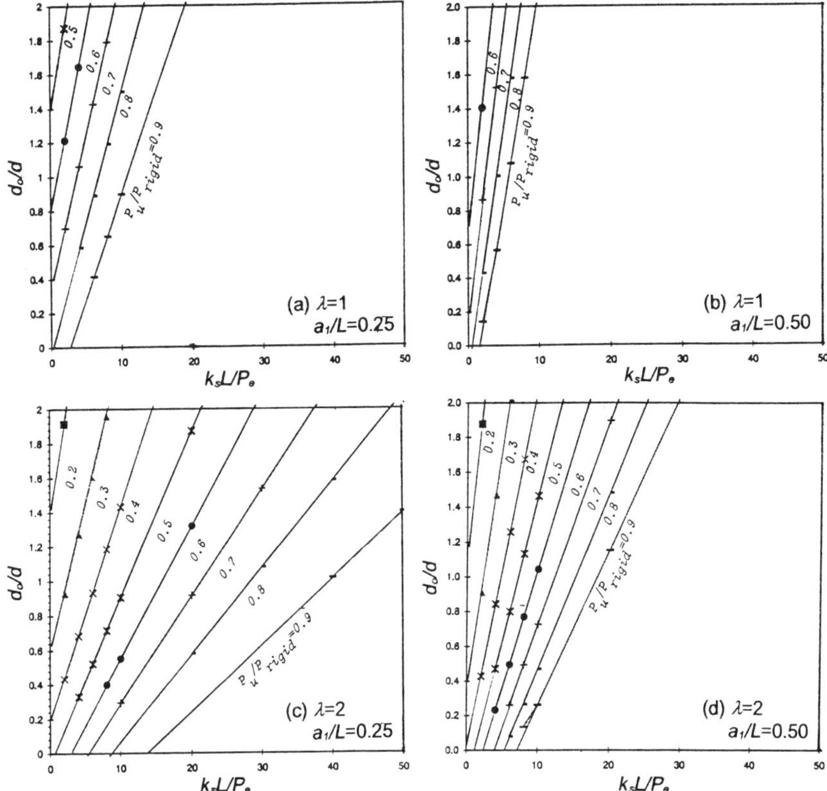

Figure 3. Effects of Bracing Stiffness and Member Strength on Deflection Ratio

References

- Ales, J.M, Jr., and Yura, J.A. (1993) "Bracing Design for Inelastic Structures," *Is Your Structure Suitably Braced*, SSRC, 29-37.
- Clarke, M.J., and Bridge R.Q. (1993) "Bracing Force and Stiffness Requirements to Develop the Design Strength of Columns," *Is Your Structure Suitably Braced*, SSRC, 175-86.
- Plaut, R.H. (1993) "Requirements for Lateral Bracing of Columns with Two Spans," *J. Struct. Engrg.*, ASCE, **119**(10), 2913-2931.
- Plaut, R.H., and Yang, J.-G. (1993) "Lateral Bracing Forces in Columns with Two Unequal Spans," *J. Struct. Engrg.*, ASCE, **119**(10), 2896-2911.
- Lui, E.M. and Zhang, C.-Y. (1990) "Nonlinear Frame Analysis by the Pseudo Load Method," *Comp. & Struct.*, **37**(5) 707-716.
- Timoshenko, S., and Gere, J.M. (1961) *Theory of Elastic Stability*, McGraw-Hill, New York.
- Winter, G. (1958) "Lateral Bracing of Columns and Beams," *J. of Struct. Div., Proc.* ASCE, **84**(2), 1561-1-1561-22.
- Zhang, H.-Y., Beliveau, J.-G., and Huston, D.R. (1993) "Minimum Lateral Stiffness for Equally Spaced Braces in Columns," *J. Engrg. Mech.*, ASCE, **119**(9), 1888-1897.

Analysis of Bridges:
What Do The Numbers Really Mean?

Dennis R. Mertz,[1] Member, ASCE

<u>Abstract</u>

Traditional analysis of bridges, the code-prescribed process of calculating force effects such as moments, shears, stresses or deformations from prescribed loads, introduces much uncertainty in the design of new bridges and the evaluation of existing bridges. Examples are cited that illustrate that the calculated force effects as per traditional designer assumptions and the *LRFD Specifications* are not necessarily accurate and precise. Two conclusions are drawn that suggest that code changes are warranted and designers' awareness should be heightened.

<u>Introduction</u>

Traditional analysis of bridges, the code-prescribed process of calculating force effects such as moments, shears, stresses or deformations from prescribed loads, introduces much uncertainty in the design of new bridges and the evaluation of existing bridges. The distribution of force effects is skewed in that the calculated force effects overestimate in-service force effects for the most part to achieve conservative upperbounds. Such conservatism may be judged prudent for new design, but is costly for the evaluation of existing bridges.

The new standard for bridge design, the *LRFD Bridge Design Specifications* (AASHTO 1998), has introduced the concept of probability-based design to the US bridge community. Therein, unreasonable design conditions, such as the simultaneous superposition of loads due to a major earthquake and the heaviest truck on the bridge in a 75-year exposure period, have been eliminated through the

---

[1]Associate Professor of Civil Engineering, University of Delaware, Newark, DE 19716

application of the theory of reliability. A similar analysis of probability could be introduced with regard to analysis to understand whether the upperbound of conservatism is prudent or merely overkill.

The uncertainties introduced in the analysis of bridges are discussed through a review of the provisions of the AASHTO *LRFD Bridge Design Specifications* which influence analysis, and traditional assumptions made by bridge engineers about which the *LRFD Specifications* are relatively silent. For this discussion, girder bridges will only be considered. The code provisions discussed, in order of appearance in the specifications, include:

- Article 2.5.3 Constructibility,

- Article 3.6.1.3 Application of Design Vehicular Live Loads,

- Article 4.5 Mathematical Modeling, and

- Article 4.6.2.2 Beam-Slab Bridges.

The traditional analysis assumptions that are discussed include:

- the assumption of unyielding piers and abutments, and

- the assumption of frictionless bearings.

Ignoring or Neglecting Construction Force Effects

Many bridge engineers ignore the effects of the construction scheme, in the case of girder bridges the order in which the cast-in-place concrete deck is placed on the girders. Many engineers assume the concrete deck is mysteriously and simultaneously placed on the bridge. This is far from the truth and can have significant effects on the dead load stresses.

Article 2.5.3 of the *LRFD Specifications*, Constructibility, states, "Bridges should be designed in such a manner that fabrication and erection can be performed without due difficulty or distress and that locked-in construction force effects are within tolerable limits." Thus, the aforementioned practice is not in accord with the specifications. Perhaps the specifications should be more explicit. In any case, the bridge engineer who does not consider the construction scheme is not analyzing the bridge properly and has not calculated accurate force effects.

<u>Superposition of Extreme Occurrences</u>

The load combinations of the *LRFD Specifications'* Table 3.4.1-1 Load Combinations and Load Factors, reflect probability. The maximum load factors for a specific transient load component are not combined with the maximum load factors for another load component in a specified load combination. For example, the maximum load factor (LL) for live load in the table is 1.75. It is not in any load combination used simultaneously with the maximum load factor for wind on structure (WS), which is 1.40. For the strength I load combination, the load factor for LL is 1.75 when the load factor for WS is 0. In other words, it is unlikely that the heaviest truck in the bridge's 75-year design life will probably not cross the bridge during a severe wind. For the strength III load combination, the load factor for WS is 1.40 when the load factor for LL is now 0. In other words, during the most severe wind to be expected, trucks will not venture onto the bridge. An intermediate combination, strength V load combination, combines less extreme values of LL and WS with load factors of 1.35 and 0.4 respectively.

While such probabilistic notions are used in combining loads, they are not used in bridge analysis where unlikely extremes are assumed to occur simultaneously. Two examples from the specifications are cited: (1) the heaviest trucks are assumed to be perfectly aligned transversely, ignoring the striped lanes, and longitudinally to produce very unlikely extreme effects, and (2) the heaviest trucks are assumed to possibly cross the bridge during a damaged condition.

In bridge design, two methods of analysis are specified: (1) refined analysis or (2) analysis by distribution factors. When using refined methods of analysis where the designer positions trucks on the bridge deck, the *LRFD Specifications* in Article 3.6.1.3, Application of Design Vehicular Live Loads, specify that, "Both the design lanes and the position of the 3000 mm loaded width in each lane shall be positioned to produce extreme force effects." When applying distribution factors, in Article 4.6.2.2 of the *LRFD Specifications*, Beam-Slab Bridges, it is specified that, "The distribution of live load, specified in Articles 4.6.2.2.2 and 4.6.2.2.3, may be used for girders, beams and stringers other than multiple steel box beams with concrete decks, . . ." The distribution factors of these articles, are derived from parametric studies of refined analyses of bridges where live load was positioned as previously specified in Article 3.6.1.3. No reduction is taken for the improbable occurrence of perfect transverse and longitudinal alignment.

In the second example, the specifications seem concern with potential damage to continuous barriers and mandate that they not be counted on for strength calculations. In Article 4.5, Mathematical Modeling, the *LRFD Specifications* state, "consideration of continuous composite barriers shall be limited to service and fatigue limit states and to structural evaluation." In actuality, continuous barriers draw moment to exterior girders and also help to resist these moments as edge beams. Ignoring their effects at the strength limit state, suggests that the heaviest truck will cross the bridge after an accident which damaged the barrier, but before the barrier can be repaired or traffic restricted. Again, this occurrence is unlikely.

Oversimplification of Boundary Conditions

Finally, in assuming supports are either fixed or expansion, the bridge engineer is again not modeling the true situation. In service, fixed supports on piers and abutments can move due to flexing of the piers and movement of the abutments. Such ability is recognized in the design of integral bridges. Also, expansion supports are not truly frictionless, especially long into the service life of the bridge.

While these effects are difficult to quantify, they are real and influence the actual force effects. Designers can at least acknowledge them in their designs as the Tennessee Department of Transportation has in the development of procedures to design integral bridges.

Summary

Examples have been cited that illustrate that the calculated force effects as per traditional designer assumptions and the *LRFD Specifications* are not necessarily accurate and precise. Whether or not such inaccuracy and imprecision, which are typically conservative, are appropriate for design is not the question.

Nonetheless, two conclusions can be reached:

- for consistency in the *LRFD Specifications*, a probability-based approach should be considered to determine reasonable analysis procedures for extreme force effects, and

- bridge engineers should understand the limitations of their traditional analyses to yield accurate and precise force effects.

References

American Association of State and Highway Officials (1998). *LRFD Bridge Design Specifications*, 2nd Edition. AASHTO, Washington, D.C.

Estimating Suspension Cable Strength

Robert J. Perry [1]

Abstract

Rehabilitation projects involving suspension bridges often require accurate estimates of the strength of existing cables. A notable example was determining strengths of the Williamsburg Bridge suspension cables by the firm of Steinman, Boynton, Gronquist & Birdsall (Steinman). This paper describes sampling and analysis methods that produce such estimates.

Introduction

Sampling wires of suspension cables to determine cable strength involves two separate issues -- engineering and statistics. The engineering issues are beyond the scope of this paper. However, proper selection of which portions of the cable are representative of the most distressed conditions is critical to success. Once these locations are selected, plans may be made for sampling and testing of individual wire elements. The wire sampling plan described here is based on that used by Steinman (1988) for the Williamsburg Bridge in New York City.

Sampling Plan

Each of the four Williamsburg Bridge cables is 19 ¼ in. in diameter and contains 7696 wires distributed among 37 strands. Cable bands are spaced every 20 ft. At each sampling location 32 wire samples were extracted. Four wires were taken at eight equally spaced circumferential locations -- one surface wire and three interior wires. Circumferential locations were 45 deg apart starting at the 12 o'clock position. Interior wires were taken at radial depths of 2-1/4, 4, and 7 in. from the surface. Each wire sample was 40 ft long. Why were these particular choices made? The 40-ft sample length was based on the physical testing program. Strength testing required a minimum of 15 ft of wire, with the balance consumed in chemical, fatigue, and corrosion testing. This length will vary with the particular testing program planned and the practical issues associated with extracting wire from a cable. The eight equally spaced circumferential locations appeared to be the minimum number needed for a reasonable description of variation in strength within the cable cross-section. Certainly, the 6 o'clock position and the two adjacent positions would be required to evaluate the bottom quadrant of the cable, where one might suspect wires to have suffered worst corrosion damage. Other schemes are possible, based on engineering judgment concerning cable condition and environmental factors that might influence cable damage.

Steinman was silent on the rationale behind the radial sampling depths chosen. Although these depths and sample-size distributions appear reasonable in context of this problem, there are other ways to view the problems of sample depth and size.

----

[1] Director, Transportation Research, New York State Department of Transportation

For instance, if cables are perpendicular to prevailing wind, more samples might be taken on the windward side. If it reasonable to assume that cores of cables are less likely to be damaged, fewer samples could be taken near the center and more in exterior portions. Lacking knowledge of interior condition of the cable on which to base reasonable assumptions, it would be reasonable to select equal annular areas and sample sizes. The engineer generally should consult a qualified statistician to discuss location and number of samples. The engineer provides the best estimate available as to cable condition and needed precision for the strength estimate, and the statistician helps develop a sampling plan to achieve these objectives at a reasonable cost.

Ten 18-in. samples were randomly selected from each 40-ft wire sample and tested in tension to determine wire strength. Length between grips in the tensile tester was 1 ft, so strength can be viewed as representing that of a random 1-ft portion of wire in a suspension cable. Each sampled section of cable produces 320 estimates of wire strength -- 32 locations and a nominal 10 tests per location, each associated with a radial depth and circumferential position. Of six locations sampled, five were used to develop wire strength data. The number of wire samples found to be broken was used to estimate the number of unbroken wires remaining in the cable. Estimated wire strength and numbers of unbroken wires were then used to estimate cable strength.

Estimating Number of Broken Wires

A subcontractor to Steinman performed an analysis to determine the expected number of broken wires to be found in a 60-ft length of cable. This length was chosen as a very conservative estimate of "effective" clamping length, which is the distance necessary for a broken wire again to share load with other cable wires. The subcontractor estimated that 96 wires would be broken, and thus cable strength would be estimated on the basis of 7600 intact wires per cable. For that figure, the subcontractor estimated the proportion of broken surface wires in the lower quadrant and interior wires, and combined them to arrive at a single estimate of broken wires using a Normal approximation to the Binomial distribution. If the purpose of the analysis was to estimate strength of the " as is " cables, this is a valid approach if done properly. These two areas are distinct populations. Under the best of circumstances one would expect a greater percentage of broken wires in the exterior of the cable. Combining the proportion of broken wires in the interior and exterior portions of the cable to obtain a single estimate of broken wires is questionable. The correct procedure would be to estimate the expected number of broken wires in each area independently and then add them. If the intent was to estimate strength of the cables after rehabilitation, then only the broken interior wires should be used, since presumably all surface/exterior breaks would be repaired. As a practical matter, the number of broken wires should have little effect on cable strength, unless the cable is composed of a small number of wires and insufficient cable protection and poor maintenance practices have resulted in significant damage.

Two interesting issues are raised in estimating broken wires. The first is what number of wires were examined? Visualize a cable split open to the center by wedges. Counting the number of exposed wires is necessarily a process subject to error, which would seem to be both unavoidable and unquantifiable. A very small value for α in the following equation may be the only way to compensate for this unknown error. The other issue is how to define "exterior" wires. Surely this involves more than surface wires. "Exterior" then becomes a matter of engineering judgment and introduces the counting problem just mentioned. Three was the greatest number of interior wire breaks found in any sampled section of the Williamsburg Bridge cables. The following equation can be used for accurate estimation of the expected number of broken wires. The bracketed quantity is the "exact" upper confidence bound on the percent of broken wires, p:

$$N * \left(1 + \frac{n - x}{(x+1) * F_{2*(x+1),\, 2*(n-x),\, \alpha}} \right)^{-1} \tag{1}$$

Where N = the number of wires in a cable as fabricated (N = 7696),
   n = the number of wires sampled to detect broken wires (n = 1057),
   x = the number of broken wires found in the sample of size n (x = 3),
   α = the percent of the time one may expect the number of broken wires to be higher than the calculated value (α = 0.0005), and
   F = the 1 - α quantile from the F distribution with degrees of freedom 2*(x + 1) and 2*(n - x) [F(8,2108,0.9995) = 3.5016].

Using the equation and the listed values results in an estimate of 101 wires expected to be broken, roughly about 1.3 percent of total cable wires. Assuming that all exterior wires are intact or have been repaired, the cable would have 7595 intact wires. The estimate of 101 broken wires may seem surprising to many, given that only three broken wires actually were found in a sample of 1057. An excellent discussion of estimating Bernoulli parameters was published by Leemis and Trivedi (1996), with comment by Klotz (1996). (It should be noted that the Normal approximation found in most elementary statistics texts is inappropriate in this situation. This approximation is appropriate only when $p \approx \frac{1}{2}$. Its use would result in a substantial underestimation of the number of broken wires.)

Estimating Wire Strength

The Weibull distribution has long been associated with problems in strength of materials. Gumbel (1958) defined the asymptotic distribution of the smallest or largest value of a sample into three types: I, II, and III. The Type III asymptotic distribution arises from distributions having a finite upper or lower bound -- of the three this is the only one that is bounded. The Weibull distribution for smallest values is a Type III distribution. Clearly wire strengths have a lower bound, which is assumed to be zero in the following discussion. Also, cable strength is based on the smallest values of wire strength -- that is the weakest-link theory. The Weibull model is the only one appropriate for describing wire strength that is consistent with the weakest-link concept (Harlow, Smith, and Taylor 1983). The general cumulative distribution function for the Weibull distribution is $F(x) = 1 - \exp\{ - [(x - u)/c]^k \}$. This distribution is determined by the three parameters c, k, and u. The first is the scale parameter or characteristic strength -- the 63.2 percentile of the distribution. The second is the shape parameter and the third is the threshold value below which probability of rupture is zero -- for this problem one may assume u = 0. These parameters should be estimated from sample values using the maximum-likelihood method. Steinman's sampling plan for the Williamsburg Bridge resulted in 160 wire samples (32 per location at 5 locations) from which the 160 smallest values from the 1600 wire-strength values (32 wires per location, 10 wire-strength specimens per wire, and 5 locations) were used to estimate the Weibull parameters. The results were c = 6377.07 psi and k = 15.74. With a model for wire strength in hand, now one can move on to estimating cable strength.

Estimating Cable Strength

The issues in estimating cable strength are complex, involving engineering, statistics, and policy. The major policy issue encountered is the overall degree of precision required in a cable strength estimate. This, of course, is influenced by the required factor of safety, anticipated loadings, and engineering judgment concerning cable condition before examination. Heavy loadings and assumed less-than-ideal cable condition would require a high degree of precision in estimating cable strength. The statistical issues primarily involve proper sampling, testing, estimation, interpretation of results, and verification of underlying assumptions. The engineering issues involve estimating current cable condition and what locations to sample. Beyond that, however, the engineer must decide on the proper model for physical interaction of wires in the cable. Steinman considered two primary models -- brittle and ductile. The brittle-wire model assumes that cable strength is the number of unbroken wires times the strength of the weakest wire. Maximum cable strength occurs when some number of the weakest wires break. This model assumes a constant external force on the cables, equal load-sharing among the wires, and elastic elongation but no plastic deformation. The ductile-wire model assumes that cable strength is the sum of all unbroken wires. Maximum cable strength occurs when all wires are intact. This model assumes wires of equal original length, uniform breaking extension, and that as wires break cable strength is reduced by the strength of the broken wires. Steinman decided, based on physical and engineering considerations, that the proper selection was the ductile wire model. However, one should be aware that model assumptions can significantly influence the estimated cable strength. A paper by Daniels (1989) illustrates this point.

Another classic paper by Daniels (1945), which led to the terms "Daniels Systems" and "Daniel Bundle," describes how to determine cable strength assuming the brittle model is correct. In that paper Daniels showed that strength tends to be Normally distributed for large N, no matter what distribution describes strength of individual wires. The mean of the expected cable strength is S = N·x·[1 - cdf(x)] and the standard deviation is x·SQRT{N·cdf(x)·[1 - cdf(x)]},   (2a, 2b)

Where N = the number of intact wires in the cable (N = 7696 - 101 = 7595),
       x = the strength of the wire which maximizes x[1 - cdf(x)][x = 5350 lb], and
   cdf(x) = the cumulative distribution function for the Weibull distribution{ 1 - exp(- [x/c]^k) }.

Performing these calculations yields S = 38.2x10^6 lb and s.d. = 11.4x10^4 lb. Although not readily apparent, the brittle model requires that 464 wires be broken, N·cdf(x), to achieve this strength. The very small standard deviation would seem to indicate that the estimate of cable strength is associated with no significant error. A better measure of variation in estimated cable strength can be obtained from the variation inherent in estimates of parameters c and k. The following equation describes the asymptotic confidence region for the parameter estimates:

$$\left(\frac{\overline{\lambda}-\lambda}{\delta}\right)^2 - \frac{2*(1-\gamma)*(\overline{\lambda}-\lambda)*(\overline{\delta}-\delta)}{\delta^2} + \left[\frac{\pi^2}{6} + (1-\gamma)^2\right]*\left(\frac{\overline{\delta}-\delta}{\delta}\right)^2 < -\frac{2*\log\alpha}{n} \qquad (3)$$

where λ = the ln transform of c [$\overline{\lambda}$ = ln(6377.07)],
   δ = the reciprocal transform of k [$\overline{\delta}$ = 1/15.74],
   γ = Euler's constant [0.57722],
   α = percent of time the calculated limits would be exceeded [0.0005] , and
   n = is the sample size used to estimate the parameters c and k [160].

Assigning values to λ and solving for δ provides the data necessary to plot the ellipsoid defining the joint variation in c and k. The whole perimeter of the ellipse must be explored to find the extremes of variation in cable strength resulting from parameter uncertainty, because of the nonlinear nature of this problem. This also can be done exactly by analytic methods. Doing this for the brittle model, one finds the lowest cable strength to be 36.4x10^6 lb, with c = 6405.52 and k = 12.00, and the highest to be 39.7x10^6 lb, with c = 6359.19 and k = 19.48. This is about a -5 and +4-percent variation, respectively -- not large, but more than the estimated standard deviation would imply. Note that magnitude of variation is controlled by 1/n in the right side of the previous equation. For small samples, variation in cable strength could be appreciable.

Cable strength for the ductile model is simply the expected wire strength times the number of unbroken wires[N·c·Γ(1 + 1/k)], which is 46.8x10^6 lb. The standard deviation for this estimate is four orders of magnitude smaller than the estimated strength, and thus not very useful. Using the parameter values on perimeter of the ellipse previously determined or the analytic method, one finds that cable strength can vary between 45.9x10^6 lb, with c = 6261.66 and k = 14.69, and 47.8x10^6 lb, with c = 6515.24 and k = 14.97, about a 2-percent variation.

The methods shown here apply to strength estimates for a short portion of cable, a bundle or clamping length of cable. However, due to the nature of this problem the strength of the bundle and the cable are virtually identical [see Smith and Phoenix (1981)]. A more detailed description of the cable strength estimation problem may be found in Perry (1998)

Recommended Practice

1.   Reach consensus on the required remaining service life of the cable system.
2.   Assess the impact of taking the structure out of service.
3.   Determine the anticipated loadings during the service life determined in Step 1.
4.   The results of Steps 1 through 3 are used in deciding the degree of precision required in the estimation of cable strength. In design of a new structure, service life would normally be assumed to exceed 100 years, Step 2 would not be a factor, and Step 3 would generally result in a safety factor in the range of 3 to 4. In a rehabilitation project the required remaining service life could be relatively short, say 10 years. In this case, it might be reasonable to consider a smaller factor of safety acceptable, consistent with assuring service throughout the remaining service life. However, such a decision requires a high degree of precision in determining strength of the cable system. As shown in the example, precision of about 3 percent was achieved for the Williamsburg cables.
5.   Select the sampling sites to encompass worst-case conditions in the cable system. In addition to examining prior inspection reports, this may involve a fairly extensive physical inspection effort.
6.   Develop and execute a sampling plan to obtain wire specimens. This sampling plan must address the problem of estimating the intact wires in the cable and estimating cable strength with required precision. Precision of the strength estimate depends on both the sample size used to estimate the Weibull parameters and the parameters themselves. Had the example sample size been 16 instead of 160, the precision associated with the ductile model would have gone from about 2% to 20%. Original test records for quality control of the wire during construction would be valuable in obtaining a preliminary estimate of the Weibull parameters, so that the sample size required for a given level of precision may be obtained. Such records would also be useful in assuring that properties of the cable wires are reasonably uniform. Non-uniform wire properties would require more complex sampling and analysis efforts. The practical problem of determining the number of wires sampled in relation to estimating the number of wires broken has been mentioned.
7.   Analyze the minimum wire strength data to determine the Weibull parameters. This step and Step 6 require participation of a statistical consultant well versed in such matters.
8.   Select a model for cable strength and estimate the strength and its precision. The ductile model used by Steinman for the Williamsburg bridge certainly appears to be acceptable. Daniels (1989) presents a more complex model, requiring knowledge of the elastic and plastic strain properties of wire. The work involved in obtaining such data and adapting the model to the cable problem would be significant. However, because of the more realistic assumptions governing wire failure, the model may significantly increase the estimate of cable strength. In some cases, such extra effort may be warranted.

Discussion

The intent of this paper has been to illustrate the issues and some suggested best practices for estimation of strength of bridge suspension cables. The first requirement, of course, is expert engineering knowledge. Knowledge of suspension cable systems; their design, construction, in-service performance characteristics; and potential failure modes is essential. Such knowledge is critical for proper selection of cable segments to be sampled. The rationale behind selecting particular segments is to produce the worst-case estimate of cable condition. Each sampling location should then be analyzed with respect to environmental conditions that might create nonuniform distress within the cable section. This information, along with estimates of required cable strength and needed precision, will provide the basis for a statistical sampling plan at each location. The plan in fact will be composed of several sampling plans, depending on wire attributes to be estimated. This paper has focused on cable strength, but other wire properties may be of interest, including metallurgy, corrosion loss, and fatigue life. Once wire samples have been obtained, specimens for study should be randomly selected and tested in random order.

The author recommends using the Weibull distribution as a model to describe minimum wire strength. This advice is based partly on experience with analysis of cable strength of the Williamsburg Bridge, where the Weibull distribution was a best fit to the data. It is also based on a growing body of literature that recommends the Weibull for a number of extreme-event phenomena, such as wind speeds and ground accelerations due to earthquakes -- events that like wire strength have physical bounds. However, in any modeling effort the analyst should take appropriate steps to check the fit of the model to the data and the distribution of residuals.

References

Daniels, H. E. (1945), "The Statistical Theory of the Strength of Bundles of Threads I," in *Proceedings of the Royal Society of London*, Series A, 183, pp. 405-435.

———— (1989), "The Maximum of a Gaussian Process Whose Mean Path has a Maximum, With an Application to the Strength of Bundles of Fibres," *Advances in Applied Probability*, 21, 313-333.

Gumbel, E. J. (1958), *Statistics of Extremes*, New York: Columbia University Press.

Harlow, D. G., Smith, R. L., and Taylor, H. M. (1983), "Lower Tail Analysis of the Distribution of the Strength of Load Sharing Systems," *Journal of Applied Probability*, 20, 358-367.

Klotz, J. H. (1996), "Leemis, L. M., and Trivedi, K. S. (1996), "A Comparison of Approximate Interval Estimators for the Bernoulli Parameter," *The American Statistician*, 50, 63-68.": Comment by Klotz and Reply.

Leemis, L. M., and Trivedi, K. S. (1996), "A Comparison of Approximate Interval Estimators for the Bernoulli Parameter," *The American Statistician*, 50, 63-68.

Perry, R. J. (1998), "Estimating Strength of the Williamsburg Bridge Suspension Cables," *The American Statistician*, 52, 211-217.

Smith, R.L., and Phoenix, S.L. (1981), "Asymptotic Distributions for the Failure of Fibrous Materials Under Series-Parallel Structure and Equal Load Sharing," *Journal of Applied Mechanics*, 48, 75-82.

Steinman, Boynton, Gronquist, and Birdsall in association with Columbia University (1988), "Williamsburg Bridge Cable Investigation Program," final report for the New York Department of Transportation and the New York City Department of Transportation.

STRENGTH EVALUATION OF A COMPLEX BRIDGE

William X. Zhang, PE, M.ASCE[1]

Abstract

Most bridges are evaluated using simplified models. For complex bridges, such as curved bridges with unusual geometry and complex configurations, the sufficiency of the structure cannot be reliably established. Therefore, special analysis methods and procedures are required to determine its capacity and provide a balance between safety and economy.

Introduction

The bridge, built in 1960's, is located in Providence, Rhode Island and carries heavy traffic exiting I-95 northbound onto the Gano Street. With a three-cell concrete box cross-section, the structure has a total length of 306 ft and a minimum radius of 300 ft. The exterior webs are non-prismatic, designed to match the arches in the main bridge. During inspection, many cracks were found extending up the majority of the web height and at the top flange in the vicinity of pier supports. 3-Dimensional finite element analysis and non-destructive truck testing were performed to evaluate this bridge.

Structure Modeling

The model of Gano street ramp consists of 1777 elements, which include plate element for top and bottom slab, shell element for webs, spandrel wall, and all of the diaphragms. The 3-D bridge model is given in Fig. 1.

Once the structural model is built and calibrated, it will become a very useful tool to perform load rating. Furthermore, the model could be used to identify the structural responses and indices that need to be monitored to evaluate any changes in the reliability of the structures during its remaining service life.

---

[1] Vanasse Hangen Brustlin, Inc., 101 Walnut Street, Watertown, MA 02471

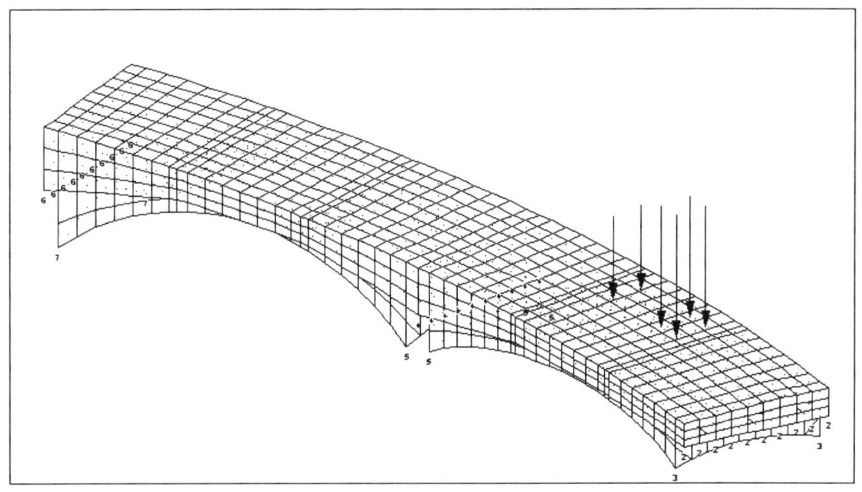

Figure 1. Structure Model

Truck Load Testing

As a major step to effectively verify 3-D modeling, a field diagnostic test was conducted and its results were compared with modeling results.

Load Rating

Live Load Distribution Factor.   The moment distribution factors from testing and modeling are given in Table 1.

Table 1. Moment Distribution Factor under Truck Load

	G1	G2	G3	G4
Testing	0.10	0.22	0.41	0.39
Modeling	0.17	0.36	0.41	0.23

Table 1 shows excellent agreement in the maximum distribution factor between testing and modeling, which means an accurate modeling can be generated by developing appropriate parameters as described in the previous sections.

Figure 2 and 3 give the details of moment and shear distribution factors.

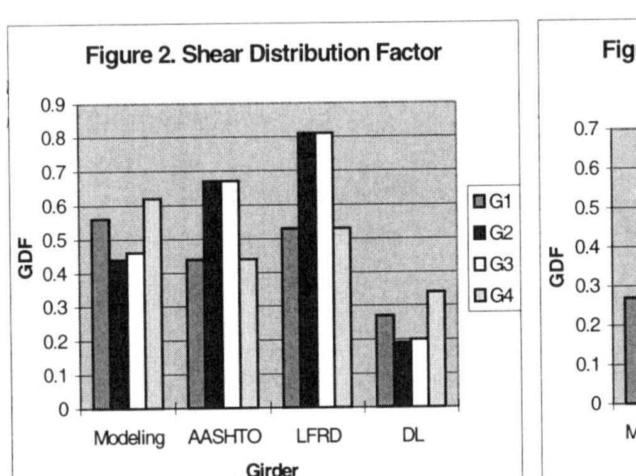

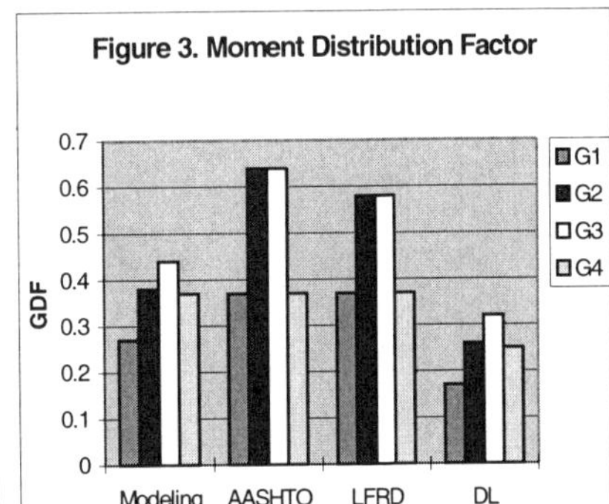

Modeling results identified two important characteristics of this bridge. First, the spandrel wall at exterior girder assisted in resisting wheel load, so bending stress was redistributed in the flange. The spandrel wall attracted about 15% more moment and 16% more shear to the exterior girders. Second, the torque induced by the horizontal curvature of bridge geometry increased shear force on the outside girders and reduced shear force on the inside girders.

Concrete Shear Capacity after Cracking. The computer program RESPONSE, developed by Michael P. Collins and Denis Mitchell[1], was used to predict shear response behavior before and after concrete cracking. The results were compared with the field cracking patterns. The computer analysis results are given in Table 2.

Table 2. Shear Behavior predicted by RESPONSE Program

Girder 2 Member	Shear Capacity Vn(kip)	Cracking Capacity Vcr(kip)	Crack Incline Angle	Tensile Stress f1(psi)	Compressive Stress f2(psi)	Crack Spacing s(in)	Pattern Width w(in)	Dead VDL kip	Load MDL kip
15	274	163	57	257	-298	13.3	0.027	-92	876
16*	292	174	32	0	-530	13.2	0.011	-97	1476
17*	325	188	32	0	-526	13.2	0.011	-103	2148
25*	281	184	35	0	-543	13.2	0.013	108	926
26	259	170	50	218	-195	13.2	0.032	104	237
27	270	158	50	218	-224	13.2	0.032	95	-371

*Member cracked under dead load

Load Rating Factor. Critical inventory and operating rating factors were computed for the bridge under HS20-44 loading. The summary of rating factors is given in Table 3.

Table 3. Summary of Rating Factors

Modeling							
Girder 2	Rating Factor				Girder 3	Rating Factor	
Member	IRF*	ORF	LRFD-RF		Member	IRF	ORF	LRFD-RF
15	2.15	3.59	3.46		15	2.31	3.85	3.72
16	2.21	3.69	3.56		16	2.34	3.90	3.77
17	2.47	4.13	3.99		17	2.58	4.30	4.16
25	1.79	2.99	2.86		25	1.80	3.01	2.87
26	1.64	2.74	2.60		26	1.73	2.89	2.74
27	2.18	3.65	3.51		27	2.35	3.93	3.78
AASHTO code							
Girder 2	Rating Factor				Girder 3	Rating Factor	
Member	IRF	ORF			Member	IRF	ORF	
15	0.71	1.18			15	0.98	1.63	
16	0.66	1.10			16	0.93	1.55	
17	0.65	1.08			17	0.92	1.53	
25	0.58	0.97			25	0.86	1.43	
26	0.58	0.97			26	0.87	1.45	
27	0.66	1.10			27	0.96	1.60	

*IRF > 1.0 means the structure is sufficient

Conclusion

Structural modeling, calibrated through load testing, can cost-effectively evaluate existing complex bridges.

References

1. M. P. Collins, D. Mitchell " *Prestressed Concrete Structures*", Pretice Hall, Englewood Cliff, NJ 07632 (1989)
2. W. X. Zhang "Nonlinear FE Analysis of Prestressed Concrete Bridge Beams", *Proceedings, Fourth National Workshop on Bridge Research in Progress*, Buffalo, NY, June 17-19, 1996

Comparisons of Structural Linear and Nonlinear Analysis of Steel Moment Frames
with Supplemental Damping

Wenshen Pong and Anna Yu

Abstract

A finite-element formulation for fluid viscous dampers is developed using the Maxwell Model to simulate the mechanical behavior of fluid viscous damper. The mathematical model for fluid dampers as incorporated into computer programs to facilitate their use in engineering practice. Comparisons of structural linear and nonlinear analysis are then made. The numerical investigations include (1) the seismic performance of buildings with and without seismic dampers, and (2) the seismic behavior of the structural linear and nonlinear analyses with nonlinear seismic dampers. The results indicate that structural behavior is quite different depending on whether one assumes that the structure is elastic or inelastic for the time-history analyses.

Analytical Model for Fluid Viscous Dampers

The fluid damper exhibits viscoelastic fluid behavior over a large frequency. The simplest model to simulate the mechanical behavior of fluid viscous dampers is the Maxwell model (Bird, et al. 1987) given by

$$P + \lambda \dot{P} = C_0 \dot{U} \tag{1}$$

where λ is the relaxation time, C_o is the damping constant at zero frequency, P is the damping force, and U is the damper position velocity.

A more general Maxwell model may also be considered where the derivatives are of fractional order (Markris, et al. 1991)

$$P + \lambda D^{\gamma} P = C_0 D^q U \qquad (2)$$

where $D^{\gamma}f(t)$ is the fractional derivative of order γ of the time dependent function $f(t)$. Eq.(2) may provide better results than Eq.(1) in simulating the mechanical behavior of complex fluid dampers. Due to the assumption that the damping coefficient is independent of the velocity over a wide range of values, the parameter q can be set equal to 1. For $q=1$, the parameter C_0 is the damping constant at zero frequency. If r is also set equal to one, Eq.(2) is equal to Eq.(1).

Finite Element Formulation for Fluid Dampers

$B=[-R,R]$ where R is a transformation matrix related to the local and global coordinate systems. Using the virtual work principle, one obtains the equivalent nodal forces, $F(t)$

$$F(t) = B^T P(t) = \frac{C_0}{1 + \dfrac{\lambda}{\Delta t}} B^T B \dot{U}(t) + \frac{\lambda}{\Delta t} \frac{1}{1 + \dfrac{\lambda}{\Delta t}} B^T P_{n-1}(t) \qquad (3)$$

Introducing a matrix, C_f, as the added damping which results from fluid dampers, Eq.(3) can be rewritten as:

$$F(t) = C_f \dot{U}(t) + \frac{\lambda}{\Delta t} \frac{1}{1 + \dfrac{\lambda}{\Delta t}} B^T P_{n-1}(t) \qquad (4)$$

where
$$C_f = \frac{C_0}{1 + \dfrac{\lambda}{\Delta t}} B^T B \qquad (5)$$

A finite element formulation for fluid viscous dampers was developed so that the damper can be easily incorporated in computer programs (Pong, et al. 1995). The structural nonlinear model based on two-surface model (Tseng and Lee, 1983) was implemented into the computer programs so that structural nonlinear analysis with dampers could be applied.

A Steel Moment Frame Building Equipped with Fluid Dampers

A 4-story steel moment frame building is designed to use supplemental damping devices to achieve a higher seismic performance during severe earthquakes. Approximately 20% of critical damping is provided using seismic dampers which are mounted on diagonal braces in selected locations on the each floor. The floor systems are 8.25 cm light weight concrete over 7.62 cm metal deck at 18 gage. Typical floor live load is 2.4 kN/m². Typical floor height is 4.3 m and floor mass is 16,400 kN.

Site Specific Ground Motion Records

Three levels of design earthquakes are defined in Table 1 below. The time histories used in spectral matching are shown in Table 2, and spectral acceleration at various damping ratios is shown in Table 3.

Design Earthquakes	Earthquake Having Probability of Exceedance	Mean Return Period (Years)
Maximum Probable Earthquake (MPE)	50%/50 year	72
Design Basis Earthquake (DBE)	10%/50 year	475
Maximum Creditable Earthquake (MCE)	10%/100 year	950

Table 1. Design Earthquakes

Design Earthquake	Earthquake	Magnitude	Time History	Dist. (km)
MPE	Loma Prieta	7.1	Santa Teresa	18
DBE	Imperial Valley	6.7	El Centro	12
	Landers	7.4	Joshua Tree	15
	Loma Prieta	7.1	Saratoga	3
MCE	Imperial Valley	6.7	El Centro	12
	Landers	7.4	Joshua Tree	15
	Kern County	7.6	Taft	56

Table 2. Time Histories Used in Spectral Matching

Damping Ratio	2%	5%	7%	10%	20%
Period (Second)					
0.00	0.444	0.444	0.444	0.444	0.444
0.05	0.666	0.607	0.583	0.555	0.492
0.10	1.155	0.929	0.843	0.750	0.565
0.50	1.061	0.865	0.792	0.703	0.503
1.00	0.725	0.591	0.540	0.488	0.380
2.00	0.369	0.300	0.276	0.249	0.197
3.00	0.236	0.200	0.187	0.173	0.131

Table 3. Spectral Acceleration (g) of 10% in 50 Years

Analysis of Frames with Seismic Dampers

Since the steel moment frames with dampers are designed to remain elastic under Design Basis Earthquake (DBE). The structural design was based upon the assumption of structural linear with damper's nonlinear behavior. The design criteria include: (1) the demand/capacity ratio of the structural member shall below unity and, (2) the story drift ratio shall be under 1% during DBE events. However, the structure will undergo some inelastic rotations under Maximum Creditable Earthquake (MCE).

Conclusions

The structural behavior is not much different during the lower demand earthquakes such as Maximum Probable Earthquake (MPE) or DBE events. The structure remains elastic under MPE and DBE events. However, the seismic behavior is quite different under MCE because some structural members start to yield and demands on the joints increase significantly. The structure contributes larger energy dissipation which reduces the role of the dampers as a major mechanism for energy dissipation under MCE events. Due to the uncertainty of the input ground motion's characteristics, conclusions for structural inelastic behavior are hard to make. But the need for structural nonlinear analysis for buildings with supplemental damping is greater in order to obtain a better understanding of structural behaviors.

References

Bird, R.B., Armstrong, R.C. and Hassager, O. (1987). "Dynamics of polymeric liquids." J. Wiley and Sons, New York.

Markris, N. and Contantiou, M.C. (1991). "Fractional derivative Maxwell model for viscous dampers." Journal of Structural Engineering, ASCE, 117(9), 2708-2724.

Pong, W.S. and Tsai, C.S. (1995). "Seismic study of buildings with viscoelastic dampers." An International Journal of Structural engineering and Mechanics, Vol. 3, Number 6, November 1995.

Tseng, N.T. and Lee, G.C. (1983). "Simple plasticity model of two-surface." Journal of Engineering Mechanics, ASCE, 109(3), 795-810.

System of Units

1 cm = .3937 inch   1 N = .2248 lbf

Easy, Concise Calculations for
Eccentrically Loaded Welded Connections

Thomas W. Hartmann, P.E.[1]

Abstract
     This paper provides a straightforward formula based upon
connection geometry to provide welded connection coefficients.
The formulas can include mixed weld sizes and skewed load cases.
Results are compared to experimental data and the Ultimate
Strength Method (USM) described in the American Institute of
Steel Construction Manual.

Introduction
     Many welded connections must be designed to resist loading
eccentric to the centroid of the weld geometry. The American
Institute of Steel Construction Manual provides tables for
standard weld configurations including load angles of 0 and 15
degree increments (AISC, 1994). Intermediate values may be
approximated or calculated through an interactive approach using
methods outlined in the AISC Manual (Butler, et.al. 1972). The
developed formula provides a general solution combining the
effects of axial and eccentric loading on a weld. Values for all
angles of loading and eccentricities are easily determined. No
iteration is required.

Development
     The strength properties of the weld are determined by
analyzing the geometric configuration of the weld.   For
simplicity of development, all weld widths are assumed equal.
First, the weld is segmented into discrete number of elements.
Six or more segments per weld leg are typically sufficient for
determining the capacity.   The length of the weld, L, is
determined by the following formula:

$$L = \sum_{i=1}^{n} A_i \qquad (1)$$

where A_i is the length of an individual segment and n is the
number of segments.

---

[1]Principal, TwHartmann, Inc. Engineering Consultants,
107 Cameron, Suite 2, Fort Collins, CO 80525.

The centroid of the connection is determined by following formulas:

$$y_c = \frac{\sum\limits_{i=1}^{n} A_i \, y_i}{\sum\limits_{i=1}^{n} A_i} = 0 \qquad (2)$$

$$x_c = \frac{\sum\limits_{i=1}^{n} A_i \, x_i}{\sum\limits_{i=1}^{n} A_i} = 0 \qquad (3)$$

where A_i is a discrete length of weld segment,
x_i is the distance from the origin to the center of the segment.
y_i is the distance from the origin to the center of the segment.
The distance, d_i, from the centroid to each segment location can now be determined:

$$d_i = \sqrt{(x_i - x_c)^2 + (y_i - y_c)^2} \qquad (4)$$

For long lap weld splices, the shear lag reduction (Bellamy, 1993) for connection length may be approximated by:

$$\gamma \approx 1 - x'/L' \qquad (5)$$

where x' is the distance from the centroid of the shape profile to the shear plane of the connection. L' is the longest leg of the weld. The capacity for long lap splices is determined by:

$$C_1 = \gamma \sum\limits_{i=1}^{n} A_i \sigma_{ult} = \gamma \, L \, \sigma_{ult} \qquad (6)$$

where n is the number of segments, L is the total weld length and σ_{ult} is the ultimate weld shearing stress. Research has shown that for loads perpendicular to the weld axis the capacity is approximately 67 percent of welds loaded parallel to the weld axis (Butler, 1972). The length reduction, γ, can be set to one.

$$\delta \sigma_{ult} \, Z = \sum\limits_{i=1}^{n} [\delta \, \sigma_{ult}] \, A_i \, |d_i| \qquad (7)$$

The rotational strength of the connection is determined by

the following formula:

where δ is given by the following formula (Lesik, 1990):

$$\delta_i = 1.00 + 0.5 \sin^{1.5}\theta_i \tag{8}$$

θ_i =   rotation of connection segment i, proportional to the distance of the connector from centroid, where $\theta_i = (d_i / d_{max})$

θ_{max} =   maximum weld deformation rotation at failure

δZ, called "zorque", is not dependent on orientation of the connection or load orientation and is based solely on connection geometry. For simplicity, δ can be taken as 1.00.

For a moment applied to weld, the capacity is determined by the following formula:

$$C_2 = \frac{\delta \ \sigma_{ult} \ Z}{M} = \frac{\delta \ \sigma_{ult} \ Z}{P \ e} \tag{9}$$

where e is the eccentricity measured from the centroid. Combining the axial formula (6) and rotational formula (9) into a circular interaction formula yields:

$$\sigma_{ult}^2 = \left(\frac{C_1}{\gamma \sum_{i=1}^{n} A_i}\right)^2 + \left(\frac{C_2 \ e}{\sum_{i=1}^{n} \delta_i A_i |d_i|}\right)^2 \tag{10}$$

Equation (10) is then normalized so that the ultimate shear stress, σ_{ult}, is set equal to 1.0, and the capacities of the two limiting cases, C_1 and C_2, are set equal to C, the capacity of the connection based on the interaction. The normalized coefficient, C, is given in Equation (11).

$$C = \frac{1}{\sqrt{\left(\frac{1}{\gamma \sum_{i=1}^{n} A_i}\right)^2 + \left(\frac{e}{\sum_{i=1}^{n} \delta_i A_i |d_i|}\right)^2}} = \frac{1}{\sqrt{\left(\frac{1}{\gamma L}\right)^2 + \left(\frac{e}{\delta Z}\right)^2}} \tag{11}$$

The following table has comparison values between the Ultimate Strength Method and Equation 11 and Butler's test results. In the table, "V" indicates a vertical weld and "C" indicates a cee shaped weld with dimensions shown.

Equation 11 can be approximated by further simplification for "high eccentricity" loads defined as connections where e is greater than L. Equation 12 shows the reduced formula:

$$C = Z/e \tag{12}$$

TABLE 1 -- SUMMARY OF WELD TEST RESULTS BY BUTLER (1972)
***ALL 1/4" WELDS NORMALIZED WITH $\gamma=1.0$ AND $\delta=1.0$**

#	e cm (in)	γL* cm (in)	Height x Width of Weld "C"	$\delta Z/4$* cm^2 in^2	TEST MPa (kips)	EQ.(11) MPa (kips)	USM MPa (kips)
V1	20.3 (8)	20.3 (8)	20.3x0 (8x0)	25.8 (4.0)	560 (126)	485 (109)	507 (114)
V2	15.2 (6)	25.4 (10)	25.4x0 (10x0)	40.3 (6.25)	1014 (228)	957 (215)	1010 (227)
V3	38.1 (15)	25.4 (10)	25.4x0 (10x0)	40.3 (6.25)	418 (94)	410 (92.1)	427 (96)
V4	15.2 (6)	30.5 (12)	30.5x0 (12x0)	58.1 (9.00)	1384 (311)	1335 (300)	1415 (318)
V5	20.3 (8)	30.5 (12)	30.5x0 (12x0)	58.1 (9.00)	1037 (233)	1050 (236)	1103 (248)
V6	30.5 (12)	30.5 (12)	30.5x0 (12x0)	58.1 (9.00)	720 (162)	725 (163)	757 (170)
V7	25.4 (10)	40.6 (16)	40.6x0 (16x0)	103 (16.00)	1450 (326)	1481 (333)	1553 (349)
V8	40.6 (16)	40.6 (16)	40.6x0 (16x0)	103 (16.00)	1059 (238)	966 (217)	1005 (226)
C1	10.8 (4.25)	30.5 (12)	15.2x7.6 (6x3)	47.6 (7.38)	1321 (297)	1500 (337)	1286 (289)
C2	25.4 (10)	45.7 (18)	15.2x15.2 (6x6)	93.7 (14.53)	1001 (225)	1380 (310)	1095 (246)
C3	27.9 (11)	40.6 (16)	20.3x10.2 (8x4)	84.3 (13.15)	1401 (315)	1144 (257)	1441 (324)
C4	30.8 (12.13)	50.8 (20)	20.3x15.2 (8x6)	123 (19.06)	1339 (301)	1495 (336)	1420 (319)
C5	38.6 (15.2)	50.8 (20)	30.5x10.2 (12x4)	141 (21.93)	1295 (291)	1380 (310)	1366 (307)

Conclusion

The developed formula provides results comparable to the Ultimate Strength Method. The calculation effort is simple compared to the interactive approach of the Ultimate Strength Method.

References

1.   American Institute of Construction, Manual of Steel Construction, Load and Resistance Factor Design, Second Edition, 1994.
2.   Lesik, D.F. and Kennedy, D.J., "Ultimate strength of fillet welded connections loaded in plane," Canadian Journal of Civil Engineering, National Research Council of Canada, Vol. 17, No. 1, 1990.
3.   Butler, L.J., Pal, S. and Kulak, G.L., "Eccentrically Loaded Welded Connections," ASCE Journal of the Structural Division, Vol. 98, ST5, Proc. Paper 8874 pp. 989-1005, May 1972.
4.   Bellamy, R.E., "A geometric approach for eccentrically loaded welded connections," Thesis for Colorado State University, Spring 1993

A Simplified Design Procedure for End- and Base-Plates of
Traffic Support Structures

Osman Hag-Elsafi[1], Sreenivas Alampalli[1], Member, ASCE, and Frank Owens[1]

Abstract

A simplified procedure is proposed for the design of end-plates and base-plates of
cantilevered traffic sign, signal, and light structures, and also for base-plates of span-
wire-mounted traffic-signal structures. This procedure is based on beam and plate
bending and torsion theories, and is intended for routine application by practicing
engineers designing these items. Results obtained using the proposed procedure
compared well with those estimated using finite-element analysis and full-scale testing.

Introduction

Although end-plates and base-plates are important structural components for traffic
signal, sign, and light support structures, no standard procedure has been established
for their design. Manufacturers of these plates normally rely on some combination of
experience and minimal analysis. A rational procedure is proposed here for designing
the typical configuration shown in Figure 1 -- square-shaped plates supported by single
bolts at the four corners, supporting a cylindrical (or multi-sided) mast or mast-arm,
load-transferring member (Owens, Hag-Elsafi, and Alampalli 1998). Design loads are
limited to dead load, ice load, and wind load, and the mast/mast arm is assumed to be
sufficiently strong to transfer these loads to the plate. The plate is assumed to be
strong enough to transfer in-plane shear and torsional forces to the bolts.

First, procedures are developed for determining load distribution to the anchor bolts
and estimating bearing forces and bearing stress distribution. Then, by investigating
plate behavior under applied loads, critical-stress states on the plate are identified, and
a simplified procedure is proposed to estimate these stresses. Plate designs based on
the proposed procedure are compared with those obtained using the procedure given
by Boulos, Fu, and Alampalli (1993), and those supplied by manufacturers.

---

[1]Transportation Research and Development Bureau, New York State Department of
Transportation, Albany, New York 12232-0869

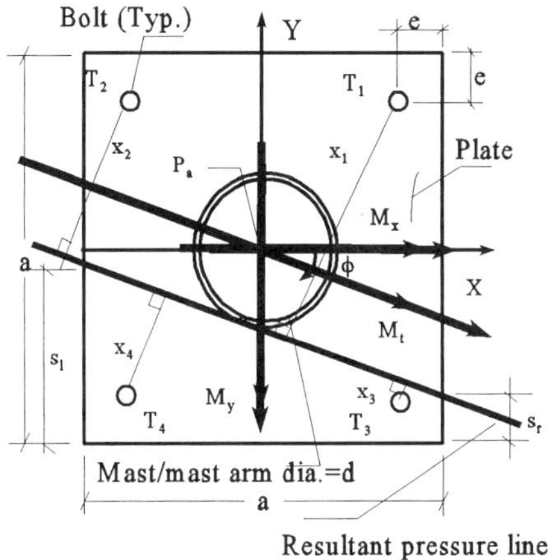

Figure 1. Typical end- or base plate.

Forces on Bolts and Plate

Axial loads on the types of traffic support structures addressed here are normally of insufficient magnitude to maintain throughout compressive bearing stresses between a base plate and foundation, and the design is typically governed by Group II or III loading (Standard Specifications 1994). Thus, for static equilibrium some bolts must carry tensile forces. An end plate is a special case of a base plate with no axial load.

For calculated design moments due to gravity load M_x and wind load M_y, one may determine the total moment on the plate M_t and its angle of incidence ϕ (as defined in Figure 1). Assuming flexibility of the plate, the resultant of compressive bearing forces may be assumed to lie at the toe of the mast/mast arm, at the location shown in Figure 1. Bolt distances x_1 to x_4 to this line may be obtained from geometry, and bolt tensions T_1 to T_4 may be calculated by taking moments about the resultant pressure line. These forces may be obtained in terms of the most-stressed bolt tension T_1 as $T_i = T_1 x_i / x_1$, where T_1 is given by $T_1 = [M_t - P_a \cdot d/2]/ [\sum x_i^2 / x_1]$ and P_a is the applied axial force on the plate.

Maximum bearing stress f_{b_max} may be obtained based on an assumed distribution (Owens, Hag-Elsafi, and Alampalli 1998) as $f_{b\_max} = (4/3) [T_{Total} + P_a]/ [a \cdot (s_l + s_r)]$, for the case shown in Figure 1, where T_{total} is the total tension force on the bolts, and s_l and s_r as defined in Figure 1. Calculated f_{b_max} should not exceed allowable bearing stress for the material the plate is bearing against (Manual 1998).

Proposed Design Procedure

By investigating plate behavior under applied loads and expected failure modes, complex stresses on the plate can be approximated by those calculated for assumed plate elements, due to estimated forces acting on those elements. Such stresses at four locations identified to be critical (Owens, Hag-Elsafi, and Alampalli 1998) are: 1) flexural stresses between tension-side bolts estimated based on a fixed-end beam approximation, 2) torsional stress at the sides of the mast/mast arm estimated using a torsional beam approximation, 3) flexural stresses at the yield line (bending and shear stresses may be calculated for an assumed yield line around the most stressed bolt), and 4) flexural stresses at the resultant pressure line estimated from an assumed bearing stress distribution. Allowable stresses should be based on Group II or III loading of the Standard Specifications (1994).

Application

The proposed procedure was used to design 23 base plates from three major suppliers of traffic poles. Estimated plate thicknesses using the proposed procedure were compared with those obtained using a procedure based on physical testing and finite-element analysis (Boulos, Fu, and Alampalli 1993), and also with those supplied by manufacturers. Based on these results (Figure 2), estimated plate thicknesses using the proposed procedure show clear agreement with Boulos, Fu, and Alampalli (1993) regarding deficiency of the investigated plates. This deficiency was observed to result primarily because of the need for thicker plates in some cases when moment components are applied at a 45-degree angle from the horizontal axis, and a plate is diagonally stressed.

Estimated stresses on base plates obtained using the proposed procedure indicated that, on average, these stresses are within 11 to 20 percent (with a standard deviation of 7 to 12 percent) of those obtained using finite-element analysis (Boulos, Fu, and Alampalli 1993).

Conclusions

A simplified procedure is proposed for design of cantilevered end- plates and base-plates of traffic sign, signal, and light structures, and for base plates of span-wire-mounted traffic-signal structures. This procedure identifies critical stress locations to estimate corresponding forces and effective plate sections. Results based on the proposed procedure showed good agreement with those obtained using finite element analysis and physical testing.

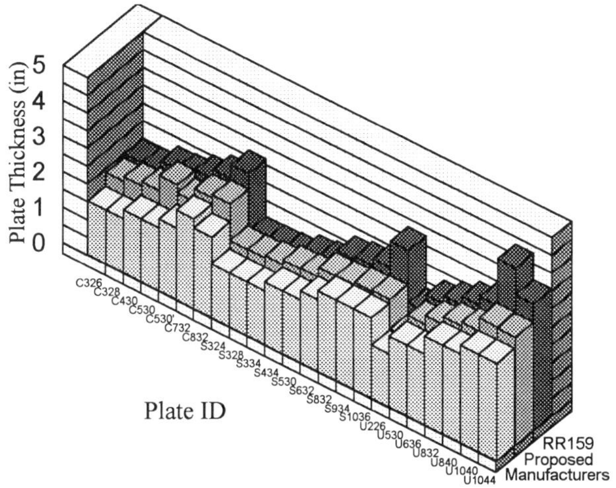

Figure 2. Comparison of plate thicknesses using the proposed procedure with those used by manufacturers and based on Boulos, Fu, and Alampalli (1993).

Acknowledgments

Several New York State Department of Transportation (NYSDOT) personnel contributed to this study. The views expressed in this paper are those of the authors and not necessarily those of the NYSDOT. This work was partially funded by the Federal Highway Administration, U.S. Department of Transportation.

References

Owens, F., Hag-Elsafi, O., and Alampalli, S. "A Simplified Procedure for Design of End-Plates and Base-Plates of Cantilevered Traffic Structures." Transportation R&D Bureau, NYSDOT, 1998 (in preparation).

"Standard Specifications for Structural Supports for Highway Signs, Luminaries and Traffic Signals." American Association of State Highway and Transportation Officials, Washington, D.C., 1994.

Boulos, S.J., Fu, G., and Alampalli, S. "Load Testing, Finite Element Analysis, and Design of Steel Traffic-Signal Poles." Research Report 159, NYSDOT, July 1993.

"Manual of Steel Construction - Allowable Stress Design." Ninth Edition, American Institute of Steel Construction, Chicago, Illinois, 1998.

Issues related to achieving qualitative improvements in reliability assessment in structural design.

Pavel Marek[1] (F. ASCE) and Le-Wu Lu[2] (M. ASCE)

Abstract

The dramatic advancement in computer technology affects nearby all areas of human activities. In structural design, considering assessment of reliability, the transition from slide rule era to computer and information freeway era leads to new approaches. In order to cross the divide from current concepts to qualitatively new methods, reengineering of the reliability assessment process is needed. Attention should be given to education, to the development of new design concepts, to restructuring the specifications, and to direct application of data and knowledge bases. Some of the main issues are addressed.

Introduction and critical comments on the current reliability assessment process

Increasing number of publications dealing with probabilistic approaches to structural reliability assessment indicate a qualitatively new trend in the concepts applicable in designer's work. The dramatic developments in the computer technology allow for considering a transition from the methods developed in the slide rule era to qualitatively new structural reliability assessment concepts.

Only some ten years ago, powerfull personal computers were not available to all designers. Today, major part of designer's activities is unmanageable without these powerful tools. How come that one of the key activities, the structural reliability assessment, is still based on methods developed in the era of primitive computational tools? What are the main issues and research needs to be taken into account while

---

[1] Prof., C.E., PhD., Institute of Theoretical and Applied Mechanics, Czech Academy of Sciences, Prosecká 76, 190 00 Prague 9., Czech Republic.
[2] Prof., C.E., PhD, Dept. of Civil and Environ. Eng., Lehigh University, 13 Packer Ave., Bethlehem, Pa, 18015, U.S.A.

attempting to cross the divide from the methods used at present to concepts of the future? Ideas and suggestions presented here should turn the attention to some of the main issues.

The concepts applied in current specifications (such as allowable stress design and partial factors method) are based on a "design point" approach, i.e., on separation of the defined maximum load effects and minimum resistance in the analysis, and on comparing these quantities. The researchers and specification writing committees earlier used the deterministic and later probabilistic approaches and calibration procedures in specifying the corresponding factors and assessment procedure presented in codes. The designer's involvement in the actual reliability check is limited to the interpretation of criteria and instructions contained in specifications.

The current design strategy according to those codes doesn't give the designer a chance to contribute directly to the actual reliability analysis.The designer does not consider variables and their interaction. He/she is applying quantities contained in the specification and he/she needs not to understand clearly the substance of criteria which are often hidden in "black boxes". From the designer's point of view all reliability assessment concepts, as applied in current specifications, are deterministic. The designer has to follow strictly the instructions given in the documents, while his/her expertise regarding the assessment are not used to any significant extent. It can be observed that the specifications are becoming more and more complex, their volume is increasing and the content is less and less manageable. Inadequate attention is given to the true role of the reliability assessment. Some other objections may also be mentioned. As already reviewed in several papers (see, e.g., *Vrouwenvelder, 1997*), the design according to current specifications based on partial factors concept does not necessairly lead to a balanced safety. Also, the common representation of loadings (characteristic values and load factors) gives only limited chance for a consistant evaluation of the load effect combinations and doesn't allow at all for an analysis of combinations of two- or more-component load effects (see, e.g., *Marek et al. 1995*).

Crossing a divide: Reeingineering of the design process and education

In order to cross the qualitative divide from current practice to an arrangement corresponding to the potential of the available computer technology, reengineering of the entire assessment procedure should be considered. Reengineering of the structural dimensioning and reliability assessment can be defined as the fundamental rethinking and radical redesign of processes to achieve dramatic improvements in critical contemporary measures of safety, durability, serviceability of structures and structural components. An important subject of this redesign is the representation and interaction of the variables involved in the analysis of the reliability function. It can be expected that the overall architecture of a future design process will differ very much from the today's routine. One of the expected main improvements seems to be the transition from semi-probabilistic to fully probabilistic reliability assessment concept. Such transition can be expressed by symbolic equations: from current criterion max S < min R , to a probabilistic check $P(R - S) < P_d$ where R is the variable resistance, S is the resulting combination of variable load effects, and P_d the specified target probability.

Attention should be given to the development of databases (containing information about variables such as material properties, imperfections, transformation models, loading etc.) and of knowledge bases. The results of the reengineering should bring about a qualitatively new level of design practice expressed by significant economical effects related to the efficient use of available computer technology. The reengineering requires education of students, designers and all others involved. Especially at the universities and university extensions the training should lead to the probabilistic way of thinking.

<u>To fully probabilistic concept</u>

'Fully' probabilistic concepts applicable in specifications can be based on various approaches, such as analytical or numerical procedures (see, e.g., *Vrouwenvelder 1997*) or on simulation techniques, see, e.g.,. SBRA documented in *Marek et al. 1995*. SBRA concept is based on Limit States philosophy, parameters generated histograms and Monte Carlo method. All input variables are expressed by bounded histograms, the reliability function RF = (R-S) is analyzed using basic Monte Carlo technique, and the reliability is expressed by probability of failure P_f. The application of the SBRA is demonstrated using two hundred examples. The application of SBRA in designer's work would require approval of criteria for assuring acceptability of the statistical input (*Galambos 1998*).

One of the important issues related to the introduction of fully probabilistic concept is the definition of the reference levels regarding safety, durability and serviceability, including the corresponding target probabilities of failure. In the partial factors concept the „ultimate" carrying capacity refers to collapse (disposal) limit state while in case of the probabilistic SBRA concept the reference levels are defined, for example, either by onset of yielding, or by a tolerable permanent deformation, or by magnitude of a "damage" in performance design.

The transition from partial factor concept to fully probabilistic concepts, (such as SBRA) requires, among others, recognition of the fully probabilistic approach in specifications. A recently revised standard, *CSN 1998,* allows already for applying SBRA in design work and specifies target probability P_d.

The concepts applied so far in specifications are restricted to the assessment of structural elements and components. As emphasized in paper by *Galambos 1998*, it is already time for considering the extension of the reliability concepts to structural systems. It can be assumed that the probabilistic concept should be introduced in the design of elements and components first and extended to systems after the fundamentals of probabilistic assessment of components are made clear to designers.

Summary and Conclusions

A significant progress in the development of fully probabilistic structural reliability assessment concepts has taken place. The modern computer technology allows for introducing general application of the Limit States philosophy using simulation technique as a powerful tool. Following issues can be considered:

(a) A successful implemenation of probabilistic design concept requires a transition from deterministic to probabilistic way of thinking of students, designers and other users of specifications. The simulation technique can serve as a tool in education and in teaching structural reliability assessment.

(b) The transition from current methods (such as partial factors concept) to fully probabilistic methods will require a reengineering of the entire assessment procedure.

(c) Considerable discussion by researchers and specifications committees should be initiated regarding the strategy of the future assessment concepts applicable in designer's work (*Galambos 1998*). One of the important issues to be evaluated is the practical applicability of fully probabilistic concepts based either on analytical approaches (such as JCSS concept, *Vrouwenwelder 1997*) or on the explicit use of simulation technique using powerful computers (*Marek et al. 1995*).

(d) Attention should be given to the basic rules of the reliability assessment including the definition of the reference level in the probabilistic analysis.

(e) The information freeway allows for considering the application of central data bases and knowledge bases on structural design.

(f) All legal aspects of the qualitatively new assessment system must be studied by specification committees in cooperation with institutions having a jurisdiction.

References

Vrouwenvelder, A.C.W.M. (1997). The JCSS probabilistic model code. *Structural Safety, Vol. 19, No.3., pp.245 - 251).*

CSN 73 1401 (1998). Design of Steel Structures. *Czech Standard Institute, Prague, Czech Republic.*

Marek, P., Guštar, M., and Anagnos, T. (1995). Simulation Based Reliability Assessment. *CRC Press Inc., Boca Raton, Florida.*

Galambos, T. (1998). Developments in LRFD in the United States of America. In: *Structural Engineering World Wide 1998, Elsevier Scienec Ltd.*

Acknowledgments

The authors would like to acknowledge the Grant Agency of the Czech Republic (Research projects Nos.103/94/0562 and 103/96/K034) and Lehigh University for their support.

The JCSS Probabilistic Model Code for new and existing Structures

Prof. Dr. Dimitris Diamantidis[1]

Abstract

The JCSS (Joint Committee on Structural Safety) is developing a model code for full probabilistic design. This note gives an overview of the set up and contents of this code.

1. History of the JCSS

In 1971 the Liaison Committee which coordinates the activities of seven international associations in Civil Engineering, CEB, CIB, ECCS, FIP, IABSE, IASS, and RILEM, created a Joint Committee on Structural Safety, JCSS, with the aim of improving the general knowledge in structural safety.Until 1984 the JCSS has had 14 meetings not included the meetings of the various subcommittees. Thereby several guidile documents have been prepared and published. In 1985 the JCSS was reorganised and had in the period 1986 to now more than 25 meetings of its Working Party. Several background papers related to structural safety and codified design were thereby produced. The long term goal of the work is the development of an operative probabilistic model code.

2. Background and Scope of a Probabilistic Model Code

Current building codes are based on the definition and the classification of limit states and on the implementation of partial safety factors for load and resistance parameters. The partial safety factors are, to some extend, derived on the basis of probabilistic methods. For this reason such codes are referred to as

---

[1] Fachhochschule Regensburg
Chairman of the Working Group of JCSS (Joint Committee on Structural Safety)

probability based codes.

There is an increasing need and consequently an increasing tendency to use probabilistic methods not only as a background for code development but as a tool for the safety assessment of special or important structures such as nuclear plants, offshore structures, bridges etc, existing as well as under design. One of the main problems which is faced with that respect is that no standardised procedures are available regarding statistical data interpretation and reliability calculation procedures.

Therefore there is an obvious need for a code which gives sufficient guidenance and information to perform a full probabilistic analysis. Ideally such a code should a) provide a complete and consistent set of models and procedures for reliability based decision making associated to structural design and reassessment and b) be operational intended for application in the context of probability based expertises.

The JCSS has considered as its basic task to produce such an operational *Probabilistic Model Code*. After a number of preliminary studies the actual drafting of several parts of the code is beeing completed.

3. Contents

The Probabilistic Model Code covers new and existing structures of various material types and consists of four main parts, which are briefly described herein.

Part 1: Basis of Design

This part includes general principles for probabilistic design which were already available from earlier work of the JCSS and are similar of those given by ISO 2394 (General Principles on Reliability of Structures). The table of contents is presented next:

1   Introduction
2   Requirements
3   Principles of limit state design
4   Basic variables and uncertainty modelling
5   Analysis procedures
6   Reliability models and methods
7   Target Reliability
ANNEX A: Robustness
ANNEX B: Durability

Part 1 provides sufficient information on reliability methods, models and tools and gives quantitative safety acceptance criteria in terms of target reliability levels based on a reliability differentiation (definition of safety classes).

Parts 2 and 3: Load and Material Models

Each type of load and material has its own chapter (load and material note) in this part. Load combinations are dealt with in the general chapter 2.0. Each note on load and material models provides all necessary information according to state-of-the art to be implemented in reliability analysis calculations. Some of the notes are almost completed and shall be published in the near future.

The scope of each note is to provide applicable stochastic models i.e. statistical distribution types and parameters, spatial and time variability, correlation functions, load spectra if applicable, etc.

However only the information that is relevant for the probabilistic modelling is presented. Mechanical calculation models such as buckling, shear etc. are not included. Each note has a length of 5 to 10 pages. Table 1 illustrates the contents of parts 2 and 3.

Table 1: Contents of Parts 2 and 3 of the Probabilistic Model Code

PART 2	PART 3
LOADS	**MATERIALS**
2.0 General	3.0 General
2.1 Selfweight	3.1 Concrete
2.2 Live Load	3.2 Reinforcement
2.3 Industrial Storage	3.3 Prestressing Steel
2.4 Cranes	3.4 Steel
2.5 Traffic Load	3.5 Timber
2.6 Car Parks	3.6 Aluminium
2.7 Silo Load	3.7 Soil
2.8 Liquids and Gasses	3.8 Masonry
2.9 Manmade Temperature	3.9 Model Uncertainties
2.10 Earth Pressure	3.10 Dimensions
2.11 Water and Groundwater	3.11 Imperfections
2.12 Snow	
2.13 Wind	
2.14 Temperature	
2.15 Waves	
2.16 Avalanches	
2.17 Earthquake	
2.18 Impact	
2.19 Explosion	
2.20 Fire	
2.21 Chemical Attack	

Part 4: Assessment of Existing Structures

The assessment of existing structures is getting more and more important, while on the other hand most codes deal explicitly with design situations only. The assessment of existing structures differs essentially from the design of new structures. One key point is that new information can become available related to the state of the existing structure. Part 4 of the Model Code is related to the reliability based reassessment of structures. The work shall be completed by the end of next year. The document is of educational type and provides reliability methods to be used in the structural reassessment. Tutorial examples and practical case studies are included as shown in the contents:

Chapter 1: Forword
Chapter 2: Guidelines
Chapter 3: Codification
Annex A: Reliability Analysis Principles
Annex B: Updating procedures
Annex C: Target Reliability
Annex D: Examples
Annex E: Case Studies

Part 4 provides relevant information on how to process specific information about an existing structure, how to update its reliability based on such information, how to base decisions regarding maintenance, strengthening, upgrading etc. This part is generally applicable for various materials and various structure types.

4. Summary and Conclusions

The development of a full probabilistic model code is currently under development by the Joint Committee on Structural Safety. The work is extremely interesting and the task is quite an ambitious one. Compared to well known Probability Based Codes the Probabilistic Model Code:

– accounts in a more consistent, direct and flexible way for the required safety level, the importance of variable uncertainty and the statistical models;
– does not hide sources of uncertainties in partial safety factors;
– is easy to use together in with associated modern reliability analysis computation tools.

PROBABILISTIC ASSESSMENT OF THE SAFETY OF HSC COLUMNS

By Sofia M. C. Diniz[1], Member ASCE, and Dan M. Frangopol[2], Fellow ASCE

Abstract

In the present study, a systematic approach for the probabilistic assessment of the safety of high-strength concrete (HSC) columns is developed. The reliability of HSC columns is assessed using a hybrid method combining both Monte Carlo simulation and Advanced First Order Reliability Methods (AFORM). An example problem is presented. In the light of the results reported herein, the major advantages of the full probabilistic approach are discussed.

Introduction

Current design codes are based on semi-probabilistic concepts. While it is agreed that the ideal would be to design a structure or structural element for a target probability of failure, the semi-probabilistic approach has been considered as a compromise between design simplicity and the need for probabilistic concepts. In spite of the large acceptance this code format has received, care shall be exercised in the case of new materials, such as high-strength concrete (HSC).

The ACI Code (1995) recommendations for column design were based mostly on concretes with strengths up to 42 MPa. Due to differences between HSC and normal-strength concrete (NSC) material and structural behavior, using these recommendations for HSC columns does not imply that the same level of safety as for NSC is obtained.

In this study, procedures for the probabilistic assessment of the safety of short and slender HSC columns are presented. The major problems that had to be dealt with are: (a) selection of the limit state; (b) definition of load and resistance; (c) selection of random variables; and (d) implementation of a reliability formulation compatible with the selected failure criterion. The main phases of the reliability assessment process are shown in Figure 1. It is shown that rather than close to uniform probabilities of failure, a wide range is obtained. Moreover, in some cases, very high probabilities of failure may be found in the case of HSC columns.

---

[1] Associate Professor, DEEs, Federal University of Minas Gerais, Belo Horizonte, 30110-060, Brazil.
[2] Professor of Civil Engineering, Department of Civil, Environmental and Architectural Engineering, University of Colorado, Boulder, CO 80309-0428

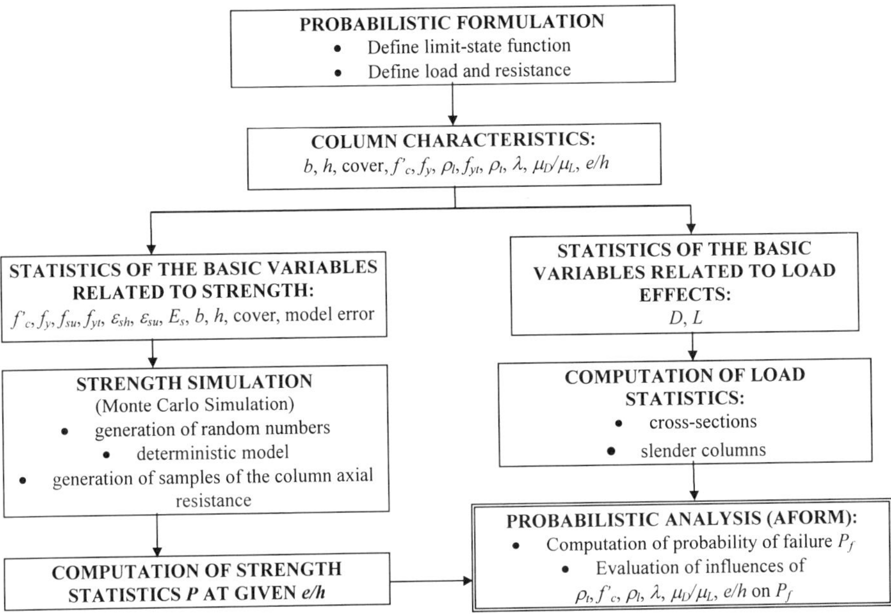

Figure 1 - Main phases of the probabilistic assessment process.

Selection of the limit state

Two major problems are faced in establishing a limit state function for reinforced concrete (RC) columns: (a) there is no unique formulation to define the limit state; and (b) there is no closed form expression to describe column strength. The following expression is used to define the limit state in the present study (Israel et al. 1987):

$$g\left(P,D,L\right)=\left[P^2+\left(\frac{P\,e}{h}\right)^2\right]^{\frac{1}{2}}-\left[\left(D+L\right)^2+\left(\frac{\left(D+L\right)e}{h}\right)^2\right]^{\frac{1}{2}}=0 \qquad (1)$$

where P, D, and L are the random variables; P = the axial resistance of the column at eccentricity e; D = the dead load; and L = the live load acting on the column. In this formulation the load eccentricity e is assumed deterministic. Although this formulation does not establish an upper bound on the "true" probability of failure of the column, it provides a good measure for comparing different design performances.

Load and resistance definition

In order to use the limit state established in Eq. (1), resistance and load effects must be defined in such a way that the external axial load and moment increase proportionally. In the case of slender columns, when the Moment Magnification Method is used, the eccentricity does not remain constant. Therefore, a more complex reliability analysis would be required. Also, the Moment Magnification Method may produce conservative or unconservative estimates of the load effects depending on the load eccentricity. These pitfalls can be avoided if loads and resistances are defined accordingly. In the present study, the load is defined as the applied axial load and the initial moment acting on the column. For each value of the slenderness ratio, a different interaction diagram is obtained.

Selection of random variables

The load statistics provided in Galambos et al. (1982) have been used to describe load variability. The computation of the load statistics for both cross-sections and slender columns was made by first assuming that the design strength exactly matches the design loads and then computing the nominal loads by assuming the mean live load to mean dead load ratio (Diniz and Frangopol 1997a). Next, the mean values were calculated from the nominal loads and the information given in Galambos et al. (1982).

The assumed random variables related to the column resistance are: concrete compressive strength, f'_c; yield strength of the longitudinal steel, f_y; ultimate strength of the longitudinal steel, f_{su}; steel modulus of elasticity, E_s; steel strain at the start of the strain-hardening, ε_{sh}; steel strain at ultimate, ε_{su}; yield strength of the tie steel, f_{yt}; lateral dimension of the column; cover thickness; and model error. A summary of the properties of these random variables is shown in Table 1 (Diniz and Frangopol 1997a).

Table 1. Statistics of the basic variables related to column strength (Diniz and Frangopol 1997a)

Variable	Mean value	St. Dev.	Type of dist.
f'_c=34.5 MPa	35.5	5.67	Lognormal
f'_c=62.1 MPa	55.6	6.67	Lognormal
f'_c=96.5 MPa	78.7	7.87	Lognormal
f_y (MPa)	460	38.2	Beta
f_{su} (MPa)	714	59.3	Beta
ε_{sh}	0.015	0.004	Normal
ε_{su}	0.150	0.030	Normal
E_s (GPa)	200	6.6	Normal
b, h (cm)[1]	+ 0.15	0.635	Normal
cover (cm)[1]	+0.84	0.422	Normal
model error	1.0	0.03-0.11	Normal

[1] Deviation from specified values.

Reliability formulation

A hybrid method using the limit state described by Eq. (1) has been selected for the reliability assessment of HSC columns. In this method, strength statistics are obtained via Monte Carlo simulation and column reliability is computed through Advanced First Order Reliability Methods. The basic requirements in Monte Carlo simulation are the generation of random numbers and a deterministic model for the computation of the column strength. The rationale and algorithm for the computation of the column strength is presented in Diniz and Frangopol (1997b).

Example problem

The probabilities of failure of three unconfined HSC columns 96-13-0, 96-13-22, and 96-13-50 (96 stands for the specified concrete strength, f'_c=96.5 MPa; 13 for longitudinal steel ratio, ρ_l=0.013; the third number to the slenderness ratio λ), at nine eccentricity ratios have been computed using the procedures outlined above. A description of the material and geometric properties of these columns is found in Diniz and Frangopol (1997b). Although all three columns comply with ACI Code-95 requirements, different probabilities of failure result in each case. As it can be seen from Table 2, rather than close to uniform probabilities of failure, a wide range is obtained. Moreover, in some cases, very high probabilities of failure may be found in the case of HSC columns. It should be pointed out that the knowledge of the probability of failure can be made only through a full probabilistic approach.

Table 2. Probability of failure P_f for several eccentricities e/h for columns 96-13-0, 96-13-22, and 96-13-50.

e/h	Probability of failure, P_f		
	96-13-0	96-13-22	96-13-50
4.0	5.41×10^{-6}	1.66×10^{-5}	2.24×10^{-4}
1.5	3.17×10^{-5}	1.65×10^{-5}	1.81×10^{-3}
1.0	5.20×10^{-5}	4.84×10^{-4}	4.53×10^{-3}
0.7	8.80×10^{-5}	1.75×10^{-3}	1.10×10^{-2}
0.5	7.80×10^{-5}	1.64×10^{-3}	1.04×10^{-2}
0.3	8.74×10^{-4}	2.98×10^{-3}	9.04×10^{-4}
0.2	1.31×10^{-3}	3.26×10^{-3}	2.16×10^{-4}
0.1	1.04×10^{-3}	1.93×10^{-3}	3.90×10^{-5}
0.05	1.47×10^{-4}	2.70×10^{-4}	4.80×10^{-5}

Conclusions

In the present study, a systematic approach for the probabilistic assessment of the safety of high-strength concrete (HSC) columns is developed. An example problem is presented. It is shown that high probabilities of failure may be obtained in the case of HSC columns. However, the computation of the probability of failure can be made only through a full probabilistic approach.

References

ACI Committee 318 (1995), "*Building Code Requirements for Reinforced Concrete (ACI 318-95) and Commentary - ACI 318R-95*", American Concrete Institute.

Diniz, S.M.C. and Frangopol, D.M. (1997a), "Reliability Bases for High-Strength Concrete Columns", *Journal of Structural Engineering*, ASCE, Vol. 123, No. 10, pp. 1375-1381.

Diniz, S.M.C. and Frangopol, D.M. (1997b), "Strength and Ductility Simulation of High-Strength Concrete Columns", *Journal of Structural Engineering*, ASCE, Vol. 123, No. 10, pp. 1365-1374.

Galambos, T., Ellingwood, B., MacGregor, J., and Cornell, A. (1982), "Probability Based Load Criteria: Assessment of Current Design Practice", *Journal of the Structural Division*, ASCE, Vol. 108, No. ST5, pp. 959-977.

Israel, M., Ellingwood, B., and Corotis, R. (1987), "Reliability-Based Code Formulation for Reinforced Concrete Buildings", *Journal of Structural Engineering*, ASCE, Vol. 113, No. 10, pp. 2235-2252.

Deterioration Model for Corrosion in Concrete Using Monte Carlo Simulation
By: Edward J Gannon[1] and Paul J. Tikalsky[2], Member ASCE

Introduction

There has been much work performed on the development of deterioration models for concrete bridges. Yet none of the models have been widely accepted by major specifications or design guidelines. All of the models reported can be divided into two groups: deterministic and probabilistic. Deterministic models are based on empirical relationships, while probabilistic models are based on the stochastic behavior of bridge populations. Both groups of models have been successfully used to predict deterioration in bridges. However, they both have inherent flaws, which make their use unwieldy.

Probabilistic models are used to develop relationships when there are a large number of contributing factors, and the contribution of each is not easily or accurately understood. The probabilistic model can include both deterministic events, such as the infiltration of chloride ions into concrete, and probabilistic events, such as the distribution of concrete strengths for a given mixture design, age and curing environment.

Monte Carlo Simulation

One solution to combining both model types is to incorporate the best of both while limiting the drawbacks of both. The use of a statistical simulation can provide the tool for such a solution. A Monte Carlo simulation is a method for generating values for an equation whose variables have a specified distribution. When the distributions are normal and the relationship relatively simple, the determination of the expected value and its distribution follow know statistical relationships. However, as the complexity of the equation increases, and the distribution changes from normal, the exact relationship of the final distribution becomes unclear. With a Monte Carlo simulation, the equation is entered along with the statistical distributions of each variable, and a pseudo-random number generator is then used to pick one value for each variable from its distribution. These values are then substituted into the equation and the result is calculated. From these results, a new distribution is calculated using specific values to develop a solution that is a distribution. One benefit to using a Monte Carlo simulation is that a variable's distribution can be assumed to have any known distribution, such as normal or exponential, or it can be based on empirical data gathered from the real system. By using the actual distribution, and not an assumed statistical distribution, the true nature of the variable can be seen in the result. Monte Carlo analysis as is often described as a series of experiments performed by the computer.

For a corrosion model, the simulator might select a values for reinforcing cover, diffusion coefficient, and equilibrium chloride content. One value from each of those distributions is chosen according to a frequency distribution or histogram . These values are then inserted in the diffusion equation and the resulting chloride

---

[1] Sear and Brown Engineers, Inc., State College, PA 16801
[2] Penn State University, University Park, PA 16803

content is calculated and compared to a threshold value. The output of this model is a distribution of chloride content at various times, the percentage of deck corroded, and various descriptive statistics developed for the results.

Corrosion of Steel in Concrete Model

Reinforcing steel in concrete will not normally corrode unless an outside element affects the system. In an environment with a high pH, like concrete, a passive film forms on the reinforcing steel that acts to prevent corrosion. In order for corrosion to occur, the pH of the concrete must drop below a value of 10, or the passive film must be attacked. The pH of the cement paste is in the range of 12.5 to 13.5 due to the presence of excess amounts of calcium hydroxide $(Ca(OH)_2)$. If the pH drops below 10, then the passive film degrades and corrosion can be initiated. The most common cause of a drop in pH is from carbonation. The rate at which carbonation occurs is controlled by the gas permeability of the concrete. More commonly, in northern climates, corrosion of reinforcing steel is the result of concrete being exposed to both moisture and deicer salts. These deicers are usually soluble chloride based compounds that penetrate the concrete as a function of the quality of the microstructure and the solution concentration. When the chlorides reach a threshold value, corrosion of the steel is initiated. Since surface concentrations of the chlorides are much higher than carbon dioxide, chloride penetration is much faster and the threshold limit may be exceeded in as little as five years. The effect of carbonation has not been reported in this paper.

The deterioration of a concrete bridge deck can be thought of as a two-part process. The first is the penetration of road deicing salts into the concrete to reach the threshold content for corrosion to begin. The second is the destruction of the concrete resulting from the expansion of corroding reinforcing steel. The method of penetration that is most often considered is the diffusion of the chloride ions through the cement matrix. A second method is the short-circuiting of the chlorides to the steel is a result of penetration through shrinkage cracks or subsidence cracks above the reinforcing bars.

The critical factor effecting subsidence cracking is the depth of cover and for shrinkage cracking is curing condition. Weyers, Conway and Cady (1982) found that the spacing of the reinforcing bars can greatly influence the probability of cracking. Dakhil, Cady and Carrier (1975) provide an equation that can be used in a monte carlo simulation model, by relating the concrete cover, the bar size, and concrete slump to the probability of occurrence of subsidence cracking. The relationship is:

$$p = \frac{1.5e^{y} - 0.5}{1 + e^{y}} \hspace{3cm} 1$$

Where: p=probability of subsidence cracking y=$1.37-0.0228x_1-0.56x_2+0.0106x_3$

x_1=concrete cover (mm) x_2=concrete cover/bar size

x_3=concrete slump (mm)

Chloride Diffusion

Most often, the diffusion of chlorides through concrete and the resulting corrosion of reinforcing is thought to be the primary cause of corrosion related deterioration to bridge decks. The penetration of chlorides through concrete is most nearly modeled as a diffusion controlled event. Fick's Second Law has been used to model this solution:

$$C_{(x,t)} = C_o\left[1 - erf\left(\frac{x}{2\sqrt{D_c t}}\right)\right] \dots\dots\dots 2$$

Where: $c_{(x,t)}$=chloride concentration at depth x after time t for an equilibrium chloride concentration, C_o at the surface.

erf=the Error Function, a standard mathematical function found in tabulated and graphical form

D_c=chloride diffusion constant.

The corrosion threshold chloride content generally varies between 0.71 and 1.07 kg of chloride per cubic meter of concrete.

Deterioration Model - Chloride Related Corrosion Damage

Typical values for the diffusion coefficient for states in the snow belt vary from 0.05 to 0.15. The equilibrium chloride content, C_o, of concrete is directly attributable to the quantity of deicer salts used. For the deterioration model, an average chloride content of 7 was used to model moderate to high chloride usage, as is typically found in northern states. For the model, a standard deviation of 1.0 was used.   With the development of distributions for the diffusion coefficient, equilibrium chloride content and depth of cover, it is possible to generate a distribution of chloride contents with time. One final distribution must be developed prior to completing the deterioration model. This distribution is for the chloride threshold beyond that corrosion of the reinforcing commences. The exact chloride level at that corrosion may start has been found to be a function of both chloride and hydroxide concentration in the cement paste.

The data collected shows a constant equilibrium chloride value of 5.3 kg of chloride per cubic meter of concrete, and a diffusion coefficient of 5.8 cm^2 per year is used.   An average threshold chloride value of 1.5 pounds per cubic yard is delineated. There is a very wide difference between the time it takes to reach the threshold value at 38 mm of cover (approximately 6 years) compared with a cover depth of 75 mm (approximately 26 years). At 26 years for a cover of 100 mm, the chloride content is only 0.47 kg per cubic meter, only half of the threshold value. Increasing the cover depth not only delays chloride penetration, but also reduces the probability of subsidence cracking, thereby providing a major role in reducing the probability of corrosion during the expected service life. For an equilibrium chloride concentration of 7 kg per cubic meter, corrosion is expected in less than 10 years.

But, corrosion will be delayed until year 27 if C_o is 2.4 kg per cubic meter. The diffusion coefficient, D_c, is a material related property of the concrete. While a relationship between the diffusion coefficient and permeability, compressive strength, and water-cement ratio is easily understood, however, the exact nature and form of this relationship is not known at this time. This is especially true of the new "high performance" concrete.

Model Results

Determination of area of the deck experiencing active corrosion was also modeled using the Monte Carlo Simulator. This was performed by comparing the chloride concentration determined by the diffusion model with the chloride threshold. To complicate this, a distribution of values of the chloride threshold limit (C_t) were used. The distribution used had an average of 0.89 and a range of 0.71 to 1.07. The chloride concentration at the reinforcing level (Cl) was calculated, and then subtracted from the chloride threshold value selected from its distribution. If Cl was less than C_t, then no corrosion would be occurring and the difference would be positive but less than 1.07. If Cl was greater than or equal to C_t, then corrosion would be occurring and the difference would be less than or equal to zero. The area corroding could then be found using zero as the point of non-exceedence. This indicates that at age 15 years, 34.11 percent of the deck has a chloride content greater than the threshold limit either now or at some time in the past.

The results show there will be some corrosion occurring at an age of only approximately 3 years and even at 40 years there will be some parts of the deck experiencing no corrosion. The median chloride concentration crosses the mean threshold line at approximately 17 years. According to the data, at 17 years, only 44 percent of the deck is corroding. The data for the deck area undergoing active corrosion show a very low rate of deterioration at early ages. The rate of deterioration increases, until it reaches a maximum value at approximately 50 percent corrosion. At this point, the rate of corrosion begins to decrease until the entire deck has corrosion activity. The rate of deterioration is lower with increasing cover. This can be attributed to the form of the diffusion equation. The maximum slope for a 50-mm cover is approximately 4.9 percent/year, this decreases to 2.8 percent/year for 75-mm of cover.

References

Cady, P.D., and Weyers, R. E. (1983), "Chloride Penetration and Deterioration of Concrete Bridge Decks," Cement, Concrete and Aggregates, CCAGDP, Volume 5, Number 2, Winter.

Cady, P.D., and Weyers, R.E. (1984), "Deterioration Rates of Concrete Bridge Decks," Journal of Transportation Engineering, ASCE, Volume 110, Number 1, January.

Dakhil, F.H., Cady, P.D. and Carrier, R.E., (1975), "Cracking of Fresh Concrete as Related to Reinforcement," ACI Journal, V 72, No. 8, August, p. 421 - 428.

Critical Review of Fully Probabilistic Design for Seismic Loadings

Jungwon Huh,[1] Ali Mehrabian,[1] Achintya Haldar,[2] Fellow, ASCE
and Alfredo Reyes Salazar[3]

Abstract

The necessity of fully probabilistic design for short duration dynamic loadings including seismic loading is critically reviewed. Implementation of fully probabilistic design for static loadings has had limited success, particularly in cases where the limit state functions are available in explicit form. However, in most cases of practical significance, the limit state functions are not available even for static loadings. For nonlinear static loadings and linear or nonlinear dynamic or seismic loadings, the limit state function is implicit, and can only be expressed in algorithmic form, e.g. using the finite element algorithm. The issues and challenges of fully probabilistic design for implicit limit state functions emphasizing seismic loading are discussed. A method is proposed and the procedure is explained with the help of an example.

Introduction

Structural reliability assessment techniques, particularly under static loadings, have evolved considerably in the last three decades. The profession is having limited success using fully probabilistic design approaches expressing performance or limit state functions in explicit form with complete distributional information on the random variables in them. The probability of failure corresponding to a limit state (strength or serviceability) can be calculated by evaluating a multidimensional integral. This approach is impractical in many cases since the joint probability density function of the basic random variables involved in the limit state may not be available. This led to the development of several reliability methods (Haldar and Mahadevan, 1999) with different

---

[1]Doctoral Student, CEEM Dept., University of Arizona, Tucson, AZ 85721
[2]Professor, CEEM Dept., University of Arizona, Tucson, AZ 85721
[3]Professor, University of Sinaloa, Mexico

levels of complexity, including the first and second order reliability methods (FORM/SORM) and many different Monte Carlo simulation schemes. In most cases, these algorithms are applicable for simple cases, e.g., the reliability of a structural element (tension or compression members, beams, etc.) considering linear behavior. The load and resistance factor design (LRFD) concept was introduced in several model building codes for this purpose. However, the available methods may not be usable for complicated structures where the presence of several members needs to be considered, or where the limit state is with respect to the overall structural behavior, e.g., lateral deflection. In general, if the structure requires an algorithmic representation, e.g., in the form of finite element, currently available methods may not be applicable, even for static loading. In short, if the limit state functions are implicit, the derivatives of the performance functions with respect to the basic random variables can not be evaluated to obtain the gradient vector required for the FORM/SORM. Thus, the applicability of the available probabilistic design methods needs to be extended to consider linear and nonlinear behavior under static and dynamic loadings. This is the subject of the paper.

Reliability Methods for Implicit Performance Function - Static Loading

To evaluate the safety of complicated structures in the presence of different sources of nonlinearity and uncertainty, a finite element-based formulation is desirable, since it is also the first step in a conventional deterministic analysis. The use of the finite element method (FEM) in the context of uncertainty leads to the concept of the stochastic finite element method (SFEM). Using iterative perturbation sensitivity analysis (the most appropriate for nonlinear analysis), the finite element method, and the FORM, the authors proposed such an SFEM algorithm for static loading only. In this approach, the value of the limit state function is determined by the deterministic finite element analysis of structures. The gradient vector is calculated using the iterative perturbation technique. The algorithm appears to be very efficient and elegant in estimating risk of any linear and nonlinear structures under static loading. It has been used to verify the LRFD concept used in the steel industry.

Reliability Methods for Implicit Performance Function - Seismic Loading

The considerable damage suffered by structures during recent earthquakes and strong winds prompted the profession to study the reliability of structures subjected to dynamic loadings. Not only are the uncertainties in dynamic loadings relatively large compared to static loadings, but the performance functions also need to be developed at the structural or system level, and they are functions of time and location. Thus, the application of fully probabilistic design approaches in the context of dynamic loadings is expected to be much more complicated than for static loading cases.

There are very few methods available to consider the uncertainties in the seismic loading and the linear and nonlinear behavior of structures. The Monte Carlo simulation method, SFEM, and response surface method (RSM) can be used for this purpose. Although a class of Monte Carlo simulation methods may be appropriate, considering their accuracy and computational efficiency, they may be too costly and cumbersome for the simulation of nonlinear dynamic systems. The RSM (Khuri and Cornell,1996) has the potential to consider the appropriate mechanical behavior of systems and the uncertainty in the load and resistance-related parameters without compromising the efficiency and accuracy to a great extent. The authors would like to explore the possibility of developing such an algorithm, in the context of RSM, SFEM, and FORM. Once the algorithm is developed, it can be verified using the Monte Carlo simulation method. Various components of the algorithm need further discussion.

Response surface methodology is an important element of the proposed study. The primary purpose of applying RSM in reliability analysis is to approximate either the original and implicit limit state function or the structural response statistics, i.e., the mean and coefficient of variation of response, using a simple and explicit polynomial. Some additional approximations are necessary to formulate the limit state function using the response statistics. The implementation potential of all the RSMs, in terms of accuracy and efficiency, depends on the selection of the center point and the sampling points around the center point.

At least a second order polynomial is necessary for nonlinear problems under seismic loading. Thus, the strategy for the selection of the center point and the sampling points needs discussion, in the context of the second order polynomials. Conceptually, the center point should be close to the most probable failure point (MPFP) so that the response surface includes most of the failure region with sufficient accuracy. Two possible schemes for the selection of the center point are a trial and error scheme, and an iterative linear interpolation scheme. The success of the trial and error scheme depends on the experience of the simulator, and may not be feasible for everyday use. Since the algorithm will be based on the FORM concept, the iterative linear interpolation scheme will be quite appropriate. It will be discussed in detail during the presentation.

Available techniques for the selection of sampling points can be divided into two categories: classical design and saturated design. In the classical design approach, responses are calculated at several sampling points and then the regression analysis is carried out to formulate the response surface. In order to fit a second order surface for k input variables, the sampling points must have at least three levels for each variable, leading to 3^k factorial design. To increase its efficiency, the central composite design is also proposed. It consists of a 2^k factorial design augmented by a center point and $2k$ axial points. The basic drawback of the central composite design is that it requires too many sampling points to fit a second order surface, particularly when a large number of random variables needs to be considered.

The saturated design consists of as many sample points as the total number of coefficients necessary to define a polynomial. The unknown coefficients are obtained by solving a set of linear equations, without any regression analysis. Without going too much in detail, it can be shown that for k random variables, the saturated design using a second-order polynomial with cross term would require $[(k+1)(k+2)/2]$ deterministic analyses. The corresponding number for the cental composite design is (2^k+2k+1). For $k = 15$, the sampling points for the two choices are 136 and 327999, respectively. To increase the efficiency, the central composite design can be used only in the final iteration to develop the response surface. In all previous iterations, saturated design using a full second order polynomial will be used.

Reliability Estimation for Both Strength and Serviceability

Since a structure can fail due to excessive lateral or interstory deflection, or due to failure of several components in strength forming either local or global mechanism, they need to be considered separately. Limit states corresponding to drift and deflection will be formulated using current code recommendations or permissible values. The response surface will be used to estimate drift and deflection due to seismic loading and then the corresponding reliability under seismic loading will be estimated.

The design criterion for a beam-column element requires that an interaction equation consisting of both the axial load and the bending moment must be satisfied. However, since the axial load effect and the bending moment are not explicit and their maximum values do not occur at the same time, it is necessary to develop the response surfaces explicitly for the axial load and bending moment separately as well as jointly. The corresponding reliabilities will be calculated using the proposed algorithm.

Examples and Conclusions

The details of the procedure can not be discussed here due to lack of space, but will be discussed during the presentation. The algorithm will be elaborated with the help of an example. Several important weaknesses in currently available methods will be identified, particularly for fully probabilistic design for seismic loadings.

Acknowledgments

This paper is based on work partly supported by the National Science Foundation under Grant CMS-9526809. Financial support for the last author from the University of Sinaloa, Mexico, is also appreciated. Any opinions, findings, conclusions, or recommendations expressed in this publication are those of the authors and do not necessarily reflect the views of the sponsor.

References

Haldar, A., and Mahadevan, S., *Reliability Methods in Engineering Design (Including Stochastic Finite Element Analysis)*, John Wiley & Sons, New York, NY, to appear in 1999.

Khuri, A. I., and Cornell, J. A.(1996) *Response Surfaces Designs and Analyses*, Marcel Dekker, Inc, New York, N.Y.

GENETIC ALGORITHM FOR MULTIOBJECTIVE OPTIMIZATION AND LIFE-CYCLE COST

Franklin Y. Cheng[1], Fellow, ASCE
A. H-S Ang[2], Honorary Member, ASCE
J.H. Lee[3] and D. Li[4], Members, ASCE

Abstract

Loss of life and property from possible future earthquakes as well as the expense and difficulty of post-earthquake rehabilitation and reconstruction strongly suggest the need for proper structural design with damage control. Design criteria should balance initial cost of the structure with expected losses from potential earthquake-induced structural damage. Life-cycle cost design addresses these issues. Such a design methodology can be developed using multiobjective and multilevel optimization techniques. These techniques are well-suited to the use of genetic algorithms (GA). GAs have the characteristic of maintaining a population of solutions, and can search in a parallel manner for many nondominated solutions. These features coincide with the requirement of seeking a Pareto optimal set in a multiobjective optimization problem. The rationale for multiobjective optimization via GAs is that at each generation, the fitness of each individual is defined according to its nondominated property. Since

[1]Curators' Professor of Civil Engineering and Senior Investigator with Intelligent Systems Center, University of Missouri-Rolla, Rolla, MO 65409-0030

[2]Professor, Department of Civil and Environmental Engineering, University of California, Irvine, CA 92697

[3]Visiting Associate Researcher, Department of Civil and Environmental Engineering, University of California, Irvine, CA 92697

[4]Structural Engineer, Ludwig Buildings, Inc., Harahan, LA 70123 (former research assistant at UMR)

nondominated individuals are assigned the highest fitness values, the convergence of a population will go to the nondominated zone: the Pareto optimal set. Based on this concept, a Pareto GA whose goal is to locate the Pareto optimal set of a multiobjective optimization problem, is developed along with Pareto-filter in order not to miss Pareto optimal points during evolutionary processes.

Life-cycle Cost Model

Recent studies have focused on loss estimation for earthquakes and other natural hazards (FEMA, 1994). Jones and Chang (1995) provide an overview of current work on the economic impact of natural disasters. Nigg (1995) presents community-level disaster preparedness planning and response as well as mitigation actions. These studies are generally conducted to determine the after-effects of an event, or to estimate gross regional losses expected in a future event, and were intended primarily for post-earthquake recovery and response planning (Cheng and Wang [eds.], 1996; Frangopol and Cheng [eds.], 1996)

The platform for the systematic integration of seismic reliability engineering and socioeconomics is the expected life-cycle cost function which can be summarized as follows

$$C_T = C_I + C_D \tag{1}$$

where C_I is the intial cost of a structure, and C_D is the damage cost over the life of the structure composed of the following (for buildings)

$$C_D = C_r + C_c + C_e + C_s + C_f \tag{2}$$

in which C_r = repair or replacement cost of the structure; C_c = loss of contents; C_e = economic impact of structural damage; C_s = cost of injuries casued by structural damage; and C_f = cost of fatalities from structural damage or collapse. Items in Eq. (2) comprising the total damage cost are functions of structural damage level, x; for example, item C_r has an expected repair cost of

$$C_F = \int_0^\infty C_F(x) f x(x) dx \tag{3}$$

where x = damage level, and fx(x) = probability density function (PDF) of X. Structural damage, x, and its PDF fx(x) are calculated for a given earthquake, from which the particular damage costs, $C_r(x)$, can be determined as a function of damage level, x.

Direct losses caused by an earthquake include human fatality and injury, as well as physical damage to structures and their contents. Concerning human fatality, Wiggins (1979) correlated the number of fatalities to building damage; Shiono and Krimgold

(1989) reported fatility data for a hospital that collapsed during the 1985 Mexico City earthquake; and Coburn *et al.* (1992) suggested a simulation model to estimate the number of fatalities in buildings of a given structural type. Besides immediate or direct losses from an earthquake, there is also secondary or indirect impact: e.g., economic loss associated with the disruption of business due to structural damage (Nigg, 1996). Ang and De Leon (1996) examined the minimum expected life-cycle cost vis-a-vis structural damage, and the reliability analysis of an RC frame building based on actual cost data from Mexico City buildings damaged in the 1985 earthquake.

The general approach described above is illustrated for a class of 5-story RC office buildings located in a soft soil site in the Los Angeles area. The structure of each model building consists of parallel frames similar to that shown in Fig. 1. Only the building response in the short direction is considered for this study. Several model buildings (SMRF frames) were designed according to UBC code for service level seismic base shear coefficients, C_b, ranging from 0.04 to 0.13. Currently, C_b is equal to 0.08 in the UBC. A finite element model is constructed for each model building and used for the nonlinear time-history analyses required for structural damage and reliability assessments.

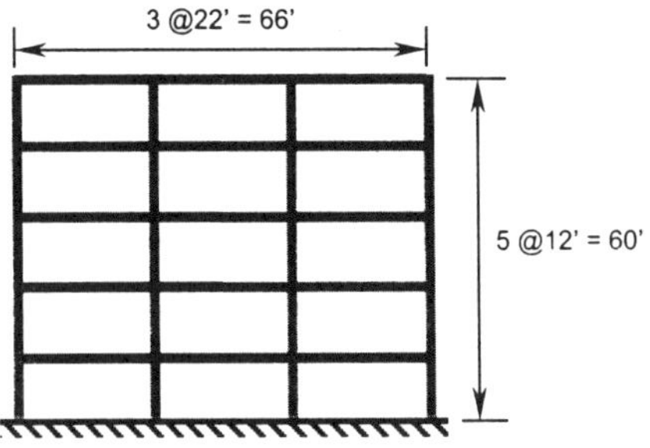

Fig. 1 Model Building

A regression relation between the computed median global damage index and the actual or estimated damage repair costs was obtained on the basis of reported damage repair costs for RC buildings damaged under previous earthquakes, together with damage assessment analyses. The regression relation for the normalized repair cost is formulated and expressed in terms of the median global damage index, leading to the following equations for the damage repair cost function

$$C_R = \alpha_1 \cdot (d_m)^{\alpha_2} ; \qquad 0 \le d_m \le d_0$$
$$C_R = C_I ; \qquad\qquad d_m > d_0 \tag{4}$$

where d_m = median global damage index of the structure and $\alpha_1 = 0.57C_I$, $\alpha_2 = 1.0$, and $d_0 = 0.5$. Value of the contents of the building, which is an office building, is assumed to be 40% of the initial building cost, and the loss of contents is assumed to reach its maximum for a median global damage index $d_m = 1.0$. Making use of available data, the loss-of-contents function for all values of the median structural damage is as follows

$$C_C = 0.4C_R; \qquad\qquad\qquad 0 \le d_m \le 0.5$$
$$= 0.114C_I + 0.572(d_m - 0.5)C_I; \quad 0.5 \le d_m \le 1.0 \qquad (5)$$
$$= 0.4C_I; \qquad\qquad\qquad d_m \ge 1.0$$

Expected life-cycle costs and structural reliabilities are obtained for seismic hazard at a site in downtown Los Angeles. Initial costs and total expected life-cycle costs for the five designs of the model building are shown in Figs. 2 and 3 as a function of the probability that a specified damage level is exceeded over the 50-year life span of the structure at a given site for an average annual discount rate q = 4%.

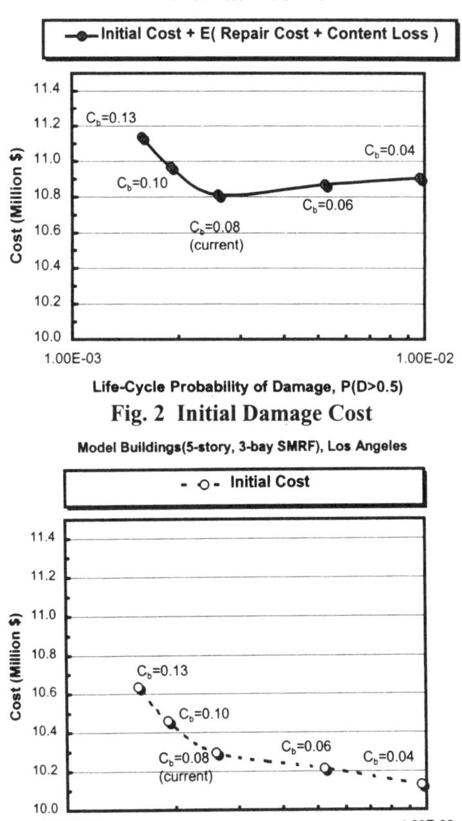

Fig. 2 Initial Damage Cost

Fig. 3 Expected Life-Cycle Cost

The work based on genetic algorithm (GA) for multiobjective optimization (MOP) with target reliability and minimum expected life-cycle cost as multilevel optimization is summarized in Fig. 4.

Acknowledgments

This paper is a portion of research results from the project supported by the National Science Foundation under grant NSF CMS-9703725 and the Manufacturing Research Training Center (MRTC) at UMR. The financial support is gratefully acknowledged.

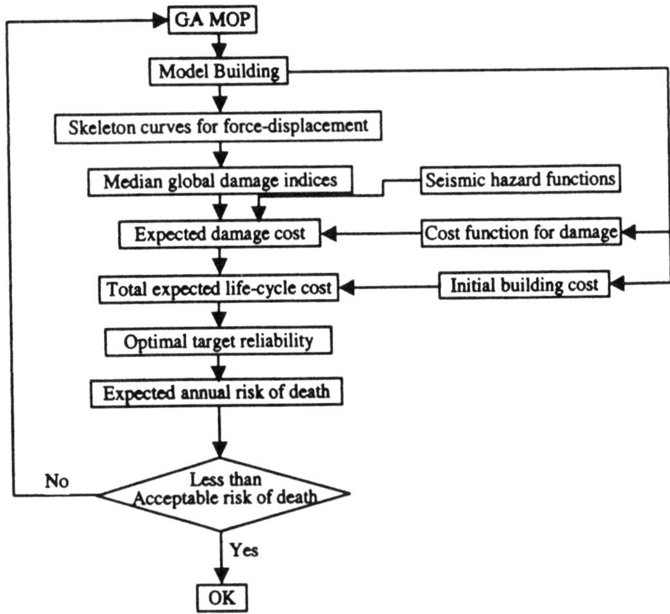

Fig. 4. Multilevel Optimization Scheme

References

Ang, A.H-S and DeLeon, D. (1996), "Basis for Cost-Effective Decision on Upgrading Existing Structures for Earthquake Protection," *Post-Earthquake Rehabilitation and Reconstruction*, eds., F.Y. Cheng and Y.Y. Wang, Elsevier Science Ltd, 69-84.

Cheng, F.Y., and Li, D. (1997), "Multiobjective Optimization Design with Pareto Genetic Algorithm," *J. Struct. Engrg.*, ASCE, 123:9, 1252-1261.

Cheng, F.Y. and Wang, Y.Y., eds. (1996), *Post-Earthquake Rehabilitation and Reconstruction*, Elsevier Science Ltd.

Coburn, A.W. Spence, R.J.S., and Pomonis, A. (1992), "Factors Determining Human Casualty Levels in Earthquakes: Mortality Prediction in Building Collapse," *Proc. 10th World Conference on Earthquake Engineering*, Madrid, 5989-5994.

FEMA-2149 (1994), "Assessment of the State-of-the-Art Earthquake Loss Estimation Methodologies," Washington, D.C.

Frangopol, D. and Cheng, F.Y., eds. (1996), *Advances in Structural Optimization*, ASCE.

Jones, B.G. and Chang, S.E. (1995), "Economic Aspects of Urban Vulnerability and Disaster Mitigation," *Urban Disaster Mitigation: The Role of Engineering and Technology*, eds., F.Y. Cheng and M.S. Sheu, Elsevier Science Ltd., 311-320.

Nigg, J.M. (1995), "Social Science Approaches in Disaster Research: Selected Research Issues and Findings on Mitigating Natural Hazards in the Urban Environment," *Urban Disaster Mitigation: The Role of Engineering and Technology*, eds., F.Y. Cheng and M.S. Sheu, Elsevier Science Ltd., 303-310.

Shiono, K. and Kirmgold, F. (1989), "A Computer Model for the Recovery of Trapped People in a Collapsed Building: Development of a Theoretical Framework and Direction for Future Data Collection," *International Workshop on Earthquake Injury Epidemiology for Mitigation and Response.*

UBC, *Uniform Building Code* (1997), International Conference of Building Officials, Whittier, Calif.

Wiggins, J.H. (1979), "Estimated Building Losses from U.S. Earthqaukes," *Proc., 2nd National Conference on Earthquake Engineering*, 253-262.

Optimal Design of Transmission Line Structures for Earthquake Loads Using a Genetic Algorithm

Fatma Y. Kocer[1] and Jasbir S. Arora[2], Member ASCE

Introduction

Major thrust of the present research is to develop realistic problem formulations for optimum design of transmission line structures subjected to dynamic loads. Although an H-Frame pole presented in ASCE (1990) is used for demonstration purposes, other types of pole structures and towers can be treated in similar ways. Formulations for optimum design of the problem are presented by increasing the number of design variables and/or by changing the cost function. The structure is subjected to an earthquake record and nonlinear time history analysis is performed to determine its. Members are allowed to deform plastically. Since all the formulations involve discrete design variables, a genetic algorithm that is suitable for solving such problems is used to search for optimal designs. From the results, it is observed that designs obtained with nonlinear time history analysis of the example structure are cheaper than the ones obtained with linear analysis. This conclusion cannot be generalized, however. Also better solutions are found as the number of design variables is increased. Details of the formulation and solutions are presented in Kocer and Arora (1998); they are only summarized here.

The overall dimensions of the example structure are given in Fig. 1. It is composed of two poles, one crossarm and an X-brace. The connection points between the crossarm and the poles are fixed. As can be seen in the figure, there are three conductors attached to the crossarm. Also there are two shield-wires, one for each pole. The reader should refer to ASCE (1990) for the material properties and the design loads. In the present examples, it will be subjected to NESC Light loading and a dynamic load simultaneously. Dynamic loading consists of an earthquake record of 15 sec. having 50 data points per second (Maison, 1992). In addition, loads due to self-weight of the structure are included.

Optimal Design Problem Formulation

Design Variables: For the structure, the design variables are: D_{PTE}, T_{Pole}, t_{Pole}, S_{Pole} = exterior diameter at the tip (mm), tapering, thickness (mm), coefficient defining cross-sectional shape of the poles; D_{CE}, $t_{Crossarm}$, $S_{Crossarm}$ = exterior diameter (mm), thickness of the crossarm (mm) and coefficient defining cross-sectional shape of the crossarm; D_{BE}, t_{Brace}, S_{Brace} = exterior diameter (mm), thickness (mm) and coefficient defining cross-sectional shape of the X-brace members; M_{Pole}, $M_{Crossarm}$, M_{Brace} = steel grade used in the pole, crossarm and X-braces (MPa). All the variables are treated as discrete variables.

Cost Function: The first cost function is minimization of the material cost (\$) $C_m = C_w \rho V$, where C_w is the cost per unit mass (0.77 \$/kg), ρ is the mass density (7,849 kg/m³) and V is the structural volume. Treatment of the cost of different grades of steel is

---

[1] Graduate Research Assistant

[2] Professor, Civil and Environmental Engineering, The University of Iowa, Iowa City, IA 52242.

explained in Kocer and Arora (1997). The second cost function is minimization of total initial cost which includes the material cost, welding cost, painting cost and galvanizing cost expressed as: $C_I = C_W \rho V + C_P A_{Surface} + C_G \rho V + C_{weld} L_{Weld}$, where C_P is the cost of painting/area (12.81 \$/m²), C_G is the unit cost of galvanizing (0.22 \$/kg), $A_{Surface}$ is the surface area. Calculation of welding cost C_{weld} is explained in Kocer and Arora (1998).

Design Constraints: Design constraints are formulated according to ASCE (1990) which are requirements for local buckling, shear stress, bending stress, and the combined stress constraints. In addition to these, deflection at the tip of the pole and the base diameter should not exceed specified limits. All the stress and the deflection constraints are time dependent.

Optimal Design Process

There are two major steps in computer implementation of the optimal design process: analysis step, and optimization step. In the present work, structural analysis is performed using ANSR, a nonlinear structural analysis program (Maison, 1992). The program calculates response histories using an implicit time integration of equations of motion. It can treat geometric nonlinearity as well as elasto-plastic material behavior. The optimization problem is solved using IDESIGN (Arora et al, 1993). IDESIGN is an interactive design optimizer, which has algorithms for both continuous and discrete design variables. For detailed explanations of these methods, Arora and Huang (1996) can be referred. Since the problem formulations involve discrete design variables, a genetic algorithm (GA) is used to obtain optimal solutions. This is a fairly simple algorithm that does not require evaluation of the gradients of the cost or constraint functions. The constraints are treated in the algorithm using a penalty function approach. An exact penalty function is proposed and implemented to treat time dependent constraints.

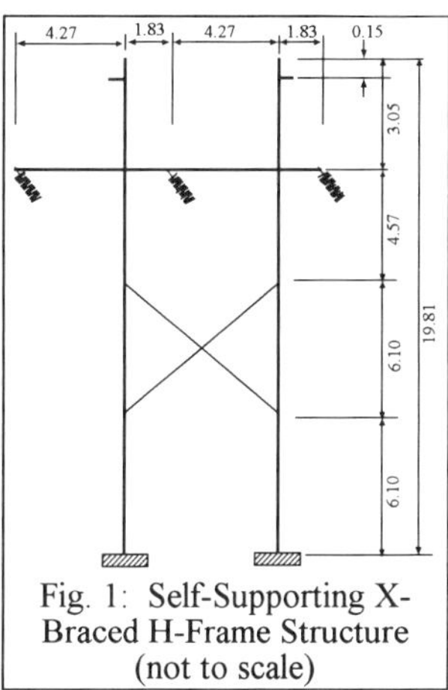

Fig. 1: Self-Supporting X-Braced H-Frame Structure (not to scale)

Numerical Results

The poles are modeled as 7 beam elements and the crossarm as 4 elements. Since there are two poles, there are 18 beam elements. Six constraints are checked at each end of the beam element and so there are 216 constraints. There are two truss elements and for each element there is one constraint. In addition to these, deflection at the tip of the pole is limited to 0.254m. and diameter at the base of the poles is limited to 0.61m. As a result there are 220 design constraints out of which 219 are time dependent. Before the problem under dynamic loads was solved, the structure was optimized for static loads to verify the computer implementation and to compare results given in ASCE (1990). The optimal design was 8.8% cheaper than the design given in that manual.

Various optimum solutions obtained for a static loading (NESC Light Loading) and a dynamic loading (an earthquake accelerogram) applied simultaneously are summarized in Table 1. 1500 time grid points are used in the analysis and the optimization process. First, the effects of analysis types on optimal designs are compared. For this purpose, the problem is analyzed and optimized using linear (Column 2), materially linear-geometrically nonlinear (Column 3) and fully nonlinear (Column 4) analyses. Design variables, design constraints and the cost function are the same as for the previous case. When Column (2) is compared with Column (3), it is seen that the pole and brace diameter, and the crossarm thickness are smaller in Column (3). The crossarm diameter, brace thickness and pole slope are larger but the pole thickness is the same. The design with materially nonlinear analysis is 1.4% lighter compared to the design with linear analysis. Design in Column (4) has larger pole and brace diameters but smaller brace thickness compared to design in Column (3). Crossarm diameter, pole and crossarm thickness and pole slope are the same in both the designs. As a result, design in Column (4) is 0.83% lighter. Overall it is seen that the structure gets lighter as the nonlinearity is included. More significantly, the final designs are quite different. In ANSR, both linear and nonlinear analyses are performed using iterative procedures. Therefore CPU time required for all three analyses are almost the same. Using modal superposition approach for linear analysis can significantly reduce the CPU time.

Table 1. Optimal Designs

(1)	Cost = C_m (2)	Cost = C_m (3)	Cost = C_m (4)	Cost = C_I (5)	Cost = C_I (6)	Ranges (7)
D_{PTE} (mm)	210.82	208.28	213.36	198.12	264.16	a
D_{CE} (mm)	251.46	266.70	266.70	248.92	233.98	b
D_{BE} (mm)	129.54	114.30	119.38	114.30	119.38	c
t_{POLE} (mm)	4.7625	4.7625	4.7625	4.7625	4.7625	d
$t_{CROSSARM}$ (mm)	6.3500	5.55625	5.55625	7.14375	6.3500	e
t_{BRACE} (mm)	4.7625	5.55625	4.7625	5.55625	5.55625	e
T_{POLE}	0.00668	0.00678	0.00658	0.00728	0.00658	f
S_{POLE}				(3.22)	3.32	g
$S_{CROSSARM}$				(3.46)	4.00	g
S_{BRACE}				(3.14)	3.14	g
M_{POLE} (MPa)				(448.16)	344.74	h
$M_{CROSSAARM}$ (MPa)				(448.16)	344.74	h
M_{BRACE} (MPa)				(241.38)	241.38	h
G (%)	0.008	0.00	0.00	0.00	0.00	-
Cost ($)	1838.07	1812.32	1797.29	3616.86	3480.54	-
Analysis Method	Linear	ML-GN	Nonlinear	Nonlinear	Nonlinear	-
CPU/an.	4.19	4.18	4.18	4.20	5.79	-
N_A (%)	0.0876	0.0844	0.0605	0.0737	0.00001	-

G is maximum constraint violation in percent, N_A = (Number of analyses performed/number of enumeration cases) × 100
a = 198.12 - 279.40; 2.54, b = 215.90 - 266.70; 2.54, c = 88.90 - 139.70; 2.54, d = 4.7625 - 7.1438; 0.79375, e = 4.7625 - 7.9375; 0.79375, f = 0.00658 - 0.00758; 0.0001, g = 3.14(round), 3.19(hexdec.), 3.22(dodec.), 3.32(octa.), 3.46(hexa.), 4.00(square), h = 241.38344.74, 413.68, 448.16

Next, the effects of problem formulation are investigated. The problem is formulated using two different cost functions and two different sets of design variables. The first cost function is the material cost and the second one is the total initial cost which is a sum of material, galvanization and painting costs. Seven design variables of the first set are diameters at the tip of the pole, crossarm and brace; pole, crossarm and brace thickness; and tapering of the pole. Second set of six additional design variables have cross-sectional shapes, and steel grades of the pole, crossarm and brace members in addition to the design variables of the first set. Allowable discrete values are listed in Column (7) of Table 1. Design with total initial cost is given in Column (5); note that the numbers in the brackets are the fixed values used for the variables. When this design is compared to the one in Column (4), one can see that the designs are quite different. Design in Column (5) has smaller diameters but thicker elements. The pole slope is also larger in Column (5). The final designs have no constraint violations. Next, we compare the designs in Column (6) and (5). Both of these designs are obtained with nonlinear analysis and the cost function is the total initial cost. However, Column (6) has 6 additional design variables. Pole and X-brace diameter, crossarm thickness are larger, brace diameter is smaller. Pole and brace thickness are the same. The shape of pole changes from dodecagonal to a octagonal, crossarm changes from hexagonal to a square, and brace remains the same. Yield strength of pole and crossarm decrease and yield strength of X-brace remains the same. The cost function value is decreased by 3.77% with the inclusion of 6 more design variables.

Discussion and Conclusions

Formulations for optimum design of transmission line structures subjected to earthquake loads were presented using an H-frame pole as an example. The structural analysis problem was treated as a nonlinear dynamics problem where geometric as well as material nonlinearities were included. An implicit time integration algorithm was used to determine the response history for the structure. Since stress and displacement constraints for the problem were time dependent, procedures to treat such constraints were developed and used in the solution process. The thickness, diameter and slope of the members were treated as discrete design variables. With this treatment, it is possible to use standard available section for fabrication of the structure (Kocer and Arora 1997). In addition to these design variables, solution cases were created to treat material properties and the cross sectional shape as design variables. Since all the formulations involved discrete design variables, a genetic algorithm (GA) was used to search for optimum solutions. This algorithm has been shown to be quite effective for discrete variable problems.

The optimal designs for the example structure obtained with nonlinear analysis were less expensive than the ones obtained with linear analysis. More importantly, the final designs were different. Problem formulations, especially definition of design variables, had an important effect on the final designs. A better optimal solution was found with a larger set of design variables but it also required a larger CPU time. The final design changed significantly as well. Similarly, with a different cost function in the formulation, the design changed significantly. Overall, implementation of different cost functions and design variables was quite straightforward with the genetic algorithm; therefore, for any given problem, several formulations can be tried to develop practical ones.

In the present work, only one earthquake load was considered. For more realistic designs, several different earthquakes should be considered as loading conditions. Inclusion of other dynamic loads will further increase the CPU time significantly but the solution process remains the same. In the present work, an entry-level workstation was used for all calculations. Use of parallel computers and supercomputers can alleviate the problem of large CPU time for optimization with genetic algorithms.

References

Arora, J.S.; Huang, M.W. (1996), "Discrete Structural Optimization with Commercially Available Sections", *J. of Str. Mech. & Earth. Eng.*, JSCE, 13 (2), 93-110.

Arora, J.S.; Lin, T.C., Elwakeil, O.A. and Huang, M.W. (1993), *IDESIGN User's Manual 4.2*, Tech. Report, ODL-93.04. Opt. Design Lab., Civil Eng., U. Iowa, Iowa City, IA.

ASCE (1990), *Design of Steel Transmission Pole Structures*, 2nd ed., ASCE, Reston, VA.

Kocer, F.Y.; Arora, J.S. (1997), "Standardization of Steel Pole Design Using Discrete Optimization.", *J. of Str. Eng.*, 123 (3), 345-350.

Kocer, F.Y.; Arora, J.S.. (1998), *Optimal Design of Transmission Poles Subjected to Earthquake Loading*, Tech. Rep. ODL-98.03. Opt. Des. Lab., Civil Eng., U. Iowa, Iowa City, IA (submitted for publication in J. of Str. Eng., ASCE, Nov. 98).

Maison, B.F. (1992), *"PC-ANSR: A Computer Program for Nonlinear Structural Analysis"*, Civil Engineering, University of California, Berkeley, CA.

Composite Frame Design Using a Genetic Algorithm

Charles Camp[1], Jifei Li[2], and Shahram Pezeshk[1]

ABSTRACT

A design procedure utilizing a genetic algorithm (GA) is developed for discrete optimization of composite structures. This procedure conforms to the load and resistance factor design (LRFD) method. The objective function considered is the cost of the structure. The objective function is minimized subjected to serviceability and strength requirements.

KEYWORDS

Optimization, genetic algorithms, composite-frames, structural engineering, and load and resistance factor design.

INTRODUCTION

The design of partial-composite steel frames is more challenging than that of steel frames because of the complexity associated with selecting the number connectors between the concrete slab and the steel beam or girder. The variation of number of shear connectors will change the stiffness of a beam, affecting the capacity of the frame.

Much work has been done in the area of composite design using the strength of the concrete floor slab and steel girders acting together by means of shear connectors. Schaffhausen and Wegmuller (1978) compared a 12-story frame design with and without consideration of composite action. Ito and Galambos (1993) presented a formulation for minimum weight design of continuous composite girders based on the American Association of State Highway and Transportation Officials (AASHTO) specifications. The economies of using LRFD in composite floor beams were discussed by Zahn (1987). A cost-based optimization model for design of composite beams was presented by Lorenz (1988). An optimum-cost design of partially-composite steel beams using LRFD was presented by Bhatti (1996).

In this study, a genetic algorithm (GA) is used to design composite steel frames in compliance with the AISC load and resistance factor design (LRFD). The objective is to find the minimum cost of the rigid frame with composite girders which satisfies strength and serviceability requirements.

---

[1]Associate Prof., Dept. of Civil Engineering., The University of Memphis, Memphis, TN 38152
[2]Graduate Asst., Dept. of Civil Engineering., The University of Memphis, Memphis, TN 38152

GENETIC ALGORITHM

The GA used in this study is a modified version of a program originally developed by David Carroll at the University of Illinois. The source code for the GA driver is free for public use and is available over the Internet. Carroll's program is a FORTRAN version of a genetic algorithm driver. The GA driver program can be used for a variety of different problems by simply designing an encoding scheme and supplying routines for estimating the fitness of a individual solution. The main advantage of using the GA drive system is modularity and code reuse. New options can be added to the GA portion of the program with little to no modifications to the fitness evaluation routines.

The GA technique is based upon theory of natural adaptation (Holland, 1975). The primary strength of the approach is that information from relatively good solutions is exchanged to create better solutions. The basis of the mechanism is patterns or schema developed in the encoded representation (Goldberg, 1989).

The GA driver initializes a random sample of individual solutions upon initiation of the algorithm. The GA driver uses binary coding for individual solutions in the population. The modified version of the GA driver has two strategies for choosing random pairs for mating: tournament selection with a shuffling technique and a partitioning scheme (Camp et al. 1998). There are several crossover techniques in the modified version of the GA driver: single-, double-, triple-point or uniform. In addition, there is an option to randomly vary the crossover method at each application. Reproduction allows for the generation of either a single child or two children from each set of parent solutions. In addition, there is an concurrent option for an elitist operation that guarantees the survival of the best solution into the next generation. Mutation is handled either at the genotype (jump mutation) or at the phenotype (creep mutation). Additional features include: a niching (sharing) operator and an option for the number of children generated per pair of parents. Each operator is designed to either enhance the convergence properties or to slow the process to ensure adequate exploration of the design search space.

OBJECTIVE FUNCTION

The objective function for the optimization problem is to minimize the total cost. For the design of frames with partially composite beams there is a trade-off between steel beam weight and number of shear connectors for a given set of design requirements. Thus, the primary design variables are the W-Shape for each beam and column in the frame and the number of shear connectors (N_s) for each composite beam. The total costs of a composite frame is the sum of steel beam and column costs and the shear connector costs. Consistent with standard practice, it will be assumed that the cost of connectors is proportional to the cost of the steel, which is called the relative cost. The objective function can be expressed as follows Bhatti (1996):

$$f = \sum_{i=1}^{n+m} W_s \, L + \sum_{j=1}^{n} N_{sj} \, C_{sm} \qquad (1)$$

where f is the objective function, n is the total number of beams, m is the total number of columns, W_s is the weight of steel (lbs/ft), L is the length of a member (ft), N_s is the total number of connectors per beam, and C_{sm} is the relative cost of connectors to the cost per pound of steel. The relative cost coefficient C_{sm} varies based on the job-size and the geographical region. Typically, the relative cost coefficient ranges between 6 and 12 (Lorenz, 1988)

LRFD DESIGN CONSTRAINTS

In this study, seven constraints are applied to composite girder design. These constraints enforce LRFD design specifications for strength (before and after the concrete cures), nominal moment capacity, live load deflection, vibration, and other practical design considerations (limiting the depth of girders). In general, the girder constraints g_i may be expressed as:

$$g_i = \begin{cases} 0 & \text{if} \quad m_i \leq 0 \\ m_i & \text{if} \quad m_i > 0 \end{cases} \tag{2}$$

where m_i is the degree of violation of constraint g_i.

The are two constraints on the design of the shear connectors. In general, the connector constraints s_i may be expressed as:

$$s_i = \begin{cases} 0 & \text{if} \quad q_i \leq 0 \\ q_i & \text{if} \quad q_i > 0 \end{cases} \tag{3}$$

where q_i is the degree of violation of constraint s_i.

There are four design constraints for column design. These constraints enforce LRFD requirements on beam-column interaction, shape compactness (for local stability), and frame stability. The column constraints c_i may be expressed as:

$$c_i = \begin{cases} 0 & \text{if} \quad n_i \leq 0 \\ n_i & \text{if} \quad n_i > 0 \end{cases} \tag{4}$$

where n_i is the degree of violation of constraint c_i.

There are several penalty function schemes proposed for structural optimization design (Camp et al. 1998). The quadratic penalty function used in this study for composite frames is:

$$\Phi = \prod_{i=1}^{7}\left(1 + g_i\right)^2 \cdot \prod_{j=1}^{4}\left(1 + c_j\right)^2 \cdot \prod_{k=1}^{2}\left(1 + s_k\right)^2 \tag{5}$$

where Φ is the penalty factor, g_i are the girder constraints, c_j are the column constraints, s_k are the shear connector constraints.

Having computed a penalty factor, the penalized objective function of a particular solution can be easily determined as:

$$F = \Phi f \tag{6}$$

EXAMPLE

Consider the design of a simple composite beam as presented by Bhatti (1996). The problem parameters are: C_{sm} is 10, F_y is 36 ksi, f_c' is 3 ksi, the live load is 250 psf, the dead load is 90 psf, the beam span is 40 feet, the beam spacing is 10 feet, a 3 inch metal deck with a 4.5 inch slab, the unit weight of concrete is 145 pcf, the maximum damping for vibration is 4%, and the strength of shear connectors is 26.1 ksi.

There are two design variables in this example: the beam size and the number of shear connectors. The member sizes and the number of connectors per beam are determined by the GA in compliance with the AISC LRFD code. In this example, 256 AISC cross-sections are considered.

The design presented in this example is developed using a population of 100 solutions run for 50 generations. A partitioning selection scheme is used where the upper 25% of the population is assigned a 50% probability of selection and the lower 75% of the population shares the remaining 50% probability. Reproduction uses uniform crossover to generating two new solutions and employs an elitist strategy. Exploration of the search space is enhanced by using a jump mutation operator.

Table 1 lists the comparison of the results from the GA with Bhatti's solution. Both methods obtained same beam size; however, the GA design required fewer connectors which results in a lower relative cost.

Table 1. Comparison of Results for Simple Composite Beam.

Analysis	W-Shapes	Connectors	Cost
GA	W27x84	33	3,660
Bhatti (1996)	W27x84	35	3,724

REFERENCES

American Institute of Steel Construction (1995). *Manual of Steel Construction, Load & Resistance Factor Design, Second Edition.*

Bhatti, M. A. (1996). "Optimum Cost Design of Partially Composite Steel Beams Using LRFD," Engineering Journal, AISC, Vol. 33, No. 1.

Camp, C. V., Pezeshk, S. and Cao, G. (1998). "Optimized Design of Two-Dimensional Structures Using A Genetic Algorithm," Journal of Structural Engineering, ASCE, Vol. 124, No. 5, 551-559.

Goldberg, D. E. (1989). *Genetic Algorithms in Search, Optimization and Machine Learning.* Addison-Wesley Publishing Company, Inc., New York, N.Y.

Holland, J. H. (1975). *Adaptation in Natural and Artificial Systems*, University of Michigan Press, Ann Arbor, Michigan.

Ito, M. and Galambos, V. T. (1993). "Minimum-Weight Design of Continuous Composite Girders," Journal of Structural Engineering, ASCE, Vol.119, No.4.

Lorenz, E. R. (1988). "Understanding Composite Beam Design Methods Using LRFD," Engineering Journal, AISC., Vol. 25, No. 2.

Schaffhausen, R. J. and Wegmuller, A. W. (1978). "Multistory Rigid Frames with Composite Girders Under Gravity and Lateral Forces," Engineering Journal, AISC, Vol. ??, No. 2.

Zahn, M. C. (1987). "The Economies of LRFD in Composite Floor Beams," Engineering Journal, AISC, Vol. 24, No. 2.

Genetic Algorithm for Design of Nonlinear Framed Structures

S. Pezeshk,[1] M. ASCE, C.V. Camp,[1] A.M. ASCE, D. Chen,[2] M. ASCE,

ABSTRACT

In this paper we present a genetic algorithm (GA) based optimization procedure for the design of geometrical nonlinear steel framed structures. The approach presented in this paper uses GAs as a tool to achieve discrete nonlinear optimal or near optimal designs. Frames are designed in accordance with the requirements of the AISC-LRFD specification.

INTRODUCTION

In this paper we present a genetic algorithm (GA) approach for optimized design of 2-D frames using discrete structural elements. GAs are efficient and broadly applicable global search procedures based on a stochastic approach which relies on a "survival of the fittest" strategy. In recent years, GAs have been used in structural optimization by many researchers (Jenkins, 1975; Grierson and Pak, 1993; and Camp et al., 1996). All these studies have shown that the GA can be a powerful design tool for discrete optimization.

OPTIMIZATION FORMULATION

The objective of our problem is to develop a design that minimizes the total structural weight W while satisfying the AISC-LRFD specifications. The objective function is to minimize weight of the structure. The AISC-LRFD specification includes strength and stability requirements. These requirements combined with displacement limits constitute the constraints for the optimization problem. Here, the displacement constraints are the allowable interstory drift. These constraints are implicit constraints because structural responses like stresses, strains, and displacements are function of design variables. Structural responses are calculated by the finite element method. In addition, practical and constructional constraints are included.

---

[1] Associate Professor, Dept. of Civil Engrg., Campus Box 526570, The Univ. of Memphis, Memphis, TN 38152.
[2] Programming Engineer, Wedgcor Inc., 6800 Hampden Ave., Denver, CO 80224

GA BASED DESIGN

The proposed design procedure involves a GA, linear and geometrically finite element analyses for fitness evaluation, enforcement of code provisions, and calculation of the penalty function. Step-by-step operations of the GA procedure used in this study can be summarized as the following:

(1)   Select the GA control parameters suitable to the given problem. These parameters include population size, string length per individual design variable, crossover, and mutation rate. The selection of these parameters may require some experimentations.

(2)   The initial population is randomly generated.

(3)   Decode the binary design variables into decimal values and generate an input file for finite element analysis (FEA).

(4)   Perform FEA using a suitable software package, check the given constraints, and calculate the value of the penalty function.

(5)   Check the convergence criteria and terminate the design process if it is satisfied; otherwise, continue.

(6)   Calculate the penalized fitness for every individual of the population and generate the next generation through reproduction, crossover, and mutation.

(7)   Repeat steps 3 through 6.

Example – Two-Bay, Three-Story Frame

The cross sections of all members are assumed to be W shapes. There are 268 available cross sections for each member according to the AISC-LRFD. For these two examples, we use the same GA control parameters which are: population of 60, binary coding length for one variable of 9, crossover probability of 0.85, mutation probability of 0.01, ratio of random generating portion of population to whole population of 0.01. Young's modulus of $E = 29,000$ ksi and a yield stress of $f_y = 36$ ksi are used. For each example, we performed three different cases of analysis and design cycles:

Case 1.   Linear analysis Ignoring the P-Δ Effects of the AISC-LRFD specification

Case 2.   Linear Analysis Following P-Δ Effects Consideration of the AISC-LRFD Specification

Case 3.   Geometrically Nonlinear Analysis in Lieu of the AISC-LRFD Specification's P-Δ Effects Magnification Factors

Figure 1 shows the topology of a 2-bay, 3-story example frame under a single load case. This frame was designed by Hall, et al. (1989) in accordance with the AISC-LRFD specification. We will also design this frame using the proposed GA. The load values indicated in Figure 1 are assumed to define a factored load level that is appropriate for direct application of the strength/stability provisions of the AISC-LRFD specification. Displacement constraints are not imposed for the design.

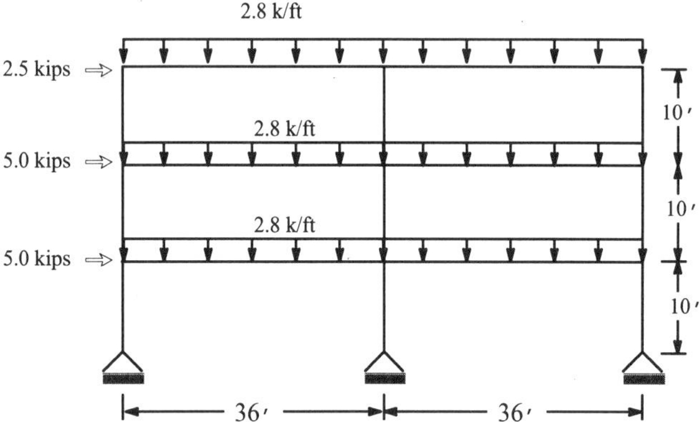

Figure 1.   Geometry and the applied loading for the two-bay, three-story frame.

Summary and Comparison of Optimal Designs

The optimal designs for different methods are summarized in Table 1. We note that the geometrically nonlinear analysis case resulted in 4% heavier structure compared to other two cases. Surprisingly, the P-Δ effects of the AISC-LRFD specification did not result in different optimized designs for cases 1 and 2. To thoroughly understand the behavioral difference between various cases, we performed pushover analyses of designs. The push-over analyses of the frames studied here are based upon the geometrically nonlinear rod model developed by Pezeshk (1992). The push-over curves are plotted in Figure 2. The load carrying capacity of the optimal design obtained using geometrically nonlinear analysis is 20% higher than that of linear analysis with about the same post-limit slope as the linear analysis. Figure 2 tells us that by providing an addition 4% weight by choosing W10X68 for column instead of W10X60, we can achieve 20% increase in strength and with the same post-limit load carrying capacity. Obviously, the design following nonlinear analysis is a much better design but more expensive.

Table 1. Summary of optimal design results for the example problem.

Analysis Procedure	Beam	Column	Weight (lb)
Case 1	W24X62	W10X60	18,792
Case 2	W24X62	W10X60	18,792
Case 3	W24X62	W10X68	19,512

CONCLUSIONS

In this paper, we presented a design procedure using a genetic algorithm for the design of 2-D framed structures. Both geometrically linear and nonlinear analysis were performed to

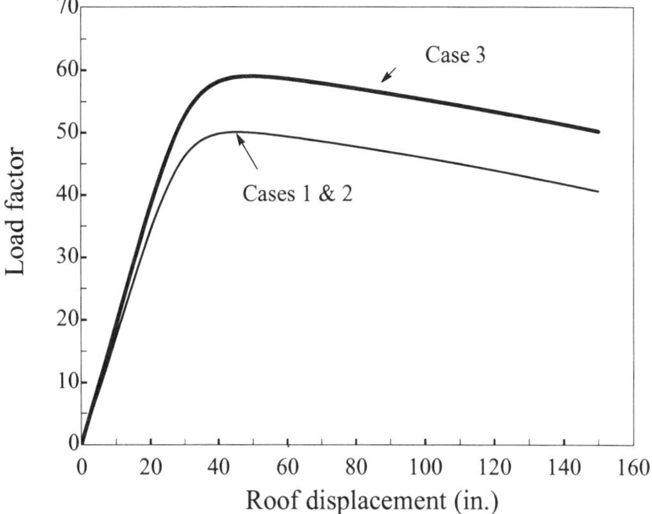

Figure 2. The Results of the Pushover Analyses.

investigate the effect of the P-Δ effects on design. The requirements of the AISC-LRFD specification were followed. We used a group selection mechanism and presented an improved adapting crossover operator. We also presented recommendations on the penalty function selection and implementation. Through one design example, we conclude that the optimized design are not affected significantly by the P-Δ effects. However, in some cases we may achieve a better design by performing nonlinear analysis instead of linear analysis.

REFERENCES

Manual of steel construction: load and resistance factor design (2nd ed.). Chicago, IL: American Institute of Steel Construction, 1994.

Camp, C.V., Pezeshk, S., and Cao, G. (1996). "Design of 3-D structures using a genetic algorithm." First U.S. - Japan Seminar on Structural Optimization, April, Chicago, Illinois.

Jenkins, W.M. (1992). "Plane frame optimum design environment based on genetic algorithms." *J,of Struct. Engrg.,* ASCE, **118**(11), 3103–3112.

Grierson, D.E., and Pak, W.H. (1993). "Optimal sizing, geometrical and topological design using a genetic algorithm." *J. of Struct. Engrg.,*ASCE, 6, 151–159.

Pezeshk, S. (1992). "Optimal Design of Structures with Kinematic Nonlinear Behavior." *J.of Engrg. Mech.,* **118**(4), pp. 702-720, April.

Structural Design

Manuscripts for some presentations were not available at time of publication.

5. *STRUCTURAL LOAD MODELS*

MODERATOR: Jeffery A. Laman
Pennsylvania State University
University Park, PA

(1) Bridge Fatigue Load for Truck Weight Limit Changes
Gongkang Fu, Harry Cohen, Fred Moses, and Husni Al-Dakkak
Wayne State University
Detroit, MI

(2) Reliability and Redundancy Under Seismic Loads
Y-K Wen, C.H. Wang, and S.H. Song
University of Illinois
Urbana-Champaign, IL

(3) Bridge Girder Distribution Factors for Live Load
Andrzej S. Nowak, Ahmet Sanli, and Junsik Eom
University of Michigan
Ann Arbor, MI

(4) Proof Load Tests of Highway Composite Bridges
Jan Bien and Pawel Rawa
Wroclaw University of Technology
Wroclaw, Poland

(5) Combining Earthquake and Other Loads in LRFD
Bruce Ellingwood
Johns Hopkins University
Baltimore, Maryland

(6) Loads and Load Combinations for Slender High-Strength Concrete Columns
Sofia Maria Carrato Diniz and Dan Frangopol
Universidade Federal de Minas Gerais
Belo Horizonte, MG
Brazil

14. *INTERNATIONAL PERSPECTIVE ON COMPOSITE CONSTRUCTION: BUILDINGS AND BRIDGES*

MODERATOR: W. Samuel Easterling
Virginia Tech
Blacksburg, VA

(1) Fire Safety of Composite Building in Germany - 3 Case Studies
Joerg Lange
Technical University of Darmstadt
Darmstadt, Germany

(2) Composite Building Design in the U.S.
Lawrence G. Griffis
Walter P. Moore and Associates
Houston, TX

(3) Composite Bridge Technique in Europe Today
Julio Martinez Calzon
Polytechnical University of Madrid
Madrid, Spain

(4) Bridge Applications of Composite Construction in the U.S.
William C. Clawson
Howard Needles Tammen & Bergendoff corporation
Kansas City, MO

23. *DESIGN OFFICE PROBLEMS*

MODERATOR: John L. Gross
National Institute of Standards and Technology

(1)Double Angle Connections with Slotted Holes in the Supported-Beam-Web-Legs
Robert Disque
Gibble Norden Champion Consulting Engineers
Old Saybrook, CT

(2)Column Economy Considering Stiffening Requirements
Charles J. Carter
American Institute of Steel Construction
Chicago, IL

(3)Design Office Rules of Thumb
Socrates Ioannides
Structural Affiliates International, Inc.

32. *AESTHETICS OF STADIA AND ARENA STRUCTURES*

MODERATOR: Kirk Mettam
Robert Silman Associates
Washington, D.C.

(1) Survey of Steel Roof Systems
Thomas Scarangello, Robert Otani, Michael Squarzini, and Steven Witkowski
Thorton-Tomassetti Engineers
New York, New York

(2) Survey of Fabric Roof Systems
Mathys Levy
Weidlinger Associates

(3) Unique Glued Laminated Timber Roof Structures for the Lisbon Multi-Use Arena
Robert Sinn
Skidmore Owings, and Merrill LLP
Chicago, IL

41. *DAMPING IN TALL BUILDINGS*

MODERATOR: Robert McNamara
McNamara/Salvia
Boston, MA

(1) Damping Estimates in Tall Buildings: A Case Study
T. Kijewski, A. Kareem, and F. Durgin
University of Notre Dame

(2) The Role of Damping on the Response of Tall Buildings to Wind Loads
Javier Horvilleur
Walter P. Moore and Associates

(3) Damping in Tall Buildings due to Foundation-Structure Interaction
Finley A. Charney
Schnabel Engineering Associates

(4) The Use of Viscous Fluid Dampers in Tall Buildings
Douglas Taylor
Taylor Devices Inc.

50. *RECENT INNOVATIONS AND FUTURE RESEARCH NEEDS IN TALL BUILDING DESIGN*

MODERATOR: B.S. Taranath
John A. Martin & Assoc.
Los Angeles, CA

(1) Rational Codified Approach for Wind and Seismic Design
P.V. Banavalkar
CBM Engineers
Houston, TX

(2) Daewoo Business Center, Shanghai, China
Richard A. Henige
LeMessurier Consultants
Cambridge, MA

(3) Composite Structural System for an 88-Story Tower
Navin R. Amin, Edward Qi, and W.H. Yang
Middlebrook+Louie Structural Engineers
San Francisco, CA

(4) Research Needs in Tall Building Design
Finley A. Charney
Schnabel Engineering Associates
Denver, CO

**(5) Seismic Isolation of the First Federal Building in
San Francisco, CA**
Anoop Mokha
Ksidmore, Owings, and Merril, LLP
San Francisco, CA

59. ADVANCED TECHNOLOGIES IN CONCEPTUAL DESIGN OF STRUCTURES

MODERATOR: Colby Swan
The University of Iowa
Iowa City, IA

(1) Simultaneous optimization of topology and nodal locations of a plane truss associated with a Bezier curve

Makoto Ohsaki and Yuji Kato
Kyoto University
Sakyo, Japan

(2) Multi-criteria conceptual design using adaptive computing paradigms
Don Grierson and S. Khajehpour
University of Waterloo
Ontario, Canada

(3)Continuum and Ground Structure Topology Methods for Concept Design of Structures
Colby C. Swan, Jasbir S. Arora and Fatma Y. Kocer
The University of Iowa
Iowa City, IA

(4) Conceptual Structural Design Through Customizable Knowledge
Nestor Gomez
Carnegie Mellon University
Pittsburgh, PA

Hugues Rivard
Concordia University
Montreal, Canada

Steven J. Fenves
Carnegie Mellon University
Pittsburgh, PA

66. *RECENT DEVELOPMENT IN EXPLOSION RESISTANT DESIGN*

MODERATOR: Sam A. Kiger
University of Missouri-Columbia
Columbia, MO

(1) Design of a Two-Story Blast Resistant Control Room
Darrell D. Barker and Johnny H. Waclawczyk
EQE International, Inc.
San Antonio, TX

(2) Investigation of Explosive Damage and Repair Costs
William H. Zehrt, Jr. and Paul M. Lahoud
US Army Engineer and Support Center
Huntsville, AL

**(3) Bomb Blast Damage to Concrete-Framed Office Building Ceylinco House:
Colombo Sri Lanka**
Tom Caldwell
Atlas Engineering, Inc.
Raleigh, NC

Bridge Fatigue Load Modeling for Truck Weight Limit Changes

Harry Cohen[1], Gongkang Fu[2], Husni Al-Dakkak[2], Fred Moses[3]

Abstract

Heavy trucks operating on highways are subject to weight limit regulations, which have significant implications to loads on highway facilities, such as bridges and pavements. This paper presents an attempt of modeling truck fatigue loads affected by weight limit increase. A model of shifting truck operating weights is proposed here. This model is applied to an example of assessment for steel bridge fatigue damage.

Introduction

It is well known that heavy trucks represent a significant load to the highway infrastructure. These loads may induce costs to highway facilities in various ways. On the other hand, highway agencies in this country constantly receive pressure for increasing truck weight limits. Truck weight here collectively refers to truck gross weight, axle weights, and axle configuration. It is apparent that such increases may induce incremental costs to the infrastructure. For bridges, for example, they may imply higher costs for new bridges for higher design load and upgrading existing bridges for deficiency and fatigue damage. TRB Special Reports 225 and 227 (1990a, 1990b) estimated the impact on bridges due to several proposed heavy trucks. These costs could exceed several billion dollars per year for the nation.

Truck weights (and sizes) have been regulated in some jurisdictions of US since 1913. The federal government first intervened in this matter with the Federal Aid Highway Act in 1956. A "grandfather clause" was also included, allowing States with higher limits to remain at the same operation for their portions of the Interstate System. In the Federal Aid Highway Amendments Act of 1974, the Federal Bridge Formula was enacted. It regulated truck weight, for the first time, explicitly including the axle

---

[1]Engineering Consultant, Ellicott City, MD 21042, [2]Associate Professor and Graduate Student, Dept. of Civil & Environmental Engg, Wayne State University, Detroit, MI48202 [3]Professor, Dept. of Civil & Environmental Engg, University of Pittsburgh, Pittsburgh, PA15261

weights, axle configuration, and gross weight. This legislation allowed States to increase the weight limits on Interstate System to 89 kN (20 kips) single axle, 151 kN (34 kips) tandem axle, and 356 kN (80 kips) GVW. Pertinent clauses allowed the issuance of special permits for overloads above the limits. After the 1982 STAA Act, States were not allowed to enact any lower weight limits than the bridge formula but were allowed higher limits if grandfather clauses were in place. By that time, all States also had regulations for special permits exceeding the weight limits. The same trend has been seen for major state highways other than Interstate System.

Overload trucks currently are accommodated by the permit systems available in all states. Needless to say, some trucks (or their trips) operate illegally hauling overweight. A clear trend has been observed of increasing number of permits. The number of overweight violation citation has also been increasing. Many highway agencies are faced with increasing pressure to allow more and heavier trucks to travel. It is thus eminent to be able to quantify the impact of possible weight limit increases on truck operation, and on bridge and pavement costs. With this capability, agencies could more rationally deal with the issue. This paper has a focus on modeling such changes in truck loading to bridges and their implications to bridge fatigue.

Truck Weight Limit vs. Load Spectra to Bridges

Truck operation is subject to pertinent regulations, which are supposed to balance trucking productivity and protection of affected lives and facilities. Changes in truck weight limits can affect the truck weight histograms operating on the highways in many ways. In turn, bridges on the system will experience different load spectra. For example, changes in GVW limits can affect the competitive balance between truck and rail for long-haul freight. They can also affect the relative cost (and hence utilization) of different types of truck configurations. In modeling truck load effects due to weight limit changes, the following concepts are proposed to be used.

A) Not all truck traffic is weight limited. For many commodities (e.g., potato chips), the cubic capacity of the truck is the limiting factor. B) For trucks operating in multiple states, the practical maximum GVW is generally controlled by limits in the most restrictive state. C) Truck weight distributions in a given state are greatly affected by the distribution of commodities carried by trucks and by levels of enforcement. D) Many trucks operate under special permits. Their reaction to weight limit changes may be affected by other factors, such as the permit fee charge system. E) It appears to be reasonable to assume that the total payload ton-miles remain the same before and after the weight limit change:

$$\text{Payload (in tons) x Vehicle-Miles of Travel (in miles)} = \text{Constant} \qquad (1)$$

But the distribution of this ton-miles over truck configurations will be altered by "shifting" some portion to different trucks. F) Weight limit changes may cause changes in longitudinal load distribution and the number of stress cycles to bridges, due to the shift of loads from one group of truck configurations to another.

Application Example, Discussion, and Conclusions

The above concepts are applied to an example here for illustration. The bridge is on the Interstate System in New York [BTML 1987]. One of the bridge's spans is considered. The cover plate's cutoff point (shown in Fig.1) is focused here for fatigue life assessment, to understand the impact of weight limit change. The GVW limit for six axle trucks is hypothetically increased from 356kN (80 kips) to 431kN (97 kips), which is referred to as Alternative Scenario compared with the Base Case (before change). The axle weight limit is assumed to remain the same. Actually this is a possibly realistic scenario, without significant impact on pavement related issues.

Fig.2 (top) shows the truck weight histogram of the Base Case, based on vehicle-miles of travel (VMT) data for New York rural interstate highways. The Base Case data were synthesized using traffic volume counts, vehicle classification counts, and weigh-in-motion data. The above concepts are applied to this set of data, with two major effects: 1) increases in truck operating weights currently carrying weight-limited traffic on the 6-axle truck (CS6), and 2) diversion of freight from 5-axle truck (3S2) to the 6-axle truck, to take advantage of the increased weight limit for CS6. The tare weights of 3S2 and CS6 are assumed to follow probability distributions with means of 133 kN (30 kips) and 145 kN (32.5 kips) respectively. Percentages of traffic to be shifted are also accordingly determined, as response to the weight limit change. The freight shifted to heavier trucks (CS6) makes new operating weights proportional to their Base Case operating weights. The proportion is equal to the ratio of the weight limits: 97/80.   The resulting truck weight histogram is given in Fig.2 (bottom), showing a separate peak of weight towards the high end. The VMT is reduced by 4.5% due to the increase of weight limit.

These two weight histograms are used to estimate the bridge's respective remaining fatigue lives, using the AASHTO approach (1994,1990). Miner's law is applied to find the effective truck weight for this purpose: $W = (\Sigma f_i W_i^3)^{1/3}$, where W_i is the GVW for interval i in the histogram, and f_i is its frequency. W is found to be increased by 5.9% from the Base Case. For a new bridge (present age=0) with ADTT = 2000 for the Base Case, it is concluded that the safe remaining life is reduced under the Alternative Scenario by 15%. It is understood that this percentage reduction will increase with the bridge's present age.

Acknowledgments

This work was funded by the National Cooperative Highway Research Program Project 12-51, FHWA Research Project DTFH61-97P00549, and CULMA of Wayne State University. This support is gratefully appreciated. However, the contents here reflect the views of the authors, not necessarily those of the sponsors.

References

AASHTO (1994) "Manual for Condition Evaluation of Bridges" 1994
AASHTO (1990)"Guide Specs for Fatigue Evaluation of Existing Steel Bridges", 1990
BTML (1987) "Effects of Truck Weights on Deterioration, Operations, and Design of Bridges and Pavements, Final Report to NYSDOT, ERDB, Oct. 1987
Fu,G. and Hag-Elsafi,O. (1996) "New Safety Based Checking Procedure for Overloads on Highway Bridges", TRB Transportation Research Record 1541, 1996, p.22
Moses,F. (1989) "Effects on Bridges of Alternative Truck Configurations and Weights" Final Report to NCHRP - TRB, 1989
TRB (1990a) "Truck Weight Limits", Special Report 225, 1990
TRB (1990b) "New Trucks for Greater Productivity and Less Road Wear - An Evaluation of the Turner Proposal", Special Report 227, 1990
Yu,C.P. and Walton,C.M (1982) "Estimating Vehicle Weight Distribution Shifts Resulting from Changes in Size and Weight Laws", TRB Trans. Res. Record 828, p.16

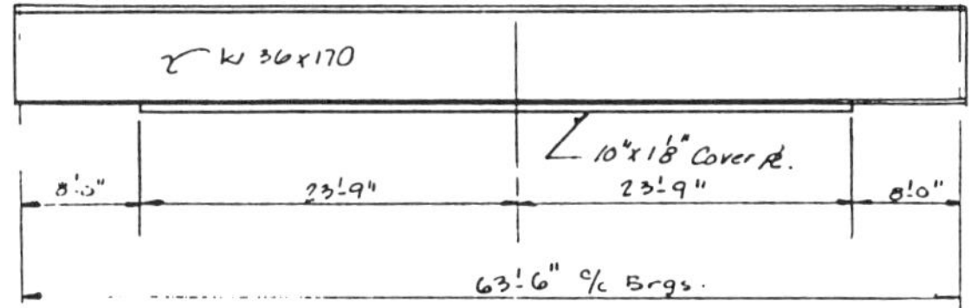

Fig.1 Example Bridge: Cover Plate Weld Fatigue Life Assessment

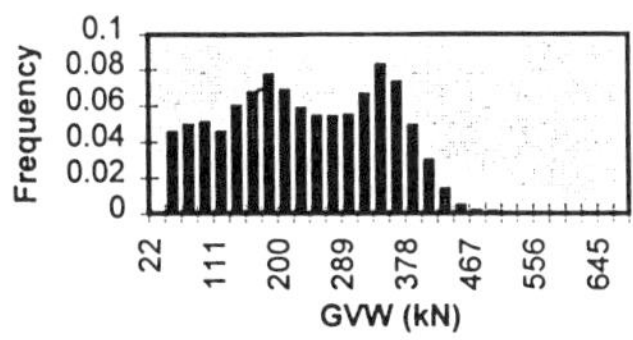

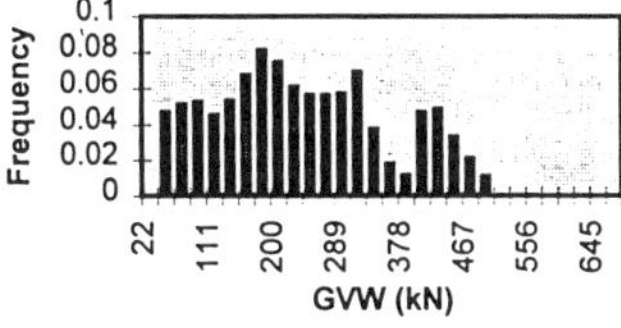

Fig.2 Truck Weight Histogram for (top) Base Case (bottom) Alternative Scenario

Reliability and Redundancy under Seismic Loads

Y. K. Wen[1], C. H. Wang[2], and S. H. Song[3]

Abstract

Structural redundancy is an important concept in structural performance and safety evaluation, yet it has not been clearly defined, and as a result, not well understood by engineers. This study investigates the various factors that have an impact on the redundancy of steel moment frames and dual (moment frame and shear wall) systems under seismic excitation. The emphasis is on effects of number of lateral force resistance components, member ductility capacity, strength variability and correlation, structural configuration, and uncertainty in the seismic excitation. A uniform-risk redundancy factor for design is proposed to achieve uniform reliability of structural systems with different degrees of redundancy.

Introduction

Redundancy of structural systems has attracted the attention of engineers after the large number of failures of structural systems observed in Northridge and Kobe earthquakes. Yet the definition and interpretation of structural redundancy vary greatly among individuals and disciplines and can lead to misunderstanding by engineers. A recent review has been given by Bertero and Bertero (1998) of redundancy and the difficulty of separation of redundancy factor from other factors such as the structural over-strength. In the 1997 UBC, there is a new reliability/redundancy factor ρ, as a function of the floor area and maximum element-story shear ratio. The allowable range of ρ can vary the seismic lateral force by as much as 50 %. In view of the large uncertainty of the seismic excitation and structural resistance, the redundancy of a structural system under seismic load can not be rigorously treated without a careful consideration of the uncertainty. For example, a simple deterministic parallel system of identical

<hr>

[1] Professor, Department of Civil Engineering, University of Illinois at Urbana-Champaign.
[2] Research Scientist, CSIRO, formerly Graduate Research Assistant, University of Illinois at Urbana-Champaign.
[3] Graduate Research Assistant, University of Illinois at Urbana-Champaign.

members under tension force will fail at the collapse load threshold regardless of the number of the parallel members since all members have the same strength and will reach the collapse threshold at the same time. Therefore, there is no advantage of having more members. The situation is drastically different if there is uncertainty in both loading and member strength as has been shown by Gollwitzer and Rackwitz (1990) and De and Cornell (1989). Under random static loads, the parallel systems have significant redundancy (much higher reliability) if there is adequate number of members, moderate degree of ductility, low strength correlation among members, and small load variability compared with that of the member resistance. It is clear that not all these factors have been considered in redundancy study of structures thus far, especial under random dynamic loads. The same is true in the case of the largely empirical redundancy factors proposed in code procedures. Considerable more research is needed before a rational and quantitative redundancy factor for design can be developed.

Parametric Study of Balanced Parallel Systems under Stochastic Loads

A parametric study is carried out of simple parallel systems with equal load distribution under random dynamic loads modeled as random processes. Factors considered include number of members, degree of ductility of member, correlation of member strength, and ratio of load variability to member strength variability. Monte-Carlo method is used in simulating the stochastic dynamic loads and resistance. The results showed that the conclusions reached for systems under random static loads by Gollwitzer and Rackwitz and De and Cornell as mentioned in the foregoing also hold under random dynamic loads. Real structural systems, however, are significantly more complex than simple parallel systems, in particular the complicated load redistribution after the failure of members including hysteretic behaviors and brittle fractures of connections observed in recent earthquakes.

Redundancy of Structural Systems under Seismic Loads

The redundancy of steel moment frames and dual systems consisting of moment frames and shear walls are studied. The nonlinear behavior of the members and connections including inelastic and brittle fracture failures are accurately modeled according to test results. The ground motions corresponding to three different probability levels (50%, 10 %, and 2% in 50 years) developed in the SAC Steel Project (Somerville et al, 1997) are used as excitation. To realistically model the load redistribution and effect of possible un-symmetric failure of members, member resistance is modeled by random variables and 3-D response analysis methods were developed which allow evaluation of the bi-axial interaction and torsional response.

<u>Steel Moment Frames</u>
Two low-rise steel moment frame buildings in the LA area, of 2 stories and 2 by 3 bays, and 3 stories and 3 by 5 bays, were designed according to current code

procedures. The diaphragms are assumed to be rigid in its own plane but flexible out of plane. Plastic hinges can form at the ends of beams and columns. Fracture failures of connections occur when the random damage index capacity (a function of ductility and cumulative energy dissipation) is exceeded. The capacity is modeled as random variable based on test results. The smooth hysteresis model (Wen 1989) was extended and used to describe the post-yielding ductile and brittle behavior of the members and connections and calibrated against test results. It reproduces the nonlinear behavior well. A 3-D response analysis method is then developed based on these element models. Response statistics under the SAC ground motions were obtained. Details of the method of modeling and response analysis can be found in Wang and Wen (1998). It was found that the brittle fracture failure of the connections have only moderate effects compared with ductile systems under uni-axial excitation. The coupling of such failures with bi-axial interaction and torsional response due to possible un-symmetric member failures, however, significantly increases the displacement demand on the structures (inter-story and global drift) that can lead to serious consequences such as instability.

Uniform-Risk Redundancy Factor R_R

To include the effects of redundancy in design with proper consideration of the uncertainty, a reliability-based redundancy factor is developed. The objective is to achieve uniform reliability for systems with different degrees of redundancy. It is therefore called a uniform-risk redundancy factor, R_R, which can be used in conjunction with the response modification factor, R, commonly used in code procedures to determine the required design force. R_R is defined as the ratio of the spectral acceleration corresponding to the probability of incipient collapse (P_{ip}) to that corresponding to the allowable value of P_{ip}. The factor is 1.0 if P_{ip} is lower than the allowable value, or in other words, the system has adequate redundancy. The design seismic force is reduced by a factor of R multiplied by R_R. Preliminary investigation shows that if an allowable P_{ip} value of 2% in 50 years is used, this factor R_R for the 3-story building varies from 1 under the assumption of un-axial excitation and ductile connections, to 0.471 under bi-axial excitation with possible brittle connection failure. The incipient collapse thresholds are assumed to be a maximum column drift ratio of 8% for ductile systems and 5% for systems with possible brittle connection failures, based on interim SAC research results by Foutch and Yun (1998). The results indicate the importance of 3-D response and effects of bi-axial interaction and torsional motions in structural redundancy.

Dual Systems of Moment Frames and Shear Walls

The investigation is being extended to dual systems of moment frames and shear walls. The interaction of the frame with the wall is important and highly dependent on the intensity of the ground excitation. Factors considered include the number and layout configuration of the walls, the nonlinear behavior (ductility) of the walls, the relative stiffness of walls versus the moment frame, the uncertainty and the correlation of the strength of the walls. An equivalent member model based on DRAIN-3DX has been developed for the shear walls.

The hysteresis compared well with experimental results. Preliminary response analyses under the SAC ground motions were carried out and the results showed the important effects of the ductility and number of shear walls (Song 1998). A bi-level, uniform- risk redundancy factor is also being developed which takes the redundancy of both shear walls and the whole system into consideration.

Conclusions

Structural redundancy under seismic load can be quantified by carefully considering the large uncertainty in the excitation and the structural resistance including the complex nonlinear behavior of the members and load redistribution. Methods of modeling the post-yielding ductile and brittle member behaviors were proposed and 3-D response analysis methods are developed. Numerical results indicated the importance of bi-axial interaction and torsional motion, especially when brittle fracture and un-symmetric member failures occur. A uniform risk redundancy factor for design based on reliability is also proposed to achieve uniform reliability for systems with different degrees of redundancy.

References

Bertero, R. D. and Bertero, V. V. " Redundancy in Earthquake-Resistance Design: How to Define It and Quantify Its Effects", Proc. 6th US National Conference on Earthquake Engineering, Seattle, Washington, June 1998.

De, R.S. , Karamchandani, A., and Cornell, C. A. " Study of Redundancy in Near-Ideal Parallel Structural Systems", Proc. ICOSSAR'89,San Francisco, Aug. 1989.

Gollwitzer, S, and Rackwitz, R," On the Reliability of Daniels Systems", Structural Safety, Vol. 7,1990, pp. 229-243.

Somerville, P., Smith, N, Punyamurthula, S, and Sun, J," Development of Ground Motion Time Histories for Phrase 2 of the FEMA/SAC Project, SAC/BD-97/04.

Song, S-H, " Structural Redundancy of Dual Systems Under Seismic Excitation", Ph D Thesis Proposal, University of Illinois, Y. K. Wen, Faculty Advisor.

Wang, C.-H. and Wen, Y. K.," Reliability and Redundancy of Buildings under Seismic Loads", Paper No. 161, Proc. 6th US National Conference on Earthquake Engineering, Seattle, Washington, June 1998.

Wen, Y. K. " Methods of Random Vibration for Inelastic Structures", Applied Mechanics Review, Vol 42, No.2, Feb.,1989.

Yun S-Y," Investigation on Performance, Prediction and Evaluation of Low Ductility Moment Frames", Ph D thesis proposal, University of Illinois at Urbana-Champaign, Oct. 1998, D. A. Foutch, Faculty Supervisor.

Bridge Girder Distribution Factors for Live Load

Andrzej S. Nowak[1], Ahmet Sanli[2], and Junsik Eom[3]

Abstract

In design and rating, load distribution to main supporting members is usually based on the AASHTO Specifications distribution factors. Prior analytical studies showed that in most cases the current code provisions are too conservative, however, for short spans and girder spacings, they can be too permissive. Therefore, this study focused on the verification of the validity of the current code specified distribution factors by field tests, particularly for short span bridges.

Introduction

A rational bridge management requires a good knowledge of the actual loads, load distribution, load effects and structural condition (load carrying capacity). An important part of the rating equation concerns the distribution of the live load to the main load-carrying members of the bridge, and to the individual components of a multi-component member. Typically, in design and rating, load distribution to main supporting members is based on the AASHTO Specifications. However, this distribution is affected by several variables, which greatly complicate the analysis. Except by field testing, it is virtually impossible to find exact values of girder distribution factors. Moreover, prior analytical studies showed that in most cases the current code provisions are too conservative. However, for short spans and girder spacings, they can be too permissive. Therefore, this study focused on these structures to verify the validity of the current code specified distribution factors, particularly for short span bridges. To accomplish this objective, field tests were carried out on five highway bridges in Michigan.

---

[1]Professor, [2]Associate Research Scientist, [3]Graduate Student Research Assistant, University of Michigan, Ann Arbor, MI48109-2125

AASHTO Girder Distribution Factors

For steel and prestressed concrete girder multi-lane bridges with a concrete deck, AASHTO Standard Specifications (1996) specifies the GDF as follows:

$$GDF = \frac{S}{1.67} \qquad (1)$$

where S = girder spacing (m). Note that, in the AASHTO Standard Specifications (1996), GDF's are specified for a wheel line load. Therefore, GDF's in Eq. (1) should be multiplied by 0.5 to make it applicable to a full truck.

For bridges with four or more girders, the AASHTO LRFD (1998) specifies the girder distribution factor (GDF) as a function of girder spacing, span length, stiffness parameters, and bridge skewness. For moment in interior girders with less than $30°$ of skew angle, the GDF for multi-lane loading is:

$$GDF = \left\{ 0.075 + \left(\frac{S}{2900}\right)^{0.6}\left(\frac{S}{L}\right)^{0.2}\left(\frac{K_g}{Lt_s^3}\right)^{0.1} \right\} \qquad (2)$$

where S = girder spacing (mm); L = span length (mm); $K_g = n(I + Ae_g^2)$; t_s = depth of concrete slab (mm); n = modular ratio between girder and slab materials; I = moment of inertia of the girder (mm^4); A = area of the girder (mm^2); e_g = distance between the center of gravity of the girder and slab (mm).

Testing

Five steel girder bridges with concrete slabs located in Michigan were selected. They are all simply supported, single span structures with two traffic lanes. Their parameters are summarized in Table 1.

Table 1 Basic Parameters of Selected Bridges

Bridge	Span Length (m)	Number of Girders	Girder Spacing (m)	Year of Construction	Speed Limit (km/h)	Skew Angle
1	13.7	10	1.32	1935	48	$30°$
2	16.8	11	1.44	1932	48	$0°$
3	9.9	12	1.36	1922	72	$10°$
4	13.7	9	1.46	1939	89	$20°$
5	11.7	10	1.42	1929	88	$0°$

Strain transducers were installed on bottom flanges of girders in the middle of span for all girders. The vehicles used were three-unit, 10-axle and 11-axle trucks, each weighing about 580 kN and 650 kN, and with a wheelbase of 14.3 m and 15.6 m, respectively. Strain data obtained from side-by-side truck tests were used to calculate load distribution factors. Results were verified by comparing the superposition of strain data from individual truck to data obtained from side-by-side truck tests. The strain measurements obtained also verified the linear elastic behavior of the bridges. Trucks were driven at a crawling speed and at a regular highway speed.

Results

Girder distribution factors obtained from crawling and regular speed truck tests are shown in Fig. 1. Superposition of girder distribution factors from single truck loadings is also shown for each bridge. In these figures, the results are taken as the maximum effect caused by the combination of different truck positions in each lane. For comparison, AASHTO (1996) and AASHTO LRFD (1998) specified girder distribution factors for one lane and two lane girder bridges are also shown. All measured GDF's are well below AASHTO Code specified GDF's.

Acknowledgements

The presented research was partially sponsored by the Michigan Department of Transportation and the Great Lakes Center for Truck and Transit Research, National Science Foundation, TRB-IDEA Program and NATO Scientific Affairs Division which gratefully acknowledged. Thanks are due to Sangjin Kim, former research assistant at the University of Michigan.

References

1. AASHTO. (1996). Standard Specifications for Highway Bridges, Bridge Design Specifications, American Association of State and Transportation Officials, Washington, D.C.

2. AASHTO. (1998). LRFD Bridge Design Specifications, American Association of State and Transportation Officials, Washington, D.C,

3. AASHTO Guide Specifications for Distribution of Loads for Highway Bridges, American Association of State Highway and Transportation Officials, Washington, D.C., 1994.

4. Kim, S., and Nowak, A. S. (1997). "Load distribution and impact factors for I-girder bridges." Journal of Bridge Engineering, ASCE. 2, 97.

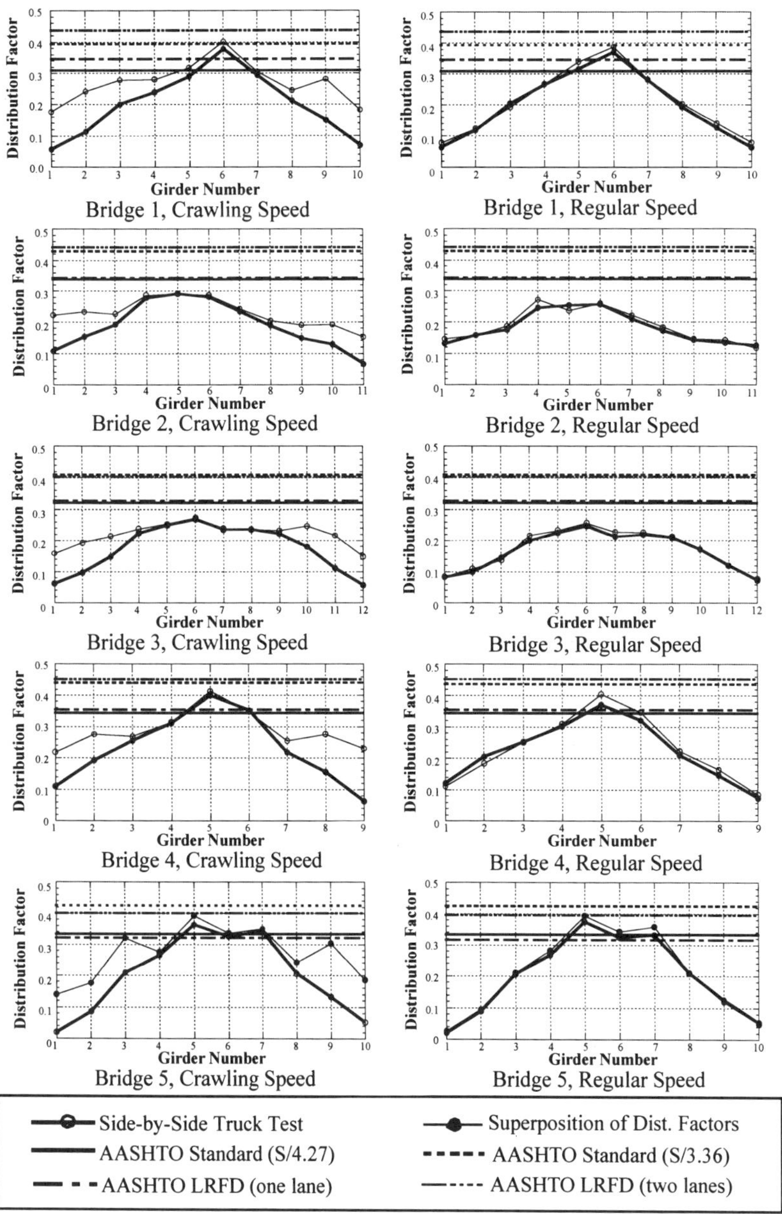

Fig.1 Girder Distribution Factors

Proof Load Tests of Highway Composite Bridges

Jan Bień[1]

Paweł Rawa[1]

Abstract

The paper describes some of the experiences of the Wrocław University of Technology (WUT) in proof loads of highway bridges in Poland and underlines needs of the unification of the testing procedures.

Introduction

In 1995 Poland started the Toll Highway Construction Program adjusted to the European Transportation System. The program involves the construction of 2,600 km of toll highways over the next 20 years. This large project of transportation infrastructure development stimulates also changes in testing of highway bridge structures. The information required for determination of vehicle loads on highway bridges and for the study of analytical assumptions used in the structural analysis can only be obtained through extensive field testing. Such an investigation should be based on the results of the unified proof loads as a background for the evaluation of structure condition and for future maintenance.

Proof loads of highway bridges

The Institute of Civil Engineering (ICE) of the WUT is involved in the supervision of the construction of thé A-1 Highway, Section Toruń. A part of the project is the preliminary study of the proof loads unification. The primary goals of the project are:
- verification of existing and application of new testing methods (Table 1);
- optimization of the computer models of loads and bridges, based on the comparison of analytical results with the measured structure's response;

----

[1] Wrocław University of Technology (WUT), Institute of Civil Engineering (ICE), Bridge Group, Wyb. Wyspiańskiego 27, 50-370 Wrocław, Poland

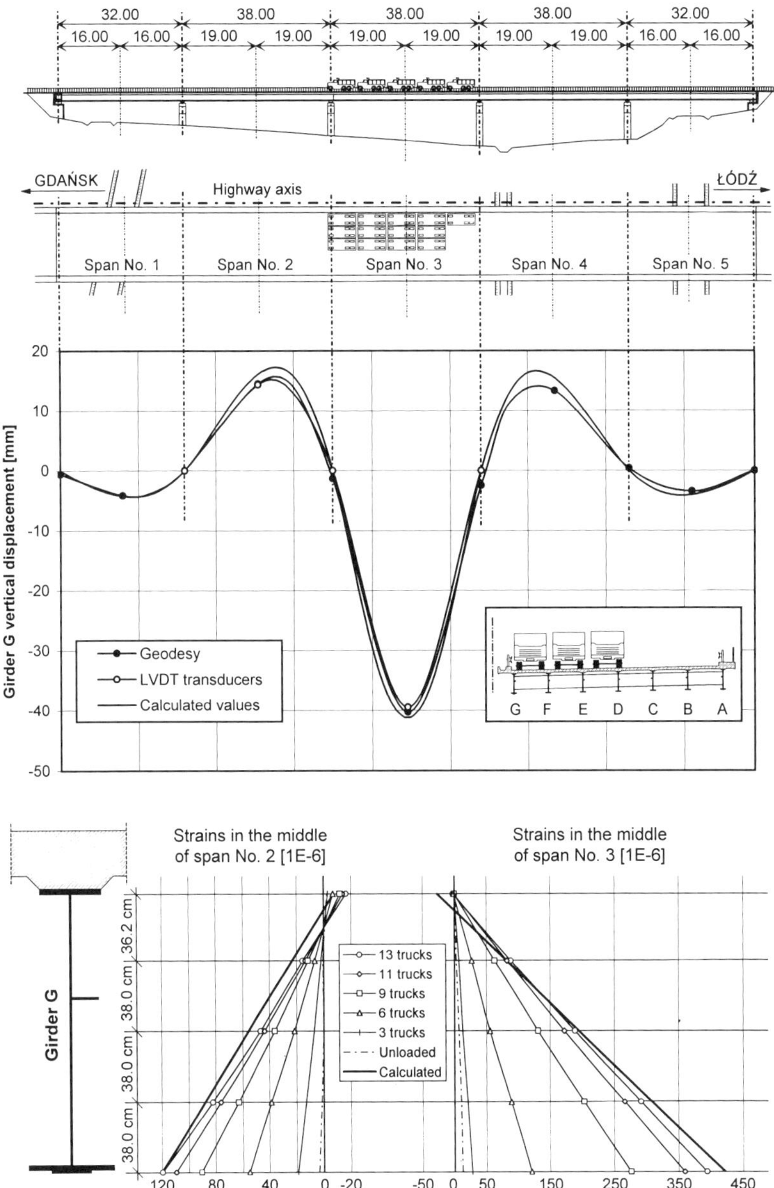

Figure 1. Ecological viaduct near Toruń: calculated and measured proof load results

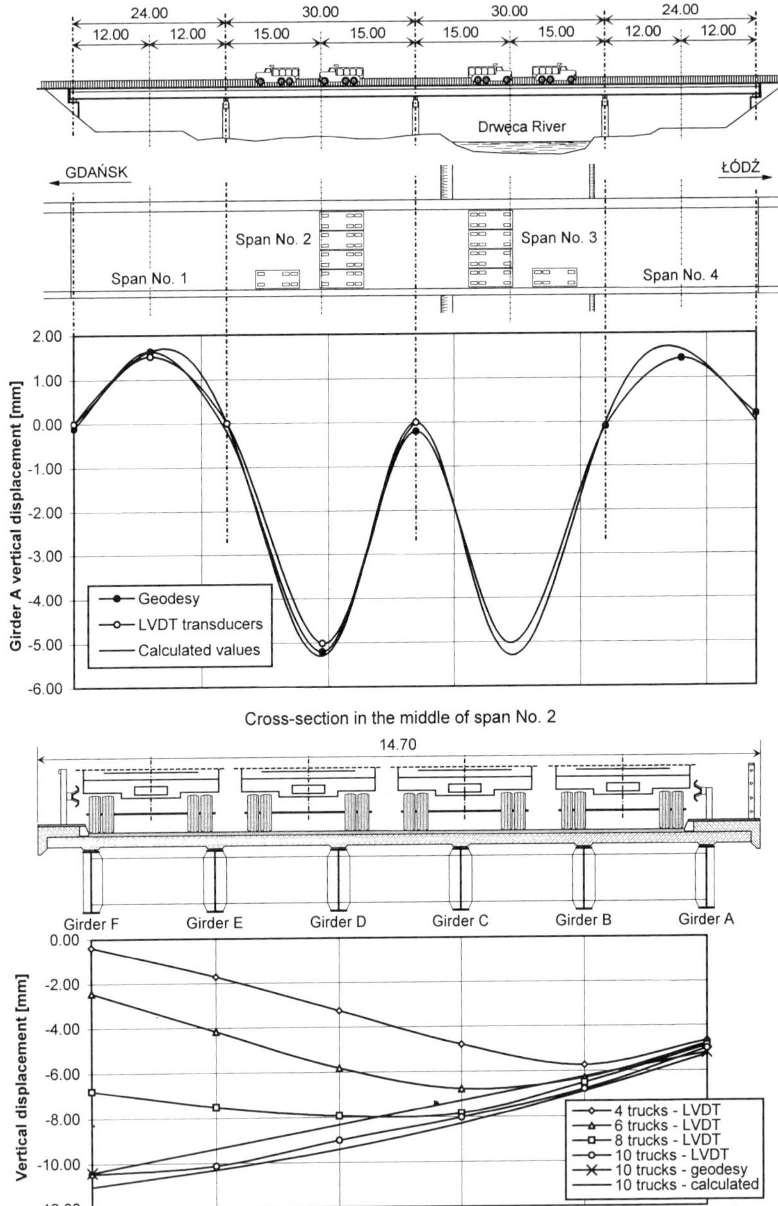

Figure 2. Bridge over Drwęca River: calculated and measured proof load results

- unification of the proof load procedures.

Measured quantities \ Measuring methods	Geodesy	Laser-based systems	Mechanical indicators	Electrical resistance gauges	Electromagnetic induction gauges	Inertial sensors	Tiltmeters	AE sensors	Thermometers
Displacement	⊕	⊕	⊕		⊕				
Rotation							⊕		
Strain			⊕	⊕	⊕	⊕			
Material effort								⊕	
Dynamic characteristics		⊕		⊕	⊕	⊕			
Temperature									⊕

Table 1. Measured quantities and measuring methods applied during proof loads

Examples of the proof load results are presented in Figure 1 (ecological viaduct near Toruń) and Figure 2 (bridge over Drwęca River). Figure 1 presents theoretical and measured values of the main girder G vertical displacement under one of the static loads (upper part) together with comparison of strains of this girder for various load levels (bottom part). Calculated and measured displacements are compared also in Figure 2: upper part - displacement of the girder A, bottom part - displacement of the cross-section in the middle of the loaded span.

In the computer analysis of the structures the finite element approach have been used. The effective moment of inertia of the structures have been calculated taking into account not only the structural members, but also the parapets, asphalt wearing surface and reinforcement.

More information regarding the proof loads is available in the reports of the ICE-WTU (Bień et al., 1996), (Bień et al., 1997).

Summary

Load testing of bridges is an efficient means of obtaining qualitative and quantitative information on the behavior of the structures. Results of the unified proof loads should be a necessary reference point for the evaluation and maintenance of the structures.

References

Bień J., Kmita J., Maliszkiewicz P., Rawa P.: Report on the proof load of the eastern construction of the road bridge over Drwęca River, WUT, Report ICE, No. SPR-41/96, 1996.

Bień J., Gładysz M., Kmita J., Rawa P., Szymkowski J., Zwolski J.: Report on the proof load of the highway A1 ecological viaduct near Toruń, WUT, Report ICE, No. SPR-100/97, 1997.

Combining Earthquake and Other Loads in LRFD

Bruce R. Ellingwood[1], M ASCE

Abstract

Load combinations in Section 2 of ASCE Standard 7-95, Minimum Design Loads for Buildings and Other Structures, are presented in a "principal action-companion action" format. This format is derived from modern probabilistic load combination analysis procedures. Yet, recent standard development activities have indicated that this approach to specifying load combinations is not well understood in the structural engineering profession at large. This paper summarizes the basis for the load combination requirements involving earthquake and other time-dependent loads that are found in ASCE Standard 7-95.

Introduction

Modern limit states codes used in structural design utilize a principal action-companion action approach to combining structural actions. This approach is based on the assumption that the maximum combined load effect during the service life of a structure occurs when one of the time-varying loads (principal action) attains its maximum value while the remaining loads (companion actions) are at their point-in-time values (Ellingwood, et al, 1982). The load combinations found in most modern limit states design standards worldwide are based on this concept.

The load combinations used in structural design for seismic effects in ASCE Standard 7-95 are,

$$U_d = 1.2 \, D_n + 1.0 \, E_n + 0.5 \, L_n \text{ or } 0.2 \, S_n \qquad (1)$$

---

[1]Professor, Department of Civil Engineering, Johns Hopkins University, Baltimore, MD 21218

$$U_d = 0.9\, D_n + 1.0\, E_n \tag{2}$$

in which D_n, L_n, S_n and E_n = nominal or code-specified dead, occupancy live, roof snow and earthquake loads. The E_n appearing in Eqns 1 and 2 is derived from an expression for base shear used in the 1994 NEHRP Recommended Provisions (NEHRP, 1995)

$$V = Cs\, W \tag{3a}$$

$$Cs = 1.2\ Cv/RT^{2/3} \tag{3b}$$

in which Cv = seismic coefficient, R = response modification factor, and T = fundamental period. The total gravity load, W, includes the dead load and a portion of the live and/or snow load.

The structural action due to earthquake is the principal action in Eqns 1 and 2. The companion action load factors on live and snow load in Eqn 1 are less than 1.0. The development and the balloting of the load combinations involving seismic effects in the last two editions of ASCE 7 (1993 and 1995) have drawn questions regarding this approach to load combination analysis. The question of appropriate companion action factors can be answered using probabilistic load combination analysis.

Probabilistic Load Modeling

In modern probabilistic load modeling, structural loads often are characterized as Poisson renewal pulse processes. The temporal characteristics and intensity of loads so modeled can be characterized by the mean rate of occurrence of the individual load events, their duration, and the probability distribution function describing the intensity of the resulting load. There now is a significant body of research data to characterize temporal characteristics and probability distributions of individual load intensities (e.g., Galambos, et al, 1982).

Live load. Live load depends on building occupancy, and includes the weight of the occupants, their possessions, and other moveable items. Numerous surveys of live loads in the "light occupancies" all have shown that the survey live load intensity is substantially less than the code value, typically on the order of 25% to 50%. However, such surveys do not measure loads from occasional transient events of very short duration (on the order of hours), such as remodeling or emergency exiting. When such loads are included in a probabilistic model, the average total load may be close to the code-specified value, but its duration is that of the transient event.

Roof Snow Load. The temporal characteristics of snow load vary widely with local climatology, and the stochastic load process that models it must reflect this variation. In temperate areas, snow loads occur intermittently during the snow season, with durations of a week or two; in the northern states or in mountainous regions, snow may accumulate more or less continuously during the snow season, with the annual extreme generally occurring in late winter.

Earthquake Load. The earthquake force depends on the ground motion parameters (acceleration or velocity) at the site and the structural characteristics, as described in Eqn 3. The uncertainty in the seismic hazard, measured by the maximum effective peak ground acceleration or spectral acceleration at the fundamental period of the building, is the largest single source of uncertainty.

Probabilistic Load Combination Analysis

Extreme loads rarely coincide. Because the duration of peak live loads, snow loads, and earthquake ground motions are very short (on the order of 1 day, 1 week, and 30 sec, respectively), the probability of a coincidence of peaks is vanishingly small. Indeed, the mean rate of occurrence of earthquake and transient live load or earthquake and significant snow can be shown to be on the order of 0.0001/yr and 0.001/yr, respectively. The controversy is over how to account for this in design: through a traditional combination factor 0.75 on all effects including dead load, or some more rational procedure.

This situation can be investigated using Monte Carlo simulation and the probabilistic load models described in the previous section (Ellingwood and Rosowsky, 1996). To this end, a series of simple steel portal frames with pinned supports that can be modelled as a single degree-of-freedom system were considered. Each roof was assumed to be flexurally rigid, with moment-resisting connections to the supporting columns. These frames were designed using Eqns 1 and 2; however, the companion action factor on snow load was allowed to vary. The stochastic column shears (or moments) due to combined (random) earthquake and snow can be written as functions of time, t,

$$V = D + E(t) + S(t) \qquad (4)$$

in which it is understood that D, E(t) and S(t) are random structural actions in the columns due to the effects of dead, earthquake and snow loads. The probability that the combined effect of snow and earthquake exceeds the design column shear, V_d, is $P[V > V_d]$. This first-passage probability (for a 50-year service period) was determined for several cases, representing extremes of snow loads found in the continental United States in terms of intensity and duration during each year. It was found that with $\gamma = 0.2$, the probability of exceeding the combined structural action was approximately 0.05 in 50 years in all cases, including those cases in which the average duration of snow during the year was significant, such as might be found at higher elevations (Ellingwood and Rosowsky, 1996). This 50-yr probability corresponds to a mean recurrence interval of approximately 975 years, which is comparable to that associated with other factored load combinations in Section 2.3 of ASCE 7-95. Similarly, the companion action factor on live load is found to vary from 0.25 to approximately 0.6. This variation depends more on the influence area for the live load than on its temporal characteristics.

Conclusions

     The hypothesis that the likelihood that an extreme event is caused by a coincidence of earthquake with snow or transient live load and earthquake is very small was confirmed. As a result, the companion action factors on live and snow load are not especially sensitive to the temporal characteristics of the individual loads. Load combinations in which the companion action is taken as the full nominal load when combined with earthquake appear to be needlessly conservative. On the other hand, adjustments to dead load are unjustified.

References

Ellingwood, B., MacGregor, J.G., Galambos, T.V. and Cornell, C.A. (1982). "Probability-based load criteria: load factors and load combinations." <u>J. Struct. Div. ASCE</u> 108(5):978-997.

Ellingwood, B. and Rosowsky, D.V. (1996). "Combining snow and earthquake loads for limit states design." <u>J. Struct. Engrg. ASCE</u> 122(11):1364-1368.

Galambos, T.V., Ellingwood, B., MacGregor, J.G. and Cornell, C.A. (1982). " "Probability-based load criteria: assessment of current design practice." <u>J. Struct. Div. ASCE</u> 108(5):959-977.

"Minimum design loads for buildings and other structures (<u>ASCE 7-95</u>)." ASCE, NY.

"NEHRP recommended provisions for seismic regulations for new buildings and other structures (1997)." FEMA Report 302, Federal Emergency Management Agency, Washington, DC.

LOADS AND LOAD COMBINATIONS FOR SLENDER HIGH-STRENGTH CONCRETE COLUMNS

By Sofia M. C. Diniz[1], Member ASCE, and Dan M. Frangopol[2], Fellow ASCE

Abstract

In this paper, procedures for the computation of the statistics of load effects for slender columns are proposed. A failure criterion is chosen, column strength and load effects are defined accordingly, and a procedure consistent with the selected failure criterion is developed. A load effect definition that does not resort on the moment magnification method is presented. The advantages of the proposed procedure are discussed.

Introduction

The ACI Code (1995) recommendations for column design were based mostly on concretes with strengths up to 42 MPa. In the case of slender columns the usual procedure for load definition is the Moment Magnification Method. Due to differences between high-strength concrete (HSC) and normal-strength concrete (NSC) material and structural behavior, the Moment Magnification Method may produce conservative or unconservative estimates of the load effects depending on the load eccentricity (Diniz and Frangopol 1997a). As a consequence, using these recommendations for slender HSC columns does not mean that the same level of safety as for NSC is obtained. Therefore, the reliability of slender HSC columns designed according to the ACI Code must be assessed.

A basic step in the reliability assessment of slender HSC columns is the computation of the statistics of the load effects due to different load combinations. This is not a simple issue. First, the column failure is largely dependent on the load path and a unique formulation of the failure criterion has not yet been established. Second, the definition of load effects must be compatible with the selected failure criterion and column strength definition. Third, strength and load effects shall be defined without resort to the moment magnification method.

---

[1] Associate Professor, DEEs, Federal University of Minas Gerais, Belo Horizonte, 30110-060, Brazil.

[2] Professor of Civil Engineering, Department of Civil, Environmental and Architectural Engineering, University of Colorado at Boulder, Boulder, CO 80309-0428

All these reliability aspects are dealt with in the present study: a failure criterion is chosen, column strength and load effects are defined accordingly, and a procedure consistent with the selected failure criterion is developed for the computation of the statistics of the load effects. The major advantages of the proposed procedure are discussed.

Formulation of the failure criterion

The computed safety of a given column depends on how the safety criterion was established, i.e., if at fixed axial load, fixed moment, fixed eccentricity, or the most general case of no perfect correlation between axial load and moment. Different formulations of the failure criterion have been proposed for the problem of eccentrically loaded reinforced concrete (RC) columns. It has been shown (Diniz and Frangopol 1997a) that in the case of the reliability assessment of HSC columns the following expression may be used to represent the failure criterion (Israel et al. 1987):

$$g(X) = [P^2 + (\frac{P.e}{h})^2]^{1/2} - [(D + L)^2 + (\frac{(D + L)e}{h})^2]^{1/2} = 0 \qquad (1)$$

where P is the axial resistance of the column at a given eccentricity e; and D and L are, respectively, the dead and live load acting on the column.

This failure criterion assumes that the moment and axial load acting on the column increase proportionally, i.e., the axial load and moment are perfectly correlated. Although this formulation does not establish an upper bound of the "true" probability of failure of the column, it is a good measure for comparing different design performances. Also, by assuming the load path, resistances and load effects can be dealt with separately.

Strength and load effects definition

In order to use the failure criterion established in Eq. (1), strength and load effects must be defined in such a way that the external axial load and moment increase proportionally. In the case of slender columns, this can be accomplished if loads and resistances are defined accordingly. Loads and resistances may be defined in the following ways:

• strength is defined as the cross-section strength; the applied external moment is magnified due to second order effects (this is the basic assumption in designing RC columns according to the moment magnification method);

• the load is defined as the applied axial load and moment; the column strength is reduced and for each value of the slenderness ratio, a different P-M interaction diagram is obtained. An algorithm for the computation of the slender column strength is presented in Diniz and Frangopol (1997b).

In the first formulation, the magnified moment is a function of the strength properties of the column; consequently, according to this formulation, resistance and load effects are correlated. Furthermore, as the axial load increases, the eccentricity does not remain constant; this implies that a failure criterion such as Eq.(1) cannot be used in conjunction with this formulation.

In the second formulation, on the other hand, the load is defined as the axial load and the initial moment acting on the column (not the magnified moment). In this case the eccentricity remains constant. Moreover, this formulation presents another advantage. Since the magnitude of the moment acting on the column does not depend on the column characteristics, resistance and load effects are uncorrelated. This procedure is adopted for computing load effects in the present study.

Variability in load effects

The load effect statistics may be computed in two different ways. In the first way, the nominal loads are known a priori. If the relationship between characteristic and mean values of a given load type is known (see Table 1), the mean value corresponding to an assumed characteristic value is computed. In the second way, the column characteristics (materials, geometry, etc.) are known a priori. In this case, the design load is assumed as identical to the design strength. Next, the nominal strength of the column section is computed and the design strength is obtained using the ACI Code understrength factors. Finally, the mean values of the load effects are computed from the design load S_d and by assuming the load ratio. For instance, if only dead load and live load act on the column, the load ratio is μ_D / μ_L, where μ_D and μ_L are mean dead and mean live load, respectively.

Whereas the first approach involves designing sections for predetermined loads, the second approach involves the calculation of loads and influence areas corresponding to a predefined column section. The advantage of the latter procedure is that by not initially restricting the influence areas and dead and live load effects, a wide range of parameters can be analyzed. This latter procedure is assumed in the present study.

Table 1 - Statistical Data on Loads (Galambos et al. 1982)

Load type	Ratio of mean to nominal load	Coefficient of variation	Probability distribution
Dead load	1.05	0.10	Normal
Live load	1.00	0.25	Extreme type I

Design loads P_u and M_u for slender columns

In the case of computing the design loads for slender columns, some remarks shall be made. First, according to the strength and load effect definition used in this study, the initial moment, and not the magnified moment, is required. Second, the columns analyzed in this study are subjected to equal moments at the ends and bent in single curvature. Moreover, it is assumed that only gravitational loads (dead and live loads) act on the column, and no sway will occur. Hence, the expression for the magnified moment M_c will be:

$$M_c = \frac{1}{1 - \dfrac{P_u}{\phi \, P_c}} \, M_u \tag{2}$$

The computation of the design loads for the slender column comprises the following steps. First, compute the design strength diagram of the column cross-section. Each point in this diagram corresponds to a pair (P_u, M_c), in which M_c is the magnified moment corresponding to the factored load P_u. Second, compute the factored initial moment M_u for a given axial load P_u. M_u and P_u define an initial eccentricity (Figure 1). Points (P_u, M_u) corresponding to several selected eccentricities are computed.

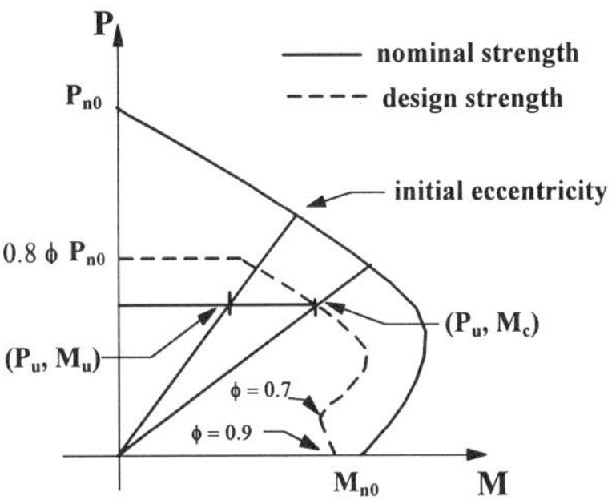

Figure 1 - Initial factored loads for slender columns (Diniz and Frangopol 1997a).

Computing mean load effects

The mean load effects corresponding to a given design may be computed as follows. According to the ACI Code, in the case only dead and live load act on the column, the design load shall be taken as $S_d = 1.4\, D_n + 1.7\, L_n$. Using the values of the ratios of mean to characteristic load (see Table 1), it follows that $S_d = 1.33\, \mu_D + 1.7\, \mu_L$. Now, the mean values μ_L and μ_D can be easily found by first computing the design load corresponding to a given column, and then assuming the ratio between μ_D and μ_L. A similar procedure may be used in the case of other load combinations.

Summary and conclusions

In this paper, procedures for the computation of the statistics of load effects for slender columns were proposed. A load effect definition that does not resort on the moment magnification method is presented, and the corresponding slender column strength was defined accordingly. It was shown that the load effect and column strength definition is compatible with the selected failure criterion. The results presented in this paper can be used in the reliability assessment of HSC columns.

References

ACI Committee 318 (1995), "*Building Code Requirements for Reinforced Concrete (ACI 318-95) and Commentary - ACI 318R-95*", American Concrete Institute.

Diniz, S.M.C. and Frangopol, D.M. (1997a), "Reliability Bases for High-Strength Concrete Columns", *Journal of Structural Engineering*, ASCE, Vol. 123, No. 10, pp. 1375-1381.

Diniz, S.M.C. and Frangopol, D.M. (1997b), "Strength and Ductility Simulation of High-Strength Concrete Columns", *Journal of Structural Engineering*, ASCE, Vol. 123, No. 10, pp. 1365-1374.

Galambos, T., Ellingwood, B., MacGregor, J., and Cornell, A. (1982), "Probability Based Load Criteria: Assessment of Current Design Practice", *Journal of the Structural Division*, ASCE, Vol. 108, No. ST5, pp. 959-977.

Israel, M., Ellingwood, B., and Corotis, R. (1987), "Reliability-Based Code Formulation for Reinforced Concrete Buildings", *Journal of Structural Engineering*, ASCE, Vol. 113, No. 10, pp. 2235-2252.

FIRE SAFETY OF COMPOSITE BUILDINGS IN GERMANY

Joerg Lange[1]

Introduction

Currently Germany is experiencing very fast development in the theoretical approach to building fire safety. Unfortunately this development is not being transferred as rapidly into practical use. One reason is that architects and structural engineers are not generally informed about the new developments. But even well informed engineers have no stimulation to use newly developed sophisticated design tools because they result in additional work for which they are not compensated. The second reason for the very slow transfer of our knowledge into daily practice is the very conservative attitude of public authorities towards fire.

A brief review of the state-of-the-art of fire-resistance-design in Europe, and especially in Germany, will be presented. Additionally a review of the codes related to this topic will be presented.

To begin with fire safety in Germany, is highly important because of historic reasons. The experiences of the narrow towns in the Middle Ages, during which a fire destroyed whole districts in tremendous speed, still live on in the collective remembrance. Much closer are the memories of the World War II firebombs, which left entire cities as ruins.

Added to this is the fact that, after the Second World War, steel was very rare in Germany. Therefore, the concrete industry had a significant head start in developing a market for its product. By the time the steel construction industry had enough material to supply the market, a large number of companies that used concrete had nationwide branches. Because concrete more easily met fire resistance requirements, it became the building material of choice until recently. During the last decade the situation changed due to two major reasons. One is that a few steel contractors finally have nationwide representation. This is a result of the reunification, which gave steel and concrete contractors equal bidding possibilities. The other reason for this change is the availability of improved design tools for newly developed composite members, such as concrete-filled hot-rolled sections.

Fire Safety in Law and Codes

Germany is a federal republic and therefore many laws and regulations are a matter of the 16 states. Culture, education, and buildings, for example, are subject to state legislation. Defense and health, on the other hand, are regulated in the nation's capital. When a law concerning buildings is developed, usually a model law is prepared by the central government, in cooperation with the states and other interested and related associations. Guided but not obliged by this, the states pass their laws and regulations. An example of this is that car parking garages in one state may have a minimum clear height of 2.05 m and in the neighboring state of 2.15 m.

---

[1] Prof. Dr.-Ing., Technische Universität Darmstadt, Institut für Stahlbau und Werkstoffmechanik Alexanderstraße 7, D-64289 Darmstadt

For fire safety of buildings, the legislative body defines three major objectives:
- to prevent the beginning and development of a fire and of smoke,
- to make possible the rescue of persons and animals in the case of a fire, and
- to enable the extinguishing of a fire.

Codes, which are usually prepared by engineers, are established by the DIN, the German Institute for Standardization. Regarding fire safety, a distinction is made between active and passive fire protection. Both areas have various codes, which evolved independently from each other. This leads in some cases to expensive solutions when both active and passive fire prevention systems have to be added without improving the fire safety.

In addition to this the European unification resulted in Eurocodes, which are a positive development of the DIN-codes. Fortunately, these Eurocodes are published and represent great improvements in fire safety design. Unfortunately, they have not yet been introduced due to various reasons. Also national codes have not been further developed because everybody is waiting for European codes. Hence there is a formal problem because the laws say to work according to old but safe codes, but the use of state-of- the-art tools is not forbidden.

Regardless of this slightly confusing code-landscape and in spite of the tight schedules of today's building projects, very innovative solutions are possible. Sir Norman Foster, an English architect whose projects are built all over the world, said at the opening of the 295 m high Commerzbank-Office-Tower in Frankfurt, that he had never before planned and built any structure as fast as this skyscraper. And this building includes many very innovative solutions.

Figure 1. Düsseldorfer Stadttor

Düsseldorfer Stadttor

General Design Considerations

Using the example of the Düsseldorfer Stadttor the ideas and possibilities of fire safety design in Germany will be presented.

The Düsseldorfer Stadttor is located above the southern entrance to the Rhineshore-Tunnel, a major four-tube highway tunnel. Two 16-story office towers have their foundations on the two outside tunnels. They are connected with a three-story tie beam that spans the two central tunnels to form a gate. The tender was open to steel and concrete design.

As with all high-rise buildings of the new generation, the design team included not only architects and structural engineers but also an expert on fire safety design. Because in this case the expert had his professional roots in structural engineering, a very fruitful cooperation took place. This lead to interesting new solutions concerning the building structure.

The special feature of this building is not only the unusual foundation but also the architect's leading idea to produce a very transparent building. To accomplish this, walls were used as little as possible. Only three concrete walls (forming n U) of the emergency stair case and the brick walls of the electrical installation were allowed. The horizontal loads are carried almost completely by three frames made of trusses with tubular members – the portal frames. The composite slabs are made of 11 cm concrete on a 4 cm steel decking. They rest on composite girders that bear on concrete-filled steel tubes.

Another result of the architect's vision of a transparent building is the loss of balustrades, which usually prevent the fire from passing from one story to the next.

The fire safety plan included the installation of a sprinkler system with a concentration of sprinkler heads in the area of the glass facade and with three independent systems alternating over three stories.

The next step was the development of an evacuation- and rescue-lane-concept. It shows two independent staircases in each semi-story. One of them is inside the concrete core and has a direct exit to the outside. Above 60 m there are 4 separate staircases.

The third major issue of the fire safety design was the installation of smoke detectors and sprinkler heads between the false ceiling and floor slab. This improves safety by reducing the spreading of fire.

These and other arrangements lead to
a)   a decrease of insurance rates, and
b)   a fire resistance rating of 90 minutes. Buildings of this size usually require 120 minutes.

Fire Protection of Members Carrying Vertical Loads - Composite Floors

The floors were designed based on the plastic hinge theory. Because their load bearing capacity under positive bending is reduced to less than 10% due to the unprotected steel, additional slab reinforcement would have been necessary. Under negative bending their capacity is still very high. Because the reduction of live loads in the case of a fire is allowed, a redistribution of moments from the weak center span cross-sections to the strong support cross sections results in an overall equilibrium. This procedure is not covered by national standards, but is included in the Eurocodes. It was only possible to use them with a special permit from the public authorities.

Beams and Girders

Beams with depths less than 270 mm were protected with a mineral fiber spray coating or silica fiber-boards. The large girders were concrete-filled I-beams. This was the most economic solution due to the large web openings and the additional stiffness, which is achieved with the concrete filling. This design method is covered by German standards.

The concrete filling is valuable because the concrete is a thermal insulation of the steel. It may also contain reinforcement to substitute for the bottom flange of the girder, which is exposed to the fire. The concrete filling is very cheap because no formwork is needed.

Columns

To simplify the interior work, the outside diameter of the columns had to be the same in all floors. This was achieved by using high-strength concrete in the lower floors. In some columns with very high loads, I-beams of high-strength steel were added to the concrete filling.

The design philosophy for these columns uses the good thermal insulating properties of the concrete. The resulting load bearing capacity if the steel tubes is reduced to less than 10% under fire. Given this and the reduction of the safety factor from 1.7 to 1.0, the design principle was to give 1.0 of the load to the concrete core (which may have additional reinforcement or steel profiles) and 0.7 of the load to the steel tube. This resulted in small columns that provided greater rentable space. The wall thickness was not reduced over the building height because this would lead to fewer orders for the smaller tubes resulting in price penalties. These penalties would sum up to a larger amount than the saving of material weight.

level	Interior columns B/3 and G/5			Exterior columns D/2-3		
	Structural steel	concrete (cube strength)	reinforcement contribution ratio	structural steel	concrete (cube strength)	reinforcement contribution ratio
	[mm]	[N/mm^2]	[%]	[mm]	[N/mm^2]	[%]
17.-19.	559/6,3	25	------	406/8,8	25	------
14.-16.	559/6,3	25	------	406/8,8	25	17
11.-13.	559/6,3	35	------	406/8,8	35	16
8.-10.	559/6,3	45	18	406/8,8	45	25
5.-7.	559/6,3	45	38	406/8,8	45	40
1.-3.	914/10	45	9	no columns		

Table 1: Interior and exterior columns (structural steel: diameter/plate thickness)
reinforcement contribution ratio = $(A_s f_{y,s})/N_{pl}$

The design of the columns is not covered by German codes. The Eurocodes were used after obtaining a special permit from the public authorities.

Fire Protection of Members Carrying Horizontal Loads

The frames are made of three-dimensional truss like columns (h = 70 m) and three trusses which cut through the three top floors. This results in a z-shaped system, as can be seen schematically in Fig. 2. The truss like columns are made of vertical steel tubes (d = 914 mm) filled with reinforced concrete and horizontal and diagonal steel tubes (d = 813 mm) which are filled with reinforced concrete below the 4th floor level only.

Although the diagonals and horizontals above the 4th floor were not filled with concrete, 90 minutes of fire resistance were achieved. It is very improbable that a fire occurs in both building wings at the same time. Therefore the framing of the "cold" building wing may withstand the horizontal loads. This saves not only concrete, but also a lot of labor because the positioning of the reinforcement in the nodal points of the trusses was very difficult.

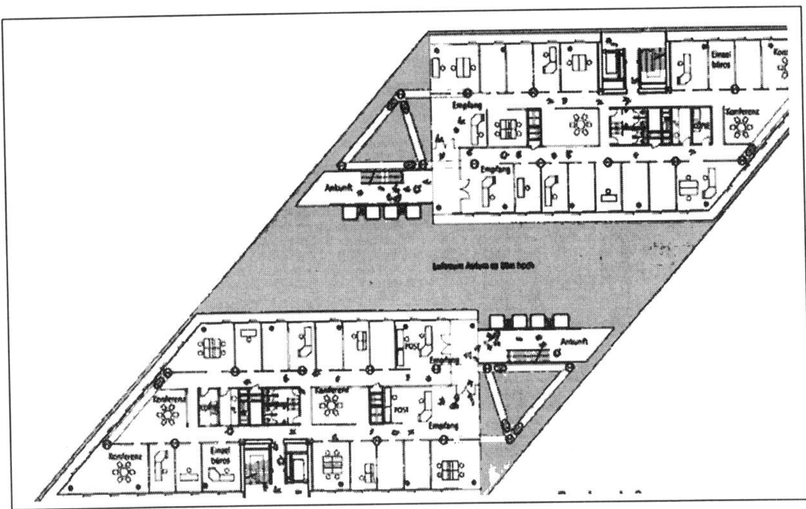

Figure 2. Schematic Plan of the Düsseldorfer Stadttor

Composite Building Design in the U.S.

Lawrence G. Griffis[1]

Abstract

This paper presents an overview of composite construction in the U.S. It traces the early development and describes its use in modern high-rise buildings.

Historical Overview of Composite Construction

Although many modern students and practitioners of structural engineering tend to think that composite construction is a product of recent design and construction practice, it actually began just prior to the start of the twentieth century.

In the USA, composite construction first appeared in the year 1894 when both a bridge and a building were constructed. The bridge was the Rock Rapids Bridge in Rock Rapids, Iowa. A Viennese engineer named Joseph Melan obtained a patent for bending steel I-beams to the curvature of an arch and then casting them in concrete. He submitted calculations to verify his composite design. The building was the Methodist Building in Pittsburgh constructed using concrete encased steel floor beams. It so happens that a fire in a nearby building in 1897 spread to this structure and destroyed the contents but not the frame of the Methodist Building. Already, one of the advantages of construction frame construction was realized - namely fire protection.

As additional buildings and bridges were constructed using steel wrapped in concrete toward the end of the nineteenth century, a need for research testing arose to better understand the behavior. A set of systematic tests for composite columns was begun at Columbia University's Civil Engineering Laboratories in 1908. This was followed by tests of composite beams in the Dominion Bridge Company's fabrication shop in Canada by Professor H.M. McKay of McGill University in 1922-4.

The first record of composite construction appearing in a US building code was in 1930 when the New York City Building Code first allowed extreme fiber stresses of 138 MPa (20 ksi) rather than the 124 MPa (18 ksi) value traditionally allowed for noncomposite beams.

[1]Senior Vice President and Director of Structural Engineering, Walter P. Moore and Associates, Inc., 3131 Eastside, Second Floor, Houston, Texas 77098

Shear connectors were also recognized in this early composite construction as an effective means to enhance the natural bond between steel and concrete. In 1903, Julius Kahn received a US patent on composite beams where shearing tabs in the beam flanges were bent upward to project into the slab. Different types of shear connectors have been proposed over the ensuing 90 years including some types still documented in the AISC *Manual of Building Construction*. It was in 1954 when welded headed metal studs were first tested at the University of Illinois. In 1956, at the completion of the tests, a formula for the design capacity of these connectors was published. The welded headed metal stud has now become the dominant method of transferring shear between steel and concrete. The first bridge to use these connectors was the Bad River Bridge in Pierre, South Dakota built in 1956. Also in 1956, IBM's Education Building in Poughkeepsie, New York became the first building to use headed stud connectors. The second floor was formed with a 38 mm deep, 0·6 mm thick steel composite deck, using wires welded to the top flutes of the deck to achieve composite action between the metal deck and the hardened concrete.

The widespread use of composite metal decks began to flourish in the 1950s in building construction. The metal deck acted as a form for the wet concrete thus reducing concrete formwork costs. The deck was shaped in such a manner as to ensure composite action so that it could serve as the positive one way reinforcement for the final hardened concrete slab. Composite action was first achieved through the use of wires welded to the deck. More recently, the standard way it is accomplished in modern composite decks is through embossments manufactured into the deck to achieve composite action with the concrete. The composite metal deck and the welded headed stud have gained such widespread popularity in modern building construction that it has become virtually the only system used in building floor construction for steel and composite frame buildings in the last 25 years. One of the first modern buildings using this technique of construction was the Federal Court House in Brooklyn, New York designed in 1960. Today almost all steel and composite framed buildings utilize this method of floor construction.

The first tall building boom occurred in the USA in the 1920s and 1930s when high-rise structures such as the Empire State Building and the Chrysler Building were built. Many of these early vintage steel frames utilized the protection that the concrete afforded the frame when it was cast around it for resistance against fire and corrosion. Only until the 1960s with the advent of modern composite frame construction have engineers actively sought rational methods to take advantage of the stiffening and strengthening effects of reinforcing bars and concrete on the capacity of the embedded steel frame. The late Dr. Fazlur Kahn, in his early discussion of structural systems for tall buildings, first proposed the concept of a composite frame system in the Control Data Building in Houston, Texas in 1970. Since that time composite frame construction has been utilized on many high-rise buildings all over the world and its usage, with a composite column as the key element, is well documented in the work of the Council on Tall Buildings and numerous other publications.

Modern Composite Construction

Over the past 25 years, numerous innovative composite floor and frame systems have developed in tall building design whereby structural steel and reinforced concrete have been combined to produce a building having the advantages of each material. The use of these systems has as its underlying principle, the combination of these two distinctive and different building materials to benefit from the advantages of both - namely, the inherent mass, stiffness and economy of reinforced concrete and the speed of construction, strength, longspan capability, and light weight of structural steel.

Composite frame construction can take on several forms. One form of composite frame construction utilizes a bare steel frame designed to carry the initial gravity, construction, and lateral loads until such time as the concrete is cast around it to form composite columns capable of resisting the total gravity and lateral loads of the completed structure. The floor number identified refers to the number of levels above which concrete has encased the erection columns. With the erection guy derrick or crane positioned on the 10th level, steel for levels 11 and 12 is being set. On levels 9 and 10 the frame is being welded or final bolt tightening is occurring and metal deck is being placed. On levels 7 and 8 studs are being welded to the top of composite beams and welded wire fabric is being laid on the floor deck. At levels 5 and 6 concrete is being poured for the floor. On levels 3 and 4 composite column reinforcing cages are being erected and tied. On levels 1 and 2 column forms are being placed and concrete is being poured for the composite columns. Finished concrete floors are needed ahead of composite column and shear wall pouring in order to have a finished surface for stacking and teeing reinforcing steel and setting the column forms.

Experience gained from this type of construction indicates that, depending on the individual contractor, there exists an optimum construction sequence and spread in the various construction activities. If this relative staging is not maintained, then problems can occur. For instance, when the gap between setting steel and placing concrete beams becomes too wide, an overload of the erection columns can occur since they have been designed for a certain number of floors of construction loading or have been sized to limit column shortening to a predetermined amount. Also, frame stability can start to be of concern. If the gap becomes too close, then construction activity becomes congested with a resulting loss of construction time and efficiency. Obviously, close coordination and control of the construction process is required for this type of construction.

Advantages of Composite Frame Construction

It is not hard to understand why the designers of tall buildings today are turning to composite frame construction. The lateral load design of high-rise buildings requires consideration of lateral deflection (drift) for prevention of cladding and partition distress and acceleration to prevent disturbing perception to lateral motion. It is interesting to compare the relative cost effectiveness of steel and composite columns in providing the necessary strength and axial stiffness for tall building design. From a strength standpoint, composite columns are approximately 11 times more cost effective in resisting axial loads than are structural steel columns. From an axial stiffness standpoint, composite columns are approximately 8.5 times more cost effective in providing resistance to axial deformation than are structural steel columns. However, for resistance to a given axial load, structural steel columns are only 25% as large in area and weigh only 80% as much as reinforced concrete columns.

In comparing composite metal deck and reinforced concrete floor systems currently in use for high-rise buildings having bays that are 9·14 m x 11·56-12·19 m, and utilizing lightweight concrete, it is interesting to note that the composite metal deck and beam floors weigh approximately 60% as much as the reinforced concrete floors (steel floors having 133 mm deck slabs, concrete floors with wide pan joists and haunch girders with a 100 mm slab). This significant difference in floor system weights for costs comparable to concrete floor costs can translate to large savings in foundation costs for tall buildings having composite metal deck floors.

Concrete can be used to economic advantage in carrying the large vertical column or shear wall loads at much lower cost per unit strength and stiffness. Steel floor construction is used to economically carry the floor loads for a reduced building mass and more economical foundation. Besides the economy of materials, composite frame structures have the advantage of speed of construction by allowing a vertical spread of construction activity so that numerous trades can engage simultaneously in the construction of the building. Structural damping from wind induced motions for composite structures can usually be justified at roughly 1.5% which is slightly greater than structural steel alone.

In evaluating the cost benefits of composite frame construction there are several factors that must be considered as follows: material cost savings in substituting concrete and reinforcing steel for structural steel; effect of time on construction schedule and the cost of construction financing; savings in fireproofing costs for the substitution of concrete for structural steel; cost and degree of reuse for the sophisticated concrete jump forming systems often used in composite frame construction; potential benefit to the owner, cost and otherwise, for earlier occupancy of the building; experience and expertise of the potential General Contractor(s) in building with composite frame construction; and potential benefit or detriment of the composite frame on the building architecture, particularly the exterior cladding system.

The Future of Composite Frame Construction

Progress in the modern design and construction of high-rise composite frame construction has been very dramatic since first introduced in the US in 1969. Unquestionably new development will occur. Several advances that could occur soon include even higher strength concrete to reduce column sizes and allow for taller buildings, development of more sophisticated forming systems to reduce construction time and perhaps most significantly, development of new and more efficient artificial damping systems that will reduce the perception to motion that usually governs tall building design today. Ultimately, the building lateral load frames may be designed for strength alone, reducing or eliminating the 'premium for height'.

The challenge for US designers and builders is to develop new components and systems that take advantage of the ductility, toughness and redundancy that are inherent with steel and concrete composites. For low-rise construction, the key may very well be in developing simple and economic connections between beams and columns or walls. These and other new connections may be able to allow composite systems to successfully compete with existing steel or composite systems in the market place.

COMPOSITE BRIDGE TECHNIQUE IN EUROPE TODAY

Julio Martínez Calzón [1]

Abstract

A brief introduction of the main considerations and techniques of modern European composite bridges in order to obtain economic, durable and aesthetic construction solutions.

Introduction

Since their introduction at the end of the 50's, there was a gentle but steady growth in the use of composite bridges in Europe right up to the mid-eighties.

From this time on there have been very important achievements and innovations and the current amount and size of bridges constructed with this technique is truly amazing.

In one way or another different European countries have helped to establish a strong body of learning based on: broad practical experience, control of techniques, testing, analysis, complex calculation programs, etc. A suitable code of practice for composite bridges is already well underway.

While German techniques were setting the guidelines at the outset –the most significant works of the first period being constructed in this country– other countries such as France, England, Belgium, Switzerland, Spain, Holland, Austria and Italy were all adding their contributions to the whole. At first, these contributions were made by outstanding individual specialists in an attempt to breach the deep-rooted traditions and technologies prevalent in bridge

---

[1] Prof. Dr. Civil Engineer, MC-2 Estudio de Ingeniería, Víctor de la Serna 21, 28016 Madrid (Spain).

construction in all of these countries. Traditions based on the use of prestressed concrete or structural steel and where it was very difficult to introduce innovations. These difficulties were further handicapped by the general lack of inertia of construction companies and Administrative bodies.

However, from the 80's, and on the introduction of a number of specific administrative measures, French technicians and companies established a set of rules which required that every bridge project considered an alternative composite solution in order to exploit the advantages of this latter to the full. This led to a remarkable impetus in composite bridge design and dramatically increased the number of composite bridges constructed. This impetus subsequently furnished the industry with new materials (high resistance welding steel grades; plates with variable thickness) and exploited intrinsic latent resources through the combination and symbiosis of materials and new ideas from technicians from very different fields.

This French awakening moved very quickly towards other countries, such as Spain, England, Belgium, Switzerland, etc. and these countries have subsequently made important contributions to the development of composite bridges. The Nordic countries are also engaged in outstanding composite works at present.

Nowadays, some of the more innovative procedures are those indicated below. These procedures may be a source of inspiration for many other new solutions:

–   Cross sections with doubled composite work with bottom concrete slabs set in the box sections in areas of strong negative bending moments close to the piers in order to favourably resist the compression of these areas and stabilise the steel elements.

–   Corrugated steel sheets in the webs of box or plate girders in order to improve all prestressing forces –external or internal– in the flanges of the cross sections, without an appreciable introduction of stress in the steel webs as a result of the easy deformation produced in these elements against longitudinal forces. This corrugation also gives the webs greater stability without requiring additional stiffeners.

–   Three dimensional truss webs. A solution of great interest for very wide bridges due to the important weight reduction achieved and the favourable resistant effect produced by the intermediate local supporting points introduced by the truss in the upper concrete deck slab: this method also offers very simple and high performance external prestressing.

–   Strict box girders. Optimum use of all materials used in the composite members, particularly when very high grade resistances are used. At the same time this solution eases site and workshop techniques, erection, transportation and assemblies of bridges of large dimensions.

–   Composite arches. Solutions designed from very different perspectives:

    • Self centering of the arch's steel member and subsequent partial or complete concrete filling
    • Active temporary prestressing diagonals for cantilever launching.
    • Bow string with composite deck in tension working as tie.
    • Tubular members
    • False arches transformed in trusses on the inclusion of slender diagonals.

–   Spatial tubular trusses. Modern designs with very high visual appeal that include advanced analytical and construction procedures for medium span bridges. With external prestressing techniques they designs can be of the utmost interest.

–   Hinges and elastic outer or internal cells. Systems with vertical ties joined to the foundations in order to achieve very favourable force diagrams under strict requirements of clearance, weight, etc., and which noticeably aid erection procedures.

–   Stay bridges. Mainly of the twin girder type, these systems are particularly favoured for the development of long span bridges, but may equally be employed in spans of more modest dimensions.

–   Unique bridges. While some individual bridges have attracted all types of criticism, others, such as those designed by the Spanish engineer, Santiago Calatrava, have been praised for their great aesthetic value. These bridges have cleared the way for other engineers to attempt highly original and bold designs and this should not always be treated with contempt.

In closing it may be said that composite bridges in Europe are currently undergoing a remarkable phase of creative activity and are showing that the combined use of materials in an intelligent way offers superior results to that provided by a single material regardless of how active and appropriate this may be.

It is interesting to note that this process has radically transformed the thinking of modern engineers. They have changed from being specialists in a precise structural line, to being experts in what could be defined as the steel-concrete structural material, with the global use of all available building materials offered by modern techniques.

BRIDGE APPLICATIONS OF COMPOSITE CONSTRUCTION IN THE U.S.

William C. Clawson[1], Ph.D., P.E., M. ASCE

Abstract

The development of innovative methods in the use of composite construction for bridges in the U.S. is an ongoing process that depends on economy, aesthetics, time constraints and other factors. Three advancements in composite bridge construction in the U.S. are discussed in this paper that are gaining increasing acceptance in the industry or are unique solutions to bridge-specific problems. First, the use of precast concrete deck panels for steel beam and girder bridges is reviewed. Secondly, the design and analysis of long-span cable-stayed steel composite bridge is discussed. Finally, an innovative composite construction system is presented for the cable anchorage system for the cable-stayed Charles River Bridge in Boston.

Introduction

Composite construction for bridges in the U.S. has been permitted by AASHTO since 1944. However, widespread usage was not achieved until the 1950's and 1960's following research of beam and shear connector behavior.

In the 1965 AASHTO specifications, allowable design capacities for individual shear connectors were specified for channels, welded studs and helical bars. In 1969, the AASHTO specifications were revised to require that shear connectors be designed for fatigue and checked for ultimate strength. Today, composite design is routinely used for most bridge designs for steel I-girders, box girders and precast concrete beams to take advantage of the inherent benefits in structural performance.

In recent years, innovations in composite construction methods have developed in response to needs for improved construction efficiencies and as structural analysis methods have evolved to allow for more comprehensive analysis. This paper discusses three recent applications and advancements in composite construction methods.

Precast Deck Panels for Girder Bridges

The use of concrete deck panels for bridges in the United States has been rather limited until recent years, although the practice has wider usage abroad. For example, the Missouri Department of Transportation now uses deck panels on the majority of its steel and

---

[1] Group Director, HNTB Corporation, 1201 Walnut, Suite 700, Kansas City, MO 64106

concrete girder bridges. Other states are following suit as it is demonstrated that bridge deck costs can be reduced.

The typical precast panel used by Missouri is a 75-mm thick concrete panel that spans between the beams or girders with a maximum girder spacing of about 3 meters. The width of the panels (longitudinal direction of the bridge) is generally limited to 2.5 m.

The panels are typically prestressed in their spanning direction with 9.5-mm diameter Grade 1860, 7-wire strands on 100 mm centers. Reinforcing in the transverse panel direction is No. 10 U-bars spaced on 150-mm centers. These bars extend from the top surface of the panel and function as lifting loops and serve to tie the panel to the cast-in-place topping concrete layer.

The panels are supported along the edges of the steel girder on a layer of expanded polystyrene bedding material, which varies in thickness to accommodate changes in top flange thickness. The deck panels generally overlap the flange by 75 mm, leaving ample space on the flange for shear studs. The girder shear studs are sized to extend at least one inch above the deck panel into the topping slab. Figure 1 shows precast panels on a bridge prior to topping slab placement.

The design of the deck panels is similar to prestressed beam design. Stresses are checked at each loading sequence and the composite slab is checked for ultimate strength. The topping slab is reinforced for negative moments over the girders in the transverse bridge direction and additional longitudinal reinforcing steel is placed in the topping slab over the piers to resist girder negative moments.

Figure 1 Precast Deck Panels on Steel Girder Bridge

The advantages of prestressed bridge deck panels include a reduction of formwork, the deck panels participate in resisting live load forces as well as dead load force (unlike steel stay-in-place forms), and the deck panels can be economically produced in precast plants. Field labor is reduced due to the lower reinforcing steel quantities and formwork requirements.

Composite Construction in Cable-Stayed Bridges

Composite construction in long-span steel cable-stayed bridges is typically used due to the cost advantages obtained. In a cable-stayed bridge, the inclined cables attached to the bridge induce axial compression forces that must be resisted by the superstructure. Since these bridges are erected usually in a balanced cantilever manner, it is advantageous to make the deck composite with the girders as earlier as possible to allow the deck to participate in resisting these axial forces.

Hague (1997) discusses the design of the Bill Emerson Memorial Bridge, a cable-stayed bridge now under construction at Cape Girardeau, Missouri, which uses large precast deck panels. Figure 2 shows the superstructure components of this design with two precast panels spanning between floorbeams.   In the balanced cantilever erection method, construction progresses outward from each tower by erecting sections of edge girders, filling in with the floorbeams and bracing, and then placing the precast panels. Gaps between the precast panels are provided for cast-in-place concrete closure strips that will enable composite action between the floorbeams and panels. After curing of the closure strips, the panels are post-tensioned in the longitudinal direction. Closure strips are then cast over the edge girders for composite action. At this point, further loading will be resisted by the composite section. A 75-mm non-structural silica fume overlay will be provided over the precast panels between the gutterlines following completion of the superstructure erection.

It can be readily seen that the erection and in-place analyses of a cable-stayed bridge can be a challenging effort due to the changing section properties and loading which locks in stresses in the various structural components.   In particular, the behavior of the slab component in a composite steel cable-stayed bridge has not been significantly studied.

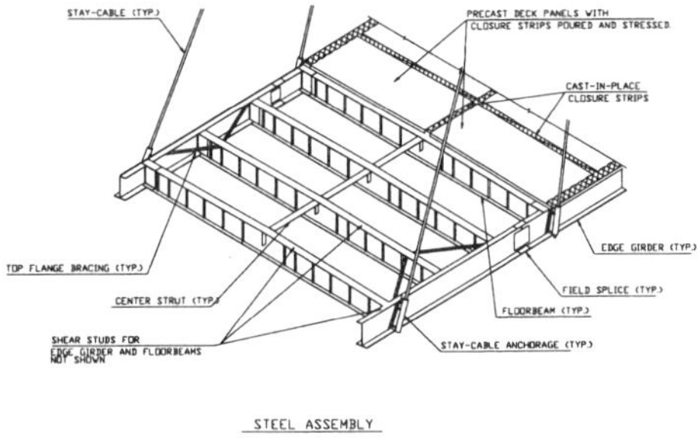

Figure 2 Superstructure Components

Studies are currently underway (Byers and McCabe 1997) to evaluate the slab participation in resisting the axial and flexural forces using finite element analysis.   Their primary findings suggest that the effective slab width throughout the span can be approximated by a simplified model. This model has been shown to provide good prediction of stresses and deflections for direct stiffness analyses.

Composite Construction in Long-Span Bridge Towers

A key component of the massive Central Artery project in Boston, Massachusetts is the Charles River bridge project. The mainline bridge, currently under construction, is a five-span cable-stayed bridge featuring a main span of 227 m and a total bridge length of 429 m. The interesting aspects of this bridge include the cable arrangement, the inverted Y-shaped towers and roadway width of almost 61 m.

The superstructure for the bridge is a hybrid design consisting of back spans of multi-cell concrete box girders while the main span is a composite steel box girder design. Due to construction considerations, the cable connections to the back spans will not be anchored along the superstructure edges but instead to a central spine along the bridge centerline. The main span anchorages are along the edge girders in the conventional fashion.   Aesthetics were a key consideration for the cable arrangement and shape of the inverted Y-shaped towers.

The top portion of the tower varies in size from 3.2 m square cross section at the top to about 4.3 m square at its base.   The limited dimensions of the tower cross section, the physical dimensions of the cables and anchorages and large cable forces led the designers to use a composite steel anchor box in the tower in lieu of conventional anchorage methods.

The anchor-box system is designed to act compositely with the concrete section by using conventional shear studs welded to the box. The cables are anchored at the interior face of structural tube sections fabricated into the anchorage boxes. Figure 3 shows the tower elevation and cross section at the anchor box. McCabe et al (1998) list the following additional advantages to this design solution as a reduction of the transverse cable spacing which reduces torsional moments, more accurate geometry control, quicker construction with no complicated forming of the tower walls, and elimination of post-tensioning.

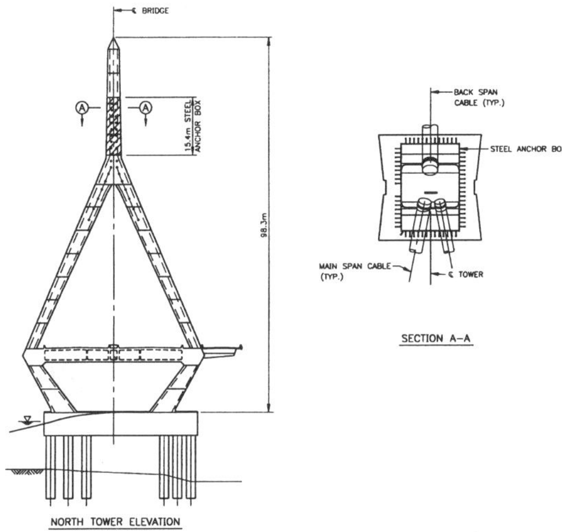

Figure 3 Charles River Tower and Anchor Box Details

The composite steel anchor-box system for the Charles River bridge towers provides an efficient solution to a problem where size limitations require the use of steel and the concrete provides additional strength and protection for the anchorages. Although similar systems have been used abroad, this is the first use of the anchor-box system in the U.S.

<u>References</u>

Hague, S.T. (1997). "Composite Design of Long Span Bridges," Proc. Structures Congress XV, Vol. 1, ASCE, Portland, OR.

Byers, D. D., and McCabe, S.L., (1997). "Effective Slab Width for Composite Cable-Stayed Bridges using Finite Elements," Proc. Structures Congress XV, Vol. 1, ASCE, Portland, OR.

McCabe, R.J. et al (1998). "The Design of Charles River Bridges," Proc. 15th Annual International Bridge Conference, Pittsburgh, PA.

DOUBLE-ANGLE CONNECTIONS WITH SLOTTED HOLES
IN THE SUPPORTED-BEAM-WEB-LEGS

Robert O. Disque, P.E.[1]

A beam that is connected to a column or girder with
a double-angle connection is considered to be simply
supported with no end moment (Fig. 1). End rotation is
usually accommodated by the flexibility of the legs
connected to the supporting member (Fig. 2). However, if
slotted holes are used in the supported-beam-web-legs,
the beam web rotation may, instead, cause sliding of the
bolts unless the bolts are made slip-critical.

If both the connection of the angles to the support
and to the supported member were designed as pins (Fig
3), the system would be similar to a catenary and in need
of thrust restraint (Fig 4). Because delta is a very
amount, the required thrust restraint becomes quite
significant and can easily exceed the available strength
that can be provided by the slab or other structural
element, if one is present at all.

To avoid the need for thrust restraint entirely, it
is recommended practice to make the beam-web-leg bolts
slip-critical and design them for eccentricity, (Fig 5).
The connection to the support can then be designed for
shear only. Alternatively, if these bolts are made
bearing-type (snug-tight or fully tensioned), the hinge
point is at or near the bolt line, and the eccentric
effect of this load must be considered in the design of
the connection angles, the connection of the angles to
the support, and the support.

[1] Consultant, Gibble Norden Champion Consulting
    Engineers, Old Saybrook, Connecticut 06475

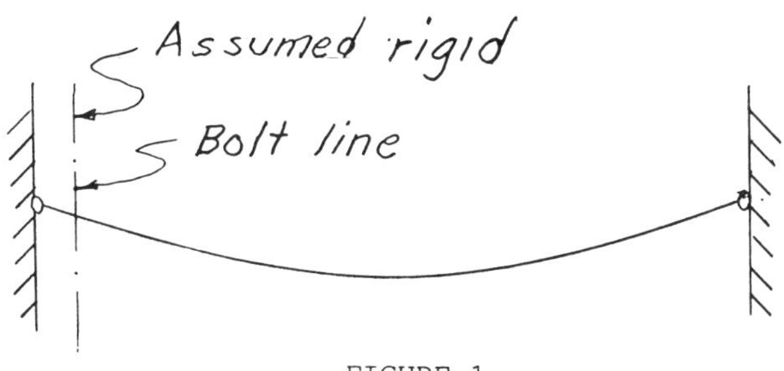

FIGURE 1

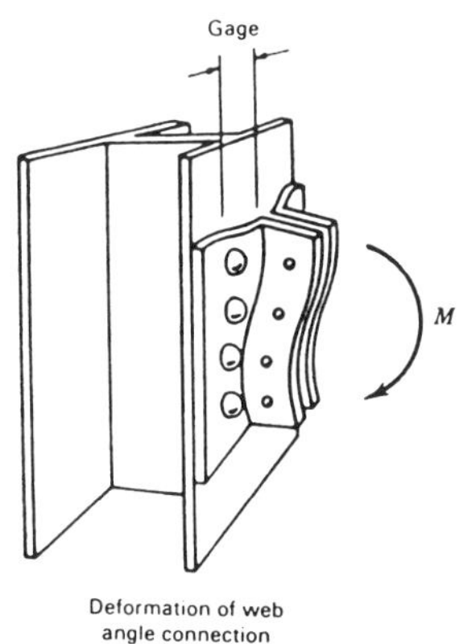

Deformation of web
angle connection

FIGURE 2

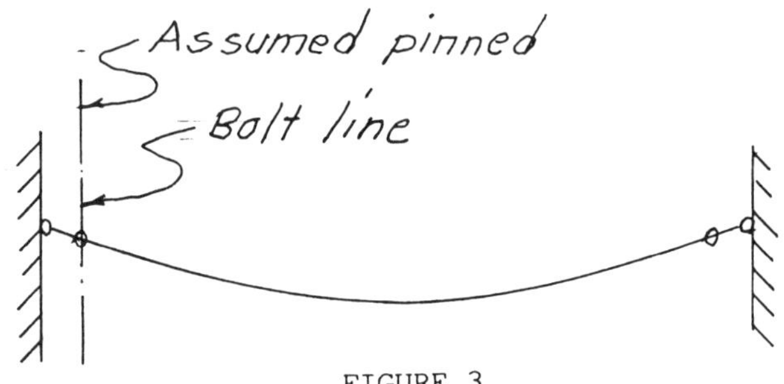

FIGURE 3

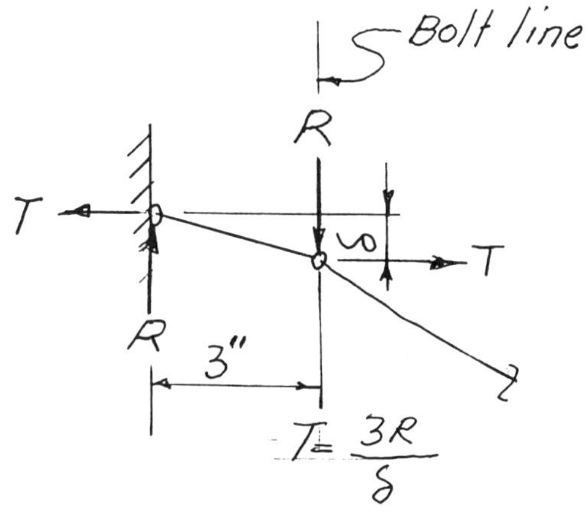

FIGURE 4

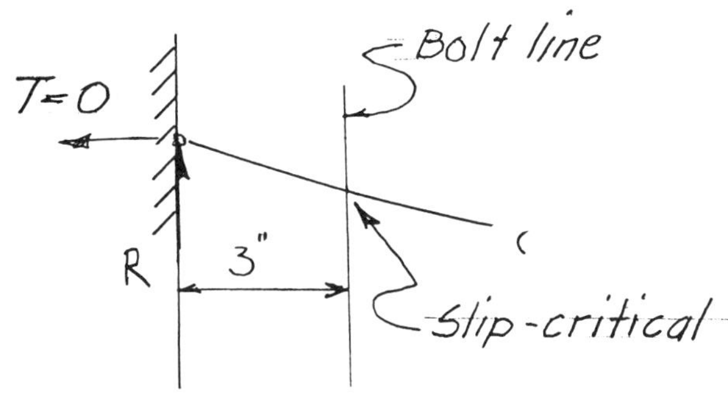

FIGURE 5

Column Economy Considering Stiffening Requirements

Charles J. Carter, PE [1]

Abstract

Transverse stiffeners and web doubler plates are extremely labor-intensive detail materials due primarily to the welding that is associated with their use. If transverse stiffeners and web doubler plates can be eliminated and an unreinforced column can be used, significant cost savings can often be realized. Additionally, the elimination of column stiffening will simplify (and thereby economize) connections that are made to the weak axis of the column.

Introduction

Transverse stiffeners and web doubler plates are extremely labor-intensive detail materials due primarily to the welding that is associated with their use. As such, they add considerable cost in spite of their disproportionately low material cost. If transverse stiffeners and web doubler plates can be eliminated and an unreinforced column can be used, significant cost savings can often be realized. Additionally, the elimination of column stiffening will simplify (and thereby economize) connections that are made to the weak axis of the column.

In low-seismic applications, the specification of transverse stiffeners as a default item is discouraged. In high-seismic applications, however, transverse stiffeners will normally be required.

In low-seismic and high-seismic applications, the specification of a web doubler plate as a default item is discouraged. Web doubler plates require significant welding into the column fillet region, which is an area of potentially lower notch toughness. The shrinkage that accompanies the cooling of these welds typically can distort the cross-section and overwelding in this region carries

---

[1] Director of Manuals, American Institute of Steel Construction, Inc., One East Wacker Drive, Suite 3100, Chicago, IL 60601-2001

the potential for cracking. Additionally, the weld joint involves a non-prequalified detail.

Eliminating Column Stiffening

There is significant potential for economic benefit when transverse stiffeners and web doubler plates can be eliminated. Therefore, the designer should consider alternatives that eliminate the need for column stiffening, when possible. Some suggestions follow.

(1)   Specify column material with yield strength of 50 ksi, such as ASTM A992 or A572 grade 50 steel. The increased minimum yield strength will increase the design strength of the column, yet there will be little or no impact on the material cost. Mill grade extras for 50-ksi wide-flange material are largely nonexistent in shapes that weigh as much as 150 lbs. per ft of length[2]. Even for W-shapes in weight ranges that have grade extras, these nominal cost differences of two or three pennies per pound are negligible when compared to the advantage gained in detail material savings. Column material with even higher yield strength, such as ASTM A913 grade 65 material, is also available; however, the associated material cost differential is greater.

(2)   Consider a different column section that has a thicker flange and/or web, as appropriate. This increase in material cost, given today's typical FOB[3] mill price for common grades[4] of steel of approximately $400 to $450 per ton, is in most cases easily offset by the savings in labor costs.

(3)   Consider a deeper cross-section for the beam that is connected to the column. Increasing the depth of the beam decreases the flange force delivered due to the increase in moment arm between the flange-force couple. If it were possible to replace a W16x50 with a W18x50, the material cost would not be increased; if a lighter, deeper shape were suitable, the material cost would in fact be decreased. Even if there were an increase in material cost, it would in most cases be easily offset by the savings in labor costs. Note that this suggestion may instead be punitive when the moment connection is designed to develop the strength of the beam because the flange force is essentially a function of the flange area.

(4)   Increase the number of moment-resisting connections and/or frames to reduce the magnitude of the moment delivered to a given connection to a level that is within the local design strength of the column section.

---

[2] Inquire with steel mills to determine the current range of shapes for which a grade extra applies.

[3] FOB stands for "free on board", which indicates that the quoted price assumes delivery to the indicated location. In the above case, the indicated location is the mill itself; subsequent shipping would incur additional cost.

[4] Common grades include ASTM A992, ASTM A572 grade 50 and A36.

Minimizing the Economic Impact of Column Stiffening Requirements

In some cases, the need for column stiffening may not be avoidable. When this is the case, the following suggestions may help minimize the cost impact for building structures in low-seismic applications:

(1) Where allowed by governing building codes, design column stiffening in response to the actual moments and resulting flange forces rather than the full moment resistance of the cross-section; the latter simply wastes money. When the Engineer of Record (EOR) delegates the responsibility to determine the column stiffening requirements, the design moments should also be provided.

(2) If designing in allowable stress design, take advantage of the allowable stress increase in wind-load applications (load combinations in LRFD inherently account for such concurrent occurrence of transient loads).

(3) Properly address reduced design strength at column-end applications. The typical beam depth is usually such that the reduced design strength provisions for column-end applications apply only at the nearer flange force.

(4) Increase the number of moment-resisting connections and/or frames to reduce the magnitude of the moment delivered to a given connection to a level that allows a more economical stiffening detail.

(5) Give preference to the use of fillet welds instead of groove welds when their strength is adequate and the application is appropriate. This is particularly true for the welds connecting transverse stiffeners to the column.

(6) When possible, use a partial-depth transverse stiffener, which is more economical than a full-depth transverse stiffener because it need not be fitted between the column flanges. Select the partial-depth transverse stiffener length to minimize the required fillet-weld size for the transverse-stiffener-to-column-web weld.

(7) While transverse stiffeners are required in pairs when the limit states of local flange bending or local web yielding are less than the required strength, a single transverse stiffener is permitted and should be considered when the limit states of web crippling and/or compression buckling of the web only are/is less than the required strength.

(8) In cases when the flange force is only compressive, allow the option to weld the transverse stiffener end or to finish it to bear on the inside flange. In most lateral load resisting frames, however, moments are reversible and the design flange force may be either tensile or compressive.

(9) Use a single web doubler plate up to a required thickness of ¾ in. If thicker web reinforcement is required consider the use of two plates, one on each side of the column web. This practice may be more economical and is likely to reduce heat input, weld shrinkage, and member distortion.

(10)   Select the web doubler plate thickness so that plug welding between the column web and web doubler plate is not required.

(11)   Recognize that, in the concentrated-flange-force design provisions in LRFD Specification Section K1, it is assumed that the connection is a directly welded flange or flange-plated moment connection, not an extended end-plate moment connection.

(12)   Limit the number of different thicknesses that are used throughout a given project for transverse stiffeners and web doubler plates. Production economy is achieved when many repetitive elements can be nested within the fewest number of source materials.

In high-seismic applications, economy suggestions 5, 6 (when a moment connection is made to one flange only), 9, 10[5], 11, and 12 above remain applicable.

---

[5] Note that this may not be possible in high-seismic applications if the column web thickness itself does not meet the seismic shear buckling criteria given in AISC Seismic Provisions Equation 9-2.

Survey of Steel Roof Systems

Thomas Z. Scarangello, P.E.[1]
Robert K. Otani, P.E.[2]
Michael S. Squarzini[3]
Steven R. Witkowski[4]

Abstract

Three arenas opening in 1999 will feature unique long-span roof systems which all meet demanding economic, construction and functional criteria.

Introduction

In 1999 three new arenas designed by Thornton-Tomasetti Engineers (T-T) for NHL and NBA teams will be opening in Atlanta, Denver and Miami. All three of these arenas have unique architectural long-span roof forms and demanding programmatic requirements. Thornton-Tomasetti's challenge in each case was to develop a safe and economical structural solution that compliments these forms as well as efficiently acts as a functional support platform for arena events. As T-T set out to achieve this goal, we were guided by some common principles.

One principle was to minimize the amount of costly field assembly/fabrication. To achieve this T-T utilized shippable 3.8-meter center-to-center deep truss segments. These elements were incorporated either as spanning elements alone or as part of a tension tied truss/arch system.

Another principle was to minimize the need for temporary shoring and, if needed, to limit shoring location to areas of the arena that would not interfere with the completion of the arena bowl construction. This approach allows for an overlap of the superstructure and roof construction, reducing the overall construction schedule, thus

----

[1]Member, Thornton-Tomasetti Engineers, 641 Avenue of the Americas, New York, NY 10011; [2]Member; [3]Member; [4]Member.

providing a significant economic benefit.

The final common principle was to ensure that the roof structure would function as an efficient, easily accessible and flexible working platform for the support of scoreboard, event rigging, maintenance functions, speaker clusters, etc.

Each facility designed accomplished these goals in a unique and efficient manner.

I. THE PEPSI CENTER

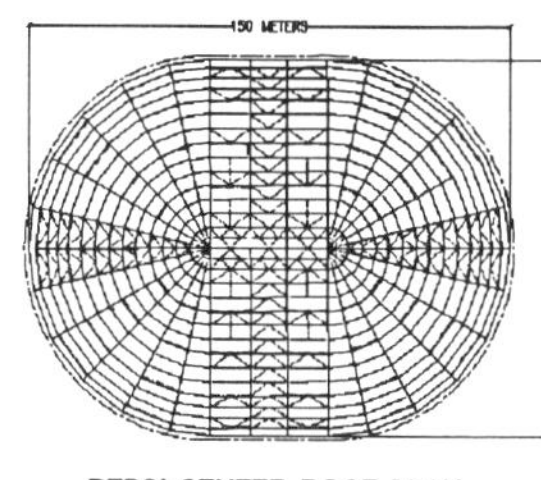

PEPSI CENTER ROOF PLAN

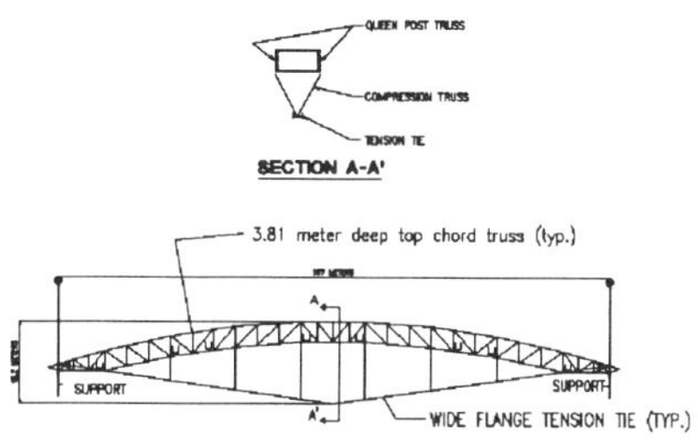

TYPICAL TENSION TIED QUEEN POST TRUSS

The Pepsi Center is a 19,000-seat arena in Denver, Colorado's surging downtown district. The arena will house the Colorado Avalanche and the Denver Nuggets beginning in the fall of 1999. The arena will satisfy all the requirements of the current sports facility market with nearly 100 private suites, and numerous restaurants and concessions.

The clear spans for the Pepsi Center roof are approximately 150m in the longitudinal direction and 107m in the lateral direction. The roof geometry consists of two semi-circles with a 53m radius separated by a spine that is 44m wide. The roof is curved at a 183m radius from the high point at the roof ridge to the supporting columns.

The 107m lateral span of the Pepsi Center roof could have been spanned by a series of simple trusses and deep tension tie elements. Due to the complexity of the radial geometry at the two semi-circled ends of the roof and the requirement to maintain clear site lines at the ends of the arena, it was determined that the roof system would use only deep queen post trusses at the center spine. At the radial ends of the roof, simple 3.81m deep by 53m long trusses would be used to span from the exterior columns to the queen post trusses. The structural system is erected using eight shoring towers in the center of the arena which do not interfere with the simultaneous construction of the superstructure.

The four queen post trusses at the center spine also have a unique aesthetic feature. Rather than providing four trusses with four tension ties, a more elegant structure could be created if two trusses shared one tension tie. Therefore, the compression struts of two queen post trusses post down to one tension tie between them, creating two pyramid supporting elements.

The catwalks and rigging beams follow the geometry and are supported by the bottom chords of the roof trusses. The catwalks serve lighting platforms as well as a scoreboard service platform located within the center box of the roof.

II.   AMERICAN AIRLINES ARENA

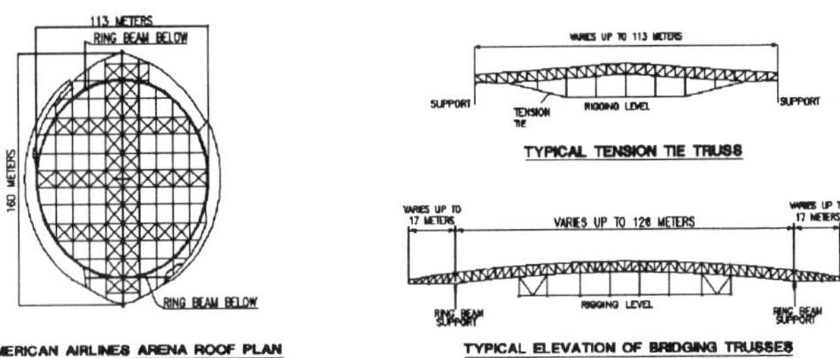

The American Airlines Arena located in downtown Miami, Florida, is a $165 Million (US) state-of-the-art sports facility for the Miami Heat basketball team with a maximum seating capacity of 20,000. The 13285 m^2 arena roof structure (see above figure) is comprised of intersecting orthogonal tied trusses and bridging trusses spaced at approximately 11m center to center. At both ends of the arena bowl, the arena roof trusses cantilever up to 17.1m beyond a concrete ring beam support and overhang well beyond the arena footprint, creating an architecturally dramatic view from below.

There are eight main tension tied trusses. Each main tension tie truss is built of a 3.81m truss section top chord, wide flange tension tie, each connected by vertical wide flange struts centered at intersecting bridging trusses. The bridging trusses consist of typical 3.81m depth trusses. Each main tied truss is erected using two cranes. Using a bridle, each crane lifts one half of a truss until the two halves are connected via a mid-air splice, negating the need for temporary shoring or erection towers.

A unique feature of the main tension tie trusses is its large (approximately 3447 m^2) rigging, lighting, catwalk, and scoreboard access support platform in which the horizontal portions of the truss tension ties, reinforced with wide flange beams, serve as structural support. Another design consideration unique to this arena located next to the oceanside is the extremely high uplift wind pressures (up to 4310 Pa). This uplift necessitated that the tension ties be laterally braced in the event that the wind uplift pressures exceeded the dead load of the roof, causing the tension ties to go into compression.

III.   ATLANTA ARENA

The new Atlanta Arena, located adjacent to the CNN Center, will be the new home of the Atlanta Hawks basketball team and the Atlanta Thrashers expansion hockey team. With a seating capacity of 20,000 and four levels of luxury suites stacked on one side, four roofs are required to cover the 18580 m^2 plan area. Three roofs are structural steel framing, while the last is concrete framing, long-span steel joists and wide flange purlins.

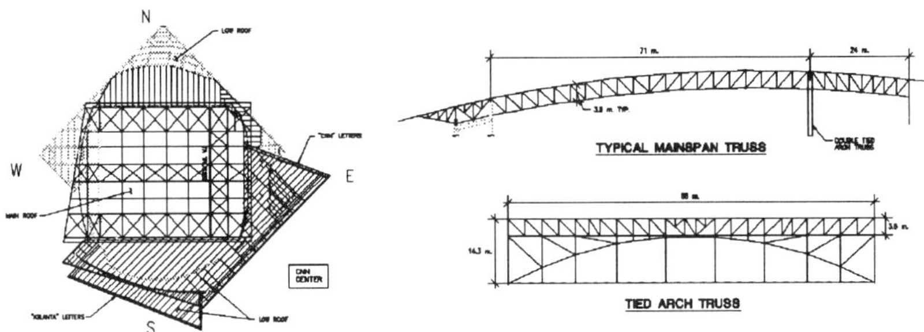

The main roof is supported by a two-way truss system composed of 3.8m deep trusses that span 95m x 85m between column supports and cantilever up to 19.8m. This 10700 m^2 roof covers the arena floor and portions of the seating bowl, and overhangs areas of the lower roofs. The 95m span is broken up by two 14.3m deep tied arch trusses, 0.9m apart, to allow the 3.8m deep trusses to span economically. A major architectural statement, the tied arch trusses, are located 24m away from the support columns on the East side and are composed of a 3.8m upper truss with a compression arch and tension tie below. This arrangement allows for an entire 85m x 24m area to be erected with the aid of only three temporary shoring towers. Once this area is completed, the towers are removed and erection can continue with the North and South perimeter trusses. With a 12.2m effective depth, these trusses act as a supporting edge for the two-way system and also support joists that frame the low roofs. This is accomplished by adding verticals, diagonals, and a 1.5m bottom chord truss below the 3.8m typical upper truss. Although two temporary shoring towers are needed for each truss, they are removed once the adjacent interior truss is erected. Since the 71m long interior trusses were designed to span as a one-way system for the dead load, they are erected without temporary shoring.

All rigging beams, catwalks, and scoreboard support are integrated within the main roof. Rigging beams are located above the staged event areas and span between bottom chord centerlines of the trusses. Catwalks are located approximately .3m above the centerline of the bottom chords. Deeper catwalk beams support the scoreboard platform.

Framing of the low roofs typically consists of joist or purlins supported by wide flange girders or 3.8m deep two-way trusses that span up to 43m. At two perimeter areas, built-up box columns, .6m to .8m square, standing as much as 27m tall, support a roof box girder. These box columns will appropriately spell out "CNN" and "ATLANTA".

Aesthetics of Fabric Structures

Matthys Levy, FASCE[1]

Abstract

Fabric structures have proliferated in the past forty years following the development of tough and durable fabrics. Although often dramatic, some fabric structures do not rise to the level of aesthetic excellence. A number of such structures are examined and compared to those that are regarded as visually pleasing as well as exciting.

Introduction

Architecture is all around us, we live in it, sleep in it, in fact almost everything we do takes place within an architectural space. Sight and touch are the senses aroused in us by our immediate surroundings. The form of a building, its proportions, the materials used, both their texture and color, all contribute to giving us a feeling of satisfaction or disturb us. I can think of a number of buildings against which I have reacted violently. Whatever our role in life, we are therefore deeply involved with architecture and the world created by the architect.

What is it about an architectural structure that generates such a strong response from us? T o understand this we must first define what we mean by an architectural structure or what are its principal attributes. First, it must fulfill a purpose, be it to house us, entertain us as in a theater, teach us as in a school, judge us as in a court house or punish us as in a jail: this is its <u>functional</u> attribute.

Secondly, it must provide a visual unity that satisfies us in a spiritual manner, or as we sometimes say, it must be beautiful or in a more general sense, ti must satisfy the spirit of man: this is its <u>aesthetic</u> attribute. Finally, and maybe this is the only area I can speak to with any authority, it must be sound. Certainly, we would not want to be within a structure built like on of the self-destruct sculptures of the Swill artist Jean Tinguely. We expect our structures to protect us from the elements, to provide shelter, to be strong: this is the <u>structural</u> attribute of an architectural structure.

It should be relatively easy to fulfill three requirements in order to recreate perfect structures. Certainly, function and structural soundness are totally rational and objective criteria that can be described clearly and set down on paper and disseminated with little argument. Unfortunately, aesthetic qualities are subjective and elusive and are responsive to the fickle mores of the times. Remember the story of the Eiffel Tower, a structure that is almost universally admired today for its

---

[1] Principal, Weidlinger Associates Inc., 375 Hudson Street, New York, NY 10014

elegance and structural correctness. When the design was first published, the artists of the day issued a manifesto stating, "We writers, painters, sculptors, architects, devoted lovers of the hitherto intact beauty of Paris, protest with all our strength and wrath…against the erection in the very heart of our capital of the useless and monstrous Eiffel Tower…" It was also called and arrogant ironmongery, black factory chimney, hollow candlestick, assemblage of ladders, disgraceful skeleton and the greatest insult of all, "..even commercial America wouldn't want it". They had forgotten that not fifty years earlier, Baron Houssmann had totally changed the city of Paris by brutally cutting broad boulevards and creating giant squares, all the elements that give the city its unique character and are admired today..

The vagaries of taste exist in our time as well: one day the International school is celebrated and the next day it is vilified; then the post modernists reign supreme only to be supplanted by the modernists. Without a guidebook it is sometimes impossible to know what you should like at any given moment. This points out the transience of our current super-industrialized society. We live in an age when change is a rapidly accelerating phenomenon…here today, gone tomorrow is a motto we hear all the time. On a personal level each generation feels about the next that they are moving too fast, growing up too fast. In the perspective of my own lifetime, I have witnessed fantastic changes: television did not exist when I was born; I solved mathematical problems in school using a slide rule which today has been totally displaced by the pocket calculator. I still call a refrigerator an icebox. When I first arrived in the US it was by boat and it took four days; the Concorde completes the same journey in less than four hours and will not cause me to become seasick. In a broader context, half of all the energy consumed by man since the beginning of time has been used in the last hundred years, a direct consequence of the phenomenal increase in technology - - cars, trains, planes, lights, machines, computers, radio, television, all new in that time and all consumers of energy.

Change

In past times the rate of change used to be so slow that it would pass unnoticed. This permitted contemplation of philosophical questions such as "what do all beautiful things have in common?" Rationalizing what we have originally defined as a subjective question, Plato answered, pure form as exemplified by geometric figures. Yet, if we apply this statement to the evaluation of a statue, its geometric aspects do not constitute its whole worth. Its value has a connection with life and feeling, and therefore its aesthetic value goes beyond the Platonic idealization of beauty. There is a similar relation between structure and architecture. A well-conceived structure contributes purity of form but is not necessarily a work of architecture. This is exemplified by comparing the delicately formed vaulted roof of Nervi's Turin Exhibition Hall seen on the interior with the bland, nondescript exterior of the building.

After Plato, Socrates defined beauty in terms of usefulness and efficiency. I applied today, this can be a very misleading criterion. There are innumerable buildings today that serve a useful purpose in an efficient way but which certainly do not have a real aesthetic value. Many office buildings fall into this category, but I hesitate to single out any one for fear of offending the designer. On the other hand sometimes the real, naked beauty is covered up, as was almost the case with the George Washington Bridge in New York for which a proposal was made to clad it in stone. Fortunately, it was never carried out.

In these classical definitions of aesthetics, the object being examined was considered to have a value entirely of itself, divorced from life and feeling. In architecture, joy and sorrow as representations of life were introduced through ornamentation and decorative structural elements: a scupper to lead water off a roof became a gargoyle, the lateral brace for a tall vault became a flying buttress. In our own time, Robert Venturi expressed the need for playfulness and variety when he deplored the "deadness that results in too great a preoccupation with tastefulness and total design".

In the Middle Ages, canons of proportion passed on from one master builder to the next, generation after generation codified taste and aesthetics. The golden section and innumerable proportional numbers were used as the rational basis for design. This was the "Harmony of the Ancients" spelled out in Alberti's Ten Books:

> And indeed I am every Day more convinced of the truth of Pythagora's saying, that Nature is sure to act consistently, and with a constant Analogy in all her Operations: From whence I conclude, that the same Numbers, by means of which the Agreement of Sounds affects our Ear with Delights, are the very same which please our Eyes and our Mind. We shall therefore borrow all our Rules for the finishing our proportions, from the Musicians, who are the greatest Masters o f this Sort of Numbers, and from those particular Things wherein Nature shews herself most excellent and compleat.

This codification of aesthetics is highly suspect if we believe that Beauty is subjective and emotional or, as we say, is in the eye of the beholder. Viollet-le-Duc, the nineteenth century French Architect, stated it best when he wrote, "Proportions in architecture are not rigid canons but a harmonic scale, a correlation of variable relationships, according to the mode adopted…The artist is always present beside the geometrician, and will be able to bend the formulas when necessary.." It seems everyone agrees what structural aesthetics is not, but then what is it?

Creativity

In a book on creativity, Arthur Koestler presented man's realm of knowledge on a graph with truth at the top left represented by pure mathematics, which is characterized, as being objective and verifiable. On the bottom right, Beauty is represented by art, which is characterized as being subjective and emotional. On this graph, structure is somewhere west of art and south of math. This is clear to anyone who has ever been involved in a forensic investigation of a structural problem. Three engineers evaluating the same structure will render opinions ranging from: it's in perfect shape, to, it will stand for now, to…,I'm surprised it's still standing. With this kind of diversity of opinion of a supposedly scientifically based design, how can we possibly expect unanimity when considering aesthetics.

The form or shape of a structure is not unique. Ask the same three forensic engineers who could not agree on the state of a damaged structure to redesign the structure to support a given load and with a given configuration, and you will no doubt obtain three differing designs. But this is perhaps where creativity and an aesthetic sense will triumph. When Robert Maillart, the Swiss bridge designer and builder, created some of the most beautiful bridges built in concrete, he was responding on many levels: as the designer, he had tom provide a structure to carry specified loads over a specified distance. As builder, he wanted the most economical structure that could be built. He obtained this result in a structure for the Salginatobel bridge completed in 19300 which was the visual expression of the envelope of the deformations under unsymmetrical loads, and a bridge which could be built with a minimum of formwork and a minimum of material and therefore cost. It is appropriate that Maillart designed a structure that was possible within the bounds of a given situation following a process which Herbert Simon, the Nobel laureate in economics called "good enough" actions. Simon, the father of artificial intelligence, created "thinking out loud' as a means of exploring the process of problem solving. Through his research into the creative process, he came to believe that human beings do not optimize but rather search for good enough solutions. This process which involves lateral thinking brings to bear on a solution ideas from divergent sources and is virtually impossible to codify. If the creative process leading to one person's aesthetically satisfying structure is not definable, then how could another person, the viewer, see it exactly the same way? I can't tell you how many times I have pointed out to my wife what I consider a really beautiful structure only to be brought up short by her one work exclamation, "That?!"

Aesthetics

Clearly we can't all expect to agree on what constitutes beauty, but a few common characteristics can be identified against which we each measure aesthetic value: **Function, Geometry, Material, Shape, Scale, Flow of forces**. In each case the most direct and simplest solution is the one that is most pleasing. Entering a room in which there is a jumble of furniture shapes and a crazy quilt of colors and materials will not yield a favorable aesthetic response. This does not imply7 that complexity is necessarily bad. The Sagrada Famiglia designed by Antonio Gaudi is a church in which the form was determined by shaping the various roof elements to the funicular of the applied loads as a rational expression of the flow of forces. The soaring filigree structures of Nervi and smooth elegance of the concrete shells of Candela are clear statements of the flow of forces and express strength through form. A simple arch does this as well, whereas a pyramid represents a structure with a constant inner stress: every lower layer of stones adds a load which must be supported by a proportionally larger stone layer below it.

Fabric Structures

The development of stronger and more durable fabrics following the Second World War introduced a new aesthetic, that of soft structures. The many fabric structures that were shown at the 1970 Osaka World's Fair demonstrated the range of possibilities from the good to the bad and the ugly. Perhaps the best was the US Pavilion designed by Davis & Brody with David Geiger as the engineer. That structure was a low profile balloon shaped like a sliced pumpkin. It was both elegant and subtle, not imposing its new technology yet exploiting its drama. Subsequent roof structures for stadiums that were designed by Geiger failed to meet the promise of elegance that was implied by the Osaka Pavilion. Instead, they were often bulbous and graceless. In another direction, the use of fabrics in tension structures led to the development of some elegant examples. The simplest canopy in the shape of a hyperbolic paraboloid is both dramatic and aesthetically pleasing. One of the best examples of such tension structures is the Riyadh Stadium roof. This structure expresses both the strength of the structure and the tension between its parts better than any other.

Fabrics as structure are important because they bring to buildings a level of translucency and softness that cannot otherwise be achieved. For that reason, it is important to exploit these qualities while using such structures appropriately. The Georgia Dome with its strong geometric pattern exploits the best aesthetic potential for large scale structure. The La Plata Dome, currently under construction, extends the same concept but with a more dramatic result and will undoubtedly be one of the premier fabric structures.

Size is not always a measure of importance. The new Millenium Dome in London is a low profile, huge roof structure covering an exhibit. Although technically well done, it does not rise to the level of aesthetic success that its name would augur.

On a smaller scale, the roof canopies over a visitor's center in Panama are delicately balanced and express the nautical origin of the site without making an overblown statement.

Using a fabric structure simply for the sake of innovation without consideration to its appropriateness is often an invitation to aesthetic failure. Nevertheless, the potential that soft architecture offers, opens the way to the development of structures that are both innovative and beautiful.

Unique Glued Laminated Timber Roof Structures
for the Lisbon Multi-Use Arena

Robert C. Sinn, MASCE[1]

Abstract

At up to 115 metres, the roof structure for the main hall of the Lisbon Multi-Use
Arena represents the longest one-way span of glued laminated timber in the world.
There are two separate roof structures framed in exposed glued laminated timber for
the project: the main hall which is 200m long by 120m wide and is oval shaped in
plan, and the auxiliary hall which is 100m long by 50m wide. While glulam
structures have been nearly non-existent in Portugal up to the present due to the lack
of suitable timber materials, glued laminated timber was chosen for the roof
structures as the material best suited to enhance the Lisbon 1998 world exposition
environmental theme as well as being cost-effective in comparison to alternative

Figure 1. Lisbon Arena Longitudinal Section

----

[1]Associate Partner, Skidmore, Owings & Merrill LLP, 224 South Michigan Ave.,
Suite 1000, Chicago, Illinois, USA, 60604.

structural steel schemes. The structural design of the roof is also noteworthy from a structural engineering standpoint as being one of the longest glued laminated timber roof spans in an active seismic zone. (See Figure 1)

Aesthetic and Programmatic Considerations

The design brief for the Lisbon arena competition called for a facility which would be a principal element of the 1998 World Expo as well as providing easy adaptation to accommodate a wide variety of indoor sporting, recreational, and cultural events for the City of Lisbon in the longer term. The stadium would be located on the Expo's 50 hectare waterfront site along the Tagus river and would have a fixed seating capacity of 12,500 and a maximum capacity of 17,500. Portugal's history is closely linked to the sea. The building's appearance and materials would naturally complement the Expo's environmentally friendly theme, "The Ocean: A Heritage for the Future." The design includes the capability to operate using natural ventilation and natural light for most of the daytime events. To promote a long term commercial viability, the playing field meets Olympic standards with additional provision for adjacent warm-up facilities in the auxiliary hall.

The shape of the arena draws inspiration from the lines of the 16th century carabelas, the ocean-going sailing ships of Vasgo da Gama's day. (See Figure 2) The same theme is further developed on the interior in the ribbed structure of the glued laminated timber roof structure, manufactured from softwood, an

Figure 2. Exterior Building Form

environmentally conscious building material produced from a sustainable source. More recent Portuguese history is reflected in the curvature of the front cantilevered elevation which is based on the nose of a PANAM Clipper, the first transatlantic aircraft that disembarked passengers near what is now the Expo site during the thirties and forties.

The choice of glulam for the roof structure was consistent with the architectural vision of the interior as suggesting a ship's hull. (See Figure 3) The primary engineering challenge was to structure the roof profile without the benefit of a pure parabolic or circular arch meeting the ground. The architectural premise was to create a unique profile without exterior abutments or piers while at the same time maintaining a tall entrance space ringing the grandstands on the interior. The portal arch form, while not as structurally efficient as a more geometrically pure arch or dome, served to create the interior and exterior architectural vision with a reasonable amount of material expended.

Structural System Components

The main exhibition hall roof structure is framed parallel to the short direction by sixteen glued laminated timber two-hinged, arched, portal truss frames spaced 9.0 metres on center with varying spans based on the oval plan up to a maximum of 115 metres between bearings. To follow the unusual shape of the plan and roof surface, each portal truss is geometrically set out with top and bottom chords along a unique

Figure 3. Interior Ceiling Surface Concept

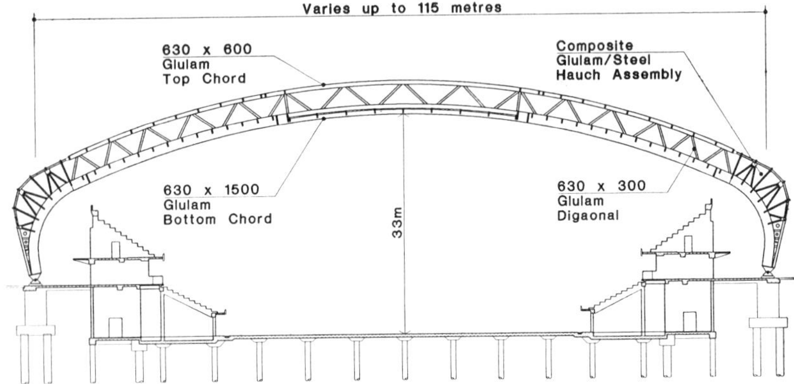

Figure 4. Typical Arched Portal Truss Elevation

circular arc meeting the haunch assembly which is in turn configured based on an identical geometry for each truss line. Typical portal truss dimensions are indicated in Figure 4. The central trussed section of each portal frame is completed by glued laminated timber diagonal members in a Warren truss configuration which are connected to the chords through a proprietary system developed by the specialty French contractors Weisrock, who were responsible for the fabrication, erection, and final design calculations for the roof structures. The trussed haunch assembly for the portals, which creates the transition from the circular central span down to the support bearings, was conceived as a composite of structural steel and glulam members using the most favorable characteristics of each material in their individual roles.

The auxiliary hall roof is also framed in glued laminated timber with nine 3.2 metre deep bowstring trusses spanning the 42 metres between supports. Transverse bridging members spanning between the primary bowstring trusses serve to stabilize the truss bottom chord under wind suction as well as reducing the span of the roof purlins. These braces take on a fan arrangement along the span of the roof trusses lending visual interest to the structural composition.

Seismic Design Considerations

Lisbon is in an area of moderate seismic activity. One of the largest earthquakes in continental Europe occurred in the year 1755; nearly leveling Lisbon and taking some 60,000 lives. From an earthquake engineering standpoint, the roof structure was completely analyzed in three dimensions through a modal response spectrum analysis in both the vertical and horizontal directions. The seismic design is based upon the structure remaining essentially elastic in a seismic event due to the lack of demonstrable ductility in the glulam construction.

Daewoo Business Center, Shanghai, China

Richard A. Henige, Jr., S.E.[1]

The centerpiece of the Daewoo Business Center is a 96-story office/hotel building to be constructed in Shanghai, China. Overall height to the top of the sloping roof is 420 meters above grade. The lower 60 floors of the building are office space with typical plan dimensions of 47.1m by 47.1m and a floor-to-floor height of 4.00m. The next 30 floors of the building are hotel space with typical plan dimensions of 37.0m by 37.0m and a floor-to-floor height of 3.25m. Hotel floor plans are rotated by 45 degrees with respect to office floor plans so that corner columns at hotel floors are aligned with mid-side columns at office floors. The upper 6 floors of the building are restaurant, observation and mechanical equipment floors with typical plan dimensions of 30.0m by 30.0m. These floors are rotated by 45 degrees with respect to hotel floors. The elevation at the 96th floor is 385m above grade. The building is capped by a 35m tall sloping roof structure.

Typical Floor Framing

Due to its economy and speed of construction, structural steel framing ($Fy=345MPa$) is used to support office and hotel floors.

Office floor construction consists of 79mm thick normalweight concrete fill on 1.2mm thick by 51mm deep composite steel deck. Typical slab span is 3.00m. Typical office floor beams are W460x60 composite steel beams with 50mm camber and a span of 12.0m. Office floor framing is designed to support 4.70 kPa of superimposed dead load and 2.00 kPa office live load.

---

[1]Vice President, LeMessurier Consultants, 675 Massachusetts Avenue, Cambridge, MA 02139

Hotel floor construction consists of 79mm thick normalweight concrete fill on 1.5mm thick by 51mm deep composite steel deck. Typical slab span is 3.40m. Typical hotel floor beams are W410x46 composite steel beams with 50mm camber and a span of 10.0m. Hotel floor framing is designed to support 4.20 kPa of superimposed dead load and 2.00 kPa hotel live load.

<u>Lateral Load System</u>

With a 385m elevation at the highest accessible floor, this is one of the tallest buildings in the world. And with an overall width of only 47.1m, this is also a very slender building: 8.2 height-to-width ratio. Local building codes specify relatively high wind loads and stringent drift criteria. Not surprisingly, the design of the lateral load system is primarily governed by control of lateral drift under wind loads. A very efficient and stiff lateral load system is required to meet this challenge: a combination of perimeter frames supplemented by reinforced concrete shear walls at the core, a 'tube-in-tube' system.

At lower office floors, perimeter frames are X-braced with steel diagonals located on a five-story module. Typical diagonal braces are W360x989 members. Perimeter braced-frames are supported by 8 reinforced concrete columns: corner columns are 3.60m by 3.60m square and mid-side columns are irregularly shaped with a cross sectional area of $12.0m^2$. Concrete cube compression strength is 60MPa and the typical vertical reinforcement ratio is 1.0%. Concrete columns provide maximum axial stiffness and dead load to resist overturning moments while steel diagonals provide stiffness and strength required to resist shear forces. A 600x600x20x20 steel box column is provided in each concrete column to transfer gravity and lateral loads from diagonal braces and to allow steel erection to proceed ahead of concrete column construction.

There is a transition from the office floor plan to the hotel floor plan between the 61st and 67th floors. Office corner columns slope inwards to support columns located at the mid-side of hotel floors above. The area of these columns is reduced from $13.0m^2$ at the 61st floor to $7.34m^2$ at the 67th floor. Office mid-side columns directly support hotel corner columns above. The area of these columns is reduced from $12.0m^2$ at the 61st floor to $7.20m^2$ at the 67th floor.

At hotel floors, perimeter frames are steel moment-frames since guest-room windows preclude the use of diagonal braces. Typical columns are W910x653 members located at 3.40m on center. Corner columns are 600x600x70x70 box columns. Typical girders are W910X488 members. One-story deep vierendeel trusses are provided at the 69th floor to transfer gravity and lateral loads from hotel moment frames above to office concrete columns below.

There is another transition from the hotel floor plan to the mechanical floor plan between the 90th and 93rd floors. Steel braced-frames with W360x744 diagonal members are provided between the 90th and 93rd floors. Steel moment frames are provided between the 93rd and 96th floors. Typical girders are W910x289 members and typical columns are W910x312 members. Corner columns are 600x600x40x40 box columns. Chevron bracing with W360x216 diagonal members is also provided in the center bays at the 94th and 95th floors.

Concrete shear walls are provided in the central core. At the office floors, these core walls provide lateral bracing for perimeter columns and diagonal braces as well as transferring lateral loads through floor diaphragms at every 5th floor to panel points of perimeter braced-frames. The overall plan dimensions of the core walls are 25.3m by 21.6m. Three 800mm thick walls are provided in the East-West direction and four 500mm thick walls are provided in the North-South direction. Openings in East-West shear walls provide access from perimeter occupied space to elevator lobbies, stairways and mechanical equipment rooms. Typical link beams located above these openings are 1200mm deep. Overall plan dimensions are reduced to 22.2m by 21.6m at the 43rd floor, since low-rise elevators do not extend above this floor.

Above the 70th floor, the overall plan dimensions of core shear walls are 15.0m by 15.0m. The alignment of hotel shear walls is rotated by 45 degrees with respect to the alignment of office shear walls below. Hotel shear walls are typically 400mm thick with 1100mm deep link beams above wall openings. Above the 90th floor, the overall plan dimensions of core walls are reduced to 6.60m by 11.0m. Concrete cube compression strength varies from 70 MPa at the base of the core walls to 40 MPa at the 96th floor.

Foundations

Shanghai is located on a flood plain near the mouth of the Yangtze River. Subsurface conditions include hundreds of meters of soft alluvial silts and clays. Foundations include 910mm diameter by 25mm thick steel pipe piles that extend 90m below grade. Typical pile spacing is 3.00m. A 4.5m thick reinforced concrete mat transfers loads from perimeter columns and core walls to the piles.

Wind Analysis

Design wind pressures from Shanghai steel and concrete design codes for tall buildings are based on a 30-year reference wind pressure of 0.55 kPa at 10m above grade. This corresponds to a 10-minute average wind speed of 29.7 m/s. For strength design of members and connections, the 100-year reference wind pressure is 20% greater, 0.66 kPa, corresponding to a wind speed of 32.5 m/s. For estimating overall drift and inter-story drift, the 50-year reference wind pressure is 0.61 kPa, corresponding to a wind speed of 31.1 m/s. The building site is consistent with Category B ground roughness for urban and suburban areas.

100-year design wind pressures vary from 0.75 kPa at grade to 3.72 kPa at 420m elevation. At Level B3, the shear force is 50,800 kN and the overturning moment is 12.8×10^6 kN-m respectively. The distribution of overturning moment is approximately 30% to core walls and 70% to perimeter columns at the base of the building. Overall building drift is H/650. Maximum inter-story drift is h/520. P-Delta effects are included in lateral load analyses, but due to the stiffness of the structure vary from less than 1% at the top of the building to a maximum of 7% at the 28th floor with an average value of 5%.

A wind tunnel test was performed to confirm the performance of the structure using a small-scale, 1:500, rigid pressure-tap model. Generalized forces for fundamental translation and torsion modes were obtained by integrating wind pressures measured on the faces of the model and weighting the resultant forces and torques along the height of the building using floor masses and corresponding mode shapes. Fundamental translation periods are 6.5 seconds and the fundamental torsion period is 2.7 seconds. Structural damping is assumed to be 1.5% of critical damping.

A probabilistic study of the Shanghai wind climate, including tropical and extra-tropical storms, estimates that 10-minute average wind speeds at gradient height are 34 m/s for 10-year and 43 m/s for 100-year return periods. For comparison, gradient wind speeds extrapolated from local code design wind pressures for a 100-year return period are approximately 33% greater, 57 m/s.

Wind tunnel estimates of base shear forces and overturning moments for a 100-year return period are 46% to 50% of corresponding local code design values. The maximum overall building drift ratio for a 50-year return period is H/1500 as compared to the code criteria of H/800. Maximum inter-story drift ratios are h/1200 as compared to the code criteria of h/750. For a 10-year return period, peak lateral acceleration at the corner of the building is estimated to be less than 1.0% of gravity. Acceleration due to torsion is less than 0.3% of gravity. Lateral and torsion accelerations are below generally accepted criteria for residential occupancy.

<u>Seismic Analysis</u>

Local seismic design codes specify a maximum earthquake base shear coefficient of: a_{max}=0.08g. As in many seismic design codes, the base shear coefficient declines with increasing fundamental period, but there is a lower-bound for long-period structures of: $0.14a_{max}$=0.0112g. Estimated building weight is 2,233,000 kN. The resultant equivalent static base shear is: V_{static} = 0.0112*2,233,000 = 25,000 kN. This is approximately 50% of the design wind shear. A response spectrum analysis was performed using a full three-dimensional computer model of the structure, which confirmed the equivalent static design forces.

The structure was also analyzed for seismic design loads in accordance with the 1994

Uniform Building Code with the following parameters: seismic zone factor Z=0.15g, importance factor I=1.0, soil site coefficient S=2.0, and response modification factor R_w=6. The resultant static base shear coefficient is 0.0191g. The equivalent static base shear is: V_{ubc} = 0.0191*2,233,000 = 43,000 kN. This is 16% lower than the wind base shear. From response spectrum analysis, the overturning moment is 7.42×10^6 kN-m. This is 42% less than the wind overturning moment, since higher modes do not contribute significantly to base overturning moment.

For design of braced-frame members and their connections, seismic design forces are amplified by a factor of $^3/_8 R_w$ = 2.25. Even with this increase, the sizes of most members are controlled by wind loads.

<u>Column Shortening Analysis</u>

Analysis of column shortening includes estimates of elastic shortening under gravity loads, concrete creep under sustained loads and concrete shrinkage, based on the recommendations of ACI 209R-82 "Prediction of Creep, Shrinkage and Temperature Effects in Concrete Structures." The following construction sequence was used in this analysis:

Building Component	Days from Start of Construction
Core Walls	1
Structural Steel	43
Floor Slabs	57
Concrete Columns	85
Exterior Facade	154
Superimposed Dead Loads	183
Superimposed Live Loads	1002

For corner columns, the maximum net elastic shortening is 31mm at the 44th floor, the maximum net shrinkage is 15mm at the 67th floor and the maximum creep shrinkage is 74mm at the 67th floor. The maximum floor-to-floor shortening is 3.3mm at lower floors. The maximum total shortening is 114mm at the 67th floor. Approximately 50% of this can be compensated for during construction. Perimeter diagonal braces and girders are designed for forces induced by perimeter column shortening.

Composite Structural System for an 88-story Tower

Navin R. Amin,[1] Edward (Xiaoxuan) Qi,[2] and W. H. Yang[3]

Abstract

This paper summarizes the analysis, design and performance of the lateral force resisting systems for an 88-story mixed use building in Southern China.

Introduction

Composite structural systems are rapidly becoming the preferred choice for high-rise buildings, due to their high strength and stiffness, large ductility, and convenient, economical construction. This is particularly true for high-rise buildings subjected to high wind forces. A successful example of such an application is Xiamen Fairwell International Center in Southern China, a 283,563 m² multi-use development with office, retail, parking and service facilities. (See Figure 1.)

Structural System

The vertical elements of the building consist of five I-shaped reinforced concrete shear walls at the core, and 18 structural steel-reinforced concrete

Figure 1 – Rendering of Tower

---

[1] Senior Project Director, Middlebrook + Louie Structural Engineers, 71 Stevenson Street, San Francisco, CA 94105
[2] Project Director, Middlebrook + Louie Structural Engineers, San Francisco, CA
[3] Senior Research Engineer, Weidlinger Assoc., Inc, 4410 El Camino Real, #110, Los Altos, CA 94022 (former structural engineer with Middlebrook + Louie, Structural Engineers, San Francisco, CA)

composite columns along the perimeter, as illustrated in Fig. 2a. The wall "web" has a constant thickness of 0.75m for the entire height of the building, whereas the thickness of the wall "flange" varies from 1.2m at the bottom to 0.75m at the top of the building. All circular columns have a constant diameter of 2.6m up to Level 74, and all square columns have constant dimensions of 2.3m by 2.3m. These elements carry all gravity loads of the building.

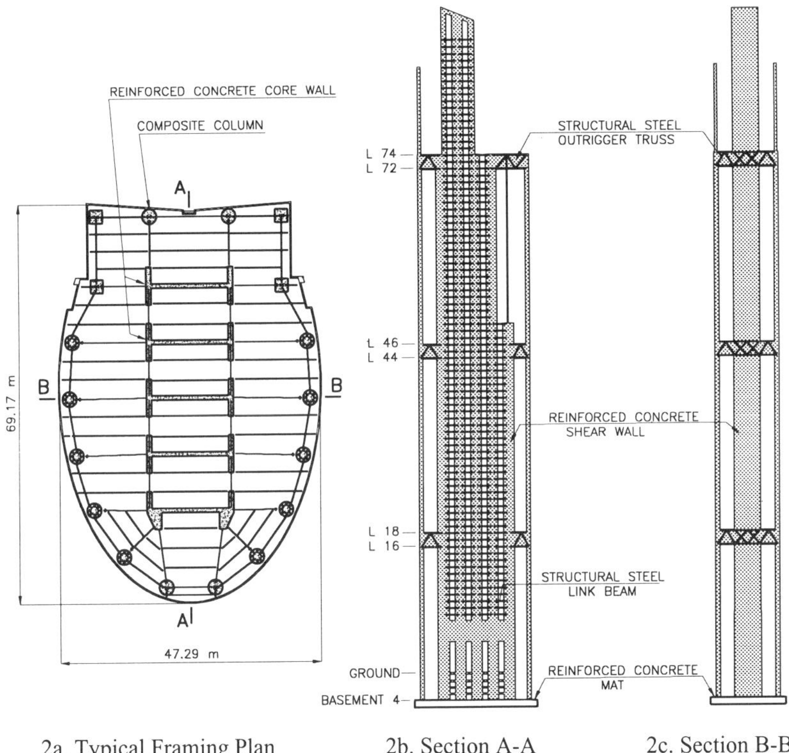

2a. Typical Framing Plan   2b. Section A-A   2c. Section B-B

Figure 2 – Structural System

The lateral force resisting system in the longitudinal direction consists of shear wall flanges interconnected with coupling beams, and further connected to the perimeter columns by three sets of outriggers (Fig. 2b). In the transverse direction, the lateral system consists of five I-shaped walls connected to exterior columns with three sets of outriggers. A typical transverse section is shown in Fig. 2c. In addition, all perimeter columns are connected by five belt trusses at refuge/mechanical floors. Such a lateral system proved to be very efficient for the given geometry of the building, because it engaged all of the vertical members in resisting the overturning moment in both principal directions. For such a slender building (with a height/width ratio of 8.3 in the transverse direction), the fundamental periods are

7.3sec, 6.7sec and 5.2sec in the transverse, longitudinal and torsional directions, respectively.

The foundations for the tower structure are hand-dug belled caissons supported on slightly weathered bedrock (granite). The depth of the caissons varies between 35 to 50 meters below grade. Each perimeter column is supported by a 4.2m shaft diameter caisson with a 4.8m bell diameter. Each I-shaped shear wall is supported by a group of three 4.4m shaft diameter caissons with a 5.2m bell diameter. A monolithic reinforced concrete mat 3 meters thick connects all caissons and the tower structure.

Wind Engineering

Two approaches were used to evaluate wind loads for the Fairwell Tower: Chinese Code formula and wind tunnel test results. The basic wind pressure defined in the Chinese Code (GBJ9-87) is 0.825 KN/m^2 for a 50-year reoccurrence event, which corresponds to a 10-min maximum mean wind speed of 36.33 m/s at 10m above ground. The building site is in a terrain of ground roughness Type B. Calculated wind pressure varies from 0.845 KN/m^2 at the bottom to 3.905 KN/m^2 at the top of the tower. The wind tunnel tests were conducted at the University of Western Ontario; London, Canada. A force-balance model was employed for the study. For a 100-year reoccurrence event, the wind tunnel test resulted in somewhat less severe structural wind loads than that from the Chinese code. The total base shear and overturning moment are summarized in Table 1.

Load Type	Results	X Direction	Y Direction
Chinese Code Wind Load with 50 years Return Period and Exposure Type B	Base Shear (KN)	54.0×10^3	69.7×10^3
	Overturning Moment (KN-M)	11.67×10^6	14.51×10^6
	Top Story Lateral Displacement (M)	0.552	0.882
	Maximum Inter-Story Drift Ratio	1/429	1/397
Wind Tunnel Test Wind Load with 100 years Return Period	Base Shear (KN)	32.2×10^3	54.4×10^3
	Overturning Moment (KN-M)	7.23×10^6	12.49×10^6
	Top Story Lateral Displacement (M)	0.330	0.766
	Maximum Inter-Story Drift Ratio	1/862	1/424
Time History Analysis with 63% Probability Earth Quake Load and 5% Damping	Base Shear (KN)	19.9×10^3	18.2×10^3
	Overturning Moment (KN-M)	3.30×10^6	3.23×10^6
	Top Story Lateral Displacement (M)	0.142	0.216
	Maximum Inter-Story Drift Ratio	1/1550	1/1352
Time History Analysis with 10% Probability Earth Quake Load and 5% Damping	Base Shear (KN)	72.9×10^3	74.6×10^3
	Overturning Moment (KN-M)	13.73×10^6	13.93×10^6
	Top Story Lateral Displacement (M)	0.602	0.860
	Maximum Inter-Story Drift Ratio	1/389	1/334
Time History Analysis with 3% Probability Earth Quake Load and 5% Damping	Base Shear (KN)	115.2×10^3	99.9×10^3
	Overturning Moment (KN-M)	19.03×10^6	19.12×10^6
	Top Story Lateral Displacement (M)	0.834	1.263
	Maximum Inter-Story Drift Ratio	1/270	1/254

Table 1 – Summary of Analysis Results under Different Load Types

Strength design for wind loads was based on the larger wind loads resulting from Chinese code for a 50-year return period event and wind tunnel test results for a

100-year return period event. The basic design guideline was a modified version of the 1997 Edition of Uniform Building Code (UBC), which incorporated applicable Chinese code requirements. Structural deformation under wind load was checked based on the wind tunnel test results for a 100-year return period event. The maximum calculated inter-story drift ratio was 1/424 , as shown in Table 1.

The mass and stiffness of the tower were studied to minimize the effects of vortex shedding (i.e. across-wind excitation) in order to achieve an optimal dynamic wind response. With the inherent damping characteristics of the composite structures, the accelerations (occupant perception) were well below the internationally accepted limits. The maximum measured wind induced acceleration and torsional velocity from the wind tunnel test for a 10-year return period event were 14.7cm/sec^2 and 0.0012rad/sec, respectively, at the top-most public occupied floor (i.e. Level 87, which is 349.8 meters from the ground).

Earthquake Engineering

Fairwell Tower was designed to resist earthquake loads based on Chinese code intensity degree 7 (comparable to UBC Zone 2A), and detailed to meet the requirements for intensity degree 8. The earthquake resistant design of the tower was conducted for three probabilities of earthquakes. A local geotechnical engineer developed the site specific ground acceleration time history for 63%, 10% and 3% probabilities of exceedance in a 50-year time period. The corresponding ground accelerations associated with these probabilities are 0.034g, 0.1239g and 0.216g, respectively. The structural elements of the lateral force resisting system were designed and detailed to meet the code requirements. A static pushover analysis was also performed along both principal axes in order to verify the post-earthquake behavior of the structure. The results indicated that even under an extreme earthquake load of 3% probability of occurrence in 50 years the majority of the tower's lateral force resisting system would remain essentially elastic, with limited regions of plastic deformations.

Conclusions

In designing Xiamen Fairwell Tower, the primary structural behavior, including the strength demand, deformation and occupant comfort, was governed by design wind loads. Moreover, the required stiffness for wind load response controlled almost all member sizes. As a result, the majority of the reinforced concrete columns and walls required only a minimum reinforcement to meet the strength requirements. The detailing requirements for reinforced concrete columns and shear walls were largely governed by Chinese codes. The composite structural system developed for the Fairwell Tower utilized reinforced concrete for its mass, stiffness, and inherent damping, and structural steel for its strength, speed of construction, and future feasibility of tenant improvements. It proved to be a very cost-effective solution to both the strength and serviceability requirements for a slender high rise structure.

Seismic Isolation of the First Federal Building
in San Francisco

Anoop Mokha, Ph.D., S.E.[1]

Abstract

Seismic retrofitting of existing historic building poses many challenges to the structural engineer in integrating structural strengthening schemes within architectural constraints. Seismic isolation in the United States is gaining wide acceptance as an attractive and alternate means to seismically upgrade historic structures without interfering with architecturally significant features. One of the important historic landmark structures currently undergoing such an upgrade is the Ninth Circuit U.S. Court of Appeals building located in San Francisco, which suffered damage during the 1989 Loma Prieta earthquake and was declared an unsafe building for further occupancy. This paper summarizes the comprehensive approach adopted for implementation of the seismic isolation technique for the seismic retrofit of this monumental structure. This historically significant structure is the first Federal Building to be retrofitted using seismic isolation.

Introduction

Seismic isolation is a design technique which reduces the demand on structures by isolating them from the damaging effects of severe ground motions. The isolation is usually achieved with specially designed bearings that provide flexibility and energy absorption capability while supporting the weight of the structure (Kelly, 1986).

The historic Ninth Circuit U.S. Court of Appeals at 7th and Mission Streets in San Francisco is owned by the General Services Administration (GSA). The building is a monumental structure of historical and architectural significance. In terms of architecture it is without peer in San Francisco, and possibly in the western United States. The building suffered damage during the Loma Prieta

---

[1] Associate, Skidmore, Owings & Merrill, LLP, One Front Street, San Francisco, CA 94111

earthquake of 1989 and was closed thereafter. The building is currently being retrofitted with seismic isolation to provide a level of life safety and damage control beyond that of conventional strengthening methods.

Building Description

The original U-shaped building, constructed in 1905, structurally survived the devastating 1906 San Francisco earthquake and fire with minimal damage. In 1933, a fourth wing was added, giving the building a rectangular shape with a central atrium. Figure 1 shows a photograph of the building. Approximate plan dimensions are 100 m by 81 m (330 feet by 265 feet) and the total floor area is about 32,500 m^2 (350,000 square feet). The building is a five-story, 24.4 m (80 feet) tall structure with steel framing, concrete slabs, unreinforced granite masonry exterior walls and hollow clay tile interior partitions. The total weight of the building (dead load + reduced live load) is about 534 MN (120,000 kips). Interior finishes are extremely ornate. They include carved marble figures, inlaid marble walls and floors, and highly articulated plaster ceilings. This Beaux Arts building is on the National Register of Historic Places.

Isolator Location

A complex matrix of costs and impacts was developed to evaluate each of the four alternate isolator locations, as shown in Fig. 2.. For each of the four locations, excavation costs, moat requirements, architectural and mechanical impact, overall constructability, and the amount of superstructure strengthening that would be required were determined.

Alternate 1 (above the foundation and below the basement slab) was selected as the optimum isolator location because it provided sufficient headroom in the basement for the architectural program, efficient installation, and minimal disruption to mechanical, electrical, and plumbing services. Alternates 2, 3, and 4 were less desirable architecturally and more disruptive to building services. Most importantly, they would have required more strengthening and involved details which would require higher costs and a longer construction schedule. To complete Alternate 1, additional tie beams would connect the footings below the plane of isolation, and a new basement floor slab would serve as a rigid diaphragm just above the isolators.

Isolation System

Given the Alternative 1 isolation location, the three location systems were then assessed in terms of superstructure response, architectural and mechanical impact, and constructability. Each isolation system was designed for three levels of base shear (15%, 17%, and 19% of seismic weight) and isolation periods in the range of 2.3 to 3.5 seconds. The three levels of base shear were selected based on engineering judgement. The design of each isolation system utilized average properties extracted from available research and practical application data.

Nonlinear dynamic analyses were conducted with computer code 3D-BASIS (Nagarajaiah et al. 1991). Friction Pendulum Isolation system was selected after a competitive bidding process.

Implementation of FPS Bearing

The Friction Pendulum System (FPS) bearing consists of an articulated slider on a concave spherical stainless steel surface (Zayas et al. 1990). A cross-sectional view of the FPS isolator of the U.S. Court of Appeals building and connection details are shown in Figure 3. Characteristics of these bearings are the polished stainless steel spherical sliding surface and the articulated slider which is faced with a PTFE-based, high bearing-capacity composite. The bearings are typically sealed and installed with the sliding surface facing down, so that contamination of the sliding interface is impossible.

The FPS bearing acts as a fuse, activated only when the earthquake force overcomes the static friction. Once in motion, the articulated slider moves along the concave spherical surface, causing the supported mass to rise, with motions equivalent to those of a simple pendulum (Zayas et al. 1990). The kinematics and operation of the bearing is the same whether the concave is facing up or down. Geometry and gravity achieve the desired seismic isolation results. The superstructure shear with isolation was limited to 16% with maximum isolator displacement of 26-8 cm for a 475 year design level earthquake (i.e. a 10% in 50 year event)

Conclusions

Systematic evaluation of isolator locations and isolator type results in considerable savings in the construction cost and time of construction. For historic building, the isolation system selection should be based on the existing constraints, desired superstructure performance and the overall economics. The structural implication of long term variation of isolation system properties and response of structure during minor and major earthquake should be accounted in the design.

REFERENCES

Kelly, J.M. (1986). Aseismic base isolation: review and bibliography. Soil Dynamics and Earthquake Engineering, 5(3), 202-216.

Nagarajaiah, S., Constantinou, M.C., and Reinhorn, A.M. (1991). 3D-BASIS non-linear dynamic analysis of three dimensional base isolated structures: Part II, Report No. NCEER-91-0005, National Center for Earthquake Engineering Research, Buffalo, NY.

Zayas, V., and Low, S. (1990). A simple pendulum technique for achieving seismic isolation, Earthquake Spectra, 6(2).

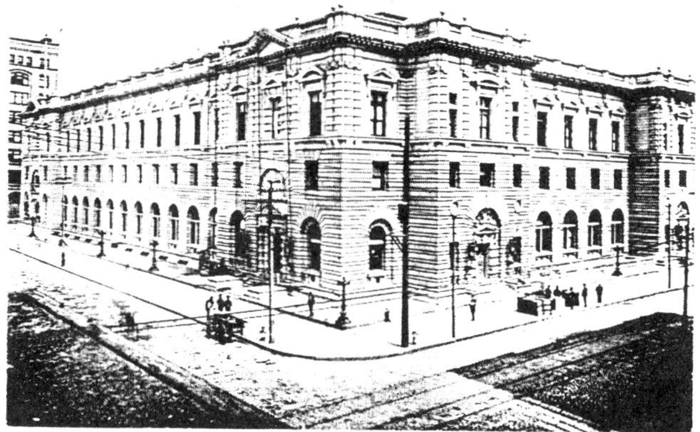

Fig. 1 Photograph of U.S. Court of Appeals

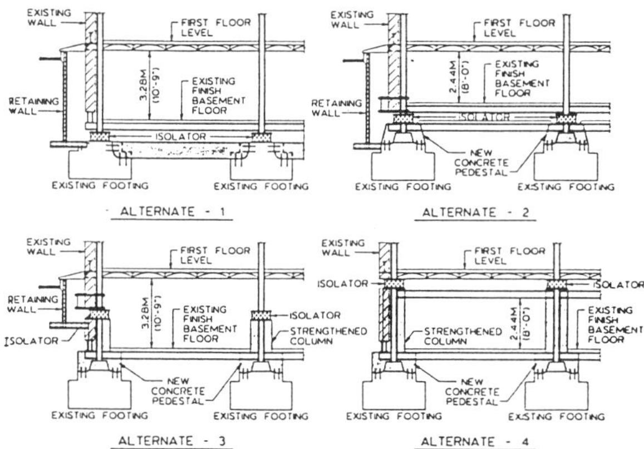

Fig. 2 Alternative Location of Isolation System

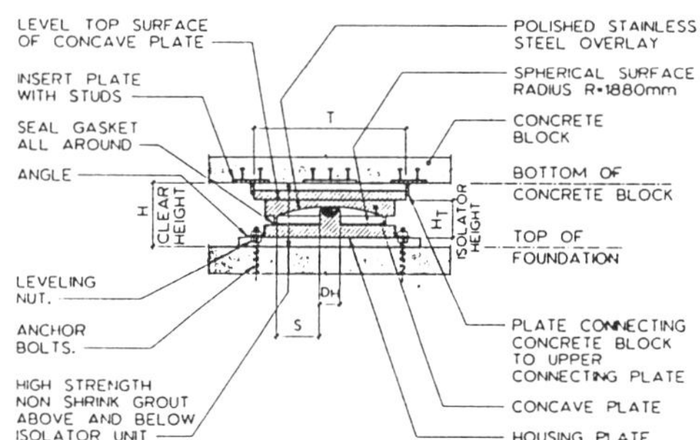

Fig. 3 Cross-Sectional View of FPS Bearing

Simultaneous Optimization of Topology and Nodal Locations of a Plane Truss
Associated with a Bézier Curve

Makoto Ohsaki[1] and Yuji Kato[2]

Abstract

A method is presented for simultaneous optimization of topology and geometry of an arch-type plane truss modeled by a Bézier curve. Explicit geometrical constraints are given to prevent existence of practically inadmissible shape, and optimal toplogies and nodal locations are found by using a genetic algorithm. It is shown that the proposed operator of mutation is effectively used for local search among the trusses with same topology, and practically admissible optimal shape is found by introducing the geometrical constraints.

Introduction

The parametric curves and surfaces developed in the field of computer aided geometric design have been successfully applied to shape optimization of distributed parameter structures (Ramm *et al.* 1993). The authors have presented several methods for geometry optimization of trusses with fixed topology by using Bézier curves and surfaces (Ohsaki *et al.* 1997, 1998b). In those methods, side constraints are given for the locations of the control points so that the optimal shape derived is practically acceptable. Ohsaki *et al.* (1998a) presented a two-level approach for shape optimization of a truss, where explicit constraints are given for the member length and the angles of the triangular units.

It is well known that optimizing the topology and the nodal locations is very difficult (Kirsch 1996). The topology optimization problem is a combinatorial optimization problem which is difficult to be solved by a gradient based

---

[1]Associate Professor, Department of Architecture and Architectural Systems, Kyoto University
[2]Graduate Student, Department of Architecture and Architectural Systems, Kyoto University

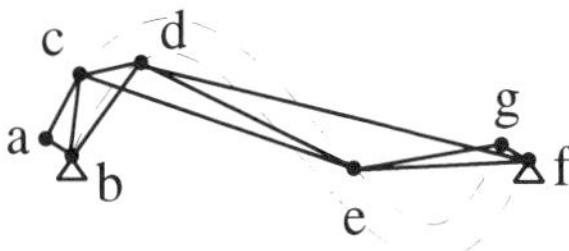

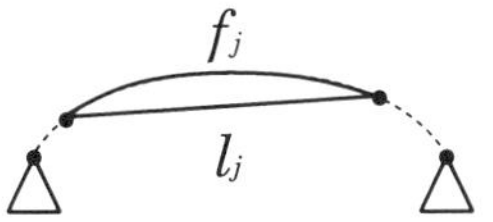

Fig. 1: Unfavorable nodal locations.

Fig. 2: A chord member and corresponding arc.

algorithm. The Genetic Algorithm (GA) has been shown to be very effective for such problems. Ohsaki and Tagawa (1997) applied GA for optimization of geometry and topology of regular plane trusses. In this study, a method based on GA is presented for optimizing a plane truss described by Bézier curves.

Plane truss model by Bézier curves

Consider an arch-type plane truss that consists of triangular units. Let $\mathbf{b}_i^L$ ($i = 0, 1, 2, 3$) denote the control points. The curve associated with the lower nodes is given as a Bézier curve of order three defined by the parameter u ($0 \le u \le 1$) as

$$\mathbf{q}^L(u) = \sum_{i=0}^{3} \binom{3}{i} u^i (1 - u)^{3-i} \mathbf{b}_i^L \tag{1}$$

The locations $\mathbf{Q}_i^L$ ($i = 1, 2, \cdots, N$) of the lower nodes including the supports are given as $\mathbf{Q}_i^L = \mathbf{q}^L(u_i^L)$, where $u_i^L = (i - 1)/(N - 1)$. The parameters u_i^U for the upper nodes are defined by $u_i^U = (u_{i-1}^L + u_i^L)/2$ ($i = 2, 3, \cdots, N$) and $u_1^U = 0$, $u_{N+1}^U = 1$. Then the locations $\mathbf{Q}_i^U$ of the upper nodes are written as $\mathbf{Q}_i^U = \mathbf{b}^L(u_i^U) + \bar{d}\mathbf{n}(u_i^U)$, where $\mathbf{n}(u)$ is the unit normal vector of $\mathbf{q}^L(u)$ and $\bar{d}$ is the prescribed depth of the truss.

Optimization problem with geometrical constraints

Let S_k denote the signed area of the kth unit such as $(\overrightarrow{\text{cb}} \times \overrightarrow{\text{cd}}) \cdot \mathbf{i}_z/2$ for the unit bcd in Fig. 1, where $\mathbf{i}_z$ is the unit vector normal to the plane of the truss. The parameter r_k is defined as $r_k = S_k/l_k^c$, where l_k^c is the length of the chord member; e.g. $\overline{\text{bd}}$ for the unit bcd. Another geometrical parameter $a_j = (l_j - f_j)/f_j$ is introduced so that the distance of a lower chord from the lower curve is sufficiently small, where f_j and l_j are defined as shown in Fig. 2. Let $\mathbf{P}$ denote the vector of applied nodal loads, and the displacements for $\mathbf{P}$ is denoted by $\mathbf{w}$. The optimization problem for minimizing the compliance of a truss with prescribed cross-sectional areas under geometrical constraints is defined as follows:

$$\text{OP}: \quad \text{Minimize} \quad C = \mathbf{P}^T \mathbf{w} \\ \text{subject to} \quad \bar{r}_L \le r_k \le \bar{r}_U, \quad \bar{a}_L \le a_l \tag{2}$$

Optimization using genetic algorithm

The locations of control points are expressed as a binary string of M^c bits. Let M^a denote the number of points for possible location of lower nodes. The existence of nodes are define by a binary bit for each existable node, and the total length of the string is $M^c + M^a$. Penalty terms for the constraints on r_k and a_j are include in the performance function. The function d_r is given as

$$d_r(r_k) = G \left(\frac{\bar{r}^* - r_k}{\bar{r}_a - \bar{r}^*} \right)^2 \quad \text{for} \quad r_k < \bar{r}_L \text{ or } \bar{r}_U < r_k \tag{3}$$

where G is the specified parameter, $\bar{r}_a = (\bar{r}_L + \bar{r}_U)/2$, $\bar{r}^* = \bar{r}_L$ or $\bar{r}_U$, and $d_r(r_k) = 0$ for $\bar{r}_L \leq r_k \leq \bar{r}_U$. Let X_i^j denote the bit string of the ith individual of the jth generation. Then the penalty term due to r_k of X_i^j is defined as

$$D_r(X_i^j) = \min\{g_r, D^L + \alpha(g_r - D^L)\} \tag{4}$$

where g_r is the sum of d_r among all the individuals, and D^L and α are the parameters introduced so that the magnitude of the penalty terms are equivalent to that of C. The penalty term $D_a(X_i^j)$ for a_j is defined in a similar manner. Then the performance function is written as

$$V(X_i^j) = C(X_i^j) + D_r(X_i^j) + D_a(X_i^j) \tag{5}$$

At the mutation process, the schema *1* is changed to 10* or *01, where $* = 0$ or 1. Since the number of nodes does not change for most of the cases, this operator is useful for local search among the trusses with same number of nodes.

Examples

Optimal shapes are found for a symmetric truss with $\bar{d} = 50.0$ cm and span length 750 cm. The load along the upper curve is 9.8 N/cm, and the self weight

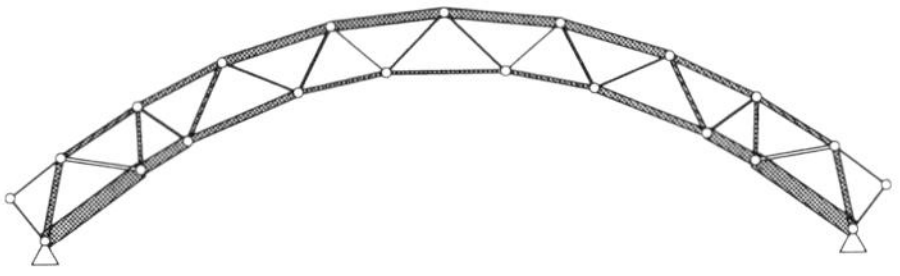

Fig. 3: Optimal shape and corresponding distribution of strains.

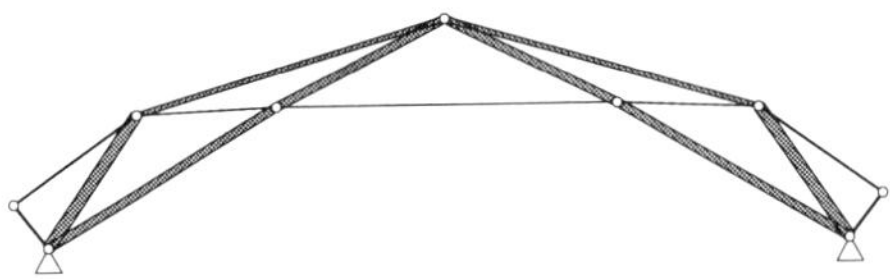

Fig. 4: Optimal shape without geometrical constraints.

is also applied at each node. The acceleration of gravity is 980.0 $(\mathrm{cm/s^2})$, and the mass density is 7.86×10^{-3} $\mathrm{kg/cm^3}$. The cross-sectional area of all members is equal to 5.0 $\mathrm{cm^2}$. The elastic modulus is 205.8 GPa, and the parameters for constraints are $\bar{r}_L = 0.4$, $\bar{r}_U = 1.4$, $\bar{a}_L = -0.05$. By using the symmetry condition, $M^c = 16$, $M^a = 50$. Other parameters are given as $D^L = 30.0$, $G = 2000.0$, $\alpha = 0.1$, probabilities of crossover and mutation are 0.6 and 0.02, respectively, and the population size is 50. The linear ranking method, two point crossover and the elite preservation strategy are used. The process is terminated if 4000 different individuals are evaluated, or the best value does not change in two consective steps. The optimal shape is sa shown in Fig. 3, where $C = 138.08$ Nm. Note that the width of each member is proportional to the absolute value of the strain, and the blank and gray members indicate the tensile and compressive strains, respectively. The optimal truss without shape constraints is as shown in Fig. 4, where $C = 113.0$ Nm.

Conclusions

An algorithm based on GA has been presented for simultaneously optimizing topology and geometry of arch-type plane trusses. The number of design variables are reduced by using Bézier functions while preserving the smoothness of the nodal locations. The topology of the truss is defined by a binary string that indicate existence of the nodes. It has been shown that local search is successfully carried out by the proposed mutation operator that exchanges the existable nodes, and that practically admissible optimal shape can be obtained by introducing constraints on geometrical properties of the units and members.

References

Kirsch, U. (1996). "Integration of reduction and expansion processes in layout optimization," *Structural Optimization*, 11, 13-18.

Ohsaki, M., Nakamura, T., and Kohiyama, M. (1997). "Shape optimization of space trusses described by parametric surfaces," *Int. J. Space Structures*, 12(2), 109-119.

Ohsaki, M. and Tagawa, H. (1997). "Genetic algorithm for simultaneous optimization of topology and geometry of a regular plane truss," *Proc. Int. Symposium on Optimization and Innovative Design (OPID97)*, Japan Soc. of Mech. Engineers, Paper #121.

Ohsaki, M., Katoh, N. and Isshiki, Y. (1998a) "Interactive multiobjective two-level optimization of dissatisfaction level of structures," *Proc. Seventh AIAA/ USAF/NASA/ISSMO Symposium on Multidisciplinary Analysis and Optimization*, AIAA, 97-106.

Ohsaki, M., Isshiki, Y. and Nakamura, T. (1998b). "Shape-size optimization of plane trusses with designer's preference," *J. Struct. Engng., ASCE*, in press.

Ramm, E., Bletzinger, K-U. and Reitinger, R. (1993). "Shape optimization of shell structures," *Bulletin of International Association for Shell and Spatial Structures*, IASS, 34(112), 103-121.

Multi-Criteria Conceptual Design Using Adaptive Computing

S. Khajehpour[1] and D. E. Grierson[2]

Abstract

This paper presents a solution to conceptual structural and architectural layout design of medium-rise office buildings, with concrete frame structural systems, using multi-criteria genetic algorithms and artificial neural networks.

Introduction

To find the optimal design of a medium-rise office building, one should include account for structural efficiency, erection cost, mechanical and electrical requirements, life-cycle cost, quality of space and comfort, rental revenue, land cost and interest on borrowed money. Significant complexity comes from the need to determine the relative benefits of all these various qualities and quantities. The different parties involved in design of an office building are 1) architect, 2) structural engineer, 3) mechanical/ electrical engineer, and 4) construction engineer. Often, the interest of these various parties are in conflict. For example, an architect wants maximum flexibility of floor space usage, a structural engineer desires a least-weight structure, and mechanical /electrical engineers want optimum energy use. It is apparent that optimum floor flexibility may conflict with having the lightest structural system and the minimum energy use.

The conceptual stage of design relies on the design team's experience and knowledge concerning the interaction between basic aspects of the design, such as between capital cost and life-cycle cost and revenue income. Of real importance is information concerning the optimal trade-offs between the different design aspects. The present paper employs a Multi-criteria Genetic Algorithm (MGA) in conjunction with Pareto optimization theory to solve this multi-criteria optimization problem for

[1]Reasearch Assistant, Civil Engineering Department, University Of Waterloo, Ontario, Canada, N2l 3G1

[2] Professor, Civil Engineering Department, University Of Waterloo, Ontario, Canada, N2l 3G1

the conceptual design of medium-rise office buildings. Artificial Neural Networks (ANNs) are employed for efficient criteria evaluation to establish design fitness.

Conceptual Layout Design

In general, the geometrical layout of a building is made prior to any member sizing. Potential construction and life-cycle cost savings, and future revenue gains, generated by layout optimization are much more significant than sizing optimization. With this in mind, the objective of this research is to determine optimal structural layouts and floor systems for a building while accounting for capital cost, life-cycle cost and revenue income. To ensure ease of construction, it is presumed that beams and slabs span over columns to form floor systems, that columns are arranged in lines in two orthogonal directions, that structural grid lines defining bay sizes are regularly spaced, and that only one type of floor layout prevails for the entire structure. For calculation of the energy consumption of a building during a one year-cycle, it is assumed that windows are installed above the task level (one meter above floor level) and are stretched to the ceiling, that the air conditioner and heating systems are each used three months a year, and that the effects of solar heat gain through windows and the energy consumption by ventilation systems are neglected. Finally, it is assumed that designs with larger spans and more window area for the same building footprint are more likely to generate higher revenue income.

Design Variables

There are five primary design variables: the distances between columns in the X and Y directions (CDX, CDY); the numbers of spans in the X and Y directions (NSX, NSY); and the ratio of the total area of windows to the maximum possible total area of windows on the perimeter of the building (WR). In addition, there are two secondary variables: the floor system type (FT) and number of stories (NS). After determining the primary variables, the most economical floor system and number of stories which provide the required floor area are found for the building. The floor-to-floor height is then found based on the floor type and the required floor to ceiling clearance

For the primary optimization, the MGA progressively investigates a population of design scenarios by considering different combinations of primary variables to find all Pareto optimal designs. In order to ensure practical designs are found for the Pareto optimal set, limits are placed on the primary and secondary variables: the building dimensions NSX x CDX and NSY x CDY are each limited to 50 meters; the number of stories NS is limited to 20.

Objective Criteria

The integrated conceptual optimization of a medium-rise building is herein conducted through optimization of the three criteria: a) Capital cost of structure, foundation, land, lighting and HVAC systems, elevators/lifts and façade; b) Life-cycle cost as defined by the amount of energy consumed per year by the HVAC and lighting systems; c) Revenue income as defined by a quantitative criterion related to architectural quality of space.

In this study, for the sake of simplicity, capital cost is calculated considering only structure, land and façade cost. The structure cost consists of floor and column costs. The floor cost varies according to bay size and type of floor system. Each column cost is defined by the total accumulated load it experiences from all stories above it. The land cost rates and size of the building footprint define the land cost. The façade cost is defined by the surface perimeter of the building. The objective function for capital cost is expressed as:

$$f_1 = Capital\ cost = C_{floor\ system} + C_{columns} + C_{land} + C_{façade}$$

Since life-cycle cost is proportional to the amount of energy used per year, this study proposes the minimization of energy consumed by the lighting and HVAC systems while considering the interaction between these two energy consumers and the effect of the window ratio WR on both. The design objective concerning life-cycle cost is expressed as:

$$f_2 = Life - cycle\ cost = C_{lighting} + C_{HVAC}$$

Regarding revenue income, this study proposes to minimize an inverse objective function in order to maximize revenue income. Improved quality of rental space is assumed to define improved revenue income, and a quantitative form of this objective function is expressed as follows:

$$f_3 = \frac{1}{revenue\ income} = \frac{\sum_j \left\{ \sum_i (TLA_i)^\alpha \right\}}{WR \times AR} \quad (j = 1,..,NF; i = 1,..,NC)$$

Where AR is the ratio of floor area benefiting from windows to the area of the building footprint, and TLA_i is the tributary area for each column i. When chosen less than unity, the exponent α is a factor which promotes longer spans and therefore better quality rental space. Similarly, both the window ratio WR and the area ratio AR ensure higher space quality for the same plan as they each increase in value.

Conceptual Design Example

Briefly presented in the following are the results of a conceptual design using the MGA approach to find building layouts that are Pareto-optimal for the three described objective criteria concerning capital cost, life-cycle cost and revenue income. The parameters governing the design were: total required floor area of 10,000m^2, clearance height of 3m, and all columns had square cross-section. The parameters used in the MGA were: population of 500 designs, one-point crossover with 100% probability, and an initial 5% mutation rate that progressively decreases as the MGA search progresses. Design fitness was calculated relative to the current

Pareto-optimal design set using an Euclidean distance norm. The convergence criteria were: the same Pareto design set is found for 15 consecutive generations; or 300 generations is reached. For given bay floor area, a neural network was used to calculate both the floor cost per square meter and the floor thickness.

The MGA was run for three different initial design populations, and all of the Pareto designs found for the three runs were combined together to form the overall Pareto design set. Figure 1(a) presents a 3-D plot of the overall Pareto-optimal set, which defines the trade-off surface for the three objective criteria. That the surface is convex in the criteria space can be seen from the 2-D plots in Figures 1(b) and (c). It remains for the designer to choose one of the Pareto designs as the conceptual layout of the building that represents an acceptable compromise for the three different criteria for the design. One such design, for example, is shown in the inset in Figure 1(a).

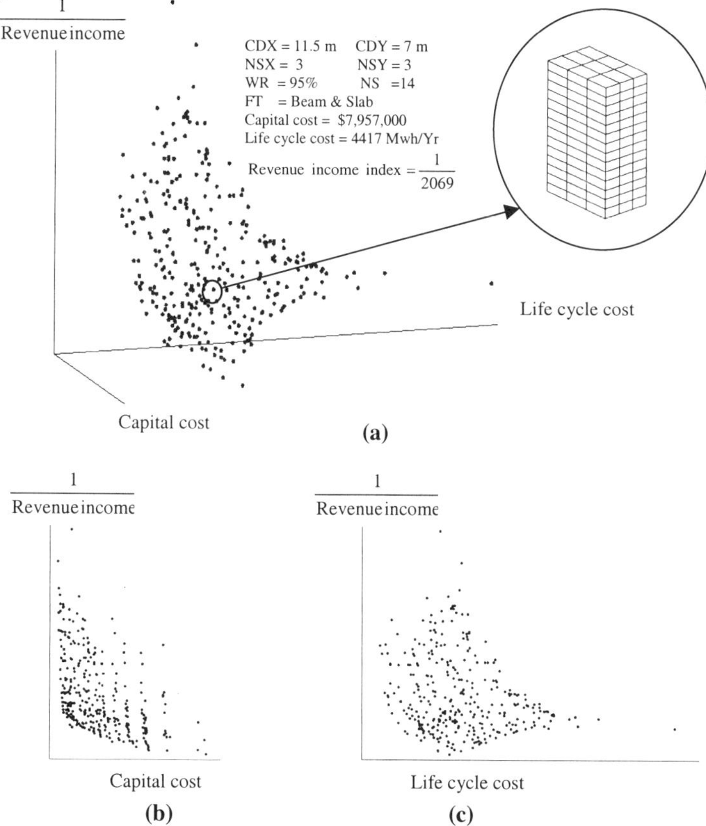

Figure 1. Pareto–optimal surface in: (a) 3–criteria space; (b) and (c) 2-criteria space

CONTINUUM AND GROUND STRUCTURE TOPOLOGY METHODS FOR CONCEPT DESIGN OF STRUCTURES

Colby C. Swan†, Jasbir S. Arora†, and Fatma Y. Kocer†

Abstract

The performance of continuum topology and ground structure topology methods are studied here for their utility and efficiency in concept design of structures. The continuum topology method is formulated using a volume-fraction/powerlaw framework and solved using sequential linear programming, whereas the ground structure method is formulated using discrete truss elements and solved using a genetic algorithm. The two alternative methods are exercised in the design of a plane truss to carry a system of static loads, and their results are compared in terms of realism and utility as well as their associated computational expenses.

Introduction and Motivation

For a given system of loads, and a given set of support conditions for a structure, an open question that often has to be answered in structural engineering, is, "what is the most natural form of the structure that achieves an appropriate blend of stiffness, lightness, economy, safety, and aesthetics?" To address this question, both continuum structural topology optimization methods, and ground structure topology optimization methods have been developed and actively investigated over the past decade. To compare the relative utility and expense of these two methods, they are tested here on the design of a planar truss.

Continuum Topology Formulation

The goal this section is to exercise a continuum topology optimization method to design a structure having the optimal arrangement of material $\mathcal{A}$ (which is steel) in a candidate spatial domain Ω_B which leads to the structural performance of minimum compliance. Once such a structural material arrangement is found, it will be interpreted as an *optimal starting concept design* which can then be further refined using shape

---

† Department of Civil and Environmental Engineering, The University of Iowa, Iowa City, Iowa 52242 USA

shape and/or sizing optimization techniques. Both to describe structural material arrangements, and to calculate the stiffness characteristics of different designs, the candidate structural domain Ω_B is discretized into a FEM mesh of NEL elements. Material arrangements, which constitute the design $\mathbf{b}$ of the structure are described using distributions of volume fractions as described in [Swan and Kosaka (1997)]:

$$\mathbf{b} = \{\phi_1, \phi_2, \ldots, \phi_{\text{NEL}}\}, \tag{1}$$

where ϕ_i represents the volume fraction of the structural material in the ith finite element of Ω_B. This system allows the structural material to be arbitrarily distributed throughout the NEL finite elements comprising the design domain Ω_B. In solving for optimal realizations of $\mathbf{b}$ that lead to high structural performance, material usage constraints are generally imposed on $\mathbf{b}$ by specifying upper limits on the global volume fraction of the structural material: $\mathcal{F}_{\text{solid}} = < \phi > \le C$ where C is a designer specified upper bound on the global volume fraction of the structural material, and $< \phi >$ the volume averaged value of ϕ over Ω_B.

In the continuum topology optimization framework, each finite element in the design domain Ω_B will generally contain a mixture of structural material $\mathcal{A}$ and a fictitious void material $\mathcal{B}$, and so it is necessary to prescribe how the stiffnesses of such mixtures will vary with the volume fractions of the materials present. Here, this is done using a powerlaw mixing rule [Bendsoe (1989)] as follows:

$$\mathbf{C}_{\text{eff}} = \phi^n \mathbf{C}_{\mathcal{A}} + (1 - \phi^n)\mathbf{C}_{\mathcal{B}} \tag{2}$$

where $\mathbf{C}_{\mathcal{A}}$ and $\mathbf{C}_{\mathcal{B}}$ are the stiffness tensors, respectively of the structural material $\mathcal{A}$ and the void material $\mathcal{B}$. To achieve controllable behavior of the powerlaw mixing rule, the exponent n can be varied during the optimization process, with $n = 1$ producing a stiff Voigt-like mixing rule, and $n \to \infty$ producing a compliant Reuss-like mixing rule [Swan and Kosaka (1997)]. It should be mentioned here that the void material $\mathcal{B}$ is assumed to be a very compliant elastic solid having $E_{\text{void}} = 10^{-6} E_{\text{solid}}$, $\nu_{\text{void}} = \nu_{\text{solid}}$, and $\rho_{\text{void}} = 10^{-6} \rho_{\text{solid}}$, where E, ν, and ρ denote Young's modulus, Poisson's ratio, and mass density, respectively.

Structural topology design optimization problems can be formulated in a number of alternative ways through varied utilization of assorted objective and constraint functionals. Here the problem is formulated to minimize the compliance of the structure subject to constraints on the structural material usage and structural equilibrium:

$$\min_{\mathbf{b}} \Pi(\mathbf{b}, \mathbf{u}) \text{ such that } \mathbf{r}(\mathbf{b}, \mathbf{u}) = \mathbf{0}; \text{ and } < \phi > -0.25 \le 0. \tag{3}$$

The specific loads and structural domain are shown in Figure 1a, and the continuum topology solution obtained is shown in Figure 1b. This problem was solved using SLP optimization methods presented in [Swan et al (1998)]. To achieve the quality design shown symmetry control [Kosaka and Swan (1998)] and filtering [Swan and Kosaka (1997)] methods were utilized.

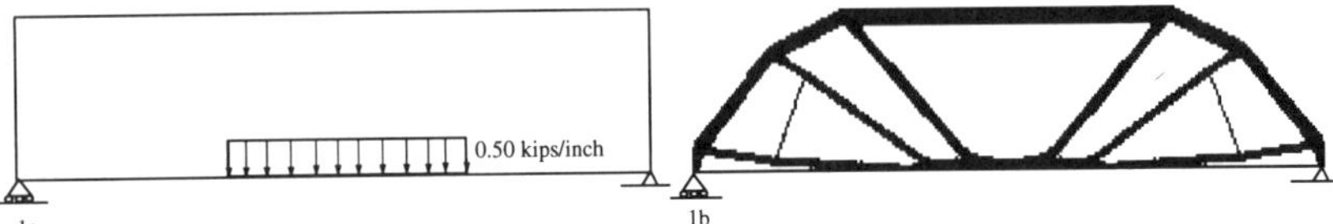

Figure 1: (a) Design domain and loading/restraint conditions; and (b) the optimal solution produced by continuum topology optimization and SLP.

Ground Structure Topology Formulation

The same structural domain treated in the above problem is now discretized using nodes and truss elements. The initial ground structure used to solve this problem is that shown in Figure 2a which features 46 structural elements. Since the objective of this optimization problem is merely to find the optimal placement of truss elements, and to eliminate those elements which are inefficient, the problem is solved to optimize element cross-section properties with only two possibilities as shown below in Table 1. With possibility #1, the cross-sectional area is vanishingly small, which represents a truss element that has effectively been removed from the structure. With possibility #2, the area is that associated with a square tube having a wall thickness of 0.01 inches and a dimension of 2.0 inches.

Value	Area (in^2)	I (in^4)	Value	Area (in^2)	I (in^4)
1	0.0001	1.00E-6	2	0.0796	5.25E-2

Table 1 Discrete design variable values and associated section properties.

Within a ground structure framework, the truss topology optimization problem solved is as follows:

$$\min_{\mathbf{b}} \quad M = \sum_i (\rho A L)_i \quad \text{such that :} \tag{4a}$$

$$\frac{\Pi}{\Pi_{\text{allowable}}} - 1 \le 0; \quad \frac{\|\sigma\|}{\sigma_{\text{allowable}}} - 1 \le 0; \quad \frac{-P}{2P_{\text{cr}}} - 1 \le 0; \quad \frac{\text{NEL}}{\text{NELMAX}} - 1 \le 0; \tag{4b}$$

In the above, Π denotes the compliance of the structure; $\Pi_{\text{allowable}}$ denotes a target compliance value (0.05 kip-inches); P_{cr} is the allowable buckling force in each element; P is the force in each element; NEL is the number of actual element in the truss and $NELMAX$ denotes the maximum desired value of truss truss elements (16). Using the genetic algorithm presented in [Huang and Arora (1997)] and the fitness function defined in [Kocer (1998)], the optimal solution obtained for this problem is that shown in Figure 2b.

Discussion and Conclusions

Both continuum structural topology optimization with SLP and discrete ground structure topology optimization with GA have been applied here to solve the concept design

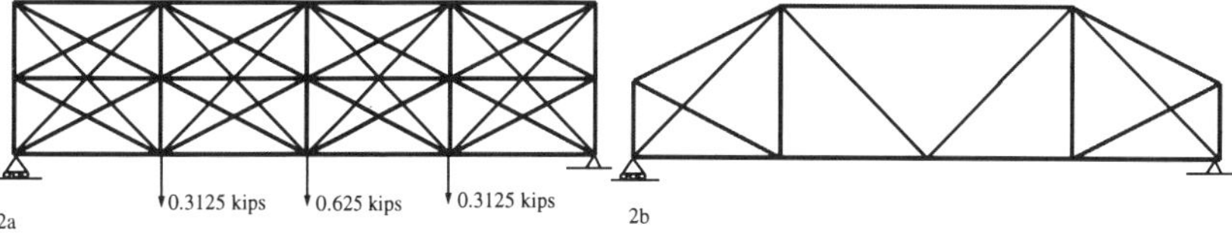

Figure 2: (a) Ground-structure truss consisting of 46 truss elements, loading case, and support conditions; and (b) the optimal solution containing 17 truss elements, produced by the genetic algorithm.

problem for a bridge type structure to carry a system of design loads. While both methods clearly provide very realistic and useful concept designs, each uses a fundamentally different method. The continuum topology approach models the structure with a highly refined continuum mesh, and this can lead to virtually any type of structure. In this method, each structural analysis can be potentially quite time consuming, but typically only a few hundred structural analyses are required to find the optimum topology. For example, the computational effort required to generate the optimal continuum topology solution shown in Figure 1b was approximately 2 cpu-hours on a Silicon Graphics Power Challenge work-station. In contrast, the ground structure topology optimization method uses a relatively coarse network of nodes which are connected by simple truss elements. For such cases, the analysis of simple truss structures is extremely rapid, taking only fraction of a cpu-second. However, since genetic algorithms require many thousands of structural analyses to find the optimal structure the overall optimization time can be considerable. For example, the optimal truss design shown in Figure 2b required approximately 1.5 cpu-hours to obtain on a HP 715 work station. Thus, for the example problems studied here, it appears that continuum topology optimization methods and ground structure topology optimization methods are competitive with respect to computational costs. Further study is required, however, before more general statements can be made about the relative efficiency of the two methods.

Since the topological designs obtained in this work are really attempts to find the best locations of the primary structural members, they are viewed strictly as good starting points for subsequent design efforts that will necessarily deal with such important details as sizing and selection of the individual structural members and design of the connections. For such efforts, optimization based on genetic algorithms such as that used in the ground structure methods may be quite useful.

References

Bendsoe, M.P., Optimal shape design as a material distribution problem. *Structural Optimization,* 1989, **1**, 193–202.

Huang, M.W. and Arora, J.S., Optimal design with discrete variables: some numerical examples. *Int. J. for Num. Meth. in Eng.,* 1997, **40**, 165–188.

Kocer, F.Y., Ph.D Dissertation, 1998. The University of Iowa, Iowa City, Iowa USA.

Kosaka, I. and Swan, C.C., A symmetry reduction method for continuum structural topology optimization, *Computers & Structures,* 1998 (in press).

Swan, C.C. and Kosaka, I., Voigt-Reuss topology optimization for structures with linear elastic material behaviors. *Int. J. for Num. Meth. in Eng.,* 1997, **40**, 3033-3057.

Swan, C.C., Arora, J.S., Kosaka, I. and Mijar, A.R., Concept design of bridge structures for stiffness and vibrations using continuum topology optimization, in *Structural Engineering World Wide 1998,* ed. Srivastava N.K., Paper T159-4.

Design of a Two-Story Blast Resistant Control Room

Darrell D. Barker, P.E.[1]
Johnny H. Waclawczyk, P.E.[2]

Abstract

Potential explosion hazards at petrochemical plants create the need for blast resistant control rooms and other occupied buildings near process units. In the case described in this paper, an existing structure was upgraded to provide substantial blast capacity. A new two-story, blast resistant shell was constructed over the top of the existing one-story masonry control building.
This paper describes the design and construction of the two-story blast resistant control room including blast load prediction and structural design.

Introduction

Chemical plants and refineries process flammable materials which create potential explosion hazards for occupied buildings. Many of these buildings are designed for conventional loads only and do not have sufficient blast capacity to provide proper personnel protection. These hazards can be produced by potential leaks from piping and vessels creating flammable clouds, which can ignite and subject buildings to potentially devastating blast loads. When possible, it is desirable to locate buildings in regions of low blast pressures. However, in many cases the building must be located very near the process unit and the structure must be designed to resist relatively high blast loads.

The project described in the paper involved an unreinforced masonry control room with plan dimensions of approximately 100 feet x 60 feet with an eave height of 15 feet. Operation requirements made it necessary to double the

---

[1] Senior Consultant, EQE International, Inc., 15600 San Pedro, Suite 400, San Antonio, TX 78232
[2] Lead Engineer, EQE International, Inc., 15600 San Pedro, Suite 400, San Antonio, TX 78232

existing floor space to accommodate additional personnel in a protected floor space. The overall building height of the completed two–story structure was 29 feet.

Design Challenge

Predicted blast loads at the existing control room were approximately 14 psi peak side-on overpressure with a load duration of 90 milliseconds. The peak reflected pressures on walls facing the potential explosion hazard were on the order of 40 psi. From the standpoint of structural design and construction cost, locating the control functions in a new building several hundred feet from the process unit was the best option. However, moving control equipment to a structure away from the unit was not an option because it would have required relocating equipment and data lines. More importantly, it would have required a process shutdown at a much higher cost than modifying the existing structure. Instead, a new blast resistant shell was constructed over the top of the existing one-story masonry control building providing a two-story facility housing a variety of support functions. A section through the building is shown in Figure 1.

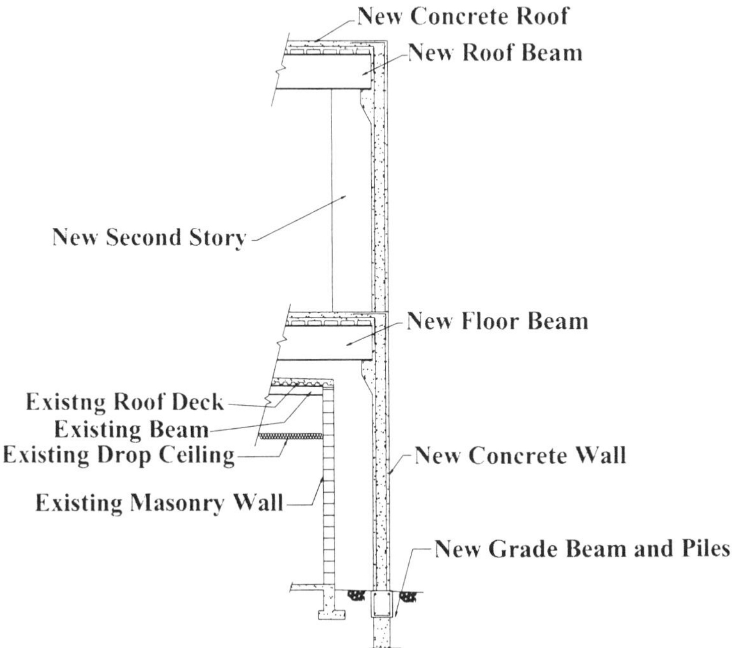

Figure 1. Two-Story Protective Shell Cross-Section

The project presented several design challenges due to the two-story configuration and high blast load. The roof system was required to span approximately 70 feet to avoid interference with the existing structural system. Interference with cable trays and piperacks combined with numerous wall penetrations required careful detailing of the reinforced concrete walls. Installation of pile foundations also required innovative methods for successful installation.

Construction of the second floor slab and support beams was accomplished in stages to protect occupants of the facility from potential accidents involving crane lifts of structural steel floor beams and other materials. Formwork and slab support systems were designed to eliminate the need for shoring while protecting the existing structure.

Design Criteria

The new structural shell was required to limit maximum response under design blast loads such that personnel would be afforded a high degree of protection. Response limit criteria were taken from the ASCE Design of Blast Resistant Structures at Petrochemical Facilities[1]. A medium response level was used which limited support rotations for concrete members to 4 degrees with stirrups or 2 degrees without shear reinforcing. A ductility ratio of 20 and a support rotation of 12 degrees were used for the limits of structural steel members.

Dynamic design techniques were used to minimize structural section requirements compared with static equivalent design used in many petrochemical designs. Members were analyzed as single-degree-of-freedom (SDOF) systems to obtain maximum midspan deflections. Dynamic end reactions were applied to supporting members tracing the load through the structures.

Structural Shell Design

A conceptual design phase evaluated several options for the structural shell design including cast-in-place and precast concrete panels. Due to the severity of the load and cable tray interference it was determined that the cast-in-place methods was the only feasible alternative.

Wall components were designed as one-way, doubly reinforced members with a thickness of 26 inches on the reflected face and 16 inches on the remaining sides. The walls spanned full height with the second floor limiting clear spans to 15 feet. Blast doors were incorporated into the design with the walls capable of resisting peak reactions from the doors. This was accomplished by designing in-wall columns at the door jambs which spanned from ground level to the second floor slab. Wall components were designed to resist not only the normal-to-plane blast loads but also the shear wall loads produced by the roof slab lateral load reaction. Openings for mechanical and electrical penetrations were accommodated with cast multi-cable transits (MCT). Openings for air intakes were configured to face away from the blast and vestibules were created to restrict the entrance of blast pressures.

Several structural configurations were analyzed during the conceptual design phase to determine the most cost-effective approach. Installation of interior columns was not feasible due to interference caused by installation of structural steel and interior foundations. The resulting clear span requirement of more than 70 feet required substantial strength in the roof support system. The final configuration selected was 27 inch deep wide flange members at 8 feet on center acting compositely with a 10 inch reinforced concrete slab. Roof beams were supported by a continuous corbel in the wall sections.

Elevated stacks were used to prevent intake of toxic gasses in conjunction with gas detection and system shutdown logic. HVAC units were mounted on the roof due to the lack of available space at ground level. The weight of the units, which was quite substantial, was accounted for in the design of the roof structure.

Foundations

Design of the structure foundation presented the greatest challenge. The soil bearing capacity of was poor requiring installation of pile foundations. Typical foundation systems for similar structures in the area utilized 12 inch square piles driven to a depth of 60 feet. A spacing of 6 feet was sufficient to resist the dead and live loads under working load conditions. A dynamic analysis was performed on response of the system under blast loads. During construction, it was determined that sufficient headroom was not available for a portion of the facility for the equipment the contractor elected to use for pile installation. A new pile configuration was designed based on the capacity of the contractor's equipment using low headroom equipment.

Conclusions

This project presented some unique challenges for design and construction of a retrofit protective structure. The structure was subjected to very high blast loads in terms of pressure and load duration. Stringent requirements for maintaining operations within the facility limited options for internal support. Dense cable tray and utility penetrations restricted access for foundation installation and required specialized framing. In spite of the challenges, this project was successfully constructed and proved the feasibility of upgrading the blast capacity of existing structures without significant interruption of operations.

References

"Design of Blast Resistant Buildings in Petrochemical Facilities," Published by American Society of Civil Engineers, 1997.

Investigation of Explosive Damage and Repair Costs

Paul M. LaHoud, P. E.[1]
Member

William H. Zehrt, Jr., P.E.[2]
Member

Abstract

On May 4, 1998, a series of explosions destroyed the Pacific Engineering Company (Pepcon) plant in Henderson, Nevada. The Pepcon plant manufactured ammonium perchlorate for rocket fuel. From damage analyses, the estimated explosive yield of the largest and most damaging single explosion was 226,800 kg. (500,000 lbs.) net explosive weight (NEW).

Following the accident, a database was developed which contained key damage claim information including the location of each damaged structure, the general type of construction for each damaged structure, and the dollar amount of each damage claim. In 1992, based on these data, insurance companies paid nearly 17,000 claims totaling $77 million. The majority of these claims were for single family residences in Henderson and Las Vegas, Nevada.

Starting in 1995, the U. S. Army Engineering and Support Center, Huntsville has supervised several studies by JWR, Inc., Albuquerque, New Mexico, which use the Pepcon database to assess structural damage and repair costs from accidental explosive detonations. In this paper, we will review the major findings of this research effort by evaluating probable repair costs from a hypothetical explosive detonation at a Department of Defense (DoD) installation. Our evaluation will be based upon the application of the current DoD minimum separation distance from the detonation to the installation boundary and will consider both residential and commercial construction.

---

1 Chief, Civil-Structures Division, U. S. Army Engineering and Support Center, Huntsville, P. O. Box 1600, Huntsville, AL 35807-4301
2 Senior Structural Engineer, U. S. Army Engineering and Support Center, Huntsville, P. O. Box 1600, Huntsville, AL 35807-4301

DoD Safety Regulations and Inhabited Building Distance

The Department of Defense Explosives Safety Board publishes and maintains safety criteria applicable to DoD ammunition and explosives. If possible, this protection is provided by requiring a minimum separation distance between potential explosive donor and acceptor structures. At a military installation's boundary, the required separation distance is termed the Inhabited Building Distance (IBD). For explosive quantities of 114,400 kg. (250,000 lbs.) or greater, the IBD is calculated with the equation: $IBD = 19.8*Q^{1/3}$ (IBD in meters, Q = NEW in kg. TNT) or $IBD = 50*W^{1/3}$ (IBD in feet, W = NEW in lbs. TNT).

According to the DoD Ammunition and Explosives Safety Standard, DoD 6055.9-STD, at IBD, "...Unstrengthened buildings can be expected to sustain damage up to about 5 percent of the replacement cost." Unfortunately, cost data from the 1988 Pepcon accident indicate that actual damage costs at this separation distance will be significantly greater.

Hypothetical Accident - Background

Fort Reagan is an active FORSCOM installation. It was originally constructed early in World War II. At that time, it was located six miles south of the small town of Clinton whose economy was based largely on agriculture. For many years, the installation was surrounded entirely by farmland. Since the early 1990's, however, Base Realignment and Closure activities have resulted in the assignment of several new missions to Ft. Reagan. This has increased requirements for ammunition storage and other facilities including troop housing. The nearby town of Clinton has grown significantly, expanding southeastward. In fact, the town limits have now reached the northwest boundary of the installation. The largest portion of the town growth near the installation has been light commercial construction and residential communities, including a new elementary school closely adjacent to the boundary.

Since its initial construction, ammunition storage activities have been concentrated in the northern end of Ft. Reagan. In accordance with the DoD Ammunition and Explosives Safety Standards, DoD 6055.9-STD, storage magazines are sited at the IBD separation distance from the installation boundary. Consequently, the incident overpressure at the boundary will not exceed 6.2 kPa (0.9 psi). Figure 1 shows a map defining the northern boundary of the installation and its relationship to the town of Clinton.

Description and Consequences of An Accidental Detonation

On April 18, 1999, at 5:00 a.m., a spontaneous detonation occurred in one of the storage magazines near the northern boundary of the installation. The resulting damage and injuries were an unexpected shock to both the installation and the community. While no one actually observed the detonation, thousands of residents were awakened by what was often described as a "sonic boom." Many also described an "earthquake" ground motion. It was later determined that the magazine which accidentally detonated had contained approximately 204,000 kg. (450,000 lbs.) of bulk explosives.

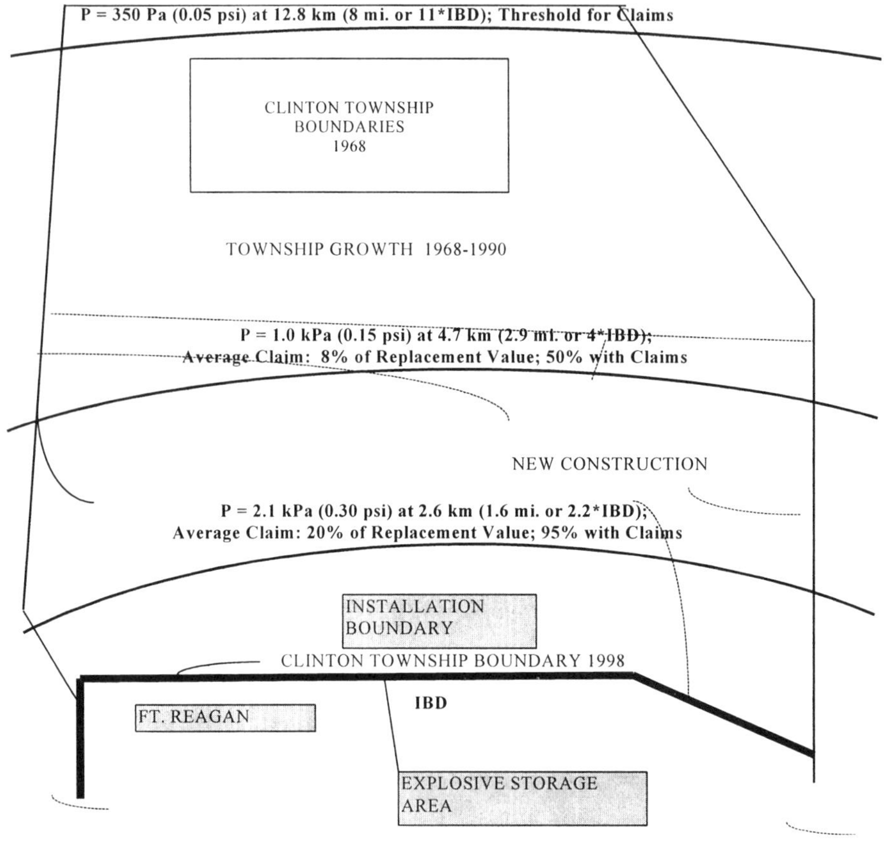

Figure 1 (not to scale)

As previously noted, DoD 6055.9-STD limits the damage expectation at IBD (0.9 psi) to mostly glass breakage and minor damage not exceeding 5% of a structure's replacement value. This damage expectation was developed through tests, performed in the 1945-1969 time frame, of sturdy, wood frame construction located at this distance from a detonation. However, in the weeks following the accident, damage claims from the Clinton community climbed into the thousands with many claims from distances of up to 6 km. (3.7 mi. or more than 5 times IBD) from the detonation. Figure 1 shows contours banding the approximate distribution of damage claims in percentage of replacement cost. Paired with these values are the associated separation distances as a ratio of IBD. As depicted on the figure, the dollar value of these claims exceeded 5% of the replacement cost at distances greatly exceeding IBD.

Total damages claimed by the community against the Army, mostly through insurance carriers, were estimated at $25 million. Fort Reagan was also faced with extensive damage to facilities on the installation. Interestingly, in areas farther from the detonation, there was a marked increase in the number and value of claims when window breakage occurred. Claims in these areas were primarily for nonstructural items such as windows, carpet, and furniture. This damage figure likely included many claims which were incorrectly attributed to the event (e.g., cracked sidewalks, walls, chimneys, plaster, etc.)

There were a significant number of injuries from falls and broken glass (cuts, abrasions, minor broken bones, etc.) Other reported injuries included numerous anxiety related complaints which, in a few cases, led to chest pains and heart attacks. There were also a few vehicle accidents caused either directly by the air blast loading or indirectly by distraction.

After the accident, there was considerable discussion of the damage guidelines presented in the DoD safety standard. The local emergency management agencies were greatly concerned about their complete lack of awareness of the possible consequences of such an event on the community. There was also a great deal of confusion in the coordination of information released to these agencies immediately after the accident.

Clearly, most Clinton citizens had not understood the possible consequences of living near an ammunition storage area. For example, although residents were thankful that the nearby school was not in session at the time of the accident, they were shocked by the extensive breakage of classroom windows and the clear potential for injuries to their children in a future accident. Similar concerns were raised at several churches which, although several miles from the accidental detonation, lost several large stained glass windows. Residents were also stunned by the extensive damage to a few lightweight, pre-engineered metal buildings adjacent to the installation boundary; these buildings were near total collapse.

Conclusions

The scenario described in this paper is hypothetical. It is intended to raise awareness in the safety community of the consequences of a maximum credible explosion at a DoD installation or private industrial plant.

Risk is the product of the probability of an event occurring and the consequences of that event. Current DoD and commercial safety regulations are just beginning to wrestle with the concept of risk based management for explosive manufacturing and storage. Since the damage consequences in this paper are extracted from an actual event, it is important to understand that the public will decide what is the likely cost of damage from an event (through insurers and lawyers), not the DoD or commercial regulations. It is also important to understand the public standard of acceptable risk may be significantly different from that defined by these regulations.

In dealing with community relations, closer coordination with local emergency management agencies, including discussion of possible event consequences and response expectations, would be a wise initiative. This coordination is already standard policy at chemical storage facilities and would benefit DoD and commercial explosive facilities as well.

Bomb Blast Damage to a Concrete-Framed Office Building
Ceylinco House - Columbo, Sri Lanka

Tom Caldwell, P.E. [1]

Introduction and Purpose

This paper briefly describes the blast damage sustained by the 16-story concrete-framed Ceylinco House office building, following a terrorist truck-bomb attack at street level. My job at this site was to evaluate the structural damage and design repairs, to return the building to its pre-blast condition, load capacity, and use. This description of the pattern and severity of the blast damage may be useful to designers of blast-resistant buildings.

Truck Bomb Attack

The Ceylinco House office building is in Colombo, the capital city of Sri Lanka. A terrorist group targeted the building across the street, the government's Central Bank building, in the heart of the city's financial district. On January 31, 1996, terrorists drove a five-ton truck partially up the steps of the Central Bank. Gunfire from bank guards brought many people to the windows of the high-rises lining the street, when the bomb detonated at 11:20 am local time. The blast killed over one hundred people. The bank and four nearby buildings suffered severe damage, and several fires broke out. The government later reported that the bomb consisted of 300 Kg of the military explosive C4.

The blast severely damaged the Ceylinco House office building, directly across the street from the targeted Central Bank building. The bomb detonated about 60 meters northwest of the northwest corner of the Ceylinco House, at a level even with the ground floor.

The Ceylinco House Office Building

Ceylinco House is a 16-story office building, constructed in 1963, and housing offices, a bank, restaurants, and several luxury apartment suites on the upper floors. Its total usable floor area is about 11,200 square meters, with overall dimensions of

----

[1]Tom Caldwell is a structural engineer and President of Atlas Engineering, Inc., of
  Raleigh, North Carolina. Member, ASCE.

about 50m by 25m. The long dimension runs east-west. The western side took the brunt of the blast forces.

The structure is of conventional reinforced cast-in-place concrete construction, bearing on a mat foundation on clay subsoils. Beams and girders are monolithic with the floor slabs, supported on concrete columns with typical bays of 7m x 5m. A central elevator and stair core provides lateral restraint at each slab level. The exterior walls are of heavy clay-block masonry supported at each level by the slab. Wide, operable windows lined the exterior, recessed from the wall and shaded with vertical concrete shade fins. The north and west exterior walls cantilever out from the building frame by about 4m, with a half-bay supported on cantilever stub beams projecting outward from the main frame.

Overall Blast Damage at Ceylinco House

The blast blew in windows and destroyed large sections of the exterior wall. Many occupants were injured by driven glass. The blast fractured a small portion of the building frame, but no part of it collapsed. Fire broke out, gutting the building.

Blast Effects on Building Envelope and Building Systems

All west-facing and most north-facing windows blew inward. The single-pane, operable windows provided no protection to occupants along the west and north elevations. Eye and head injuries were especially severe, since many people came to the windows at the sound of the gunfire just before detonation.

Many of the concrete shade fins broke inward, especially at the 4th floor. Large areas of the masonry exterior wall fractured inward on the west and north elevations, but the heavy, clay-block masonry walls did afford some protection. The masonry fractured and bulged inward but remained largely in place, without disintegrating.

It was reported that electrical power failed and fire broke out very soon after the blast. The central core elevators quit and the stairways were unusable due to darkness and smoke. Fortunately, a retrofit fire stair had been added to the building's south elevation a few years before. This unusual, exterior-mounted spiral steel staircase apparently saved many lives during the fire evacuation and rescue of blast victims. The destruction and chaos on the street below and at nearby buildings delayed and limited fire-fighting efforts. Fire spread through floors 2 to 11.

Blast and Fire Effects on Main Structural Frame

I inspected the building about three months after the blast. All finishes and interior partition walls had been removed, exposing almost all the concrete slab panels, girders, beams, and columns.

Fire effects were widespread. Concrete spalled in the high temperatures of the fire, exposing the bottom reinforcing steel at over 2,000 sq. m of slab, and along 3,700 m of beams and girders. Many columns were heavily spalled, and at 14 columns, longitudinal steel broke free and bowed outward. The original concrete cover was observed to be thin, on the order of 1 to 3 cm. Also, column ties were widely spaced, giving less column containment than is mandated by modern codes, and leading to increased spalling and movement of the longitudinal bars.

Direct blast effects on the main building frame appeared limited. I found no signs of gross deformation of the overall building frame. The "drag beam" lateral restraint connections between the central core and each floor system were intact, and the core itself appeared straight and sound. At frame members shielded from the blast (most were) and untouched by fire-spalling (about 60% were not exposed to fire), the concrete appeared sound, uncracked, of good workmanship, and in good condition.

The exterior walls and the concrete shade fins, on the west elevation, showed more damage at the 4th floor than at any other. It appeared clear from the exterior that the blast force hit the building with special violence and focus at the 4th level, for reasons I cannot explain.

Direct blast effects were concentrated at the western row of columns on the 5th floor, and the 5th floor slab. These effects were unusual and spectacular. The western column line was closest to the blast, yet was protected from direct lateral forces by the outer, cantilevered half-bay and the heavy masonry walls. No columns showed damage from lateral displacement. However, four columns in the western row were damaged by compressive overload, and one failed completely, forming the 45 degree failure cones so familiar in concrete cylinder tests. The concrete between the failure cones was completely crushed, and the longitudinal bars buckled outward several inches.

Observations and measurements at the interior showed that the western portion of the 5th floor slab had heaved upward, apparently from over-pressure between the 4th and 5th slab levels. The upward heave had cracked the 5th floor slab beams and overloaded four 5th floor columns in compression. The floor to ceiling distance near these columns had shortened, by 4 cm near the failed column, and 2 - 3 cm near the other columns along the western row.

Dead loads evidently redistributed themselves around the failed and damaged columns, and no slab panels or beams along the west side collapsed or appeared particularly unstable. The frame's survival after the loss of one or more columns showed the redundancy and toughness of this conventional, reinforced concrete structural system.

<u>Lessons from the Blast</u>

Although not an expert on blast effects, I offer the following observations:

- The windows failed immediately and flying glass injured many occupants.
- The heavy, frangible masonry walls absorbed blast energy and protected occupants and building columns from direct blast damage.
- The building safety systems were vulnerable. The building lost power, ventilation, light, and escape and rescue routes. Their loss could have caused more casualties than the blast, but for the recent installation of the second fire escape.
- The blast forces were evidently uneven and focused in an unusual way, delivering a concentrated punch to the 4th floor level.
- The 5th floor slab "collected" over-pressure and delivered concentrated force at columns and beams, causing localized, intense structural damage.
- With close-spaced small columns, the frame was better able to absorb blast damage without progressive collapse.

<u>In Conclusion</u>

The truck-bomb attack on this conventional concrete-framed office high-rise caused severe damage to exterior walls and windows on two sides of the building. Blast forces entering the window openings injured many occupants, destroyed interior rooms, and started a massive fire. Building safety systems failed, including power, lighting, and access to primary escape routes. Loss of these systems also hampered fire-fighters and rescuers.

The building frame remained largely intact, although several columns were damaged and at least one column failed completely. The column damage resulted from an unusual focusing of blast force at the 4th floor level. Apparently, over-pressure at this level heaved the 5th floor slab upwards, overloading several columns on the 5th floor in compression. The frame survived these losses without collapse or excessive deflection.

This brief paper is presented to ASCE to aid researchers and designers of blast-resistant buildings. I would like to thank my client, Evans International, Ltd., for its kind permission to present this paper.

Wood Engineering

Manuscripts for some presentations were not available at time of publication.

6. *LRFD IN WOOD: IS IT LOADED? IS IT RESISTED? ARE WE FACTORING IT INTO DESIGN?*

MODERATOR: Dan L. Wheat
University of Texas at Austin
Austin, Texas

(1) LRFD for Wood from an Industry Perspective: How Loaded Is it?
D. Gromala
Weyerhaeuser Company
Tacoma, WA

G. Robak
Trus Joist McMillan
Boise, ID

D. Soderquist
Willamette Industries Inc.
Woodburn, IL

(2) LRFD in the Structural Engineering Classroom
M. Criswell
Colorado State University
Fort Collins, Colorado

(3) Where Can I Find Wood LRFD Instruction?
D. Wheat
University of Texas
Austin, TX

S. Cramer
University of Wisconsin
Madison, WI

(4) LRFD in Structural Steel: Trials and Tribulations, Accomplishments, and the Future
N. Iwankiw
American Institute of Steel Construction, Inc.
Chicago, IL

15. *WOOD CONNECTION DESIGN AND RESEARCH*

MODERATOR: Philip Line
American Forest and Paper Association
Washington, D.C.

(1) Bolt Bearing Behavior of Engineered Wood Composites
Stephen Carstens and David Pollock
Washington State University
Pullman, WA

**(2) Development of a Novel Nail Plate Connection System for
Ready-To-Assemble Wood Structures**
Joseph R. Loferski
Virginia Tech Blue Ridge
Blacksburg, VA

R. Terry Platt
Timberwrights, Inc.
Christiansburg, VA

(3) Evaluation of Nail Dowel Bearing Strength
Douglas R. Rammer
USDA Forest Products Laboratory
Madison, WI

(4)Cyclic Response of Wood Dowel Connections
J. Daniel Dolan
Virginia Tech
Blacksburg, VA

24. *CHANGES IN BUILDING AND DESIGN CODES FOR WOOD CONSTRUCTION*

MODERATOR: J. Daniel Dolan
Virginia Polytechnic Institute and State University
Blacksburg, VA

**(1) Changes and Future Trends in the NDS® for Wood Construction and
LFRD Specifications**
Buddy Showalter
American Forestry & Paper Association
Washington, D.C.

**(2) Changes and Future Direction of the NEHRP Provisions for
Wood Construction**
J. Daniel Dolan
Building Seismic Safety Council
Blacksburg, VA

**(3) The IBC and IRC and Their Impact on Wood-Frame Design
and Construction**
Dennis Pitt and David Tyree
American Forestry & Paper Association
Washington, D.C.

**(4) The Future Direction of Performance Based Codes for Structural
and Fire**
Rodney McPhee
Canadian Wood Council

33. *NEW LIFE FOR FORGOTTEN MATERIALS, BUILDING WITH RECYCLED TIMBERS*

MODERATOR: Richard J. Schmidt
University of Wyoming
Laramie, WY

(1) Stress Grading of Recycled Lumber and Timber
R. H. Falk and David Green
USDA Forest Products Laboratory
Madison, WI

(2) Analysis of Wood-Peg Connected Timber Frames
W. M. Bulleit
Michigan Tech. University
Houghton, MI

Matthew W. Drewek
Cleveland-Cliffs, Inc.
Ishpeming, MI

(3) Design Considerations for Timber Frame Connections
R. J. Schmidt, C. E. Daniels, and G. F. Scholl
University of Wyoming
Laramie, WY

(4) The Painful Realities and Lovely Results of Using Recycled Timbers
R. L. Brungraber
Benson Woodworking Co.
Alstead Center, NH

42. *PERFORMANCE, SAFETY, AND RELIABILITY-BASED DESIGN OF WOOD STRUCTURES*

MODERATOR: David Rosowsky and Ken Fridley
Clemson University
Clemson, SC

LRFD for Wood from an Industry Perspective:
How Loaded is it?

David S. Gromala, P.E.[1], M. ASCE
Glen D. Robak, P.E.[2], M. ASCE
David C. Soderquist, P.E.[3], M. ASCE

Abstract

Is load and resistance factor design (LRFD) a loaded gun pointed at the captive structural engineers of the country? Is it a loaded cannon that will provide a long overdue "explosion" that will blow apart the engineering conventions of the past, making room for a system that simply works better? Is it a set of loaded dice – in which the end result can be manipulated to someone's advantage?

The authors, each with a different perspective on this topic, provide several differing answers to these questions. As this paper shows, we disagree on whether LRFD, in the short term, serves any useful purpose for engineered wood construction. We agree that LRFD, if implemented poorly, has the potential to blow apart a system that is working well, only to replace it with an unproven and unstable substitute. We also agree that LRFD, if implemented properly, will provide a unique set of advantages that will likely serve as the foundation for engineering design in the next century. Finally, we agree that acceptance of LRFD does not, and should not, imply acceptance of reliability analysis as a design tool.

Introduction

Load and resistance factor design (LRFD) has been promoted as being far superior to allowable stress design (ASD). A major advantage claimed for LRFD is its ability to more completely quantify the reliability implications of all major

---

[1] Senior Engineering Specialist, Weyerhaeuser Company, Tacoma, WA 98477

[2] Manager of Product Performance, Trus Joist MacMillan, Boise, ID 83707

[3] Engineering Manager, Willamette Industries Inc., Woodburn, OR 97071

parameters that impact structural safety. Obviously, if this claim is true, we as structural engineers will rush to apply it every day in every design. Why hasn't this occurred? Is it simply because engineers are too "set in their ways" to embrace change, as some LRFD proponents contend? While resistance to change is certainly one factor, we don't think it's the biggest factor. We believe that the reluctance to fully embrace LRFD is largely centered in a "communications gap" between most practicing engineers and most reliability theorists. This paper identifies several areas in which the gap is particularly troublesome. For clarification, when the authors refer to ASD, we are referring to the National Design Specification (AF&PA, 1997). LRFD refers to the AF&PA/ASCE 16-95 (ASCE, 1995) standard. LSD refers to Canadian limit states design, CSA O86.1-1994 (CSA, 1994) – similar to LRFD in the U.S., but based on several different underlying assumptions.

There are hundreds of structural engineers in North America who, like the authors of this paper, are employed by manufacturers of engineered products. Our duties include the search for the delicate balance in which our products are used safely, while providing economical design solutions. Because our primary responsibility as professional engineers is to protect the public safety, we often bias our judgments more toward safety than economy. In our litigious society, this bias toward safety also helps to minimize the liability exposure of our companies. This bias, coupled with our experience and confidence using ASD, makes us wary of some of the promises of LRFD (*i.e., "has higher reliability AND spans further!"*).

The authors of this paper are not unified in our opinions regarding the implementation of LRFD. At least one of us sees compelling advantages to LRFD, and believes that the wood products industry must stay "ahead of the wave" by aggressively promoting its use. At least one of us sees LRFD as a means to find design advantages that benefit us in an increasingly competitive market. At least one of us believes that LRFD, like death and taxes, is probably inevitable – to be faced with similar joy and anticipation. While this description may paint our opinions in a rather extreme manner, it indicates our own diversity of opinion regarding the costs versus the benefits of LRFD implementation.

LRFD Implementation Issues

All of us are wary of LRFD because we have already seen evidence of the problems that it will cause as it reaches widespread implementation. This paper provides four examples, each addressing a specific implementation issue. The first two are general in nature – addressing issues of communicating simple design information to our customers and of the mis-use of reliability procedures as design tools rather than as research tools. The last two examples focus on specific design considerations for specific products – discussing discrepancies between design systems that have yet to be resolved.

Example 1: Presenting LRFD design information to other professionals or to customers without confusion is a challenge.

Some aspects of LRFD implementation will be particularly painful during the "transition period" in which any person in the design/build process is still working (or even just thinking) in terms of ASD. Transition issues start with simple communications. Let's say that my beam design is complete, and I show a maximum reaction of 8.5 kips. In a pure ASD world, that's all the information I need to supply to the steel column designer. In a mixed ASD/LRFD world, I must specify whether that reaction is due to factored or unfactored loads. In addition, since relative magnitudes of various load types are important in LRFD, I must also specify the percentages of that reaction that are due to each type of load. And, of course, I must specify the unfactored load as well, just in case any of the subsequent design calculations much perform a deflection check. Try to perform this "bookkeeping" exercise for a multi-story building with multi-span members (requiring analysis under various types of balanced and unbalanced loading). Just for fun, don't forget to check load combinations that include lateral loads.

The communications issues aren't limited to tracking loads within the building. Assume that the design calls for a joist with $F_b = 3000$. If interpreted as an LRFD value, a low-grade joist will work fine. Conversely, if interpreted as an ASD value, few joists have properties that will meet this very high requirement. (So, in LRFD, if somebody asks for a 2400f glulam beam, they actually want a beam with a reference strength of 6.1 ksi – the LRFD equivalent of 2400f.)

Other implementation challenges involve the presentation of simple load (or plf) tables. Current tables (ASD-based) indicate the maximum load based on a deflection limit and another maximum load based on a stress limit. In LRFD, the deflection column stays the same (using unfactored loads), but the stress-based column (using factored loads) is much higher. In such a table, the user cannot determine which case will control simply by inspection.

Example 2: Reliability analysis is still being promoted as a project/product specific design tool (rather than simply as a specification development aid).

Examine the engineering literature. You'll find no shortage of articles discussing the use of reliability analysis as a tool to be used to establish design values for a particular product. You'll find similar articles discussing reliability calculations as they apply to the design of specific building types or specific assembly types. It is our belief that, while such articles provide intriguing concepts for philosophical discussions, those that promote applications to day-to-day engineering design are misguided. We believe that reliability analysis results provide useful insights in the calibration process for code requirements and design specifications. But we believe that the usefulness of reliability analysis stops there. Our experience is that the tool called reliability analysis requires a skilled and experienced operator to provide meaningful results. In the hands of anyone else, reliability results can provide virtually any answer that one desires – depending only on the subtle limitations of various assumptions in the analysis. We as authors differ on our recommendations for solving this problem. Our opinions range from the belief that LRFD should not be fully-implemented until these problems are resolved, to the contention that LRFD as a format is fine – as long as it is differentiated from reliability analysis, which is not suitable for use in design.

Example 3: Prefabricated wood I-joist design using LSD does not match our judgment and experience.

Design a wood I-joist using 2400f-2.0E flanges (MSR lumber) using three design formats (ASD, LRFD, LSD). The allowable design values (ASD) for the flanges are 1975 psi (compression) and 1925 psi (tension). On this basis, the design of any joist with full lateral support of the compression flange will be controlled by the tension flange capacity. Converting these design values to a reference strength basis under LRFD using ASTM D5457 (ASTM, 1993) format conversion procedures is accomplished by multiplying the ASD values by 2.16/phi. This yields 4740 psi (compression) and 5198 psi (tension) – apparently reversing the design control from tension to compression. When multiplied by their respective phi-factors for a specific design, the flange capacities revert to their proper relationships (tension controls). Unfortunately designs using LSD (CSA O86, Supplement 1) do not exhibit similar behavior. The conversion from ASD to LSD characteristic values is accomplished by multiplying tension values by 2.1 and compression values by 1.9. This difference is sufficient to flip the LSD calculation to a compression-controlled design. Even after multiplication by normalization factors and resistance factors, the problem remains.

Example 4: Pre-engineered connector design using LSD not match our judgment and experience.

Design a joist hanger for a residential floor load using three design formats (ASD, LRFD, LSD). Choose a standard face-mount hanger with an ASD capacity of 1150 pounds. The LRFD format conversion procedure for this hanger is complex – because the original ASD derivation procedure was also complex. Without delving into the depths of hanger design, let it suffice to say that hanger design values tabulated in manufacturers' catalogs include consideration of multiple possible stress modes and/or failure modes. Using rules established by the building codes and product standards, the minimum of these multiple calculations is used to provide a sufficiently safe design value. The conversion to LRFD values can be accomplished by the manufacturer by retracing each calculation using LRFD rules and again taking the minimum value. For users who have ASD design catalogs, this process is more difficult. The best approach available at this time for users is to follow the steps given in the Pre-Engineered Metal Connector Guideline to AF&PA's Manual of Wood Construction: LRFD. These steps use tabulated information from the manufacturer's catalog to make inferences regarding the limiting ASD mode. Armed with this information, the appropriate conversion factor can be used. Conversion to LSD using CSA O86 proceeds along a somewhat different path. The O86 committee chose to generate LSD values for pre-engineered metal connectors on a fundamentally different basis than prior working stress values. While the prior values were benchmarked against an arbitrary limit state (called 5% offset yield point), the new values are based on maximum load carrying capacity. Because the basis for calculation is completely different between the two methods, there is no practical way to compare one against the other. For example, in reliability-lingo, what is the value of saying that the old system had a safety factor of 3.2 on the yield point while the new system has a reliability index of 2.9 computed on the ultimate load?

One method for computing the relative safety levels provided by each system is to compute how many square feet of floor load can be supported by this joist hanger under the various systems. If the same hanger supports more floor area, the authors contend that its safety margin in service has been reduced. Our analysis shows that this hanger will carry 23 square feet of floor area under ASD, 25 square feet under LRFD, and 42 square feet under LSD. This near doubling of effective design capacity is a concern because it raises serious questions – Are we overdesigning the hanger in ASD? If not, how can we claim adequate safety under LSD carrying nearly twice the floor area? We don't claim to have all of the answers to these questions – but we would like to see these issues clearly stated and debated in our public engineering forums.

Summary and Recommendations

So, it's time to answer the questions posed earlier. Are practicing engineers being held hostage by the "loaded gun" of load and resistance factor design? Our answer is "yes" in those areas in which regulators are either dropping ASD as an option or are adopting provisions that give preference to the LRFD format. We believe that both ASD and LRFD should be maintained in parallel for the foreseeable future. May the better format win!

Should we fire the LRFD "cannon" into the ASD camp to blow apart the excuses by which "old timers" cling to outdated ASD procedures? Our answer is a resounding "no." While we fully appreciate the complications we will endure by maintaining dual formats, we are not yet convinced that the promoters of LRFD have properly transferred the depth and breadth of judgment and experience embedded in our ASD procedures into their new format. If and when this transfer occurs, we will be better prepared to support a shift from the old format to the new.

How can we stop the practice of "loading the dice" by using LRFD or reliability analysis to achieve a preferred result? First, let us clarify the issue -- we do not believe that professional engineers will knowingly mis-use these tools to achieve predetermined results. However, we have two implementation concerns in this regard. First, now that both the ASD and LRFD systems are accepted by the building codes, we shouldn't be surprised to see some projects in which some parts are designed using one system and some parts using the other. We strongly encourage regulatory interpretation in this area. Second, we are concerned that reliability analysis will be interpreted as being "equivalent" to LRFD. This is not the case! Real product or project design applications should follow consensus-based and code accepted LRFD standards. These standards present a very specific (and very deterministic) design format. They DO NOT permit the use of reliability calculations for product or project engineering. Leave the reliability analyses to the professors – and leave the interpretation of their results to the standards-writing committees. Finally, we urge practicing engineers to become active in those standards-writing committees – your input and your perspective are a critical ingredient to the process.

References

American Society of Civil Engineers. (1995). "Standard for Load and Resistance Factor Design (LRFD) for Engineered Wood Construction." AF&PA/ASCE 16-95. ASCE, New York, NY.

American Society for Testing and Materials. (1993). "Standard Specification for Computing the Reference Resistance of Wood-Based Materials and Structural Connections for Load and Resistance Factor Design." ASTM D5457-93. ASTM, Philadelphia, PA.

Canadian Standards Association. (1994). "Engineering Design in Wood (Limit States Design)." CAN/CSA-O86.1-1994, CSA, Etobicoke, Ontario, Canada.

American Forest and Paper Association. (1997). "National Design Specification for Wood Construction." AF&PA, Washington, D.C.

Where Can I Find Wood LRFD Instruction?

Steven M. Cramer[1], M. ASCE
Dan L. Wheat[2], M. ASCE

Abstract

A survey was conducted in order to determine how many colleges and universities are teaching the load and resistance factor design, or LRFD, for wood structures. It was found that there are many departments that teach wood courses, but the number of departments adopting LRFD for classroom usage is small. In the paper, we identify from the survey results some of the reasons why the adoption of LRFD may be so slow. Also, we make an effort to discern general perceptions of LRFD and other issues that need to be discussed by that part of the civil engineering community with a clear interest in wood.

Introduction

In the summer of 1997, the first complete load and resistance factor design (LRFD) manual with reference stresses (AF&PA 1996) became available for wood structural design. College and university instructors were faced with, and continue to be faced with, a decision whether to convert to the LRFD design procedures or to continue instruction with the well-established allowable stress design (ASD) procedures contained in the 1997 National Design Specification® for Wood Construction and previous versions.

This effort in wood design followed the development of LRFD in steel design during the 1970's leading to the adoption of the first steel design LRFD specification in 1986 (Galambos 1995). The steel industry has followed a path whereby LRFD and ASD formats have been maintained in parallel and wood design is proceeding similarly. Estimates vary on the degree to which steel designers have adopted the

---

[1] Professor, Department of Civil and Environmental Engineering, University of Wisconsin, Madison, WI 53706.
[2] Associate Professor, Civil Engineering Department, University of Texas, Austin, TX 78712

LRFD specification, but Galambos (1998) claims there are no structural engineering programs in the United States that do not teach steel design based on the LRFD specification.

Load and resistance factor design (LRFD) is a form of limit states design. Factors are applied to loads and to structural member resistances to determine structural adequacy at the "limit of structural usefulness" (Galambos 1995). The criteria in LRFD wood design require the effect of forces from extreme event factored loads to be less than or equal to factored member adjusted resistance, or

$$R_u \leq \lambda \phi R'$$

where

R_u = factored load effect;
λ = time effect factor;
ϕ = resistance (strength reduction) factor; and
R' = resistance adjusted for conditions such as moisture content.

The decision to change the teaching of design to a new design philosophy is not trivial. First and foremost, there must be a clear motivation for changing course content. The continuation of the existing ASD format has made the conversion to LRFD slow in steel design and a similar situation exists with wood design. Assuming such a motivation exists, the instructor must retool his/her own understanding, notes, homework problems and exams. The availability of books and problem manuals to guide these efforts is also a critical factor in the decision. Our objective in this paper was to establish the level of usage of LRFD in higher education and to determine the key issues shaping the decision whether to continue emphasizing ASD or to accelerate the conversion of instruction to LRFD.

Survey

During the Fall of 1998, we sent out 360 surveys to faculty in 213 colleges and universities in the United States. Because wood design is taught at various levels to various student majors, departments including civil engineering, agricultural engineering, architecture, engineering technology, industrial technology and construction management were included in the survey. Some schools received multiple surveys to cover the possibility of wood design being taught in one department versus another. The survey was very short consisting of only several questions related to whether one or more wood design courses were offered at the institution and, if so, whether the course was taught using ASD or LRFD formats, and why one format was chosen over the other. We felt that a brief survey would yield a higher response rate. In addition, rather than risk "leading" the person being surveyed through a long questionnaire, we left one question open for general comments, thus allowing the respondent a free choice in providing commentary about any aspect of the ASD/LRFD issue. The survey is meant to provide a benchmark of the earliest teaching of the LRFD specification in wood design.

We received responses from approximately 55 percent, or 196, of the 360 surveys mailed. We received responses from 170 different institutions and found that 68 percent of those institutions responding offer at least a portion of a course devoted to wood design. When wood design is offered as part of a course, a frequent combination is wood along with masonry; wood and steel combinations were reported also.

During the Fall of 1998, the new LRFD specification for wood design has been or is about to be adopted in the first--and perhaps only--course in 17 percent of the institutions that have at least part of a course devoted to wood design. Approximately 1/3rd of those who have at least part of a course devoted to wood also offer a second course at the graduate level. And approximately 40 percent of those offering a second course teach LRFD in at least part of the course. The survey did not establish how frequently these courses actually are offered. Comments suggested a first course in wood design could be offered as infrequently as once every two years.

As expected, the number of schools using the LRFD specification at this early stage is not large. Twenty schools offer LRFD instruction in wood design at the undergraduate level and approximately 18 at the graduate level out of 170 schools responding. After having steel LRFD specification for 13 years, there appear to be no structural engineering programs in U.S. universities that do not teach steel design using LRFD, according to Galambos (1998). We have no data on the number of institutions that offered steel design using LRFD within a year or two of its appearance.

Discussion

The survey revealed many reasons for a reluctance to adopt the LRFD specification in wood design education, but two reasons were prevalent. First the lack of adoption of LRFD by practicing designers who will employ graduates was cited by many as a primary reason to continue to teach ASD rather than LRFD. The second most commonly cited reason was the lack of teaching materials such textbooks, problem manuals, and other teaching aids.

A dearth of practicing engineers who have adopted LRFD in steel seems to have had no impact on the design philosophy being taught. Thus, the question that arises is who should be the driving force for the changes in design philosophy: the design profession or academia? Should academia force LRFD on an unwilling design community by supplying only LRFD-trained engineers; or should the design community say to academia, "we're the employer and you're there to provide what we want, not what you think we should have? The role of universities and colleges to educate and train as well as to lead in technical innovation seems to be in question.

The second primary reason cited retarding the adoption of LRFD in wood is lack of textbooks and teaching aids, but this likely will obviate itself with time. Approximately ten textbooks on LRFD steel design are available and form a foundation for LRFD-based steel design education (Galambos 1998).

Many other reasons were cited for not teaching LRFD, but many fall under the category of education related to LRFD. Some were not aware that the LRFD specification was available and some expected the National Design Specification to convert to LRFD. As mentioned above the National Design Specification for Wood Construction is the ASD procedure for wood design and will remain so. There was also a perception by some that LRFD is more complex than ASD. With limited course time and an objective to provide wood design awareness rather than expertise, some educators in non-engineering disciplines noted that the perceived additional complexity of LRFD can not be accommodated.

LRFD design procedures have now been established and implemented for steel, including steel buildings, cold-formed steel, stainless steel; aluminum; concrete; bridges; and wood structures (Galambos 1998). In addition, masonry now has a strength design procedure (International Conference of Building Officials 1997). Despite the fact that LRFD design procedures exist for all of these commonly used materials, some educators feel that ASD still should be a component of the curriculum and that component often is a wood course.

The Case for LRFD

LRFD is based on the notion of a uniform reliability of structures to meet specific needs. Uncertainties in loads and material resistances are accounted for separately in LRFD with factors that distinguish the levels and nature of uncertainty, rather than lumping uncertainty in ASD safety factors that often have little in the way of a rational basis. While smaller sections may be achieved with LRFD in some cases, it is the more rational factoring of loads and resistances and the uniformity across material types in load treatment that are the primary motivations in adopting LRFD.

Both practicing design engineers and academics are concerned about a growing chasm between structural engineering research conducted in universities and that actually implemented into practice. This concern was indeed reflected in the survey with many academics unwilling to be out of step with the majority of design practice. Yet the role of universities and colleges to foster innovation must continue if the structural engineering profession is to advance.

Summary

With release of the wood design manual for load and resistance factor design in 1997 and the adoption of this standard by the major building codes, educators have become faced with the dilemma of which design philosophy to teach; meanwhile, the traditional allowable stress design procedures continue to be accepted and used. A survey conducted by the authors reveals that of those institutions that offer a single course or only a portion of a course, 17 percent teach LRFD. Primary reasons for the lack of adoption are the lack of adoption by design firms and the lack of teaching aids such as textbooks and problem manuals.

References

American Forest and Paper Association. 1996. *Load and Resistance Factor Design (LRFD) – Manual for Engineered Wood Construction.* Washington DC.

American Forest and Paper Association. *1997 National Design Specification® for Wood Construction.* Washington, DC.

Galambos, T. 1995. "The Future of LRFD". *Restructuring: America and Beyond, Proceedings of Structures Congress XIII.* Edited by M. Sanayei. American Society of Civil Engineers, 734-737.

Galambos, T. 1998. "Developments of LRFD in the United States of America". Proceedings SEWC '98. Edited by Srivastava. Elsevier Science. Paper T104-1 on CD ROM.

International Conference of Building Officials 1997. *Uniform Building Code, Vol.2.* Whittier, California.

LRFD in Structural Steel: Status and the Future

Nestor R. Iwankiw, AISC[1]

Abstract

The development and widespread utilization of any new product or technique often requires much time and effort. The most recent example of such in structural engineering is occurring in the evolution of limit states design, which is also known as Load and Resistance Factor Design (LRFD) for steel buildings. The current status and acceptance of this second edition AISC-LRFD Specification for structural steel in the US will be presented from the perspective of its overall origins, public reaction to date, progress, and future prospects. While the road for steel LRFD development and acceptance has not been short or easy, these experiences do parallel similar limit states design transitions in other building materials, both here and abroad. The excellent technological basis for LRFD provides the best consensus design methodology whose full day is soon approaching.

Introduction and Background

LRFD for structural steel is conceptually similar to the LFD (load factor design)/LSD (limit states design)/USD (ultimate strength design) standards that have been developed for various building materials in the US and other industrialized countries. Steel LRFD astutely combines plastic design/ultimate strength principles with modern structural reliability theory to result in a broad and powerful design method. Years of industry sponsored and cooperative research by Prof. T.V. Galambos and his graduate students at both Washington University and the University of Minnesota has led to the fundamentals of today's LRFD criteria. A milestone in this development was the series of papers published in the September, 1978 ASCE Structural Journal and in the AISC Engineering Journal on LRFD by Dr. Galambos and committee members involved in its research and development. Further Committee work in the early 1980's on trial building designs and additional refinements led to an extended public review period of LRFD and ultimately, its historical release in 1986. In the near future and distant decades to come, this 1986 LRFD 1st Edition will be identified as

[1]American Institute of Steel Construction, Inc.
Nestor R. Iwankiw, Vice President/ Engineering & Research
1 E. Wacker Drive Ste. 3100
Chicago, IL 60601-2001

a major breakthrough, much as the original predecessor 1923 AISC-ASD 1st Edition served as a benchmark for several generations of successful design practice. This introduction of LRFD was probably the single most comprehensive design change undertaken by AISC in its long history.

In 1984, AISC released a public review copy of this first LRFD Specification. Two years later, the 1986 AISC-LRFD 1st Edition officially appeared as a landmark document. It marked the beginning of the U.S. change from traditional Allowable Stress Design (ASD) to the new LRFD in structural steel. Since that time, AISC has maintained both design methods as alternatives.

The 2nd Edition of AISC-LRFD Specification was produced in late 1993, (AISC, 1993) and is now in use, while a 3rd Edition is anticipated for late 1999.

Due to the participatory open nature of the building code process in the U.S., major changes in design standards are often rather difficult to quickly implement into practice. In contrast, many other countries have single national building codes, which are developed and promulgated, at the central government level. In addition to the usual time lag delays through the many levels of code authorities, the U.S. consensus system itself tends to slow the rapid acceptance of new technology when existing standards already provide a ready source. This fact combined with the general user inertia to change leads to the observation that, LRFD acceptance by the broad design profession is moving at a slow pace. This is understandably disheartening to the LRFD advocates and developers. Many practitioners still favor the convenience of the familiar ASD over the potential of LRFD, and demonstrative support for maintaining ASD still remains among the consulting establishment. Complaints are expressed about the difficulty of understanding and using LRFD, the time/cost necessary to learn it and accordingly change office practice, and lack of a clear economic incentive for the change. (LRFD was intentionally calibrated by its writers to provide, on average, similar designs to ASD.) While these may be valid transitional reservations, the underlying basic obstacle seems to be the proverbial user resistance to change.

Status and Prognosis

The current AISC LRFD and ASD dual design standards requires additional staff effort and committee oversight to maintain. It also tended to obscure AISC's long-run commitment for this new approach. In order to offer a more explicit signal to the public relative to the available ASD vs. LRFD design choice, the AISC Board of Directors, based upon the recommendations of its Committee on Specifications, adopted the following resolution in early 1995. As published in the June 1995 issue of AISC *Modern Steel Construction*, (AISC 1995) this resolution reads:

"Based upon expert input from its Committee on Specifications, the Board of Directors of AISC affirms that the *1993 Load and Resistance Factor Design (LRFD) Specification for Structural Steel Buildings* is the preferred Specification for the fabricated structural steel industry. LRFD is a modern and technologically superior steel design specification. Its direct representation of ultimate structural behavior is especially relevant for seismic design, design of frames with partially restrained connections, and composite systems design. It offers engineers the opportunity to innovate in the analysis and design of highly reliable and competitive steel structures by encouraging the consideration of strength and serviceability criteria under appropriate combinations of gravity and lateral loads. In this way, LRFD is consistent with the prevailing trend toward limit-states design in all materials, both domestically and internationally."

Several features/advantages of LRFD identified in this resolution deserve further highlighting:

1. Technical superiority: LRFD is state-of-the-art in its representation of ultimate strength limit states, expected loading demand, and reliability.

2. Advancement potential: Because of these inherent capabilities, LRFD is best suited for new applications in seismic design, partially restrained connections, and composite systems, wherein non-linearity's and inelasticity must be handled not just for structural performance accuracy, but also for building efficiency and economy. In contrast, the long-standing ASD is quite limited by its linear elastic basis.

3. Consistency: The clear global trend in all building materials is moving toward limit state design. Canada, Europe, Japan, and Australia either already have mandatory LSD national codes in place, or are well along in this process. US structural steel design and construction must keep in step through LRFD or risk falling behind, or becoming isolated, from the mainstream international and domestic practices.

Furthermore, at the AISC joint Board and Specification Committee leadership meeting in June 1997, this resolution was further reinforced. It was agreed that the final 1989, 9th Edition ASD Specification and Manual will be maintained, and remain publicly available, in their current form for the foreseeable future, subject only to changes by addendum for reasons of safety and/or major discrepancy. However, the very critical decision was made that there would not be a new 10th Edition of either the ASD Specification or Manual.

The expected result of these recent official AISC directives is to more forcefully influence the domestic growth and acceptance of LRFD, with the gradual voluntary phase-out of ASD practice. LRFD is now the integral part of design courses in most universities, and the emerging generation of young structural engineers will be proficient in its use and understanding. Numerous textbooks, publications, software, and other design aids also exist for steel LRFD, thereby removing another possible obstacle to its office use.

During the fall of 1994, AISC commissioned the Gallup Organization to conduct a steel design survey (Gallup, 1994). A random sample of about 450 engineers was interviewed to assess their perceptions on LRFD and ASD. Overwhelmingly, structural steel design in ASD was still the method of choice, about 80%. This has been intermittently substantiated by the author during informal surveys at selected technical presentations, and likewise confirmed by the still continuing external pressures to keep ASD "alive". Familiarity with ASD was by far the biggest reason given in the Gallup study to favor ASD, and 60% wanted AISC to continue supporting ASD. Nevertheless, in a key concession, 64% of this same sample admitted that LRFD represents the future, and 44% vs. 31% thought that LRFD is the technically superior method.

The general observation from this Gallup survey that has been reflected in the AISC resolutions emphasizing LRFD is that while the majority of the profession clings to the familiar ASD, the evolution toward LRFD predominance is well underway. There are no fundamental or serious technical barriers to its use. With sufficient transition time and natural ASD attrition, steel LRFD will become the primary design method, probably paralleling the historic ACI experiences in reinforced concrete ultimate strength design over the last 25-30 years.

Conclusions

After about ten years of neutrality since the introduction of LRFD in 1986, AISC has taken a proactive public position clearly favoring LRFD over ASD in structural steel. The professional transition and conversion to LRFD from the traditional ASD is continuing, but now with more of a sense of urgency and commitment from the engineering and industry sides. The academic sector is firmly behind this design change and is helping to facilitate it with proper educational fundamentals for its student graduates. The strong leadership posture of AISC, and other similar standards organizations, is necessary to continue, and accelerate, the application uses of limit states design.

REFERENCES

1.	AISC, *Load and Resistance Factor Design Specification for Structural Steel Buildings*, 2nd Ed. Dec. 1, 1993.
2.	AISC, *Modern Steel Construction*, "Essentials of LRFD," June, 1995.
3.	Gallup Organization, *Steel Design and Specification Study Report to AISC*, November, 1994

BOLT BEARING BEHAVIOR OF ENGINEERED WOOD COMPOSITES

Stephen J. Carstens, M. ASCE [1]
David G. Pollock, Jr., Ph. D., P.E., M. ASCE [2]

Abstract

The goal of this research was to gain a better understanding of the bolt bearing behavior of engineered wood composites made from yellow poplar lumber. Lumber specimens included in this study were laminated veneer lumber, strand-based lumber, yellow poplar lumber, and Douglas-fir larch lumber. Testing followed the half-hole and full-hole configuration as set forth in ASTM Standard D5764 (1998).

In a previous study by Wilkinson (1991), a strong correlation was shown between bearing strength perpendicular to grain and bolt diameter. This study supports Wilkinson's finding for bearing strength perpendicular-to-grain based on the half hole test configuration. Other findings in this study indicate there may be a correlation between bolt diameter and bearing strength parallel-to-grain for the half-hole test configuration as well as a correlation between bolt diameter and bearing strength both perpendicular- and parallel-to-grain for the full-hole test configuration.

In general, half-hole tests resulted in a greater dowel-bearing strength than full-hole tests, especially for 12.7mm (½ in) diameter bolts. Also, engineered wood composites generally provided equivalent or greater dowel-bearing strength in the half-hole configuration and greater dowel-bearing strength in the full-hole configuration when compared to lumber from the same species.

----

[1]Graduate Research Assistant, Washington State University, Department of Civil and Environmental Engineering, Pullman, WA 99164
[2]Assistant Professor, Washington State University, Department of Civil and Environmental Engineering, Pullman, WA 99164

Introduction

The scope of this study was specifically focused toward quantifying the bolt bearing strength (F_e) of lumber and wood-based products commonly used in wood-frame construction. Bolt sizes and grain orientation used for the half-hole (HH) test configuration are given in Table 1 and bolt sizes and grain orientation for the full-hole (FH) test configuration can be found in Table 2. Structural grade Douglas fir-larch lumber (DF), laminated veneer lumber (LVL), strand-based lumber (SBL), and yellow poplar lumber (YP) were selected for this study. The LVL and SBL specimens were fabricated entirely from YP material. The YP and DF lumber specimens were selected as a basis of reference. Physical testing was used to determine F_e in accordance with the provisions of ASTM Standard D5764 (1998).

Testing Procedure

The purpose of the HH configuration of the F_e test was to determine the load resistance and displacement characteristics for fasteners in wood-based products. This was obtained through application of a force/load on a bolt placed in a drilled hole. Each rectangular wood specimen had a hole drilled in the 'width' face and was used to evaluate the resistance to fastener embedment in the hole without bending the fastener. Crushing strength of the wood under the fastener can be used to establish connection design values. The effects of bolt diameter, moisture content, specific gravity, and grain direction on F_e can also be evaluated using these tests. 5% offset and maximum load of the specimen were determined from the test data.

The original intent of the FH configuration was to facilitate completion of the test for specimens that tend to fail prior to reaching the 5% offset load in the HH configuration. A goal of this study was to provide a direct comparison between the HH and FH tests configurations for lumber and engineered wood composite specimens.

Twelve replications were conducted for each wood product, angle to grain, and bolt size. The bolt sizes used in the HH tests were 12.7 mm (½ in), 19.05 mm (¾ in), and 25.4 mm (1 in) diameter. 12.7 mm (½ in) and 19.05 mm (¾ in) diameter bolts were used in the FH tests. Matched specimens were used for the FH and HH tests in this study.

Analysis of Results

In evaluating the HH configuration, it is interesting to note that the engineered wood composites always exhibited higher F_e values than the YP lumber for the perpendicular-to-grain loading orientation. However, for parallel-to-grain loading the results were mixed, with YP lumber sometimes exhibiting higher F_e values than either LVL or SBL. In all cases, SBL had higher F_e values than LVL. Specifically, SBL exhibited an 11% higher F_e value for 12.7mm (½ in) diameter tests and a 15% higher F_e value for 19.05mm (¾ in) diameter tests, versus LVL parallel-to-grain. Similarly, SBL F_e values perpendicular-to-grain for the 12.7mm (½ in) diameter tests were 93% higher than LVL F_e values, and were 43% higher for 19.05mm (¾ in) diameter tests of LVL. These F_e values are shown in Table 1.

Diameter effects were observed for all products in both perpendicular and parallel-to-grain orientations. Wilkinson's (1991) study showed F_e decreasing as bolt diameter increased for specimens loaded perpendicular-to-grain. However, the results of this study indicate somewhat of a "reverse effect", with F_e increasing as the bolt diameter increased for LVL and YP lumber specimens loaded perpendicular-to-grain.

Table 1.   Average Dowel-Bearing Strengths for the HH Configuration, MPa (psi).

Bolt Diameter		12.7mm (½ in)	12.7mm (½ in)	19.05 mm (¾ in)	19.05 mm (¾in)	25.4 mm (1 in)
Grain Orientation		Parallel	Perpendicular	Parallel	Perpendicular	Perpendicular
Material	Avg SG	MPa (psi)	MPa (psi)	MPa (psi)	MPa (psi)	MPa (psi)
DF	0.498	50.5 (7322)	20.5 (2985)	36.1 (5234)	N/A	N/A
YP	0.492	47.1 (6830)	18.1 (2624)	46.2 (6710)	19.7 (2870)	20.2 (2944)
LVL	0.516	43.6 (6326)	27.7 (3304)	36.6 (5313)	23.7 (3448)	N/A
SBL	0.629	48.8 (7084)	43.9 (6379)	42.2 (6124)	34.8 (5047)	40.0 (5806)

In evaluating the FH test, both LVL and SBL always exhibited higher F_e values than the YP lumber, and SBL F_e values were always greater than LVL. Specifically, SBL had a 1% higher F_e than LVL and a 22% higher F_e than YP lumber in the 12.7mm (½ in) parallel-to-grain test.   In the 12.7mm (½ in) perpendicular test, SBL had a 27% and 80% higher F_e than LVL and YP lumber, respectively. In the 19.05mm (¾ in) tests, SBL F_e values were 0.3% higher than LVL and 23% higher than YP lumber in the parallel tests, and 72% higher than LVL and 123% higher than YP lumber in the perpendicular tests. These F_e values are shown in Table 2.

Reversed diameter effects were observed for all products in the FH tests loaded parallel-to-grain. F_e values for SBL specimens loaded perpendicular-to-grain also increased with increasing bolt diameter. However, F_e values for LVL decreased approximately 8% for 19.05mm (¾ in) versus 12.7mm (½ in) diameter bolts loaded perpendicular-to-grain, and F_e values for YP lumber load perpendicular-to-grain remained constant.

F_e values from the FH test were lower than F_e values from the HH test for all test scenarios except SBL specimens with 19.05mm (¾ in) diameter bolts. The largest decrease in F_e is a 48% drop with LVL 12.7mm (½ in) perpendicular-to-grain. The disparity in F_e values for FH versus HH tests was greatest for the 12.7mm (½ in) diameter bolt specimens. This was possibly due to elastic bolt deflection within the FH specimen, causing a non-uniform stress distribution through the thickness of the specimen. Differences between the FH and HH were less pronounced for 19.05mm (¾ in) bolts.

Table 2. Average Dowel-Bearing Strengths for the FH Configuration, MPa (psi).

Bolt Diameter		12.7mm (½ in)	12.7mm (½ in)	19.05 mm (¾ in)	19.05 mm (¾in)
Grain Orientation		Parallel	Perpendicular	Parallel	Perpendicular
Material	Avg SG	MPa (psi)	MPa (psi)	MPa (psi)	MPa (psi)
YP	0.491	33.2 (4828)	18.7 (2713)	43.5 (6310)	18.8 (2727)
LVL	0.517	33.1 (4799)	23.1 (3359)	45.4 (6579)	21.4 (3097)
SBL	0.627	35.8 (5191)	29.5 (4283)	46.9 (6805)	36.7 (5324)

Conclusions

While this study validates some attributes of a previous F_e and bolt diameter study, it also suggests that other relationships exist that warrant further testing. Specifically, a "reverse" diameter effect for two bolt sizes loaded perpendicular-to-grain, and the existence of a correlation between F_e and bolt diameter for specimens loaded parallel-to-grain, were observed. This research also indicates that YP-based engineered composite lumber exhibits higher F_e values when compared to YP dimension lumber using the FH test configuration. Engineered wood products also exhibit higher F_e values than lumber when loaded perpendicular-to-grain using either test configuration. Though it was initially assumed that the two test configurations would give similar results, the FH test resulted in equivalent or increased Fe values for 19.05mm (¾ in) diameter bolts only. The initial results indicate that further investigation of the FH versus HH configurations in ASTM Standard D5764 (1998) may be warranted for a wider range of fastener diameters and structural wood-based products.

Acknowledgement

This study was funded through Project G854001 of the U.S. Department of Energy Idaho National Engineering and Environmental Laboratory (INEEL) University Research Consortium.

References

American Society for Testing and Materials (ASTM). 1998. Standard Test Method for Evaluating Dowel-Bearing Strength of Wood and Wood-Based Products. Standard 5764-97a. Philadelphia, PA: ASTM.

Wilkinson, T. L. 1991. Dowel-Bearing Strength. Research Paper RP-505. Madison, WI: U.S. Forest Products laboratory.

Development of a Novel Nail Plate Connection System
for Ready-To-Assemble Wood Structures[1]
by
Joseph R. Loferski[2], R. Terry Platt[3]

Abstract
Despite many advances in building construction, affordable housing is still in short supply in many countries, including the USA, and is critically needed. This paper discusses the development of a ready-to-assemble construction system for wood framed structures that are assembled with novel multiple-nail connector plates. These nail connector plates are used to produce built-up structural elements from solid wood or wood-based composites for use in a variety of structural applications. The RTA system design concept has potential for worldwide use because it provides a method for rapid assembly of low cost buildings using minimal on-site labor and simple tools.

A variety of connector plates have been designed and tested in various modes including lateral translation and moment rotation. Design equations to predict the strength of these connectors in lateral loading have been developed for specific structural conditions.

Introduction
In today's world, there is an increasing need for buildings for housing, industrial, and agricultural uses. This situation offers an opportunity for the wood products industry, especially if new and innovative methods are developed to produce affordable, efficient, and effective buildings. Recently, a novel nail plate connector system has been developed to produce built-up members for beams, columns, trusses, and rafters. The system, under development, will produce the built-up elements in factories in standardized sizes for shipment to a building site for bolt-together assembly. The perceived advantages of this system are rapid assembly and disassembly of structures using relatively unskilled labor. The system has potential uses for low cost housing, disaster or emergency relief buildings, and innovative structures such as space frames and space trusses.

The objective of this paper is to present an overview of the RTA construction system, and some of the methods used to test the strength and stiffness of these connectors.

Background
The RTA connection system was inspired by the work of Y. Piskunov (Piskunov 1993) of Vyatka Polytechnic Institute in Kirov, Russia. He developed connectors to transmit shear between layers of mechanically laminated beams. His connectors consist of multiple double pointed nails that are spot welded at their centers to the perimeter of a steel plate. The connectors are pressed into the surfaces of sawn timbers to produce built-up elements.

---

[1] Partial funding for this research was provided by USDA SBIR Project # 97-00022.
[2] Associate Professor, Dept. of Wood Science & Forest Products, Virginia Tech, Blacksburg, VA, USA
[3] Director of Product Development, Blue Ridge Timberwrights, Christiansburg, VA, USA

The RTA construction system which is under development uses nail plate connectors which consist of 5 or more double pointed nails inserted through a steel plate. The nail plates are pressed into the wood elements at strategic locations to produce a built-up structural member ready for assembly into a building (photograph 1).

Photograph 1: Example of the sequence of element fabrication (from left to right) A) connector prior to assembly. B) connector pressed into one-half of a wood element. C) completed element.

The steel plates have holes bored through them to allow bolting together of individual elements to form the structure (Photograph 2). The elements can be produced from solid wood or wood-based composites such as parallel strand lumber or laminated veneer lumber. Photograph 3 shows an RTA frame without sheathing.

Photograph 2: Column-beam-rafter connection: A) joint layout prior to assembly B) joints bolted together.

<u>Experimental</u>
In this project, four RTA connector designs were tested in six different loading conditions to measure the strength and stiffness in translation (i.e. lateral loading), and moment rotation situations. Lateral loads were applied at three angles to the grain (0°, 45°, and 90°). All tests were designed to simulate the in-service conditions of joints in a full-scale RTA structure. Figure 1 shows the average load vs. deformation of one of the tested RTA joint conditions. The figure shows the average curve of 15 replications and the range of the data.

The yield theory for dowel type connectors was applied to determine if it was suitable for predicting the lateral strength of these connectors. After adjusting the experimental data for load duration, safety factors, and calibration to NDS (AFPA 1991) yield equations, it was concluded that the yield theory can be used to accurately predict the strength of these connections in lateral loading. Figure 1 shows a proposed design value for this particular connection, computed from NDS (AFPA 1991) yield equations for three member, double shear joints with a steel main member.

Photograph 3: Example of an assembled RTA frame.

Currently, the authors are developing a method to predict the stiffness of these connectors for use in structural modeling of full-size buildings. Validation of the computer model will be accomplished through full-scale testing.

To date, more than 500 tests of connectors in seven different materials have been conducted for use in a database to assist in obtaining building code approval. Various nail patterns have also been tested. More details of these experiments and the results of some of these tests are in Platt (1998).

<u>Summary</u>
A novel connection system for ready-to-assemble wood buildings has been developed and tested in various configurations. The experimental work shows that yield theory can be used to predict the strength of these connections in lateral loading. The system concept is to produce factory fabricated elements ready for on-site assembly. Future work will include methods to predict the stiffness of these connections for use in finite element modeling of full size structures.

<u>References</u>

American Forest and Paper Association. 1991. <u>National Design Specification for Wood Construction</u>. NOV American Forest & Paper Association. pp 125

Piskunov, Y.V. 1989. "Development of Design Specification for Timber Structures made of Built-up bars with use of Nail Connections." Proc. CIB/W18A/TG9. Timber Structures, Berlin, Germany.

Piskunov. 1993. "Scientific and Technical Problems of the Nail Plates used for Manufacturing and Building Timber structures." CIB/W18A/TG6, Intl. Coun. For Bldg. Res. St. and Doc. Proceedings of "Design and construction of Timber Structures Assembled with Metal Connector Plates and Nail Connectors." Kirov, Russia. pp36-47.

Platt, R.T. 1998. "Development of a Ready-To-Assemble Construction System" Ph.D Dissertation. Virginia Tech. Blacksburg, VA USA.

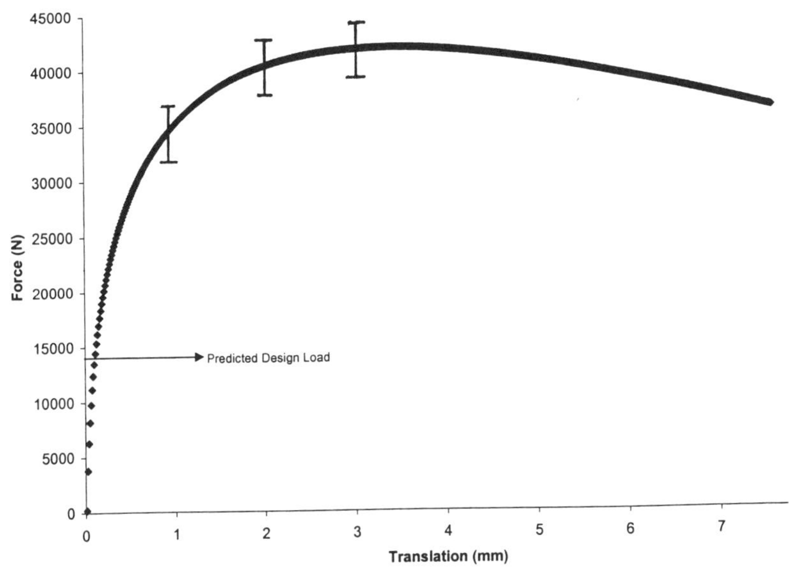

Figure 1: Force vs. deformation diagram (average curve of 15 specimens) of an RTA joint loaded in tension parallel to the grain of the wood elements and the predicted design load.

Evaluation of Nail Dowel Bearing Strength Expression

Douglas R. Rammer, ASCE Member[1]

Abstract

An experimental study was conducted to improve the dowel bearing specific gravity relationship for nails. Dowel-bearing strength tests were performed on Maple, Southern Pine, Spruce–Pine–Fir, and Yellow Cedar specimens using 4.11-mm nails. New derived expressions are compared to current design equations.

Introduction

In the United States, structural connections utilizing wood under lateral loading are currently designed by the yield theory (*National* 1997, "Standard" 1996), which relates the connection load to geometry and dowel bearing and bending strength. European communities developed the yield model and have generated much information on joint strength and dowel yielding and bearing strength. Using a method that loads the dowel at the ends, Whale and Smith (1986) determined that nail bearing strength, defined by maximum load or deformation of 2 mm, is a function of both specific gravity and nail diameter. They noted a difference between parallel and perpendicular to grain orientation, but since the difference was small they combined data for regression analysis.

Since an ASTM standard ("Standard" 1998) for evaluating dowel bearing strength has been developed and accepted only recently, dowel bearing strength data generated according to this new standard are limited. Wilkinson (1991) determined the nail bearing strength of 4.11-mm nails for wood species with specific gravity ranging between 0.37 and 0.50. For nails driven into side grain, he developed the following expression relating specific gravity and dowel bearing strength:

$$F_e = 126.8 G_{12}^{1.84}$$

(1)

where F_e is expressed in N/mm and G_{12} is specific gravity using volume at 12% moisture content. After converting expression (1) to a specific gravity based on oven-dry volume, this expression was accepted into the current U.S. wood construction standards for wood dowel bearing strength (*National* 1997, "Standard" 1996). Expression (1) is based on a limited data set that contained Douglas Fir

---

[1] Research Engineer, USDA Forest Service, Forest Products Laboratory, One Gifford Pinchot Drive, Madison, WI 53705–2398

material with a specific gravity value lower than published norms and two species groupings with moisture content values below 6%.

In comparing the current nail dowel bearing design expression with two Guatemalan hardwoods with high specific gravity (0.70 and 0.76), Rammer (submitted for publication) showed that the Wilkinson (1991) dowel bearing–specific gravity relationship, based on 0.37 to 0.50 specific gravity range, overpredicts the experimental values. The difference between the relationship and experimental data was greater for the wood with higher specific gravity. Several researchers have investigated the effect of moisture content on bolt dowel bearing strength (Fahlbusch 1949; Koponen 1991, Winistorfer, in press). Expressions developed by these researchers predict an 8% to 30% increase in bearing strength for a 6% decrease in moisture content.

My objective was to improve the empirical relationship between specific gravity and dowel bearing strength for nails used for connection design (*National* 1997, "Standard" 1996). This objective is part of a larger study on the effects of specific gravity, nail diameter, and moisture content on nail dowel bearing strength.

Research Methods

The methods described here address only the 4.11-mm nail size and 12% moisture content condition. Dowel bearing strength of four species groupings—Southern Pine, Yellow Cedar, Spruce–Pine–Fir, and Maple—was investigated both parallel and perpendicular to the grain. Specimens were conditioned in a 21°C–65% relative humidity environment to achieve 12% moisture content. Pilot holes approximately 75% the diameter of the shank were drilled prior to nailing in accordance with ASTM D5764 ("Standard" 1998) and specimens were tested between 24 and 32 h after nailing. During the 24-h waiting period, specimens remained in the 21°C–65% relative humidity environment.

Results

Table 1 shows total number of tests, average specific gravity and moisture content, average and coefficient of variation (COV) values for stiffness (calculated between 20% and 40% of maximum load), and dowel bearing strength parallel and perpendicular to grain (calculated using 5%-diameter offset load ("Standard" 1998)).

Data Analysis

T-tests indicated that grain orientation has a significant ($p = 0.05$) effect on 5%-diameter dowel bearing strength for all species except Maple. Comparison of mean parallel- and perpendicular-to-grain values revealed the greatest difference for the lower specific gravity woods (Spruce–Pine–Fir and Yellow Cedar).

In this study, the specific gravity range (0.4 to 0.7) was greater and the moisture content variation (10.8% to 14.4%) was less than that in Wilkinson's dowel bearing study. Therefore, a regression analysis of the combined parallel and perpendicular to grain results led to a better relationship between specific gravity and dowel bearing strength.

Table 1. Dowel bearing strength test results[a]

Species	Grain[b]	Tests (no.)	Dry SG	MC (%)	Stiffness Mean (kN/mm)	COV (%)	Bearing strength Mean (MPa)	COV (%)
SPF	P	49	0.42	11.2	7.75	40.9	27.68	26.0
	R	49	0.42	11.4	2.68	42.4	20.45	26.8
Yellow Cedar	P	34	0.51	12.4	16.00	26.5	42.38	15.4
	R	26	0.50	12.7	4.28	39.7	27.62	23.2
Southern Pine	P	43	0.60	11.0	16.62	34.9	46.59	19.4
	R	42	0.60	10.8	7.43	19.6	40.08	17.0
Maple	P	30	0.69	14.3	13.36	32.8	48.09	16.4
	R	30	0.70	14.4	10.13	35.0	46.99	20.4

[a]MC is moisture content; SG, specific gravity; SPF, Spruce–Pine–Fir.
[b]P is loading parallel to grain; R, loading perpendicular to grain.

A least-square fit of a power function resulted in the following expression:

$$F_e = 91.15 G_{12}^{1.39} \tag{2}$$

where F_e is expressed in N/mm and G_{12} is specific gravity based on volume before drying. Figure 1 shows experimental results of the relationship of dowel bearing strength, both parallel and perpendicular to grain, to specific gravity along with Equations (1) and (2). Over the entire specific gravity range, parallel-to-grain strength values were higher than perpendicular-to-grain values, as confirmed by statistical comparison for each species. The fitted expression predicts the middle response of the data, with a range of percent deviation (PD) between the fitted expression and experimental values of 57.7% to −48.5% and an average PD of −0.52. For Wilkinson's expression, the PD range was 88.8% to −51.1% with an average of −2.00. Based on PD and analysis of the actual residuals, Equation (1) overpredicts dowel bearing strength for higher specific gravity values and slightly underpredicts at the lower extreme. These results were not compared to Whale and Smith's expression because of the difference in the definition of failure load between the two test procedures.

Equations (1) and (2) are equivalent when specific gravity is 0.48, but for values lower than 0.39 and higher than 0.6, the expressions differ by more than 10% (Fig. 1). For typical wood species for construction, the difference between best fit and current practice is less than 10%.

Conclusions

The results indicate that grain orientation significantly affects dowel bearing strength in wood with specific gravity of less than 0.7. The current nail dowel bearing design expression is 10% different from the best-fit power function for specific gravity less than 0.4 and greater than 0.6. This design expression overpredicts experimental values by 10% or greater for specific gravity greater than 0.6.

References

Fahlbusch, H. (1949). A contribution to the problem of the bearing strength of bolts in wood under static loads. Report No. 49–09, Institute for Mechanical Construction and Carpentry, Technical School Braunschweig.

Koponen, S. (1991). Embedding characteristics of wood in the grain direction. Report 25, Helsinki University of Technology, Laboratory of Structural Engineering and Building Physics. Espoo, Finland.

National design specification for wood construction. (1997). ANSI/NFPA NDS 1997. American Forest & Paper Association, Washington, DC.

Rammer, D.R. Dowel bearing strength of danto and ramon hardwoods. Submitted to Forest Products Journal.

"Standard for load and resistance factor design for engineered wood construction." (1996). *AF&PA/ASCE –16–95*, ASCE, New York, NY.

"Standard test method for evaluating dowel-bearing strength of wood and wood-base materials." (1998). *ASTM D5764–95*, ASTM, West Conshohocken, PA.

Whale, L.R.J. and Smith, I. (1986). The derivation of design classes for nailed and bolted joints in EUROCODE 5. International Council for Building Research Studies and Documentation, Working Commission W18, Timber Structures, Florence, Italy.

Wilkinson, T.L. (1991). Dowel bearing strength. Res. Pap. FPL–RP–505. USDA Forest Service, Forest Products Laboratory, Madison, WI.

Winistorfer, S.G. Effect of moisture content on dowel-bearing strength. Forest Products Journal (in press).

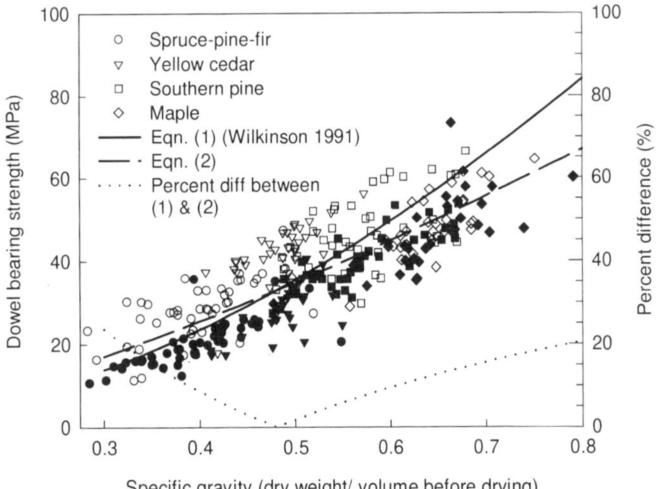

Figure 1. Dowel bearing strength versus specific gravity for 4.11-mm nails (Open symbols = parallel-to-grain strength, filled symbols = perpendicular-to-grain strength.)

The *International Building Code* and *International Residential Code* and
Their Impact on Wood-Frame Design and Construction

David P. Tyree[1], P.E., C.B.O.
Dennis L. Pitts[2]

ABSTRACT

This paper presents the evolutionary history of development of a new single set of uniform
building codes in the United States, and outlines the impact of these new provisions on wood
design and construction. The first *International Building Code* (IBC) and *International
Residential Code* (IRC) are scheduled for publication in the year 2000. Both are being developed
by the International Codes Council (ICC), a new entity formed by the three U.S. model building
code agencies. The IBC and IRC will contain revisions to existing code requirements and new
provisions that will impact wood design and construction. Possible benefits of the new codes will
also be discussed.

INTRODUCTION

The purpose of a building code is to serve as a regulatory document aimed at protecting the
public's life, health, and welfare in the built environment. In the U.S. this is usually
accomplished by the adoption of a model set of codes through state statute or local government
ordinance with enforcement being provided by state agencies and local governments having
specific jurisdiction. Model building codes are most frequently used because cities and counties
lack individual resources to write and maintain a comprehensive regulatory document. Model
building codes represent a partnership of cities, counties, states, industries, laboratories,
educational and research institutions.

BACKGROUND

In the early 1900s, it became clear that better regulation of the built environment was required in
the U.S. Code practitioners banded together to promote the concept of professionalism and to
promulgate 'model' building codes. Three groups came to preeminence: the Building Officials
and Code Administrators (BOCA), International was founded in 1915; the International
Conference of Building Officials (ICBO) was founded in 1922; and the Southern Building Code

---

[1]Southwest Regional Manager, American Wood Council, American Forest & Paper
Association, 1111 19th St., NW, #800, Washington, D.C., USA

[2]South Central Regional Manager, American Wood Council, American Forest & Paper
Association, 1111 19th St., NW, #800, Washington, D.C., USA

Congress, International (SBCCI) was founded in 1940. Because these model codes reflected the characteristics of construction and the environmental conditions that were prevalent in their region, they differed in their format, content, and appearance, despite the fact that they imposed very similar regulations on many construction aspects, Francis and Stone, 1998.

In 1972 the Council of American Building Officials (CABO) was created by the three model code groups as an organization that would, among other things, provide the means to address matters of common concern among BOCA, ICBO, and SBCCI. CABO, in turn, established the Board for the Coordination of the Model Codes (BCMC) which was intended to identify conflicts between the three model codes and recommend revisions to address those conflicts.

One of the early joint code-writing efforts by the three code groups occurred under the CABO umbrella. It was the development of the *One- & Two-Family Dwelling Code* (OTFDC). The OTFDC is a specification code which is applicable to one- and two-family dwellings and townhouses, as defined in the code, and addresses not only the construction of the structural elements of the building, but also the plumbing, mechanical, electrical, and fuel gas systems. Although all of these elements are addressed in the larger building codes and their related codes, the OTFDC provides a single source of primarily prescriptive requirements and contains little performance language. Although it is recognized by all three code organizations, the OTFDC is not a mandatory document unless specifically adopted by a jurisdiction.

Around the time that BOCA, ICBO, and SBCCI reorganized their code books to reflect the BCMC format recommendations, the three model code groups formed a new umbrella organization called the International Code Council (ICC). Subsequently, the code groups rolled CABO into the ICC. ICC was incorporated as a not-for-profit corporation in 1994, dedicated to developing a single set of comprehensive and coordinated. This bold and ambitious program called for a new set of model codes to be completely developed and published by the first quarter of the year 2000, in time for the new millennium.

The ICC code development process quickly dispatched several elements of its family of codes. The first to be promulgated was the *International Mechanical Code* (IMC) in 1995. Quickly following was the promulgation of the *International Plumbing Code* (IPC), the *International Private Sewage Disposal Code* (IPSDC), and an *International Zoning Code* (IZC). Finally, in late 1996, Building Code Development Committees were established by ICC with the task of melding three distinctly different regional building codes into a single model code.

INTERNATIONAL BUILDING CODE
The task of developing a single model code was broken up into five technical code development committees: Structural, Occupancies, Fire Safety, Means of Egress, and General. These code development committees were formed and tasked with developing model code provisions from materials already available on each subject within the existing model codes, BCMC reports, or in rare circumstances, the work of other groups such as the Building Seismic Safety Council (BSSC). Each of the three model code groups sent a three-member delegation and one staff member to each of the committees. The time schedule set forth by the ICC Steering Committee resulted in the committees meeting between six and ten times to develop their respective draft chapters. These were then consolidated into an initial working draft of a new *International Building Code* (IBC).

As is typical in legislative processes, each committee developed its own personality and thus its own philosophy during developmental activity. The Structural Committee expressed an intention to incorporate the most current and generally accepted engineering standards for conditions such as environmental loads (wind, seismic and snow), new maps for wind and snow, and the latest in design methodologies for various materials. The Occupancy committee decided to develop use groups and height and area limits which would embrace all of the limits then existing in the various codes. The Fire Safety Committee wanted to utilize the "most restrictive" provisions of the model codes. The Egress and the General Committees developed similar independent personalities.

Structural Committee
The result of the Structural Committee's work is embodied in various chapters including: Chapter 16, "General Design Requirements;" Chapter 17, "Structural Tests and Inspections;" Chapter 18, "Soils and Foundations;" and Chapters 19 through 23 which address the particular building materials, i.e., concrete, aluminum, masonry, steel, and wood. Early on, the committee turned to the various materials groups, such as the American Forest & Paper Association (AF&PA), for guidance and input in development of related provisions of the code. The Committee also worked closely with BSSC on new seismic provisions and with engineering associations such as the Structural Engineers Association of California (SEAOC) and other groups to develop the loads contained in Chapter 16.

Prescriptive wood construction provisions, known as "conventional construction," were included in Chapter 23. Requirements of the conventional construction section were taken primarily from the UBC. The NBC contained a limited range of prescriptive material, and while the SBC conventional construction requirements were similar to those in the UBC, the committee considered the UBC version to be more complete. Various amendments were made to the UBC provisions to address issues which were a concern in portions of the country which did not use the UBC. Since the UBC requirements address issues relating to resistance of seismic forces, most revisions were made to address wind-related issues not adequately addressed in the UBC.

As in current editions of the three model codes, Chapter 23 of the *International Building Code* governs materials, design, construction, and quality of wood members and their fasteners. Aside from prescriptive conventional construction provisions, Chapter 23 is performance-based and relies in large part on references to the design standards of AF&PA: the *National Design Specification® (NDS®) for Wood Construction, Load and Resistance Factor Design (LRFD) for Engineered Wood Construction*, and *Wood Frame Construction Manual (WFCM) for One- and Two-Family Dwellings, SBC High Wind Edition*.

The format of the chapter is different from that of any of the three existing codes. The new format was suggested by AF&PA, and it placed the requirements into a more logical and user-friendly arrangement. Chapter 23 is now divided into sections that address definitions of terms particular to the chapter, minimum quality standards, general requirements applicable throughout the chapter, lateral force resisting systems, allowable stress design, load and resistance factor design, and conventional construction.

The section on conventional construction defines those situations in which prescriptive requirements of the section apply. It also contains requirements for lateral bracing and continous load paths, similar to what exists in the current UBC. Engineered precalculated span tables for joists and rafters and for girders and headers have been included in the conventional construction section. Prescriptive requirements for high-seismic areas, based on the provisions of the National Earthquake Hazards Reduction Program (NEHRP) and recommended by the Code Resources Development Committee (for BSSC), have been included.

Occupancies Committee
The Occupancies Committee had what some observers believed to be the most difficult task of the group. They had to find a way to meld widely divergent use group categories and extremely dissimilar height and area (H&A) limits into a single set of requirements and definitions. The H&A limits were generally regarded as the single biggest stumbling block to the whole IBC process.

The Committee consciously decided to take the approach that the H&A table ought to allow construction of any building currently permitted by any one of the three model codes. Essentially, the philosophy was for the H&A table to reflect the least restrictive provisions of the three model codes. This is a very important concept and a fairly innovative approach to codes, where the usual reaction is to use the most restrictive provision. It means that for any given occupancy, the IBC H&A limit would be almost identical to one of the existing codes. But that also means that the H&A table would permit larger buildings than presently permitted by one or two of the current model codes! In some cases, the increase from that which is currently permitted is quite significant.

Another effort by the Occupancies Committee involved combining definitions of types of construction from the three model codes and creating definitions of use groups and types of occupancies from the model codes. With few exceptions, definitions include an expanded role permitted for wood or wood products. For example, Type III construction, commonly referred to as ordinary construction with its non-combustible walls and combustible floor and roof/ceiling assemblies, allows use of fire retardant treated wood for exterior walls, notwithstanding the normal requirement for noncombustible construction.

INTERNATIONAL RESIDENTIAL CODE
In 1996, the ICC Board of Directors established the ICC/NAHB Task Force. This group, composed of ICC members and members of the National Association of Home Builders, was created to review the OTFDC and determine its place in the ICC family of codes. The task force was also charged with identifying other issues that would affect use of the code under the ICC, including whether the CABO process of updating the OTFDC should be retained. The task force recommended that a new stand-alone residential code be developed, called the *International Residential Code.*

ICC also decided that use of the IRC would be mandatory, unlike the OTFDC. All material related only to one- and two-family dwellings and to townhouses, again, as defined in the IRC, will be deleted from the IBC and its related codes. Code provisions within that scope will only be found in the IRC.

A drafting committee was formed by ICC which contained one code official from each of the three code groups, three home builders, and three industry individuals nominated by the ICC Industry Advisory Committee. The industry members of the original drafting committee were all architects.

The IRC drafting committee conducted its first meeting in September of 1997 and met monthly through April of 1998, at which time a draft of the IRC was published for public comment. Although the 1995 edition of the OTFDC was used as a basis, the committee made substantial revisions to that document. An administrative section was added, as was a section stipulating the structural basis for the code. In many instances the technical content of the OTFDC was updated and expanded to be more complete. New provisions were taken from the IBC as well as the existing model codes, although relevant outside technical material was also considered.

The existing format of the OTFDC -- chapters based on building elements rather than on materials, as is done in the IBC -- was retained. The existing OTFDC requirements relating to wood-frame construction were retained, but updated.

The requirements in the IRC draft for high load areas are somewhat of an odd mixture. There are prescriptive requirements for light-gauge steel framing in both high-wind and high-seismic areas of the country. There are limited high-seismic requirements for masonry and concrete, and for construction in high-wind areas these buildings must either be designed to comply with the latest version of ASCE 7 or must comply with prescriptive requirements of SBCCI's SSTD 10, *Standard for Hurricane Resistant Residential Construction.* Wood frame construction has prescriptive provisions for high-seismic areas. Buildings in high-wind areas must comply with SSTD 10, AF&PA's *Wood Frame Construction Manual for One- & Two-Family Dwellings (High Wind Edition),* or be designed in accordance with ASCE 7.

The seismic portions of the IRC draft reflect the methodology and terminology used in the IBC, but the material is somewhat simplified. The provisions of the IRC draft apply to Seismic Design Categories A, B, C, and D. A major undertaking was the inclusion of new seismic provisions from NEHRP, including new seismic maps. These proposed revisions from the BSSC Code Resource Development Committee corrected the seismic risk maps and brought them into compliance with the 1997 NEHRP provisions.

ICC CODE DEVELOPMENT PROCESS
Like the democratic processes of the three model code organizations, any interested party may submit a code change proposal to ICC and participate in the proceedings in which it and all other such proposals are considered. This open debate and broad participation before a committee comprising representatives from across the building regulatory industry, including code regulators and construction industry representatives, ensures a fair hearing in the decision-making process. A new feature of ICC's code development process is that it will allow both the ICC code development committees and eligible voting members (code enforcement professionals) at code change hearings to participate in establishing the results of each proposal. Voting members may either ratify a committee's recommendation or make their own recommendation, with the members' recommendation being treated as a challenge to the action of the committee. The results of all votes will be published in the report of the ICC code development hearings.

Eligible voting members of the three model code groups will review recommendations of the respective ICC code development committee at their annual conference and determine final action. Following consideration of all public comments, each proposal will be individually balloted by the eligible voters. This process is intended to ensure that the International Codes reflect the latest technical advances and address concerns of those throughout industry in a fair and equitable manner.

CONCLUSION

The development of the new International Codes is part of an evolutionary process to improve protection of the public's life, health and welfare in the built environment. Although, it is doubtful that a single set of codes will successfully discourage states and localities from attempting to amend the International Codes or adopting the International Codes with technical amendments, it is hoped that uniform adoption will lead to consistent code enforcement and higher quality construction. By combining the efforts of the three model code organizations to produce a single set of codes, architects, engineers, designers and contractors will be better able to expand and market their services to broader geographical areas, rather than being limited to a small region. Manufacturers of building products will also be better able to place their efforts into research and development rather than designing products to meet differing sets of regional standards. This, in turn, should enable manufacturers to improve their competitiveness in worldwide markets. Further, uniform education and certification programs should benefit the mobility of code enforcement professionals to market their skills nationally, and uniform adoption should improve mutual aid disaster relief efforts of code enforcement professionals.

REFERENCES

American Forest & Paper Association (1996), *Load and Resistance Factor Design (LRFD) Manual for Engineered Wood Construction*, Washington, DC.

American Forest & Paper Association (1997), *National Design Specification (NDS) for Wood Construction*, Washington, DC.

American Forest & Paper Association (1996), *Wood Frame Construction Manual (WFCM) for One- and Two-Family Dwellings, SBC High Wind Edition*, Washington, DC.

Building Officials and Code Administrators, International (1996), *National Building Code*, 1996, Country Club Hills, IL.

Francis, S.W, and Stone, J.B., *The* International Building Code *and Its Impact on Wood-Frame Design and Construction*, Presented at the July 12-16, 1998 ASAE Annual International Meeting. Paper No.984007, ASAE, 2950 Niles Rd., St. Joseph, MI 49085-9659 USA.

International Codes Council (1998), *International Building Code, Final Draft*, 1998, Falls Church, VA.

International Codes Council (1996), *International Mechanical Code*, 1996, Falls Church, VA.

International Codes Council (1997), *International Plumbing Code*, 1997, Falls Church, VA.

International Codes Council (1997), *International Private Sewage Disposal Code*, 1997, Falls Church, VA.

International Codes Council (1998), *International Residential Code, Second Draft*, 1998, Falls Church, VA.

International Codes Council (1998), *International Zoning Code,* 1998, Falls Church, VA.

International Codes Council (1995), *One- and Two-Family Dwelling Code,* 1995, Falls Church, VA.

International Conference of Building Officials (1997), *Uniform Building Code*, 1997, Whittier, CA.

Southern Building Code Congress, International (1997), *Standard Building Code*, 1997, Birmingham, AL.

Southern Building Code Congress, International (1997), *Standard for Hurricane Resistant Residential Construction, SSTD-10*, 1997, Birmingham, AL.

**Changes and Future Trends in the *NDS® for Wood Construction*
and *LRFD* Specifications**

John "Buddy" Showalter, P.E.[1]

ABSTRACT

This paper explores the latest changes to both the *National Design Specification®
(NDS) for Wood Construction* and the *Load and Resistance Factor Design (LRFD)
Manual for Engineered Wood Construction*. Also included is an examination of
allowable stress design (ASD) as provided in the NDS compared to LRFD. Finally, a
discussion of future trends for both methodologies is provided.

INTRODUCTION

Traditionally in the United States, allowable stress design (ASD) for most
structural members is based on the *National Design Specification (NDS) for Wood
Construction*, AF&PA (1997). The 1997 edition of the NDS incorporates the latest state-
of-the-art design information for wood products. In structural engineering, however, there
is a growing trend to replace existing ASD methodologies with load and resistance factor
design (LRFD) procedures, Pollock, *et al.* (1994). The wood industry responded to this
emerging preference by publishing the *Load and Resistance Factor Design (LRFD)
Manual for Engineered Wood Construction*, AF&PA (1996). To support practicing
designers, the wood industry will maintain both methodologies for the foreseeable future.
Future changes to wood design will be developed in both formats.

NDS for Wood Construction

The *1997 Edition of the National Design Specification (NDS) for Wood
Construction* was approved as an American National Standard on August 7, 1997 with
the designation ANSI/AF&PA NDS-1997.

The *NDS Supplement: Design Values for Wood Construction*, AF&PA (1997), an

---

[1]Director, Technology Transfer, American Wood Council, American Forest & Paper
Association, 1111 19th St., NW, #800, Washington, D.C., USA

integral part of the NDS, has also been updated to provide the latest design values for lumber and glued laminated timber.

Highlights of changes to the NDS are as follows:

* new provisions for timber rivets;
* new provisions for wood-to-concrete connections;
* new provisions for designing notches on compression face of bending member;
* new incising adjustment factors;
* simplified wet service factors for connectors;
* new combined lateral/withdrawal equation for nails;
* modification of nail clinching provisions;
* clarification of built-up column provisions;
* revised upper limit on load duration factors for pressure-preservative and fire retardant treated wood;
* update of lumber and glued laminated timber design values (*NDS Supplement*);
* permissive language replaced with mandatory language.

More details regarding these changes are found in Showalter and Line, 1998.

In 1998, AF&PA also began development of an *Allowable Stress Design (ASD) Manual for Engineered Wood Construction*, AF&PA (1999). The *ASD Manual* will parallel the format of AF&PA's *LRFD Manual*, building on the industry-wide effort to bring together all design information in single source documents for the benefit of wood specifiers and designers.

The package will incorporate the following design information:

Allowable Stress Design Manual for Engineered Wood Construction
National Design Specification (NDS) for Wood Construction, 1997 Edition
NDS Supplement, Design Values for Wood Construction
NDS Commentary
Structural Lumber Supplement
Structural Glued Laminated Timber Supplement
Structural Panels Supplement
Timber Poles and Piles Supplement
Structural Connections Supplement
Wood I-Joist Guideline
Structural Composite Lumber Guideline
Metal Plate Connected Wood Truss Guideline
Pre-engineered Metal Connector Guideline

Publication is anticipated in the second quarter of 1999.

LRFD for Engineered Wood Construction

An integral part of the *LRFD Manual* is the *AF&PA/ASCE 16-95 Standard for Load and Resistance Factor Design (LRFD) for Engineered Wood Construction,* which was approved in November 1995 as a consensus standard through the ASCE standards development process. It is currently approved as a recognized standard by the ICBO *Uniform Building Code.* It was adopted as a reference standard by SBCCI's *Standard Building Code* and BOCA's *National Building Code.* It is also included as a reference standard in the International Code Council's proposed *International Building Code.*

The wood industry chose to adopt two stages out of a three-stage process to transition from deterministic (ASD) design to a reliability (LRFD) basis. LRFD for wood adopts code accepted load factors (stage 1) that smooth reliability across load cases. It also adopts the concept of a target reliability index (stage 2) for a reference design case and adjusts designs to match that target. The final stage (stage 3), that permits additional adjustment of designs based on individual data set analysis, is not currently being employed (even though it is permitted in the underlying standards). The wood industry does not believe that there is sufficient consensus regarding techniques for adopting this level of refinement in a uniform and equitable manner across construction materials.

The initial audience for LRFD primarily will be academia. Marx (1996) indicated that,

> ...most engineering colleges today teach students LRFD for designing with steel and concrete. And, with LRFD, much of the design process is actually material-independent. This means that students who learn LRFD already understand most of the concepts needed to design wood structures using LRFD.

The wood industry also recognizes that there are a great number of designers unfamiliar with the various nuances of wood design. LRFD is becoming a universally recognized structural design methodology. This should enable the designer coming into wood design for the first time to pick up quickly on the requirements, thus reducing the learning curve and reluctance to "learn a new material."

There are a number of benefits related to changing to Reliability Based Design (RBD) that include: 1) Applying partial factors of safety to the load and strength terms, making the sources of design uncertainty more apparent and better accounting for variability; 2) LRFD has narrowed the range of reliabilities in practice and, as a result, economies in design can be realized without sacrificing safety and performance; 3) Finally, LRFD is easier to teach, Gromala (1996).

LRFD vs. ASD

Case studies comparing designs using the LRFD methodology to those using ASD procedures as prescribed in the *NDS* indicate similar results for beams and connections with live to dead load ratios of 3:1. Showalter, et al (1998) indicated that a 15% smaller cross-section was achieved for certain column design using LRFD versus ASD, due primarily to load factoring for combinations of wind, snow, and crane loads on the column. Additional calculations indicate as much as 30% smaller cross sections for structural members carrying multiple transient loads (roof live and occupancy) - again due to load factoring.

Based on input from users, coupled with observation of prior experience of the steel and concrete industries, ASD provisions of the *NDS for Wood Construction* will be supported for the foreseeable future. However, the worldwide trend toward LRFD is driven by factors including its subtle "smoothing" of reliability across a range of designs and improvements in "teachability" of structural design across a range of materials. While differences in design results using LRFD versus ASD are not exceptionally large, they are sufficient to provide significant economic benefits for some projects.

Future Trends

As noted, ASD and LRFD will co-exist for the foreseeable future. Therefore, new research and changes to structural design will need to consider both methodologies. The following topics represent current thinking about possible changes to design in the near future.

Energy-Based Seismic Design

Current activity in the building code arena shows a growing need to characterize wood members and connections with objective, performance-based means rather than subjective, prescriptive-based methods. To that end, AF&PA has awarded a grant to Washington State University for a research proposal entitled "Energy Approach for the Seismic Design of Wood Structures." It is hoped that results of this research can be used to counter subjective limits in codes and standards for high seismic areas of the country. It should also provide a more rational approach for future development of seismic design provisions for wood products.

Design criteria for wood structural components, including shearwalls and connections, are increasingly limited by anecdotal damage observations rather than engineering criteria. Examples of this include the Federal Emergency Management Agency's (FEMA) National Earthquake Hazard Reduction Program (NEHRP) *Recommended Provisions for Seismic Regulations of New Buildings and Other* Structures, ICBO's *Uniform Building Code*, and the proposed *International Building Code*.

Building codes are also shifting from a narrow "minimum life-safety" performance level toward a seismic design philosophy in which multiple performance levels are considered (i.e., performance-based seismic design). This new design approach will allow better characterization of various performance levels.

Connection Design

Connection ultimate capacity design is also emerging as a primary point of concentration for both ASD and LRFD enhancements. The yield limit equations specified in the *NDS* and *LRFD* for bolt, lag screw, wood screw, nail, spike and drift pin connections represent a mechanics-based approach for connection design. This approach, which was originally incorporated in the *NDS for Wood Construction* in 1991, and LRFD in 1995, permits the designer to determine effects of member thickness, member strength, fastener size, and fastener strength on lateral connection values for the majority of connections found in wood construction. A report developed by AF&PA entitled *Calculation of Lateral Connection Values Using General Dowel Equations*, AF&PA 1999, covers calculation of lateral values for single dowel type fastener connections using a generalized and expanded form of the yield limit equations. These general dowel equations apply to NDS/LRFD connection conditions, but also permit rational and consistent treatment of gaps and fastener moment resistance, and consideration of various connection limit states. Nominal lateral connection values at various limit states such as proportional limit, 5% offset, and ultimate (maximum) load can be determined.

Shearwall Design
Wood-frame shearwalls are traditionally designed using full-height shearwall segments restrained against overturning by structural weight and/or tension anchorage. Traditional methods require shearwalls that contain openings to be considered as multiple shearwall segments, increasing the requirements for tension anchorage within the wall. The perforated shearwall design method, described in the *Wood Frame Construction Manual for One- and Two-Family Dwellings*, AF&PA (1995), and published in the *Standard Building Code*, reduces the capacity of shearwalls with openings when intermediate tension anchorage is omitted. While giving designers an additional design method and analytical tool, it also provides verification of conventional construction techniques, allows for fewer uplift restraints, and more efficient use of construction materials.

AF&PA and APA-The Engineered Wood Association have also performed additional shearwall tests to verify a mechanics-based design procedure to account for the combined uplift and shear resistance of walls sheathed with structural sheathing. This proposed procedure would allow designers to replace labor-intensive uplift connectors with exterior structural sheathing.

CONCLUSION
ASD and LRFD methodologies will coexist for the foreseeable future. Current research to incorporate new design criteria will need to consider both. Research in the areas of energy-based seismic design, ultimate connection capacities, and mechanics-based shearwall design may eventually be incorporated in the *NDS for Wood Construction* and *LRFD Standard for Engineered Wood Construction*. ASD is currently the predominant methodology used for structural wood design, with LRFD as an alternate approach. Eventually, it is expected that these roles will reverse.

REFERENCES
American Forest & Paper Association, *Allowable Stress Design (ASD) Manual for*

Engineered Wood Construction (1999 publication pending), Washington, DC.

American Forest & Paper Association, *Commentary on the National Design Specification (NDS) for Wood Construction*, 1999 Edition, Washington, DC.

American Forest & Paper Association, *Calculation of Lateral Connection Values Using General Dowel Equation, Technical Report 12*, 1999, Washington, DC.

American Forest & Paper Association, *Load and Resistance Factor Design (LRFD) Manual for Engineered Wood Construction*, 1996, Washington, DC.

American Forest & Paper Association, *National Design Specification (NDS) for Wood Construction*, 1997 Edition, Washington, DC.

American Forest & Paper Association, *NDS Supplement, Design Values for Wood Construction*, 1997 Edition, Washington, DC.

American Forest & Paper Association, *Wood Frame Construction Manual (WFCM) for One- and Two-Family Dwellings, SBC High Wind Edition*, 1995 Edition, Washington, DC.

Building Officials and Code Administrators, International, *National Building Code*, 1996, Country Club Hills, IL.

Gromala, D.S., *Implementation of Load and Resistance Factor Design in the United States*, 1996 International Wood Engineering Conference Proceedings, Vol. 3, p. 45

International Codes Council, *International Building Code, Final Draft*, 1998, Falls Church, VA.

International Conference of Building Officials, *Uniform Building Code*, 1997, Whittier, CA.

Marx, C. M., *Wood Engineering Education in the U.S.*, Proceedings of the International Wood Engineering Conference, October 28-31, 1996, Vol. 3, 3-10.

Pollock, D.G. and Williamson, T.G., LRFD vs. ASD for Wood Structures, *Spatial, Lattice and Tension Structures*, Proceedings of the IASS-ASCE International Symposium, 1994, 76-85.

Showalter, J.H., and Line, P., *1997 NDS Provisions*, Frame Building News Magazine, January 1998, pp.66-69.

Showalter, J.H., Manbeck, H.B., and Pollock, D.G., *LRFD versus ASD for Wood Design*, Presented at the July 12-16, 1998 ASAE Annual International Meeting. Paper No.984006, ASAE, 2950 Niles Rd., St. Joseph, MI 49085-9659 USA.

Southern Building Code Congress, International, *Standard Building Code*, 1997, Birmingham, AL.

Stress Grading of Recycled Lumber and Timber

Robert H. Falk and David Green[1]

Abstract

This paper presents an overview of selected research at the Forest Products Laboratory (FPL) to characterize the grade distribution and engineering properties of lumber and timber recycled from deconstructed buildings on U.S. Army installations. The effects of splits on timber beam and column strength and the effects of damage on lumber grade yield are reported here.

Introduction

For decades, the preferred method of disposal for buildings has been to demolish them mechanically and place the debris in a landfill, saving little material for reuse. Over the last decade, the demand for old timbers has grown significantly, making it worthwhile to salvage this material from buildings slated for disposal. This is particularly the case for large old-growth softwood timbers, which can be resawn into structural members or millwork. While larger timbers command a high price and are regularly recycled, dimensional lumber is not often reused. However, recent studies suggest the feasibility of deconstructing buildings and salvaging and reusing the dimensional lumber stock (NAHB 1997, FORA 1997). Ongoing research at FPL is characterizing the grade distributions and engineering properties of lumber and timber recycled from deconstructed buildings (Falk et al. 1998a,b,c; Green et al. 1998). To date, more than 1,700 pieces of lumber and timber have been collected from the U.S. Army's Twin Cities Army Ammunition Plant (TCAAP) in Minnesota and Fort Ord in California. At TCAAP alone, this cooperative program has resulted in recycling of more than 4,700 m^3 (2 million board feet) of lumber and timber (Falk et. al 1995, Lantz and Falk 1996). This paper highlights the results of testing timbers in bending and columns in compression and grading dimension lumber.

---

[1]Research Engineers, USDA Forest Service, Forest Products Laboratory, One Gifford Pinchot Drive, Madison, WI 53705–2398. The Forest Products Laboratory is maintained in cooperation with the University of Wisconsin. This article was written and prepared by U.S. Government employees on official time, and it is therefore in the public domain and not subject to copyright.

Materials and Methods

Material from both military bases ranges from nominal 2 by 4 in. (standard 38 by 89 mm) to nominal 10 by 18 in. (standard 241 by 445 mm). A sample of lumber and timber was collected from a 59,000-m^2 (548,000-ft^2) dismantled building at TAACP in 1995. The FPL research staff and U.S. Army facility engineers and demolition contractors selected approximately 82.6 m^3 (35,000 board feet) of lumber and timber for testing (Falk et al. 1998b,c; Green et al. 1998).

The 1994 closure of the Fort Ord U.S. Army Military Reservation in Marina, California left more than 1,200 buildings that either did not meet current building code requirements or contained remnant hazardous materials requiring abatement. The Fort Ord Reuse Authority (FORA) developed a deconstruction project focused on distinct building types and monitored the cost, timing, and work involved in building disassembly and material collection and reuse (FORA 1997). The FPL developed a cooperative research agreement with FORA and the West Coast Lumber Inspection Bureau (WCLIB) to develop information on the grades of lumber reclaimed from the deconstructed buildings.

The lumber and timber collected from both military bases was primarily Douglas Fir and was visually assessed for structural grade by a WCLIB grading supervisor according to standard no. 17 in *Grading Rules for West Coast Lumber* (WCLIB 1996). Particular attention was paid to damage, defined as holes resulting from nails or bolts, splits caused by factors other than drying, saw cuts, notches, decay, and mechanical damage (such as gouges, broken ends, and missing sections resulting from splits). If a bolt and/or nail hole or holes were present in the piece, the grader estimated an equivalent knot size for determining the grade.

Results

Ninety nominal 6- by 8-in. (standard 140- by 191-mm) timbers were collected from TCAAP and shipped to FPL for testing (Green et al. 1998). Thirty timbers with heart checks (boxheart splits), characteristic of old timbers installed in dry locations, and 60 "unchecked" timbers were selected for testing. Most beams were Select Structural Beams and Stringers grade by current grading rules. Bending tests were performed according to ASTM D198 methods (ASTM 1996). Analyses of bending strength data indicated that the mean modulus of rupture of beams with heart checks was about 15% lower than that of beams without heart checks.

Nominal 8- by 8-in. (standard 191- by 191-mm) Douglas Fir columns were collected at TCAAP and sent to FPL for grading and testing (Falk et al. 1998b). Columns were tested in direct compression with no intermediate lateral support (ASTM 1996). The ends were laterally supported to prevent slippage, although no attempt was made to stabilize them. An inspection of the building indicated that the timber had been installed green and many members had developed significant drying checks and/or splits. In spite of being in service for 55 years and containing many in-service defects, 75% of columns were graded as No. 2 or higher and 40% as

Select Structural. In-service defects, such as checks, splits, and mechanical damage, resulted in downgrading of approximately one-third of the columns. To study the effect of defects on column strength, "checked" and "unchecked" members were selected on site. Checks had little effect on column compressive strength (Fig. 1). All columns were found to be higher in strength than expected by current design procedures (Fig. 2).

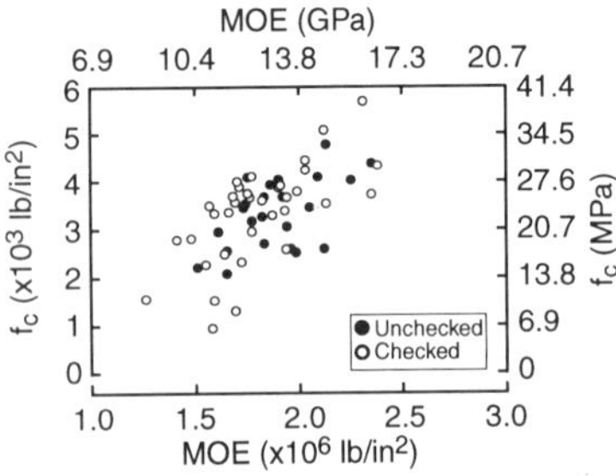

Figure 1. Effect of checks on strength of 8 by 8 columns.

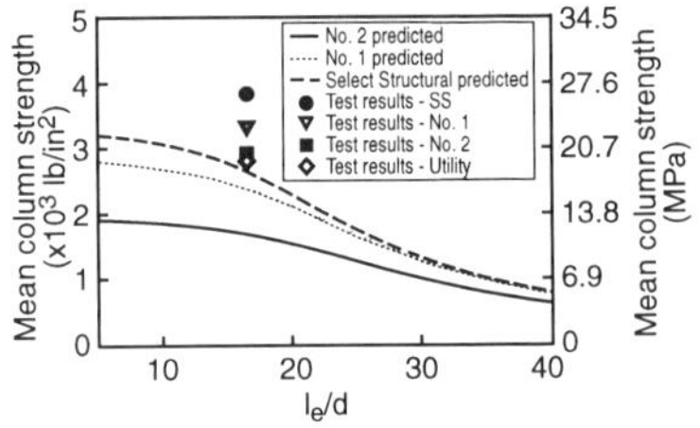

Figure 2. Comparison of actual and predicted strength of columns.

More than 900 pieces of 2 by 4, 2 by 6, 2 by 8, and 2 by 10 lumber (standard 38 by 89, 38 by 140, 38 by 191, and 38 by 235 mm lumber) were collected from four deconstructed buildings at Fort Ord (Falk et al. 1998a). Most pieces graded as Structural Joists and Planks qualified for No. 2 grade (47%); most 2 by 4 pieces were graded as Standard (68%). As expected, Douglas Fir was the predominate species group (92%), although Hem–Fir (6%) and sugar pine (2%) were also present. From the standpoint of structural use, the most distinguishing feature of the recycled lumber compared to freshly sawn lumber was the presence of damage, which may have been a result of the original construction process (for example, nail holes, bolt holes, saw cuts, notches), building use (drying defects, decay, termite damage), and/or the deconstruction process (edge damage, end splitting, gouges). Damage reduced the average grade of the lumber (Fig. 3).

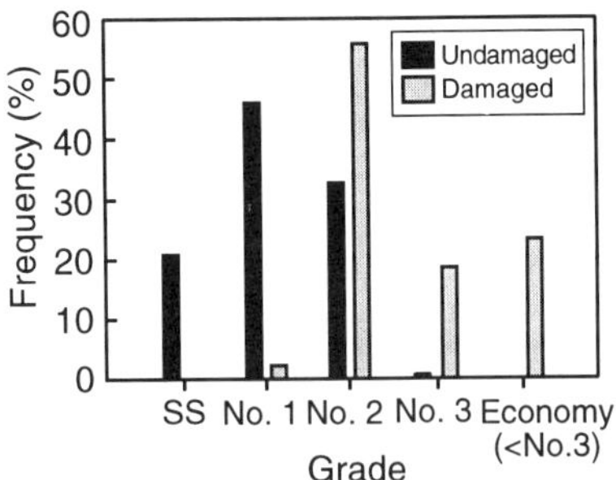

Figure 3. Effect of damage on grade of 2 by 8 lumber. n = 504.

Conclusions

Results to date indicate that heart checks lower the modulus of rupture in recycled timber beams but have little effect on the strength of recycled timber columns. The quality of dimensional lumber from the reconstructed buildings is on average one grade lower than that of freshly sawn lumber as a result of damage incurred during deconstruction. Because the value of lumber is tied directly to its quality, evaluation of the grades of lumber from these buildings will help determine reuse options and market value.

References

ASTM (1996) Standard methods of static tests of lumber in structural sizes. Annual Book of Standards, vol. 04.10, ASTM Standard D198, American Society for Testing and Materials.

Falk, R.H., DeVisser, D., Cook, S., and Stansbury, D. (1998a) Military building deconstruction: lumber grade yield from recycling. Submitted to Forest Prod. J.

Falk, R.H., Green, D.W., Rammer, D, and Lantz, S.F. (1998b) Engineering evaluation of 55 year old 8×8 timber columns recycled from an industrial military building. Submitted to Forest Prod. J.

Falk, R.H., Green, D.W., and Lantz, S.F. (1998c) An evaluation of lumber recycled from an industrial military building. Submitted to Forest Prod. J.

Falk, R.H., Green, D.W., Lantz, S.F., and Fix, M.R. (1995) Recycled lumber and timbers. In: Proceedings, 1995 ASCE Structures Congress XIII, vol. I, April 2–5, Boston, MA: 1065–1068.

FORA (1997) Pilot deconstruction project. Final Report, Fort Ord Reuse Authority, Marina, CA.

Green, D.W., Falk, R.H., and Lantz, S.F. (1998) Effect of heart checks on the flexural properties of recycled Douglas-fir 6×8 timbers. Submitted to Forest Prod. J.

Lantz, S.F. and Falk, R.H. (1996) Feasibility of recycling timbers from military industrial buildings. In: Proceedings, Conference on use of recycled wood and paper in building applications, September 9–11, Forest Products Society, Madison, WI: 41–48.

NAHB (1997) Deconstruction: building disassembly and material salvage. The Riverdale case study. NAHB Research Center, National Association of HomeBuilders, Upper Marlboro, MD.

WCLIB (1996) Standard no. 17, Grading rules for West Coast lumber. Revised January 1, 1996. West Coast Lumber Inspection Bureau, Portland, OR.

Analysis of Wood-Peg Connected Timber Frames

By William M. Bulleit[1] and Matthew W. Drewek[2]

Abstract

Wood-peg connected timber frames require the use of an analysis procedure that accounts for the connection behavior. The peg stiffness can be modeled using short elements whose axial stiffness is determined assuming that the peg acts as a simply supported beam with a concentrated load at mid span. This approach has been shown to work when compared to test data on the behavior of a single connection. This modeling approach has been extended to a fairly large frame to examine the sensitivity of the frame to various modeling assumptions, particularly the effects of member-to-member contact at a joint. If the effects of contact are not accounted for, estimations of some load effects may be underestimated by more than 100 percent, although most underestimations are 25 percent or less.

Introduction

The need to understand the structural behavior of wood-pegged timber connections has become more pressing over the last decade or so. This need has been driven by the trend toward renovation and rehabilitation of historic wood structures and the significant increase in the use of traditional methods in the construction of new structures. The objective of this paper is to summarize the findings related to analysis of timber frames.

The analysis of frames is first addressed by Brungraber (1985). He performed 2–D finite element analyses on some joint details and proposed a 3–spring joint model for frame analysis as well as suggesting the use of pinned connections. His 3-spring model would have required testing of many different

1. Professor, Dept. of Civil & Env. Engrg., Michigan Technological Univ., 1400 Townsend Dr. , Houghton, MI 49931-1295
2. Structural Engineer, Cleveland-Cliffs, Inc., Ishpeming, MI

joint configurations in order to determine the spring constants for each joint type. Kessel et al. (1988) reported on the reconstruction of an eight-story timber frame building in Germany. They modeled the joints as pinned connections, but no tests were performed to examine the adequacy of the model. Kessel and Augustin (1995a, 1995b) have published experimental studies on the load capacity of wood–pegged joints. Some more recent work on wood-pegged joints has be performed by Schmidt and MacKay (1997). A summary of work on testing and analysis performed by the first author and his colleagues can be found in Bulleit et al. (1999).

Modeling Guidelines

Observations during testing and some confirming analyses (Bulleit et al. 1999), have led to basic guidelines for modeling traditionally connected timber frames. The guidelines are: 1. Frame members should be modeled as beam-columns with the effects of shear included. 2. Joints should be assumed to carry no moment, i.e. free to rotate. 3. Eccentricity of force should be included. For instance, when the pegs are not located at the centroid of a column or beam, the eccentricity between the pegs and the column or beam centroid should be included in the analysis. 4. Where contact between one member and another may have a significant effect on the behavior of the frame, the effect of contact should be examined. This fourth guideline generally requires that the frame be analyzed more than once for each load case. First, an analysis must be performed to discern which members are in tension. Members in tension should have the axial stiffness of the connections on that member based on the peg behavior only, i.e. no contact effects. This is particularly important for members that act as braces, e.g. knee braces and collar ties. Second, braces in compression should be included in the analysis by accounting for the possiblity that they may or may not be in contact with the adjacent member. This last guideline is a direct result of the behavior seen in simulated gravity load tests of the knee brace specimens where contact between the beam and the knee brace had a significant effect on the behavior of the frame and was not consistently present until fairly high load levels.

Modeling

To show how a joint is modeled, consider the column-beam connection shown in Fig. 1a. The elements framing into the joint from the top and bottom, column elements 5 & 2, and the beam element, no. 8, should be modeled as beam-column elements, including the effects of shear in the stiffness matrix. The elements labeled 3 & 4 are beam-column elements that model the reduced cross section at the mortise. Some recent work (Drewek 1997) indicates that it may be adequate to continue the full column section properties into this region. Element 6 accounts for eccentricity of force. This is a very stiff element, on the order of 10,000 times stiffer than the mortise element, i.e. element 3 or 4, framing into it. The last element, no. 7, with nodes 19 & 20, models the section of the beam tenon from the face of the column to the pegs. At the pegs, node 19, there should be a moment release on this element that allows that end of the element to act as a hinge thus preventing the pegs from carrying a moment. The axial stiffness of this element is controlled by the state of stress in element 8. If element 8 is in compression, then contact may be assumed and the axial stiffness becomes the same as the the beam, i.e. element 8. If the beam is in tension, then the axial stiffness of element 7 is controlled by the pegs. The pegs are modeled as short beams with a concentrated load acting at the center of the tenon. The span of the short beam is assumed to be the distance between the middle of the mortise walls on each side of the column. The axial stiffness of element 7 would then be calculated using $n(48EI/L^3)$ where n is the number of pegs, E is the modulus of elasticity of the peg, I is the moment of inertia of the peg, and L is the peg span described above. The other properties of this element should be based on the tenon dimensions. Again, there are indications (Drewek 1997) that the properties, other than axial stiffness, may be based on the full beam cross section dimensions.

Next, consider the knee brace/beam connection shown in Fig. 1b. The two beam elements, numbers 8 & 12 and the short element, number 11, have their section properties determined in a manner similar to that described for the column-beam connection. The short element with nodes 12 and 13 is used to model the peg effects. There is a moment release at node 12. The axial stiffness is based on the cross sectional area of the knee brace when the knee brace is in compression and is in contact with the beam and column. It is based on the single peg in bending, as described above, when the knee brace is in tension or if the knee brace is in compression and not in contact with the beam or column. The other properties are based on the knee brace cross section.

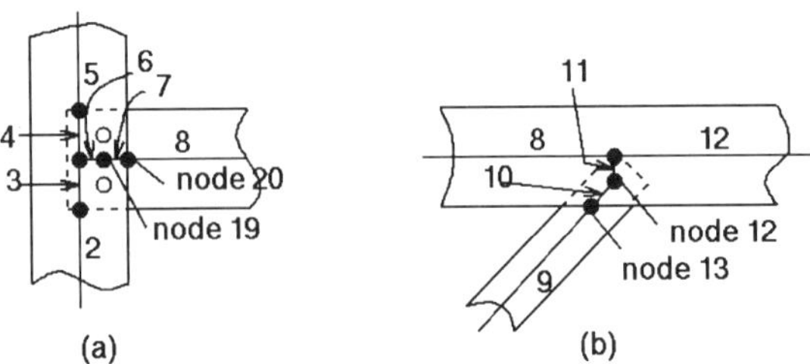

Figure 1 Representative Joint Models

As an example of the effect of the knee brace, consider positive moment in a beam supported by the knee brace. For floor load on the beam, the positive moment in the beam when the knee brace is in compression but not in contact is more than 100 percent of the moment when the knee brace is in contact. This is a significant difference. In our tests, loads at service load levels did not always produce contact. This indicates that until better knowledge is available about contact behavior in actual frames, bracketing of load effects is important.

Now, consider wind loads that produce tensile forces in the knee braces. The force if contact is assumed for all members in compression is 80 percent of the force if knee braces in compression are assumed not to be in contact. This difference is potentially significant, particularly in the way it affects the pegs connecting the knee brace to the post and beam. Remember that knee braces in tension should always have their axial stiffnees based on the peg.

Further details can be found in Bulleit et al. (1999).

Conclusions

A technique for analyzing traditional timber frames using commerical structural analysis software has been developed. Initial studies indicate that multiple analyses are required for each load case in order to bracket load effects. This is necessary because of the possible effects of contact on the behavior of the joints. If contact is not accounted for, some load effects may be underestimated by more than 100 percent, although most estimation differences are 25 percent or less. Further research is required to better model peg behavior and to better understand the effects of contact on the frame behavior. Potential simplifications to the analysis procedure, such as removal of the short stiff elements, need to be examined. Comparison of the analysis results to full-scale frame tests is also necessary.

References

Brungraber, R. L. (1985). <u>Traditional Timber Joinery: A Modern Analysis.</u> Ph.D. dissertation, Stanford University, Palo Alto, CA.

Bulleit, W. M., Sandberg, L. B., Drewek, M. W., O'Bryant, T, L. (1999) "Behavior and Modeling of Wood-Pegged Timber Joints". <u>Journal of Structrual Engineering</u>, ASCE, 129(1), to be published.

Bulleit, W. M., Sandberg, L. B., O'Bryant, T. L., Weaver, D. A., Pattison, W. E. (1996). "Analysis of frames with traditional timber connections". <u>Proceedings of the 1996 International Wood Engineering Conference</u>; Vol. 4, Omnipress, Madison, WI, 232-239.

Drewek, M. W. (1997). <u>Modeling the Behavior of Traditional Timber Frames.</u> Masters thesis, Michigan Technological University, Houghton, MI.

Kessel, M. H., Speich, M., Hinkes, F.-J. (1988). "The reconstruction of an eight floor timber frame house at Hildesheim (FRG)". <u>Proceedings of the 1988 International Timber Engineering Conference</u>; Forest Products Research Society, Madison, WI, 415-421.

Kessel, M. H., Augustin, R. (1995). "Load behavior of connections with pegs I". Trans. by Peavy, M. and Schmidt, R. <u>Timber Framing</u>, Timber Framers Guild of North America, 38, 6-9.

Kessel, M. H., Augustin, R. (1995). "Load behavior of connections with pegs II". Trans. by Peavy, M. and Schmidt, R. <u>Timber Framing</u>, Timber Framers Guild of North America, 39, 8-10.

Schmidt, R. J., and MacKay, R. B. (1997). <u>Timber Frame Tension Joinery</u>. University of Wyoming, Laramie, WY.

Design Considerations for Timber Frame Connections

Daniels, C. E., Schmidt, R. J., and Scholl, G. F.[1]

Introduction

The objective of this research is to determine strength and stiffness characteristics of recycled Douglas fir timber-frame connections in tension. Timber-frame connections use wood pegs to secure the tenon (main member) within the mortise (side members). The design of these connections is currently beyond the scope of building codes, including the National Design Specification for Wood (NDS). Full size joints in double shear were tested to determine possible failure modes and to establish minimum edge and end distances to prevent abrupt failures of the main or side members. The results of the tests utilizing recycled Douglas fir are compared to the characteristics of similar new growth species.

General Overview

Timber framed mortise and tenon connections have been used for centuries in some form or another throughout the world. The connections are constructed entirely of wood utilizing wood pegs to secure the joint. Figure 1 shows a typical timber frame mortise and tenon connection. This method of construction became rare in the United States during the 19th century when the use of wire nails became common and the country had a need for buildings that could be constructed quickly by relatively unskilled labor. Timber frame structures have demonstrated their longevity throughout the world, with structures that have survived hundreds of years of service. The durability of the structures, coupled with their beauty and convenient use with holistic building schemes, has led to their renewed popularity.

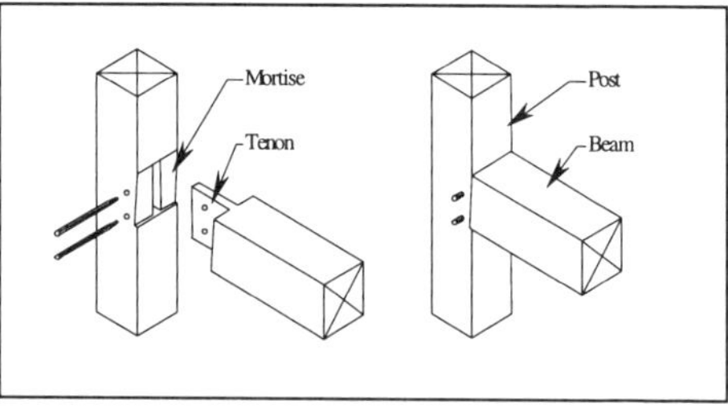

Figure 1-Mortise and Tenon Connection

[1] Department of Civil and Architectural Engineering, University of Wyoming, Laramie, WY 82071

Recycled materials are thought to provide a more stable structure and help preserve the forest resource. Heavy timber buildings are usually constructed with green material. Shrinkage and drying of this material creates checking and twisting of the timbers that can be a problem for builders and homeowners alike.

The use of recycled material allows builders to start with a timber that has been dried through years of previous service and the unknowns of seasoning defects are eliminated. Use of recycled materials also conserves resources by utilizing existing material supplies, hence reducing harvest demands.

Test Procedure and Results

Full-size joints were tested using a steel load frame and a hydraulic ram operated by a hand pump. Data acquisition software was used to record data from a pressure transducer, and two linear potentiometers. The potentiometers measured displacement between the mortised and tenoned. The test apparatus and a typical joint detail are shown in Figure 2[2].

The joints were secured with two 25 mm diameter white oak pegs. Pegs for all tests were obtained from the same stand of trees in the North Eastern United States. For correlation to results from the full-size joint tests, the shear/bending strengths of the pegs were determined using a test fixture designed to closely represent the behavior of a wood peg in a mortise and tenon joint. The yield point for both the peg shear/bending tests and the full-size joint tests were calculated using the 5% offset method (ASTM D 5652-95).

Twelve tests were conducted on recycled coastal Douglas fir joints. Ten of the twelve joints were loaded monotonically to failure. The remaining two joints were cyclically loaded (with increments of 1.3 mm in joint displacement) to determine hysteresis characteristics. The timbers were ungraded and had an average moisture content of 9.8%.

Strength of the monotonically loaded recycled Douglas fir (RDF) joints was dominated by peg shear/bending failure (Figure 3). The average strength for the monotonically loaded joints with a peg shear/bending failure was 25.4 kN. The COV was 10.2% and the 5% exclusion value for yield load was 19.8 kN. Table 1 shows the tabulated strengths for the 10 tests and the geometry of each connection. The end distance describes the

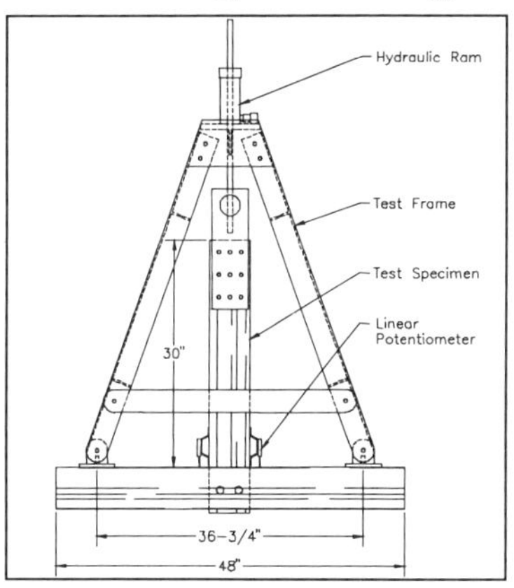

Figure 2-Test Set-up

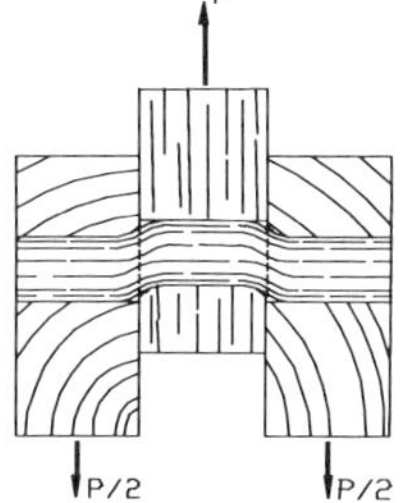

Figure 3 - Peg Shear/Bending

---

[2] Used with permission from MacKay 1997

measurement from the center of the wooden dowel to the end of the tenon member parallel to grain, and edge distance is the measurement from the center of the dowel to the loaded edge of the mortise member perpendicular to grain. The minimum required values found from these tests are much lower than those presently specified by the NDS for connections with steel bolts.

Test	End Dist. (Iv)	Edge Dist. (Ie)	Spacing Dist. (Is)	Yield Disp. (mm)	Yield Load (kN)	Yield Stress (MPa)	Stiffness (kN/mm)	Ultim. Disp. (mm)	Ultim. Load (kN)	Failure Type @ Yield	Failure Type @ Ultimate
DF1	4D	2D	3D	3.2	26.2	18.0	12.7	4.1	28.9	Peg Shear/Bending	Mortise Split
DF2	4D	2D	3D				14.2	1.6	23.0		Simult. Peg & Mortise
DF3	3D	2.5D	3D	2.3	23.9	16.3	23.8	5.3	27.0	Peg Shear/Bending	Mortise Split
DF4	3D	2.5D	3D	3.0	23.8	16.5	11.4	4.6	27.9	Peg Shear/Bending	Peg Shear/Bending
DF5	2D	2.5D	3D	1.9	21.3	15.1	32.6	3.9	28.2	Peg Shear/Bending	Simult. Tenon & Mortise
DF6	2D	2.5D	3D	2.1	28.7	20.0	29.1	2.8	29.9	Peg Shear/Bending	Peg Shear/Bending
DF7	2D	2.5D	2.5D	2.3	25.0	16.9	23.0	4.4	29.1	Peg Shear/Bending	Peg Shear/Bending
DF8	2D	2.5D	2.5D	2.4	25.2	17.1	17.9	3.7	28.8	Peg Shear/Bending	Peg Shear/Bending
DF9	1.5D	2D	3D				26.0	0.9	13.1		Tenon Split
DF10	2D	2.5D	2.5D	3.2	28.8	19.7	15.2	5.7	31.6	Peg Shear/Bending	Mortise Split

Mean= 25.4 17.4
Std. Dev.= 2.5 1.7   [1]From ASTM D 2915
5% Exclusion= 19.8 13.8
COV= 0.10 0.10
k^1= 2.189 2.189

Table 1-Monotonic Test Results

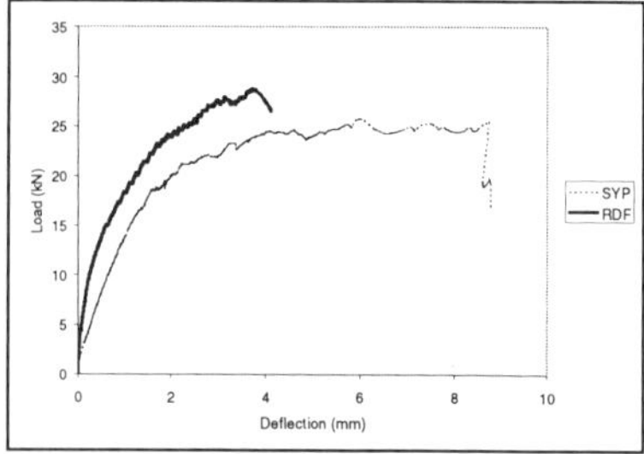

Figure 4-Monotonic Load Tests

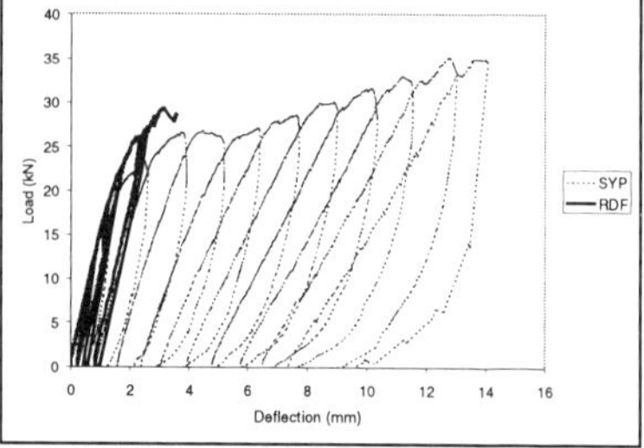

Figure 5-Cyclic Load Tests

The behavior of the recycled Douglas fir joints was not as ductile as other softwood joint tests. Figures 4 and 5 show a monotonic RDF test and a cyclic test compared to southern yellow pine joints (SYP) with the same geometry and white oak pegs. The tests with the SYP material had pegs from the same population but the pegs were at fiber saturation point. The mean 5% offset yield value for the RDF joints was 13% higher than for the SYP joints. However, while the RDF joints had an average ultimate displacement of only 17% of the peg diameter, whereas the SYP joints consistently reached 40% of the fastener diameter at ultimate displacement. This behavior is shown in Figure 4. Behavior under cyclic loading for the two species was consistent with that for the monotonically loaded joints. Each of the two SYP joints withstood 11 cycles, while each of the two RDF joints failed after the fourth cycle. The RDF joints used for the cyclic tests had defects that could have affected the results. One joint had a severely checked mortise member and developed a tension failure perpendicular to grain. The other RDF joint had a large check parallel to the orientation of the pegs through the center of the tenon. During testing of this joint, the check closed and the tenon formed two individual splits behind each peg.

Design Values

A total of 56 tests was conducted on 42 individual joint specimens. Three different wood species were tested: southern yellow pine, recycled Douglas fir and red oak. The results of these tests have been compared to similar research by Kessel (1988) in Germany. Recommendations for design strength of these connections are under development. These recommendations will be based the 5% offset yield point from the tests and analytical procedures based on the European Yield Model.

Current end distance, edge distance, and peg spacing requirements are based the NDS specifications for steel bolted connections. A method for determining end and edge distances for wood pegs, proposed by MacKay (1997), was compared to the results of this work. The method, in which end and edge distances are chosen for a steel bolt with equivalent strength to that of the wood peg, has been shown to be conservative. That is, end and edge distances selected by the method assure that connection failure occurs in the fastener, rather than in the main or side member.

Acknowledgements

Funding for this research was provided by the U. S. Department of Agriculture, NRI Competitive Grants Program, under contract #9702896 as well as the Timber Frame Business Council. Individual timber frame companies provided materials, facilities, labor, and shipping for the joint tests. In particular, the Southern Pine joints were supplied by Red Suspenders Timber Frames, Nacogdoches, Texas, the Douglas Fir joints were supplied by Big Timberworks, Inc, Gallatin Gateway, Montana, and the Red Oak joints were supplied by Sunset Structures, Elkview, WV. Scott Northcott generously supplied the White Oak pegs used in this work.

References

American Forest & Paper Association (1997). "National Design Specification for Wood Construction," American Forest & Paper Association (AFPA), Washington, DC.

ASTM, (1995a). *1995 Annual Book of ASTM Standards, 04.10 Wood,* Philadelphia, PA.

Kessel, M. H. & Augustin, R. (1996). "Load Behavior of Connections with Pegs II," *Timber Framing, Journal of the Timber Framers Guild,* No. 39, pp. 8-10, March

MacKay, R. B. (1997). "Timber Frame Tension Joinery," M.S. Thesis, Department of Civil and Architectural Engineering, University of Wyoming, August.

The Painful Realities and Lovely Results
of Using Recycled Timbers

Robert Lyman ("Ben") Brungraber, Ph.D., PE[1]
Member, A.S.C.E.

Abstract

Modern timber framed structures are fully exposed to the interior and are connected with traditional joinery methods, often executed to cabinetry standards. The search for dry and stable, but solid sawn, timbers has led many timber framing companies to use timbers recycled from dismantled old mill buildings. The results of using these timbers have been rewarding, but the challenges included increased costs and elevated risks. The costs are involved in both the material processing and the labor to repair and make best use of the more valuable material. Grading the recycled timbers is feasible and a growing issue.

Introduction

Benson Woodworking has been designing and building timber framed buildings since the early 1970's. Several hundred timber framing companies have formed in North America since then. The entire North American industry only builds a couple thousand homes and public structures a year, but their market has been solid and growing. Timber framed structures can be characterized as having heavy timbers exposed to the interior of the completed building, relying on knee braces for lateral strength, and being connected with traditional joinery methods. Variations on the pegged mortise and tenon joint are the most commonly used connections in timber framed buildings, both new and old.

Benson Woodworking first built timber frames with fresh sawn and boxed heart timbers from logs available in New England: northern red oak, eastern white pine, and hemlock. The search for higher quality, large and solid sawn timbers led us to fresh-sawn, free-of-heart-center Douglas fir and export quality southern yellow

---

[1] Operations Director, Benson Woodworking Company, Inc., Box 224 Pratt Road, Alstead Center, NH 03602

pine. Those noble timbers were a significant improvement and got us accustomed to paying quite a bit more for our raw material. But even the free-of-heart-center timbers still shrank, distorted, and checked as they dried toward equilibrium moisture content in the completed frame. This movement has both structural and aesthetic impacts. Shifting timbers within a connection which relies on bearing faces can mean the load transfer mechanisms change as the timbers dry. An experienced timber framer craftsman can achieve cabinetry quality in the fits of the assembled frame connections. This initial quality, however, can be almost totally obscured as the timbers dry and shift in the first years after the craftsman's work is installed.

A straightforward source of large structural timbers that are dry and stable is glue laminated wood. Timber framers are largely convinced that their clients are looking for the more traditional look of solid sawn timbers. Our first approach to getting solid sawn timbers that were already at nearly equilibrium moisture content was to experiment with kiln-drying them. We tried hard, and even ruined some timbers by trying too hard, but we finally convinced ourselves that the minor stabilization that we could achieve was not worth the time and money investment involved. Very recently, we have been able to buy timbers which have been dried in radio frequency kilns; an entirely different drying mechanism and one which offers new hope for rapid (if not at all cheap) and high quality timber drying.

We so wanted to use stable timbers for our frames that we started looking for old, already dry timbers which were large enough for us to resaw down to the sizes we need. Obsolete mill buildings, being demolished all over the country, have proven a steady (if unpredictable) source of large and dry timbers. Douglas fir and southern yellow pine are the most commonly available species. Buildings west of the Mississippi were generally framed with Douglas fir; eastern ones from southern yellow pine. We have also used redwood from wine vat staves and highway bridges, first growth eastern white pine from an early Connecticut mill, and various first growth softwoods that had been hand hewn into barn timbers. This paper deals with some of the realities confronted when recycling timbers into new structures.

The benefits of using recycled timbers

There are two reasons to use recycled timbers: enhanced aesthetics and heightened karma. Combined, these two improvements can increase the sales appeal of timber framed buildings; the main inspiration for undertaking the risk and investment involved. We started using recycled timbers because they were already largely dried and would, therefore, not shrink, twist, and check after the frame was assembled. This has been justified; the five-year-old frames we fabricated from recycled timbers look exactly as great as they did when we erected them.

An unanticipated benefit of using recycled timbers has been the generally extraordinary high quality of the raw material itself. Many of these timbers were cut to build mill buildings more than a hundred years ago. Some of the barn timbers we have recycled came from barns more than twice as old. The logs that were yielding these timbers were often from first growth forests. While the highest quality lumber was cut for other uses from the outside of the log, even then, the trees themselves were remarkable specimens, no longer readily available.   When we resaw the timbers, in order to reveal fresh wood and to get rid of the mill building patina and the demolition damage, we are often confronted with fabulous wood. Since the large timbers were cut from the centers of the logs, the southern yellow pine timbers are often largely heartwood on all surfaces; some of the loveliest and most desirable wood on the market today. We often saw one face off the hand hewn beams we recycle from barns. While the client wants only to enjoy the patina of the originally hewn faces, the fabricator and builder enjoy the predictably flat surface; at least where other finishes are to be installed. These faces are not exposed in the finished structure, but they reveal to us just how fine were the trees the American colonists were cutting down to clear the land.

Recycling timbers offers some significantly improved karma; enough to add sales appeal, and even in the face of heightened costs. No forests were cut down (directly) to provide the timber. Furthermore, the waste going into our landfills is reduced, at least to some extent. Finally, the timbers' previous lifestyle adds great interest for a lot of our clients. They all seem to want to know the location and history of the buildings that had housed the timbers now in their building. While these benefits may seem a bit esoteric for use in sales, our clients tend to be at the high end, and are already sufficiently interested in the qualities of their finished building as to be building with exposed and finely crafted structure.

The challenges in using recycled timbers

Timber frames built from recycled timbers tend to be (and stay) so much lovelier than those built from fresh sawn timbers, that there must be some reasons why we are not building all of our structures with them. Predictably, these reasons involve money; specifically the increased costs of using recycled timbers. The increased costs are tied up in both the material, by the time it is on the framers' bunks, and in extra labor required to use recycled timbers. These costs can be passed on to the buyer, but there are also increased risks inherent to using innovative products, such as recycled timbers, the costs of which are tougher to allocate.

Buying raw timbers from either a demolition contractor, or one of the brokers who specializes in recycled timber, can seem deceptively inexpensive. Then you start investing in the various processes that prepare the raw timber for use in a finished building. Many of these steps not only cost money; they directly reduce the amount of finished material yielded by the process; a double whammy when it comes to the unit cost of the finished product.

Depending on timber sizes and lengths, the quantities involved, timber species and condition, and their availability and location; I have been offered raw recycled timbers for costs ranging from $0.30 to $2.00 per board foot. Then the timbers have to be shipped from the demolition site to a processing facility. That can add from $0.05 to $0.34 per board foot. We have been successful in having the team resawing the timber also responsible for removing the miscellaneous metal. This adds about $0.12 per board foot. Now, we start spending money to get rid of wood: "sawing the jacket" off the outside faces costs about $0.16 per board foot; based on the size of the timber as it goes into the sawmill. We saw anywhere from half an inch to an inch and a half off all faces, depending on how straight the raw timbers are and whether we are trying to saw away particularly deep damage. Clearly, the reduction in the timber quantity is variable, but a range of 25% to 40% is to be expected. Then, we generally plane and square the sawn faces with a timber sizer, for another $0.11 per board foot and 2-5% reduction in wood quantity. Since we pay for the timbers "as is," we almost always have some end cutoffs after we have cut our as-needed timber. The losses associated with end cuts can vary, but another 6-12% is not unlikely. Finally, there are always a few timbers which are rejected for poor quality; either too badly damaged or undue spiral grain being the most likely flaws. After all this, the timbers can be worth from $2.50 to $3.50 per board foot by the time they get into the shop. This cost for dry, solid sawn timber is comparable with our cost for glue laminated timbers.

Fabricating timber frames from recycled timbers also takes more labor than to do so with fresh sawn timbers. First, the usual extra length means an extra decision has to be made as how best to locate the finished timber in the raw material; or, how much of which end to cut off. The recycled timbers have already checked and we like to minimize their appearance in the completed structure, so the framer is making location and orientation decisions that are not required with fresh sawn, unchecked timbers. Recycled timbers almost always need some remedial work, before we set them loose in a finished structure. Some of the work undoes the modifications made by the original builders; filling bolt holes and notches. Other work is required to repair the signs of demolition workers, not the most dainty members of the construction industry. Finally, after all these decisions and repairs are made, we are left with wood that is simply harder to cut than are fresh sawn and fully moisturized timbers.

Fabricating structures with recycled timbers

The major risks inherent to the user of recycled timbers can be divided into the pre and post fabrication phases. It can be difficult to predict what the raw timbers will contain when they arrive from the demolition site. It can be even tougher to describe the finished product, both to building authorities and to clients. We try to preview salvaged timbers, before buying and shipping them. We look for wood quality; tight grain and heart wood, and condition; not too drilled and preferably unpainted. Ideally, at least some timbers will have already been removed so that we can assess the likely damage we can expect from the demolition process. We like to design our structures in response to the available timber sizes. Clearly, we can make any timber from another one, so long as the raw material is at least larger and longer than the desired product. But this blind adherence to an original design can mean brutishly inefficient use of available sizes. We prefer to adjust the spacings, sizes, and schemes to make better use of the timber we have; or can get.

Once we have the timbers, we can repair most damage done by the builders or the dismantlers. We can plug bolt holes, glue and lag splits, and even implant splints and patches. Much of the remedial work is done in an effort to improve the finished appearance. The clients' expectations play an enormous role in the amount of effort required on this repair work. Educating clients as to reasonable levels of stains, holes, and repaired dings is a crucial part of the building process. One thing which consistently concerns potential buyers is the dripping stains from metal oxides in old nail and bolt holes, running down the exposed faces after being soaked by the weather. We mitigate these problems by plugging the bigger holes, filling the smaller ones with epoxy or urethane finish, coating the entire timber with a durable finish, and getting the timbers protected by an enclosure system as quickly as we can.

Another risk we have only had to confront officially a few times is building department approvals of recycled timbers. Usually, I am able to assuage any concerns with my assurance that our frames are fabricated under my supervision and, therefore, the engineer of record is monitoring timber quality. We have, on occasion, had our recycled timbers graded after fabricating them. The timbers usually have no trouble meeting guidelines for #2 and better and can often be described as dense #1 and better. The checking is an issue in these inspections, but one which we have been able to handle to date. The entire issue of grading any timbers which are going to be mortised and drilled is one which timber framers expect to address more often, as the popularity of their craft continues to spread.

The results of using recycled timbers

We have built more than one hundred structures from recycled timbers. We have used hundreds of thousands of board feet of potential landfill to craft some of our loveliest homes and public buildings. The supply of, and demand for, these timbers shows no sign of drying up, as our economy booms. I expect that the economic and quality issues involved with using recycled timbers are ones which will continue to engage us for the rest of my career.

Inverse Reliability Methods in the Design of Wood Structures

Ricardo O. Foschi[1], Member ASCE; Felix Yao[2], Hong Li[3] and Meho Karalic[4]

Abstract

Inverse reliability methods allow the direct determination of design parameters corresponding to a specified reliability level for each of the limit states considered in the design. This paper illustrates an application of these methods to the direct calculation of the span L of a timber floor (the "design parameter"), satisfying pre-set reliability levels β for different limit states and criteria (static deflection limits, vibration control, human acceptance of vibrations and bending strength).

Introduction

Consider a floor with joists and nailed (or glued) subfloor sheathing. For a given joist type (lumber or engineered I-joists), and for given sheathing and nailing characteristics, a floor span L can be calculated which will satisfy all design criteria. These usually include: 1) that the maximum bending stress in the joists does not exceed a tolerable level for the combination of a permanent and a uniformly distributed "design" live load, and 2) that the static deflection under the same combined loads does not exceed a fraction of the span (L/K). In most codes, K = 240. Also, the deflection under only the design live load must not exceed a fraction of the span L/K. In this case, K is normally 360. The calculation of deflections and stresses are normally done for the joist alone (no composite behavior), even if the structure is

[1] Professor, Department of Civil Engineering, University of British Columbia, Vancouver, B.C. Canada V6T 1Z4.
[2] Research Engineer, Department of Civil Engineering, University of British Columbia, Vancouver, B.C. Canada V6T 1Z4
[3] Graduate Student, Department of Civil Engineering, University of British Columbia, Vancouver, B.C. Canada V6T 1Z4
[4] President, Matrix Timber Ltd., 24611 Fraser Highway, Langley, B.C. Canada V2Z 2L2

a system of several interacting components. To compensate for this simplified approach, codes have introduced "system factors". The calculation of these is not a simple matter since, obviously, they depend on the specific configuration. Furthermore, in a reliability-based code, the factors must be calculated on a reliability basis, such that the intended reliability is maintained when changing floor configuration (Foschi et al.,1989). In the codes, factors are normally associated with the limit state of bending strength. For example, the Canadian code CSA-086 specifies a factor 1.40 as a low common denominator for all practical floor configurations. There are no factors applied to the deflection limit states. Furthermore, the bare joist deflection calculation is performed using the mean modulus of elasticity. Under this condition, the span satisfying L/360 will obviously result in the tolerable deflection being exceeded, under the design loads, approximately 50% of the time if, in fact, *one had a bare joist*. The deflection of a floor system of the same span will exceed L/360 with a much lower probability. In fact, it can be verified that the Canadian design loads (maxima over a 30 year period), and the simplified code rule lead to spans with a floor deflection exceedence probability in the range of 5-10%, depending on the configuration. In using the simple design procedure, we then implicitly accept this level of non-compliance in Canada. This implies that the span for the floor could also be calculated with a floor analysis (not a bare joist), looking for that L to which corresponds a deflection exceedence probability of around 5% (or β = 1.65). It can be similarly verified that the code rule, together with the 50-year design window used in the United States, result in an exceedence probability approximately corresponding to β = 1.30. For bending strength, the corresponding levels are 2.60 in Canada and 2.40 in the USA.

However, the limiting condition for spans is normally *vibration control*. Although the intention of the L/360 deflection limit is to control vibration, it has long been recognized that it is not adequate. In fact, floors spans recommended for engineered joists are usually obtained using more stringent criteria (e.g., L/480). Still, this may not be sufficient. A practical alternative is the use of effective, engineered load-sharing stiffeners able to improve floor vibration response. A proper calculation of floor deflection and stresses, when including stiffeners, requires computer analysis. Specialized static and dynamic software has been developed by the authors on the basis of a finite strip model (FAP) by the first author (Foschi, 1982). Several vibration controls have also been proposed independent of the L/360 criterion. For example, 1) that the natural frequency of a floor should exceed 15 Hz, or 2) that the rms of the acceleration for a specified impact should be less than a threshold (Smith and Chui, 1988), or 3) that the probability of an occupant assigning to a floor a "strongly perceived vibration" rating (Wiss and Parmelee, 1974; Foschi et al., 1995) should be less than a threshold, or 4) that the static deflection for a concentrated, 1kN load at midfloor should be less than a span-dependent limit. This concentrated load criterion was developed in Canada on the basis of surveys of occupants' reactions to their floors, and the deflection limits were set to discriminate "good" from "bad" floors in an approximately 50-50 split, when the floor had average properties (Onysko, 1986). This criterion has been adopted by the National Building Code of Canada (NBCC).

Table 1: Reliability-based spans, in inches, 2x10 No.2 Douglas fir floors.

Criterion	Reliability β	No Stiffeners	1 Stiff. Row[5]	2 Stiff. Rows
1 kN (NBCC)	0.00	165.33	195.10	200.09
L/360 (30 yrs)[6]	1.65	212.55	217.66	218.66
L/480 (30 yrs)	1.65	191.57	194.71	195.64
L/240 (30 yrs)	1.65	230.10	236.56	237.59
Stress (30 yrs)	2.60	243.12	250.09	251.04
L/360 (50 yrs)	1.30	216.02	221.24	222.23
L/480 (50 yrs)	1.30	194.72	197.93	198.87
L/240 (50 yrs)	1.30	233.26	239.77	240.80
Stress (50 yrs)	2.40	253.15	262.00	262.90
f = 15 Hz	1.65	209.94	208.69	212.25
Rms(a)	1.65	181.02	198.85	212.21
W-P (50%)[7]	1.65	156.49	225.28	243.80
W-P (75%)	1.65	95.91	120.19	135.67
W-P, NBCC[8]	1.65	46%	57%	60%
A = area ratio[9]		1.0	0.53	0.57

Table 2: Reliability-based spans, in inches, 12" I-Joist floors.

Criterion	Reliability β	No Stiffeners	1 Stiff. Row	2 Stiff. Rows
1 kN (NBCC)	0.00	191.63	244.52	255.92
L/360 (30 yrs)	1.65	268.95	273.02	273.94
L/480 (30 yrs)	1.65	240.67	243.68	244.36
L/240 (30 yrs)	1.65	292.44	297.27	298.36
Stress (30 yrs)	2.60	413.01	428.93	430.12
L/360 (50 yrs)	1.30	273.25	277.45	278.41
L/480 (50 yrs)	1.30	244.55	247.71	248.43
L/240 (50 yrs)	1.30	296.28	301.21	302.33
Stress (50 yrs)	2.40	420.16	436.71	437.86
f = 15 Hz	1.65	295.90	295.02	296.54
Rms(a)	1.65	263.81	314.90	316.60
W-P (50%)	1.65	229.59	367.44	441.18
W-P (75%)	1.65	158.34	202.67	224.68
W-P, NBCC	1.65	63%	69%	71%
A = area ratio		1.0	0.40	0.41

---

[5] IBS2000-type load sharing stiffeners, Luxor Industrial Corp., Vancouver, B.C.
1 row at L/2, 2 rows at L/3.
[6] Either 30-years maximum design load (Canada) or 50-years design load (USA).
[7] Wiss-Parmelee criterion, with acceptance rate either >50% or >75% for 95% of the floors.
[8] Wiss-Parmelee acceptance rate achieved by the 1kN criterion in the NBCC, for 95% of the floors.
[9] Ratios of areas under free vibration displacement response after an impact, for spans obtained with the NBCC (1kN load). Ratios are with respect to the no-stiffener configuration for an average floor.

Results

The software IRELAN (Li and Foschi, 1998) was used for the inverse analysis, determining directly the value of the span L corresponding to the target reliabilities. For this, IRELAN was linked to the extensions of FAP. If a forward procedure had been used, the reliability level would have had to be estimated for different spans, interpolating the L for the desired β . Table 1 presents spans for a floor with 2x10 joists, Douglas fir No.2, and Table 2 shows the corresponding spans for 12in. I-Joists. The sheathing is 5/8in plywood attached with nominal nailing at 12in o.c. The shear stiffness of the nails is 8000 lb/in, and that of the stiffeners is 5000 lb/in. The floors have 12 joists at 12in o.c. for lumber and at 16in o.c. for I-Joists, with no side supports. Tables 1 and 2 also show an assessment of the NBCC spans using either the Wiss-Parmelee criterion or evaluating the area under the displacement response to a specified impact. This area is smaller when amplitudes are smaller and the damping greater, terminating vibrations sooner, and can also be used as a sensitive vibration control when calibrated to either the 1 kN load or the Wiss-Parmelee criteria.

Conclusions

Inverse reliability is a very useful and efficient tool to find, directly, the design parameters corresponding to target reliabilities. In this paper, the span of a floor was calculated to satisfy different criteria related to vibration, deflections and bending strength. The 1kN load procedure used in the NBCC can be evaluated with the benchmark Wiss-Parmelee criterion. As shown, performance can be *substantially improved* by the use of engineered load-sharing stiffeners, particularly for I-Joists. Note that not only spans are longer, but that the behavior is improved (as shown by greater acceptance rates, and decreases in area A, showing smaller amplitudes and improved damping). Note that criteria based on deflections or frequency do not perform adequately, since they are quite insensitive to the usefulness of stiffeners. The tables also show reliability levels when comparing Canadian and US calibrations.

References

Foschi, R.O. (1982). "Structural Analysis of Wood Floor Systems", *ASCE Journal of the Structural Division*, 108, 1557-1574.

Foschi, R.O., Folz, B. and Yao, F. (1989). "Reliability-Based Design of Wood Structures", *Structural Research Series Report No. 34*, Department of Civil Engineering, Univ. of British Columbia, Vancouver, B.C. Canada.

Foschi, R.O., Neumann, G., Yao, F. and Folz, B. (1995). "Floor Vibration due to Occupants and Reliability-Based Design Guidelines", *Canadian J. of Civil Engineering*, 22, 471-479.

Li, H. and Foschi, R.O. (1998). "An Inverse Reliability Method and Applications", *Structural Safety* (in press).

Onysko, D. (1986). "Serviceability Criteria for Residential Floors Based on a Field Study of Consumer Response", *Forintek Canada Corp.*, Vancouver, B.C. Canada.

Smith, I. and Chui, Y.H. (1988). "Design of Lighweight Wooden Floors to Avoid Human Discomfort", *Canadian J. of Civil Engineering*, 15, 254-262.

Wiss, J.F. and Parmalee, R.A. (1974). "Human Perception of Transient Vibrations", *ASCE Journal of the Structural Division*, 100, 773-787.

Performance Reliability of Timber Fender Systems

Steven A. Davidow[1] and Kenneth J. Fridley[2]

Abstract

Due to the deterioration of existing timber fender systems in waterfront facilities, high performance composite replacement components for these structures are in demand. The performance criteria for these replacement components must meet or exceed the performance of existing timber systems. For this reason, a reliability analysis of timber fender systems has been performed to (1) establish the level of performance in existing timber fender systems and (2) establish target performance levels for new component replacements.

Introduction

The requirements of fender systems in marine facilities are becoming increasingly more sophisticated in order to accommodate the range of load demands created by a wide variety ship sizes and shapes. As we approach the new millenium, many of the existing timber fender systems built both during and after the World War II era are in need of rehabilitation and/or replacement components. With the advent of high performance fenders and the use of new composite materials, the structural demand placed on these replacement components can be met in an economical fashion. The performance of these replacement components will be required to meet or exceed the current performance level provided by existing timber elements. With the wide range of variability and correlations that exist between material properties in timber elements, the current structural performance level in the existing timber fender components must be assessed using a reliability based evaluation.

Performance Criteria

The main purpose of the fender system is to act as the interface between the berthing ship and the pier while reducing the load transferred by the berthing process to both the pier and the ship hull. The key issues that define the performance criteria for the fender

---

[1] Graduate Research Assistant, Department of Civil and Environmental Engineering, Washington State University, Pullman, WA 99164-2910
[2] Associate Professor, Department of Civil and Environmental Engineering, Washington State University, Pullman, WA 99164-2910

systems are energy dissipation, and reaction force. The reaction force created by the berthing ship is a function of the flexibility or energy dissipation capability of the fendering system design. Current fender system design is typically based on the *Naval Facilities Engineering Command (NAVFACENGCOM), Military Handbook 1025/1 Piers and Wharfs (1987)*. The design of waterfront fender systems is based on applied energy rather that applied loads. The most widely used and accepted method for determining the applied energy created by the berthing ship is the kinetic energy method. The design vessel has a given mass and approach velocity, which are used to determine its kinetic energy, K, through the equation

$$K = \frac{1}{2}M V^2 \tag{1}$$

where,

M = mass or water displacement of the berthing vessel
V = approach velocity of the berthing ship at the time of impact with the fender

The fendering system is then required to dissipate this energy through the deflection of several individual elements, such as, wales, chocks, piles, and various rubber or pneumatic fendering units. The summation of the energy dissipated by each of these elements must equal the applied kinetic energy of the berthing vessel. Since the design of the fendering system is based on energy requirements, the design limit state can be written in terms of the potential energy of an individual element, E_w, and a portion of the kinetic energy applied by the berthing ship,

$$g(x) = E_w - \Omega K \leq 0 \tag{2}$$

where the factor Ω represents the fraction of the applied berthing energy that an individual element is required to dissipate. Failure of the system is defined by this equation, when the above equation is less than or equal to zero. The potential energy of an element in the fender system is given by,

$$E_w = \frac{1}{2}k_w \Delta^2_w \tag{3}$$

where,

k_w = stiffness of the timber element
Δ_w = deflection of the timber element

By focusing on an individual element in the fender system and making assumptions regarding the location of loading and support conditions, the limit state can be written incorporating ratios of the design variables to their respective nominal design values (shown with the subscript n),

$$g(x) = \left(\frac{F_b}{F_{bn}}\right)^2 - \frac{E}{E_n}\frac{M}{M_n}\left(\frac{V}{V_n}\right)^2 \tag{4}$$

where,

F_b = actual material bending stress (MPa)
E = actual material modulus of elasticity (MPa)
M = mass of the berthing vessel (Kg)
V = approach velocity of berthing ship (cm/s)

In evaluating the level of performance in existing fender designs with equation (4), we make the assumption that the nominal design requirements have been met in the derivation of the limit state, and thus limit this equation to the evaluation of systems conforming to current design standards.

Reliability Analysis and Results

Advanced first-order second-moment (AFOSM) reliability analysis was conducted to determine the performance reliability of the timber fender components. Several cases were investigated based upon assumptions made about the loading and material variables. Statistical data on the berthing vessel mass and approach velocity is extremely limited. For this reason, and investigation of performance reliability was based upon both a deterministic applied kinetic energy and an applied kinetic energy with vessel mass and approach velocity as random variables. The other defining factor in the analysis was the existence of a correlation between modulus of elasticity and modulus of rupture in timber components. The analyses investigated to what extent this correlation affected the performance of the system. For all analyses, Douglas Fir was assumed to be the design material.

The effect of a correlation existing between the modulus of rupture and modulus of elasticity in the timber elements is illustrated in Fig 1. As expected, by inducing a correlation between the modulus of rupture and modulus of elasticity, the performance reliability increases. Typical mean-to-nominal ratios of applied kinetic energy range from 0.4 to 0.8. When assuming no correlation, the reliability index ranges from 2.3 to 3.3 for this given range of applied energy. As the correlation increases, the range of the reliability index increases to approximately 2.8 to 3.8, depending upon the assumed correlation.

To examine the impact of loading variability on the performance reliability of the fender components, the applied berthing vessel mass and approach velocity were assumed to have some variability. Both loading variables were assumed to following log-normal distributions. A correlation between the timber material properties of 0.6 was also assumed. The same analysis procedure as the correlated case was used, where two new variables were now transformed and used in the analysis. This analysis was the most realistic approach, as the loading variables will have some variability. The vessel approach velocity is the most critical variable in the kinetic energy term, as its value is squared, and will most likely have more variability than the vessel mass. The performance reliability was investigated for mean-to-nominal values of velocity varying from 0.1 to 1.5 and COV of velocity varying from 10% to 50%. The vessel mass was assumed to have a mean-to-nominal value of 1.0 and a COV of 10% through all analyses. The results of this analysis are summarized in Fig 2.

This analysis was conducted to illustrate the effect of a randomized approach velocity on the performance reliability of the fender system. By again looking at a mean-to-nominal velocity range between 0.3 and 0.8, the variability in approach velocity will change the performance reliability from 5 to 3.5 for a conservative estimate of velocity, and from 3.1 to 1.8 for a more liberal estimate of velocity. Statistical data on the berthing vessel approach velocity must be gathered in the future before any meaningful results can be drawn from this analysis.

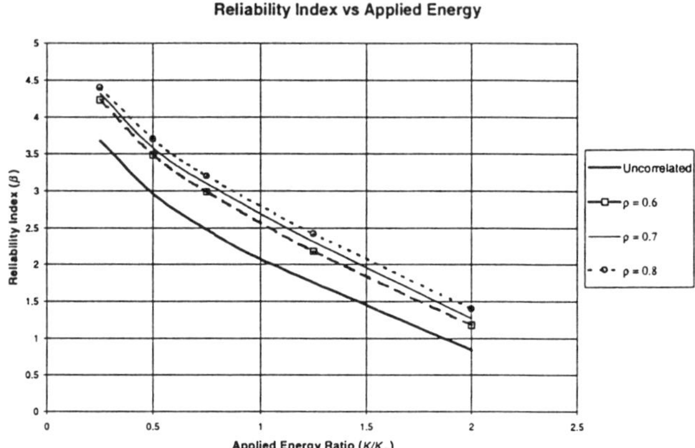

Figure 1. Reliability Index vs Mean-to-Nominal Applied Energy

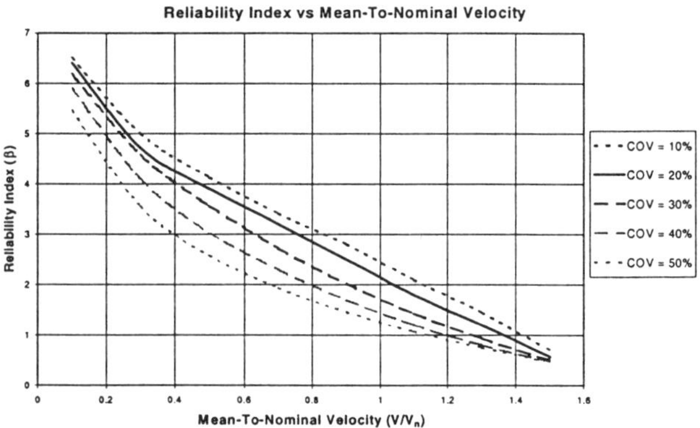

Figure 2. Reliability Index vs Mean-to-Nominal Velocity

References

Military Handbook: Piers and Warfs, MIL-HDBK-1025/1, Naval Facilities Engineering Command, U.S. Department of Defense, 1987.

Correlation Structure of Two Lumber Properties

By William M. Bulleit[1], Member, ASCE

Abstract

Spatially averaged data from the literature for along-the-length correlation of lumber flatwise bending modulus of elasticity and tensile strength was analyzed to determine the correlation structure of the base process. The exponential correlation function, equivalent to a first-order autoregressive process, was the best. This function can be used to generate correlation information for spatially averaged processes averaged over lengths different than the lengths used to obtain the test data.

Introduction

Some past research has been directed toward obtaining information about the along-the-length correlation of lumber properties (Kline et al. 1986, Richburg and Bender 1992, Taylor and Bender 1991). The two primary properties of interest so far have been flatwise bending modulus of elasticity, E, and ultimate tensile strength, T. These two properties are important in modeling the length effect for lumber in tension and for probability modeling of glulam members. The data from the testing was generally fit to autoregressive models. There is one drawback to these models. Although they model the segment property correlations well, they are only able to model segment properties for segments with the length used in the tests. It is not possible to use different lengths due to spatial averaging effects that cause the between segment correlation to be different for different lengths. This drawback can be dealt with using random field models (Vanmarcke 1983).

---

[1] Professor, Dept. of Civil and Environmental Engg., Michigan Technological University, 1400 Townsend Drive, Houghton, MI   49931-1295.

The objective of this paper is to show how random fields can be used to find the correlation function from spatially averaged data.

Random Fields

The data from the above literature are spatially averaged.   Consider a realization of a random process, say modulus of elasticity, E, along the length of a piece of lumber. The tests for E are performed by deflecting a specific length of the board. Thus, the value of E for that specific length is an averaged value which depends on the length chosen. The values for E obtained by testing a specific length are a realization from a process that is different than the base process. The new process is a spatially averaged process. The spatial averaging retains the mean of the base process but reduces the variance. A longer length of averaging of the base process produces a smaller variance for the averaged process.

Consider a continuous parameter stationary random process, such as E along a board. Denote this base process E(r) with mean M and variance σ^2. A family of moving average processes $E_R(r)$ can be obtained from (Vanmarcke 1983)

$$E_R(r) = \frac{1}{R} \int_{r-R/2}^{r+R/2} E(x)dx \tag{1}$$

where R denotes the length over which the process is averaged. The mean of the process is not affected by the averaging. The variance of the averaged process is (Vanmarcke 1983)

$$Var\ (E_R) = \sigma_R^2 = \gamma(R)\sigma^2 \tag{2}$$

where $\gamma(R)$ is the variance function. The variance function is related to the correlation function by (Vanmarcke 1983)

$$\gamma(R) = \frac{2}{R} \int_o^R (1 - \frac{r}{R})\ \rho(r)dr \tag{3}$$

where $\rho(r)$ is the correlation function. It is the variance function which allows a relatively straightforward determination of the autocorrelation between segment lengths other than those used in testing. Once the correlation function, $\rho(r)$, is known, the correlation between different segment lengths can be determined. Consider multiple equal length segments as shown in Figure 1. Two segments of length L are separated by a distance, S_o. If the variance function for the base random process is known, then the correlation between the process averaged over the two segments can be determined (Der Kiureghian 1985).

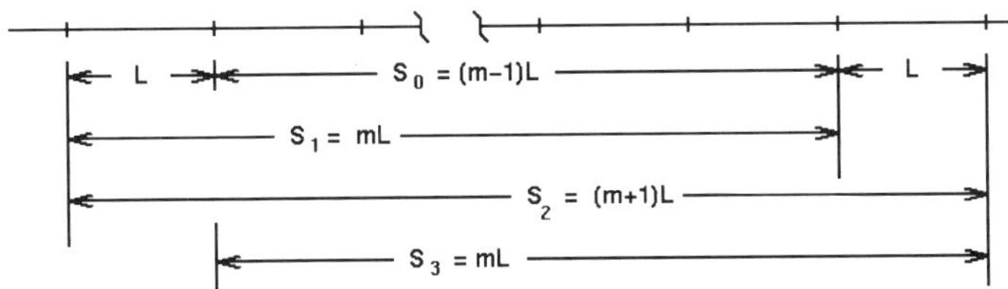

Figure 1. Equal Length Segments

$$\rho_m = \frac{(m-1)^2 \ \gamma((m-1) \ L) \ - \ 2m^2 \ \gamma(mL) \ + \ (m+1)^2 \ \gamma((m+1) \ L)}{2\gamma(L)} \tag{4}$$

Eq. 4 can be used to determine the best fit correlation function for the data from the literature, where the variables are shown in Figure 1. Eq. 4 in combination with Eq. 3 can be used to find the correlation coefficients between the averaged properties from the data. The correlation function, $\rho(r)$, will be determined from the spatially averaged data. This correlation function can then be used to determine correlation coefficients when the process is averaged over lengths other than those used in testing. The exponential correlation function, associated with a first-order auto-regressive process is (Vanmarcke 1983):

$$\rho(r) = e^{-|r|/a}$$

and its variance function is: $\tag{5}$

$$\gamma(R) = 2 \left(\frac{a}{R} \right)^2 \left(\frac{R}{a} - 1 + e^{-R/a} \right)$$

In the above equation, a is the correlation function parameter whose value must be obtained from data. Other correlation functions were considered, including the triangular correlation function and the correlation function associated with a second-order autoregressive process. The approach used in this study was to find the value of the correlation function parameter which minimized the indicator function

$$I(a) = \sum_{m=1}^{n} \left(\hat{\rho}_m - \rho_m(a) \right)^2 \tag{6}$$

where $\hat{\rho}_m$ is the correlation coefficient from the data for the mth segment and $\rho_m(a)$ is the correlation coefficient determined from Eq. 4 using a given variance function with a specific parameter value, a. The function, I(a), is summed over the number of segment separation distances available from the data, in this case, n=3, for many values of a. The value of a that minimizes I(a) is the value chosen for that particular correlation function. The chosen value will be denoted a*. The minimization process is performed considering selected correlation functions and the function which gives the smallest value of I(a*) is chosen.

Results and Conclusions

Taylor and Bender (1991), working with 2 x 6, 302-24 lumber, obtained 2-, 3-, and 4-segment (1-lag, 2-lag, 3-lag, respectively) correlation coefficients using 610 mm (24 in) segments. Eq. 6 was minimized for n = 3 considering each of the selected correlation functions. For both E and T, the exponential correlation function gave the minimum value of I(a) with a* = 13,106 mm (516 in) for E and a* = 2,286 mm (90 in) for T. The test versus predicted correlation coefficients are shown below. The fit is good.

Lag	E-data	E-pred.	T-data	T-pred
1	0.9581	0.9695	0.8008	0.8405
2	0.9277	0.9255	0.6413	0.6437
3	0.8902	0.8834	0.5136	0.4931

The parameter a* is related to the correlation structure. A large value of a* means the correlation function decreases slowly with increasing r indicating high spatial correlation. A small value indicates the opposite.

The exponential correlation function, equivalent to a first-order autoregressive process, was the best for all the literature data, except in one case the triangular correlation function was best by a very small margin. For the exponential correlation function, the value a* ranged from 1,500 mm (59 in) to 13,700 mm (539 in) for E and from 740 mm (29 in) to 2,500 mm (98 in) for T.

References

Der Kiureghian, A. (1985) "Finite element methods in structural safety studies," Structural Safety Studies, ASCE, New York, NY, pp. 40-52.

Kline, D. E., Woeste, F. E., and Bendtsen, B. A. (1986) "Stochastic model for modulus of elasticity of lumber," Wood and Fiber Science, 18(2), pp. 228-238.

Richburg, B. A. and Bender, D. A. (1992) "Localized tensile strength and modulus of elasticity of E-rated laminating grades of lumber," Wood and Fiber Science, 24(2), pp. 225-232.

Taylor, S. E. and Bender, D. A. (1991) "Stochastic model for localized tensile strength and modulus of elasticity in lumber," Wood and Fiber Science, 23(4), pp. 501-519.

Vanmarcke, E. H. (1983) Random Fields: Analysis and Synthesis, The MIT Press, Cambridge, MA.

Simplified Reliability Analysis of Light-Frame Roofs Subject to Uplift Loads

David V. Rosowsky[1] and Ningqiang Cheng[2]

Abstract

In this study, a simplified reliability analysis is conducted for a simple gable-end roof. Specific limit states considered include loss of sheathing and failure of the roof-to-wall connection. The results from the simplified analysis are in good agreement with those from a more advanced first-order second-moment analysis.

Introduction

Light-frame structures comprise the majority of the building stock most vulnerable to high wind hazards. One of the most significant performance issues is maintaining the integrity of the roof system. This includes both preventing loss of sheathing and ensuring adequate connections between the roof rafter (or truss) and the top-plate of the wall.

This paper reports on one small aspect of a larger project to evaluate the relative safety of light-frame roof components on typical residential structures located along the Carolina coast. Specific limit states considered were loss of sheathing and failure of the roof-to-wall connection, both associated with uplift loads caused by high winds. For more details, see [Cheng, 1998]. The focus of the present paper is on the use of a simplified, two-variable (R-S) reliability analysis as a closed-form alternative to the more complex second-moment techniques.

Wind Load Statistics for Reliability Analysis

ASCE 7-95 was used to calculate both the nominal *and* the actual wind loads acting on individual components. (Other standards were also considered.) To describe the 50-year maximum wind load, information on the 50-year maximum wind speed and statistics on the code factors (exposure factors, gust factors, pressure coefficients, etc.) taken from the recently complete Delphi study [Ellingwood and Tekie, 1997] were also used. Simulation was used to determine the mean-to-nominal wind load and COV for a set of coastal NC/SC sites which were then aggregated to develop a composite set of wind load statistics suitable for use in a reliability analysis. For more details, see [Cheng, 1998].

---

[1] Assoc. Prof., Dept. of Civil Engrg., Clemson University, Clemson, SC 29634-0911.
[2] Formerly, Grad. Res. Asst., Dept. of Civil Engrg., Clemson University, Clemson, SC 29634-0911.

Reliability Analysis

The limit state function for a roof component subject to uplift load can be written as $g(x) = R - (W - D)$ where x = vector of basic random variables, R = capacity of the sheathing panel or connection, W = wind load acting on the component, and D = dead load. The probability of failure is found by integrating the joint probability density function (PDF) of the basic variables $x = (R, W, D)$ over the failure region $g(x) < 0$. In most cases, the exact form of the joint PDF is unknown since complete probability information on x is seldom available. Alternatives to integration of the joint PDF include first-order second-moment (FOSM) techniques and simulation. A second-moment reliability index β is defined as a function of the failure probability as $\beta = \Phi^{-1}(1 - P_f)$ where Φ^{-1} = inverse standard normal CDF.

Statistics for the resistance of the individual roof components (sheathing attachment, roof-to-wall connection) are taken from previous testing programs [e.g., Mizzell, 1995; Schiff et al., 1996; Tyner, 1996]. The uplift capacity statistics for the roof sheathing panels and roof-to-wall connections considered in the present study are summarized in Table 1.

The wind load can be modeled as a Lognormal random variable with mean-to-nominal = 0.41 and COV = 0.41 [Cheng, 1998]. The nominal wind load values can be determined using the appropriate load standard. Load distribution is determined by the code-specified roof zones and the assumed effective tributary areas. The dead load statistics are also shown in Table 1.

Table 1. Summary of Resistance and Dead Load Statistics for the Present Study

Resistance (Uplift capacity)	Mean	COV	CDF
4 ft × 4 ft (1.22 m × 1.22 m) sheathing	73.3 psf (3.51 kN/m^2)	0.20	Normal
4 ft × 8 ft (1.22 m × 2.44 m) sheathing	57.7 psf (2.76 kN/m^2)	0.20	Normal
Connection built with Simpson H2.5 strap	1312 lbs (5.84 kN)	0.10	Normal
Dead Load	Mean	COV	CDF
Dead load on sheathing panel	3.5 psf (0.17 kN/m^2)	0.10	Normal
Dead load on roof-to-wall Connections	15 psf (0.72 kN/m^2)	0.10	Normal

Since an "exact" solution to the limit state probability of failure (or reliability) cannot be provided in closed-form, advanced first-order second-moment (AFOSM) techniques were used. The sheathing panel reliabilities were shown to be lowest at or near the gable end. The lowest roof-to-wall connection reliability indices were generally found at the second and third connections from the gable end. A more complete discussion of the results may be found in [Cheng, 1998].

Simplified Reliability Analysis

The wind load was shown to have a Lognormal distribution. For the sheathing and roof-to-wall connection capacities, both the Normal and Lognormal distributions were found to fit equally well, in part due to the relatively small sample sizes. Therefore, the Lognormal distribution could be assumed for the convenience of developing a simplified closed-form solution. The COV of the dead load is very small, therefore it may be assumed to be deterministic in the simplified analysis. Since the dead load acts in an opposite direction from the wind uplift load, it can simply be added to the uplift capacity. Under these simplifications, the limit state function can be reduced to a two-variable R-S problem in which both the resistance (R) and the load (S) are Lognormal random variables, and the second-moment reliability index can be expressed as:

$$\beta = \frac{\lambda_R - \lambda_S}{\sqrt{\xi_R^2 + \xi_S^2}} = \frac{\ln\left(\dfrac{\mu_R \sqrt{1+V_S^2}}{\mu_S \sqrt{1+V_R^2}}\right)}{\sqrt{\ln\left[(1+V_R^2)(1+V_S^2)\right]}} \tag{1}$$

in which λ_R, ξ_R = Lognormal parameters of R, λ_S, ξ_S = Lognormal parameters of S; and μ_R, V_R = mean and COV of R and μ_S, V_S = mean and COV of S.

For comparative purposes, reliability indices were calculated using eqn. (1) and compared with those using advanced first-order second-moment techniques. One comparison is provided herein; a more complete presentation may be found in [Cheng, 1998]. The safety indices calculated using eqn. (1) are shown in Figure 1 for one case and set of design assumptions. Also shown on this figure (in parentheses) are the results from the AFOSM analysis. The simplified closed-form approach yields very similar safety indices to those obtained using the more complete AFOSM method. Note that for the edge zone full-size sheathing panels, the safety indices from the closed-form analysis are very close to those obtained using AFOSM methods. This is due to the large degree of overlap of the PDF's of sheathing panel capacity and wind load, i.e., large failure probabilities. In these cases, simplifications affecting the distribution tails have little effect on the estimated reliabilities.

Summary and Conclusions

A study was conducted to investigate the relative safety (reliability) of roof system components in light-frame construction subject to high wind uplift [Cheng, 1998]. This was achieved through reliability analysis of roof sheathing panels and roof-to-wall connections in three "typical" residential structures assumed to be located along the NC/SC coast. The baseline houses were all assumed to have gable-end roofs. Specific limit states considered in the reliability analysis of the roof system included loss of sheathing and failure of the roof-to-wall connections. Resistance statistics were obtained from a combination of experimental and analytical investigations conducted previously. Statistics of the wind loads acting on the different components were determined using the nominal wind loads and the composite wind load statistics described in [Cheng, 1998].

In the present paper, a simplified two-variable (R-S) reliability analysis was presented as a closed-form alternative to the more complete second-moment analysis. Within the context of the problem being investigated (i.e., for the purposes of providing some measure of relative risk of individual roof components), the simplified analysis was shown to provide reasonably accurate results. As a closed-form analysis, it is also very easily incorporated into a spreadsheet.

References

Cheng, N. (1998). "Reliability of Light-Frame Roof Systems Subject to Wind Uplift," MS Thesis, Department of Civil Engineering, Clemson University, Clemson, SC.

Ellingwood, B., Galambos, T.V., MacGregor, J.G. and Cornell, C.A. (1980). *Development of a Probability Based Load Criterion for American National Standard A58*, NBS Special Publication SP577, U.S. Department of Commerce, National Bureau of Standards, Washington, DC.

Ellingwood, B. and Tekie, P.B. (1997). *Wind Load Statistics for Probability-Based Structural Design*, NAHB Research Center, Upper Marlboro, MD.

Mizzell, D. P. (1994). "Wind Resistance of Sheathing for Residential Roofs," MS Thesis, Department of Civil Engineering, Clemson University, Clemson, SC.

Schiff, S.D., Rosowsky, D.V. and Lee, A.W.C. (1996). "Uplift Capacity of Nailed Roof Sheathing Panels," *Proceedings*: International Wood Engineering Conference, New Orleans, LA, 4:446-470.

Tyner, K.G. (1996). "Uplift Capacity of Rafter-to-Wall Connections in Light-Frame Construction," MS Thesis, Department of Civil Engineering, Clemson University, Clemson, SC.

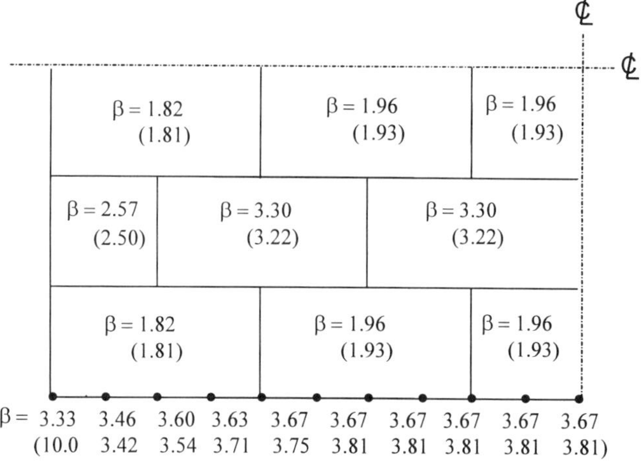

Figure 1. Component-level reliabilities (AFOSM results in parentheses) for one-story house, 22.6 × 40 ft (6.9 × 12.2. m), gable roof, 4:12 slope, framing 24 in (610 mm) o.c., no overhang (ASCE 7-95, closed building, components & cladding provisions)

PROBABILISTIC NONLINEAR COMPUTER MODEL FOR REINFORCED GLULAMS

H.J. Dagher, Ph.D., P.E.
Robert F. Lindyberg, M.S.C.E., P.E.
Advanced Engineered Wood Composites Center
University of Maine

1. Introduction

In 1997 the University of Maine completed the laboratory testing of ninety FRP-reinforced glulam beams, 22 feet long, in conjunction with Willamette Industries, Georgia-Pacific Resins Inc., Strongwell, and APA-the Engineered Wood Association. This test showed that FRP reinforcement in the order of 3% can increase the allowable strength of the beams by over 100%. In addition, the physical testing was used to verify a nonlinear stochastic computer model for reinforced glulam, called ReLAM.

ReLAM uses Monte Carlo simulation techniques to predict the strength and stiffness of a population of reinforced glulams. ReLAM requires as input distributions of lumber MOE, Ultimate Tensile Strength (UTS), and Ultimate Compression Strength (UCS). The objective of this study is to verify ReLAM using the 90 tested reinforced glulams.

2. Full Scale Beam Tests

102 twenty-two foot long, 5 1/8inch x12 inch beams were fabricated by Willamette Industries. Half were western hemlock and the other half were douglas fir. Three reinforcement ratios (0%, 1.2%, and 3.6%) were used for each species. The FRP was placed directly above the lowest wood lamination in the beam (herein called bumper-lam). The size of this lowest wood lamination was adjusted so that the total depth of each beam remains at 12 inches. A random layup of special L2/L3 lamstock grade was used in all beams.

- Ninety (90) of the one-hundred and two (102) beams were tested under static load to failure. The static flexure tests were performed in accordance with ASTM D198: Standard Methods of Static Tests of Lumber in Structural Sizes (4-point bending).
- The twelve (12) remaining beams are currently undergoing an accelerated creep test at the University of Maine, using a load equivalent to 1.25 x (design allowable).

3. Full-Scale Beam Test Results

The full-scale beam tests were used to calculate the following beam properties: For each beam, the Modulus of Elasticity (MOE) was calculated from the initial linear-elastic slope of the load deflection curve. Two values of the Modulus of Rupture were calculated: One value is based on the gross cross section at bumper strip failure (MOR_b), and the other value is based on the reduced section modulus (without the bumper-lam) at ultimate failure (MOR_{ult}). This was done to see what the performance gains would be had the bumper strip of wood been omitted. Table 1 summarizes the beam test results.

Using the mean ultimate bending stresses and the COVs from Table 1, the allowable bending stresses are determined using methods in ASTM D3737. The allowable bending stress is the 5% lower tolerance limit (LTL) with 75% confidence, divided by 2.1. The 5% LTL with 75% confidence was calculated as based on a sample size of 15 beams following ASTM D2915:

Test 5% LTL MOR $= \mu_{MOR(test)} - 1.991 * \sigma_{MOR(test)}$ Eqn. (1)

F_b = Test 5% LTL MOR/2.1   Eqn. (2)

Where 2.1 is a combination load duration factor (1.6) and safety factor (1.3).

Table 1: Test Results: Ultimate Strength and MOE

Test Group		MOE	MOR$_b$ Gross Section		MOR$_{ult}$ Reduced Section	
Beam Designation	# Specimens	(x 10^6 psi)	(psi)	COV (%)	(psi)	COV (%)
W Hem Control	15	1.32	4,338	15.4	N/A	N/A
1.2% W Hem	15	1.41	4,794	15.3	6251	17.0
3.6% W Hem	15	1.51	5,521	10.9	8357	12.7
Doug-fir Control	15	1.49	4,309	11.5	N/A	N/A
1.2% Doug-fir	15	1.58	5,044	19.2	6,656	9.8
3.6% Doug-fir	15	1.74	5,755	13.4	8,844	8.9

The allowable stresses were calculated based on the gross section at bumper strip failure ($F_{b\text{-}bumper}$), and based on the reduced section at ultimate failure ($F_{b\text{-}ult}$). Table 2 summarizes the allowable stress results:

Table 2: Test Results: Allowable Stresses, MOE, and Performance Gains

Test Group		Average MOE		$F_{b\text{-}bump}$ Gross Section		$F_{b\text{-}ult}$ Reduced Section	
Beam Designation	# Specimens	(x 10^6 psi)	% Gain	(psi)	% Gain	(psi)	% Gain
W Hem Control	15	1.32	-	1,433	-	N/A	-
1.2% W Hem	15	1.41	6.8%	1587	10.7%	1,970	37.5%
3.6% W Hem	15	1.51	14.4%	2,058	43.6%	2,975	108%
Doug-fir Control	15	1.49	-	1,581	-	N/A	-
1.2% Doug-fir	15	1.58	6.0%	1,483	-6.2%	2,553	61.5%
3.6% Doug-fir	15	1.74	16.8%	2,010	27.1%	3,466	119%

4.   In-Grade Properties of Laminating Stock

Approximately 1500 board feet (BF) were selected at random out of the L2/L3 lamstock used to fabricate the test beams. The lamstock had been MOE tested with a Continuous Lumber Tester (CLT) by Willamette Industries and the CLT-MOE was recorded for each lamstock specimen. The majority of the specimens consisted of 13-foot long 2x6's. In total, there were (50) Douglas Fir specimens, and fifty (50) Western Hemlock specimens.

Compression Testing: The lamstock specimens were shipped to the APA, where one foot was cut off one end (see Figure 1 below) for compression testing. The 100 one-foot long specimens were shipped to the University of Maine for testing in compression parallel to grain. The gage length for the displacement measurements was 9 inches. The tests were performed in accordance with ASTM D198.

Tension Testing: The 100 twelve (12) foot long specimens were tested at the APA for tension parallel to grain, using a 7-ft gage length. The tests were performed in accordance with ASTM D198.

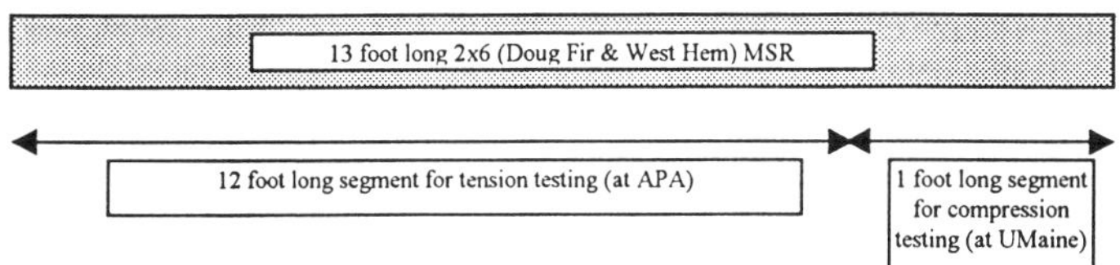

Figure 1: Lamstock Test Specimens

The following mechanical properties were obtained:

1. **MOE$_{CLT}$**: Modulus of Elasticity obtained from the Continuous Lumber Tester (Table 3)
2. **UTS**: The Ultimate Tensile Stress parallel to grain (Table 3).
3. **UCS, Compression stress-strain curve**: The Ultimate Compression Stress parallel to grain, the full stress-strain curve in compression parallel to grain. The data were obtained by testing one-foot long lamstock specimens to failure in compression parallel to grain. Using this data, the falling slope of an approximated bilinear stress-strain curve was obtained (Table 3).
4. **The correlations between MOE$_{CLT}$, UTS, and UCS.** See Table 4.

Table 3: Mechanical Properties of Lamstock in Test Beams

	Douglas Fir			Western Hemlock		
	Data points (number)	Mean (psi)	COV (%)	Data points (number)	Mean (psi)	COV (%)
MOE$_{CLT}$	50	1.54×10^6	10.2	50	1.41×10^6	10.0
UTS$_{1\,foot}$	50	4,067	24.4	50	4,712	31.8
UCS	30	6,447	13.2	30	5,982	14.2

Table 4: Correlation Matrix of MOE$_{CLT}$, UTS & UCS

	Doug Fir			Western Hemlock		
	MOE$_{CLT}$	UTS	UCS	MOE$_{CLT}$	UTS	UCS
MOE$_{CLT}$	1.0	0.22	0.11	1.0	0.50	0.46
UTS	0.22	1.0	0.20	0.50	1.0	-0.13
UCS	0.11	0.20	1.0	0.46	-0.13	1.0

5. Simulation of Glulam Beams

The first step of ReLAM is to simulate a glulam beam layup. ReLAM simulates a single cross section of a layup as specified by the user (see Figure 2 below). The lumber properties used in the simulation were generated based on the means and the COVs listed in Tables 3 and 4 above. Following are the input parameters currently required by ReLAM:

1. **MOE**: Mean and COV for each species and grade of wood, and the reinforcement.
2. **UTS:** Mean and COV for each species and grade of wood, and the reinforcement.
3. **UCS:** Mean and COV for each species and grade of wood.
4. **Correlations between MOE, UTS, &UCS:** for each species and grade of wood.
5. **Descending Slope of the Bilinear Stress-Strain Curve:** for wood in compression.
6. **Ultimate Compression Strain:** for wood at extreme compression fiber of beam at beam failure.

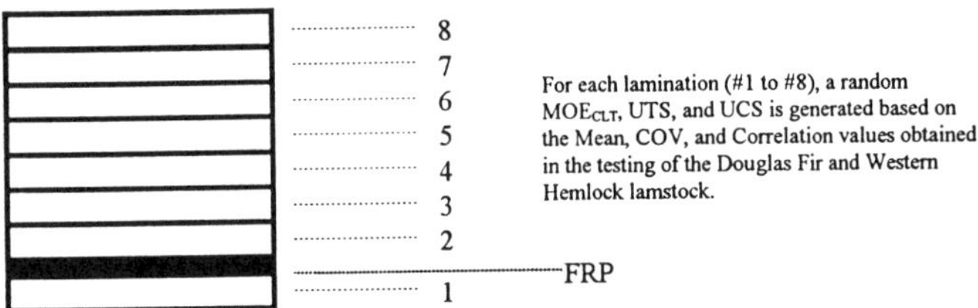

Figure 2: Simulation of Beam Cross-Section

After simulating the mechanical properties for the section shown in Figure 2, the computer program uses nonlinear approach to analyze the section in bending. This method of analysis does not assume a tension or compression failure, but rather calculates the real bending stresses through the section. This analysis method thus calculates the load deflection curve for a beam through progressive tension failure and compression yielding of laminations, to complete failure of the beam.

Using ReLAM, 500 beam simulations were performed for each beam category tested. The mean and COV of MOE and Ultimate Modulus of Rupture (MOR$_{ult}$) were calculated for the 500 simulated beams and compared to the data from the 90 tested beams (15 samples for each of six beam types). The results are given in Tables 5 and 6.

From the property distribution listed in Tables 5 and 6, the MOE and allowable bending stress can be calculated for the simulated beams. The design MOE for the simulated beams is simply the mean MOE. The allowable bending stress for the simulated beams can be found by calculating the 5% LTL of the beam MOR, and dividing by a factor of 2.1. From ASTM D2915 Table 3, the 5% LTL with 75% confidence in a data set of 30 (the total number of compression lamstock samples tested) is:

Test 5% LTL MOR = $\mu_{MOR(test)}$ - 1.869 * $\sigma_{MOR(test)}$ Eqn. (3)

Equation (2) is then used to calculate F_b for the simulated beams. Table 7 shows a comparison between the test values for MOE and $F_{b\text{-}ult}$ and those values predicted by the computer model. Table 7 shows that ReLAM predicted the beam MOE within 5.7% and in most cases within 1%. With the exception of the control beams, ReLAM predicted the allowable bending stress with an error of −5.5% to 6.7%. It should be noted that the current version of ReLAM is only intended for the analysis of <u>reinforced</u> glulams. This data shows that ReLAM is very accurate in predicting the MOE and F_b for reinforced glulam beams.

Table 5: Comparison of Experimental & ReLAM Results for Doug-fir

	Doug-fir Control		1.2% Doug-fir		3.6% Doug-fir	
	Test	ReLAM	Test	ReLAM	Test	ReLAM
Mean MOE ($\times 10^6$ psi)	1.49	1.50	1.58	1.59	1.74	1.74
COV MOE (%)	4.75	6.11	6.71	5.30	8.02	4.57
Mean MOR_{ult} (psi)	4309	4035	6656	6261	8844	9062
COV MOR_{ult}	11.53	19.53	9.78	10.19	8.90	7.67

Table 6: Comparison of Experimental & ReLAM Results for Western Hemlock

	W Hem Control		1.2% W Hem		3.6% W Hem	
	Test	ReLAM	Test	ReLAM	Test	ReLAM
Mean MOE ($\times 10^6$ psi)	1.32	1.32	1.41	1.33	1.51	1.50
COV MOE (%)	6.10	6.29	5.74	5.85	8.11	6.01
Mean MOR_{ult} (psi)	4338	4467	6251	5853	8357	7854
COV MOR_{ult}	15.39	18.46	16.98	15.52	12.68	10.71

Table 7: ReLAM Predictions versus Test Results

Test Group			Mean MOE ($\times 10^6$ psi)			$F_{b\text{-}ult}$ Reduced Section (psi)		
Beam Designation	Number of Lab Test Samples	Number of ReLAM Simulations	Lab Test	ReLAM	% Diff	Lab Test	ReLAM	% Diff
Control WH	15	500	1.32	1.32	0.0	1433	1345	-6.1%
1.2% WH	15	500	1.41	1.33	-5.7%	1970	1926	-2.2%
3.6% WH	15	500	1.51	1.50	-0.7%	2975	2942	-1.1%
Control DF	15	500	1.49	1.50	0.7%	1581	1220	-22.8
1.2% DF	15	500	1.58	1.59	-0.6%	2553	2413	-5.5%
3.6% DF	15	500	1.74	1.74	0%	3466	3697	6.7%

6. Conclusions

The probabilistic model developed at the University of Maine was able to predict the mean MOE for the reinforced Doug-fir and Hem-fir beams within 6%, and the allowable strength for the reinforced Doug-fir and Hem-fir beams within 7%. This demonstrates that the probabilistic nonlinear approach used by the University of Maine is an accurate method to analyze CR-glulam beams. As envisioned, the model requires simple input data and is expected to be very valuable tool for industry.

Structural Reliability of Plank Decks

Andrzej S. Nowak[1], Chris Eamon[2] and Michael A. Ritter[3].

Abstract

The structural reliability of plank decks designed by the AASHTO codes are determined. Inadequacies in load distribution and plank resistance in the current Specifications are identified.

Introduction

The objective of this study is to evaluate the adequacy of the current AASHTO Specifications' design criteria for plank decks for highway bridges. This evaluation will be based on the reliability index, β, the probabilistic measure of the structural performance. It has been observed that the current AASHTO code (1996) provisions for load distribution for plank decks are not realistic. The problem was identified by the AASHTO Committee on Timber Bridges as a priority item requiring an urgent solution.

Plank decks are classified into one of two categories, depending on the direction of planks relative to the direction of traffic; transverse decks and longitudinal decks.

----

[1]Professor of Civil Engineering, University of Michigan, Ann Arbor, Mich. 48109
[2]Graduate Student, University of Michigan, Ann Arbor, Mich. 48109
[3]Forest Products Lab, Madison, WI 53705

For transverse plank decks, the span of stringers is usually 5-6m, while in older structures it can be up to 11m. Stringers are spaced between 300-450mm center-to-center, and not more than 600mm. They are made of sawn lumber. Typical Southern Pine size is 150x450mm, and Douglas-Fir can have larger sizes. Planks are typically 100x250mm or 100x300mm, with a length of 3.5-11m, and are nailed to stringers. Longitudinal plank decks are similar. For longitudinal decks, the major design parameter determined by the designer is the spacing between stringers.

In this paper, it is assumed that stringers have an adequate load carrying capacity and that they provide a sufficient support for planks. The design of stringers is not considered.

The study is focused on distribution of the truck load to plank decks and plank resistance. Material properties, in particular modulus of rupture (MOR), are based on actual test data. The load model is based on available weigh-in-motion measurements data, and the contact pressure between truck tire and road surface is modeled using the available literature.

Load and Resistance Models

The live load model is developed on the basis of the actual truck measurements. Extensive weigh-in-motion (WIM) measurements were carried out by researchers at the University of Michigan (Nowak et al. 1994). The study provided statistical data on gross vehicle weights (GVW), axle weights and axle spacing. The mean maximum weight for a two-tire wheel is 200 kN, or 100 kN for a single tire, and the coefficient of variation is 0.25. Actual tire contact area is based on work by Pezo et al. (1989), and is taken as 200x250mm for a single tire, where 250mm is in the direction of traffic. For a two-tire wheel unit, the area is taken as 500x250mm, with a 100mm space between tires.

The wheel load is applied as a uniform pressure on the planks. For transverse planks, if the plank width is larger than the length of contact area (250mm), then it is assumed that the load is distributed over the whole plank width. If the plank width is less than 250mm, then the plank takes only a portion of the wheel load proportional to the ratio of plank width and 250mm. Longitudinal planks are treated similarly. If the plank width is larger than the contract area width (200mm), then the load is distributed over the whole plank width. If the plank width is less than 200mm, then the load is reduced proportionally. Planks are modeled as continuous beams on elastic supports.

The major parameter which determines the structural performance of wood components is the modulus of rupture (MOR). The statistical model of MOR is based on actual in-grade tests carried out by researchers in Canada (Madsen and Nielsen 1978) and the test data was processed by Nowak (1983). The flatwise use factor, the ratio of MOR for edge-wise and flat-wise loading, is based on work by Stankiewicz and Nowak (1997).

The reliability analysis is carried our using the procedure developed for calibration of the AASHTO LRFD (Nowak 1995). Load and resistance are treated as lognormal random variables. Reliability is measured in terms of the reliability index, β. The analysis is performed for plank decks designed using AASHTO (1996) and AASHTO LRFD (1994). The reliability index is calculated using the following equation:

$$\beta = [\ln (m_R) - 0.5 \ln (V_R^2 + 1) - \ln (m_Q) + 0.5 \ln (V_Q^2 + 1)] /$$
$$[\ln (V_R^2 + 1) + \ln (V_Q^2 + 1)]^{1/2} \qquad (1)$$

Where m_R = mean resistance, m_Q = mean load, V_R = coefficient of variation of resistance, V_Q = coefficient of variation of load.

The corresponding β's are representative of current design practice. Calculation results are presented in Table 1 for AASHTO (1996) and AASHTO LRFD (1998).

Table 1. Reliability Indices for Douglas-Fir Planks, AASHTO LRFD (1998).

| | Transverse Planks | | | | Longitudinal Planks | | | |
| | Select | | Grade 1&2 | | Select | | Grade 1&2 | |
Size	1996	1998	1996	1998	1996	1998	1996	1998
100 x 150	5.3	5.3	6.1	6.1	*	6.4	*	*
100 x 200	4.6	4.6	5.3	5.3	*	5.6	*	6.6
100 x 250	3.6	3.6	4.9	4.9	7.6	5.2	*	6.5
100 x 300	4.2	4.2	6.1	5.4	7.9	5.9	*	7.0

The results indicate that there are considerable differences in the reliability indices. The lowest values of β are obtained for 100x150 and 100x200. On the other hand, from the system reliability point of view, 100x150 planks are better because they allow for a better load sharing between the planks.

Conclusions

Given that the LRFD code target reliability index is 3.5, it is clear that in most cases the codes are overly conservative. This is primarily the result of two factors: an unrealistic load distribution model, which assumes the entire wheel load is carried by a single plank, regardless of plank width, and flat-use factors which do not adequately predict plank capacity (current values in the code were significantly lower than those found in testing). Simple changes in the Specifications, such as adopting a more reasonable load distribution and analysis model, and more accurate flat-use factors, may dramatically improve results and establish code reliability index consistency.

REFERENCES

AASHTO, "Standard Specifications for Highway Bridges", Washington, D.C., 1996.

AASHTO, "LRFD Design Code for Highway Bridges", Washington, D.C., 1994.

Nowak, A.S., "Calibration of LRFD Bridge Design Code", ASCE Journal of Structural Engineering, Vol. 121, No. 8, pp. 1245-1251.

Nowak, A.S., "Statistical Analysis of Timber", Report UMCE 83-12, Department of Civil Engineering, University of Michigan, Ann Arbor, MI, 1983.

Pezo, R. F., Marshek, K. M., and Hudson, W. R., "Truck Tire Pavement Contact Pressure Distribution Characteristics for the Bias Goodyear 18-22.5, the Radial Michelin 275/80R/24.5, and the Radial Michelin 255/70R/22.5, and the Radial Goodyear 11R24.5 Tires", Research Report Number FHWA/TX-90+1190-2F, Center for Transportation Research, University of Texas, Austin, September 1989.

Stankiewicz, P.R. and Nowak, A.S., "Bending Tests of Bridge Deck Planks", Report UMCEE 97-10, University of Michigan, Department of Civil and Environmental Engineering, Ann Arbor, MI, May 1997.

Structural Instrumentation

Manuscripts for some presentations were not available at time of publication.

51. *DESIGN OF PRECISION EXPERIMENTS*

MODERATOR: Michael Giltrud
Defense Threat Reduction Agency
Alexandria, VA

(1) Sequential Experiments for Test Minimization
Thomas F. Curry, Deborah C. Parker, and William E. Skeith
Logicon RDA
Colorado Springs, CO

(2)Precision Experiments for Evaluating FE Models
Frank D. Dallriva and James L. Davis
USAE Waterways Experiment Station
Vicksburg, MS

(3) Load Calculations and Comparison with PWT Data
Charles E. Needham, Kara Peterson and Barry L. Bingham
Applied Research Associates, Inc.
Albuquerque, NM

(4) The Choice of Test Data for Validation of Analysis Codes
Brett A. Lewis and Barbra B. Lewis
APTEK, Inc.
Colorado Springs, CO

**(5) Validation and Verification of a Nonlinear Reinforced Concrete
Structure Model Using Precision Test Data**
Mark C. Anderson and Timothy K. Hasselman
ACTA Inc.
Torrance, CA

John E. Crawford
Karagozian & Case Structural Engineers
Glendale, CA

60. *METHODS OF MONITORING LARGE SCALE STRUCTURAL BEHAVIOR*

MODERATORS: Dr. Dryver R. Huston
University of Vermont
Burlington, VT

John DeWolf
University of Connecticut
Storrs, CT

1) Shoring Methods at Museum Towers
Dr. Dryver R. Huston, Peter Fuhr
University of Vermont
Burlington, VT

David Rosowsky
Clemson University
Clemson, SC

Wai-Fah Chen
Purdue University
West Lafayette, IN

Minhaj Kirmani
Weidlinger Associates, Inc.
Cambridge, MA

(2) Strain and Acceleration Assessment of Bridge Performance
John DeWolf
University of Conneticut
Storrs, CN

(3) Precursor Transformation Method for Health Monitoring Systems
Armin Mehrabi and Habib Tabatabai
Construction Technology Laboratories, Inc.
Skokie, IL

(4) Bridge Monitoring Systems
Harold Bosch
Federal Highway Administration
McLean, VA

(5) Remote Roof Stability Monitoring for Underground Nonmetal Mines
A. Iannocchione, R. Grau, and L. Prosser
National Institute for Occupational Safety & Health
Pittsburgh, PA

Sequential Experiments for Test Minimization

Thomas F. Curry, Deborah C. Parker, and William E. Skeith[1]

Characterization and modeling of complex physical processes such as the structural response of a hardened target to detonation of a conventional weapon requires destructive testing. Due to the expense of such tests, it is essential to keep the number of required tests to a minimum. The technique of Design of Experiments associates each experiment with a term in a polynomial that predicts damage as a function of the inputs. Thus, the terms (but not their coefficients) are known prior to obtaining the results of the experiments. This paper presents an overview of Design of Experiments as a tool for systematic planning and efficient analysis of physical tests. The methodology is used to determine the effect of a single bomb casing fragment on a concrete slab. A sequential experimental design approach decreased the number of required tests from 27 to 17 over more traditional experimental designs. Repeated experiments allowed estimation of natural process variability which increases the statistical validity of the prediction interval of the polynomial. The polynomial can be used directly in a simulation, or if the simulation already has a predictive algorithm, the algorithm can be checked to see if its result falls within the prediction interval of the polynomial.

With defense dollars constantly shrinking, more emphasis is being placed on obtaining the most information from the fewest tests. Design of Experiments (DOE) offers an efficient, systematic method for planning and evaluating many types of physical tests, whenever the purpose is to characterize a process by determining which factors affect the outcome. This process is often combined with simulation in a model-test-model paradigm where the results of the physical tests are used to validate the results of the simulation.

One such effort is currently underway to investigate the physical response of a hardened structure to detonation of a conventional weapon. The dynamic loading produced when the weapon detonates can include impulse from both the airblast and bomb casing fragments. In addition to the global structural response of a concrete slab (floor, walls, or ceiling), the concrete near the fragment impact location may be damaged and/or ejected. An understanding of the loading and material damage resulting from bomb-fragment impact is an essential input to structural response analyses. In support of the Conventional Weapons Effects Program sponsored by the Defense Threat Reduction Agency, the US Army Corps of Engineers' Waterways Experiment Station (WES), along with Logicon RDA, developed and conducted a set of experiments to evaluate the effects of bomb fragments impacting reinforced concrete walls.

The purpose of the first phase of the project was to determine how a fragment's mass, impact velocity, impact angle, and their interactions affect percent of momentum transferred (PMT) to a concrete slab suspended from a ballistic pendulum. An objective was to produce a high fidelity mathematical model from the fewest tests. Using a sequential experimental design approach, analysis conducted after each set of tests determined the follow-on tests required. This analysis (1) identified key factors and their influence on the response variables, (2) determined sources of variation, (3) produced

---

[1] Logicon RDA, 2864 S. Circle Dr., Suite 1100, Colorado Springs, CO 80906

mathematical models that can be used to predict percent of momentum transferred as a function of fragment mass, impact angle, and fragment velocity, and (4) demonstrated the methodology and effectiveness of sequential experimentation.

OVERVIEW OF EXPERIMENTAL DESIGN

Modern designed experiments use purposeful changes to the inputs of a process and efficient analysis of the corresponding changes in the outputs to predict how the inputs affect the mean and variability of the outputs. The process is viewed as a "black box" into which inputs (factors) are injected and outputs (responses) are measured (*Figure 1*).

From this process-oriented view, an application of experimental design methods involves (1) choosing the output variable to measure, (2) choosing input factors, and their settings with which to experiment, (3) selecting the appropriate experimental design, (4) conducting the experiment, (5) prioritizing the factors by the magnitude of their effect per unit change on the output,

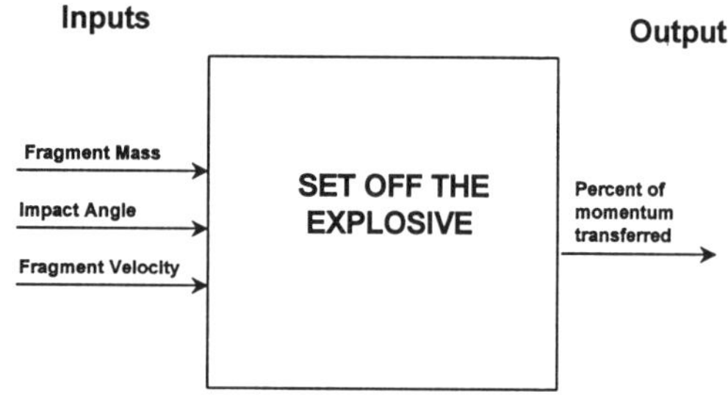

Figure 1. **Single Fragment Black Box**

(6) developing the mathematical model, and (7) applying the results of the analysis to meet the experiment's objectives.

In a designed experiment, all inputs are varied simultaneously in a controlled manner. If analysis of the test results indicates that a more detailed model is needed, one makes specific additional runs and combines them with previous results to obtain greater fidelity.

Poor initial planning and lack of test discipline are the primary causes of disappointing experimental results. The overall objective is to gather information about a process. An organized approach to choosing experimental factors, determining their settings for the experiment, conducting the experiment, collecting the data and analyzing the results is critical to success.

Since testing always comes at a price (whether measured in time, money, resources, or lost opportunity), it is important to structure the experiments into a carefully ordered sequence that can be terminated when objectives are met. A good sequential approach conserves resources and provides information in phases to aid in planning future experiments, thereby increasing the chances of success.

SINGLE FRAGMENT EXPERIMENTATION

For each experiment, a cylindrically shaped mass of steel, representative of a bomb casing fragment, was shot from a 50 caliber gun at a concrete slab. The .6096m x .6096m x .1542m slab was suspended from a ballistic pendulum in order to measure horizontal displacement (lateral movement of the slab). Horizontal displacement is then used to compute the percent of momentum transferred to the slab. It was concluded that, in order to realistically represent the characteristics of a bomb fragment, mass should vary from 5.8329g to 9.7215g, impact velocity should vary from 1066.8 to 1432.56 meters per second, and impact angle should vary from 45 to 90 degrees.

Analysis of Main Effects

Experimental designs are often presented as "test matrices" that list the combinations of input factor settings for each test. The test matrix for the first set of experiments is given in *Table 1*. In design of experiments, input factors are generally orthogonalized, and scaled from -1 to 1. (*Table 1* also gives the non-scaled factor settings.) There are three main factors to be analyzed. Consequently, four runs are required to screen for just the main

Test #	Fragment Mass (g)		Fragment Velocity (m/sec)		Impact Angle (deg)	
	Setting	Actual	Setting	Actual (target)	Setting	Actual
1-3	+	9.7215	+	1432.56	+	90
4-6	+	9.7215	−	1066.80	−	45
7-9	−	5.8329	+	1432.56	−	45
10-12	−	5.8329	−	1066.80	+	90

***Table 1.* Factor Settings, Initial Runs**

effects. If one pictures these factors in three-space (see *Figure* 2), then the parameter space of the experiment is a cube with 0 at the center, and its corners represent the three factors at combinations of their low (−1) and high (+1) settings. The four runs in *Table 1* correspond to 4 corners of this parameter space (one-half of the 8 corners, hence the name half-fractional factorial). Three experiments were conducted at each of the four combinations of settings to allow estimation of process variability.

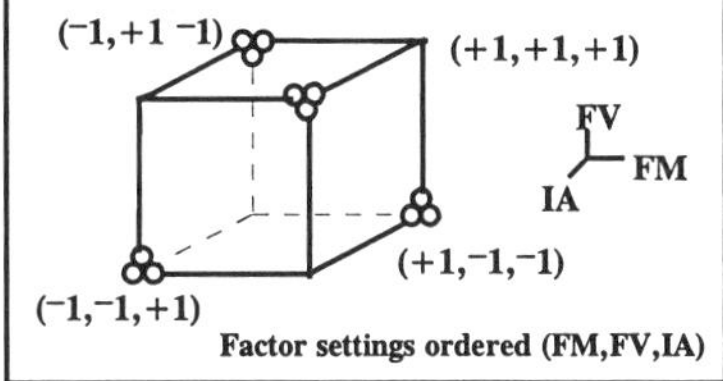

***Figure 2.* Orthogonalized Factors**

The actual data matrix for the first 12 experiments is in *Table 2*. Multiple linear regression was used to analyze these first 12 runs for main effects. The results are in *Table 3*. At this point in the analysis, the math model is

$$\mathbf{PMT} = 172.54 - 5.08*\mathbf{FM} + 3.08*\mathbf{FV} - 5.91*\mathbf{IA}$$

where levels for all factors are first orthogonalized (placed on their respective −1 to +1 scale).

It is impossible to tell whether impact angle is truly significant, or if the interaction with which it is aliased (FM*FV) is significant. Additional runs are required at the other four corners of the parameter space to allow estimation of the interaction terms.

After adding those additional 4 runs to the test data, analysis determined the interaction terms were not significant. The coefficients for the main effects were recomputed using 16 runs which resulted in the following prediction equation

Test #	FM	FV	IA	(PMT)
1	9.7215	1381.628	90	162.1
2	9.7215	1345.479	90	159.9
3	9.7215	1389.827	90	168.9
4	9.7215	1068.812	45	167.8
5	9.7215	1019.800	45	173.1
6	9.7215	1047.293	45	168.8
7	5.8329	1419.941	45	190.3
8	5.8329	1428.354	45	172.1
9	5.8329	1418.783	45	197.0
10	5.8329	1067.958	90	166.4
11	5.8329	1091.946	90	159.9
12	5.8329	1093.500	90	180.4

***Table 2.* Results of Tests 1-12**

$$\mathbf{PMT} = 174.71 - 4.80*\mathbf{FM} + 4.90*\mathbf{FV} - 4.80*\mathbf{IA}.$$

Model Confirmation

The math model was used to predict the response at the center of the parameter space, where all factors are set at their mid-levels (zero when orthogonalized). Fragment velocity, because it can't be controlled exactly, was somewhat

Factor	PMT	
	Coefficient	Significance
Constant	172.5374	.0001
~ Fragment Mass (FM)	-5.0831	.0812
~ Fragment Velocity (FV)	3.0767	.2986
~ Impact Angle (IA)	-5.9126	.0474

~ indicates factors are scaled

***Table 3.* Main Effect Coefficients (first 12 runs)**

below center. Results of this confirmation run are summarized in *Table 4.* Models are said to "confirm" when subsequent observations fall within the confidence interval. The four term linear model for PMT confirmed at the center point with 90% confidence.

Input Parameters			Prediction and 90% Confidence Interval				
FM	FV	IA	Lower 90% CI	Predicted	Upper 90% CI	Actual	Confirm?
7.7772	1200	68	157.6	173.7	189.8	173.4	Yes

Table 4. **Predictions and 90% Confidence Intervals vs. Actual Results at the Center Point**

Confirmation indicates that the linear approximations for PMT are adequate over this range of input values. If the model had failed to confirm, a quadratic effect would have been suspected, since only linear effects can be modeled with two level designs.

CONCLUSIONS

Experimental design techniques enabled efficient planning and analysis of tests to characterize the loading and structural responses associated with bomb fragment impact. These single fragment tests allowed estimation of the effect of fragment mass, fragment impact velocity and impact angle on the percent of momentum transferred when the fragment impacts a concrete slab. Tests using multiple fragments will build on the results obtained here.

Analysis of these designed experiments indicates that fragment mass, fragment velocity, and impact angle all have a significant effect on PMT.

These results were obtained more economically (fewer tests) using a specific experimental design approach called sequential experimentation. In sequential experimentation, analysis is conducted at specified points in the testing process to determine the additional tests required. Starting with a full factorial experimental design to estimate all three main effects and their interactions (without prior knowledge of process variability) would have required three replications each of $2^3=8$ experiments, for a total of 24 experiments. Three additional confirmation runs would have brought the total to 27 runs. By conducting only the first half of the replicated tests (12 runs), followed by analysis of main effects and process variability, it was determined that all effects appeared significant and process variability was constant. Since all main effects tested as significant, it was necessary to conduct four additional tests to allow estimation of interactions separately from main effects. However, since the initial tests also showed that process variability was constant, the additional four tests and the confirmation test did not require replication. So, only 5 additional tests (instead of 15) were required to complete the analysis. This savings of 10 tests using a sequential experimental design was in addition to the immeasurable benefits of using an experimental design in the first place.

ACKNOWLEDGEMENT: The analysis presented herein was performed by Logicon RDA in cooperation with the U.S. Army Engineer Waterways Experiment Station. We would especially like to thank Mr. Vince Chiarito and Mr. Frank Dallriva for their role in conducting the tests that produced the data used in this analysis, and for their assistance throughout all phases of the study.

Precision Experiments for Evaluating FE Models

Frank D. Dallriva[1] and James L. Davis[1]

Abstract

Experiments were conducted to evaluate the capability of structural analysts, using nonlinear finite-element methods, to predict the response of reinforced concrete walls subjected to the airblast from the detonation of bare explosive charges. Each experiment involved a detonation in the center bay of a three-bay structure to produce damage to the two interior walls. In each structure, the two interior walls were of identical design for evaluating the repeatability of the structural response. Instrumentation was installed to provide measurements of the airblast environment and the structural response.

Introduction

The finite-element method is a valuable tool for computing the response of reinforced concrete structures subjected to the dynamic loading associated with the detonation of explosives in or near the structure. It is useful for predicting the response associated with a particular explosive scenario of interest, as well as for developing response surfaces for a set of variable parameters for use in fast-running PC-based algorithms. The finite-element models must, however, be validated with experimental data to provide confidence in the response-surface fits. By conducting precision explosive experiments, the U.S. Army Engineer Waterways Experiment Station (WES) has supported the Defense Threat Reduction Agency's effort to improve and validate finite-element models for predicting structural response to explosive loading.

Experiment/Analysis Comparison Issues

When using data from an explosive experiment to validate a structural-response analysis, certain information must be provided to the structural analyst for input to the analytical model. This information includes the structural details, the

---

[1] US Army Engineer Waterways Experiment Station, 3909 Halls Ferry Road, Vicksburg, Mississippi 39180-6199

material properties, and the dynamic loading on the structure. The accuracy of this information and the way in which it is incorporated into the finite-element model will contribute to how well the experimental and analytical responses compare. If the information provided to the analyst were 100 percent accurate with respect to the as-built structure, then any deviation from the experimental results could be attributed to some problem in the calculated results. All measurements, however, contain inaccuracies. It is important, therefore, if the analytical results are to be judged based on how well they compare with the experimental data, that one understands the effects of the inaccuracies in the information provided to the analyst. If possible, the experiment should be designed to minimize their effects. For instance, it is undesirable to design an experiment in which the response mode changes or the degree of damage significantly varies within the range of accuracy of the measured load. Hence, pretest calculations to identify potential problems such as these may be necessary.   Also, it may be necessary for structural analysts to conduct several analyses as a parameter study, to determine the effects of measurement inaccuracies and uncertainties on their results.

Due to construction tolerances, there will always be inherent differences between two supposed identical structures. Therefore, replicate test specimens may be necessary to evaluate the repeatability of the response under loading conditions as nearly identical as possible.   The structural analysis results should not be expected to compare any better with the experiment than the response can be repeated experimentally under the same loading conditions.

Model Structure Description

As shown in Figure 1, each 1/4-scale structure was one-story high and consisted of three rooms, including one center room and two end rooms. The two interior walls, which were the responding walls of interest, were replicates in their respective structures for the purpose of evaluating the repeatability of the wall response under essentially the same loading conditions.

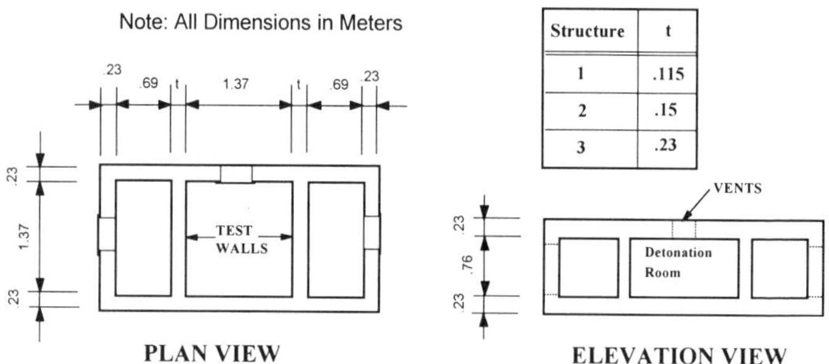

Figure 1. Structure dimensions.

Instrumentation

The structures were instrumented with transducers to record airblast pressure, structural velocities, and reinforcing bar strains. Airblast pressures were measured on the exterior walls in the detonation room, rather than on the interior walls, since the installation of gage mounts in the test walls could result in an effect on the wall response. In addition to shock pressure measurements on the walls, quasi-static pressure measurements were recorded on the floors of interior and exterior rooms. Measurement of the pressure in the exterior rooms was to provide information on venting from the interior to the exterior rooms in case of failure of the test walls. Accelerometers were mounted on the back face of the interior walls to provide measurements of velocity and displacement at key locations. Strain gages were installed in the test walls on both vertical and horizontal reinforcing bars at mid-span and near the supports.

Test Configuration and Execution

After construction and instrumentation-installation were complete, the structure was placed on a layer of sand in a steel cylindrical container. Instrumentation cables were connected to the transducers, high-speed film cameras were mounted in protective enclosures, and the structures were backfilled as shown in Figure 2. In each experiment, a bare, cylindrically shaped explosive charge of Composition C-4 was placed in the center of the detonation room at mid-height. After installation of the explosive charge, the steel top to the container was set in place, and the container was rolled into a large concrete reaction structure for experiment execution.

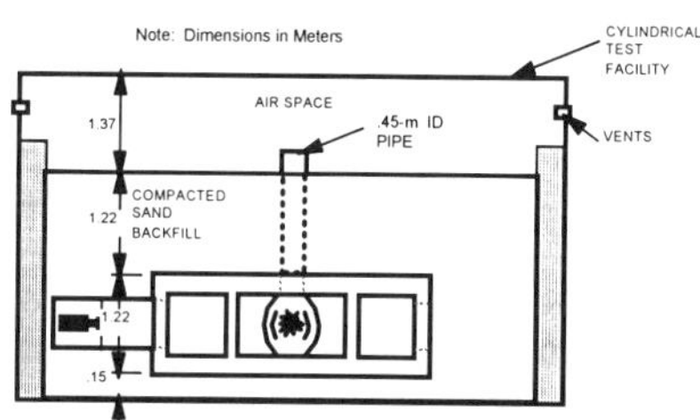

Figure 2. Test configuration.

Sample of Results

The structural analysts required plots of airblast-pressure vs. time at many locations on their grid, to provide a non-uniform loading representative of that in the experiment. It was not feasible to measure the airblast on the interior walls because placing gage mounts in the walls would have significantly affected the wall response. Also, measuring the airblast at the number of locations required by the analyst would have been not only cost prohibitive, but difficult to install due to the physical size of the mounts in a small-scale structure. For these reasons, the airblast was measured at selected locations on the exterior sidewalls, and the data were compared with airblast from hydrocode calculations of the experiments.   Adjustments were made to the airblast calculations, when necessary, to provide good correlation with the experimental airblast data.   The computed airblast pressure histories were then provided to the structural analysts for use as input loading to the structural response analysis.

The interior walls in Test 1 experienced light-to-moderate damage, responding primarily in flexure. The interior walls in Test 2 experienced severe damage due to combined flexure and shearing at the floor. The shearing at the floor in Test 2 appeared to have resulted due to the construction joint at the floor/wall intersection. The evaluation of how well analytical and experimental results compared involved comparing the time-dependent structural velocities and displacements, and reinforcing-bar strains, as well as comparing the final condition of the structure. Figure 3 shows an example of early-time impulse and displacements recorded along the vertical centerline of a test wall.

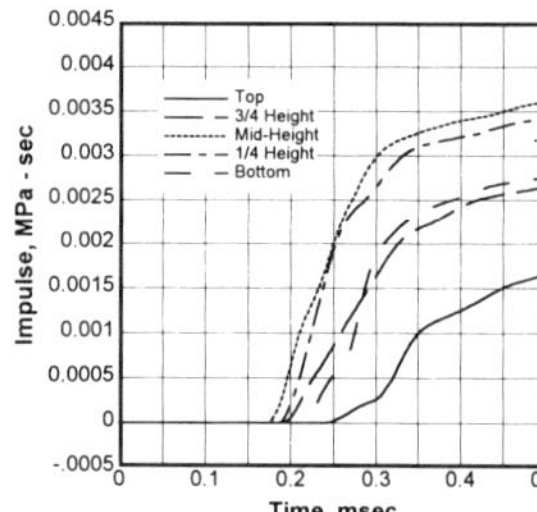

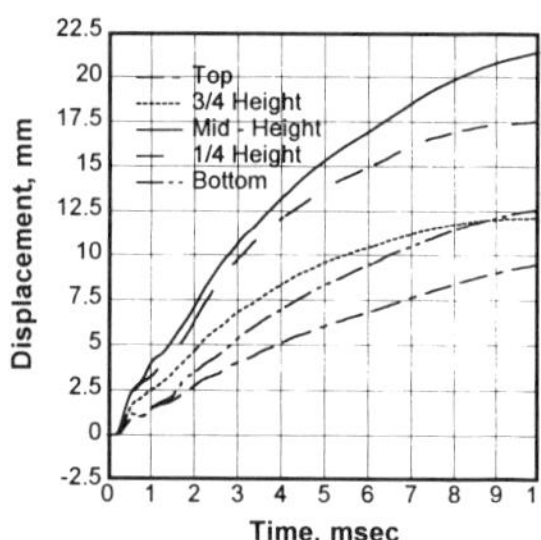

Figure 3. Early-time impulse and displacements along vertical centerline of wall.

Summary

When conducting precision explosive experiments for validating analytical models, there should be close coordination between the experimenter and the analysts. The experiment should be designed, if possible, such that relatively small differences in the loading, as compared with the design load, do not cause major changes in the structural response. The instrumentation layout should be designed to capture the response mode of the structure and to provide for direct comparison of individual records with analytical output at locations of interest.   If deemed necessary, the structural analyst should perform a parameter study to evaluate the effects of inaccuracies or uncertainties in the information used as input to the structural analysis.

Acknowledgments

The Defense Threat Reduction Agency (DTRA), Washington, DC, sponsored this research under the direction of Mr. Mike Giltrud. Permission to publish this paper was granted by DTRA and the Office, Chief of Engineers.

Load Calculations and Comparisons with PWT Data

Charles E. Needham[1], Kara Peterson[1], and Barry Bingham[1]

Abstract

A series of tests were conducted at Waterways Experiment Station (WES) under the sponsorship of the Defense Special Weapons Agency (DSWA). One-sixth scale reinforced concrete structures were instrumented and tested at various pressure levels and wall thickness ratios. The tests, designed to provide high quality experimental data for comparisons with results of numerical structural response codes, were called Precision Wall Tests (PWT). ARA made full 3-D hydrodynamic calculations of the airblast loads on the walls, ceiling, and floor of the structures for three tests and the calculated overpressure vs. time histories were provided to the structure response calculation contractors. These data were provided at hundreds of nodes on each wall and permitted the structure finite element codes to have forcing functions defined at each node. While the airblast calculation was resolved to 5 mm, the data files were provided at approximately 0.08 m resolution.

PWT Test Configuration

The PWT tests were performed in a one-room concrete structure 1.3716 m by 1.3716 m and 0.762 m high. The ceiling had two circular vents 0.2128 m in radius placed symmetrically along the axis of the room. The vents extended to 2.5 m above the floor of the structure and emptied into a large plenum chamber. The C-4 charge was placed in the center of the room and detonated from the top. Experiment PWT1a used 1.6 kg of C-4 formed into a cylindrical charge with a length-to-diameter ratio of approximately 5. PWT2 used 2.7 kg of C-4 in a geometrically similar charge in the same sized room and PWT3 used 3.4 kg of C-4 in the same geometry.

ARA SHARC 2-D Detonation Calculation

The first calculation, a 2-D axisymmetric calculation, modeled C-4 burn and initial shock propagation. The grid used in this calculation contained 450 zones in

---

[1] Applied Research Associates, Inc., 4300 San Mateo Blvd., NE, Suite A220, Albuquerque, NM, 87110 (505) 883-3636.

the x direction and 508 zones in the y direction, with a zone size of approximately 1.5 mm. The grid boundaries were perfectly reflecting, with their locations corresponding to the positions of the ceiling, floor, and walls in the PWT structure. The axis of the grid corresponded to the axis of the charge. The detonation location was set at 32.1 mm from the top of the cylinder. When the burn was complete and before the shock had reached the position of the circular vent at the level of the ceiling, approximately 130.2 mm from the center of the charge, the 2-D calculation results were mapped into the 3-D calculation.

ARA SHARC 3-D Shock Propagation Calculation

In the 3-D SHARC calculation, the concrete walls were modeled with perfectly reflecting zones, and the boundaries along the axis of symmetry were made to be reflective. In both horizontal directions the active grid extended for .6858 m, half the length and width of the room. The x axis was parallel to the room bisector plane with the exterior walls at x = +/- .6858 m. In the vertical direction reflective zones representing the roof of the structure were located 0.762 m above the floor. The vertical grid extended above the roof to model the two vent pipes leading from the circular openings. The vent pipes extended 2.5 m above the floor and the grid extended an additional 0.5 m above the top of the vent pipes ending at a transmissive boundary. The grid used for all calculations was identical. The zone sizes in the mesh along the x and y directions were between 10 and 5 mm, with the smallest zones against the walls. Because the grid was longest in the z direction, ending at 3 m above the floor, the zones in this direction increased from approximately 10 mm to 80 mm at the top. This allowed for an efficient run with the finest zones located near the gage locations.

The 3-D calculation provided loading histories for all walls, floor, and ceiling. The measurements on the exterior walls were used to validate the calculated loads. Calculated airblast loads at all locations were provided to the structural response calculators.

Comparison of Calculated and Experimental Waveforms

The gages in PWT3 used protective screens over the gage heads. These gage protectors reduced the measured peak overpressure and was so severe that the early impulse was also reduced.

At location P3, other than the low measured peak and resultant loss of impulse, the calculation agreed very well with the experimental data in magnitude and reasonably well with the secondary wave arrival times. At P4 there was apparently a small signal arriving prior to the main wave in the experimental data. The first measured peak was less than half of the calculated peak. The second peak was also lower than the calculated value. The impulse after the first two peaks was very close and the magnitude at 3 ms was very good. At location P6 the same comments hold as for P4. It was interesting to note that at position P7 the arrivals of all calculated secondary shocks are early by nearly the same amount. After the first reflection, the shock transit times appeared to be correct. The impulses, after the first peak, remained parallel to each other.

Position P8, in the upper right quarter panel, showed excellent agreement in both arrival time and magnitude of the second peak. The calculated impulse remains parallel to the experimental data, indicating very close agreement in full waveform detail after the first peak. At P9, the closest gage to the vent openings, showed the effects of the perturbation on the first peak. Multiple shocks arrived in the first 200 microseconds. The experimental shock front was somewhat rounded and the shocks had more peaks than the calculated waveforms. Even the secondary shocks, which the calculation indicated had higher peaks than the first arrivals, showed a rounded front and a slow rise. These were also indications of the effects of the gage protectors.

PWT1a

At the center of the wall, P1, the first shock arrival was about 30 microseconds earlier than calculated. This early arrival was not consistent over all waveforms but was observed on gages P1, P4, and P5. The early arrival was not as great on gages P6, P10, and P12. This indicated a slight asymmetry during the charge detonation; especially evident by comparison with the close agreement in arrival on the opposite wall at P12. The overall agreement at P1 was excellent and the impulse was within a percent or two.

At location P4, a double shock structure was seen in the experimental data for the second and third arrivals. This was also an indication of a jet or asymmetry in the initial expansion of the charge. The impulse, while agreeing so well at location P1, here showed significantly higher values experimentally and the differences continued to grow with time. Location P5 showed good agreement except that after about 1.2 ms the impulse difference grew significantly. The rise at this location was not consistent with the comparisons at location P6. At this location the detailed agreement was excellent and the overall impulse was within a percent or two.

At the center of the upper right quarter panel, the agreement was again quite good. The impulse agreed well throughout the waveform with a slight upward drift in the experimental data. The agreement in impulse over the first 3 ms was within 10 percent. Station P9 was the closest to the vent openings and showed the effects in the first set of shock arrivals. While the calculation and the experiment showed complex structure of the waveforms and multiple shock arrivals in the first pulse, the phasing of the shocks was different and resulted in a higher calculated peak overpressure. The reflected wave arriving at about 0.8 ms showed a single peak and good agreement in arrival time but the calculated peak was somewhat larger than measured. After that time the impulse appeared to agree very well.

Location P10 (near the floor and least influenced by the vents) showed overall excellent agreement between calculation and experiment. The phenomena here included a strong influence of the detonation products and formation of a complex Mach reflection prior to striking the wall. The wave shapes, arrival times and peaks were all in excellent agreement and the impulse was very close over the majority of the extent of the waveform.

Gage P12 was on the opposite wall and was within 1.6 cm of perfect symmetry with position P4. The calculated peaks at these two locations were nearly identical but the measured first peaks differed by nearly a factor of two, a strong indication of a jet or asymmetry from the charge. The experimental impulse grew at a higher rate than calculated which may have been caused by a slight baseline shift in the gage.

Conclusions

Calculations were completed for three different yields in "identical" structures (differing only in the thickness of certain walls). The airblast load calculations reported here assumed that the walls were non-responding and therefore construction details were not important. The agreement between the 3-D calculations and the experimental data was a graphic example of the current state of the art in detonation and airblast computational capability.

The fact that there was very little divergence of the measured and calculated impulse values after the first reflected shock was a strong indication that wall motion was not an important factor during the first 3 ms on any of the events. On event PWT2 there appeared to be a slight but consistent reduction in the measured impulse after a time of about 4 ms. The effect was about 12 percent on the total impulse at 20 ms. From the point of divergence to the 20 ms time, the peak pressure was less than 20 percent of the peak. It was not plausible that the effect of the movement of the wall had any significant effect on the wall loading.

For PWT3, the highest yield event which generally gave the greatest loading, we had no observable difference in impulse between calculation and experiment during the first 3 ms after second shock arrival. We saw a slight difference between the SHARC calculation and experiment after a time of about 10 ms. In fact, after a time of 15 ms the experimental data were higher than the calculated data. This slight pressure difference occurs well after the wall had received the airblast impulse and had a very small influence by the wall motion.

The quality of the PWT airblast data provided a set of high precision waveforms for validation of hydrodynamic codes and shock models. The data included arrival times, representative loading pressure as a function of time, and impulse as a function of time. The placement of the gages provided symmetry checks on the experiments and the effects of Mach reflections from walls, floor, and ceiling. The high quality and consistency of the PWT data provide a benchmark against which modern hydrodynamic computer code models can be validated.

The Choice of Test Data for Validation of Analysis Codes

Brett A. Lewis, Ph.D.[1]
Barbara B. Lewis, Member ASCE[1]

Abstract

When using precision test data to validate non-linear dynamic structural response analysis computer codes, the main question is what data should be used for comparison. In this paper, we will discuss the types of measurements usually taken in a precision test and their relationship to validation of analysis tools and models. We will describe the test data used for comparison to blind predictions for validation of models and discuss how well the data represents the response.

Introduction

Structural test data generally falls into two categories: time-dependent and static. Both types are necessary for validation of dynamic structural response computer codes and material models. Time-dependent data includes accelerations, strains, and pressures. It also includes other quantities not measured directly, but derived from time-dependent measurements such as velocities and displacements derived from accelerations. High-speed video is another form of time-dependent data. If done properly, high-speed video can provide quantity measurements as well as quality of the response. Static assessment is done on the post-test structure. It includes measurements of residual displacement, determination of failure mechanisms, level of damage, and amount of spall.

In testing of reinforced concrete structures, the accelerometer data is the most useful although accelerations themselves are not easily compared. The test accelerations are usually filtered to remove high frequency ringing while analysis results are not. By integrating the data to obtain velocity, the high frequencies effects are minimized. The use of velocities as a comparative measure also has the advantage that structure rebound is most clearly shown by reversal of the velocities. A second integration of the accelerations gives displacements, which are easy to visualize and can be used as a check against photopole or residual displacement data. One drawback is that the accelerometer measurements do not always return to zero after the test so the data must be baseline

---

[1]Research Engineer, APTEK, Inc., 1257 Lake Plaza Drive, Colorado Springs, CO 80906. This work sponsored by the Defense Threat Reduction Agency.

shifted. Errors made in shifting the accelerations will be magnified in the velocities and displacements.

Strain histories can also be useful, but can have a great amount of local variation. For example, the strains in reinforcing bars in a concrete wall may vary considerably based on where cracks or breakage have occurred. Pressures are commonly recorded for explosive tests, which give a check on predicted airblast loading. Thermocouples are sometimes used to record temperature, but the structural analyst does not use this data.

Post-test inspection of the structure is one of the most important means of obtaining data for comparison with analysis results. The analysis must be carried out to a point that material damage is over and the residual shape can be estimated. The analysis results can be compared to the damage mechanisms and locations and the final shape of the test structure. Comparison of failure mechanisms is critical. For example, if the analysis predicts a bending failure but the test structure fails in shear, the analysis has not captured the correct response even though single measurements such as peak displacement may correlate well.

Examples

The Defense Threat Reduction Agency (formerly the Defense Special Weapons Agency) sponsored a series of tests and accompanying analyses of reinforced concrete walls. One specific test was of two interior walls of 533 mm (21 in.) and 381 mm (15 in.) thickness. Both walls were 3.66 m (12 ft.) wide by 2.44 m (8 ft.) high. Loading was airblast pressure and fragment impacts from a cased weapon detonated at the center of the room. Damage was light for the thicker wall and moderate for the thinner wall.

APTEK made blind predictions of the test results using the model shown in Figure 1. We explicitly modeled the steel reinforcement and used half symmetry. The analysis was run using DYNA3D, an explicit dynamic finite element code. The concrete material model was a three-invariant cap model with damage and plasticity. We modeled the steel reinforcement using a strain-rate dependent material model with Von Mises plasticity. Measurements taken on the tests included accelerations, rebar strains, pressures and thermocouples recorded as a function of time; high-speed video of photopoles on the back side of each wall; and post-test measurements and observation.

Much of the significant test data used for comparison with the analysis came from integration of the accelerometer data to get velocities and displacements. Figure 2 shows comparisons of the predicted and measured center displacements for the two walls. These plots show that the magnitudes of the displacements correlate very well, indicating that the predicted strengths of the walls are close to the actual. The period of both walls is slightly longer than predicted, indicating that the stiffness of the analysis model is higher than the stiffness of the test structure. The predicted residual displacements, estimated from the late-time response, also agreed well with the measured residual displacements of 71 mm (2.8 in.) for the thinner wall and 13 mm (0.5 in.)for the thicker wall. This shows that the overall damage in the concrete and plasticity in the rebar are predicted very well. Although not shown, comparisons of velocities gave similar results.

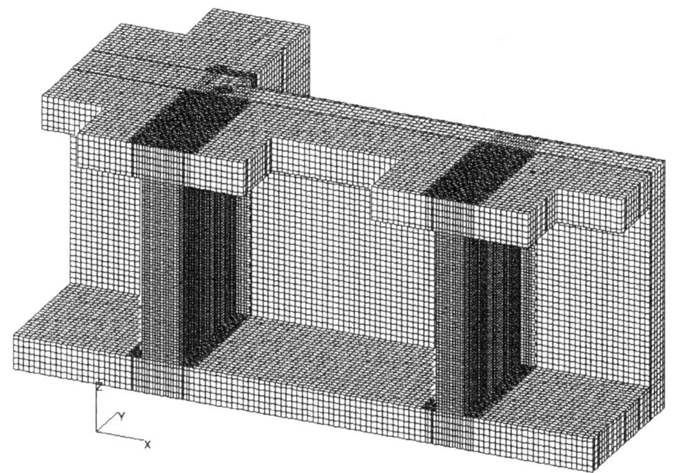

Figure 1. Analysis model for test predictions

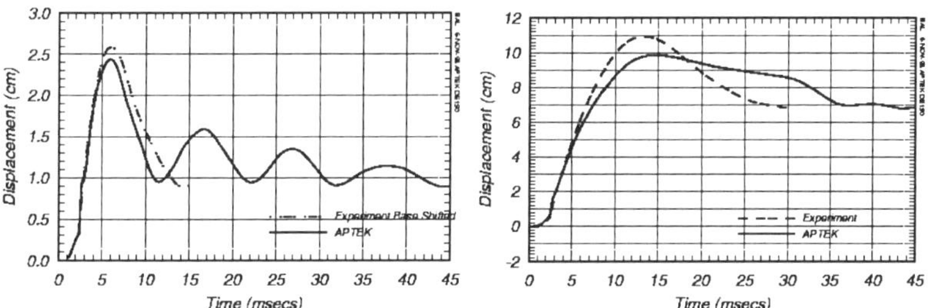

Figure 2. Calculated and measured displacements for 21" (left) and 15" (right) walls

Figure 3 shows comparisons of analytical predictions of rebar strain with measurements for front and back vertical rebar at the center of the symmetry plane in the thicker wall. The comparison is good for the measured and predicted front rebar strains. Both test and analysis show compression in the front bars and tension in the back at early times indicating a bending response. Both also show the front bars going into tension indicating membrane action by the wall. The analysis shows strains on the same order for front and back, as would be expected for a purely membrane response. However, the measured rear strain increases dramatically at about the time of peak velocities, indicative of a bending response in addition to the membrane response. Some of the other pairs of strain measurements showed the same characteristics and it is difficult to explain the discrepancy. Given that the analysis predicts other measurements so well, we have to question whether we understand what these few strain measurements represent.

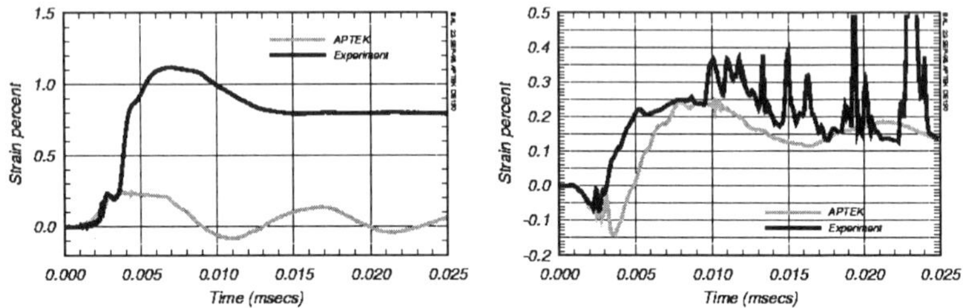

Figure 3. Calculated and measured rebar strains for back (left) and front (right) center vertical bars

Figure 4 shows the comparison of predicted and observed damage to the 15-inch wall. John Higgins of RDA/Logicon determined the post-test damage and developed the plot shown on the right. Both show shearing of the wall near the floor as the incipient failure mechanism. Observed damage patterns due to the fragment impacts also matched well.

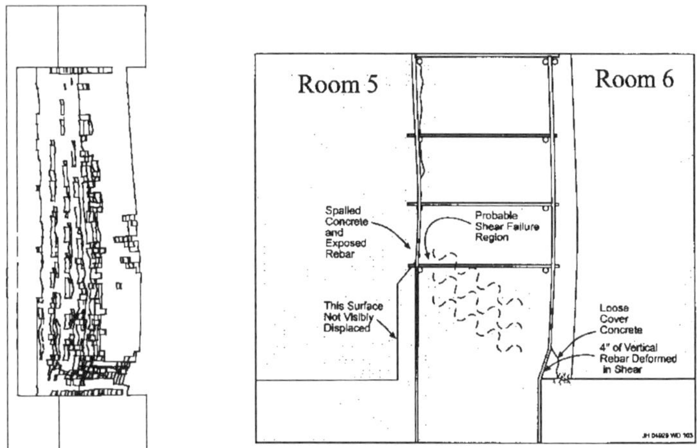

Figure 4. Calculated and observed damage mechanism

Conclusions

The comparison of test and analysis gives a good picture of how the model is performing. In this particular example, we know from post-test observation that the model gives the correct failure mechanism and predicts damage and deformations very well. Velocities and displacements derived from accelerometer data correlate well with the predictions, but also indicate that improvements could be made to the model to better simulate the stiffness of the wall. We also conclude that we need to better understand the rebar strain measurements to determine whether the differences in test and analysis strains represents a real deficiency in the analysis model or a lack of understanding of the test data. No single measurement or type of measurement gives an overall picture; all of the data contributes to the validation of the model.

Validation and Verification of a Nonlinear Reinforced Concrete Structure Model Using Precision Test Data

Mark C. Anderson[1], Timothy K. Hasselman[2], and John E. Crawford[3]

Abstract

With the advent of sophisticated, nonlinear finite element codes and ever higher performance computers, there is an increasing trend to supplant much of the testing currently performed to characterize the high strain rate response of reinforced concrete structures with more economical analyses using high fidelity, physics-based finite element models. However, these models have proven difficult to validate via traditional means, even when direct measurements are available. This paper outlines a methodology for systematic validation and verification of nonlinear structural dynamics models and applies this methodology to a model of a reinforced concrete structure subjected to internal blast loading.

Introduction

Validation and verification of structural dynamic models is an iterative process which includes the correlation of model predictions with available test results, qualitative adjustment of model physics to match observed behavior, parameter estimation to minimize the quantitative differences between predicted and measured response, and uncertainty analysis to estimate the predictive accuracy of the model. Previously developed tools for model-test correlation, parameter estimation, and predictive accuracy evaluation are based on the linearity of the equations of motion, the ability to characterize the structure in terms of its normal modes and evaluate structural response to arbitrary excitation by modal superposition.

These tools are not directly applicable to nonlinear models. In principle it may be possible to discretize the response time-histories, generate a response surface at each temporal point, interpolate these response surfaces in the time domain, and apply some systematic tools. If the functional dependencies are simple enough to permit such an approach, the number of curve fits is directly proportional to the temporal resolution and can quickly become untenable. The usual approach involves either ad hoc comparisons or generation of response surfaces for certain response variables in terms of the physical parameters of the model. At best, only a few response variables (for example: initial velocity, peak displacement, time to peak, etc.) are treated systematically. This does not permit rigorous characterization of the correlation of the effects of the physical parameters on the response parameters or of the residual modeling error.

An approach that overcomes these limitations has been developed (Anderson et al. 1997; Hasselman et al. 1998). This method effectively separates the temporal dependency from the spatial and parameter dependency and yields a set of "modes" which facilitates the extension of parameter estimation and uncertainty analysis tools to nonlinear models.

---

[1] Senior Scientist, ACTA Inc., 2790 Skypark Dr., Suite 310, Torrance, CA, 90505
[2] Director, Engineering Mechanics Div., ACTA Inc., Skypark Dr., Suite 310, Torrance, CA, 90505
[3] Principal, Karagozian & Case Structural Engineers, 625 North Maryland Ave., Glendale, CA, 91206

Outline of the Methodology

The methodology uses a technique referred to as Principal Components (PC) analysis. PC analysis is based on the singular value decomposition of a collection of time-histories. Let $x(t)$ denote a response time-history, where x may be any time-dependent quantity of interest. A response matrix, X, is a collection of discretized time-histories,

$$X = \begin{bmatrix} x_1(t_1) & \cdots & x_1(t_n) \\ \vdots & \ddots & \vdots \\ x_m(t_1) & \cdots & x_m(t_n) \end{bmatrix} \tag{1}$$

where each row corresponds to either a different measurement location or set of physical parameters, and each column corresponds to response at a specific time. Typically, $m \neq n$, so X is generally rectangular. The response matrix in (1) may be factored as

$$X = \phi D \eta, \quad \phi^T \phi = \eta \eta^T = I_p \tag{2}$$

where D is the diagonal matrix of nonzero singular values, d_i $(i = 1, \ldots, p \leq \min(m,n))$, ϕ and η are the matrices of left and right singular vectors, respectively, corresponding to the nonzero singular values, and I_p is the p-dimensional identity matrix. The columns (rows) of ϕ (η) are called the left (right) principal vectors of X. The factorization (2) is called the Principal Components Decomposition (PCD) of the response matrix.

The PCD furnishes a compact representation of the response predictions of a nonlinear model, or the corresponding test data (if available). The right principal row vectors represent the normalized basic shapes comprising the response time-histories. The squared singular values correspond to the energy content of these shape functions. Each row of the left principal vector matrix denotes the specific linear combination of the shape functions which reproduces the response time-history at the corresponding value of the physical parameter vector, θ, and spatial location of the response.

The matrix parameters comprising the PCD representation may be used to generate metrics for model-test correlation, uncertainty analysis, and predictive accuracy estimation. For example, the difference between model predictions and data may be characterized by

$$\psi = {}^0\phi^T\phi, \quad \Delta\tilde{D} = {}^0D^{-1}(D - {}^0D), \quad v = \eta^0\eta^T \tag{3}$$

where ψ is the left principal vector cross-orthogonality matrix, $\Delta\tilde{D}$ is the (scaled) difference of the singular values, and v is the cross-orthogonality matrix of the right principal vectors. In (3), ${}^0\phi$, ${}^0D$, and ${}^0\eta$ represent nonlinear "modal" parameters derived from analysis for comparison with the corresponding modal parameters ϕ, D, and η derived from test data.

Alternatively, assume that all rows of an analytical response matrix correspond to the response at a single location, but each row corresponds to a unique set of parameters. Then each left singular vector can be considered as a function of the parameter vector only. If so, then ${}^0X$, the response matrix of analytical predictions, may be approximated locally by

$$ {}^0\hat{x}(\theta) = {}^0\hat{\phi}(\theta){}^0D^0\eta \tag{4}$$

where the columns of ${}^0\phi(\theta)$ are represented by individual response surfaces. Note that (4) may be further reduced by truncation based on the number of dominant singular values. The representation in (4) is referred to as a fast-running model (FRM) and can be used in lieu of the computationally intensive finite element model for parameter estimation.

Application to Concrete Structures

This methodology was applied to a high fidelity, physics-based (HFPB) finite element model of a dimensionally scaled, three-room, buried bunker test structure. Figure 1a depicts a quarter-symmetry model of the structure. Figure 1b shows a finite element model of an interior wall of this structure. Internal blast tests were conducted by exploding a charge in the center of the center room, and measuring the blast overpressures inside the room and the corresponding accelerations of the two interior walls. Two accelerometer measurements were recorded at the center of each interior wall.

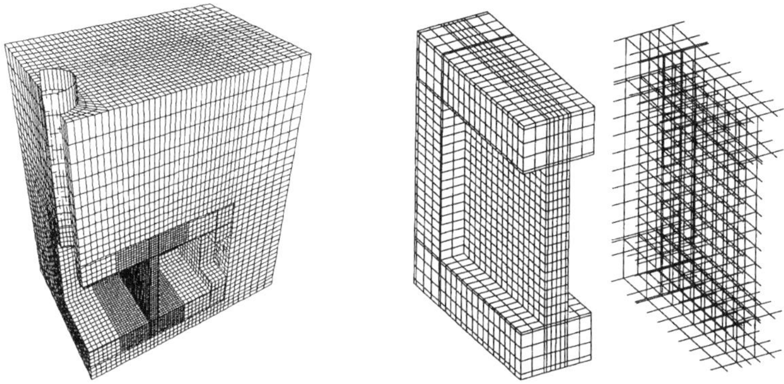

(a) Full (quarter symmetry) model   (b) Coarse mesh stub wall model

Figure 1. HFPB structure models.

A sensitivity analysis was performed to identify the parameters most affecting center-of-wall response. The most sensitive parameters included the concrete strength, charge weight, concrete tensile rate enhancement parameters (initial slope and "knee point"), cold joint friction parameters (static friction, kinetic friction, and decay factor), and concrete dilatancy fraction. The two concrete tensile rate enhancement parameters and the three cold joint friction parameters were assumed to be perfectly correlated, so that only five parameters were estimated. Statistical design of experiments techniques were used to investigate systematic parameter effects and interactions and reduce the set of significant parameters for estimation to include only the charge weight, concrete strain rate enhancement, and shear dilatancy parameters.

Bayesian estimation using a PC-based FRM produced the results given in Table 1. The effect of the model revision is shown in Figure 2a. The first column of the table (following the parameter name) lists the initial parameter values, θ_i. The second column lists the initial standard deviations, σ_i, of these estimates. The third column lists the revised parameter estimates, θ_i^*. The fourth and fifth columns, respectively, list the ratio of parameter changes to the initial standard deviation, $\Delta\theta_i/\sigma_i$, and the ratio of revised-to-initial standard derivations, σ_i^*/σ_i. With Bayesian estimation, the latter ratio is always less than or equal to unity.

Table 1. Parameter estimation results.

Parameter	θ_i	σ_i	θ_i^*	$\Delta\theta_i/\sigma_i$	σ_i^*/σ_i
1. Charge weight (N)	16.02	2.425	15.17	-0.349	. 0.438
2. Rate enhancement	0.029	0.012	0.021	-0.667	0.443
3. Concrete dilatancy	0.500	0.121	0.318	-1.504	0.609

The statistics produced by Bayesian parameter estimation can be used to evaluate the quality of the estimates, i.e. their statistical significance and consistency (Hasselman 1983). The ratios listed in Table 1 are used for this purpose. Figure 2b presents a graphic interpretation of these statistics. The quantity $(\sigma_i/\sigma_i^* - 1)\times100\%$ is plotted as a function of $\Delta\theta_i/\sigma_i$ for each of the three parameters. The vertical axis represents the statistical significance of the estimates, while the horizontal axis represents statistical consistency. An increase in confidence of 100%, for example, means that the information gain of a parameter is 100%, i.e., the information content doubled. Conversely, a confidence increase of 0% means that no new information was gained from the data.

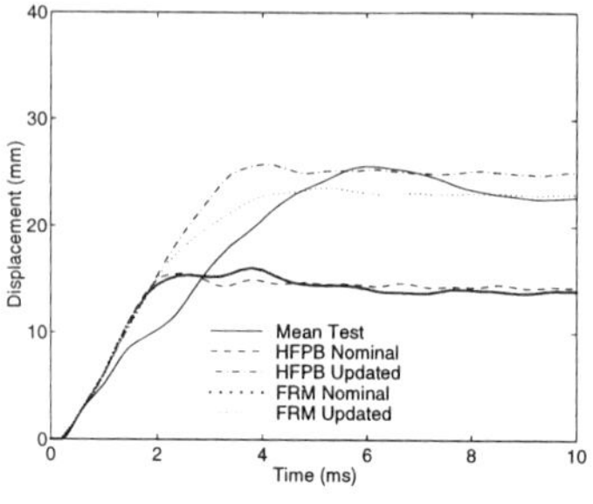

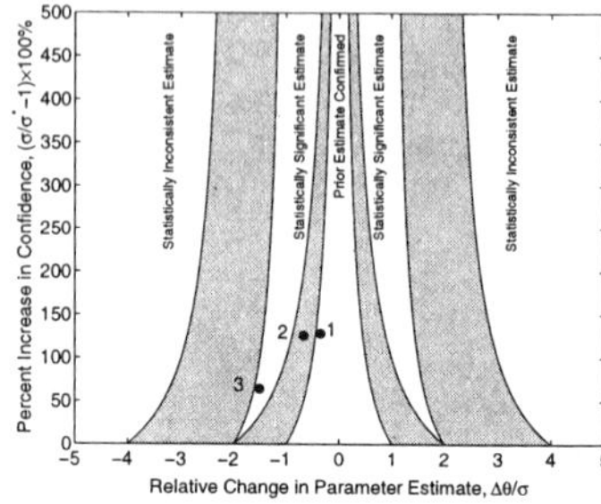

(a) Nominal and updated results   (b) Statistics of parameter estimates

Figure 2. Model updating results.

Figure 3 shows the response of the bias-corrected nominal and updated models relative to all four test measurements, with $\pm 1\sigma$ and $\pm 2\sigma$ uncertainty bands on the residual differences between the models and the data. The predictive accuracy of the updated model shown in Figure 3b may be compared with that of the original model shown in Figure 3a, indicating a progressive improvement of predictive accuracy with parameter estimation.

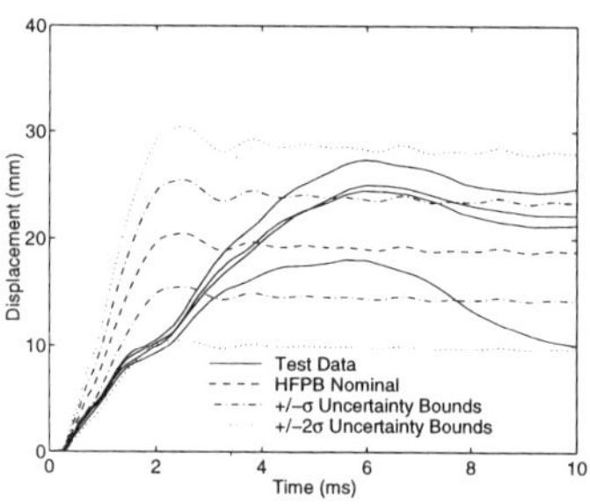

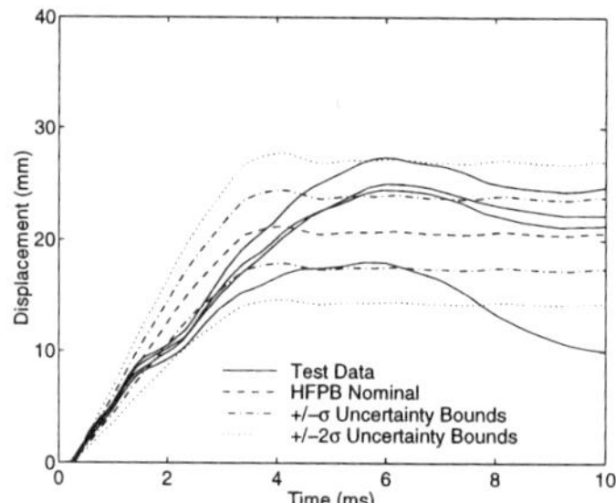

(a) Nominal model with bias correction   (b) Updated model with bias correction

Figure 3. Improvement of predictive accuracy through model updating.

Acknowledgements

The work reported in this paper was performed under contract to the Defense Special Weapons Agency. The authors wish to acknowledge the encouragement and support of the COTR, Michael Giltrud.

References

Anderson, M. C., T. K. Hasselman and J. E. Crawford (1997), "Model Validation and Verification of Reinforced Concrete Walls Subjected to Blast Loading," *Proceedings of the 68th Shock & Vibration Symposium*, Hunt Valley, MD.

Hasselman, T. K., "A Perspective on Dynamic Model Verification (1983)," *Modal Testing and Model Refinement*, AMD Vol. 59, ASME.

Hasselman, T. K., M. C. Anderson and W. Gan (1998), "Principal Components Analysis for Nonlinear Model Correlation, Updating and Uncertainty Evaluation," *Proceedings of the 16th International Modal Analysis Conference*, Santa Barbara, CA.

Shoring Measurements at Museum Towers

Dryver Huston[1], Peter Fuhr[1], David Rosowsky[2], Wai-Fah Chen[3], Minhaj Kirmani[4]

Abstract

Temporary shoring supports for slab-style concrete building construction
have a history of collapses. Hadipriono and Wang (1986) reported 85 collapses
in 25 years. In order to establish better shoring procedures, shore loads were
measured simultaneously in 10 shores in a 26-story building as it was built. The
building was part of the Museum Towers at North Point complex in Cambridge,
MA. The construction sequence was moderately fast track. Up to five stories of
shoring were in place at one time. Loads were measured at 10 locations that
spanned 2 columns of shoring cycles during pouring, curing and form removal
activities.

Introduction

In general, the primary causes of formwork failures are: 1) excessive
loads, 2) premature removal of forms or shores, and 3) inadequate lateral support
for the shoring members (Chen and Mossallam, 1991). Hadipriono and Wang
(1986) determined that 48% of the recorded failures were due to inadequate
vertical shores. Ayyub and Eldukair (1989) report that errors in design and
construction procedures are responsible for about 51% and 57% of construction
failures in the United States, respectively. A detailed discussion of various
shoring failures appears in ACI SCM-19(89) (ACI, 1989).

[1]College of Engineering, University of Vermont, Burlington VT 05405
[2]Dept. of Civil Engineering, Clemson University, Clemson SC 29634
[3]School of Civil Engineering, Purdue University, W. Lafayette IN
[3]Weidlinger Associates, Inc., One Broadway, Cambridge MA 02142

Agarwal and Gardner (1974), Fattal (1983) and Karshenas (1989) have reported previous field measurements and estimates of shoring loads. One of the distinctive features of the present study is the measurement of horizontal as well as vertical loads.

Boston Museum Towers Building Test Site

In the summer of 1997 a major concrete construction site, referred to as the Boston Museum Towers, in Cambridge, MA became available for shore load measurements. This site was a multiple building complex with two 25-story, one 5-story and one 7-story reinforced concrete buildings. Figure 1 shows one of the Boston Museum Towers as it is about two thirds complete.

Shore load data were taken during the construction of the South 25-story tower. The slabs were 8-in thick. In order to reduce costs and to avoid cold-weather concreting, the building was constructed at a fairly fast pace. A floor was poured every 4 or 5 working days. In order to sustain this pace without resorting to the use of rapidly curing concrete, a multilevel shoring system was used. The shoring system, from Peri, used a rather unique drophead feature that enabled the removal of the formwork without requiring the removal the shores and a subsequent reshoring operation, Figure 2.

Figure 1 Boston Museum Tower.   Figure 2 Shores with load cells.

Shore load data were taken at the site from June 2, 1997 to August 29, 1997. Since multilevel shoring was used, the load cell configuration was set up so as to measure the loads through cycles of multiple level shoring for the largest bays. A measurement cycle consisted of placing two load cells at locations in the bay and recording the loads during pouring and curing operations on that floor. On the next floor two more load cells were placed on the shores at locations directly above the lower load cells. This cycle was continued for up to five floors, then the cycle was restarted. Load data were collected starting on the 8th floor and continued up to the 22nd floor. The data were collected fairly continuously over this time period. However, certain interruptions due to power losses, missed opportunities for load cell placement and the shorting of load cell cables, resulted in lost data. The data collection rates were set at once every second during pouring and once every 10 minutes during curing.

Data Analysis

The amount of data collected was quite large. A first pass reduction involved scanning for spurious data sets, such as that caused by water temporarily shorting cables, and calculating local load maxima. Table 1 is a typical vertical load maximum data set for the June 1997.

Date	Comments	1st story	2nd story	3rd story	4th story	5th story
6/2/97	8th Story Pouring					
6/5/97		3433				
6/6/97	9th Story Pouring	3425				
6/6/97		4066	5501	NA	NA	
6/7/97		4123	5964	NA	NA	
6/8/97		4304	6258	NA	NA	
6/9/97		4626	6772	NA	NA	
6/10/97		3415	4836	NA	NA	
6/11/97		2593	4096	NA	NA	
6/12/97	10th Story Pouring	2226	3566	NA	NA	
6/12/97		NA	3805	4331	NA	
6/16/97		2600	2957	7309	NA	
6/17/97		1463	2548	4806	NA	
6/18/97	11th Story Pouring	1478	2461	4716	NA	
6/18/97		NA	3485	3354	5387	
6/19/97		NA	NA	NA	NA	
6/20/97		4675	3025	3407	6827	
6/21/97		4946	3205	3657	7309	
6/25/97	12th Story Pouring	NA	NA	NA	NA	
6/30/97		2940	4460	2890	NA	NA

Table 1 Maximum Vertical Shore Loads in June 1997

Conclusions

Vertical and horizontal load data were taken in a multistory shore system for a 25-story building. The load data form a useful set for comparison with design codes and mechanical models.

Acknowledgements

This project was supported by grant 5 R01OH03157 from the National Institute for Occupational Safety and Health, Weidlinger Associates and Beacon/Skanska Construction. The authors would like to thank Dr. James Becker, Walter Turgeon, Suresh Babu, Juan Pedro, Matt Nelson, Hongchu Qiu, and Mario Guerra for their assistance on this project.

References

Agarwal RK, Gardner NJ. (1974). "Form and Shore Requirements for Multistory Flat Slab Type Buildings." *ACI Journal*, Vol. 71, No. 11,559-569,.

Chen WF, Mosallam KH. (1991). *Concrete Buildings: Analysis for Safe Construction*. CRC Press.

Fattal SG, (1983) "Evaluation of Construction Loads in Multistory Concrete Buildings," NBS BSS 146, National Bureau of Standards, Washington, DC,.

Hadipriono FC, Wang HK. (1986) "Analysis of Causes of Formwork Failures in Concrete Structures." *Journal of Construction Engineering and Management*, ASCE, Vol. 112, No. 1,112-121

ACI. (1989) *Avoiding Failures in Concrete Construction*, Seminar Course Manual/SCM-19(89), American Concrete Institute, Detroit.

Ayyub, B.M., and Eldukair, Z.A. (1989), "Impact of Errors on Safety During Construction," Proc. of Sessions Related to Design, Analysis and Testing, ASCE Structures Congress '89, San Francisco, CA, May 1-5, pp. 835-843.

Karshenas, S., and Ayoub, H. (1989), "An Investigation of Live Loads on Concrete Structures During Construction," Proc. of the 5th International Conf. on Struc. Safety and Reliability, San Francisco, CA, August.

Keywords

Shoring, load, measurement, construction, concrete, multistory

Strain and Acceleration Assessment of Bridge Performance

John T. DeWolf, Fellow, ASCE[1]

Abstract

Work has been underway at the University of Connecticut during the past 15 years to develop effective and efficient assessment techniques for bridge monitoring. This work has been based on use of strain sensors and accelerometers for local and global evaluations. The has included the development and application of different field monitoring equipment, the evaluation of data collected from a variety of bridges and the study and development of techniques for the evaluation of the performance of both the overall structure and the components.

Introduction

Bridges are generally evaluated with visual techniques, requiring extensive time and man power. On occasion, specialized monitoring equipment may be used for evaluation of localized details. However, these approaches only provide information during inspection, i.e. they do not give any information during the periods between inspections. Additionally, it is possible to miss areas or problems with the present inspection procedures.

Today, advances in computer technology along with extensive development of sensors and equipment for use on bridges in all kinds of weather provides new opportunities to monitor bridges. A number of promising approaches are available to determine how critical parts are contributing to the overall bridge performance. It is increasingly possible to have a global picture of how a bridge is performing and to provide warning of major changes which can lead to serious consequences.

Use of monitoring equipment can provide engineers with better information on

---

[1]

Professor of Civil and Environmental Engineering, University of Connecticut, Storrs, CT 06269

how different structural systems are actually resisting loads and how changes due to varying weather conditions effect the bridge. Typically, design of different parts of bridges are based on application of theory to practice, with emphasis on how the structure is presumed to behave. In even the most carefully conceived design, it is not always possible to consider all variables. Analytical tools are not always adequate in evaluating different members and connection elements. This is particularly true in localized areas. Even finite element analyses, which have great potential for determining strains and stresses in specific parts, do not provide exact answers. The development of the analytical basis of the method is based on assumptions, and any application must rely on simplifications, particularly when localized areas are involved.

Field monitoring at the University of Connecticut has been used to assist the Connecticut and Rhode Island Departments of Transportation in the evaluation of some of their critical bridges. This work has involved both strain measurements and vibrational behavior. Both short and long term studies have been conducted. Short term monitoring studies have typically been based on strain monitoring, with some effort to integrate vibrational information with the strain data. These studies have generated data which has been used in localized determinations which allow for evaluation of specific members or connections. They have also provided information on the overall bridge performance. Long term monitoring has involved use of accelerometers to determine vibrational information, both to evaluate the causes of what was perceived as excessive vibrations and to evaluate the overall structural integrity.

Strain Monitoring

The strain monitoring studies have involved the development and application of portable monitoring systems for the determination of strains and stresses. They have normally involved testing over one to three days. While most studies have involved use of 8 strain gages or less, as many as 100 gages, not all read simultaneously, have been used. Strain data has been collected under normal traffic for most bridges. In some, test vehicles were used, and in the case of movable bridges, data was collected during opening and closing of the bridge structure. Extensive software has been written to use during both the strain monitoring and in the postprocessing of the data. One of the goals has been automatization of the data analysis for the evaluation of the remaining fatigue life.

This work reported has involved field monitoring of fifteen different steel bridges and two reinforced concrete bridges. The studies have treated a number of different concerns and requirements for information:

1. Fatigue problems - This has included evaluation of the causes, determination of whether repairs should be made, and if needed, how they can be carried out economically. This has included diaphragm connections, connections between girders, weld crack problems and evaluation of cracks at changes in beam cross-sections.

2. Corrosion evaluation - This has involved the determination of stress levels in aged, corroded major load carrying members.

3. Live load rating - Evaluation of load distributions to different girders in a multi-girder bridge has been used to determine if replacement or strengthening are required.

4. Ice Flow - Monitoring was set up to provide evaluation of the consequence of problems which could arise with ice flows.

5. Information for renovations - The evaluation of the main drive shafts in one bascule bridge and the determination of the load distribution to different members in another bascule bridge were used in the design of renovations.

6. Safety evaluation - A field review of a newly constructed curved bridge in which the deflections were larger than expected confirmed that the deflections were acceptable, allowing the bridge to be opened as planned.

7. Historic truss bridge - The determination of how loads are transferred to the main supporting apparatus and the evaluation of how different truss elements are performing assisted in design of renovations in a historic bridge.

8. Older reinforced concrete bridges - The determination of the load carrying capacity of a bridge with box girders and the evaluation of the columns supporting the deck in a bridge with arch spans provided guidance in the renovations of these bridges.

Vibration Monitoring

The vibrational monitoring work began with the evaluation of a continuous four span, non-prismatic steel plate girder bridge across the Connecticut River. The Department of Transportation continually received complaints from the public about the vibrations. Researchers at the University of Connecticut conducted a field study to evaluate the cause of the vibrations, to determine if they would be detrimental to the structural integrity and to suggest methods for alterations, if needed. An extensive experimental study was carried out to determine the lowest natural frequencies, the corresponding mode shapes and the acceleration levels. A finite element analysis was used to correlate the field data with the actual vibrational behavior. The results demonstrated that while the vibrational magnitudes were higher than desirable in terms of comfort levels, they were not causing structural problems. Continued analytical studies have provided additional information on the causes of vibrations, with information ways to reduce their magnitude.

Vibrational monitoring of four additional bridges has been used to develop techniques for global bridge monitoring. This work was initially used in the design of a prototype monitoring system for continuous, long-term installation on bridges, involving support from a State of Connecticut in conjunction with a Connecticut company specializing in the manufacture of vibrational equipment. The system was successfully used on two different bridges during all weather conditions. The first confirmed that it is possible to statistically determine a vibrational signature which does not change under different vehicle loading patterns. Placement of the prototype system on an older bridge was then used to demonstrate how the vibrational signature changes when changes in the structural system occur due to temperature changes. This was confirmed with an extensive finite element analysis based vibrational study.

An additional study as used to evaluate the vibrational behavior in an existing bridge when a major crack was introduced into the exterior girder. This study demonstrated that changes in the overall vibrational signature can be used to show major changes in the structural integrity.

Design of Continuous Monitoring Systems

Currently, work is underway to design and implement continuous monitoring systems on different bridges in the State of Connecticut. Both the strain monitoring and the vibrational monitoring studies have demonstrated the feasibility of using field data collected from normal traffic loading to evaluate different aspects of bridge performance. These efforts have been used to design and develop new permanent monitoring systems for placement on different in Connecticut. This work has begun with the installation of the first system, using 50 different sensors, on a curved, cast in place, three span continuous box girder bridge with post-tensioned cables. The sensors include strain gages, accelerometers, tiltmeters and temperature gages. Other systems under design are being developed for a segmental, multi span post tensioned concrete box girder bridge, a curved composite multi span, steel box section bridge, a large multi span truss bridge, a multi girder bridge with high traffic volume, a steel plate girder bridge which is regularly subjected to overweight loads and a prestressed concrete bridge with multiple I-beam girders.

Conclusions

The studies at the University of Connecticut have involved development and implementation of different bridge monitoring systems on a variety of bridges, with a variety of sensors and development of different techniques for evaluation. This has included use of both portable monitoring systems and operation of a remotely controlled monitoring systems for the collection and evaluation of the data during normal usage. This work has resulted in savings to the States of Connecticut and Rhode Island by providing information needed in their continued efforts to maintain and renew the bridge infrastructure.

Damage Detection Using Precursor Transformation Method

Armin B. Mehrabi[1], Associate Member, ASCE, and Habib Tabatabai[2]

Abstract

In this paper, a new concept for damage detection and health monitoring of structures is presented. The Precursor Transformation Method (PTM) is based on determining the causes (precursors) of change in the measured state of the structure under non-variable loading conditions. In this method, the measured changes in the state of a structure are related to damage precursors through a transformation matrix. This matrix is formed by determining the patterns of change in the state of structure associated with externally imposed strains or displacements representing possible damage scenarios.

Introduction

Civil engineering structures can be exposed to varying environmental conditions, and undergo changes in stiffness, material properties and boundary conditions over time. They can experience damage from various sources. A systematic and rational method is required to relate measured changes in structural parameters reflected by individual sensor outputs to damage sources and locations. These parameters may include member forces, geometry profiles (deflections), support reaction forces, structural strains, support settlements, etc.

Several investigations have been performed on damage detection of structures including vibration-based modal analysis (Aktan et al., 1997), dissipated energy density method (Mast et al., 1994), and parameter estimation methods (Banan et al., 1994; Sanayei and Saletnik, 1996). These methods usually require a relatively large computational effort and a knowledge of exact loading configuration. In this paper, a new analytical procedure, Precursor Transformation Method (PTM), is presented that identifies the location(s) and relative significance of possible damage sources based on measured changes in structural response parameters over time. This method offers advantages in sensitivity and cost efficiency when compared to other available methods.

[1] Engineer, Construction Technology Laboratories, Inc., Skokie, IL 60077-1030
[2] Principal Engineer, Construction Technology Laboratories, Inc., Skokie, IL 60077-1030

Concept and Verification

In this method, changes in the state of the structure are experimentally assessed through measurement of structural response parameters (displacement, strain, or internal forces) at discrete points on the structure at a reference time and later at any desired time. The external loading state at different measurement times should be constant.   This loading could be the dead load of the structure alone or augmented with additional live load. To uncouple the effects of different sources of damage and determine their locations and relative significance based on the experimental data, an analytically-determined transformation matrix is utilized.

From an analytical standpoint, the sources of damage can be characterized as precursor events (or damage precursors) that precipitate changes in the state of the structure. Precursors are externally imposed, and are therefore independent of the structure and the subsequent changes in the state of the structure. Examples of damage sources that can be modeled as precursor events include loss of material or stiffness, joint slippage, support settlements, loosening of bolts, etc.   In analytical modeling of structures, most precursors can be introduced as a change in temperature (or initial strains). It has been shown (Tabatabai et al. 1998) that the change in a vector of response parameters ($\Delta\mathbf{P}$) due to changes in temperature, displacements, etc. ($\Delta\mathbf{T}$) in an structure can be expressed as:

$$\Delta\mathbf{P} = \mathbf{C}\ \Delta\mathbf{T} \qquad \text{or} \qquad \Delta\mathbf{T} = \mathbf{C}^{-1}\ \Delta\mathbf{P} \tag{1}$$

where $\mathbf{C}$ is termed "precursor transformation matrix".   The component C_{ij} of matrix $\mathbf{C}$ is the change of i-th selected response parameter due to a unit temperature (or other precursors) change in location j. This matrix can be formed by analytical modeling (e.g., finite element modeling) of the intact structure. The singularity and other conditions for Matrix $\mathbf{C}$ is discussed by Tabatabai et al. (1998). It has also been shown that there exist a direct relationship, linear in the case of relatively small damages, between the precursors ($\Delta\mathbf{T}$) and the damage or change of stiffness.   In any case, precursors calculated based on measured response parameter changes ($\Delta\mathbf{P}$) using Eq. 1 are indicators of location and relative significance of damage.

PTM has been examined through several simulation examples by Tabatabai et al. (1998). In one of these examples, a damage scenario was simulated for the Weirton-Steubenville bridge (West Virginia). A finite element model of the bridge was generated (Fig. 1). It was assumed that three cables in this bridge had experienced damage equivalent to 10, 15, and 20 percent of cross-sectional area reduction and support settlements at two locations. Using the finite element model (in presence of bridge dead weight), force changes in stay cables due to the simulated damages were calculated. To demonstrate the applicability of PTM in the presence of measurement errors, additional random errors up to $\pm$ 5 percent of the initial cable forces were added to force change "measurements". Figure 2 shows the simulated measured changes of cable forces, $\Delta\mathbf{P}$ (including "measurement" errors). To generate the transformation matrix $\mathbf{C}$, temperature changes (on each cable) and support settlements were individually applied within the finite element model. These information were then processed (as measured data) using the PTM. Figure 3 show damage precursors resulting from this analysis. For solution of this problem, a linear optimization program was used to minimize the error effects. The analysis has clearly indicated the location and relative significance of simulated damages.

Summary

A new damage detection method (PTM) for structural health monitoring was introduced. This method can effectively identify locations and relative significance of damage sources based on the measured changes in structural response parameters over time. Finite element method is utilized to derive a behavioral matrix that transforms the structural state changes to corresponding causes of damage. The method originally was developed for the case of structures whose critical members are axial elements such as cable-stayed bridges. However, the concepts discussed here can be adopted for other types of structures as well. The main advantage of the PTM is that it can determine the location and relative significance of damages under certain loading without knowledge of the magnitude and configuration of the loads, as long as the loading remains constant at measurement times. The proposed method is able to detect damages with a single set of measurements at each time, however, the accuracy will be improved if multiple measurements are conducted and/or averaging and optimization methods can be employed.

Acknowledgment

The study presented in this paper was supported by Federal Highway Administration under Contract No. DTFH-61-96-C-00029. Opinions expressed in this paper are those of the writers and do not necessarily represent those of the Federal Highway Administration.

References

Aktan, E., Brown, D., Farrar, C., Helmicki, A., Hunt, V., and Yao, J., 1997, "Objective Global Condition Assessment," *Proc., 15th Int. Modal Analysis Conf.*, Orlando, FL, Vol. 1, 364-373.

Mast, P.W., Michopoulos, J.G., Badaliance, R., and Chaskelis, H., 1994, "Dissipated Energy as the Means for Health Monitoring of Smart Structure," *Proc., Smart Structures and Material* , SPIE, Orlando, FL, 199-207.

Banan, M.R. Banan, M.R., and Hjelmstad, K.D., 1994, "Parameter Estimation of Structures from Static Response I: Computational Aspects," *J. Struct. Engrg.*, ASCE, 120(11), 3243-3258.

Sanayei, M., and Saletnik, M.J., 1996, "Parameter Estimation of Structures from Static Strain Measurements. I: Formulation," *J. Struct. Engrg.*, ASCE, 122(5), 555-562.

Tabatabai, H., Mehrabi, A.B., Morgan, B.J., and Lotfi, H.R., 1998, "Non-Destructive Bridge Evaluation Technology: Bridge Stay Cable Condition Assessment," Report Submitted to the Federal Highway Administration, Construction Technology Laboratories, Inc., Skokie, IL.

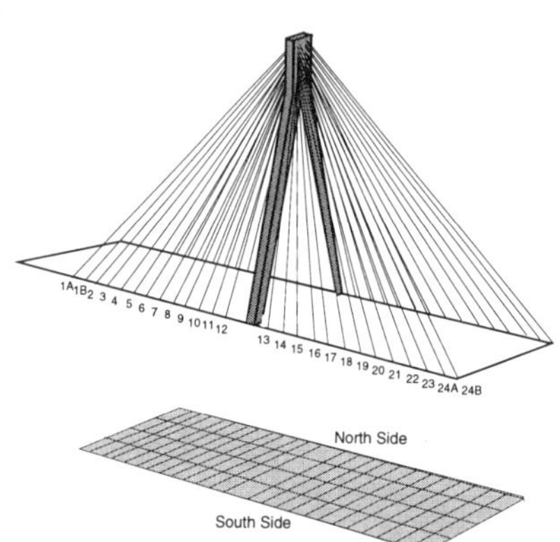

Figure 1. A schematic of the Weirton-Steubenville Bridge

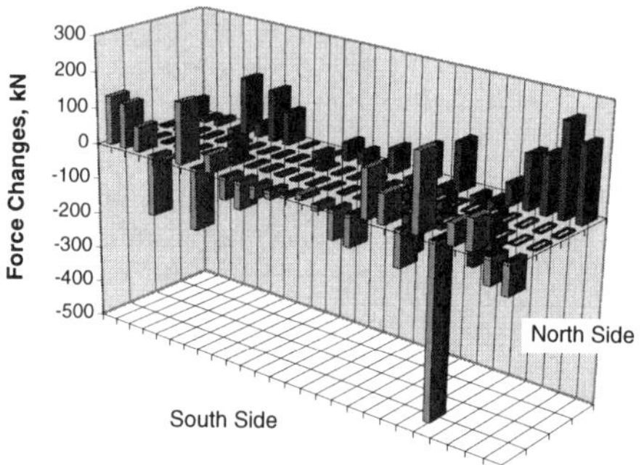

Figure 2. Force changes due to damage in bridge plus up to 5% error

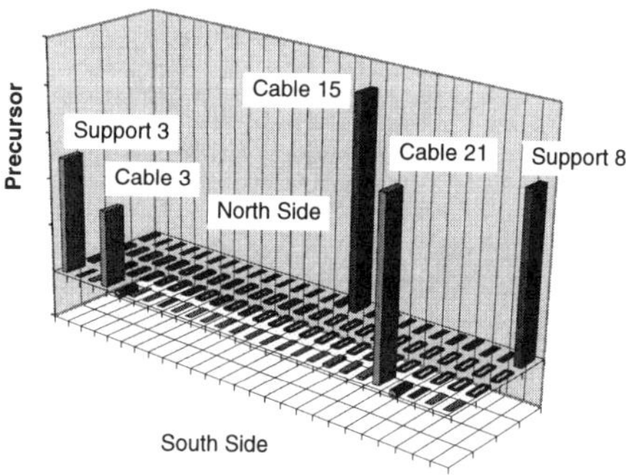

Figure 3. Damage precursors

REMOTE ROOF STABILITY MONITORING FOR UNDERGROUND NONMETAL MINES

by: R. H. Grau III, A. T. Iannacchione, and L. J. Prosser
National Institute for Occupational Safety and Health
Pittsburgh, PA

INTRODUCTION AND CURRENT ROOF MONITORING TECHNOLOGY

Unrecognized roof beam failure is responsible for most of the falls of ground fatalities and injuries occurring in U.S. underground stone mines. Observational techniques to inspect roof integrity, such as sounding the rock, observing drilling performance, etc., have always been available to miners. These techniques can be enhanced by monitoring mine roof movement on a regular basis. Monitors can be divided into two basic types: 1) roof-to-floor convergence monitors, and 2) roof and rib extensometer monitors. Because most stone mine development rooms average 7.1 m (23 ft) and bench rooms average 16.4 m (54 ft) in height, roof-to-floor convergence monitors are difficult to install, maintain, and analyze. Roof and rib extensometers have enjoyed wider use than convergence monitors; however, they are difficult to read because of their location on the roof line or back.

In its simplest form, roof and rib extensometer monitoring can be accomplished with a scratch tool. This device can detect separations and provide an indication of loose rock layers or roof beam deflection. Information on the location and size of the separation can be marked on the roof and used to assess potential future roof degradation.

Extensometers permanently installed in drill holes have been used for many years in underground mines to detect ground fall hazards. Sonic probe extensometers have been widely used in the United States, the United Kingdom, and Australia. The probe is temporarily inserted when measurements are made. Homemade mechanical extensometers have been used for decades in metal mines in Michigan, Missouri, and Idaho. For example, in the Missouri lead belt district a deflection rate of 0.007 in/month is considered a good warning of strata failure. Extensometers have been used as real-time hazard warning devices. Parker 1973 discussed an ingenious method to alert miners of strata movement by adding a warning light to an extensometer.

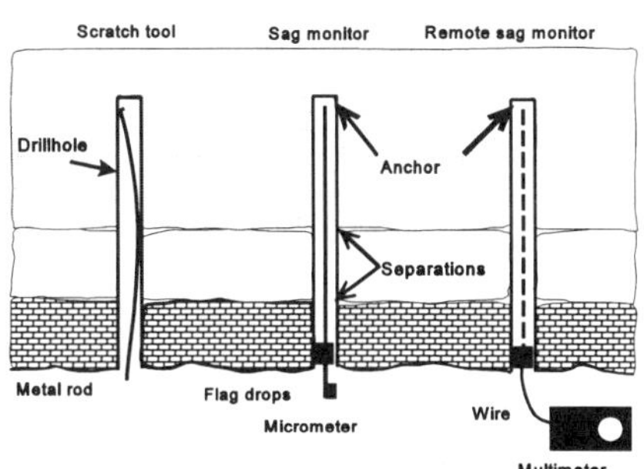

Figure 1 - Comparison of three roof monitoring techniques.

The most common commercially available mechanical extensometer monitoring devices are the Miners Helper, the Guardian Angel, and the Dual Height Telltale.

These monitors generally have one or two anchors points that measure the overall separation of rock layers in the immediate roof. If roof deflection is detected by the Miners Helper and the Guardian Angel, a reflecting flag drops from the roof line, signaling the potential for imminent roof failure. In some cases, this information has been used to indicate a need to add roof support, remove roof rock, or danger off affected areas.

Coal mines have used telltales for decades to warn miners of strata movement. The telltale is a rigid bar, possibly a roof bolt, anchored into the roof. A small section of rod protruding from the borehole is covered with three bands of reflective tape. The portion of the bar closest to the roof is generally green, followed downward by yellow and then red. As the roof deflects downward, the roof line can easily be seen to move through the green, yellow, and red tape zones. Recently, Bay Tech, a Canadian company, produced an electronic telltale. In the United Kingdom, coal mines use telltales every 20 m (65 ft) with action bands from 0.4 to 2 cm (0.2 to 1 in). Between 1990 and 1995, falls of ground were reduced from 267 to 6, partially because of the use of telltales [Altounyan et al., 1997]. Figure 1 illustrates the advantage and convenience of the electro-mechanical RMSS over other existing techniques.

NIOSH REMOTE MONITORING SAFETY SYSTEM

Although considerable advancements have been made in roof monitoring, existing instruments have several limitations that are minimizing the impact of this technology on underground stone mining. These limitations include: 1) difficulty in taking readings in high roof or back areas, 2) exposure to dangerous ground while reading the monitors, 3) complexity in making repetitive readings, 4) difficulty in seeing warning devices in the dusty and foggy production face areas, and 5) expense of most commercial monitors.

A Remote Monitoring Safety System (RMSS) developed by NIOSH improves on the existing methods for determining roof stability. This electro mechanical roof monitor includes the following features: 1) can be fabricated in most standard mine shops, 2) is relatively inexpensive (single-point RMSS costs less than $40/unit to fabricate in-house), 3) instrument damage from face blast because of the in-hole positioning is minimized (single-point RMSS only), 4) compatible with existing commercial data measurement devices, 5) capability to observe monitor reading at ground level (allows miner to take readings away from potentially unstable roof conditions) 6) remotely monitored with either multimeter or data acquisition systems (figure 2) and can accommodate as many as six anchor points (multipoint RMSS only).

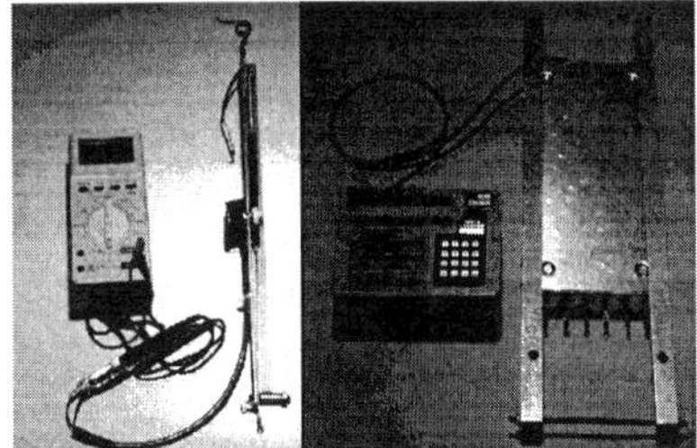

Figure 2 - Single-point RMSS with multimeter and multipoint RMSS with data acquisition.

RMSS FABRICATION, INSTALLATION, AND DATA ANALYSIS

An instruction booklet providing detailed information on how to fabricate, assemble, install, operate, and analyze data of the single-point RMSS has been prepared [Iannacchione, et al., 1997]; a parts schematic is provided in figure 3. It is also worth noting that the top-anchor can be inserted into the monitoring holes with insertion tools made of pipe 3/4-in galvanized pipe or schedule 80 PVC with a 3/8- in wide by 5/8-in deep notch cut into the end of the rod.

Monitor installation takes about 20 min and requires an open 2-in hole drilled into the roof to a depth of 14 ft or more. An electrical cable from monitor to a multimeter is installed with enough length so that readings can be made safely while on the floor. To read the RMSS, a multimeter or a data acquisition unit with high resolution and accuracy is required. The minimum meter resolution and accuracy acceptable for use with the RMSS are 1 ohm and 0.2%, respectively. The instruction manual provides examples of data sheets and graphs to record and analyze beam deflection with time. The two graphs consist of a Resistance Conversion Chart (converts multimeter resistance readings to beam deflection) and a beam deflection graph (plot of deflection versus time).

SUMMARY AND CONCLUSIONS

This research is intended to serve as a catalyst to develop better engineering tools and strategies that will improve safety by better understanding roof behavior. To help address these problems and to provide a better means of collecting and sharing roof deflection data, NIOSH has developed the RMSS. The RMSS has the following advantages: 1) local fabrication, 2) inexpensive, 3) placed in boreholes protected from blast damage, 4) read remotely, 5) compatible with many kinds of data acquisition systems, and both the single- and multiplc-point RMSS can be incorporated into a mine-wide monitoring system.

Because hazardous roof beam deflection is dependent on site-specific geologic, stress, and mining characteristics, any roof monitoring technique must be calibrated for local conditions. The use of the RMSS, as well as other observational and monitoring techniques by the underground stone mines, is expected to enhance miners' understanding of roof behavior and provide a tool for proactive intervention when hazardous ground conditions exist.

REFERENCES

Anon, "Guidance on the use of rockbolts to support roadways in coal mines, HSE Books, 1996, 35 pp.

Altounyan, P.F.R., D.N. Bigby, K.G. Hurt, and H.V. Peake, "Instrumentation and Procedures for Routine Monitoring of Reinforced Mine Roadways to Prevent Falls of Ground, 27th Int. Conf. Of Safety in Mines Research Inst., 1997, New Delhi, India, pp. 759-766.

Iannacchione, A.T., D. R. Dolinar, L. J. Prosser, T. E. Marshall, D. C. Oyler, and C. S. Compton, "Controlling Roof Beam Failures from High Horizontal Stresses in Underground Stone Mines", 17th Int. Conf. On Ground Control in Mining, Morgantown, WV, Aug. 4-6, 1998, pp. 102-112.

Iannacchione, A.T., L. J. Prosser, T.E. Marshall, C.S. Compton, D. R. Dolinar, D. C. Oyler, and D.M. Pappas, "An Instruction Booklet for the Remote Monitoring Safety System (RMSS)," National Institute for Occupational Safety and Health Internal Publication, Dec. 10, 1997, 21 pp.

Parker, J., "How Convergence Measurements Can Save Money", Engineering and Mining Journal, August, 1973, pp. 92-97.

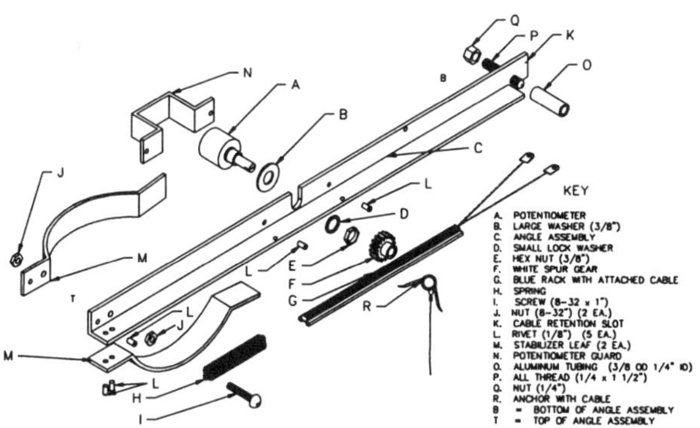

Figure 3 - Overall view of RMSS components.

Steel Structures

Manuscripts for some presentations were not available at time of publication.

7. *SELECTION CRITERIA FOR STRUCTURAL STEEL*

MODERATOR: Robert J. Dexter
University of Minnesota
Minneapolis, MN

(1) Steel Material Property Considerations
Reidar Bjorhovde
The Bjorhovde Group
Tucson, AZ

(2) Purchasing Steel For Structural Fabrication
Larry A. Kloiber
LeJeune Steel Company
Minneapolis, MN

(3) Selection Criteria for Structural Steel: Design Considerations
James O. Malley
Degenkolb Engineers
San Francisco, CA

(4) Steel Producer Considerations
Michael F. Engestrom
Nucor-Yamato Steel Company
Blytheville, AR

16. *FHWA'S HIGH PERFORMANCE STEEL (HPS) NATIONAL INITIATIVE*

MODERATOR: Joseph Hartman
Federal Highway Administration

(1) Recent Developments in HPS Research
Bill Wright
FHWA Research

(2) HPS Design Issues
Dennis Mertz
University of Delaware
Newark, DE

(3) Funding Mechanisms for HPS Demonstration Projects
John Hooks
FHWA Office of Technology Applications

(4) HPS Demonstration Project: The Ford City Bridge
Thomas Macioce
Pennsylvania Department of Transportation
Harrisburg, PA

25. *TENSILE MEMBRANE STRUCTURES*

MODERATOR: R. E. Shaeffer
Florida Agricultural and Mechanical University

(1) Materials for the Next Millennium
Nicholas Goldsmith
FTL Associates
New York, NY

**(2) The Watts Tower Cultural Crescent: Tensile Design Amidst
a National Landmark**
Mic Patterson and Will Shepphird
Advanced Tensile Structures, Inc.
Santa Monica, CA

(3) The Technology of Final Tensioning Fabric Structures
Harry Daugherty
Harry Daugherty Consulting Engineers
Whitehouse, OH

(4) Tensile Sculpture
Geoffrey Bruce
G. H. Bruce Designers
Scottsdale, AZ

(5) Fabrication and Construction of the Millennium Dome
Martin L. Brown
Birdair, Inc.
Amherst, NY

34. *COMPARISON OF AUSTRALIAN, EUROPEAN, AND NORTH AMERICAN CONNECTION DESIGN METHODS*

MODERATOR: Thomas M. Murray
Virginia Polytechnic Institute
Blacksburg, VA

(1) US Connection Design Methods
William A. Thornton
Cives Steel Company
Roswell, GA

2) UK Connection Design
D. A. Nethercot
University of Nottingham
University Park, Nottingham, UK

A.S. Malik
The Steel Construction Institute
Ascot, UK

(3) Australian Connection Design Methods
Arun Syam
Australian Institute of Steel Construction
North Sidney, NSW, Australia

43. *UPDATE ON SEISMIC PERFORMANCE OF STEEL MOMENT CONNECTIONS*

MODERATOR: Chia-Ming Uang
University of California
San Diego, CA

(1) The SAC Database: Understanding and Using Steel Frame Test Results
David Bonowitz

(2) Effect of Connection Fractures on Seismic Performance of Steel Moment Frames
Allin Cornell and Nicolas Luco
Terman Engineering

(3) Through-thickness Properties of Column Flanges
Robert J. Dexter
University of Minnesota
Minneapolis, MN

(4) Welded Steel Moment Connections - Old and New
Bozidar Stojadinovic, Subash Goel, and K.H. Lee
University of Michigan
Ann Arbor, MI

52. *DESIGN OF STEEL BUILDINGS AFTER NORTHRIDGE*

MODERATOR: Y. Henry Huang
Alhambra, CA

(1) Practical Implications of Research on Seismic Performance of Structural Systems
Y. Henry Huang
Alhambra, CA

(2) Update on Seismic Performance of Steel Frame Connections
James Malley
Degenkolb Engineers
San Francisco, CA

Charles W. Roeder
University of Washington
Seattle, Washington

(3) Analysis and Behavior of Steel Frame Buildings
Douglas A. Foutch
University of Illinois
Urbana, IL

(4) Effect of Material Properties on Steel Frame Building Behavior
Karl H. Frank
University of Texas at Austin
Austin, TX

61. *FATIGUE EVALUATION OF EXISTING STEEL BRIDGES*

MODERATOR: Y. Edward Zhou
URS Greiner, Inc.
Hunt Valley, MD

(1) The Basis for the Fatigue-Evaluation Methodology in the Proposed *Manual for Condition Evaluation and Load and Resistance Factor Rating of Highway Bridges*
Dennis R. Mertz
University of Delware
Newark, DE

(2) Fatigue Behavior of Intermittently Welded Diaphragm-to-Beam Connection
Amy S. Barth
SAI Consulting Engineers
Westover, WV

Mark D. Bowman
Purdue University
West Lafayette, IN

(3) Fatigue Evaluation Through Field Testing
J. Preston Halstead
TVGA Engineering
Elma, NY

Jerome S. O'Conner
NYSDOT
Hornell, NY

Peter W. Szustak
TVGA Engineering
Elma, NY

(4) Fatigue Strength Assessment of I-95 Bridge Over James River
Peter J. Massarelli
Virginia Transportation Research Council

Y. Edward Zhou
URS Greiner, Inc.
Hunt Valley, MD

Jose P. Gomez
Virginia Transportation Research Council

67. *USE OF NDE IN FATIGUE AND FRACTURE RELIABILITY*

MODERATOR: Richard A. Walther
Wiss, Janney, Elsmer Associates, Inc.
Northbrook, IL

(1) Application of Field Data in Fatigue Life Estimation of Highway Bridges
Jamshid Mohammadi
Illinois Institute of Technology
Chicago, IL

Ramakrishna Poleteddi
Packer Engineering
Naperville, IL

(2) Fatigue Reliability Updating Using NDE
Achintya Haldar
University of Arizona
Tuscon, AZ

Zhengwei Zhao
Department of Research and Development
New York, New York

3) Integration of NDE in Life-Cycle Cost of Highway Bridges
Dan M. Frangopol
University of Colorado
Boulder, Colorado

Michael P. Enright
University of Colorado
Boulder, Colorado

Allen Estes
U.S. Army Corps of Engineers
Fort Leonard Wood, Missouri

Kai-Yung Lin
Chung-Sehn Institute of Science and Technology
Taiwan, Republic of China

(4) FHWA Approach for Reliability Testing of the Visual Inspection Method
Richard A. Walther
Wiss, Janney, Elstner Associates, Inc.
Northbrook, IL

Glenn Washer
Federal Highway Administration
McLean, VA

Mark E. Moore
Wiss, Janney, Elstner Associates, Inc.
Atlanta, GA

(5) Evaluating UT as a Tool for Reliability Assessment of Welded Steel Moment Frame Buildings
Terrance Paret
Wiss, Janney, Elstner Associates, Inc.
Emeryville, CA

(6) Missing Elements in Crack Detection Procedures
Richard Seals
Richard I. Seals Consultant
Berkeley, CA

Steel Material Property Considerations

Reidar Bjorhovde[1], F.ASCE

Abstract

For the optimum use of steel in a structure, it is essential to understand and assess all relevant characteristics of the material. The paper presents a review of the properties that need to be addressed by designers and fabricators. Comments are given on mechanical, chemical and other properties, as well as on testing and other issues that are necessary to ensure satisfactory material performance.

Introduction

In selecting and using steel for structures, it is essential to consider all aspects of the material, not just the mechanical properties. A number of structural steels arc available, and important advances in steel production have taken place over the past ten to fifteen years. In addition, the complexity of structures and advances in analysis methods and tools have led to increasingly fine-tuned designs. Loads and load transfer mechanisms in connections and other parts of the structure affect the response of the structure. Finally, fabrication operations such as welding and flame cutting impose significant, albeit localized demands on the steel, and lead to changes in the microstructure and deformation characteristics of the material.

Mechanical Properties of Structural Steel

The major characteristics of structural steel include mechanical and other properties; all are critical to the structural response and performance during fabrication, construction and service. For the structural engineer, the mechanical properties are the most important, since they enter directly into the design equations for strength and stiffness. The yield stress represents the boundary between elastic and inelastic behavior; it is also the one material property that is the easiest to

---

[1] President, *The Bjorhovde Group*, P. O. Box 37168, Tucson, AZ 85740-7168

control in the steelmaking process, relatively speaking. For connection regions, on the other hand, the yield stress, the ultimate tensile strength and the ductility of the steel play equally important roles, since all of them control the performance of the assembly of elements. Under low-temperature and high restraint conditions, material toughness is critical. On the whole, therefore, the engineer must appreciate the interaction between micro and macro mechanical responses.

The Certified Mill Test Report

The certified mill test report (CMTR) gives the results of the steel mill production line tests, as prescribed by the ASTM A6 standard. The data are used to ascertain that the material meets the requirements of the relevant ASTM specification. It is important to recognize that these properties do not necessarily reflect the response of the steel in the structure. The tests that are used reflect standard (ASTM) practice, as developed on a consensus basis by producers, users and general interest individuals, and the results have to meet the minimum requirements in order for the steel to be deemed acceptable.

Testing for Mechanical Properties

For mechanical properties the tension test is the common one, using samples with either 200 or 50 mm gage lengths. ASTM gives detailed criteria for tension testing. It is important to know where the samples are taken from shapes and plates, and how they are oriented within the members. For example, since 1997 tension specimens for wide-flange shapes have been taken from the flange; previously web samples were used. Furthermore, certain supplementary tests may be specified by the user, as allowed under some of the individual ASTM specifications. For example, fracture toughness is a common requirement for steel intended for low-temperature and other brittle fracture applications. The usual toughness test is the Charpy V-Notch (CVN), the results of which will be shown in the CMTR when the steel order has specified such testing. Finally, hardness and bend testing is sometimes used, but such data are only marginally related to structural response, and should not be used for reliable property assessment.

Steel Chemistry

The major types of structural steel are classified according to chemical composition and processing characteristics, as follows:

1.   Carbon or carbon-manganese ("mild") steels
2.   High strength, low alloy (HSLA) steels
3.   High strength, quenched and tempered (Q&T) alloy steels
4.   High strength, quenched and self-tempered (QST) alloy steels

The primary chemical element in all steels is carbon (C). Depending on the type and strength grade of the steel, the maximum specified amount of C in rolled shapes varies between 0.26 and 0.12 percent; the actual amount also depends on product size and thickness, among other things. The limits are similar for plates and other products. It is important to note that current US shape production practices generally result in steels with carbon contents less than 0.10 percent. As a rule, higher strength steels have smaller amounts of carbon; the higher strength is obtained through suitable alloying elements such as manganese, columbium, molybdenum and vanadium. Chromium, copper and nickel enhance corrosion resistance; nickel also is significant for increasing fracture toughness.

Silicon and aluminum are so-called de-oxidizing elements; they reduce or eliminate the oxygen that is released during the solidification process of the molten steel. This process is known as killing of the steel. The less oxygen that is retained, the more uniform and finegrained the crystalline structure of the material will be. In past mill practice steel could be rimmed, semi-killed or fully killed; with today's continuous casting processes all steel is fully killed. American mills use silicon killing for structural shapes; aluminum is used for plates. Foreign producers utilize aluminum for all products.

Sulfur and phosphorus are detrimental to steel strength, ductility and weldability. These elements are therefore restricted to no more than 0.04 to 0.05 percent of the steel matrix. Current US steel shapes are produced by continuous casting with silicon killing; this results in sulfur contents of 0.02 to 0.03 percent.

Weldability

For welding, it is essential that the steel has a chemical composition that promotes the fusion of the base metal and the filler (weld electrode) metal, without the formation of cracks and other imperfections. This is called the weldability of the steel. The most common measure is the carbon equivalent (CE); this is a number that reflects the influence of the most important chemical elements insofar as weld quality is concerned. Several CE equations can be used; the most common is that of the International Institute of Welding (IIW). All currently available structural steel grades are weldable, although the highest strength grades require specially developed weld procedures. For example, welding for Q&T steels is more restrictive than for A36. For dual certified material, welding should be done according to the needs of the higher strength steel; this also applies when different steels are to be joined.

Grain Structures of Steel

The grain structures of steel are a function of chemistry, production method, heat treatment, cooling protocol and localized heat input from welding and flame cutting. Generally, steels with higher strength and ductility tend to have finer and more uniform grain structure. However, a very fine-grained structure such as martensite is not desirable, since it has high hardness and low ductility. Ingot-based steel products tend to exhibit more variability in grain size; continuous cast steel are more uniform. For welding, carefully controlled preheat and post-fabrication cooling are essential features of successful fabrication.

Grades and Types of Structural Steel

Many grades of steel are acceptable for construction today, focusing on the grades for wide-flange and other shapes, tubes and plates. The key characteristics reflect strength, ductility, and other properties, including weldability. The availability of the various grades in the form of rolled shapes, tubes in different forms, and plates must be addressed by the designer in consultation with the fabricator and the steel supplier. Similarly, appropriate uses for the various grades are paramount considerations in the selection process.

Last, but certainly not least, economy is paramount. Current prices for A36 and A572 (50) steels are essentially equal, although for most purposes it will be better to select the higher grade. For improved performance under all conditions, including for high-demand events such as earthquakes, the new (1998) A992 is the steel of choice for wide-flange shapes.

Summary: The Properties of Steel

Mechanical, chemical and metallurgical properties of steel that are of direct interest to structural engineers have been examined. It is demonstrated that all of these must be understood and fully considered when steel structures are designed and fabricated.

Bibliography

AISC/ASTM (1997), *"Selected ASTM Standards for Structural Steel Fabrication"*, AISC, Chicago, IL/ASTM, West Conshohocken, PA.

Barsom, J. M. (1987), "Material Considerations in Structural Steel Design", *Engineering Journal*, Vol. 24, No. 3, AISC, Chicago, IL.

Bjorhovde, Reidar, Engestrom, M. F., Griffis, L. G., Kloiber, L. A., and Malley, J. O. (1998), *"Selection Criteria for Structural Steel - A Primer"*, publication in progress.

Dowling, P. J., Harding, J. E., and Bjorhovde, Reidar, Editors and Authors (1992), *"Constructional Steel Design - An International Guide"*, Elsevier Applied Science, London.

Fruehan, R. J., Editor (1998), *"The Making, Shaping and Treating of Steel: Steelmaking and Refining"*, 11th Ed., AISE Steel Foundation, Pittsburgh, PA.

Lay, M. G. (1982), *"Structural Steel Fundamentals"*, Australian Road Research Board, Vermont South, Victoria, Australia.

Purchasing Steel for Structural Fabrication

Lawrence A. Kloiber [1], F. ASCE

Abstract
The structural steel industry has under gone major changes in the methods of producing and selling structural steel. This paper outlines current purchasing requirements, indicates what products are availiable and how to best specify and purchase them for project needs and economy.

Introduction

Structural steel production in the U.S. has under gone revolutionary changes in the last decade. While purchase terms and procedures vary with each producer and fabricator it is still possible to establish general guidelines on how the system operates. The basic primer for anyone specifying, buying, or selling structural steel is the ASTM A6 Standard Specification It covers terminology, manufacturing information, quality requirements including surface conditioning and tolerances, test methods, reports, identification, and supplementary requirements. All structural steel must comply with A6 and the specifications for physical and chemical requirements.

Order Preparation

Typically material orders are prepared using the structural bid drawings. These may not be final release for construction documents but they need to show, size, and dimension all structural members. The column orientation must be indicated and the elevation of all framing shown. Framing for mechanical openings can be added later provided the material can be purchased from stock and will not change the size or length of the members already shown. Any special metallurgical, or testing requirements should be clearly given. These should be project specific and limited to the members where these properties are needed.   Special material requirements such as toughness for the core zone of heavy shapes must be given.

---

[1]Vice President Engineering, Lejeune Steel Co. 118 W 60th St. Mpls. MN. 55419

Wide Flange Sections

The price of A572 Gr 50 wide flange shapes today is approximately 30% less then it was 10 years ago. The important thing to know here is that at least a part of the efficiency that produced these cost savings involves some restrictions on order size and length. The U.S. shape producers are gradually changing over to a minimum yield material of 50 KSI. Currently there is no increase in price for ASTM A572 Gr 50 or AISC "Improved Properties"(See AISC Tech Bulletin #3 March 1997) material for shapes weighing up to approximately 90 lb. per ft. Above 90 lb. per ft. there is a slight premium over ASTM A36. Even with this slight additional charge if the designer can gain any efficiency such as reducing the size or eliminating stiffeners or doublers A572 Gr 50 is usually cost effective.

Order entry requirements have changed. Mills now require a minimum order of 20 tons and a minimum item quantity of one bundle. A bundle ranges from slightly over one ton to as much as 9 ton depending on the footweight and length of the pieces. The number of pieces of the same length needed to make up a bundle ranges a from one to 10 pieces. Cutting tolerances have also changed. Mills will now only provide a length tolerance of -0 to +4 inches. These changes have obviously made it desirable to standardize sizes to meet bundling minimums.

All mills publish a rolling schedules and the cycle for each section series varies based on demand and the ability to produce and sell stock. Most heavy sections are rolled every 4 to 6 weeks. Some of the light beam sections are only rolled on a 3 month cycle because they are generally sold in stock lengths. If the rolling is sold out the oder time will be increased by the cycle time to an open rolling.

Normally it is possible to order any grade of material the mill produces when the section series rolls. The exception might be a lack of an adequate quantity of beam blanks if a large amount of material such as ASTM A588 is required. A588 material may also be subject to a minimum item requirement of 20 to 80 tons.

Mills maintain a substantial amount of stock in common sizes in length increments of 5 ft. from 30 to 65 ft. Availability varies but this is often a good option to speed delivery at a modest increase in scrap cost. The same bundle and minimum order requirements apply here.

While stock lengths are limited it is possible to order lengths of 90 ft or more from rollings. Handling and transportation may be a problem so the fabricator should have the option of splicing if this more economical on extra long lengths.

U.S. mills can produce sections up to a maximum of 400 lb. per ft. and up to 40 inches deep. Any material over this size must be made by an foreign producer. There is a substantial premium for this material and the delivery is extended due to shipping and the rolling cycle.

Service centers are an alternate to mill purchases especially for small projects and rush deliveries. They usually stock only popular footweights and depths sometimes described as "First run sections" such as W18X35 and not W18X46.

Standard or "I" beams are available from mill but because of item limitations and limited application they are usually purchased from a service center.

Steel Plate

Structural steel building frames use plate primarily for base plates, stiffeners, and connections. This material is usually ordered in stock sizes from which multiple pieces are cut. Most fabricators maintain a limited stock of common thickness plates in A36 and may also stock some A572 Gr 50. These same grades are readily available from service centers at prices only slightly above mill cost. Stock lengths for plate are typically 20, 30 and 40 ft. Thickness less then 5/8 inch are often cut from coil stock so other lengths are available if needed. This material is often available in an ASTM A529 grade thereby providing a ready source of 50 ksi yield material for thiner plates.

The maximum size plate available is determined by the width of the rolls and the yield from a 20 ton slab. Usually handling capabilities limit the practical maximum length to about 60 ft. The fabricator should normally be given the option of splicing anything over 40 ft. Plates are normally available in thickness of 1/16 inch increments to 1/2 inch thickness and 1/8 inch to 2 inch thickness, above this an inquiry should be made.

Plate material meeting A588 and A709 specifications is typically available only from mill or from a specialty service center at a premium price especially if CVN values are required.

Mill rollings are the best source for large quantities or orders with special material requirements. Delivery schedules at plate mills are not determined as much by rolling cycles as by the demand on the mill and any post processing that may be required. The addition of supplementary requirements for standard CVN values or ultrasonic testing will normally not add significantly to the delivery time. As plate thickness increase to over 2 inches it becomes more difficult however to provide toughness without heat treating the material. This is a post processing operation and will require more time and more cost.

Any through thickness ductility requirement will require special material with low suflur and other processing requirements.

It is important that any required special inspection or testing be specified at the time of purchase and done at the mill. Any testing after purchase that goes beyond standard A6 requirements will not normally be grounds for replacement of the material.

Angles and Bars

Angles and bars, like plate, are typically used for connection material in building applications. The exceptions being bracing, truss members and deck angles. .Angles are typically rolled in A36 and A529 material but most mills and service centers and mills stock primarily A36. Because angles are typically used in short lengths they are cut from stock lengths of 20 and 40 ft. Mills require a minimum item of 5 tons to produce any special length. For truss and bracing designs that require special sizes and lengths and possibly 50 ksi material it is important the designer standardize on as few sizes as possible to enable the fabricator to meet minimum order requirements. For bracing members when only a few pieces of long lengths are required it may be cost effective to allow some splicing to effectively use stock lengths.

Service centers are the only option when the item quantities are not sufficient to order from mill. This limits the order to 20 or 40 ft lengths and usually A36 material. There are some specialty centers that that stock A588 material at a premium.

Flat bars in thickness up to 1 inch and widths up 12 inches are a very economical alternate to plate for detail material. Widths are available in 1/2 inch increments to 5 inches and 1 inch increments above that. are similar to angles. Cost is approximately 70% of equivalent plate size and this does not include the savings in cutting.

Hollow Structural Sections

Fabricated HSS tonnage has been gradually increasing. The majority of these are rectangular sections produced to an A500 Gr B specification with a 46 ksi yield. There is a limited of amount of round HSS produced to the same specification but at 42 ksi yield. Most producers do not make a full range of products. The sizes listed in the AISC manual are not all readily available. The list published yearly in Modern Steel Construction is somewhat more accurate. Some sections are produced only by one mill on an infrequent or demand only basis. Designers should if possible avoid sections made by only one producer and where possible 5/8 inch wall material.

HSS is typically produced in stock lengths of 24, 40, or 48 ft. With a 20 ton minimum order and some minimum item requirements it is possible to get cut lengths up to 60 ft. Service centers typically stock 44 ft lengths and are a source for items that do not meet mill item minimums. Service centers are the most available source for most 5/8 inch wall material since many of the sections are infrequently rolled and often by only one producer. The price per lb. for common HSS sizes is about one third higher then wide flange shapes ans the price for 5/8 inch material and some of the specialty sizes can easily be double that of wide flange shapes.

Selection Criteria for Structural Steel: Design Considerations

James O. Malley[1]

Abstract

Selection and specification of structural steels to be used on building projects is done by the Engineer-of-Record. In the design of members and connections, the engineer must properly consider the material properties and capacities. In addition to typical design considerations, the engineer must be aware of demands that result from special loads, environmental conditions, and residual stresses.

Introduction

Selection of the structural steel to be used on a specific project is done by the Engineer-of-Record in an effort to most efficiently meet the design requirements. Considerations to be used in determining the most efficient application of the material includes the following:

1) Structural efficiency, in terms of least weight of material
2) Simplicity and ease of connection to other structural members
3) Relative unit cost of the shape or plate
4) Availability of the shape material

In the past, many engineers have attempted to rely mainly on providing maximum structural efficiency in their designs, but in many cases this will not result in the least cost solution. The engineer must be aware of all cost factors of the constructed project in the proper selection of structural steel systems, members and connections.

Specifying Structural Steel

Steel material selection is delineated in the Project Specifications by the different ASTM Grades (A36, A572 Gr. 50, etc.) that are appropriate for the members and connection elements. Where necessary to meet special project design demands, additional requirements beyond those contained in the standard ASTM specifications may be included in the specifications. Such demands could include minimum material toughness, adequate through-thickness properties, etc. Materials testing

---

[1]Senior Principal, Degenkolb Engineers, 25 Bush Street, San Francisco, CA 94104-4207

beyond standard tension tests and/or specific chemical composition requirements may be necessary to meet these demands. Where special requirements are specified, they must be clearly delineated, using other ASTM specifications wherever possible, in order to ensure complete understanding of the designers' intent.

Member Shape and Plate Selection

In many building applications, structural wide flange shapes are the predominant section for beams and columns. For beams, this is largely because of the shape efficiency and the availability of a large range of sizes. For columns, the wide flange shape is most widely used because its' open shape allows for relatively simple connection to the beams, and because of their availability to very large sizes. Where wide flange shapes are selected, they are typically specified to meet the requirements of ASTM A36, A572 Gr. 42 and 50, A588, A913, and now A992.

Other shapes, such as channels, angles and WT sections are also widely used in structural applications. For example, channels are used for short span beams and miscellaneous framing members. Many short span trusses are composed of single or double angle members. For longer spans, the truss chords may be WT sections. Typically, channels and angles are available as A36 material. Since they are cut from wide flange shapes, WT sections would be the same material as wide flange sections.

In addition to the common open sections, structural steel members with closed cross sections are also commonly used, such as round or rectangular tubes, designated as HSS (hollow structural shapes). These shapes are used for columns because of their efficiency in resisting axial compression or torsion. Also, for architectural reasons, exposed structural elements often use these shapes. Hot formed circular tube members are most commonly ASTM A53, Grade B material, whereas cold-formed circular or rectangular tubes are typically A500 material of various grades.

In addition to rolled shapes of various cross sections, structural steel construction makes extensive use of plate material. Examples include built-up plate girder or box column shapes, column base plates and plates used as part of the connection between various members. In addition to all grades used to produce structural shapes, plate material is available in a number of additional grades that may have special properties, such as very high yield strength levels. In addition to plate material, bar stock material is widely used for connecting plates. Bar stock is available up to one inch thick, and up to one foot wide, generally in the A36 grade.

Member Design - Strength Considerations

Calculation of the member loads in structural frames is typically determined through one dimensional analysis (simple beams, e.g.). For lateral force resisting members, linear elastic two dimensional frame analyses are most commonly used. Three dimensional analysis is generally only used for space frame structures. Such analyses generate the axial, bending, shear and torsion force distribution along the member. The three dimensional aspects of the cross sections are not explicitly considered. Prior to the 1986 publication of the AISC LRFD Specification, structural steel member design was based on Allowable Stress Design (ASD), where members are considered to remain in the linear elastic range for all loading conditions. Using the assumption that "plane sections remain plane," simple equations based on mechanics are used to determine the member stresses for expected loading from various sources (dead, live, wind seismic, snow, etc.). In the LRFD Specification, the element design is based on stress resultants and limit state procedures. Member fiber strains and deformations are not directly addressed in a transparent manner. In fact, only Fy, Fu and E are needed to apply any of the requirements in the LRFD Specification. While this approach leads to relatively simple design procedures, actual member and connection performance is closely related to the strain distributions imposed by the member loads, and the ability of the steel to achieve these distributions.

Member Design - Stiffness/Serviceability Considerations

In addition to designing members for minimum strength levels, stiffness and serviceability considerations must be considered. Examples of where serviceability concerns are important include excessive deflection of floor beams, lateral drift in moment frames subject to wind or seismic actions, and vibration control. Since increases in the yield strength of modern steels has not also had a commensurate increase in modulus of elasticity, stiffness becomes the controlling limit state on member design in many cases. In fact, for moment frames the stiffness considerations control by such a wide margin over strength that designers can specify lower yield strength material and still meet all of the design requirements. It is noted that specification of lower strength materials in such cases may not result in significant Design for vibration control is becoming common in modern structures. Occupant sensitivity in standard office type settings has been a concern for a number of years. Unique occupancies such as aerobics studios and dance floors have also received special attention. More recently, the extreme vibration sensitivity of modern equipment has led to the need for explicit design for vibration control. Since this is a dynamic problem, the mass, stiffness, damping, loading function and occupant sensitivity (either human or mechanical) all interact. The designer has some control over the mass and damping, but little or no control over the loading function and the occupant sensitivity. The most direct design means to control vibration is through the structural stiffness.

Connection Design - Strength Considerations

Steel connection design focuses almost exclusively on strength considerations. As with member design, connection designs are typically based on simplifying assumptions that utilize stress resultants. It is important for designers to recognize that the actual distribution of stresses is likely different from that used in the design assumptions. Inherent in these procedures is the assumption that the steel has enough ductility to redistribute the member forces to reach a state of equilibrium.

It is important to remember St. Venant's principle that plane sections do <u>not</u> remain in the vicinity of connections. The strains at connections are substantially affected by the three dimensional aspects of the members being joined. This must be recognized, if not directly considered in design. Stress concentrations result that may lead to unanticipated cracking if the material is not allowed to yield. Caution should be exercised whenever high restraint is proximate to significant stress concentrations.

The mechanical properties of structural shapes are not uniform over the cross section, as a result of the production, rolling, straightening and/or cold forming processes. Examples are the web-to-flange fillet area (also known as the "k-area") of wide flange shapes, and the corners of HSS shapes. This needs to be recognized and considered in connection designs, especially those that result in highly restrained conditions. It is important to recognize that in many instances, the greatest local strains on steel connections are from welding and subsequent cooling during fabrication. Recent fabrication cracking in a number of projects in the "k-area" of wide flange columns has generated substantial concern in the design community and a substantial research effort to address this issue is underway. In each connection design, consideration of the design requirements, the connection demands on the material, and the ability of the steel to meet those demands, must be made to produce a satisfactory result.

Environmental and Other Loading Considerations in Structural Steel Design

Structural steel in buildings is typically in enclosed, environmentally controlled space. In other instances, such as in bridges, the steel is subjected to environmental conditions that must be addressed. Extreme temperature effects in arctic regions can impose special material demands, such as increasing the susceptibility to fracture. The stress and strain effects of large temperature change can also substantially affect the design of supports and members of long span structures. Exposure to corrosive environments, from de-icing salts, industrial applications, close proximity to the ocean, etc., may require the selection of corrosion resistant steel, such as ASTM A588.

In addition to the standard loads considered in design (dead, live, wind, seismic, etc.), some structures are subjected to special loads. Moving loads on bridges creates fatigue loads (high cycle, low strain loads) that increases susceptibility to fracture. Special design and detailing considerations are, therefore, included in the standards for bridge design. The AISC Specifications include special requirements for the members subjected to impact loads (crane girders, e.g.).

Impact of Residual Stresses on Member Performance

The varied cooling rate that occurs along the length and across the section combines with stresses induced by the straightening process and the rest of the production to induce a set of internal stresses. Since there are no external member loads, residual stresses are self-equilibrating at any section. Residual stresses are considered in the AISC LRFD Specifications. The Specification assumes that member capacities can be reached without explicitly considering residual stresses, implying that there will be local fiber yielding and redistribution of stresses under loading.

After the members are fabricated and shipped to the site of the constructed project, another set of internal stresses are applied during erection. In most applications, these stresses are considered to be negligible. But, there are instances where significant stresses can be generated during erection. One such case would be a long span truss that is erected in segments or a multi-bay frame with welded connections. Such conditions should be addressed in developing the project specifications.

High-Performance Steel Bridge Design Issues

Dennis R. Mertz,[1] Member, ASCE

Abstract

The introduction of high-performance steel for bridge applications has raised several issues with regard to bridge design. The design issues which must be addressed relate to limitations inherent to the *LRFD Specifications* for steels with yield strengths in excess of 345 MPa (50 ksi), and for all steel bridges. AASHTO is considering revisions to the *LRFD Specifications* allowing HPS485W flexural resistance to be as great as M_p. The fatigue design provisions and the live-load deflection criteria will also be examined for possible revision. The proposed and future possible revisions are discussed herein.

Introduction

The introduction of high-performance steel for bridge applications has raised several issues with regard to bridge design. The first such grade of steel, designated ASTM A709, Grade HPS485W, has a yield strength of 485 MPa (70 ksi) and toughness far in excess of the specified toughness which is equal to the highest toughness specified in the *LRFD Bridge Design Specifications* (AASHTO 1998) for any bridge applications. The design issues which must be addressed relate to:

- limitations inherent to the *LRFD Specifications* for steels with yield strengths in excess of 345 MPa (50 ksi), and

- limitations inherent to the *LRFD Specifications* for all steel bridges.

------

[1]Associate Professor of Civil Engineering, University of Delaware, Newark, DE 19716

Limitations for Steels with Yield Strengths Greater than 345 MPa (50 ksi)

The design process for flexure of I-shaped girders is summarized in Figures C6.10.4-1 and C6.10.4-2 of the commentary to the *LRFD Specifications*. An examination of these figures reveals that the specifications limit the flexural resistance of I-shaped girders of yield strength greater than 345 MPa (50 ksi) to that which results in yielding of the extreme fiber or M_y. In the design of modern steel bridges, this limitation represents a significant barrier to full utilization of high-performance steel.

The *LRFD Specifications* allow I-shaped girders of yield strengths of 250 and 345 MPa (36 and 50 ksi) to go beyond M_y using the Q Formula all of the way to M_p, and beyond through inelastic analysis procedures.

Research (Fahnestock and Sause 1998, White, Ramirez and Barth 1997 and Yakel, Mans and Azizinamini 1998) suggests that the limitations in the *LRFD Specifications* for 485 and 690 MPa (70 and 100 ksi) can be relaxed. Based upon this research, the AASHTO Technical Committee on Steel Bridges, T-14, voted to remove most of the limitations on flexural resistance for grade 485 steels, and will submit the a revision to the *LRFD Specifications* as a agenda item at the Spring 1999 meeting of the AASHTO Highway Subcommittee on Bridges and Structures.

The proposed revision removes ASTM A709, Grade 485W from Table 6.4.1-1, replacing it with ASTM A709, Grade HPS70W. It allows the flexural resistance of grade 485 steels to be as great as M_p, based upon geometric requirements already established for lower grades of steel. The revision also extends the use of the Q Formula of Article 6.10.4.1.8 to grade 485 steels. The inelastic analysis procedures of Article 6.10.10 of the *LRFD Specifications* have not been extended to grade 485 steels as the traditionally required rotational capacity has not been demonstrated for such steels. Further, the 10% redistribution of moment after elastic analysis of Article 6.10.4.4 has not been extended to grade 485 steels for similar reasons.

Limitations Inherent to All Steel Bridges

The *LRFD Specifications'* limitations to all steel bridges which limit the effective application of HPS are:

- the fatigue design provisions of Article 6.6.1.2, and

- the live-load deflection criteria of Article 2.5.2.6.2.

The enhanced toughness of HPS is utilized to simplify welding and eliminate special toughness requirements for fracture-critical members (FCM). In terms of design, no benefit of the observed toughness enhancement of HPS has been taken. The typical observed toughness enhancement of HPS485W is far in excess of that required to eliminate the FCM toughness requirements.

An investigation is required to ascertain whether or not this toughness enhancement can be utilized to liberalize the fatigue design requirements which have been observed to limit the full utilization of the increased strength of HPS. The investigation shall consist of an analytical fracture-mechanics study of bridge-girder cross sections to determine the theoretical level of toughness required to realize the proposed liberalization.

Three possible outcomes are envisioned:

1   If the resultant toughness is less than or equal to the observed enhancement, a proposal shall be made to (1) revise ASTM A709 Grade HPS485W to mandate the required toughness, and (2) revise the AASHTO LRFD fatigue provisions to utilize the toughness enhancement,

2   If the resultant toughness is greater than the observed enhancement but seemingly attainable, a proposal shall be made to research steel production procedures to routinely achieve the required toughness to allow the fatigue provisions to be liberalized, or

3   If the resultant toughness is far beyond that currently observed in HPS70W, the concept of the liberalization of the fatigue provisions shall be "put to rest."

Such a study has been proposed by the Design Advisory Group of the American Iron and Steel Institute/Federal Highway Administration/U.S. Navy's joint research effort on HPS.

The issue of live-load deflection criteria is important and will be investigated shortly. The criteria in Article 2.5.2.6.2 of the *LRFD Specifications* are intended to "avoid undesirable structural and psychological effects due to . . . deformations." These criteria have little scientific basis. No undesirable structural effects have been attributed to live-load deflection, and psychological effects are not a function of live-load deflection.

Summary

AASHTO is considering revisions to the provisions of the *LRFD Specifications* allowing HPS485W flexural resistance to exceed M_y and to be as great as M_p. Moment redistribution of any kind, either 10% of the elastic moments or inelastic redistribution, is not proposed to be allowed for HPS.

The fatigue design provisions and the live-load deflection criteria will be examined for possible revision due to the enhanced toughness of HPS and the unscientific current nature of the live-load deflection criteria, respectively.

References

American Association of State and Highway Officials (1998). *LRFD Bridge Design Specifications*, 2nd Edition. AASHTO, Washington, D.C.

Fahnestock, Larry A. and Sause, Richard (1998). *Flexural Strength and Ductility of HPS-100W Steel I-Girders*. ATLSS Report No. 98-05, Lehigh University, Bethlehem, PA.

White, Donald W., Ramirez, Julio A. and Barth, Karl E. (1997). *Moment-Rotation Relationships for Unified Autostress Designs of Continuous-Span Bridge Beams and Girders*. FHWA/IN/JTRP-97/8, Purdue University, West Lafayette, IN.

Yakel, Aaron, Mans, Patrick and Azizinamini, Atorod (1998). *Inelastic Behavior of Steel Plate Girders with 70 ksi Steels*. AISI Draft Report, University of Nebraska, Lincoln, NE.

Acknowledgments

The Design Advisory Group (DAG) which is investigating the HPS issues relating to design was established by the American Iron and Steel Institute (AISI), the Federal Highway Administration (FHWA) and the U.S. Navy as a part of their joint research into HPS for highway bridges. The author is the chair of the Group which consists of Professor Karl H. Frank of the University of Texas at Austin, Professor Theodore V. Galambos of the University of Minnesota, Dr. John M. Kulicki of Modjeski and Masters, Inc., Mr. Robert L.Nickerson of NBE, Ltd., Mr. Camile G. Rubeiz of AISI, Mr. Krishna K. Verma of FHWA, Mr. Edward Wasserman of the Tennessee Department of Transportation, Mr. Alex Wilson of Bethlehem Lukens Plate and Mr. William J. Wright of FHWA.

HPS Demonstration Project: The Ford City Bridge

[1]Thomas P. Macioce
Member, ASCE

Abstract

The Ford City Bridge is a 3 span continuous steel plate girder structure with spans of 98m – 127m – 98m over the Allegheny River. The girders are designed with HPS485W steel in the negative moment region and Grade 345W is positive moment regions. An extensive research program is being conducted to study the feasibility of using weld metals used to fabricate Gr. 345W steel to fabricate HPS485W steel. The research program undertaken will try to demonstrate that undermatched weld metals can successfully be used to provide bridge members of acceptable strength and fatigue resistance using 485 MPa yield strength design criteria. The more conventional materials used for welding Gr. 345W are well known and fabrication costs are known to be acceptable. The research program consists of two components: 1) wide plate tension tests to characterize the strength and ductility of closely or undermatched HPS485W steel weldment and 2) full scale girder tests to characterize their fatigue resistance. The results of the wide plate specimens indicate sufficient strength and ductility. The results showed that if the undermatching of the weld metal strength to the plate was no more than 12-15%, the weldment achieved at least the yield strength of the plate. The fatigue tests have not been finalized but results indicate no decrease in fatigue performance due to undermatching.

Introduction

The Ford City Bridge is a 3 span continuous steel plate girder structure that carries state route 128 over the Allegheny River. This bridge replaces an existing 3 span camel back truss bridge with an open steel grid deck and wooden sidewalk. The existing bridge was built in 1914, and, it is structurally deficient and functionally obsolete. The replacement bridge is located approximately 38.5m upstream from the existing bridge. The bridge site is located approximately 60 km north east of the city of Pittsburgh, Pa. This project was designed using metric units and load factored design method.

[1]Assistant Chief Bridge Engineer, Pennsylvania Department of Transportation, Bureau of Design P.O. Box 3560, Harrisburg, PA 17105-3560

Some of the key features of the bridge are:
- 3 span continuous steel plate girder with spans of 98m – 127m – 98m
- Horizontally curved structure with a 155m radius in span 1, other spans on tangent alignment
- Constant depth webs 4.26m, longitudinally and transversely stiffened
- 4 girder cross section with 4.1m girder spacing
- HPS485W in negative moment region, Grade 345W elsewhere

Substructure Design

The two river piers are constructed of reinforced concrete. The piers were designed for the typical American Association of State Highway and Transportation Officials (AASHTO) Standard Specifications for Highway Bridges load groups. In addition, due to the barge traffic on the Allegheny River, the piers were designed for barge collision forces. Method I from the AASHTO Guide Specification and Commentary for Vessel Collision Design of Highway Bridges was used to determine the design vessel for collision impact.

The two river piers are founded on sandstone bedrock located approximately 15m below the river normal pool water elevation. The river piers were designed for an allowable bearing pressure of 0.958 MPa. Abutment 1 consisted of a spread footing on sandstone designed for an allowable bearing pressure of 0.958 MPa. Abutment 2 is founded on steel HPiles driven to rock with an allowable pile load of 837 kN.

Superstructure Design

The superstructure was analyzed and designed using a 3D finite element bridge design software program from Bridge Software Development International, Ltd. A 3D analysis was required due to the horizontal curved girders in span 1. The severe curvature in span 1 effected the moments and shears in the remaining spans of the structure. To accommodate the future ½ width redecking, the design drawings showed an assumed redecking scheme including details for a concrete counterweight at abutment 1.

The HPS485W steel was used in the negative moment regions for the flanges and the webs. Grade 345W was used in the remaining portions of the bridge including the longitudinal and transverse stiffeners. Some factors that must be considerations when designing using HPS485W is:
 Limit plate lengths to approximately 15,200 mm
 Limit plate thickness to 80 mm
 These plate limitations affect the location and number of shop splices in a field segment.

Constructability of steel plate girders is an important apsect that needs addressed during the girder design. For the deep girders of the Ford City bridge, web bend buckling during construction and in service where critical design issues.

Construction

The winning general contractor is Trumbull Corporation, Pittsburgh, Pa. The fabricator is PDM Bridge, Eau Claire, WI. The erector is Abate-Irwin, Pittsburgh, Pa. The plate material is being supplied by Bethlehem Lukens. The bid cost for the bridge was $13.5 million. The cost for fabricated structural steel, delivered and in place, was bid at $3.09/kg. The quantity of fabricated structural steel for the bridge consisted of 1,768,175 kg of Grade 345W and 474,738 kg of Grade HPS485W.

The bridge is currently under construction. The contractor is working on constructing the substructure units. The HPS485W plates have been produced. These plates are being independently tested Lehigh University for chemical analysis and mechanical properties to obtain information on the variability and quality of manufactured HPS485W plates. Fabrication has begun with erection to begin in the spring of 1999. The girders are being fabricated with conventional materials using Grade 345W material.

Research Program

The research program undertaken will try to demonstrate that undermatched weld metals can successfully be used to provide bridge members of acceptable strength and fatigue resistance using 485 MPa yield strength design criteria. The more conventional materials used for welding Gr. 345W are well known and fabrication costs are known to be acceptable. The participants in conducting this research are:
- Federal Highway Administration
- Pennsylvania Department of Transportation
- ATLSS Center at Lehigh University
- American Iron and Steel Institute
- High Steel Structures, Inc.

The cost of conducting this research program is $225,000.00.

Wide Plate Tension Tests – The wide plate tension tests consist of a specimen with typical dimensions are 610mm wide and 1500 mm in test length. The specimen size was selected to provide constraint conditions for the steel. A full penetration groove weld in the center of the specimen is used to fabricate the specimen. These full scale tests will demonstrate the overall strength and ductility of the weldments compared to the base metal. Four wide plate tests were performed, 2 specimens were fabricated with the L61 AXXX-10 (undermatched) and one specimen for each of the LA100 + M800-H (overmatched) and E7018 (undermatched) weld material.

Full Size Girder Fatigue Tests – The full size girder fatigue tests are designed to correspond to prior fatigue tests. This will allow comparison of these data to the existing database. The girders will include two each of the two undermatched conditions, L61 AXXX-10 submerged arc weld metal and E7018 shielded metal arc (4 fatigue tests).

RESULTS TO DATE

The wide plate tension tests have been completed. Examination of the results shows that all weldments had virtually identical yield and maximum tensile stresses as seen below:

Weld Metal used to Fabricate Wide Plate Tension Specimen	Wide Plate Specimen Yield Strength (MPa)	Wide Plate Specimen Tensile Strength (MPa)
L61 AXXX-10	572	651
E7018	580	654
LA100 + M800-H	558	648

Based on the mill test report for the HPS485W plate yield strength is 575 MPa and its tensile strength is 649 MPa. The wide plate test results and the mill test report for the HP485W are consistent and indicate feasibility of using undermatched welds to fabricate girders with HPS485W material.

The elongation and failure mode for the undermatched and overmatched wide plate tests are not the same. The overmatched weld combination failed in the plate with an overall strain of about 12% with considerable necking deformation located in a diagonal line in the base metal. The undermatched weld combination failed in the weld metal, with smaller strains. The undermatched wide plate specimens separated into two pieces. The plastic elongation, which occurs after yielding in these undermatched specimens, is concentrated in the weld metal. This small region cannot support the same elongation, thus fracture occurs.

The first two fatigue tests of the girders fabricated with L61 AXXX-10 are complete. The test was terminated due to fatigue cracks in the full penetration butt weld in the flange and in the flange to web fillet weld. The results are consistent with Category B weld detail, thus no decrease in fatigue performance due to undermatching.

The Watts Towers Cultural Crescent:
Tensile Design Amidst A National Landmark

Mic Patterson / Will Shepphird

Introduction

The Watts Towers Cultural Crescent is part of a newly designed linear park in Watts, a South Los Angeles community. Anchoring the park to the Watts Towers National Landmark, a permanent amphitheater has been planned. A removable tensile canopy shades the amphitheater. The canopy design required sensitivity to a host of issues both social and technological. Primary to the design objective was the desire to control the visual and physical impact on this historically significant site while satisfying the predetermined functional requirements of the design program.

Historical / Cultural Context

Watts is linked to the history of the City of Los Angeles in both development and disenfranchisement. Located along a railway corridor linking downtown Los Angles to the port of Los Angeles, the community of Watts developed servicing the commercial and industrial interests of early twentieth century America. The community found a namesake through a local developer, Charles H. Watts, who purchased the remaining portion of the original Spanish land grant, Rancho Tajuata. In 1902 the Pacific Electric Railway determined that the area would become Watts with the development of a railway junction. The City of Los Angeles annexed Watts in 1926. The Watts Depot, a vestige to the Red Car era of the Los Angeles, is being redeveloped as gallery space for the new linear park.

As trains and time passed over the railway lines, the residents of Watts took part in the cultural development of their community. One man, Simon Rodia (an Italian immigrant) used his life's earnings and 34 years of blood, sweat and tears to perfect a singular masterpiece of urban art known as the Watts Towers. Built with shards of glass and pottery, steel tendons and cement mortar from 1921 to 1955, a remarkable construct of five towers and adjacent walls forms a whimsical skyline at the Watts Towers site today.

The 1950's found Watts on the losing side of Los Angeles' love affair with the automobile. The abandoned Red Car junction linking Los Angeles with outlying Counties, a massive highway construction boom, as well as families leaving Watts for the suburbs slowly eroded the community. Watts, once a small and peaceable working man's community was lost in the ensuing years. Owing to numerous social dysfunctions, the civil unrest that erupted in 1965 cast a stigma on Watts that in some quarters remains today. In 1992, a ray of hope emerged from a desperate situation. The Rodney King civil unrest provided the impetus for civic renewal in many South Central Los Angeles communities. The Community Redevelopment Agency of Los Angeles, acting in concert with the Watts community and City and State governments has undertaken the task of knitting together the past and present communities of Watts in the construction of the Watts Towers Cultural Crescent. The desire to showcase the unique cultural aspects of Watts will be demonstrated in the completion of this civic park.

Project Participants

The design for the park incorporated a wide range of skills from landscape architects, artists, historians, various engineers and, of course, politicians. Of interest to this paper, however, is the interaction of the designers responsible for the fabric canopy and amphitheater. Owing to the political nature of the project, the park design team was selected from pre-qualified firms. Takata Associates, a landscape architecture firm, were responsible for the overall park design. Local architect Joe Addo, who developed the original concept with University of Southern California professor of architecture Dr. G.G. Schierle, headed the canopy design team. Once the overall park design was finalized, including a schematic design for the amphitheater canopy, Advanced Structures Incorporated (ASI), was hired to develop the design of the fabric canopy through contract documents. Englekirk & Sabol, a structural engineering firm, were responsible for the foundation design.

A Removable Tensile Fabric Canopy

The semi-circular geometry of the amphitheater provided a basis for a radial and modular fabric canopy. Early attempts to shade both stage and seating with independent structures were incorporated into a single fabric canopy. The final design incorporated a series of prestressed cable trusses that spring from simple pipe and cable pylons that reinforce the amphitheater geometry and form a closely spaced network of support points for the fabric panels.

The primary function of the canopy is to provide shade to the stage and seating areas of the amphitheater. The challenge was to provide adequate shade while minimizing the overall visual impact of the canopy on the site. The canopy design attempts to meet this challenge by providing a highly transparent structure (minimized visual mass). A computer analysis was performed to study the issues of visual mass from multiple vantage points. Gaps and breaks provided by the curved forms of the fabric panels serve to reduce visual mass in the final design. Transparency is further enhanced through the use of an 85% mesh for the fabric panels.

The issue of minimizing the visual mass of the canopy drove a decision to develop a demountable structure, providing the option for complete removal from the site. In addition, the fabric panels are independent of the cable trusses and may be partially or completely removed without affecting the overall canopy stability. The amphitheater may thus utilize the cable structure with or without fabric panels, or the cable truss structure may be removed in its entirety. This feature adds to the overall functional and visual flexibility of the system. Connection detailing for both the fabric terminations and the pylon support points allows for ease of assembly and disassembly. Residents from the local trades will be trained in the procedures for raising the canopy structure and prestressing the component parts.

The canopy can be classified as a hybrid structure utilizing both cable trusses and fabric panels. The fabric panels require double curvature to resist out of plane loads. Varying the elevation of the fabric terminations provides the anticlastic shape of each panel. Each fabric panel is edge bound by galvanized wire ropes or nylon webbing depending on the tensile load to be resisted. The design of the fabric and edge cables was performed using a "dynamic relaxation" program developed by ASI called SoftSpace. The program facilitated the initial form-finding and structural analysis, as well as ultimately generating the cutting patterns as required for fabrication of the final membrane. The cable trusses were modeled with the fabric panels to determine cable size and foundation loads. The pipe pylons were designed under AISC ASD prescriptions for steel members. While the analysis showed that the fabric terminations never develop the cable capacity, each termination was designed using an allowable load defined as the cable breaking strength divided by a factor of safety.

Other Design Challenges

The context of a large design team comprised of multiple players with overlapping responsibilities and highly varied concerns within a political and bureaucratic environment presents a distinct challenge to any tensile design project. Projects utilizing tension fabric structures demand clear and effective lines of communication and sound project management. Tensile structures are design and engineering intensive and reliant upon high tolerance prefabrication in the factory and relatively high tolerances in the field location of supporting structures. The design of a tension structure needs to be frozen at a predetermined point in the design process. The significance of this must be effectively communicated to all players in the design team. Equally importantly, the frozen design itself must be communicated to all players in the design team. Subsequent changes, even small ones, to a frozen design are frustrating, disruptive and costly. On this project, several small changes occurring late in the design process required redesign and analysis of the canopy.

Summary and Project Status

Subsequent to the design phase, delays occurred to this project that were political in nature. The implementation of the project was threatened by interdepartmental feuding, but the City of Los Angles is now moving ahead in construction of the Watts linear park. The amphitheater foundations have been cast and the infrastructure is in place. The construction of the canopy will occur in 1999. The design team believes that the new construction will be successfully implemented, and that the amphitheater canopy will fully satisfy the design criteria and provide an elegant and effective addition to the unique visual vocabulary of the Watts Towers.

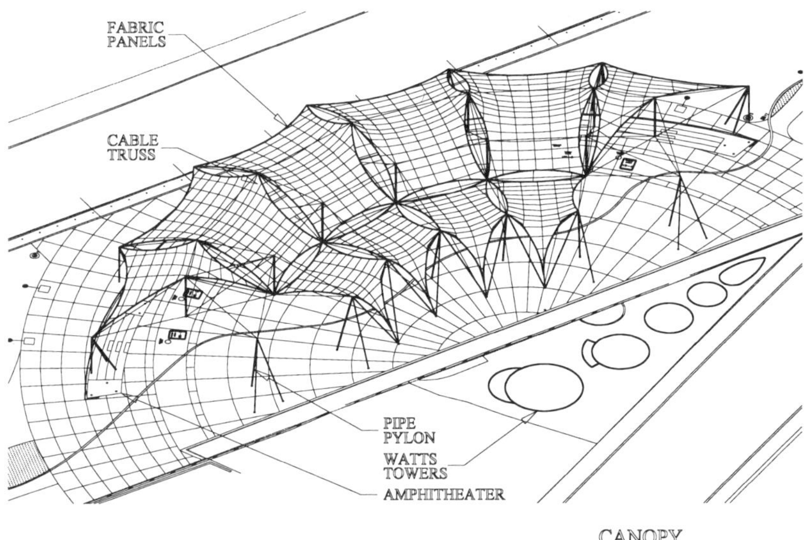

THE TECHNOLOGY OF FINAL TENSIONING FABRIC STRUCTURES

H. B.Daugherty, P.E., Member[1]

Abstract

This paper deals with a typical example of a modern, engineered fabric structure, including the cable system that typically helps support it, and the corresponding technical considerations during final tensioning in the field that affect the achievement of the desired pre-stresses and final surface geometry in the resulting structural system. The example to be used in this paper is the fabric pavilion at Caras Park in Missoula, Montana.

Text

The Caras Park pavilion is representative of the state of art of current tensile membrane structure projects. It is a clear-span permanent roof system made of fabric and cables supported on a perimeter frame. It is 148 feet long by 68 feet wide, symmetrical, and is open on the sides. The center zone of fabric is supported by "flying" masts. The technology has now been developed to the point where an engineer experienced in this field can accurately define surface shapes and stresses.

This author makes the case for a better-performing, longer-lasting, safer installation when the structural designer of the project is providing advice during the installation process.

Fabric, cables, and connection steel for a modern tensile membrane structure are normally fabricated in the shop, taking into account the desired finished surface geometry at the desired pre-stress level and the expected stretch of fabric and cables at that combination of shape and stress. Inevitably, discrepancies in the locations of

----

[1]Principal, HBDaugherty Consulting Engineers, 6776 Providence Street, Whitehouse, OH 43571

workpoint interfaces, fabrication tolerances, etc., lead to deviations from the idealized design model. The installer is generally left to his own devices to deal with these factors, and more.

Some of the other important factors are:

> History dependent stress/strain.
> When to stop tensioning?
> How to deal with wrinkles?
> How to justify backing up?
> Importance of a survey of workpoints
> Weather implications
> System adjustability
> Material sensitivity

A tensile membrane structure must be tightly controlled dimensionally. Fabric patterns are sized to anticipate stretch as small as a few tenths of one percent and cable sags are very sensitive to end-to-end lengths. In other words, small incongruities between the design dimensions and those encountered in the field can make the difference between a structure that tends to be too tight, too loose, or just right. A tight structure or a loose structure will generally cause an installer to react – and well he should. Unfortunately, he does not have the benefit of anything other than common sense to guide his response. A fabric structure project should incorporate a survey to verify the exact locations of workpoints. Without knowing whether the interfacing workpoints are in the right place there is no way to anticipate what kinds of potential problems may arise in the fitting out of the structure.

Obviously there is potential for tensioning problems if fabricated fabric, fabricated cables and/or steel fabrications are not precisely sized. Cables are generally cut precisely to length after pre-stretching them; this is an expense that some contractors try to avoid, but, it is important.

One of the remarkable idiosyncrasies of tensile membrane structures is their history-dependent stress/strain behavior. In effect, the stress/strain condition produced on the structure varies, to some degree, depending on what is pulled first, second or third, and how much. This can actually be simulated on the computer, but probably what is more important for the purpose of this paper is that it actually has been demonstrated in the field! Therefore, if an installer responds to a local condition by, for instance, taking up on a turnbuckle, raising the mast top, and so forth, he may be inadvertently simply moving the problem to another area of the structure. What's worse, he probably doesn't have a good idea where the problem went, so, he probably won't address it. In fact, it is likely that he will feel good about the problem that he has "solved". The structural engineer who designed that project is likely to be the only one who is in a strong position to respond to such field problems.

Even if interface workpoints have been installed reasonably accurately, and everything seems to be proceeding in the field in an orderly manner, the installer is generally ultimately faced with the question of when to stop tensioning, i.e., when to lock the structure off. My experience has been that, without close input from the engineer, the installer will respond to this in one of two ways. If the installer is inexperienced at this kind of work he will tend to proceed with abandon, and probably over-tighten or ignore warning signs. If the installer is experienced he tends to be more conservative and usually leaves the fabric under-tensioned – too slack. The structural engineer is in the best position to evaluate all the field deviations from the design and determine whether the final tension position should be adjusted.

What would a discussion of fabric structures be without a discussion of wrinkles? Inasmuch as this kind of roof system is at least novel, and generally chosen for its architectural effect, it draws the eye to itself. Wrinkles are unsightly, and basically unacceptable. Wrinkles occur when the fabric stress in one direction is much greater than the corresponding stress in the other direction. Thus, it is important to realize that a wrinkle is a technical event – not simply a visual distraction. When a wrinkle occurs in a tensile membrane roof system it is an indication that the structure needs some specific attention in the way of either reducing overstress in the direction parallel to the axis of the wrinkle, or increasing the stress in the direction perpendicular to the wrinkle. These types of choices can be often made effectively by the installer if it is at an early stage in the tensioning process. If wrinkles exist in the latter stages of tensioning, the response is probably better left to the engineer.

Sometimes it can be necessary to reverse the tensioning process in order to address a problem condition. Ordinarily, the installer is under a time constraint to finish the project. Under those conditions, it is very difficult to get an installer to "back up". In fact, my experience has been that only the responsible engineer has enough "clout" to order a "back up", where the tension is actually temporarily reduced in order to try another load path.

Tensile membrane roofs are big sails. Thus they are susceptible to wind at all times, including during installation, perhaps especially during installation. During the installation process, workers may be on the roof, the edges of the fabric/cable envelope may not be securely tied down, and the fabric is probably loose and susceptible to buffeting. This presents a potential human safety issue as well as a potential property damage issue. There should be some understanding between the engineer and the installer as to how the structure should be stabilized overnight or while it is between work shifts. Some types of fabrics are easily damaged if left slack even in moderate winds; such fabrics may appear to be normal, in other words, not torn, frayed, etc., but actually internally broken. The engineer should know of this in advance and should be made part of the installation process for this reason.

In unusual cases temperature can affect the tensioning process or the level of pre-stress. Extreme ambient temperatures, both high and low, have been known to impact the ability of the installer to achieve his objectives of properly stretching out the fabric and cables. Very high ambient temperatures can cause fabric to shrink, making it tighter on the structure, or cause fabric joints to fail. High frictional resistance can develop when fabric is installed over metal framing, requiring larger forces to pull out the fabric and giving the impression that the fabric is too small or the framing is too large. Very low cold temperatures can cause fabric to expand; a structure which appears to have been properly pre-tensioned at very cold temperatures may become too tight as ambient temperature rises.

The incorporation of adjustment, such as turnbuckles, drawbolts, and other such features in a tensile membrane roof system must be sufficient to address field problems , but otherwise minimized. In the first place adjustment is expensive in terms of fabrication cost. More to the point of this paper, however, is the opportunity to "play" with the geometry of the structure and end up hurting more than helping. Adjustment from nominal settings is inadvisable unless one has a clear plan on how to proceed systematically and knows whether the effect of an adjustment solves a problem or aggravates it. Over the long run, the engineer is likely to offer the best insight in this matter. Proper use of available adjustment is not unlike the tuning of a quality musical stringed instrument, such as violin or guitar. In the case of a guitar, say, the string (cable) is roughly adjusted until the string (cable) reaches the correct approximate pitch (stress); this is analogous to the cable and fabric in a tensile membrane roof system being tensioned and adjusted until they reach the approximate desired stress. Tuning a single string on a guitar to the correct pitch throws the strings on either side of it out of pitch due to deflection of the neck. Some people never get to be any good at tuning their instrument and end up making noise instead of music, primarily because of a lack of understanding of basic physics. In the same general way, the installer needs a technical understanding of the general physics of the fabric structure or else he is probably just fiddling with the structure and has little chance to achieve the results intended by the designer. This technical understanding can and should be made available by securing the services of the engineer.

A tensile membrane roof structure that has been installed with the direct participation of the structural designer is likely to perform better. Peak stresses in the system are likely to remain within an acceptable range, pre-stress levels will likely be high enough that the fabric/cables are not beat up by weather-induced vibrations and movement- but not too high that creep occurs. Supporting structures such as frames, beams, columns, footings, and so forth, will perform better because the reaction forces imposed on them by the fabric/cable system are accurately defined. The structure is likely to last longer, too, because incorrect stress levels and distortion brought about by incorrect final tensioning has a debilitating effect on fabric and cables.

TENSILE SCULPTURE

Geoffrey H. Bruce[1]

My mentor, the late C. William (Bill) Moss, used to say "...our work bridges art and function." Since beginning my career in tensile structures in September of 1994, I've made a distinct change in the philosophical notion of who I am and what I do.

Figure 1. Embracing in the Wind, Ewing Irrigation

I used to call myself the "Principal" of G.H. BRUCE, and I thought of us as a "design firm" specializing in Shade Structures. Now I'm the lead sculptor of G.H. BRUCE and we create "tensioned fabric sculpture." Our work, to a certain extent, is the same... we "protect" some aspect of an environment from the sun. Though the assignments are similar today, our clients see us differently. They no longer lump us in with the "trades," nor ask us to quote by the square foot.

----

[1]Geoffrey H. Bruce, G.H. Bruce, 5815 Canal Bank Road, Scottsdale, AZ 85250-6106, 602-948-2086

The first project I designed for BILL MOSS, INC., Embracing in the Wind (Figure 1), is an example of pure freedom of form, for the sake of function, for a client with plenty of vision. Our largest project to date is a 3,800 square foot, three canopy composition for a large MicroElectronics plant in Chandler, Arizona (Figure 2). Although the majority of our work is high-end residential, we're ever more frequently being invited to concept larger and more unusual commercial and civic projects. A number of city "slide banks" keep our work on file, and we're frequently being invited to enter competitions for various public art projects. Currently, we have a piece at the Phoenix Art Museum (Figure 3), and at the Escalante Community Center (Figure 4) in Tempe, Arizona.

Our designs are one-of-a-kind (Figures 5 and 6) and we're fortunate to work with clients/architects/designers who invite our creativity to blend with theirs. We enjoy the freedom to develop new and different canopy point supports and connection hardware. Each project offers opportunities to expand our thinking, and perfect our details. The end product is never "static," the fabric panels of any given sculpture are expected to move. They undulate gracefully and shadow patterns dance on both vertical and horizontal surfaces. Patterns and movement are as integral as sun protection, and sometimes a major reason for creating the piece. The fabric we're most familiar with is a knitted polyethylene originally manufactured to shade crops. It has a predictable stretch and a memory. It also twists and warps beautifully and lasts a long time in the Arizona sun.

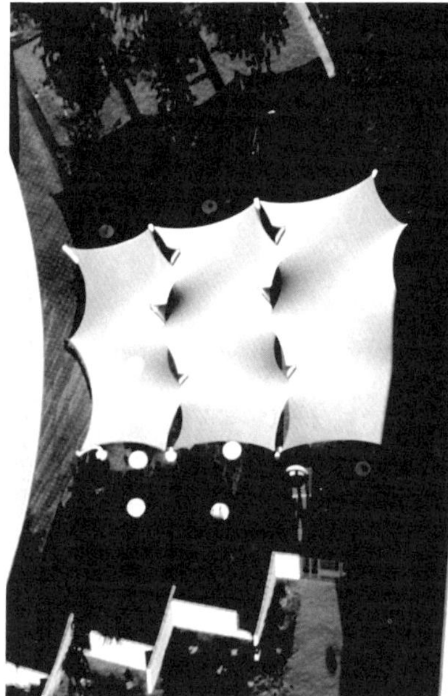

Figure 2. Solar Membrane Microelectronics Plant

The general visual impression of the majority of our pieces is "light," and many compositions are purposely minimal in elevation. We strive to maintain a balance between structure and canopy which enhances that "light" quality.

Figure 3. Cloud Memories, Phoenix Art Museum

We rely on webbing, not cable, as perimeter reinforcement of a panel. In some designs we look to "structure flexibility" in partnership with fabric, to respond to general wind loads and even withstand the inevitable micro bursts.

Figure 4. Aerial Cascade, Escalante Community Center

G.H. BRUCE is still a relative newcomer to this exciting and rapidly expanding field of tensile structures. We intend to continue to explore light weight forms and

Figure 5. Desert Flower, Desert Botanical Garden

designs, tensile products and pure sculpture. Though we may never take on the challenge of an airport or a stadium, we will always be expanding our horizons to include larger and more elaborate compositions as well as singular pieces of kinetic art.

Wherever possible we'll invite our clients to give us the latitude to use "fabric forms under tension" to solve functional and/or aesthetic needs. From our point of view, it is

Figure 6. Sky Bridge, Linthicum Constructors, Inc.

imperative that we continue to work with enlightened architects and engineers who are aware and in tune with the unique possibilities of the medium... people with vision and skills, who become allies in introducing "Tensile Sculpture" (Figure 7) to clients and agencies which are currently unfamiliar with this art form.

Figure 7. Arizona Pinwheel, Concept Model

U.S. Connection Design Methods
W.A. Thornton[1]
Fellow, ASCE

Abstract

Connection design practice in the U.S. is explained and applied to three connections. Limit states that must be considered are identified for the connections and completely detailed designs are presented.

Introduction

Connection design proceeds from an assumed load path and a set of limit states for each element of the load path to a completed connection. The limit states insure that the assumed load path has sufficient capacity in terms of strength, stability, and ductility to sustain the applied load. The limit states used are listed in Table 1. Not all of those will be used for any one connection, but all will be used at some point in the design of the three connections presented here. There is not sufficient space here to reproduce the calculations involved in these connections, but detailed calculations for similar connections can be found in the American Institute of Steel Construction's Manual of Steel Construction, AISC (1993).

Table 1 Limit States For Connection Design

Type	Number	Type	Number
Coped Section Yield	1	Bearing At Bolt Holes	9
Coped Section Buckling	2	Fillet Weld Fracture	10
Shear Yielding On Gross Area	3	Fillet Weld Ductility	11
Tension Yielding on Gross Area	4	Bolt Shear/Tension Fracture	12
Shear Fracture On Net Area	5	Plate Bending/Bolt Prying Action	13
Tension Fracture on Net Area	6	Whitmore Section Yield	14
Block Shear Fracture	7	Whitmore Section Buckling	15
Bolt Shear Fracture	8		

[1] President, Cives Engineering Corporation, 411 Rouse Lane, Suite 200, Roswell, Georgia 30076

Single Plate Shear Connection

Fig. 1 shows a completed design for a factored shear load V_u = 11 kips. The limit states used are numbers 1, 2, 3, 7, 8, 9, 10, and 11 of Table 1. The load path is from the full section of the W10x12, through the coped section to the bolts, thence into the shear plate and through the fillet welds to the web of the W14x311. The fillet weld of the shear plate to the W14x311 web is sized for ductility (Limit State 11) by using a weld $^3/_4$ of the plate thickness. This insures that the shear plate will yield before the weld fractures. The controlling Limit State is Number 1 and the design strength is 11.2 kips > 11 kips OK. This assumes a rigid support. If the W14x311 can rotate, a flexible support must be assumed and the design strength is then 8.6 kips, with Limit State 1 still controlling.

The method used to design the connection of Fig. 1 involves an empirical factor to determine eccentricity which was derived from physical tests. This limits the method to a single column of from two to nine bolts, and to a bolt to weld distance of from $2^1/_2$ to $3^1/_2$ inches.

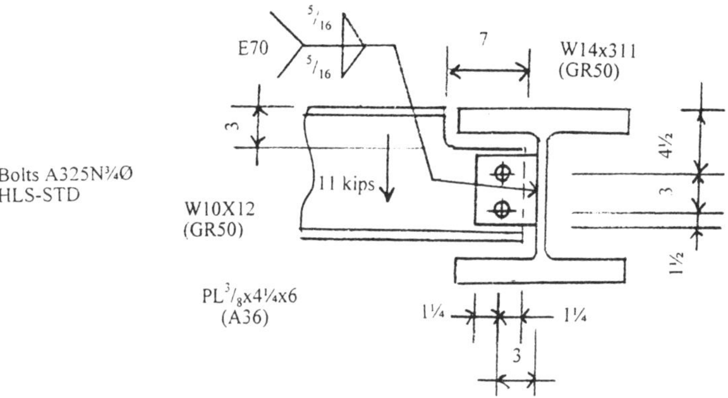

Fig. 1 Single Plate Framing Connection

End Plate Moment Connection

Fig. 2 shows a completed design for a moment of 260 kip feet and a shear of 52 kips, both factored. The limit states considered are numbers 3, 5, 7, 8, 9, 10, 11, and 12 of Table 1. The load path is such that the load travels from the W24x55 to the end plate through the welds, thence into the bolts and the column flange. Each of these components is checked using the appropriate limit states. In addition to those listed above, the column flange and web have limit states of flange bending, web yielding, web crippling, and web buckling which need to be checked. The large $^5/_{16}$ fillet welds of the beam web to the end plate are required for ductility

(Limit State 11). The method used to design this connection, AISC (1993), involves several empirical factors to account for Limit State 13 and thus has certain limitations, including that is it limited to static (non-cyclic) loadings.

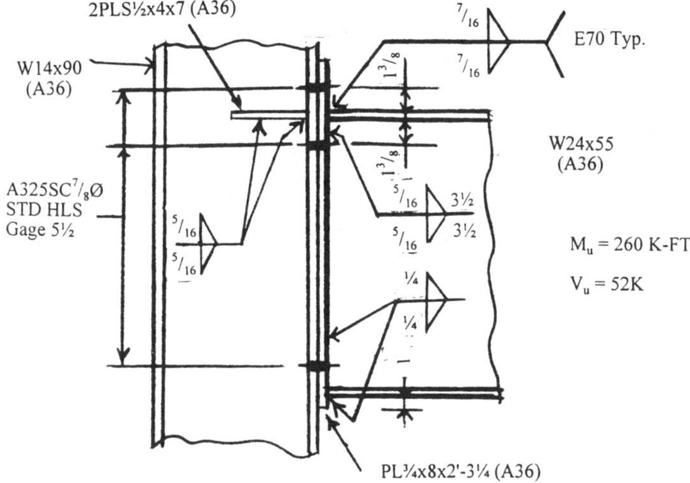

Bracing Connection   Fig. 2 End Plate Moment Connection

Fig. 3 shows a completed design for a brace load of 840 kips, a beam shear of 57 kips, and a transfer or drag force of 100 kips. All of the Limit States of Table 1 (except numbers 1 and 2) are used. Because this connection is highly statically indeterminate, there are many possible assumed load paths associated with admissible internal force distributions. The connection of Fig. 3 was achieved using the admissible force distribution recommended by the American Institute of Steel Construction, AISC (1993), and called the "uniform force method." Ductility (Limit State 11) is enhanced by increasing the size of fillet welds by a factor of 1.4 when the welds join two rigid elements, such as the gusset to beam flange fillet weld. This factor is the ratio of the peak to average stresses as determined by Richard (1986). Bracing connections consist of four connections, labeled A, B, C, and D in Fig. 3. For connection A, Limit States 8 and 15 control; for B, Limit States 3, 4, 10, and 11 control; for C and D, Limit States 3, 4, 10, 12, 13 and control. The column flange is also controlled by Limit State 13. The admissible force distribution of the method used to design the connection of Fig. 3 involves no empirical factors and is thus not limited to any particular connection configuration. For instance, instead of the end plate of Fig. 3, a shear plate similar to that of Fig. 1 could be welded to the column flange and bolted to the gusset and the beam web. In this case, the beam flange would coped or blocked to clear the shear plate and Limit States 1 and 2 may need to be considered.

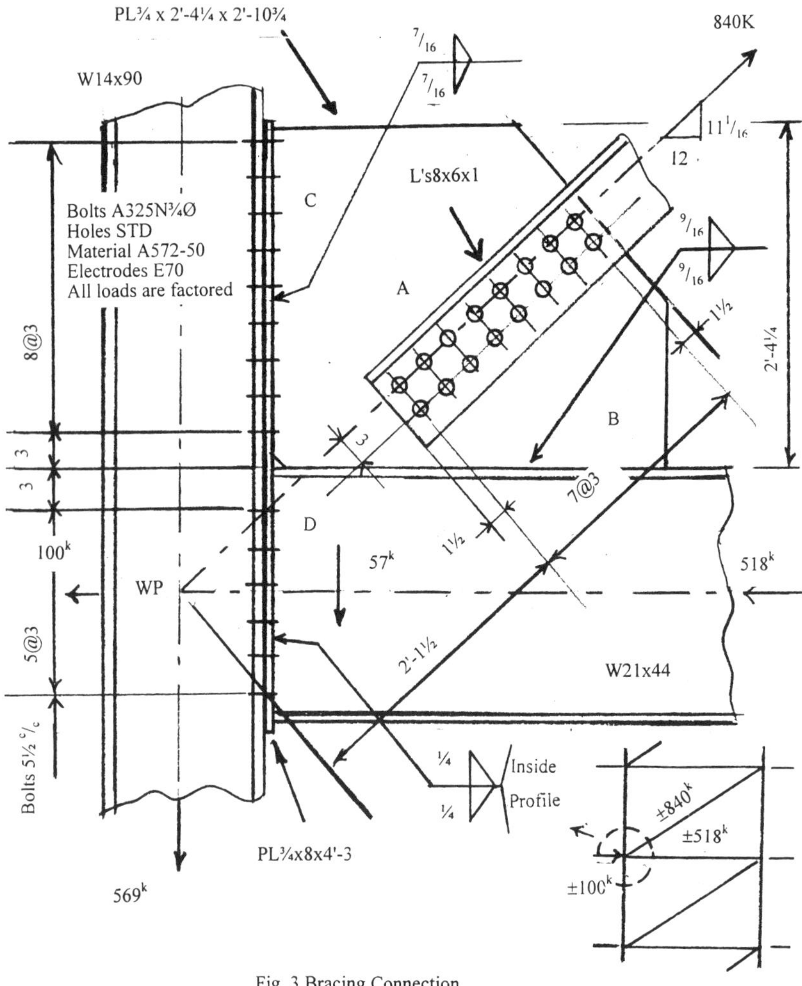

Fig. 3 Bracing Connection

Typical Bracing Elev.

References

1. American Institute of Steel Construction, 1993, Manual of Steel Construction, Load and Resistance Factor Design, Vol. II, AISC, Chicago, IL.
2. Richard, Ralph, 1986, Analysis of Large Bracing Connection Designs for Heavy Construction, National Steel Construction Conference Proceedings, pp. 31.1-31.24, AISC, Chicago, IL.

UK Connection Design

D A Nethercot[1] and A S Malik[2]

Abstract

The current UK approach to the design of shear tabs, moment transmitting endplates and bracing connections is reviewed.

Introduction

Although UK Steel Design Codes have traditionally included a section entitled "Connections", coverage is restricted to detailed rules on bolt strength, edge distances etc. The Codes do not provide design models for particular connection types. Some ten years ago a review was undertaken to investigate attitudes towards a greater degree of standardisation of connection design methods and/or the actual details themselves. Overwhelming support was evident for the former, with not inconsiderable support for a range of standard connections. As a result, the two organisations that together broadly cover the AISC remit in the UK – BCSA and SCI – jointly established a group representing designers, fabricators, researchers and regulators to produce a series of industry standards.

Shear Tab Connections

The basic design model (1,2) assumes the line of shear transfer to be at the face of the supporting element. Because this is a reasonable assumption for the connection to "flexible support" but may not be the case for connection to a "rigid support", the bolt group and the tab is checked for vertical shear plus the eccentric moment. The weld is then made 20% stronger than the plate. Ductility is ensured by making bearing on the plate or the beam web critical.

The following detailing requirements ensure that tab connections can be classified as a "simple connection" and have sufficient rotation capacity:

1. The tab is positioned close to the top;
2. The tab depth is at least 0.6 times the supported beam depth;

---

[1]Professor of Civil Engineering, University of Nottingham, UK
[2]Principal Engineer, The Steel Construction Institute, Ascot, UK

3.   End projection, $t_l \geq 10mm$;
4.   Tab thickness = 10mm;
5.   The thickness of the tab or the beam web is: $\leq 0.42d$ (S355 steel) or $\leq$ 0.50d (S275 steel);
6.   Grade 8.8 bolts are used untorqued and in clearance holes;
7.   All end and edge distances are at least 2 x (nominal bolt diameter).

Designs must satisfy the following structural checks:

1.   Recommended detailing requirement.
2.   Bolt group (for shear + moment due to eccentricity [Qa or Q(a+g/2)].
3.   Shear and bending of tab
4.   Shear and bending of supported beam at the connection (shear is the smaller of plain and block shear).
5.   Shear and bending of supported beam at the notch.
6.   Local stability of notched beam.
7.   Overall stability of unrestrained, notched beam.
8.   Weld design.
9.   Local shear and punching shear of supporting member (web of I-Section, RHS, CHS).
10.   Structural integrity checks.

Moment Transmitting Endplate

The design model adopted and the checks required (3) are shown in Figure 1. The approach is to calculate the moment capacity of a trial arrangement.

For the tension zone each potential "weak link" is examined in turn, starting with the "top" row of bolts. The potential resistance of the bolt row is taken as the capacity of the least capable component. End plate and column flange bending are based on yield line analysis. However, the complex pattern of yield lines around a pair of bolts is simplified by transforming them into an "equivalent T-stub" from which the resistance is easily computed. Subsequent rows are checked as individual rows and also in combination with previous rows.

The compression zone capacity may be limited by column web bearing or buckling or the capacity of the beam compression flange. The latter is checked against a capacity of 1.4 x p_y x flange area. The factor 1.4 is a convenient device to allow all the force to be transferred through the flange along. In practice, the web will also contribute to the transfer of the compression.

The shear capacity of the column web (= 0.6 p_y t_c D_c) is checked against the local shear force, as this may be a limiting feature, especially in one-sided connections. The compression and shear capacities must be at least equal to the sum of the potential resistances in the tension zone. If this is not the case, then stiffeners may be added or the forces in the tension zone reduced to maintain equilibrium. The moment capacity is the sum of the bolt row forces multiplied by their respective lever arms, measured from the centre of the beam compression flange.

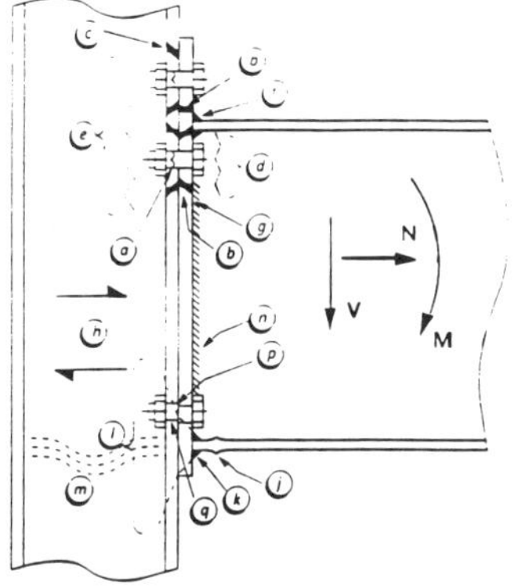

ZONE	REF	CHECKLIST ITEM
TENSION	a	Bolt tension
	b	End plate bending
	c	Column flange bending
	d	Beam web tension
	e	Column web tension
	f	Flange to end plate weld
	g	Web to end plate weld
HORIZONTAL SHEAR	h	Column web panel shear
COMPRESSION	j	Beam flange compression
	k	Beam flange weld
	l	Column web crushing
	m	Column web buckling
VERTICAL SHEAR	n	Web to end plate weld
	p	Bolt shear
	q	Bolt bearing (plate or flange)

Figure 1.   End plate moment connection – Design model and checks required

Bracing Connection

The particular example is an arrangement in which a beam to column end plate connection must also accommodate a diagonal angle bracing member.   Such a detail is not, presently, accommodated within the UK's "Green Book" series.   The following therefore represents a synthesis of views obtained from the Design Departments of a number of UK fabricators.

The key step is to decide on an appropriate load path, such that the set of forces for which the fastening at each of the 5 interfaces must be designed can be identified. In addition, loads carried by the gusset plate, end plate and regions of the members themselves in the immediate vicinity of the connection need to be determined.

Angle to gusset – The consensus view is that the set out would be the centroid of the bracing, no account would be taken of eccentricity for the angle bracing and all bolts would be designed to take an equal load.

Gusset to beam – Bracing force F must be resolved into horizontal F_H and vertical F_V components.   Whether allowance needs to be made for the bending effects introduced as a result of the horizontal eccentricity in the line of action of the bracing force F seems unclear.   If no allowance is made, then the gusset to beam weld is designed simply for the force F_H .   Recognising the existence of the eccentricity and the resulting bending requires this force to be carried by the gusset to beam weld since it is, almost certainly, the stiffest part of the connection.

<u>Gusset to column</u> – The simple approach is to design this for the vertical component of the bracing load F_V. However, if it is assumed that the eccentric bending in the beam equal to $F_H.e_b$ is taken on the bolt group, then a resultant weld load should be determined following a similar procedure to that outlined above.

<u>Beam to end plate weld</u> – This is designed for the shear in the beam due to the end reaction but may also carry tension as a result of the eccentric bending in the beam.

<u>End plate to column bolts</u> – Bolt shear is taken as the value of the beam shear plus or minus the bracing shear. An allowance may need to be made for bolt tension if the beam cannot carry the eccentric bending. In addition, there may be some direct transfer of tension across the column width.

<u>Plate components</u> – The end plate will need to be designed in the normal way for a shear only connection. If the bracing force can reverse, then the exposed portion of the gusset plate should be checked for its ability to transfer a compressive force.

It should be stated that the above does not represent any form of "standard procedure" – at least partly because enquiries amongst fabricators revealed wide diversity in the types of bracing connections they were expected to design e.g. frequently unsymmetrical arrangements of bracing were provided at both the top and bottom of the beam, bracing members were either flats or angles, different detailing arrangements were employed etc. A rather fundamental point concerns the extent to which accommodating the diagonal bracing force (F) requires sufficient modification to the basic "simple" end plate arrangement that the assumption of an essentially simply supported beam for gravity load remains valid.

Conclusions

The procedures outlined herein will be elaborated upon with the aid of specific worked examples in the Conference presentation.

References

1.   The SCI and BCSA, Joints in Simple Construction, Vol. 1: Design Methods (2nd Ed.) 1993, SCI-P215.

2.   The SCI and BCSA, Joints in Simple Construction, Vol. 2: Practical Applications 1992, SCI-P206/92.

3.   The SCI and BCSA, Joints in Steel Construction, Moment Connections 1995, SCI-P207/95

Australian Connection Design Methods

Arun Syam[1]

Abstract

Three types of structural steel connections are considered – the web side plate, bolted moment (extended) end plate and bracing connections. This paper considers the design of these connections from an Australian perspective.

Introduction

Three types of structural steel connections are considered from an Australian structural design and detailing perspective – the web side plate, bolted moment (extended) end plate and the (extended end plate) brace connection. For each of the joints, the paper considers USA detailing practice (eg plate/member strengths, bolt spacing, etc) though uses Australian bolt types/sizes, bolt hole sizes and standard width/thickness flat bar elements. All the design loads and capacities used in the paper are for the (factored) strength limit state and readers should be wary of the differing load factors and combinations used in these countries.

Web Side Plate (WSP) Connection

The WSP connection configuration noted in Figure 1(a) is considered to be structurally adequate in the Australian context. A design shear force (V^*) of 11 kips (48.9kN) acts on the connection. A standard design model of this connection is considered in Hogan & Thomas (1994) based on provisions in AS 4100 (1998). Due to the complex evaluation of support rotational stiffness a design shear force eccentricity, e ($= s_{g1}$), conservatively applies to all connection components – unless otherwise noted. V^* must be greater than or equal to the minimum of [V_a, V_b, V_c, V_d, V_e, V_f, V_g] where $V_a....V_g$ are design capacities for a specific connection failure mode.
- Design capacity of Fillet Weld due to eccentric shear force (V_a): This failure mode includes a check of fracture failure on the most critical portion of the fillet weld group.
$$V_a = \textbf{94.2 kN (21.2 kips)} > V^*$$
- Design capacity of Bolt Group (V_b): The check on the bolt group includes consideration of failures involving: bolt shear (82.8 kN (18.6 kips)); ply (local) bearing failure of the beam web (111 kN (25.0 kips)) and side plate (131 kN (29.4 kips)); horizontal bolt tearout of the

---
[1] National Manager – Technology, Australian Institute of Steel Construction, P.O. Box 6366, North Sydney NSW 2059, Australia.

side plate (72.6 kN (16.3 kips)) and beam web (61.8 kN (13.9 kips)); and vertical bolt tearout of the side plate (174 kN (39.1 kips)) and beam web (148 kN (33.3 kips)). From this, the critical bolt group failure mode is beam web horizontal tearout failure.

$$V_b = \textbf{61.8 kN (13.9 kips)} > V^*$$

- <u>Design capacity of Side Plate – Shear (V_c):</u> $V_c = \textbf{108 kN (24.3 kips)} > V^*$
- <u>Design capacity of Side Plate – Moment (V_d):</u>Checks on shear-bending interaction are not required for such connections with practical configurations. $V_d = \textbf{108 kN (24.3 kips)} > V^*$
- <u>Design capacity of Coped Supported (Purlin) Member - Shear (V_e):</u> Due to the presence of the top cope, the member is considered as a tee section and the design capacity is determined for the section on a non-uniform shear stress basis. $V_e = \textbf{137 kN (30.8 kips)} > V^*$
- <u>Design capacity of Coped Supported (Purlin) Member - Moment (V_f):</u> Again, the member is designed as a tee section and the eccentric design shear force is evaluated to develop the section moment capacity based on the effective portions of the section. The shear force eccentricity in this instance is taken from the bolt line to the end of the cope (222 mm (8¾")). $V_f = \textbf{19.1 kN (4.29 kips)} < V^*$

The connection fails due to AS 4100 (1998) being conservative in determining the local buckling behaviour of the tee section (it is categorised as "slender"). In this instance, it is assumed the tee section is longer than the length of the cope. American practice allows for the presence of the I-section at the end of the cope and (presumably) the inability of the tee section to develop a half-buckle wavelength. This permits a higher effective section modulus (Z_{ex}) for flexure and is now being considered in Australia. Assuming local buckling does not control due to these effects (i.e. it is a compact section and $Z_e = 2.26$ in$^3 = 37000$ mm^3 (AISC(US) (1994)) then: $V_f = \textbf{51.7 kN (11.6 kips)} > V^*$

- <u>Design capacity of Coped Supported (Purlin) Member – Block Shear (V_g):</u> This method is based on US design practice (AISC(US) (1994)). $V_g = \textbf{142 kN (31.9 kips)} > V^*$
- <u>Additional design and detailing requirements:</u> The nominal moment capacity of the weld (9.46 kNm (6.98 ft-kips)) is greater than the nominal capacity of the plate (9.15 kNm (6.75 ft-kips)) as a plastic hinge must form in the plate instead of the weld. To observe rotational demand requirements, the thickness of the side plate must be less than the thickness of the supported web plus 1 mm and less than 10 mm. This is not satisfied. However, this provision was generally considered applicable for sections 360-530 mm (14.2" – 20.9") deep. Other detailing requirements such as depth of plate to depth of purlin web, and placement of side plate along the purlin web are satisfied.

Bolted Moment End Plate (BMEP) Connection

The BMEP connection configuration noted in Figure 1(b) is considered to be structurally adequate in the Australian context. A standard design model of this connection is considered in Hogan & Thomas (1994) based on provisions in AS 4100 (1998). Unlike the above WSP connection, the designer cannot readily compare each failure mode load with a common design load as each component transmits a different design load. Checks are made on connection components and are based on their position in the load path.

- <u>Beam to End Plate Welds</u> *Flange welds* are checked against the design load in each flange. For fillet welds, AS 4100 (1998) only requires the vector resultant of the design load – i.e. there is no requirement to resolve the weld load into components. Fillet welds are selected due to their economy and the "thin" beam flanges, which do not lend themselves to edge preparation if a complete penetration butt weld is used. *Web welds* are designed for a proportion of the bending moment and the total shear force.
- <u>Bolts</u> *Bolts placed about the tension flange* are designed to take the total tension

component of the beam bending moment – this is generally higher than the design tension load used for the weld connecting the beam flange to end plate. The design tension load is applied uniformly to these bolts with a prying factor included – i.e. between 20-30% with 30% being used in the check of the connection considered. These bolts are not designed to take the connection shear load. *Bolts not placed about the tension flange* are designed only to resist the total shear load at the connection. Only two bolts are required in the bottom flange region in this instance. Ply bearing and tearout checks are done for the end plate and column flange, though these generally do not govern.

• Design of End Plate A simple single dimensional yield model is used for design of the end plate, which is assumed to be undergoing double curvature (hence the bolt prying force) at the tension flange. This is considered to give conservative results, which are acceptable as the design model remains simple. Additionally, end plates are generally cut from flat bars and the common range of Australian thicknesses are 20, 25 and 32 mm for these sections. Consequently, in comparison, the Australian BMEP connections may have thicker end plates than those in the USA and UK. Based on this design method, the 25 and 28 mm thickness end plates were not adequate for the design loads and connection configuration – 32 mm being used. A transverse shear check is also made of the end plate cross section.

• Column Stiffening – tension stiffener The column flanges connected to the end plate are susceptible to flexing outwards in the tension region. An initial check is made on the adequacy of the unstiffened flanges. A tension stiffener is required if the design flange pullout load (same as tension bolt loads) is greater than the flange flexural capacity. The stiffener is then designed to resist the residual force between design load and flange capacity. The calculations in this instance indicated that no tension stiffener was required as the unstiffened flange capacity equalled design load – which is very close.

• Column Stiffening – compression stiffener These stiffeners help to stabilise the column web from crushing and buckling due to the compression flange bearing loads. No stiffener was required in this instance.

• Column Stiffening – check of overlapping stress regions in web This check indicated that there is an interaction of the overlapping tension/compression stress regions and a stiffener is required. A check on column web shear stiffener requirements will assist in determining the placement of the stiffener.

• Column Stiffening – Shear stiffener A check of the adequacy of the unstiffened column web panel indicates that it cannot resist the design shear loads in this region. A diagonal shear stiffener (which tends to be more common in Australia than web doubler plates) is required and is designed to resist the difference in load between the unstiffened web panel shear capacity, the storey shear load and the flange loads acting on the web. The calculations used in this exercise assumed that there was no storey shear force in the column web. It has been recognised in Australia that the assessment criteria for this stiffening method is conservative and requires revision. The requirement of a diagonal shear stiffener necessitated the use of tension and compression column web stiffeners (Figure 1(b)).

Bracing Connection

Within Australia, a standard design model for single brace connections is considered in Hogan & Thomas (1994), Syam & Chapman (1996) and based on provisions in AS 4100 (1998). However, the configuration of this brace connection is limited to single members connected to a flat bar or plate component with one edge welded to a support and two edges parallel to the brace. There is currently no standardised Australian design model, or common views on the bracing connection configuration noted in Figure 1(c).

It has been recognised that a standardised design model is required for this connection type and that specific elements of current Australian connection design methodologies can be utilised to this end. There is, however, one significant element missing from a potentially common Australian design model – that is the resolution of design actions for each of the connection elements in what is a statically indeterminate connection. Based on available material, one feasible option is to use the "uniform force method" as noted in AISC (US) (1994). Thereafter, either the relevant Australian or USA provisions (e.g. Whitmore criteria, block shear, etc) can be applied for the connection components being designed. This situation is being further reviewed in Australia.

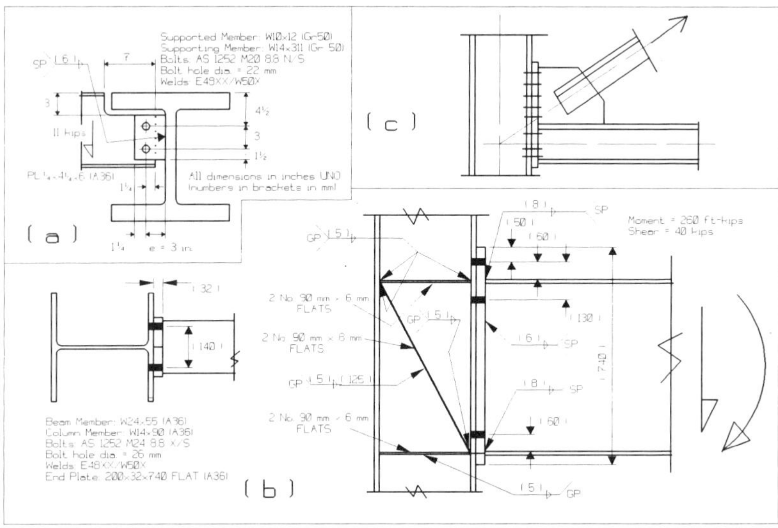

Figure 1: Structural steel connections considered (a) web side plate; (b) bolted moment end plate; (c) bracing connection. (Note: Number in brackets are in mm)

Conclusion

A summary of Australian connection design methodologies has been presented on three different types of structural steel connections. This summary will be further described during the Conference presentation.

References

AISC(US) (1994), "Load & Resistance Factor Design – Volume II: Connections", 2nd edition, Manual of Steel Construction, American Institute of Steel Construction.
Hogan, T.J. & Thomas, I.R. (1994), "Design of Structural Connections", 4th edition, Australian Institute of Steel Construction.
AS 4100 (1998), "AS 4100:1998 Steel Structures", Standards Australia.
Syam, A.A. & Chapman, B.G. (1996), "Design of Structural Steel Hollow Section Connections", Australian Institute of Steel Construction.

The SAC Database:
Understanding and Using Steel Frame Test Results

David Bonowitz, S.E.[1]

Abstract

The SAC Database of WSMF Connection Tests (SAC, 1998) is an online, searchable tabulation of recent full-scale tests. This paper discusses the test data in terms of their implications for building evaluation and design. Key issues include the distinction between connection capacity and reliability, and the appropriate use of existing test results to qualify proposed new details.

The Database

Building code provisions for steel frames changed after the Northridge earthquake. Connection designs must now be supported by specific tests or calculations. In response to this change, over 300 full scale specimens have been tested in the last four years. It's a bonanza of useful data, but it puts a new burden on engineers and building officials. Recognizing the need for a catalog of test results, the SEAONC Steel Subcommittee began compiling test data in 1996. Later, the SAC Joint Venture extended the list into an online database sortable by connection type, member size, and other test parameters.

The database is available at the SAC web site (SAC, 1998). As of late 1998, FEMA also expected to publish a bound volume of one-page test summaries. A full description of the database, including summaries by connection type and preliminary analysis of data trends, is provided in the SAC subcontractor's reports (Bonowitz, 1998).

Test Results by Connection Type

Selected results are presented here to demonstrate broad trends only. Conclusions about specific tests or details should not be drawn without consulting original test reports.

---

[1]887 Bush Street No. 610, San Francisco, CA 94108, 415-771-3227, dbonowitz@mindspring.com.

Acceptance criteria typically call for a minimum plastic rotation capacity and limited strength degradation (SAC, 1995, 1997; AISC, 1997). Because plastic rotations are reported inconsistently, the database emphasizes equivalent story drift, which includes both elastic and inelastic deformation in all connection components.

The Pre-Northridge Connection

Figure 1 shows the tested capacities of pre-Northridge (PreNR) connections. It also shows how many tests developed local flange buckling (LFB) and the number for which detailed behavior was not reported (nr). LFB is the preferred failure mode in PreNR connections.

FIGURE 1. Pre-Northridge Connections

Results confirm an inverse relationship between beam depth and inelastic capacity; most of the tests at the left side of Figure 1 are W36 beams, and most of the right side comes from W24's. Regardless of depth, Figure 1 suggests that good performance, controlled by LFB, is associated with a threshold story drift of about 0.025 radians. The database also shows that notch-tough weld metal improves PreNR performance.

Cover-plated Beams

Figure 2 shows the distribution of drift capacity and LFB in cover plate tests. While various beam sizes and plate details are represented, the general performance clearly surpasses that of PreNR specimens. Some brittle fracture patterns did occur, however. Of the six tests at the left end of Figure 2, four failed by fracture of the groove weld at the column face.

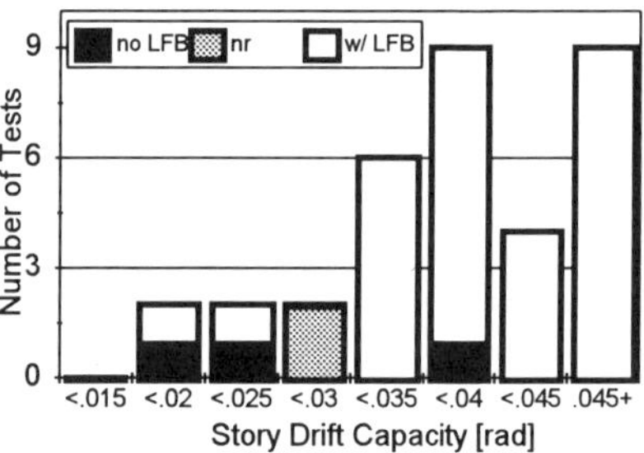

FIGURE 2. Cover Plate Connections

Reduced Beam Sections

Various reduced beam section (RBS) details have been tested. Figure 3 shows the distribution of drift in specimens with cut flanges (some of these may also have small cover plates or rib plates). LFB often can not develop in RBS specimens, so Figure 3 notes those governed by hinging at the reduced section.

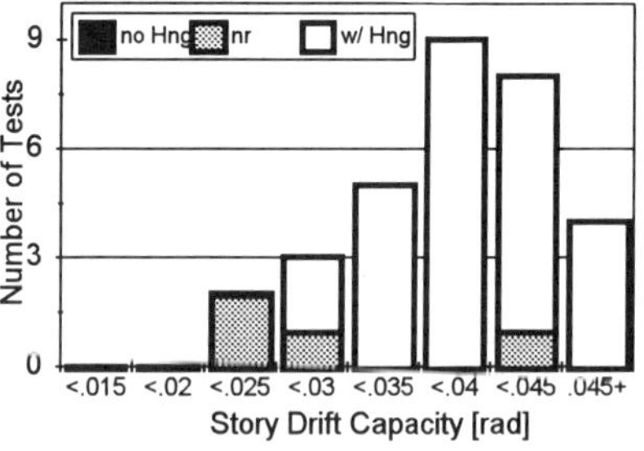

FIGURE 3. Flange Cut+ Connections

While Figure 3 suggests consistently good

performance, RBS details may be prone to lateral torsional buckling and strength degradation. These effects are not yet reliably quantified. Also, cuts at the bottom flange only (an economical retrofit option) have not tested well.

Implications for Evaluation and Design

Capacity and Reliability
While a few tests can establish the potential capacity of a given detail, they can not establish that such capacity will be achieved, say, 90% of the time. Simply stated, there are not enough test results of any connection detail to justify high reliability. Reliability can be gaged by the proportion of tests governed by ductile failure modes such as LFB and hinging. Figure 1, for example, suggests a reliability of about 40%, while Figure 3 suggests reliability approaching 100%. But these are estimates based on small samples. The reliability that can be assumed with confidence is lower than the nominal estimate. Furthermore, the number of existing tests that are comparable to a proposed design (see AISC, 1997) is generally quite small, rarely more than five or six.

Current criteria by SAC (1995, 1997) and AISC (1997) allow design based on just two or three tests. The reliability implications are significant. Consider a building with 100 connections highly stressed by a design ground motion. Based on three tests, can all 100 be expected to perform? Binomial statistics can estimate the answer: going 3-for-3 in the lab gives us only 50% confidence that the overall success rate will be 80% or better. Can the building survive with only 80% of its critical connections intact? Maybe. Is 50% confidence acceptable? If not, we can have 90% confidence from the same three tests, but only if we assume a success rate as low as 45%. Design criteria should focus less on plastic rotation and more on failure mode, less on capacity and more on reliability.

Without high reliability, evaluation and design must consider potential connection failure. One design approach is simply to provide additional or redundant connections so that performance objectives can be met even if some connections fracture. A policy approach is to adopt lower performance objectives suitable for existing buildings. From the analysis side, a probabilistic approach based on simulations can take advantage of known variability. Such an approach has been described by Bonowitz and Maison (1998).

Regulatory Issues
Currently, most new designs are being qualified by comparison to existing test results, although a number of key issues are still unresolved. First, nobody yet knows what makes a sufficiently comparable test. If a test beam had a yield stress of 42 ksi, can you use that test to qualify a detail with Dual Cert material? If you expect panel zone yielding in your frame but the closest available tests were one-sided specimens (less likely to exhibit panel zone yielding), can you use those tests? The current database is not sufficiently robust to offer any clear rules. In fact, such details as the shape of a weld access hole or the nature of welds at the tip of a cover plate have affected performance, so the idea of "comparability" based on gross properties seems somewhat strained.

Second, the reliability question still must be addressed. As shown above, a handful of good results can not justify an assumption of perfectly repeatable behavior. A related issue concerns failed tests, about which current acceptance criteria say little. Current criteria ask only for submission of two or three successful tests, effectively asking for proof of *potential* performance, as opposed to *likely* performance. If, for example, ten comparable tests can be found, but two were failures, what should this mean to your design? Is 8-for-10 better or worse than 3-for-3? (It depends on the confidence you need.) As discussed above, binomial statistics and probabilistic analysis can answer these questions, but this rational approach will not be taken unless building officials require it.

Third, if qualification by comparison is tightly restricted, engineers will use the same qualified member sizes and frame dimensions over and over, regardless of cost or structural inefficiencies, just to avoid new testing. Ultimately, this will be harmful to the design and construction community as a whole.

Finally, while additional testing could resolve some technical uncertainties, whose responsibility is it? Even if an owner is required to pay for testing, can she be required to share the results without compensation? And if two tests mean so little in terms of reliability, how can any project-specific testing requirements be justified? Legally and politically, it is difficult for building departments to use fees or incentives to encourage more testing, so without institutional funding, testing is likely to dry up, and all future designs will be based on the data available today.

References

AISC (1997). *Seismic Provisions for Structural Steel Buildings*. AISC, April 15, 1997.
Bonowitz, D. (1998). "SAC Database of WSMF Connection Tests." January, 1998. Awaiting publication by the SAC Joint Venture. To be updated in early 1999.
Bonowitz, D. and Maison, B. (1998). "Are Steel Frames Safe? Probabilistic Seismic Evaluation of Existing WSMF Buildings." *Sixth U.S. National Conference on Earthquake Engineering*, Seattle, June, 1998. Published by EERI, 1998.
SAC (1995). *Interim Guidelines* (FEMA 267). The SAC Joint Venture, August, 1995.
SAC (1997). *Interim Guidelines Advisory No. 1* (FEMA 267A). The SAC Joint Venture, March, 1997.
SAC (1998). http://quiver.eerc.berkeley.edu:8080/design/conndbase/. This site offers a list of over 80 references on specific tests.

Acknowledgments

Development and maintenance of the SAC Database of WSMF Connection Tests is funded by FEMA through the SAC Joint Venture. Credit is also due to the engineers and researchers who made test results available. The opinions and conclusions presented in this paper are entirely those of the author. The author assumes no legal liability or responsibility for the use of this paper's contents by others.

Through-thickness Properties of Column Flanges

Robert J. Dexter[1], M. ASCE

Abstract

An investigation is described which is intended to determine the through-thickness strength of the column flanges under constrained conditions similar to those of the connection. Large-scale "T" joints were tested in tension with high-strength (690 MPa yield strength) pull plates welded transversely to heavy column sections. Results show that the apparent through-thickness strength of the column flange exceeded 690 MPa. This result can be explained by the existence of triaxial constraint of the column flange material, which creates hydrostatic tension stresses, raising the apparent through-thickness strength.

Introduction

There are few, if any, documented instances of fractures from the 1994 Northridge Earthquake that can be attributed to the strength or toughness of the column base metal. Some of these fractures curved into the column base metal and then back toward the surface of the column flange, forming a "divot" in the column flange. These fractures have often mistakenly been associated with the through-thickness strength of the column flanges. All of these fractures initiated in the weld metal, and the fact that the fractures propagated into the column material is not related to low strength or toughness of the column material (Fisher, Dexter and Kaufmann, 1997, and Kaufmann, et al, 1997).

However, there are lingering questions about the through-thickness strength of the column sections. This lingering doubt may be due to the results of uniaxial through-thickness tensile tests performed in accord with ASTM A770. These results often show: 1) much less reduction in area less than is typical for

[1]Department of Civil Engineering, University of Minnesota,
500 Pillsbury Dr. S.E. Minneapolis, MN 55455, USA, dexter@tc.umn.edu

longitudinal tests; and, 2) a high degree of scatter including some low strength values that may even be below the minimum specified yield strength (MSYS) in the longitudinal direction (Barsom and Korvink, 1998).

To explain the effect of constraint, consider a "scoop" of column flange material that is postulated to pull away with the beam flange. This scoop of yielding column material is surrounded by column flange material that is not yielding, which prevents shrinkage in the column transverse and longitudinal directions to accommodate stretching in the through-thickness direction. The scoop of column material will begin to develop tension in both the transverse and longitudinal directions, in addition to tension in the through-thickness direction due to the applied load. According to Von-Mises' yield theory, yielding depends on the differences between the principal stresses, so the tension in the transverse and longitudinal directions elevates the stress in the through-thickness direction required to cause yielding (Ferrel and Dexter, 1995).

Large-scale "T-joint" Tests

Thirty-four large-scale "T" joints were loaded in tension through high-strength (690 MPa MSYS) pull plates that were welded transversely to heavy column sections (Figure 1). Minimal weld reinforcement was used. High strength and high toughness pull-plates and welds were used to attempt to induce failure in the through-thickness direction of the column flange. Column sections were obtained from all the major steel mills.

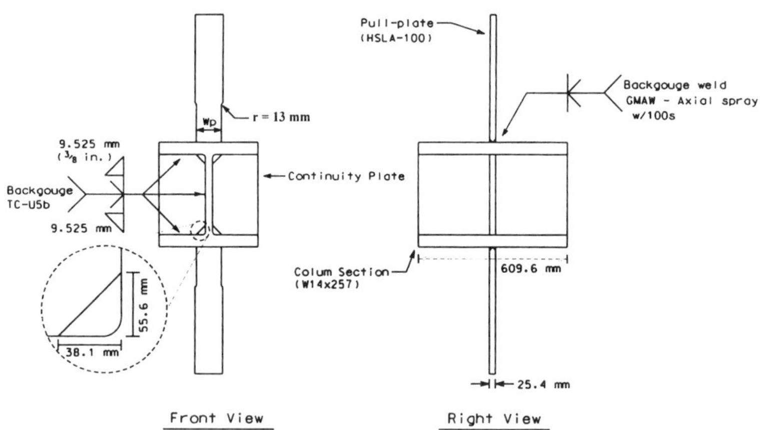

Figure 1: "T-joint" test specimen

As shown in Table 1, all specimens (except the last three in the table) broke in the pull plates at stress levels exceeding the MSYS of 690 MPa. The strength in the T-joint tests was not dependent on the width of the specimens or the strain rate. The last three specimens in the table were attempts to get a column flange failure by using poor welds or details and grinding so that there was absolutely no reinforcement. Use of E70-T4 weld root pass and the use of high-heat input (which degraded toughness of the 100S weld filler) caused fracture in the weld. The last specimen was tested without continuity plates, which resulted in a divot-type failure of the column flange material, albeit at very high nominal stress levels. No unusual features such as lamellar defects were observed on the fracture surface of this specimen.

Table 1: Selected T-joint Test Results

Column Size	Pull-plate stress (MPa)	Weld stress** (MPa)	Type of "T-joint" test
W14x257	798	572	4 in. wide pull-plate
W14x176	778	427	High strain rate
W14x176	800	413	
W14x257	797	381	
W14x257	802	422	
W14x176	799	412	
W14x176	790	510	4 in. wide pull-plate Quasi-static strain rate
W14x257	701	493	
W14x257	777	542	
W14x176	770	485	12 in. wide pull-plate
W14x455	779	574	Quasi-static strain rate
W14x176	788	462	
W14x257	764	448	
W14x257	654	654	12 in. wide pull-plate w/E70-T4 root pass & backing bar
W14x257	762	762	4 in. wide pull-plate w/high heat input weld
W14x257	698	698	12 in. wide pull-plate w/out continuity plates

**1.6mm away from the column

In addition, seven tests were performed with pull-plates 204 mm wide with another plate lapped onto the pull plate giving 25 mm of eccentricity. In these plates, the bending moment exceeded the gross-section plastic moment and the plates essentially straightened out but did not fail. Despite the large plastic rotation, bending and prying did not induce column failures.

Conclusions

Through-thickness failure of the column section typically could not be induced in the T-joint test, despite applying stress greater than 690 MPa, i.e. well above the strength of any structural steel. The creation of tensile stresses in the transverse and longitudinal direction as well as in the through-thickness direction elevated the apparent yield strength in the through thickness direction in the T-joint test. Through-thickness failures were not induced by high-strain rate loading, bending and prying, using high-heat input welds, or using E70-T4 weld root pass with a backing bar in place. Therefore, it is concluded that the through-thickness failure of column sections is very unlikely in these types of T-joints. It is recommended that the through-thickness strength does not need to be explicitly checked in the design of welded beam-to-column connections.

Acknowledgements and Disclaimer

This work was sponsored by the FEMA-funded SAC Joint Venture. The author appreciates the guidance of project director James Malley and topical team leader Karl Frank. Testing was performed at Lehigh University by research assistant Minerva Melendrez with the guidance of Eric Kaufmann. The information contained herein is not for design use and is not applicable to specific building projects. These organizations and individuals do not assume any legal liability or responsibility for the accuracy, completeness, or usefulness of any of the information or processes described in this paper.

References

Barsom, J.M. and S.A. Korvink, "Through-thickness Properties of Structural Steels", ASCE Journal of Structural Engineering, 1998.

Ferrell, M. and R.J. Dexter "Tensile and Shear Behavior of Undermatched Welded Joints", Proceedings of the Fifth (1995) International Offshore and Polar Engineering Conference, The Hague, Netherlands, 11-16 June 1995, Vol. IV, pp.1-5, 1995.

Fisher, J.W., R.J. Dexter, and E.J. Kaufmann, "Fracture Mechanics of Welded Structural Steel Connections", <u>Background Reports: Metallurgy, Fracture Mechanics, Welding, Moment Connections, and Frame Systems Behavior</u>, Report No. SAC 95-09, FEMA-288, March 1997

Kaufmann, E.J., J.W. Fisher, R.M. Di Julio, Jr., and J.L. Gross, Failure Analysis of Welded Steel Moment Frames Damaged in the Northridge Earthquake, NISTIR 5944, National Institute of Standards and Technology, Gaithersburg, MD, January 1997.

Welded Steel Moment Connections – Old and New

Božidar Stojadinović[1]   Subhash C. Goel[2]   Kyoung-Hyeog Lee[3]

Abstract

Fully restrained steel moment connections have been traditionally designed using a prescriptive design procedure. This procedure was invalidated by brittle fractures in these connections discovered after the 1994 Northridge earthquake. Until today, a widely accepted new design that satisfies the requirements for earthquake resistance has not been developed.

This paper presents a summary of the test results on a post-Northridge steel moment connection design proposed by the SAC Joint Venture. This connection design deploys some fracture mitigation measures, such as the use of notch-tough weld metal and improved welding practices. This connection is capable of achieving a mean plastic rotation of 1.5% radian, averaged over a wide range of connection parameters.

The results of these tests show that the use of fracture mitigation measures alone can not produce a good connection design. An independent study at the University of Michigan demonstrated the effect of the boundary conditions on the stress flow in the connection and the importance of considering the resulting over-stress in the flanges. A new, earthquake-resistant connection design must be based on a combination of fracture and over-stress mitigation measures. Several candidate designs implement such combinations of mitigation measures and are close to achieving satisfactory seismic behavior.

Pre-Northridge Design

Before the 1994 Northridge earthquake fully restrained steel moment connections were designed following the AISC Manual of Steel Construction (Figure 1). The flanges of the beam were welded to the column using a complete joint penetration weld. The beam web was bolted to the shear tab with the intent to transfer the shear generated at the nominal moment capacity of the beam. This prescriptive design was developed

[1] Assistant Professor, Department of Civil and Environmental Engineering, 2380 G. G.Brown Building, The University of Michigan, Ann Arbor, MI, 48109-2125, phone: 734-764-6576, fax: 734-764-4292, e-mail: boza@umich.edu

[2] Professor, Department of Civil and Environmental Engineering, 2376 G. G.Brown Building, The University of Michigan, e-mail: subhash@umich.edu

[3] Postdoctoral Fellow, Department of Civil and Environmental Engineering, 1368 G. G.Brown Building, The University of Michigan, e-mail: younglee@umich.edu

in the late 1960's and early 1970's and was widely used in steel moment-resisting frame structures.

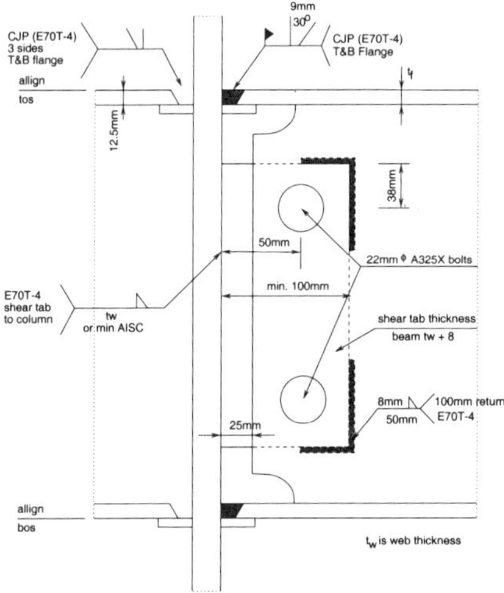

Figure 1: Pre-Northridge connection detail.

Steel moment connections suffered serious damage during the Northridge earthquake. They exhibited a multitude of failure modes. All of these modes were reproduced in laboratory conditions and most of them were shown to be very brittle: failures occurred before any significant yielding in the connection. Even though the engineering community was shocked by this outcome, the small plastic rotation capacity of the pre-Northridge connection design is not surprising. A statistical analysis of the 1970–1993 cyclic tests data shows that these connections have a mean plastic rotation capacity of 0.9% radian (Fry, 1998).

Post-Northridge Designs

The search for a new, earthquake-resistant steel moment connection proceeded in two ways: through individual proof-tests and through the SAC Joint Venture. The SAC Joint Venture effort focused on the connection welds and mitigation of brittle fracture. Different measures, such as the use of notch-tough electrodes, weld root reinforcement, improved welding practices, and the use of design assumptions consistent with the statistical variation of the properties of today's steel, were implemented in a post-Northridge connection design (Figure 2).

This connection design was extensively tested. The parameters under investigation were: beam size strength; beam yield strength; and panel zone strength. The list of specimens is shown in Table 3.

The outcomes of the tests are shown in Figure 4 in terms of the total rotation and the total plastic rotation of the specimens. The total plastic rotation is further divided into beam and panel zone components. The average total connection rotation was 2.6%

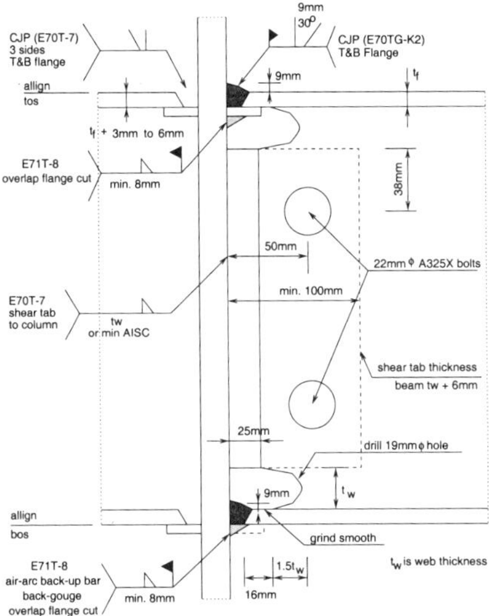

Figure 2: Post-Northridge connection detail.

radian, while the average total plastic rotation was 1.5% radian. The average beam contribution to the total plastic rotation was 0.9% radian, regardless of the beam size or strength. The remainder was provided by the plastic deformation of the panel zone. This contribution depends on the panel zone strength.

Failure of the post-Northridge specimens occurs in the base metal of the flanges, in the heat affected zone. The crack is located in the area between the edge of the weld on the outside flange surface and the end of the access hole on the inside flange surface. This crack propagates by ductile tearing in the middle of the flange first, followed by brittle fracturing towards the sides.

Conclusion

The use of fracture mitigation measures tested in these post-Northridge specimens is beneficial. These measures improve the seismic response of the unreinforced fully restrained steel moment connection. However, the newly detailed connection is not capable of generating a plastic rotation of 3% radian in two consecutive load cycles. Thus, it does not pass the FEMA 267/267A acceptance test.

The reason for such performance lies in the redistribution of forces in the connection caused by the boundary effects. A recent study at the University of Michigan (Lee, 1997) shows that the restraint of beam shear warping and transverse deformation due to the Poisson effect introduces additional self-equilibrating stresses in the beam. The redistributed normal and shear stresses concentrate in the flanges, resulting in a gross over-stress, while most of the beam web is virtually stress-free.

A new, earthquake-resistant, steel moment connection must incorporate both fracture and over-stress mitigation measures. The search for such connection, that is also

Specimen	W36x150	W30x99	W24x68	W14x257	W14x176	W14x145	W14x120
3.1			365/475				358/482
3.2			365/475				358/482
4.1		407/496				358/496	
4.2		317/462				358/496	
5.1		407/496			386/503		
5.2		317/462			386/503		
6.1		400/503		386/524			
6.2		310/427		386/524			
7.1	310/427			393/503			
7.2	310/427			393/503			

Figure 3: Post-Northridge specimens (mill test σ_y/σ_u [MPa]).

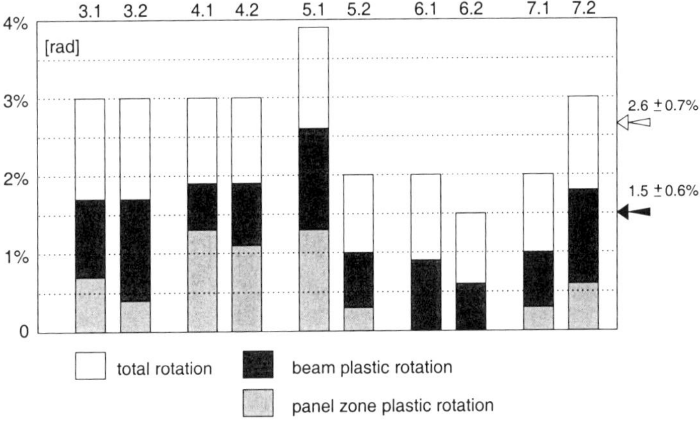

Figure 4: Post-Northridge connection detail.

economical and submits to rational analysis, is ongoing. To date, several good connection configurations have been proposed. One connection configuration, based on an indirect connection between the beam and the column using cover plates and vertical ribs, has been developed and successfully tested at the University of Michigan (Goel, 1997). Other new connection configurations, such as the slotted web connection or the reduced beam section connection, can also provide satisfactory seismic response.

References

Gary T. Fry "Reduced Beam Section Connections", presentation at the Meeting of SAC Steel Project Participants, Los Angeles, September 15–18, 1998

S. C. Goel, B. Stojadinović and K.-H. Lee, "Truss Analogy for Steel Moment Connections", *AISC Engineering Journal*, pp. 43–53, 2nd Quarter, 1997

K.-H. Lee, S. C. Goel and B. Stojadinović, "Boundary Effects in Welded Steel Moment Connections", Technical Report UMCEE 97-20, Department of Civil and Environmental Engineering, University of Michigan, Ann Arbor, December 1997

Update on Seismic Performance of Steel Frame Connections

James O. Malley[1] M. ASCE
and
Charles W. Roeder[2] M. ASCE

Abstract

During the January 17, 1994, Northridge Earthquake, many steel frame buildings
sustained cracking and other damage. The SAC Steel Project was funded by FEMA
to determine possible causes of the damage, develop of repair and retrofit methods
for pre-Northridge connections, and establish a suite of connections that could be
used for seismic design of new steel frame buildings. To achieve these goals models
that predict the strength, stiffness and ductility of connections that are to be used for
seismic design are needed. Experimental and detailed analytical studies are in
progress to develop the understanding needed to develop these engineering models.
This paper provides an update to this work.

Introduction

Steel frame buildings sustained cracking damage during the January 17, 1994,
Northridge Earthquake. The fractures all occurred in the pre-Northridge welded
flange-bolted web connection. The damage was not expected, and the SAC Steel
Project was funded by FEMA to provide practical guidance to the structural
engineers designing, evaluating, and repairing these connections. The SAC Phase 1
Program provided immediate, preliminary guidance regarding these connections
and potential repairs for damaged connections. The SAC Phase 2 program is
intended to provide a more in depth examination of the issues. Phase 2 must
identify the causes of the problem, develop an improved understanding of
connection behavior, provide a range of alternative connections, and develop the
necessary knowledge to use these alternative connections.

Considerable Phase 2 effort is being applied to experimental investigations of
connection behavior. The experiments are closely coordinated to assure that all
issues are addressed without repetition or critical gaps, and all experimental studies
include a component of analysis to assure that the maximum benefit is achieved
with the limited funds. The analysis may consist of finite element analyses and
parameter studies which are used to extend the range of applicability of the test

---

[1] Structural Engineer, Degenkolb Engineers, 225 Bush Street, Suite 1000, San Francisco, CA 94104

[2] Professor of Civil Engineering, 233B More Hall, University of Washington, Seattle 98195-2700.

results and analysis of past experiments so that these past results can be combined with test results from the SAC Program.   Separate analytical studies are also coordinated with these experimental studies through Task 5.3. The Connection Performance Technical Advisory Panel (Connection Performance TAP) planned and coordinated these analytical and experimental studies.

Scope of Phase 2 Program

The global objectives for the SAC Program require a series of moment connections which provide satisfactory seismic performance, evaluation and analysis techniques for connections, and methods of repairing or retrofitting existing structures. These objectives require consideration of many different aspects of connection behavior. The Northridge Earthquake has shown that different failure modes have a dramatic effect on the seismic behavior. The Northridge Earthquake also has shown that it is inappropriate to believe that undesirable failure modes can be avoided through design rules and specifications, since unexpected changes in the engineering and construction practice can negate these provisions. Therefore, every yield mechanism and failure mode that can be achieved with each of the connections and their effects the strength, stiffness, energy dissipation, and rotation or deformation limits of the connection must be evaluated. The net effect of the different yield mechanisms, failure modes and connection types is that a very broad research program was needed. This broad program was planned and coordinated by the Connection Performance Technical Advisory Panel (TAP).

Description of Connection Performance Projects

Task 5.3.1 is an analytical study which is divided into two parts. Task 5.3.1a was directed toward analysis of fracture critical connections and was performed by Prof. Deierlein at Cornell University. The goal was to establish whether analysis tools can accurately predict resistance and mode of failure. The work focused on pre-Northridge connections and it attempted to predict different crack directions noted in SAC Phase I Testing. The study included elastic and inelastic 2D and elastic 3D crack propagation analyses. The analyses showed good correlation between the measured and calculated force deflection behavior, and it indicated that beam yielding shields the critical regions of weld until redistribution of stress and strain hardening occur. This suggests the possible use of overmatched weld metal, and it also has potential impact in the selection of the materials for the beam and the column. The work also indicated that continuous supplemental fillet weld on the underside of the backing bar may benefit existing pre-Northridge welded connections, since an external initial crack becomes a less critical internal flaw of similar size.

Task 5.3.1b is a companion study performed by Professors El-Tawil and Kunnath at the University of Central Florida. They approach the cracking problem from a different direction, since inelastic deformation of the connection is predicted while examining cracking potential. The ABAQUS Computer Program was used to evaluate a range of parameters which may affect yielding and crack potential. These parameters include panel zone deformation, flange thickness, beam depth, column web and continuity plate thickness, weld cope holes, relative yield and tensile strength of steel, shear tab thickness, and span to depth ratio. The results suggest that panel zone yielding increases the potential for cracking. The connection ductility is insensitive to yield/tensile ratios less than 0.8 but larger ratios reduce ductility.

Task 5.3.2 is a coordination study of the connection research program. It develops simple design, analytical models for all connection types. This work evaluates hundreds of past connection tests in addition to the analysis and testing included in the SAC Project. The work is being done by Prof. Roeder of University of Washington, it will establish relatively simple but accurate models for establishing the strength, stiffness and ductility of a wide range of connection types. Task 5.3.3 is a follow-up study to Task 5.3.1, and is performed by Prof. Deierlein at Stanford University. It will perform fracture analysis of connections, and it will coordinate results from a number of test programs.

Task 7.01 developed the test program, test procedures and instrumentation plans. Task 7.02 evaluates welded flange bolted web connections with and without improved welding procedures. This study helps to understand pre-Northridge connections, to establish the extent that improved welding will improve connection performance, and to examine the effect of panel zone yield, beam depth and material properties on the connection performance. The work is done by Professors Goel and Stojadinovic at the University of Michigan. Task 7.03 started at a similar time and evaluates the use of bolted connections (T-Stub connections) for seismic design. This is being done by Professor Leon at Georgia Institute of Technology. He will address the complex behavior of bolted connections including prying forces, connection stiffness, the large number of failure modes and variations in connection type.

Task 7.04 focuses on simple connections including the effects of the composite slab, and it is being done by Prof. Astaneh-Asl at the University of California at Berkeley. Pre-Northridge buildings used simple beam-column connections at most gridline intersections. These single erection plates were bolted to the beam web and welded to the column. These connections provide some resistance and stiffness but were not included in the seismic design. This study will determine if and when these composite PR connections can be used as an aid to resist seismic loading and reduce the cost of retrofit. Task 7.05 is a follow-up to Task 7.02 since it will examine further effects of improved welding on connection performance. Cope details, continuity plates, web welds and some possible ecomonmic improvements are being evaluated. Professors Ricles and Lu at Lehigh University are doing this work.

Task 7.06 will develop a reliable design model for the reduced beam section connections. The work is done by Professors Engelhardt and Fry at the University of Texas at Austin and Texas A & M.   The work will examine all failure modes not considered in past testing, and it will focus on the effects of the composite slab, lateral-torsional buckling and panel zone yielding. Task 7.07 will be done by

Professor Anderson at University of Southern California. It examines economical methods for retrofitting modest sized pre-Northridge connections such as fins and weld overlays. Task 7.08 focuses on welded flange plate and coverplated connections. The work is being done by Professor Bertero and Dr. Whittaker at University of California, Berkeley, Earthquake Engineering Research Center. They will develop design procedures for welded flange plate connections, and they will address several failure modes which have recently been observed in coverplate connection test.

Task 7.09 is being performed by Professor Schneider at the University of Illinois. He addresses bolted flange plate connections and considers the possibility of supplementing bolted connections with friction dampers. This study will examine failure modes of bolted flange connections and will improve models for predicting connection behavior. Task 7.10 focuses on the bolted extended end plate connections and is being performed by Professor Murray at Virginia Tech. Extended End Plates appear to be one of the more promising bolted connection alternatives for seismic design, but failure modes are complex and methods for predicting the behavior need improvement. This study will attempt to fill that gap.

Task 7.11 will explore other connection issues such as loading history and frame behavior and will be done by Professor Uang at the University of California, San Diego. Task 7.12 includes supplemental weld tests in support of the overall connection test program. This work is being performed by Dr. Johnson of the Edison Welding Institute. Material tests are being performed on welded connections identical to those used in test studies so that analytical predictions and experimental results can be compared.

Summary and Conclusions

Work on the SAC Phase 2 program is nearing completion. However, considerable addition evaluation will be needed before the final results can be presented. This paper has illustrated the concerns, focus and scope of this work.

Acknowledgments

The work forming the basis for this paper was conducted pursuant to a contract with the Federal Emergency Management Agency. The substance of such work is dedicated to the public. The funding is provided through the SAC Joint Venture. The authors are solely responsible for the accuracy of statements or interpretations contained in this publication. No warranty is offered with regard to the results, findings and recommendations contained herein, either by the Federal Emergency Management Agency, the SAC Joint Venture, the individual joint partners, their directors, members or employees. These organizations and individuals do not assume any legal liability or responsibility for the accuracy, completeness or usefulness of any of the information, product or processes included in this publication.

Analysis and Behavior of Steel Frame Buildings

Douglas A. Foutch[1] Seung-Yul Yun[2] and Kihak Lee[2]

Abstract

A performance based design procedure for design of steel moment frames is being developed for the SAC Phase 2 project. The three performance objectives are considered. The procedure is developed within the reliability framework established for the project. Several analysis procedures are recognized and bias factors will be developed so that each method will produce buildings with approximately the same probability of satisfying the design objective. A dynamic pushover analysis is used to determine the maximum drift allowed to prevent global instability. Special, Intermediate and Ordinary moment frames will be investigated.

Introduction

The Northridge Earthquake revealed that the current method of making beam-to-column connections was inadequate when fractured connections were found in over 200 buildings. The common method of welding the flanges of the beams directly to the flange of the column using a full penetration grove weld was found to be inadequate. Several factors contributed to the problem including brittle weld metal. However, tests of beam-column subassemblies performed after the earthquake revealed that, even if the fabrication is done very carefully using a notch-tough electrode, inadequate ductility is obtained.

The SAC Phase 2 project is evaluating a number of possible methods of making beam-to-column-connections in steel moment connections. These include fully restrained and partially restrained connections. Analytical models of the moment-rotation behavior of each connection type are being developed. The maximum plastic rotation capacities are also being evaluated. In addition to these tests, a number of studies are being done to investigate system performance,

---

1 Professor, Civil and Environmental Department, University of Illinois at Urbana-Champaign, 3129 Newmark Lab, 205 N. Mathews, Urbana, IL 61801.
2Research Assistant, Department of Civil and Environmental Engineering, University of Illinois at Urbana-Champaign.

material behavior and performance prediction and evaluation. The results of all of these tests and other studies are being used to develop performance based design and evaluation procedures.

Performance Based Design and Evaluation

The performance based procedures are currently being developed. These procedures are based on reliability concepts. Three performance levels are currently being investigated. These are named Collapse Prevention (CP), Immediate Occupancy (IO) and Fully Operational (FO). For the CP performance level no local or global instabilities would be allowed. Local instability will occur if the shear connection of any beam can no longer carry gravity loads. This would include fracture of the shear tab or the bolts. This behavior will be evaluated for ground motions that have a 2% probability of being exceeded in 50 years (2500 year return period). A 95% confidence level of not exceeding the performance requirements will also be required.

For the IO performance level only slight structural damage will be allowed including some connection fractures and some beam local flange buckling. No structural damage will be allowed for the FO performance level. The hazard level to be paired with each of these performance levels has not yet been decided. One option being considered is to couple the IO performance level with the 50%-in-50-year hazard and the FO performance level with the 50%-in-30-year hazard. It has not been decided if a confidence level will be required for each of these.

Acceptance Criteria

For the CP performance level the rotation capacity of the gravity connection will be determined from test data. The actual design limits will be given in terms of story drift.

Global instability will be determined using a dynamic pushover analysis (dpa) developed by Luco and Cornell (1998). An earthquake accelerogram is selected and scaled so that no yielding occurs in the frame when subjected to it. The intensity of the ground motion is increased slightly and the analysis is performed again. This process is continued until instability of the frame occurs or the drift reaches 10% when local instability is expected to occur. The 10% drift or the drift at which instability occurs is considered to be to drift limit. The process is repeated for another accelerogram. The number of accelerograms to be used is 20.

The nine-story frame shown in Figure 1 is used to demonstrate the results for a dpa analysis using 10 accelerograms. The white circles in Figure 2 show the collapse drift for each accelerogram. For this example the average collapse drift is 9.4%.

As mentioned above, the story drift will be used as the main compliance parameter. This is easily computed by the design professional. The actual critical parameter for beams is the plastic rotation which is not calculated for elastic analysis procedures. Story drift can only be used if there is a predictable relationship between story drift and plastic rotation. Figure 3 shows the plastic rotation, panel zone rotation and story drift. These results indicate that the drift is approximately equal to the panel zone deformation and the beam plastic rotation. Additional analyses revealed that the plastic rotation of the beam is related to the ratio of the panel zone strength and the plastic strength of the beam. As a result, story drift will be an effective acceptance parameter.

Bias Factors

Several analysis procedures are being evaluated. These are the 1997 NEHRP equivalent lateral force and linear modal analysis procedures, the FEMA linear static, nonlinear static, elastic and inelastic time history, and capacity spectrum procedures. Bias factors are being developed for each procedure so that the buildings designed with each will have approximately the same reliability when subjected to the design ground motion.

Acknowledgments

This study is supported by a grant from the SAC Joint Venture through a grant from the Federal Emergency Management Agency. All findings, opinions and conclusions expressed in this paper are those of the authors and do not necessarily represent those of the sponsors.

References

Luco, N. and Cornell, C.A., "Effects of Random Connection Fractures on the Demands and Reliability for a 3-Story Pre-Northridge SMRF Structure", Proc. 6th U.S. National Conference on Earthquake Engineering, Seattle, Washington, June 1998.

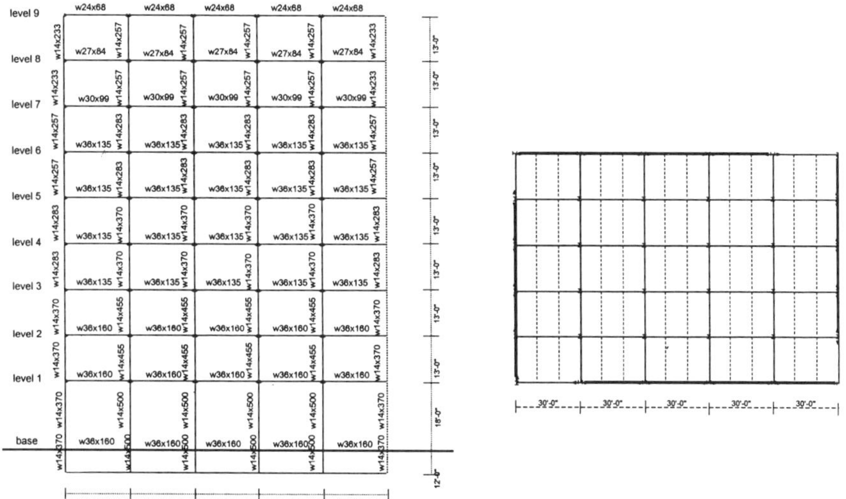

Figure 1. Elevation and plan view of Pre-Northridge Nine-story building

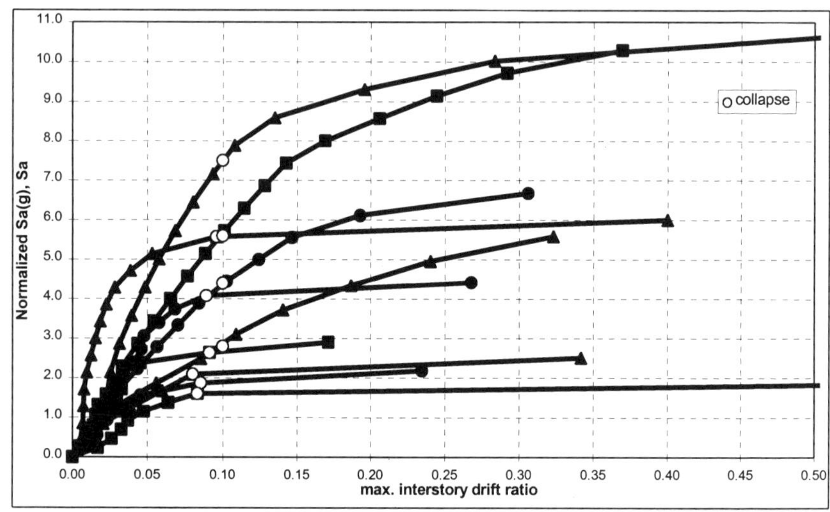

Figure 2. Dynamic Pushover of Pre-Northridge Nine-story building

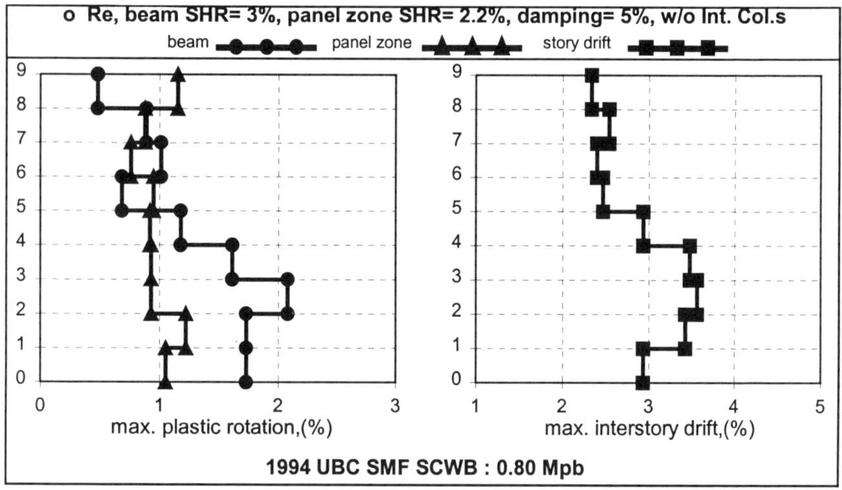

Figure 3. Plastic deformations and story drifts of Pre-Northridge Nine-story building

The Basis for the Fatigue-Evaluation Methodology in the Proposed *Manual for Condition Evaluation and Load and Resistance Factor Rating of Highway Bridges*

Dennis R. Mertz,[1] Member, ASCE

Abstract

A new fatigue-life evaluation methodology is being developed for the new *Manual for Condition Evaluation and Load and Resistance Factor Rating of Highway Bridges*. It is based upon the *Guide Specifications for Fatigue Evaluation of Existing Steel Bridges* and the fatigue design procedures of the *LRFD Bridge Design Specifications*. The basis for the new methodology is described herein.

Introduction

A new *Manual for Condition Evaluation and Load and Resistance Factor Rating of Highway Bridges* is being developed as the National Cooperative Hgihway Research Program (NCHRP) Project 12-46. New fatigue-life evaluation procedures for existing steel bridges are specified in Section 7 of the Manual which combine aspects of the *Guide Specifications for Fatigue Evaluation of Existing Steel Bridges* (AASHTO 1990) and the fatigue design procedures of the *LRFD Bridge Design Specifications* (AASHTO 1998). The basis of these fatigue-life evaluation procedures are described herein.

A new evaluation fatigue life is defined through new reliability-based resistance factors, resulting in a more reasonable estimate of remaining life, yet still taking into account the extreme variability of fatigue resistance of welded steel bridge details. These resistance factors allow the evaluator to estimate mean and design fatigue life for comparison purposes also.

Partial load factors, adapted from the *Guide Specifications*, model variability and uncertainty of the estimated stress range used in the evaluation.

---

[1]Associate Professor of Civil Engineering, University of Delaware, Newark, DE 19716

Table 1 - Partial Load Factors; R_{sa}, R_{st} and R_s

column 1	column 2	column 3	column 4
Fatigue-life Evaluation Methods	Analysis Partial Load Factor, R_{sa}	Truck-weight Partial Load Factor, R_{st}	Stress-range Estimate Partial Load Factor, R_s (In general, R_s equals R_{sa} times R_{st})
For Evaluation or Minimum Fatigue Life			
stress range by simplified analysis, and truck weight per Article 3.6.1.4 of the *LRFD Specifications*	1.0	1.0	1.0
stress range by simplified analysis, and truck weight estimated through weight-in-motion study	1.0	0.95	0.95
stress range by refined analysis, and truck weight per Article 3.6.1.4 of the *LRFD Specifications*	0.95	1.0	0.95
stress range by refined analysis, and truck weight by weight-in-motion study	0.95	0.95	0.90
stress range by field-measured strains	na	na	0.85
For Mean Fatigue Life			
all methods	na	na	1.00

The Load Side of the Load and Resistance Factor Evaluation (LRFE) Methodology

The estimate of stress range for use in the fatigue-life evaluation procedures utilizes partial load factors which are a function of the uncertainty of both the analysis procedure and the estimate of the effective fatigue truck weight. The effective stress range is estimated by:

$$(\Delta f)_{eff} = R_s(\Delta f)$$

where R_s is given in column 4 of Table 1 and, in general represents the product of columns 2 and 3, and Δf is the calculated or measured stress range. These partial load factors are adapted from the *Guide Specifications*, with a more simplified presentation.

The Resistance Side of the LEFE Methodology

The resistance side of the LRFE methodology for estimating load-induced fatigue life is divided into two levels: the infinite-life check of Article 7.2.4 of the Manual and the finite fatigue life estimate of Article 7.2.5 of the Manual.

In the infinite-life check, a detail is assumed to have infinite fatigue life if the maximum stress range, defined as twice the effective stress range from the load side of the methodlogy, is less than the constant-amplitude fatigue threshold of the *LRFD Specifications*. No variability is introduced for the fatigue threshold, as it has not presently been adequately quantified.

If the detail under evaluation fails the infinite-life check, the finite fatigue life is calculated using load and category related variables defined in the *LRFD Specifications* and probability-based resistance factors in Table 7.2.5.1-1 of the Manual, given as Table 2, herein. The observed variability of the resistance of welded steel details is embodied in these resistance factors. A new evaluation fatigue life has been defined as one standard deviation above the mean life, and the fatigue life is calculated as:

$$Y = R_R A / [365 n (ADTT)_{SL} ((\Delta f)_{eff})^3]$$

where R_R is given in Table 2, with the remainder of the variables are given in the fatigue provisions of the *LRFD Specifications*. The minimum life represents the design condition of the *LRFD Specifications*.

Summary

The principles of reliability employed in the *LRFD Specifications* for design, which originated in the *Guide Specifications*, have been extended to fatigue-life evaluation in the new Manual. The basis for partial load factors and resistance factors is briefly discussed herein.

Table 2 - Resistance Factor for Evaluation, Minimum or Mean Fatigue Life, R_R

Detail Category (from Table 6.6.1.2.3-1 and Figure 6.6.1.2.3-1 of the *LRFD Specifications*)	R_R		
	Evaluation Life	Minimum Life	Mean Life
A	1.7	1.0	2.8
B	1.4	1.0	2.0
B'	1.5	1.0	2.4
C	1.2	1.0	1.3
C'	1.2	1.0	1.3
D	1.3	1.0	1.6
E	1.3	1.0	1.6
E'	1.6	1.0	2.5

References

American Association of State and Highway Officials (1998). *LRFD Bridge Design Specifications*, 2nd Edition. AASHTO, Washington, D.C.

American Association of State and Highway Officials (1990). *Guide Specifications for Fatigue Evaluation of Existing Steel Bridges*. AASHTO, Washington, D.C.

Acknowledgment

The development of the proposed Manual, and the fatigue evaluation procedures therein, is sponsored by the American Association of State Highway and Transportation Officials, in cooperation with the Federal Highway Administration, and is conducted in the National Cooperative Highway Research Program which is administered by the Transportation Research Board of the National Research Council.

The research agency developing the Manual as NCHRP Project 12-46 is A. G. Lichtenstein & Associates, Inc. with Charles Minervino and Bala Sivakumar as Principal Investigators. David Beal is the NCHRP's program officer responsible for this project.

Fatigue Behavior of Intermittently Welded Diaphragm-to-Beam Connections

Amy S. Barth[1] and Mark D. Bowman[2]

Abstract

Fatigue tests were conducted to examine the fatigue resistance of a bridge diaphragm detail that is directly welded to the web of the supporting beam. Additional tests were also conducted to evaluate possible repair procedures. The testing demonstrated that, although some fatigue cracking might occur at certain locations, the detail is not likely to reduce the service life of the bridge.

Introduction

Diaphragms in multigirder steel bridges are frequently used to provide lateral stability and to distribute loads laterally among bridge girders. Structural diaphragms range from rolled beam shapes to built-up cross frames. Usually, diaphragm members are connected to steel girders through vertical connection plates. The connection plates are welded to the web, sometimes welded to the compression flange, and often cut short of the tension flange thus creating a small unstiffened gap. Various types of cracking have occurred in the gap region due to displacement-induced fatigue loading (Grider and Bowman, 1998).

A common diaphragm detail in Indiana highway bridges involves a hot-rolled beam member that is directly welded to the web of the longitudinal beam members at mid-height. An intermittent fillet weld is used to connect the diaphragm web to the web of the supporting beam member. A short fillet weld is also used to attach the top side of the top and bottom flanges of the diaphragm to the beam web. This detail is no longer used in new bridges, but many bridges with this detail are still in service.

Fatigue cracks have been observed in the welds that connect the diaphragm to the beam web. The cracks, which were found during visual inspection, have been detected primarily in the intermittent web welds and the lower flange welds. The cracks have not been observed to propagate into the beam web. A series of fatigue tests were conducted to evaluate the severity of the cracking that was observed. Also, a few tests were conducted to evaluate the effectiveness of retrofit and repair procedures.

[1] SAI Consulting Engineers, Inc., 26 Commerce Dr., Westover, WV 26505-3874
[2] School of Civil Engineering, Purdue University, West Lafayette, IN 47907-1284

Experimental Program

Nine steel beams with welded diaphragms were tested under constant amplitude cyclic loading. Six of the beams were loaded cyclically with no repairs being performed. Three additional tests were conducted to determine the effectiveness of hole drilling and air-hammer peening to repair/retrofit the diaphragm detail. A more detailed discussion of the testing program is provided by Grider (1998).

The test configuration consisted of three longitudinal beam members that were interconnected with diaphragm members. The load was applied at midspan of the middle beam, which was the primary member of the test set-up, with a servo-hydraulic actuator. The beam members were W24x55 hot-rolled shapes and the diaphragm members were W14x26 sections. ASTM A572 Grade 50 steel was used for all of the members. The diaphragm members were attached to the primary test beam member with 6.4 mm (0.25-in.) fillet welds produced by Shielded Metal Arc welding with 3.2 mm (0.125-in.) diameter E6010 electrodes. The diaphragm members were attached to the outside beam members with a bolted connection so that the outside beam members could be reused.

To simulate different possible field conditions, two different test configuration were used for the experimental program. In one case, the diaphragms were directly opposite of each other, while in the other configuration the diaphragms were staggered. The staggered diaphragm configuration is typical when the bridge is skewed. To evaluate the influence of the diaphragm configuration relative to the load, the loading actuator was positioned at two different locations for the staggered diaphragm specimens.

The test procedure involved the application of cyclic load along with periodic interruptions to monitor the response of the system. The beam was subjected to a sine-wave cyclic loading, with a stress ratio of R = 0.1 and a frequency of 1-2 Hz. At regular intervals the cyclic loading was stopped and static load measurements were performed. Visual inspections were performed at various times throughout the cyclic loading to detect the onset of fatigue cracking in the welds connecting the diaphragms to the beam web.

Test Results

Static loading measurements reveal that the two diaphragm configurations behave quite differently. When the diaphragms are directly opposite, they act like a rigid connection and transfer load to the adjacent beams. As long as the welded connection remains intact, the diaphragm transfers load through the system. The staggered diaphragm, however, does not stiffen the web, but instead pulls the beam-web out-of-plane. The measurements showed that very little load was transferred through the staggered diaphragms, since the system is able to deform with the load. These experimental results were consistent with observations obtained through field measurements of strains developed in both staggered and adjacent diaphragms taken by Canna (1996).

Due to the low stresses induced in the staggered diaphragm members, the welds for the staggered diaphragms seldom fractured. The welds for the adjacent diaphragms, however, were more susceptible to cracking since significant forces were carried by these members. In fact, after a few thousand cycles of loading, the bottom flange welds and some of the intermittent web welds on one side fractured. After this occurred, then the load transfer through the uncracked connection decreased, the out-of-plane bending increased, and the adjacent diaphragm configuration was found to behave like a staggered diaphragm set-up.

The cyclic test results for the experimental program are shown in Figure 1. Several observations can be drawn from the S-N plot shown. Firstly, it can be observed that many of the tests did not fail at the time the cyclic loading was terminated. These tests are denoted in the figure with an arrow. In fact, only two beam members actually failed.   Both of the beam failures occurred in members with a staggered diaphragm configuration; both diaphragm locations for one beam member failed, and one end only for the other beam failed. The cracks that led to failure were observed to consistently develop at the very end of the bottom flange diaphragm weld.

Secondly, for the beam members that failed, the cyclic life was observed to fall very near Category D. Several other tests, however, had developed significant cracks but had not failed at the time the cyclic loading was stopped. As shown in Figure 1, most of these data fell above Category D but below Category C.

Thirdly, the repair techniques were quite effective in repairing and retro-fitting the connection.   In two specimens (staggered and adjacent diaphragm configurations), peening was performed before cracks were present in the beam. The tests were then subjected to many more cycles of loading after the repair without any evidence of cracking at the weld toe at the end of the bottom flange weld where the critical cracking had been observed in the other tests. Also, in one of the staggered specimens, holes were drilled to arrest a fatigue crack after it had grown to 60 mm in length. The drilled holes were effective in arresting the cracks that were repaired.

Lastly, the diaphragm details were not found to be particularly susceptible to significant fatigue damage. The computed stress range at common diaphragm details for typical HS truck loading was found to be considerably lower than the stress range utilized in the cyclic tests. The primary reason for this behavior is that the diaphragm members are attached to the bridge beams near mid-height of the beam members, which tends to be in a region of lower primary bending stress. Moreover, the beam web gap between the diaphragm connection and the lower beam flange is large enough that the secondary bending stresses are not too significant. Consequently, it is not expected that a significant reduction in the service life of the bridges will occur due to fatigue cracking at the diaphragm welds.

Conclusions
Fatigue tests were conducted to examine the fatigue resistance of welded diaphragm details and to evaluate possible repair procedures. The testing demonstrated that, although some fatigue cracking might occur at certain locations, the detail is not likely to reduce the service life of the bridge due to the loading experienced at the critical location for the diaphragm detail.

References
Canna, T. (1996) "Field Evaluation of Bridge Diaphragm Members with Welded Connections." MSCE Thesis, Purdue University.

Grider, A.S. (1998) "Fatigue Behavior of Welded Diaphragm-To-Beam Connections." Ph.D. Thesis, Purdue University.

Grider, A.S. and Bowman, M.D. (1998) "Displacement-Induced Fatigue Cracking of Structural Details," *Proceedings*, Structural Engineers World Congress, San Francisco, CA, Paper T103-4, 8 pp.

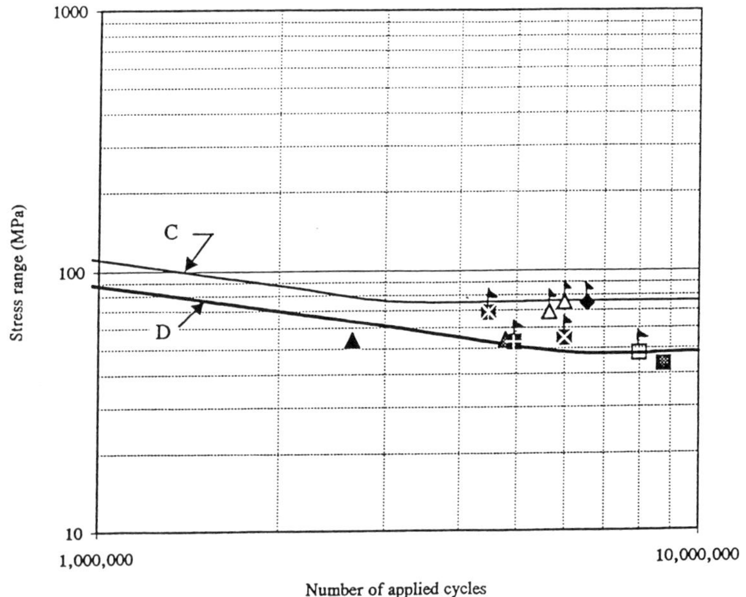

Figure 1   Cyclic Test Results of Diaphragm Members

Fatigue Evaluation Through Field Testing

J. Preston Halstead, P.E.[1], Jerome S. O'Connor, P.E.[2], Peter W. Szustak, E.I.T.[3]

Abstract

To address the inadequate fatigue life of a riveted steel truss bridge, field testing was undertaken to determine actual stresses in the critical members under traffic loading. Data was collected using magnetically attached acoustic strain gauges, and was processed on-site. Following post-processing and verification, the remaining safe fatigue life was determined to be +32 years; the bridge was determined to be safe from fatigue problems for its remaining desired service life.

Background

The bridge carrying U.S. Route 15 over the Cowanesque River in Lindley, NY is a 51.8m single-span riveted steel Camelback Warren Truss constructed in 1942. A fatigue analysis on the subject bridge was completed by TVGA Engineering, Surveying, P.C. (TVGA) in June of 1997. This study utilized a theoretical analysis to determine the effects of the predefined 240 kN fatigue truck, in accordance with the General Procedure outlined in Section 2.1 of the Guide Specifications for Fatigue Evaluation of Existing Steel Bridges [AASHTO, 1990]. Significant results for the controlling members are presented in Table 1. All other members were found to have "infinite" remaining fatigue life.

Table 1 - Theoretical Fatigue Analysis Results

Truss	Members	Remaining Safe Fatigue Life
East	U3-L4, L4-U5	-49.6 years
East	L2-U3, U5-L6	-47.6 years
West	U3-L4, L4-U5	-48.9 years

[1]Proj. Manager, TVGA Engineering, Surveying, P.C., 1000 Maple Rd., Elma, NY 14059
[2]Bridge Management Engineer, NYSDOT Region 6, 107 Broadway, Hornell, NY 14843
[3]Engineer, TVGA Engineering, Surveying, P.C., 1000 Maple Rd., Elma, NY 14059

The following options for addressing truss members with inadequate remaining fatigue life were examined for feasibility and cost-effectiveness:

1. *Inspection:* This option involves annual inspections of the fatigue sensitive details to assure adequate safety. While this option would not change the computed remaining fatigue life, it would ensure that any fatigue crack initiation and propagation would be detected early.

2. *Retrofit:* This option would increase the computed remaining fatigue life by replacing the existing rivets with high-strength bolts.

3. *Instrumentation:* This option would provide a more accurate measure of stress levels and fatigue cycles, by measuring strains in members under normal traffic loads. This would reflect the actual effects of truck loading and the built-in redundancies of the structural system. Therefore, an improved fatigue rating would likely be achieved.

A cost-benefit analysis was conducted for each option, including the cost of traffic delays to businesses and commuters. The cost of Option 1 was estimated at $300,000 over a 25-year period, the cost of Option 2 was estimated at $250,000, and the cost of Option 3 was estimated at $15,000. In hopes of determining that the structure actually has an adequate remaining safe fatigue life, Option 3 was recommended and adopted.

Procedures

"Alternative 1" of Article 2.1 of the AASHTO Guide Specifications provides for the calculation of the effective stress range to be used in the fatigue evaluation "through field measurements while the bridge is under normal traffic." This approach takes into account the actual wheel load distribution, the end conditions of the truss members, the multiple load paths inherent within the structure, and the impact of each vehicle, as related to its own speed and load configuration.

The load distribution of trucks observed during the field testing was assumed to be representative of "normal" conditions. Truck weight and classification data was collected during the field testing by "weigh-in-motion" sensors located approximately 20 km north of the bridge. The truck load distribution was also assumed to be "normal" for the time from original construction of the bridge to present day. This assumption is inherently conservative.

Strain data was collected using gauges placed on the critical members. The gauges were Acoustic Strain Gauges (ASG) manufactured by SonicForce, LLC. The ASG is magnetically attached, utilizing a non-contact ultrasonic technology to measure applied axial strain in structural steel members. Measurements were taken at L2-U3 and U3-L4 of the east truss, and at U3-L4 of the west truss. Gauges were placed at locations which had no section loss. Strain gauges were not placed on members L4-U5 and U5-L6 of the east truss, or on L4-U5 of the west truss, as these results can be inferred due to the symmetry of the truss. A sample time-history is shown in Figure 1, presenting recorded microstrain (μstrain) versus time, and shows the effects of a number of trucks, with both small and large magnitude cycles.

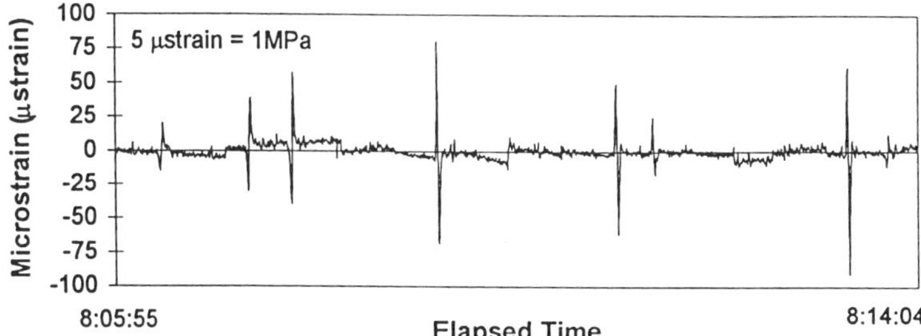

Figure 1 - Sample Strain Time History Data
(Member L2-U3 of the East Truss)

Cycles of fatigue loading were calculated from the recorded strain time-history data using a rainflow algorithm [Downing and Socie, 1982]. This algorithm analyzes the strain time-histories and determines the range of individual strain cycles in order to generate cycle count histograms. From the cycle count histograms, it is possible, using Miner's Law, to determine an "effective stress range" [AASHTO, 1990],

$$S_r = \left(\sum f_i S_{ri}^3 \right)^{1/3} \tag{1}$$

where S_r is the calculated effective stress range, f_i is the fraction of recorded stress cycles occurring within a given interval, and S_{ri} is the midwidth value of the interval. The effective stress ranges for the critical members (after accounting for section loss in the members) are presented in Table 2.

Table 2 - Effective Stress Ranges by Miner's Law

Truss	Location	Effective Stress Range
East	U3-L4, L4-U5	17.2 MPa
East	L2-U3, U5-L6	17.9 MPa
West	U3-L4, L4-U5	17.9 MPa

The safe remaining fatigue life was calculated from the effective stress ranges, in accordance with Section 3 of the AASHTO Guide Specifications, using:

$$Y_f = \frac{fK \times 10^6}{T_a C (R_s S_r)^3} - a \tag{2}$$

where Y_f is the remaining fatigue life in years, f is 1.0 for calculating safe life, K is the detail constant from Article 3.3 of the Guide Specifications, T_a is the estimated lifetime average daily truck volume in the outer lane, C is the stress cycles per truck passage, and a is the present age of the bridge in years. The calculated values of remaining safe fatigue life of the critical members using Equation (2) are presented in Table 3.

Table 3 - Fatigue Analysis Results

| | | Remaining Safe Fatigue Life | |
Truss	Location	Theoretical	Instrumentation
East	U3-L4, L4-U5	-49.6 years	+77.8 years
East	L2-U3, U5-L6	-47.6 years	+32.2 years
West	U3-L4, L4-U5	-48.9 years	+79.6 years

Conclusions

This paper summarizes the results of a fatigue evaluation performed on a riveted steel truss bridge built in 1942, using both "theoretical" and "instrumentation" analyses. The "theoretical" analysis yielded a safe remaining fatigue life of -50 years. This analysis utilized assumed truck traffic weights and wheel load distribution, and a structural model which ignored the beneficial effects of multiple load paths and member end effects.

The "instrumentation" analysis utilized acoustic strain gauges to obtain a more accurate determination of stress levels in the critical members under actual truck traffic loading. Recorded stresses were used to determine an effective stress range based on normal truck traffic and actual load distribution. Results of the "instrumentation" analysis demonstrated that the structure's remaining safe fatigue life of 32 years exceeds its remaining desired service life.

The cost of the instrumentation and engineering evaluation was $15,000, while the alternative measures would have cost approximately $250,000. This project demonstrated that condition analysis by instrumentation can provide a significantly more cost-effective solution, and a more accurate estimation of remaining fatigue life.

References

American Association of State Highway Transportation Officials (AASHTO), *Guide Specifications for Fatigue Evaluation of Existing Steel Bridges*, 1990.

S.D. Downing & D.F. Socie, *Simple Rainflow Counting Algorithms*, International Journal of Fatigue, Butterworth & Co., Ltd., pp. 31-40, 1982.

Fatigue Strength Assessment of I-95 Bridge over James River

Peter J. Massarelli[1]
Y. Edward Zhou,[2] M. ASCE
Jose P. Gomez,[3] M. ASCE

ABSTRACT

An assessment of the remaining life of the I-95 bridge over the James River in Richmond, Virginia was conducted using two procedures outlined in the AASHTO Guide Specifications for Fatigue Evaluation of Existing Steel Bridges (1990). Strain data were recorded for five fatigue critical details which were previously determined to have finite remaining fatigue lives, based upon an analysis conducted using the AASHTO fatigue truck. The effective stress ranges calculated from the field data were considerably lower than the stress ranges based upon the fatigue truck analysis, and the details were determined to have infinite remaining safe fatigue lives, as defined by the AASHTO criterion.

INTRODUCTION

The bridge carries six lanes of traffic, and was built in the mid 1950's with sections widened in the 1970's. It is comprised of 102 spans, consisting of four different types of steel superstructure elements; rolled wide flange beams, riveted plate girders, built-up floor beams, and a deck truss span. All spans are simply supported with deck joints at each end. The rolled beams are built composite with a 178 mm (7 in) concrete deck. The riveted plate girders and deck truss are non-composite with a 178 mm (7 in) and 165 mm (6.5 in) concrete deck, respectively.

There are four transverse riveted, built-up, I-shaped floor beams (designated FB13, FB14, FB15, and FB16) that span both north and southbound lanes between single column piers. Floor Beam 13 is shown in Figure 1. These floor beams support the ends of two beam spans and two riveted plate girder spans. All the elements were fabricated with ASTM A7

---

[1] Faculty Research Associate, Virginia Transportation Research Council, 530 Edgemont Road, Charlottesville, VA 22903

[2] Senior Bridge Engineer, formerly of URS Greiner, Inc., 2219 York Road, Suite 200, Timonium, MD 21093.

[3] Senior Research Scientist, Virginia Transportation Research Council, 530 Edgemont Road, Charlottesville, VA 22903

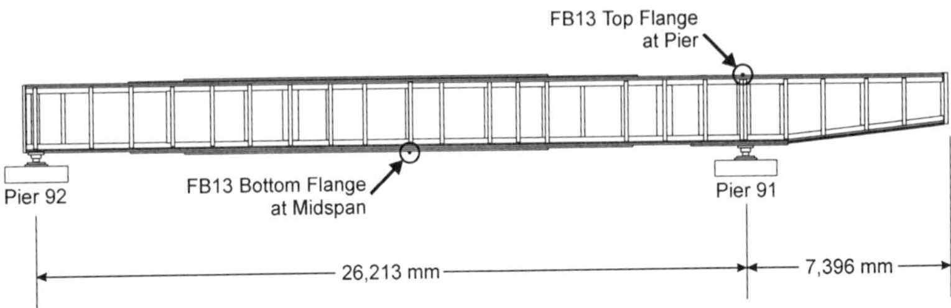

Figure 1.   Location of strain gages on FB13 (oriented longitudinally along the floor beam).

structural steel with a yield stress of 228 MPa (33 ksi). A category D fatigue detail is formed at the riveted locations. The floor beams are considered to be fracture critical elements.

In 1995, the Virginia Department of Transportation determined that the deck must be replaced. A detailed inspection indicated very little deterioration of the floor beams which were found to be structurally adequate to support a 216 mm (8.5 in) lightweight concrete deck.

URS Greiner conducted a fatigue evaluation in accordance with the AASHTO Standard Specifications for Highway Bridges (AASHTO, 1996) and the AASHTO Guide Specifications for Fatigue Evaluation of Existing Steel Bridges (AASHTO, 1990). The AASHTO guidelines stipulate that the remaining fatigue life of a category D structural detail is infinite if

$$R_S\, S_r < 17.9 \text{ MPa } (2.6 \text{ ksi})\tag{1}$$

where S_r is the calculated (or measured) stress range and R_S is the reliability factor.

The results of the evaluation indicated that the allowable fatigue stress range was exceeded according to the design specifications and the limiting stress range for infinite fatigue life was exceeded using the fatigue truck based on the fatigue guide specifications. It was recommended that either the four floor beams be replaced or a fatigue evaluation based upon field stress measurements be investigated. Since replacing the floor beams would add significant cost to the rehabilitation project, a field test was conducted (as allowed in the AASHTO guidelines) to determine the actual stress ranges at the critical locations.

FIELD TEST INSTRUMENTATION

Floor beams 13 and 14 and an interior longitudinal girder (designated G4) were instrumented with conventional quarter-bridge electrical resistance strain gages. Longitudinal girder, G4, was selected because it was determined to be closest to a wheel line of truck traffic. A total of five gages were employed for the fatigue study. Figure 1 shows the location of the two gages on FB13 -- one on the bottom surface of the bottom flange at the centerline of the web to record the tensile strain and the other on the side of the top flange plate at the centerline of pier 91 to measure the maximum negative moment strains. Floor beams 13 and 14 were instrumented identically. The fifth gage was placed at the midspan of G4 on the bottom flange.

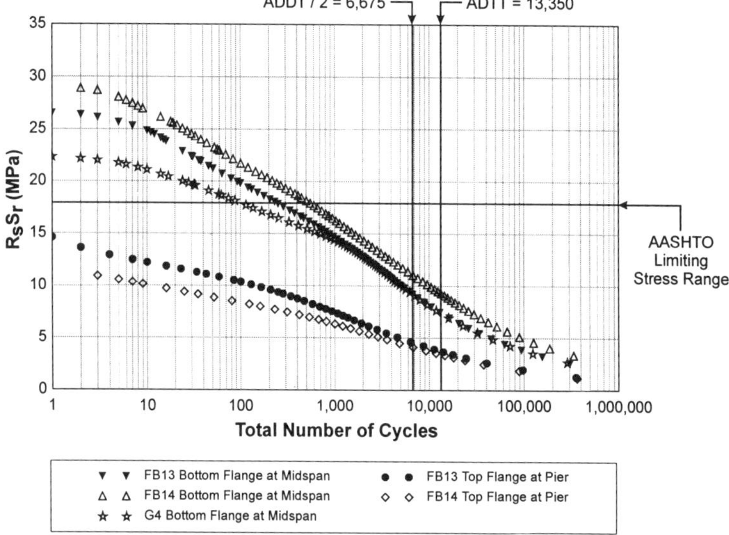

Figure 2.   **AASHTO factored stress range as a function of total number of cycles used in effective stress range calculation. The data presented here was recorded over a 24-hour period on April 2, 1998.**

The data were recorded with a Campbell Scientific CR9000 data acquisition system connected to a laptop computer. The measurements were recorded continuously at a sampling rate of 50 Hz from March 31 to April 7, 1998.

RESULTS

The data records were digitally filtered with a low-pass frequency algorithm (Carleton, 1970) to remove unwanted noise. Stress ranges were then calculated using rainflow cycle counting. Histograms were then constructed for each of the gage locations. No measured stress range exceeded 20.7 MPa (3 ksi) which is 42% of the constant amplitude fatigue threshold for a Category D fatigue detail (AASHTO, 1996).

The AASHTO guidelines provide the following equation for the effective stress range, S_r:

$$S_r = \left(\sum f_i S_{ri}^{\ 3} \right)^{1/3} \tag{2}$$

where f_i is the fraction of stress ranges within an interval and S_{ri} is the midwidth of the stress range interval. The effective stress range is highly dependant upon the number of total cycles used in the calculation. This is shown in the semi-log plot of Figure 2 where the effective

Table 1. Comparison of factored stress range results based upon field measurements and AASHTO fatigue truck analyses.

	Limiting Stress Range		AASHTO Fatigue Truck Analysis R_sS_r		Field Data Calculation R_sS_r	
	(MPa)	(ksi)	(MPa)	(ksi)	(MPa)	(ksi)
Floor Beam 13, Top Flange at Pier	17.9	2.60	32.5	4.72	3.7	0.54
Floor Beam 13, Bottom Flange at Midspan	17.9	2.60	28.7	4.16	7.5	1.09
Floor Beam 14, Top Flange at Pier	17.9	2.60	30.3	4.40	3.4	0.49
Floor Beam 14, Bottom Flange at Midspan	17.9	2.60	23.9	3.46	9.2	1.33
Girder 4, Bottom Flange at Midspan	17.9	2.60	21.0	3.04	9.2	1.33

stress range for a single day of data is shown as a function of the total number of cycles. As shown in the plot, the inclusion of all the rainflow cycles in the effective stress range calculation results in a drastically unconservative estimate.

It was decided that a reasonable value could be computed by limiting the total number of cycles to a value corresponding to the Average Daily Truck Traffic (ADTT) which was estimated to be 13,350 (VDOT, 1990). The floor beams respond to vehicles crossing the bridge in both directions, but Girder 4 is only subjected to stresses caused by traffic travelling in one direction. Using this rationale, the effective stress ranges were calculated and applied to the infinite remaining life criterion given in Equation 1. The reliability factor, R_S, for non-redundant members is given in the AASHTO guidelines as $0.85 \times 1.75 = 1.4875$.

All of the details tested were determined to have an infinite remaining fatigue life. Table 1 offers a comparison of the results from the field test and fatigue truck analysis.

REFERENCES

AASHTO (1996). "Standard Specifications for Highway Bridges," Sixteenth Edition, American Association of State Highway and Transportation Officials, Washington, D.C.

AASHTO (1990). "Guide Specifications for Fatigue Evaluation of Existing Steel Bridges," American Association of State Highway and Transportation Officials, Washington, D.C.

Carleton, H. D. (1970). "Digital Filters for Routine Data Reduction," U.S. Army Engineer Waterways Experiment Station, Vicksburg, Mississippi, Miscellaneous Paper N-70-1.

VDOT (1990). "Average Daily Traffic Volumes on Interstate, Arterial and Primary Routes," Commonwealth of Virginia Department of Transportation, Traffic Engineering Division, in cooperation with the U.S. Dept. of Transportation Federal Highway Administration.

Application of Field Data in Fatigue Life Estimation of Highway Bridges

Jamshid Mohammadi[1] and Ramakrishna Polepeddi[2] Members ASCE

Abstract

Non-destructive evaluation of highway bridges can effectively be conducted by monitoring and use of field stress data. This study focuses on the use of stress data compiled for bridges in the estimation of the remaining useful lives of highway bridges. The study examined several bridges located in the State of Illinois for damage evaluation purposes. The example bridges are all made up of steel girders with reinforced concrete deck slabs. For each bridge, the study examined the superstructure and investigated the potential for fatigue damage to the critical structural details of the main steel girders. The critical components investigated are steel girders with welded cover plates. The field stress range data compiled for each bridge was used along with a probabilistic method to estimate the bridge fatigue life.

Introduction

The increase in the load and the growth in the truck traffic in the past four decades have caused a rapid fatigue damage accumulation in many steel girder bridges. Fatigue damage in steel girder bridges depends on: (i) the severity of the stress range level; (ii) the number of stress cycles; and (iii) the type of structural details. Several categories of fatigue-critical components are specified in the current design guidelines (such as AAHTO) and have been utilized in bridge fatigue design. In general, the guidelines provide for a safe maximum stress range level based on an expected fatigue life. Bridges usually undergo changes during their service lives wherein the stress history, stress range level, previous damage history, and volume of traffic change; and, as such, fatigue damage may dramatically change also.

----

[1] Professor and Chairman, Civil and Architectural Engineering Department, Illinois Institute of Technology, Chicago, IL 60616

[2] Senior Engineer, Packer Engineering, 1950 Washington St., Naperville, IL 60566

To account for these factors, more refined procedures for fatigue evaluation of bridges have been proposed (Moses, et al, 1987). To properly include the effect of nearly all factors that would affect fatigue life of an existing bridge, it is more appropriate to obtain a realistic set of data on the stress history which depends upon bridge traffic. This can be achieved by conducting a comprehensive program of traffic monitoring and by compiling data on the resulting stress history of the bridge. To avoid the formidable costs associated with a comprehensive and continuous monitoring program, the strain history of a bridge at only a few critical locations may be obtained and used in the fatigue evaluation purposes. This paper presents the results of an investigation conducted on several highway bridges in the State of Illinois in which strain data was compiled and used together with roadway traffic data to estimate remaining life. The results of the study were then used to investigate the significance of truck weight increase and traffic growth on fatigue life of the example bridges.

Bridge Stress Range Data

The stress range frequency data may be developed by means of a data compression/reduction process, called "cycle counting". Various methods for cycle counting are available. One widely-used method is the "rainflow" process which is described in standards of the American Society for Testing and Materials (ASTM). A major issue in compiling data to construct stress range frequency statistics is the duration of the data acquisition process. A sufficiently long period of time is needed to capture a realistic representation of all stress ranges and their respective frequencies. In highway bridges, a two or three-day period seems to be satisfactory (Mohammadi, et al, 1992). It is expected that during this period nearly all stress ranges, that may apply to a critical component in a bridge, are captured. Preferably, the data acquisition device should be left running continuously to capture the effect of day and night traffic, and empty and full trucks. To include any seasonal change in the traffic pattern, the data acquisition process may have to be conducted in several two or three-day sessions at various times during the year. In the investigation reported herein, the example bridges were monitored with assistance provided by the Bureau of Materials of Illinois Department of Transportation (IDOT) in the State of Illinois. These example bridges are steel girder structures with reinforced concrete deck slabs. The maximum stress range values were generally quite low and within the 27-34 Mpa (4-5 ksi) range. It is also noted that the stress range data gathered correspond to the live loads on the bridge. The fatigue damage estimated in this study ignored the effect of the dead load stress (which turned out to be small compared with the live load stress ranges).

Fatigue Damage Estimation

Bridge fatigue damage was estimated using a probabilistic model. The Miner damage rule was used to compute the expected value of damage. This was done utilizing the stress ranges compiled for the example bridges. To estimate the remaining fatigue life of a bridge component, the procedure outlined by Moses et al (1987) or Ang and Munse (1975) can be used. In this study the latter was adopted. It was further assumed that failure of a critical component will cause either the failure of a bridge or a severe reduction in its serviceability. Thus the estimated fatigue life of a component was assumed to be identical with the life of the bridge.

One purpose of this study was to investigate the significance of truck weight increase and traffic volume growth to the service life of highway bridges. The field data compiled for the bridges provided the history of bridge stress ranges under current conditions. The investigation into the condition of these bridges was intended to specifically consider the following two factors and their effect on the remaining life of the bridges: (1) Truck weight increase by about 10% , from about 320 kN (72 kips) gross weight to 356 kN (80 kips); and (2) Traffic volume growth on rural and urban highways. Both factors were expected to shorten the fatigue life of these bridges. The weight increase was considered to be a critical factor, especially for the many bridges in Illinois that were designed for a lower gross truck weight. In addition to the weight increase, urban roadways experience an average of 5% traffic increase per year. The growth for rural roadways is about 5.6% per year (Mohammadi, Guralnick, Polepeddi, 1992). This growth will be compounded yearly and will accelerate the rate of fatigue damage growth.

Considering the aforementioned two factors, the study followed the steps described below to arrive at the fatigue life of any given bridge:

(1) A target reliability level equal to 97.7% for non-redundant and 99.9% for redundant bridges was selected.

(2) Based on the stress range data complied for a bridge, and an appropriate relation describing the number of stress cycles to failure versus the stress range, and the equation describing the reliability versus fatigue life, as stipulated by Ang and Munse (1975) and Byers, et al (1997), the remaining fatigue life was computed for two conditions; namely, without the truck weight increase and with the 10% weight increase.

(3) The rate of traffic volume increase and the average daily traffic for the roadway were used to convert the fatigue life into years. Appropriate factors to distribute the traffic volume to roadway lanes were used for this purpose. This was done for all example bridges.

Upon estimation of their lives, bridges were divided into three groups, those with lives shorter than 25 years, those with lives between 25 and 75 years and those with lives longer than 75 years. An examination of the results reveals that with the 72-kip weight and no traffic growth, nearly 80% of the bridges have a fatigue life longer than 75 years. However, with the 80-kip gross truck weight and with predicted traffic growth, only 40% of the bridges will have a life of 75 years or longer. The results indicate that 60% of these bridges will have a life shorter than 50 years and, in particular 20% will have a life shorter than 25 years. From these, it was concluded that an annual traffic growth of 5-5.6% is a more influential parameter in reducing bridge life than is an increase in gross truck weight from 72 kips to 80 kips.

Conclusions

This paper presents a non-destructive test method for the condition assessment and estimation of fatigue lives of steel girder bridges based on bridge field stress range data. The stress range data compiled for a bridge is used in a probabilistic study to estimate the extend of fatigue damage in a bridge critical component and the remaining useful life of the bridge. Based on the results of this study, it is concluded that:

- Field strain range data can be used effectively and easily to estimate fatigue life of bridges. Both deterministic and probabilistic methods can be used for this purpose.

- The duration of the data acquisition period may be limited to only a few days to capture the most significant variations in the loads experienced by a bridge.

- The method described herein for fatigue life estimation can be useful in evaluating effects upon bridge life when changes in legal truck weight and/or other factors such as traffic increases are expected.

References

Ang, A. H.-S. and Munse, W.H. (1975). "Practical reliability basis for structural fatigue." *Preprint No.2494, Proc., Nat. Struct. Engrg. Conf.,* ASCE, New York, N.Y.

Byers, W.G., Marley, M.J., Mohammadi, J., Nielsen, R.J., and Sarkani, S. (1997). "Fatigue reliability reassessment applications: state-of-the-art paper." *J. Struct. Engrg.,* ASCE, 123(3), 277-285.

Mohammadi, J., Guralnick, S.A., and Polepeddi, R. (1992). "Effect of increased truck weight upon Illinois highway bridges." *Rep. No. FHWA/IL/RC-013,* Dept. of Civ. Engrg., Illinois Inst. Of Techol., Chicago, Ill.

Moses, F., Schilling, C.G., and Raju, K.S. (1987). "Fatigue evaluation procedures for steel bridges." *Nat. Cooperative Hwy. Res. Program, Rep. No. 299,* Transp. Res. Board, Nat. Res. Council, Washington, D.C.

Fatigue Reliability Updating Using NDE

Achintya Haldar,[1] Fellow, ASCE and Zhengwei Zhao,[2] Member, ASCE

Abstract

Using a linear elastic fracture mechanics (LEFM) approach, the fatigue reliability is estimated considering all major sources of uncertainty, including uncertainties in nondestructive inspection outcomes. The fatigue reliability is then updated using uncertainty-filled information from nondestructive inspections, as it becomes available. The updated information on the reliability is used as a decision making tool regarding what to do next, in terms of whether to do nothing, reschedule the next inspection to an earlier date, or immediately repair or replace the damaged element. The procedure is explained with the help of an example.

Introduction

Nondestructive inspections (NDI) are becoming an important and essential part of the rehabilitation and maintenance of existing structures. Because of the high cost of replacement and a lack of funds, the possible extension of the service life of existing structures using NDI is a very attractive option. Fatigue is a very slow process, and its risk management using NDI is appropriate and is suggested by many regulatory agencies. All aspects of inspections, i.e., objectives, method, acceptance criteria, and interval, affect the risk management of a fatigue-sensitive structure. Furthermore, the decision on what to do after an inspection, such as doing nothing, inspecting more frequently, or repairing or replacing the damaged structural element, needs to be made based on the risk level and economic considerations. The challenge to the profession is to develop a comprehensive risk-based fatigue

---

[1]Professor, Department of Civil Engineering and Engineering Mechanics, University of Arizona, Tucson, AZ 85721

[2]Department of Research and Development, American Bureau of Shipping, 2 World Trade Center, 106th Floor, New York, NY 10048

evaluation model using the current and previous imperfect NDI results. This concept is the subject of this paper.

Fatigue Damage Assessment Methods

The deterministic models available to predict and evaluate fatigue damage can be classified into three groups: (1) the stress-life (S-N) curve-based approach, (2) the strain-life curve-based approach, and (3) studies based on crack propagation theory. The limitations of the first two approaches are: (1) they do not use crack size information even it is available, (2) it is not possible to estimate the remaining life of cracked structures, and (3) the modeling updating technique can not be used. The third approach is superior to the other two methods and can incorporate more information to describe complicated fatigue damage accumulation. This approach is emphasized in this paper.

The general form of the crack growth theory based on fracture mechanics approach can be expressed as:

$$\frac{d\alpha}{dN} = f(Q,\ \Delta K,\ \alpha,\ R) \tag{1}$$

where α is the crack size, N is the number of stress cycles, Q is the vector of material constant, ΔK is the range of stress intensity, and R is the stress ratio. In this approach, failure can be defined as the maximum crack size that will lead to fracture or a tolerance level that is unacceptable for practical designs. Paris (1964) suggested a model for predicting crack growth under cyclic tensile loading which is routinely used in the profession.

Fatigue Reliability Analysis

A considerable amount of uncertainty is expected in all aspects of the fatigue damage evaluation process. Using the first-order reliability method (FORM) (Haldar and Mahadevan, 1999), and the LEFM approach, the authors proposed a fatigue reliability method. For a limit state of $\alpha_c - \alpha_N \leq 0$, where α_c is the critical or permissible crack size and α_N is the in-service crack size after N stress cycles, they calculated the reliability index β considering all major sources of uncertainty. This will be discussed in detail during the presentation. Zhao and Haldar (1997) also introduced the concept of target reliability β_T, the reliability index below which the structure will fail to safely perform its intended function. They suggested a target reliability index of 3.0.

Non-Destructive Inspection

Non-destructive inspection (NDI) is an important and economical tool for controlling and improving structural reliability under in-service conditions. Various NDI techniques have been used for the purpose of detecting cracks, including visual inspection, ultrasonic inspection, liquid penetrant inspection, magnetic particle inspection, magnetic field inspection, acoustic emission inspection, eddy current inspection, radiographic inspection, and dynamic measurement. For a particular NDI technique, several factors are expected to affect inspection results, including modeling effects, human factors, and inspection factors. No single NDI technique is unquestionably superior to the others in all cases. They may fail to consistently produce the same results under the same conditions. For any given NDI technique, there is always a critical crack size below which a crack can not be detected.

Fatigue Reliability using NDI

A structure is expected to be inspected several times during its lifetime. The possible outcomes of an inspection can be classified into one of the following cases: (1) crack detection without size measurement; (2) crack non-detection, and (3) crack detection with measurement of crack size. The problem can be stated in its simplest form as follows. A fatigue-sensitive structure is inspected for the N*th* times using an imperfect NDI technique. Considering the three inspection outcomes just mentioned, Zhao and Haldar (1997) proposed an efficient method to estimate fatigue reliability. It will be discussed in detail during the presentation.

Updating Fatigue Reliability using NDI

Whether or not any cracks are detected, each inspection provides additional information and results in changes in the prior estimated reliability and the uncertainty in the basic variables. Since an inspection reflects the current level of integrity of the structure at the time of inspection, after the inspection the structure's limit state function and the distributions of all the random variables need to be updated for the three crack detection events discussed earlier. Using the Bayesian updating technique, the authors proposed such techniques (Zhao and Haldar, 1997). If the updated reliability index is unacceptable, a repair or replacement should be made immediately or within a short time period. The mathematical procedure to calculate the underlying reliability after a repair is very similar to the case before a repair, but the limit state function needs to be modified.

Risk-Based Criteria For Repair and Replacement

The ultimate goal of reliability-based fatigue inspection and maintenance is to ensure an appropriate level of safety for a structure throughout its lifetime. Just after an NDI, the updated information can be used to make maintenance decisions regarding what to do next in terms of doing nothing, inspecting more frequently, or repairing or replacing. The discussion is made simpler by introducing the concept of the initial reliability index (β_i), the updated reliability index (β_{up}), the target reliability index (β_T), and the permissible maintainability index (β_c). Zhao and Haldar (1997) proposed that the maintenance decision can be made based on the ratio of β_{up}/β_c. If the ratio is greater than 1.0, the structural integrity is within the acceptable level. However, if $\beta_i < \beta_{up} \leq \beta_T$, the structure needs vigorous monitoring, and the inspection interval needs to be reduced. If β_{up}/β_c is less than 1.0, the structural detail under consideration should be repaired immediately. In some cases, it is not easy to repair due to accessibility problems or because the repair process would cause additional damage. In other cases, the intended function of a structure may not be recovered entirely after the repair. Considering the high cost of repairs and other associated risks, a complete replacement may be warranted. In rare cases, the ratio β_{up}/β_T may be less than 0.6. This can be regarded as an indication of imminent structural failure. In this case, the structure should be closed until the danger of structural failure can be removed by rehabilitation.

If reducing the inspection interval is an option, the following equation can be used to estimate the next inspection interval:

$$\Delta T_{inp, \, up} = Min \left[\frac{\beta_{up} - \beta_c}{\beta_i - \beta_c} \, \Delta T_{inp}, \, \Delta T_{inp} \right] \tag{3}$$

where ΔT_{inp} is the constant inspection interval, and $\Delta T_{inp, \, up}$ is the updated inspection interval.

The procedure will be explained with the help of an example during the presentation.

References

Haldar, A., and Mahadevan, S., *Reliability Methods in Engineering Designs (Including Stochastic Finite Element Analysis)*, John Wiley & Sons, to appear in 1999.

Paris, P.C., *Fatigue an Interdisciplinary Approach*, Syracuse University Press, 1964, pp. 107-132.

Zhao, Z., and Haldar, A., "Reliability-Based Structural Fatigue Damage Evaluation and Maintenance Using Non-Destructive Inspections," *Uncertainty Modeling in Finite Element, Fatigue, and Stability of Systems*, edited by Haldar, Guran, and Ayyub, World Scientific Publishing Co., 1997, pp. 159-214.

INTEGRATION OF NDE IN LIFE-CYCLE COST OF HIGHWAY BRIDGES

Dan M. Frangopol, [1] Fellow, ASCE, Michael P. Enright, [2] Associate Member, ASCE, Allen C. Estes, [3] Member, ASCE, and Kai-Yung Lin [4]

Abstract

The integration of quantitative non-destructive evaluation (NDE) data into bridge management systems is instrumental to the development of optimal maintenance planning strategies for the U.S. bridge inventory. This study presents recent progress in (a) integration of NDE in life-cycle cost of highway bridges, and (b) identification of optimum lifetime reliability-based maintenance strategies for bridges using inspection data and life-cycle cost methods. The overall goals are to predict bridge condition quantitatively, incorporate inspection results into existing bridge management systems, and use the results to efficiently plan and prioritize bridge maintenance efforts.

Introduction

As the U.S. bridge inventory continues to deteriorate, the need for reliable quantitative assessment technologies continues to increase. Although several new NDE methods are emerging, many state transportation departments continue to rely on visual inspections for bridge condition assessments. In fact, all of the data that feed into bridge management systems today are based on visual inspection and subjective condition assessment (Chase and Washer 1997). Subjective or inaccurate condition assessment has been identified as the most critical barrier to effective management of highway bridges (Aktan *et al.* 1996).

---

[1]Professor, Department of Civil, Environmental, and Architectural Engineering, University of Colorado, Campus Box 428, Boulder, CO 80309-0428.

[2]Graduate Research Assistant, Department of Civil, Environmental, and Architectural Engineering, University of Colorado, Campus Box 428, Boulder, CO 80309-0428.

[3]Lieutenant Colonel, U.S. Army Corps of Engineers, 169th Engineer Battalion, Fort Leonard Wood, MO 65473; formerly, Department of Civil, Environmental, and Architectural Engineering, University of Colorado, Boulder, CO 80309-0428.

[4]Project Manager, Chung-Shen Inst. of Sci. and Technology, Taiwan, Republic of China; formerly, Department of Civil, Environmental, and Architectural Engineering, University of Colorado, Boulder, CO 80309-0428.

Life-cycle cost methods have been successfully applied in many industries, and are becoming a requirement for bridge maintenance planning decisions (Frangopol and Hearn eds. 1996). Recently, methods have been developed for the optimal inspection/repair planning for bridges based on life-cycle cost and lifetime system reliability. These methods rely on quantitative condition assessment data (e.g., NDE results) and can be used to identify maintenance planning strategies which minimize life-cycle cost while maintaining a target reliability level for bridges.

This study presents recent progress in (a) integration of NDE in life-cycle cost of highway bridges, and (b) identification of optimum lifetime reliability-based maintenance strategies for bridges using inspection data and life-cycle cost methods. The overall goals are to predict bridge condition quantitatively, incorporate inspection results into existing bridge management systems, and use the results to efficiently plan and prioritize bridge maintenance efforts. Applications of integrating NDE in bridge life-cycle cost and determining the optimum lifetime inspection/repair planning strategy are briefly presented. These applications illustrate a comprehensive life-cycle cost analysis for bridges using both reliability and optimization techniques developed at the University of Colorado (Lin 1995; Estes 1997; Frangopol *et al.* 1997; Enright 1998).

Selection of Optimum Number of Lifetime Inspections/Repairs

Using the minimum expected life-cycle cost criterion, Frangopol *et al.* (1997) developed a methodology for identifying the optimum number of lifetime inspections/repairs for deteriorating concrete bridges. The methodology takes into account the quality of NDE methods, the effects of deterioration and subsequent

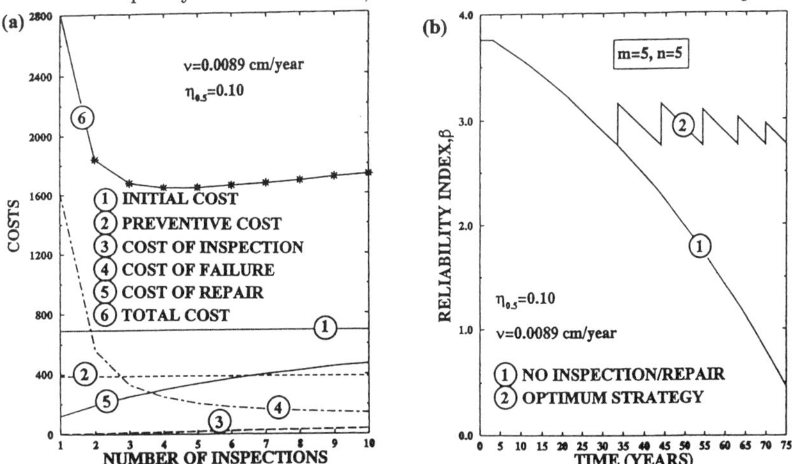

Figure 1: (a) Costs as Function of Number of Nonuniform Interval Inspections; and (b) Optimum Inspection/Repair Strategy (Frangopol *et al.* 1997)

repair on structural reliability, and the time value of money. A typical example of an optimum lifetime strategy for the maintenance planning of a deteriorating reinforced concrete girder bridge is shown in Fig. 1 (Frangopol *et al.* 1997). The notations in Fig. 1 are as follows: ν = uniform corrosion rate of steel reinforcement, $\eta_{0.5}$ = damage intensity at which the NDE method has a 50% probability of detection, m = number of lifetime inspections, and n = number of lifetime repairs.

Selection of Optimum Lifetime NDE Inspection Method

Lin (1995) and Estes (1997) showed that it is possible to identify from a group of possible NDE methods characterized by different detection abilities and costs the method that will produce the best lifetime solution by minimizing the expected life-cycle cost. As an illustrative example, Fig. 2 (Estes 1997) shows the probability of detection of four NDE techniques denoted as A, B, C, and D.

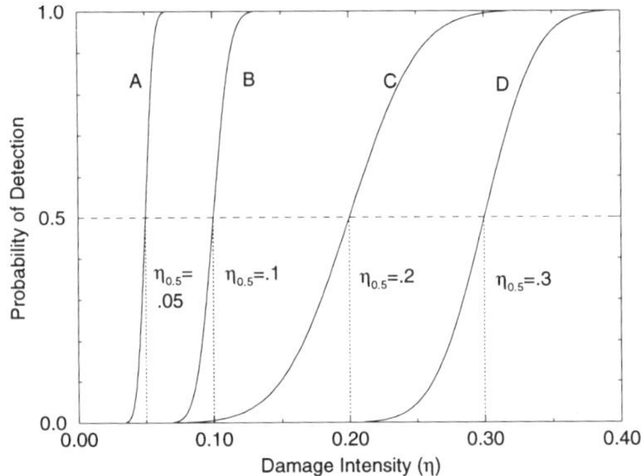

Figure 2: Probability of Detection of Four Inspection Techniques

Depending on the bridge type, deterioration process, ratio between cost of inspection and cost of repair, failure cost, net discount rate of money, and service-life of the bridge, each of the four NDE techniques shown in Fig. 2 might be the one minimizing the total expected life-cycle cost (see Estes 1997).

Effect of Inspection/Repair on Time-Variant Reliability

In a recent study, Enright (1998) quantifies the effect of inspection/repair on time-variant reliability of deteriorating concrete bridges. As an example, Fig. 3 shows the effects of the optimum inspection/replacement time t_R associated with different cost of failure coefficients C_F on the cumulative-time failure probability of a deteriorating girder.

Concluding Remarks

As illustrated in this study, the integration of NDE in life-cycle cost and life-time reliability of highway bridges is a practical possibility. The FHWA's program

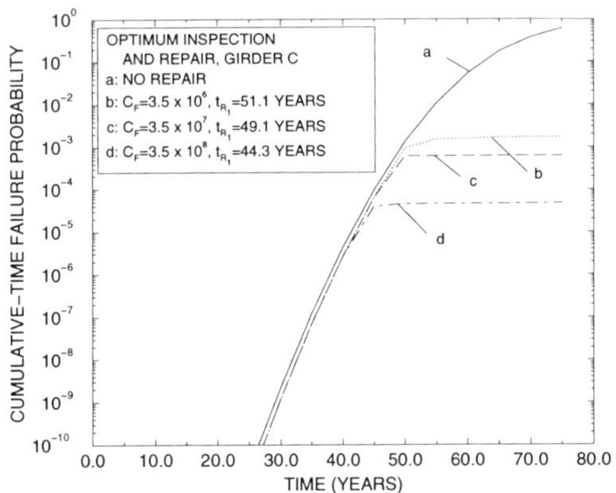

Figure 3: Effect of Single Inspection/Replacement on the Cumulative-Time Failure Probability of a Deteriorating Girder

of research and development in validating existing and new NDE technologies (Chase and Washer 1997) is helping this integration process. Optimum strategies for bridge inspection and repair can be identified using reliability-based life-cycle cost methods. These strategies rely on the availability of quantitative inspection data from nondestructive evaluations.

References

Aktan, A.E., Farhey, D.N., Brown, D.L., Dalal, V., Helmicki, A.J., Hunt, V.J., and Shelley, S.J. (1996). "Condition assessment for bridge management." *Journal of Infrastructure Systems*, ASCE, 2 (3), 108-117.

Chase, S.B. and Washer, G. (1997). "Nondestructive evaluation for bridge management in the next century." *Public Roads*, July/August, 16-25.

Enright, M.P. (1998). "Time-variant reliability of reinforced concrete bridges under environmental attack." PhD Thesis, Dept. of Civ., Envir., and Arch. Engr., University of Colorado, Boulder, Colo.

Estes, A.C. (1997). "A System Reliability Approach to the Lifetime Optimization of Inspection and Repair of Highway Bridges." PhD Thesis, Dept. of Civ., Envir., and Arch. Engr., University of Colorado, Boulder, Colo.

Frangopol, D.M., and Hearn, G., eds. (1996). *Structural Reliability in Bridge Engineering*, McGraw-Hill, New York.

Frangopol, D.M., Lin, K-Y., and Estes, A.C. (1997). "Life-cycle cost design of deteriorating structures," *J. Struct. Engrg.*, ASCE, 123 (10), 1390-1401.

Lin, K-Y. (1995). "Reliability-based minimum life-cycle cost design of reinforced concrete girder bridges." PhD Thesis, Dept. of Civ., Envir., and Arch. Engr., University of Colorado, Boulder, Colo.

FHWA Approach For Reliability Testing of the Visual Inspection Method

Glenn Washer[1], P.E., Mark Moore[2], P.E., and Richard Walther[3], P.E.,
Members ASCE

Abstract

The state-of-the-practice for bridge inspection relies almost exclusively on
the Visual Inspection (VI) Method to assess the condition of our nation's bridges.
Visual inspection, especially by well trained and experienced inspectors, can be a
very effective tool for rapidly and economically detecting deterioration or distress.
However, the visual inspection method has inherent limitations that affect its
reliability. The Federal Highway Administration (FHWA), through the newly
established FHWA NDE Validation Center (NDEVC), has initiated a comprehensive
study to define the reliability of the visual inspection method, as it is currently
practiced in the United States. This study will include a comprehensive literature
search and questionnaire survey of transportation agencies and consultants to define
the current state-of-the-practice of bridge inspection. A series of performance and
capability trials will be carried out under laboratory and field conditions to assess
individual components of the visual inspection method. The overall study and
preliminary scientific approach are described, with study completion in early 2000.

Background

The FHWA's NDEVC, which officially opened in the fall of 1998, was
established with the goal of providing a national resource for the evaluation and
validation of NDE technologies with highway applications. Specifically, the
NDEVC is designed to evaluate the reliability and performance of NDE technologies
and methods currently used or proposed for use in normal bridge inspections. The

---

[1]   Research Engineer, Turner Fairbank Highway Research Center, Special Projects
and Engineering Division, 6300 Georgetown Pike, McLean, Virginia 22101

[2]   Senior Consultant, Wiss, Janney, Elstner Associates, Inc., 225 Peachtree Street,
N.E., Suite 1600, Atlanta, GA 30303-1730

[3]   Senior Structural Engineer, Wiss, Janney, Elstner Associates, Inc. 208 Carlton
Lane, North Andover, MA 01845

nucleus of the NDEVC is located at the Turner Fairbank Highway Research Center in McLean, Virginia and includes 439 m^2 of newly renovated state-of-the-art laboratory space. An inventory of bridge component specimen containing representative bridge defects or deterioration is maintained by the NDEVC for conducting control trials of NDE technologies and methods. Seven test bridges located within 150 miles of the center are also supported and permit field testing of NDE technologies and methods under "real world" conditions.

One of the first initiatives for the NDEVC is to undertake a comprehensive study of the visual inspection method to quantify its reliability as a bridge inspection tool. The visual inspection method is by far the predominant nondestructive evaluation technique used in bridge inspections. Furthermore, the visual inspection method serves as the standard by which all other nondestructive evaluation technologies and methods are measured. However, since the inception of the National Bridge Inspection Standard in 1971, a complete study of the reliability of this method has not been undertaken.

This proposed FHWA study and the NDEVC represents a significant contribution to the field of nondestructive evaluation. Findings generated through a comprehensive literature search and survey questionnaire of state departments of transportation, consultants and other bridge inspectors will define the current state-of-the-practice of bridge inspection. Laboratory and field trials will define the capability and performance characteristics of the overall method, as well as its individual components. The study was recently initiated, with completion expected in early 2000. The project scope and preliminary scientific approach are presented.

Scientific Approach

The following model describes the reliability of a nondestructive method:

$$R = f(\text{IC}) - g(\text{SF}) - h(\text{HF}),$$

where R is the method reliability, $f(\text{IC})$ is the intrinsic capability function, $g(\text{SF})$ is the situational function, and $h(\text{HF})$ is the human factors function. The intrinsic capability of the method being tested represents a theoretical limit for the method based on ideal conditions and is generally determined analytically. The situational function describes the capability of the method void of human factors and under a given set of controls. Situational functions are generally evaluated through capability trials and ultimately serve to reduce the reliability of the method from its theoretical limit. The human factor element further reduces the method reliability, and can impact method reliability significantly. Performance trials are used to evaluate the human factors function and provide data for given situational conditions.

The intrinsic capability of the visual inspection method is fairly straightforward. Visual inspection achieves ultimate reliability only when a defect appears on the surface and is discernable to the human eye. However, what is and is not a defect is defined arbitrarily. In terms of bridge inspections, defects are generally defined with some consideration to the consequence the defect ultimately has on bridge performance. Setting the defect size to a sufficiently high level will result in nearly 100% detection reliability, but has the negative impact of increasing the likelihood that an existing defect will go undetected and cause failure. Alternatively, setting the defect level to a small level may unnecessarily increase the cost of bridge inspections. While this study will address some of these issues, the research required to fully evaluate these concepts is beyond the scope of this study. Beyond the intrinsic capability of the visual inspection method, the individual factors that contribute to method reliability are numerous; however, preliminary analysis has produced several broad-based categories of factors for study. They are Human Factors and Situational Factors, of which Environmental and Managerial Factors are the most significant.

Human factors to be investigated include physical, mental and team/individual characteristics of the bridge inspector. Physical characteristics pertinent to this study include visual, auditory, and olfactory acuity, touch, conditioning/stamina, age, and aliments. Mental characteristics considered include attitude, inspection habits, mental conditioning, fear, psychological, circadian rhythm, home/family problems, and substance abuse. Finally, team/individual characteristics to be investigated include team composition and interaction, knowledge and background, and use of available resources. In general, the effect of human factors on the reliability of the visual inspection method will be determined through performance trials on realistic specimen under realistic conditions.

Situational factors include environmental factors such as ambient light, both quantity and quality of direct or reflected light, ambient noise, and weather conditions. Specific attributes of the specimen being inspected and the inspection techniques are also included within the situational factor category. For example, structural attributes might include surface conditions, accessibility and viewing angle, complexity, and maintenance. Inspection techniques considered might include visual enhancements, magnification, platform stability, and cleaning methods.

Managerial factors include financial, team attributes, and inspection procedures. Within the realm of financial factors, subset factors to be considered include budget, QA/QC, schedule, training, and scope and frequency of inspection. The team composition and supervision must also be evaluated. Inspection procedures dictated by management also play an important role in visual inspection reliability. How the inspection team prepares, conducts and reports findings have bearing on method reliability. Management's approach to safety and results are also considered.

Proposed Scope of Study

This study will include an exhaustive literature search regarding the state of visual inspection as it applies to highway bridges. The literature search will include studies on bridge inspection practices, studies on facility and structural inspections in related fields such as the aerospace, nuclear, and military industries, and studies on human factors in the ergonomics and psychology fields. Available documentation on bridge inspection procedures authored by transportation agencies or consultants will also be obtained. Documents pertinent to this study will be obtained for deposition in the NDEVC technical library. The technical library, like the NDEVC itself, was established to serve as a national resource for nondestructive testing.

A questionnaire survey will be prepared and submitted to transportation agencies and consultants. The survey objectives include the compilation of a state-of-the-practice for bridge inspection, especially as it pertains to visual inspection and the general use of nondestructive evaluation techniques and methods. In addition, the survey will be used to gather information regarding inspection management, and how management influences the reliability of visual inspection. Topics addressed by the survey will include commonly used nondestructive evaluation techniques and methods, typical composition and experience of bridge inspection teams, the ratio of inspections performed by transportation agency staff versus inspections performed by consultants, and time allotted per average inspection.

A test plan will then be developed to examine the significant factors affecting reliability of the visual inspection method. It is envisioned that test subjects will consist of practicing bridge inspectors who will perform laboratory and field trials. In developing the test plan, the use of the NDEVC component specimen inventory and seven test bridges will be maximized. The specific tests to be conducted and the procedures for such tests have not be fully defined; however, information collected through the literature search and survey should prove to be useful. Test designs might include controlling such factors as distance to target, access, lighting and team composition and experience.

Upon approval of the test plan, the NDEVC will coordinate and execute the various field and laboratory trials necessary to define the reliability of the visual inspection method. Results obtained will be synthesized and reported. The FHWA through technical meetings and symposiums will keep the bridge inspection community up-to-date on the progress of this important study and present conclusion findings at the end of the study.

Conclusion

The visual inspection method is the primary tool used in the inspection of our nation's bridges and serves as the benchmark to all other nondestructive evaluation techniques and methods. However, since the inception of the 1971 National Bridge Inspection Standards, a complete study of the reliability of this method has not been undertaken. This study will provide a crucial benchmark in the effort to enhance bridge inspections through the use of nondestructive technologies and methods. Furthermore, this study will represent a significant contribution by the NDEVC to the bridge inspection industry, which will benefit by being able to refine inspection efforts to enhance reliability and improve productivity.

EVALUATING UT AS A TOOL FOR RELIABILITY ASSESSMENT

Terrence F. Paret[1]

Introduction

Following the Northridge earthquake, an unprecedented UT assessment of existing welded steel moment frames took place in and around Los Angeles, California. The assessments were prompted by weld fracturing observed after the earthquake and by a mandatory inspection ordinance adopted by the City of Los Angeles. The records of over 200 of these assessments are maintained in the files of the Los Angeles Department of Building and Safety and comprise the largest extant repository of post-earthquake UT evaluations of existing buildings.

Most of the assessments were tailored to comply with the SAC Interim Guidelines [SAC, 1995], and made note of the weld fractures and the rejectable and non-rejectable defects found. Well over one thousand weld fractures were observed with only a few of the many SAC categories of damage comprising the bulk of these, but nearly all the rejectable defects found consisted of large planar ultrasonic indications in the weld root on the column fusion face. These types of planar root indications were termed W1a and W1b by SAC. Nearly 15% of the connections inspected and subsequently documented in the City of Los Angeles files were found to have W1's --- twice as many connections as were found to have all other categories of damage combined. The vast majority of the W1's documented conform to W1b, i.e., greater than 3/16-inches or $t_f/4$ in height or greater than $b_f/4$ in length.

Not surprisingly in a post-earthquake environment, these large planar indications were typically treated as earthquake damage by the engineering community. However, a number of samples of W1b's trepanned from welded connections and examined in the laboratory were determined to contain only areas of nonfusion and slag, without any crack extension or other potentially earthquake related conditions. In addition, for the SAC fracture types most commonly observed after the earthquake (through-weld fractures), nearly all of the fracture surfaces

---

[1] Consultant, Wiss, Janney, Elstner Associates, Inc., 2200 Powell Street, Suite 925, Emeryville, CA 94608, as subcontractor to SAC Joint Venture, a partnership of the Structural Engineers Association of California, the Applied Technology Council, and California Universities for Earthquake Engineering.

contain large visible planar defects along the root adjacent to the column flange that bring to mind what W1b's might look like if they were not completely hidden by adjacent weld metal and parent material. If W1's are construction-related defects as the sum of the evidence seems to suggest, then the actual frequency of occurrence of W1-type defects is substantially more than the 15% found by ultrasonics.

The presence of such a large number of planar indications in the inventory of buildings in Los Angeles raises numerous questions whose answers bear directly on the subject at hand. If W1's are construction-related conditions, for example, how did so many escape detection during normal construction-related inspection activities. Given the great frequency of occurrence of these W1's, considerable effort has been devoted to explaining this phenomenon. This paper presents some of the findings of this work, concentrating on issues related to the use of UT as a tool for reliability assessment.

An Early Investigation

One study in particular led early researchers to conclude that the occurrence of W1's is related to earthquake ground motions and that study speaks loudly to the reliability of UT. In that study, full-scale beam-column assemblies were subjected to cyclic load reversals after having been UT'd. Subsequent UT examination conducted after the load cycling was reported in the SAC Interim Guidelines as having located rejectable indications that were not detected prior to the load tests, and this was taken to validate the notion that the Northridge earthquake might have caused the W1's so widely observed. Unfortunately, long after the publication of the SAC Interim Guidelines, a critical evaluation of the UT examinations performed could not confirm this validation.

Review of the specifications for UT in the tests reported by SAC indicated that all inspections were to be performed by ASNT Level II qualified individuals in accordance with AWS D1.1-94. Therefore, in principal and to the same degree that might be expected on any project involving ultrasonics, the UT technicians involved were equally qualified and were expected to employ comparable testing methods. The records show that there were three UT technicians involved: one of the three (T1), UT'd all three connection specimens prior to load cycling; one technician (T2) performed the post-load cycling UT on two of the connections; and the third technician (T3) only examined one specimen after load cycling. T1 was employed by one testing lab and T2 and T3 by another. Each inspector employed similar, but not identical recordkeeping.

A superficial comparison of the pre load cycling UT reports with the post load cycling UT reports clearly implies that there was an increase in the number of discontinuities due to load cycling; however, a detailed comparison suggests that in addition to having initiated during load cycling, many of the discontinuities identified by UT seem to have either shifted, grown, shrunk, or disappeared. (Table 1) The differences between the pre and post load cycling UT findings are not minor.

The data in Table 1 suggests such tremendous variations between the findings of the different technicians that one is forced to seriously question the use of UT as an assessment tool. For all three specimens, the locations and sizes of every discontinuity observed suggests that the pre and post load cycling UT technicians were not even looking at the same connection.

Table 1. Location and size of indications found in specimens prior to and after load cycling. Dimensions provided are as noted on UT inspection reports.

Indication #	X	Y	Y Converted*	Depth	Length
Specimen #1 t_f = .95"					
Pre-Load Test (T1)					
#1	3/8	2½	8½	13/16	3¼
#2	3/8	0	6	13/16	2
Post-Load Test (T2)					
#1	0	0		.96	2.0
#2	0	6		.91	.75
#3	0	9.5		1.0	3.75
#4	+.15	9.0		.64/.31	2.0
Specimen #2 t_f = .95"					
Pre-Load Test (T1)					
#1	3/8	+3	+9	13/16	3
#2	3/8	0	6	13/16	2¼
#3	3/8	3½	+2½	13/16	2
Post-Load Test (T2)					
#1	+.1	0		.89	1.7
#2	0	6.2		.84	1.2
#3	+.1	9.0		.94	2.5
Specimen #3 t_f = .900"					
Pre-Load Test (T1)					
None	None	None	None	None	None
Post-Load Test (T3)**					
#1	+.03-.07	0-1.75		.868	1.750
#2	-.350	2.3-4.8		.661	2.500
#3	+.123	5.5-7.5		.834	2.00
#4	-.250	7.5-9.0		.609	1.50
#5	+.05	9.5-12.0		.817	2.50

*The pre and post load-cycling inspection companies used different Y origins. Y converted provides a basis for comparison.

**T3 used the acceptance/rejection criteria in Table 9.3, as opposed to T1 and T2 who are believed to have used Table 8.2.

Evaluation of the Reliability of UT for Identifying Root Defects

A separate study was performed to evaluate the reliability of UT for identifying W1 type root defects. The study included visual confirmation of the UT findings and utilized a 1970's vintage 17-story building known to have been 100% ultrasonically inspected during construction. The building's lateral load resisting system is a complete moment resisting space frame with supplemental concentric bracing in one direction. Typical column sizes range from W14 x 426 to W14 x 87. Typical beam sizes range from W27 x 94 to W24 x 68 in one direction and from W16 x 78 to W16 x 45 in the other. The building is located in a part of California that has not experienced ground shaking in excess of 0.2g since the date of construction and is therefore believed to have not experienced significant earthquake induced damage.

The study requirements specified ASNT Level II inspectors and AWS D1.1. A testing protocol was developed within the constraints provided by the building owner, which allowed for verifying the ability of Face A scans and Face B scans to identify W1 type defects. To circumvent biasing the testing results, at each connection tested, different inspectors were used for Face A and Face B scanning. When UT scanning was completed at a given connection, yet another inspector mapped visually observable root defects after air-arc removal of the backing bar and measured the maximum height of backgouging required to remove the defects. Reporting forms were specially developed for use by the inspectors. While the use of separate inspectors for each task shielded them from knowledge developed about a particular connection by other inspectors, the protocol introduced unknowns due to operator variability. Furthermore, because the UT technicians knew that their work would later be confirmed by destructive examination, the UT staff may have been more vigilant than usual. Ultimately, however, because the destructive testing found so many more large defects than either UT scan could find, the protocol did not appear to have influenced the outcome of the testing program significantly.

Evaluation of Data

In all, 37 welded bottom flange connections were subjected to two or more of the three specified testing procedures; Face A scans, Face B scans or destructive investigation. The distribution of connections subjected to the various testing regimens and the distribution of connections inspected by floor framing level is shown on Table 2.

Table 2. Distribution of connections by inspection type and by framing level.

Inspection Type	UT-Face A & Destructive	UT-Face B & Destructive	UT Faces A & B & Destructive	
# of Connections	15	12	10	

Framing Level	8th	10th	16th	Roof
# of Connections inspected	7	10	10	10

All 37 of the inspected connections were subjected to destructive testing. Of these 37 connections, 36 were found via destructive testing to have substantial occurrences of nonfusion, incomplete fusion, slag and porosity in the weld root. The vast majority of these defects were located along the interface of the weld and the column flange, and the cumulative length of these defects accounted for approximately one-fourth of the flange width ($b_f/4$) or greater in 34 of these 36.

Therefore, on the basis of the lateral extent of the observed defects only 2 of the 36 connections with defects could have been classified as W1a ($\leq b_f/4$), with the balance being W1b. The final measured depths of the backgouges after removal of the defects exposed after the backing was removed are tabulated in Table 3. On the basis of the maximum height of the defects (assuming the backgouge depth equals the defect height) none of the 36 connections were W1a, and all were W1b.

Table 3. Distribution of connections by depth of backgouge.

Measured Depth of Backgouge (inches)	3/16	1/4	$\leq 5/16$	$\leq 3/8$	$\leq 7/16$	$\leq 1/2$	$\leq 5/8$	$> 5/8$
Number of Connections	4	5	9	3	1	5	4	4

Of the 15 flanges that were subjected to only Face A UT scans and destructive testing, 14 were found by destructive testing to have root defects (primarily non-fusion and incomplete fusion) adjacent to the column flange with cumulative length in excess of $b_f/4$. Of these same 15 flanges, 13 were identified by Face A UT scans as being _free_ of rejectable defects. The few rejectable defects identified by UT were significantly smaller than the visually observed defects.

Of the 12 flanges that were subjected to only Face B UT scans and destructive testing, 11 were found by destructive testing to have root defects (primarily continuous and intermittent zones of nonfusion and incomplete fusion) adjacent to the column flange with a cumulative length in excess of $b_f/4$. (The remaining flange had a 1" long root defect.) Of these 12 flanges, 8 were identified by Face B UT scans as being _free_ of rejectable defects. The rejectable indications found by UT were always substantially smaller than the visually observed defects.

Of the 10 flanges that were subjected to both Face A and Face B UT scans and to destructive testing, 9 were found by destructive testing to have root defects (primarily continuous and intermittent zones of non-fusion, incomplete fusion and/or slag) adjacent to the column flange with cumulative lengths in excess of $b_f/4$. Of these same 10 flanges, 4 were identified by Face A UT scans as being _free_ of rejectable defects and 8 were identified by Face B UT scans as being _free_ of rejectable objects. The rejectable indications found by Face A UT scans were, in 5 of the 6 flanges, less than ½" long. Of the two flanges found by Face B UT scans to have rejectable indications, one was also identified by Face A scans as having a rejectable indication. In the other instance, the rejectable indication identified by the Face B scan was not confirmed by either Face A scans or destructive investigation.

Conclusions

- The collected data clearly shows that nearly every flange tested had large root defects that at times extend almost completely across the width of the flange. These defects consisted primarily of nonfusion and incomplete fusion to the column flange, as well as of slag and pinholes.

- The ultrasonic testing performed, though specified to be in accordance with AWS, was for the most part unable to identify the presence of most visually observed root defects. Where defects were identified via UT, they were typically noted to be much smaller than were later observed during destructive testing.
- No clear differences in the relative reliability of FACE A versus FACE B scans are apparent in the data. Both appear on the basis of this study to be inadequate for identification of defective weld roots in full penetration T-joints with backing.
- It is not known whether ultrasonic testing as practiced in Los Angeles after Northridge would have more successfully identified the presence of large root defects, although experience of the inspectors, the time available for each inspection, and the different circumstances presumably affected the inspection process to some degree.
- The location and orientation of the root defects found during destructive testing, although clearly not earthquake related, appear to be similar if not identical to the location and orientation of conditions identified throughout the Los Angeles area after Northridge as W1a and W1b.
- The nearly ubiquitous presence of the root defects raises the question of whether the studied building, which was 100% UT'd during original construction, is unusual or is typical of steel frame construction.
- Based upon the height and extent of the typical defects, it appears as if a standard construction specification for backgouging to a depth of 5/8-inch and re-welding would have eliminated 90% of the poor weld roots.

References

SAC Joint Venture, 1995, <u>Interim Guidelines: Evaluation, Repair, Modification and Design of Welded Steel Moment Frame Structures</u>, FEMA 267, Report No. SAC-95-02, August 1995.

Acknowledgements And Disclosures

A portion of the work forming the basis for this publication was conducted pursuant to a contract with the Federal Emergency Management Agency. The substance of such work is dedicated to the public.

The author is solely responsible for the accuracy of statements or interpretations contained in this publication. No warranty is offered with regard to the results, findings, and recommendations contained herein, either by the Federal Emergency Management Agency, the SAC Joint Venture, the individual joint venture partners, their directors, members, or employees. These organizations and individuals do not assume any legal liability or responsibility for the accuracy, completeness, or usefulness of any of the information, product, or processes included in this publication.

Special Structures

Manuscripts for some presentations were not available at time of publication.

8. DESIGN ISSUES FOR OFFSHORE STRUCTURES

MODERATORS: Guillermo D. Hahn
Vanderbilt University
Nashville, TN

Stephen P. Schneider
University of Illinois at Urbana-Champaign
Urbana, IL

(1) Structural Challenges in Deepwater Oil Production
Jose M. Roesset
Offshore Technology Research Center
College Station, TX

(2) Human and Organizational Factors in the Reliability and Engineering of Offshore
Robert G. Bea
University of California at Berkeley
Berkeley, CA

(3)Compliant Towers: Structures that Benefit from Dynamics
Larry Light and Cheng-Yo Chen
J. Ray McDermott Engineering, LLC
Houston, TX

(4) Practical Approaches in Offshore Engineering
G. D. Hahn and Q. Abughazleh
Vanderbilt University
Nashville, TN

17. SEMI-ACTIVE CONTROL SYSTEM IMPLEMENTATION ISSUES

MODERATOR: Shirley Dyke
Washington University
St. Louis, MO

(1) Variable Stiffness and Instantaneous Frequency
Satish Nagarajaiah, Sanjay Sahasrabudhe, and Varadarajan Nadathur
Rice University
Houston, TX

2) Semi-Active Hybrid Seismic Isolation Systems: Addressing the Limitations of Passive Isolation Systems
Michael D. Symans, Glenn J. Madden, and Nat Wongprasert
Washington State University
Pullman, WA

(3) Variable Property Devices for Structural Control
Henri P. Gavin and Nitin S. Doke
Duke University
Durham, NC

(4) Stability and Performance of Controlled Civil Engineering Structures Using MR Dampers
S. J. Dyke
Washington University
St. Louis, MO

26. *STRUCTURAL REHABILITATION OF OPERATING INDUSTRIAL FACILITIES*

MODERATOR: Merle E. Brander
Brander Construction Technology, Inc.
Green Bay, WI

(1) Structural Rehabilitation of Operating Industrial Facilities: An Introduction
Merle E. Brander
Brander Construction Technology, Inc.
Green Bay, WI

(2) Temporary Structures: Role of the Design Engineer
John Frauenhoffer
Frauenhoffer & Associates
Champaign, IL

(3) Rehabilitation Projects in Operating Facilities are in the Hands of the Structural Engineer
Thomas Rewerts
Construction Technology Laboratories, Inc.
Skokie, IL

(4)Demolition of a Pulp Mill in an Operating Paper Facility
John J. Jeanquart and Merle E. Brander
Brander Construction Technology, Inc.
Green Bay, WI

(5) The Challenges Created by the Hierarchy in Facility Rehabilitation Projects
Kenneth B. Simons
Damage Consultants, Inc.
Mercer Island, WA

MODERATOR: Randy Kissell,
       The TGB Partnership

(1) Investigation of Turn-of Nut Method for Slip-Critical Joints of Aluminum Using A325 Bolts
C. R. Luttrell,
Oak Ridge National Laboratory
Oak Ridge, TN

(2) Effect of Thermal Cycling on Slip-Critical Connections of Aluminum
C. R. Luttrell,
Oak Ridge National Laboratory
Oak Ridge, TN

(3) More Efficient Aluminum IBeam Shapes
Randy Kissell,
The TGB Partnership
Hillsborough, NC

(4) Aluminum Gusset Plates in Tension
Craig C. Menzemer
The University of Akron
Akron, OH

(5) A Review of Design Codes For Aluminum
Cedric Marsh,
Concordia University
Montreal, Canada

Structural Challenges in Deepwater Oil Production

Jose M. Roesset[1]

Introduction

In just 50 years, oil production has gone from 20 ft of water to over 3,000 ft with surface piercing structures and over 5,000 ft with subsea completion systems. The possibility of extracting oil from water depths of 10,000 ft is already being considered. In this process, the offshore industry has designed, fabricated, and installed platforms which have defied the imagination of structural engineers and which have repeatedly earned the award for civil engineering accomplishments of the year (Cognac, Bullwinkle, and Auger platforms are examples). These achievements have been made possible through a great deal of imagination, sound engineering principles and judgment, and a considerable amount of research and development work conducted both in-house by some oil companies, and at research institutions.

The purpose of this paper is to review some of the challenges that deep water oil production is posing in the areas of Structural Mechanics. These challenges are related to the definition of the environmental loads acting on the different components of the structures (the hull, risers, tethers and mooring lines), to their dynamic response and detailed stress analysis, the characterization of new materials and their behavior in the marine environment, and the analysis and design of the foundations or anchoring elements.

Structural Types

The traditional platforms in the Gulf of Mexico were the steel jackets. For relatively shallow waters, the fundamental period of these towers was well below the range of periods where most of the wave energy is concentrated. Dynamic effects were thus considered negligible and dynamic analyses were not deemed necessary. With increasing water depth, however, as the flexibility of the structures increased, dynamic amplification became more significant. For water depths of 3,000 ft or more the steel jacket concept is no longer applicable with the normal geometric proportions because the natural period of the platform would be close to resonance with the design waves. The solution is then to design very flexible structures with fundamental periods much longer than those of the waves. Included in these new types of structures are compliant, very slender towers and large floaters held in place by

---

[1] Director, Offshore Technology Research Center, Texas A&M University, 1200 Mariner Drive, College Station, TX 77845

vertical tethers and/or mooring lines. The former can still have important dynamic amplifications in higher modes and only one platform of this type has been designed and fabricated to date. The latter include tension leg platforms (TLPs), Spars, semisubmersibles (FPO's), and tanker-based FPSO's (Floating Production Storage and Offloading Systems). Several of these systems have been installed and are being installed in the Gulf of Mexico, off the coast of Brazil, in the North Sea, the West coast of Africa, and the Pacific Rim.

Hydrodynamic Forces and Dynamic Response

A first characteristic of these compliant systems due to their flexibility and the very small amount of available damping, is that the amplitude of the free vibration terms can be much larger than that of the steady state response and that it can take a very long time to reach the latter. This is important when performing nonlinear dynamic analyses in the time domain under a harmonic excitation, or when performing tests in a wave basin for the same conditions. Since in reality the maximum wave will not occur suddenly with the structure at rest, but the excitation (corresponding to an irregular wave with many different harmonic components) will grow slowly in intensity, the very large responses associated with the free vibrations are meaningless. To reach the steady state conditions fast, it is necessary to apply a gradual ramp of sufficient duration to the forcing function. On the other hand, if a wave of much larger amplitude than the others were to arrive suddenly, it would give rise to important free vibration terms which would take a long time to dissipate.

A second and more important characteristic is that the linear dynamic response in the range of frequencies of the waves will normally be very small. Nonlinear effects will give rise, however, to small components of excitation at the natural frequencies of the structure which will be highly amplified due to the small amount of damping in these systems. The nonlinear response will be important relative to the linear one. The nonlinear excitation at the natural frequencies may result from sum frequency terms (vertical vibrations of tension leg platforms), or from difference frequency terms (horizontal vibrations of TLPs and for all components of motion of typical moored spars). The response at the natural frequencies may be of a steady state or of a transient nature (ringing of TLPs).

There are a number of sources of nonlinear behavior, such as: the nonlinear wave kinematics resulting from imposing the free surface boundary condition on the actual free surface of the water; the nonlinearity resulting from the computation of the hydrodynamic forces up to the instantaneous free water surface (wetted surface) and in the displaced position of the structure; the nonlinearities associated with the convective and temporal acceleration terms when using potential theory; the nonlinearities associated with viscous flow around slender members and the resulting drag forces; the coupling between axial and transverse vibrations of risers and tethers (nonlinear geometry).

Many of these effects are of similar magnitude but opposite signs and may cancel each other in some cases. Solutions that neglect them entirely may then provide good agreement with experimental data. It is important, however, to have a proper understanding of their nature to guarantee that they are not omitted in other cases where they may be significant. It is also common to approximate these effects using perturbation theory and keeping only up to second order terms (second order waves, second order diffraction). More has to be known, however, about the potential importance of higher order terms using complete nonlinear solutions. Much has to be learnt also about vortex-induced vibrations of slender members and drag forces for the ranges of values of Reynold's number and Keulegan Carpenter's numbers encountered in practice in the ocean. Even if Morison's equation were an accurate representation of the phenomena involved, the selection of the values of the inertia and drag coefficients on the basis of experimental data with much smaller Reynold's numbers than those of interest, is not very satisfactory. The present lack of connection between analysis techniques of a purely empirical nature, which are only valid for the conditions of the tests on which they are based, and the very sophisticated and expensive numerical solutions of the complete Navier Stokes equations, must be resolved.

The third important characteristic is the combination of large diameter bodies (the floaters), which may disturb the wave field significantly and require nonlinear diffraction theory, and slender bodies (tethers, risers, mooring lines), where viscous effects are important and which must be analyzed using some form of Morison's equation. It is common to perform uncoupled analyses simulating the effect of the tethers, risers, and mooring lines on the response of the floater using equivalent linear or nonlinear springs. A limited number of coupled analyses have indicated that it may be important to account properly for the complete coupling. This is another area in which much work remains to be done, particularly when dealing with more than one large body (an FPSO and an offloading tanker) and more than one system of slender elements (various riser systems, mooring lines, etc.).

Materials

For some of the new types of structures (such as the TLPs) weight is an important factor affecting the cost, and the use of lighter, but sufficiently strong materials (such as phenolic compounds) in the superstructure (for deck gratings, stairs, etc.) can result in substantial savings. Even larger savings could be achieved if tethers, risers, or mooring lines could also be made of new and lighter materials. One possible solution is the development of riserless drilling procedures (an alternative which is being seriously studied by industry) and the elimination of long production risers by transporting the well products by pipelines to structures in shallower depths. Another alternative is to use composite materials (made of resin matrix with graphite and/or glass fibers) for the risers (and tethers). The main advantage of these composites is that they can be designed to have any desired characteristics through appropriate selection of the components, the number and angle of the plies, etc. Their main problem is that there is a lack of sufficient knowledge of their long term behavior in the marine environment, particularly for the types of shapes and load combinations of interest in offshore applications. While drilling risers can be extracted periodically (say once a year) for inspection, production risers must remain in service underwater for the life of the platform (their extraction, interrupting operations would represent a significant cost) and the replacement of the tethers would also be a major operation.

While there has been a substantial amount of work on crack propagation and the nonlinear behavior in general of composite tubulars subjected to various load combinations (including high pressures) under water, much more research remains to be done on the aging and degradation of these materials and their time-dependent nonlinear constitutive equations. It is also necessary to develop accurate analytical models for the dynamic response of the elements made of these materials, incorporating their nonlinear constitutive behavior and accounting for possible delaminations or other damage. A combined effort, experimental, analytical and computational is needed. Yet, without a long-term experimental database, it will be hard to get these materials accepted and certified unless one can monitor their performance in a continuous way and detect with sufficient warning time potential damage (which may be internal and hard to observe visually). It is thus necessary to develop at the same time the instrumentation and nondestructive testing techniques needed for this purpose. Both acoustic emissions and stress waves are being investigated as possible solutions. (It should be noticed that instrumentation for performance monitoring and damage detection is important not only for the elements made of composite materials but for the complete structures due to the lack of accumulated experience on their behavior).

Fiber rope is also being considered and used in some parts of the world as a substitute for chain in mooring lines, particularly when dealing with taut moorings, in ultra deep waters. The preferred material at present is polyester because its long-term behavior is better known, but more research should be conducted on the properties and behavior of other fibers.

Foundations

In the Gulf of Mexico, vertically tethered platforms such as TLPs use traditional piles for foundations, although in this case the piles are subjected to important tensile forces in addition to the lateral loads. Apparently, the design of these piles does not present any major difficulties at the present time, yet our basic understanding of their behavior and our ability to accurately predict their response and the stresses in the surrounding soil is far from satisfactory. There are two distinct and unrelated approaches for dynamic analysis of pile foundations. One is based on the use of equivalent, nonlinear springs (P-y and T-z curves) to reproduce the nonlinear behavior of the soil around the pile. The other relies on more accurate elasto-dynamic solutions assuming linear elastic material behavior and perfect bonding between the piles and the soil. Dynamic effects are sometimes approximated in the first approach by adding viscous dashpots to simulate the radiation damping associated with the linear dynamic solution. Nonlinear effects are approximated in the second approach assuming a softer cylinder of soil surrounding the pile or reducing the soil properties on the basis of estimated levels of strain (both crude approximations which are qualitatively reasonable but offer little quantitative accuracy). A proper combination of the two approaches reproducing with similar degrees of sophistication nonlinear behavior and dynamic effects and accounting properly for the state of effective stresses in the soil, is still lacking, although a number of nonlinear finite analyses have been conducted.

A different, innovative foundation type was developed for North Sea platforms. It is referred to as "suction piles," "suction caissons," or "skirted foundations." In the normal applications, these are unusual piles with diameter to depth of penetration ratios larger than unity, which are driven into the soil with the help of suction. The penetration process and the ensuing behavior of these foundations under lateral and vertical loads has been studied experimentally (through full scale as well as laboratory tests) and numerically (using various nonlinear soil models and analysis procedures). Suction piles, or suction-assisted piles with depth to diameter ratios larger than unity (by opposition to the previous cases) and of the order of 6 or more, are being considered and used now as anchors for taut mooring lines. More experimental and computational work is needed to properly understand the behavior of these anchor-foundations in different soils and to be able to predict their response.

Human & Organizational Factors:
Engineering and Reliability of Offshore Structures

Robert Bea[1], Fellow, ASCE

Abstract

The past record of problems with offshore platforms clearly indicates that a significant contributor to critical deficiencies in the quality of these systems are human and organizational factors that are neglected during design. This paper proposes approaches to reduce the incidence and severity of these problems.

Human and Organizational Factors

Any activity that involves human activities is subject to the results of human influences. These are influences that have both positive and negative impacts of varying degrees and involve (Bea 1994): 1) System operators (design engineers, constructors, users), 2) Organizations that they work for and with, 3) Procedures (formal, informal, software) used, 4) Structures and equipment (hardware) involved, 5) Environments (external, internal, social) in which the tasks occur, and 6) Interfaces among these five elements

These six elements comprise offshore platform systems. The Human and Organizational Factors (HOF) involved in these elements have the major influence on the quality of the engineered structure system during its life-cycle (design, construction, operation, maintenance).

Operator Malfunctions

In the quest to assure quality (serviceability, safety, durability, computability) in design of offshore structures, the engineer is concerned primarily with malfunctions, errors, and defects that develop in the six elements cited above. There are many different ways to define, classify and describe design engineering malfunctions. Design engineering malfunctions can be defined as actions taken by individuals that result in realizing a lower quality than intended. These are malfunctions of commission (Reason 1990). Design engineering malfunctions also include actions not taken that result in realizing a lower quality than intended. These are malfunctions of omission. Designer malfunctions might best be described as actions and inactions that result in lower than acceptable quality.

---

[1] Professor, 212 McLaughlin Hall, Dept. of Civil & Environmental Engrg., University of California, Berkeley, 94720-1712, (510) 642-0967, e-mail bea@ce.berkeley.edu

A taxonomy to characterize design engineering malfunctions during the life-cycle of structure systems (Bea 1994) includes: communications, selection and training, mistakes, ignorance, violations, planning and preparations, and limitations and impairments.

Organizational Malfunctions

Analysis of the history of failures of structure systems provides many examples in which organizational malfunctions have been primarily responsible for compromises in quality. An organizational malfunction is defined as a departure from acceptable or desirable practice on the part of a group of individuals that produces unacceptable or undesirable results. A taxonomy of organizational malfunctions includes: communications, culture, structure and organization, monitoring and controlling, ignorance, violations, planning and preparations, and mistakes (Bea 1994).

Procedure Malfunctions

Procedure malfunctions can be embedded in engineering design guidelines and computer programs, construction specifications, and operations manuals. They can be embedded in how people are taught to do things. With the advent of computers and their integration into many aspects of the design, construction, and operation of structures, software errors are of particular concern because computers can only function within the limitations of their designers. A taxonomy of procedure / software malfunctions includes procedures that are: incorrect, inaccurate, incomplete, excessively complex, poorly organized, poorly documented, and have inadequate provisions for checking and verifications (Bea 1994).

Structure Malfunctions

Human malfunctions can be initiated by or exacerbated by poorly engineered structure systems. Such systems (hardware and/or structure) are difficult to construct, operate, and maintain. A classification of hardware / structure malfunctions would include inadequate or insufficient serviceability, safety, durability, and compatibility. New technologies compound the problems of latent system flaws. Complex design, close coupling (failure of one component leads to failure of other components) and severe performance demands on systems increase the difficulty in controlling the impact of human malfunctions even in well operated systems.

The field of *ergonomics* has largely developed to address the human - machine or system interfaces. Specific guidelines have been developed to facilitate the development of people friendly systems. Structure design engineers need to be versed in the ergonomic aspects of their systems to help reduce worker injuries and accidents and to help promote quality in the resulting structure systems.

The issues of system robustness (defect or damage tolerance), design for constructability, and design for IMR (Inspection, Maintenance, Repair) are critical aspects of engineering structure systems that will be able to deliver acceptable quality (Bea 1992). Design of the structure system to assure robustness is intended to combine the beneficial aspects of three primary attributes:

1) *redundancy* (degree of structure indeterminacy, alternative primary load carrying elements and components in the structure),

2) *ductility* (ability to deform plastically without substantial loss in load carrying capacity and redistribute loadings to other elements and components in the structure), and

3) *excess capacity* (intentional excess load carrying abilities in the structure elements and components to safety carry overloadings and redistributed loadings).

Environmental Malfunctions

Environmental influences can have important effects on the performance characteristics of individuals, organizations, hardware, and software. Environmental influences include external (e.g. wind, temperature, rain, fog, time of day), internal (lighting, ventilation, noise, motions) and sociological factors (e.g. values, beliefs, mores). Environmental malfunctions are those that promote or invite human and organizational malfunctions. Because of their pervasive nature, the importance of sociological factors should not be under estimated. Sociological factors on both local and national levels can have major influences on the quality of structure systems. Unnecessary litigation, use of win - loose processes, and acceptance of behavior that does not promote integrity are examples.

Interfaces

Research on structure design errors clearly indicates the importance of interfaces. Important breakdowns that result in compromises in the quality of the structure system frequently develop at these interfaces. These interfaces develop between individuals on design teams, between the design teams and their organizations, and between their organizations and those that represent the public. Most importantly, interface breakdowns develop between structure design engineering, construction, operation, and maintenance. Lack of adequate continuity, communications, incentives, and resources for the design engineer to see the structure system through its life-cycle seems to be prevalent. It appears that often the engineer and design team are only concerned with configuration and analysis of the structure system. The engineer and design team seems not to be similarly concerned with the other life-cycle phases and activities that have important influences on the quality of the structure system. Inadequate feed-back to the engineering process results, mistakes are repeated, and learning is dramatically slowed. Perhaps, the current education and training processes for many structure design engineers actually promotes these interface breakdowns.

Engineering Defenses

The structure design process should be viewed in the context of the multiplicity of organizations that influence the quality of that process. The organizations and their activities form a 'mega-system' that should be recognized and addressed. These mega-systems and their organizational components must be understood as 'organisms, living systems that relate to each other.'

The implementation of Total Quality Management (TQM) in design organizations is a current example of efforts directed at the organization aspects associated with design of structures. The International Standards Organization (ISO) design Quality Standards can compliment the general efforts to achieve quality in design organizations. Critical flaws to avoid in implementing these approaches is development of 'minimum compliance mentalities' and making them an unnecessarily burdensome 'paper chase'.

Studies of HRO (High Reliability Organizations) (Roberts 1992) has shed some light on the factors that contribute to risk mitigation in HRO (Roberts, 1989). HRO are those organizations that have operated nearly 'error free' over long periods of time. A variety of HRO ranging from the U. S. Navy nuclear aircraft carriers to the Federal Aviation Administration Air Traffic Control System have been studied. The HRO research has been directed to define what these organizations do to reduce the probabilities of serious errors. Reduction in error occurrence is accomplished by the following: 1) command by exception or negation, 2) redundancy, 3) procedures and rules, 4) training, 5) appropriate rewards and punishment, and 6) the ability of management to 'see the big picture', detect at early stages degradation in quality, and then implement measures to achieve the desirable quality.

There are two primary lines of defense to prevent and / or detect and correct structure design engineering malfunctions. The first line of defense is centered in the individuals performing the design analyses; the design team. The second line of defense is identified as Quality Assurance and Quality Control (QA / QC). QA are those measures and procedures that are put in place before the design process is initiated. QC are those measures and procedures that are implemented during and after the design process to assure that the desired quality is achieved.

The first line of defense is associated with prevention and minimization of errors made and not corrected by the individuals that perform the design processes. The quality of the structural design is a direct function of the quality of the design team that performs the design. The qualities cited earlier for HRO are those that should be developed in high reliability design teams.

A clear conclusion from the research on which this paper is based is that HOF should be reflected in 'next generation' structure design procedures and criteria (Bea 1994). This then leads to the question of how it should be reflected. The results of this research indicates that there are three major ways in which considerations of HOF should be reflected in structure design procedures. The first two ways have been discussed in the preceding sections. The third way is directed at helping achieve adequate and acceptable quality in the design procedures and processes themselves. There are three strategies that should be considered:

• *Strategy 1* - QA / QC the design procedures and processes (fault avoidance),

• *Strategy 2* - QA / QC is integrated as a requirement directly in the design procedures and processes (fault detection and correction), and

• *Strategy 3* - Measures are introduced into the design procedures and processes that will minimize the effects of HOF on the quality of the structure (fault tolerance through structure robustness).

Conclusions

Some organizations have successfully applied this knowledge. Their records of quality, reliability, productivity, and profitability indicate that applying this knowledge can make good sense and business. The primary challenge to implementation rests squarely on the shoulders of offshore structures engineers: to admit that consideration of HOF is an integral part of their responsibilities, and to take definitive and aggressive steps to learn how to apply what is known about HOF in the engineering and reliability of structures.

References

Bea, R. G. (1992). Marine Structural Integrity Programs (MSIP), Ship Structure Committee, Report SSC-365, Washington, DC.

Bea, R. G. (1994). The Role of Human Error in Design, Construction, and Reliability of Marine Structures, Ship Structure Committee, Report SSC-378, NTIS # PB95-126827, Washington, DC.

Reason, J. (1990). Human Error, Cambridge University Press, New York, NY.

Roberts, K. H. (1989). "New Challenges in Organizational Research: High Reliability Organizations," Industrial Crisis Quarterly, Vol. 3, Elsevier Science Publishers B. V. Amsterdam - Netherlands.

Compliant Towers: Structures that Benefit from Dynamics

Larry Light and Cheng-Yo Chen[1]

Abstract

Designs of recent compliant towers take advantage of platform dynamics to minimize the storm loadings on the platforms and minimize the moments in the structures.

Introduction

This paper offers a brief, non-technical introduction to compliant towers and how dynamic behavior is used to their benefit. The compliant tower concept is described and reasons for its use are given. Mechanisms to make a tower compliant are presented and recent platforms using these mechanisms are identified. Tuning towers to gain benefit from dynamics is described.

Definition and Description

Compliant towers are relatively slender, bottom-founded platforms designed for use in water too deep for conventional fixed platforms. Compliant towers are designed to sway, or comply, with storm waves rather than to rigidly resist the waves. A platform that sways doesn't sound desirable. So, why make deep water platforms compliant? The short answer is: "to reduce the forces on the platform and thus to reduce the cost." This is easiest explained by describing problems that occur with conventional, fixed platforms as their design water depths increase and explaining how compliant towers overcome these problems.

---

[1] Manager of Engineering and Staff Consultant, respectively, at J. Ray McDermott Engineering, LLC; 801 North Eldridge; Houston, TX 77079

Conventional platforms resist wave and storm forces by having sufficient strength to stand rigidly against them. In shallow water, less than roughly 150 meters, platforms rigidly resist waves efficiently. In deeper water, the platform strength required to resist the waves becomes large and expensive. In addition, because of their height, the deeper platforms tend to be less stiff and have sway periods large enough (5 to 6 seconds) to allow the platforms to react dynamically with the waves. Compared to wave forces calculated statically, the forces on a conventional, fixed platform in deep water are amplified by the platform's dynamic interaction with waves. Building adequate fatigue life into these deeper fixed platforms also becomes expensive since the seastates with periods matching the platforms' periods, 5-6 seconds, occur frequently and are large enough, 1.5 to 2.5 meter significant, to control fatigue life.

Compliant towers, unlike fixed platforms, resist storm waves with their inertia. Because a compliant tower is sufficiently flexible to have sway periods roughly twice as long as the design wave periods (tower periods of 30 seconds), the tower will sway, or comply, with approaching waves thus allowing its inertia to help resist storm wave forces. The inertia of the tower and the water entrapped and enclosed by tower members reduces the strength that must be built into the tower. The motion of the tower affects the relative velocities between the members and the waves and produces damping that helps reduce harmonic vibrations.

What Makes the Tower Compliant?

A tower can be made compliant if it is free to sway about the seabed and is held upright with anchor lines or guy wires attached near the water surface and anchored into the seabed well away from the platform. One such structure has been built, Exxon's Platform Lena.

A structure can be made compliant by reducing its cross section to reduce its stiffness. Considering that a typical drilling and production platform topsides is likely to weigh in excess of 10,000 tons, the tower supporting it must be stiff enough to support the topsides without buckling. A small cross section also leads to large pile reactions. A tower flexible enough to comply with the waves and still be stiff enough to support a topsides can be made in depths exceeding about 700 meters.

A structure can be made compliant with sufficient foundation flexibility to allow the structure to be flexible yet still stable. Normal tubular steel piles attached to the base of a platform are too stiff to provide platform wave compliance. Pile area is set by the forces applied to it, but pile length can vary depending on how far above the seabed the foundation piling are attached to the tower legs. Please refer to Figure 1. The sketches are out of scale to show some features more clearly. The length of piling from the seabed to the point of attachment to the tower legs serves as a spring to reduce the foundation stiffness. Attaching the foundation piling to the tower well above the seabed not only provides foundation flexibility, it allows the framing to be lighter because the tower is supported nearer to its center rather than cantilevered off the seabed. This type tower was used for the Texaco Petronius tower installed in 535 meter deep water in November of '98.

Another method to make a structure compliant is to build a "hinge" or articulation point in the lower part of the tower and provide long tubular members to act as "springs" connecting the upper and lower parts of the tower legs. These springs support the top of the tower much as the piles supported the tower described in the paragraph above. This concept allows a tower to sit on a fixed base with a broader foundation pile spacing. Please refer to Figure 2. This tower concept was used for the Hess Baldpate tower installed in 500 meter deep water in May of '98.

Design for Dynamics

Because a compliant tower sways or complies with waves, the engineer must control or tune its sway and bending periods to produce the optimum structure. The tower's sway period must be long enough compared to the storm waves' periods to allow the tower to comply to take full advantage of inertial effects and its bending period must be short enough to avoid fatigue problems. Initial tower configuration is set using parametric studies considering effects on installed cost as platform base width and pile attachment height are varied.

Compliant towers are designed for their maximum responses or motions in selected storm return intervals. Determination of a tower's responses must be based on trains or series of random waves. The trains of waves must be selected to give the appropriate responses for the characteristics of the tower being designed. One train of waves might give the maximum design shear responses in the top of a tower while another train might give the maximum design bending response. Selecting appropriate wave trains requires determining responses of each level of the tower for many seastates, statistically determining the design level responses at each level, then defining design seastates that produce the required responses. The analyses determining the level responses are time-domain, random wave analyses considering the effects of the tower weights, including the topsides loads, on the tower tilt (P-Delta effects) and the relative velocities between the tower and the waves.

A properly tuned compliant tower is an efficient structure with a long life because its design takes advantage of the benefits from its dynamic behavior with storm waves.

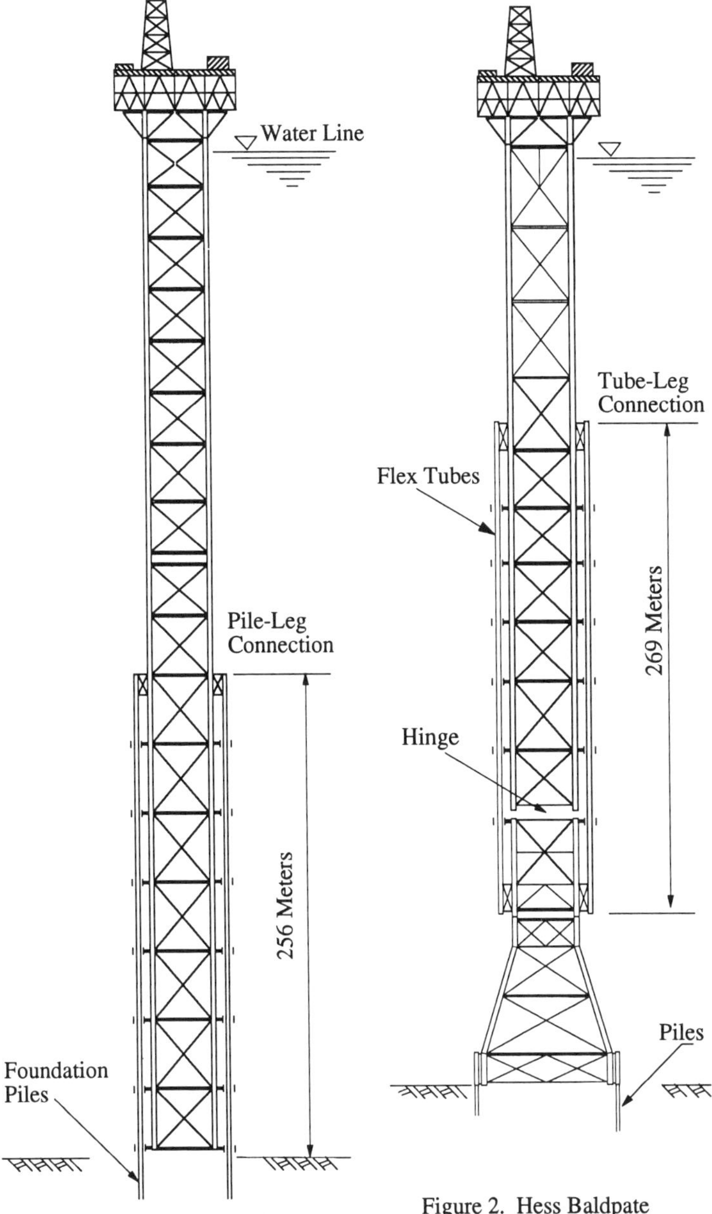

Figure 1. Texaco Petronius

Figure 2. Hess Baldpate

Practical Approaches in Offshore Engineering

G. D. Hahn[1] and Q. Abughazleh[1]

Abstract

A presentation is first made to illustrate that the inertia and drag forces which govern the overall response of bottom-supported, deep-water offshore structures are controlled by the sea surface kinematics. This presentation is then extended to discuss the definition of the amount of hydrodynamic damping resulting from the interaction of such structures with the fluid. The damping is also shown to be dominated by the characteristics of the sea surface kinematics.

Introduction

Although many advances have been made in the area of the dynamics of offshore structures under wave forces, there remains a need for improved appreciations of the main effects involved in the wave excitation and the influences they have on the resulting structural response. These appreciations can be accomplished only by critical examinations of the fundamentals of the problem, and they are of considerable importance in order to develop efficient design methods.

The main objectives of the present paper are: (1) to discuss practical approaches for the understanding of dominant features of the excitation induced by waves on bottom supported, deep-water offshore structures; and (2) to provide information that can be useful to make proper interpretations of the main effects of the excitation on the resulting structural response. The reported results apply to the type of offshore structure that consists of a continuous frame directly supported by the foundation (bottom-supported), operating in a water depth exceeding about 300 m. The wave motion is defined by using the deep-water approximation of the linear wave theory, and the corresponding wave forces are specified by the Morison equation (Sarpkaya and Issacson 1981), in terms of inertia and drag force components.

---

[1]Department of Civil and Environmental Engineering, Vanderbilt University, P.O. Box 124-B, Nashville, TN 37235

854

System and Excitation Considered

It is of interest to examine the lateral response of offshore structures with uniform properties. Let y represent vertical distance measured from the still water level (SWL), and t be time. Also, let the volume per unit of length and the projected area per unit of length of the structure be V and A, respectively. Furthermore, the associated inertia and drag force coefficients are respectively denoted by c_M and c_D, and the mass density of the fluid is denoted by ρ. Using the mode superposition method, the lateral displacements of the structure can be expressed as

$$x(y,t) = \sum_{j=1}^{n} \phi_j(y)\, q_j(t) \tag{1}$$

where $\phi_j(y)$ is the jth natural mode of vibration of the structure; n is the number of modes to be considered; and $q_j(t)$ is the jth modal response, which can be obtained from the solution of

$$\ddot{q}_j(t) + 2\,\zeta_j\,\omega_j\,\dot{q}_j(t) + \omega_j^2\, q_j(t) = \frac{1}{m_j^*}\, P_j^*(t) \tag{2}$$

in which m_j^*, ω_j, ζ_j, and $P_j^*(t)$ are the jth modal mass, natural frequency, damping ratio, and force, respectively. In particular, the modal force is given by

$$P_j^*(t) = P_{Mj}^*(t) + P_{Dj}^*(t) \tag{3}$$

where $P_{Mj}^*(t)$ and $P_{Dj}^*(t)$ are the modal inertia and drag forces, respectively, defined by

$$P_{Mj}^*(t) = \int_0^H \phi_j(y)\, p_M(y,t)\, dy \; ; \qquad P_{Dj}^*(t) = \int_0^H \phi_j(y)\, p_D(y,t)\, dy \tag{4,5}$$

in which H is the total water depth, and $p_M(y,t)$ and $p_D(y,t)$, the inertia and drag force intensities, are given by

$$p_M(y,t) = B_M\, \ddot{u}(y,t) \; ; \qquad p_D(y,t) = B_D\, |\dot{u}(y,t)|\, \dot{u}(y,t) \tag{6,7}$$

where $B_M = c_M\,\rho V$, $B_D = 0.5\,c_D\,\rho A$, and the quantities $\dot{u}(y,t)$ and $\ddot{u}(y,t)$, the horizontal fluid particle velocity and acceleration, are be defined by

$$\dot{u}(y,t) = \sum_{i=1}^{N} \bar{\omega}_i\, a_i\, \exp\!\left(-\frac{\bar{\omega}_i^2\, y}{g}\right) \cos(\bar{\omega}_i - \theta_i) \tag{8}$$

$$\ddot{u}(y,t) = -\sum_{i=1}^{N} \bar{\omega}_i^2\, a_i\, \exp\!\left(-\frac{\bar{\omega}_i^2\, y}{g}\right) \sin(\bar{\omega}_i - \theta_i) \tag{9}$$

in which the $\bar{\omega}_i$'s are frequencies in rad/sec; the a_i's are the corresponding wave displacement amplitudes; the θ_i's are random phase angles uniformly distributed between 0 and 2π; and g is the gravitational acceleration. Both $\bar{\omega}_i$ and a_i can be obtained by sampling an applicable wave spectrum at N points.

Inertia and Drag Force Resultants

Before discussing the modal forces $P^*_{Mj}(t)$ and $P^*_{Dj}(t)$ further, it is desirable to address the definition of the related inertia and drag force resultants, given by

$$R_M(t) = \int_0^H p_M(y,t)\,dy \; ; \qquad R_D(t) = \int_0^H p_D(y,t)\,dy \qquad (10,11)$$

As demonstrated by Hahn (1995), evaluations of the integrals in (10,11) lead to

$$R_M(t) = -B_M\,g\,u(0,t) \; ; \qquad R_D(t) = 0.46\,B_D\,g\,|v(0,t)|\,v(0,t) \qquad (12,13)$$

where $u(0,t)$ and $v(0,t)$ are, respectively, the horizontal and vertical components of water particle displacements computed at the SWL (i.e., the sea surface). Note that $v(0,t)$ is usually denoted by $\eta(t)$ in the technical literature. The quantities $u(0,t)$ and $v(0,t)$ can be expressed as

$$u(0,t) = \sum_{i=1}^N a_i\,\sin(\bar{\omega}_i t - \theta_i) \; ; \qquad v(0,t) = \sum_{i=1}^N a_i\,\cos(\bar{\omega}_i t - \theta_i) \qquad (14,15)$$

Modal Forces

Considering a particular modal shape $\phi_j(y)$, it is possible to obtain a linear approximation for it as

$$\phi_j(y) \approx 1 - \beta_j \frac{y}{H} \qquad (16)$$

where β_j is a parameter chosen to minimize the square of the error between the linear approximation of (16) and the exact variation for $\phi_j(y)$, as reported by Hahn (1995).

For the linearly varying modal shape, it has been shown (Hahn 1995) that, for deep waters, the inertia and drag modal forces $P_{Mj}(t)$ and $P_{Dj}(t)$ in (4,5) are given with a good degree of approximation by

$$P_{Mj}(t) = R_M(t)\left[1 - \beta_j \frac{y_M(t)}{H}\right] \; ; \qquad P_D(t) = R_D(t)\left[1 - \beta_j \frac{y_D(t)}{H}\right] \qquad (17,18)$$

where $y_M(t)$ and $y_D(t)$ are the positions, measured from the SWL, of the lines of actions of $R_M(t)$ and $R_D(t)$, respectively. The quantities $y_M(t)$ and $y_D(t)$ have also been defined by Hahn (1995), but they do not have a significant effect on the modal forces of (17,18) because in deep waters they are small in comparison with H.

Fluid-Structure Interaction and Hydrodynamic Damping

To account for fluid-structure interaction, the drag force intensity of (7) is re-expressed as

$$\hat{p}_D(y,t) = B_D \left| \dot{u}(y,t) - \dot{x}(y,t) \right| \left[\dot{u}(y,t) - \dot{x}(y,t) \right] \tag{19}$$

Considering the contribution of the jth mode only, $\hat{p}_D(y,t)$ reduces to

$$\hat{p}_D(y,t) = B_D \left| \dot{u}(y,t) - \phi_j(y)\dot{q}_j(t) \right| \left[\dot{u}(y,t) - \phi_j(y)\dot{q}_j(t) \right] \tag{20}$$

Note that if the interaction effects are ignored, then $\hat{p}_D(y,t) = p_D(y,t)$. By making a number of assumptions that are consistent with the nature of the problem, it can be shown that the effects of fluid-structure interaction can be reduced to an added damping ratio approximately given by

$$\hat{\zeta}_j \approx \frac{0.5\ \phi_j^2(0)\ B_D}{\omega_j\ \bar{\omega}_o\ m_j^*} \left(1 - \beta_j\ \frac{g}{H\ \bar{\omega}_o^2} \right) g\, u_o \tag{21}$$

where $\bar{\omega}_o$ is the mean wave frequency in rad/sec, and $u_o = |u(0,t)|_{max}$. By replacing ζ_j with $\zeta_j + \hat{\zeta}_j$ in (2) and using (5) in (3) to define the excitation (no interaction), the actions of fluid-structure interaction can be accounted for approximately.

Conclusion

Examinations have been made of the forces induced by waves on offshore structures, and the effects of fluid-structure interaction have been accounted for by defining additional damping ratios for the individual modes. This technique ignores the coupling of modal responses due to cross-modal damping ratios resulting from the actions of fluid-structure interaction. The information presented can be useful for interpreting the effects on the response of relevant factors involved.

References

Hahn, G. D. (1995). Concepts for wave-excited offshore structures, *J. of Structural Engineering*, ASCE, 121(10), 1427-1435

Sarpkaya, T., and Issacson, M. (1981). *Mechanics of wave forces on offshore structures*, Van Nostrand, New York

VARIABLE STIFFNESS AND INSTANTANEOUS FREQUENCY

Satish Nagarajaiah[1], M. ASCE, Vardarajan Nadathur [2] and Sanjay Sahasrabudhe[2]

Abstract

This paper presents a new variable stiffness control system, which can switch the stiffness continuously and smoothly. The new and innovative semi-active variable stiffness control (SAIVS) device and tests performed to study the effectiveness of the device are presented. A new control algorithm based on instantaneous frequency estimation using Hilbert transform is developed and implemented in real time digital signal processing system and controller. Shake table test results of a single degree of freedom system with SAIVS device is presented.

Introduction

In semi-active stiffness systems the control forces are developed due to variation in stiffness of the system and not due to direct application of control force as in fully active systems; hence, these systems need nominal power. Variable stiffness devices have been developed by Kobori et al. (1993) and Sack et al. (1993). The system developed by Kobori et al. (1993) maintains the structure in a non-resonant state under seismic excitations by altering the building stiffness based on the nature of the earthquake. The system has been implemented in an actual building and has performed to expectations in real earthquakes. The system developed by Sack et al. (1993) is based on a variable stiffness and damping device and has been implemented in a bridge; the system has performed to expectations. The main limitation of these devices is that the stiffness is switched abruptly. To overcome the limitations of existing devices, a new semi-active continuously variable stiffness (SAIVS) device has been developed. This paper presents the results of the experimental and analytical study of the SAIVS system.

SAIVS Device

The SAIVS device consists of four sets of spring elements and telescoping tube elements. Varying the position of the springs produces the variable stiffness. The telescoping tubes develop frictional forces, which are beneficial due to the resulting energy dissipation. A DC servomotor with a rack and pinion assembly controls the position of the device between the fully closed and open configurations. The DC servomotor is controlled by the digital signal processor based dSPACE controller. The power required by the DC servomotor to position the device is nominal.

---

1. Associate Professor, 2. Graduate Research Assistant, Dept. of Civil Engineering, Rice University, Houston, TX 77005.

The SAIVS device has maximum stiffness in its fully closed position and minimum stiffness in its fully open position. The device can be positioned in any configuration between closed and open positions; thus, the device is capable of varying the stiffness continuously and smoothly between the maximum and minimum. In a typical framed structure, the SAIVS device is designed to connect the floor beam and inverted V bracing system and transfer forces between them. The bracing is connected with variable stiffness depending on the device position.

SAIVS Device Test Results

In order to evaluate the device, dynamic tests were conducted. These tests were performed with harmonic excitation imposed to the SAIVS device by a servo-hydraulic actuator. A series of tests were performed on the SAIVS device. The tests involved different frequencies and amplitudes of harmonic excitation imposed by the actuator and different frequencies and switching positions of the SAIVS device imposed by the DC motor and dSPACE real time DSP controller. The test results were recorded in the form of time history of the force in the SAIVS device, measured by a load cell, the displacement, and the position of the SAIVS device. The force-displacement loop of the SAIVS device for 1.0 Hz harmonic excitation of amplitude 0.25 inch with the device switched by a sinusoidal signal of 0.325 Hz is shown in Fig. 1. The minimum stiffness of the SAIVS device in the open position is 7 lb./inch and the maximum stiffness is 38 lb./inch. It is clearly evident from Fig. 1 that device can vary the stiffness continuously and smoothly. The hysteretic behavior due to friction in the telescoping tubes and other connections is evident in the force-displacement loop

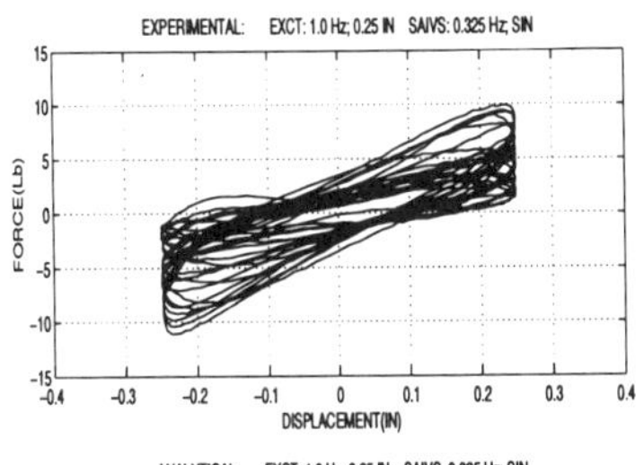

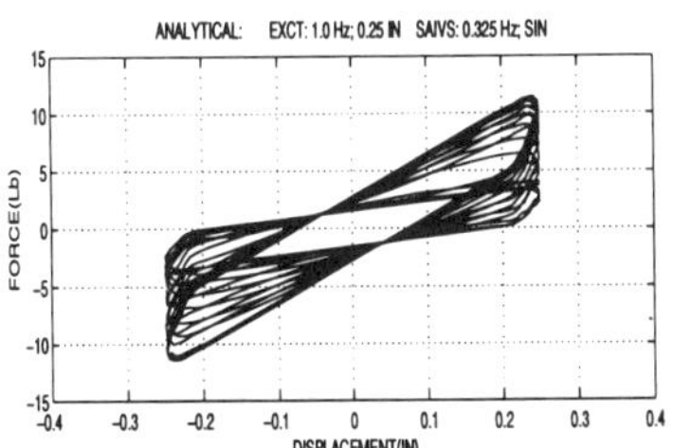

Fig. 1. Device Switched with a Sinusoidal Signal of 0.325 Hz

(shown in Fig. 1). The results of the analytical model are also shown in Fig. 1. The analytical model is not presented due to lack of space. Comparing the analytical and experimental results, shown in Fig. 1, it is evident that the analytical model is capable of capturing the features of the device satisfactorily. The low force level is due to the small scale SAIVS device. The device can easily be scaled due to its mechanical nature.

Instantaneous Frequency using Hilbert Transform

The analytical signal $y(t)$ associated with $x(t)$ is defined by

$$y(t) = x(t) + j\tilde{x}(t) \tag{1}$$

which can also be written as

$$y(t) = A(t)e^{j\phi(t)} \qquad (2)$$

where instantaneous amplitude

$$A(t) = (x^2(t) + \tilde{x}^2(t))^{1/2} \text{ and}$$

instantaneous phase

$$\phi(t) = tan^{-1}(\frac{\tilde{x}(t)}{x(t)}) = 2\pi f t.$$

The instantaneous frequency f (Cohen 1995) is given by

$$f = (\frac{1}{2\pi})\frac{d\phi(t)}{dt} \qquad (3)$$

Switching Control Algorithm

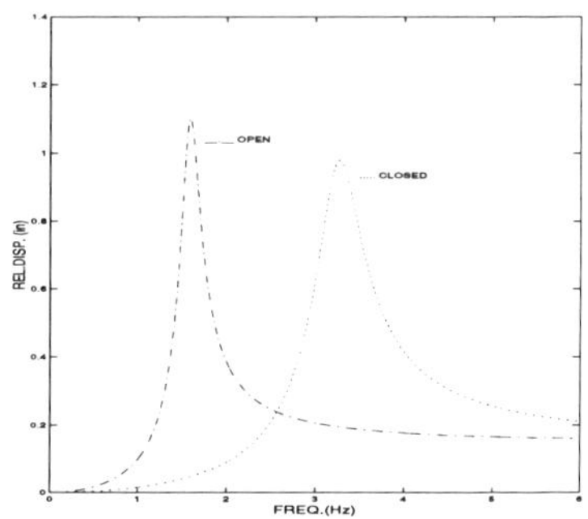

Fig. 2. Frequency Response Curves for the SDOF system with SAIVS device in the Fixed Open and Fixed Closed Position

The control algorithm developed involves (1) tracking the instantaneous frequency of the excitation using the Hilbert transform of shake table displacement, (2) switching smoothly from open to closed position or vice versa once the intersection frequency is detected. The value of the frequency at the intersection of the frequency response curves for fixed open and fixed closed conditions, shown in Fig. 2, can be determined using an approach similar to that presented by Nagarajaiah (1997).

Shake Table Test Results

A SDOF system with the SAIVS device was tested on the shaking table. The SDOF system was subjected to sine sweep excitation of amplitude $\pm$ 0.25 inch, with frequency range 5.0 Hz to 0 Hz (shown in Fig. 3). The system was instrumented to measure acceleration and displacement of the model and the shake table. The dSPACE digital signal processing system was used for data acquisition and real time control. The tests were performed with the SAIVS device in the fixed open, fixed closed and controlled condition. The measured time history response results of the system for the three cases are shown in Fig. 3. It is evident from Fig. 3 that the controlled case maintains a nonresonant state and produces the least displacement response amongst the three cases. The resonant peaks in the fixed open and fixed closed positions are evident in Fig. 2 and Fig. 3. The instantaneous frequency tracking using Hilbert transform, shown in Fig. 3 is effective.

Conclusions

The test results presented demonstrate that the SAIVS device can switch the stiffness continuously and smoothly. It is evident from the shake table test results of the SDOF system with the SAIVS device can produce a non-resonant system. The developed switching control algorithm based on instantaneous frequency tracking using Hilbert transform is effective.

Acknowledgments

Funding for this project provided by the National Science Foundation Grant (CMS-9625979), with Dr. E. Sabadell and Dr. S. C. Liu as Program directors gratefully acknowledged.

References

Cohen L. (1995). <u>Time-Frequency Analysis</u>, 1st Ed., Prentice –Hall, New Jersey.

Kobori, T., Takahashi, M., and Niwa, N. (1993). "Seismic response controlled structure with active variable stiffness system," *Earthquake Eng. Struct. Dyn.*, Vol. 22, No. 12, 925-941.

Nagarajaiah, S. (1997). "Semi-active control of structures," *Proc. Struct. Congress,* ASCE, Portland, Oregon, 1574-1578.

Nagarajaiah, S., and Mate, D. (1998). "Semi-active control of continuously variable stiffness system," *Proc. Second World Conf. on Structural Control,* Kyoto, Japan.

Sack, R. L., and Patten, W. N. (1993). "Semi-active hydraulic structural control," *Proc. International Workshop on Structural Control*, Hawaii, 417-431.

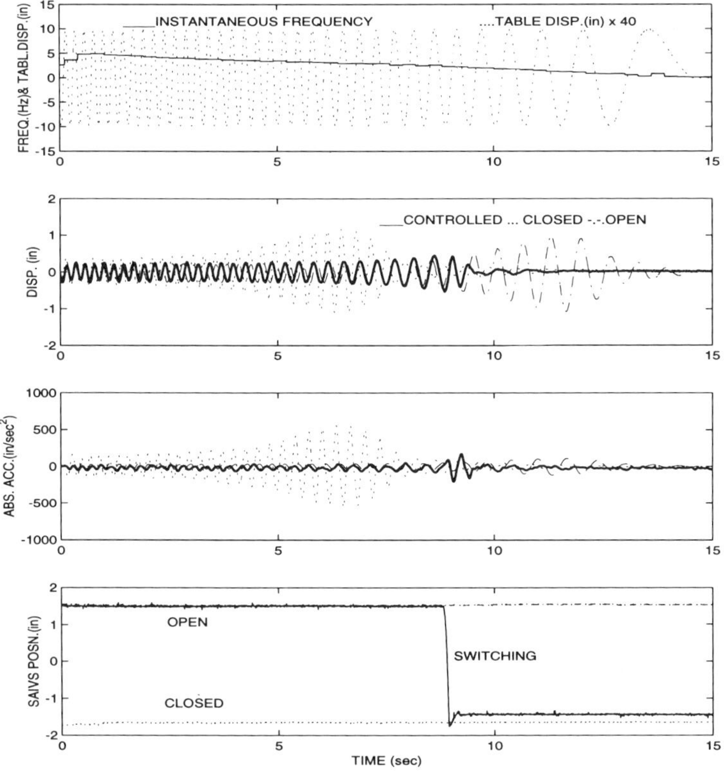

Fig. 3.   Time History Response of SDOF System to Sine Sweep Excitation

Semi-Active Hybrid Seismic Isolation Systems: Addressing the Limitations of Passive Isolation Systems

Michael D. Symans[1], Glenn J. Madden[2] and Nat Wongprasert[2]

Abstract

Passive seismic isolation systems consisting of passive isolation bearings and passive damping devices have been utilized in at least five building structures located in regions of potentially strong seismic activity. As the application of performance-based seismic design specifications becomes more widespread, it is expected that the number of such applications of hybrid structural control technology will increase. The performance of such hybrid control systems, however, may be further enhanced by modifying the passive damping devices to behave as semi-active control elements. This paper discusses the features of semi-active hybrid seismic isolation systems that address some of the limitations of passive hybrid isolation systems. In addition, some of the implementation issues related to the use of semi-active hybrid seismic isolation systems are discussed. Finally, the development of a semi-active hybrid seismic isolation system consisting of passive sliding isolation bearings and semi-active fluid damping devices is presented.

Introduction

As earthquake engineering professionals move toward the acceptance of performance-based seismic design specifications, the expected performance level of structures will become more readily apparent to building owners. As a result, building owners will be more likely to consider the application of advanced technologies for protecting their structures. To a limited degree, this has already occurred for passive structural control systems. In particular, a number of structures utilize passive fluid dampers for absorbing seismic energy either within the framing of the structure or within a base isolation system (Soong and Constantinou, 1993). A passive hybrid isolation system consisting of passive base isolation bearings and passive dampers offers a very reliable and cost-effective approach to mitigating the

---

[1] Assistant Professor (A.M. ASCE) and [2] Graduate Research Assistant (S.M. ASCE), Dept. of Civil and Env. Engineering, Washington State University, Pullman, WA 99164-2910

effects of strong earthquake-induced ground motion. However, there are limitations to the performance of a passive hybrid isolation system. In particular, such systems may not perform well under a variety of earthquake ground motion characteristics. For example, as noted by Johnson et. al. (1998), when isolated structures are designed for extreme earthquake events, the isolation system may not be very effective for more frequent moderate earthquakes. One approach to addressing the limitations of a passive hybrid isolation system is to replace the passive dampers with semi-active dampers. This paper explores the potential for implementation of semi-active dampers within the base-isolation system of a structure.

Passive and Semi-Active Hybrid Seismic Isolation Systems
Passive hybrid seismic isolation systems consist of passive isolation bearings combined with a passive damping system (see Figure 1). Such systems are generally very effective for controlling the response of structures since the bearings increase the fundamental period of vibration, resulting in a reduction of spectral acceleration, while the dampers limit the displacement response at the isolation level. Recent applications of passive hybrid seismic isolation systems have utilized either elastomeric bearings or friction pendulum sliding bearings combined with fluid dampers. The fluid dampers that are utilized in such applications are typically inertial force dampers that dissipate energy via the orificing of a fluid. The force-velocity relation in the damper may be linear or nonlinear but, in either case, is a fixed function.

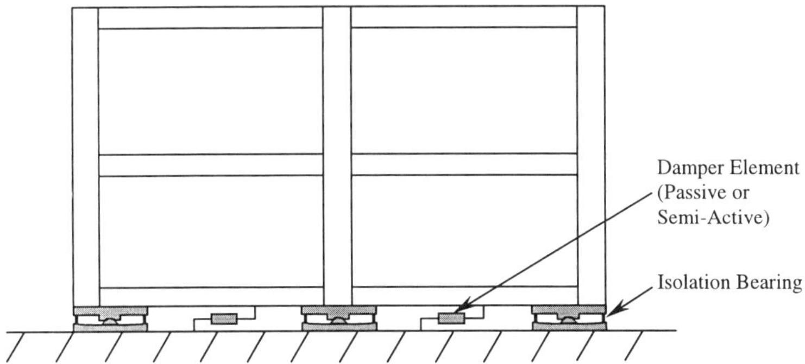

Figure 1 Building Structure Containing Hybrid Seismic Isolation System

Semi-active hybrid seismic isolation systems consist of passive isolation bearings combined with a semi-active damping system (see Figure 1). There are a variety of semi-active control elements for application within hybrid seismic isolation systems (e.g., see Symans and Constantinou (1998)). One type of semi-active damper dissipates energy via the orificing of a fluid. The size of the orifice can be modulated based on the measured response of the structure. As a result, the force-velocity relation in the damper is a variable function that can be controlled in real-time.

Implementation Issues

There are a wide variety of issues that must be addressed to achieve successful implementation of a semi-active hybrid seismic isolation system. In the author's opinion, the most important implementation issue is the need for clear demonstrations of significant performance enhancements over that which can be achieved by passive hybrid isolation systems. Passive hybrid isolation systems are generally capable of significant response reductions over conventional seismic design approaches. In addition, they may be regarded as reliable in the sense that they have no hydraulic or electrical power requirements. As a result of these features, passive hybrid isolation systems have received some acceptance within the structural engineering community as evidenced by a few applications within buildings. Once the performance levels of semi-active hybrid seismic isolation systems are shown to clearly exceed that of passive hybrid isolation systems, a number of other implementation issues must be resolved. Examples of some of these issues include: 1) Development of simple yet reasonably accurate models of semi-active isolation systems; 2) Development of robust control algorithms for operation of the semi-active portion of the isolation system; 3) System integration issues wherein the structure, the semi-active isolation system, the response sensors, and the control hardware/software are considered as a whole in the design process; 4) Reliable power sources for operation of the semi-active control elements; and 5) Development of robust semi-active control elements that will respond when called upon after remaining dormant for extended periods of time.

As to the performance issue, it is apparent that passive hybrid isolation systems are limited in their ability to respond in an optimal fashion to a wide variety of earthquake ground motion characteristics. For example, a structure may be prone to a typical short-period ground motion from a far-field source and to a long-period pulse-like ground motion from a near-field source. These two ground motions induce significantly different demands on the seismic isolation system. An isolation system that can adapt in an optimal fashion to both types of ground motions may exhibit superior performance as compared to a passive isolation system that has been designed with a particular type of ground motion in mind. Recent studies by Makris (1997) have demonstrated the potential advantages of a semi-active seismic isolation system for near-field pulse-like ground motion. In addition to the near-fault problem, recent studies by Sadek and Mohraz (1998) have indicated that the acceleration response of multi-story structures with semi-active hybrid isolation systems can be reduced while simultaneously limiting the displacement response.

Development and Testing of a Semi-Active Hybrid Seismic Isolation System

As of this writing, the authors are currently performing an analytical and experimental study on the seismic response of a three-story, ¼-scale model, moment-resisting frame containing a semi-active hybrid seismic isolation system. The hybrid isolation system consists of friction pendulum sliders and semi-active fluid viscous dampers (see Figure 1). The semi-active dampers behave as linear viscous dashpots in which the damping coefficient may be continuously modulated between an upper and lower bound. The dynamic response of the building structure is monitored and utilized within a feedback control system to determine the optimal value of the damping coefficient. The structure is excited at the base by a uni-axial seismic simulator that has recently been constructed at Washington State University. The simulator is being used to subject the test structure to both far-field and near-field ground motions. Some of the implementation issues which have been identified and characterized through these studies include analytical model development, sensor dynamics, real-time operation, fail-safe operation, power requirements and structure/control system interaction.

Summary

Passive hybrid seismic isolation systems are available for protecting structures subjected to strong earthquakes. These systems have already been accepted to some degree by practicing engineers and it is expected that the number of applications will continue to grow due to their ability to significantly reduce seismic response and their inherent reliability. In contrast, semi-active hybrid isolation systems face a variety of obstacles before they will be accepted for implementation. However, it is readily apparent that a semi-active isolation system offers adaptability characteristics that may prove particularly beneficial in the case of disparate ground motions. As building owners continue to demand higher performance levels for their structures, it is inevitable that semi-active hybrid isolation systems will continue to move forward toward implementation.

Acknowledgements

This material is based upon work supported by the National Science Foundation (NSF) under Grant No. CMS-9624227. This support is gratefully acknowledged. Any opinions, findings, and conclusions or recommendations expressed in this material are those of the authors and do not necessarily reflect the views of the National Science Foundation.

References

Johnson, E.A., Ramallo, J.C., Spencer, B.F. Jr. and Sain, M.K. (1998). "Intelligent Base Isolation Systems," Proc. of Second World Conf. on Structural Control, Kyoto, Japan, June (in press).

Makris, N. (1997). "Rigidity-Plasticity-Viscosity: Can Electrorheological Dampers Protect Base-Isolated Structures from Near-Source Ground Motions?," *Earthquake Engineering and Structural Dynamics*, 26, 571-591.

Sadek, F. and Mohraz, B. (1998). "Semi-Active Control Algorithms for Structures with Variable Dampers," *J. of Engineering Mechanics*, ASCE, 124(9), 981-990.

Soong, T.T. and Constantinou, M.C. (Editors) (1993). Passive and Active Structural Vibration Control in Civil Engineering, Springer-Verlag, Wien-New York.

Symans, M.D. and Constantinou, M.C. (1998). "Semi-Active Control Systems for Seismic Protection of Structures: A State-of-the-Art Review," *Engineering Structures*, 21(6), 469-487.

Variable Property Devices for Structural Control

Henri P. Gavin[1], A.M., Nitin S. Doke[2]

Abstract

Base isolation reduces the the base shear, inter-story drift, and absolute accelerations of structures at the expense of potentially large deflections of the isolators. By incorporating variable damping and stiffness devices into the isolation system, the isolator deformations can be reduced with out significantly increasing the base shear. In this paper, the relative importance of the variable damping and variable stiffness mechanisms to this control system is analyzed.

Introduction

Base isolation systems protect building structures from earthquakes by lengthening the natural period of the structure (through the compliance of the bearings), by adding damping to the base isolation interface (through the inherent damping of the isolator bearings, or through supplemental energy dissipation), and ultimately by limiting the base shear (Imbimbo and Kelly 1998). Both compliance and dissipation are necessary mechanisms in a base-isolated structure. The deformations of compliant isolators can become large if the ground motion has a coherent pulse close to the isolated structure's period. Under these conditions, the energy dissipation demand on supplemental dampers in the isolation system can be very large.

As an alternative to passive dampers, devices with variable properties, controlled through a very simple feedback control rule, can control isolator deflections without increasing the absolute accelerations and the base shear of the structure. Furthermore, variable-property devices can reduce isolator deformations without dissipating large amounts of energy. These devices allow a single (adaptive) structural system to perform well under a wide range of excitations.

Variable-property devices can be made using hydraulic dampers with a controllable orifice (Patten 1994), with mechanical devices (Nagarajiah 1998), or with

---

[1]Assistant Professor, Dept. Civil & Environ. Eng., Duke Univ., Durham, NC 27708–0287
[2]Graduate Research Assistant, Dept. Civil & Environ. Eng., Duke Univ., Durham, NC 27708–0287

controllable fluids, such as electro-rheological (ER) (Gavin 1998b) and magneto-rheological (MR) materials (Dyke 1998). These devices can be designed such that their properties can be modified over a finite range. This paper addresses the design problem of sizing the range of devices properties for a given structure.

When controllable fluids are subjected to high fields, their yield stress and stiffness increase by orders of magnitude. Experimental studies on a small scale building model equipped with a controllable ER device illustrates two important points (Gavin 1998a):

1. When the voltage to the ER device is turned on, the resonant frequencies of the structure increase.

2. When the voltage to the ER device is controlled through a feedback circuit, the structure has no resonances at all.

This paper shows that this controllable solid-like behavior of energized ER materials is essential to the observed shift in resonances.

Variable Stiffness or Non-proportional Damping?

Figure 1 illustrates an idealized three-story base isolated building frame. The variable property device (illustrated with vertical hash marks) is installed in parallel with low-stiffness base isolation bearings (illustrated with horizontal hash marks). A very simple control rule calls for the force in the device to mimic the force in a *fictitious damper* installed between the isolation level and the inertial reference frame. In other words, the device stiffness and damping are high (ON) when its force has the same sign as the absolute velocity $(\dot{x} + \dot{z})$ of the isolation level. Otherwise, the device should have low damping and stiffness (OFF). This control rule can be implemented with a small integrated circuit and sensors located on the device itself. Figure 2 illustrates qualitative results which are characteristic of many simulations and experiments implementing this control method with ER devices. When the device is turned ON, the resonant frequencies increase. When the device is controlled, the spectral accelerations follow the lower bound of the two constant cases. This paper illustrates the effects of variable isolator stiffness and damping on the resonant frequencies of a building structure. The imaginary part of the eigenvalues of the system dynamics matrix,

$$A = \begin{bmatrix} 0 & I \\ -M^{-1}K & -M^{-1}C \end{bmatrix}, \tag{1}$$

are plotted against the isolator stiffness and the isolator damping in Figures 3 and 4. The resulting damped natural frequencies are non-dimensionalized by $\sqrt{(k/m)}$, where k is the inter-story stiffness and m is the mass of each story. The isolator stiffness is non-dimensionalized by k, and the isolator damping is non-dimensionalized by the (light) inter-story damping. These generalized results show that a shift in

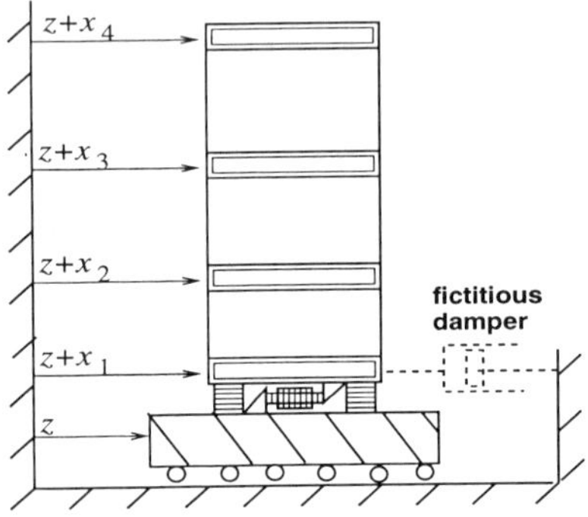

Figure 1: Base isolated building model with a controllable fluid device.

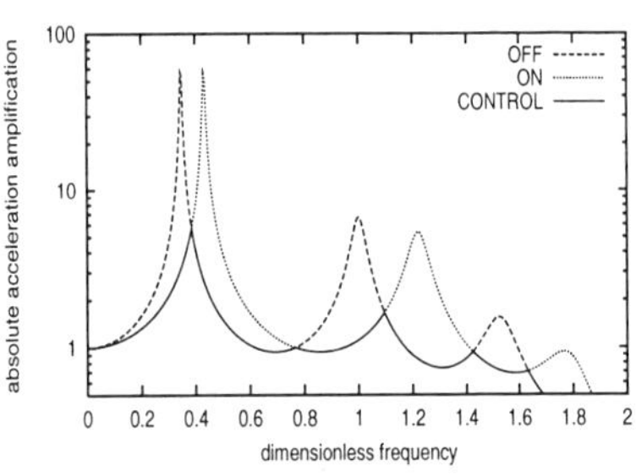

Figure 2: Spectral accelerations when the device is ON, OFF, and CONTROLLED.

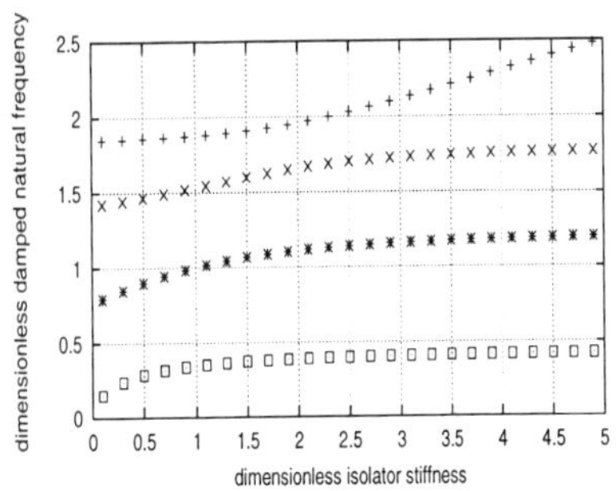

Figure 3: Effect of isolator stiffness on the damped natural frequencies.

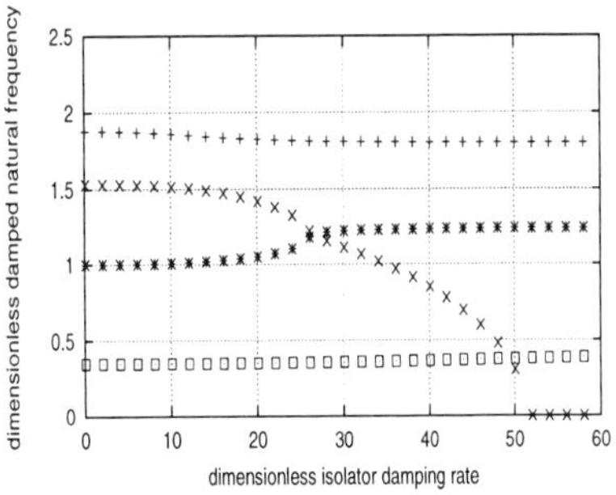

Figure 4: Effect of isolator damping on the damped natural frequencies.

resonant frequencies in the low modes can occur only through a modification in the isolator stiffness. The isolator damping does not significantly affect the lower mode frequencies. While the isolator damping can shift the second and third mode frequencies, the damping must be over twenty percent critical in those modes.

Response Times of Variable Stiffness Fluidic Devices

Variable stiffness devices can be designed using hydraulic cylinders and controllable by-pass valves. When the valve is closed, the oil column acts as a fluid spring. Figure 5 illustrates the importance of valves that can open quickly. The solid line represents the typical hysteresis of a controllable stiffness device with a fast (five millisecond) valve response time. If the valve is slow to open, the variable stiffness device does not accurately represent the ideal (fast) behavior. On the other hand, the valve can be slow to close without diminishing the device behavior.

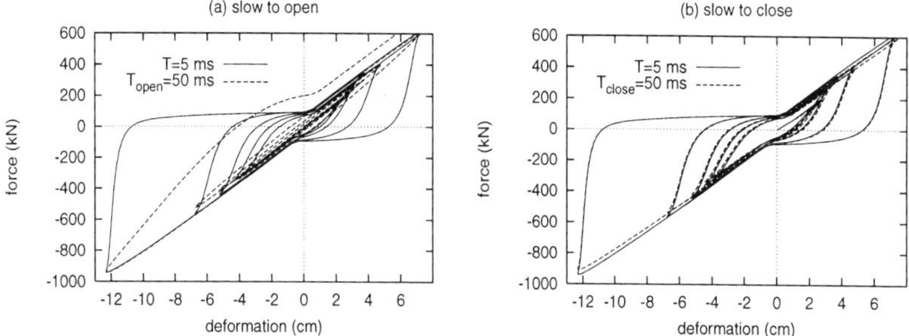

Figure 5: Controllable stiffness device hysteresis for a fast device and devices with valves that are slow to open (a) or close (b).

Acknowledgments

This material is based on work supported by the National Science Foundation, Earthquake Hazard Mitigation Program, under Award No.s CMS-9622177 and CMS-9624949, and by the Ford Motor Corporation. Any opinions, findings, and conclusions or recommendations expressed in this publication are those of the author and do not necessarily reflect the views of the National Science Foundation or the Ford Motor Corporation.

References

1. Dyke, S.J., B.F. Spencer, M.K. Sain, and J.D. Carlson (1998) "An experimental study of MR dampers for seismic protection," *Smart Materials and Structures*, vol. 7, pp. 693-703.

2. Gavin, H.P. (1998a) "Energy-based stiffness control for hysteretic structures," *Proc. 6th Nat'l Conf. on Earthquake Engineering*, Seattle WA, June 1998.

3. Gavin, H. P., and D. Hoang (1998b) "Construction of multiduct electrorheological dampers *Proc. SPIE*, Passive damping and isolation, San Diego, CA, March 1998, Issue 3327-19, pp. 214-225.

4. Imbimbo, Maura, and James M. Kelly, (1998) "Influence of material stiffening on stability of elastomeric bearings at large displacements," *J. Eng. Mech.* vol. 124, pp. 1045-1049.

5. Patten, William N., Q. He, C.C. Kuo, L. Liu, and R.L. Sack (1994) "Seismic structural control via hydraulic semi-active vibration dampers," *Proc. 1st World Conf. on Structural Control*, Pasadena, CA, pp. 83–89.

6. Nagarajiah, S., and D. Mate (1998) "Development of a semiactive continuously variable stiffness control device," *Proc. 12th Eng. Mech. Conf.* La Jolla, CA 17-20 May 1998, pp. 257-260.

STABILITY AND PERFORMANCE OF CONTROLLED
CIVIL ENGINEERING STRUCTURES USING MR DAMPERS

S.J. Dyke[*]

Abstract

One issue brought forth regarding the use of active control systems is in their ability to perform adequately when unmodeled or mismodeled dynamics are present in the system. In fact, instabilities may result when the control system is not robust enough to deal with such effects. Semi-active devices such as the magnetorheological (MR) damper are considered to be inherently stable because they do not have the ability to add energy to the structural system. However the performance of semi-active devices in such situations must be considered. Here the ability of the semi-active system to perform adequately is examined when various types of modeling errors are present.

Introduction

The development of effective and implementable structural control techniques has attracted a great deal of attention in recent years. Extensive research has been done in the area of active structural control, and these systems have been installed in over twenty commercial buildings and more than ten bridges (during construction). However, questions have been brought forth regarding the stability, cost effectiveness, reliability, power requirements, *etc* of active systems. For instance, active systems have the ability to input mechanical energy into the structural system, making them capable of generating instabilities. For actively controlled structures, researchers have shown that the controlled system may become unstable if a pure time delay is introduced into the feedback loop. Additionally Dyke, et al. (1995) demonstrated that instabilities may also arise if the dynamics of the control device, or the interaction between the device and the structure, are not properly accounted for in the control design. Furthermore, unmodeled dynamics or nonlinearities in a structure that has been damaged may result in instabilities.

Semi-active systems offer another alternative in structural control. Devices in this class have the ability to dynamically vary their properties, indicating they will be effective for a wide variety of disturbances. Additionally, they typically have low power requirements, eliminating the need for a large external power source. Moreover, semi-active devices are considered to be stable because they do not have the ability to input energy into the structural system (including the control device and the structure).

In this paper the stability and performance of a semi-actively controlled structure is examined. Through simulations, the controlled responses are evaluated for a variety of situations in which modeling errors are introduced. The types of modeling errors considered herein include: a pure time delay, stiffness reduction simulating damage, and an inverted feedback signal. A magnetorheological (MR) damper is used in these studies as the semi-active control device.

---

*. Assistant Prof., Dept. of Civil Engrg., Washington University, St. Louis, Missouri 63130.

870

MR Damper Modeling and Control

Prior studies using MR dampers have indicated that this device has a great deal of promise for civil engineering applications. A series of analytical and experimental studies was conducted to demonstrate the potential of MR dampers (Yi et al., 1998; Dyke et al, 1996b, 1998). The performance of the semi-active control system surpassed that of comparable passive systems. Further, experimental and analytical studies have demonstrated that MR dampers can provide forces of the magnitude required for full-scale structural control applications.

The phenomenological model of the MR damper developed by Spencer et al. (1996) is used in this study. The model was developed based on the experimentally measured responses of a prototype MR damper. The simple mechanical model is based on a Bouc-Wen model and accurately portrays the behavior of the device for time-varying displacement and control inputs. Through comparison with experimental results for a broad range of inputs, the performance of this model was found to be superior to that of various other models available in the literature.

In this study, to determine the control input to the MR damper, the clipped-optimal control algorithm proposed by Dyke, *et al.* (1996a-b, 1998) is used. The approach is to design a linear optimal controller that calculates a desired control force f_c based on the measured structural responses and the measured force f applied to the structure. Because the force generated in the MR damper is dependent on the responses of the structural system, the desired optimal control force f_c cannot always be produced by the MR damper. The voltage v applied to the current driver for the MR damper is determined based on the following control law

$$v = V_{\max} H(\{f_c - f\}f) \tag{1}$$

where $V_{\max}$ is the voltage to the current driver associated with saturation of the MR effect in the tested device, and $H(\cdot)$ is the Heaviside step function.

Types of Modeling Errors Considered

In this study three types of modeling errors are considered. These are described in the following paragraphs.

• <u>Pure Time Delay:</u> To examine the stability characteristics of the system, a pure time delay is introduced into the feedback loop. Thus the calculated control input voltage $v(t)$ is applied at $t + \tau$. The time delay was selected to be 0.046 sec, which is one quarter the fundamental period of the structure. For the first mode response, this means that the control signal determined when the displacement is a maximum is actually applied when the displacement is zero. For semi-active systems this will result in a worst case scenario, unlike active systems where the worst case time delay would be one half of the fundamental period (180 degrees out of phase with the response).

• <u>Structural Damage:</u> A second type of study is performed to assess the performance of the controlled system when degradation of the structure has occurred. For civil applications, a control system should be sufficiently robust to perform adequately in such a situation. To examine the performance of the controller after degradation has occurred, the controller designed using the original structure is applied to a model of the structure in which the stiffnesses are reduced to emulate damage, or softening, of the structure. Here, the stiffnesses was reduced by 25%.

• <u>Inverted Control Signal:</u> In this case the control signal is assumed to be inverted. Thus, the control signal applied to the MR damper is exactly the opposite of that determined by the control algorithm (e.g., when the calculated command signal is $v = V_{\max}$, the control signal applied is $v = 0$).

Numerical Example

The performance of the MR damper when various types of unmodeled dynamics are introduced is examined through a numerical example. Consider the three-story building controlled with a single MR damper, as shown in Fig. 1. The MR damper is rigidly connected between the ground and the first floor.

The structural model considered here is a simple model of the scaled, three-story, test structure, that has been used in various control studies at the Structural Dynamics and Control / Earthquake Engineering Laboratory at the University of Notre Dame. The parameters of the structural model and the MR damper are provided in Dyke et al. (1996a).

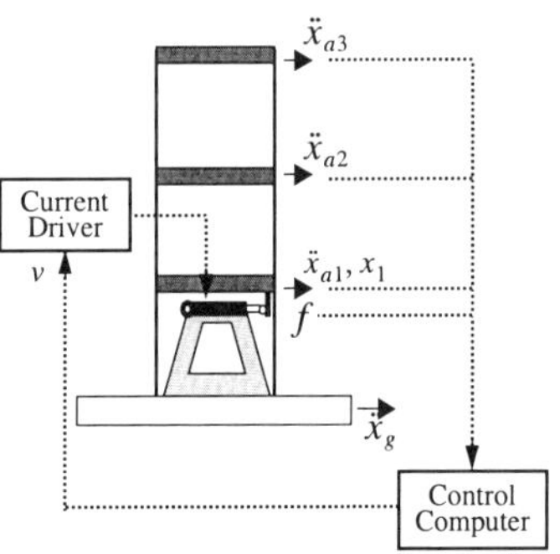

Figure 1. Diagram of MR Damper Implementation.

In simulation, the model of the structure is subjected to the NS component of the 1940 El Centro earthquake. Simulations were performed in MATLAB (1997). Because the structure under consideration is a scaled model, the earthquake was reproduced at five times the recorded rate. The maximum structural responses in each case are presented in Table 1. Here, x_i is the displacement of the ith floor relative to the ground, d_i is the interstory drift (*i.e.*, $x_i - x_{i-1}$), $\ddot{x}_{ai}$ is the absolute acceleration of the ith floor, and f is the applied control force.

Table 1: Peak Responses Due to a Scaled El Centro Earthquake.

Control Strategy	Uncontrolled	Delay (0.046)	Damaged (25%)	Inverted Command
x_i	1.883	0.701	0.694	0.554
(cm)	2.869	1.217	1.002	1.004
	3.369	1.574	1.241	1.250
d_i	1.883	0.701	0.694	0.554
(cm)	1.116	0.532	0.442	0.509
	0.704	0.361	0.336	0.275
$\ddot{x}_{ai}$	2992	1998	1640	1499
(cm/sec^2)	3608	2023	2186	1852
	4893	2505	1753	1910
f (N)	–	1706	1431	1607

Based on the results in Table 1, the semi-active system performs significantly better than the uncontrolled structure, even in the presence of significant modeling errors. All of the cases considered achieve a 55–65% reduction in relative displacements, a 32–64% reduction in absolute accelerations, and a 49–65% reduction in interstory displacements. The largest displacements and accelerations resulted when a pure time delay of 0.046 sec (one quarter the fundamental period) was introduced into the feedback loop. Even in this case the largest floor displacement is reduced by 55% (from 3.369 cm in the uncontrolled case to 1.574 cm), the largest interstory displacement is reduced by 58% (from 1.883 cm to 0.701 cm) and the largest acceleration is reduced by 48% (from 4893 cm/sec^2 to 2505 cm/sec^2). Thus, the semi-active control system using the MR damper appears to achieve a significant response reduction when unmodeled dynamics are present in the control system.

Conclusion

Through a series of numerical simulations the performance of the magnetorheological damper has been examined when various types of unmodeled dynamics are introduced into the system. A single MR damper was used to control a three story model structure and the responses were found for an El Centro earthquake excitation. In the first study a pure time delay was introduced into the feedback loop. Next, the semi-active control system was applied to a model of the structure in which the stiffnesses were reduced to simulate damage. In the final study, an inverted control signal was applied to the structural system, making the control input exactly the opposite of what the algorithm determined it should be. These numerical studies demonstrated that, even in the presence of significant unmodeled dynamics, the semi-actively controlled system performs significantly better than the uncontrolled structure.

The study presented herein represents some preliminary results on the stability of semi-active control devices. For further results, email Dr. S. J. Dyke at *sdyke@seas.wustl.edu*, or visit the web site at:

www.seas.wustl.edu/research/quake/.

Acknowledgment

This research is partially supported by National Science Foundation Grant No. CMS 97–33272.

References

Carlson, J.D. and Spencer Jr., B.F. (1996). "Magneto-Rheological Fluid Dampers for Semi-Active Seismic Control," *Proc. 3rd Int. Conf. on Motion and Vibr. Control*, Chiba, Japan, Vol. III, pp. 35–40.

Dyke, S.J., Spencer Jr., B.F., Quast, P. and Sain, M.K. (1995). "Role of Control-Structure Interaction in Protective System Design," *J. of Engrg. Mech, ASCE*, Vol. 121, No. 2, pp. 322–38.

Dyke, S.J., B.F. Spencer, Jr., M.K. Sain and J.D. Carlson (1996a). "Modeling and Control of Magnetorheological Dampers for Seismic Response Reduction," *Smart Materials and Structures*, Vol. 5, pp. 565–575.

Dyke, S.J., B.F. Spencer, Jr., M.K. Sain and J.D. Carlson (1996b). "Experimental Verification of Semi-Active Structural Control Strategies Using Acceleration Feedback," *Proc. of the 3rd Int. Conf. on Motion and Vibr. Control*, Chiba, Japan, Vol. III, pp. 291–296.

Dyke, S.J., B.F. Spencer, Jr., M.K. Sain and J.D. Carlson (1998). "An Experimental Study of MR Dampers for Seismic Protection," *Smart Materials and Structures: Special Issue on Large Civil Structures*, Vol. 7, pp. 693–703.

MATLAB. The Math Works, Inc. Natick, Massachusetts (1997).

Spencer Jr., B.F (1996). "Recent Trends in Vibration Control in the U.S.A.," *Proc., 3rd Int. Conf. on Motion and Vibr. Control*, Chiba, Japan, Vol. II, pp. K1–K6.

Spencer Jr., B.F., Dyke, S.J., Sain, M.K. and Carlson, J.D. (1996). "Phenomenological Model of a Magnetorheological Damper." *J. Engrg. Mech., ASCE*, Vol. 123, No. 3, pp. 230–238.

Yi, F., Dyke, S.J., Frech, S., and Carlson, JD. (1998). "Investigation of Magnetorheological Dampers for Earthquake Hazard Mitigation" *Proc. of the 2nd Intl. Conf. on Structural Control*, Kyoto, JAPAN, June 30–July 2.

Temporary Structures: Role of the Design Engineer

By: John A. Frauenhoffer, MASCE

Introduction

In 1995, a major improvement to a wastewater treatment plant located on the banks of the Ohio River in Pennsylvania was planned. Flows to this plant could exceed 200 million gallons per day, and proper operation of the plant has a large impact upon the Ohio River ecology and the use of the river as a raw water source for downstream communities. The improvements required deep excavations in very difficult subsurface conditions. Because of the nature of the plant site materials and the critical importance of the plant infrastructure relative to the Ohio River ecology, the civil engineering team retained to create the Capital Improvement Plans made a careful reassessment of the Role of the Design Civil Engineer regarding Temporary Structures.

Plant Site Geology

The plant site is essentially a landfill which lies on the banks of the Ohio River with high bluffs on the opposite side of the river. The site was fairly level and the surface is partially covered by grass, with the majority covered with impervious surfaces such as concrete and asphalt. The normal pool of the Ohio River was approximately eight meters below the plant site ground surface, controlled by a downstream Corps of Engineers operated dam. Several times in a year, the river had risen to within a meter of the plant site, and the plant site had been flooded in the past.

[1]President, Frauenhoffer and Associates, Champaign IL

Underlying the plant site and the river was bedrock, primarily composed of limestone and claystone. Above the rock was a layer of river gravel overlain with an alluvial clay layer. On top of the clay layer was approximately seven meters of cinder and rubble fill, the cinders being the product of past steel mills previously operating in the area. There were old foundations below the surface, as the site was previously occupied by foundries and residences. Traces of bricks, wood and metal were found in many of the boring samples. Utilities consisting of power lines, telephone lines, sewer lines, gas lines, pressurized force mains, surcharged process piping, surcharged reinforced concrete conduits, and storm drains ran throughout the plant site. Most of the plant buildings were constructed on piles. Tanks were constructed on one meter thick slag fills. The cinder fill was very loose. Standard Penetration Test blowcounts of zero or one occurred frequently.

The Design Dilemma

Historically, Design Civil Engineers have preferred to not participate or interfere with contractors' choices of temporary structures, such as sheeting and shoring for excavations. There have been many motivations for not participating, including not wishing to dampen a contractor's economic advantages in search of the lowest price for the Owner, eliminating the design cost of temporary structures from the design engineer's fee, and admonitions to abstain from temporary structure design from professional liability insurance underwriters. As a practical matter, often the contractor's choice of sheeting or shoring has sufficient capacity or excess capacity to reduce lateral loads.

The geotechnical engineering problem in this project was the magnitude and variability of the lateral pressures which could be applied to temporary excavation structures. Due to the loose condition of the cinder fill, lateral pressures exerted from the cinder fills alone would be relatively high. Further, groundwater levels could rise overnight.

The Ohio River Pool Stage dictated plant site groundwater levels, since the cinder fill was very permeable. Installed sheeting and shoring could reasonably expect groundwater level lateral pressures ranging from zero to seven meters of head during their short life.

The Analysis of the Engineering Problem

The engineering team defined the following problems with contractor designed sheeting and shoring:

1. There was no good data on the condition of the cinder fill. It was extremely variable. Standard Penetration Test blowcounts on fill material can be misleading. There were numerous indications of buried debris, which can influence blowcounts. Should the contractor or the contractor's shoring designer rely on data from the soil borings given to the bidders, the design may not be adequate.

2. The shoring was expected to experience groundwater shock loads, which is anything but common. Deep excavations would be located within seventy meters of the Ohio River. Groundwater levels on the plant site essentially rose and fell with the River Stage with little to no time lag. Sheeting and shoring, installed to allow construction to proceed within its boundaries, could fail overnight due to swift river stage rises.

3. Failure of deep sheeting and shoring would likely damage buried power ductbanks, process piping and conduits, natural gas lines, and adjacent foundations. Such damage could easily lead to the forced bypass of raw sewage to the Ohio River, which would not have been palatable to any regulating governing body, downstream municipalities, or the public.

4. The potential economic loss from failed sheeting and shoring was huge, likely causing severe financial distress to the Owner and the Engineer after exhausting insurance coverage limits.

The Role of the Design Civil Engineer

The engineering team chose to reject the see no evil, hear no evil, speak no evil engineering management philosophy relative to the design of sheeting and shoring. It was the opinion of the civil engineers that due to the special nature of the site and due to their superior knowledge of the risks of excavating on the site, the best choice was to design and detail the temporary structures and include them as a part of the construction contract. With this approach, arguments with a successful bidder regarding the actual lateral pressures would be eliminated, the Owner would have the same reliability with the temporary structures as with the permanent structures designed and detailed in the contract documents, and the ecology of the Ohio River would be protected to a reasonable degree of professional engineering certainty.

Recommendations: Role of the Design Civil Engineer

Based upon the author's experience as the Geotechnical Engineer of Record for this project, the following recommendations for the Role of the Design Civil Engineer in projects which include deep excavations are presented:

1. Assess the commonality of the subsurface conditions. Odd conditions should prompt more detailed involvement of the Design Civil Engineer in temporary structures.

2. Assess the load action against temporary structures. If the loads can abruptly increase by multiples of two or more, more detailed involvement by the Design Civil Engineer in temporary structures is warranted.

3. Assess the potential economic and ecological consequences and increase involvement appropriately.

Structural Rehabilitation of Operating Industrial Facilities:
An Introduction

Merle E. Brander, P.E., ASCE

Overview

Unlike construction of new facilities that are unoccupied until substantial completion, structural modifications often require that work be done while operations continue in the facility. While there are risks associated with any building project, in new construction or rehabilitation of an unoccupied structure, the on-site workers are responsible for, aware of, and prepared for safe operation in a construction environment.

The risks associated with rehabilitation of operating industrial facilities, however, are greater than normal because the occupants are people driving lift trucks, delivering papers, meeting, eating lunch, etc., in the facility. These people prepare for the risks associated with their specific jobs, but not necessarily for the risks associated with construction. For that reason, rehabilitation work must be done with the assumption that the building is always ready for final occupancy, with personnel and equipment in place and operating.

Approaching the Operating Industrial Facility

The building sciences have evolved over the centuries, allowing people trained in construction related activities to complete greater and more complex construction projects, and the structural rehabilitation of an operating industrial facility may be the most demanding task faced by today's construction community. Such rehabilitation requires continual accommodation of the non-construction workers within the facility and, frequently, management of harsh physical conditions.

When a building owner decides to rehabilitate an industrial facility, through modification, repair, or demolition, his primary concern will understandably be the operations of the facility. Certainly he will recognize the importance of maintaining a sound structure both during and after the rehabilitation, but he will generally not realize the number of variables that can affect structural integrity, rehabilitation costs, facility service life, and the safety of personnel both during and after rehabilitation. With rehabilitation of operating facilities, variables not clearly defined and addressed can be the origins of failures of existing structures, temporary structures, or new structural components.

Planning Facility Rehabilitation

During each stage of a rehabilitation project, the strength, stability, and durability of all affected structures must be understood and accounted for if the facility is to remain safe and fully operational. Ultimately, it is the owner who will bear the consequences of the project, both positive and negative, but the owner generally does not have the knowledge or training needed to plan a rehabilitation project in detail. He needs someone to take responsibility for ensuring that buildings, platforms within buildings, and access structures are free of hazards; when new building configurations are being proposed, he needs someone who can design to account for the required strength, stability, and durability; and when the components of the structure are removed or altered to make way for new building configurations, he needs someone who understands how loads are redistributed.

Effective rehabilitation of operating industrial facilities requires the supervision of a Qualified Structural Engineer at all phases. It is generally understood that Competent people can construct a project, choosing the means and methods of construction and consulting a Qualified person when unknown loading conditions arise. However, in the case of rehabilitation of operating industrial facilities, the level of analysis required to achieve acceptable levels of predictability and safety increases dramatically. The structural and operational integration of the work site with other parts of the facility increases the base level of risk associated with the work, and the rational, moral response of the owner would be to proportionately augment risk-reduction measures by placing a Qualified Structural Engineer in charge from beginning to end.

Risk and Responsibility

All too often, though he shoulders the increased risk, the owner does not commensurately enhance his strategy to prevent accidents and failures. He may attempt to do so by paying a high price for rehabilitation services, trusting that the high cost of the services will lead to an adequate factor of safety, but unless the *strength*, *stability*, and *durability* of each structure is understood for each stage of the project, the risk of damage or injury remains high. Alternatively, a contractor performing the rehabilitation without the benefit of specifications for Means and

Methods from a structural engineer may provide an unnecessarily high degree of support for some parts of the building, leading to higher costs and lost time for the owner.

Planning for the redistribution of load, and designing the new structure accordingly, is the most critical aspect of the rehabilitation of an operating industrial facility. The owner may not understand the forces at work within the structural members of his building, but still he bears the responsibility of their performance. If he unwittingly authorizes the modification of his facility structures by a person who does not fully understand load redistribution, he sets himself up for costly failures. The owner can not safely redirect his attention to facility operations until he transfers both the responsibility and the authority for structural modification to a party that can properly accept them.

The Structural Engineer has the professional liability insurance as well as the professional training that enable him to confidently direct the rehabilitation of an operating industrial facility. With so many more variables to consider during the work, as compared with new construction, this is an excellent example of a project requiring the input of a Qualified Structural Engineer regarding Means and Methods.

Conclusion

Building science has evolved to the point where ordinarily high-risk, high-cost rehabilitation projects can be executed quickly, safely, and without interruption of on-site industrial operations. An increased number of relevant calculations and structural analyses, resulting from an expanded role for the Structural Engineer in structural modification, leads directly to less downtime, fewer accidents, and lower overall costs.

Demolition of a Pulp Mill in an Operating Paper Facility

Merle E. Brander and John J. Jeanquart

Introduction

If you would ask people on the street what a structural engineer does, the most frequent response would probably be, "I don't know". A few people may know that structural engineers design buildings or additions or modifications to buildings; but you would seldom hear that structural engineers design for demolition of buildings. Yet, it is crucial that demolition, particularly demolition of a structure within an operating facility, include structural engineering to establish the loading and proper sequence of demolition to prevent unplanned collapse that could endanger workers, damage adjacent structures and equipment, or interrupt mill operations.

Setting The Stage

A major integrated papermaking facility, consisting of a pulp mill, a paper mill, and converting operations, was constructed on a site in various stages beginning in 1916. As the output of the company increased over time, additional structures were added to the site, resulting in a facility with partially intermingled processes and operations. By the time the company ceased on-site pulp operations several years ago, several pulp mill buildings also were serving secondary functions unrelated to the pulp operations. For example, one of the buildings housed the facility's main water pump and piping; some retired pulp mill buildings contained components of the electrical system that served other still active buildings. In addition, some structural members of the pulp mill buildings still served a primary function, as they were shared with other still active buildings.

When demolition was being considered, the size and complexity of the project, the number and variety of equipment inside the buildings, the interior location of the buildings, the lack of physical separation, and the incomplete operational separation convinced the Owner that the demolition proposed required the attention of a structural engineer. The costs associated with shutdown of operations in the vicinity of the proposed demolition would have been prohibitive;

the only reasonable course of action was to conduct the demolition without interrupting the operations in surrounding buildings.

Defining The Problems

The demolition project proposed involved 10 buildings for a total of approximately 39,000 square feet. More than half of the perimeter of the demolition area involved walls that were shared with other structures that were to remain and continue operating. When demolition was to be undertaken, not only was it necessary to assure the structural integrity of all the other buildings sharing those walls, but also to assure that the environmental conditions within those buildings would remain appropriate for personnel functioning in those buildings, including specifically maintenance of fire protection and escape and firefighting access. In addition, with the pulp mill housing electrical systems and water systems on which operations in other buildings were relying, those had to be adequately addressed if operations were to continue uninterrupted.

Because the pulp mill buildings were constructed before hazards associated with various materials were known, consideration had to be given to the likely possibility that the buildings would contain lead paint, asbestos, and other materials commonly in use at the time of construction. It was also necessary to account for other hazardous materials, such as sulfur, ammonia, and lead in mortar inside digesters and accumulators, known to be common in industrial facilities, and for other existing conditions, such as the presence of underground piping, that could pose a hazard or interrupt operations if encountered.

Pre-Demolition Activities

Before demolition activity began, we met with the Owner representative to discuss the project scope and likely cost. A demolition sequence had been proposed as part of Condition Survey of the Pulp Mill buildings done by Brander in 1996. That involved repairs that had already been done, as well as the following:

- A survey to determine if hazardous materials had been used in any of the pipes, tanks, or other equipment on the property or building, and tests and purging to eliminate any hazards found or suspected.

- A checklist of items within the pulp mills to be removed or relocated prior to demolition, such as fire suppression, electrical, automation, and water piping systems, and development of a plan for the removal or relocation.

- Drawings and specifications for demolition of the existing buildings and structural components and for severing the buildings scheduled to be demolished from the adjacent buildings that were to remain intact.

- Demolition of the Pulp Mill buildings and structural components, following all applicable OSHA safety requirements.

The 1996 Condition Survey of the pulp mill buildings served as the OSHA required survey by a competent person. Since mill personnel had already taken action to assure the stability of the structures shortly after issuance of the Condition Survey report, it was only necessary to verify that conditions remained as previously noted to be assured of the structural integrity of the buildings being addressed.

Specifying The Demolition

It was determined by the Owner's engineer and Brander that the demolition project would be done under a single prime contract with a general contractor, with one of three listed prequalified demolition contractors serving as subcontractor to the general contractor. In the specifications, the project was defined as follows:

1. Demolition of existing Pulp Mill Buildings, involving numerous buildings, most of which are steel-framed structures with masonry walls, cast-in-place concrete slabs, and precast concrete, steel, or wood roof decks and smooth surface asbestos-containing builtup roofing systems. The buildings are at various levels and collectively contain digesters, pressure accumulators, stone towers, and numerous other structures and equipment, such as chests, tanks, chip bins, elevators, conveyors, mechanical equipment, conduit, piping, etc.

2. Demolition of peripheral concrete tanks, tank foundations, containment structures, and peripheral buildings constructed primarily of masonry walls with concrete foundations, all of which are to be assumed to be deep foundations.

3. Detachment of buildings being demolished from existing buildings that are to remain, and reinforcement and closure of those existing buildings, with closure to include construction of a new wall, construction of new roof framing and decking, installation of bracing for existing masonry walls, infilling of all penetrations through masonry walls, installation of new doors, and installation of platforms.

The Owner accepted responsibility, either in-house or under separate contract, for state-required asbestos assessment and removal, for identification and relocation of all active electrical or mechanical lines or equipment, and for emptying all chests and tanks. Contractors were advised to include in their bids the costs for dealing with lead paint. The specifications were prepared with disclosure of all known hazardous materials and contractors were offered the opportunity to tour the buildings to determine the types and numbers of items requiring special handling and disposal. Since even with the best prior planning, there are things that cannot be known until they are uncovered, there was a provision in the specifications for unit price payments for truly hidden conditions.

Each set of drawings included 15 demolition drawings and 20 structural drawings. The demolition drawings included plans of each floor to be demolished, as well as a cross section to identify the structures, equipment, and elevation changes, based on which volumes and areas could be calculated for bidding. The structural drawings included a sequence of structural reinforcement for walls of buildings that were to remain. Structural drawings also included details and plans for new roofs and walls to be installed after demolition, door and wall opening infill schedules, and windows for exterior walls.

Demolition

To facilitate coordination of the project and keep the Owner informed about the project, weekly meetings were held. During the meetings, questions could be answered as they arose so that work could proceed uninterrupted. The demolition work proceeded safely and on schedule, with only minor delays resulting in a demolition project that turned out to be less troublesome than a routine construction project.

Cost Considerations

The owner had initially requested funding for the demolition based on a cost estimate from a local contractor, provided without the benefit of drawings and specifications. After drawings and specifications were prepared and were used in the preparation of lump sum bids, it was found that the project cost was approximately 40% lower than originally expected, a substantial savings considering the size of the project.

The Role Of The Structural Engineer

The institutional role of the structural engineer developed when architects could no longer rely on a craftsman or builder to use and understand the properties and performance capabilities of the materials that were being produced. To have a builder take a building down and reconstruct adjacent buildings without being qualified to evaluate the support conditions and understand the integration of the materials needed for reconstruction exposes the building Owner to unnecessary risks and costs. A qualified structural engineer can reduce the risk and cost, allowing for demolition and rehabilitation even as the Owner's facility continues to operate.

CHALLENGES CREATED BY THE HIERARCHY IN FACILITY REHABILITATION PROJECTS

By Kenneth B. Simons[1], Member ASCE

Introduction

During the latter portion of December, 1996, heavy snowfall accumulated in western Washington state. The snow changed to rain as temperatures rose on December 29, 1996, at which time there was almost two inches of rainfall. This rain became confined within the snow and ponded on deflected roof surfaces. Most susceptible were flat (low sloping) roofs where the water was confined and/or obstructed to the roof drains or the perimeters. One of the structures our office investigated is a covered moorage facility.

Description of Covered Moorage Facility

The subject 56' wide (east to west) by 268' long (north to south) covered moorage facility was entirely covered by a roof structure constructed in 1981. The roof structure is supported by 16" diameter timber pilings. Glued-laminated wood beams extend across the tops of the pilings, and additional glued-laminated beams extend between every other piling from east to west. The glued-laminated wood beams at the east and west sides support pre-fabricated wood trusses. The roof trusses are covered with 3/4" thick plywood, which in turn is covered with a fluid-applied roofing membrane. The bottoms of the trusses are located approximately 20' above the water level. A short fascia (facade) surrounds the roof of the subject pier, which is braced at the bottom by 2 x 4 wood members extended horizontally beneath the glued-laminated wood beams.

Repair Design

Our office was initially retained by an insurance company to determine the cause and extent of the collapse. We were asked by the covered moorage facility owner to provide the necessary repair design/drawings to repair the structure. The design replaced the existing collapsed and damaged trusses with new open web joists that consist of tubular steel members extended diagonally up and down to connect the wood top and bottom chords of the trusses. The exterior fascia was shown to be reconstructed in general conformance with the original construction with 1

1   Principal Engineer, Damage Consultants, Inc.,
3003 Island Crest Way, PO Box 1336, Mercer Island, WA 98040-1336

x 8 cedar clapboard siding over building paper and plywood. The roof sheathing was to consist of 3/4" thick tongue and groove plywood, which in turn, was covered by a new built-up roofing. Roof drainage was to be collected by an external gutter at the downhill side of the roof. A 4 x 6 was shown to be provided above the existing east/west glued-laminated wood beams, tapered ("ripped") as required to conform to the camber of the new 33" deep open web truss joists.

The Hierarchy

The moorage facility was managed by a condominium association of boat owners. Our office met with the condominium association and discussed the proposed repairs and recommended that the repairs commence on the northern (land) side of the pier, such that a work platform could be constructed by erecting several trusses at a time, then installing the braces and continuing to the south. It was also recommended that the short wood framed fascias could be constructed on the roof surface once the roof was constructed and then hung from the ends of the trusses as necessary. Questions concerning bird infestation were presented and it was recommended by our office that the individual trusses could be wrapped with netting on the surface of the roof prior to being installed in order prohibit birds from nesting in the trusses. The contractor selected by the covered moorage facility was experienced in construction of marinas and covered moorage facilities. The covered moorage facility also retained a roofing consultant and a separate consulting engineer to provide input.

The Challenges

1) The first challenge was the high cost of the repairs which although the engineer of record had no financial interest in keeping the cost of the repairs down, it was our opinion that the repair cost was excessive for this type of roof. When discussed with the contractor and owner about commencing the work from the northern land side and constructing outward to the south, it was stated by the contractor that they could not nail the bottom bracing of the trusses from above, and thus would have to work from below from a barge, and since they already had the barge mounted crane available to do the work, this would be simpler and thus they would work from the south to the north. Construction commenced in October, 1997 and was not completed until February, 1998.

2) The second challenge was that on the repair drawings, it states that, "The contractor and truss fabricator shall verify all dimensions of the existing covered moorage prior to fabrication of the trusses." Subsequently, we received a request by the truss manufacturer to, "Please specify exact dimension for the clear span of the trusses". We reiterated the requirement on the drawings and it was our understanding from the reports from the contractor that the dimensions shown by the truss manufacturer were correct.

3) The third challenge was that the contractor discovered after commencing construction that the locations of the pilings from one side to the other were skewed slightly. Our office stated that this did not matter because we were not using the beams to carry roof loads, they were only intended to connect (tie) the two opposing piers to each other and thus, could remain in a skewed condition. We had anticipated that the tops of the beams would not conform to the cambered condition of the new open web truss joists. Thus, the drawings showed a continuous 4 x 6 above the existing glued-laminated beam ripped to conform to the camber of the truss joist. The contractor proposed providing 2 x 4 wood members (struts) between the new trusses to span the existing east to west glued-laminated beams. The contractor stated that this system would not cost any more than installing the 4 x 6 on top of the glued-laminated beam as shown on our drawings.

4) The fourth challenge was that an internal gutter system was constructed by the repair contractor which required a 4" high curb (cant strip) at the low end of the roof structure. Since the roof structure has a downward slope to the east of 1/2" in 1', we warned the owner that this will result in water ponding on the lower portion of the roof, between 4' and 8' west of the low portion of the roof if there was any malfunction in the drains, including snow or ice dams (similar to what previously happened). Ponding water on a roof membrane is discouraged and typically roofing manufacturers void the warranty if water ponds on the roof. In addition, the weight of the water (five pounds per square foot per each inch in depth) which may pond on the roof had not been included in our design.

5) The fifth challenge, although it did not present a concern for the support of the roof structure, was that we stated that care should be exercised in the selection of the metal cladding system such that it can withstand winds in addition to the application of an appropriate barrier

behind the siding to allow water (which is driven by the wind between the joints of the metal cladding and condensation which forms on the back side of the cladding) to drain at the bottom and not infiltrate the plywood and framing. The metal siding was installed as described by the repair contractor at the owner's request, without review from our office.

The Aftermath

A follow-up discussion with the facility owner indicated that the contractor's proposal to install the 2 x 4 struts across the beams instead of providing the blocking above the beams as shown on our drawings, resulted in an approximate $40,000 change order. This additional cost was inconsistent with the contractor's statement that there would not be an additional cost to provide the struts. The owner also stated that water was ponding adjacent to the curb at the low end of the roof and that their roofing consultant concurred with our opinion as to the problems inherent to the internal gutter system constructed by the contractor.

Conclusion

The Engineer of Record is the only entity in the project who legally has a duty to protect the safety and welfare of the public. This not only means the individuals who may be using the covered moorage facility, but also the contractor's workmen. The majority of thought that is exercised by the Engineer prior to the selection of design of the repair or rehabilitation includes the anticipated means and methods of the reconstruction of the facility for the contractor in order to facilitate construction and limit the risk to the workmen involved and reduce the risk of potential failure during construction. Allowing the contractors to conduct business as usual only increases the cost of construction and may be increasing the risk of safety to the personnel. The passivity of the engineer of record on rehabilitation projects relegates him to a low (if not the lowest) position on the hierarchical order of the construction project. The insistence of liability insurance carriers to have designers avoid the proposed means and methods of construction also relegates the engineer to a lower position.

In order to attain higher quality, safer buildings with less risk to the public, that is, both workmen and occupants of rehabilitation projects, it is imperative that the engineer of record insist upon an active role in the means and methods, material selection, the bidding process and the safety aspects of the project. This can only be done through education of the clients or the public to elevate the stature of the engineer.

espace

More Efficient Aluminum I Beam Shapes

by

Randy Kissell, P.E., Member ASCE[1]

Abstract

Designers of aluminum I beams have tended to use the same cross sectional shape as steel I beams, that is, wide flange, or W shapes, and American Standard, or S shapes. Aluminum beams are usually extruded, however, while steel beams are hot-rolled, a method of production that limits variations on the shape much more so than extruding. Even so, designers have been slow to take advantage of the variety of extruded shapes that are possible. Around 1970 flat flanges began to replace tapered thickness flanges for aluminum I beams, but standard aluminum I shapes have not changed since then.

This paper investigates alternative shapes that may be more efficient than current standard aluminum I beams. One of the obstacles in designing such shapes is the difficulty in analyzing them, since they tend to be more complex than standard shapes composed of simple rectangular elements. To overcome this, the finite strip method was used to determine the elastic local buckling strength of the alternative shapes.

Background

When aluminum first came into use in civil construction, aluminum was treated as if it were steel that just happened to have a low modulus of elasticity. Aluminum I beam and channel shapes were dimensionally identical to their steel counterparts, and had sloped flanges just like hot-rolled steel shapes. The sloped flanges complicated detailing and design and made bolted connections inconvenient at best. They were a necessary evil,

---

[1]Partner, The TGB Partnership, 1325 Farmview Rd., Hillsborough, NC 27278

however, a by-product of the hot-rolling production process by which the release of the rolls from the part was made possible.

A major difference in the two metals, however, is that aluminum can be extruded. Extrusions are made by heating a cylinder of aluminum and then pushing it through an opening called a die which outlines the shape of the desired cross section. A wide variety of shapes can be very economically produced by the process, including hollow shapes and shapes with curved elements, and no draft or taper is necessary on the parts. It was not until around 1970, however, that the Aluminum Association, an association of aluminum producers, introduced standard extruded I beam and channel shapes. These shapes had constant thickness flanges and constant thickness webs and were specifically designed to optimize structural properties when made of 6061-T6 aluminum, a popular alloy/temper for structural applications. No changes have been made to these shapes since their introduction nearly 30 years ago.

The Aluminum Association's *Specifications for Aluminum Structures* provides rules for the structural design of aluminum members. Since extruded members can have such a great variety of cross sections, designers cannot assume a given extruded shape is compact (using the terminology of the AISC *Specification for Structural Steel Buildings*) for which local buckling will not limit the strength of the member. Thus the design flexibility associated with extrusions comes at the price of additional required design checks. Not only that, but while the Aluminum *Specifications* include rules for determining the local buckling strength of shapes when they are subjected to bending or axial compression, these rules are necessarily limited to a finite number of cases such as constant thickness elements supported on one edge (flanges) or both edges (webs). No rules are provided for elements with varying thickness or multiple intermediate stiffeners, to name just a few. Finally, the buckling strengths of elements in the Aluminum *Specifications* are based on idealized plate buckling equations, which do not account for interaction between elements and assume pin-ended support conditions at the elements' edges.

The Finite Strip Method

It is precisely the ability to extrude customized shapes, however, that gives aluminum designers the potential to use more efficient shapes than can be provided in any other metal. This potential could be fully realized if simple tools were available to calculate the local buckling strength of arbitrary shapes.

While there are a number of methods that can be used to analyze shapes, including the finite element method and testing, the finite strip method has the advantage of providing fast results, given the limitation that only simple supports can be modeled at the ends of the member. To perform a finite strip analysis, the member is divided into a collection of strips which run the length of the member; the results give elastic local, distortional (if any), and overall member buckling strengths. The finite strip method as used here only predicts elastic buckling strengths and does not consider residual stresses and imperfections that would be factors in determining inelastic buckling strength, nor does it account for post-buckling strength. The webs of the larger Aluminum Association standard I-beams, however, are slender enough that they buckle elastically. Also, the finite strip method can be used to compare the elastic buckling strength of various shapes on an equal basis.

A More Efficient I-Beam Shape

Except for short, deep beams carrying shear loads, I-beam shapes benefit from placing as little material in the web as possible. By moving material away from the midheight of the web, both axial and bending compression performance is enhanced: for axial compression, the shape has a larger major axis radius of gyration, tending to increase column buckling strength, and for bending compression, the shape has larger section moduli. The approach used for the Aluminum Association standard I-beams (and channels) has been to make the web's aspect ratio as slender as possible within certain limitations. One constraint is that very slender elements are difficult and thus more costly to extrude; none of the standard I-beams has a height to thickness ratio greater than approximately 40, since it is difficult to push metal through such a narrow opening in the extrusion process. Also, a certain minimum thickness and maximum slenderness ratio are desirable to avoid shipping and handling damage.

Within these constraints, however, more efficient shapes can be achieved by varying the thickness of the web over its height, using a thinner web section near the neutral axis and thicker portions near the flanges. The variation of web thickness can be smooth or stepped. A stepped variation was used on the Aluminum Association standard I-beam designated I 12 x 11.7, 12" deep weighing 11.7 lb/ft, with a 7 in. wide, 0.47 in. thick flange and 0.29 in. thick web. The standard shape has a web slenderness ratio of 38.1, neglecting the effect of the web-to-flange radii for simplicity. In the revised shape, the web was divided into three parts over its height: two parts nearest the flanges 1.29 in. long and 0.47 in. thick, and one part at the midheight 8.47 in. long and 0.235 in. thick. To dimension the revised web, three conditions were set: 1) the area of the web of the revised shape is equal to the area of the original web; 2) the thin portion of the web is one half as thick as the thick portion of the web, and 3) the thick portions of the web are the same thickness as the flanges. The flanges are the same for both shapes.

The Aluminum Association standard shape and the revised shape are compared in the table below. Local buckling strengths are for axial compression and determined by the finite strip method.

Parameter	AA Standard Shape	Revised Shape	Improvement
Major Axis I (in^4)	251.5	260	3%
Major Axis r (in.)	5.07	5.15	2%
Local Buckling Strength (ksi)	36.8	37.9	3%

The local buckling strength computed by the Aluminum *Specifications* is 24.6 ksi and assumes the web's edge supports are pinned. The finite strip-calculated strength is about 50% greater (36.8 ksi) because it models the interaction between the web and flange; since the flange is considerably thicker than the web, the flange actually provides greater edge support to the web, somewhere between pinned and fixed (74% greater).

Greater improvements are possible by making the thin portion of the web thinner and the thick portion of the web thicker. If the thickness transition is too abrupt, however, extrusion costs will increase. By limiting the transition to 2:1, the transition is no greater than that employed between flanges and webs in standard shapes.

Additional Considerations

This example considered only the elastic buckling strength. Further investigation including consideration of inelastic and possibly postbuckling strength would be useful, especially since many aluminum shapes have webs that buckle inelastically. Testing would be particularly helpful in establishing inelastic buckling strength because this may be the only way to determine the dimensional imperfections and residual stresses due to unequal cooling associated with thickness transitions so that their effect can be evaluated. Another consideration is bending compression; here the flange usually buckles before the web. Even so, concentrating web material near the flange increased the flange local buckling strength by 6% in this example by increasing the flange edge fixity at the flange-web juncture. Finally, varying the thickness of webs could complicate bolted connections there, possibly requiring filler plates. Limiting the variation to a step relatively near the flanges, however, mitigates this concern.

Summary

The local buckling strength and section properties of standard extruded aluminum I-beams and channels can be improved by varying the thickness of their webs over the depth of the shape, using thinner web portions near the neutral axis. The finite strip method can be used to determine the local buckling strength of such members.

Aluminum Gusset Plates in Tension

Dr. Craig C. Menzemer[1]

Dr. Tirumalai S. Srivatsan[2]

Abstract

An experimental and analytical program was conducted as a means to study block shear failure of aluminum gusset plates in tension. A total of twenty-three plates representing four different bolt patterns were mechanically deformed. Mathematical models used to estimate joint capacity were examined and compared with experimental results. In this study, finite element models were developed and scanning electron micrographs of failure surfaces recorded. The highlights of the model are presented and discussed.

Introduction

Plates are often used as connection elements in aluminum structures. Examples include railcars, latticed roofs, intermodal containers, bridges and automotive systems. Many connections utilize mechanical fasteners. Reliable and efficient connection design necessitates an examination of bolt or rivet failure, bearing distress of connected material, net-section tensile rupture and tear out of the fastener group

There is an acute paucity of scientific research on mechanical connection block failures made from aluminum alloys. Sharp in 1993 provided and rationalized data from a restricted study on the behavior of extruded angles attached by a single leg [1]. Extruded 6061 (Al-Mg-Si)-T6 (maximum strength temper) angles were connected at the ends by varying numbers of 2024 (Al-Cu-Mg)-T4 bolts and mechanically deformed in tension to failure. As expected, the failure mode changed as a function of the number of fasteners. Angles connected by a single bolt failed by combination of shear and bearing of the angle. Those connected by two bolts failed by block shear, while those utilizing three or more fasteners failed through the net section by tensile rupture.

Marsh has reported on one of the larger bodies of experimental data on the tear out of

[1] Assistant Professor Civil Engineering, [2] Professor, Mechanical Engineering.
The University of Akron, Akron, OH 44325-3905

bolt groups [2]. A model was developed and put-forth to estimate the capacity of bolt groups. The model was based upon data obtained from tests of double lap joints fabricated from 6063 (Al-Mg-Si)-T6 (maximum strength) sheet. Sixteen tests were conducted on samples having different geometry. The extrinsic variables included: (i) edge distance, (ii) gage spacing, and (iii) pitch. The resulting mathematical model to estimate joint capacity was a function of: (a) bolt group periphery, (b) sheet/plate thickness, and (c) ultimate tensile strength of the base material. Strength predictions generally accorded well within several percentage points of test values. The only significant variation from model prediction was those joints that failed along a single fastener line.

Experimental Program and Results

A total of twenty-three gusset plates were fabricated. Four different specimen types were evaluated, and are exemplified in Figure 1. The variables included: (a) joint length, (b) fastener gage spacing, and (c) alloy chemistry (composition). Twenty samples were manufactured from 6.25 mm thick plate of aluminum alloy 6061-T6 plate. Three specimens were fabricated from 6.25 mm thick plate of alloy 5083-H321. All samples were evaluated using 15.9-mm diameter, high strength steel bolts. Grade 8 fasteners were used instead of standard A7 or A325 bolts since the SAE variety provides for longer thread lengths and greater flexibility in test set-up. Threads were excluded from the shear plane.

Figure 2 shows the general load – displacement behavior for the gusset plate samples. Bolts were brought to a snug-fit condition prior to the start of each test. Faying surfaces were left in the as-fabricated condition. As such, bearing was the predominant mechanism for load transfer. A majority of the load – displacement records revealed no horizontal plateau during loading, characteristic of a sudden slip into bearing. Rather, the test records showed a gradual concave shape during the early stages of testing. This behavior is indicative of slack removal from the load train and a progressive transition into bearing.

All samples exhibited tensile ligament failure. Rupture occurred between the bolt holes in the uppermost fastener row. Further inspection revealed a thinning of the plate between the fastener lines. Necking may be construed as an orange peel roughness effect on the surface of the specimen. Bolt holes along both the tensile and shear planes were elongated. Holes located near the bottom edge of the sample were elongated and rotated with respect to the primary loading direction. In some instances, the rotation was as high as 30°. Shear rupture at locations of the bottom most fastener holes occurred in samples having larger spacing between the fastener lines. This is ascribed to the larger gage spacing which facilitates the development of shear along the bolt line. Table 1 summarizes the results of all connection plate tests.

Block Shear Model Development

Four predictive models were examined, all of which considered the transfer of load to be a combination of tension along the ligament between the uppermost row of bolts and shear along the vertical bolt lines. Variations included tensile rupture and shear yield on combinations of net or gross areas. A comprehensive evaluation of the developed models was based upon previously documented methodologies [3,4]. A professional factor, defined as the ratio of the ultimate test load to the calculated value, was evaluated. Agreement between the test results and prediction values when the professional factor is close to 1. A conservative result is interpreted as a factor greater than 1.

While the analysis of the various models have been described in detail in depth elsewhere [5], the final form follows

$$P = A_{nt} * \sigma_u + A_{gv} * 0.6 * \sigma_y$$

Where:
P = predicted load; A_{nt} = net tensile area; A_{gv} = gross shear area;
σ_u = ultimate strength of base material; σ_y = yield strength of base material
Figure 3 shows the results of the tests compared to model predictions based upon guaranteed minimum mechanical properties.

Conclusions

- A series of tests were conducted on gusset plates fabricated from aluminum alloys 6061-T6 and 5083-H321
- Block shear failure is a potential limit state for aluminum alloy connection plates with mechanical fasteners
- Failure occurred by rupture of the tensile ligament between the uppermost row of fasteners and yielding in shear along the bolt lines
- A predictive model governing failure under load is: $P = A_{nt} * \sigma_u + A_{gv} * 0.6 * \sigma_y$

References

1) Sharp, M. L., "Behavior and Design of Aluminum Structures", McGraw-Hill, New York, 1993.
2) Marsh, C., "Tear-Out Failures of Bolt Groups", American Society of Civil Engineers, ASCE Structural Journal, No. ST10, October, 1979.
3) Haradash, S. and Bjorhoude, R., "New Design Criteria for Gusset Plates in Tension", AISC Engineering Journal, 2nd Qtr, 1985.
4) Galambos, T., "Load and Resistance Factor Design", AISC Engineering Journal, 3rd Qtr, 1981.
5) Menzemer, C. and Fei, L., "Design Criteria for Bolted Connection Elements in Aluminum, Report to the Aluminum Association, 1998.

Table 1: Test Result Summary

Sample #	Material	Average Failure Load (KN)
ST1	6061-T6	513
ST2	6061-T6	605
ST3	6061-T6	863
ST4	6061-T6	953
SH1	5083-H321	403
SH2	5083-H321	529
SH3	5083-H321	727

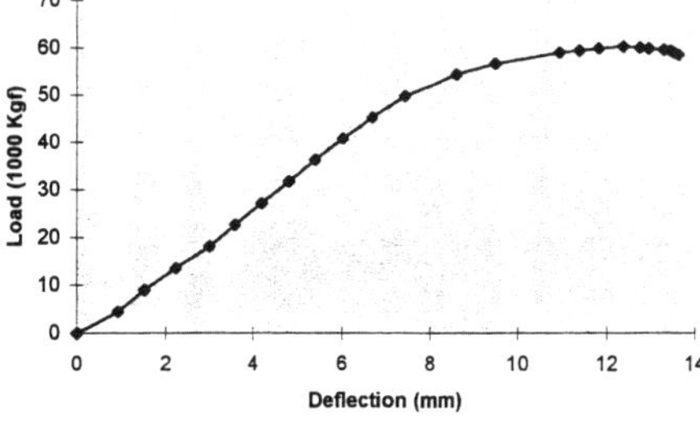

Figure 2: Load – Displacement Plot

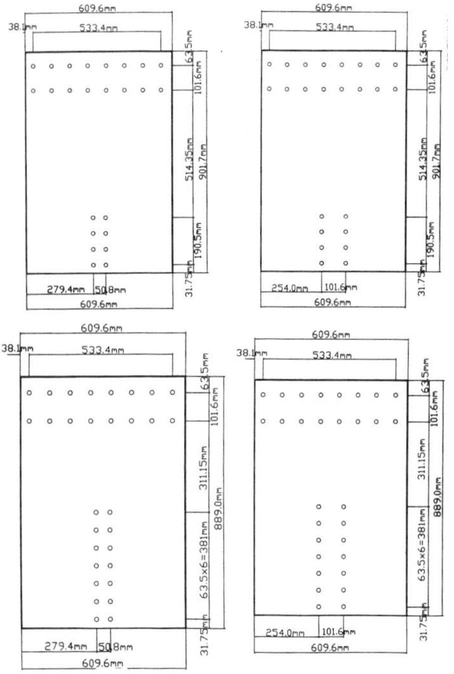

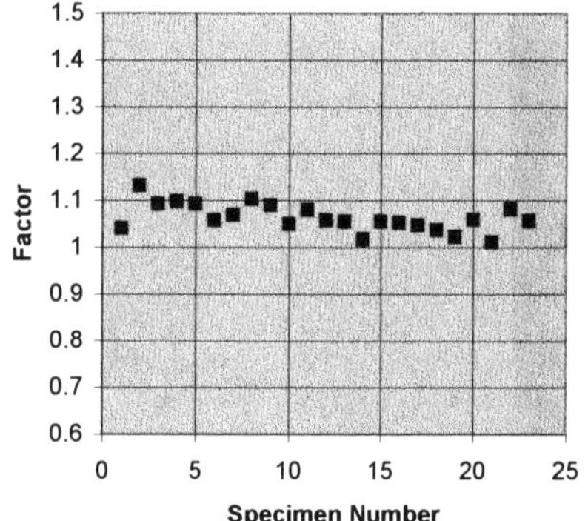

Figure 3: Model and Test Results

Figure 1: Specimen Geometry

(a)   Sample ST1.
(b)   Sample ST2.
(c)   Sample ST3.
(d)   Sample ST4.

A Review of Design Codes for Aluminum

Cedric Marsh[1], MASCE

Abstract

Some of the differences between the codes for strength design in aluminum are discussed in general, without specific references, to provide the basis for a wider review aimed at the drafting of a Guide which will take the most suitable procedures and present them in a uniform and coherent fashion.

Introduction

Each country has its own standard for the strength design of aluminum components, but standards are becoming available, such as Eurocode 9 and ISOTR1069, that aim towards a universal acceptance. Codes are in continuous revue, and, as the committees often have members in common, the major shifts such as the change to SI units and the adoption of Limit States Design (LRFD) are found in most of the documents, but a few fundamental concepts and many individual items still receive a variety of treatments. This paper looks at some of the area where differences occur.

Load and resistance factor design

Most codes are now cast in the LRFD format, or LRFD versions are available, and the concept of a calculable reliability as the basis for design is generally accepted.. The limit states of interest are collapse, which is a matter of public safety, and serviceability, which is the concern of the owner. Thus a code for the strength of components will deal primarily with the ultimate load capacity, and need pay little attention to service requirements. This distinction is not always expressed.

---

[1]Professor Emeritus, Dept of Building , Civil and Environmental Engineering, Concordia University, Montreal, QC. Canada.

To move from working stress design to LRFD demands a study of the relationships between the existing design procedures and the reliability of the end product. However, the factors that have been adopted were calibrated against the traditional procedures to ensure that the designs were in reasonable agreement, and little attention was paid to the design procedures themselves, thus the opportunity was missed to rationalize the overall design practices. The very process of calibration carried forward many of the past inconsistences in the codes.

The usefulness of a code depends on how the procedures are presented. There is much talk of simplicity, but the clauses tend to become more complex. There can be no great precision in the predicted strength, and even less in determining the loads that will be encountered. Two significant figures in the coefficients adopted provides an accuracy that is greater than that of the predictors used. This limit is seldom observed.

Mechanical properties

Only the yield and ultimate strengths in tension are guaranteed. For a given alloy and temper, the properties vary with thickness and product. To facilitate design a value is sometimes given representing the product most commonly used. This leads to some conservatism but the economic cost of the simplification is small.

Yield in compression is usually taken to be equal to that in tension, although some codes recognize the reduction caused by the Bausinger effect in work hardened alloys. Shear strength is a derived property. Using von Mises, the strength is usually taken to be 0.6 times the tensile properties. This is valid for the yield strength but cannot be applied to the ultimate strength, which is an artificial stress. However , in general, it gives strengths within 15% with the measured values. Bearing strength is related to the ultimate tensile strength. "Bearing yield strength" is sometimes quoted but finds little utility as it correlates poorly with the either the tensile yield or the true bearing strength.

Weld properties are not guaranteed and there is no consistent correlation with the properties of the parent metal. Values are preferably those specified in the qualifying weld procedures, but as the strength is different for the weld bead and the heat affected zone, and varies with the amount of filler wire deposited, the codes differ in the value chosen, while some give strengths that vary with the welding process. In reality, the designer may not know the weld method to be used and requires a value that can be realized by any weld procedure.

Tension

All codes agree that overall yielding of the gross cross section of a tension member limits its utility. The use of the yield or ultimate strength at net sections and at transverse welds, is not as consistent. Tension failure across a net section, tear out of a bolt group, and bearing failure are all controlled by the ultimate strength of the connected material, but they are not so viewed in most codes.

Compression

All design curves for buckling fall between the yield strength and the theoretical elastic buckling stress, and there is an almost universal move to normalized buckling curves, but the shapes vary widely from code to code and within the same code for different cases. Individual treatments for flexural, torsional, lateral-torsional, local buckling and postbuckling may be provided, while other codes adopt a single normalized buckling curve as the standard for all modes of buckling for all shapes, whether prismatic, built-up, or lattice. The basis for the buckling curves may be theoretic analysis for nonlinear material or for columns with imperfections, by FEM analysis for columns with residual stresses, or from test results. Ideally the methods will be combined to provide formulas that predict values that lie at about two standard deviations below the mean of test results. Some codes use mean values while others use lower bounds. Coupled instability, between local and overall buckling, may be handled by a simple cut off at the local buckling stress or by a more realistic model for the interaction.

Angles failing by torsional buckling and outstanding flanges that fail by local buckling lead to collapse, and some codes treat them equally. An element supported on both long edges possesses postbuckling strength. How this is determined varies widely.

.Beams

To carry the fully plastic moment the required compact section is usually defined. The influence of holes in the tension zone and local buckling in the compression zone receive various treatments. Lateral-torsional buckling for unrestrained beams is uniform but the case of a restrained tension flange is not always included.

In shear webs, formulas for the initial elastic buckling stress are the classic ones, but the postbuckling capacity may be attributed to a diagonal tension field, or to a redistribution of shear stress along the boundaries. The limited strength of the boundaries in riveted or welded aluminum plate girders is not always recognized.

Beam-columns

Laterally loaded and eccentrically loaded columns may fail in the plane of bending when the extreme fibre stress reaches a limiting value, or by lateral-torsional buckling. These two modes are not always differentiated, and in some codes the lateral buckling of an I-beam section receives the same treatment as that used for latticed masts, in spite of the clearly distinct failure modes.

Connections

The treatment of mechanical fasteners themselves usually follows the practices in other metal structures, but, for the connected material, bearing stresses vary widely and tear-out failures may not be covered. For eccentrically loaded groups of fasteners some codes assume that elastic behavior limits the capacity while others make use of the nonlinear load/displacement as the ultimate force is approached and methods comparable to those adopted in North America for steel members are used,

Design stresses for simple butt welds vary between codes, but there is even less agreement on fillet welds for which the strength varies with the three direction of loading. Eccentrically loaded joints, as with bolt groups, are analyzed elastically in some codes while in others, following North American steel practice, a plastic limiting condition is permitted.

Conclusions

There is a sufficiently wide variation between the national standards for structural design in aluminum that designs that satisfy a particular code may be unacceptable in another region. There is evidently a clear need for the extraction of the best practices from the various codes and the adoption of a universally accepted document, as is being attempted in the new Eurocode 9, although that draft version contains much that is not acceptable in North America.

Documents reviewed include:

Aluminum Association, Inc., Washington. DC. "Aluminum Design Manual: Specifications for Aluminum Structures", 1994.

British Standards Institute, London. BS 8118 "Structural use of aluminium. Part 1 Code of practice for design".1991

Canadian Standards Association, Rexdale, ON. CAN3-S157-M83 "Strength Design in Aluminum", 1993. CAN/CSA-S157-M "Strength Design in Aluminum" (Draft for 1999)

Comité Europeén de Normalisation, CEN/TC250/SC9, Brussels. Eurocode 9 "Design of aluminium structures-Part 1-1: General rules- General rules and rules for buildings", (Draft)

European Convention for Constructional Steelwork, Committee T2 "European recommendations for aluminium alloy structures", 1978.

International Organization for Standardization, Geneva, ISO TR 11069, "Aluminium structures-Material and design-Ultimate limit state under static loading", 1995.

Construction

Manuscripts for some presentations were not available at time of publication.

35. *EVALUATION OF PERFORMANCE OF EXISTING STRUCTURES DURING CONSTRUCTION EXCAVATION*

MODERATOR: Brian Brenner
Bechtel/Parsons Brinckerhoff
Boston, MA

(1) Evaluation of Construction Impacts on the Central Artery/Tunnel Project
Huang Ni, Dave Druss, and Brian Brenner
Bechtel/Parsons Brinckerhoff
Boston, MA

(2) A Soil-Structure Interaction Analysis Model for Diaphragm Walls
I. Steven Varga and George Aristorenas
Weidlinger Associates Consulting Engineers
Cambridge, MA

(3) Analysis for Impacts During Excavation
Bob Palermo and Terese Kwiatkowsky
GZA
Newton Upper Falls, MA

(4) Construction Excavation Excavation Impacts on the LA Metro
Amanda Elioff
Engineering Management Consultants
Los Angeles, CA

44. *DESIGNING TEMPORARY STRUCTURES FOR CONSTRUCTION: WHO? HOW? ON WHAT BASIS? WITH WHAT CONSEQUENCES?*

MODERATOR: Robert T. Ratay
Polytechnic University

(1) ASD- The Design Method of Choice for Engineered Temporary Structures
Alan D. Fisher
CIANBRO
Pittsfield, ME

(2) A Multimedia CD ROM-based Instructional Tool for the Design of Temporary Structures
Bob McCullouch
Purdue University
West Lafayette, IN

**(3) Design Issues for Temporary Containment Structures in the Ship Building
and Repair Industry**
Brian Peterson
Puget Sound Naval Shipyard
Bremerton, WA

**(4) The Role of the Construction Engineer in the Design of Temporary Structures
for Construction in the Power, Generation, and Petrochemical Industries**
Dennis S. Fedock
Babcock & Wilcox Construction Company, Inc.
Barberton, OH

(5) Design Responsibilities for Temporary Structures in Construction
Robert T. Ratay
Polytechnic University

53. *DESIGN LOADS ON STRUCTURES DURING CONSTRUCTION*

MODERATOR: John F. Duntemann
Wiss, Janney, Elstner Associates, Inc.
Northbrook, IL

(1) The Equivalent Shore Method
Sandoval Rodriques Jr.
Pontifical Catholic University of Rio de Janeiro
Rio de Janeiro, Brazil

José Filho
University of Alberta
Edmonton, Canada

**(2) Construction Loads Produced During Heavy Lifting, Rigging,
and Handling Operations**
Dennis S. Fedock
The Babcock & Wilcox Construction Co., Inc.
Barberton, OH

(3) Conflict in Codes Applicable to the Job Site
John S. Deerkoski
Deerkoski and Associates
Warwick, NY

(4) Design Wind Loads on Structures During Construction
John F. Duntemann
Wiss, Janney, Elstner Associates, Inc.
Northbrook, IL

**(5) Application and Design Ramifications of ASCE Seismic Provisions
on Recent Caltrans and AASHTO Designed Based Projects**
Cris D. Subrizi
Parsons, Brickerhoff, Quade, and Douglas
San Francisco, CA

A SOIL-STRUCTURE INTERACTION ANALYSIS MODEL FOR DIAPHRAGM WALLS

George Aristorenas, Associate Member [1]
I. Steven Varga, Member [2]

Abstract

A method to analyze diaphragm (slurry) walls is presented. The method utilizes two-dimensional plane strain finite elements to model the soil. Such a method allows the soil to interact with the wall during various stages of excavation and therefore produces more realistic results than conventional analytical approaches. In addition, there are other meaningful results which can be obtained from such a method. The modeling technique is presented using a hypothetical test case.

Introduction

Components of an infrastructure project involving excavations have traditionally been modeled and analyzed by various techniques by different designers. For instance, excavation-induced ground movement was analyzed by geotechnical engineers who employed semi-empirical and finite element methods to estimate effects of excavations on adjacent structures. The geotechnical engineers would, in turn, provide structural engineers with lateral pressures and soil parameters to design the structural members supporting the excavation, such as diaphragm walls and struts. Thus, in terms of displacements, shear forces and bending moments, the same structure supporting the excavation could potentially have two sets of totally different results due to the different approaches adopted by different engineers.

---

[1] Engineer, Weidlinger Associates Consulting Engineers, Cambridge, Massachusetts
[2] Principal, Weidlinger Associates Consulting Engineers, Cambridge, Massachusetts

With the Central Artery/Third Harbor Tunnel (CA/T) Project in Boston, there has been an effort to unify the method of analysis by providing the structural model with a more realistic way to interact with the surrounding soil. The resulting model is more sophisticated than the conventional methods using "free earth support" and "beam on elastic foundation" methods (Xanthakos, 1994; Bowles, 1996) but does not possess the complexity of geotechnical models involving coupled displacement-pore pressure formulations and general effective stress material models (Whittle, 1997).

This paper presents the methodology of the proposed modeling technique and the results of analysis for a hypothetical excavation supported by diaphragm walls. A discussion of the model's advantages and limitations is also included.

Methodology

The proposed modeling technique uses the Finite Element Method. The soil is discretized by plane strain four-noded quadrilateral finite elements with two displacement degrees of freedom at each node. Walls, Roof and Invert Slabs are modeled by two-noded beam elements, having a nodal rotational degree of freedom in addition to two displacement degrees of freedom. Struts are modeled by so-called link, or truss, elements which only take axial loads.

Structural elements are assumed to be linearly elastic. That is, the only material properties required are the Young's Modulus, E, and Poisson's Ratio, v. Sectional properties (e.g., cross sectional area, moment of inertia) are also required.

Soil material non-linearities are considered. In general, cohesionless soils (e.g., fill, sand and glacial till) are modeled using the Drucker-Prager criterion which is basically an extension of the more familiar Mohr-Coulomb criterion. Clay is modeled using a hyperbolic stress-strain relationship simulating undrained behavior, similar to that developed by Duncan and Chang (1970).

Analogous to most project design criteria in which loads on underground structures consist of soil and water components, loads on walls and slabs in the proposed finite element models consist of: (a) soil pressure which is automatically applied by the soil elements, and (b) water pressure which is assumed to be hydrostatic and directly applied as a distributed load on structural elements. The soil pressure is thus generated from the soil's buoyant (or effective) unit weight.

A more detailed description of the modeling technique is presented in ASCE (1999).

Example

In this section, an example, or base model, will be presented. Figure 1 shows the finite element model of a 12.2m wide by 12.2m deep excavation. Due to symmetry, only half the excavation is modeled. The soil profile consists of Fill, Clay, and Rock. The key features of the model are summarized in Table 1, while Table 2 describes the stages of excavation.

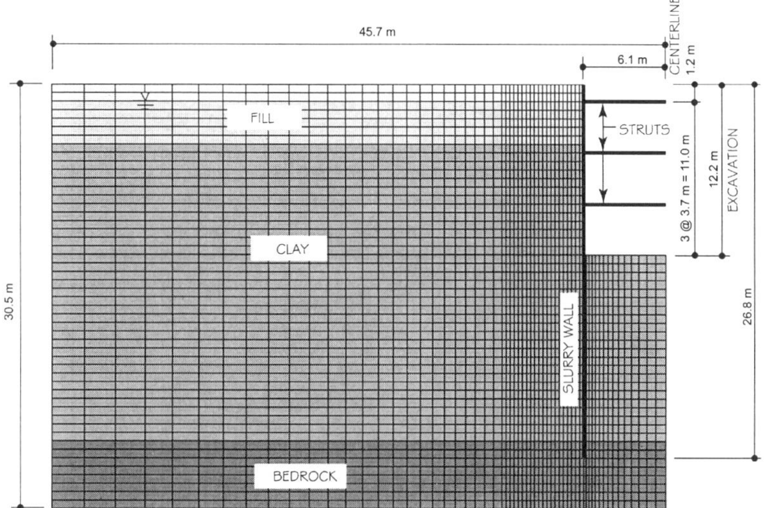

Figure 1: Finite Element Model

Table 1: Characteristics of the Base Model

ITEM	DESCRIPTION
Tunnel Width	12.2 m
Depth to Bottom of Excavation (Tunnel Depth)	12.2 m
Soil Profile	Fill, Clay, Bedrock Depth to Water Table: 1.2 m Depth to top of Clay: 4.3 m Depth to top of Bedrock: 25.6 m
Wall Embedment	1.2 m into bedrock
Wall Stiffness	$EI = 1.48 \times 10^6$ kN-m^2/m (1.1 m thick SPTC wall with W36x300 @ 1.8 m)
Struts	$A = 97.55$ cm^2/m (91.4 cm ϕ x 19.1 mm @ 5.5 m Hor.)
Strut Vertical Spacing	3.7 m

Table 2: Stages of Excavation

STAGE	DESCRIPTION
1	Excavate to a depth of 1.8 m below ground surface.
2	Install first strut level at a depth of 1.2 m below ground surface. Excavate to a depth of 5.5 m below ground surface.
3	Install second strut level at a depth of 4.9 m below ground surface. Excavate to a depth of 9.1 m below ground surface.
4	Install third strut level at a depth of 8.5 m below ground surface. Excavate to a final depth of 12.2 m below ground surface.

Soil properties are defined in Table 3 below. Fill is modeled using the Drucker-Prager criterion, and the Clay is described by a hyperbolic stress-strain elasto-plastic curve, similar to that proposed by Duncan and Chang (1970). The parameter R_f is, thus, part of the hyperbolic model. The bedrock is modeled as a linear elastic material.

Table 3: Soil Properties

PROPERTY	FILL	CLAY	BEDROCK
Total Unit Weight (kN/m^3)	19.6	18.5	22.0
Cohesion (kN/m^2)	0	-	-
Friction Angle (degrees)	32	-	-
Undrained Shear Strength (kN/m^2)	-	76.6	-
Young's Modulus (kN/m^2)	95.8	93.4	478.8
Poisson's Ratio	0.33	0.47	0.43
R_f	-	0.92	-

Results:

Figures 2, 3, and 4 present results which are useful for structural design of the diaphragm wall, walers and struts. Figure 2 shows that the wall displacements progressively increase as excavation advances, with a maximum value of 13 mm (0.5") occurring during the last excavation stage.

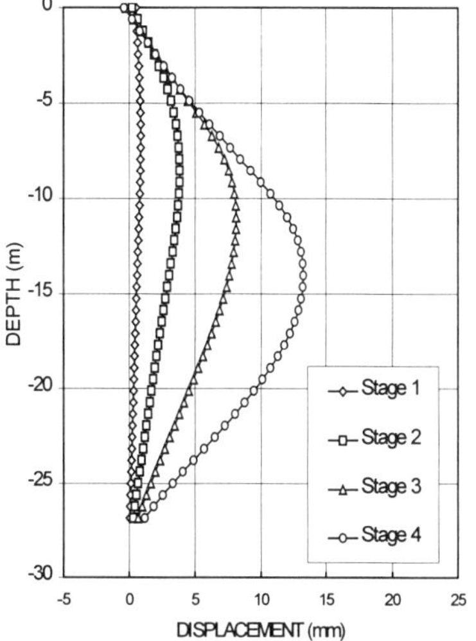

Figure 2: Wall Displacements

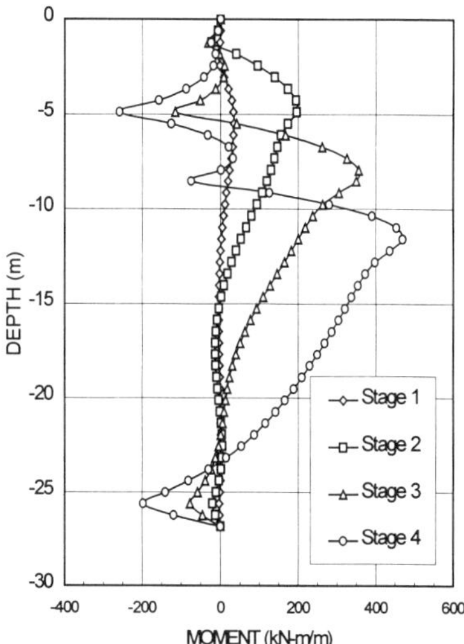

Figure 3: Wall Bending Moment

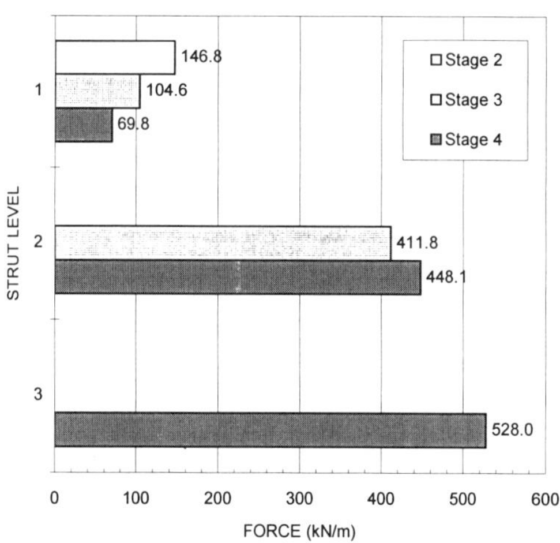

Figure 4: Strut Loads

The evolution of the wall's bending moments is shown in Figure 3. Negative moments are produced near the strut locatons, while positive bending moments exist between lateral supports. Figure 4 shows that, as expected, the maximum strut load increases with depth.

Information regarding ground movement may also be obtained from the analysis, as presented in Figures 5 and 6. In Figure 5, horizontal displacement contours suggest that the maximum horizontal displacement occurs near the bottom of excavation.

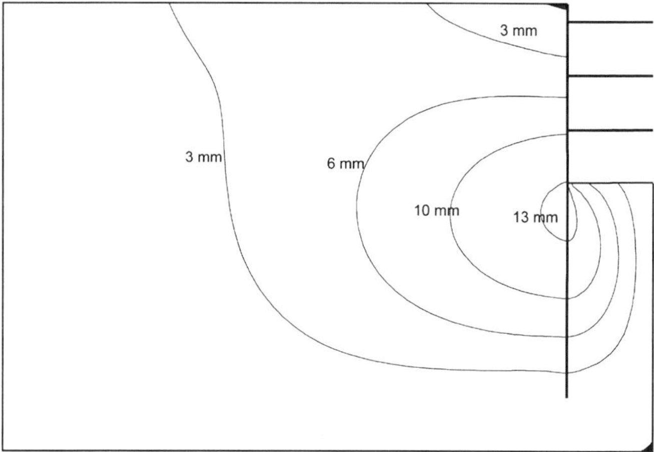

Figure 5: Horizontal Displacement after Stage 4

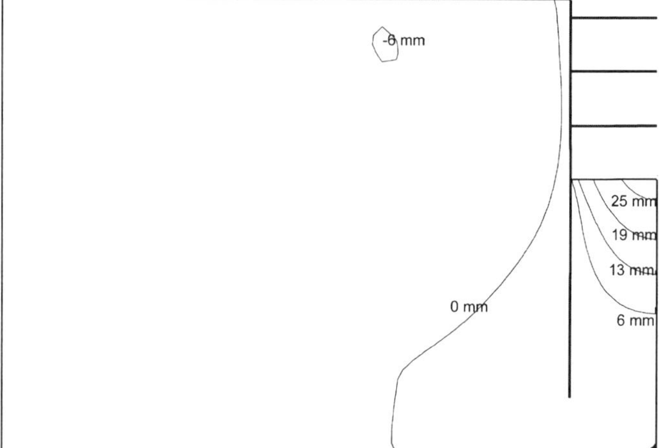

Figure 6: Vertical Displacement after Stage 4

Vertical displacement contours, shown in Figure 6, reveal that the maximum settlement actually occurs below the ground surface at a distance of approximately 12.2 m (40') away from the excavation. The information shown in Figures 5 and 6 may be used to evaluate how the excavation affects adjacent structures.

Advantages and Limitations

The main advantage of the proposed modeling method is that, relative to conventional methods such as *beam on elastic foundation* analyses, it allows one to perform a more realistic analysis of an excavation problem involving diaphragm walls over a reasonable amount of time. Also, additional factors intrinsic to the excavation, such as excavation width, unsymmetrical loading and soil profile, may be considered in such an analysis. Modeling the soil in two dimensions is conceptually appealing since it takes into consideration the shearing behavior of the soil as well as the interaction between soil elements themselves.

Simplifying assumptions are introduced by the proposed method compared to geotechnical models employing fully coupled displacements and pore pressures. The resulting limitation is twofold: (a) excess pore pressures resulting from the excavation cannot be obtained, and there may be uncertainties regarding the stress paths of different soil elements around the excavation; and (b) time-dependent deformations resulting from consolidation cannot be modeled. In terms of wall displacements, bending moments and strut loads, the effect of the first limitation is minimal (ASCE, 1999). Therefore, the proposed method is acceptable for structural design. Due to the second limitation, the time history of displacements during the transient phase is unknown. However, because the soil material model uses properties during an undrained state, the proposed method produces results which are acceptable for the short term excavation and construction stages. Often, an additional analysis using drained soil properties to model the long term permanent conditions is also performed.

Acknowledgements

Part of this work was funded by the American Society of Civil Engineers. The authors wish to thank the members of the ASCE Technical Committee on Performance of Structures During Construction for their contribution.

References

ASCE (1999). "Effective Analysis of Diaphragm Walls." Technical Committee on Performance of Structures During Construction.

Bowles, J. E. (1996). "Foundation Analysis and Design, 5e." The McGraw-Hill Companies, Inc.

Duncan, J.M. and Chang, C.Y. (1970). "Nonlinear Analysis of Stress and Strain in Soils." J. Soil Mech. Found. Div., ASCE. Vol.96. No.SM5.

Whittle, A. J. (1997). "Prediction of Excavation Performance in Clays." Civil Engineering Practice, Journal of the Boston Society of Civil Engineers Section/ASCE. Fall/Winter 1997, Vol. 12, No. 2.

Xanthakos, P. P. (1994). "Slurry Walls as Structural Systems, 2e." McGraw-Hill, Inc.

ASD
The Design Method of Choice for Engineered Temporary Structures

Alan D. Fisher, P.E., Member ASCE[1]

Abstract

Allowable Stress Design (ASD) is the traditional method of analysis and design for engineered temporary structures. The reasons for this go beyond tradition. ASD must remain in a paramount position in all standards regulating such structures.

Introduction

The types of structures which now are grouped together under the heading of Temporary Structures have been until recently the orphans of the structural engineering field. Long unappreciated by the structural engineering profession, temporary structures are now being recognized for the complex works they are. This new attention has brought about the development of the <u>ASCE Design Loads on Structures During Construction Standard</u>, (ASCE, 1999). Originally conceived as a Load and Resistance Factor Design (LRFD) based standard, it now maintains Allowable Stress Design (ASD) as an equal design methodology to LRFD. The purpose of this paper is to support this philosophy of equality.

History

Before discussing which design methodology a designer should choose, it is worthwhile to review the history behind the use of ASD, LRFD and the precursors to LRFD, Ultimate Strength Design (USD) and Plastic Design. At the start of this century, elastic design became the first generally accepted structural design methodology. Elastic design is the basis of ASD. However, even in the early

----

[1]Manager, Temporary Design Section, CIANBRO, Pittsfield, Maine 04967

"

1900's, there were strong proponents for using ultimate strength design theory for the design of reinforced concrete. In 1956 the American Concrete Institute's <u>Building Code Requirements for Reinforced Concrete</u> (ACI 318-56) included USD as an alternate design method. In 1963, ACI 318-63 made ASD and USD equal methods. In 1971, ACI 318-71 made USD the preferred method and ASD an alternate method. The reason for this change from ASD to USD is that reinforced concrete behaves inelastically as it nears its failure capacity. This inelastic behavior results in strengths 20% greater than those predicted by elastic design theory. This strength advantage gives USD designed concrete structures an economic advantage in material costs over ASD designed structures.

The American Institute of Steel Construction (AISC) formally introduced USD concepts in a limited way with the plastic design method in the 1963 AISC Specification for Structural Steel Buildings. Plastic design utilized load factors, but not resistance factors, making it not directly compatible with the USD methods of reinforced concrete. LRFD was developed by AISC during the 1970's to achieve a common design format with ACI 318 and to obtain the increased capacity that plastic design can offer. AISC formally introduced LRFD in the 1986 AISC LRFD Specification, and stated AISC's intention to discontinue support for the AISC ASD Specification after the Ninth Edition was published in 1989.

The trend from ASD to LRFD by ACI and AISC may make some believe that ASD will become obsolete. This is an incorrect perception. As will be shown below, ASD is still needed for the design of temporary structures.

Basis for Use of ASD

LRFD and ASD each have beneficial aspects. Galambos and Ravindra in a paper presented in the First Quarter 1978 AISC Engineering Journal concluded that LRFD had the following advantages:
1. It is a rational and consistent design methodology.
2. It can be updated and modified as more data becomes available from research.
3. It may result in more economical design, especially when dead load dominates.
4. It permits an integrated treatment of loads and load effects where two material specifications, i.e. steel and reinforced concrete, interface.

It is not the intention of the writer to denigrate the LRFD approach but rather to propose that ASD offers the following benefits:
1. It uses loads directly for structure analysis and calculates strengths directly.
2. It requires only a single analysis for both the evaluation of strength and deflections.
3. It results in moments, shears and reactions which are directly comparable to the ratings used by manufactured systems common to temporary structures.
4. It can be used with all structural materials.

ASD's use of loads and forces directly rather than multiplied by modifying factors as in LRFD, simplifies the analysis process. There are almost twenty different LRFD load categories in the <u>ASCE Design Loads on Structures During Construction Standard</u>. Adding to the potential confusion this may cause, some loads may change categories from one construction stage to the next. ASD's combination of loads directly, eliminates this potential for confusion. Mental calculations are therefore easier with ASD than with LRFD. Mental calculators are sometimes the only ones available.

ASD uses one analysis for both stress calculations and deflection calculations. LRFD requires two analyses, one for strength and a second for service load deflections. Many temporary structures are deflection controlled, requiring that a deflection analysis be included as a part of their design. Analyzing the structure only once is another effort saving aspect of ASD.

ASD analysis results in moments, shears and reactions which can be used directly with the ratings used by manufactured systems common to temporary structures. Components of off-the-shelf shoring systems, formwork systems, rigging systems, jacking systems, and personnel access systems are almost entirely rated in terms of safe working loads, shears and moments. Use of these components with an LRFD based design requires the unfactoring of the results of the LRFD analysis.

LRFD design is promoted on the basis of reduced construction cost with structures whose dead load is large compared to its live load. This savings may be slight to nonexistent with temporary structures whose dead load is usually small compared to its live load. Further, this reduced cost results from reduced material sizes based on strength needs which may not apply if deflection controls or if construction expediency dictates the material sizes used.

ASD and Specific Temporary Structure Types

A review of several of the primary types of temporary structures may help illustrate these points. Cofferdams, concrete formwork, and rigging are widely used types of temporary structures.

Cofferdams are often constructed with used materials and are constructed in environments with unpredictable loadings. Steel sheet piling is often used in one cofferdam and then pulled for later reuse. Horizontal support members, may likewise be fabricated, installed, used, removed and refabricated several times before being scrapped. Construction with used materials encourages the use of ASD since the materials of choice are apt to be kept to a few common sizes to minimize inventory issues. An ASD analysis is accurate enough to permit selection between these few sizes and a more accurate, but longer, LRFD analysis will not result in a significant cost savings.

Installation control of cofferdams may be less than perfect resulting in a high degree of variability between theoretical forces and actual forces. Construction methods may cause incidental impacts which cannot be specifically provided for. These factors support the use of ASD for the higher inherent factor of safety it affords.

A second very common temporary structure type is concrete formwork and shoring. Concrete shoring and formwork are heavily based on component systems. These systems are exclusively rated in terms of working loads or of allowable shears and moments. Use of an ASD analysis which results in actual loads, shears and moments is preferable. Concrete shoring and formwork design is based on deflection criteria, another benefit of the ASD approach. Formwork design based on a LRFD approach can easily involve ten or more load categories each with its own load factor, resulting in a more elaborate design than with ASD. ASD is therefore preferable for concrete shoring and formwork design.

A third common group of temporary structures are designs involving jacking and rigging elements. Here again these tasks are based largely on system components such as slings, shackles, jacks and other devices. These components have a long history of being exclusively rated by allowable capacities. Loads on these systems are almost entirely live loads and therefore no benefit is gained from the lower safety factors provided for dead loads in LRFD. Rigging almost always involves crane usage, and cranes are all exclusively rated in terms of allowable loads. ASD is therefore preferable for jacking and rigging designs.

Conclusion

Design of temporary structures is driven more by "turn around" time than by structural refinement. The temporary structures on many projects are the first order of business when the project is awarded. Overhead and equipment costs can run to tens of thousands of dollars per week on even a moderate sized project. If a temporary structure is on the critical path of a project the cost of refining a design can easily exceed any savings offered by the refined design. The design and construction of the temporary structure must therefore be carried forward at as fast a rate as is safely possible. Simple analysis, reduced amount of analysis, conformity with prevailing sources of component systems and little to no negative cost impact make ASD the preferable design methodology for the design of temporary construction structures.

A Multimedia CD-ROM based Instructional Tool for the Design of Temporary Structures

Bob McCullouch [1] P.E., Ph.D.

Abstract

Professional practicing engineers and construction professionals receive very little if any formal training on the subject of temporary structures. From both an economic and humanitarian standpoint, there is a significant need to be met by developing a training tool for professionals involved in the design and supervision of constructed facilities.

A CD-ROM continuing engineering education credit course on temporary structures is being developed (completion date is May 1999) at Purdue University. The CD-ROM approach is appropriate because the student audience will most likely have access to computer resources and the topic requires the use of different media forms. The student will work through the content by forcing a quiz/response requirement on them. To receive continuing education credit an electronic submittal will be required. This paper describes this course and demonstrates why multimedia technologies should be used with this topic.

Approach

Multimedia is a good fit for Temporary Structures because it opens a new data set to work with. By adding pictures, sounds, animations, and video; temporary structures can be described and design concepts explained more effectively. Presenting the information in electronic form provides a resource that is easier to disseminate and reaches a broader audience. Also by taking a computer-based training approach the learner becomes more engaged and involved in the learning process which in turn creates a better learning environment. The vision is to package and present the

[1] Research Scientist, School of Civil Engineering, Purdue University, 1284 Civil Engineering Building, West Lafayette, IN 47907, email: bgm@ecn.purdue.edu.

information in a way that makes it useful in a college course as well as a training package for practicing professionals.

Temporary structures can be divided into three general areas: Vertical Above Ground Temporary work, Below Ground Temporary work, and Horizontal Above Ground Temporary work like that used in bridges. The Vertical Above Ground Temporary is the subject of this CD. An overall plan is to produce modules for each of these areas with this being the first one.

Another reason for taking a modular approach is to use it in existing courses in the civil engineering, construction engineering, and building construction management curriculum. The module could be used by students to support existing courses. Also when all three modules have been completed a complete course on the subject of temporary structures can be offered electronically.

Topics covered in the area of Vertical Above Ground Temporary are the following:

>   Legal Aspects
>   Codes and Design Standards
>   Construction Loads
>   Formwork Design
>   Shoring and Scaffolding Design

A brief content description helps to explain the instructional design plan that is being followed during development.

Legal Aspects covers; insurance requirements, legal requirements, legal responsibilities, liabilities, and submittals. This section is text information. The information will be organized into a menu structure and presented in popups and scrollable boxes through hyperlinks. Court cases and decisions will be cited as well as any regulatory and statutory law requirements.

The Codes and Standards section describes the applicable building code sections and standards that apply in design and construction. The American Society of Civil Engineers (ASCE) is developing a standard for "Design Loads on Structures During Construction." An overview description of this proposed standard is included. Suggested references from ASCE and ACI are found on the CD as well.

Examples of construction loads, their source, magnitudes, and combinations will be organized in a menu structure and explained through a variety of media forms.

Formwork Design will be divided into vertical and horizontal forms. Vertical forms include walls, columns, and slip forming. Horizontal forms are used for floors, beams and girders. The different variations of these types will be described. This section will contain design examples, pictures, videos of forming operations, and formwork supplier product info.

The last major section is Shoring and Scaffolding.   The Shoring section will describe the shoring and reshoring process and analysis encountered in multistory construction. A sound and animation clip is used to describe this technique. The Scaffolding section provides information on the different systems, safety issues, load capacities, and support options. Supplier information will be solicited and placed into the appropriate categories.

To ensure the continuing education requirement is met, the student will be required to respond to a series of checklists and quizzes during the review of the CD-ROM. These responses as well and student info will be recorded electronically and returned for course validation and continuing education credit.

A course evaluation survey is included on the CD-ROM. Every student completing the course will be asked to print the form, answer the questions and return it to the office of Continuing Engineering Education. The information will be compiled and presented to the course developers as a means of providing insight into future course module development and for revising future editions of the current course. The course developers will also produce a short report of "lessons learned" that may be beneficial to others proposing to develop material using similar delivery technology.

At this stage of development (November 1998) only a skeletal layout of the content exists. A considerable number of screens have been developed and the testing component is just starting to take shape. Numerous multimedia files are developed and some animations and video are being created. By Congress, the application will be completed and a demo can be performed during the session. Below is the main menu screen. The user mouse clicks the icon to branch into a particular topic. One icon contains instructions on how the tool operates.

The user navigates with mouse clicks and the only time the keyboard will be needed is during the quiz phase. The application is developed to run on Windows 95, 98, and NT. An install program comes on the CD that sets up the application on the student's computer and creates an icon for launching the application. Bookmarking will be available so the student will have a record of what parts have been completed. The application requires a computer with a minimum CD-ROM speed of 6X and a sound card with speakers.

Design Issues for Temporary Containment Structures in the Ship Building and Repair Industries

Brian A. Peterson, P.E., M.ASCE[1]

Abstract

This paper examines current practices related to the implementation of outdoor temporary containment in the ship building and repair industries. The results of a shipyard survey on containment practices is examined from a design engineer's perspective.

Introduction

A shipyard's proximity to open bodies of water and nearby public/private development results in environmental scrutiny of work practices. To maintain environmental compliance, the ship building and repair industries have gradually come to terms with the necessity to "contain" abrasive blasting and/or painting work-sites. This is accomplished by enclosing the areas with temporary structures.

Until recently, these temporary structures were not a necessity of the work process. Many facility owners therefore view containment investment solely as an increase in the cost of doing business to be passed on to the customer. Therefore, to maintain a competitive edge the owners often rely on poorly designed structures, using marginal materials, which usually lack engineering input. Failures, although not typically life-threatening, are not uncommon. They have resulted in production delays, damage to the supporting structures, damage to the item being contained, and questionable environmental compliance.

---

[1] Structural Engineer, Puget Sound Naval Shipyard, 1400 Farragut Ave., Bremerton, WA 98314

Containment Practices

A survey was conducted to examine current shipyard containment practices. A total of (16) facilities replied to the survey (46% response rate), ranging in size from some of the largest domestic facilities to smaller regional facilities. The results pertinent to this paper are listed below:

- 88% use some sort of temporary outdoor containment.

- 25% utilize some engineered containment designs.

- 85% monitor the weather and disassemble a containment when needed to minimize the impact of severe weather.

- A response of "fair" to "poor" was given for the overall satisfaction with containment performance.

- 50% identified 1-4 weeks as the duration of containment installation.

From the low percentage of engineered designs identified, it becomes apparent that shop personnel, relying on their experience, are responsible for the majority of containment designs.   Problems with these shop-designed containments have occurred. For example, a dry docked vessel undergoing re-painting had scaffolding installed along the mast and yardarms extending 20 m above the ship's deck. Portions of the scaffold were sheathed with heat-shrunk plastic to contain the work. A winter storm with winds in excess of 100 km/hr caused the top portion of the containment to displace approximately 1.5 m back and forth as it oscillated in the wind. Due to the unstable condition of the scaffold, personnel were unable to cut open the plastic sheathing to relieve the load. Although the scaffold did not collapse, the yardarms were damaged. Approximately one month later, this same ship had a partially completed containment on its adjacent mast and yardarm.   The shop containment designer increased the amount of sheathed surface area with the false impression that it would make the structure more stable.   With gale force wind warnings in effect, a production supervisor took no action to "open" the structure with the feeling that "the other containment survived just fine, what's wrong with this one?" When winds of 100 km/hr arrived that afternoon, the only thing that prevented a catastrophic scaffold collapse was the fact that the structure had not yet been fully sheathed.

Other examples abound of flawed containment design. Ship to dock "curtains" (tarps/cables) pulled the dockside handrails, to which they were secured, from their supports during high winds. A facility utilizing tarps without back-up reinforcement experiences frequent failures, with replacement costs in material alone exceeding $500,000 annually. A cable and tarp containment for crane refurbishment, destroyed by heavy snow, cost $100,000 to replace and resulted in a 7 month project delay.

Design Engineer Issues

The previous examples came from only a few facilities. When the other facilities and their increased reliance on containments are considered, the total number of problems will undoubtedly increase.

Many of these problems could be minimized or prevented with design involvement by a structural engineer. However, reluctance to utilize engineered input exists on many levels and must be overcome. The engineer's design of a temporary structure often is critically scrutinized since it is perceived as unnecessary and cost excessive. It is imperative that the engineer be proactive when it comes to these structures. The engineers must actively seek out situations where their services should be utilized, rather than wait for last minute calls seeking advice (i.e., just as a storm is approaching). In essence, it is necessary to repeatedly convey the dangers of neglecting allowances for the severity of wind loading.

Past reliance on lightweight materials often necessitates a reverse engineering approach. When confronted with the "preferred" lightweight materials utilized by shop personnel, the engineer may wish to impose a level of allowed wind load beyond which "opening" the containment is required. This allows the shop to utilize materials with which they are most familiar, and a more economical structure can be designed based on lower wind loads.

While a stringently engineered design is impractical for many situations, engineering involvement should be utilized as much as possible. The engineer should recognize several issues relating to design. The engineer must first be cognizant of the load which may be transmitted to the structure being contained. Superstructure which is contained has a substantial increase in surface area susceptible to wind loading. Similarly, large "curtains" with cables secured to dock and ship structure transmit significant cable loads to their anchor points. A routine check of these conditions should be made to ensure acceptability.

Another priority is to minimize the chance of catastrophic containment failure. The severity of wind loading on these structures stems from their exposure to open bodies of water (e.g., exposure category "D"), the height at which many of these structures exist, and the excessive "sail" that is created by the containment. Unlike building containment, ship related temporary structures may span great distances without the aid of bearing on adjacent surfaces. It therefore may be necessary to again consider a "wind release" contingency. Large "curtains" spanning 20 m were used on the bow of an aircraft carrier for paint overspray containment. Velcro seams incorporated into the curtain allowed separation of the curtains when wind speeds were excessive. The lighter weight materials performed well and resulted in lower design loads for the supporting cables and anchorage.

The final design consideration is identifying applicable codes and standards. From ANSI A10.8, 4.30 - "When scaffolds are to be partially or fully enclosed, precautions shall be taken to assure the adequacy of the number, placement, and strength of ties attaching the scaffolding to a structure due to the increased load conditions resulting from the effects of wind and weather. The scaffolding components to which the ties are attached shall be strong enough to sustain, without failure, the additional loads imposed upon them." Meeting those objectives, as the minimum, is crucial to the designer.

While OSHA regulates scaffold use, a sheathed scaffold is given only minor consideration - "Wind screens shall not be used unless the scaffold is secured against the anticipated wind forces imposed." Care must be taken to consider overturning and uplift on the structure, as well as localized loads. OSHA stability requirements of 4 to 1 are entirely inadequate for a sheathed scaffold, yet some designers quote this as a governing requirement.

ASCE 7-95 is the sole available source for design loads, although its applicability is questionable for structures of this nature. ASCE's forthcoming guide on loads for temporary structures may prove more useful. The guide's load reduction factors, based on time in place of the structure, are viable when compared with the duration of installations identified in the survey. Inclusion of temporary structures within published guides can legitimize these structures in shipyards, providing a much better appreciation for the importance of their design, and greater respect for the role of the engineer.

Conclusions

This paper presented issues related to Shipyard containment, which differ in many ways from that of traditional construction containment. Until shipyard processes can mitigate a need for "enclosing" work areas, containment will be necessary, and very likely will increase in use, due to environmental regulatory pressures. With minimal design input by an engineer, greater serviceability and reliability of the structures can be obtained. The forthcoming ASCE guide on temporary structures may legitimize a need for bona fide engineered designs of temporary containments in shipyards.

Acknowledgment

The facility survey was performed with funding from the National Shipbuilding Research Program in conjunction with the project "Open Area Painting – Overspray Containment."

THE ROLE OF THE CONSTRUCTION ENGINEER IN THE DESIGN OF TEMPORARY STRUCTURES FOR CONSTRUCTION IN THE POWER GENERATION, PROCESS, AND PETROCHEMICAL INDUSTRIES

Dennis S. Fedock, P.E., Member ASCE [1]

Abstract

With an ever increasing awareness of construction jobsite safety and the cost of workmen's compensation claims, owners, contractors, and architect/engineers are realizing the benefits of planning construction work early in the life of a project, and supporting that plan with a dedicated project constructability program and an engineered approach to project execution. The construction engineers assigned to the project play critical roles in the implementation of these programs.

Temporary structures, including scaffolding, personnel protection systems, work platforms, trestles, temporary support systems for existing infrastructure, and lifting, rigging, and material handling systems are critical elements of the overall construction plan. This paper will emphasize the need for the construction engineer's involvement, and ultimate responsibility, in the design of these temporary structures for construction projects in the power generation, process, and petrochemical industries.

Historical Perspectives

The construction industry is changing. Many tales have been told about the way things were in the "old days". There are tales such as the one that explains a crawler crane is not being used to its maximum capability if during the course of a lift it isn't up on at least seven of its rollers. Or it isn't being used effectively if it isn't necessary to tie a "D9 Caterpillar" to the crane's back end to hold it down. Then there is the tale

---

[1] Senior Engineer & Manager of Construction Technology, Babcock & Wilcox Construction Company, Inc., 90 East Tuscarawas Avenue, Barberton, Ohio.

about the methods used to size spreader beams. If a rigging superintendent required a spreader beam to handle a particular load, he found a beam on the jobsite, attached the beam to the load, and started the lift. If the beam bowed excessively, it was probably undersized and as such, it was time to find a bigger beam. Never were any calculations made! Finally, there were many tales told about ironworkers and boilermakers who never wore safety belts or lanyards, and were never tied off. The need to tie off was perceived as a restriction to their required mobility in performing the work of "hanging iron". Only a minimal amount of scaffolding was ever erected. It wasn't needed; the ironworkers climbed to many precarious precipices to hang a load block or a piece of structural steel. In order to access their work area, they rode the headache ball on the crane.

By today's standards, the practices described above are strictly prohibited and cannot be considered as a means of performing construction work. Construction safety is governed by OSHA, 29 CFR, Part 1926 [2], wherein such issues as scaffolding, fall protection, cranes and derricks, rigging and material handling systems, and structural steel erection are specifically addressed.

Domestically, the nature of the construction industry has changed with time. In the power, process, and petrochemical industries, the mega-projects of new plant construction undertaken during the 1960's, 1970's, and early 1980's have been completed. These projects represented an extremely large amount of work and were performed in large part on a cost plus basis with a seemingly endless schedule. These projects served as good training areas for apprentice craftspeople, engineers, and technicians entering the construction industry, and provided a comfortable living for many people.

Unfortunately, the trend has not continued. New plant construction has been at a relatively low level in the United States throughout the 1980's and 1990's. In that time frame, construction efforts have been focused on retrofit, repair, and maintenance projects within the existing plants. Unlike the new plant construction, retrofit projects have been undertaken on a fixed price basis, with short schedules fit into critical outages, with liquidated damages attached for any schedule overruns. The competition for these projects has been extremely fierce, and low price and short schedule have dictated the contract award; and while schedules and budgets have been constantly reduced, the emphasis on construction safety and quality has constantly increased.

In the past year, construction work in the power, process, and petrochemical industries has returned to a relatively high level. But prior to this recent surge, it was sporadic at best. During this time frame, union and merit shop labor organizations underwent considerable changes. Because of the uncertain market and nature of the work, young people did not view the trades as an attractive profession. Similarly, in the absence of steady employment, the older craftspeople who were eligible for their pension, chose to retire. As such, the base of knowledge within the trades has been diminished.

In light of these changes, the construction industry has been challenged. In order to remain in business, contractors have recognized that they must work harder, work smarter, work safer, and work more efficiently via the use of existing technology.

An Engineered Approach to Project Execution

The recognition of the need to work harder, smarter, safer, and more efficiently dictates the need for an engineered approach to project execution. This approach defines a four step process for incorporating construction technology into specific project execution, and for the development of historical data for future projects. First, as early in the project as possible, and based on the project scope of work, influence the material supply with regard to the size and configuration of shipping assemblies, the number of shipping assemblies, and the overall shipping sequence. Second, plan the construction work to identify methods, techniques, tooling, equipment, temporary structures, required resources and manpower, and work sequences, based on the material supply, project constraints, and project opportunities. Third, upon project completion, conduct a "lessons learned" session to evaluate the effectiveness of both the material supply for constructability, and the associated methods, techniques, tooling, equipment, and temporary structures used. Fourth, and finally, develop and maintain construction standards on the basis of the evaluation.

It should be noted that the techniques, methods, tooling, equipment, and temporary structures used on a specific project are determined via a careful evaluation of the project scope of work and existing site conditions. Each project presents its own unique set of constraints and opportunities. Many schemes have been tried and proven over the years. The challenge to the project team is to choose the scheme that will be most effective in completing erection work in a safe, economic, and timely manner, subject to the specific constraints of the project.

Project Responsibilities of the Construction Engineer

The construction engineering function plays a critical role in the implementation of the engineered approach to project execution. The construction engineer's mission is to provide technical services to construction operations, with the goals of (1.) reducing costs of field operations; (2.) minimizing project span time; (3.) constructing a quality product; and (4.) providing a safe working environment, all through the effective use of methods, techniques, tooling, and equipment.

Specifically, with regard to temporary structures, the construction engineer is responsible for the design of all (1.) personnel access, scaffolding, and work platforms; (2.) material handling, heavy lifting, rigging, and hauling systems; (3.) crane and derrick installations; (4.) support systems for the existing infrastructure; (5.) construction rigs, jigs, fixtures, devices, and mechanisms; (6.) earthwork and earth retaining structures; (7.) bridges, ramps, and trestles; (8.) barge landings and moorings; and for the evaluation of the existing infrastructure with regard to imposed construction loads and effects. This design work may be performed directly by the construction engineer, or depending upon the complexity of the installation, subassigned to a specialist with greater relative design experience.

Background and Qualifications of the Construction Engineer

In order to perform the functions described in the previous sections, the construction engineer shall preferably possess a background in civil, structural, and / or mechanical engineering, and be well versed in construction field operations and structural analysis and design. Minimum degree requirements include a Bachelor of Science, but a Master of Science is preferred. In the recent past, certain project specifications have required the inclusion of a Professional Engineer's stamp, for the state in which the work is being performed, on all drawings and calculations for all temporary structures. The construction engineer should be computer literate and be able to readily produce drawings in a CAD environment such as AutoCad or MicroStation.

Of equal importance to the academic and professional requirements identified above, the construction engineer must be able to readily deal and communicate with field labor and supervision on a mutual basis. All the drawings and calculations that can be produced are ineffective if they do not represent what the field really wants to do. Finally, the construction engineer must possess the acute ability to assess the potential for things to go wrong, and to develop contingency plans to maintain the progress of the work.

Conclusion

Evolving changes have dictated a critical role by the construction engineer in the design of temporary structures for construction in the power, process, and petrochemical industries. Construction drawings and calculations must be produced for temporary structures to ensure that construction will proceed in a safe and expeditious manner, consistent with the project plan, resulting in a high quality installation upon its completion.

References

[1.]   Construction Industry Institute (1986). *Constructability - A Primer,* Bureau of Engineering Research, The University of Texas at Austin.

[2.]   Occupational Safety and Health Administration, Department of Labor (1995). *Code of Federal Regulations CFR 29, Part 1926,* U. S. Government Printing Office, Washington, D. C.

The Equivalent Shore Method

Sandoval Rodrigues[1] Jr. and José Napoleão Filho[2]

Abstract

This paper proposes a simple method to calculate the construction loads on slabs of ultistory reinforced concrete buildings. The method adopts some simplifying assumptions pertinent to the Simplified Method (Grundy and Kabaila, 1963). However, the method considers the shore and reshore flexibilities. Others factors taken into account are the construction cycle, concrete age, shore-slab interaction and the effect of concentrated loads on the slabs. The formulation of the method is developed considering equivalent shores acting in between slabs. A practical example is provided for comparison and performance purposes. The method can be easily coded into a pocket programmable calculator and may fit the needs of a structural design office.

Introduction

The slabs of a reinforced concrete building are normally designed considering permanent and live loads that would attain their full values after the phase of construction and by the time of occupancy of the building. At this stage, concrete compressive strength has reached its conventional 28 days value. During construction, however, *construction loads* due to slab self-weight, stored material, pouring of concrete, movable equipment and workers, are effective on building slabs with strengths much lower than the traditional 28-days strength. In fact, measured construction loads on slabs are as high as twice the slab dead load (Moragues et al, 1996). Considering the low strength of the fresh poured concrete, construction loads are the primary cause (71%) of multistory building collapses (Hadipriono, 1985).

---

[1]Assistant Professor, Department of Civil Engineering, Federal University of Pará, Belém, PA, Brasil, 66000. (Formerly a M.Sc. graduate at Department of Civil Engineering of PUC-Rio)

[2] Assistant Professor, Department of Civil Engineering, Pontifical Catholic University of Rio de Janeiro, PUC-Rio, Brasil, R. Mq. S. Vicente, 225, Rio de Janeiro, Brasil, 22453-900. E-mail: napoleao@civ.puc-rio.br (Currently on leave of absence at Department of Civil Engineering, The University of Alberta, Edmonton, AB, Canada, T6G 2G7, E-mail: jnapoleao@civil.ualberta.ca).

This paper aims to present a simple method to calculate the construction loads on slabs during the construction of multistory reinforced concrete buildings (Rodrigues Jr., 1996). The major asset of the method is the replacement of a group of shores by an equivalent shore. As a consequence, the treatment of loads and deflections is unidimensional. This artifice simplifies considerably the algebraic formulation. The method is developed in three phases. In the first phase, the loads and deflections for the slabs as well as the loads for the equivalent shores – after an operation of concreting or shore removal - are calculated following closely the Simplified Method proposed by (Grundy and Kabaila, 1963). The basic assumptions in this phase are: slabs show a linearly elastic behavior; floors have the same geometry and dimensions; shores and reshores are rigid; slab stiffness is proportional to concrete age and foundations are considered rigid. In the second phase, the method takes into account the flexibilities of the equivalent shores. This causes a load redistribution on the slabs. In the third phase, after calculating the new loads and deflections on the varying age slabs, the principle of superposition is applied to furnish the final loads and deflections of the slabs. The final loads on the equivalent shores in between floors are then computed from equilibrium considerations.

The method can be easily implemented in a pocket programmable calculator for use in structural offices or at construction sites. Therefore, the equivalent shore method may be considered as a simpler alternative to more sophisticated 3D refined methods already available (Liu et al, 1985).

The Equivalent Shore Method

The formulation of the method is initially presented in its three phases considering a three-story building for the sake of simplicity.

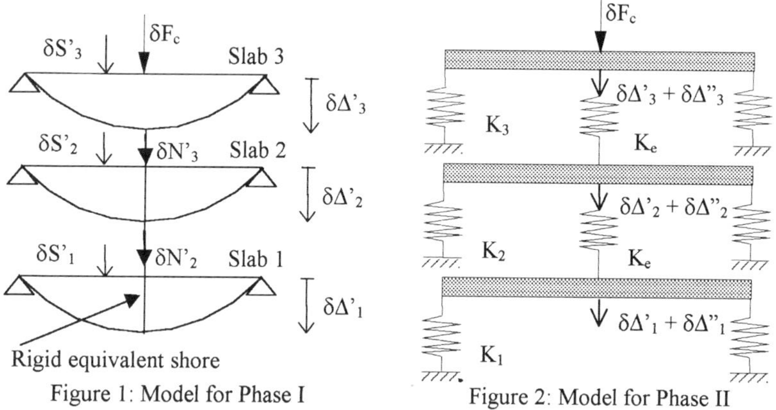

Figure 1: Model for Phase I

Figure 2: Model for Phase II

Phase I:

In this phase, the equivalent shores are considered rigid. As a result of an operation of concreting or shore removal, an increment of total load δF_c is introduced into the system (slabs + shores). The slab deflections (Figure 1) attain the same value $\delta\Delta'$, given by:

$$\delta\Delta' = \delta\Delta'_1 = \delta\Delta'_2 = \delta\Delta'_3 = \frac{\delta F_c}{\sum\limits_{j=1}^{3} K_j} \qquad (01)$$

where K_j is the flexural stiffness of slab j and is evaluated at t days after concreting. K_j is given in function of the moduli of elasticity $E_{c(28)}$ at (28) days and $E_{c(t)}$ at (t) days (ACI Committee 347, 1978) and a reference stiffness K as:

$$K_j = [E_{c(t)} / E_{c(28)}]\, K \qquad (02)$$

The construction load on each slab is proportional to its stiffness, that is,

$$\delta S'_j = \frac{K_j}{\sum\limits_{i=1}^{3} K_i} \cdot \delta F_c \qquad ; j=1,2,3 \qquad (03)$$

whereas the load in each equivalent shore is computed from equilibrium. It yields:

$$\delta N'_j = \delta F_c - \sum\limits_{i=j}^{3} \delta S'_i \quad ; j=2,3 \ : \text{after an operation of concreting} \qquad (04)$$

or

$$\delta N'_j = -\sum\limits_{i=j}^{3} \delta S'_i \quad ; j=2,3 \ : \text{after an operation of shore removal} \qquad (05)$$

Phase II:

In this phase each equivalent shore, subjected to a load $\Delta N'_j$, is allowed to deform through a deflection $\delta\Delta''_j$ (Figure 1). Therefore, the resulting deflection of each slab is given by the sum $\delta\Delta'_j + \delta\Delta''_j$ (Figure 2). Following the model in Figure 2, the equilibrium equations for slabs 3, 2 and 1 can be written respectively as:

$$\begin{cases} \delta F_c = K_3 \cdot (\delta\Delta''_3 + \delta\Delta'_3) + K_e \cdot [(\delta\Delta'_3 + \delta\Delta''_3) - (\delta\Delta'_2 + \delta\Delta''_2)] \\[2ex] \delta F_c = K_3 \cdot (\delta\Delta''_3 + \delta\Delta'_3) + K_2 \cdot (\delta\Delta''_2 + \delta\Delta'_2) + K_e \cdot [(\delta\Delta'_2 + \delta\Delta''_2) - (\delta\Delta'_1 + \delta\Delta''_1)] \\[2ex] \delta F_c = K_3 \cdot (\delta\Delta''_3 + \delta\Delta'_3) + K_2 \cdot (\delta\Delta''_2 + \delta\Delta'_2) + K_1 \cdot (\delta\Delta''_1 + \delta\Delta'_1) \end{cases} \quad (06)$$

After reorganizing the above equations, it yields:

$$\begin{cases} \delta F_c - K_3 \cdot \delta\Delta'_3 = K_3 \cdot \delta\Delta''_3 + K_e \cdot [(\delta\Delta'_3 - \delta\Delta'_2) + (\delta\Delta''_3 - \delta\Delta''_2)] \\[2ex] \delta F_c - (K_3 \cdot \delta\Delta'_3 + K_2 \cdot \delta\Delta'_2) = K_3 \delta\Delta''_3 + K_2 \delta\Delta''_2 + K_e [(\delta\Delta'_2 - \delta\Delta'_1) + (\delta\Delta''_2 - \delta\Delta''_1)] \\[2ex] \delta F_c - (K_3 \cdot \delta\Delta'_3 + K_2 \cdot \delta\Delta'_2 + K_1 \cdot \delta\Delta'_1) = K_3 \cdot \delta\Delta''_3 + K_2 \cdot \delta\Delta''_2 + K_1 \cdot \delta\Delta''_1 \end{cases} \quad (07)$$

From the first phase, it is known that:

$$\delta\Delta' = \delta\Delta'_1 = \delta\Delta'_2 = \delta\Delta'_3$$

$$\delta F_c = K_3 \cdot \delta\Delta_3' + K_2 \cdot \delta\Delta_2' + K_1 \cdot \delta\Delta_1' \tag{08}$$

$$\delta N_3' = \delta F_c - K_3 \cdot \delta\Delta_3'$$

$$\delta N_2' = \delta F_c - K_3 \cdot \delta\Delta_3' - K_2 \cdot \delta\Delta_2'$$

Introducing the last three relations in (08) into (07), it results:

$$\begin{cases} \delta N_3' = K_3.\delta\Delta_3'' + K_e.[\delta\Delta_3'' - \delta\Delta_2''] \\[2em] \delta N_2' = K_3.\delta\Delta_3'' + K_2.\delta\Delta_2'' + K_e.[\delta\Delta_2'' - \delta\Delta_1''] \\[2em] 0 = K_3.\delta\Delta_3'' + K_2.\delta\Delta_2'' + K_1.\delta\Delta_1'' \end{cases} \tag{09}$$

The equations in (09) are given in function of the unknowns $\delta\Delta''_j$, namely, the deflections of the slabs that result from a redistribution of loads caused by the deformation within the equivalent shores. The solution for the deflections is possible with the convenient incorporation of the following constants:

$$D_1 = 1 \qquad C_1 = K_1 \qquad D_2 = 1 + K_e \cdot \frac{D_1}{C_1}$$

$$C_2 = K_2 \cdot D_2 + K_e \qquad D_3 = 1 + K_e \cdot \frac{D_2}{C_2} \qquad C_3 = K_3.D_3 + K_e \tag{10}$$

After solving (09) for the deflections, it results:

$$\delta\Delta_3'' = \frac{\delta N_3'}{C_3} + K_e \cdot \frac{\delta N_2'}{C_3.C_2}$$

$$\delta\Delta_2'' = \frac{\delta N_2'}{C_2} - [K_3.\delta\Delta_3''].\frac{D_2}{C_2} \tag{11}$$

$$\delta\Delta_1'' = -[K_3.\delta\Delta_3'' + K_2.\delta\Delta_2''].\frac{D_1}{C_1}$$

Finally, the current construction loads on the slabs are given by:

$$\delta S_j'' = K_j.\delta\Delta_j'' \; ; j=1,2,3 \tag{12}$$

Phase III:

Applying the principle of superposition, the final values for deflections and loads on the slabs are given respectively by:

$$\delta\Delta_j = \delta\Delta_j' + \delta\Delta_j'' \qquad ; j=1,2,3 \tag{13}$$

$$\delta S_j = \delta S_j' + \delta S_j'' \qquad ; j=1,2,3 \tag{14}$$

From equilibrium, the final loads in the equivalent shores are computed as:

$$\delta N_j = \delta F - \sum_{i=j}^{3} \delta S_i \qquad ; j=1,2 \quad : \text{after an operation of concreting} \tag{15}$$

$$\delta N_j = -\sum_{i=j}^{3} \delta S_i \qquad ; j=1,2 \quad : \text{after an operation of shore removal} \tag{16}$$

Although the present application of the method is carried out for a simple construction system of three slabs, the formulation can easily be generalized for n floors, considering the construction cycle of the building and different support conditions for the slabs. The complete formulation is in (Rodrigues Jr., 1996).

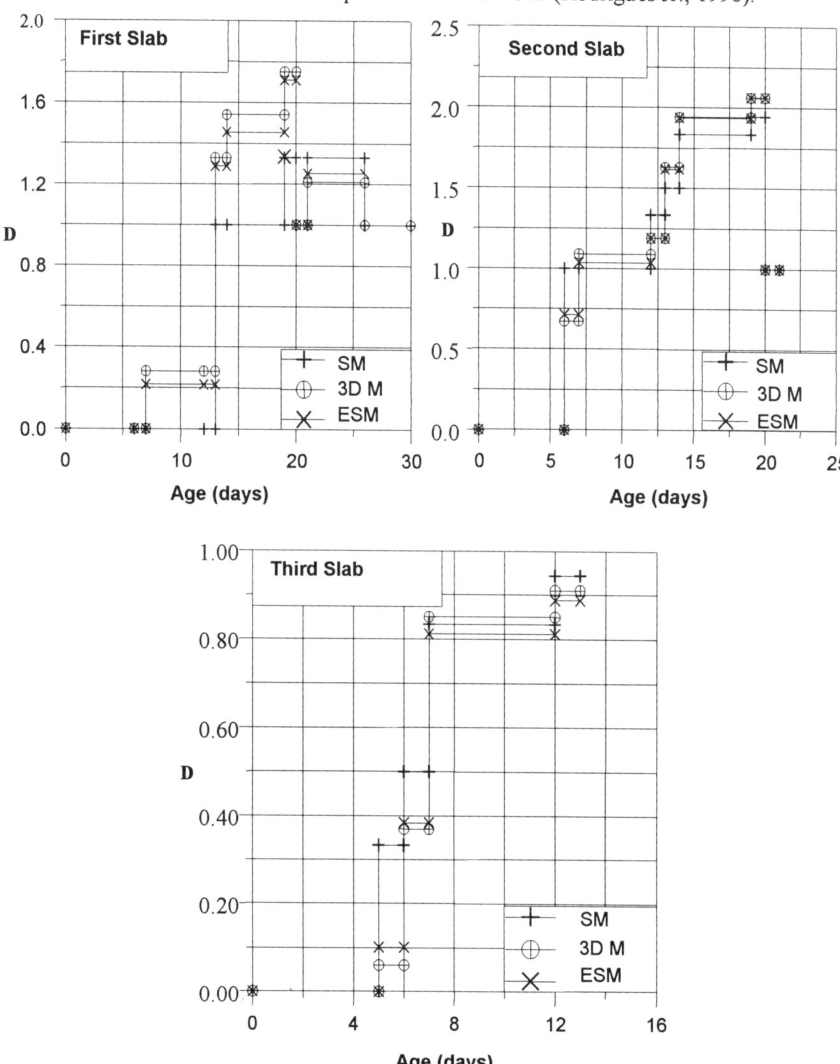

Figure 3: Evolution of Construction Loads

<u>Application</u>

An example studied previously by (Liu et al, 1985) serves as a comparison to the equivalent shore method (**ESM**). It refers to a building with four slabs. The construction procedure consists of two shores of forming and one floor of reshoring. The construction rate is of one floor per week. The slabs have dimensions 6 x 3.6 x 0.18m. The shores and reshores are spaced 1.5m in the direction of the larger span and 0.9m in the direction of the smaller span. The properties of concrete are: E_c = 35GPa, f '$_{c(28)}$ = 41MPa and v_c = 0.2. The wooden shores and reshores have the properties: E_w =7.75GPa, compressive strength f_w = 5.6MPa and Poisson's ratio v_w = 0.3. The analysis proceeded by (Liu et al, 1985) is based on a refined 3D finite element model (**3D M**) for the slabs and shores. The results are also compared with the Simplified Method (**SM**) (Grundy and Kabaila, 1963). When comparing the loads on slabs and shores, it is customary to indicate their values in terms of D, the dead load per unit area of the slab.

Figure 3 shows the evolution of the construction loads with time (days) for the first, second and third slabs. The maximum load on each slab given by the ESM is considerably close to the values computed using the 3D M and the SM. In fact, the maximum relative difference between values is about 2%. The maximum load (2.06D) takes place on the second slab at the age of 20 days, after the reshores beneath the second slab are removed.

<u>Conclusions</u>

The equivalent shore method has shown a reasonable performance in the computation of construction loads on slabs of reinforced concrete buildings during construction. The method is based on the concept on a single equivalent shore that interacts with the slabs in between two consecutive floors. The method is easily programmable into a pocket calculator and may be used in practice in structural engineering offices or even in construction sites. The comparison of the values of the construction loads provided by the ESM with respect to the ones computed through other available methods (3D M and SM) is considerably satisfactory and in the order of a 2% relative difference.

<u>References</u>
- Grundy, P., and Kabaila, A., 1963, *Construction Loads on Slabs with Shored Formwork in Multistory Buildings*, ACI Journal, V. 60, No. 12, December, pp. 1729-1738.
- Moragues, J.J., Catalá, J. and Pellicer, E., 1996, *An Analysis of Concrete Framed Structures During the Construction Process*, Concrete International, ACI, November, pp. 44-48.
- Hadipriono, F.C., 1985, *Analysis of Events in Recent Structural Failures*, J. of Structural Engineering, ASCE, V. 111, No. 7, July, pp.1468-1481.
- ACI Committee 347, 1978, *Recommended Practice For Concrete Formwork*, (ACI 347R-78), American Concrete Institute, Detroit, 37p.
- Rodrigues Jr., S., 1996, *Construction Loads on Reinforced Concrete Slabs*, M.Sc. Dissertation, Department of Civil Engineering, PUC-Rio, Rio de Janeiro, Brazil, 158p.
- Liu, X. L., Chen, W. F., and Bowman, M. D., 1985, *Construction Load Analysis for Concrete Structures*, Journal of Structural Engineering, ASCE, Vol. 111, No. 5, May, pp. 1019-1036.

CONSTRUCTION LOADS DURING LIFTING, RIGGING, AND HANDLING

Dennis S. Fedock, P.E., Member ASCE [1]

Abstract

Construction loads produced during the handling operations of heavy component erection govern the design of all elements of the rigging and handling system, and affect both the existing infrastructure and the respective component being erected. These loads are a function of the gross rigging weight of the component, its center of gravity location, required handling operations, and the configuration of the rigging and handling system being used. This paper presents a detailed description of the construction loads to be considered in the analysis and design of handling systems for heavy component erection.

Introduction

As used in the context of this paper, the rigging and handling system is defined as all of the mechanical equipment and structural installations specifically made to move and handle components from location "A" to location "B". In most cases, the rigging and handling system is a temporary installation, and is removed following the completion of heavy component erection. However, in certain cases, systems have been left in place as permanent plant equipment, for use in the ongoing maintenance or future replacement of the erected components.

The existing infrastructure refers to all existing installations that are subject to the imposition of construction loads produced by heavy lifting, rigging, and handling operations. The existing infrastructure further refers to existing installations that create access limitations and define the route to be traversed between locations "A" and "B".

---

[1] Senior Engineer & Manager of Construction Technology, Babcock & Wilcox Construction Company, Inc., 90 East Tuscarawas Avenue, Barberton, Ohio.

Characterization of Loads

Construction loads imposed on the existing infrastructure are not necessarily the service loads for which the infrastructure was designed. Variations in load magnitude, load type, and point of application are most likely to occur. In order to evaluate the effects of the imposed construction loads, the infrastructure design loads must be thoroughly understood.

Infrastructure Design Loads

Infrastructure design loads are established by governing building codes for the particular installation. In the case of buildings, minimum design loads are specified in ANSI/ASCE 7 [4.] and the Uniform Building Code [9.]. Per ANSI/ASCE 7, buildings shall be designed to accommodate the effects of dead load; live load including uniformly distributed, partially distributed, and concentrated loads produced by the occupancy and use of the building; soil pressure, hydrostatic pressure, and flood loads; wind loads; snow, ice, and rain loads; and seismic loads. Further, the building design must account for any self-straining forces arising from differential settlements of foundations and from restrained dimensional changes due to temperature, moisture, shrinkage, creep, and similar effects. In the particular case of heavy industrial installations, live loading includes the weight of water for hydrostatic testing, the weight of any product accumulations such as fly ash, slag, and dust, and the weight of any vessel contents such as coal in a bunker. Any pressures, friction loading, and impact that are produced by operating equipment must be accounted for in the design of the equipments' supporting structure.

Imposed Construction Loads During Heavy Lifting, Rigging, and Handling Operations

Gravity loading produced during lifting, rigging, and handling operations includes construction dead loads and live loads. Construction dead loads are defined as the distributed loads associated with the weight of all immovable or fixed elements of the handling system. Primary construction live loads are defined as the net rigging weight of the component being handled, plus the distributed loads associated with the weight of the movable or traveling elements of the handling system. Secondary construction live loads that may or may not exist in concurrence with handling operations include snow and ice loads, and live loads associated with construction personnel and staged material, tools, and equipment in the vicinity of the work area.

Non-gravity loads combine with gravity loads to act on the handled component and all elements of the handling system. However, the non-gravity loads that govern the design of the handling system are generally not the same non-gravity loads that govern the design of the infrastructure upon which the handling system is founded.

In the context of heavy lifting, rigging, and handling, Shapiro [10.] defines impact as the increase in load effect due to dynamic causes. These causes relate to three conditions that may occur in a component handling application: moving a component suddenly from a condition of rest, stopping a component suddenly from a condition of movement at constant velocity, and sudden release of load.

Friction loading is encountered in a variety of forms in the design of the handling system. For example, Whenever a wire rope passes over a sheave in a multiple part rigging system, a loss of mechanical advantage occurs. This loss of mechanical advantage is associated with the friction encountered at the sheave's bearings. In a hauling type application, wherein a component is skidded or rolled from point "A" to point "B", the handling system must first overcome static friction to cause the component to move. Following the start of movement, the handling system must overcome kinetic friction to maintain movement of the component. These friction forces act in a direction opposite to the direction of motion.

Over the course of construction, design wind loading, calculated in accordance with ANSI/ASCE 7, may act on the elements of the handling system for the period of time that the system remains in place. Wind load may also act on the handled component as the component is being erected. However, the magnitude of that wind load will be appreciably less than the design wind loading.

Seismic loading is considered to act only on the permanent elements of the handling system. The weight of these elements adds dead load to the supporting infrastructure. Because the seismic shear forces are a function of the total vertical gravity loading, the increased dead load represents an increase in seismic shear forces. The chance of occurrence of an earthquake at the same time a heavy component is being moved is remote. As such, seismic loading is not calculated in conjunction with the worst case of moving a handled component.

Similar to the infrastructure design loading, good engineering practice dictates the inclusion of contingency in the imposed construction loading. Contingency on the construction loading is intended to account for unknowns that may transpire over the course of handling operations. The magnitude of this contingency loading is not currently specified in any building code and hence will vary from designer to designer. A contingency load equal to 10% of the total combined gravity plus non-gravity loading acting on the handling system has been successfully applied by the author.

Design Loading for Heavy Lifting, Rigging, and Handling Operations

With construction loads being as characterized in the preceding section, a design loading condition, imposed at the center of gravity of the handled component, is determined as the worst of:

(Component Rigging Weight + Impact + Wind @ 20 Miles per Hour) x
Contingency Factor @ 10%;

(Component Rigging Weight + Friction + Wind @ 20 Miles per Hour) x
Contingency Factor @ 10%.

Note that in the identification of the design loading condition, impact and friction loading conditions are non-concurrent events.

Distribution of Construction Loads

The design loading condition is taken at the center of gravity of the handled component. The distribution of that loading into the elements of the handling system, and to the supporting infrastructure, is dependent on a number of variables. These variables include the geometry and orientation of the handled component, the three-dimensional location of the center of gravity of the handled component, and the handling operations to be performed.

As an example of how handling operations affect the distribution of construction loads, consider the steam drum [6.] shown in Figure No. #1 below. In its final position, it is suspended from boiler support steel that is erected prior to the steam drum. The drum is raised from grade to its final elevation via two independent sets of hoist, load lines, and load blocks suspended from the boiler support steel. Each set of load blocks is pinned to lifting lugs equally spaced a distance L/2 outboard of the center of gravity of the drum. The lifting lugs are shop attached to the steam drum shell such that the centerline of their pin holes is positioned a distance d above the steam drum's center of gravity.

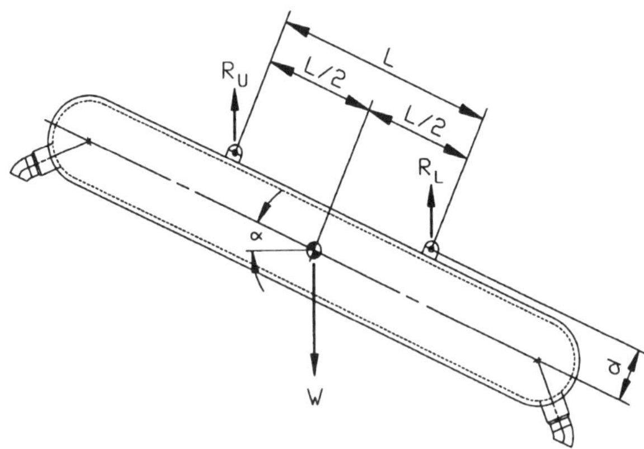

Figure 1. Steam drum raising.

In many cases, the length of the steam drum is greater than the available clear dimension between the main column rows of the support structure. As such, in order to gain the necessary raising clearance, the steam drum is raised at incremental angles α from the horizontal. Because each set of load blocks is independent of the other, this is readily accomplished by spooling more rope through one set of blocks than the other.

Initially, suspended in a horizontal orientation from two sets of rigging lines, the steam drum rigging weight, W, is equally distributed to each lifting lug and to each set of load blocks and lines. That is

$$(1.) \qquad R_U = R_L = \frac{W}{2} .$$

However, it can be shown by alternately summing moments about each lifting lug, that as the drum assumes an angle α from the horizontal, the rigging weight is respectively distributed to the upper and lower lifting lugs as

$$(2.) \qquad R_U = \frac{W}{2} + \frac{Wd}{l}\tan\alpha ,$$

$$(3.) \qquad R_L = \frac{W}{2} - \frac{Wd}{l}\tan\alpha .$$

It should be noted that besides producing a change in the magnitude of the loading distributed to each lug and set of load blocks, the change in orientation of the steam drum produces a change in the nature of the load imposed on the lifting lugs. In the horizontal orientation of the drum, the lifting lugs and their connections to the drum shell are subjected to a case of pure tension. At the angle α, the lifting lugs and their connections to the drum shell are subjected to combined shear and tension. As the angle α approaches $90°$, the lugs and their connections approach a case of pure shear.

A change in the orientation of the handled component, produced by a change in the mode of handling, results in a change of magnitude and nature of loads imposed on the component, the handling system, and the supporting infrastructure.

The distribution of construction loads imposed on the supporting infrastructure is governed by the configuration of the handling system. For the same magnitude of lifted load, the nature and magnitude of the support reactions will vary with the configuration of the handling system. Hence, the loads imposed on the supporting infrastructure will likewise differ.

Design Considerations

As discussed in the preceding sections, the magnitude and distribution of the design loading, and the nature of that loading, govern the design of all elements of the handling system, and form the basis for the structural evaluation of the handled component and the existing infrastructure. While a detailed discussion of that evaluation is beyond the scope of this paper, certain considerations should be made in its performance. These considerations include current codes and specifications, allowable stress design, safety factors, sustained vs. transient conditions, construction sequence, and component orientation variations.

Conclusion

The character and magnitude of loads produced during heavy lifting, rigging, and handling operations are significantly different from the service loads for which a particular installation was designed. These differences mandate the necessity for an evaluation of their effects to ensure that construction will proceed in a safe and expeditious manner, consistent with the project plan, and will result in a high quality installation upon its completion.

References
(for full text version, BR-1670, available from author)

[1.]   American Institute of Steel Construction, Inc. (1974). *Guide for the Analysis of Guy and Stiffleg Derricks*, AISC, Chicago, Illinois.

[2.]   American Institute of Steel Construction, Inc. (1989). *Manual of Steel Construction - Allowable Stress Design*, Ninth Edition, AISC, Chicago, Illinois.

[3.]   American Institute of Steel Construction, Inc. (1994). *Manual of Steel Construction - Load & Resistance Factor Design - Vols. I & II*, Second Edition, AISC, Chicago, Illinois.

[4.]   American Society of Civil Engineers (1996). *ANSI/ASCE 7-95 - Minimum Design Loads for Buildings and Other Structures*, ASCE, New York, New York.

[5.]   Avallone, Eugene A. And Theodore Baumeister III, ed. (1987). *Marks' Standard Handbook for Mechanical Engineers*, Ninth Edition, McGraw-Hill, Inc., New York, New York.

[6.]   The Babcock & Wilcox Company (1992). *Steam - Its Generation and Use*, 40th Edition, S. C. Stultz and J. B. Kitto, ed., The Babcock & Wilcox Company, Barberton, Ohio, Chapter 39, "Construction".

[7.]   The Crosby Group, Inc. (1998). *General Catalog - Blocks & Fittings for Wire Rope & Chain*, Tulsa, Oklahoma.

[8.]   Federation Europeene de la Manutention (1970). *Rules for the Design of Hoisting Appliances*, Second Edition, Paris, Section I, "Heavy Lifting Equipment".

[9.]   International Conference of Building Officials (1996). *Uniform Building Code*, International Conference of Building Officials, Whittier, California.

[10.]   Shapiro, Howard I. (1980). *Cranes and Derricks*, McGraw-Hill, Inc., New York, New York.

"CONFLICTS IN CODES APPLICABLE TO THE JOB SITE"

John S. Deerkoski, P.E.[1]
Fellow ASCE

ABSTRACT:

The structural designer of the temporary works at the job site is faced with conflicting code requirements. When design loads are specified but not enumerated the designer bears an unreasonable fiduciary responsibility. Misinterpretation of conflicting safety factors can lead to fiduciary and criminal exposure since the temporary works are part of the contractors "means and methods".

There are many major conflicts in codes applicable to the job site, a sampling of the most obvious are as follows:

A. OSHA

> Part 1910 Subpart D is titled, "Walking-Working Surfaces"
> What is a walking-working surface?
> This could be a scaffold, formwork, the floor or the roof of the permanent structure being constructed. Scaffolds require a 4 to 1 safety factor on failure permanent structures approximately 2 to 1 in ASD.

B. CABLES (Wire Rope)

> Safety factors required could vary from 3 to 1 for a static line to 6 to 1 for a running line to even higher for elevators.
> Rigging materials such as, turnbuckles and shackles are rated with a safety factor of 5 to 1.
> Can you adjust the hardware safety factor down to 3 to 1 for static applications such as guy cables?

---

[1] President, John S. Deerkoski, P.E. & Associates, 56 Forester Avenue, Warwick, New York 10990

C.   LEVEL OF SAFETY TO BE PROVIDED TO THE GENERAL PUBLIC

Some authorities having jurisdiction, such as New York City, specify a design load for sidewalk sheds of 300 lbs. per square foot.
What do you do when the jurisdiction has not specified a level of safety to be afforded to the general public and merely states, "provide a shed" to protect the general public.

D.   SAFETY FACTOR

Manufactured scaffolding and shoring system are loaded to failure and then safety factor is applied to that failure load.

Allowable stress design safety factor is placed on the yield stress.

What is a safety factor in an LRFD design?

E.   SCAFFOLDS

By OSHA, scaffolds are required to "support without failure at least 4 times the maximum intended load."
Is the floor of the structure you are working on required to have a 4 to 1 safety for the intended load during construction? Are you required to take into account construction loads when you are designing the permanent structure? Should you take into account construction loads when you are designing the permanent structure?

OSHA

The title in the noted section in OSHA is "Working Surfaces." Scaffolding is a separate section within OSHA, with its associated safety factors. OSHA is silent on what the safety factor should be on a working surface, if that working surface is part of the permanent structure.

The author is aware of projects that have been shut down by OSHA inspectors when work being performed on a working surface which was part of the permanent structure and was within design capacity.   In one particular case a brick laying operation, which would normally require a 75 lbs. per square foot loading with a 4 to 1 factor of safety if it were a scaffold. The structure in question had a design-working load per code in excess of 75 lbs. per square foot, however the safety factor with that permanent structure design load was only 2 to 1. Applying the 4 to 1 safety factor made that surface inadequate.

OSHA's requires that the 4 to 1 safety factors be applied to failure. A steel structure designed by an allowable stress design to a 2 to 1 safety factor, has in fact close to a 4 to 1 safety factor based on ultimate. So it is arguable that a permanent steel structure does meet the 4 to 1 criteria of OSHA.

CABLES

When cables are used for a standing line, (such as a guy cable) it is appropriate to use a 3 to 1 safety factor. The published safe working load for most rigging hardware is based on a safety factor of 5 or 6 to 1, depending on the manufacturer. You end up with a situation where a ¾ inch diameter steel cable (wire rope) loaded to a 3 to 1 safety factor (approximately 16 kips) on its breaking load is matched with an 1 3/8" turnbuckle, which has a matching published safe working load of 17.4 kips. In this case, it is more appropriate to adjust the safe working load published by the manufacturer to a 3 to 1 safety factor so that more reasonable match is obtained between the turnbuckles and cables (1 1/8" turnbuckles).

This adjustment of the safe working load should be clearly stated on all drawings and calculations so that there is no misunderstanding by any authorities have jurisdiction that would be reviewing the documents. If nothing else, it demonstrates to the authorities having jurisdiction, that you are aware of and deliberately adjusted the safe working load of the rigging materials in order that the system is appropriately matched in terms of safety factor.

LEVEL OF SAFETY TO BE PROVIDED BY THE GENERAL PUBLIC

Many job or project specifications as well as codes require that sidewalk or roadway sheds or shields shall be provided by and designed for the "anticipated loads". Or merely state that they should be provided. In the absence of a specified design load such requirements imply absolute protection of the general public. This is an unreasonable fiduciary burden to put on the designer.

The authority having jurisdiction should specify the level of protection to be provided to the general public. This is done for all permanent structures in order to shield the designer and the owners of those permanent structures from fiduciary liability. The designer or provider of the temporary structures should be afforded the same level of protection.

When you are dealing with a multiple story building, debris falling from the structure or a load being raised by a crane can drop from a substantial height. It is not practical to provide protection for the general public from all such incidents. Authorities having jurisdiction but not specifying a safety factor and design load make the designer and supplier of sidewalk sheds liable for such protection against incidents.

BOCA specifically mentions sidewalk sheds for the protection of the general public, they are silent on the design load requirements. The New York City Building Code clearly specifies a design load for sidewalk sheds of 300 pounds per square foot. Providing a sidewalk shed with that capacity places a requirement on the contractor and designer and at the same time offers them protection from unforeseen incidents.

In the absence of design load and safety factor requirements an unscrupulous designer and/or contractor can use an unreasonable low design load and state that, that is the anticipated load since they are going to be provided adequate quality control on the job site. This argument can be taken to an extreme and provide nothing more then a canvass awning. This not only defeats the purpose of protection of the public but also gives the unscrupulous designer and contractors an opportunity to circumvent the reasonable level of safety that should be provided to the general public.

An ethical professional designer should insist on the level of protection to be afforded to the public be provided by the authority having jurisdiction. It is equivocally their responsibility.

SAFETY FACTOR

The term safety factor without defining what is applied to is really meaningless. Trying to compare safety factors between different codes, specifications and OSHA can also be misleading and lead to dangerous situations.

Applying a safety factor of 2 to 1 on ultimate for ductile materials such as common construction steels such as A36 brings your allowable load very close to yield or may even exceed yield depending on the material. While a designer may feel that he still has the reserve between yield and ultimate that may not be true since the structure in distorting may lost its geometric stability thus magnifying P-delta effects which will cause a failure.

An additional problem that the designer is faced with is where one acknowledged and specified safety factor ends and where another applies. The author is currently involved in designing a suspended scaffold to be attached beneath a major bridge. The scaffold is a fixed type not a rolling type. The owner has specified the design load for the scaffold. Being defined as a scaffold rather then a shield a safety factor of 4 to 1 is being utilized for all of the scaffold components. A problem occurs when the scaffold is attached to the existing structure. The attachment devices are designed with a safety factor of 4 to 1 but the existing structure components on the permanent structure that where specified as the attaching points can not sustain a 4 to 1 safety factor for the scaffold load being applied. The existing structure therefore is analyzed for AASHTO design stress levels, which are approximately 2 to 1.

This is a common problem with varying safety factors. For example a major scaffold which is supported on grade reaches the point where we have to deal with the allowable soil bearing pressure. Should the soil bearing pressure be modified?

The scaffold safety factor of 4 to 1 is applied to "failure", the physically constructed and supplied components for the scaffold are attached to a permanent structure. The analysis of the permanent structure should reasonable be based on a safety factor of 2 to 1 based on yield, which results in approximately a 4 to 1 safety factor on ultimate or "failure" as defined in OSHA.

DESIGN WIND LOADS ON
STRUCTURES DURING CONSTRUCTION

John F. Duntemann, P.E., S.E., Member ASCE [*]

Abstract

Experience indicates that the occurrence of the failures during construction is problematic. In many cases, these failures are related to wind loading and resultant instability. The objective of this paper is to identify the significant features of current standards as they relate to wind loads on structures during construction. Design velocity for short-term exposure, drag and shielding coefficients for lattice-type structures or partially completed structures, and other variables will be discussed.

Introduction

The determination of design wind loads for permanent structures is well-established in current design practice. However, until recently there has been little guidance with respect to design wind loads on partially constructed facilities or temporary works. This paper identifies some of the existing standards and their application to structures during construction, including ASCE 7-95 *Minimum Design Loads for Buildings and Other Structures*, British Standard 5975 *Code of Practice for Falsework*, and the AASHTO *Guide Design Specification for Bridge Temporary Works*.[1,2,3] While the ASCE 7 wind provisions correspond to long-term exposure periods, ASCE 7 serves as the basic reference document for a proposed ASCE Standard.[4] The proposed ASCE standard prescribes a simplified notion of reliability, similar to British Standard 5975, where failure due to wind is determined as a function of several parameters including the wind speed probability distribution, factor of safety, mean recurrence interval of the design wind speed, and the exposure period.

---

[*] Consultant, Wiss, Janney, Elstner Associates, Inc., 330 Pfingsten Road, Northbrook, Illinois 60062

Existing Standards

ASCE 7-95, Minimum Design Loads for Buildings and Other Structures

In ASCE 7, wind pressures are determined using the following two formulae:

$$q = 0.00256K\,(IV)^2 \qquad (1)$$
$$p = qGC \qquad (2)$$

where p: design pressure in psf,
       q: velocity pressure in psf,
       K: velocity pressure exposure coefficient,
       I: importance factor,
       G: gust response factor,
       V: basic wind speed in mph, and
       C: pressure or force coefficient.

The basic wind speed, V, is the fastest-mile speed at 33 ft above ground for flat or open terrain and is associated with an annual probability of 0.02 (50-year mean recurrence interval). The importance factor, I, adjusts the design wind speed to annual probabilities of being exceeded other than the value of 0.02 (50-year mean recurrence interval). The velocity pressure exposure coefficient, K, adjusts the basic wind speed for heights above ground and terrain roughness.

In Eqn. 2, the gust response factor, G, accounts for the additional loading effects due to wind turbulence over the fastest-mile wind speed. The pressure or force coefficient, C, translates wind flow into wind pressures over the entire structure or over a component. The coefficient depends on the shape and size of the structure, as well as the location of the component on the structure. For trussed towers and lattice frameworks, C can vary anywhere between 0.8 and 4.0.

British Standard

The *British Standards Code of Practice* in CP3: Chapter V: Part 2: *Wind Loads*, is similar to the ASCE 7 and the pressure p exerted at any point on the surface of a building is given by the equation

$$p = C_p\, q \qquad (3)$$

where C_p is the pressure coefficient (C_{pe}-C_{pi}) and q is the dynamic pressure provided by

$$q = k\, V^2 \qquad (4)$$

in which k = 0.00256 with imperial units or k = 0.613 in SI units, and V^2 is the design wind speed calculated from

$$V_s = V\, S_1\, S_2\, S_3 \qquad (5)$$

The basic wind speed V is the 3-second gust speed estimated to be exceeded on the average once in 50 years.[5] The basic wind speed corresponds to open level country with no obstructions and is adjusted by factors S_1, S_2 and S_3. The topography factor S_1 can very from 0.9 to 1.1, but is generally assumed to be 1.0 unless special topographical features are present. The factor S_2 is derived from a nomograph based upon ground surface conditions and height above ground. The statistical factor S_3 is normally equal to 1.0 with the exception of temporary structures.

British Standards BS5975, *Code of Practice for Falsework*, prescribes the following statistical factors for exposure periods less than 10 years:

<u>Life of Falsework</u>	<u>Wind Speed Factor</u>
Less than 2 years	0.77
2 to 5 years	0.83
5 to 10 years	0.88
More than 10 years	1.00

The factor is for a probability of 0.63, based on a 50-year wind return period and is a function of the degree of required security and of the period of time the structure is exposed to wind.

AASHTO Guide Design Specifications for Bridge Temporary Works

The AASHTO *Guide Design Specification for Bridge Temporary Works* adopts wind pressure values developed by Caltrans.[6] The minimum horizontal wind load is the sum of the products of the wind impact area and the applicable wind pressure value for each height zone listed in Table 1. The basic wind speed used in the determination of design wind loads is derived from ASCE 7.

Table 1 - Wind Pressure Values ($P = C_e C_q q_s I$)

Height Zone (ft above ground)	Pressure, psf for Indicated Wind Velocity, mph			
	70	80	90	100
0 to 30	1.5 Q	2.0 Q	2.5 Q	3.0 Q
50 to 100	2.0 Q	2.5 Q	3.0 Q	3.5 Q
50 to 100	2.5 Q	3.0 Q	3.5 Q	4.0 Q
over 100	3.0 Q	3.5 Q	4.0 Q	4.5 Q

Note: $Q = 1 + 0.2\,W$, but not more than 10. In the preceding formula, W is the width of the false work system, in ft, measured in the direction of the wind force being considered.

Proposed ASCE Standard

The basic reference for computation of wind loads in the proposed ASCE Design Loads on Structures During Construction Standard is the 1995 edition of

ASCE 7. The design wind speed is the basic wind speed in ASCE 7 modified by the following duration factors for the period of exposure:

Construction Period	Factor
less than six weeks	0.75
from six weeks to one year	0.8
from one to two years	0.85
from two to five years	0.9

The importance factor is 1.0 for all environmental loads during construction, regardless of the occupancy after construction.

The objective of this standard is to provide for a level of safety during construction that is comparable to that of the completed structure. To achieve this, the probability of a load exceeding the factored nominal construction load during the construction period should be roughly the same as that of a load exceeding the factored nominal design load during the projected life of the completed structure. The reduced construction period velocity factors have been developed to achieve this objective.[7]

Closure

Experience with both the British and AASHTO standards indicates reasonable design results with respect to wind loads on partially constructed facilities or temporary works. The proposed ASCE standard provides a rational method for developing design loads on structures during construction subject to shorter exposure periods. Further analysis is required to determine what impact this new standard will have on current practice.

References

1. ASCE 7-95, "Minimum Design Loads for Buildings and Other Structures," American Society of Civil Engineers, 1995.
2. British Standards Institution, "Code of Practice for Falsework," (BS 5975:1982), London, 1982.
3. American Association of State Highway and Transportation Officials, "Guide Design Specification for Bridge Temporary Works," Washington, DC, 1995.
4. American Society of Civil Engineers, "Design Loads on Structures During Construction (Draft)," June, 1998.
5. British Standard Code of Practice, CP3: Ch.V: Part 2, British Standards Institution, London, 1972.
6. California Department of Transportation, "California Falsework Manual," Sacramento, CA, 1988.
7. Boggs, D.W. and Peterka J.A., "Wind Speeds for Design of Temporary Structures," Proceedings of the ASCE Tenth Structures Congress, San Antonio, TX, April 13-15, 1992, pp. 800-802.

Application and Design Ramifications of ASCE Seismic Provisions on Current Caltrans and AASHTO Designed Based Projects

Cris D. Subrizi P.E. Member ASCE[1]

Abstract

This paper compares and evaluates the impact that the seismic design provisions in the proposed ASCE Standard, "Design Loads on Structures During Construction" have on current Caltrans and AASHTO based projects. The proposed ASCE Standard is compared against the current AASHTO and CALTRANS recommendations for seismic design loads during staged construction. The comparison focuses on several areas; the design philosophy behind the derivation of seismic demand and the context of usage. The comparison is followed by recommendations for future work.

Introduction

The proposed ASCE Standard, "Design Loads on Structures During Construction" (herein referred to as the proposed ASCE standard) contains a series of recommendations aimed at providing a minimum degree of seismic safety for structures during construction. Though presently geared to building structures this living document is expected to mature into a comprehensive general structure's document.

The goal of this paper is to examine the impact that this ASCE standard will have on the stage construction, stage demolition, and maintenance operations of highway bridges. The proposed standard will be evaluated against the current design load recommendations published in the United States; namely, Section 3.12 in the AASHTO Standard Specification for Highway Bridges (16th ed., c.1995) and CALTRANS' Memo to Designers, 20-2 (c.1989).

Another standard that provides guidance in the selection of the design earthquake during the construction of a bridge is EUROCODE 8, Annex A (c.1994), though not of primary focus in this document it is referenced periodically to provide a broader picture of current design practice in the world.

---

[1] Lead Engineer, Major Bridge Engineering Service Center, Parsons Brinckerhoff Quade and Douglas, 303 Second Street, San Francisco California 94107

Application of the proposed ASCE standard

For what type of bridge related structures can this proposed ASCE standard be used?

The proposed ASCE standard can be applied to provide a measure of seismic safety for the designs of temporary works as well as partially completed permanent bridge works. The scenarios that these works could be used are; staged construction, staged demolition and maintenance operations such as replacement of bearings or defective members. Structural configurations could include; falsework, temporary supports, bracing, frameworks of fixed equipment, stockpile restraints, temporary bearings, permanent substructures supporting portions of superstructures, etc.

When is the proposed ASCE standard to be used?

The proposed ASCE standard requires that seismic loads during construction only be considered when a seismic risk exists; i.e. the effective peak acceleration (A_a) of 0.15 is exceeded.

Where is the proposed ASCE standard to be used?

The following table lists the affected states and territories under US jurisdiction.

Locations where Proposed ASCE Standard Seismic Provisions are to be Considered			
California	Wyoming	Tennessee	
Oregon	Idaho	Mississippi	Puerto Rico
Washington	Montana	Arkansas	US Virgin Islands
Nevada	Kentucky		
Arizona	Illinois	Alaska	Guam
Utah	Missouri	Hawaii	American Samoa

(Based on the "Effective Peak Acceleration Contour Map" published in ASCE 7-95. Note: California, Nevada, Oregon Arizona and Washington have developed their own Seismic Hazard Maps.)

The fact that the seismic load case must be considered does not mean that it will govern. The proposed ASCE standard has developed an additional threshold which states that; If non-seismic lateral loads are found to be greater than $0.2A_a$ multiplied by the weight to be braced, seismic loads need not be considered. This threshold would be equivalent to seismic response coefficients, C_s range of approximately 0.03 to 0.09.

How does this parameter refine our understanding of the applicability of these provisions?

Remember that seismic demand is derived from:

$V = C_s \times W$

where, W is the seismic weight.

Let's compare the threshold with the peak values of C_s that are limited by the equation:

$$C_s = 2.5\, C_a\, /\, R$$

where, C_a is the seismic coefficient that is a function of soil type and A_a.

The proposed ASCE standard imposes a maximum limit on the response modification coefficient, R. This limit corresponds to temporary bracing systems common in construction, R = 2.5.

The table below summarizes the C_s for the maximum (assuming some ductility) and the minimum (assuming elastic response) R-values.

Seismic Response Coefficient ,C_s For the Temporary Bracing Systems

Soil Type	Aa=0.20g	Aa=0.30g	Aa=0.40g *
Hard Rock :	0.03	0.05	0.06
R = 2.5	0.08	0.12	0.16
R = 1.0			
Rock	0.04	0.06	0.08
R = 2.5	0.1	0.15	0.2
R = 1.0			
Dense Soil / Soft Rock	0.05	0.07	0.08
R = 2.5	0.12	0.17	0.2
R = 1.0			
Stiff Soil	0.06	0.07	0.09
R = 2.5	0.14	0.18	0.22
R = 1.0			
Soil	0.07	0.07	0.07
R = 2.5	0.17	0.18	0.18
R = 1.0			

(This value would be increased by a factor of 2 if the project site is along a major fault. The revised values are 0.13, 0.16, 0.16, 0.18, & 0.15 for R=2.5)*

As can be seen from the above table if the temporary structure were expected to respond elastically or if the project site were along a major fault seismic loads would have to be considered. Remember the values shown are the peak values of an earthquake that is more likely to occur than the design earthquake for the permanent structure. This design philosophy is shared by ASCE, AASHTO and CALTRANS and the context of its application is examined subsequent section.

Do AASHTO & CALTRANS provisions place similar limits on their applicability?

To start, it must be reiterated that AASHTO and CALTRANS provisions apply only to normal highway bridges. If the structure is of irregular geometry, exceptional height or type; i.e. not a girder bridge of steel or concrete construction, the designer must develop a project specific seismic criteria with the owner. AASHTO and CALTRANS do not recommend a specific threshold that limit the application of their seismic load provisions during construction. These codes do, however, establish the context in which seismic loads during construction should be checked. AASHTO and CALTRANS recommend that checks be performed for stage construction that puts the public at risk. That is the bridge construction passes over traffic or the bridge carries traffic during the staged construction. The designer is responsible for determining whether or not the seismic load case governs.

AASHTO seems to have its roots in the provisions of Annex A, of Eurocode 8 (E8), which recommends that a seismic event associated with a minimum return period of 100-yrs be considered during the construction of the bridge. The design acceleration level during construction is approximately 50% that of the design acceleration used for the permanent design. AASHTO & E8 share the same goal for their seismic design check; to prevent collapse of the bridge.

Caltrans takes a different approach, due to its deterministic design philosophy. Caltrans' Memo to Designers reminds the engineer that earthquakes of 5.0 magnitude have produced peak bedrock accelerations (A) greater than 0.25g within 2 miles of an active fault. As a result the seismic demand is a function of proximity to an active fault. These provisions fall more into the seismic load definition category that AASHTO and ASCE do not address with directly and recommend special study.

The table below compares the key characteristics of the three codes.

Derivation and Application of Seismic Demand			
	ASCE	AASHTO	CALTRANS
Philosophy	Probabilistic	Probabilistic	Deterministic
Duration considered temporary	6-months	Up to 5-yrs	Not Defined
Applicability	Areas where Aa is greater than 0.15	Where bridge carries &/or spans over traffic	Where bridge carries &/or spans over traffic
Design EQ Return Period: Permanent structure	475-yrs	475-yrs	Maximum Credible EQ (MCE)with mean attenuation spectra

Derivation and Application of Seismic Demand

	ASCE	AASHTO	CALTRANS
Risk Map	NEHRP 1994	NEHRP 1988 & State specific Hazard Maps	CDMG
EQ Demand Permanent structure	Design EQ Elastic Response Spectra	Design EQ Elastic Response Spectra	MCE ARS, Elastic Response Spectra
Peak Design Acceleration (*sites close to active fault require special study)	Max contour = 0.40g* Min contour = 0.01g	Max contour = 0.80g* Min contour = 0.01g	Miles from fault / A: 0-1 → 0.4 g; 1-2 → 0.3 g; 2-12 → 0.2 g; >12 → 0.1 g
Design EQ Return Period: Temporary structure	Not Defined, but < 475-yrs	Not Defined, but < 475-yrs	Not Defined But < MCE
EQ Demand: Design Coefficient for Temporary structure	Minimum of 0.2 C_s except within 0.4g contour where minimum is 0.4 C_s (values shown on previous table)	Minimum of 0.2 C_s, At 0.8g countour Max Peak C_s = 1.0g Min Peak C_s = 0.8g	ARS $A = 0.4g$ Max Peak ARS = 1.45g Min Peak ARS = 0.9g $A = 0.1g$ Max Peak ARS = 0.70g Min Peak ARS = 0.25g
Modification to " R-Factor "	R is limited to 2.5	R can be increased by up to 50% Except at connections	R can be increased by up to 50% Except at shear keys or hinge restrainers

Abbreviations:
EQ: Earthquake
R Factor: Inelastic Force Reduction Factor
ARS: Seismic design coefficient equivalent to C_s

NEHRP: National Earthquake Hazard Reduction Program
CDMG: California Division of Mines and Geology
A: Peak Bedrock Acceleration, R: Normalized Rock Spectra, S: Soil Amplification Factor

Conclusions: Ramifications of using the ASCE standard

The primary ramification is a legal one, the proposed ACSE standard will be a consensus standard that will potentially carry more weight than the current non-consensus design recommendations published by AASHTO and CALTRANS. Currently the proposed ASCE standard defers detailed recommendations for bridges to the AASHTO &CALTRANS provisions. It seems logical that these documents should evolve into a single document. However, until that happens the standard should continue to refer to local practice.

AASHTO and CALTRANS provisions limit the applicability of a seismic load during the construction phase on projects that place the general public at risk. This implies that projects, in remote but seismically active areas, would not need to be evaluated for seismic safety during construction. The proposed ASCE standard does not make such a distinction and therefore would require seismic safety to be assessed.

AASHTO and ASCE provisions view seismic risk probabilistically. Currently Caltrans Bridge Design Specifications use deterministic design parameters. In some cases deterministic values exceed probabilistic design values. This issue is a current subject of debate in maturation process of the permanent structure code. An effective interim mechanism to resolve this disparity is the use of a consensus peer group.

ASCE limits the R-factor based on the assumption that temporary works have limited ductility. This assumes that the bracing is the weakest link in the chain of the partially completed structure. This assumption may be felt to be too conservative when checking a partially completed bridge with a well confined reinforced concrete substructure. AASHTO and CALTRANS tackle this issue from the perspective of the partially completed structure, however, these codes allow up to 50% increases in the R-factor that is applicable to the substructure intended for permanent use. While the R-factors for reinforced concrete columns and pier walls continue to evolve as a result of extensive testing since the Loma Prieta Earthquake, the inelastic behavior of common substructure and bracing configurations made from steel and timber have not reached comparable levels of understanding. As a result, the AASHTO & CALTRANS recommendation to increase the R-factor should be implemented with the utmost caution. A thorough understanding of the meaning of the R-factor for the permanent structure is needed before it can be modified. Remember high R values suggest extensive deformation, thus necessitating checks on the stability of the structure.

Recommendations for Future Work

The current levels of maturity of the proposed ASCE standard as well as the current AASHTO and CALTRANS provisions do not provide designers with a clear path to follow. However, the lack of clarity is mainly due to the rapidly change design provisions for permanent construction. As a result it is important that any standard for temporary works evolve with the standards of the permanent works.

Further refinement is necessary to keep designs realistic and construction costs and schedules in check. This will be possible once we obtain a better understanding of:
- seismic resistant design
- the ductility of patent shoring systems and traditional falsework construction
- the response of patent shoring systems and traditional falsework construction
- the needs of the Standards Consensus Peer Review Group,i.e. you the reader

As with any design the establishment of a clear load path is the paramount ingredient to a successfully performing structural design.

Communication and Transmission Structures

Manuscripts for some presentations were not available at time of publication.

9. *INTERNATIONAL TRENDS IN THE DESIGN OF TELECOMMUNICATION STRUCTURES*

MODERATORS: Bruce Sparling
University of Saskatchewan
Saskatoon, Saskatchewan, Canada

Ghyslaine McClure
McGill University
Montreal, Quebec, Canada

(1) The Reliability of Communication Towers in Strong Wind
Alan Davenport
Boundary Layer Wind Tunnel Laboratory
London, Ontario, Canada

(2) Experience with New British and European Design Codes for Masts and Towers
Brian Smith
Flint and Neill Partnership
London, England

(3) Aesthetic Aspects of Telecommunication Tower Design
Ulrik Stottrup-Andersen
RAMBOLL
Virum, Denmark

(4) Proposed Revision to the ANSI/TIA/EIA-222 Design Standard
David Brinker
Rohn Engineering
Peoria, IL

18. *PRESENTATION OF THE ASCE GUIDE ON DYNAMIC REPONSE OF LATTICE TOWERS*

MODERATOR: Murty K. S. Madugula
University of Windsor
Windsor, Ontario, Canada

(1) Introduction to the ASCE Guide for the Dynamic Response of Lattice Towers
Murty K. S. Madugula
University of Windsor
Windsor, Ontario, Canada

(2) Dynamic Analysis of Lattice Towers
George E. Blandford
University of Kentucky
Lexington, KY

Bruce Sparling
University of Saskatchewan
Saskatoon, Saskatchewan, Canada

Markus Ostendorp
EPRI
Haslet, TX

(3) Wind Loads and the Response to Wind
Bruce F. Sparling
University of Saskatchewan
Saskatoon, Saskatchewan, Canada

(4) Seismic Loads and the Response to Earthquakes
G. McClure
McGill University
Montreal, Quebec, Canada

(5) Vibration Control in Lattice Communication Towers
Celina U. Penalba
California Polytechnic State University
San Luis Obispo, CA

27. *GUIDELINES FOR THE DESIGN OF ELECTRICAL SUBSTATION STRUCTURES*

MODERATOR AND PANELIST: George T. Watson
Houston Lighting & Power Co.
Houston, TX

(1) Background Information
George T. Watson
Houston Lighting & Power Co.
Houston, TX

(2) Electrical Substation Structure Design Guide: A Panel Presentation
Leon Kempner, Jr.
Bonneville Power Administration

(3) Deflections
William Magee
PECO Energy Co.

(4) Design
Terry G. Burley
Western Area Power Administration

(5) Connections to Foundation
Jim Hogan
Burns & McDonnell

62. *SPECIAL ISSUES FOR ELECTRICAL TRANSMISSION STRUCTURES*

MODERATOR: Mark Ostendorp
EPRI – Energy Delivery and Utilization Center
Haslet, TX

(1) Performance of Electrical Transmission Structures During Earthquakes

Leon Kempner
Bonneville Power Administration

Gina Gobo and Wendelin H. Mueller, III
Portland State University
Portland, OR

(2) Structural Loading Tests of Pre-Cast, Pre-Stresses, Tapered Concrete Poles and Frames with Steel Cross-Arms
Harry Durden and David Mayo
Alabama Power Company
Birmingham, AL

Mark Ostendorp
EPRI – Energy Delivery and Utilization Center
Haslet, TX

(3) Ice Storm '98: Characteristics, Events, and Interpretations
Mark Ostendorp
EPRI – Energy Delivery and Utilization Center
Haslet, TX

(4)Performance of Electrical Transmission Structures During the 1998 Ice Storm
H. Brian White
Consultant

(5)Cascading Failure Risk Assessment of an Electric Transmission Line
Thien Do
Bonneville Power Administration

Mark Ostendorp
EPRI – Energy Delivery and Utilization Center
Haslet, TX

Experience with the New British and European
Design Codes for Masts and Towers

Brian W Smith[1]

Abstract

The new British Standards, BS 8100[1][2][3] and the recently drafted Eurocode, ENV 1993-3-1[4] for the design of steel towers and masts have both embraced the latest theoretical and research studies undertaken in this field. Initial experiences of the use of both these documents are provided.

Introduction

The British Standards Institution (BSI) published the Standard for the loading of Lattice Towers (BS 8100: Part 1)[1] in 1986 and for the loading of Guyed Masts (BS 8100: Part 4)[2] in 1995. The complementary Design Code, providing the strength criteria for members, their connections and guys, was published as a Draft for Public Comment in 1997[3]. A final version, taking due account of the comments received, is presently with BSI and publication of the Standard is expected in early 1999.

As part of the Eurocode programme a draft Eurocode (ENV 1993-3-1)[4] has been published, covering the design of lattice towers and guyed masts and incorporating the treatment to be used for their response under wind and ice.

Whilst these Drafts are not mandatory documents at this stage, they are being used on a trial basis against existing Codes, thereby providing some indication of the implications they will have on current design practice. This paper highlights the results of some of these calibration exercises.

Loading of Lattice Towers

The procedure adopted for the loading of self-supporting lattice towers in both BS 8100: Part 1 and ENV 1993-3-1 is to use a gust response factor in

---

[1] Consultant, Flint & Neill Partnership, 21 Dartmouth Street, London SW1H 9BP, England

conjunction with a mean wind speed, whereas existing practice is to use a gust wind speed as a static load. Direct comparisons between the procedures are not possible due to the different parameters adopted, but generally for a given site the results are similar. The most significant differences occur when:

(a) the site is on a hill where BS 8100 and ENV 1993-3-1 take proper account of topographic and altitude effects, leading to higher loading;
(b) the structure is mounted with significant aerials where BS 8100 and ENV 1993-3-1 show a reduced drag from existing practice. This is a direct result of the parametric wind tunnel tests undertaken on structures with varying aerial complements; and
(c) diagonal bracing members in highly eiffelated towers are more onerously loaded by the new Codes, which take due account of unbalanced gusts applied to the tower.

Loading of Guyed Masts

Both BS 8100 and ENV 1993-3-1 adopt the static patch loading technique to simulate the dynamic response of these wind sensitive structures. Previous publications[5][6] have provided comparisons with existing practice, which has shown that the new Codes produce a more realistic assessment of response.

A further example is shown in Figure 1 for an existing 225m high mast, which was designed some 30 years ago. It can be seen that the gust loading bending moment is in fact very unbalanced – probably a reflection of the procedures available when the mast was originally designed. The BS 8100 results also reflect this unbalance but provide an envelope to the gust results in the bottom part of the mast and reduced values in the upper part. Overall the new procedures are more robust and, using appropriate computer software, structures are readily analysed and assessed.

Strength of Lattice Members

Predictions of the strengths of lattice members in towers and masts have developed principally from work undertaken in the transmission tower industry, where towers are full-scale tested and the results used to provide appropriate strut curves associated with the relevant end fixities and forms of buckling.

Unfortunately such strut curves are limited primarily to bolted angle-section members and in drafting both BS 8100 and ENV 1993-3-1 due care had to be taken to ensure that the philosophy of the strength rules was consistent for all member types. The principal findings in comparative exercises undertaken are:-

(a) Angle legs with symmetric bracing – the new Codes provide as good as or better results than existing practice – in this case taken to be BS 5950[7], the United Kingdom (UK) Standard for the design of building structures, as may be seen from Figure 2. The reduction in the new Codes above a slenderness ratio of about 75 is academic as efficient leg sections are unlikely to have such high slenderness. Angle sections with high $^b/_t$ ratios however are likely to suffer a 10% shortfall compared with existing practice.

(b) Angle legs with asymmetric bracing – interpretation of BS 5950 for such situations is suspect and most designers have taken due account of the problems of applying this Standard in such cases. Consequently, inappropriate application of BS 5950 provides higher capacities than use of the new Codes.

(c) Tubular legs with symmetric or asymmetric bracing – the new Codes do not allow any reduction in effective length if the primary bracing is single-bolted. When double-bolted, a reduction factor of 0.9 may be used in BS 8100 but in ENV 1993-3-1 any reduction has to be justified by analysis. BS 5950 may be interpreted to allow for such fixity providing a higher capacity than the new Codes, as may be seen in Figure 3.

(d) Bracing members – generally the new Codes provide higher strengths for bracing members, particularly in the range of practical slendernesses. A typical comparison is shown in Figure 4.

Other Aspects Covered in the New Codes

Ice loading – BS 8100 provides quantitative data on ice thicknesses and weights to be used for the design of towers and masts in the UK. Conditions of wind with ice as well as ice alone are covered. ENV 1993-3-1 provides guidance on ice loading but refers to an ISO document, available now as a Committee Draft[8]. However, at present, the ISO document does not provide quantifiable data and it is hoped that work will be undertaken to redress this.

Guy rupture – Masts of the highest reliability in ENV 1993-3-1 have to be designed to withstand the rupture of one guy without collapsing. To satisfy this criteria two conditions need to be considered (which generally are met):

(a) a dynamic load applied to the mast in still air conditions, to simulate the rupture of the guy; and

(b) a static analysis under 50% of the characteristic mean wind loading in the absence of the ruptured guy.

Fatigue – Both of the new Standards deal with fatigue arising from vortex-excited cross wind vibrations for structures mounted with cylindrical aerials or are of cylindrical form themselves.

In addition, consideration is given to the likelihood of fatigue damage under gust response. Whilst the criteria differ slightly between BS 8100 and ENV 1993-3-1 both Codes show that for structures constructed entirely of bolted connections, the fatigue life due to this effect is likely to be over 50 years. However, for welded structures, fatigue checks are needed if relatively poor class structural details are used.

Assessment of existing guyed masts, which have been in use for some 30 years, has started to show that welded leg details in sections of the mast high up in the structure may be exhibiting signs of fatigue damage; these structures are being carefully monitored and, in some cases, by-pass leg clamps have been installed to provide an alternative load path.

Serviceability – ENV 1993-3-1 provides guidance on calculating deflections, which are not generally critical as far as structural performance is concerned, but can be a governing criteria for the performance of directionally sensitive antennae. BS 8100 provides more specific data in that graphs are included to show the exceedance (in hours per year) of different mean wind speeds for sites in the UK.

Guy strengths – BS 8100 and ENV 1993-3-1 provide information on the behaviour and criteria for the prediction of strength of guys and their connections, which had previously been based on variable practice and empirical rules.

Conclusions

Limited use of the two drafts of BS 8100: Part 3 and ENV 1993-3-1 has shown that whilst they may not provide significant savings in material, compared with current practice, they should provide a more uniform reliability. They are certainly more complex than existing procedures but as these structures are invariably designed and analysed using computer software, this should be acceptable. For small light structures, however, use of these Codes may well cause problems and only by experience will satisfactory design guides become available, which should enable deemed-to-satisfy criteria to be developed to minimise the computational effort.

References

[1] British Standards Institution. BS 8100: Lattice Towers and Masts, Part 1: Code of Practice for Loading, London, BSI 1986.

[2] British Standards Institution. BS 8100: Lattice Towers and Masts, Part 4: Code of Practice for Loading of Guyed Masts. London, BSI, 1995.

[3] British Standards Institution. Draft BS 8100: Lattice Towers and Masts, Part 3: Code of Practice for Strength Assessment of Members of Lattice Towers and Masts. London, BSI, 1997.

[4] Comité Européen de Normalisation. Eurocode 3: Design of Steel Structures – Towers and Masts – ENV 1993-3-1. Brussels, CEN, 1997.

[5] Sparling, B. F., Smith, B. W. and Davenport, A. G. Simplified Dynamic Analysis Methods for Guyed Masts in Turbulent Winds. Journal of the International Association for Shell and Spatial Structures (IASS), Volume 37 No. 2. Madrid, August 1996.

[6] Smith, B. W. Towers and Masts: A General View. Proceedings of the IASS International Symposium on Shell and Spatial Structures. Singapore, November 1997.

[7] British Standards Institution. BS 5950: Structural Use of Steelwork in Building, Part 1: Code of Practice for Design in Simple and Continuous Construction: Hot Rolled Sections. London, BSI, 1990.

[8] International Standards Organisation. TC98/SC3/WG6 Committee Draft: Atmospheric Icing on Structures, ISO/CD/12494. 1998.

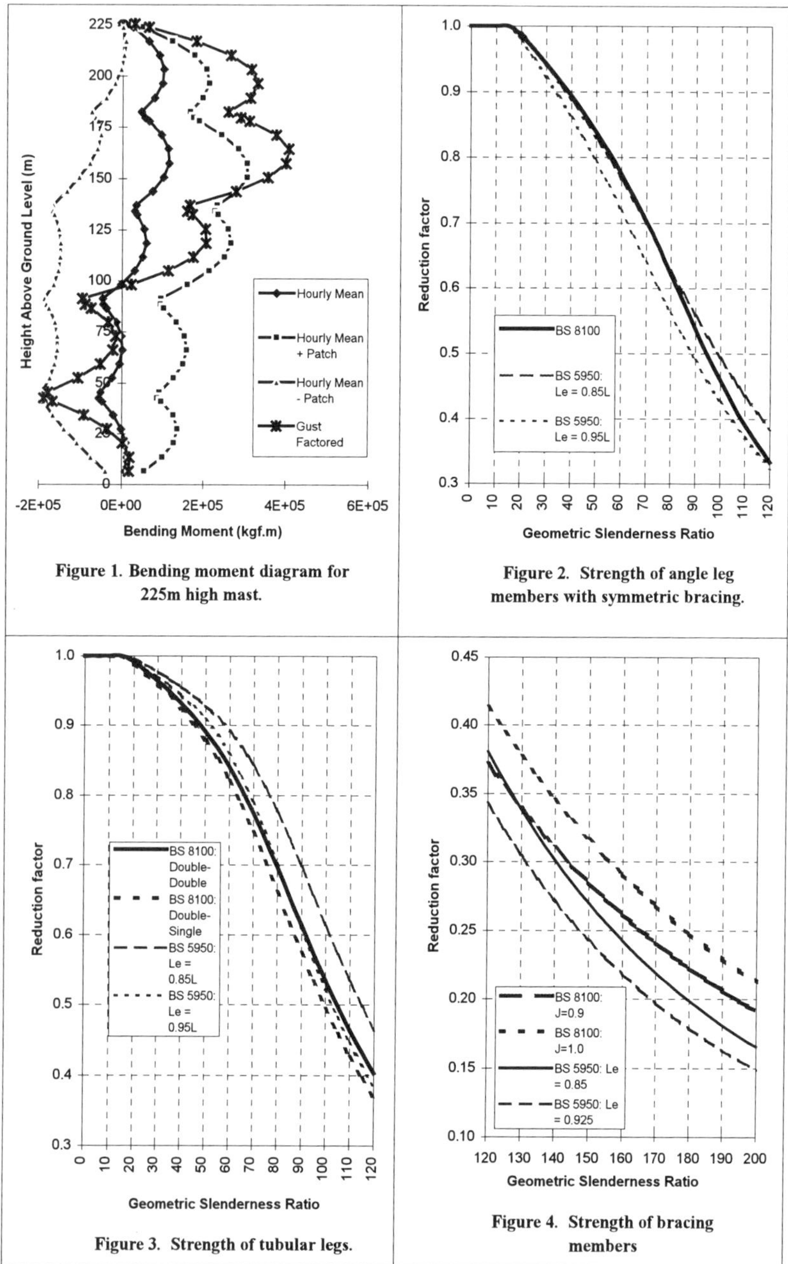

Figure 1. Bending moment diagram for 225m high mast.

Figure 2. Strength of angle leg members with symmetric bracing.

Figure 3. Strength of tubular legs.

Figure 4. Strength of bracing members

Aesthetics in Mast and Tower Design

U Støttrup-Andersen [1]

ABSTRACT

The very rapidly increase in the need for telecommunication also heavily influence the required number of masts and towers. Especially the exponential growth in the celluar telephone area give rise to numerous new masts and towers spread all over the landscape and in the cities.

Both local authorities, nature conservancy organizations, etc. set forward requirements for reducing the impact on the landscape from new antenna masts and towers. This have already resulted in more attention and concern from the telecommunication companies towards an aesthetic approach in the layout and design of new antenna supporting structures.

In the paper is given a brief introduction to the problem, and a discussion of the new approach to design based on aesthetics more than the engineering functional and economical optimization principles that normally govern the design.

INTRODUCTION

Only a few years ago many experts thought that the need for masts and towers for communication would decrease due to the new technique with satellites, small parabolics, hybrid cables, sensitive receivers etc. However until now the need for new masts and towers has in fact increased.

The rapid extension of local and commercial radio and TV channels combined with the explosion in the use of cellular telephones calls for more and more masts and towers. This considerable increase in the number of new structures, especially for cellular telephone

---

[1] Head of Department of Advanced Steel Structures, RAMBØLL, DK-2830 Virum, Denmark

nets, has given increased importance in many countries to the aesthetic appearance of these structures, and their impact on the landscape, and the aesthetic qualities are now often an important issue for new structures. This is also a new challenge to the designers and today a new mast or tower design often involves a close co-operation between architects and engineers while in the past engineers/fabricators traditionally had undertaken the total design.

A NEW IDENDITY OF LATTICE TOWERS

Lattice masts and towers are traditionally the most popular structural configuration used for telecommunication masts and towers. The lattice outline is the optimal solution when regarding the economics and the functional requirements, and this kind of structures is therefore dominant in almost all countries.

In the beginning of the implementation of the network for the cellular telephones in Denmark all the new antenna masts and towers were of the latticed type. The lattice towers were designed purely to optimize the economic aspect satisfying of course the technical and functional requirements of the antennas. As mentioned are lattice masts and towers very cost-effective however do not all find that the traditional lattice antenna tower is a beautiful structure that fit into the landscape. This attitude has become more and more common as the number of masts and towers increase, and now it is more a rule than an exception that a new antenna tower is not accepted by the communities and the authorities.

Eventhough the network for the various systems of cellular telephones is relatively well established throughout Denmark, the first operator Tele Denmark A/S, still needs relatively many new antenna towers over the country. Faced with this fact and combined with the increasing difficulties of acceptance of new structures Tele Denmark A/S initiated a development of a new design of a series of lattice towers. International acknowledged architects and consulting engineers were ask to come up with a new design where the aesthetic parameters were the most important. However it was also a clear requirement that the functionally of the new tower should be maintained as well as the overall cost of the new series inclusive foundations, erection, maintenance, etc. should be within reasonable limits. It is acceptable that the costs for the "aesthetic tower" is higher than for the traditional layout, but as for everything else the extra price should be relatively low.

After a detailed analysis of the various elements and possibilities of lattice towers the group of architects and engineers proposed a quite new design of a standard antenna tower to Tele Denmark A/S. The tower shall express the actual function and not be like any "standard tower", that is that the tower shall express the new age the mobile telephones are a part of. The tower shall be an identity creator, it shall be a landmark at the same time as it express its high technology function.

The new lattice tower has a triangular cross-section. One of the three legs is formed as a bended steel plate in which the many cables and feeders are hidden, as well as the climbing ladder is integrated as a part of the "main leg". This leg is arranged vertically while the two other legs are sloping towards the top of the tower. In this way is the vertical element accentuated as well as the sloping elements add suppleness and direction to the structure. The legs are connected by a very light cross-bracing system of solid round bars.

| The new lattice tower | Erection of the tower | Antennas on the top |

As a transparent landmark in the landscape is the tower divided horizontally by platform elements with equal mutual distance of 3.6 m. These elements serve visually as a scale and functionally as resting/working platforms. An other advantage of this vertical scale is that despite the height of the towers they look to be within the same family of design.

A very important aspect of the visual impression of the structures is the antennas and their arrangement and support on the tower structure. Quite often give the placing and location of the antennas on the tower an impression of randomness and they dominate the visual impression of the structure. It has therefore been an important part of the new design to analyze the various possible antenna configurations, including not only the antennas of Tele Denmark A/S, but also antennas for other operators and services, as there might be requirements from the authorities in the nearest future that new masts and towers should be capable to support antennas from a number of operators.

To be sure that the visual impression of the new tower design is maintained in the future it is necessary that the installation of new antennas is planned and that there are certain restrictions in the placing of the various antennas. It might of course in the future show that the layout and design of new antennas differ dramatically from what is known to day, but in this circumstances a new analysis and design of new supporting systems may be necessary.

The first tower of the series of the new standard lattice antenna tower has been taken into service in during summer 1998.

A view up in the tower

Erection

CONCLUDING REMARKS

The Scandinavian countries have today a very close and fine coverage by cellular telephones, and Denmark is probably the country with the best coverage in the World today. As a relatively small country with rather sensitive landscape and a population relatively evenly distributed over the country, the density of masts and towers is already now high, and the lobby towards new structures is becoming more and more strong. However if the population also want the modern services offered compromises have to be found, and careful designs taking the aesthetics aspects serious may be a way ahead.

There are no doubts that the same problems will soon arise in many other countries, and maybe they will be inspired by the present development.

INVOLVED PARTIES

The new design of the series of lattice towers with heights between 30 m and 63 m is developed for

Tele Denmark A/S	by:	RAMBØLL	and:	KHR AS
Sportorno Allé 12		Bredevej 2		Teknikerbyen 7
DK 2630 Taastrup		DK 2830 Virum		DK 2830 Virum
Phone: + 45 43 58 02 02		Phone: + 45 45 98 60 00		Phone: + 45 45 85 44 44

Proposed Revision to the ANSI/TIA/EIA-222 Standard

David G. Brinker, P.E.,[1] Member ASCE

Abstract

The next revision of the ANSI/TIA/EIA-222-F standard "Structural Standards for Steel Antenna Towers and Antenna Supporting Structures," will represent the most drastic change to the standard since the first publication in 1949. The purpose of the paper is to discuss some of the major changes being considered and the reasons the changes are necessary.

Introduction

The proposed changes to the standard are not being driven by a rise in the number of failures or increased complaints from the users of the standard. There are two major trends that are driving the need for a major change, specifically, the methods used to measure and record wind speeds and the use of limit states structural steel design criteria. Also, significant research has been undertaken regarding ice loading which may lead to the inclusion of mandatory minimum ice loading requirements in the standard.

Basic Wind Speed

The current TIA/EIA standard is based on fastest-mile basic wind speeds. A fastest-mile wind speed represents a wind speed averaged over the shortest time for a mile of wind to pass an anemometer. For example, if the shortest time for a mile of wind to pass an anemometer was 60 seconds, the fastest-mile wind speed would be 1 mile per 60 seconds or 60 mph. The averaging period is different for each wind speed, the higher the wind speed, the shorter the averaging period. The wind speeds listed in the current TIA/EIA standard were based on the fastest-mile wind map published in the1988 edition of the ASCE 7 Standard "Minimum Design Loads for Buildings and Other Structures."

[1] Vice President, Engineering, ROHN Industries, PO Box 2000, Peoria, IL

The trend in measuring and recording wind speeds has been to measure and record gust wind speeds as opposed to fastest-mile wind speeds. The gust wind speeds are generally averaged over 3 to 5 second time periods and are therefore higher than fastest-mile wind speeds. Gust wind speeds are generally the wind speeds reported by the media and are not easily converted to fastest-mile wind speeds for comparison purposes.

Based on new wind studies and the phasing out of the fastest-mile wind speed, the most recent ASCE 7-95 Standard "Minimum Design Loads for Buildings and Other Structures" has replaced the fastest-mile wind map with a new map based on a higher, 3-second gust reference wind speed. Since most building codes have traditionally adopted the wind maps published by ASCE 7, it is anticipated that the 3-second gust wind speed map will also be adopted. The new map will therefore have to be adopted by the TIA/EIA standard in order to be compatible with the major building codes.

An entirely new model for wind loading is required when changing to a different reference wind speed. Specifically, different gust and height escalation factors will be required. Since the fastest-mile wind speed is an averaged wind speed, a gust factor greater than or equal to one, dependent upon tower height, is currently used to account for gusts of shorter duration. For a 3-second gust wind speed, a gust factor of 1.0 or less will most likely be adopted.

Wind speeds are escalated with height up to a gradient wind speed where the wind speed is considered to be independent of the roughness of the ground surface. The gradient wind height is considered the same by ASCE 7 for both the fastest-mile and the 3-second basic wind speeds. The 3-second wind speed model, however, does not reduce towards the earth's surface as rapidly as the fastest-mile wind speed. The height escalation factors for a 3-second gust basic wind speed will therefore be less than the height escalation factors used with the current standard.

Structural Steel Design

The current TIA/EIA standard utilizes the 9th edition of the "Manual of Steel Construction" published by the American Institute of Steel Construction (AISC). This AISC standard is based on an allowable stress design (ASD) approach. For ASD, the anticipated loads for a structure are considered in the structural analysis without modification. Safety factors are applied to the capacities of the structural components and compared to the results of the analysis.

AISC has published a specification for structural steel design, referred to as LRFD (load and resistance factor design). The first LRFD AISC specification was published in 1986. With this approach, load factors are applied to the anticipated loads to obtain a limit state or ultimate loading condition for structural analysis. Resistance factors are applied to the capacities of the structural components and compared to the results of the analysis.

Different resistance factors are used for different types of capacities (tension, compression, bending, etc.) in accordance with the accuracy or confidence level of the methods used to determine the various capacities. The load factor can be thought of as applying the main portion of the safety factor to the loads rather than to capacities as done with ASD. A small portion of the safety factor remains with the capacity as a resistance factor.

More flexibility can be achieved with LRFD accounting for different types of loading. For example, a higher load factor could be applied to wind loads compared to dead loads which are more accurately determined relative to wind loads. Using ASD, it is not possible to vary safety factors with the type of loading since safety factors are considered after analysis.

The LRFD approach has been widely adopted for many types of structures. In fact, it is unlikely that AISC will publish another edition of the ASD specification. Due to the popularity of the LRFD standard, it is inevitable that the TIA/EIA standard will have to abandon the use of the ASD approach.

Ice Loading

A large portion of the United States has been mapped using an ice load model for freezing rain. The modeling utilized past weather records for temperature, wind speed, and precipitation type and rate along with damage reports from past severe icing events to verify the model. An extreme value analyses was used to generate 50-year extreme ice thicknesses and extreme wind-on-ice loads. The extreme ice thickness values were mapped along with calculated concurrent wind speeds, which when applied to an extreme ice thickness, results in the extreme wind-on-ice load. This methodology results in one loading condition that represents both the extreme ice weight and the extreme lateral load with ice.

Extensions of the mapped regions may be possible based on local observations and other data allowing mandatory minimum glaze ice loading criteria to be included in the next revision. Rime ice, however, from in-cloud icing, will likely remain a site-specific issue as in the current standard.

Other Issues

Probably the most controversial issue to be resolved is how to model the dynamic effects of wind loading. In the current standard, one gust factor is applied to the entire structure which does not account for the effects of local gusts acting on a structure. A guyed mast, where a local gust may occur entirely within one span, is a good example of why this effect needs to be considered. Since the mast behaves as a continuous beam, a change in load in one span can increase the member forces in other spans.

The controversy concerns the best method to account for the dynamic effects of wind. These methods vary from very complicated dynamic analyses to prescriptive design approaches. Patch loading and empirical design approaches based on guy stiffness, span lengths, and other factors have also been proposed. The prescriptive design approach, due to it's simplicity and similarity with existing design practices, is being evaluated as perhaps the best method to be developed for the next revision.

A prescriptive design approach would involve establishing design rules believed to result in reliable structures. These rules would be used in conjunction with a conventional analysis using an escalated wind speed with a constant gust factor. Many studies are available regarding the dynamic effects of wind which would form the basis of the rules, but perhaps more importantly, the rules would document the practice of many engineers who have long recognized the unpredictable nature of wind loading and the inevitable limitations of all manual and computer analyses.

An example of a prescriptive approach would be to require minimum leg and brace loads relative to the maximum values within a span for a guyed tower. Another approach would be to modify the mast's shear and moment diagrams. The goal would be to place additional strength at the mast locations known to be affected by local gusts and perhaps reduce requirements at locations known to be over designed.

Other issues that have long been considered for inclusion into the standard will most likely be included in the next revision. Many of these considerations have already been addressed by ASCE 7, such as reliability classes, considerations for structures placed on hills, ridges etc.; and height escalation equations for terrain conditions other than open flat terrain. It is anticipated that ASCE 7 criteria will be adopted for these issues, modified as appropriate for telecommunication structures.

Introduction to the ASCE Guide for the Dynamic Response of Lattice Towers

Murty K. S. Madugula[1], M. ASCE.

Abstract

The information regarding the dynamic response of lattice towers is scattered throughout the literature, making it difficult for the practicing engineers to obtain the necessary information for design purposes. The Task Committee prepared the Guide by compiling and clarifying current methodologies in a single source. Both self-supporting lattice towers and guyed lattice masts are included in the Guide which is targeted to practicing design engineers. The topics covered in the Guide include, among others, dynamic analysis, response to wind loads, ice shedding, seismic input, and vibration control. Extensive bibliography, divided into six topics, is appended.

Introduction

The Task Committee on the Dynamic Response of Lattice Towers is under the jurisdiction of the Technical Committee on Special Structures of the Technical Administrative Committee on Metals. It was approved by the Executive Committee of the Structural Engineering Institute on April 14, 1996. The Committee consists of 24 members distributed over three continents - Asia, Europe, and North America. The list of members is given in the Appendix. The Committee held five meetings – in Chicago, Illinois (April 1996 and September 1997), Haslet, Texas (September 1996), Portland, Oregon (April 1997), and Toronto, Ontario, Canada (May 1998). In addition, an informal meeting of the Task Committee was held in San Francisco, California, on July 19, 1998 during the Structural Engineers World Congress. A draft was put together in the first week of November 1998 for review by the Task Committee members. After their comments are received, the draft will be revised and released for public comment. The document will be finalized after consideration of these comments.

[1]Professor of Civil Engineering, University of Windsor, Windsor, Ontario, Canada N9B 3P4.

Scope

Communications are playing an ever-increasing role in our society, and as the information superhighway expands, there is a need to better utilize the capacity of existing towers and improve the capacity of new towers. In order to use the existing and new towers more efficiently, a better understanding of their dynamic response is required. The analysis of lattice towers has far lagged behind state-of-the art dynamic analysis methods used for the design of buildings. Currently, dynamic loads on lattice towers, such as wind and earthquake, are analyzed by equivalent static methods even for the tallest guyed towers. Analysis methods suitable to calculate the wind-induced response of lattice towers have not been defined until recently. Additionally, there is a belief that the seismic response of lattice towers does not govern their design, because of their light weight.

The control of the vibration of lattice towers subjected to wind has been implemented successfully only in recent years. Most of the vibration control of structures has been achieved in the past by using the concept of ductility to dissipate the energy caused by the dynamic vibrations. However, the application of ductility often requires complex and costly connections and may develop permanent deformations, which are unsuitable for towers. New approaches for the control of vibrations consist in providing the structure with special devices that may generate forces, displacements, or dissipate energy to restrain excessive vibrations.

The information regarding the dynamic response of lattice towers is scattered throughout the literature, making it difficult for the practicing engineers to obtain the necessary information for design purposes. The Task Committee prepared the Guide by compiling and clarifying current methodologies in a single source. Both self-supporting lattice towers and guyed lattice masts are included in the Guide which is targeted to practicing design engineers. North American Standards for the design of communication towers do not directly address the dynamic response of such towers and it is hoped that the Guide will fill that void. The Guide is divided into five chapters. Extensive bibliography divided into six sections – (a) Static and dynamic analysis of lattice towers, (b) Dynamics of cables, (c) experimental investigation, (d) Response to wind and ice loading, (e) Guides and Standards, and (f) Tower failures – is appended to the Guide. The following sections briefly introduce the contents of the various chapters.

Chapter 1 - Introduction

Historical perspectives, description of the self-supporting and guyed lattice towers, and Standards related to latticed communication towers are presented in this Chapter.

Chapter 2 – Time-varying Loads

Chapter 2 deals with wind loads, ice-shedding loads, seismic input, and foundations. Since tower gravity loads are typically quite light, wind can account for a significant portion of the critical load effects. A reliable tower design therefore requires an accurate assessment of the wind induced response of the structure that accurately reflects the dynamic characteristics of the wind.

Another source of transient loads which is particular to a guyed mast is the sudden fall of ice from its guy wires. Normally these effects are treated quasi-statically and the tower is analyzed under several unbalanced scenarios. However, in cases of induced ice-shedding, where all the ice on a guy wire is forced to shed almost instantaneously, or in sudden natural ice-shedding situations, the tower responds dynamically.

Little work has been done to document the performance of towers in earthquakes. Only recently have there been any significant analytical studies. Evidence from the 1989 Loma Prieta Earthquake (Magnitude 7.1) and the 1994 Northridge Earthquake (Magnitude 6.8) demonstrated that towers are affected by earthquakes and sometimes catastrophically.

Whether the tower is guyed or self-supporting, the local soil conditions should be investigated for their influence on earthquake ground motions, and the susceptibility for vibration induced settlement and liquefaction.

Chapter 3 – Dynamic Analysis

Dynamic analysis is dealt with in Chapter 3. Both linear and non-linear dynamic analyses are included. Linear dynamic analysis methods, generally used for the dynamic analysis of building structures, can be used to analyze self-supporting towers. Modal, response spectrum, and time history strategies are included in the linear dynamic analysis. However, such methods are not adequate for analyzing guyed towers. Guyed towers are geometrically nonlinear and require nonlinear methods of dynamic analysis. Nonlinear stiffness models and second-order nonlinear time history analyses are considered.

Chapter 4- Wind-induced Response

. The dynamic response to wind is discussed in Chapter 4. To overcome the deficiencies of existing static approaches while avoiding the complexity of full dynamic analysis, a pseudo-static patch load method of analysis was included. As its name suggests, the patch load method employs a series of static load patterns to reproduce the effect of gusty winds. Since only static loads are involved, the method can be implemented on standard commercial static analysis software.

Chapter 5 – Vibration Control

Vibration control is the topic of Chapter 5. Depending on the mechanism used to activate these devices, they are classified as Active Control Systems or Passive Control Systems. In the Active systems, the control is achieved by automated devices composed of sensors, controllers, and actuators. Sensors measure the response of the structure and send signals to the controllers to process. They then compute the forces required to control the vibrations, and send them to the actuators, which in turn, impose the control forces or displacements to the structure. The motion of the structure is resisted by applying external forces or by modifying the structural properties of active elements as in the case of tuned mass or liquid dampers that add damping to the critical modes of vibration.

In the passive control systems, energy dissipation is part of the structure. Control forces are developed from the motion of the structure subjected to dynamic excitations. The amplitude and direction of these forces depend on the relative motion of the attachment points of the energy devices.

Both systems are currently used to control induced wind vibration on lattice towers. The active systems may be more effective on guyed towers, but are more expensive and difficult to implement and require continued supervision.

Appendix

Membership of the ASCE Task Committee
on the Dynamic Response of Lattice Towers

Andrew Allsop, U. K.	Ramesh B. Malla, U. S. A.
T. V. S. R. Apparao, India	Cedric Marsh, Canada
Madison J. Batt, U. S. A.	Ghyslaine McClure, Canada
George E. Blandford, U. S. A.	Amir Mirmiran, U. S. A.
David Brinker, U. S. A.	Mogens Gunhard Nielsen, Denmark
James S. Cohen, U. S. A.	Markus Ostendorp, U. S. A., Secretary
Allan G. Davenport, Canada	Celina U. Penalba, U. S. A.
Mark W. Fantozzi, U. S. A.	Bruce Sparling, Canada
Badri Guru, Canada	Ulrik Stottrup-Andersen, Denmark
Mehran Keshavarzian, U. S. A.	John Tyrrell, U. K.
J. L. Lilien, Belgium	Yohanna M. F. Wahba, Canada
Murty K. S. Madugula, Canada, Chair	Simon Weisman, Canada

Dynamic Analysis of Lattice Towers

George E. Blandford[1], *M. ASCE* and Bruce Sparling[2]

ABSTRACT – A review of techniques for the dynamics analysis of self-supporting and guyed lattice towers is presented. Both linear and nonlinear response is considered. For linear analysis, mode superposition using eigenvalue decomposition or load dependent Ritz vector approaches is considered along with response spectrum and time history analyses. In the nonlinear case, nonlinear stiffness models are presented for truss members, guy cables, frame members, and flexible connections. Dynamic analysis for the nonlinear response of lattice towers is restricted to time history analysis.

INTRODUCTION – The primary objective of a deterministic structural dynamic analysis is the evaluation of the displacement and member force time histories for a given lattice tower subjected to the time-varying loads. In most cases, an approximate analysis involving a limited number of degrees of freedom will provide sufficient accuracy. Solving the equations of motion in terms of these degrees of freedom provides the desired time history response of a lattice tower structure subjected to the prescribed time-varying loads.

This paper presents a descriptive overview of Chapter 3 in the *ASCE Guide on Dynamic Response of Lattice Towers*. Chapter 3 is divided into three main sections: structural modeling, linear dynamic analysis, and nonlinear dynamic analysis. This guide is being prepared by the Task Committee on Dynamic Analysis and Design of Lattice Towers, which is a subcommittee of the Technical Committee on Special Structures.

STRUCTURAL MODELING – There are two general procedures for representing the discrete structural members of a lattice tower structure: (1) analytical member representation and (2) interpolation models in terms of member displacement degrees of freedom (typically use a finite element representation). Chapter 3 of the guide focuses on the interpolation models. For such systems, the algebraic equations are expressed in terms of the structure mass, damping, and stiffness matrices.

---

[1] Dept. of Civil Engineering, University of Kentucky, Lexington, KY 40506-0281
[2] Dept. of Civil Engineering, University of Saskatchewan, Saskatoon, Saskatchewan, CANADA

Element stiffness equations are expressed in terms of the principal cross section co-ordinates that include flexible rotational behavior at the ends of the members to model joint connection slippage and rotation behavior. A closed-formed stiffness matrix is provided in an appendix that is based on a flexibility-stiffness transformation procedure. Emphasis is placed on the use of principal cross section coordinates since numerous tower structures are composed of singly symmetric angle members. For such members, bending moments, axial forces, and the associated displacements are expressed in terms of the cross section centroid; whereas torsion and transverse force variables are expressed in terms of the cross section shear center. Details are presented in the guide to transform the force and displacement degrees of freedom evaluated at the shear center to the centroid. This transformation provides a consistent representation of the forces and displacements.

Element mass is represented in terms of a consistent mass representation. A lumped mass contribution is also included to represent the additional mass introduced by the connection details at the tower structure joints.

Two different damping models are considered. The first model is based on a linear viscous representation in which the damping is frequency-dependent and proportional to the structure velocity. Only the Rayleigh damping approximation is considered for the frequency-dependent case due to its simplicity and universal acceptance by the profession. A least-squares procedure is presented to calculate the Rayleigh damping constants when the damping ratio is specified in more than two modes.

A frequency-independent model is also considered, referred to as structural damping herein but is also commonly referred to as solid or hysteretic damping. Structural damping is not proportional to the velocity and hence is not a viscous damping model.

LINEAR DYNAMIC ANALYSIS – Three different strategies are considered: (1) modal, (2) response spectrum, and (3) time history.

Modal analysis involves transforming the discrete, dynamic equilibrium equations into an alternative form that should be easier and less expensive to analyze. In theory, many different techniques can be used to construct a transformation matrix. In practice, only two techniques are commonly used: (1) exact transformation based on an eigenvalue decomposition, and (2) an approximate transformation based on load-dependent Ritz vectors.

The eigensolution procedure for calculating the transformation matrix is large and computationally expensive compared to the use of orthogonal Ritz vectors, which were popular before the early '70's (Wilson 1995). It was presumed that exact eigensolutions would produce solutions that are more accurate. Wilson *et al.* (1982) presented a method of mode superposition analysis and showed that a set of m load-dependent Ritz (LDR) vectors produced more accurate results than those produced using m exact eigenvectors for some typical structural engineering problems. Bayo and Wilson (1984) further illustrated the efficiency and accuracy of the LDR method for the solution of wave propagation problems.

Mode superposition analysis produces the complete time history response of joint displacements and member forces due to specified dynamic loading. Two disadvantages of this approach are (Wilson 1996): (1) large volume of output for each dynamic load, and (2) analysis must be repeated for each dynamic load.

Response spectrum seismic analysis provides computational advantages for the earthquake analysis of tower structures. The method only involves the calculation of maximum values of the displacements and member forces in each mode using smooth design spectra that are the average of several earthquake records. Since the maximum displacements and member forces in each mode occur at different points in time, some averaging technique must be used to superimpose the modal maximums.

One common approach to superimpose the modal maximums is the Square Root of the Sum of Squares (SRSS) technique that has produced excellent results for two-dimensional structures. The SRSS method assumes that all the maximum modal values are statistically independent. This assumption can produce significant errors in the analysis of three-dimensional structures (Wilson *et al.* 1981).

A more recent development is the Complete Quadratic Combination (CQC) method (Wilson *et al.* 1981). This method has been shown to reduce errors in combining modal maxima. Furthermore, the CQC method degenerates into the SRSS method for systems with well-spaced natural frequencies.

Frequently, the linear dynamic equilibrium equations are evaluated using a direct integration procedure rather than performing the matrix transformations required in both the modal and response spectrum analyses. This is the case when the desired time interval is relatively small or simply to avoid the transformation computations. Direct integration can also be used for the solution of the modal equations. Direct time integration involves expressing the dynamic equilibrium equations at discrete points in time, i.e. **time history analysis**).

Numerous time integration schemes are available for the recursive evaluation of the dynamic equilibrium equations (e.g., Bathe 1996). Only two schemes are considered in the guide: central difference method and the Hilber-Hughes-Taylor α-method (Hilber *et al.* 1977). Both schemes are second-order accurate in time. However, the central difference method is a conditionally stable, explicit time integration scheme. The α-method is an unconditionally stable, implicit time integration scheme that numerically dampens high frequency responses for $\alpha < 0$, a plus for multiple degree of freedom systems. For $\alpha = 0$, the α-method is simply the trapezoidal rule or Newmark's constant-average acceleration method. Recall (e.g., Bathe 1996) that conditionally stable time integration schemes must utilize a discrete time increment $\Delta t \leq \Delta t_{cr}$ to maintain stability, though this may still be much too large to achieve accuracy; $\Delta t_{cr} \equiv$ critical time increment. No such restriction is imposed on unconditionally stable schemes, though the selected time increment must be small enough to achieve accuracy. If both the mass and damping matrices are diagonal, the recursive solution provided by the central difference method is trivial. However, if the mass matrix or damping matrix is not diagonal, then the α-method should be used.

Furthermore, since the α-method introduces numerical damping for $\alpha < 0$, it may not be necessary to introduce a damping approximation into the structural system.

NONLINEAR DYNAMIC ANALYSIS – Geometric nonlinear behavior is restricted to first-order, i.e. large displacements with small rotations and strains. A finite element model is given to represent this behavior for a truss member. Analytic expressions are presented for the guy wires for both static and dynamic behavior. Analytic expressions are also presented for frame members that include flexible connection response at the ends of the members.

Material nonlinear response is restricted to the truss members (failed member is assumed to have zero stiffness) and connections. A nonlinear moment-rotation model is presented that includes unloading modifications and can model joint rotational plasticity.

Solution of the nonlinear equations is expressed in terms of an incremental – iterative time history analysis. A modified Newton-Raphson solution scheme is discussed.

REFERENCES

1. Bathe, K.-J. (1996). *Finite Element Procedures*, Prentice Hall, Englewood Cliffs, NJ, Chapter 9.

2. Bayo, E.P. and Wilson, E.L. (1984). "Use of Ritz vectors in wave propagation and foundation response," *Earth. Engrg. Struct. Dyn.*,**12**, 499-505.

3. Hilber, H.M., Hughes, T.J.R. and Taylor, R.L. (1977). "Improved numerical dissipation for time integration algorithms in structural dynamics." *Earth. Engrg. Struct. Dyn.*, **5**, 283-292.

4. Wilson, E.L., Der Kiureghian, A. and Bayo, E.P. (1981). "A replacement for the SRSS method in seismic analysis." *Earth. Engrg. Struct. Dyn.*, **9**, 187-194.

5. Wilson, E.L. and Button, M.R. (1982). "Three-dimensional dynamic analysis for multi-component earthquake spectra." *Earth. Engrg. Struct. Dyn.*, **10**, 471-476.

6. Wilson, E.L., Yuan, M.-W. and Dickens, J.M. (1982). "Dynamic analysis by direct superposition of Ritz vectors." *Earth. Engrg. Struct. Dyn.*, **10**, 813-821.

7. Wilson, E.L. (1995). "A complete subspace basis for a dynamic mode superposition analysis." *Proceedings of the Symposia Honoring Professor Stanley Dong*, ASME, AMD-MD '95, University of California, Los Angeles, June 28-30, 21 pp.

8. Wilson, E.L. (1996). *Three Dimensional Dynamic Analysis of Structures – with Emphasis on Earthquake Engineering*. Computers and Structures, Inc., Berkeley, CA.

Wind Loads and the Response to Wind

B. F. Sparling[1]

Abstract

This paper summarizes the sections in Chapters 2 and 4 of the ASCE Guide on the Dynamic Response of Lattice Towers (ASCE 1998) dealing with dynamic loading due to gusty winds and the resulting dynamic response. A brief overview is presented of the most common statistical measures used to describe the wind for the purposes of analysis and design. In addition, various methods for estimating dynamic response are reviewed, along with the potential advantages and disadvantages of each approach.

Overview of Gusty Winds

Due to the light weight of lattice towers, wind is a major contributor to many of the critical load effects governing the design of such structures. The dynamic nature of wind gusts can also effectively excite a significant degree of rather complex vibrations. It is therefore essential to obtain reliable estimates of wind loads and wind loading effects.

Friction generated between a moving air mass and the ground surface results in the formation of a turbulent *atmospheric boundary layer* that can vary in depth between several hundred meters and several kilometers. The fundamental characteristics of the flow within this boundary layer have a direct impact on tall, dynamically sensitive structures such as lattice towers. First of all, the frictional drag force impedes the movement of air near the ground, causing a gradual increase in the average, or *mean*, wind speed with increasing elevation. Secondly, turbulent eddies, or gusts, result in wind speed fluctuations that vary in a random fashion over time, as well as spatially.

The instantaneous wind speed $U(z,t)$ at some elevation z can be characterized by its time-averaged mean value, $\overline{U}(z)$, and a turbulent component $u(z,t)$ that fluctuates randomly about that mean value.

---

[1] Associate Professor, Department of Civil Engineering, University of Saskatchewan, 57 Campus Drive, Saskatoon, SK, Canada, S7N 5A9

Mean Wind Speeds

For a given wind record, the magnitude of the mean wind speed $\overline{U}$ is not unique, but will depend on the duration T of the averaging period used. Shorter averaging periods tend to accentuate the influence of strong gusts, thereby increasing the value of $\overline{U}$. Since the definition of the mean wind speed varies widely in different jurisdictions, and can also change over time, care must be exercised when interpreting design requirements. In the United States, for example, the most recent ASCE/ANSI 7-95 Standard (ASCE 1996) has adopted a 3 second gust speed to replace the traditional *fastest mile of wind* value as the basis for design. Hourly and 10 min mean wind speeds are widely used in other countries.

Because of the variations in wind speeds with height, mean wind speed values must be specified at some reference elevation, typically 10 m above the ground. The increase in mean wind speeds with increasing elevation can be described in a number of ways, including the familiar power law form

$$\overline{U}(z) = \overline{U}_{ref}\left(z/z_{ref}\right)^{\alpha} \qquad [1]$$

in which $\overline{U}_{ref}$ is the reference wind speed specified at elevation z_{ref}. The exponent α defines the rate of increase of $\overline{U}(z)$, depending on such factors as the ground roughness at the site and the duration of the averaging period, T.

Descriptions of Turbulence

Although wind gusts are random, they can be characterized by specific statistical properties. Among the attributes to be considered in the design of lattice towers are the intensity of the turbulence, the spatial organization and frequency distribution of wind gusts, and the probability distribution of gust wind speeds. Chapter 2 of the Guide on the Dynamic Response of Lattice Towers provides expressions describing these properties.

The intensity of turbulence is related primarily to the strength of the wind storm ($\overline{U}$) and the ground roughness at the tower site. The turbulence intensity is generally quantified in terms of the *rms* (root-mean-square) value of the fluctuating component of the wind speed, $\tilde{u}$.

Starting with the formation of large, unstable eddies, turbulence is generated as these eddies are broken down into progressively smaller sizes. The resulting distribution of gust sizes can be related to the *power spectral density* function, or spectrum, which describes how $\tilde{u}^2$ is distributed with frequency. Although a number of expressions have been proposed to model the wind spectrum, all suggest that the majority of the turbulent energy lies within a range extending roughly from 0.01 Hz to 2.0 Hz, with very little left above 5.0 Hz.

The overall effectiveness of the gusty wind in producing a given response is also dependent on the spatial organization of the turbulent eddies. Correlation functions, expressed in terms of the spatial separation between any two points on a structure, are used to quantify the expected degree of similarity between the fluctuating wind speeds at those two points. While a lack of correlation in wind speeds acts to reduce the effective load on tall cantilevered structures such as self-supporting towers, it can have the effect of actually increasing the dynamic response of guyed towers whose masts are continuous across several guy support levels.

Dynamic Response to Gusty Winds

The typical form of response in a flexible structure subjected to gusty winds is illustrated in Figure 1. As shown, the total response $r(t)$ may be broken up into a static time-averaged mean component $\bar{r}$ and a fluctuating component. The fluctuating response can be further sub-divided into a quasi-static background response r_B to large, slowly varying gusts, and the more rapid resonant component r_R that includes the response at all participating natural frequencies of the system. The power response spectrum $S_r(f)$, shown here in its logarithmic form, illustrates how the dynamic response is distributed with frequency.

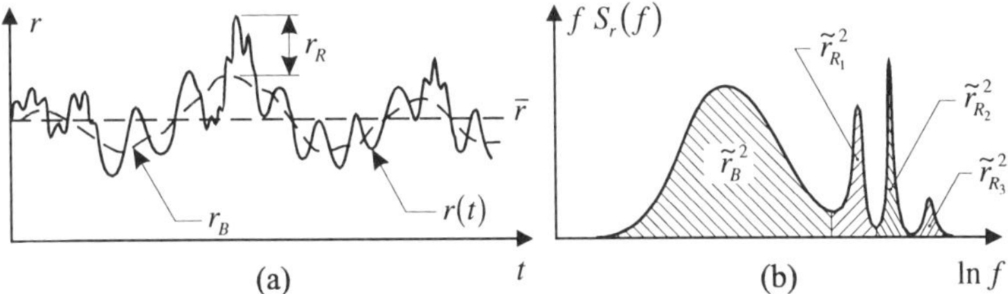

Figure 1. Dynamic response to wind: (a) Time history; (b) Power spectrum.

The peak design response $\hat{r}$ can be expressed in the following form:

$$\hat{r} = \bar{r} \pm g_p\, \tilde{r} \qquad\qquad [2]$$

where $\tilde{r}$ is the total *rms* dynamic response (background plus resonant) and g_p is a statistical peak factor that varies over a narrow range from about 3.5 to 4.5.

Dynamic Analysis Methods

Traditionally, peak dynamic responses have been estimated using the *Gust Factor* method, in which the static basic (mean) wind loads are simply increased by a constant factor to account for all gust-related effects (TIA/EIA. 1996). While this approach is adequate for self-supporting towers that respond primarily in their lowest vibration mode, it has been found to be potentially unconservative for guyed towers (Sparling et al. 1996). In particular, the gust factor method tends to underestimate mid span dynamic bending moments and shear in the mast. Also, it does not define the full envelope of possible response magnitudes generated by vibrations about the mean position; stress reversals resulting from these vibrations can be critical in the design of leg splices and in considerations of fatigue performance.

A rigorous dynamic analysis of guyed towers is complicated by the complex interaction between the mast and the guy cables, as well as by the random nature of the wind loads. In general, dynamic analysis approaches can be classified as one of two types: frequency domain methods utilizing modal analysis techniques, or time domain methods using a time-marching scheme. In the modal analysis of guyed towers, a large number of modes must be considered to fully described the motion of the mast and the guys. In the time domain approach, explicit wind load time histories must be defined for each location on the tower; however, nonlinear dynamic response can be readily accounted for. In both methods, it is imperative to properly model the spatial structure of wind turbulence, as well as the inertial and damping characteristics of the guys.

Due the deficiencies of the gust factor method and the difficulty associated with a full dynamic analysis, a number of approximate dynamic analysis methods have been proposed. "Patch load" methods, which use a series of static load patterns to replicate the effects of gusty winds, have been adopted by the British design standard BS-8100 Part 4 (BSI 1994), as well as by others. The British standard also includes a "simple" method for short guyed towers, in which designated values taken from mean response diagrams are used as a basis for estimating dynamic response values.

Summary

This paper presents an overview of material contained in Chapters 2 and 4 of the ASCE Guide on the Dynamic Response of Lattice Towers dealing with wind loads and wind-induced dynamic response.

References

ASCE. (1996). "ASCE/ANSI 7-95 Standard: Minimum Design Loads for Buildings and Other Structures", American Society of Civil Engineers, NY.

BSI. (1994). "British Standard BS 8100 Part 4 - Lattice Towers and Masts; Code of Practice for Lattice Masts", British Standards Institute, London, UK.

ASCE. (1998). "Guide on the Dynamic Response of Lattice Towers - Draft Edition", Task Force on the Dynamic Response of Lattice Structures, American Society of Civil Engineers, NY.

Sparling, B.F., Smith B.W., and Davenport A.G. 1996. "Simplified Dynamic Analysis Methods for Guyed Masts in Turbulent Winds". Journal of the IASS, Vol. 37, No. 2, pp. 89-106.

TIA/EIA. (1996). "TIA/EIA-222-F: Structural Standards for Steel Antenna Towers and Antenna Supporting Structures", Telecommunications Industry Association, Arlington, VA.

Vibration Control in Lattice Communication Towers

Celina U. Penalba[1]

ABSTRACT

Lattice communication towers are sensitive to dynamic environment generated by wind, ice, earthquakes, impact, blast loads and mechanical failures, such as guy rupture. The vibrations excited by these environments cover an ample spectrum of frequencies which affects the towers in different ways, ranging from serviceability problems, to fatigue or collapse. The control of such vibrations by supplemental damping devices in the categories of Active and Passive Control Systems have been implemented in lattice towers. The active system may be more effective in guy towers, but is more expensive, difficult to implement and requires continued supervision. The paper will focus on passive control devices for self supporting towers, tower masts and guys.

INTRODUCTION

The vibration control of lattice towers requires the assessment of the total damping provided by the structural, environmental, soil and supplemental damping. For linear elastic behavior, the total fraction of the critical damping can be determined by the sum of the damping ratios from all sources, that is:

$$\xi = \xi_s + \xi_e + \xi_S + \xi_d \tag{1}$$

ξ_s, ξ_e, ξ_S, and ξ_d = structural, environmental, soil and supplemental damping ratios.

Supplemental damping provided by damping devices is an effective, low cost and practical source of additional damping for lattice towers and is the subject of this presentation. The selection of a damper is governed by the source of excitation, dominant vibrations, efficiency, compactness, cost, maintenance requirements and safety.

PASSIVE CONTROL SYSTEMS

Passive control systems modify the response of the tower or guy by dissipating the energy induced by dynamic excitations reducing the corresponding inertia forces and relative displacements. They may be classified as displacement-dependent, velocity-dependent and others.

<u>**Displacement-Dependent Devices.**</u> Modeling of these dissipators is based on their force-displacement responses, axial-shear-flexure interaction or bilateral deformation. For the purpose of design, the force-deformation response may be expressed as:

---

[1] 570 Peach St. #10, San Luis Obispo, CA 93401

$$F = k_{eff} \, d \tag{2}$$

$$k_{eff} = \frac{|F^+| + |F^-|}{|d^+| + |d^-|} \tag{3}$$

where k_{eff} = effective stiffness of the device, F^+ and F^-, the forces corresponding to the displacements d^+ and d^-, respectively. The energy dissipation provided by the damper, should be determined from the area contained within the histeresis curve.

(a) *Friction Devices*. Examples of friction dampers are: The Pall Damper and the Slotted Bolted Connection (SBC).

The **Pall Friction Damper**, Fig 1(a,b), is composed of a series of metal plates clamped together with high strength bolts, treated to develop a predictable and reliable friction. The damper can fit at the intersection of the different bracing systems currently used in self-supporting towers. Its high energy dissipation properties under seismic excitation, low cost, little or no maintenance, a behavior not affected by temperature, velocity or stiffness degradation, makes the Pall Damper very effective for the seismic-protection of new and retrofitted lattice towers.

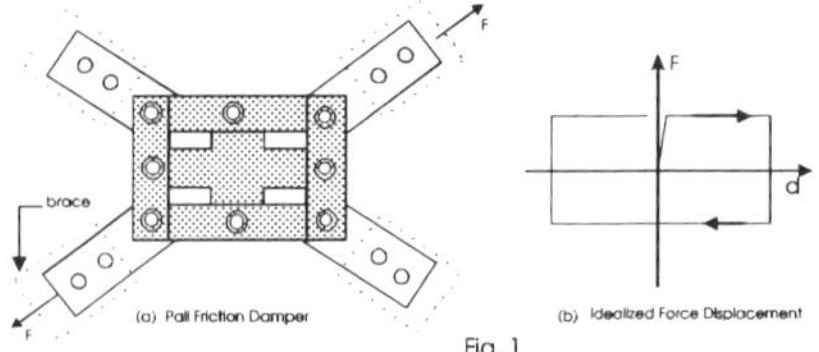

Fig. 1

(b) *Yielding Devices* Yielding dampers that utilize flexure, shear or extension deformation into the plastic range, has been proposed for structural energy dissipation. Among them, the Added Damping and Stiffness devices (ADAS). uses X-shaped bolted steel plates that deform plastically during moderate to severe earthquakes. As the Pall Damper, it can be mounted in different bracing systems.

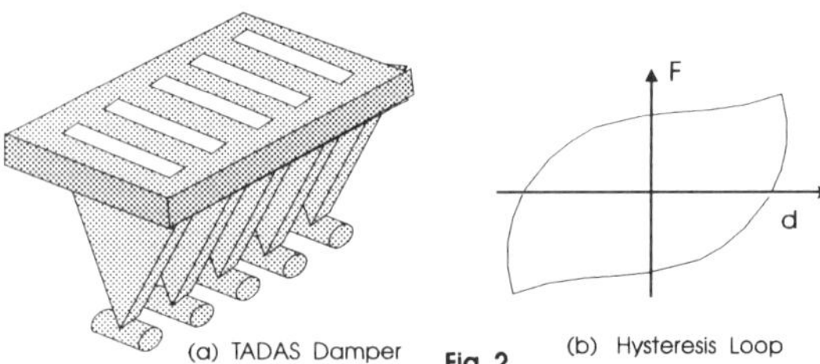

Fig. 2

Modification of the ADAS device, that replaced the X-shaped bolted plates with steel triangular ones and later with triangular welded plates, evolved in the **Triangular Added Damping and Stiffness Damper (TADAS)**, Fig. 2 (a,b), designed to overcome the stiffness sensitivity of the tightness of the bolts and increase its capacity to dissipate the energy generated during a severe earthquake.

The application of the triangular plate concept is based on the fact, that the bending curvature of the plate with an applied force at its end, is uniform over its height , and thus, it can deform into de plastic range without curvature concentrations. The TADAS device is usually installed in structures with Chevron bracing providing an alternative to reduce seismic vibrations in self-supporting towers.

Velocity-Dependent Devices

(a) **Solid Viscoelastic Dampers** (VE). These devices dissipate energy primarily by shear deformations of viscoelastic polymer layers constrained by metallic

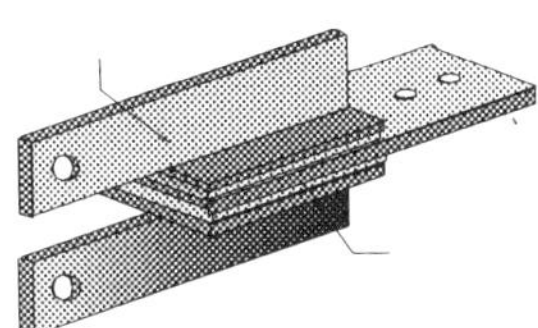

members that transmit axial loads. These metallic members provide the stiffness of the damper and function as paths to dissipate the heat generated by the viscoelastic material.

The force-deformation of a typical VE damper, subjected to a sinusoidal motion with circular frequency ω, may be expressed as:

$$F = k_{eff}\, d + C\dot{d} \qquad (4)$$

where C is the damping coefficient of the viscoelastic device; d and $\dot{d}$ are the relative displacement and relative velocity between each end of the device; and k_{eff} is the effective stiffness given by expression (3). The VE damper is effective for the control of seismic and wind oscillations, easy to design and install and could be used for the retrofit, and construction of self-supporting towers and masts protected with enclosures.

(b) **Fluid Viscoelastic Devices (FVED).** These devices exhibit a behavior similar to the solid ones, except they have zero effective stiffness under static loading conditions For practical applications, they can be represented by the Maxwell model.

(c) **Fluid Viscous Devices (FVD).** A prototype of a damper that exhibit purely viscous behavior, Fig. 4, consists in a stainless steel piston which travels through chambers filled with silicone oil, which flows through an orifice in and around the piston head. These devices are very effective for the control of induced seismic vibrations by transforming the seismic energy into heat that is dissipated into the atmosphere.

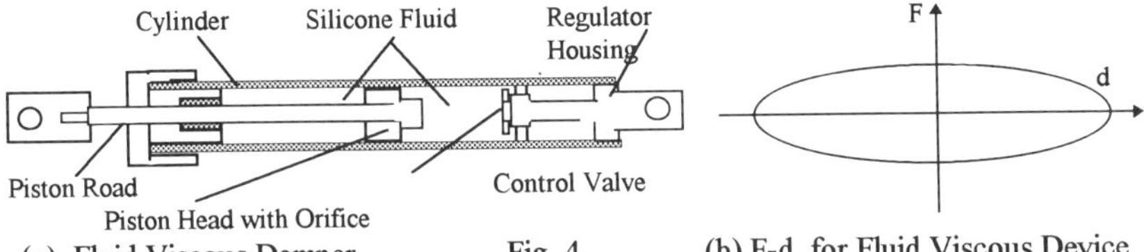

(a) Fluid Viscous Damper   Fig. 4   (b) F-d for Fluid Viscous Device

<u>Other Type of Devices</u> Energy dissipation devices that can not be classified as displacement-dependent or velocity-dependent require different modeling techniques.

(a) Tuned Mass Dampers. A tuned mass damper (TMD) provides indirect damping by the tuning of its frequency to that of tower. A typical passive TMD consists of an inertia mass attached to the tower by a spring-dashpot system, at the location of maximum motion, usually the top. The vibrating energy is dissipated by the dashpot when relative motion occurs between the tower and the TMD.

(b) Tuned Liquid Dampers- Tuned Liquid Column Dampers. The Tuned Liquid Dampers (TLD), have proven effective in reducing the dynamic-induced vibrations of towers, by tuning its frequency to that of the tower. Examples are:

The Sloshing Damper (TSD) in which the liquid in a large container serves as the tuned mass. It dissipates energy through the viscous action of the sloshing liquid.

The Tuned Liquid Column Damper (TLCD), dissipates energy by restricting the flow of the vibration liquid in a U shape container, through one or multiple orifices.

(c) Hanging Chain Dampers. Hanging chain dampers (HCD) can be used to control the dynamic vibrations of tall flexible structures, such as towers and masts, when the oscillations occur primarily in one plane. The HCD consists of a chain enclosed

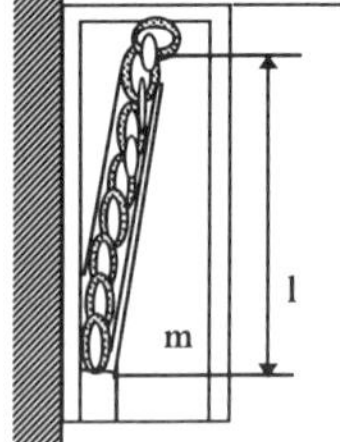

in a cylinder hung at the top of the tower that increases the damping of the tower by: (a) impact of its free end against the enclosing cylinder and (b) sliding friction between the chain links. Impact accounts for most of the energy dissipated; the effective damping coefficient due to impact can be ten times larger than the coefficient by friction of the chain links. Dissipation of energy by impact is achieved, when as a result of the movement of the structure, the free end of the chain moves with sufficient amplitude to impact the wall of the enclosing cylinder.

GUY DAMPERS

Guys may vibrate due to excitation generated in the mast or the guys themselves. Wind-induced oscillations result from gust excitations, vortex shedding or galloping.

Vortex shedding generates high frequency low amplitude vibrations and when they coincide with the natural frequency of the guy, it renders heavy vibrations. In general, the higher the guy initial tension, the more vulnerable is to vortex shedding. IASS, recommends the mounting of dampers in towers where the initial tension is greater than 10% of the breaking strength of the guy and the use of more than one damper in each guy, in planes 90^0 apart. Among the dampers in use to suppress or reduce vortex shedding excitations are, the **Rope Loops**, clamped to the guys, **the Heavy Chain** suspended from a guy and the **Stockbridge Damper,** a practical spring vibration device.

Galloping may occur with changes of the cable shape by ice or grease, generating very low and erratic frequencies, that endanger primarily the anchorage system. Among dampers to reduce them are the **Hem Rope** attached from guy to guy to the points of maximum amplitude and the **Sandamper** containing a closed cylinder partially filled with dry sand such that, when the cylinder rotates energy is dissipated.

A brief outline of a general procedure that may be used for the analysis and design of damped self-supporting towers is given in the Guide (Reference).

CONCLUSION

The feasibility of applying passive dampers to control dynamic vibrations in lattice towers has proven to be effective and the technologies required for their development and implementation are evolving, specially for the control of wind vibrations.

REFERENCE
Task Committee on Dynamic Response of Lattice Towers, "Guide to the Dynamic Response of Lattice Towers", ASCE (In preparation)

ELECTRICAL SUBSTATION STRUCTURE DESIGN GUIDE: A PANEL PRESENTATION

Moderator: L. Kempner, Jr., Ph.D., P.E, M. ASCE[1]

Panelists: Terry G. Burley, P.E., Western Area Power Admin., Golden, CO
Jim Hogan, P.E., Burns and McDonnell, Kansas City, MO
William Magee, P.E., PECO Energy Co., Berwyn, PA
George T. Watson, P.E., Houston Lighting & Power Co., Houston, TX

ABSTRACT

This technical session is sponsored by the Subcommittee on Electrical Substation Structures, of the Committee on Electrical Transmission Structures. The panel members will present the latest draft of the Electrical Substation Structures subcommittee's efforts. Session summaries are provide within this preprint that represent the information included in the subcommittee's draft titled "ASCE Substation Structure Design Guide."

INTRODUCTION

The purpose of the design guide is to provide a comprehensive document for the design of outdoor electrical transmission substation structures. The recommendations of this document apply to substation structures that support electrical equipment, rigid bus, and cables. The electrical equipment can be of significant weight and have attachments of brittle porcelain components. Specific guidelines for structural loads, deflection limits, analysis, design, fabrication, maintenance and construction of substation structures are recommended. Guidelines for the design of the structure to foundation connections and special foundation design considerations are presented. The guide addresses steel, concrete, wood, and aluminum used for the design of substation structures. Design equations are provided when references to existing structural design documents are not appropriate or convenient. The intent of the design guide is for structures that are located inside the fenced area of a substation yard. These guidelines may be appropriate for structures that serve the purpose of supporting electrical equipment located outside the substation fence, but must satisfy electrical substation design criteria.

---

[1] Structural Engineer, Bonneville Power Admin., PO Box 3612 (TNF-3), Portland, Oregon, 97208

ELECTRICAL EQUIPMENT & STRUCTURE TYPES

This section is used to establish common terminology and definitions for the different types of substation and switchyard structures used to support the above-grade components and electrical equipment. Substation structures support conductors, switches, buses, lightning arrestors, insulators, and other equipment. Structures can be fabricated from latticed angles forming chords and trusses, wide flanges, tubes (round, square, and rectangular), pipes, polygonal tubes (straight or tapered). This section shows photographs and provides an overview of typical types of substation support structures and gives brief descriptions of the electrical equipment they are required to support. An understanding of the function, operation, and relationship of electrical equipment and their support structures is a prerequisite to good structural support design.

LOADING CRITERIA FOR SUBSTATION STRUCTURES

All substation structures should be designed to withstand applicable loads transmitted by wind, ice, wire line tensions, seismic, construction and maintenance, electrical equipment loads, and other specified or unusual service conditions. Equipment manufacturers should be consulted with regard to operational loads and deflection requirements specific to the particular equipment. The loading section discusses guidelines for development of structure loading criteria.

DEFLECTION CRITERIA

Deflection and rotation of substation structures and members can affect the mechanical operation of supported electrical equipment, reduce electrical clearances, and cause unpredicted stress in structures, insulators, connectors, and rigid bus. For these reasons, structural deflections should be investigated and limited to magnitudes that are not detrimental to the mechanical and electrical operation of the substation. The sensitivity of equipment to deflection of supporting structures varies considerably. Disconnect switches, with complex mechanical operating mechanisms, are highly susceptible to binding if the structure distorts from the installed geometry. Conversely, structures supporting only stranded conductor bus or line dead-ends could grossly deflect without any impact on operation. Accordingly, structures are classified for the purpose of applying deflection limitations in accordance with the potential sensitivity of the supported equipment.

METHOD OF ANALYSIS

The design of substation structures requires knowledge of the equipment being supported by the structure, its operation and, electrical and safety codes. Analysis, as used herein, is defined as the mathematical formulation of the behavior of a structure under loads. The solution yields the calculated displacements, support reactions and internal forces or stresses. The analysis of a structure begins by developing a model that defines the structural configuration, connection characteristics, support boundary conditions, and loading cases. These items are discussed in this section.

DESIGN

The guide refers to other documents for design guidelines and will note any exceptions, to the referenced documents. Load factors and deflection criteria specified in the Loading and Deflection Section must be used rather than the load factors and deflection criteria specified in the referenced documents. Factored loads must be used with ultimate strength design. Unfactored loads should be used with allowable stress design. There is no intention to exclude any material or section types. If the material or section type is not addressed in the guide, the structural engineer should use the appropriate design code or document. Ultimate strength design and allowable stress design are both acceptable for design of substation structures. Ultimate strength design is recommended, since the structural design engineering trend is toward ultimate strength design. Structures that support conductors and overhead groundwires that extend outside the boundaries of the substation should meet or exceed the load and strength requirements of the NESC code.

CONNECTIONS USED IN FOUNDATIONS

The variety of structures used in electrical substations has a wide range of groundline reactions. The foundations used depend on the different types of soil present as well as individual preferences. These foundations can be slabs on grade, spread footings, drilled shafts or piling with pile caps. With these variables, many different types of anchorage are used to connect substation structures to their foundations. The most common means of transferring structure reactions to the foundations are by anchor bolts and welding to embedded plates. These types of anchorage provide good transfer of load. Anchor bolts can be headed bolts or a straight length of deformed reinforcing bar. Cast-in-place headed bolts are the recommended anchor bolt types. Stub angles and direct embedded structures can also be used in substations. The approach for the design of anchor bolts in this section is based on ultimate strength design (USD). Design loads must include applicable overload factors.

QUALITY CONTROL AND QUALITY ASSURANCE

In order to assure product quality, a good quality control (QC) and quality assurance (QA) program should be instituted by both the fabricator and purchaser throughout the entire production process. A QC/QA program will insure the purchaser that the fabricator has the personnel, organization, experience, procedures, knowledge, equipment, capability, and commitment to produce the specified structure. Quality control is the responsibility of the fabricator while quality assurance is the responsibility of the purchaser. Quality control guidelines used by the fabricator should be clearly defined and available for review and approval by the purchaser. The purchaser should also specify any additional requirements in order to achieve the desired degree of structure quality. It is necessary that the QC/QA programs be agreed upon between the fabricator and the purchaser prior to the start of any fabrication. The extent of QC/QA programs may vary based on initial investigations, the purchaser's experience, the fabricator's experience, and past performance, and the degree of reliability required for the specific job.

TESTING

Full-scale structural proof tests are rarely performed on substation non-electrical equipment support structures, (wire-support structures). It is not cost effective to perform a full-scale test because substation structures are not fabricated in large quantities (similar to a structure used in a transmission line). Full-scale testing should be considered if a particular substation structure is a standard and will be used in large quantities or if the structure uses a unique structural system not typical of current practice. Component testing (a section of the tower, connections, etc.) may be cost effective for substation structures. Electrical equipment support structures are typically simple cantilever structures that are generally not proof tested for static loads. Seismic response (dynamic loading) requires that the support structure and equipment be seismically tested/evaluated as a system. Seismic tests are performed in accordance with IEEE's Std. 693 (1997).

CONSTRUCTION AND MAINTENANCE

Designers of substation structures should anticipate construction loads imposed on the structure. These loads could be locally higher than others, especially on pull-off or other strain-type structures. Maintenance and/or operation loads should be considered. Equipment should have provisions for access; i.e., cross-over platforms or working platforms. These will need to be considered, especially around large transformers with coolant and fire protection piping. Rigid bus system maintenance is addressed as a special issue. There should be a systematic maintenance routine to keep the entire rigid bus system in good operating condition. Some inspections may have to be done during scheduled outages. Thermographic inspections can be used to detect excessive temperatures before discoloration is visible. Infrared scanning inspections can be routinely made in accordance with the international electrical commission charts. Worker safety is also discussed as a separate issue. All structures and equipment, which will be inaccessible with bucket trucks or small ladders, should be considered for climbing devices (ladders) mounted to the structure with a fall protection device. High areas that require movement, either during construction or for maintenance, must have fall protection devices; i.e., safety cables for attachment to workers. In all cases, OSHA and local codes must be adhered to, especially in energized substations.

SUMMARY

The sections of the ASCE Substation Structures Design Guide discussed, form the basis of the subcommittee document. A copy of the latest draft can be obtained for the session moderator (lkempnerjr@bpa.gov). The objective of the subcommittee is to submit this document for ASCE publication in 1999.

REFERENCE

IEEE Std 693, Recommended Practices for Seismic Design of Substations, Institute of Electrical and Electronics Engineers, Pascatawy, NJ 1997

PERFORMANCE OF ELECTRIC TRANSMISSION STRUCTURES DURING EARTHQUAKES

Gina T. Gobo, M. ASCE[1]
Leon Kempner, Jr., Ph.D., P.E., M. ASCE[2]
Wendelin H. Mueller, Ph.D., P.E., M. ASCE[3]

ABSTRACT

This paper describes ongoing research to determine the significance of seismic loading on a high-voltage electric transmission line tower/conductor structural system. The purpose of the research is to compare the seismic member forces to member forces obtained using standard wind, ice, and unbalanced longitudinal tension load criteria. A review of the historical earthquake performance of electric transmission towers is presented along with a summary of the results of a literature search.

INTRODUCTION

An electric transmission line system consists of a generation source, transmission lines, substations, and distribution lines. The generation and substation components are designed to specific seismic loading criteria and will not be discussed in this paper. In the US and typically throughout the world, electric transmission and distribution lines/structures are not designed to specified seismic loads. The structures used to support the wires are designed for combinations of wind, ice, and unbalanced longitudinal tension loads that are specified in industry guidelines (ASCE 1991) and State legislative documents (NESC 1997, GO95 1998). The loading criteria specified in these documents have provided adequate structural capacity to resist the loads generated by seismic ground motions. This statement is supported by the historical earthquake performance of both electric transmission and distribution wire-supporting structures (FEMA 202).

Electric transmission and distribution lines are distributed systems. Therefore, when an earthquake event occurs there is a high probability that a significant number of the wire-support structures will experience the resulting seismic ground motions. The wire-support structures typically consist of lattice steel or aluminum members,

---

[1] Graduate Student, Portland State University, Portland, Oregon, 97207
[2] Structural Engineer, Bonneville Power Admin., PO Box 3612 (TNF-3), Portland, Oregon, 97208
[3] Professor of Civil Engineering, Portland State University, Portland, Oregon, 97207

tubular steel members/poles, wood structures, and concrete poles. The structural systems used are space-truss, planar-frame, and single-pole structures. Structural problems reported with these structural systems after major earthquakes have primarily been the result of foundation issues, this excludes the problem of buildings falling into distribution structures. Reported foundation problems (NIST 1998) have consisted of fault separation, ridge shatter, liquefaction, landslides, lateral spreading, and differential footing displacement.

Long (1973) and Agrawal (1976), evaluated the seismic load effect on lattice steel and wood pole structures and determined that the industry specified wind loads control the design of these structure types. Long (1973) and Alt (1984) reported that the conductor seismic mass loads, particularly in the case of suspension towers, have no significant effect on the resulting member design forces.

Recent studies in China and Japan (Li 1991, 1996, 1997, Ozono 1992, 1998, and Suzuki 1992) have investigated the seismic effects on lattice transmission towers. These studies were initiated to investigate the seismic effects of long-span river-crossing and ultra-high voltage (1100kV) transmission line structures. The concern with ultra-high voltage transmission line structures is the higher bundle conductor to tower mass ratio than that for conventional high-voltage transmission line systems. These investigators studied both the in-plane and out-of-plane coupled tower/conductor system response. The coupling mechanism of the in-plane and out-of-plane tower/conductor system is different. The in-plane coupling restrains deformations of the tower and the out-of-plane coupling amplifies the deformations. The stiffness matrix of the tower/conductor system for the in-plane seismic motion is uncoupled and the mass matrix is coupled, whereas, for the out-of-plane seismic motion the stiffness matrix is coupled and the mass matrix is uncoupled. The conclusion of these studies is that the conductor seismic mass loads are significant and need to be included in the analysis/design of lattice transmission towers. None of these studies presented a comparison of the seismic loads to the standard design loads of combined wind, ice, and unbalanced longitudinal tension loads.

ONGOING INVESTIGATION

An ASCE special project grant was obtained by Portland State University to investigate the significance of the seismic loads in comparison to standard design wind, ice, and unbalance longitudinal tension loads for a long-span 500kV double circuit lattice steel electric transmission tower/conductor river-crossing system. This section of the paper describes the development of a computer model to evaluate the seismic response of this coupled transmission tower/conductor system. The model includes two transmission towers supporting three-bundle conductor phases, a total of 18 sub-conductors. The transmission tower system is modeled using the SAP2000 (1996) structural analysis computer program. A seismic response analysis of the coupled tower/conductor model will be used to determine the forces of the transmission tower members. The member seismic forces will be compared to wind, ice, and unbalance longitudinal tension design forces in an effort to determine which forces control the design of the tower. The need to understand the seismic response of transmission towers is essential in determining whether or not current design practices should include seismic loads.

Table 1 shows the configuration of the tower and conductor used for the investigation. The initial SAP2000 model consisted of three level spans and two complete suspension towers. The base of the towers is fixed and the boundary condition of the conductor end-spans is pinned. This configuration was chosen to simulate a river-crossing tower/conductor system.

Table 1 - Tower and Conductor Information

Tower		Conductor	
Type:	525/550 kV Double Circuit	Type:	1.752" Special AACSR
Weight:	102 kN	Weight:	36 kN
Height:	141 m	Span:	1050 m

This model was used to obtain the tower/conductor system natural frequencies and vibration mode shapes. A preliminary Response-Spectrum analysis was performed to determine if there would be any problems with SAP2000 using the developed tower/conductor system model. The results of this analysis showed dominant (significant mass contribution) conductor natural frequencies of 0.15 Hz with the mass contribution in the vertical direction, 0.21 Hz and 0.41 Hz with the mass contribution in the in-plane direction, and 0.22 Hz and 0.44 Hz with the mass contribution in the out-of-plane direction. The dominant tower natural frequencies were 1.17 Hz in the out-of-plane direction and 1.20 Hz in the in-plane direction. The participating mass ratios of these vibration modes accounted for 75 percent in the in-plane (x) direction, 57 percent in the out-of-plane (y) direction, and 65 percent in the vertical (z) direction.

The next phase of the investigation will be to perform seismic response analysis using selected earthquake time histories/spectra. The member forces obtained from these analyzes will be compared to the member forces obtained from standard wind, ice, and unbalanced tension load combinations. Results will be presented at the 1998 ASCE Structures Congress.

CONCLUSION

The purpose of this paper is to present the latest information on the seismic analysis/design performance of electric transmission line and distribution line structures. There is a renewed interest by the design code/standard groups on how these structure types should be evaluated for seismic loading. The latest edition of the NEHRP (1997) seismic guidelines includes a chapter titled "Nonbuilding Structure Design Requirements," in which an attempt was made to provide a "red-flag" provision for transmission and distribution line structures. A simple static seismic force calculation is recommended to compare to standard design load criteria. The ongoing research, funded by an ASCE grant, will be used to assess the seismic load effect of a long-span river-crossing. Results from this investigation will be presented in a subsequent ASCE report.

REFERENCES

Agrawal P.K., Kramer J.M., "Analysis of Transmission Structures and Substation Structures and Equipment for Seismic Loading," Transmission and Substation Conference, 1976.

Alt K., Bauer K.H., Boos K.V., Jurdens C., and Paschen R., "Dynamic Effects on Transmission Lines," International Conference on Large High Voltage Electric Systems, 1984.

ASCE, Guidelines for Electrical Transmission Line Structural Loading, 1991.

FEMA 202, Earthquake Resistant Construction of Electric Transmission and Telecommunication Facilities serving the Federal Government Report, Federal Emergency Management Agency, 1990.

GO95, Rules for Overhead Electric Line Construction, California, 1997.

NEHRP, Recommended Provisions for Seismic Regulations for New Buildings and other Structures, FEMA 302, Federal Emergency Management Agency, 1997.

Li H.N., Wang S., and Wang Q.X., "Response Of Transmission Tower System To Horizontal And Rocking Earthquake Excitations," Earthquake Engineering and Engineering Vibration, Dec. 1997.

Li H.N., Wang Q.X., Li M., and Singh M.P., "Aseismic Calculations for Transmission Towers," Technical Council on Lifeline Earthquake Engineering, Monograph No. 4,1991.

Li H.N., Wang Q.X., and Singh M.P., "Seismic Response analysis Method for Coupled System of Transmission Lines and Towers, Part I: Out-of-Plane" Earthquake Engineering and Engineering Vibration, Dec. 1996.

Li H.N., Wang Q.X., and Singh M.P., "Seismic Response analysis Method for Coupled System of Transmission Lines and Towers, Part II: In-Plane" Earthquake Engineering and Engineering Vibration, Dec. 1996.

Li H.N., Suarez L.E., and Singh M.P., "Seismic Effects on High-Voltage Transmission Tower and Cable Systems," Earthquake Engineering and Engineering Vibration, Dec. 1997.

Long L.W., "Analysis of Seismic Effects on Transmission Structures," IEEE PES Summer Meeting and EHV/UHV Conference, 1973.

NESC, National Electrical Safety Code, IEEE Inc., Pascataway, NJ, 1997.

NIST, Guide to Improved Earthquake Performance of Electrical Power Systems, National Institute of Standards and Technology, GCR 98-757, 1998.

Ozono S., Maeda J., and Makino M., "Characteristics of In-plane Free Vibration of Transmission Line Systems," Engineering Structures, 1998.

Ozono S., Maeda J., "In-plane Dynamic Interaction Between a Tower and Conductors at Lower Frequencies," Engineering Structures, 1992.

SAP2000, Version 6.0, Computers and Structures, Inc., Berkeley, California, 1996.

Suzuki T., Tamamatus K., and Fukasawa T., "Seismic Response Characteristics of Transmission Towers," Tenth Conference on Earthquake Engineering, 1992.

Structural Loading Tests of
Pre-Cast, Pre-Stressed, Tapered Concrete Poles and Frames with Steel Cross-Arms

Harry Durden[1], David Mayo[2], Mark Ostendorp, PhD[3]

Abstract

Twenty-four full-scale load tests were performed on six different pole and frame configurations. Poles and frames consisted of pre-cast, pre-stressed, solid and hollow structural members of varying height, embedment depth, cross-section, and wall thickness. Selected pre-stressing strands and structural members were strain gauged to compare measured strains with predicted values. Applied loads and deflections were monitored continuously during each test. Deflections and strain gauge readings were analyzed and evaluated.

Introduction

Twenty-four full-scale load tests were performed on six different pole and frame configurations at the Energy Delivery and Utilization Center in Haslet, TX. Poles and frames consisted of pre-cast, pre-stressed, solid and hollow structural members of varying height, embedment depth, cross-section, and wall thickness. Test structures consisted of six two-pole frames with and without steel cross-arms and tension back braces, and two single pole structures. Three of the two-pole structures were assembled with tapered, round, hollow, spun-cast, pre-stressed, concrete members, while the other three frames were composed of tapered, square, solid, static-cast, pre-stressed concrete members.

Similarly, one of the single pole test structures was a tapered, round, hollow, spun-cast, pre-stressed, concrete pole, while the other was a tapered, square, static-cast, pre-stressed concrete pole. Both of the single pole tests were used to define the base line performance of the concrete poles of the frames.

---

[1]Project Manager, Alabama Power Company, Birmingham, AL
[2]Senior Engineer, Alabama Power Company, Birmingham, AL
[3]Project Manager, Electric Power Research Institute, TX

Concrete and steel strain gauges were used to measure strains at selected locations on various components of the test structures. Strain gauge measurements were used to compare the predicted response of the structures to measured strains on the concrete poles, steel cross-arms, and steel tension braces. Additionally, strain gauges were placed on the pre-stressing strands to monitor pre-stressing losses and strains during manufacturing, and prior to and during the test of the control poles.

Test Configuration

Test structure TS5 was a single, tapered, round, hollow, pre-cast (i.e., spun), pre-stressed, concrete pole. Figure 1 shows a sketch of the test item including the foundation condition, strain gauge locations, and the maximum load applied at failure. The pole was directly emdedded in a steel pipe back-filled with clean gravel to approximate actual field installation conditions. The embedment depth was determined in accordance with industry standards.

Multiple strain gauges were applied at each location within the pole cross-section. Six concrete gauges were applied to locations B, C, and C' as shown in Figure 1. Pre-stressing strand gauges were located at location E (i.e., 3 strain gauges for each selected pre-stressing strand). Concrete strain gauges used in the tests were wired in a half-bridge configuration and were not temperature compensated.

A unique method was developed to strain gauge the pre-stressing strands of the control poles prior to the manufacturing of the concrete poles. Strain gauged hollow aluminum sleeves of appropriate diameter were compressed on the pre-stressing wire. Each aluminum sleeve was strain gauged via three half-bridge resistance gauges located at 120-degree angles within the cross-section of the sleeve. Following the installation of the sleeve on the wire, strain gauged pre-stressing strands were calibrated to 75 percent of their ultimate strength. Strain measurements on the pre-stressing strands were recorded prior to the manufacturing of the poles and before the each test.

Deflection targets of an electronic survey station were attached at equal increments along the height of the pole above the ground line. Transverse, longitudinal, and vertical deflections were measured at 25%, 50%, 75%, 95%, and 100% of the applied load at each of the three targets. Applied loads were measured continuously in the test via a calibrated load cell. Similarly, strains were measured continuously at each gauge location.

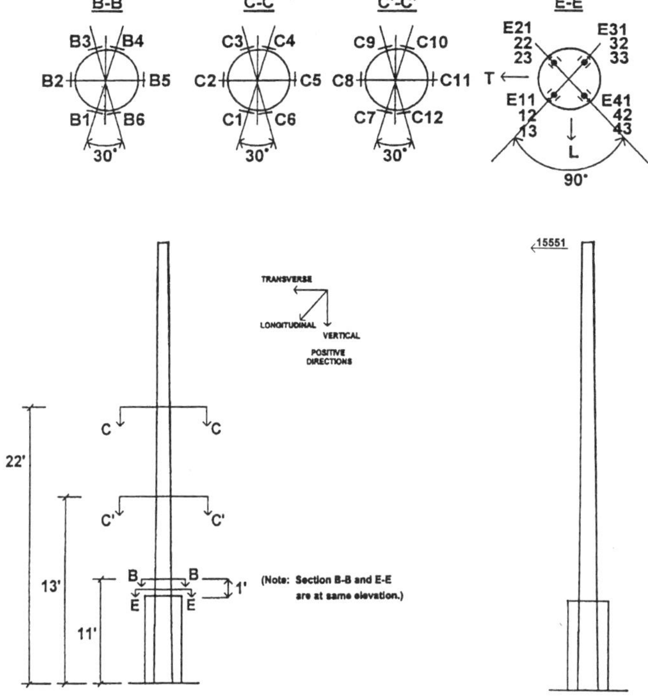

Figure 1. TS5 Test Structure

Test Results

Strain gage data was collected real-time via the data acquisition and analysis system. Each record contained the scan number, percent load applied to the test item, and the data collected on each active channel.

Figure 2 shows strains measured at six locations within the cross-section shown in Section B-B (Figure 1). Measurements are concrete strains recorded at the surface of the concrete. Figure 3 shows strains measured at the surface of the aluminum sleeves that were attached to the pre-stressing wires. Strains were recorded by gauges on each sleeve at four different locations within the cross-section shown in Section E-E.

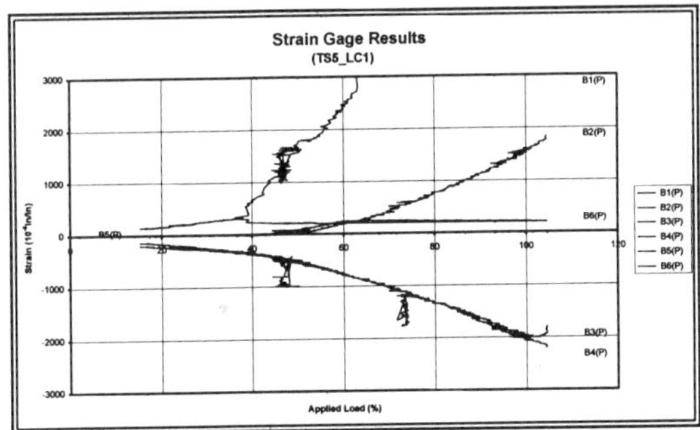

Figure 2. Strain Gage Measurements - Location B

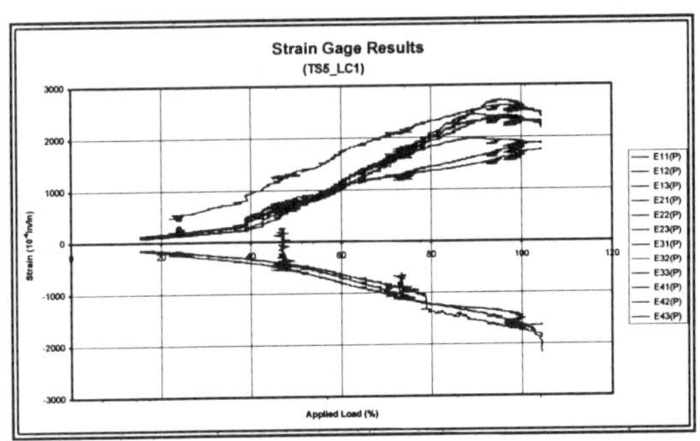

Figure 3. Strain Gage Measurements - Location E

Conclusions

Aluminum sleeves were used to measure the strains on pre-stressing strands embedded in high strength concrete. The sleeves were prepared, instrumented, calibrated, and installed in the form prior to the manufacturing of the pre-stressed, spun-cast concrete pole. Strain gauge measurements of the pre-stressing strands were recorded during the test. Measured strains will be compared with values from a non-linear analysis of the poles.

Acknowledgment

This study was performed under a research grant from the Alabama Power Company and the Electric Power Research Institute (EPRI). The authors wish to thank Paul Lyons of EPRI and Fouad Fouad of the University of Alabama at Birmingham for their guidance and support.

Ice Storm '98
Characteristics, Events, and Interpretation

Mark Ostendorp, Ph.D., PE[1]

Abstract

The 1998 northeastern ice storm constituted a series of three distinguishable icing events. Temperatures remained below freezing, allowing ice on conductors and ground wires to remain in place throughout the event. Wind speeds during the storm were light to moderate and average values ranged from 8 to 25 km/hr (5 to 15 mph) with gusts not exceeding 40 km/hr (25 mph). The return period of the ice storm within the most critically affected area ranged from 50 to 150 years while return periods for other areas ranged from 5 to 35 years. Radial ice thicknesses were calculated (i.e., based on water equivalent of all precipitation) for conductors parallel and perpendicular to the wind direction. Maximum radial ice thicknesses ranged from 25 to 33 mm (1 to 1.3 inches) for the Ontario area, 25 to 45 mm (1 to 1.8 inches) for the Quebec area, 23 to 50 mm (0.9 to 2.0 inches) for the New York region, and 23 to 38 mm (0.9 to 1.5 inches) for the New England region, respectively.

Introduction

Ice storm '98 impacted significant portions of the northeastern Unites States and southeastern Canada. It was an exceptional event for the severity, geographic extent, and the amount of freezing precipitation, and persisted for nearly five days from January 5 through January 9, 1998, producing three separate icing events.

At the peak of the storm, nearly 4 million people were left without power for periods ranging from hours to weeks. Delivery systems experienced numerous outages and devastating failures, nearly resulting in the blackout of Montreal, Quebec. Estimates of the economic damages incurred as a result of the January '98 ice storm range from 1 to 1.25 Billion US dollars.

---

[1] Project Manager, Energy Delivery & Utilization Center, Electric Power Research Institute, Haslet, TX, 76052.

Objectives

The accumulation of ice on power lines constitutes one of the most severe loading criteria that each line component is likely to experience. The appropriate selection of a design radial ice thickness significantly influences the severity of this frequently governing load condition. Comparisons of observed and predicted values are likely to improve the accuracy of the models. The objectives were:

- Collect information from state, federal agencies, electric utilities, and other organizational entities to identify characteristics of the '98 ice storm.
- Analyze ice storm data to identify relevant parameters such as the radial ice thickness, predicted return period, and other significant statistical parameters.
- Analyze line failures to evaluate the performance of delivery systems and design methods and models.

Results

Ice accretion measurements were analyzed via a model to calculate the corresponding return periods. Figure 1 shows return periods calculated based on the observations made at a selected number of the weather stations. Values are based on the water equivalent of all precipitation (i.e., rain, snow, freezing rain) on January 4-10, 1998. Consequently, return periods are larger than values calculated based on frozen precipitation only.

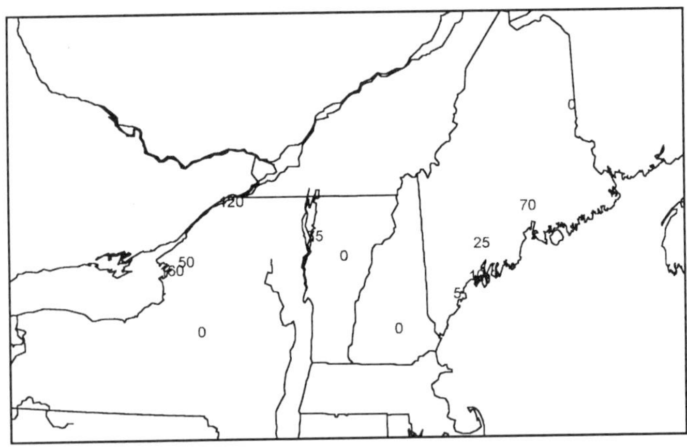

Figure 1. Maximum Probable Return Periods (Based on Water Equivalent of all '98 Storm Precipitation)

Climatic data have been used to predict radial ice thickness accumulations during the '98 storm. Values were compared to field observations collected during the ice storm. Radial ice thicknesses are reported for power lines parallel and perpendicular to the wind direction. Actual radial ice thicknesses for lines oriented at other angles are bracketed by these limiting values.

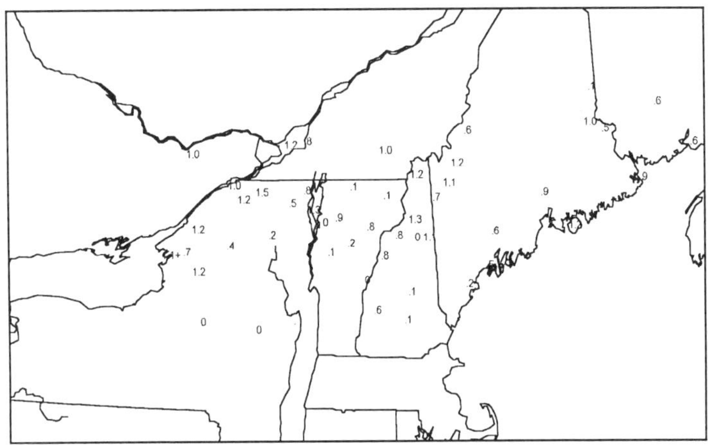

Figure 2. Cumulative Radial Ice Thickness (mm/25)
(Wind Parallel to 25 mm (1 inch) Wire)

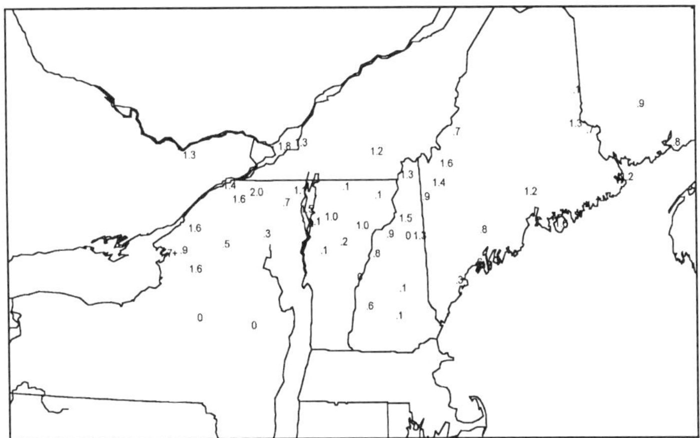

Figure 3. Cumulative Radial Ice Thickness (mm/25)
(Wind Perpendicular to 25 mm (1 inch) Wire)

Figures 2 and 3 show maximum cumulative radial ice thicknesses for conductors and ground wires parallel and perpendicular to the predominant wind direction, respectively. Cumulative radial ice thicknesses were calculated (25 mm (1 inch) diameter conductor at 10 m (33 feet) standard elevation) based on water equivalent precipitation records (i.e., rain, snow, freezing rain, freezing drizzle) for each 24 hours period. As evidenced by temperature records, it was assumed that none of the accreted ice melted for the duration of the storm.

Significant variations of the cumulative radial ice thickness were observed in the predictions developed using weather station data from northern Vermont, New Hampshire, and northeastern Maine. These variations were largely contributed to the differences in elevation, geographic features, and weather conditions at each of the observation stations.

Conclusions

Weather station records show temperatures to remain below freezing during the event, allowing significant accumulations of ice on the power lines. Wind speeds were light to moderate ranging from 8 to 25 km/hr (5 to 15 mph) with gusts not exceeding 40 km/hr (25 mph). Data indicates a predominant wind direction during icing events (i.e., based on 50 years records). Total water equivalent precipitation (i.e., rain, snow, freezing rain, freezing drizzle) ranged from 25 to 100 mm (1 to 4 inches) within the affected region.

Return periods within critically affected areas ranged from 50 to 150 years while return periods for other areas ranged from 5 to 35 years. Maximum radial ice thickness (i.e., based on water equivalent of all precipitation) ranged from 25 to 33 mm (1 to 1.3 inches) for the Ontario area, 25 to 45 mm (1 to 1.8 inches) for the Quebec area, 23 to 50 mm (0.9 to 2.0 inches) for the New York region, and 23 to 38 mm (0.9 to 1.5 inches) for the New England region, respectively. Minimum values correspond to lines parallel to the wind while maximum values are for lines perpendicular to the wind.

Ice loads observed during the '98 ice storm constituted a `real world' load test that quickly and efficiently identified any weak areas in the power delivery systems. Failures of wires, insulators, and suspension components degraded the overall resistance of some of the systems causing failures to occur at ice loads well below the design criteria.

Tomoka River Transmission Line Crossing
Just Another Wood Pole Replacement Project

C. Jerry Wong,[1] M. ASCE

Abstract

Inspection reports for the Bunnell – Ormond Beach 115 kV transmission line revealed serious deterioration to the three (3) wood H-frame structures and their wooden pile foundations in the Tomoka River. Preliminary investigations determined that the replacement of these three (3) structures in the river was cost prohibitive due to the inaccessibility around pristine wetland areas. A decision was made to utilize steel-concrete hybrid structures of 65.5 meters (215 ft) in length to span over the Tomoka River. These are the tallest single pole transmission structures in the US. Many technical and scheduling obstacles of this unique approach had to be overcome. With the cooperative efforts from all parties involved, Florida Power and Light Company was able to develop a timely, cost effective solution that ensure the reliability of the electrical transmission system, satisfy all regulatory requirements, and minimized the environmental impact.

Introduction

The Bunnell – Ormond Beach 115 kV transmission line provides electrical service to Florida Power and Light's (FPL) customers in the Daytona Beach Area. Wood poles were utilized when this forty (40) kilometer line was originally built in 1951. Many of the original structures were replaced due to deterioration over the years.

Three (3) of the wood H-frames were located in the Tomoka River, north of Ormond Beach, Florida. These structures were supported by wooden pile clusters and galvanized steel guy wires. Average water depth in the river crossing exceeds 3 meters (10 ft). Therefore, all maintenance and replacement work on this line section requires marine construction equipment.

---

[1] Principal Engineer, Florida Power & Light, Transmission Line Design, 700 Universe Blvd., Juno Beach, Florida 33408-0420

A recent inspection report indicated that deterioration of these wood structures and their pile foundation systems had degraded their structural integrity beyond acceptable limits.   The structural capacity of this line also needed to be upgraded in the near future due to the electricity needs from the population growth on the East Coast of Florida. To avoid any potential problems, FPL's Transmission Engineering Group determined that it was necessary to replace these structures before the peak of the 1998 hurricane season. This limited the structure design and manufacturing schedule to a maximum of six (6) weeks.

Preliminary negotiations, with utility contractors, indicated that the replacement of these three (3) structures in the river would be cost prohibitive. The heavy construction equipment needed to install the traditional concrete pole and concrete pile system increased the initial cost significantly. The aquatic life (such as manatees), bird inhabitants (Osprey nest), boat traffic, and residences in the nearby area also increased the complexity of construction. Impacts on these environmental elements had to be minimized. Certain pristine wetland areas inside the right-of-way made equipment and material accessibility a major issue. In order to satisfy all regulatory requirements and meet FPL's aggressive internal cost and schedule targets, it became obvious that an innovative new approach was required.

<u>Feasibility Study</u>

To minimize the construction duration and to ensure electrical transmission system reliability, the designer was forced to investigate the possibility of spanning the entire river.

By installing the new supporting structures only in upland locations, the environmental impacts to the wetland could be drastically reduced. The proposed structures also eliminated the need for a sectional river barge, which was estimated at a cost of $190,000 for the duration of the construction. As the accessibility improved, the costs of matting over environmentally sensitive areas would be reduced. This approach also reduced the future maintenance and inspection costs since a barge or boat would not be needed to accomplish these tasks.

However, designing overhead facilities that could span over the 586.7 meters (1,925 ft) wide riverbed presented other technical challenges. The chosen conductor (795 kcmil, 26/7, ACSR/AW conductor) would sag up to 36.5 meters (120 ft) under the controlling design criteria. Unfavorable soil conditions yielded a setting depth of 11.3 meters (37 ft) on the north bank of the river. The minimum structure assembled length required to provide the clearance specified by the National Electrical Safety Code over waterways subject to sailboats and launch traffic was calculated at 65.5 meters (215 ft).

Although lattice transmission structures at this height and greater are not uncommon, there was not enough space on either sides of the river to physically locate structures that require such a large footprint area. Also, lattice towers would not be aesthetically acceptable. From the discussion with the local agencies, it would be difficult to obtain the necessary permits and get agreements from the surrounding neighborhood within the short time frame desired for this project.

The underground alternative was also investigated. However, the project site did not provide enough room to setup the equipment needed for directional drilling. It was also estimated that the cost of an underground facility would be approximately 10 times as high as an overhead transmission line. It would not be possible to complete the underground construction on schedule.

When all factors were considered, it became obvious that a single pole transmission structure was the only feasible solution for the Tomoka river crossing. FPL and Newmark International, a utility pole manufacturer, jointly developed the final solution for this project, and quickly made if happen.

Engineering, Manufacturing, and Construction

The river crossing structure located on the south bank of the river is a double dead-end structure with a 105° line deflection angle. It has an above ground height of over 57.3 meters (188 ft) and multiple guy wires in all directions to support conductor tensions and to prevent buckling. This pole has four (4) sections. The bottom section, a spun concrete pole section, is directly embedded and has a reverse taper. This arrangement shortened the construction time considerably compared to a traditional drilled pier foundation. The upper sections are made of 12-sided, tapered tubular steel poles. Due to the heavy axial load, flanged joints were specified.

The river crossing structure located on the north bank of the river is a self supported tangent structure with minimal deflection angle. It has an above ground height of 54.3 meters (178 ft). It is composed of four (4) sections. The bottom section is a directly embedded spun concrete pole that functions like an open caisson foundation. The upper sections are made of 12-sided, tapered steel tubes. Slip splices were selected for the top two splices, and a flanged joint was selected for the mating of tubular steel to the spun concrete bottom section. This pole is a cantilever structure with a ground line moment capacity that required the spun concrete bottom section to have over 160 prestressed strands in a unique double cage construction inside a 17.8 centimeters (7 in) thick wall. This bottom section is the largest (capacity wise) spun concrete pole ever made in the US.

Both river crossing structures have an overall assembled length of 65.5 meters (215 ft) in length. The steel-concrete hybrid pole concept is not new to the industry. However, the capacity requirement for this application complicated the connection design. Special attention was given to the steel-concrete connections to ensure that proper strength was developed.

The adjacent structures were also replaced to ensure adequate strength is provided. The overall length of these spun concrete structures was increased to 38.1 meters (125 ft) and to 50.3 meter (165 ft) to account for the uplift forces created by the river crossing span. The maximum diameter for poles is just over 1.5 meters (60 in) so that the impact to the surrounding area was minimized. The total project replaced eleven (11) wood H-frame structures with three (3) concrete and two (2) hybrid single poles. All pole sections were manufactured and delivered on time.

Construction started as originally scheduled. However, it was delayed due to one of the most severe forest fires in the history of Florida. All structures were manufactured in multiple sections to minimize the impacts to the surrounding heavily wooded and residential area. No trees were cut and not even a single mailbox was knocked down during the construction. The construction activities were never as uneventful as they could possibly be. Yet, it went smooth enough so that it did not cause any unplanned outage.

Conclusion

The new Tomoka river-crossing structures are unique in both the problem solving approach taken and in the final structural design. Almost every new pole in the project represents a milestone in the pole manufacturing industry. Two structures set a new record for total pole assembled length of a single pole electrical transmission structure. Some other poles pack extremely high moment capacity into a rather small maximum diameter limitation. The high capacity hybrid concept and the associated connection designs provide new options for future applications. Many details, structural elements, and the their combinations have never been used before.

The project also provided plenty of challenge in the electrical engineering field. These poles are the tallest structures in the nearby area and could function like a lightning rod. Precautionary measures were incorporated to ensure adequate lightning performance. By being so tall, the foreign interference from the boats or the trees will be significantly reduced. In fact, since the construction completion, no interruptions have occurred on the newly built line section.

The design allowed FPL to complete the line rebuild project without adversely affecting the environment. Input from all applicable State and Federal Agencies, City and County Government, the Sierra Club, the Audubon Society, the concerned citizens and local businesses were incorporated to make this project successful. The area residents have actually commented that river and neighborhood aesthetics have been improved by the use of these taller single poles and by spanning over the entire river.

Wind Engineering

(3) Assessment of Pedestrian Thermal Comfort
Michael J. Soligo, Hanqing Wu, and Peter Irwin
Rowan Williams Davies & Irwin Inc.
Guelph, Ontario, Canada

(4) Microclimate Design Features for Buildings and Landscaping
Leighton Cochran
Cermak Peterka Peterson Inc.
Fort Collins, CO

Peter Irwin
Rowan Williams Davies & Irwin Inc.
Guelph, Ontario, Canada

54. *WIND UPLIFT ON BUILDINGS AND OTHER STRUCTURES*

MODERATOR: Peter A. Irwin,
Rowan Williams Davies & Irwin, Inc.,
Guelph, Ontario, Canada

(1) Effect of Roofing Membrane on Wind Uplift Structures
A. Baskaran, U. Vilaipornsawai, M.G. Savage, and K. R. Cooper
National Research Council Canada
Ottawa, Canada

**(2) Simulation Requirements for Roof Wind Loads near the Corners
of Low Buildings with Low-slope Roofs**
D. Surry T.C.E. Ho, and G.R. Lythe
The University of Western Ontario,
London, Ontario, Canada

(3) Wind-Induced Internal Pressures on Sports Arenas
G. Conly, S.L. Gamble, P.A. Irwin, and M. Soligo
Rowan Williams Davies & Irwin, Inc.,
Guelph, Ontario, Canada

(4) Simultaneous Measurement of Roof Pressures and Application
Noriaki Hosoya,
Colorado State University
Fort Collins, CO

Jack Cermak,
Cermak Peterka Peterson Inc.
Fort Collins, CO

63. *DESIGNING FOR WIND AND COASTAL FLOODING HAZARDS*

MODERATOR: Arthur N. L. Chiu
University of Hawaii at Manoa
Honolulu, HI

(1) Using Wind Design Criteria to Gauge Seismic Performance
Edwin T. Dean
Dean Engineering
Portland, OR

Jeffrey R. Soulages
Degenkolb Engineers
Portland, OR

(2) Designing for Hazards in an Uncertain World
Charles H. Thornton and Leonard M. Joseph
Thornton-Tomasetti Engineers
New York, NY

(3) Applied Technology Council's Role in Wind and Coastal Flood Hazard Mitigation
Christopher Rojahn
Applied Technology Council
Redwood City, CA

(4) Design Considerations for Coastal Zones Exposed to Hurricane-Induced Wave Action
Robert G. Dean
University of Florida
Gainesville, FL

(5) Design Considerations for Effects of Complex Topography on Wind Characteristics
Arthur N. L. Chiu
University of Hawaii at Manoa
Honolulu, HI

69. *VIBRATION AND FATIGUE OF SIGN, SIGNAL, AND LIGHT SUPPORT STRUCTURES*

MODERATOR: Mark Kaczinski,
     The D. S. Brown Company
     North Baltimore, OH

(1) Mechanical Damping System for Mast Arm Traffic Signal Structures
Ronald A. Cook, David Bloomquist, Michael A. Kalajian,
University of Florida
Gainesville, FL

(2) Truck-Induced Wind Loads on Highway Sign Support Structures
Kevin W. Johns
Modjeski and Masters, Inc.
Harrisburg, PA

Robert J. Dexter,
University of Minnesota
Minneapolis, MN

Investigation of Engineered Building Wind Failures

Joseph E. Minor[1], P.E., F. ASCE

Abstract

In the wake of recent disasters, investigations of building failures have become important to building code officials, building trade associations, the insurance industry, engineers and architects, and the legal profession. The paper outlines four general types of wind damage investigations and discusses their respective roles in reporting on and influencing construction practice. Principles to guide the investigation of wind-damaged engineered buildings are outlined.

Introduction

The writer has been involved with the investigation of more than sixty damaging windstorm events dating from 1970 through 1995 (Minor and Mehta, 1979). Investigations early in this time period related failure modes to building classifications (Minor, Mehta and McDonald, 1972). Insights into building failure modes that emerged from the damage data correlated well with then new building code initiatives such as those offered in early drafts of ANSI A58.1 (ANSI, 1972). Overall pressure, local pressure, internal pressure and windborne debris were identified as the principal wind effects to be addressed in designing buildings. Field investigations in more recent years served to reinforce these prior observations.

In 1992 the damage caused by Hurricane Andrew stimulated a flurry of damage surveys by many types of organizations. The insurance industry, the housing industry, and building trade associations were especially active. These investigations broadened the general understanding of building failure mechanisms, but this plethora of damage surveys revealed some weaknesses in documentation and reporting processes.

---

[1] Research Professor, Graduate Center for Materials Research, University of Missouri-Rolla, P.O. Box 603, Rockport, TX 78381.

Wind Damage Investigations

Four general types of damage documentation have been conducted following windstorms: (1) "windshield" surveys, (2) walk-through documentation, (3) detailed documentation, and (4) the analysis of engineering buildings. Surveys through the windshield of an automobile or aircraft may have visual impact, but contribute little to the understanding of causes of building failures. Walk-through documentation, commonly accomplished only with a camera, can preserve data and illustrate behavior. However, this type of survey is often incomplete or produces misleading observations. Care must be exercised to not prejudge causes of failure because unknown factors such as material properties and connection strengths can mislead the investigation. If dimensions, material samples and other data are not taken in a walk-through investigation, critical calculations cannot be performed subsequently.

The most useful surveys involve detailed documentation, including the recording of important dimensions, the securing of samples for later analysis and the recording of pertinent data about the building site. Detailed documentation can be used to establish the cause of building failures. This approach is essential if the investigation is to be used to influence design practice, building codes, the marketplace, insurance issues or expert opinion. The detailed documentation process is tedious and time consuming both in the field and subsequently in the office. Investigators will find that they will be restricted to documenting only a few buildings of interest in a given windstorm and that large amounts of office time will be required to fully develop the investigation. The detailed documentation of non-engineered building failures in windstorms is addressed in a companion paper in this conference (Phang, 1999). Principles guiding the detailed analysis of engineered buildings are outlined below.

Assessment of Surveys

Following Hurricane Andrew, Saffir (1993) offered an opinion that "Investigations (which) rely on photographic surveys, which become the basis for reports prepared by investigating engineers and professionals from other disciplines, . . . are wholly unsatisfactory for practicing structural engineers." He recommends using competent structural engineers, analyzing design plans and computing the apparent loads which caused structural distress or failure. The present writer has been asked to perform engineering analyses of damaged buildings. In some cases the buildings were previously documented in a "walk-through" survey, while in others the damaged buildings were examined in detail during first-time visits. In both situations, oftentimes it was found that initial judgments were not correct, or there was more to the assessment of building response than was immediately apparent.

The Analysis of Engineered Buildings

If the results of a survey are to be used to influence design practice, change building codes, promote a product, resolve insurance disputes, pursue litigation, contribute to research, or support some other worthwhile objective, a detailed engineering analysis is the only way to proceed. Saffir (1993) notes that "No benefits can accrue to an individual structural engineer or code-writing authority without such rigorous and detailed stress analyses." The following procedure outlines principles for the analysis of engineered buildings following a damaging windstorm.

1. *Set Objectives:* It is important to have an understanding of why a team is being committed to the field and what is to be done with the survey data upon return. Many field surveys have gone unreported because this commitment was not made in advance. After a few initial surveys by the writer and his colleagues, it became a part of their standard procedure to set objectives for field surveys before leaving the office. Field teams were assigned specific topics (e.g., manufactured housing, metal buildings, performance of window glass) and work commenced only if there was a commitment to finalizing the analysis upon returning to the office.

2. *Plan Field Work:* Sufficient time must be allocated to conduct full, complete and detailed surveys. Operating in the field under post-storm conditions is difficult, at best, and investigators often underestimate the amount of work that can be accomplished. Documentation time should be measured in days, not hours. If a topical analysis is being pursued (e.g., the analysis of a specific type of roofing system or the performance of a specific type of material) planning should include the selection of sample buildings for study from the population of available damaged buildings. If a single engineered building presents a unique opportunity for analysis, a systematic documentation plan should be scheduled to assure completion. If a detailed documentation is to be performed, it is essential that permission be obtained from the building owner or manager.

3. *Perform Field Work:* Field work must be systematically executed according to plan. The investigations should proceed with several working theories as to building performance, while exercising care not to prejudge failure modes. Critical dimensions should be recorded, material samples may be taken, and pertinent data regarding the building site must be obtained. It is often easier to obtain information on the building code in force at time of construction while in the field. Damage to the surrounding area should also be recorded.

4. *Perform Office Work:* Analysis done in the office must use established engineering principles. Plans and specifications should be obtained to confirm construction details and establish material properties. A record of the wind field (wind speeds and directions) that affected buildings being studied is an essential part of the analysis process. A consulting meteorologist or wind engineer should prepare this record. Loads that affected the building and building components should be calculated.

5. *Report Results:* Results of a field investigation are often offered as justification for changing a building code, for using one product over another, for defining failures for insurance purposes and for forming expert opinion. Hence, it is essential that windstorm damage reports be factual and complete. Reports must use statistics and mathematical calculations as may be appropriate to establish the validity of the sample of buildings studied and the accuracy of the technical conclusions reached. Most importantly, it should be published in a forum that permits peer review.

Conclusion

Experience suggests that while "windshield" surveys and walk-through documentation may be useful for illustration purposes, meaningful conclusions cannot be advanced without accompanying detailed analyses. Experience also indicates that detailed documentation surveys require considerable time and effort to fulfill requirements for completeness and accuracy. It is recommended that organizations or individuals that contemplate the conduct of field surveys following a damaging wind event (1) carefully consider the objectives of the survey in advance, (2) plan field activities before leaving the office, (3) budget for the necessary office time following field activities and (4) publish reports in peer reviewed forums.

References

ANSI, "Building Code Requirements for Minimum Design Loads in Buildings and Other Structures," ANSI A58.1, American National Standards Institute, New York, NY, 1972.

Minor, J.E. and Mehta, K.C., "Wind Damage Observations and Implications," Journal of the Structural Division, ASCE, Vol. 105, No. ST11, Proc. Paper 14980, November 1979, pp. 2279-2291.

Minor, J.E., Mehta, K.C. and McDonald, J.R., "Failures of Structures due to Extreme Winds," Journal of the Structural Division, ASCE, Vol. 98, No. ST11, Proc. Paper 9324, November 1972, pp. 2455-2471.

Phang, M., "Investigation of Non-engineered Building Wind Failures," Proceedings, 1999 ASCE Structures Congress (New Orleans, LA, April 18-21, 1999), ASCE, Reston, VA, 1999.

Saffir, H.S., "Evaluation of Structural Damage Caused by Hurricanes," Phase 1 Final Report, SBIR/Division of Industrial Innovation Interface 01/93-09/93, National Science Foundation, Washington, DC, 1993.

WIND DAMAGE INVESTIGATION OF LOW-RISE BUILDINGS

Michael K. Phang[1], F. ASCE

Abstract:

This paper is based on findings gathered as a result of an in-depth investigation of low-rise buildings following Hurricane Andrew. Damage caused by high magnitude winds is assessed from an engineering perspective, and the performance of various types of buildings was analyzed. By means of pertinent illustrations, the paper will allow readers to visualize failures in critical building elements and results of construction plagued by flaws and poor judgement.

Introduction:

Hurricane Andrew swept across South Florida August 24, 1992. The National Hurricane Center classified Hurricane Andrew as a category 4 storm with maximum sustained winds of approximately 225.3 km/h (140 mph). Satellite imagery indicated that it was a well organized storm with an eye diameter of approximately 24.1 km (15 miles) and a translational speed of approximately 32.2 km/h (20 mph). In its wake, Andrew devastated an area of approximately 311 km^2 (120 sq. miles) of developed land, killed 20 people, destroyed or damaged nearly 80,000 homes, displaced nearly 250,000 people, and caused over 30 billion dollars in damages, making this storm one of the costliest disasters in U.S. history.

Building Damage and Performance Assessment:

The impact Hurricane Andrew had on low-rise buildings is of extreme importance due to the simple fact that it is where most people are concentrated during a storm. Obviously, something went seriously wrong. The following sections will point out these shortcomings and analyze them from an engineering point of view.

---

[1] Professor and Director, Architectural Engineering, Department of Civil, Architectural, and Environmental Engineering, University of Miami, Coral Gables, Florida 33124-0630

1. Roof Sheathing

Sheathing plays a critical role in roof construction. It contributes to preserve the structural integrity of the home by providing lateral support to roof trusses by means of its diaphragm action. Sheathing also protects the interior of the structure from the elements because it is where the impermeable roof membrane is attached. If sheathing is lost during a storm, the interior of the structure will suffer from extensive damages due to water and wind penetration. Wind penetration can be especially damaging since it pressurizes the interior and could therefore cause wall failure, particularly in wood frame construction.

Two types of sheathing material were found in South Florida homes. Some homes had plywood sheathing, while others had pressed board sheathing. Homes with plywood sheathing performed remarkably better than their less expensive counterparts.

Sheathing loss was due to the sheathing fasteners failing to properly penetrate the top chord of the trusses. This is obviously a field problem and can be attributed to workmanship. When the roofers are nailing down the sheathing, the roof trusses are not easily located, making it a difficult task to identify their location. Secondly, when using power-assisted nail driving equipment, it is somewhat difficult for the roofers to perceive if the nail is penetrating the truss properly. However, if nails are hand-driven, the penetration can be easily sensed. Because of these reasons, many strewn sheathing panels exhibited attached nails that served no anchoring purpose.

2. Bracing and Fastening of Roof Trusses

The bracing and fastening of roof trusses is crucial if they are expected to weather a strong wing (Fig. 1). This became a big factor during Hurricane Andrew. A countless number of homes exhibited roof truss collapse, an occurrence which became known as the "domino" effect. When the wind struck the gable end, the trusses collapsed like dominoes (Fig. 2).

Roof sheathing, through its diaphragm action, provides lateral support for roof trusses. However, this may not be enough to prevent collapse. There are two basic types of bracing; horizontal and diagonal bracing.

Fig. 1- Collapse of the roof trusses due to lack of horizontal, diagonal, or cross-bracing.

Fig. 2 - "Domino" effect on the gable end collapse.

Horizontal bracing extends along the entire roof span, from gable end to gable end. This type of bracing should be present along the bottom chord plane, as well as along a plane which is between the top and bottom chords of the trusses.

Diagonal bracing provides a connection between the top of one truss and adjacent trusses. This type of bracing is especially critical at gables ends, where the gable truss needs the additional support in order to resist the wind force. By supplying adequate diagonal bracing, the forces on the gable truss are distributed to adjacent interior trusses.

The fastening of trusses to the wall by means of proper hurricane straps is also of prime concern. Hurricane straps are found at either ends of the trusses and serve to provide resistance against vertical wind forces. It is fair to say that hurricane straps performed quite well when properly installed, and the bulk of the problems were not due to truss uplift.

3. <u>Gable and Hip Roofs</u>

Gable Roofs suffered from structural damage in a disproportionate manner, as compared to Hip Roofs. In fact, extremely few hip roofs suffered structural damage, and those, which did, did not exhibit as serious damage as Gable Roofs. The Hip Roof allows the wind loads to be transferred and distributed to the walls and throughout the roof in a more equitable manner. This is due to the fact that the hip roof has four sloping sides. Also, the very shape of the Hip Roof tends to deflect wind forces significantly better than Gable Roofs. Another reason is that hip roofs do not have vertical members at their ends, unlike Gable Roofs, which must withstand great negative and positive pressures. Finally, bracing of trusses in Hip Roof construction is essential to the way the trusses are assembled. In Hip Roofs, to a certain extent, trusses are braced by adjoining trusses that frame in, especially at the corners of the building.

Gable Roof failure was observed in two manners. The first is that the gable-end collapsed outward. This type of failure took place due to the combination of negative pressure forming on the leeward side of the structure and internal pressurization of the structure due to penetration of its exterior envelope. Since both of these forces act in the same direction, the resultant force is quite large, and hence, causes the failure. Another typical failure of gable ends is an inward collapse, causing the already mentioned "domino effect". This failure took place when direct positive pressure on the windward side of the structure overcame the strength of the gable's lateral supports. Most often this process was initiated when the roof sheathing, providing lateral support along the top chord of the trusses, was lost due to wind action. When the sheathing along the roof edge was removed, the gable end had no other support, except any diagonal bracing that may have been installed.

4. Wood Framing and CBS Construction

Residential structures of CBS construction generally outperformed those made of wood framing. When the structural shell was penetrated and internal pressurization began to take place, wood frame exterior walls tended to suffer structural damage in the form of collapse with much greater frequency than their masonry counterparts (Fig. 3). However, this fact does not mean that masonry walls are hurricane proof.

Fig. 3 - Collapse of the exterior frame walls from window loss and failure of roof.

CBS construction performed relatively well when put together in an appropriate manner. Most of the buildings inspected, where failure or collapse of exterior walls occurred, clearly showed that there was no vertical steel reinforcement running through the block cells. The South Florida Building Code requires that a #5 reinforcing steel bar be present in a concrete-filled cell at all wall corners and at maximum intervals of 6.1m (20ft.) thereafter for single story homes. This vertical reinforcement is intended to provide continuity between roof, walls and foundation since the steel is meant to extend from the footings all the way to the tie beam. Without this critical reinforcement, the wall cannot act as a monolithic structural unit and, under intense wind loading, a collapse is imminent.

It was also not uncommon to find collapsed walls with the exposed reinforcing bars still extending from the foundation slab. This was apparently due to the failure of filling or grouting the masonry cells, containing the reinforcement, with concrete. Obviously, if the cells are not filled with concrete, the reinforcement serves no purpose.

Another common problem associated with CBS construction, which is also related to vertical reinforcement, is the anchoring of the tie beams to the wall. In this type of failure, little or no vertical reinforcement was used to anchor the tie beam. Most probably the steel extending from the foundation did not reach the tie beam, and, if it did, it did not penetrate the tie beam sufficiently. This type of failure is especially critical and endangers the occupants of a structure. Since the roof is anchored to the tie beam, the displacement of the unanchored beam will most likely cause a near-total collapse of the structure.

In regards to wood framing, if properly built, wood frame homes could be resistant to hurricane related wind loading. However, it is also important to mention that there are many more interdependent details involved in wood construction, therefore, quality assurance becomes a very important factor.

The way wood framing transmits loads from roof to foundation is by attaching the trusses to the studs below by means of metal straps. These metal straps are anchored by providing a minimum of three 16-d nails around the truss and three 16-d nails along the top of the stud below. Furthermore, the bottom of this stud is attached to a bottom wood plate that is bolted to the foundation. In the case of two story wood frame construction, the bottoms of the second story studs are attached to a sole plate using two 16-d endnails and metal straps that extend from the bottom of the sole plate to the lower portion of the stud. The sole plate, in turn, is anchored to the floor joists using staggered 16-d nails at 40.6 cm (16 in.) on center. In addition, the second story studs are attached to the top of the first floor studs using the same straps mentioned above.

Wood frame wall construction did not, however, live up to the expectations of the South Florida community. Several problems seemed to plague this type of construction. First of all, there was an apparent problem with connection details. This was especially common in homes in which the first floor was constructed of masonry block and the second floor of wood framing (Fig. 4). Upon penetration of the exterior shell, the resulting internal pressurization seemed to have been sufficient to cause the collapse of the second floor walls. Close inspection seemed to indicate that this was due to poor connection of the sole plate to the floor joists, poor connection of the studs to the sole plate, or, in many cases, a lack of adequate fastening that ensures continuity between intersecting walls. In most situations, collapse was caused by a combination of these.

Fig. 4 - Failure of the second story wood frame due to poor connections between studs and sole plate.

Problems also arose when wood frame construction was combined with masonry block construction. This was common in townhouse complexes where the end walls were built of masonry blocks, while front and rear walls were wood frame. Apparently, the problem was in the connection between the heterogeneous walls. Failure to provide continuity in these intersecting walls permitted collapse in both wall types. If appropriate fastening would have been present at the corners, the walls would have contributed symbiotic support that could have prevented each other's destruction.

Since wood frame construction undoubtedly involves more details that concrete block construction to ensure a sound structure, errors in workmanship can prove quite severe. In addition, the more massive concrete block offers more natural resistance against collapse than wood frame, and the block's rigidity provides added protection against penetration of windborne debris. All in all, if quality of construction is upheld, both systems should perform in an adequate manner.

5. <u>Roofing Paper, Shingles, and Clay/Concrete Tiles</u>

Loss of roofing paper caused significant damage to the interior of homes. Even if the sheathing remained intact, once this impermeable layer was removed, water was introduced inside the structure causing material losses to mount. Inspection revealed that roofing paper remained in place more often when it was fastened using tin caps instead of staples. When staples were used, it was noticed that the staples remained while the felt paper tore loose. Another problem with the installation of roofing paper is at the overlaps of the two felt layers and at roof edges. In both places, care should be taken to hot-mop these critical spots since they have a tendency to be uplifted by the wind. Failure to ensure this bonding led to a great portion of a roof's felt to be peeled off. In addition to these, the loss of felt paper was sometimes attributed to the poor quality of the material itself.

With respect to asphalt and fiberglass shingles, they performed well when fastened correctly. Nonetheless, there were ample examples of total loss of shingles. In most cases, the manufacturer's installation procedure were apparently ignored. It was found repeatedly that the film covering the shingles adhesive surface was not removed, thereby, reducing the shingle's capacity to resist uplift. As in the case of roofing paper, shingles that were fastened using staples performed poorly, while those using nails remained in place more often.

Not surprisingly, the performance of clay and lightweight concrete roof tiles was determined by how well they were fastened. One type of fastening system involves the nailing of the roof tile to 2.54 cm by 5.08 cm (1 in. by 2 in.) batten strips. Extensive damage resulted when the connections between the battens and the roof failed and when not enough nails were used to secure the roof tiles to the batten strips. The other type of roof tile fastening system uses mortar beds to secure the tile. When sufficient mortar was used for the setting, the roof tiles remained in place and minimal damage was incurred. However, when mortar was applied sparingly, extensive damage could be readily detected. It should also be mentioned that concrete tiles seemed to perform slightly better than clay tiles. This is due to concrete being a stronger and more rigid material than clay and the concrete's natural ability to establish a better bond with the mortar than clay. In addition, the recommendation that clay tile should be dipped in water before installing it in the mortar bed, to improve bonding, is hardly ever heeded.

Anchorage of clay and concrete roof tiles is of utmost importance since they can become massive flying projectiles. Many homes suffered from window breakage and penetration of their wall cladding when tiles from neighboring homes impacted.

6. Residential Garage Door Failure

Garage door failure is quite critical since it covers a relatively large area. Should the garage door fail, this large opening enables the structure to quickly become

Fig. 5 - Typical garage door failure.

pressurized. Typical effects of this pressurization can result in wall collapse, loss of roof sheathing, and failure of roof trusses (Fig. 5).

Most garage door failures were apparently due to inadequate bracing of the door itself and weak connections between the door frame and the structure. Manufacturer's installation requirements should be carefully followed to prevent future losses in this vital part of a home's exterior envelope.

Conclusions:

In the author's opinion the failure of many residential structures could have been avoided if their design and construction would have been carried out according to the prescriptive requirements of the South Florida Building Code. The overall performance of masonry walls was superior to that of walls constructed of wood framing, and gable-end roof construction experienced much more damage than that of hipped roof design. Wood frame structures performed especially poor when there was a breach of the structures' exterior protective envelopes, as in the case of roofing material, door and/or window failure. These failures led to weather penetration and consequently internal pressurization of the structures, which resulted in extensive interior damage. The need to preserve the integrity of the structural envelope calls for these critical construction details to be carefully analyzed and explicitly addressed by the code in order to prevent future catastrophes. In addition, concern should also be directed toward enforcement of the code requirements by building department officials, since it does not matter how precise and detailed the code is written if there is no certainty that the construction industry will base its practices on these high building standards.

GUIDELINES TO CONDUCT A WIND DAMAGE INVESTIGATION OF MANUFACTURED HOUSING
John Mehnert, ASCE Life Member, P. E. [1]

Abstract: The highly destructive effects of wind on manufactured homes has acted as a mandate for change in the Federal Manufactured Home Construction and Safety Standard (FMHCSS). Since June 15, 1976, manufactured homes were required to be built in accordance with the FMHCSS as administered by the U. S. Department of Housing and Urban Development (DHUD). Now, there are proposals to further adjust and raise this standard.

Manufactured Housing can be divided into four major groups: precut, panelized, modular and HUD-Code, previously known as mobile. HUD-Code housing composes 31 percent of all housing in America,and since the Code is national in scope, it preempts all local building codesfor single family manufactured dwellings.

This paper is a discussion about HUD-Code housing and it's susceptibility to wind damage, the history and evolution of the name, and the acceptability/affordability and proposed changes to the Code. A discussion of this information should assist the practicing engineer in performing a wind damage investigation.

Hurricane Andrew struck south of Miami, Florida on August 24, 1992, and subjected commercial and residential housing to participate in a full scale testing laboratory experiment. The large scale devastation of over 8000 HUD-Code housing units proved that the structural strength requirements found in the Manufactured Housing HUD-Code unit was inadequate. The design wind loads in the standard are actually less than typically experienced. Also poor structural connections provide a weak link in the integrity of these homes allowing for progressive failure.

Further analysis also revealed the (FMHCSS) design specifications in the HUD-Code

---

[1]Regional Supervisory Structural Engineer for the United States Department of Housing and Urban Development (Retired); Consulting Engineer, JM Residential Structural, 12689 West 82nd Terrace, Lenexa, KS 66216-2644

was equivalent to an 80 mph wind velocity in hurricane zones, whereas the local building code used by the engineering profession recommends 115-120 mph for other residential   and commercial construction.

Preliminary observations of the Hurricane Andrews effects by the Wind Engineering Research Council indicated the wind speed to be between 110 and 125 miles per hour. The Council also stated that "mobile (HUD-Code) homes and other forms of manufactured housing performed poorly as currently designed and constructed. The homes cannot be relied upon to provide for the safety and welfare of inhabitants."

BACKGROUND:

HUD-Code housing is extremely popular throughout the United States in both rural and urban settings, comprising 31 per cent of all housing. The Manufactured Housing term is in itself a generic name that refers to housing built in a factory rather than on-site. The name evolved from House Trailer to Mobile Home when the industry sought to improve it's image. In 1989 the term Mobile Home was discarded when the U. S. Congress officially changed the name to Manufactured Home.

A Manufactured Home is a permanent residence and is seldom moved from it's origi nal site. The wheels and axles are not for continuous use, but a built in means of transporta-tion to the site.

The HUD label affixed to every manufactured home certifies that the home has been factory constructed, tested and inspected to comply with the HUD Code that includes specifi-cations on basic dimensions, emergency egress, lighting and ventilation, plumbing, electrical, heating/cooling, thermal protection and fire and safety. The HUD label is very much like a cer-tificate of occupancy.

HUD-Code housing is designed with single and double-wide floor plans. A single wide is manufactured in 12, 14 and 16 foot widths, whereas the doublewide is 24, 28 and 32 feet. A double wide home usually contains more amenities and costs more than its narrower coun-terpart. It is designed, engineered and assembled in a factory and then towed to a site. At this point, it can be placed on a permanent foundation to resist the wind or seismic lateral force. Federal law prohibits the removal of any part of the frame of a manufactured home's perma-nent chassis, since it is essential to provide support to the home during transportation an on-site. Parts not essential to the homes structural support, such as running gear assembly, lights, coupling mechanisms and the draw bar, may be removed after delivery to the site. I will refer to this category of manufactured home as HUD-Code for brevity in the following discussions since it is synonymous.

AFFORDABILITY:

Although manufactured housing has advantages over site construction in terms of con-struction time, decreased material waste, labor force and quality control, the major difference is its affordability. This factor along with low maintenance cost is certainly a reason for it's pop-ularity in the United States. Current typical retail costs place HUD-Code homes per square foot cost less than half as much a site built homes: $25.19 compared to $54.65. This equates to an average cost for a singlewide of $19,800.00 to more than $33,600.00 for a multi-wide. HUD-Code housing fills a niche and now composes 31 percent of all housing in America.

DAMAGE STATISTICS:

Statistical data is necessary, particularly when surveying a wind impacted area for the purpose of damage assessment --local, state and national governments normally require the statistics immediately to determine the need for emergency housing and funding for recovery.

The following methodology was developed by Vann and McDonald in 1978 for estimating the extent of damage for a statistical study. A chart titled Descriptions of Wind Damage Classes for manufactured Housing is included and describes the seriousness of damage varies from a class of 1 to 4. Class 1 and 2 are assigned to units that are still habitable, but have an increasing extent of damage. In damage class 3, the unit is judged uninhabitable,but repairable, Class 4units are not repairable. See table 1 for further explanation of the classes.

TYPICAL INSTALLATION DEFICIENCIES:

The installation of a HUD-Code home is not covered in the Federal Standards. It is the responsibility of the organization setting up the home to assure that the unit is set up in accordance with the manufacturers instructions and state and local regulations. The Federal manufactured home program managed by the Department of Housing and Urban Development regulates only the design and production of homes at the factories. When the home produced under the Federal manufactured home programs leaves the factory, it is sold by dealers and transported and installed either by the dealer or subcontractor on the selected site. The Federal Standards 24 CFR, Ch. XX, Part 3280.306 specifies the minimum performance requirements for windstorm protection to support and anchor the unit. Periodic monitoring by National Conference of States on Building Codes and Standards (NCSBCS) of the installation of HUD-Code homes using requirements formulated by states and local governments has discovered several typical deficiencies in the installation deficiencies summarized into several categories:

1. Site preparation, (drainage, spacing or placement of footings, vapor barrier).

2. Foundation, footings, piers and shims.

3. Tie-downs (ground anchors) against sliding, uplift and overturning movement caused by wind.

4. Marriage wall connections (roof, floor, insulation caulking and alignment between the two halves).

The lack of ground anchors or installation of those of inadequate strength is the most critical deficiency. HUD-Code homes are especially vulnerable in wind storms. Minor damage ranging in loss up to $1000.00 occurs routinely at wind speeds between 50 and70 MPH and major damage which includes total destruction can occur at wind speeds in excess of 70MPH. The NCSBCS monitoring report drew the following conclusions with regard to faulty installation of the critical ground anchors.

1. Test data from manufacturers indicates that ground anchors are generally tested for pull-out, using a force applied along the axis of the ground anchor. Therefore,applications of the ground anchors which allow diagonal struts to be aligned tin directions other than those along the axis of the ground anchors will not carry the load.

2. If a ground anchor is installed by drilling a hole in the ground and then backfilling the hole after installation of the ground anchor with compacted soil, a reduced capacity will result.

3. Since different types of ground anchors are required for various classifications of soil and their depth required for various classifications of soil and their depth requirement may also

depend upon the quality of the soil to develop the desired capacity, a uniform industry wide method to classify soils is needed. In addition to soil classification, charts need to be developed to establish capacities for anchors with variables such as depths and angle of anchor to ground.

4. Test data is not available for performance of ground anchors under cyclic loads caused by wind gusts.

5. The combination of an over the roof tie and diagonal frame tie sharing a common ground anchor necessitates an anchor capacity capable of resisting this loading.

6. Straps and buckles must comply with the variations in manufacturers instructions to be assured of the desired capacity.

INVESTIGATION OF WIND EFFECTS:

The mechanisms for wind damage to a HUD-Code home are:

1. High Aerodynamics Pressures resulting from air flowing over and around the home

The investigator should be aware that many times the failure of the walls and roof give the appearance of exploding, but in fact are due to the inward pressure on the windward side. Outward pressures are on all other sides. HUD-Code homes are relatively small in comparison to many structures and are sometimes engulfed in gusts of wind. As a result, they are frequently damaged because of the effects of turbulence (gusts) embedded in straight line winds. Damage from such turbulence is sometimes erroneously attributed to small to tornadoes. Then actual frequency of a HUD-Code home is relatively high compared to typical gust frequencies. However, the house components such as the roof covering, the siding, or a long unsupported wall may experience resonance effect. Resonance occurs when the gust frequency and the frequency of the HUD-Code house components are very nearly the same. Relatively high deformation scan then take place. (See figure1.)

2. Impact of Windborne missiles

Missiles are defined for this discussion from gravel blown off built up roofs to other HUD-Code homes becoming airborne. Where the wind velocity in greater than 100 MPH, the denser and larger missiles become extremely dangerous to personal safety and personal property. HUD-Code homes are typically vulnerable to sliding off the support piers, overturning and rolling or tumbling until destroyed. This, of course, is due to inadequate anchorage to resist the horizontal and uplift forces on the walls and roof. It has been reported that when tipping up or starting to rotate because of wind pressures, The HUD-Code home exposes its under floor at an angle to the wind and the resulting uplift action can cause an airborne missile.

3. Change in Atmospheric Pressure

This mechanism is limited to tornadoes. Although the drop in pressure from an intense tornado may approach 285 psf, the norm is much less. However, recent research suggests that a erodynamic pressures are responsible for the explosive pattern of damage patterns. The high velocity tornadic winds normally strike the home, first breaking windows, opening doors and causing wall openings that equalize the inside and outside pressures. Therefore, atmospheric pressure change causing the damage is unlikely.

FAILURE PROFILES:

1. Roof damage or loss:

Roofs on HUD-Code homes are normally constructed with metal or a flexible waterproof membrane. Once an impact by missiles or repetitive flexing causes a tear, the roofing begins to progressively peel away. The removal of the roof can result from exposure of the sheathing and its anchorage to long term moisture of localized high wind velocity pressures. (See Figure 2.) Removal of the rafters of a HUD-Code home roof also leaves exterior walls unsupported along the top. Walls of large rooms which have long spaces between transverse interior walls are the most susceptible to being damaged in this manner.

2. Sliding off the piers:

   Sliding off the piers can cause major manufactured home damage. The sudden impact with the ground will rack the framing system, leaving it permanently deformed. Table 2 shows calculated values of wind speed required to initiate sliding or overturning of unanchored manufactured homes which have different total widths, weights and runner widths. The narrower and lighter homes represent older models with wider and heavier homes are common to current construction.

3. Overturning:

   Major damage such as racking of the structural frame often occurs after sliding off the piers and the leeward edge of the floor towards the ground. As the home continues to rotate, damage becomes so extensive that it is likely to be classified as a total loss. Table 2 demonstrates the following:

   a) Weight of home strongly influences the wind speed that will produce either sliding or overturning. Sliding and overturning wind speeds increase in proportion to the square root of the average width.

   b) Spreading runners further apart increases the overturning wind speed to only a very moderate degree.

   c) Damaging movement will be likely to occur in homes without tiedowns at wind speeds as low as 47mph.

4. Separation from underframe:

   The entire super structure can be sheared or lifted off the underframe if the lag bolts that are connecting the underframe outriggers to the perimeter flooring system are too small or few in number. HUD-Code home construction techniques such a 1 by 4 base plates toenailing wall studs and allowing carpeting to be placed between the floor joists and base plate affects the resistance to large shear and uplift forces. (See figure 3.)

5. Separation of walls from floor:

   The connection made by nailing the walls wooden base plate to the edge of the floor system can fail in tension or shear. Metal straps are sometimes used to make a connection from the stud to the floor joist.

6. Support and anchoring systems:

   Footings, piers and ground anchors work in concert to form and stabilize the overall system. Failures often are due to lack of compliance to state or local codes with regard to allowable soil bearing capacity, footing size and depth, height of unreinforced piers and installation of ground anchors. Activities such as site preparation and installation are regulated by the state and local government. Reviews of the state and local regulations by the National Conference of States on Building Codes and Standards (NCSBCS) has indicated less than satisfactory regulations and enforcement. The adoption of a National Installation Standard such as NCSBCS A225.1 by the states would benefit uniformity of enforcement and compliance by the manufacturers. When ground anchor failure has been identified by the investigator as the weak link, an excellent source for further study is   <u>NBS Studies of Mobile Home Foundations.</u>

A commonality exists through the previously presented failure profiles. The HUD-Code homes have all been subjected to some degree of lifting and have experienced a connection failure.

HUD-Code homes are fragile structures as presently codified and constructed. Therefore, the exact failure profile maybe obscure as it may be a co-mingling of factors. However,there are three basic mechanisms for severe damage to the superstructure or HUD-Code unit. First the ground anchors fail and the unit is overturned and racked. Second, the light construction permitted wind borne debris damage. Third, minimal framing connections acted upon by localized wind velocity pressures permitted superstructure elements to pull apart.

CONCLUSION:

In a paper titled "Standards of Practice for Manufactured Housing" and published in the Journal of Aerospace Engineering,ASCE, Vol. 2, April 1989, the authors analyzed the design criteria for wind pressure on walls and roofs required by the FMHCSS as published in 24 C.F.R., parts3289,3282 and 3283. This standard defines two geographical zones--Standard and Hurricane. The wind pressures are as follows:

	<u>Walls</u>	<u>Roof</u>
Standard Zone	15 psf	9 psf
Hurricane Zone	25 psf	15 psf

Using the procedures of the current national wind load standard ASCE 7-88, Minimum Design Loads for Buildings and Other Structures, the wind pressures are equivalent to a design wind speed of 65 mph and 80 mph for the Standard Zone and Hurricane Zone respectively. It is apparent that FMHCSS is substantially lower in pressure and velocity when compared to the wind speed charts in the major building codes throughout the country. (See figure 4 and figure 5.)

In 1977, DHUD in conjunction with the National bureau of Standards conducted a study titled "The Measurement of Wind Loads on a Full Scale Mobile Home." The findings determined that the FMHCSS wind load requirements issued in 1976 are not adequate to provide wind safety for mobile homes. Refer to Design Load Table 3, option 1, a loading table which lists forces that the mobile homes must withstand in wind zones 1 (70 MPH) and 2 (90 MPH) to prevent structural failure. These recommendations were never implemented by the industry.

Dr. James R. McDonald stated in his paper to the Subcommittee on Housing and Community Development, October, 1992, that the present HUD-Code designs cannot be upgraded to resist winds much above 80 mph. It appears that, rather than simply increasing the wind pressures, a new strategy needs to be developed for new HUD-Code designs utilizing materials and construction techniques to allow for redundancy and ductility. He suggests the following basic concepts of manufactured home design should be changed.

1. The idea that manufactured homes are mobile should be abandoned.

2. The long, narrow, single wide shape should be eliminated.

3. The use of ground anchors should be stopped in favor of a permanent concrete or masonry foundation.

4. The HUD-Code home structure should be made more redundant and ductile.

Obviously, HUD-Code homes as presently constructed are unsafe in strong wind storms, even if they are placed on stable foundations and their floors or undercarriages are tied down. They must be considered expendable in storms of great intensity, such as Hurricane Andrew.

The solutions presented by Dr. McDonald could remedy HUD-Code problems and thus resolve the safety hazards now confronting the industry and HUD-Code investigators.

APPENDIX 1. --- REFERENCES

Defense Civil Preparedness Agency (1974).   Protecting Mobile Homes from High Winds ,   U. S. Government Printing Office, Washington, D. C.

Federal Register. 24 CFR, Ch. XX, Part 3280.   Manufactured Home Construction and Safety Standards, Office of Assistant Secretary for Housing, HUD. Washington, D. C.

McDonald,J. R., (1992). "Strategies for Mitigating Damage to Manufactured Homes Subjected to High Winds," Institute for Disaster Research, Texas Tech University, Lubbock, Texas, 1-6

National Conference of States on building codes and Standards, Inc. (1989).   Preliminary Study of the Installation of Manufactured Homes and Installation of Manufactured Homes and instal - lation Monitoring by States and Local Governments , NCSBCS, Washington.

Vann, W. P./McDonald, J. R. (1979).   An Engineering Analysis: Mobile Homes in Windstorms , Institute for Disaster Research, Texas Tech University, Lubbock, Texas. 1-15, 70-95

Waldrip, Travis G. (1976).   Mobile Home Anchoring Systems and Related Construction , Institute for Disaster Research, Texas Tech University, Lubbock, Texas. 13-31.

Walters, Jeffrey L. (1980). Wind Safety Analysis , Technology Plus Economics, Inc., prepared for: U. S. Department of Housing and Urban Development, Office of Policy and Research Contract, No. H2514, Washington, D. C.

Wind engineering Research Council, Inc. (1992).   "Hurricane Andrew - Preliminary Observations of WERC Post-Disaster Team," 1-5

Yokel, R. Y./Chung, R. M./Yancey, C. W. C. A. (1981).   NBS Studies of Mobile Home Foundations, U. S. Department of Commerce, National bureau of Standards, Washington, D. C., 4-10

DESCRIPTIONS OF WIND DAMAGE CLASSES FOR MANUFACTURED HOUSES

Damage Class	Description of Damage
0	**No Damage or Very Minor Damage** Little or no visible damage from the outside. Slight shifting on the blocks that would suggest relevelling, but not off the blocks. Some Cracked windows, but no resulting water damage.
1	**Minor Damage** Shifting off the blocks or so that blocks press up on the floor; relevelling required. Walls, doors, etc. buckled slightly, but able to be corrected by relevelling. Minor eave and upper wall damage, with slight water damage, but roof not pulled all the way back. Minor pulling away of siding, with slight water damage. Minor missile and/or tree damage. Slight window breakage and attendant water damage.
2	**Moderate Damage (Still Livable)** Severe shifting off blocks with some attendant floor and superstructure damage (punching, racking, etc.) Roof removed over a portion or all of the home, but joists intact, walls not collapsed. Missile and/or tree damage to a section of the wall or roof, including deep dents or punctures. Serious water damage from holes in roof, walls, windows, doors or floors.
3	**Severe Damage (Not Livable but Repairable)** Unit rolled onto side but frame intact. Extreme shifting causing severe racking and separations in the superstructure. Roof off, joists damaged or removed, walls damaged from lack of lateral support at top. Severe tree damage, including crushing of one wall or roof section. Superstructure partially separated from underframe.
4	**Destruction (Not Repairable)** Unit rolled onto top or rolled several times. Unit tossed or vaulted through the air. Superstructure separated from underframe or collapsed to side on the underframe. Roof off, joists removed and walls collapsed. Destruction of a major section by a falling tree.

TABLE 1

Broadside Windspeeds at 30 Feet in Open Terrain to Cause
Sliding or Overturning of Unanchored Mobile Homes

Nominal Width (feet)	Average Weight (psf)	Sliding Windspeed* (mph)	Overturning Windspeed (mph) for a Runner Spread, w_r (in feet) of:				
			4.5'	5.0	6.0	7.0	10.0
10'	15	46.6	47.6	49.6	53.2	56.3	63.5
	20	53.8	55.0	57.3	61.4	65.0	73.3
	25	60.2	61.5	64.1	68.7	72.6	81.9
	30	65.9	67.3	70.2	75.2	79.6	89.8
	35	71.2	72.7	75.8	81.2	85.9	96.9
	40	76.1	77.7	81.0	86.9	91.9	103.6
			4.5	5.0	6.0	8.0	12.0
12'	15	50.1	51.1	53.2	56.8	62.6	70.7
	20	57.8	59.0	61.4	65.6	72.3	81.6
	25	64.6	66.0	68.6	73.3	80.8	91.2
	30	70.8	72.3	75.2	80.3	88.5	99.9
	35	76.5	78.1	81.2	86.8	95.6	107.9
	40	81.8	83.4	86.8	92.8	102.2	115.4
			6.0	7.0	8.0	10.0	14.0
14'	15	53.0	59.7	62.8	65.5	69.9	76.3
	20	61.2	68.9	72.5	75.6	80.7	88.1
	25	68.4	77.0	81.1	84.5	90.3	98.5
	30	74.9	84.4	88.8	92.6	98.9	107.9
	35	80.9	91.1	95.9	100.0	106.8	116.6
	40	86.5	97.4	102.5	106.9	114.2	124.6

*With friction coefficient of 0.4 (shaded portions are most representative).

TABLE 2

TABLE 3

Design Loads for Standard and Hurricane Wind Zones

		Option 1				Option 2	
		Standard Wind Zone	Hurricane Zone				
		70 MPH		90 MPH		105 MPH	
		Single-wide	Double-wide	S. Wide	D. Wide	S. Wide	D. Wide
ROOF	Design of trusses, roof membrane and fasteners (except as noted below)	-23	-18	-40	-30	-54	-41
	Roof membrane and fasteners on strip 2 feet wide extending around perimeter of roof	-36	-36	-61	-61	-83	-83
	Overhangs (net uplift)	45	45	77	77	105	105
SIDE WALLS	Design of studs, doors, windows, exterior wall covering and fasteners (except as noted below)	15 -12	15 -12	26 -21	26 -21	35 -29	35 -29
	Exterior wall covering and fasteners at ends of sidewalls on vertical strips 6 feet wide	-24	-24	-40	-40	-54	-54
END WALLS	Design of studs, windows, exterior wall covering and fasteners	15 -32	15 -24	26 -56	26 -40	35 -76	35 -54
FLOOR	Design of joists, floor panels and fasteners (occupancy load excluded).	6	6	10	10	14	14
DRAG LOAD	Load acting on horizontally projected area of structure and used for design of structural subsystems to resist racking (See Note 3) End sections (1/4 length).	17	15	29	24	39	33
	Central section (1/2 length)	15	15	24	24	33	33
UPLIFT LOAD	Load acting vertically upward on plan area of structure and used for design of structural subsystems to resist bending in vertical plane (See Note 4)	16	16	28	28	38	41

NOTES:

1. All pressures in pounds per square foot.
2. Negative sign indicates pressures acting outward.
3. Resultant to be applied at 0.6h above ground level.
   h is height of roof-wall intersection.
4. Resultant to be applied at 0.4W from windward edge of roof.
   W is width of mobile homes.

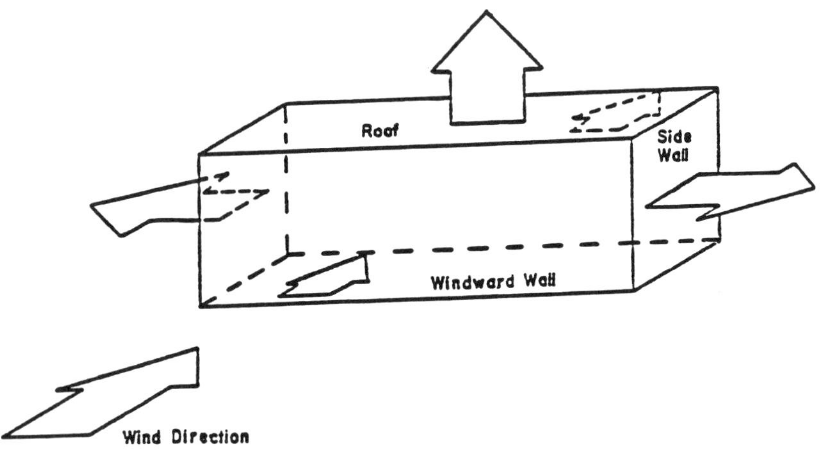

OVERALL EXTERNAL PRESSURES ACTING
ON A MOBILE HOME

FIG. 1

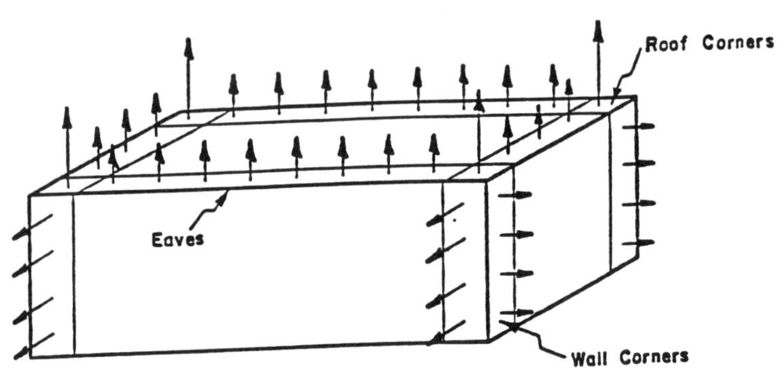

LOCAL WIND PRESSURES

FIG. 2

LAG BOLT
12" CHANNEL
2" x 6" HEADER
12" RUNNER BEAMS
3" CHANNEL
2" x 6" FLOOR JOISTS
SIDE PLATE
LAG SCREW
OUTRIGGER
5/8" FLOOR DECKING
2" x 6" FLOOR JOISTS
CONSTRUCTION DETAILS OF UNDERFRAME AND FLOOR SYSTEM

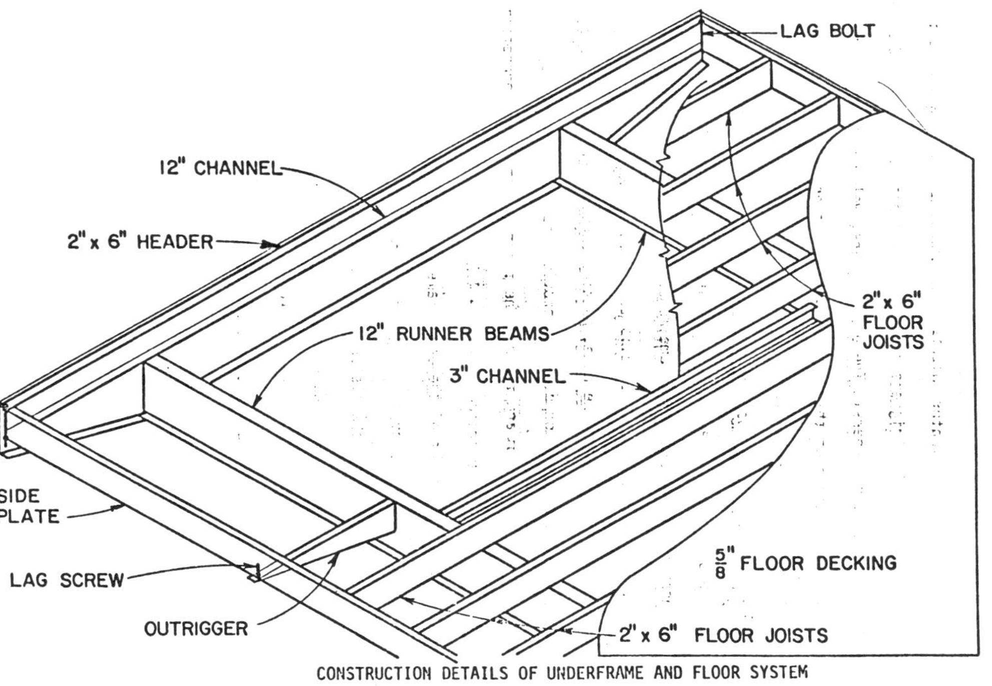

FIG. 3

Office of Asst. Sec. for Housing, HUD §3280.305

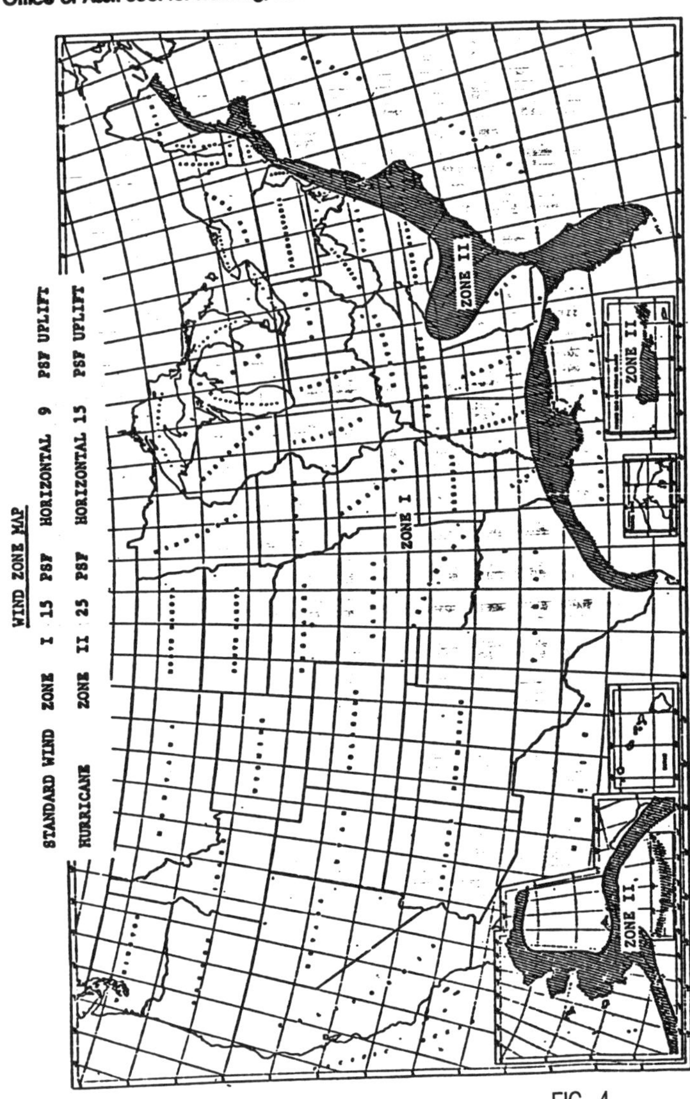

FIG. 4

AMERICAN NATIONAL STANDARD A58.1-1982

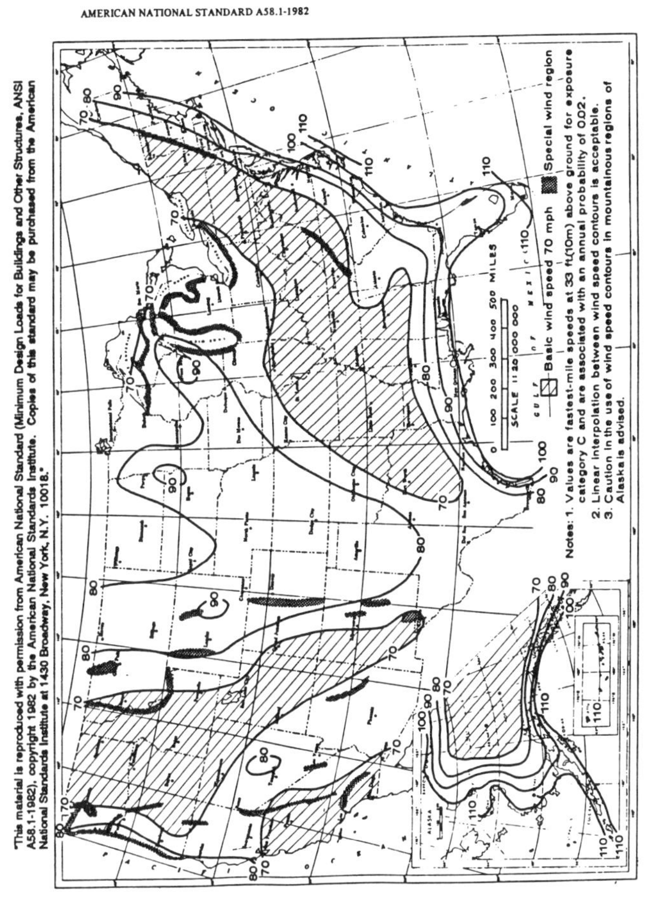

FIG. 5

Upgrading Building Code Requirements
In Hurricane-Prone Areas

Herbert S. Saffir[1], F. ASCE, P.E.

Abstract:

This paper reviews some of the history in improving building codes in hurricane-prone areas. Most of the improvements or advancements can be traced to past hurricanes and structural and flood damage resulting from those hurricanes. This paper reviews and recommends certain specific structural requirements required in building codes to provide hurricane-resistant building requirements.

Historical Records:

Historically, descriptions of hurricanes go back to 1494 when Columbus experienced his first hurricane. Other subsequent hurricanes were recorded in 1513 by Ponce de Leon, and later by Hernando Cortes. Of special note was the Great Hurricane of 1780 that killed about 22,000 people in the Eastern Caribbean: St. Lucia, Barbados, Martinique and other islands to the north.

The September 8th and 9th 1900 Hurricane that struck Galveston, Texas was the first storm in the present century that was noteworthy; some estimates indicated that over 7,000 people died in the Galveston area in the hurricane. The storm surge was the basic reason for the loss of life; buildings collapsed or were washed away by wind and water. After this storm, a well-designed concrete sea wall was constructed along the Galveston Gulf front by the Corps of Engineers.

---

[1]Herbert Saffir Consulting Engineers, 350 Sevilla Avenue
Coral Gables, Florida 33134

The September 1926 hurricane that struck the Miami and South Florida area resulted in an economic collapse in the Miami area, in addition to the extensive structural collapses. Buildings were severely damaged by the wind; buildings collapsed under the storm surge; many buildings lost their roofs. The storm continued across the Gulf of Mexico to the Pensacola, Florida area and caused considerable damage there.

1937 Code:

The 1926 hurricane resulted in probably the first extensive building code for the City of Miami. The wind load section of this code is rather meagre; the 1937 version specified design wind loads for walls of buildings to meet this table:

"Height above ground	Wind Pressure over entire building
25 feet	25 pounds per sq. ft.
75 feet	35 pounds per sq. ft.
over 75 feet	45 pounds per sq. ft.

Tanks, signs, exposed structures and their supports are to be designed for 50 pounds per sq. ft. on the gross area of the plane surface, acting in any direction."

In the early 1930's, these loads exceeded wind loads called for in other codes used in the U.S.

The chief value of the 1937 code was that it established prescriptive requirements for buildings of one and two stories, and, by detailed drawing, indicated these requirements as a mandatory part of the code. Details were given for the mandatory continuous reinforced concrete footings, and for reinforced concrete tie beams, used in masonry buildings. (C.B.S. construction).

See typical wall-section:

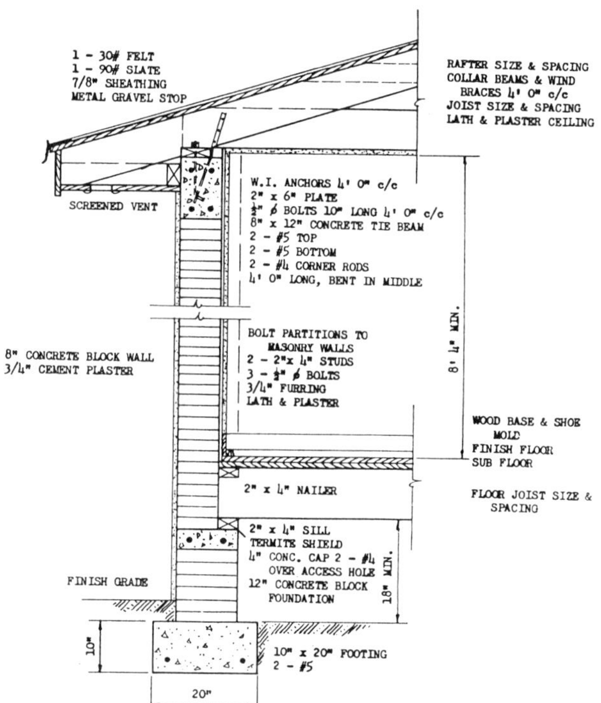

TYPICAL C.B.S. WALL SECTIONS-RESIDENTIAL

1960 Hurricane Donna and Uniform Code:

Subsequent hurricanes, such as Donna (1960), resulted in various improvements to the building code. Previous to Donna, in 1957, a uniform building code was adopted throughout South Florida. (South Florida Building Code). This code is now used by a permanent population of over four million people. The code had improved the earlier codes; the code now called for more detailed and rational wind load requirements, which requirements came up to the "state-of-the-art" at that time. This code was the first that set forth wind loading and installation requirements for sliding glass doors, windows, doors, glass and curtain-wall.

1992 Hurricane Andrew:

Hurricane Andrew which made landfall 30 miles south of Miami in August 1992 caused estimated damage of 30 billion dollars; it was probably the worst natural disaster that has hit the U.S. This hurricane had sustained one-minute winds of 145 miles/hour, with gusts reaching at least 180 miles/hour and probably higher. The South Florida Building Code was improved on, as a result of this hurricane. The tremendous structural damage caused by wind was attributed to various factors, including poor construction practices, lack of attention to the building code, poor inspection procedures by the various Building Departments, deficiencies in the building code and, foremost - the magnitude of the hurricane winds. In the opinion of the author, the building code, even with its deficiencies, prevented even much worse structural damage then did occur.

An investigation and analysis by the author, of the failed new Alert Shelter Complex building at Homestead Air Force Base - designed to the ANSI A58-1982 wind requirements - indicate the magnitude and power of Hurricane Andrew as exceeding 178 miles/hour in gusts.

The present revision of the South Florida Building Code incorporated by reference part of the wind load section of ASCE 7-88 Standard.

Recommended Requirements for New Codes or for
Upgrading Hurricane-Resistant Codes:

1. A determination of wind load requirements is one of the important steps to take in formulating this code. Both the ASCE 7-93 Standard (based on a sustained "fastest-mile" of wind) and the ASCE 7-95 Standard (based on a 3-second gust) are adequate in the opinion of the author. The basis for each of the Standards is a <u>50 year frequency of occurrence</u>, and neither standard provides for the Hurricane Andrew-type storm. ASCE 7-98 is currently in preparation and will supplant 7-95 when finished; 7-98 is also based on 3-second gust velocities, and is also based on a 50-year frequency of occurrence. The owner or the designer must decide whether the importance of the structure being designed warrants an increase to a 100 year frequency of occurrence. FEMA, in its establishment of flood elevations for floor elevations for insured construction, uses a 100 year

storm to decide on required floor elevations; the structure itself may only be designed for a 50 year wind event. (This is inconsistent!)

2. The entire subject of shop drawings must be clarified and enforced in the building code. Shop drawings are, in many cases, design drawings (not just fabrication details for the shop). For instance, the subject of residential roof truss design must be fully enforced as to responsibility. Many failures that occurred during Hurricane Andrew were due to a lack of responsibility by the building architect-of-record, who refused responsibility for the truss design and installation and appurtenant bracing. Bracing items were omitted or were poor because the residential truss designer took no responsibility for this part of the building system. Sometimes, this was left up to the builder.

    This applies to other structural roof systems, also. This applies to curtain-wall installation. The code requirements for the design and inspection of these systems must be clarified and fully enforced in the building code, as to full professional responsibility.

3. Mandatory protection of building envelopes in new construction must be adopted by requiring that the components of exterior walls such as glazing, doors, and windows of enclosed buildings be specifically designed and constructed to preserve the enclosed building envelope against wind pressure <u>and impact loads</u> from wind-borne debris or be protected by fixed, operable, or portable demountable storm shutters.

4. Mandatory testing of shingles and tiles as a system at 49 m/s (110 mi/hr) must be required and a testing standard for incorporation must be adopted into the building code.

5. The use of wall systems other than concrete blocks in multistory residential and commercial buildings must be reexamined, and prefabricated systems must be banned if necessary.

6. The role of the structural engineer in the design of residential and commercial structures must be strengthened.

7. The architect or the engineer of record must be required to participate in

the structural inspection process and sign an affidavit prior to the issuance of the Certificate of Completion by the Building Official, certifying that to the best of his/her knowledge the building construction fully complies with the design intent and the building code, and the approved permit plans reflect the structural as-built conditions.

8. It must be required as part of the permit set of plans that all drawings and specifications for building component items, such as windows, doors, and roof trusses, that have received product control approval, be included so that thorough conforming inspections can be performed by the building inspectors.

9. Design responsibility must be clarified for loads transmitted to the structure by mechanical equipment mounted on the roof.

10. For pre-engineered metal buildings, the following requirements must be part of the code:

   a. Roof and wall lateral and cross bracing, in addition to panel and deck diaphragm action for wind bracing, must be required.

   b. Doors must be anchored as part of the frame in the closed position, and <u>all doors must meet the performance and strength requirements of the code</u>.

   c. A heavier metal siding (minimum gauge of 24) (0.06 cm) must be used.

   d. A quality-control program must be instituted for all steel fabrication and installation.

   e. The use of self-tapping screws for fastening siding and roof panels must be reexamined and banned, if necessary.

   f. Metal siding must be designed continuously, allowable deflection must be decreased, and siding must be subjected to an impact load test consistent with the criteria established for the protection of building envelopes.

g. As in the case of "shop" drawings, an engineer-of-record shall take full responsibility for the entire process for pre-engineered metal buildings.

11. Standard details for typical one and two story "non-engineered" buildings must be included in the building code as a mandatory, not advisory feature. These details shall cover residential details for (a) typical masonry residences and (b) typical wood frame structures. Details shall include but not be limited to reinforced concrete continuous foundations, reinforced concrete tie (bond) beams, typical anchorage details for roof trusses, typical wood framing details, etc.

Key Words:

Building codes
Hurricanes
Hurricane-Resistant Construction
Wind Load Design
Structural Engineering
South Florida Building Code

Assessment of the Wind Environment around Buildings

Hanqing Wu[1] and Theodore Stathopoulos[2], F. ASCE

Abstract

Pedestrian-level wind environment can be assessed by wind-tunnel testing, field measurement, and more recently, by numerical computation. The wind speed data obtained through these methods can be combined with local meteorological records to determine the occurrence frequencies of wind speeds below or exceeding the threshold values for specific pedestrian activities. This short paper illustrates the basic procedures for the assessment of pedestrian-level wind environment around buildings through wind-tunnel testing.

Introduction

Wind flow induced by buildings and structures may create uncomfortable or even dangerous conditions for pedestrian activities. While both field measurement and numerical computation are valuable tools, wind tunnel testing remains the most reliable and efficient means for the evaluation of wind flow at the pedestrian level (Lawson and Penwarden, 1975). Artificial intelligence techniques by utilizing knowledge-based expert systems have also been suggested at least for preliminary assessment - see Wu and Stathopoulos (1996).

Wind Tunnel Testing Methodologies

Boundary layer wind tunnels can be used to model the building of interest, its surroundings and upstream terrains. Over the years, various experimental techniques have been developed for flow measurements. Compared to the widely-used hot wire (film) thermal anemometers, the pressure sensor proposed by Irwin (1981) has shown

---

[1] Specialist, RWDI, Inc., 650 Woodlawn Road West, Guelph, Ontario, Canada N1K 1B8

[2] Professor and Director, Centre for Building Studies, Concordia University, Montreal, Quebec, Canada H3G 1M8

remarkable advantages in pedestrian-level wind studies. It is very sturdy, easy and inexpensive to fabricate. A large number of sensors could be installed on a model site provided they do not interfere with each other. Their outputs can be read by pressure transducers as in the case of pressure measurements. Both the hot wire method and the Irwin method can run into difficulty in measuring winds accurately when there is a downwash vertical component. However, generally this happens only at low wind locations and thus, is not a serious limitation in practice. The particle scour technique, surface flow visualization and smoke injection have also been used for assisting in understanding the general flow pattern and identifying the windy spots around buildings.

Meteorological and Wind Tunnel Data

Meteorological data are obtained from a local weather station for a typical period of 30 years. The hourly records of wind speeds can be simulated by Weibull distributions, usually for 16 equally incremented wind directions:

$$P_i(>V_o) = A_i e^{-(\frac{V_o}{c_i})^{k_i}} \qquad i = 1,2,...,16 \tag{1}$$

where A_i is the wind probability in sector i. A large c_i or a small k_i indicates a high likelihood of strong winds in the sector. These constants vary with the wind direction and the season or time period selected and, to be more accurate in the regression, a separate set of Weibull constants may be necessary for the strong winds. In addition, the value of c_i, in km/h for instance, has to be converted from that at the measurement height (usually 10 m) to the gradient height by using an appropriate scaling factor for each sector. Note that at the gradient height wind speeds and their distributions are assumed to be the same over the local weather station and the study site. Table 1 shows Weibull constants at the gradient height for the summer and winter seasons for a particular site, with the highest wind frequency from the West in the summer and from the WNW in the winter.

A given threshold speed (V_T) at the pedestrian level corresponds to the value of V_T/R_i at gradient height, where R_i is the speed ratio referred to the gradient height. The total frequency of wind speeds exceeding the threshold value at a test location can be calculated by combining the speed ratios and Weibull constants by using:

$$P(>V_T) = \sum_{i=1}^{16} P_i(>\frac{V_T}{R_i}) = \sum_{i=1}^{16} A_i e^{-(\frac{V_T}{R_i c_i})^{k_i}} \tag{2}$$

In the wind tunnel, mean and gust speeds can be measured and reduced to the form of speed ratios (R), by dividing each measured speed at an equivalent full-scale height of 1.5-2.0 m by the reference wind speed at the top of the simulated boundary layer. Table 1 also lists typical wind tunnel results at 3 selected locations. In general, both the mean and gust ratios at Location 3 are higher than those at the other two locations.

Table 1. Weibull Constants and Speed Ratios with Reference to Gradient Height.

| | Weibull Constants | | | | | | Speed Ratios (R) | | | | | |
| | Summer (May - Oct.) | | | Winter (Nov. - Apr.) | | | Location 1 | | Location 2 | | Location 3 | |
Sector	A(%)	c(km/h)	k	A(%)	c(km/h)	k	Mean	Gust	Mean	Gust	Mean	Gust
NNE	3.2	28.6	2.35	2.8	29.4	2.13	0.17	0.47	0.24	0.58	0.28	0.60
NE	4.5	29.1	1.83	4.4	29.1	2.32	0.20	0.61	0.28	0.67	0.26	0.61
ENE	4.1	30.6	2.59	5.1	31.2	2.16	0.37	0.72	0.34	0.71	0.22	0.55
E	5.2	28.5	2.68	5.5	28.5	2.35	0.37	0.72	0.33	0.63	0.28	0.68
ESE	2.9	26.1	2.43	2.5	24.5	2.32	0.38	0.61	0.41	0.64	0.52	0.79
SE	4.1	26.8	2.42	3.5	27.4	2.18	0.41	0.68	0.47	0.70	0.70	0.94
SSE	5.1	29.3	2.58	3.5	29.9	2.08	0.24	0.51	0.46	0.80	0.75	1.04
S	8.5	27.1	2.28	4.9	27.1	2.13	0.22	0.55	0.44	0.85	0.61	0.96
SSW	5.9	28.9	2.36	4.5	32.4	2.19	0.21	0.53	0.27	0.69	0.37	0.82
SW	6.7	28.7	2.23	5.8	31.8	2.18	0.30	0.59	0.32	0.69	0.27	0.66
WSW	6.8	27.4	2.42	5.3	29.1	2.16	0.34	0.70	0.51	0.90	0.46	0.86
W	14.5	25.3	2.29	14.1	32.3	2.06	0.43	0.77	0.50	0.84	0.53	0.87
WNW	8.7	31.4	2.28	14.3	45.5	2.19	0.28	0.63	0.40	0.75	0.46	0.81
NW	6.4	38.0	2.43	9.7	47.7	2.61	0.16	0.47	0.26	0.62	0.42	0.74
NNW	4.0	35.2	2.60	5.0	41.6	2.51	0.15	0.33	0.15	0.41	0.42	0.72
N	5.2	29.2	2.27	5.0	30.9	2.18	0.14	0.31	0.16	0.43	0.32	0.58
Calm	4.2	-	-	4.1	-	-	-	-	-	-	-	-

Wind Comfort Criteria

There is no full consensus among existing wind comfort criteria, due to differences in the selection of wind speed, occurrence frequency, activity category, presentation format and other aspects (Ratcliff and Peterka, 1989). For instance, Lawson and Penwarden (1975) use the Beaufort Scale as the limit of mean wind speeds, while RWDI criteria (Soligo et al, 1997) use the Gust Equivalent Mean (*GEM*) defined as *GEM* = max (*Mean*, *Gust*/1.85). Details about these two criteria are listed in Table 2.

Table 2. Wind Criteria Proposed by Lawson & Penwarden (1975) and RWDI (1997).

Category	Lawson &Penwarden's Criteria Mean Speed (km/h) and Frequency	RWDI Criteria GEM (km/h) and Frequency
Sitting	> 12.0 for < 4.0%	≤ 9.5 for ≥ 80%
Standing	> 19.6 for < 4.0%	≤ 14 for ≥ 80%
Walking	> 28.6 for < 4.0%	≤ 18 for ≥ 80%
Uncomfortable	All other cases	> 18 for > 20%
Unacceptable/Severe	> 49.9 for > 2.0%	> 53 for ≥ 0.1%

Assessment Results

The assessment results can be presented by an easy-to-read tabulation-see Tables 3a and 3b. Both criteria predict Location 1 to have wind conditions comfortable for standing in the summer. In the winter, however, one predicts walking is the appropriate category and the other predicts standing. Assessments on the other two locations are identical. Note that uncomfortable wind conditions are predicted during the winter at Location 3, where mitigative measures are usually required. All locations pass the unacceptable or safety-related criteria.

Table 3a. Wind Comfort and Safety Assessed by Lawson & Penwarden Criteria.

COMFORT CATEGORY		Sitting	Standing	Walking		SAFETY CATEGORY	
Mean Speed (km/h)		>12.0	>19.6	>28.6		>49.9	
Category Limit		<4.0%	<4.0%	<4.0%		>2.0%	
Location	Season	%	%	%	**RATING**	%	**RATING**
1	Summer	12.2	1.2	0.1	Standing	0.0	PASS
	Winter	20.5	4.5	0.6	Walking	0.0	PASS
2	Summer	26.5	4.1	0.4	Walking	0.0	PASS
	Winter	35.1	11.6	2.3	Walking	0.0	PASS
3	Summer	37.8	11.1	1.9	Walking	0.0	PASS
	Winter	46.3	20.3	6.1	Uncomfortable	0.1	PASS

Table 3b. Pedestrian Level Wind Comfort and Safety Categories (RWDI Criteria).

COMFORT CATEGORY		Sitting	Standing	Walking		SAFETY CATEGORY	
GEM Range (km/h)		0 - 9.5	0 - 14	0 - 18		≥ 53	
Category Limit		≥80%	≥80%	≥80%		> 0.1% of the Time	
Location	Season	%	%	%	**RATING**	%	**RATING**
1	Summer	66	90	97	Standing	0.00	PASS
	Winter	55	80	90	Standing	0.00	PASS
2	Summer	50	79	92	Walking	0.00	PASS
	Winter	43	69	83	Walking	0.01	PASS
3	Summer	42	74	84	Walking	0.01	PASS
	Winter	36	59	74	Uncomfortable	0.05	PASS

References

Irwin, P.A. (1981) "A simple omnidirectional sensor for wind-tunnel studies of pedestrian level winds", *J. Wind Engrg & Ind. Aerodyn.*, 7, 219-239.

Lawson, T.V. and Penwarden, A.D. (1975) "The effects of wind on people in the vicinity of buildings", *Proc. 4th Int. Conf. on Wind Effects on Buildings and Struct.*, London, UK.

Ratcliff, M.A. and Peterka, J.A. (1989) "Comparison of pedestrian wind acceptability criteria", *Proc. 6th US Nat. Conf. on Wind Engrg.*, Univ. of Houston, TX.

Soligo, M.J. et al (1997) "A comprehensive assessment of pedestrian comfort including thermal effects", *8th US Nat. Conf. on Wind Engrg.*, Baltimore, MD.

Wu, H and Stathopoulos, T (1996) "Computer-aided prediction of pedestrian-level wind environment around buildings", *Proc. 3rd Canadian Conf. on Computing in Civil & Building Engrg.*, Montreal, Canada.

Assessment of Pedestrian Thermal Comfort

Michael J. Soligo[1], M. ASCE, Hanqing Wu[2] and Peter A. Irwin[1], M. ASCE

Abstract

Pedestrian comfort is affected not only by wind speed, but also by air temperature, solar radiation and other parameters. A comprehensive model has been developed at RWDI for the assessment of the pedestrian comfort by considering three major components, i.e. wind force, thermal comfort and wind chill. This paper illustrates the application of the model through three recent consulting projects carried out for different climates. It demonstrates that these three components may be of equal importance and should be studied concurrently in determining the true comfort level of the outdoor microclimate.

Introduction

Pedestrian comfort may be affected by wind speed, air temperature, relative humidity, solar radiation, human activity, clothing level, exposure time and other parameters. Traditional pedestrian-level wind studies only dealt with the mechanical effect of winds. One of the earlier studies by Lawson and Penwarden (1975) concluded that, within certain limitations, the criteria of acceptability for thermal comfort would automatically be satisfied, providing the criteria for mechanical effects were met. This might be true for the moderate climate in England, considering people's ability to adapt to the local weather conditions by adjusting clothing and activity accordingly.

The climate in North America varies in a much wider range, which may alter people's perception on the wind effects. On a hot, humid, sunny day in Miami, for instance, a moderate breeze is often considered pleasant for its cooling effect. On a typical winter day in Canada, however, a breeze of the same magnitude multiplies the undesirable heat loss from human body, which has to be compensated by increasing solar heat, clothing and/or activity level in order to maintain the overall comfort.

----

[1] Principal, [2] Specialist, Rowan Williams Davies & Irwin, Inc. (RWDI), 650 Woodlawn Road West, Guelph, Ontario, Canada N1K 1B8

Outdoor Thermal Comfort Model

For evaluating the combined impact of outdoor environment on pedestrian comfort, a comprehensive model has been established at RWDI to consider three major components. (1) <u>Wind Force</u> is assessed by the Gust Equivalent Mean (GEM) which accounts for both the mean and gust wind speeds. The acceptable ranges of GEM are derived from literature information and authors' experience; (2) <u>Thermal Comfort</u> is measured by the inner body temperature using the Pierce Two-Node Model and the thermal sensation index (TSENS); and (3) <u>Wind Chill</u> combines wind speed and air temperature to determine the chilling effect on the exposed skin.

Note that these three components are studied concurrently in the model. In order to pass the overall comfort in any given hour, all three individual comfort components must pass for that hour. The overall conditions are considered comfortable for a particular activity, if they pass all criteria for that activity at least 80% of the time. More details on the comfort model and its development can be found in Soligo et al (1993 & 1997). The following examples illustrate its application.

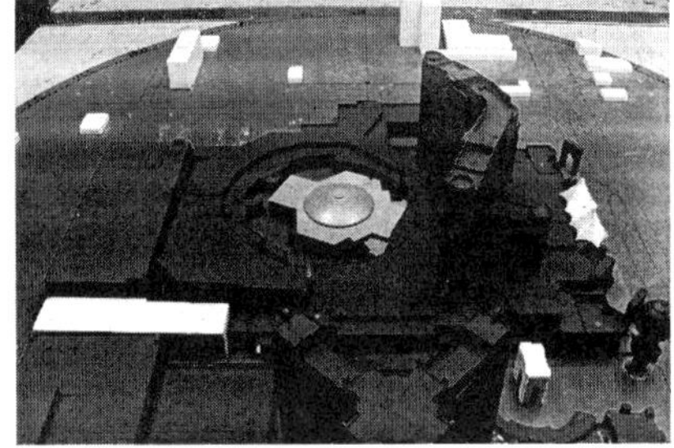

Figure 1. Hotel & Casino in Las Vegas.

Case Study and Discussion

(1) <u>Hotel & Casino Tower</u>

A total of 119 sensors are installed on the model of a Hotel & Casino Tower in Las Vegas (Figure 1) to measure mean and gust wind speeds at a full-scale height of about 1.5 m for 16 equally incremented wind directions. Solar conditions are also determined at the same 119 locations using a computerized sun/shade analysis. These data are then combined with long-term meteorological data recorded at the McCarran International Airport.

In general, winter days in Las Vegas are pleasant with mild temperatures and clear sky. The spring and fall seasons are considered by many to be ideal, although rather sharp temperature changes can occur. Wind chill is not a factor in Las Vegas for any seasons. In the spring, fall and winter seasons, the uncomfortable conditions caused solely by failing the thermal comfort criteria are found to be less than 3% of the time.

During the summer, however, the uncomfortable conditions are more likely to be experienced due to the high temperatures in the 100 °F (38 °C) range. Table 1 summarizes the lowest and highest percentages of time of comfort conditions for the 119 locations evaluated in the summer (May 16 through September 15). The percentage of comfort wind conditions changes significantly with the activity category. Within the same category there also exists a remarkable difference between the low and the high, depending upon the sensor location and wind direction. The percentages of thermal comfort, however, vary within a relatively narrow range for all tested locations. This indicates that the direct solar radiation is not an overwhelming factor alone and the thermal discomfort is mainly caused by the extremely high temperature during early afternoons. Since the highest percentage for overall comfort is only 77.9%, no location is considered to have comfortable conditions in the summer and prolonged outdoor activities should be avoided, if possible.

Table 1. Lowest/Highest Time Percentages for Comfort Conditions in the Summer.

	ACTIVITY		
COMPONENT	**Sitting**	**Standing**	**Walking**
Wind Force	28.2 / 86.4	55.0 / 92.7	71.8 / 99.6
Thermal Comfort	72.2 / 81.8	72.3 / 82.0	72.9 / 82.7
Overall Comfort	21.8 / 62.7	43.5 / 74.7	58.8 / 77.9

Note: Wind chill criterion is met 100% of the time at all locations.

(2) Sports Stadium in Seattle

Figure 2 shows the 1:400 scale model of a sports stadium in Seattle, where the climate can be characterized by mild temperatures and considerable cloudiness. Due to the high cloudiness, a large portion of the solar heat comes from diffuse radiation, and the test location relative to the sun or shade becomes less important in calculating the solar contribution to the overall

Figure 2. Model of Sports Stadium.

comfort. Between 10 am and 10 pm when the stadium is expected to be used, the wind chill factor is negligible (less than 0.4% of the time is unsatisfactory). The predicted conditions at all 131 test locations are comfortable for their intended use with the summer and fall seasons being more pleasant than the winter and spring.

In the thermal comfort evaluation, the most appropriate clothing values are assumed for acclimatized local residents. The computer program used for the calculation allows changes of clothing three times a day following the changes of temperature and other weather conditions during the day. Some unexpected cold or hot conditions may be detected by this procedure when the clothing change cannot accommodate the rapid changes in weather conditions. This is more likely to occur between March and October when the average daily temperature difference is higher than that in other months. Thermal discomfort, either too hot or too cold for standing, for example, occurs for 4.1 to 10.0% of the time in May and June, and 1.3 to 3.5 % in December and January.

(3) <u>Medical Center in Minnesota</u>

In contrast to the other two cases, the wind chill factor in Minnesota plays a critical role in the evaluation of overall comfort in the winter season with low temperatures and strong winds. Figure 3 shows a wind tunnel model of medical center in Rochester, Minnesota. At the windiest location, the passing rate for wind chill is only 70.0%, compared to 95.2% at a calmer location. The

Figure 3. Medical Center in Wind Tunnel.

combination of strong winds and low temperatures in the winter also causes low passing rates for the thermal comfort (70.3 to 77.5% for sitting, 80.5 to 86.5% for standing and 97.6 to 99.3% for walking). The percentage difference between these categories is induced by the different metabolic rates for these activities. Overall, there are only 12 locations, out of a total of 92 tested, satisfying the overall comfort criteria during the winter season.

Therefore, activities of long exposure should be limited and two groups of mitigative measures were recommended. These include the soft landscaping in the form of coniferous trees and plants, and the hard landscaping in the form of tall wind screens. With the reduced wind activity around the building, the overall outdoor comfort in the winter is expected to be improved considerably.

Concluding Remarks

Three recent consulting projects are highlighted in this paper to demonstrate the important contributions of wind force, thermal comfort and wind chill to the overall pedestrian comfort. The comprehensive model developed for pedestrian comfort, including new principles, methodology and criteria, represents a significant advance over other criteria based solely on wind force.

References

Lawson, T.V. and Penwarden, A.D. (1975) "The effects of wind on people in the vicinity of buildings." *Proc. 4th Int. Conf. on Wind Effects on Bldgs. & Struct.*, London, UK.

Soligo, M.J, Irwin, P.A. and Williams, C.J. (1993) "Pedestrian comfort including wind and thermal effects." *Proc. 3rd Asia-Pacific Symp. on Wind Engrg.*, Hong Kong.

Soligo, M.J, Irwin, P.A., Williams, C.J. and Schuyler, G.D. (1997) "A comprehensive assessment of pedestrian comfort including thermal effects." *Proc. 8th US Nat. Conf. on Wind Engrg.*, Baltimore, MD.

Microclimate Design Features for Buildings and Landscaping.

Leighton Cochran[1] (Member ASCE) and Peter Irwin[2] (Member ASCE)

Introduction

A windy environment around the base of a building, particularly near a main entrance or plaza area, will detract from the appeal of the site and perhaps discourage clients and shoppers from visiting the area. Many examples exist of outdoor restaurants and cafes that have failed at the base of tall buildings as a result of a windy environment (Cochran, 1979). Similarly, an outdoor pedestrian space, such as a recreational pool area of a residential condominium, should be protected from strong winds. Thus, there is a direct financial motivation for ameliorating the wind environment if it is going to affect the appeal of a tenanted building to the users and customers of that building. In the extreme case a site may be dangerous; particularly to the infirm. Penwarden and Wise (1975) discuss the case of two elderly women who were killed when a gust of wind at the base of a tall building blew them over.

Many factors will have an impact on the wind conditions around a building. Some parameters, such as the ambient wind statistics, local topography, or whether the building is surrounded by similarly tall structures, will influence the resulting winds around the base of a new building. It is for this reason that many new-building designers evaluate their project in a boundary-layer wind tunnel with the building both installed and removed from the turntable. In this way, the project's impact on the local environment may be assessed.

Building Massing and Orientation

It is well known that the design of a building will influence the quality of the ambient wind environment at its base. A shear curtainwall to ground level with a rectilinear floor plan (circular shapes typically do not cause flows of this type) is often a design which can aggravate street-level winds by allowing the high-elevation,

- - - - - - - - - -

[1]Associates, Cermak Peterka Petersen Inc., 1415 Blue Spruce Drive, Fort Collins, CO 80524, USA.
[2]Principal, Rowan Williams Davies & Irwin Inc., 650 Woodlawn Road West, Guelph, N1K 1B8, Ontario, Canada.

faster winds to flow down the face of the structure. The mechanism is called downwash (see Figure 1) which is then accelerated around the ground-level corners (see Figures 2 and 3). The flow may be interrupted by a large canopy over the building entrances or a podium to set the tower back from the street (Figures 4 and 5, respectively). Large canopies are a common feature over main entrances. The podium approach is also a common architectural feature of many major projects in recent years, but it is counterproductive if the architect wishes to use the podium roof for long-term pedestrian activities.

Another massing issue which may be a cause of strong ground-level winds is an arcade or thoroughfare opening from one side of the building to the other. This effectively connects a positive-pressure region on the windward side with a negative-pressure region on the lee side. A strong flow through the opening often results as illustrated in Figure 6. A similar phenomenon occurs with a high-rise building raised up on columns; a design popular in the 1960's (Penwarden and Wise, 1975). The uninvitingly windy nature of these open areas is a contributing reason behind the rarity of this type of architectural massing in modern high-rise buildings. One exception is in calm, tropical climates where the extra breeze creates a desirable feature. For example, several structures in Singapore use this approach to provide shaded cooler areas at the ground-level entrances.

An entrance alcove behind the building line will generally produce a calmer entrance area (Figure 7) at a mid-building location. In many cases a canopy may not be necessary with this scenario. The same approach at a building corner is usually quite unsuccessful (Figure 3).

Horizontally accelerated flows between two tall towers (see Figure 8) is another cause of a windy ground-level pedestrian environment, which may also be locally aggravated by ground topography. By inspection of the available wind data, the designer may find a dominant wind direction that can be used to align the building on the site so as to minimize these accelerated flows in highly trafficked pedestrian areas.

The way in which a building's vertical line is broken up may also have an impact. For example, if the floor plans have a decreasing area with height the flow down the windward face may be greatly diminished. To a lesser extent the presence of many balconies can have a similar impact on ground-level winds. However, designs with many elevated balconies and deck areas near a building corner may attract a windy environment to those locations. Mid-building balconies are usually a lot calmer. Corner balconies are generally a lot windier.

In summary, there are two principal types of flow that adversely effect the pedestrian environment: (i) downwash flows bring higher energy wind to lower elevations (usually best diminished by a podium or large canopy), and (ii) horizontally accelerated flows (often ameliorated by porous screens or plantings). The latter approach is discussed below.

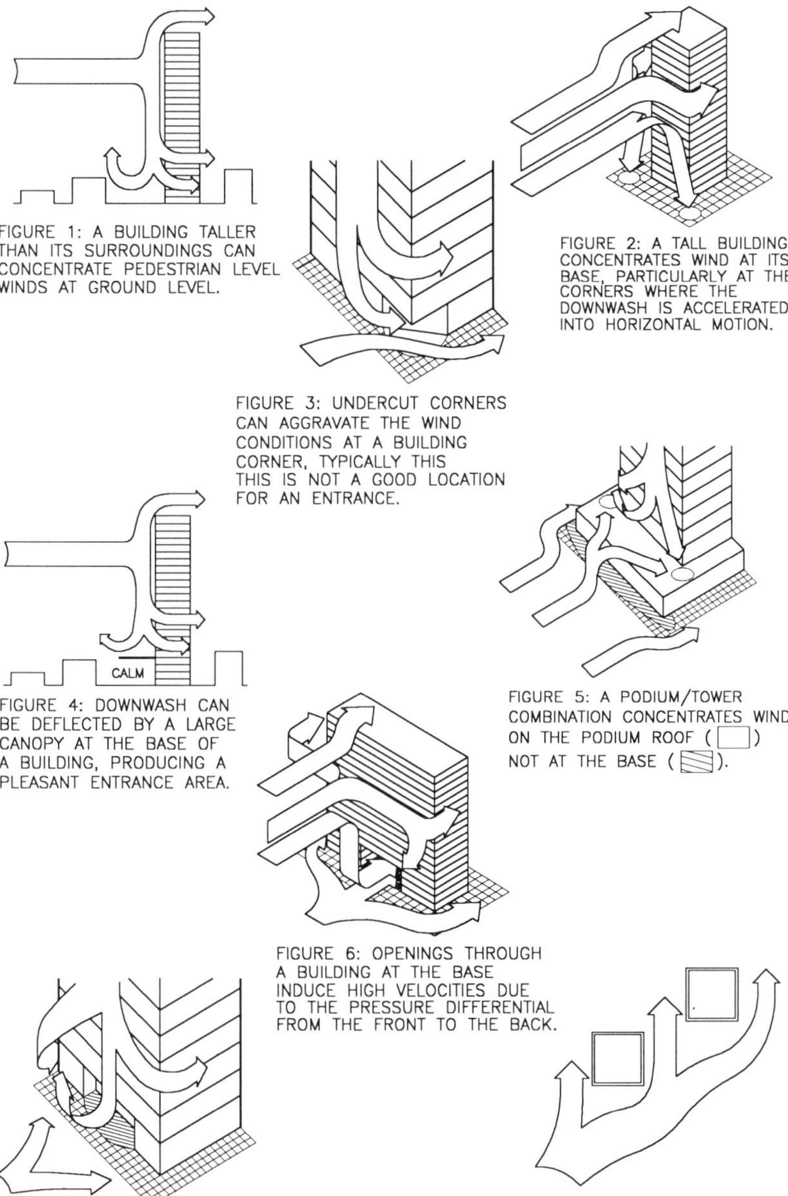

FIGURE 1: A BUILDING TALLER THAN ITS SURROUNDINGS CAN CONCENTRATE PEDESTRIAN LEVEL WINDS AT GROUND LEVEL.

FIGURE 2: A TALL BUILDING CONCENTRATES WIND AT ITS BASE, PARTICULARLY AT THE CORNERS WHERE THE DOWNWASH IS ACCELERATED INTO HORIZONTAL MOTION.

FIGURE 3: UNDERCUT CORNERS CAN AGGRAVATE THE WIND CONDITIONS AT A BUILDING CORNER, TYPICALLY THIS THIS IS NOT A GOOD LOCATION FOR AN ENTRANCE.

FIGURE 4: DOWNWASH CAN BE DEFLECTED BY A LARGE CANOPY AT THE BASE OF A BUILDING, PRODUCING A PLEASANT ENTRANCE AREA.

FIGURE 5: A PODIUM/TOWER COMBINATION CONCENTRATES WIND ON THE PODIUM ROOF ( ) NOT AT THE BASE ( ).

FIGURE 6: OPENINGS THROUGH A BUILDING AT THE BASE INDUCE HIGH VELOCITIES DUE TO THE PRESSURE DIFFERENTIAL FROM THE FRONT TO THE BACK.

FIGURE 7: RECESSED ENTRY PROVIDES LOW WINDS AT DOOR LOCATIONS.

FIGURE 8: ADJACENT BUILDING PLACEMENT CAN CAUSE A COMPRESSION OF THE MEAN STREAMLINES, RESULTING IN HORIZONTALLY ACCELERATED FLOWS AT GROUND LEVEL.

Use of Landscaping, Screens and Canopies

Once the flow mechanism at a problem location has been established for the critical wind directions the remedial solutions may be explored for effectiveness in a boundary-layer wind tunnel. As noted earlier, downwash off a tower may be deflected away from pedestrian areas by large canopies or podium blocks. The downwash then effectively impacts the podium roof rather than the public areas at the base of the tower (see Figure 5). Provided that this podium roof area is not intended for recreational use (e.g. swimming pool, tennis court or putting green), this massing method is typically quite successful.

Horizontally accelerated flows that create a windy environment are best dealt with by using porous screens or substantial landscaping. Large hedges, bushes or other porous media serve to retard the flow and consume the kinetic energy available to cause a windy environment. A solidity ratio (ie proportion of solid area to total area) of about 50-70% has been shown to be most effective in reducing the flow's momentum (Rouse, 1950). These physical changes to the pedestrian areas are most easily evaluated by a model study in a boundary-layer wind tunnel. A comparative study showing the impact on the site with and without the ameliorative additions is a useful method of defining their effectiveness, and it also allows the architect to define the extent and usefulness of the pedestrian space. When these studies are done as part of the design process the architect can be more assured of the success of open space and entrances around a new project.

Conclusions

The wind comfort and safety of pedestrian areas around a new development may be assessed by a properly conducted wind-tunnel study, combined with direct interaction with the architect. Physical modeling of the wind flow used in conjunction with the statistical description of the ambient site winds yields a powerful predictor of how a new project will be judged by the public, from the perspective of wind comfort, before construction commences. The knowledge gained from previous studies (Figures 1 to 8) provides useful massing guidance to the architect in the design phase.

References

Cochran, L.S., "Full-Scale Ground Level Wind Study of the AMP Building, Brisbane", Baccalaureate Thesis, University of Queensland, Australia, 1979.

Penwarden, A.D. and Wise, A.F.E., "Wind Environment Around Buildings", Building Research Establishment Report, Number 9(E7), Her Majesty's Stationery Office, 1975.

Rouse, H., "Form Drag of Composite Surfaces", Selected Writings of Hunter Rouse, Dover Publications, pages 248-253, 1950.

Effect of Roofing Membrane on Wind Uplift Pressures

A. Baskaran[1], U. Vilaipornsawai, M.G. Savage and K.R. Cooper

Abstract

Wind Standards and Codes of Practice derive pressure coefficients, mostly, from the wind tunnel studies. Conventionally, the wind tunnel roof models were fabricated using rigid material. This paper presents the benefits of wind tunnel studies on models with full-scale roof components.

Preamble

Wind uplift resistance of a roofing system is one of the critical features of the building envelope design. In a conventional roofing system, normally, there are four components: membrane, insulation, barriers and deck. Functional requirements of these components are respectively waterproofing, thermal resistance, air/fire/vapor control and structural support. As shown in Figure 1, fasteners (or ballasts) are used to integrate these components to form a roofing assembly that can withstand environmental forces. Field observations clearly indicate that the membrane, located at the top, balloons and flutters between the fastener attachments during high wind events. The dynamic response of the membrane also varies based on its elasticity and attachment spacing. Building codes and wind standards specify design loads mainly based on data obtained from wind tunnel studies. However, wind tunnel studies utilize rigid material such as plexiglas to fabricate scale models to represent roof assemblies as shown in the Figure 1. Thus, it is clear that the existing design data on wind uplift pressure on roofing systems were obtained only by simulating the turbulent nature of the approaching wind and not accounting for the dynamic effects of the roofing membrane. The main focus of the paper is to identify the effect of roofing membrane on the wind uplift pressures.

Wind Tunnel Experiments

A systematic research effort investigating roofing systems, 3 m by 3 m in size, consisting of full-scale components was recently completed. Wind induced pressures and forces were collected, under simulated wind flow conditions in the 9 m by 9 m wind tunnel of the National Research Council. Two types of roofing membranes, *Poly (Vinyl Chloride) and Ethylene Propylene Diene Monomer (PVC and EPDM)* were used for the wind tunnel modeling. PVC membrane is stiffer than the EPDM. About 80 different

---

[1] *Research officer, National Research Council Canada, Ottawa, Ontario, Canada K1A 0R6, e-mail : Bas.baskaran@nrc.ca*

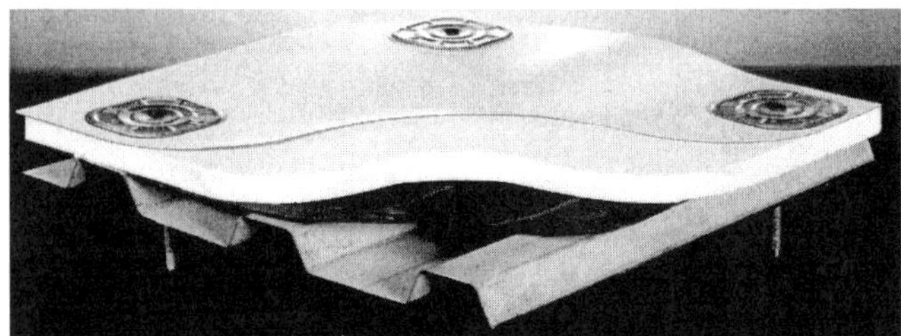

Figure 1. Typical components of a conventional roofing system

configurations were tested for a variation in building heights, wind speeds and wind angles {Savage et al. (1996) and Baskaran *et al.* (1996)}.

The large-scale model offered the opportunity to measure unsteady surface pressures on the deflecting membrane, a task that would be difficult to perform on a small-scale model. It also offered the opportunity to measure loads on fasteners for the same reason of size. By utilizing full-scale materials, it also offered a unique insight into the behavior of a prototype roof system in a controlled, gusty, wind environment. However, due to the use of the large model built from full-scale components, the simulation requirement was unusual. It was decided to produce the largest flow simulation that the tunnel could provide in the 18.3 m development length available upstream of the model using the spire technique. Normally, spires require 6 spire-heights for the flow to fully mix, but this distance was unavailable with the 4.6 m tall spires used, so the number of spires was increased from five to eight to accelerate mixing, at the cost of reduced length scales. The wind flow condition chosen for the study was that over open terrain. The spires were designed following the technique developed by Irwin (1979), with the exception that the number of spires was increased to enhance mixing and so shorten the distance for full mixing to occur. The simulated flow is characterized by its measured mean and turbulence velocity profiles, and by the velocity spectra of the flow. The required turbulence length scale for the model can be obtained by matching the measured power spectral density of the longitudinal turbulence component of wind speed in the wind tunnel. The length scale at the roof height (1.37 m) was found to be 1.20 m. Selecting unity velocity scaling, the time scale was found to be 56:1. For Further details of the wind simulation can be found from Savage et al (1997).

Results and Discussion

Typical mean pressure distributions over the roof membrane for the PVC configuration are presented as contour plots in Figure 2. The data show symmetry about the wind vector, with a leading-edge separation evident at 0 degrees wind direction and a pair of leading-edge vortices present at 45 degrees wind direction.

The largest mean negative pressure coefficients occur near the leading edges of the roof at the 45-degree wind direction due to the presence of the vortex pair formed from the upstream apex of the roof. The pressures over the remainder of the roof were small and negative. Similar trends can be seen in the mean surface pressures at 0 degrees wind direction. The pressures at 0 degrees are not as negative as the pressures under the oblique vortices, but the region over which the pressure coefficients are of order -1.0 is much larger than for the 45-degree flow.

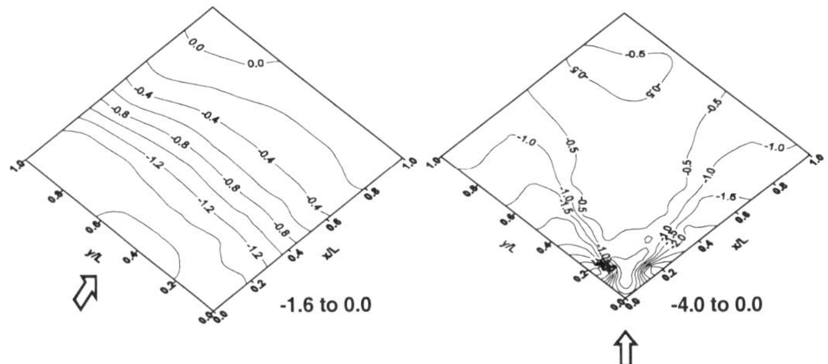

Figure 2. Pressure coefficient distribution on the PVC roof assembly

To evaluate the membrane effect, the mean pressure coefficients measured from the two roofs are compared in Figure 3.

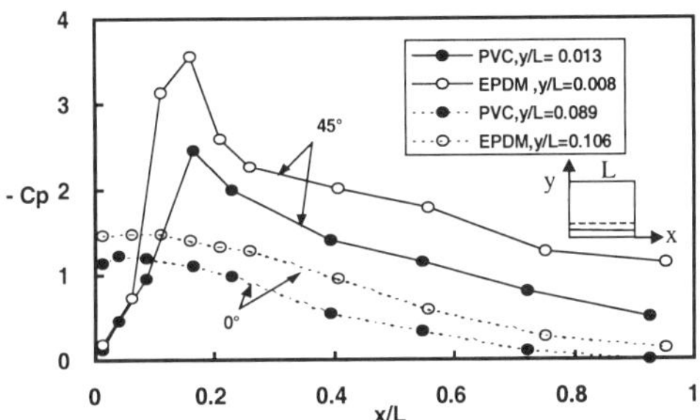

Figure 3. Effect of roofing membrane on the mean pressure coefficient

Irrespective of the roofing membrane, it is clear that the mean pressures are higher for oblique wind direction compared to wind approaching normal to the building width. Thus, it is clear that the mean pressure distribution measured on the wind tunnel models with roofing membranes are similar to that of the rigid models.

Examination of the wind induced fastener forces revealed the contrast as discussed below. Mean force coefficients are plotted in Figure 4. The tested PVC membrane was attached with 40 spot fasteners and it had two perimeter sheets and two field sheets. Bar attachment was used for the EPDM system. By instrumenting selected fasteners, induced forces were measured and force coefficients (Fc) are calculated.

$$Fc = F/q^*A$$

Where:

F is the mean measured force,

q is the dynamic pressure at roof height and

A is the tributary area indicated in Figure 5.

For the PVC system forces on the fastener F7 are shown. Forces of F5 and F6 were combined and plotted for the EPDM system representing the uplift force on the bar attachment. It is evident from Figure 4 that irrespective of the roofing system the induced forces due to the normal wind are significantly higher than that derived from the oblique wind. The difference is more pronounced for the EPDM assembly than the PVC system. Note that the PVC membrane is stiffer than the EPDM membrane.

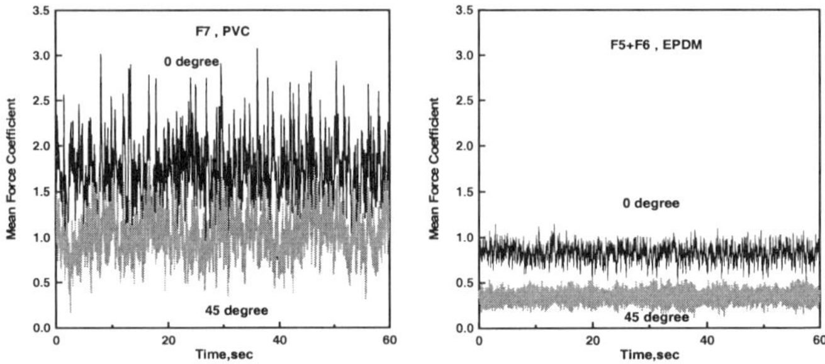

Figure 4. Measured fastener forces for two approaching wind angles

To further investigate this observation, the wind tunnel experiments were repeated at five different wind speeds. Calculated mean force coefficients are plotted in Figure 5. As shown, the Figure 5 data not only confirmed that the force coefficients are higher for normal wind as compared to 45-wind but as well revealed the following:

- Bar attachment of the EPDM system experiences less forces compared to the PVC spot attachments. This is due to the availability of larger area to distribute the forces in the case of the EPDM system.
- For the EPDM system, normal wind force coefficients are about 3 times higher than those computed for the 45-wind. In the case of the PVC system the force coefficients varied by a factor 2 between the data of the two wind angles.
- Force coefficient magnitudes are significantly high, irrespective of the wind speed, for the PVC system. The EPDM membrane's elastic nature facilitates the membrane to absorb a certain portion of the wind-induced forces. The rigid PVC membrane transfers most of the forces directly to the structural deck through the fasteners.

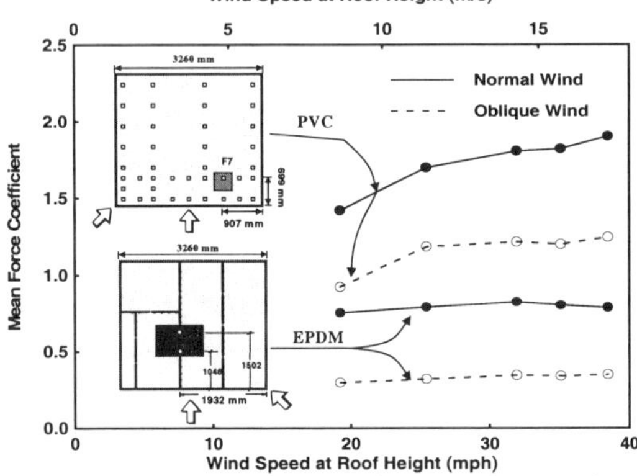

Figure 5. Effect of roofing membrane on the mean fastener forces

Data of this nature are critical in the design of the membrane roof systems. As mentioned before, only experiments with full-scale roof components can provide the opportunity to identify such new information about the membrane roofing system response. These finding revealed two critical factors for the wind design of the membrane roof systems. First, the membrane layouts and its stiffness should be taken in to account. Second, the distribution and type of fastener attachment can alter the wind-induced forces on the membrane roofs. Considering the available options for these two factors in the roofing industry, it is rather misleading to develop design guidelines only based on the wind tunnel pressures measured from the rigid models. In other words, provisions are needed to account for the membrane response during the wind design of the roof systems with membranes.

Conclusions

From the presented summary paper, it can be concluded that the overall mean pressure distribution patterns are not affected by the presence of the roofing membrane. Mean pressures are higher for the oblique wind direction compared to wind approaching normal to the building width. This is similar to conventional wind tunnel testing on rigid models. In contrast, the induced forces on the fasteners are less dependent on the approaching wind direction and more influenced by the membrane width and fastener layouts. Future wind tunnel studies on a rigid roof model with identical size and flow simulation are needed to isolate the membrane influence from the wind-induced effects.

References

Baskaran, A, Savage, M.G., Alfawakhiri, F and Cooper, K.R, 1996. Pressure Distribution Data Measured During the October 1995 Wind Tunnel Tests on a Mechanically Attached EPDM Single Ply Roofing Systems Internal Report LTR A 004, Institute for Aerospace Research, National Research Council Canada

Irwin, H.P.A.H., 1979, Design and Use of Spires for Natural Wind Simulation, National Research Council of Canada, NAE, Report LTR-LA-233, August 1979.

Savage, M.G, Baskaran, A., Cooper, K.R and Lei, W., 1996. "Pressure Distribution Data Measured During the November 1994 Wind Tunnel Tests on a Mechanically Attached, PVC Single-Ply Roofing System", Internal Report LTR A 003, Institute for Aerospace Research, National Research Council Canada

Savage, M.G, Cooper, K.R and Baskaran, A., 1997. "Wind Tunnel Investigation of the Wind Loads on a Single Ply Mechanically Attached, PVC Roofing System", Internal Report LTR A 012, Institute for Aerospace Research, National Research Council Canada

Acknowledgments SIGDERS consortium is carrying out the presented research. It consists of:

Manufacturers: Canadian General Tower Ltd., Carlisle SynTec Inc, GAF Materials Cooperation, Firestone Building Products Co., IKO industries Canada, JPS Elastomerics Corp. - Construction Products Group, Soprema Canada, Vicwest Steel. **Building owners:** Canada Post Corporation, Department of National Defense, Public Works and Government Services Canada. **Associations:** Canadian Roofing Contractors' Association, Industrial Risk Insurers, National Roofing Contractors' Association and Roof Consultants Institute.

SIMULATION REQUIREMENTS FOR ROOF WIND LOADS NEAR THE CORNERS OF LOW BUILDINGS WITH LOW-SLOPE ROOFS

D. Surry[1], T.C.E. Ho[2] and G.R. Lythe[3]

Abstract

The wind loads that occur near the corners of low-slope roofs on low buildings are problematic in two significant aspects:

1) recent research has shown that wind tunnel models tend to underestimate the peak suction loads acting on very small areas in these regions, and

2) the loads that do occur illustrate most dramatically the severity of spatial and temporal gradients that the wind can produce. These gradients cannot, in general, be reproduced in any full-scale component test machine available today.

The purpose of this paper is to highlight these two current areas of difficulty and, in particular, to illustrate the types of loading imposed by the wind and the related shortcomings of current full-scale component test protocols. Although various approaches are being pursued to solve the primary problem (number 2, above), a quick resolution is unlikely; indicating the need for further research.

---

[1] Research Director, Boundary Layer Wind Tunnel Laboratory, The University of Western Ontario, London, Ontario, Canada N6A 5B9

[2] Associate Research Director, Boundary Layer Wind Tunnel Laboratory, The University of Western Ontario, London, Ontario, Canada N6A 5B9

[3] Research Engineer, Boundary Layer Wind Tunnel Laboratory, The University of Western Ontario, London, Ontario, Canada N6A 5B9

Wind Tunnel/Full-Scale Comparisons

The widely-documented full-scale experiments at Texas Tech University (TTU) have generated considerable data on the wind loading of a nearly flat-roofed low building. This information has been very useful in, among other things, testing the credibility of wind tunnel simulations. On the whole, the comparisons have been excellent; however, the TTU experiments have also shown up a region close to the roof corner which is influenced by strong vortex flows for cornering wind directions, and for which the wind tunnel simulations of local loads are not as good [1,2]. Typically, although mean pressure coefficients are in excellent agreement, the instantaneous extreme suction coefficients measured in the wind tunnel in this region of the roof fall short by about 20%, and the more robust statistic offered by the root-mean-square of the fluctuations can fall short by even more. Research into potential shortcomings of model simulation is ongoing; however, fundamental scaling of these vortex phenomena may be involved and may not easily be overcome. More important though, is the fact that the influence of these extreme peaks on the structure is not understood.

The full-scale pressure tap location that has received the most attention (50501) is roughly 5' from one edge and 1' from the other (and hence on a ray from the corner of about 11° from the edge). At this single point location, extreme pressure coefficients, relative to the 13' mean roof height dynamic pressure, have been recorded of about -12 [1]. For a Florida coastline gust speed of 140 mph at 33 feet, the equivalent mean hourly speed at 13 feet would be expected to be about 75 to 80 mph, giving a mean roof height dynamic pressure of about 15 psf and hence implying a peak local suction at the 50501 location of about 180 psf. This, of course, would be even larger for taller buildings or more severe storm conditions. Furthermore, it has been shown by wind tunnel experiments [3], and has recently been verified in the field, that the local suction loads continue to increase as the corner is approached on rays about 15° from the edges. Field measurements of pressure coefficients of -25 have already been recorded.

These extreme suction loads affect only a very small area and are only sustained for a very short period of time. Based on a wind tunnel model, a typical time history of pressure at the most highly loaded instrumented point of a 32' high building with a low-slope roof, (discussed in more detail below), is illustrated in full-scale in the uppermost trace of Fig. 1 along with pressures averaged over successively larger areas, all of which contain the original point. The location of the areas is further illustrated in Fig. 2. Clearly, increasing areas rapidly ameliorate the loads. The operative question, then, is what are the consequences to the structure of these short duration very local loads? This question is really only the most pressing form of a more general question, since wind loads everywhere on a structure exhibit similar characteristics, albeit not so severely.

Full-Scale Test Protocols

Testing of full-scale roof or wall component assemblies for wind loading has traditionally been carried out using controlled static air pressures applied to the subject assembly as it forms one wall of a closed box. For example, in the UL 580 tests [4], specified constant pressures are applied to a sample for a fixed period, including one sequence of slow (10-second) oscillations. The magnitudes involved in a successful sequence in which the component did not fail determines the rating. Clearly, such tests apply uniform static loads to the specimen, with only a limited attention to duration of loading. More recently, concern over fatigue of components has led to the suggestion of a number of tests in which cyclic loads of different amplitudes are applied in specific sequences and durations. Perhaps the most sophisticated currently-available test process is that offered by BRERWULF [5], a pressure chamber that can be driven at pressure levels that follow an arbitrary pre-set time history. Thus, the time history of the loading averaged over the entire panel can be determined from full-scale or wind tunnel tests and used as the indexing time history.

All of these test procedures suffer from two inherent shortcomings:

1)   the applied pressures are spatially uniform over the specimen at any instant;

2)   the structural boundary conditions of the specimen play a crucial role in the structural performance of the specimen.

To some extent, the boundary condition problem can be addressed through the design of the manner in which the specimen is mounted in the test box, and through analysis that can relate the test box situation to the actual field mounting. Boundary conditions can be especially difficult for very flexible systems whose load/deflection characteristics can be strongly non-linear (eg, [5]).

The uniformity of the test pressure is inherent in these tests involving essentially no air flow. Since air pressures propogate at the speed of sound, gradients can only be sustained if flows are occurring or if physical barriers are introduced. Thus, even in BRERWULF, an 'a priori' decision must be made as to what is the relevant loading time history to be modelled.

A novel alternative to air pressures is currently being explored at Mississippi State University (MSU) under the direction of Dr. Ralph Sinno. Electromagnets are being used to apply local loads to metal roofs. Given a sufficient number of magnetic nodes and a sophisticated control system, any arbitrary time and spatially varying pressure pattern can be simulated.

The work reported below has been carried out for the Metal Building Manufacturers Association (MBMA) in support of the MSU program.

Examples of Wind Pressure Loads Near Corners

To illustrate the nature of the simulation problem near the corner of a low building roof, experimental data are presented below as determined from wind tunnel tests [7] on a model of a 12' by 25' panel which was placed at various locations on the roofs of model buildings of various heights and roof slopes. The panel was instrumented with 128 pressure taps that were all measured simultaneously. Each of the 128 time histories were recorded so that they could then be integrated over larger component areas. This was accomplished by dividing the 12' x 25' area into 6" x 6" squares and creating an associated pressure time history for each, by suitable combinations of the most relevant pressure tap data. Near the corners and edges, the taps were placed with very high resolution, with full-scale spacings of only a few inches. Further from the edges, the spacings were wider apart. Thus, some 6" x 6" squares did not actually contain a tap, but the associated time histories were formulated through interpolation of data from nearby taps. Note that such finite integration techniques tend to be conservative, and add high frequency "noise" to integrated loads since they inherently assume that even the highest frequencies in the local pressures are uniform over the 6" x 6" component areas.

The results illustrated in Figs. 1 to 3 are for a 32' eave height gable-roofed building with a 1/4 on 12 roof slope, subjected to a mean hourly roof height dynamic pressure of 18 psf approaching from the most severe wind loading direction for the corner roof panel (35° from an edge). The value of 18 psf is roughly comparable to the previously quoted 140 mph gust speed case. Within the 128 individual pressure taps, the largest local suction recorded was about 370 psf (for the -25 full-scale value reported earlier, this becomes 450 psf). Figs. 2a and 2b show the corresponding data for 1' x 1' areas and for 4' x 4' areas over the 8' x 14' panel closest to the corner. (This panel size represents the proposed test area at MSU). Each square contains 3 numbers: the most positive peak pressure, the mean pressure, and the most negative peak pressure associated with the pressure time history averaged over that area. These data have been extracted from time histories of averaged pressure versus full-scale time such as those plotted for selected areas in Fig. 1. The peak values from different squares do not occur simultaneously but all are common to the same continuous 6-minute storm from the specified wind direction. The largest 1' x 1' area suction is already ameliorated to about 260 psf and does not actually include the largest local suction, which occurred in area 2:A; whereas the largest 4' x 4' area suction is less than 140 psf and the entire 12' x 25' panel load is below 70 psf. Thus, there is a ratio of more than a factor of 5 between the peak load averaged over the corner 12' x 25' panel and the largest local load recorded.

It is particularly instructive to examine the time histories of the 1' x 1' loads around the time that the worst peaks were recorded. These are illustrated in Figs. 3a and 3b. The upper four traces in Fig. 3a are for the 4 corner 1' x 1' squares, where 2:A includes the largest local point pressure of Fig. 1. The lowest trace is for the 8' x 14' panel. At the time when the most severely loaded 1' x 1' square area exceeds 250 psf, another adjacent 1' x 1' area is less than the total panel load of about 70 psf. In Fig. 3b, the individual data points are shown (about 1/30 of a second apart) in an expanded-time view of the peak occurrence, indicating the peak lasted for about 1/20 second, and was less than 60% of the peak value just 1/5 second earlier. Clearly, both the spatial and temporal gradients are severe; and their effects need to be correctly assessed for various structural systems. It is also likely to be excessively conservative to take the worst measured local pressure and apply it over larger areas without first applying suitable reduction factors for area averaging (eg [8]) and, potentially, for the inability of the structure to follow the highest frequencies. Unfortunately, the key word is "suitable", and what is suitable cannot currently be defined with confidence.

Towards Improved Full-Scale Tests

The primary objective should be to evolve tests that are as simple as possible but that can be calibrated in some reliable manner to full-scale performance. There seems to be three different ways of pursuing this end: directly through correct full-scale tests; indirectly through analysis; and indirectly through modelling approaches.

Full-scale testing can be accomplished through field tests, through very large test facilities (exemplified by the current "Wall of Wind" facility proposed by Clemson University and INEL researchers [9]) through specialized techniques such as the MSU magnetic loading system, through a servo system as suggested by Kareem at the University of Notre Dame for structural loads [9], or through other novel approaches. Each approach has its advantages and disadvantages, and only the first two escape the test box boundary condition problems.

Analytical approaches, particularly using finite element techniques designed to follow large deflections, clearly have a role to play, both directly in predicting response of complex systems to load and indirectly by helping to interpret test sequences for the effects of boundary conditions.

Model scale testing can also play a role. The authors are initiating a research program using models designed to fail under wind tunnel loads, hence maintaining all of the spatial and temporal non-uniformities. The same model components will be tested under uniform loading conditions, ie in model representations of uniform load tests. In this way, correlations can be found between the full representative, controlled loading in the wind tunnel simulation and the non-representative simplified test process, albeit for simplified structural models.

Note that it is not sufficient to test full-scale components in large wind tunnels where correct length scaling cannot be maintained, as it is important to maintain the relative length scale of the wind-induced pressure non-uniformities to the size of the components under study.

Concluding Remarks

The spatial and temporal gradients of pressure produced by wind are particularly severe near the roof corners of low buildings. The spatial scales can involve factors of two in load over a foot in distance, and over temporal scales of a fraction of a second. These severe non-uniformities are not represented in current test procedures, and their implications for different types of structural systems are not well understood. Furthermore, the relationship between typical uniform pressure test results and the response of actual structural systems to the true non-uniform loading has not been established.

Acknowledgements

Financial support for this work has been received from the Metal Building Manufacturers Association, the National Sciences and Engineering Research Council of Canada, and the Boundary Layer Wind Tunnel Laboratory, U.W.O.

References

1) H.W. Tieleman, D. Surry and K. Mehta, "Full/Model Scale Comparison of Surface Pressures on the Texas Tech Experimental Building", J. Wind Eng.& Ind. Aero., 61(1996), pp. 1-23.

2) L.S. Cochran, "Wind Tunnel Modelling of Low-Rise Structures", Ph.D. Thesis, Colorado State U., 1992.

3) J.-X. Lin, D. Surry and H.W. Tieleman, "The Distribution of Pressure Near Roof Corners of Flat Roof Low Buildings", J. wind Eng. & Ind. Aero., 56(1995), pp. 235-265.

4) UL580 Standards For Safety, Test For Wind-Uplift Resistance of Roof Assemblies, Underwriters Laboratories, Inc., 1989.

5) N.J. Cook, A.P. Keevil and R.K. Stobart, "BRERWULF - the Big Bad Wolf", J. Wind Eng. & Ind. Aero., 29 (1980), pp. 99-107.

6) D.O. Prevatt, S.D. Schiff and P.R. Sparks, "A Technique to Assess Wind Uplift Performance of Standing Seam Metal Roofs", Proc. of 11th Conference on Roofing Technology, Gaithersburg, Maryland, Sept. 1995.

7) A.G. Davenport, D. Surry and T.C.E. Ho, "Wind Uplift Loading on Low Building Roofs", Engineering Science Research Report, The University of Western Ontario, BLWT-SS33-1993.

8) J.-X. Lin and D. Surry, "The Variation of Peak Loads with Tributary Area Near Corners on Flat Low Building Roofs", Accepted for publication in J. Wind Eng. & Ind. Aero.; also [9].

9) Proceedings Eighth U.S. Nat. Conf. on Wind Engineering, Baltimore, Md., June 1997 (on CD-ROM).

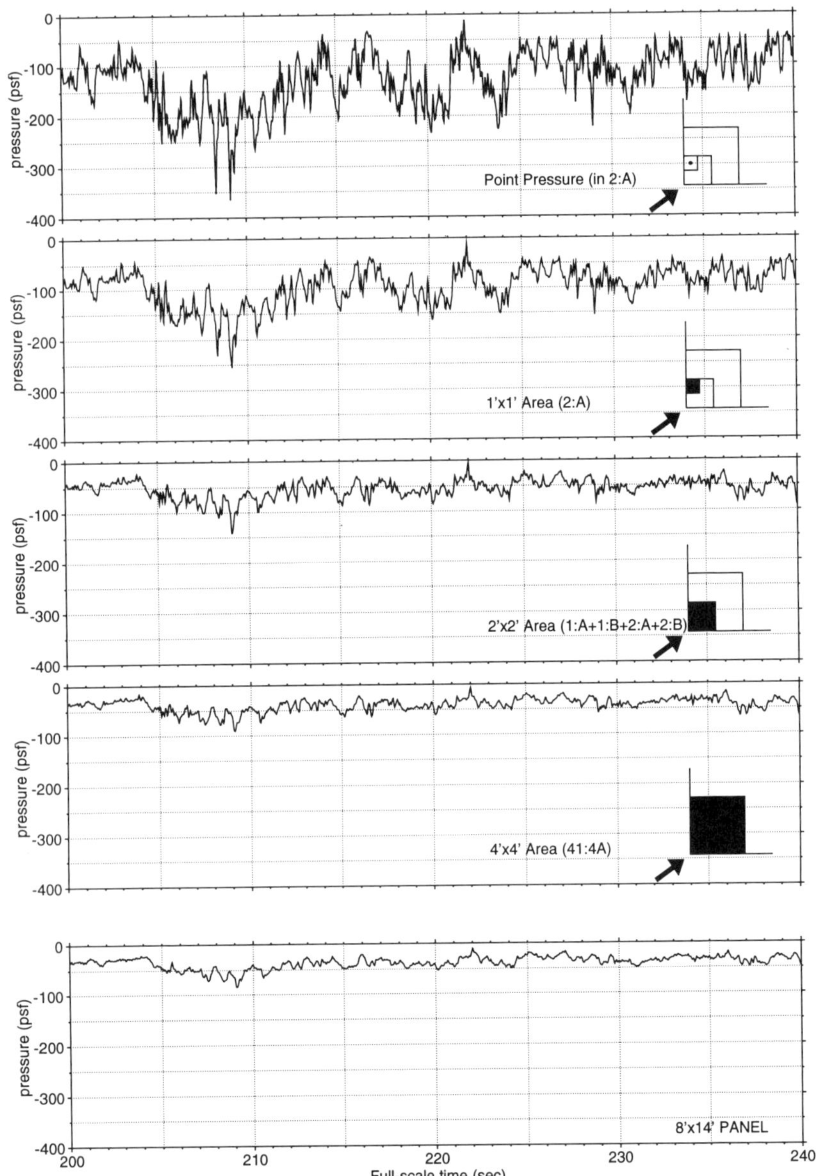

Figure 1   Time Histories of Loads On Different Tributary Areas Under the Corner of a 32' High Building (1/4 on 12 roof slope, $\bar{q}_H = 18$ psf)

a)

Legend: (psf)
Maximum
Mean
Minimum

Row	Stat	A	B	C	D	E	F	G	H
14	Maximum	-5.9	-5.5	-3.4	-3.4	7.2	7.2	7.2	11.9
	Mean	-38.9	-39.8	-46.1	-46.1	-26.2	-26.2	-26.2	-6.4
	Minimum	-102.9	-104.6	-121.9	-121.9	-124.9	-124.9	-124.9	-57.2
13	Maximum	-8.0	-7.2	-2.1	6.8	5.9	4.7	7.2	11.9
	Mean	-43.6	-45.7	-55.9	-47.4	-41.5	-28.8	-26.2	-6.4
	Minimum	-108.4	-115.2	-160.9	-154.9	-135.5	-128.7	-124.9	-57.2
12	Maximum	-5.1	-5.1	-2.1	6.8	6.8	6.8	1.7	5.5
	Mean	-41.9	-47.8	-55.9	-47.4	-47.4	-47.4	-24.1	-6.8
	Minimum	-119.4	-132.9	-160.9	-154.9	-154.9	-154.9	-88.1	-35.1
11	Maximum	-7.2	-6.4	-0.4	6.8	6.8	3.8	13.1	7.2
	Mean	-47.0	-52.1	-58.0	-41.9	-41.9	-24.6	-7.2	-7.2
	Minimum	-127.9	-135.9	-164.7	-163.0	-163.0	-92.7	-70.7	-22.0
10	Maximum	-8.5	-7.6	3.4	10.6	7.2	13.1	13.1	7.2
	Mean	-48.7	-55.9	-60.1	-34.7	-24.1	-7.2	-7.2	-7.2
	Minimum	-123.6	-140.6	-171.9	-159.6	-105.4	-70.7	-70.7	-22.0
9	Maximum	-9.3	-5.9	6.4	10.6	5.1	13.1	13.1	7.2
	Mean	-52.5	-60.5	-56.7	-24.1	-15.7	-7.2	-7.2	-7.2
	Minimum	-138.9	-154.9	-175.3	-148.6	-91.4	-70.7	-70.7	-22.0
8	Maximum	-8.0	-3.8	8.5	11.4	8.0	13.1	13.1	7.2
	Mean	-55.5	-63.9	-48.7	-16.9	-13.1	-7.2	-7.2	-7.2
	Minimum	-141.4	-168.5	-188.8	-143.1	-98.6	-70.7	-70.7	-22.0
7	Maximum	-11.0	-4.7	6.4	13.1	5.9	9.3	9.3	6.8
	Mean	-59.7	-69.4	-35.6	-12.3	-11.0	-8.9	-8.9	-8.0
	Minimum	-162.6	-194.3	-186.7	-107.5	-66.9	-33.4	-33.4	-27.1
6	Maximum	-8.9	3.0	8.9	10.6	4.7	9.3	9.3	6.8
	Mean	-65.2	-73.7	-21.6	-10.2	-9.3	-8.9	-8.9	-8.0
	Minimum	-173.6	-218.5	-166.4	-86.4	-55.5	-33.4	-33.4	-27.1
5	Maximum	-5.5	5.5	10.6	7.6	2.5	2.5	1.7	3.4
	Mean	-73.2	-65.2	-14.0	-9.7	-9.7	-9.3	-8.9	-8.5
	Minimum	-199.4	-240.5	-132.1	-74.5	-47.0	-31.8	-29.6	-42.8
4	Maximum	-8.5	6.4	10.6	11.9	3.4	8.5	5.9	4.2
	Mean	-80.4	-33.0	-12.3	-10.6	-10.2	-9.7	-9.3	-8.9
	Minimum	-229.0	-229.5	-118.1	-58.0	-38.9	-35.6	-50.4	-68.6
3	Maximum	-3.8	5.9	5.9	5.1	4.7	5.1	4.7	3.8
	Mean	-88.9	-16.5	-13.5	-11.9	-10.6	-10.6	-11.9	-15.2
	Minimum	-261.6	-146.1	-101.6	-69.4	-56.3	-88.9	-96.1	-105.0
2	Maximum	1.7	3.0	3.0	3.0	3.4	4.2	2.1	0.0
	Mean	-72.0	-17.4	-14.8	-16.9	-22.4	-30.5	-37.7	-39.4
	Minimum	-253.6	-117.7	-66.0	-117.7	-128.3	-137.2	-138.9	-130.0
1	Maximum	-5.5	0.0	-3.0	-1.7	-1.7	-0.8	-1.7	-2.5
	Mean	-49.5	-38.9	-50.8	-46.6	-45.3	-40.6	-38.1	-35.6
	Minimum	-157.5	-158.8	-181.6	-151.1	-154.9	-130.0	-118.5	-109.2

8x14 Panel: -4.2 / -30.5 / -81.7

Entire 12x25 Panel: -4.2 / -25.4 / -67.7

b)

Row	Stat	4A	4B
44	Maximum	-5.1	4.2
	Mean	-44.0	-24.6
	Minimum	-111.8	-112.6
43	Maximum	-5.1	1.3
	Mean	-49.1	-18.2
	Minimum	-132.5	-65.2
42	Maximum	-4.2	1.7
	Mean	-45.3	-8.9
	Minimum	-138.4	-34.3
41	Maximum	-3.8	-1.3
	Mean	-36.0	-23.3
	Minimum	-97.8	-68.2

Figure 2   Mean and Extreme Pressures Averaged Over 1'x1' and 4'x4' Areas For a Continuous 6 Minute Storm Duration (32' eave height; ¼ in 12 roof slope; open exposure; $\bar{q}_H = 18$ psf)

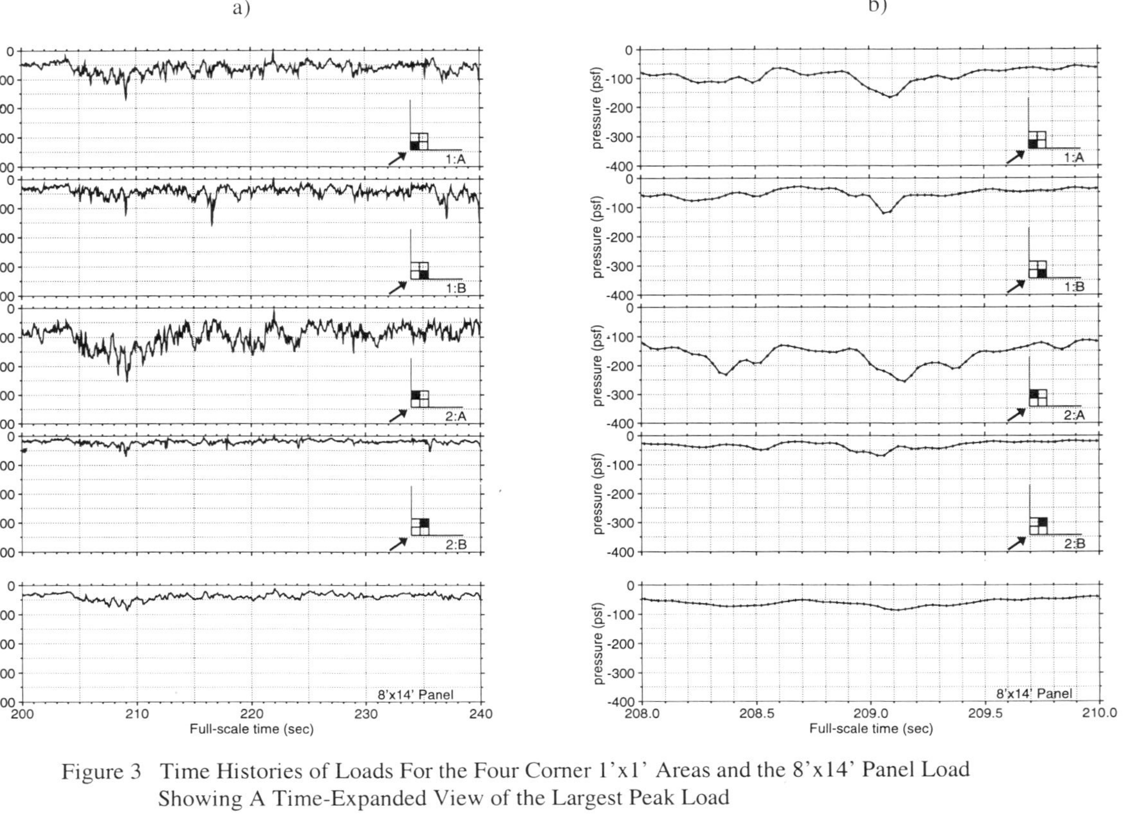

Figure 3   Time Histories of Loads For the Four Corner 1'x1' Areas and the 8'x14' Panel Load Showing A Time-Expanded View of the Largest Peak Load

Wind-Induced Internal Pressures on Sports Arenas

Greg Conley[1], Scott Gamble[2], Peter Irwin[2], M. ASCE, and
Michael Soligo[2], M. ASCE

Abstract

During a major storm event, a breach in the exterior envelope of an arena is possible. Review of these events in the past has indicated that a breach is most likely due to the impact of projectiles from the area surrounding the site unless impact resistant materials are used. A large enough opening could lead to high internal pressures which would add to the values considered for the design of structure and cladding elements. Building codes and standards generally have provisions for calculating internal pressures through simple formulae. However, wind loads can vary greatly for different shapes and orientations of the exterior envelope. Therefore, it is a very difficult task for simple methods to address all of the combinations of conditions that can occur. Additionally, the large interior volume of an arena reduces the response of the internal pressure to wind gusts on the exterior. This paper illustrates the use of wind tunnel tests to estimate the internal pressures in such cases, more precisely.

Introduction

Internal pressures on an arena are caused by: i) mechanical systems; ii) stack effect; and iii) leakage of exterior air into and out of the building. Of these, the most important component of internal pressure is the leakage of exterior air when envelope breaches and mechanical system louvers are taken into consideration. The size and location of leakages in an arena's envelope and the time for the internal volume of the arena to react to wind gusts acting through the leakages greatly influence the magnitude of the internal pressure. General gaps in the envelope not related to damage can also play a small role.

A breach in the exterior envelope caused by the impact of debris can occur at various locations on an arena producing different sizes of openings. Projectiles which

---

[1]Senior Engineer, [2] Principal, RWDI Inc., 650 Woodlawn Road West, Guelph, Ontario, Canada N1K 1B8

can cause a breach in the envelope range in size from small stones to tree branches. Larger projectiles impacting the building would be a rare event. The most likely location for a breach on an arena would be glazing near the ground. The size of the breach would depend on the size of glass panes, their impact resistance, projectile size and the spacing of mullions or exterior columns which may limit the damage in large areas of glazing. Loading doors inadvertently left open in a storm would have a similar effect.

Behaviour of Internal Pressures

The location of the breach determines the sign (positive or negative) and the magnitude of the wind pressure acting on the internal space. For example, a breach on the windward side of an arena would produce a positive internal pressure causing the arena's envelope to push out. A breach on the leeward side of an arena in a negative wind pressure region, would produce a negative internal pressure causing the arena's envelope to pull in. In the case of multiple openings on various sides, the effects would tend to average out and thus cancel each other. Multiple openings produce another ameliorating effect. The correlation between peak gust pressures acting over the total area of all openings reduces, causing the internal pressure to be more related to the mean pressures than the peak pressures.

Internal partitioning in an arena will also influence the response of the internal pressure. If a breach in the envelope has occurred in a room that is well sealed from the remainder of the arena, the internal pressure created by this breach will be limited to the exterior envelope of this room (and the internal partition walls). However an arena is generally open, except for office and restaurant space, and a breach in the exterior envelope will generally affect the internal pressure of the entire arena.

After a breach in the envelope of an arena has been suddenly created, there is a time lag before the internal volume of an arena can react and equalize to the exterior pressure at the breach. The time constant for response of the internal volume of the arena has been estimated by Irwin (1994)

$$\tau \approx \frac{1}{695}\frac{V_0}{A}\left(1 + 1.42x10^5\frac{A_S}{KV_0}\right) \tag{1}$$

where τ is the response time in seconds, V_0 is the internal volume in m^3, A is the total area of opening in m^2, A_S is the total surface area in m^2, and K is a measure of the stiffness of the walls and roof against deflections caused by internal pressures in N/m^3.

The wind-induced internal pressure of an arena has a mean and a dynamic component. The mean internal pressure will be equal to the mean external pressure at the breach. An estimate of the internal dynamic component modified from Irwin (1994), taking into account attenuation of the external pressure fluctuations, is given by the following:

$$g_{pi} = \frac{g_p}{\sqrt{\dfrac{\tau}{10} + 1}} \tag{2}$$

where g_{pi} and g_p are the internal and external peak factors respectively. An estimate of the peak internal pressure $\hat{p}$ of the arena can then made by:

$$\hat{p} = \overline{p} \pm g_{pi} p_{rms} \tag{3}$$

where $\overline{p}$ is the mean external pressure and p_{rms} is the root mean square of the external pressure.

Case Study and Discussion

A recent study at RWDI involved wind tunnel testing of the Broward County Civic Arena to determine cladding and structural wind loads (Figure 1). The arena is situated in South Florida in a hurricane zone. Therefore, the potential for glass breakage during a major storm event warranted a wind tunnel test to determine internal pressures in the arena.

Figure 1 Wind Tunnel Model of Arena

Pressure taps were installed on the scale model of the arena in locations where breaches in the envelope would more likely appear and where mechanical ventilation intakes and exhausts were present. The time series of wind pressure at each of the taps were recorded for a range of wind directions. This approach allowed multiple breach scenarios (i.e., areas involving more than one tap) to be examined on an instantaneous basis through post-processing. Each breach in the envelope was assumed to be approximately 20m^2 and was based on the size of glass panes, spacing of mullions and structural columns, and the possible size of projectiles at the site. The internal pressure of the arena was estimated using equation (3).

The wind tunnel study predicted that the worst case internal pressure for the arena would occur from a single large breach in the envelope. The ASCE 7-93 and 7-95 Standards were used as a comparison to the measured internal pressure from the wind tunnel tests. This comparison is presented in Table 1. Cases II and III are comparable to the RWDI wind tunnel results. ASCE 7-93 Case II positive pressure agrees exactly, the wind tunnel indicated higher negative pressures and the ASCE 7-95 Case III produced higher values than the wind tunnel tests.

Table 1:   Comparison Between ASCE and Internal Pressures Obtained from Wind Tunnel Tests (kPa)

ASCE 7-93		ASCE 7-95		RWDI Wind Tunnel Tests
Case I	Case II	Case I	Case III	
+/- 0.4	+1.3, -0.4	+/- 0.6	+2.5, -1.4	+1.3, -1.0

Case I assumes uniformly distributed leakage

Case II assumes a dominant opening

Case III - assumes a hurricane zone and no impact resistant glass in the lower 18 m

An investigation was initiated by Ellerbe Becket of Kansas City, the structural design engineers for the arena, to determine whether the internal pressures required for design purposes could be reduced. This involved investigating the use of existing ventilation louver locations as internal pressure relief. This approach utilized the arena's existing design for smoke purge and building pressure control. Louvers located near the main roof on two sides of the arena, which are directly accessible to the main bowl of the arena were designed to be fail-safe with operable dampers in the open position to relieve internal pressures in the event of a storm.

Through several iterations of louver positioning and sizing tests and fail-safe condition assumptions the relief system design was optimized. It was found that the internal pressures of the arena were reduced to approximately +0.5 and -0.8 kPa.

Concluding Remarks

It has been found that appropriate internal pressure allowances for the design of buildings with large internal volumes, such as sports arenas, are the result of a complex combination of physical parameters and relative probabilities. Building codes and standards provide for internal pressures in a necessarily simplified manner which may not reflect the situation for arenas very well. More sophisticated wind tunnel approaches can be used to advantage in order to fine-tune the internal pressure allowance and as illustrated, to provide tools required for the refined design of ventilation systems with a view to optimizing the structural and cladding design.

References

ASCE 7-93 Standard (1993) "Minimum Design Loads for Building and Other Structures" *The American Society of Civil Engineers*, New York.

ASCE 7-95 Standard (1995) "Minimum Design Loads for Building and Other Structures" *The American Society of Civil Engineers*, New York.

Conley, G.J., Soligo, M.J., Irwin, P.A. (1996) "Wind Loading Study - Broward County Civic Arena" *RWDI Final Report 97-109*, Guelph, Ontario Canada.

Irwin, P.A. (1994) "Review of Internal Pressures on Low Rise Buildings" *RWDI Final Report 93-270*, Guelph, Ontario Canada.

Simultaneous Measurement of Roof Pressures and Applications

Noriaki Hosoya[1], Member, ASCE and Jack E. Cermak[2], Honorary Member, ASCE

Abstract

Simultaneously sampled pressure data provide a variety of wind load information useful for designers of a structure. A number of locally measured pressures can be combined, in time series, to obtain, for example, area-averaged loads on structural elements and total loads on a main wind resisting system. Two practical applications of simultaneous pressure measurements are presented.

Introduction

Recent advancement in electronic architectures significantly modernized data acquisition strategies involving measurement of pressures in wind tunnels on buildings and other structures. A data-acquisition system, referred to here as a multi-pressure system (MPS), allows instantaneous pressures at a number of locations to be sampled simultaneously. Data obtained by an MPS can be used to analyze not only local cladding pressures but also effective pressures averaged over larger areas as well as loads on structural frames. These load effects are primarily derived by a linear combination of local pressures with appropriate weight factors which may specify local tributary areas, load influence factors, or a combination of both and others. In addition, local loads may be accumulated to represent a modal load from which resonance response loads are evaluated.

Cermak Peterka Petersen, Inc. (CPP) developed and has been operating a unique MPS since 1993. This summary paper describes two selected applications from studies at Colorado State University (CSU) and CPP -- loads on a flat roof of a low rectangular building and structural loads on a truss system of a large sports arena roof. Other examples of MPS applications are described by Irwin and Kochanski (1995) and Case and Isyumov (1997).

---

[1] Graduate Research Assistant, Department of Civil Engineering, Fluid Mechanics and Wind Engineering Program, Colorado State University, Fort Collins, CO 80523; and Senior Project Engineer, Cermak Peterka Petersen, Inc., 1415 Blue Spruce Drive, Fort Collins, CO 80524.

[2] University Distinguished Professor, Department of Civil Engineering, Fluid Mechanics and Wind Engineering Program, Colorado State University, Fort Collins, CO 80523; and President, Cermak Peterka Petersen, Inc., 1415 Blue Spruce Drive, Fort Collins, CO 80524.

Flat Roof of a Low Rectangular Building

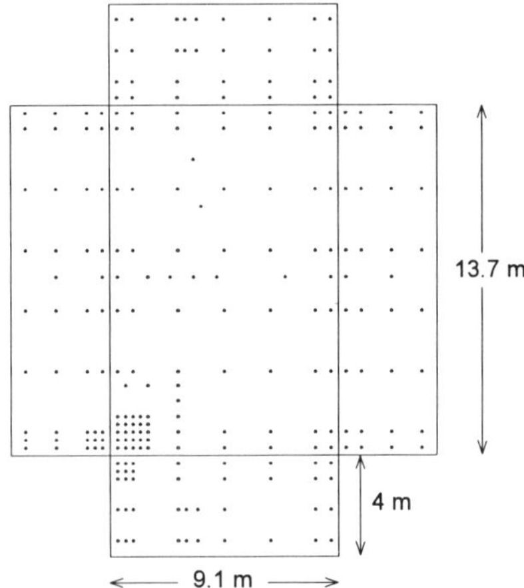

Figure 1. Flat roof building and pressure tap locations

Figure 1 illustrates an unfolded view of a low building featuring a flat roof. The small circles represent locations of pressure taps on the wind-tunnel model constructed at a scale of 1:50. The model is essentially a replica of the flat-roof building at the Wind Engineering Research Field Laboratory (WERFL) at Texas Tech University (TTU). A total of 236 pressure taps (91 on the roof) were available, and pressures were measured simultaneously for all pressure taps. The tests were conducted on the isolated model building placed in a simulated open-country exposure for various wind directions.

Figure 2 shows overall uplift coefficients on the roof as a function of wind direction, normalized by hourly-mean dynamic pressure at the roof height. The wind directions 0 and 180 degrees correspond to wind approaching normal to the short walls of the building. The resultant fluctuating uplift force is negative, upward acting, for all wind directions. The measured peak roof uplift force coefficients agree reasonably well with those specified in ASCE 7-95 (1995) and converted for use with hourly-mean reference pressure.

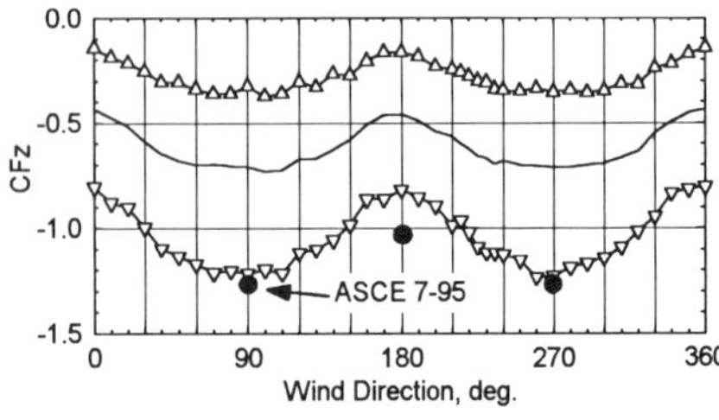

Figure 2. Uplift force coefficient on flat roof

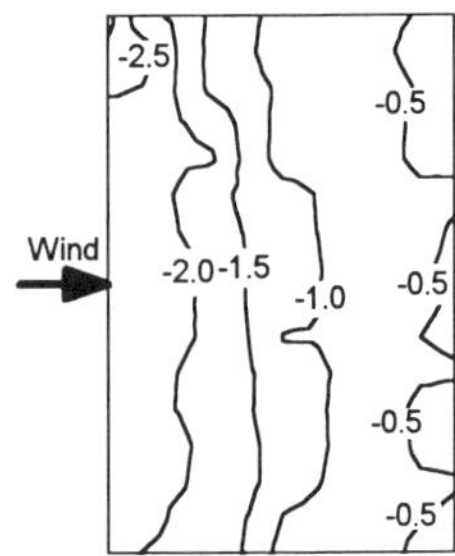

Figure 3. Instantaneous roof pressure coefficients

One of the advantages in using the MPS is that it readily provides distributions of instantaneous pressures in detail for load conditions which might be considered critical for a structural design. Figure 3 illustrates an instantaneous snap-shot of pressure coefficients which resulted in the peak minimum roof uplift shown in Figure 2.

Roof Truss Load of a Sports Arena in New Orleans, Louisiana

For designs of larger structures, it is sensible to identify not only overall wind loads but also distributed loads which individual structural frames must resist. A frame system of a sports arena roof, for example, can consist of several primary and a number of secondary trusses and girders. For those frame members, lateral and axial shears and bending moments exerted by winds are of a major concern. In addition, effects of inertial loads due to resonance response should not be blindly dismissed, especially for long-span frame members. All of these load effects can be determined by a use of the MPS in conjunction with a proper structural analysis performed on the frame system.

At a request of Walter P. Moore and Associates (WPMA) at Houston, Texas, a wind-tunnel test at a scale of 1:400 was conducted for the proposed New Orleans Sports Arena, schematically shown in Figure 4. Over the roof alone, 170 pressure taps were installed for simultaneous measurements of pressures. The wind-tunnel test included modeling of surrounding buildings near the sports arena planned for construction. The roof is primarily supported by two concrete box trusses, indicated by heavy lines, and 109 secondary joists. The box trusses span 98 meters across the short dimension of the arena, 15 meters deep, located at approximately the 1/3 points along the long dimension. The configuration of the box truss is also given in Figure 4, showing relevant element members.

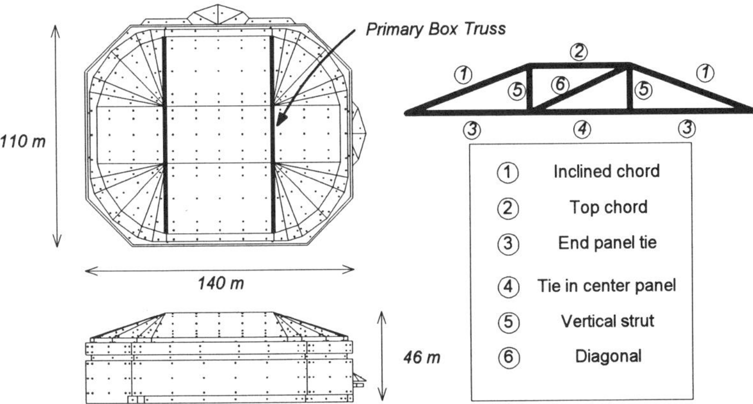

Figure 4. Sports arena and main truss system

The load effects on each truss members were determined in cascading steps. First, local concentrated load distributions were obtained from measured pressures and associated tributary areas. Contribution of the concentrated loads were allocated according to a simple beam theory to result in vertical end-shears for all secondary joists. The end-shears were then collectively transferred to key nodes on the box trusses to form concentrated nodal loads. Using influence coefficients supplied by WPMA, following a finite-element analysis of the box truss, the axial loads of the truss members were determined. The nodal loads on the truss were also used to estimate loads due to resonance response by conventional random vibration theory and modal analysis. The resonance response loads were subsequently added to the previously obtained background dynamic loads.

This test provided critical design wind loads on the box truss members as well as on representative secondary joists for the structural design. Test results, not shown here due to space constraint, revealed that the effect of resonance response was much smaller than anticipated because of a lack of pressure correlation over significant areas of the roof and the relatively high fundamental frequency of the frame system.

Concluding Remarks

The two examples briefly illustrate the versatility of MPS applications. Multi-pressure systems provide comprehensive pressure database, allowing a variety of load analyses be performed in both time and frequency domain. As an added benefit, MPS can reduce time for wind-tunnel tests, otherwise required using a conventional pressure measurement system.

Acknowledgements

The study on the low building was conducted as a part of CSU/TTU Cooperative Program in Wind Engineering supported by National Science Foundation Grant CMS-9411147. The authors thank WPMA for permission to present information on the New Orleans Sports Arena wind-tunnel study.

References

American Society of Civil Engineers (1995), *Minimum Design Loads for Buildings and Structures*, ASCE Standard ASCE 7-95, New York.

Case, P. C. and Isyumov, N. (1997), "Wind Loads on Low Buildings with 4:12 Gable Roofs in Open Country and Suburban Exposures," Proc. Eighth U.S. Nat. Conf. On Wind Engineering, Johns Hopkins University, Baltimore, MD, 10p.

Irwin, P. A. and Kochanski, W. W. (1995), "Measurements of Structural Wind Loads Using the High Frequency Pressure integration Method," Proc. of ASCE Structures Congress XIII, Vol. 2, pp. 1631-1634.

Using Wind Design Criteria to Gauge Seismic Performance

Edwin T. Dean, M. ASCE[1] and Jeffrey R. Soulages[2]

Abstract

When performing seismic evaluations of structures in low seismic regions it is not uncommon to find that the structures have been primarily designed for wind forces, with little or no consideration for earthquake performance. To more quickly evaluate structures for seismic hazards in these regions, parameters may be established that permit engineers to determine when wind load criteria will govern over seismic criteria for the primary structural system. The evaluation parameter, if easily defined, could enable an evaluation of a building designed for known wind load criteria to be quickly assessed to identify its potential seismic vulnerability. The objective would be to develop threshold "Quick Check" parameters that could be used to gauge when wind design criteria will control the performance of a structure's primary lateral load-resisting system. This procedure would be particularly valuable in regions of the country where mapped seismic hazards have increased in recent years and prior prevailing design practices did not include seismic provisions.

Introduction

Wind force loading is introduced into a building structure through pressure gradients applied over the surface of a building's exterior, while seismic forces are generated by the inertial response of a building to an earthquake's ground motion. In evaluating buildings for seismic hazards in regions of low seismicity or in regions where the seismic hazard has only recently been recognized as high it is not uncommon to encounter buildings that were only designed for wind forces. This paper explores the differences between these two fundamentally different environmental loadings, in an attempt to ascertain what parameters can be used to quickly judge the controlling influence on a building's performance.

---

[1] Principal, Dean Engineering, Inc., 319 SW Washington Street, Suite 812, Portland, OR 97204
[2] Associate, Degenkolb Engineers, 620 SW 5th Avenue, Suite 1100, Portland, OR 97204

Evaluation Criteria

In order to perform a comparison between wind and earthquake forces on a building, a common basis for analysis needs to be established. The most basic parameter to use for comparison is base shear, or the total accumulated lateral force imposed on a building structure. Building wind and earthquake design criteria is generally drawn from one of the three model building codes used in various regions of the United States. Wind provisions of the Standard Building Code (SBCC, 1993) and the National Building Code (BOCA, 1993) are taken from the ANSI/ASCE 7-93 (ASCE, 1993) provisions (formerly known as ANSI A58.1). The Uniform Building Code (UBC, 1994) wind loads have a similar basis, but uses a different formulation.

The ASCE 7 wind force formulation for primary structural systems is as follows:

$$p = q\, G_h\, C_p - q_h\, (G\, C_{pi}) \tag{1a}$$

where,

p = Design wind pressure.
q = Velocity pressure.
G_h = Gust response factor.
C_p = External pressure coefficient.
C_{pi} = Internal pressure coefficient.

The total wind load on the building can be determined by summing the wind pressure on the times the projected surface area of the building:

$$V_{wind} = \Sigma p_z\, A \tag{1b}$$

where,

V_{wind} = Wind base shear.
A = Project area.

The computation of seismic base shear (pseudo lateral force) is developed from equation (3-1) of Section 3.5.2 of FEMA 310 (FEMA, 1998) for seismic is:

$$V = C\, S_a\, W \tag{2a}$$

Where,

V = Pseudo lateral seismic force. In a displacement-based analytical approach this is the force required to achieve "actual" inelastic earthquake displacements in a structure analyzed elastically.

C = Modification factor to relate expected maximum inelastic displacements to displacements calculated for linear elastic response. The factor varies between 1.4 to 1.0 short to tall structures respectively (one story to four or greater, depending on the structural system).

S_a = Response spectral acceleration at the fundamental period of the building in the direction under consideration. For the purposes of this evaluation the S_{DS} values from table 2-1 are used to envelop the spectral accelerations being considered.

W = Total dead load and anticipated live load.

The first thing that must be examined is the fundamentally different response that each of these load conditions represent. Specifically, wind design is based on a linearly elastic response while seismic design anticipates a nonlinear, inelastic response. Before any comparison can be made between these to different design parameters they need to be aligned to represent roughly equivalent response characteristics. This alignment, while not necessarily rational, can be accomplished empirically by reducing the seismic base shear by a factor that represents the nonlinear response reduction experienced in the structure. The reduction or R factor (ATC, 1995) that generally representative of the global reduction could be taken as equivalent to an R = 6. The equivalent, effective linear static base shear would be expressed as:

$$V_{seismic} = V/R \qquad (2b)$$

where,

$V_{seismic}$ = Effective lateral seismic force.

R = Nonlinear response (ductility) factor. In the FEMA 310 evaluation procedure, unlike its predecessor FEMA 178 and model codes for new construction, ductility response coefficients (m-values) are applied at a component level. Conventional seismic analysis has relied on global ductility response coefficients.

Discussion

Using these "equivalent" base shear parameters a series of parametric analysis can be made of model building types can be made. The model building types would be defined by the type of construction, mass and projected area. The graphs would be developed with increasing building height along the abscissa and increasing base shear along the ordinate. The illustration documents where the earthquake base shear clearly exceeds wind. By varying the parameters a suite of graphs can be developed to aid engineers to quickly determine when wind design criteria will control the performance of a structure's primary lateral load-resisting system. Additionally, graphs can be developed for a given model building of fixed size, plotting the varying wind base as a function of year constructed and prevailing Code requirements, along with a constant earthquake threshold base shear. Several graphs will be presented to illustrate these relationships.

Conclusions

While it is convenient at times to develop simplified tools to permit insight into the controlling structural characteristics of a complex structural system, such simplifications should always be exercised with good deal of measured engineering judgement and caution. Structural response to wind and earthquake ground motion are inherently quite different. It is also important to recognize that the behavior of the building's components will be quite different from the primary building system -- so much so, that where the primary design is governed by earthquake criteria, the components design may be governed by wind load or vice versa. Nonetheless "Quick Check" parameters such as those presented here offer engineers a rule-of-thumb as an indicator of potential controlling building response.

References

American Society of Civil Engineers (ASCE, 1993), ASCE 7-93, *Minimum Design Loads for Buildings and Other Structures*, 1993.

Applied Technology Council (ATC), *ATC-34 A Critical Review of Current Approaches to Earthquake-resistant Design*, 1995.

Applied Technology Council (ATC), *ATC-19 Structural Response Modification Factors*, 1995.

Building Officials Code of America (BOCA), *National Building Code*, 11th Edition, 1993.

International Council of Building Officials (ICBO), *Uniform Building Code*, 1994 Edition.

Federal Emergency Management Agency (FEMA), FEMA 310, *The Handbook for the Seismic Evaluation of Existing Buildings -- A Prestandard*, January 1998.

Southern Building Code Council International (SBCCI), *Standard Building Code*, 1993.

Designing for a World of Multiple Hazards

Charles H. Thornton, Ph.D., P.E.[1]
Leonard M. Joseph, P.E.[2]

Abstract

Building response to multiple hazards such as wind, flood, earthquake and explosion can be greatly improved through consistent use of structural design features that provide ductility, toughness, continuity and redundancy in the event of unanticipated loadings.

Introduction

Earth, air, fire, water -- the 'elemental' materials of the ancient Greeks, are also the tools of civil engineering. In foundations, windmills, furnaces and reservoirs, they were tamed to do our bidding -- or so we thought. When these forces confounded expectations and led to failures, it was an 'act of God.' Today we should know better.

Structural designers must recognize the dual nature of forces. Wind tunnels, computational fluid dynamics, geomechanics and fire engineering thermodynamics improve our understanding. Databases on winds and hydrology define probabilities for different loading conditions. However, 'outliers,' less-probable but devastating events, are important. Events in Kobe, Northridge and Loma Prieta suggest that subtle effects can lead to earthquake responses more severe than anticipated. The 'greenhouse effect' seems to yield stronger storms, from tornadoes to hurricanes. El Niño rains trigger floods and landslides. Volcanoes bring ashfalls and lahars. And terrorists use the explosive effects of rapid combustion to tragic effect. These forces are ubiquitous. The public mentally assigns particular hazards to particular locations, but regions can experience tornadoes and earthquakes, hurricanes and landslides, floods and bombings. What's an engineer to do?

---

1 Member, Thornton-Tomasetti Engineers, 641 Ave. of the Americas, NY, NY 10011
2 Member

It is neither economically feasible nor technically practical to design most buildings for all possible hazards in optimal fashion. That's reserved for nuclear power plants and missile silos. However, many of the design principles for specific loading conditions are useful across a broad range of hazards. Practical multi-hazard mitigation needn't require designing the same building multiple times; a well-considered design approach with ductility, connectivity and redundancy can cover the demands of many loading types. Specific hazards will require special design features only rarely. The challenge now is political and economic: getting the basic features included in building codes, especially in areas currently perceived as not having 'that kind of problem.'

Philosophy of Multi-hazard Mitigation

Consider the nature and range of hazards external to the structural frame:
- High winds bring roof suction, wall pressure and debris missiles.
- Earthquakes input energy through shaking.
- Floods involve buoyancy, wall pressure, lateral drag, scour and debris impact.
- Explosions, accidental or malevolent, generate high pressures and shatter members.
- Vehicular impacts laterally load individual members.
- Fires reduce member capacities.
- Foundations settle from bearing failure, consolidation, pile failure or erosion.
- Landslides, mudslides and lahars undermine and laterally load structures.
- Excessive snow, storage loads and waterlogged paper can be unexpected loads.
- Structures interact with loadings, causing ponding and vortex shedding.

There are also hazards related to the frame itself:
- Understrength members or connections fail due to material inadequacies.
- Members and connections can be sensitive to minor force changes, where the net force is the small difference between large opposing forces (like gravity and uplift).
- Sensitivity to overload exists where failure is by buckling or rupture.
- Unanticipated structural behavior can occur due to design lapses.

Consider the options for handling such hazards:
- Handle all hazards within allowable stresses or factors of safety? This is economically impractical and counterproductive. Energy absorption, essential in seismic and explosive events, is greater for yielding members than elastic ones.
- Count ultimate strength of members and connections when considering extreme loadings? This is practical and economical. The challenge is determining the strength needed.
- Provide alternative load paths? This improves economy and reliability, as redundancy can spread demand over more members and connections.
- Allow for unconventional load-resisting mechanisms? This can provide economical capacity against catastrophic failure by accepting large local deformations. One example is using slabs as membranes and axially connected beams as catenaries.

Examples

Ductility (toughness), connectivity and redundancy are the keys to multi-hazard mitigation. Consider the Alfred P. Murrah Federal Building in Oklahoma City, site of an infamous 1995 terror bombing which collapsed a large portion of the reinforced concrete, beam-and-slab structure and killed 168 occupants. Four design features of the building, following prevailing codes and practices, controlled its response. First, every alternate column was interrupted between the first and the third floors, and its load was shifted to neighboring columns by a transfer girder. Thus a loss of one ground floor column affected three upper columns and four bays. Second, the columns had limited shear strength. Thus failures were of a sudden and brittle nature when laterally loaded by explosive pressures. While one column would have shattered in any case, two others probably would have survived if well confined by transverse reinforcing. Third, the continuous transfer girders and spandrel beams had negligible rebar continuity where compression was expected. Once failure of ground floor columns caused flexural forces to reverse, there was no way to span load to adjacent columns. Fourth, one-way floor slabs had bottom rebars interrupted at beam lines and top bars located just over supporting beams. Stress reversals due to upward explosive pressures 'broke the back' of floor slabs at unreinforced top surfaces and hinged at supports. Also, the potentially helpful catenary or membrane action of a sagging floor acting like a suspension bridge could not be maintained due to insufficient tensile continuity.

If the Murrah Building were designed and detailed using current requirements for an active seismic zone, such as in UBC 1997, the weaknesses would have been addressed. The code discourages omitting ground floor columns and creating a 'soft first story.' Beam and column shear reinforcement is sized, spaced and detailed so flexure, not shear, controls. Continuity of main girder reinforcing is required, with compression zone rebar equal to 50% of tension reinforcing carried through joints. This provides some load redistribution ability. Slab bottom rebar continuity is required. Under 'back-breaking' uplift forces, this could help resist failure at slab ends. And continuous bottom bars could resist the tensile forces from catenary action if required.

A non-seismic design considering load paths and energy absorption is the Skybridge of the Petronas Twin Towers in Malaysia. This 55m long steel-framed double-deck pedestrian bridge linking the 41st and 42nd floors of the Towers has a steel pipe arch propping its midspan. See Fig. 1. The design reconciled sliding end supports, two-span main girders which flex during building movements, visually light upper deck and roof framing with joints to minimize cladding motion, and 'fail-safe' behavior in the very unlikely event of arch failure. Energy absorption was 'built in' through redundant yielding systems. For an arch suddenly 'missing,' the midspan reaction force would generate kinetic energy as the girder sags downward. Sagging would stop when the girder and upper framing absorb this energy. Since elastic bending of the girder offers only a minor energy contribution, formation of a midspan plastic hinge would be permitted. See Fig. 2. Because this would not absorb energy fast enough to offset the kinetic energy input, deck joints would close as 'bumpers' meet, and the steel pipe columns of upper deck framing would yield in bending. See Fig. 3. With this additional energy absorption mechanism, all the kinetic energy would be absorbed at one meter of midspan sag - an acceptable situation, considering the extremely small chance of occurrence. See Fig. 4. Through activation of many elements, a catastrophic failure of one part of the system could be handled while maintaining reasonable member sizes.

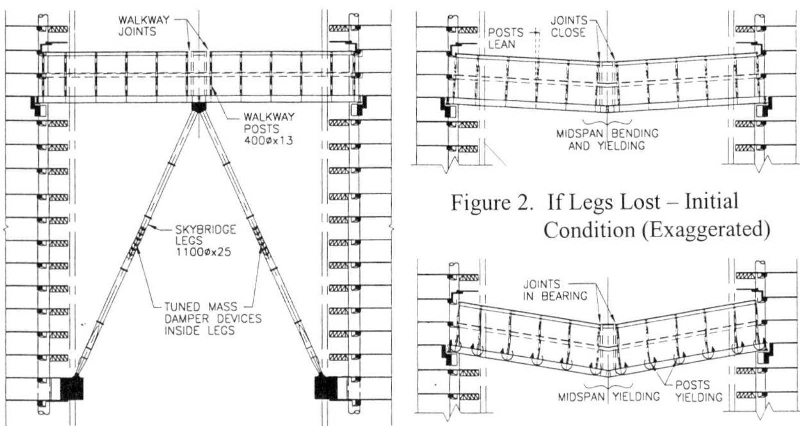

Figure 2. If Legs Lost – Initial
Condition (Exaggerated)

Figure 3. If Legs Lost - Final
Condition (Exaggerated)

Figure 1. Skybridge Elevation

Figure 4. Schematic Energy
Balance Graph

Conclusion - A Call for Action

Engineers understand that building performance and safety can be improved with only a small increase in building frame cost, through the use of ductility, connectivity/continuity and redundancy. Modern seismic codes address these aspects of design. Such features make sense even in 'non-seismic' areas to handle exceptional wind, flood, explosion and fire events, but where they are 'beyond code' they are rarely incorporated. As a profession, we must help code writers, building department officials and politicians understand the tremendous potential benefit such provisions offer. The fact that they were not historically applied is no excuse; we have all learned from sad recent experience. Let's put that expensive education to use and reduce future damage from the severe, exceptional but still probable hazardous events to come.

Applied Technology Council's Role in Wind and Coastal Flood Hazard Mitigation

Christopher Rojahn,[1] Member

Abstract

The recently commenced Applied Technology Council (ATC) program in wind and coastal flood hazard mitigation is being designed to build on technologies, concepts, and approaches already developed for earthquake hazard mitigation. The program includes reconnaissance efforts to document the effects of hurricanes on man-made structures, and in-depth efforts to develop engineering applications for wind and coastal flood hazard mitigation. ATC has signed a Memorandum of Understanding with the American Association for Wind Engineering to seek and perform wind engineering research and application projects jointly. ATC will also draw on its extensive experience over the last several decades in developing methods for evaluating and rehabilitating buildings and other structures to resist earthquakes.

Introduction

Since its founding in the early 1970s, the Applied Technology Council (ATC) has been charged with the responsibility of developing resource documents that translate and summarize useful research information for practicing engineers. To date, most of ATC's projects have been concerned with earthquake engineering, and many reports have gained national prominence. Examples include: ATC-3-06, *Tentative Provisions for the Development of Seismic Regulations for Buildings* (ATC, 1978); ATC-6, *Seismic Design Guidelines for Highway Bridges* (ATC, 1981); ATC-13, *Earthquake Damage Evaluation Data for California* (ATC, 1985); ATC-14, *Evaluating the Seismic Resistance of Existing Buildings* (ATC, 1987); ATC-20, *Procedures for Postearthquake Safety Evaluation of Buildings* (ATC, 1989); ATC-21, *A Handbook for Rapid Visual Screening of Buildings for Potential Seismic Hazards* (ATC, 1988), and the *Guidelines for the Seismic Rehabilitation of Buildings (FEMA 273 Report)*, which were prepared under the recently completed ATC-33 project (ATC, 1997a).

In 1996, recognizing that structural damage from wind and wave action has much in common with earthquake damage, the ATC Board of Directors extended ATC's mandate to address wind and coastal flood hazard mitigation. Concurrently

---

[1] Executive Director, Applied Technology Council, 555 Twin Dolphin Drive, Suite 550, Redwood City, California 94065

the Board was expanded to include two nationally recognized specialists in wind and coastal engineering, and a Board committee was established to identify projects that could appropriately be pursued by ATC, alone and in collaboration with other organizations. Initial projects included (1) a reconnaissance study of Hurricane Fran, which damaged structures on the North Carolina coast in September 1996 (ATC-44 Project); and (2) development of procedures for assessing the safety of buildings damaged by wind storm and coastal flooding (ATC-45 Project). In 1998 ATC signed a Memorandum of Understanding with the American Association for Wind Engineering that calls for ATC and AAWE to jointly seek and conduct projects to reduce wind storm and coastal flood hazards. Other projects are under consideration, including methods and techniques for evaluating and retrofitting buildings and other structures to reduce wind and coastal flood hazards. Training programs have also been identified.

Hurricane Fran, South Carolina, September 5, 1996: Reconnaissance Report

Following Hurricane Fran, a Category 3 Hurricane that made landfall on the North Carolina coast on September 5, 1996, newly-appointed ATC Director Robert Dean led an ATC-organized team consisting of Dan Cuoco of Thornton-Tomasetti Engineers, New York, New York, and Michael J. Griffin of EQE International, Aiken, South Carolina, on a two-day mission to investigate the hurricane's effects. The primary objectives of the reconnaissance investigation were to (1) examine the types and extent of damage, (2) collect data, (3) interpret damage causes, and (4) identify means of damage reduction in the landfall area.

Results of the reconnaissance investigation are reported in the ATC-44 Report, *Hurricane Fran, South Carolina, September 5, 1996: Reconnaissance Report* (ATC, 1997b). Included in this report are: (1) a summary of coastal impacts (e.g., storm surges and waves, role of erosion, and wave forces); (2) an overview of building code requirements; (3) observations and interpretation of damage to building foundations, breakaway walls, exterior walls, and roofing; (4) an overview of the performance of telephone, power, water and wastewater systems; and (5) conclusions pertaining to the relative and absolute magnitudes of Hurricane Fran, the role of beach nourishment, the effect of debris acting as missiles, and damage resulting from breaches in the Barrier Island.

Procedures for Post-Hurricane Safety Evaluation of Buildings

The ATC-45 Procedures for Post-Hurricane Safety Evaluation of Buildings (ATC, in preparation) have evolved from procedures for post flood and wind storm safety evaluation of postal service facilities developed by ATC in the early 1990's with funding from the U. S. Postal Service (ATC, 1991, 1992). The procedures are based on similar procedures developed for earthquakes under the ATC-20 project (ATC, 1989), which have been widely used in California since the Loma Prieta earthquake of 1989.

The ATC-45 Procedures are being developed to provide specific guidance on rapid and detailed inspections, posting criteria, selection and size of inspection teams, field safety, human behavior following natural disasters, and other aspects of hurricane safety evaluation. The procedures specify criteria for three recommended posting placards: INSPECTED, for buildings with no apparent hazard found (repairs may be required) that have no restrictions on use or occupancy (green placard), RESTRICTED USE, for buildings where restrictions are placed on use or occupancy of a specified portion (of the building) because a dangerous condition is believed to be present (yellow placard specifying restriction), and UNSAFE, for buildings that have collapsed, partially collapsed or are in eminent danger of collapse and that can not be occupied except by local regulatory authorities for emergency-related reasons. The intended audience for the document includes public works agency employees, fire fighters, police officers, military personnel, facility managers, and other disaster workers, as well as those individuals normally charged with post-hurricane safety evaluation of buildings: civil and structural engineers, architects, and building safety officials.

<u>ATC-AAWE Memorandum of Understanding (MOU)</u>

In August 1998 the American Association for Wind Engineering (AAWE) and ATC agreed in principal to seek and perform wind engineering research application projects jointly, including wind storm reconnaissance investigations. The intent of the MOU is to create a partnership that can respond to the urgent public need for improved techniques and procedures in wind engineering, and for systematic documentation of the effects of wind storms on the built environment, including correlations of wind speed, building system attributes, and building performance.

Each organization brings to the partnership special and unique capabilities. The purpose of AAWE is to promote and perform wind engineering research and education, and AAWE membership consists of researchers and design professionals. ATC identifies and develops state-of-the-art engineering resources and applications for use by design professionals, develops consensus opinions on natural disaster mitigation issues, and has extensive experience in managing complex technical projects involving the nation's leading structural engineering design professionals and researchers.

The MOU was signed by Michael P. Gaus, AAWE President, and Charles H. Thornton, ATC President.

<u>Other Projects In Wind and Coastal Flood Hazard Mitigation</u>

Based on input provided by the ATC Board Committee on Multi-Hazard Mitigation, ATC is planning to develop other needed engineering applications in wind and coastal flood hazard mitigation by building on the technology and concepts already developed for earthquake hazard mitigation. This approach has already been implemented in the development of procedures for post-hurricane safety evaluation of buildings, as described above. Other applications that could be readily adapted from existing earthquake-related procedures include:

1. Development of a comprehensive, agreed-upon methodology for estimation of damage and loss resulting from hurricanes.

2. Development of a handbook for rapid visual screening of buildings for potential wind hazards.

3. Development of a handbook for wind evaluation of existing buildings.

4. Development of guidelines for strengthening buildings to reduce wind hazards.

Currently these projects are in the planning and conceptual stage. They will commence when adequate funding and other needed resources have been obtained.

<u>References</u>

ATC, 1978, *Tentative Provisions for the Development of Seismic Regulations for Buildings*, Applied Technology Council, ATC-3-06 Report, Redwood City, California.

ATC, 1981, *Seismic Design Guidelines for Highway Bridges*, Applied Technology Council, ATC-6 Report, Redwood City, California.

ATC, 1985, *Earthquake Damage Evaluation Data for California*, Applied Technology Council, ATC-13 Report, Redwood City, California.

ATC, 1987, *Evaluating the Seismic Resistance of Existing Buildings*, Applied Technology Council, ATC-14 Report, Redwood City, California.

ATC, 1988, *Rapid Visual Screening of Buildings for Potential Seismic Hazards: A Handbook*, Applied Technology Council, ATC-21 Report, Redwood City, California.

ATC, 1989, *Procedures for Postearthquake Safety Evaluation of Buildings*, Applied Technology Council, ATC-20 Report, Redwood City, California.

ATC, 1991, *Procedures for Postdisaster Safety Evaluation of Postal Service Facilities (Interim)*, Applied Technology Council, ATC-26-2 Report, Redwood City, California.

ATC, 1992, *Field Manual: Post Flood and Wind Storm Evaluation of Postal Buildings (Interim)*, Applied Technology Council, ATC-26-3A Report, Redwood City, California.

ATC, 1997, *Guidelines for the Seismic Rehabilitation of Buildings*, Prepared by the Applied Technology Council for the Building Seismic Safety Council, published by the Federal Emergency Management Agency, FEMA 273 Report, Washington, DC.

ATC, in preparation, *Procedures for Post-Hurricane Safety Evaluation of Buildings*, Applied Technology Council, ATC-45 Report, Redwood City, California.

ATC, 1997b, *Hurricane Fran, South Carolina, September 5, 1996: Reconnaissance Report*, Applied Technology Council, ATC-44 Report, Redwood City, Calif.

DESIGN CONSIDERATIONS FOR COASTAL ZONES EXPOSED
TO HURRICANE-INDUCED WAVE ACTION

by

Robert G. Dean
Department of Coastal and Oceanographic Engineering
University of Florida

INTRODUCTION

It is well known that a wide beach fronting a structure provides an excellent dissipator of wave and storm surge energies during major storms. Thus, the vulnerability of areas subject to beach erosion gradually increases with time due to the increasing proximity to the ocean. Not only is the ocean closer, but the potential vertical scour increases which increases the bending moment due to the wave and current velocities and decreases the pull-out and overturning resistance of the piling. One approach to maintaining the integrity of structures located in eroding areas is to nourish the beach thereby restoring approximately the conditions that existed at some previous time. To date there has not been a rational approach presented to provide an appropriate benefit/cost analysis of beach nourishment in such areas. It is the objective of this paper to initiate the development of such an approach.

THE PROBLEM

Many of the U. S. shorelines are eroding. Various estimates are that some one-half to two-thirds of the sandy beach shoreline is subject to erosion. With the increasing utilization of the coastal zone for residential, commercial and industrial purposes, the risk to a structure located along an eroding coast increases with time. Those responsible for coastal property are faced with three choices: (1) Retreat from the coast as it recedes, (2) Armor the coastline, and (3) Nourish the coastline. Each of these three options may be the most appropriate for a particular situation. Of these three options, nourishment is the only alternative which satisfies the dual objective of allowing the structure to remain in its original position while reducing the risk and retaining the natural character of the shoreline. Nourishment is expensive and there must be a substantial economic base associated with the upland development in order to warrant the costs of this approach. Moreover, in some areas, for example southeast Florida, the availability of good quality sand for beach nourishment purposes is limited. With the value of shorefront property rising, the economic basis for nourishment is increasing in almost all coastal areas. Finally, it is important to recognize that the economic justification of beach nourishment is not the same in all areas. For example, the longevity of beach nourishment projects is inversely proportional to the wave height to the 2.5 power. Thus projects constructed in areas with generally mild waves will perform much better that those located in areas with more energetic climates. Additionally, the background erosion rate is a factor in the longevity of the nourishment project. If the background erosion rate is 30 cm per year, the performance of the project can be expected to be much better than if the background rate is 1 meter per year.

This paper will illustrate the loss of structural integrity due to a reduction in land elevation through consideration of the increases in forces and moments on a single vertical piling.

ANALYSIS BASIS

Figure 1 presents a definition sketch of the system to be considered. The land elevation at the piling of interest is z_G relative to mean sea level, the storm surge at the piling is S, and the wave setup is $\bar{\eta}$, the wave crest elevation above the local still water level $(S +\bar{\eta})$ is η_c and the piling diameter is D. The sensitivity to the ground elevation of the drag forces and moments on this piling will be investigated. It is noted that the wave forces on a piling include a drag force component which is dependent on the square of the wave induced velocities and an inertia force component which is in phase with the wave induced accelerations. These two force components are 90° out of phase. It can be shown that in shallow water and for piling diameters of the size used to support residences, the drag forces will be dominant. Since the maximum water particle velocities are proportional to the square of the water particle velocities which are maximum when the wave induced water level is the highest (ie at the crest phase position), it is this condition that will be examined.

Drag Force

The maximum drag force on a piling, F_D, as the wave propagates past the piling is:

$$F_D = \frac{C_D \rho D}{2} \int_{z_G}^{S+\bar{\eta}+\eta_c} u^2(z)\, dz \tag{1}$$

where C_D is the drag coefficient and is approximately unity, $u(z)$ is the wave-induced water particle velocity, ρ is the mass density of water and the other variables have been defined. Eq.(1) can be put in a more convenient non-dimensional form through definition of the dummy variable, v, for the vertical elevation.

$$v = \frac{z - z_G}{S+\bar{\eta} - z_G} \tag{2}$$

which yields upon substitution into Eq. (1)

$$F_D = \frac{C_D \rho D}{2} (S+\bar{\eta} - z_G) C^2 \int_0^{1+\eta_c/(S+\bar{\eta}-z_G)} \left(\frac{u}{C}\right)^2 dv \tag{3}$$

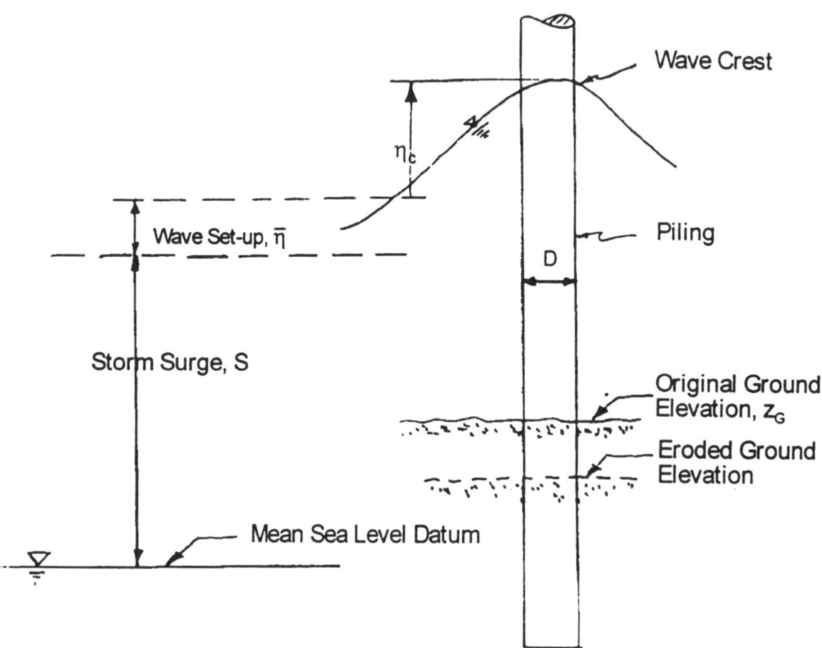

Figure 1. Definition Sketch

where C is the wave celerity, ie the speed at which the wave propagates and is given approximately by

$$C = \sqrt{g(S + \bar{\eta} - z_G)} \tag{4}$$

and in the wave breaking zone, the ratio of maximum water particle velocity to wave celerity is approximately unity, yielding

$$F_D \approx \frac{C_D \rho g D}{2} (S + \bar{\eta} - z_G)(S + \bar{\eta} - z_G + \eta_c) \tag{5}$$

To establish the changes that occur due to a vertical change in ground elevation, Δz_G, we differentiate the drag force with respect to the ground elevation by applying the chain rule since several of the other variables depend on the ground elevation, z_G. In particular, the wave set-up, η, wave height, H, and crest elevation, η_c all vary with ground elevation.

The wave set-up, $\bar{\eta}$, is given by

$$\bar{\eta} = -\frac{H_b}{16} + 0.186(1.28 H_b - S + z_G) \tag{6}$$

where H_b is the breaking wave height and the ratio of local wave height, H, to local still water depth $(= S + \bar{\eta} - z_G)$ is approximately 0.78 and the ratio of crest elevation to local wave height is approximately 0.8.

The chain rule yields:

$$\frac{d F_D}{d z_G} = \frac{\partial F_D}{\partial z_G} + \frac{\partial F_D}{\partial \bar{\eta}} \frac{\partial \bar{\eta}}{\partial H} \frac{\partial H}{\partial z_G} + \frac{\partial F_D}{\partial \eta_c} \frac{\partial \eta_c}{\partial H} \tag{7}$$

inserting the approximations previously discussed, and putting in a somewhat more meaningful form

$$\frac{\dfrac{dF_D}{F_D}}{\dfrac{dz_G}{z_G}} = \frac{-1.01 - 0.814/(S + \bar{\eta} - z_G)}{\dfrac{1}{z_G}(S + \bar{\eta} - z_G + \eta_c)} \qquad (8)$$

in which the left hand side represents the proportional increase in drag force due to a proportional increase in ground elevation. Note that Eq.(8) is negative resulting in an <u>increase</u> in drag force for a <u>decrease</u> in ground elevation. Because of the several variables appearing in Eq. (8), it is most useful to evaluate the relationship by considering an example. The following conditions will be considered

$S + \bar{\eta} = 3.5m$

$z_G = 2.5$ m

For these conditions, it can be shown that for each percent reduction in ground elevation, there is a corresponding increase in total drag force, F_D, equal to 2.8 percent. As an example, if the change in ground level were -0.5 m, (-20%) which is considered realistic, the corresponding <u>increase</u> in drag force would be 56%. A decrease in ground elevation of 1 m (40%) would cause an increase in drag force of 112%.

Maximum Drag Moment

The treatment of maximum drag moment about the pile base will be abbreviated as it follows the same path as illustrated for the maximum drag force. The moment is given by

$$M_D = \frac{C_D \rho D}{2} \int_{z_G}^{S + \bar{\eta} + \eta_c} (z - z_G) u^2(z)\, dz \qquad (9)$$

and employing the same dummy variable, v and other approximations as for the drag force

$$M_D \approx \frac{C_D \rho g D}{4} (S + \bar{\eta} - z_G)^2 (S + \bar{\eta} - z_G + \eta_c) \qquad (10)$$

For the same example conditions as considered for the drag force, for each percent reduction in ground elevation, there is a corresponding increase in total drag moment, M_D, equal to 4.3 percent.

As an example, if the change in ground level were -0.5 m, (-20%), the corresponding <u>increase</u> in drag moment would be + 86%. A decrease in ground elevation of 1 m (40%) would cause an increase in drag moment of 172%.

SUMMARY AND CONCLUSION

The reduction in structural integrity due to a lowering of the ground elevation associated with coastal erosion results in an increased loading due to larger storm wave heights and more exposed piling. The proportionate increases in drag forces and moments can be substantially greater than the lowering of the ground elevation. For the example conditions considered here, the proportionate increases in drag forces and moments are 2.8 and 4.3 times larger than the proportionate decrease in ground elevation. In addition to the increased loading associated with erosion of the ground surface, there would be a decreased resistance to pull out or overturning of the piling, effects which are not considered here.

The maintenance of structural integrity of pile supported or other coastal structures is possible through beach nourishment and in many locations, the economics are probably justified.

Design Considerations for Effects of Complex Topography on Wind Characteristics

Arthur N. L. Chiu[1]

Abstract

Complex topography can cause pronounced changes to the wind patterns which in turn can exacerbate the damaging effects on the built environment. Examples of this phenomenon include: funneling effects caused by gaps in mountain ranges; escalating effects on upward sloping terrain; sudden changes in direction and high turbulence over cliff edges; turbulence around the edges of mountains or ranges; and downslope winds. Much damage has been inflicted on residences and other structures from strong wind situations in such locales. Various Codes and Standards have attempted to provide procedures for estimating the escalation of wind speeds in simple topography. However, effects from complex topography cannot be easily quantified and the designer must be aware of these conditions.

Introduction

Hurricanes have inflicted severe damage to the built environment; reported insured losses from property damaged by Hurricane Andrew amounted to more than \$22 billion, and Hurricane Iniki inflicted more than \$2 billion of losses. In addition, there are long-term economic losses to business and industry that are not measurable immediately, but which linger on for many years after the event.

The damaging effects from tropical cyclones to the built environment have also been reported for many areas in the Carribeans, South Pacific and Asia: Guam, Saipan, Philippines, Korea, Japan, Taiwan, eastern coast of China, Hong Kong, and the east coast of India. Where the terrain changes rapidly from flat land to cliffs and mountains, the complex topography will affect the wind characteristics that impinge the built environment. These topographic features can exacerbate the wind conditions: increased speeds and turbulence. For strong winds approaching land, this situation becomes crucial for structures that are located on top of the

---

[1] Hon. Member, Dept. of Civil Engineering, University of Hawaii at Manoa, Honolulu HI 96822

ridges at the coastlines. However, although there are procedures for estimating basic wind speed profiles for various exposure categories, there is less definitive information for such complex topography. These environments are not easily quantifiable for inclusion in design procedures.

Hurricanes Iwa and Iniki

The damage caused by Hurricanes Iwa (1982) and Iniki (1992) served as clarion calls to the need for studying topographical effects on wind characteristics. Severe damage to structures and buildings was observed in many areas of complex topography. Hurricane Iwa (NRC 1983) was the first major severe windstorm on record to hit Hawaii and caused more than $203 million of property damage. The reported sustained wind speed at the Honolulu International Airport was 49 mph (22 m/s) and a peak gust of 84 mph (38 m/s) from 200 deg. However, on Kauai, which bore the brunt of the storm, the reported sustained wind speed was 73 mph (33 m/s) and a peak gust of 91 mph (41 m/s) from 180 deg. Severe damage was inflicted on buildings and structures, especially in locales of complex topography because of the resulting increased local wind speeds and turbulence. Although Iwa did not hit the island of Oahu directly, the strong winds damaged the transmission lines which then caused a total power blackout on Oahu. The long-span lines traverse over complex topography from ridge to ridge.

During the peak of Iniki, as it crossed the island of Kauai in a northerly direction, the reported sustained wind speeds and peak gusts, respectively, at the Lihue Airport (Chiu et al. 1995) were:

63 mph (28 m/s), 128 mph (57 m/s) from 130 deg at 1502 hrs
96 mph (43 m/s), 114 mph (51 m/s) from 150 deg at 1552 hrs

The hurricane moved northward along the path of the Waimea Canyon. Most of the newly-built residences in the vicinity of the top edge of the canyon walls were destroyed or were severely damaged. This complex topography created a highly turbulent wind environment and the canyon walls also funneled the wind; the combined effects resulted in higher wind speeds.

The northern side of Kauai, specifically the Princeville area, was severely impacted by Iwa in 1982 and again by Iniki in 1992. Within that area, the coastal topography changes suddenly from sea-level to more than 100 feet (30 m), and this profile caused increased turbulence and speed of the onshore winds over the cliffs. Much of the damage caused by Iniki were similar, and in some cases almost identical, to those caused by Iwa. Because of this repeated pattern of damage, a model study ensued to study the wind flow patterns over escarpments and the distribution of wind loads on low-rise structures (Means 1995). The results showed that wind loads on buildings located on escarpments can be substantially higher than those estimated by building codes, particularly for the middle portions of the roofs.

Estimating Wind Speed-Up

The ASCE 7-95 Standard (ASCE 1995) has added provisions to estimate wind speed-up over isolated hills and escarpments. Wind pressures can be estimated from Bernoulli's equation:

$$q = 0.5mV^2 \tag{1}$$

in which m is the air mass density, and V is the wind speed. To include the variation of wind speed with height, the exposure category in which the structure is located, the importance of the structure, and the mass density of air, the equation is modified:

$$q_t = 0.00256K_zK_{zt}V^2\,I \quad \text{(psf)} \tag{2a}$$
$$q_t = 0.613K_zK_{zt}V^2I \quad \text{(N/m}^2) \tag{2b}$$

in which V is the basic wind speed, I is the importance factor, K_z is the factor for exposure condition and elevation above the standard elevation of 10 m above ground level, and K_{zt} is a factor to account for wind speed-up over hills and escarpments but shall not be less than 1.0. The numerical coefficient of 0.00256 (0.613 for SI units) is to be used unless there are sufficient climatic data available to justify selecting a different value for a specific design application. The provision from ASCE7-95 Standard, Section 6.5.5, for wind speed-up over hills and escarpments states:

"6.5.5 Wind Speed-up over Hills and Escarpments The provisions of this section shall apply to isolated hills or escarpments located in Exposure B, C, or D where the upwind terrain is free of such topographic features for a distance equal to 50H or 1 mile, whichever is smaller, as measured from the point at which H is determined. Wind speed-up over isolated hills and escarpments that constitute abrupt changes in the general topography shall be considered for buildings and other structures sited on the upper half of hills and ridges or near the edges of escarpments by using factor K_{zt}:

$$K_{zt} = (1 + K_1K_2K_3)^2 \tag{3}$$

where K_1, K_2 and K_3 are given in the ASCE Standard. The effect of wind speed-up shall not be required to be considered when $H/L_h<0.2$, or when H<15 ft (4.5m) for Exposure D, or <30 ft (9m) for Exposure C, or <60 ft (18m) for all other exposures."

Detailed procedures for estimating wind speed-up can also be found in the codes of Australia (ASA 1989) and Canada (NRCC 1990).

Field Measurements

Experience with measuring behavior of wind in Hong Kong addresses the questions associated with predicting flow of air over uneven terrain. Measurements high above the city center have shown that there are several apparent deviations from currently accepted theories as to how the wind should behave when crossing a mountainous area (Jeary 1997). These measurements have shown that the effects of transition from an over-sea profile of horizontal gustiness to that occurring over a city can be seen much closer to the coast than had previously been assumed. Additionally, in Hong Kong and Japan, measurements of turbulence intensity have shown there is a gradual increase as a storm passes a measurement point. The intensity of turbulence is modified by the flow over rough terrain in a way that is currently predictable only in the most generalized form. These effects are symptomatic of the fact that the current description of wind behavior is imperfect. When these deviations from the currently accepted norms are considered together, then, the overall effects on the calculation of the wind force dominates the precision of the estimate. The need is for a full-scale three-dimensional measurement to permit improvement of the precision for understanding the windfield in a complex terrain.

Closure

The damage that continues to be inflicted by strong winds on structures located in complex terrain invites the question of the adequacy of the current design methodology. Data from field studies are meager. Some experimental data (on low-rise structures) show that the wind loads caused by the highly-turbulent and high-speed winds, at such locales, are larger than those estimated from standards and codes. A coordinated program, encompassing both model studies and field measurements, is needed to investigate this phenomenon in detail. The data from these studies will lead to a better understanding of the effects of complex topography on wind characteristics and thereby lead to an improvement in the design of structures to better resist strong winds.

Acknowledgment

The author acknowledges the contributions from Dr. Timothy A. Reinhold, Clemson Univ. and Dr. Alan P. Jeary, Univ. of Sydney.

References

American Society of Civil Engineers (1995). Minimum Design Loads for Buildings and Other Structures, ASCE, New York, NY.

Chiu, A. N. L., Chiu, G. L. F., Fletcher, C. H., Krock, H-J, Mitchell, J. K. Schroeder, T.A. (1995). Hurricane Iniki's Impact on Kauai, National Technical Information Service, Springfield, VA.

Jeary, A. P. (1997). Personal communication with A. P. Jeary, Univ. of Sydney.

Means, B. E. (1995). Wind Loads for Low-Rise Structures on Escarpments, Master of Science in Civil Engineering thesis, Clemson Univ., Clemson, SC.

National Research Council (1983). Hurricane Iwa, Hawaii, November 23, 1982, National Academy Press, Washington, D. C.

National Research Council of Canada (1990). National Building Code of Canada, NRC, Ottawa, Canada.

Standards Association of Australia (1989). Minimum Design Loads on Structures, Australian Standard, ASA, North Sydney, Australia.

Mechanical Damping System For
Mast Arm Traffic Signal Structures

Ronald A. Cook[1], David Bloomquist[1], and Michael A. Kalajian[2]

Abstract

Failures of mast arm traffic signal structures have resulted from wind-induced vibrations. This paper describes the development and testing of several different damping devices for cantilevered mast arm structures. Prototype damping devices were tested on a 11.3 m mast arm constructed in the Structures Laboratory at the University of Florida. Successful devices were then field tested on several different lengths of mast arms (11.0 m to 21.4 m) to determine the effectiveness over the full range of installations. The selected device developed was relatively insensitive to the natural frequency of the pole and consists of a tapered steel tube, weight, and spring system. The device has the potential to limit the magnitude and duration of vibrations resulting in an increase in the design life of these structures.

Introduction

Mast arm failures have been observed in Florida and other states from fatigue caused by wind-induced vibrations. These wind-induced vibrations are typically caused by vortex shedding and galloping of the cantilevered structures. An economical method to mitigate the effects of wind induced vibration on these types of structures is to install mechanical damping systems directly on the arm. These devices will limit the magnitude and duration of vibrations resulting in an increase in the design life of the structure.

---

[1] Associate Professor, Department of Civil Engineering, University of Florida, Gainesville, Florida, 32611

[2] Structural Engineer, AJT & Associates, Cape Canaveral, FL 32920, former Graduate Research Assistant, Department of Civil Engineering, University of Florida

Several devices were developed and tested in the structural laboratory at the University of Florida. The devices consisted of common damping devices for structures and modified devices specifically designed for mast arms. After successful devices were identified in the laboratory tests, the damping devices were tested in the field. The selected device provided three percent of critical damping on the upper and lower range of typical mast arms.

Damping Devices

Damping is the rate at which the vibrational amplitude decreases relative to time. The effectiveness of damping devices was measured in terms of percent of critical damping. Although there are many successful damping devices and theories on damping available for structures, most of these are related to viscous elastic damping. The most successful devices in this study did not follow conventional theory since they relied on highly non-linear impact damping. This precluded the use of analytical techniques for many of the devices tested.

The following damping devices were tested, and improved versions were developed specifically for the mast arm. The damping devices are categorized as follows:

1. Damping at Arm-Pole Connection (Disc Springs and Neoprene Pads)
2. Stockbridge Type Damper
3. Liquid Tuned Damper
4. Tuned Mass Damper
5. Spring/Mass Friction Damper
6. Spring/Mass Tapered Impact Damper

Laboratory Setup

The objective of the laboratory test setup was to evaluate the response of a mast arm under laboratory conditions and to develop and test the characteristics of damping devices to mitigate vibrations. The 11.3 m mast arm test structure was located in the Structures Laboratory at the University of Florida. The objectives of installing and testing a mast arm in the laboratory were as follows:

1. Determine natural frequencies and damping characteristics of typical traffic signal structures.
2. Reproduce large displacements and vibrations similar to those which might be encountered in the field.
3. Develop and test various damping devices prior to field tests.

Although the laboratory test setup was excellent for comparing the relative damping characteristics of different devices, it was limited in that the laboratory mast arm had a fixed length. For this reason, field tests of devices that proved to be successful in the laboratory were also performed on various lengths of mast arm structures. The testing technique used in both the laboratory and field tests was to introduce an initial vertical displacement at the tip of the mast arm and measure the free vibration response. Although this is not the same forcing function as that produced by wind, the objective was to identify devices that were relatively independent of the forcing function and primarily dependent on tip displacement. If a device was proven to be effective in damping vibrations in the laboratory, then this was a reasonable indication that the device would work in the field when subjected to wind-induced vibrations.

Field Experiments

The objective of the field experiments was to check the damping characteristics of several different poles with the different damping devices. These devices had been previously proved effective on the mast arm in the lab. The lengths of arms tested in the field ranged from 11 m to 21.4 m. The first field-test was performed at the beginning of the research to determine the natural frequency of various length arms in the field. The remaining field-tests determined if the devices tested in the lab would work properly on various length arms. After the field tests were completed, the data was analyzed and the percent critical damping was calculated for each device on each mast arm. Up to three percent critical damping was achieved on various lengths of arms in the field.

Selected Damping Device

The device that was chosen to be the most suitable in damping the vibrations of the arm was the tapered impact device. A schematic of the device is shown in Figure 1.

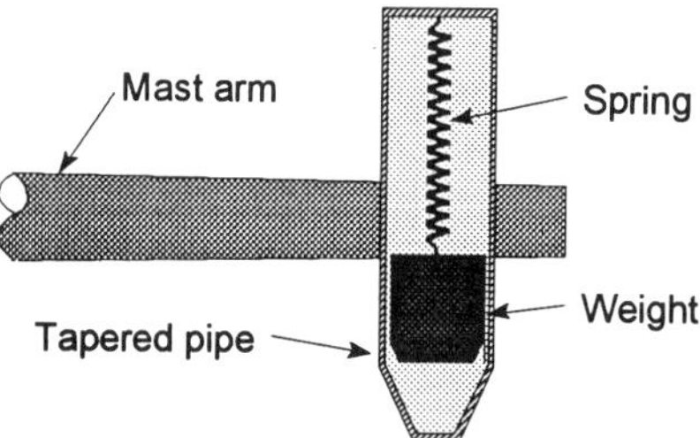

Figure 1. Schematic Diagram of Tapered Impact Damping Device

This device had 4 percent of critical damping in laboratory tests and was a size small enough so that it would not be a distraction to passing motorists. The device was composed of 102 mm steel galvanized pipe, 686 mm total length, a spring stiffness of 0.025 kg/mm and a mass of 6.8 kg. Figure 2 shows the free vibration response of the 11.3 m mast arm and the response after installation of the tapered impact device.

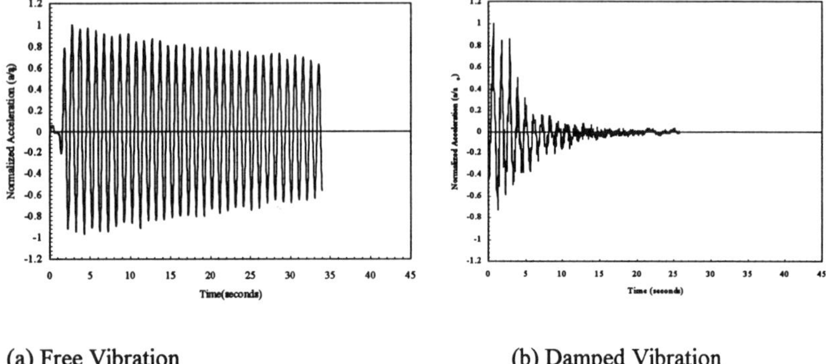

(a) Free Vibration   (b) Damped Vibration

Figure 2. Free and Damped Vibration with Selected Device on 11.3 m Mast Arm

Conclusion

Various mechanical damping devices were tested in both the laboratory and the field in order to develop a damping device that was visually unobtrusive yet provided a significant reduction in the amplitude and number of cycles associated with the vibration of typical mast arm traffic signal support structures. The selected device utilizes a tapered galvanized steel pipe (10.2 cm to 5.1 cm tapered diameter x 68.6 cm high) enclosing a mass (6.8 kg) suspended on a spring (0.25 kg/cm). The dominant damping mechanism is the suspended mass impacting the tapered section of pipe. Both laboratory and field testing were necessary since the damping action of the selected device is highly non-linear and does not lend itself to conventional analytical models. Laboratory and field testing of the device on mast arms ranging from 11.0 m to 21.4 m indicated that the device was relatively independent of typical mast arm frequencies. The device produced 3.6% critical damping on an 11.0 m mast arm (0.7 HZ) and 2.8% critical damping on a 21.4 m, mast arm (1.4 HZ) compared to 0.2% and 0.6% critical damping respectively without the device.

Truck-Induced Wind Loads on Highway Sign Support Structures

Kevin W. Johns[1]
Associate Member

Robert J. Dexter, P.E.[2]
Member

Abstract

Recent research has given more accurate fatigue design loads for truck-induced wind loads on cantilevered highway sign support structures. This research consisted of short and long-term field testing on a variable message sign over an interstate highway. The results of the research indicated that a worst case truck-induced wind load of 1760 Pa with a reduction dependant upon the height above the road results in an adequate fatigue design load.

Introduction

Cantilevered sign and signal support structures are used extensively on major interstate highways and at local intersections for the purposes of traffic control. There are some obvious reasons for the sensitivity to vibration of cantilevered support structures. The single support significantly increases the flexibility of the cantilevered structures relative to overhead structures. The ratio of stiffness to mass consistently gives these structures a low natural frequency of about 1.0 Hz. These cantilevered support structures also have extremely low critical damping ratios, typically less than one percent of critical damping. These conditions make cantilevered support structures particularly susceptible to large-amplitude vibration and/or fatigue cracking due to wind loading [Dexter, 98].

Variable message signs (VMS) are particularly susceptible to truck-induced wind gusts, because of their width in the direction of traffic and the large soffit area. Recent research at Lehigh University, sponsored by the New Jersey DOT has shown that the VMS are affected by truck-induced wind gusts. This paper will describe the test procedures and conclusions from this research. Conclusions from this research will be compared to truck-induced wind gusts recommended in NCHRP 10-38 [Kaczinski, 96]. To accomplish this objective, a cantilevered VMS was instrumented and monitored. Short-term and long-term field tests were conducted. These tests and the results will be summarized in the following sections.

Field Testing

The first step of the short-term test was to establish the stiffness, natural frequency, and damping ratio of the sign support structure. The structure was deflected and released in a method

---

[1] Engineer, Modjeski and Masters, Inc., P.O. Box 2345, Harrisburg, PA 17105. kjohns3559@aol.com

[2] Associate Professor, University of Minnesota, 122 Civil Engineering Bldg., 500 Pillsbury SE Dr., Minneapolis, MN 55455-0116, dexter@tc.umn.edu

that would allow the structure to oscillate at its natural frequency and dampen at its critical damping ratio.

This test was performed with two different configurations. One the sign was pulled straight down and the second the sign was pulled at a 45-degree angle. This would enable determination of the dynamic characteristics in the two primary modes of vibration, the "twisting mode" and the "hatchet mode". The "twisting mode" describes the sign rotating about the column. The "hatchet mode" describes an up and down in plane movement.

The second phase of the test was to determine the effect of the truck-induced wind gusts on the structure. Two trucks were hired to drive under the sign and the upward gusts from these trucks would be measured with pressure transducers. The drivers were instructed to note their speed and radio it back so a correlation could be made with truck-induced wind gusts and truck speed. The two traffic lanes under the VMS box were closed so the trucks could easily get up to speed and position themselves correctly in their lane.

Long-term testing was intended to collect data from any significant event on the VMS for three months from June to August 1997. The long-term test monitored instrumentation that was determined to be critical from the results of the short-term test. The data logger was programmed to monitor these gages continuously, however it was triggered to record events only when predetermined thresholds were exceeded. The data that was collected in the long-term testing was to be transmitted back to the lab through a cellular phone link. Not only could data be received from the logger, new programs could be sent to make any necessary changes. Two 20-watt solar panels were connected to two deep-cycle marine batteries that helped support the data logger's battery.

Dynamic Characteristics and Static Test Results

The short-term test successfully determined the dynamic characteristics of the VMS support structure. Data indicated that the stiffness was 0.24 kN/mm. The log-decrement equation was used to determine the percent of critical damping in the sign support structure. The percent of critical damping in the "twisting mode" and "hatchet mode" was 0.57 percent and 0.25 percent respectively. The larger damping in the "twisting mode" is from the large frontal area of the sign box moving through the air. The sign has an easier time slicing through the air in the "hatchet mode". A Fast Fourier Transform was performed on data from a strain gage while the structure oscillated at its natural frequencies. The natural frequency in the "twisting mode" and "hatchet mode" was 0.87 cycles/s and 1.22 cycles/s respectively [Johns, 98].

Discussion of Field Test Results

The data that was collected during field testing enabled a decision to be made on the appropriate fatigue design pressures for truck-induced gusts. The appropriate truck-induced gust pressure was calculated using strain gage data rather than the pressure transducers. Based on data taken from the short-term testing, the magnitude of pressure indicated by the pressure transducers was in error in most cases. This was probably due to the extreme turbulence in the truck gust.

Strain gages on the column measured stress ranges from random truck-induced wind gusts reached as high as 10 MPa with an accompanying natural wind speed of 2.5 m/s. It is reasonable to assume that with any significant headwind this stress range in the column could have easily reached 14 MPa. The equivalent static pressure on the horizontal area of the sign and truss that would cause the stress range of 14 MPa in the column is 525 Pa. This pressure is the worst case equivalent static pressure that was deduced from the long-term monitoring.

In the NCHRP 10-38 project DeSantis' model was recommended for truck-gust loading. Assuming a drag coefficient of 1.45 the Desantis model would imply an equivalent static pressure of 2550 Pa [DeSantis, 96]. Note that this is almost five times bigger than the maximum equivalent static pressure from strain gage data in the New Jersey research. However, during the monitoring in New Jersey there were no significant headwinds and it is not known what synergetic effect this may have. The back calculated pressure of 525 Pa is about 3.3 time less than Desantis' recommendation of 1760 Pa. The exact reason for this is not known, however there are some plausible explanations for it. The structure that Desantis modeled may have been closer to the trucks. The structure could have been more flexible than the one monitored by Lehigh. If there was more ambient wind this could have a synergetic effect that is still not fully quantifiable.

Although the pressure transducers were not relied upon to give a magnitude of truck-gust pressure they were used to try to quantify a pressure gradient at different elevations above the truck. Figure 1 shows pressure transducer data from above each of the lanes in the short-term test. The pressure from the truck gust above lane one ranged from 4 Pa to 2.5 Pa at heights of 0.3 m to 2.67 m up from the bottom of the sign respectively. The pressure from the truck gust above lane two ranged from 6 Pa to 3 Pa at heights of 0.3 m to 2.67 m up from the bottom of the sign respectively. This data indicates that at a height of 4.5 m above the bottom of the existing VMS box the pressure would go to essentially zero. This gradient will be incorporated into the design recommendations for truck-induced wind gusts.

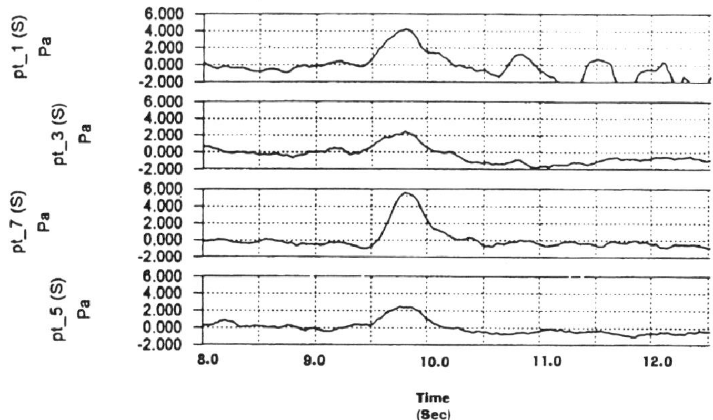

Figure 1 – Pressure Gradient along Height of VMS

Conclusions

The following conclusions were drawn from the findings of monitoring the cantilevered VMS during the short and long-term field testing.

The recommended design loads from NCHRP 10-38 were modified based on the knowledge obtained in this research. These modified design loads were recommended for use in the design of cantilevered VMS support structures.

Truck-induced gusts were measured during the monitoring of the VMS and have been determined to be a potential concern for cantilevered VMS support structures. Truck-induced gusts could be designed for with the following loads. Sign and signal support structures could be designed for truck-induced gust loads using an equivalent static pressure of "TG" Pa times the drag coefficient, which is applied vertically to the horizontally-projected area of the structural members and any attachments mounted to the horizontal mast-arm along a length of the mast arm which is the greater of the length of the sign or 3.7 m (12 ft). Equation 1 indicates the appropriate pressure from truck-induced wind gusts

$$P_{TG} = (TG)C_d I_F \quad (Pa) \hspace{4cm} \text{(Equation 1)}$$

where TG is the appropriate value from Table 4-1, C_d is the appropriate drag coefficient from table 1.2.5C of the AASHTO specification and I_F is the importance factor based on the environment of the VMS. There is no lane-load reduction factor to be considered with truck-induced gusts [Johns, 98].

Elevation Above Road Surface	Value of "TG"
6 m or less	1760 Pa
6.1 m to 7 m	1530 Pa
7.1 m to 8 m	1150 Pa
8.1 m to 9 m	690 Pa
9.1 m to 10 m	380 Pa
10.1 m or above	0 Pa

Table 4-1: Truck-Induced Gust Values

Based on the field test data and the previous research, it was recommended that the fatigue design load range suggested in Project 10-38 be used for design with only slight modification to the truck-induced gusts, as mentioned above. The recommended design value of 2550 Pa (the drag coefficient of 1.45 has already been included) for truck-induced gusts is about 5 above what was measured on the VMS in the field. Given that there has been at least one cantilevered VMS support structure in Florida, which was reported to have failed due to truck-induced gusts, it is not clear if during the monitoring period the worst case loading was observed [Johns, 98]. It could be that the worst case loading occurs when there are significant natural wind gusts acting together with the truck induced wind gusts. Large natural wind gusts did not occur during the period the structure was monitored. Therefore, it is prudent to stay with the larger design loads recommended in NCHRP 10-38.

References

Dexter, R.J., et al, <u>Fatigue-Resistance Design of Cantilevered Signal, Sign, and Light Supports,</u> National Cooperative Highway Research Program, Interim Report - NCHRP Project 10-38. Transportation Research Board, Washington D.C., 1998.

DeSantis, P.V. and Haig, P., "Unanticipated Loading Causes Highway Sign Failure," Proceedings of ANSYS Convention, 1996.

Johns, K.W., Dexter, R.J., <u>Fatigue Related Wind Loads on Highway Support Structures,</u> ATLSS Report No. 98-03, Lehigh University, Bethlehem, PA, 1998

Kaczinski, M.R., et al., <u>Fatigue-Resistance Design of Cantilevered Signal, Sign, and Light Supports,</u> National Cooperative Highway Research Program, Final Report - NCHRP Project 10-38, Transportation Research Board, Washington D.C., 1996.

Traffic Signal Structure Research
Univ. of Wyoming

Gray, B, Wang, P., Hamilton, H.R., Puckett, J.A.[1]

Introduction

Two traffic signal poles recently failed in Wyoming. These failures occurred at the connection between the mast arm and post and were due to fatigue cracking in the post near the base of the weld. The fatigue cracking is due to the mast arm vibrations caused by wind. Previous research has shown that the vertical movement of the mast arm ("in-plane" motion) is caused primarily by galloping and that the horizontal movement ("out-of-plane" motion) is caused by natural and truck induced gusts (South, May 1994). It has also been reported that galloping is the primary cause of the damage. The crack patterns indicate that out-of-plane motion may also be contributing significantly to the damage.

This paper summarizes the research work currently underway at the University of Wyoming. The research includes field monitoring, finite element analysis (FEA), and laboratory testing of full-scale specimens.

Field Testing

An in-service traffic signal pole was instrumented and monitored to observe wind effects to determine the relative significance of the in-plane and out-of-plan movements. Field monitoring consists of measuring wind speed and direction along with strain in the mast arm associated with both in-plane and out-of-plane movement. The instrumentation is shown schematically in Figure 1. The number and magnitude of cycles experienced by the pole under wind conditions greater than 10 mph were recorded using a datalogger. The wind anemometer was placed approximately 30 feet above the ground.

The first, of several, traffic signal poles instrumented is located on the southeast corner of Curtis St. and McCue St. in Laramie, Wyoming. The mast arm points north over Curtis St. This location was chosen because it had been observed

---

[1] Civil Engineering Department, University of Wyoming, 16th & Gibbon, Laramie, WY 82070

by WYDOT engineers to gallop. Initially the back planes around the lights were removed prior to testing. They were reinstalled after several weeks of data collection for comparison purposes. The standard signal detail in Wyoming includes back planes.

<u>Results</u>

Figure 2 shows the nominal stress range at the weld, in the box connection, for both in-plane and out-of-plane movements. The stress ranges are slightly different for the two motions. This is due to the method of reducing the data. Note that there are significant numbers of cycles at the lower stress levels. Further monitoring will include wind speeds below 10 mph to compare the cycles experienced at the lower wind speeds.

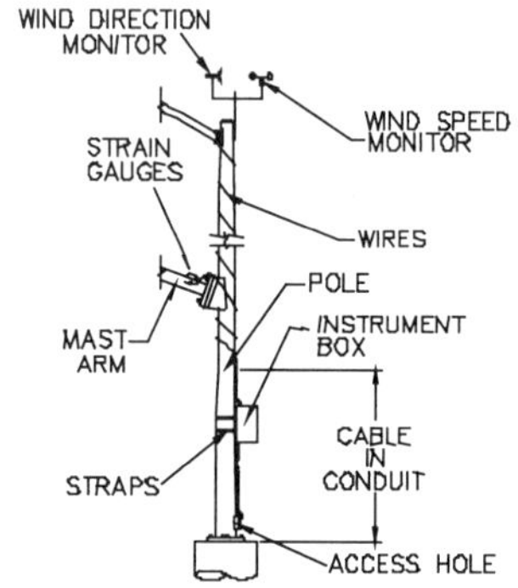

Figure 1-Field Instrumentation

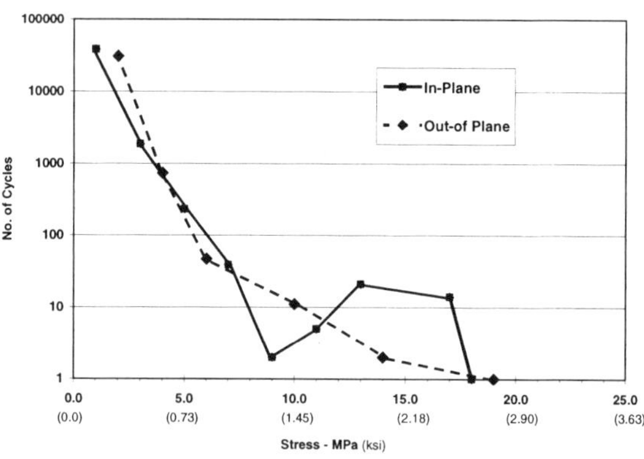

Figure 2 - Cumulative Cycles over six-month period

Finite Element Analysis (FEA)

A typical pole connection, shown in Figure 3, was analyzed using MSC/PATRAN7 to study the influence of the out-of-plane moment and in-plane moment on the maximum stresses at the corner of the box connections. Eight-node quadratic and six-node triangular thin-shell elements were used. For this analysis, an 18.48 kN-m (163.6 kip-in) moment was calculated as wind load moment based on the current proposed specification for signal pole structures. The moment was assumed to be acting as both in-plane moment and out-of-plane moment as shown in Figure 4.

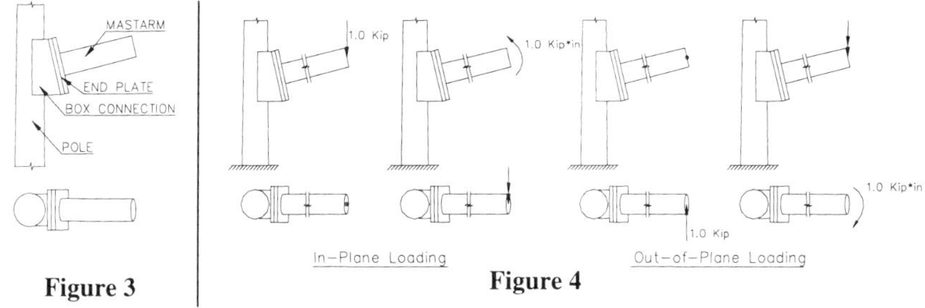

Figure 3   **Figure 4**

It is observed from the deformation plot (see Figure 5) that subjected to in-plane moment, the top plate and bottom plate of the box connection can distribute the load more uniformly in the pole and thus cause less stress concentration. Out-of-plane moment causes the box connection to twist the pole and the corners of the box connection cause significant stress concentration in pole. According to the FEA results, the out-of-plane moment can cause stress as much as 73.3Mpa (10.5 ksi) in circumference direction and 46.2Mpa (6.62 ksi) in longitudinal direction. The maximum stress caused by the in-plane moment is 16.4Mpa (2.35 ksi) in circumference direction and 20.3MPa (2.91 ksi) in longitudinal direction. The FEA results indicate that out-of–plane moment may be a major consideration in fatigue crack initiation and growth. Note that the out-of-plane moments create much higher stresses.

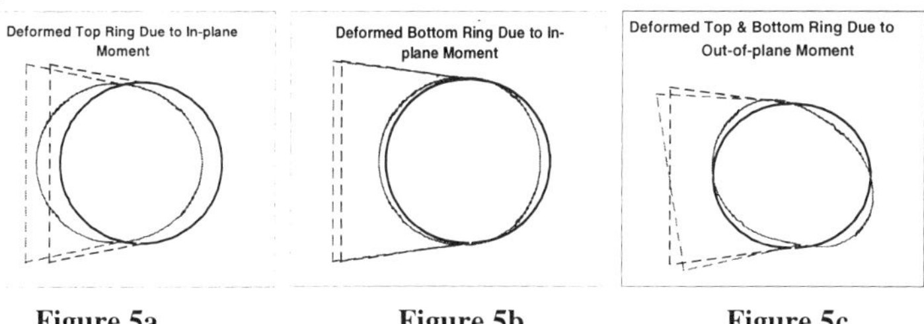

Figure 5a   **Figure 5b**   **Figure 5c**

Full-Scale Laboratory Testing – Static Tests

A sample pole was attached to a reinforced concrete base and mounted to the laboratory floor. An actuator was positioned between the pole mast arm and a reinforced concrete strong wall. The layouts of strain gauges are illustrated in Figure 6. The numbers in brackets indicate that the strain gages are placed on the inside face of the

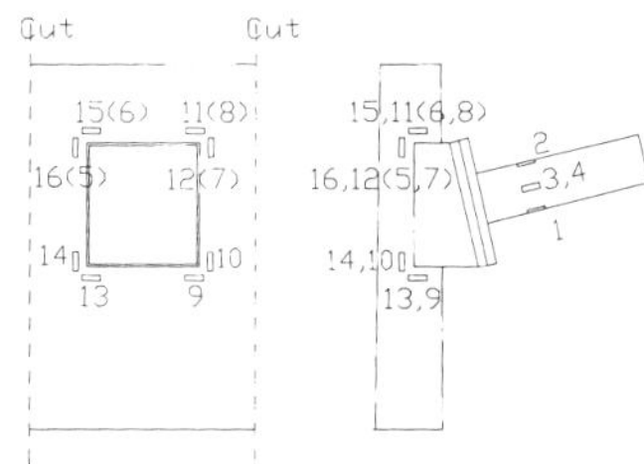

Figure 6

pole. The strain gages are approximately 38mm (1.5in) away from the box corners.

An in-plane static test was conducted and it was observed that the load-deflection curve was linear except for some initial slip. Note that the maximum stresses are well below the yield stresses of the steel, see Table 1. The slopes of the

Table 1 Summary of FEA stresses and static test stresses (MPa)

stress	M (kN-m)	Load (Kip)	S.G. 1	S.G. 2	S.G. 3	S.G. 4	S.G. 5	S.G. 6	S.G. 7	S.G. 8
test (pull)	4.91	4.45	-18.8	22.1	-0.9	-1.4	-0.6	14.6	-1.0	14.1
test (push)	4.91	4.45	-23.2	18.8	3.9	2.8	-0.6	13.2	-1.1	13.0
FEA	4.91	4.45	-20.8	20.8	0.0	0.0	-2.3	8.4	-5.9	7.4

stress	M (kN-m)	Load (Kip)	S.G. 9	S.G. 10	S.G. 11	S.G. 12	S.G. 13	S.G. 14	S.G. 15	S.G. 16
test (pull)	4.91	4.45	1.4	21.6	-12.5	-0.9	3.9	15.5	-11.2	-0.1
test (push)	4.91	4.45	2.9	23.2	-10.2	-0.9	5.7	18.9	-9.6	-0.3
FEA	4.91	4.45	9.9	5.1	-8.7	-5.2	8.5	0.6	-7.7	-1.7

load vs. stress experimental relationships were used in subsequent computations eliminating any start-up effects.

FEA results were compared with stress readings at the corresponding strain gages locations shown in Table 1. Some agreement between the FEA and test was achieved. The strain gages are located very close to the concrete support, which may affect the stress distribution significantly. Considering the imperfection of the pole cylinder and the welding along the cylinder, certain disagreement between test results and FEA results are understandable.

References:

1. AASHTO Standard Specifications for Structural Supports for Highway Signs, Luminaries, and Traffic Signals. 1994.
2. Dexter, R.J., Kaczinski, M.R., Van Dien, J.P., Fatigue-Resistant Design of Cantilevered Signal Sign and Light Supports. Prepared for: NCHRP. Report by: Leigh University. February 1996.
3. McDonald, J.R., Mehta, K.C., Oler, W., Pulipaka, N. Wind Load Effects on Signs, Luminaries, and Traffic Signal Structures. Texas Tech University. Prepared for: Texas Department of Transportation. February 1995.
4. South, Jeffery M. P.E. Fatigue Analysis of Overhead Sign and Signal Structures. Illinois Department of Transportation. Prepared for: Illinois Department of Transportation. May 1994.

Acknowledgement

The financial and technical support of the Wyoming Department of Transportation is appreciated.

Disclaimer

The information presented herein is the sole responsibility of the authors and has not been approved by the Wyoming Department of Transportation.

Strain Measurements on Traffic Signal Mast Arms

Bryan A. Hartnagel[1] and Michael G. Barker[2], Members, ASCE

Abstract

The Missouri Department of Transportation (MoDOT) has had over a dozen traffic signal mast arms fail in the past six years. The failures occurred at the weld joining the mast arm tube to the connection plate. In order to investigate these failures a study was designed to measure the in-service strains experienced by two traffic signal mast arms.

Introduction

Over the past few years the Missouri Department of Transportation (MoDOT) has had several failures out of the over 6000 traffic signal mast arms in the state. The failures occurred at the weld joining the mast arm tube to the connection plate. This prompted a design revision of the MoDOT standard plans. Joint failures appeared to be caused by fatigue due to truck-induced wind and ambient wind. The design revision included a "fatigue resistant weld". This weld, shown in Figure 1, has a shallow angle on each of the legs and is intended to lessen the stress concentration of the weld. After making the standard specification change MoDOT wanted to determine if the new specification was indeed better than the old and proposed a research program.

Research Program

The proposed research program consisted of several components. The first component was to instrument two mast arms in the field to determine the in-service range of stresses (strains) to which they were exposed. Truck-induced

---

[1] Assistant Professor of Civil Engineering, Colorado State University, Fort Collins, CO 80523
[2] Associate Professor of Civil Engineering, University of Missouri - Columbia, Columbia, MO 65211

wind and ambient wind are the cause of in-service strains. The second component of the test was to take the in-service strain data and determine the equivalent constant amplitude strain that would represent the fatigue loading applied to the mast arms in service. This constant amplitude loading would then be applied to mast arm samples in the laboratory. Laboratory tests were the third component of the research program. Four mast arm samples made with the old design specification and four mast arm samples with the new design specification were made to test in the laboratory. The laboratory tests were designed to apply a bending moment to simulate the in-service constant amplitude strain experienced by the mast arm in the field. A comparison of the two design specifications could be made after the laboratory fatigue tests.

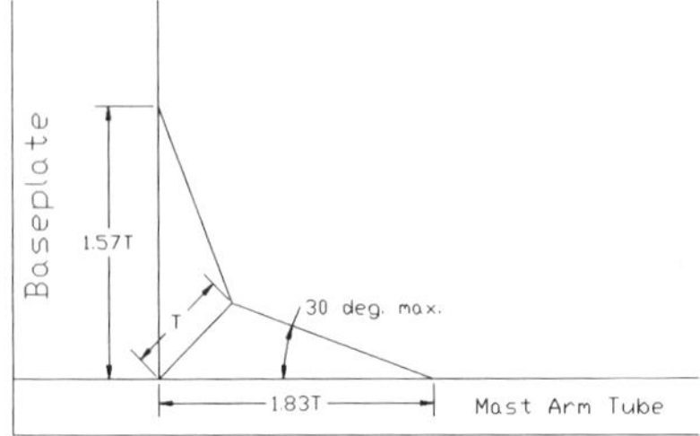

Figure 1 Fatigue Resistant Weld

Field Tests

Field tests were conducted on two mast arms in Columbia, Missouri. The mast arms were under the jurisdiction of MoDOT on state route 763 and 740. The two locations were chosen for the following reasons. Both locations have two lanes of high-speed (speed limit > 64 kmph (40 mph)) traffic which pass under the arm. Each location has some probability of trucks and buses passing under the arm. Neither location is shielded from ambient wind. The location on route 763 has three signal heads and two signs. It accommodates two through lanes and one left turning lane. The cross section is octagonal and the arm is 12.8 m (42 ft) long. The mast arm on route 740 has four signal heads and two signs. Two through lanes and two left turn lanes of traffic pass below the arm. The cross section of the arm is circular and it is 16.5 m (54 ft) long. The first arm on 763 was installed in approximately 1986 and the second arm on 740 was installed in December of 1997. The new arm was designed using the revised standard specifications.

Instrumentation

Twelve strain gages were installed on each of the traffic signal mast arms to measure the in-service strains. Eight of the strain gages were located 102 mm

(4 in) from the mast arm connection plate. Four were orientated along the length of the tube to measure bending strains at the top, bottom, front, and back of the arm. The other four were orientated to measure shear strains at the connection. Four remaining strain gages were placed a distance away from the connection plate (4.57 m (15 ft) on the 12.8 m (42 ft) long arm and 5.49 m (18 ft) on the 16.5 m (54 ft) long arm) to measure bending strains. At the tip of the 12.8 m long arm an accelerometer was attached to measure the vertical acceleration. Two accelerometers were placed on the tip of the 16.5 m long arm to measure vertical and horizontal acceleration. Cables from all the strain gages were connected to a junction box mounted near the mast arm connection to the post. In order to collect data, the data acquisition vehicle was parked below the mast arm and a cable was connected to the junction box that powered the strain gages and also returned strain gage signals to the acquisition vehicle.

Data Collection

Once the strain gages were ready to collect data a video camera was set up to record the traffic that passed under the mast arm while data was collected. This required an operator for the data acquisition computer and an operator for the video camera. The two operators were able to communicate when to start and stop an acquisition sample with the use of a wireless communications system. Each data sample usually consisted of one or two complete cycles of the traffic signal. Once the data was collected it could be compared to the video to determine what type of traffic caused a particular strain event.

A flatbed truck that was outfitted with an air dam on the cab was also used to collect some data samples. The truck was driven below the mast arm at various speeds to try to determine the effect of truck induced winds.

Results

The results of each acquisition sample consisted of the twelve strain gage signals and the accelerometer signals versus time. As mentioned earlier these signals could be compared to video of the traffic passing below the mast arm during the acquisition period. Information obtained from the strain gage signals was maximum strain experienced, natural frequency of the vibrations and damping of the system.

Bending strains were measured at 102 mm (4 in) from the mast arm connection plate. For the data samples collected it was seen that bending back and forth was more severe than bending up and down of the mast arm. A sample of the bending strains for a truck passing below the mast arm is shown in Figure 2. In Figure 2 strain 1 and strain 5 are the up and down bending strains and strain 3 and strain 7 are the back and forth bending strains.

Summary

Two traffic signal mast arms were instrumented with strain gages and accelerometers to determine the effect of truck-induced wind and ambient wind on the weld between the mast arm and the connection plate. Data was collected under various weather and traffic conditions. The instrumentation of the two arms was a portion of a larger project to determine why there have been about a dozen failures of traffic signal mast arms in Missouri during the past six years.

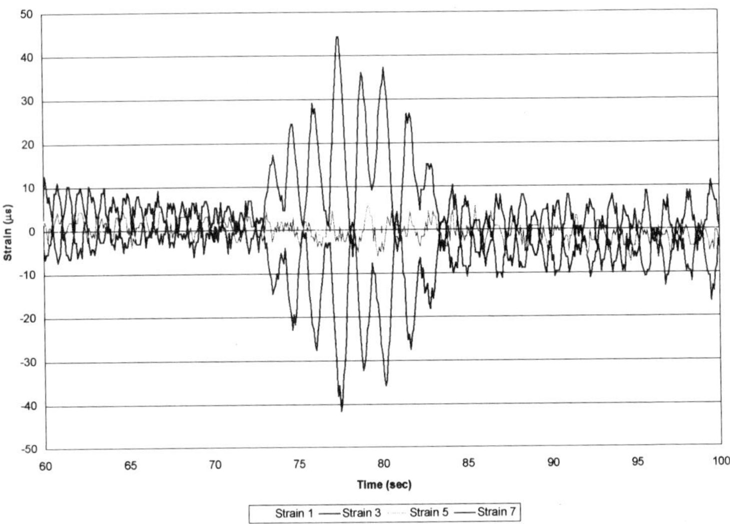

Figure 2 Strain History for Truck Passing Event

Acknowledgements

The authors would like to thank the Missouri Department of Transportation for funding this project. Also deserving recognition are Genda Chen, Lokesh Dharani, and Chris Ramsay all faculty of the University of Missouri - Rolla who are assisting with other portions of the project.

Analysis of the Catastrophic Failure of the Support Structure of a Changeable Message Sign

Lawrence Kashar, M. Russell Nester, M. ASCE, James W. Jones
Mohammad Hariri, M. ASCE, and Sanford Friezner[1]

INTRODUCTION AND BACKGROUND

On November 27, 1995, the support structure for an electronic changeable message sign (CMS) failed, crushing a passing vehicle on a California interstate freeway. Erected just 18 months prior, the sign had been subjected to high seasonal winds. Winds averaged 51 kph (31 mph) during the previous 24 hours, with gusts to 98 kph (60 mph). This was the windiest day of the previous six months.

The 13,000-kg (6300-pound) CMS was 8.05-meters (25.4 ft) high and cantilevered 9.98 meters (32.8 ft) from its single column support base. The failed support column was a 457-mm (18-inch) diameter by 12.7-mm (½ inch) thick galvanized carbon steel pipe. The steel base plate was 914x914x70 mm (36x36x2.75 inch) octagonal, supported on eight 35-mm (1.375 inch) diameter double-nutted anchor bolts. The column/plate joint was a socket design with 13-mm (½ inch) fillet welds at both upper (outside) and lower (inside). There is an electrical service access hole in the column above the base. No discrepancies in quality assurance were found despite a thorough review of fabrication, construction and inspection records.

VISUAL EXAMINATION OF THE FAILURE

Failure resulted from a fracture of the pipe column just above the upper fillet weld to the base plate. A visual examination revealed four distinct stages of the failure:

- **Stage 1** – A smooth, flat, rusty, horizontal fatigue crack extending 560 mm (22 inches) across the side away from the sign.
- **Stage 2** – Two 127-mm (5-inch) long unrusted fatigue cracks at the ends of the Stage 1 crack. Irregular and with areas of ductile yield.
- **Stage 3** – A smooth, clean, partial-thickness fatigue crack extending 305 mm (12 inches) across the side closest to the sign.
- **Stage 4** – Final ductile overload fracture of the remaining pipe section.

---

1. Dr. Lawrence Kashar, et. al., are members of The Consultants Bureau, a division of Kashar Technical Services, Inc., 6305 Arizona Place, Los Angeles, California 90045.

DOCUMENTS SEARCH

Contract drawings, specifications, and records were reviewed, revealing that the CMS weight, height, and cantilever length all violated California Department of Transportation (CalTrans) Standard Plans limitations. CalTrans Standard Plans were the basis of contract documents. Further, the Standard Plans called for a solid base plate with 254-mm (10-inch) diameter center hole for electrical conduits, with a butt-welded connection to the column. The socket configuration with a 457-mm (18-inch) diameter hole was a field-approved alternate not found in the Standard Plans.

Studies at Lehigh University (Fisher, Kaczinski) on cantilevered steel lighting and sign poles found that fatigue strengths near the bases of poles were much lower than expected. Complicating this situation, fatigue cracks were very found to be difficult to detect in galvanized materials. Design procedure changes were recommended.

Numerous contemporary references have documented the vulnerability of cantilevered single steel pole sign and lighting structures to wind-induced fatigue failure (DeSantis).

MECHANICAL PROPERTIES OF PIPE AND WELD

Testing found the yield strength of the failed pipe to be 65 ksi (448 mpa) and tensile strength to be 460 mpa (67 ksi). Elongation was 36% with 63% area reduction. These values exceed those of the specified API 5L-X42 material. Charpy impact values at $0°C$ $(32°F)$ were 270 J (200 ft-lb), 41 J (30 ft-lb), and 200 J (150 ft-lb) for the base material, weld, and heat affected zone (HAZ) respectively. Corresponding Rockwell hardnesses were B82-87, B93-94, and B79-81, indicating that the HAZ strength also exceeded API 5L-X42 requirements. Metallurgical and mechanical investigations revealed no likely causes to the short fatigue life of the subject CMS.

ESTIMATE OF STRESSES USING SOLID MECHANICS

Initial stress analyses employed closed-form solid mechanics solutions (hand calculated). Base plate bending stiffness was assumed to be much greater than either pipe wall or anchor bolt stiffnesses. Loadings included dead and wind loads per the 1994 Uniform Building Code (80 mph, Exposure D). Per UBC, the applied wind pressure was 34.4 psf on the full exposed area.

Case 1 (lower bounds stress) assumed that grout provided full fixity against base plate rotation outside the anchor bolt circle. Rourke and Young equation 24-1e estimates the base plate rotation at the pipe, and equation 30-10 estimates the resultant local pipe bending stress. Case 2 (upper bounds stress) assumed no grout fixity. Rourke and Young equation 24-1 estimates the base plate rotation, and equation 30-10 again estimates the local pipe bending stress. As seen in Table 1, local stresses are increased significantly over the stresses calculated for gross section properties. Increases of 45%

for Case 1 and 331% for Case 2 were calculated. For Case 2, total stresses with the UBC wind load exceed material ultimate strength regardless of wind direction.

Table 1. Closed Form Calculation Results (Maximum Tensile Stress)

	DEAD LOAD STRESS		DL + WL TOTAL** STRESSES	
	GROSS*	TOTAL**	WIND ON AXIS	WIND NORMAL
CASE 1	13.7 ksi	19.8 ksi	23.6 ksi	25.2 ksi
CASE 2	13.7 ksi	59.1 ksi	70.4 ksi	75.6 ksi

*** from gross section properties   ** total including local stresses**

STRESS ANALYSIS USING AASHTO-1994 DESIGN METHODS

A design stress analysis per AASHTO Standard Specifications (1994 edition) was performed. Design stress at the column base was found to be 10% over the AASHTO allowable. AASHTO methodology neglected both local stresses due to base plate flexibility and the effects of dynamic wind-structure interaction.

A fatigue design check due to truck-induced loadings was also performed. Per AASHTO standards, the failed structure would exhibit "infinite life" at 2,000,000 cycles. Actual traffic counts estimate approximately 1,000,000 truck passings at the time of failure. Either (a) passing trucks induce multiple stress cycles per passing, or (b) stresses were much higher than predicted by AASHTO methodology, or (c) some other source of cyclic stress was present to account for the fatigue failure.

FEA AND DYNAMIC ANALYSES

FEA models using the ANSYS program were developed of the base plate, anchor bolts, and lower pipe column for both the butt-welded and for the socketed configurations. Models employed a mapped mesh of 3D brick elements. The remaining structure was represented by pipe elements for column and mast, and by shell elements for the sign. The electrical service access hole was modeled. Various construction sequences were considered.

The socket configuration showed higher local stresses than the butt-welded design due to increased base plate warping. At the toe of the upper fillet weld, local stresses were 4.5 to 6.0 times those calculated using gross pipe section properties only. This agreed with the closed form solution results (Case 2). Location of highest stress coincided closely with the observed crack locations. Results were not sensitive to weld size. The electrical access hole had negligible effect on stress at the base.

Fundamental cantilever mode frequency was found to be 1.0 Hz. The sign vortex-shedding frequency was calculated to be 0.91 Hz for a 49 kph (30 mph) wind. Remedial designs considered included the butt-welded configuration, heavier materials, and gussets. Gussets were the most effective in reducing stress.

STRAIN GAGE DATA ON "SISTER" CMS STRUCTURE

A similar CMS was instrumented with strain gages, and data taken at another (less windy) site. This CMS was then moved to the failed CMS location and data taken again. Dead load stresses measured 83 kpa (12 ksi) at 0.9 meters (3 ft) above the base plate, and 160-190 kpa (24-28 ksi) just above the weld. These data confirmed calculated local stress increases, and indicate that grout provided significant restraint.

Mild to strong winds and typical truck passings produced peak-to-peak stresses of about 40 kpa (6 ksi) at the 3-foot height, and 63-96 kpa (12-14 ksi) just above the weld. Natural frequency was 1.05 Hz, also confirming FEA modal calculations. Moderate to severe wind-structure interaction ("galloping") was observed, with vertical excursions estimated to be up to 0.3 meters (12 inches).

SUMMARY AND CONCLUSIONS

The catastrophic failure of a CMS just 18 months after installation was initiated by a low-cycle fatigue crack just above the column/base plate weld. The fracture was within the pipe's base material, and the initial crack was over 12 months old at the time of failure. Materials and quality assurance were found to be not at issue.

Both closed form and FEA calculations predict high local stresses at the crack location. Base plate configuration and grout were found to have large effects on local stresses. Weld size and service access hole were found to have negligible effect. Field measurements confirmed both the high local stresses and the predicted "wind galloping". Measured stresses indicate that grout may contribute significantly to the base plate stiffness.

AASHTO design methodology was found to be inadequate, neglecting to address dynamic wind-structure interaction, truck-induced oscillations, and local stresses. Gussets would be effective in reducing local stresses, and are readily retrofit.

SELECTED REFERENCES

DeSantis, P. V. And Haig, P., "Unanticipated Loading Causes Highway Sign Failure" Proceedings of ANSYS Convention, 1966.

Fisher, J. W. Et. Al., <u>Fatigue Behavior of Steel Light Poles,</u> Report No. FHWA/CA/SD-81-82, California Department of Transportation, Sacramento, CA, 1981.

Kaczinski, M. R., Dexter, R. J. And Van Dien, J. P., <u>Fatigue-Resistant Design of Cantilevered Signal, Sign and Light Supports, Final Report for National Cooperative Highway Research Program</u>, ATLSS Engineering Research Center, LeHigh University, Bethlehem, PA, July, 1996.

Rourke and Young, <u>Formulas for Stress and Strain, 5th Edition,</u> McGraw-Hill, Inc., 1975.

Subject Index

Page number refers to the first page of paper

Cracking, 129, 1107
Cracks, 276, 448
Creep, 761
Critical load, 430
Curved beams, 231, 239
Cyclic loads, 426, 805
Cyclic tests, 207

Damage, 154, 187, 251, 336, 508,
 598, 793, 817, 1011, 1015, 1095
Damage assessment, 357, 602, 695,
 723
Damping, 15, 66, 103, 361, 452, 854,
 866, 870, 983
Data collection, 340
Databases, 781
Decision making, 821
Deep water, 854
Deformation, 369, 893
Degradation, 357
Demolition, 878, 881, 948
Design, 87, 113, 129, 133, 137, 183,
 235, 243, 377, 381, 422, 434, 460,
 468, 494, 516, 532, 548, 630, 634,
 638, 737, 757, 765, 769, 773, 777,
 789, 842, 846, 874, 889, 897, 911,
 915, 919, 923, 958, 963, 979, 987,
 991, 1051, 1055, 1099
Design criteria, 125, 141, 158, 464,
 586, 611, 658, 667, 671, 745, 749,
 753, 797, 1077, 1089, 1095
Design improvements, 498
Design wind speed, 944
Deterioration, 154, 243, 251, 476,
 671
Development, 560
Diaphragm wall, 903
Diaphragms, mechanics, 805
Disasters, 1011
Displacement, 332
Distribution patterns, 516
Dowels, 626, 634
Drag, 854

Ductility, 79, 215, 259, 785
Durability, 203, 881
Dynamic analysis, 975, 979, 1055
Dynamic loads, 490, 979
Dynamic models, 365
Dynamic response, 971, 979, 983
Dynamic tests, 320, 324, 328
Dynamics, 850, 854, 870

Earthquake damage, 33, 45, 95, 484,
 833
Earthquake excitation, 75, 512
Earthquake loads, 187, 361, 480,
 490, 512, 524, 948, 991
Earthquake resistant structures, 33,
 61, 83
Earthquakes, 11, 15, 41, 49, 57, 79,
 91, 99, 103, 308, 406, 1077, 1081,
 1085
Economic analysis, 552, 556
Education, 464, 617
Efficiency, 889
Eigenvalues, 430
Electric power transmission, 991
Electrical equipment, 987, 999
Electronic equipment, 1115
Elevated structures, 33
Embedded foundations, 61
Energy dissipation, 452
Engineering, 846, 923
Engineers, 915
Environmental factors, 268
Errors, 162
Esthetics, 560, 963
Estimating, 373, 377, 381, 442
Estimation, 162, 817
Europe, 133, 541, 958
Evaluation, 288, 292, 296, 300, 448,
 781, 801, 809
Excavation, 874, 903
Expansion, 27
Experimentation, 191, 695, 699
Explosions, 598, 695, 699, 1081

Author Index

Page number refers to the first page of paper